Engineering Sciences and Fundamentals

Presentations at the 2009 AIChE Annual Meeting

Nashville, Tennessee, USA
8-13 November 2009

Volume 1 of 2

ISBN: 978-1-61567-925-6

Printed from e-media with permission by:

Curran Associates, Inc.
57 Morehouse Lane
Red Hook, NY 12571

Printed by Curran Associates, Inc. (2010)

For permission requests, please contact AIChE
at the address below.

AIChE
3 Park Avenue
New York, NY 10016-5991

Phone: (203) 702-7660
Fax: (203) 775-5177

www.aiche.org

Additional copies of this publication are available from:

Curran Associates, Inc.
57 Morehouse Lane
Red Hook, NY 12571 USA
Phone: 845-758-0400
Fax: 845-758-2634
Email: curran@proceedings.com
Web: www.proceedings.com

TABLE OF CONTENTS

VOLUME 1

VOLUME 2

Atomistic Modeling Approach to Structural Reorganization of Self-Assembled Monolayer On H-Si (111) Surface

Seung Soon Jang, School of Materials Science and Engineering, Georgia Institute of Technology, Atlanta, GA

Won Sang Koh, School of Physics, Georgia Institute of Technology, Atlanta, GA

In this talk, we present our study on amphiphilic self-assembled monolayers (SAM) on Si (111) surface which has the capability of adapting its configuration to minimize the interfacial tension for both hydrophilic and hydrophobic solvent. To simulate such amphiphilicity of SAM, we attached Y-shaped amphiphilic molecules on Si (111) surface through covalent bonds and then equilibrated the systems in the presence or absence of solvent molecules using molecular dynamics simulations. From these simulations, we found that indeed, the amphiphilic molecules attached on the Si (111) surface reorganize themselves depending on the solvent condition. This structural reorganization was also confirmed from the corresponding change of interfacial tension calculated by Kirkwood-Buff theory. We believe the importance of this study is not only because of its promising potential for various technological applications but also because of fundamental knowledge on molecular interface regarding to how it reorganizes its configuration as a function of hydrophobicity/hydrophilicity of solvent.

See more of this Session: Computational Modeling of Surfaces and Surface Phenomena
See more of this Group/Topical: Engineering Sciences and Fundamentals

The adsorption of CO_2 on a rumpled CaO (0 0 1) surface and the effect of water vapor investigated by density functional theory

Bo Gyeong Kim and Paul Blowers

Department of Chemical and Environmental Engineering

The University of Arizona

Tucson, AZ 85721-0011

Ph: 520-626-5319 Fax: 520-621-6048

Calcium oxide (CaO) is converted to calcium carbonate ($CaCO_3$) by adsorbing carbon dioxide (CO_2) to undergo carbonation in a reversible calcination process. There is enhanced adsorption in the presence of water vapor, which has been proven by experiment. While the CaO surface is widely used in industry as a highly reactive adsorbent and has been studied in many experiments, there is lack of theoretical studies to investigate the mechanism of CO_2 adsorption on the CaO. In this study, the CaO surface was investigated for CO_2 adsorption including the role of water molecules. The reaction rate constants and equilibrium constants for a CO_2 gas molecule adsorption on a solid CaO surface were calculated.

The adsorption of CO_2 onto rumpled CaO (0 0 1) surfaces was investigated with density functional theory (DFT) methods. The surface was modeled using a periodic slab designed with a 5×5×3 unit cell. CO_2 coverage for the periodic slab models was 12 % since the surface was designed with a 5×5 unit cell. This low coverage was used to approximate pure component adsorption on clean adsorbents. Structures were optimized with the Perdew-Wang (PWC) functional with the local density approximation (LDA). Adsorption energies of CO_2 binding on the CaO surface were calculated with the gradient corrected (GGA) method with the BLYP functional using the optimized lower level geometries. All atomistic adsorption models on periodic CaO surfaces used the DFT semi-core pseudopotentials method and double-numeric polarization basis set which is comparable with Gaussian 6-31G** basis set. The spin polarization effect was included in the calculations for higher quality results. Furthermore, the adsorption of CO_2 on rumpled CaO surface in the presence of water vapor was studied to investigate the effect of water vapor on the reaction. Results show how the carbonate was formed on the CaO surface by adsorbing carbon dioxide and the strong adsorption energy on the CaO surface. Water vapor as a catalyst will increase carbonation performance of the CaO surface.

Current-Driven Surface Morphological Evolution and Pattern Formation in Stressed Elastic Solid Conductors

Vivek Tomar, Department of Chemical Engineering, University of Massachusetts, Amherst, MA

M. Rauf Gungor, Department of Chemical Engineering, University of Massachusetts, Amherst, MA

Dimitrios Maroudas, Department of Chemical Engineering, University of Massachusetts, Amherst, MA

Surface morphological instability underlies various reliability problems of technologically important materials, which have a broad range of applications from aerospace engineering to microelectronics and nanotechnology. Applied mechanical stress has been known to induce surface morphological instabilities in crystalline solid materials. Specifically, linear stability analyses have demonstrated that the competition between elastic strain energy and surface energy can cause the growth of perturbations from a planar surface morphology and specified the conditions for the onset of the so-called Asaro-Tiller or Grinfeld (ATG) instability. In addition, numerical simulations accounting for all the important nonlinearities have addressed the surface dynamics beyond the instability onset and revealed instability mechanisms. In particular, it has been demonstrated that a planar surface of a stressed elastic solid can evolve rapidly into a cusped surface, with smooth tops and deep crack-like grooves by surface diffusion, in agreement with experimental observations. These theoretical predictions are consistent with experimental findings over a broad class of materials. However, the effects of the simultaneous action of an electric field on the surface morphological response of a conducting stressed solid have not been explored systematically. To address this need, this study focuses on the current-driven morphological evolution and pattern formation in stressed elastic solid conductors.

In this presentation, we report a detailed analysis of morphological stability of planar surfaces of electrically conducting, stressed elastic crystalline solids under the simultaneous action of an electric field using linear stability theory (LST) and numerical simulation. Our analysis is based on a fully nonlinear surface transport model that accounts for curvature-driven surface diffusion, surface electromigration, and stress-driven surface diffusion along with surface diffusional anisotropy. Our self-consistent dynamical simulations combine Galerkin boundary-integral computations of elastic displacement fields and electrostatic potentials with a front tracking method for monitoring surface morphological evolution. We present a detailed analysis of the effects of material properties (surface diffusional anisotropy parameters) on the morphological stability of planar surfaces of face-centered cubic (fcc) metals under the simultaneous action of mechanical stress and an electric field. We demonstrate that (a) there exists a preferred direction for the applied electric field that optimizes the surface morphological response; (b) under given electromechanical conditions, the surface morphological response of <111>-oriented surfaces of fcc metals is superior to that of <100>- and <110>- oriented surfaces; and (c) increasing the strength of the surface diffusional anisotropy, by lowering the temperature, has beneficial effects on the surface morphological stability of the stressed conducting solid under the simultaneous action of an electric field. These results have been used to derive systematic surface engineering rules, expressed by the dependence of the critical electric-field strength for stabilization of the surface morphology of the stressed solid as a function of the surface diffusional anisotropy parameters.

Based on self-consistent dynamical simulations according to the fully nonlinear model of driven surface morphological evolution, we have discovered complex aspects of surface morphological response and pattern formation that are not accounted for by LST. Specifically, we have found that, in addition to the ATG instability, a very-long-wavelength tip-splitting instability may be triggered forming a pattern of secondary ripples on the surface; such rippling occurs in the absence of electric field application, as well as for weaker-than-critical applied electric fields. Such shorter-wavelength ripples on an initially long-wavelength perturbation of a planar solid surface can accelerate surface cracking acting cooperatively with an ATG instability or they may induce surface cracking under conditions that do not trigger ATG instabilities according to LST. The critical wavelength for the onset of this secondary rippling is not predicted by LST, but it has been computed numerically. Most importantly, a stronger-than-critical electric field inhibits both the ATG and the rippling instability. Furthermore, we have characterized the morphology of the surfaces that undergo rippling instability. We have found that the number of ripples formed increases linearly with the wavelength of the initial perturbation scaled with the maximally unstable wavelength. Our analysis adds this new rippling instability to the mechanisms of stress relaxation through surface pattern formation that may lead to surface cracking.

See more of this Session: Computational Modeling of Surfaces and Surface Phenomena
See more of this Group/Topical: Engineering Sciences and Fundamentals

Electromechanically-Driven Oscillatory and Chaotic Dynamics of Void Surfaces in Metallic Thin Films

Vivek Tomar, Department of Chemical Engineering, University of Massachusetts, Amherst, MA

Jaeseol Cho, Department of Chemical Engineering and Materials Science, University of Minnesota, Minneapolis, MN

M. Rauf Gungor, Department of Chemical Engineering, University of Massachusetts, Amherst, MA

Dimitrios Maroudas, Department of Chemical Engineering, University of Massachusetts, Amherst, MA

Electromigration-driven void dynamics in metallic thin films is significant toward a fundamental understanding of interconnect failure that is responsible for materials reliability problems in integrated circuits. Recent theoretical work in this area has demonstrated extremely rich nonlinear dynamics for the electromigration-induced morphological evolution of void surfaces due to surface diffusional anisotropy and pronounced current crowding effects. In this presentation, we report a systematic comparison of the electromechanically driven void dynamics in thin films of face-centered cubic metals with two-fold and four-fold symmetry of surface diffusional anisotropy. We report oscillatory void dynamics in both cases and, most importantly, for two-fold symmetry, complex shape evolution that sets the material system on a route to chaos.

Our nonlinear analysis is based on self-consistent numerical simulations of driven migration and morphological evolution of void surfaces in metallic thin films according to a well-validated fully nonlinear model of surface mass transport in its two-dimensional implementation. The simulations account rigorously for current crowding and stress concentration effects that become particularly important in narrow metallic films, as well as surface curvature effects that are particularly strong due to the strong anisotropy of adatom diffusion on void surfaces. The mass transport problem on the void surface is solved coupled with the electrostatic and mechanical deformation problems in the conducting film that contains the morphologically evolving void, assuming that the metallic material responds to stress elastically.

Under the simultaneous action of the electric field and mechanical stress, our analysis predicts the range of parameters over which a void present at the edge of a metallic thin film is morphologically stable and translates through the film with a steady shape and at a constant speed, i.e., the moving void corresponds to a soliton or a steady state in the frame of reference moving at the void velocity. A systematic exploration of parameter space has predicted that variations in the electric-field strength, or the mechanical stress level, or the void size, or the strength of surface diffusional anisotropy past certain critical values leads to morphological transition from the above steady state to time-periodic states through Hopf bifurcations; these time-periodic sates are characterized by wave propagation on the void surface with the void migrating along the film at a constant speed. For two-fold and four-fold symmetry of surface diffusional anisotropy, these Hopf bifurcations are sub-critical and super-critical, respectively. Focusing on the effects of mechanical stress on the driven void dynamics, we have found that, in the case of two-fold symmetry, increasing the applied stress level past the Hopf point results in increased complexity in the void dynamics through a sequence of sub-critical period doubling bifurcations. More importantly, we show that, for two-fold symmetry, this period-doubling bifurcation sequence sets the system on a route to chaos and, as the stress level keeps increasing, the void dynamics becomes chaotic. For four-fold symmetry, however, increasing the stress level past the original Hopf point leads eventually to a transition to steady state through a parameter range characterized by single-period time-periodic states. Finally, we present a systematic characterization of the void asymptotic states, including strange attractors, over the region of parameter space examined.

See more of this Session: Computational Modeling of Surfaces and Surface Phenomena
See more of this Group/Topical: Engineering Sciences and Fundamentals

Organic Heterojunctions for Photovoltaic Applications: C60 Growth On Pentacene

Rebecca A. Cantrell, Chemical and Biomolecular Engineering, Cornell University, Ithaca, NY

Paulette Clancy, Cornell University, Ithaca, NY

Making reliable organic electronic devices from carbon-rich (organic) semiconductors challenges researchers worldwide. There are many hurdles, one of which is making highly crystalline thin films out of organic materials. The focus in this proposal is on organic-on-organic thin film growth, which is less studied than organic-on-metal thin film growth. Specifically, the C_{60}/Pentacene organic heterojunction is a model system due to its applications for photovoltaic devices. In addition, C_{60} and pentacene both have relatively high crystallinities, and hence relatively large charge carrier mobilities. Experimental evidence shows promising electrical properties of photovoltaic devices made from C_{60} films grown on pentacene films. One of the main areas for improvement is in making the C_{60} films grow in a crystalline, two dimensional manner on pentacene. Molecular Dynamics has shown the diffusive behavior of up to a few C_{60} molecules on the bulk-phase pentacene surface; the surface diffusion coefficients of C_{60} on pentacene are relatively liquid-like. Molecular dynamics has also shown a very interesting anisotropic diffusion behavior of C_{60} on bulk-phase pentacene around room temperature. Kinetic Monte Carlo in combination with a continuum coalescence theory aid in the understanding of how multiple C_{60} molecules grow on pentacene. Evidence of interesting faceted grain shapes bring understanding to the C_{60}/pentacene system. There are very little previous simulation studies of what happens at organic interfaces, and this paper will bring insight into the properties of such a system that would be otherwise hard to see experimentally or predict theoretically.

See more of this Session: Computational Modeling of Surfaces and Surface Phenomena
See more of this Group/Topical: Engineering Sciences and Fundamentals

Water Ordering at Low-Index Alumina Surfaces: Structure and Reactivity

Amy H. Fletcher, Energy, Environmental and Chemical Engineering, Washington University in St. Louis, St. Louis, MO

K. Alexandria Cassady, Energy, Environmental and Chemical Engineering, Washington University in St. Louis, St. Louis, MO

Cynthia S. Lo, Energy, Environmental and Chemical Engineering, Washington University in St. Louis, St. Louis, MO

The interaction of water with mineral surfaces is expected to play a major role in controlling chemical processes in biological, chemical, and environmental systems; for example, the water-aluminum oxide (Al_2O_3) interface governs reactions on biological implants, heterogeneous catalysts, and atmospheric nanoparticles. The clean and hydrated Al_2O_3 (1-102) and (11-20) surface structures and energetics are calculated using pseudopotential plane wave density functional theory. Since both alumina and hematite (Fe_2O_3) share the corundum structure, their surface terminations are expected to be similar. The hydrated Al_2O_3 (1-102) surface is found to be isostructural to the hydrated Fe_2O_3 (1-102) surface, but the nature of the energetically preferred surface termination is dependent on temperature. Initial structural models of the hydrated Al_2O_3 (11-20) surface are also compared to experimental data on the hydrated Fe_2O_3 (11-20) surface. These results of water ordering at alumina surfaces are an important step towards a molecular scale understanding of the factors that dictate surface structure and reactivity, which would greatly aid in the development of robust models of interfacial processes in the environment.

See more of this Session: Computational Modeling of Surfaces and Surface Phenomena
See more of this Group/Topical: Engineering Sciences and Fundamentals

First-Principles Study of Small Molecule Adsorption On $SrTiO_3$ Surfaces

Beverly Brooks Hinojosa, Department of Chemical Engineering, University of Florida, Gainesville, FL

Aravind R. Asthagiri, Department of Chemical Engineering, University of Florida, Gainesville, FL

The adsorption and reactivity of low-coordinated sites on oxide surfaces is of interest in a range of catalyst applications. In this contribution, we examine representative probe molecules, H_2O and NH_3, on the perovskite $SrTiO_3$ surfaces using density functional theory (DFT). We have characterized adsorption on the low-Miller index surfaces to examine the effect of polar versus non-polar surfaces. We will report on DFT calculations of H_2O and NH_3 adsorption and disassociation on all terminations of the $SrTiO_3(100)$, $SrTiO_3(110)$ and $SrTiO_3(111)$ surfaces. We have examined the effects of surface coverage and termination on the favorability of molecular or dissociative adsorption and the resulting adsorption structures. Where possible, our DFT results will be compared to available ultra high vacuum experimental data to provide further insight into adsorption of small molecules on various $SrTiO_3$ surfaces.

See more of this Session: Computational Modeling of Surfaces and Surface Phenomena
See more of this Group/Topical: Engineering Sciences and Fundamentals

Interfacial Phenomena in Miniemulsions and Applications

Mohamed S. El-Aasser, Department of Chemical Engineering, Lehigh University, Bethlehem, PA

Miniemulsions provide an alternate route to preparing a wide array of polymer colloids (latexes) for applications ranging from large volume usage in paints, adhesives and coatings to small volume, high value products such as in medical diagnostics and therapeutics, and drug delivery systems. They are particularly useful for preparing materials that cannot be made by conventional emulsion polymerization processes. Miniemulsions differ from conventional emulsions in their droplet size (50 - 500 nm vs. 1 - 10 microns) and their increased stability brought about by the addition of a costabilizer, a low molecular weight and low water solubility compound. Two types of costabilizer, namely a fatty alcohol (cetyl alcohol) or an alkane (hexadecane), have been successfully used in preparing miniemulsions. The former system acts in a complex manner to bring about stability against both diffusional degradation (Ostwald ripening) and aggregation by organizing with the surfactant (sodium dodecyl sulfate) primarily at the oil/water interface and much effort has been made to characterize this interfacial behavior. The latter acts primarily to retard Ostwald ripening owing to its presence within the droplets. Polymerization occurs by direct entry of free radicals into the small monomer droplets and efforts to explain the polymerization kinetics both without (reduced rate) and with a small amount of added polymer in the droplets (increased rate referred to as enhanced nucleation) have been made. Further fundamental efforts are underway to better understand the mechanism of droplet formation; adsorption at the interface is the prime suspect being investigated in terms of the rate of surfactant adsorption, desorption, and energetics. Applications of miniemulsion technology are being advanced in many areas to prepare unconventional polymer-based colloids such as synthetic and hybrid latexes, high solids latexes, polymerization of highly water-insoluble monomers and macromonomers, controlled molecular weight via living free radical polymerization, controlled polymer microstructure and morphology, and encapsulation and entrapment of inorganic particles, metal particles, dyes, drugs, and perfumes. Our studies have shown that interfacial behaviors are key to the success of the encapsulation process of nanoparticles via the miniemulsion process.

See more of this Session: Interfacial Phenomena Plenary Session
See more of this Group/Topical: Engineering Sciences and Fundamentals

Directed Self-Oriented Self-Assembly of Block Copolymers: Bottom-up Meeting Top-Down

Thomas P. Russell, Polymer Science and Engineering Department, University of Massachusetts, Amherst, MA

As the size scale of device features becomes increasingly smaller, conventional lithographic processes are limited. Alternative routes need to be developed to circumvent this hard stop. Ideally, if existing technological processes can be used with novel materials, significant advances can be made. Block copolymers (BCP), two polymer chains covalently linked together at one end, provide one solution. BCP self-assemble into a range of highly-ordered morphologies and by controlling the orientation and lateral ordering of the nanoscopic microdomains, numerous applications will emerge. By combining the "bottom-up" self-assembly of BCPs with "top-down" micro-fabrication processes faster, better and cheaper devices can be generated in very simple, yet robust, ways.

See more of this Session: Interfacial Phenomena Plenary Session
See more of this Group/Topical: Engineering Sciences and Fundamentals

Breathing New Life Into Classical Colloid Science

Joesph A. Zasadzinski, Department of Chemical Engineering, University of California, Santa Barbara, CA

Lung surfactant is a mixture of lipids and specific proteins that forms monolayer films at the air-liquid interface in the alveoli that reduce surface tension, allowing easy breathing. Surfactant deficiency results in serious health problems, including Acute respiratory distress syndrome, which affects about 150,000 in the US each year with a 35% mortality rate. ARDS causes capillaries to leak, resulting in higher than normal concentrations of blood proteins in the alveolar fluids. Many of these proteins are surface-active - for example, albumin acts like a simple surfactant, lowering the surface tension by about 20 mN/m at milligram concentrations. Lung surfactant adsorption is strongly inhibited by the competitive adsorption of these proteins to the interface. This reduction in lung surfactant adsorption leads to an increased surface tension in the lung and an increased work of breathing, which is likely an important factor in the development and severity of ARDS.

The factors influencing surfactant adsorption in the presence of a protein film are surprisingly similar to those that determine colloid stability. Albumin adsorbed to the interface induces an electrostatic energy barrier to surfactant diffusion of order 5-6 kBT, leading to a reduction in adsorption equivalent to reducing the surfactant concentration by a factor of more than 100! Adding hydrophilic, non-adsorbing polymers such as polyethylene glycol to the subphase provides a "depletion attraction" between the surfactant aggregates and the interface that eliminates the energy barrier; the same polymers are used to flocculate colloidal particles. Surfactant adsorption increases exponentially with polymer concentration as predicted by the simple Asakura and Oosawa model of depletion attraction. Surfactant adsorption is also strongly affected by the ionic strength and valence (z) of added electrolytes: the ion concentration needed to restore surfactant adsorption scales as z-6, the same scaling as the classic Schulz-Hardy rule for colloid stability. Polyelectrolytes also show the same behavior in enhancing competitive adsorption as in colloid stability; oppositely charged polyelectrolytes first enhance surfactant adsorption, then as the polyelectrolyte concentration is increased, reduce adsorption, consistent with charge overcompensation.

See more of this Session: Interfacial Phenomena Plenary Session
See more of this Group/Topical: Engineering Sciences and Fundamentals

Delta-Shock, a New Type of Composition Front in Nonlinear Chromatography

Marco Mazzotti, ETH Zurich, Zurich, Switzerland

The local equilibrium theory of chromatography is applied to study binary systems subject to the mixed generalized Langmuir isotherm:

$$n_i = \frac{H_i c_i}{1 - K_1 c_1 + K_2 c_2} \qquad (i = 1, 2)$$

(with $H_2 > H_1$), where component 1 and 2 are anti-Langmuir-like and Langmuir-like, respectively. Riemann problems, where a stream of a given composition is fed to a column initially saturated at a different composition, are analyzed. Within the hyperbolic region of the hodograph plane close to the origin it is shown that such Riemann problems admit not only classical solutions, consisting of constant states separated by simple wave and shock wave transitions, but also non-classical solutions, namely including a continuous non-simple wave transition and a delta-shock [1]. The latter can be viewed as a travelling spike of growing strength superimposed to the discontinuity separating the initial and the feed states. Exact conditions for the occurrence of classical and non-classical solutions are derived. Thus, this work adds two new types of composition fronts to the classical ones already known in nonlinear chromatography.

In the case of the delta-shock, the most striking new transition, it is possible to derive exact relationships for its properties, including the slope of its image in the physical plane, i.e. the reciprocal of its propagation speed:

$$\frac{dt}{dz} = \frac{1}{V}\left(1 + \nu \frac{H_1 K_2 [n_2] - H_2 K_1 [n_1]}{H_1 K_2 [c_2] - H_2 K_1 [c_1]}\right)$$

where the symbol $[\cdot]$ denotes the jump across the discontinuity of the quantity enclosed. These analytical expressions match the results obtained through numerical simulations. The physical legitimacy of the model considered in this work is demonstrated theoretically, and is further supported by the experimental evidence of the existence of delta-shocks that has been recently provided [2].

[1] Mazzotti M., Non-classical composition fronts in nonlinear chromatography - Delta-shock, *Ind. Eng. Chem. Res.* (2009) submitted.

[2] Mazzotti M., Tarafder A., Gritti F., Guiochon G., Experimental evidence of a delta-shock in nonlinear chromatography, *J. Chromatogr. A* (2009) submitted.

See more of this Session: Mathematical Modeling of Transport Processes
See more of this Group/Topical: Engineering Sciences and Fundamentals

Semi-Analytical Solution of Nonlinear Diffusion Equations

Simon L. Marshall
CSIRO Mathematical and Information Sciences
Gate 5, Normanby Road
Clayton, Victoria, 3168
Australia

Abstract

The nonlinear diffusion equations that describe many practically-important phenomena offer very limited scope for the development of analytical solutions. While application of the well known Kirchhoff integral transformation allows the corresponding steady state concentration profiles to be expressed in terms of harmonic functions, nonlinearities remain in the transient diffusion equation. A pseudotime integral transformation reduces the nonlinear equation for the Kirchhoff transformation variable into a linear diffusion equation. The purpose of this paper is to show how the Kirchhoff and pseudotime transformations can be used to relate the desired nonlinear solution to the solution of the corresponding linear equation, for which many analytical results are available in domains of simple geometry. The concentration-dependence of the diffusivity gives rise to a local distortion of the time scale, which can be used to establish a nonlinear mapping between the solutions of the nonlinear equation and the corresponding linear equation. If an analytical solution is available for the linear equation, the solution of the nonlinear equation thus reduces to a one-dimensional quadrature and solution of a single nonlinear algebraic equation, thereby providing a highly efficient and accurate computational method that does not rely on any finite-difference approximations.

1 Introduction

The diffusive phenomena that occur in many branches of the chemical process industries are usually nonlinear. This nonlinearity could result from the dependence of the relevant diffusivity on the concentration of the diffusing species, from a fundamental in applicability of the linear constitutive relationship between the flux and concentration gradient, or perhaps a combination of both factors. Whatever the cause, construction of mathematical models of such processes requires application of numerical techniques such as finite-difference or finite element algorithms. Such calculations produce typically large sets of numerical data, from which it is sometimes difficult to extract the information of interest.

Interest in the analytical solution of nonlinear diffusion equations has persisted in spite of the current availability of computers and software capable of accurate and efficient numerical solutions. The first major advances towards this end were made by Kirchhoff [1894], who developed a way of transforming steady nonlinear equations to Laplace's equation, and Boltzmann [1894], who discovered a transformation that converts the nonlinear PDE in space and time coordinates to an ODE in the similarity variable. The Kirchhoff transformation was used in the pioneering analog numerical solutions by Eyres et al. [1946] of the transient heat-conduction equation, and forms the basis for the finite-difference discretization scheme described in *Numerical Recipes* [Press et al., 1992], and boundary element methods developed by Wrobel and Brebbia [1987] and Satravaha and Zhu [1997]. An account of similarity solutions for variety of nonlinear diffusion models in one dimensional semiinfinite regions (to which such approaches are, in fact, restricted) is presented by Crank [1956, chapter 9]. A numerical method developed more recently by Lee [1972a,b] involves application of the Kirchhoff transformation to the ODE resulting from the similarity transformation. Application of Lie group theory was first described by Ovsyannikov [1959], and inspired many subsequent works (such as that of Ivanova and Sophocleous [2006] and references cited therein) directed towards the identification of nonlinearities for which analytically soluble transformation groups can be constructed. A variety of intriguing alternative approaches such as antireduction [Fushchych and Zhdanov, 1994], generalized separation of variables [Doyle and Vassiliou, 1998; Polyanin and Zhurov, 2002; Vyaz'mina and Polyanin, 2006] and operator expansion methods [Adomian, 1988; Wazwaz, 2000, 2001] have also been described.

The imposition of boundary conditions on the solutions determined by these analytical transformations usually undoes the advantages realised by linearising the equation, thereby greatly limiting the usefulness of such techniques in describing practically important phenomena. These restrictions are further compounded by the general inapplicability of the superposition principle underlying the solution of the corresponding problems for the linear equation.

Although the neglect of nonlinear effects cannot, in general, be justified on either mathematical or physical grounds, many solutions of linear diffusion models offer compelling advantages of simplicity and ease of use. For example, data on the penetration of chlorides into concrete have been fitted to the classical error-function solution, in which time dependence of diffusivity on time and/or concentration is introduced after the fact [Costa and Appleton, 1999a,b]. It would be highly desirable if there were a more rigorous way of determining solutions of nonlinear diffusion equations from solutions of the corresponding linear problem. It is the purpose of this paper to describe such a scheme.

2 Analysis

This paper considers solution of nonlinear diffusion equations of the form

$$\frac{\partial c}{\partial t} = \nabla \cdot [D(c)\nabla c], \tag{2-1}$$

with a concentration dependent diffusivity of the form $D(c) = D_0 g(c)$, where D_0 is a reference diffusivity and $g(c)$ is a dimensionless function of concentration determined by the dependence of the activity coefficient of the diffusing species on its concentration. Specification of a reference length L allows the introduction of dimensionless spatial coordinates $X_i = x_i/L$, and a dimensionless time $\tau = D_0 t/L^2$, in terms of which the equation becomes

$$\frac{\partial c}{\partial \tau} = \nabla \cdot (g(c)\nabla c), \tag{2-2}$$

in which the gradients are taken with respect to the X_i. The changes in the spatial and temporal variables that convert this to the linear diffusion equation

$$\frac{\partial y}{\partial v} = \nabla^2 y \equiv \nabla \cdot \nabla y \tag{2-3}$$

can be derived as shown in the following paragraphs.

2.1 Kirchhoff Transformation

As first shown by Kirchhoff [1894] (see also Carslaw and Jaeger [1959, page 11]), the substitution

$$\nabla y = g(c)\nabla c \Longrightarrow y = \int_0^c g(c')dc' \tag{2-4}$$

converts the equation to

$$g(c)\frac{\partial c}{\partial v} = \nabla \cdot [g(c)\nabla c]. \tag{2-5}$$

Although the steady-state solutions of this nonlinear equation can be obtained without difficulty, transient solutions require application of a suitable numerical method.

2.2 Pseudotime Transformation

If a new time scale τ is introduced such that

$$g(c) = \frac{dv}{d\tau} \Longrightarrow v = \int_0^\tau g(c)d\tau', \tag{2-6}$$

application of the chain rule in the form

$$g(c)\frac{\partial c}{\partial v} = \frac{\partial c}{\partial v}\frac{dv}{d\tau} = \frac{\partial c}{\partial \tau}, \tag{2-7}$$

results in the nonlinear diffusion equation 2-2.

2.3 Inverse Transformations

Aternatively, proceeding in the opposite direction, application of the Kirchhoff transformation to this equation results in

$$\frac{1}{g(c)}\frac{\partial y}{\partial \tau} = \frac{1}{f(y)}\frac{\partial y}{\partial \tau} = \nabla^2 y, \tag{2-8}$$

where the new function $f(y)$ results from rewriting the concentration dependence function $g(c)$ in terms of the Kirchhoff transform variable y: $f(y) = g(c)$. The resulting quasilinear diffusion equation

$$\frac{1}{f(y)}\frac{\partial y}{\partial \tau} = \nabla^2 y \tag{2-9}$$

can now be linearized by introduction of the pseudotime variable defined by the equivalent definitions

$$v = \int_0^{\tau} f(y) d\tau' \qquad \text{or} \qquad \frac{d\tau}{dv} = \frac{1}{f(y)} \tag{2-10}$$

If the function f were independent of y, v would be proportional to τ. Another limiting case, in which the diffusivity is independent of concentration but depends explicitly on time, is mentioned briefly in the monograph by Crank [1956, section 9.1]. But in the most general case, the above manipulations show that the dependence of the diffusivity on concentration or pseudoconcentration y causes a distortion of the time scale, which depends on position. To emphasise this point, the integrand should preferably be written as $f(y(\mathbf{X}, \tau))$; as a result, the determination of the value of τ corresponding to a given value of v requires solution of a system of nonlinear equations [Wrobel and Brebbia, 1987].

2.4 A Novel Implementation

The relation between the actual time and pseudotime can also be expressed by equivalent definition

$$\tau = \int_0^{v} \frac{1}{f(y)} du \qquad \frac{d\tau}{dv} = \frac{1}{f(y)}, \tag{2-11}$$

which can be regarded as a nonlinear equation for v in terms of τ. This proves to be much more computationally convenient. Thus, if the boundary value problem for y is solved as a function of position and pseudotime (either analytically or by a finite difference or finite element algorithm), the value of τ corresponding to each position and value of y can be determined by numerical quadrature. Alternatively, if the value of τ is specified, the corresponding value of the pseudotime follows by numerical solution of a nonlinear algebraic equation for v. If an analytical solution of the linear boundary value problem is available, the solution of this nonlinear equation is especially well suited to Newton-Raphson iteration, since the derivative required for this purpose is simply the integrand. If the Kirchhoff transformation can be inverted analytically, solution of the original PDE for c thus reduces to a numerical quadrature and solution of a single nonlinear equation for the pseudotime, at each spatial point of interest.

3 Thermodynamic Considerations

While the formal manipulations in the preceding paragraphs are successful in linearizing the nonlinear diffusion equation 2-2, these transformations are useful only if they can be inverted. It is evident from equation 2-4 that the transformation is invertible if $g(c) > 0$, since the integral in this equation increases with increasing c, and each value of y corresponds to a unique c. While such considerations are of obvious fundamental importance, they are not particularly helpful in the identification of a physically meaningful form for the concentration dependence of the diffusivity. This can be achieved more easily by application of relevant principles of thermodynamics. In these terms, the driving force for diffusive transport of matter is identified as the negative gradient of the chemical potential, μ. This gradient provides, in effect, a force acting on each mole of solute, which can also be regarded as the product of a velocity $\mathbf{v}$ and a drag coefficient K. Thus, in a binary mixture,

$$\mathbf{v}K = -\nabla\mu, \tag{3-1}$$

and assuming that the overall molar density of the mixture is uniform, the flux of solute is

$$\mathbf{J} = c\mathbf{v} = -\frac{c}{K}\nabla\mu. \tag{3-2}$$

The chemical potential gradient can be converted into a concentration gradient by differentiation of the chemical potential with respect to c, making use of the definition

$$\mu = \mu^0 + RT\ln fc, \tag{3-3}$$

where f is the molar-scale activity coefficient. The result is the generalized Fick Law

$$\mathbf{J} = -\frac{RT}{K}\left[1 + c\frac{d\ln f}{dc}\right]\nabla c, \tag{3-4}$$

from which it follows that the function $g(c)$ appearing in the earlier derivation is

$$g(c) = 1 + \frac{d\ln f}{dc}, \tag{3-5}$$

and that $D_0 = RT/K$. Thus, although the Kirchhoff transformation as defined in the preceding section is valid and invertible for a very wide range of functions g, the above result shows that g is ultimately determined by the equation of state of the system, and is therefore subject to the constraints imposed on f, as a function of c, by the Gibbs-Duhem equation.

Application of the Kirchhoff transformation to the above general form of concentration dependence leads to the establishment of some intriguing relationships between the transform variable y and the properties of the solute or the solvent in a binary solution. For example, in

$$y = \int_0^c \left[1 + c\frac{d\ln f}{dc}\right] dc, \tag{3-6}$$

integration by parts leads to

$$y = (1 + \ln f)c - \int_0^c \ln f dc. \tag{3-7}$$

From this, it is also clear that for an ideal solution, with $\ln f = 0$, $y = c$, so the Kirchhoff transformation reduces to an identity.

These ideas can be made somewhat more definite by consideration of possibly the simplest model for a nonideal solution, for which

$$\ln f = e_1 c, \tag{3-8}$$

where e_1 is a constant that could be determined from direct vapor pressure measurements or isopiestic equilibration experiments. For a real solute, such a relationship is expected to be valid over a rather limited concentration range. Applying the definition, it is found without difficulty that

$$g(c) = 1 + e_1 c \Longrightarrow y = \int_0^c (1 + e_1 c')dc' = c + \frac{1}{2}e_1 c^2. \tag{3-9}$$

Inversion of the Kirchhoff transformation requires solution of the equation

$$\frac{1}{2}e_1 c^2 + c - y = 0 \Longrightarrow c = \frac{-1 \pm \sqrt{1 + 2e_1 y}}{e_1}, \tag{3-10}$$

from which it is evident that only the positive root is physically significant. Thus,

$$c = \frac{1}{e_1}\left[\sqrt{1 + 2e_1 y} - 1\right], \tag{3-11}$$

in terms of which the concentration dependence of the diffusivity is

$$f(y) = \sqrt{1 + 2e_1 y}. \tag{3-12}$$

Does this make physical sense? The fact that the activity coefficient, and the chemical potential, increase with concentration means that the solute will be energetically 'happier' in a dilute solution than in a concentrated solution. This implies an overall repulsive effect exerted by the solute molecules on each other, which would be intuitively associated with an enhanced tendency to diffuse away from each other.

4 Nonlinear Diffusion into a Cylinder

An example of a practically significant problem to which this approach could be applied is the diffusion of sucrose into a porous cylindrical foodstuff from a solution of concentration c_1 into which it is immersed. The concentration profile of sucrose as a function of radial coordinate and time is described by the following nonlinear boundary value problem

$$\frac{\partial c}{\partial t} = \frac{1}{r}\frac{\partial}{\partial r}\left[rD(c)\frac{\partial c}{\partial r}\right]: \quad c(a,t) = c_1, \quad c(r,0) = 0, \tag{4-1}$$

in the infinite circular cylindrical region $r \in [0, a]$. From the isopiestic measurements by Robinson and Stokes [1961] and densities reported by Cerdeirina et al. [1997], equation 3-8 is found to be followed reasonably well up to c=1 mol/L, with e_1=0.45 L/mol, so with c_1=1 mol/L, β=0.45. The concentration dependence of diffusivity is $D(c) = D_0(1 + e_1 c)$, where the diffusivity D_0 is the infinite-dilution diffusivity of sucrose in water, multiplied by an empirical correction factor determined by the porosity or permeability of the foodstuff. This problem can be transformed by introduction of the dimensionless variables $C = c/c_1$, $R = r/a$, and $\tau = D_0 t/a^2$ to

$$\frac{\partial C}{\partial \tau} = \frac{1}{R}\frac{\partial}{\partial R}\left[R(1+\beta C)\frac{\partial C}{\partial R}\right]: \qquad C(1,\tau) = 1, \quad C(R,0) = 0, \tag{4-2}$$

where $\beta = e_1 c_1$. Introduction of the Kirchhoff transformation variable

$$y = \int_0^C (1+\beta C)dC = C + \frac{1}{2}\beta C^2, \qquad C = \frac{1}{\beta}\left[\sqrt{1+2\beta y} - 1\right], \tag{4-3}$$

for which

$$y(1,\tau) = y_1 = 1 + \frac{\beta}{2} \quad \text{and} \quad f(y) = \sqrt{1+2\beta y}, \tag{4-4}$$

gives

$$\frac{1}{\sqrt{1+2\beta y}}\frac{\partial y}{\partial \tau} = \frac{1}{R}\frac{\partial}{\partial R}\left[R\frac{\partial y}{\partial R}\right]: \qquad y(1,\tau) = y_1, \quad y(R,0) = 0, \tag{4-5}$$

and in terms of the pseudotime defined by

$$\frac{d\tau}{dv} = \frac{1}{\sqrt{1+2\beta y}} \Longrightarrow \int_0^v \frac{du}{\sqrt{1+2\beta y(R,u)}} = \tau, \tag{4-6}$$

the transformed boundary value problem is

$$\frac{\partial y}{\partial v} = \frac{1}{R}\frac{\partial}{\partial R}\left[R\frac{\partial y}{\partial R}\right]: \qquad y(1,v) = y_1, \quad y(R,0) = 0. \tag{4-7}$$

From Carslaw and Jaeger [1959, page 199, equation 8], the solution is

$$y(R,v) = y_1\left[1 - 2\sum_{n=1}^{\infty} e^{-\alpha_n^2 v}\frac{J_0(\alpha_n R)}{\alpha_n J_1(\alpha_n)}\right] \tag{4-8}$$

where J_0 and J_1 are Bessel functions of the first kind, of order zero and one, respectively, and α_n is the nth positive root of J_0. The solution of the nonlinear problem for C is therefore

$$C = \frac{1}{\beta}\left[\sqrt{1 + 2\beta y} - 1\right], \tag{4-9}$$

where

$$y(R,v) = \left(1 + \frac{\beta}{2}\right)\left[1 - 2\sum_{n=1}^{\infty} e^{-\alpha_n^2 v}\frac{J_0(\alpha_n R)}{\alpha_n J_1(\alpha_n)}\right], \tag{4-10}$$

and v satisfies

$$\int_0^v \left\{1 + (2+\beta)\left[1 - 2\sum_{n=1}^{\infty} e^{-\alpha_n^2 u}\frac{J_0(\alpha_n R)}{\alpha_n J_1(\alpha_n)}\right]\right\}^{-1/2} du = \tau. \tag{4-11}$$

That this approaches the correct limiting value can be verified by observing first that $y \to 1 + \beta/2$ as $v \to \infty$. When y is set equal to this value, $\sqrt{1 + 2\beta y} = \sqrt{1 + 2\beta + \beta^2} = 1 + \beta$, so that the corresponding limiting value of C is $[1 + \beta - 1]/\beta = 1$, as required.

The numerical calculations in the above scheme can all be performed with a cubic spline subroutine, such as that given in *Numerical Recipes* [Press et al., 1992], extended to provide interpolated estimates of the integral of a function defined by a sequence of discrete numerical values. For example, the exact solution equation 4-8 and the integrand of equation 4-11 are evaluated for $R = 0$ and $v_k = 0.01(k-1)$, where k=1, ... 100, and the cubic spline integration used to determine corresponding values I_k of the integral. Numerical results are shown in Table 1. Next, the coefficients are calculated for the data points (I_k, v_k), and solution of equation 4-11 follows by evaluation of the spline at points $\tau_k = 0.01k$, k=1, 2,..., up to an upper limit determined by I_{100} (which in this example is about 0.74.) Results are shown in Table 2.

Table 1: Numerical Integration of $1/f(y(0,v)$ with respect to v, by cubic spline quadrature.

k	v_k	$y(0, v_k)$	$f(y(0, v_k))$	$\int_0^{v_k} 1/f(y(0,u)\,du$
1	0.00	0.000000	1.000000	0.000000
2	0.01	0.000000	1.000000	0.010000
3	0.02	0.000009	0.999996	0.020000
4	0.03	0.000573	0.999742	0.029999
5	0.04	0.004564	0.997953	0.039990
6	0.05	0.015803	0.992963	0.049947
7	0.06	0.036087	0.984146	0.059836
8	0.07	0.064963	0.971989	0.069619
9	0.08	0.100798	0.957511	0.079268
10	0.09	0.141660	0.941766	0.088765
11	0.10	0.185765	0.925613	0.098102
12	0.11	0.231644	0.909663	0.107278
13	0.12	0.278160	0.894304	0.116297
14	0.13	0.324471	0.879761	0.125167
15	0.14	0.369975	0.866141	0.133896
16	0.15	0.414253	0.853477	0.142493
17	0.16	0.457029	0.841756	0.150968
18	0.17	0.498130	0.830937	0.159331
19	0.18	0.537459	0.820966	0.167590
20	0.19	0.574976	0.811781	0.175753
21	0.20	0.610679	0.803320	0.183828
...	...	...	...	...
31	0.30	0.878953	0.747214	0.260996
41	0.40	1.030851	0.720233	0.334215
51	0.50	1.116110	0.706313	0.405469
61	0.60	1.163930	0.698850	0.475689
71	0.70	1.190749	0.694767	0.545350
81	0.80	1.205791	0.692508	0.614703
91	0.90	1.214227	0.691251	0.683885
100	0.99	1.218598	0.690602	0.746065

Table 2: Values of the Pseudotime determined by inverse cubic spline interpolation.

τ	$\upsilon(\tau)$	τ	$\upsilon(\tau)$	τ	$\upsilon(\tau)$
0.01	0.010000	0.31	0.366537	0.61	0.793210
0.02	0.020000	0.32	0.380316	0.62	0.807650
0.03	0.030001	0.33	0.394152	0.63	0.822095
0.04	0.040010	0.34	0.408041	0.64	0.836545
0.05	0.050053	0.35	0.421977	0.65	0.850999
0.06	0.060167	0.36	0.435958	0.66	0.865456
0.07	0.070392	0.37	0.449980	0.67	0.879917
0.08	0.080765	0.38	0.464040	0.68	0.894381
0.09	0.091313	0.39	0.478135	0.69	0.908847
0.10	0.102054	0.40	0.492262	0.70	0.923317
0.11	0.113000	0.41	0.506418	0.71	0.937789
0.12	0.124155	0.42	0.520602	0.72	0.952263
0.13	0.135518	0.43	0.534811	0.73	0.966739
0.14	0.147085	0.44	0.549043	0.74	0.981217
0.15	0.158850	0.45	0.563297	-	-
0.16	0.170805	0.46	0.577570	-	-
0.17	0.182941	0.47	0.591862	-	-
0.18	0.195246	0.48	0.606170	-	-
0.19	0.207713	0.49	0.620493	-	-
0.20	0.220329	0.50	0.634831	-	-
0.21	0.233087	0.51	0.649181	-	-
0.22	0.245975	0.52	0.663544	-	-
0.23	0.258986	0.53	0.677917	-	-
0.24	0.272110	0.54	0.692301	-	-
0.25	0.285340	0.55	0.706694	-	-
0.26	0.298668	0.56	0.721095	-	-
0.27	0.312086	0.57	0.735505	-	-
0.28	0.325588	0.58	0.749922	-	-
0.29	0.339168	0.59	0.764345	-	-
0.30	0.352819	0.60	0.778775	-	-

The nonlinearity resulting from the concentration dependence of the diffusivity has the effect of causing the pseudotime to differ from the actual time. In this regard, it is particularly interesting to observe that the deviations between these two quantities are insignificant at very early times, becoming appreciable (greater than 1%) for τ greater than about 0.10.

The above calculations can be performed in a variety of ways. For example, it would be possible to evaluate the integral in the first step by application of Simpson's rule to consecutive groups of three points, and determine the pseudotime by inverse parabolic interpolation. Further improvements of accuracy could be achieved by application of a superior integration rule, such as a Gaussian Quadrature formula, in combination with a Newton-Raphson iteration to solve the equation for the pseudotime. In principle, equations in the form of 4-11 are especially well suited to treatment in this manner, since the required derivative is just the integrand, and the rate of convergence can be greatly improved by use of an inverse interpolation scheme to provide an initial estimate of the solution.

A particularly important observation is that the method does not depend on the availability of an analytical solution for the corresponding linear problem. Thus, the solutions of the linear problem could just as easily have been obtained as the output of a finite-difference or finite-element program, as would be required in a region of complicated geometry. Such cases would appear to be better treated by the cubic spline method described above; application of a Gauss Quadrature rule would be less convenient, since the required values of the function at the node points are less easily obtained.

5 Conclusions

The scheme described in this paper allows simple and efficient solutions of nonlinear diffusion equations to be generated from a known concentration dependence of the diffusivity and an analytical or numerical solution of the corresponding linear problem. Since no finite difference approximations are required, in either the temporal or spatial dimensions, higher accuracy is achievable by the propsed method than by the finite-difference or finite-element methods that are commonly applied to such problems. The nonlinearities specific to the physics of the problem are, in effect, decoupled from the solution of the partial differential equation, in a way that is not permitted by conventional approaches.

Acknowledgments

Financial support of the CSIRO Future Manufacturing National Research Flagship is gratefully acknowledged.

References

G. Adomian. A review of the decomposition method in applied mathematics. *Journal of Mathematical Analysis and Applications*, 135:501–544, 1988.

Ludwig Boltzmann. Zur Integration der Diffusionsgleichung bei variablen Diffusionskoeffizienten. *Wiedemanns Annalen der Physik*, 53:959–964, 1894.

H.S. Carslaw and J.C. Jaeger. *Conduction of Heat in Solids.* Oxford University Press, New York, 2nd edition, 1959.

C.A. Cerdeirina, E. Carballo, C.A. Tovar, and L. Romani. Thermodynamic properties of aqueous carbohydrate solutions. *Journal of Chemical and Engineering Data*, 42(1):124–127, 1997.

A. Costa and J. Appleton. Chloride penetration into concrete in marine environment - part I: main parameters affecting chloride penetration. *Materials and Structures / Matériaux et Constructions*, 32, 252–259 1999a.

A. Costa and J. Appleton. Chloride penetration into concrete in marine environment - part II: prediction of long term chloride penetration. *Materials and Structures / Matériaux et Constructions*, 32:354–359, 1999b.

J. Crank. *The Mathematics of Diffusion.* Oxford University Press, London, 1956.

Philip W. Doyle and Peter J. Vassiliou. Separation of variables for the 1-dimensional nonlinear diffusion equation. *International Journal of Non-linear Mechanics*, 33(2):315–326, 1998.

N.R. Eyres, D.R. Hartree, J. Ingham, R. Jackson, R.J. Sarjant, and J.B. Wagstaff. The calculation of variable heat flow in solids. *Philosophical Transactions of the Royal Society of London A*, 240:1–57, 1946.

W.I. Fushchych and R.Z. Zhdanov. Antireduction and exact solutions of nonlinear heat equations. *Journal of Nonlinear Mathematical Physics*, 1 (1):60–64, 1994.

N.M. Ivanova and C. Sophocleous. On the group classification of variable-coefficient diffusion-convection equations. *Journal of Computational and Applied Mathematics*, 197:322–344, 2006.

Gustav Robert Kirchhoff. *Vorlesungen über die Theorie der Wärme*. B.G. Teubner, Leipzig, 1894.

C.F. Lee. On the solution of some diffusion equations with concentration-dependent diffusion coefficients - II. *Journal of the Institute of Mathematics and its Applications*, 10:129–133, 1972a.

C.F. Lee. On the existence of solutions of a nonlinear differential diffusion system. *Proceedings of the Royal Society of Edinburgh A*, 71:1–7, 1972b.

L.V. Ovsyannikov. Gruppovye svoistva uravneniya nelineinoi teploprovodnosti. *Doklady Akademii Nauk SSSR*, 125(3):492–495, 1959.

A.D. Polyanin and A.I. Zhurov. Methods of generalized and functional separation of variables in hydrodynamic and heat- and mass-transfer equations. *Theoretical Foundations of Chemical Engineering*, 36(3):201–213, 2002.

William H. Press, Saul A. Teukolsky, William T. Vetterling, and Brian P. Flannery. *Numerical Recipes in FORTRAN 77: The Art of Scientific Computing*. Cambridge University Press, Cambridge, 2 edition, 1992.

R.A. Robinson and R.H. Stokes. Activity coefficients in aqueous solutions of sucrose, mannitol and their mixtures at 25°. *Journal of Physical Chemistry*, 65:1954–1958, 1961.

Pornchai Satravaha and Songping Zhu. An application of the LTDRM to transient diffusion problems with nonlinear material properties and nonlinear boundary conditions. *Applied Mathematics and Computation*, 87: 127–160, 1997.

E.A. Vyaz'mina and A.D. Polyanin. New classes of exact solutions to general nonlinear diffusion-kinetic equations. *Theoretical Foundations of Chemical Engineering*, 40(6):555–563, 2006.

Abdul-Majid Wazwaz. A new algorithm for calculating adomian polynomials for nonlinear operators. *Applied Mathematics and Computation*, 111:53–69, 2000.

Abdul-Majid Wazwaz. Exact solutions to nonlinear diffusion equations obtained by the decomposition method. *Applied Mathematics and Computation*, 123:109–122, 2001.

L.C. Wrobel and C.A. Brebbia. The dual reciprocity boundary element formulation for nonlinear diffusion problems. *Computer Methods in Applied Mechanics and Engineering*, 65:147–164, 1987.

Low-Dimensional Models for Taylor Dispersion

Ram R. Ratnakar, Department of Chemical and Biomolecular Engineering, University of Houston, Houston, TX

Vemuri Balakotaiah, Department of Chemical and Biomolecular Engineering, University of Houston, Houston, TX

We use the Liapunov-Schmidt technique of classical bifurcation theory to derive a low-dimensional model for describing the time evolution of a non-reactive tracer in laminar flow through a tube. Unlike the other averaging techniques, the Liapunov Schmidt formalism leads to an exact averaged model which is valid for all times and converges for all values of parameters with any arbitrary initial or inlet conditions, including points sources in space and time, respectively. This model is consistent with the physics of system as the single mode combined model is parabolic in terms of cross-sectional average concentration and hyperbolic in terms of cup-mixing concentration after truncating up to the second order spatial and temporal derivatives. We also analyze the temporal evolution of spatial moments for the general initial release of a tracer and show that it does not have the centroid displacement or variance deficits predicted by other methods. The moment analysis suggests the existence of a critical time τ_{CD} named as convection-diffusion time which characterizes the dominance of diffusion over convection processes.

See more of this Session: Mathematical Modeling of Transport Processes
See more of this Group/Topical: Engineering Sciences and Fundamentals

MATHEMATICAL MODELING OF DRUG RELEASE: ANALYSIS OF THE EFFECT OF ABSORPTION RATE AND FINITE DISSOLUTION RATE ON DRUG RELEASE.

Ramana Susarla, Norman Loney, Department of chemical, biological and pharmaceutical engineering, New Jersey Institute of Technology, NJ

INTRODUCTION

The significance of controlled release drug delivery systems lies in their ability to deliver the drug at a steady rate thus reducing the dosage interval and providing a prolonged pharmacodynamic effect. But despite the steadily increasing practical importance of these devices, little is known regarding their underlying drug release mechanisms [1].

Mathematical modeling of these drug delivery systems could help us understand the underlying mass transport mechanisms involved in the control of drug release. Mathematical modeling also plays an important role in providing us with valuable information such as the amount of drug released during a certain period of time or when the next dosage needs to be administered. Thus, potentially reducing the number of in-vitro and in-vivo experiments which in some cases are infeasible.

There is a large spectrum of published mathematical models and approaches describing drug release from various types of controlled drug delivery devices [1-8]. Most of these models focus on drug release from either diffusion controlled and/or degradation controlled devices which are developed taking into account in vitro conditions wherein a perfect sink condition is maintained. However in a real system this may not be true. Physiochemical properties such as solubility, diffusion, particle size, crystal form etc... and physiological factors such as gastrointestinal tract (GI) pH, stomach emptying, (GI) motility, presence of food etc., may effect the rate of drug release [9,10].As the physiological parameters are not taken into consideration during model development discrepancies arise in the drug release behavior in vitro and in vivo. Ouruemchi et al., [11] developed a numerical solution taking into account the pharmacokinetic parameters like rate of absorption and rate of elimination and in vitro data to predict the concentration of a given drug in the plasma compartment. Ainaoui et al.,[12] (2000) combined a mathematical model developed for diffusional release systems of various geometries with a compartmental modeling approach to calculate the plasma drug concentration.

However, there are very few models which take into account the effect of dissolution rate on the rate of drug release [13, 14].The solubility of drugs is known to significantly effect the rate of absorption in the GI tract [10,15-16]. Doo-Man Oh et.al., [10] developed a mathematical model for estimating the fraction of dose absorbed in the intestine for poorly soluble compounds. But there are no mathematical models in the literature as per our knowledge that consider the effect of absorption rate on drug release from slowly dissolving drug- polymer matrices.

In this work a mathematical model is developed for drug release from spherical matrices consisting of slowly dissolving drug particles. Also included in this model is the notion of the effect of absorption rate in the GI tract. The inclusion of absorption in the GI tract has not been modeled before. The model is developed by assuming that the drug is release into a finite volume (as opposed to a perfect sink condition) and is thereby absorbed through the GI tract following first order kinetics

PROBLEM DESCRIPTION

Consider a spherical polymer matrix system wherein the solid drug particles are uniformly dispersed within the matrix. When the drug reaches the GI tract the fluid from the surroundings diffuses into the sphere and dissolves some of the drug. Consider that the drug is dissolved slowly and it is not instantly soluble. Therefore at one point there is both dissolved and undissolved drug present within the polymer matrix. The dissolved drug then enters into the GI tract where it is eventually absorbed. Following absorption in the GI tract the drug thereby enters into the plasma compartment. The rate of absorption depends on various factors such as the amount of food present in the stomach, type of food taken, pH etc., [9]. This situation can be interpreted as an encapsulated drug matrix system of radius R and volume V_s placed in a liquid compartment of finite volume V_l .The concentration of the solid drug within the matrix is taken as S . The concentration of dissolved drug in the matrix is taken as C .The drug enters the liquid phase by the mass transfer process diffusion at time $t = 0$. The concentration of the drug in the liquid phase is taken as C_l . A well- stirred tank assumption is made such that C_l is independent of position. Thereby the drug is consumed by a first order absorption k_r (Figure 1).

Assumptions:

1. A matrix type of a system is considered wherein the drug particles are dispersed uniformly within the matrix system. The entire drug is assumed to be present in solid form initially.
2. As the liquid enters the matrix, some of the drug is dissolved and release into the compartment (simulating the GI tract) through the mass transfer process of diffusion. Therefore at some point there is both dissolved and solid drug present within the matrix.
3. The drug releases from the polymer matrix into the GI tract. Only absorption of the drug in the GI tract is considered and the transport of the drug into the plasma compartment is not considered. Thus the effect of absorption rate on the rate of drug release in the GI tract is the only consideration
4. Initially there is no drug present in the liquid compartment.
5. The rate of absorption is assumed to be first order rate constant.

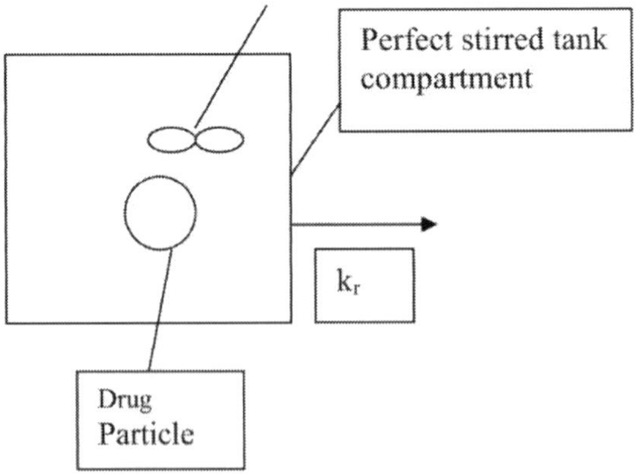

Figure 1. Schematic Representation of Drug Release in the GI Tract.

Model Development

The basic equations governing the system are given by Equations (1) – (3)

$$\frac{\partial C}{\partial t} = D_{AB}\left(\frac{\partial^2 C}{\partial r^2} + \frac{2}{r}\frac{\partial C}{\partial r}\right) + kS^{\frac{2}{3}}(C_s - C) \tag{1}$$

$$\frac{\partial S}{\partial t} = -kS^{\frac{2}{3}}(C_s - C) \tag{2}$$

Where k is the dissolution rate constant.

Subject to absorption in the GI tract is treated here as a surface boundary condition:

$$-(\frac{3}{B})D_{AB}\left(\frac{\partial C}{\partial r}\right)_{r=R} = V_l\frac{dC_l}{dt} + V_l k_r C_l \tag{3}$$

Boundary conditions governing equations (1) - (3) are:

$$C(r,0) = 0 \tag{4}$$

$$C_l(0) = 0 \tag{5}$$

$$S(r,0) \text{ is finite} \tag{6}$$

$$C = K_P C_l \tag{7}$$

Where K_P is the partition coefficient between drug and the solvent. Equations (1) and (2) were proposed by Frenning [14] for the dissolution without absorption in the GI tract.
Using dimensionless variables Equations (1)-(7) become:

$$\frac{\partial \theta}{\partial \tau} = \left(\frac{\partial^2 \theta}{\partial \xi^2} + \frac{2}{\xi}\frac{\partial \theta}{\partial \xi}\right) + M\theta_s^{\frac{2}{3}}(\overline{\theta} - \theta) \tag{8}$$

$$\frac{\partial \theta_s}{\partial \tau} = -M\theta_s^{\frac{2}{3}}(\overline{\theta} - \theta) \tag{9}$$

$$-\left(\frac{3}{B}\right)\left(\frac{\partial \theta}{\partial \xi}\right)_{r=R} = \frac{d\theta_l}{d\tau} + N\theta_l \tag{10}$$

The boundary conditions become:

$$\theta(\xi,0) = 0 \tag{11}$$

$$\theta(1,\tau)=\theta_l(\tau) \tag{12}$$

$$\theta(0,\tau) \text{ is finite} \tag{13}$$

$$\theta_l(0)=0 \tag{14}$$

The following approximation was introduced by Frenning [14].

$$U=1-\theta_S \tag{15}$$

Assuming U<<1 during early stages of release we get:

$$\theta_s^{\frac{2}{3}}=(1-U)^{\frac{2}{3}} \approx 1-\frac{2}{3}U \tag{16}$$

Therefore equations (1) and (2) become:

$$\frac{\partial\theta}{\partial\tau}=\left(\frac{\partial^2\theta}{\partial\xi^2}+\frac{2}{\xi}\frac{\partial\theta}{\partial\xi}\right)+(1-\frac{2}{3}U)(\bar{\theta}-\theta) \tag{17}$$

$$\frac{\partial U}{\partial\tau}=-M(\bar{\theta}-\theta)(1-\frac{2}{3}U)$$

$$=-M(\bar{\theta}-\theta)+M\bar{\theta}\frac{2}{3}U+M\theta\frac{2}{3}U \tag{18}$$

Keeping only the linear terms we get:

$$\frac{\partial U}{\partial\tau}=M(\bar{\theta}-\theta)-(\alpha U) \tag{19}$$

Where

$$\alpha=\frac{2}{3}M\bar{\theta} \tag{20}$$

Equations (8) and (9) become:

$$\frac{\partial\theta}{\partial\tau}=\left(\frac{\partial^2\theta}{\partial\xi^2}+\frac{2}{\xi}\frac{\partial\theta}{\partial\xi}\right)+M(\bar{\theta}-\theta)-\alpha U \tag{21}$$

$$\frac{\partial U}{\partial \tau} = M(\overline{\theta} - \theta) - (\alpha U) \tag{22}$$

Applying Laplace transform method to Equations (8) and (9) respectively we get

$$s.u(\xi,s) - u(\xi,0) = \left(\frac{d^2u(\xi,s)}{d\xi^2} + \frac{2}{\xi}\frac{du(\xi,s)}{d\xi} \right) + M\left(\frac{\overline{\theta}}{s} - u(\xi,s) - \alpha\widetilde{U} \right) \tag{23}$$

$$s\widetilde{U} - \widetilde{U}(\xi,0) = M\left(\frac{\overline{\theta}}{s} - u(\xi,s) - (\alpha\widetilde{U}) \right) \tag{24}$$

On rearranging and collecting like terms results in

$$\left(\frac{d^2u}{d\xi^2} + \frac{2}{\xi}\frac{du}{d\xi} \right) - u\left(\frac{s(s+\alpha+M)}{s+\alpha} \right) = -M\left(\frac{\overline{\theta}}{s+\alpha} \right) \tag{25}$$

Or

$$\left(\frac{d^2u}{d\xi^2} + \frac{2}{\xi}\frac{du}{d\xi} \right) - u\left(\lambda^2 \right) = -M\left(\frac{\overline{\theta}}{s+\alpha} \right) \tag{26}$$

Where $\lambda^2 = \dfrac{s(s+\alpha+M)}{s+\alpha}$ (27)

Using the usual substituting $q = \xi.u(\xi,s)$ for spherical geometry Equation (26) reduces to

$$\left(\frac{d^2q}{d\xi^2} \right) - q\left(\lambda^2 \right) = -M\left(\frac{\overline{\theta}}{s+\alpha} \right)\xi \tag{28}$$

Or

$$\left(\frac{d^2q}{d\xi^2} \right) - q\left(\lambda^2 \right) = A\xi \tag{20}$$

Where $A = -M\left(\dfrac{\overline{\theta}}{s+\alpha} \right)$ (30)

The characteristic equation for Equation (28) is

$$m^2(m^2 - \lambda^2) = 0 \tag{31}$$

Such that

$$q(\xi) = a_1 + a_2\xi + a_3 \cosh \lambda\xi + a_4 \sinh \lambda\xi \quad (32)$$

Where $a_1 + a_2\xi = q_p$ (particular solution) and $a_3 \cosh \lambda\xi + a_4 \sinh \lambda\xi = q_c$ (complimentary solution).

Therefore $$u(\xi, s) = \frac{q}{\xi} = -\frac{A}{\lambda^2} + a_3 \frac{\cosh \lambda\xi}{\xi} + a_4 \frac{\sinh \lambda}{\xi} \quad (33)$$

By substituting the boundary condition $u(0, s) = 0$ we must choose $a_3 = 0$ for boundedness which results in:

$$u(\xi, s) = -\frac{A}{\lambda^2} + a_4 \frac{\sinh \lambda}{\xi} \quad (34)$$

By applying the boundary condition $u(1, s) = v(s)$ we get

$$v(s) = a_4 \, Sinh \, \lambda - \frac{A}{\lambda^2} \quad (35)$$

Therefore $$a_4 = \frac{v(s) + \frac{A}{\lambda^2}}{\sinh \lambda} \quad (36)$$

Also Equation (10) becomes

$$-\left(\frac{3}{B}\right) \frac{du}{d\xi}_{\xi=1} = v(s)(N + s) \quad (37)$$

Such that the transformed Equation (10) becomes:

$$-\left(\frac{B}{3}\right) v(s)(N + s) = \left[\frac{(v(s))(\lambda \cosh \lambda - \sinh \lambda)}{\sinh \lambda} + \frac{A[\lambda \cosh \lambda - \sinh \lambda]}{\lambda^2 \sinh \lambda}\right] \quad (38)$$

Such that $v(s)$ is obtained as:

$$v(s) = -\frac{A(\lambda \cosh \lambda - \sinh \lambda)}{\lambda^2 \sinh \lambda \left[\frac{(\lambda \cosh \lambda - \sinh \lambda)}{\sinh \lambda} + \left(\frac{B}{3}\right)(N + s)\right]} \quad (39)$$

$$v(s) = -\frac{A(\lambda\cosh\lambda - \sinh\lambda)}{\lambda^2\left[(\lambda\cosh\lambda - \sinh\lambda) + \left(\frac{B}{3}\right)(N+s)\sinh\lambda\right]} \tag{40}$$

Where $\lambda^2 = \dfrac{s(s+\alpha+k)}{s+\alpha}$ (41)

$$v(s) = -\frac{A(\lambda\cosh\lambda - \sinh\lambda)}{\lambda^2\left[(\lambda\cosh) + \left(\frac{B}{3}\right)(N+s-1)\sinh\lambda\right]} \tag{42}$$

$$v(s) = \frac{-A\left[\left(\frac{1}{2!} - \frac{1}{3!}\right)\lambda^2 + \left(\frac{1}{4!} - \frac{1}{5!}\right)\lambda^4 + ...\right]}{\left\{\lambda^2\left[\left(1 + \frac{\lambda^2}{2!} + \frac{\lambda^4}{4!} + ...\right) + \left(\frac{B}{3}\right)(N+s-1)\left(1 + \frac{\lambda^2}{3!} + \frac{\lambda^4}{5!} + ...\right)\right]\right\}} \tag{43}$$

The residue theorem for performing Inverse Laplace Transform is given as [17]:

$$f(s) = L^{-1}\{F(s)\} = \sum_{n=1}^{n=\infty} \rho_n(t) \tag{44}$$

Where

$$F(s) = \frac{P(s)}{Q(s)} \tag{45}$$

If S_n is a simple pole of $F(s)$ then $\rho_n(t)$ is given by:

$$\rho_n(t) = \frac{P(s_n)}{Q'(s_n)} e^{s_n t} \tag{47}$$

Therefore $\theta_i(\tau) = L^{-1}\{v(s)\} = \sum_1^{\infty} \rho_n(t)$

Where

$$v(s) = \frac{P(s)}{Q(s)} \frac{M\bar{\theta}\lambda^3}{(s+\alpha)(3+B(N+s-1)} + \frac{3\left\{\frac{M\bar{\theta}}{10(s+\alpha)} - \frac{M\bar{\theta}}{(s+\alpha)(3+B(N+s-1))}\left(\frac{9+B(N+s-1)}{6}\right)\right\}}{(3+B(N+s-1)} \tag{48}$$

Results and Discussions

Future work would include obtaining a semi analytical solution for fraction of drug released within the GI tract by performing Inverse Laplace Transform for Equation (48). The obtained Equation would also include the effect of absorption rate on drug release rate. All the parameter values needed for calculation of fraction of drug released will be taken from the literature. The model results will be validated by making comparisons with dissolution model developed by Polakovic [13] and diffusion-dissolution model developed by Frenning [14] which do not take into account the absorption kinetics and are developed for perfect sink conditions.

Nomenclature

Dimensionless concentration of solid drug is given as: $\theta_s = \frac{S}{S_o}$

Dimensionless concentration of dissolved drug present in the liquid compartment is given as: $\theta_l = \frac{C_l K_p}{S_o}$

Dimensionless solubility: $\bar{\theta} = \frac{C_s}{S_o}$

Dimensionless time: $\tau = \frac{D_{AB} t}{R^2}$

Dimensionless radial coordinate: $\xi = \frac{r}{R}$

Dimensionless parameters $B = \frac{V_l}{V_s K_P}$ $M = \frac{kR^2}{D_{AB}}$ $N = \frac{k_r R^2}{D_{AB}}$

References

1. Siepmann J, Gopferich A. , Mathematical modeling of bioerodible drug delivery systems, Advanced drug delivery reviews, 2001, 48, 229-247.

2. Charlier A., Leclerc B., Couarraze G., Release of Mifepristone from Biodegradable Matrices: experiment and theoretical evaluations, International Journal of Pharmaceutics, 2000, 200, pp.115-120.

3. Crank, J., The mathematics of diffusion (1976) 2nd edition, New York: oxford University.

4. Abdekhodaie, M.J., Diffusional release of a solute from a spherical reservoir into a finite volume, Journal of Pharmaceutical Sciences, 2002, 91(8), pp. 1803-1809.

5. Lao Luciana Lisa, Venkatraman S, Peppas N.A, Modeling of drug release from biodegradable polymer blends, European Journal of Pharmaceutics and Biopharmaceutics, 2008, 70, pp.796-803.

6. Siepmann, J., Siepmann, F., Mathematical modeling of drug delivery, International Journal of Pharmaceutics, 2008, 364, pp. 328-343.

7. Raman, C., Berkland, C., Kyekyoon, K., Pack, D.W., Modeling small-molecule release from PLG microspheres: effects of polymer degradation and non uniform drug distribution, Journal of Controlled Release, 2005, 103, pp. 149-158.

8. Faisant, N., Siepmann, J., Benoit, J.P., PLGA-based microspheres: elucidation of mechanisms and a new simple mathematical model quantifying drug release, European Journal of Pharmaceutical Sciences, 2002, 15 (4), pp. 355-366.

9. Shargel, L., Wu-Pong, S., Yu, A., Applied Biopharmaceutics and Pharmacokinetics, 5th edition, 2005 McGraw-Hill.

10. Estimating the fraction dose absorbed from suspensions of poorly soluble compounds in humans: A mathematical model, Pharmaceutical Research, 1993, 10 (2), pp. 264-270.

11. Ouriemchi, E.M., Bouzon, J., Vergnaud, J.M., Modelling the process of controlled release of drug in in vitro and in vivo tests, International Journal of Pharmaceutics, 1995, 113, pp. 231-240.

12. Ainaoui, A., Siepmann, J., Bodmeier R., Vergnaud, J.M., Calculation of the dimensions of dosage forms with release controlled by diffusion for in vivo use, European Journal of Pharmaceutics and Biopharmaceutics, 2001, 51, pp. 17-24.

13. Polakovic, M., Gorner, T., Gref, R., Dellacherie, E., Lidocaine loaded biodegradable nanospheres II.Modeling of drug release, Journal of controlled release, 1999, 60(2-3), pp. 169-177.

14. Frenning, G., Theoretical analysis of the release of slowly dissolving drug from spherical matrix systems, Journal of Controlled Release, 2004, 95, pp. 109-117.

15. A theoretical basis for a biopharmaceutic drug classification: The correlation of in vitro drug product dissolution and in vivo bioavailability, Pharmaceutical Research, 12 (3), pp. 413-420.

16. The effect of physical barriers and properties on the oral absorption of particulates, Advance Drug Delivery Review, 1998, 34, pp.135-154.

17. Loney, N.W., Applied mathematical methods for chemical engineers, 2nd edition, 2006, Boca Raton: Taylor and Francis

Ciliary Transport Strategies of Spherical Particles

O. Berk Usta,Chemical and Petroleum Engineering, University of Pittsburgh, Pittsburgh, PA

We investigate the transport of particulates on a ciliated surface using a continuum modeling approach. The elastic cilia can beat due to an applied external field which is apllied in a periodic elliptical fashion. We have recently shown [1] that this beating pattern combined with an adhesive interaction of the cilia tips with the particles result in a net velocity of the particles. Here we investigate the effect of particle size and different surface arrangements on the transport velocity of the particles. Consequently we also discuss the feasibility of seperating particles based on their size using this ciliary transport system.

[1] G. A. Buxton, A. Alexeev, O. B. Usta, A. C. Balazs, submitted to Phys. Rev. Lett., 2005

See more of this Session: Mathematical Modeling of Transport Processes
See more of this Group/Topical: Engineering Sciences and Fundamentals

Application of An Energy-Conserving Electron-Ion Interaction Model to Radiation Damage in a Lennard-Jones and Binary Lennard-Jones Crystal

Carolyn L. Phillips, Applied Physics, University of Michigan, Ann Arbor, MI

Paul S. Crozier, Multiscale Dynamic Material Modeling, Sandia National Laboraotry, DOE, Albuquerque, NM

Two temperature models are used to model the interaction between the electronic subsystem and the atomic subsystem during thermal transients such as radiation damage, laser heating, and cascade simulations. In this paper, we introduce an energy-conserving version of an inhomogeneous finite reservoir two temperature model using a Langevin thermostat to communicate energy between the two systems. This energy-conserving modification allows the inhomogeneous two temperature model to be used for longer and larger simulations and simulations of small energy phenomena, without introducing non-physical energy fluctuations that may a ect the conclusions of the simulation. We test this model on the annealing of a Frenkel defect in a Lennard-Jones crystal. We then consider two idealized models of radiation damage due to a local deposition of heat. We first consider radiation damage in a large Lennard-Jones crystal that readily re-crystallizes. Second we consider radiation damage in a large binary glass-forming Lennard-Jones crystal that retains permanent damage. We find that the electronic subsystem parameters can influence the way heat is transported through the system and have a significant impact on the number of defects after the heat deposition event. We also find that the two model systems have di erent responses to the electronic subsystem.

See more of this Session: Mathematical Modeling of Transport Processes
See more of this Group/Topical: Engineering Sciences and Fundamentals

Higher Order Galilean Invariant Lattice Boltzmann Model and Entropic Gauss-Hermite Quadrature

Wahyu P. Yudistiawan, Division of Chemical and Biomolecular Engineering, Nanyang Technological University, Singapore, Singapore

Sang Kyu Kwak, Division of Chemical and Biomolecular Engineering, Nanyang Technological University, Singapore, Singapore

Ansumali Santosh, Division of Chemical and Biomolecular Engineering, Nanyang Technological University, Singapore, Singapore

We introduce a novel scheme obtained by distorting energy level of the velocity shells, which give rise to additional degree of freedom for the same number of discrete velocities. We claim that the proposed model is the smallest known 3-dimensional system in terms of the number of velocities for the accuracy of full sixth-order moments. Velocity profile is solved for the current model, which contains the Knudsen layers terms, and the Knudsen minima exists for the flowrate.

See more of this Session: Mathematical Modeling of Transport Processes
See more of this Group/Topical: Engineering Sciences and Fundamentals

A Three-Dimensional Computational Fluid Dynamics Study of Ozone Absorption in the Respiratory Tract

Banafsheh Keshavarzi, Chemical Engineering, The Pennsylvania State University, University Park, PA

James Ultman, Chemical Engineering, The Pennsylvania State University, University Park, PA

Ali Borhan, Chemical Engineering, The Pennsylvania State University, University Park, PA

We present the results of numerical simulations of ozone transport and uptake in an anatomically-correct model of the respiratory tract of a Rhesus monkey. The model geometry was created using three-dimensional reconstruction of MRI images of the respiratory tract, including the nasal passages, the larynx, and the first thirteen generations of the tracheobronchial tree. An unstructured mesh was generated for the resulting structure, and three-dimensional flow and concentration distributions were obtained through numerical solution of the Navier-Stokes, continuity, and species convection-diffusion equations. A quasi-steady diffusion-reaction model was used to account for the interaction between O3 and endogenous substrates in the respiratory tract lining fluid. The total rate of ozone uptake within each section of the respiratory tract was determined, and hot spots of ozone flux on the walls were identified. For steady inspiratory flow under quiet breathing conditions, the predicted locations of spikes in ozone flux are consistent with existing experimental observations of the focal sites of epithelial damage in rats [Postlethwait et al., Am. J. Respir. Cell Molec. Biol. 22:191-199, 2000].

See more of this Session: Mathematical Modeling of Transport Processes
See more of this Group/Topical: Engineering Sciences and Fundamentals

Swimming with a Crowd: Collective Behavior of Active Suspensions

M. Cristina Marchetti, Physics, Syracuse University, Syracuse, NY

Bacterial suspensions and extracts of cytoskeletal filaments and motor proteins are examples of assemblies of interacting self-driven units that form active complex fluids. These systems exhibit fascinating collective behavior, including nonequilibrium phase transitions, novel long-range correlations, and pattern formation on mesoscopic scales. In this talk I will review our work on using nonequilibrium statistical physics to derive a continuum description of active suspensions from specific models of single particle dynamics. Specifically I will highlight the role of physical interactions, such as excluded volume and medium-mediated hydrodynamic couplings, in controlling the large-scale behavior.

See more of this Session: Novel Flows
See more of this Group/Topical: Engineering Sciences and Fundamentals

The Hydrodynamics of Active Suspensions

Michael J. Shelley, Courant Institute of Mathematical Sciences, New York University, New York, NY

David Saintillan, Department of Mechanical Science and Engineering, University of Illinois at Urbana-Champaign, Urbana, IL

Locomotion of an organism through a fluid is one of the most

fascinating instances of fluid-structure interaction. Just as fascinating are the large-scale flows that result when there are not one but many locomoting bodies moving and interacting with each other through the fluid. Aside from their intrinsic importance to biology, such systems offer new approaches to micro-scale mixing and transport, as well as extraction of mechanical work, from active fluids. I will overview recent mathematical and simulational work in describing "active suspensions" here composed of many self-locomoting particles moving through a Stokesian fluid. This work ranges from using large-scale particle-based simulations of hydrodynamically interacting swimmers, to the development and application of continuum kinetic theories. Our analyses and simulations of these models demonstrate the existence of large-scale mixing and transport, dependent upon micro-mechanical swimming mechanisms and upon system scales, with these features associated with large-scale flow instability and the dynamics of quasi-periodic coherent structures.

See more of this Session: Novel Flows
See more of this Group/Topical: Engineering Sciences and Fundamentals

Reduction of Viscosity in Suspension of Swimming Bacteria and Algae

Dr. Igor S. Aranson, Dr. Andrey Sokolov, Dr. Liaohai. Chen, Dr. Qiaoling Jin,
Argonne National Laboratory, Argonne, Illinois, USA

Introduction

Rheological and physical properties of suspensions of motile microorganisms, such as bacteria, sperm cells, swimming unicellular algae, continue attracting enormous attention in the scientific community [1-9]. Theoretical studies have predicted a plethora of dynamic instabilities and non-trivial stress-strain relations in such active non-equilibrium systems [10, 11]. A phenomenological model of active particle suspensions [12] hinted that the viscosity becomes a function of the particles' motility. Recent analytical studies [13, 14] indicate that the shear viscosity in a dilute suspension of model swimmers can be smaller than the viscosity of the ambient liquid.

Here we present experimental studies of the viscosity of suspensions of two motile microorganisms, *Bacillus subtilis*, swimming aerobic bacteria, and swimming unicellular algae *Chlamydomonas reinhardtii*. The experiments are performed in free standing liquid films. Depending on the concentration of bacteria and their activity (typical swimming speed), up to a seven-fold reduction in the viscosity was observed compared to that of the same suspension of non-motile bacteria. Similar, but smaller, of the order of 10-20%, viscosity reduction is also observed for the algae suspensions. The effect is interpreted as a transformation by the swimming microorganisms of chemical energy from nutrients into mechanical energy of fluid motion, thus counterbalancing the energy loss due to viscous dissipation.

Materials and Methods

Experiments were conducted on strain 1085 of *Bacillus subtilis*, a rod-shaped bacterium ~4-5 μm long and ~0.7 μm in diameter see Fig. 1a. The bacteria were grown in a Terrific Broth (TB) medium (Sigma T5574), concentrated by centrifuging, and then placed in a fresh growth medium. In a typical experiment, the average concentration of bacteria in TB medium was $\sim 2\times10^{10}$ cm^{-3}, which is approximately 20 times higher than in the stationary phase of growing.

Chlamydomonas reinhardtii algae wild type strain was obtained from American Type Culture Collection. A single colony was inoculated into 15 ml of Tris-acetate-phosphate (TAP) medium [15] and grown to mid-logarithmic phase (concentration $2\text{-}5\times10^{6}$ cm^{-3}) at about 25°C with continuous shaking at 100 rpm. During growth, the culture was continuously illuminated with cool-white fluorescent light (~ 90 $\mu Einstein\cdot m^{-2}\cdot s^{-1}$). The algae have almost spherical body, diameter about 4-8 μm, propelled by two flagella, see Fig. 1b.

The computer-controlled experimental setups, see Fig.2 were based on an earlier design [4] with a number of important modifications, discussed below. A 10 μl drop of suspension was placed between two supporting crossed pairs of fibers, which formed a small square window. The drop was stretched to a thickness of ≈200 μm by moving a supporting platform attached to one crossed pair of the fibers. The size of the square window in most of our experiments was ~7 mm×7 mm. The speed of aerobic bacteria depends on the concentration of dissolved Oxygen: lack of Oxygen suppresses the motility of bacteria. To control the swimming speed, we replaced the air in the experimental chamber with Nitrogen over a period of approximately 2 minutes. During this time, the bacteria gradually reduced their swimming speed to zero. The swimming speed of bacteria in relatively thin films was estimated using particle image velocimetry, or from the trajectories of small fluorescent markers (0.53 μm in diameter, Red Nile, Spherotech) in thicker films where the direct tracking of bacteria becomes difficult. In thin films we observed that there is an almost linear relation between the speed of bacteria V_b and the marker's velocity V_m.

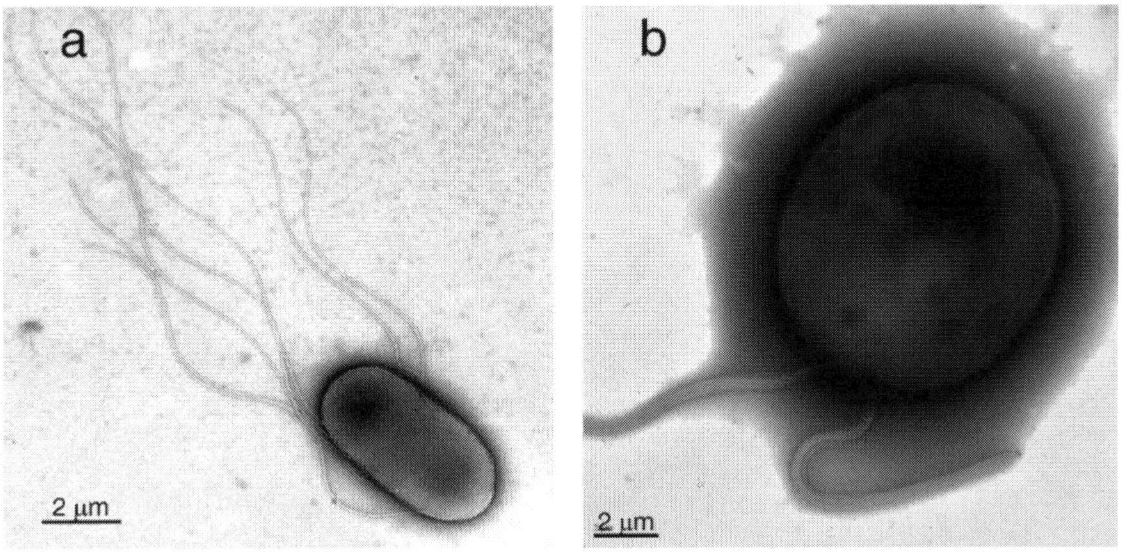

Figure 1. Transmission electron micrograms. (a) bacterium *Bacillus subtilis*, from J.J. Cardamone Jr./Biological Photo Service; (b) algae *Chlamydomonas reinhardtii,* L. Chen, Argonne National Laboratory.

Experimental Results

Vortex decay experiment. To initialize a vortex flow in a film, we used a custom-made micromanipulator built from the optical actuator taken from a commercial DVD-ROM, Fig. 2a. Two pairs of magnetic biasing coils control the horizontal and vertical position of the lens holder. The lens was replaced with a microprobe (0.5 mm wire) to produce a vortex through short (0.5 mm) and fast (0.1 seconds) horizontal movement of the probe. Fluorescent 2.54 μm tracer particles (Spherotech) were added to the suspension in order to visualize the flow.

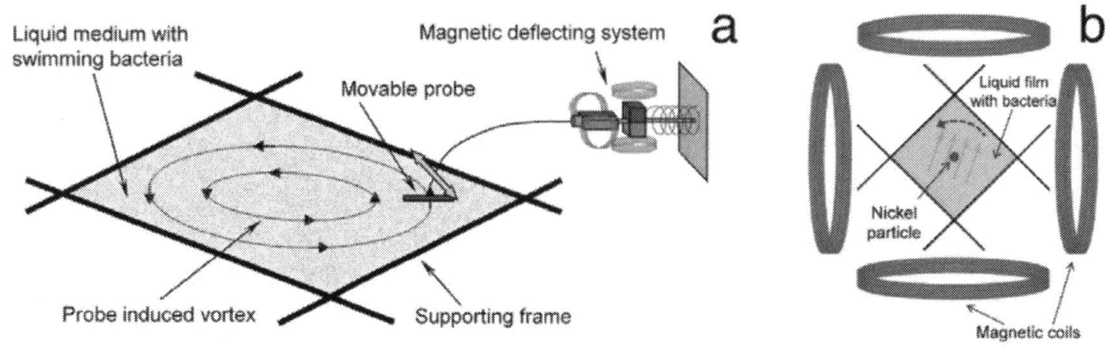

Figure 2. Two experimental setups. (a) Vortex decay experiment: a liquid film with bacteria or algae spans between four movable fibers. A micromanipulator with a magnetic deflecting system is used to initiate a large vortex through movement of the probe. (b) Magnetic particle rotation experiment: a thin liquid film containing a bacterial suspension and submersed Ni particle spanning between four movable fibers. Magnetic coils create a rotating magnetic field (four green arrows).

The effective viscosity is inferred from the time dependence of the tracer particles velocity during the decay of the probe-induced vortex. We assume that the liquid is incompressible and the flow is purely two-dimensional. The decay of the vortex of a characteristic size L occurs due to friction between the liquid and the container wall. For simplicity, we assume that the viscosity ν is practically constant for small shear strain rates and begins to depend on the strain rate only above a certain critical value. To satisfy this assumption, the initial rapid stage of relaxation which typically last 0.2-1 seconds is excluded. During this stage, the characteristic velocity of fluorescent tracers drops abruptly from 5 mm/sec to 200-500 μm/sec. During the following slower stage, the velocity decreases from 200-500 μm/sec during 1-3 seconds to a value defined by a chaotic motion of fluorescent tracers. For this stage the viscosity ν is practically constant and can be extracted from the following simple expression

$$V(t) = V_0 \exp\left(-\frac{a_0 \nu t}{L^2}\right)$$

where $V(t)$ is the flow velocity in the field of view of the microscope, and constant a_0 depends on the geometry of the experimental cell. The combination L^2/a_0 can be found from additional experiment where the viscosity is known (liquid without bacteria or with dead bacteria). We measured the viscosity for 6 different concentrations of bacteria, see Fig. 3a. At high concentrations the increase of viscosity occurs due to two effects: reduction of the bacterial motility [4]; in passive suspensions, the viscosity increases with volume fraction of particles.

Rotation of a magnetic particle experiment. While the vortex decay technique offers a simple and intuitive way to estimate the viscosity, it is indirect and based on a number of simplifying assumptions. To probe the viscosity directly, a technique based on measuring the viscous drag exerted on rotating magnetic particle in a rotating magnetic field was employed, see Fig. 2b. Orthogonal pairs of magnetic coils were connected through current amplifiers to synchronized function generators in order to create a magnetic field with constant amplitude of 0-10 Gs rotating horizontally in the plane of the liquid film. A magnetized 100 μm Nickel particle was placed in the center of the liquid film and was held in place through the gravitational depression of the film. First, the direction of particle's internal magnetic moment was determined by applying a relatively strong constant magnetic field (10 Gs) of known direction, forcing the particle to align its magnetic moment with the direction of the applied field. Then, using custom made MatLab software, the image for an arbitrary position and orientation of the particle with the reference image was compared and the direction of the magnetic moment of our particle α was extracted. Secondly, a rotating magnetic field H with amplitude of 3 Gs and a rotational frequency of Ω=0.5 Hz was applied.

The viscosity ν is extracted from the balance of rotational viscous drag $T_v \sim \nu\, \partial_t\, \alpha$ with the magnetic torque $T_m \sim \mu_0\, H \sin\varphi$, where φ is the angle between the magnetic moment and the external field and μ_0 is the particle's magnetic moment. In the case of synchronous rotation of the particle and field, $\partial_t\, \alpha = \Omega$, the viscosity can be extracted from the angle φ, since $\nu \sim (\sin\varphi)/\Omega$. The orientation of the particle's internal magnetic moment in each recorded frame was measured while filling the chamber with nitrogen N_2. Due to the change in the viscosity, the mean value of φ, and correspondingly, magnetic torque increases with time T_m as the N_2 /O_2 ratio increases (hence the bacterial motility decreases). The particle stops rotating when the motility of the bacteria drops below some critical value: the viscosity of the suspension becomes so high that the torque required for rotation of the particle at constant rate Ω is larger than the maximum magnetic torque. Since the magnetic torque is proportional to the viscosity ν, we obtained that the viscosity of the suspension at maximum swimming speed (which occurs when the chamber filled with air) was roughly 6-7 times lower than the viscosity of the same suspension of immobilized bacteria, see Fig. 3b. Compared to our vortex decay experiment, we observed a similar trend: the viscosity of a bacterial suspension at the same concentration, in the presence of sufficient amount of Oxygen, is one seventh the viscosity of the liquid without bacteria. However, since the suspension of immobilized bacteria has a slightly higher viscosity than the fluid, the relative viscosity reduction was even higher in the magnetic torque experiment.

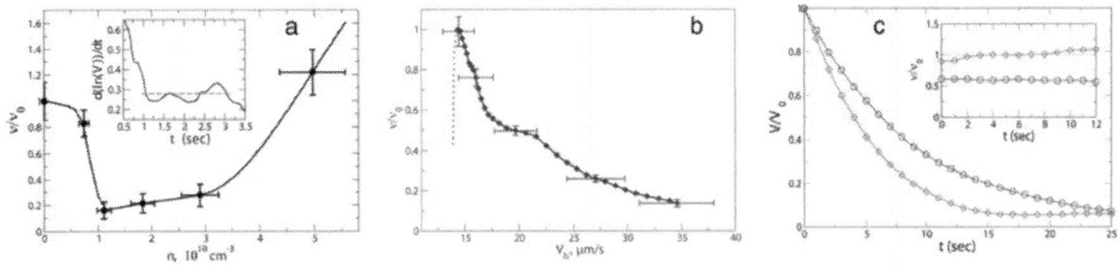

Figure 3. Summary of experimental results. (a) Vortex decay experiment: viscosity for 6 different concentrations of bacteria, ν_0 is viscosity of the liquid without bacteria. Inset: instant viscosity versus time during decay of the vortex for the concentration n=2.9×10^{10} cm^{-3}.The dashed line is the average value of the viscosity during the slow phase of decay. (b) Magnetic particle rotation experiment: viscosity versus speed of the bacteria V_b for the concentration n=1.8×10^{10} cm^{-3}. (c) Experimental results for algae, concentration n≈10^7 cm^{-3}. Main plot shows vortex velocity versus time, black line

with circles corresponds to living cells; red line with diamonds corresponds to dead cells. Inset: instant viscosity versus time for living cells (circles) and dead cells (diamonds).

The reversibility of the process was confirmed in additional experiments. After the chamber was fully filled with nitrogen, the concentration of oxygen was increased, resulting in an increase in the motility of the bacteria and a decrease of the viscosity almost to the initial value. The difference between the initial and final levels of the viscosity likely stemmed from a change in the concentration of chemicals in the fluid due to the metabolism of the bacteria and from the evaporation of liquid.

We conducted the vortex decay experiments with swimming suspensions of unicellular algae *Chlamydomonas reinhardtii.* The viscosity of the algae suspension was compared with the suspension of algae cells killed by a drop of alcohol, see Figure 3c. We observed a similar trend; the viscosity of living suspension is smaller than that of the dead cells. However, in contrast to the bacteria, the overall decrease is relatively small, of the order of 10-20% only. We attribute such a small effect to two major factors. Firstly, it was impossible to obtain highly concentrated suspensions of swimming algae: at the concentrations about 10^8 cm^{-3} the algae either died or stopped swimming. Secondly, since the algae have an almost spherical body, they have much weaker tendency to align with the external shear flow, which, in turn, according to [13, 14], produces much smaller decrease of the viscosity due to swimming of microorganisms.

Conclusions

From our experiments with algae and bacteria we observed that the effective viscosity of the suspension strongly depends on the concentration and swimming speed. In particular, the effective viscosity of a suspension of bacteria at concentrations ~$1\text{-}3\times10^{10}$ cm^{-3} is seven times lower than the viscosity of the liquid without bacteria. Similar, but smaller decrease is observed also for the suspensions of algae. The mechanism for the viscosity reduction is transformation by swimmers of chemical energy of the nutrient into kinetic energy of fluid motion and thus compensating energy loss due to viscous dissipation. This work was supported by the DOE contract DE-AC02-06CH11357.

References

1. Wu, X.-L. and Libchaber, A. (2000), "Particle Diffusion in a Quasi-Two-Dimensional Bacterial Bath", *Physical Review Letter*, **84**, 3017
2. Dombrowski, C., Cisneros, L., Chatkaew, S., Goldstein, R. E., and Kessler, J. O. (2004), "Self-Concentration and Large-Scale Coherence in Bacterial Dynamics", *Physical Review Letter*, **93**, 098103
3. Saintillan, D. and Shelley, M.J. (2007), "Orientational Order and Instabilities in Suspensions of Self-Locomoting Rods", *Physical Review Letter*, **99**, 058102.
4. Sokolov, A., Aranson, I.S., Kessler, J.O, and Goldstein, R.E. (2007), "Concentration Dependence of the Collective Dynamics of Swimming Bacteria", *Physical Review Letter*, **98**, 158102
5. Berke A.P., Turner, L., Berg, H.C. and Lauga, E. (2008), "Hydrodynamic Attraction of Swimming Microorganisms by Surfaces", *Physical Review Letter*, **101**, 038102

6. Hernandez-Ortiz, J.P., Stoltz, C.G., and Graham, M.D. (2005), "Transport and Collective Dynamics in Suspensions of Confined Swimming Particles", *Physical Review Letter*, **95**, 204501
7. Drescher K.,, Leptos, K., Ishikawa, T., Pedley, T. J., and Goldstein, R. E. (2009), Dancing *Volvox*: Hydrodynamic Bound States of Swimming Algae, *Physical Review Letter*, **102**, 168101
8. Riedel, I.H., Kruse, K., and Howard, J. (2005), "A Self-Organized Vortex Array of Hydrodynamically Entrained Sperm Cells", *Science*, **309**, 300
9. Couzin I.D. and Krause, J. (2003), "Self-organization and collective behavior in vertebrates", *Adv. Study Behavior*, **32**, 1
10. Simha, R.A. and Ramaswamy, S. (2002), "Hydrodynamic Fluctuations and Instabilities in Ordered Suspensions of Self-Propelled Particles", *Physical Review Letter*, **89**, 058101
11. Aranson, I. S., Sokolov, A. Kessler, J.O.and Goldstein, R. E. (2007), "Model for dynamical coherence in thin films of self-propelled microorganisms", *Physical Review E*, **75**, 040901(R)
12. Hatwalne, Y., Ramaswamy, S., Rao, M., and Simha, R.A. (2004), "Rheology of Active-Particle Suspensions," *Physical Review Letter*, **92**, 118101
13. Haines, B. M., Aranson, I. S., Berlyand, L., and Karpeev, D. A. (2008), "Effective Viscosity of Dilute Bacterial Suspensions: A Two-Dimensional Model", *Physical Biology*, **5**, 046003 (2008).
14. Haines, B. M., Sokolov, A., Aranson, I.S., Berlyand, L., Karpeev, D.A. (2009), A three-dimensional model for the effective viscosity of bacterial suspensions", submitted to *Physical Review E.*
15. Gorman, D.S., and Levine, R.P. (1965), "Cytochrome f and plastocyanin: their sequence in the photosynthetic electron transport chain of *Chlamydomonas reinhardi*", *Proc. Natl. Acad. Sci. USA* **54**, 1665

Cell Quakes: Mechanics and Microrheology in Active Gels and Living Cells

Alex J. Levine, Chemistry & Biochemistry, UCLA, Los Angeles, CA

The mechanics of the in vivo cytoskeleton is controlled in part by the details of its non-equilibrium steady-state. In this ``active'' material, molecular motors (e.g. myosin) exert transient contractile stresses on the F-actin filament network. Since microrheology traditionally relies of the linear response properties of the soft materials in thermal equilibrium, this departure from equilibrium has profound implications for the interpretation of microrheological data from the interior of living cells and in vitro active networks. In active networks, such as the in vitro systems of Mizuno et al. [Science 315 (5810) pp. 370-373 (2007).] and in living cells, the underlying theoretical foundation of the interpretation of microrheology -- the Fluctuation-Dissipation theorem -- does not apply. New ideas are needed. In this talk, I review microrheology, and then discuss a new theoretical interpretation of microrheology in active (i.e. molecular motor driven) networks. I also explore how molecular motor activity can reversibly control the elastic properties of these active gels. The cytoskeleton points towards the development of new biomimetic materials whose elastic properties can be tuned by controlling the material's non-equilibrium steady-state.

See more of this Session: Novel Flows
See more of this Group/Topical: Engineering Sciences and Fundamentals

Designing Synthetic Cilia to Regulate Fluid Flow and Particle Transport in Microfluidic Devices

Anna Balazs, Chemical Engineering, Univ. of Pittsburgh, Pittsburgh, PA

O. Berk Usta, Chemical and Petroleum Engineering, University of Pittsburgh, Pittsburgh, PA

Alexander Alexeev, Georgia Institute of Technology, Atlanta, GA

Gavin A. Buxton, Case Western Reserve Univ., Cleveland, OH

We model the transport of a neutrally buoyant, microscopic particle via a regular array of beating elastic filaments, which are anchored to a substrate. These synthetic cilia are actuated by an external periodic force and thus, undergo a cyclic motion. The tips of the beating cilia experience an attractive interaction with the particle's surface and thus, can transfer momentum to the suspended particle. We find that the concerted motion of the cilia causes the particle to be effectively passed from one cilium to the next, and thereby transported along the top of the layer. We isolate the parameter space that provides the most effective particle transport. These are the first simulations to show that the motility of artificial cilia can lead to the net directed motion of microscopic particulates, and thus, indicate that such filaments could be used to control the trafficking of particles in microfluidic devices.

See more of this Session: Novel Flows
See more of this Group/Topical: Engineering Sciences and Fundamentals

Area 1A Keynote Address: Direct Simulation-Based Evidence for Fluid-Solid Transition in Nanoconfined Non-Polar Fluids

Peter T. Cummings, Department of Chemical and Biomolecular Engineering, Vanderbilt University, Nashville, TN

It is well known that the behavior of fluids confined to the order of a few nanometers may differ greatly from that of the corresponding bulk fluid [1]. For example, in the 1980s and early 1990s a large number of studies were reported on a variety of ultrathin liquid films confined between mica surfaces [2]. In all cases, irrespective of the fluid being studied, a common feature was noted: When the confinement reached the order of several molecular diameters, a rapid many-orders-of-magnitude increase in the viscosity of the confined fluid was observed, together with behavior typical of the stick-slip response of a crystalline solid structure [3]. Unfortunately, as attention shifted to the nature of the transition from fluid-like to solid-like behavior a similar consensus was not achieved and, for over a decade, there has been intense debate as to whether it is a first-order (crystallization) or second-order (vitrification) order phase transition. Whilst experimental attempts have been made to clarify this issue, and end this debate, they have been severely hampered by the extreme difficulty of observing directly what occurs at this scale [4].

In contrast to the difficulties involved with experimental observation, molecular simulations techniques have an inherent atomic resolution and, as such, are ideally suited to the study of nanoscale phenomena. Because of this, a number of simulation studies have been undertaken with the aim of elucidating the nature of nanoconfinement induced phase transitions. Unfortunately, these studies have typically made use of fairly simplistic models (E.g. mica represented by an fcc lattice of Lennard-Jones spheres) [5.6], or have performed the simulations in a manner that may be criticized as biased towards the formation of such structures, leading to concerns that they fail to accurately represent the experimental systems under investigation. Thus, with the aim of elucidating the nature of any change that may occur upon nanoconfinement, here we present unprecedentedly detailed molecular dynamics simulation results for dodecane, and cyclohexane confined between molecularly-smooth, but fully atomistically-detailed mica surfaces.

1. Van Alsten, J. and Granick, S., Langmuir 6, 876 (1990).

2. Alba-Simionesco, C. et al., Journal of Physics-Condensed Matter 18, R15 (2006).

3. Klein, J. and Kumacheva, E., Journal of Chemical Physics 108, 6996 (1998).

4. Bae, S. C. et al., Philosophical Transactions of the Royal Society a-Mathematical Physical and Engineering Sciences 366, 1443 (2008).

5. Cui, S. T., Cummings, P. T., and Cochran, H. D., Molecular simulation of the transition from liquidlike to solidlike behavior in complex fluids confined to nanoscale gaps. Journal of Chemical Physics 114 (16), 7189 (2001).

6. Jabbarzadeh, A., Harrowell, P., and Tanner, I., Low friction lubrication between amorphous walls: unraveling the contributions of surface roughness and in-plane disorder. Journal of Chemical Physics 125, 034703 (2006).

See more of this Session: Thermodynamic Properties and Phase Behavior I
See more of this Group/Topical: Engineering Sciences and Fundamentals

Improving the Efficiency of Virial-Coefficient Calculations: a Hybrid Approach Employing Integral-Equation Theories and Mayer-Sampling Monte Carlo

Katherine R. Schadel, Chemical & Biological Engineering, University at Buffalo, The State University of New York, Buffalo, NY

Andrew J. Schultz, Chemical & Biological Engineering, University at Buffalo, The State University of New York, Buffalo, NY

David A. Kofke, Chemical and Biological Engineering, University at Buffalo, The State University of New York, Buffalo, NY

Mayer-sampling Monte Carlo (MSMC) has enabled calculation of high-order virial coefficients for a variety of potential models, including coefficients up to B_6 for several water models and B_8 for the Lennard-Jones (LJ) potential. A particularly appealing application of such results is the estimation of the critical point from the virial equation of state (VEOS), but virial coefficients of even higher order are required to apply the VEOS at densities approaching the critical density. Calculation of these higher-order coefficients remains impractical with the current method.

We present two improved virial-coefficient calculation methods appropriate for spherically symmetric potentials. In the first method, approximations to the virial coefficients, as defined by an integral-equation theory (IET), are computed quickly via a deterministic algorithm. Only the corrections to these approximations are computed by MSMC. For the LJ potential, we demonstrate that this hybrid approach is faster than MSMC alone for computing fourth and fifth virial coefficients.

However, the quality of the approximate virial coefficients diminishes with increasing order. In the deterministic method, an approximate virial coefficient is computed from an approximate density-expansion coefficient of the direct correlation function. In accordance with the IET, this approximate coefficient of the direct correlation function is computed from approximate coefficients of lower-order, such that the approximations compound. As the order increases, so does the magnitude of the required correction relative to that of the full virial coefficient, reducing the benefit of computing only the correction by MSMC.

We present another hybrid approach which can prevent the rapid deterioration of the quality of the approximations, and thus prove more useful for calculation of higher-order virial coefficients. In this second hybrid method, MSMC is used to correct the coefficient of the direct correlation function prior to computing the corresponding virial coefficient and to approximating a higher-order coefficient of the direct correlation function via the IET.

Extension of this hybrid method to more complicated potential models of practical interest is especially desirable as pure MSMC calculations for such models require considerably more time than those for the LJ potential. Water, for example, is often described with a multisite potential, in which interactions between sites on different molecules are spherically symmetric but the composite molecular interactions are not. Generalized integral-equation theories have been developed to compute approximate correlation functions for such potentials. We describe how we can employ these theories to generalize the hybrid method for virial-coefficient calculation.

See more of this Session: Thermodynamic Properties and Phase Behavior I
See more of this Group/Topical: Engineering Sciences and Fundamentals

Influence of An Applied Electric Field On Phase Behavior and Microheterogeneity

J. Ilja Siepmann, University of Minnesota, Minneapolis, MN

Katie A. Maerzke, University of Minnesota, Minneapolis, MN

Previous experimental studies [1] indicate that o-nitrotoluene/n-decane and m-nitrotoluene/n-decane mixtures are of particular interest as they exhibit anomalous behavior in the non-linear dielectric effect near the upper critical solution point, and it has been surmised that these effects are due to an increase in cluster formation as each mixture approaches its critical point. In this work, Monte Carlo simulations are performed using the TraPPE force field at multiple temperatures, concentrations, and electric field strengths in order to examine the structural microheterogeneity of these mixtures and determine the dielectric constants and non-linear dielectric effect. At low field strength, the simulations indicate substantial structural microheterogeneities but these persist over a wide range of conditions. At high field strength (an order of magnitude higher than in the experimental work), the simulations indicate a field induced segregation.

In addition, Gibbs ensemble Monte Carlo simulations are used to explore the influence of applied external field on the vapor-liquid phase equilibria of neat water, methanol, and dimethyl ether.

[1] M. Sliwinska-Bartkowiak, B. Szurkowski, and T. Hilczer, Chem. Phys. Lett. 94, 609 (1983).

See more of this Session: Thermodynamic Properties and Phase Behavior I
See more of this Group/Topical: Engineering Sciences and Fundamentals

An Adaptable Method to Calculate the Solubility Limit of Solids by Molecular Simulation

Andrew S. Paluch, Department of Chemical and Biomolecular Engineering, University of Notre Dame, Notre Dame, IN

Saivenkataraman Jayaraman, Department of Chemical and Biomolecular Engineering, University of Notre Dame, Notre Dame, IN

Jindal K. Shah, Department of Chemical and Biomolecular Engineering, University of Notre Dame, Notre Dame, IN

Edward J. Maginn, Department of Chemical and Biomolecular Engineering, University of Notre Dame, Notre Dame, IN

Despite the development of sophisticated methods to study vapor-liquid equilibrium, the development of methodologies to study solid-liquid equilibrium has received much less attention due to the inherent complexity involved with the calculation of the free energy of solids. In this talk, we will demonstrate a robust method to calculate the solubility limit of solids in different solvents. The method combines pseudo-path integration for the free energy calculations of the solid phase, and recently developed Wang-Landau Transition-Matrix Monte Carlo for phase equilibria calculations. We validate the method by computing the solubility limits of sodium chloride in water.

See more of this Session: Thermodynamic Properties and Phase Behavior I
See more of this Group/Topical: Engineering Sciences and Fundamentals

Blowing Bubbles in Lennard-Jonesium along the Saturation Curve

Hank Ashbaugh, Department of Chemical and Biomolecular Engineering, Tulane University, New Orleans, LA

Solvation plays a crucial role in multi-scale processes from gas solubility, to the mesoscopic organization of self-assembled structures, to macroscopic phase separation. We have performed extensive molecular simulations of the Lennard-Jones fluid to determine its liquid-vapor coexistence properties, and solvent contact densities with cavity solutes up to 10 times the diameter of the solvent from the triple point to the critical point. These simulations are analyzed using a revised scaled-particle theory to evaluate the thermodynamics of cavity solvation and curvature dependent interfacial properties along the saturation curve. While the thermodynamic signatures of cavity solvation are distinct from those in water, exhibiting a chemical potential dominated by a large temperature independent enthalpy, the solvent dewets cavities of increasing size similar with water near coexistence. The interfacial tension for forming a liquid-wall interface is found to be consistently greater than the liquid-vapor surface tension of the Lennard-Jones fluid by up to 10%, and potentially reflects the suppression of high amplitude fluctuations at the cavity surface. The first-order curvature correction for the surface tension is negative and appears to diverge to negative infinity at temperatures approaching the critical point. Our results point to the success of the revised scaled-particle theory at bridging molecular and macroscopic descriptions of cavity solvation.

See more of this Session: Thermodynamic Properties and Phase Behavior I
See more of this Group/Topical: Engineering Sciences and Fundamentals

Two-Phase Molecular Dynamics Simulations of n-Propane Vapor-Liquid Equilibria

Sonal Patel, Chemical Engineering, Brigham Young University, Provo, UT

W. Vincent Wilding, Chemical Engineering, Brigham Young University, Provo, UT

Richard L. Rowley, Chemical Engineering, Brigham Young University, Provo, UT

A molecular dynamics (MD) method [Fern et al., J. Phys. Chem. B 2007, 111, 3469.] recently developed for determination of phase equilibrium in a single, two-phase simulation has been shown to be accurate even near the critical point because complete phase separation and no particle insertions are required. In this work, we modify the method for structured molecules by employing a Monte Carlo (MC) method to determine the associated volume for every particle in the simulation cell. Bulk liquid and vapor densities are then determined by comparing the molecular volume distributions of liquid and vapor in the two-phase simulations to those obtained in single-phase simulations. Two-phase molecular dynamic (MD) simulations were performed on n-Propane using a Transferable Potential for Phase Equilibria (TraPPE) model. The coexisting liquid and vapor densities obtained agreed well with literature values from Gibbs Ensemble Monte Carlo (GEMC) simulations.

See more of this Session: Thermodynamic Properties and Phase Behavior I
See more of this Group/Topical: Engineering Sciences and Fundamentals

Bi-Velocity Hydrodynamics

Howard Brenner, MIT, Cambridge, MA

This talk presents experimental and theoretical evidence that ordinary viscous fluid mechanics requires the use of two velocities, rather than one, in order to properly describe mass, momentum, and energy transport processes in continua. In addition to the usual velocity representing a flux of mass, the second velocity represents a flux of volume. The difference between the two constitutes a diffuse flux of volume arising from molecular motion. This independent flux is addition to the usual diffuse fluxes of momentum (Newton) and heat (Fourier). This presentation updates a recent publication: H. Brenner, Bi-velocity hydrodynamics, Physica A 388, 3391-3398 (2009).

See more of this Session: Fundamental Research in Transport Processes
See more of this Group/Topical: Engineering Sciences and Fundamentals

The Constrained Vapor Bubble Experiment for ISS - Earth's Gravity Results

Arya Chatterjee, The Isermann Department of Chemical and Biological Engineering, Rensselaer Polytechnic Institute, Troy, NY

Joel L. Plawsky, The Isermann Department of Chemical and Biological Engineering, Rensselaer Polytechnic Institute, Troy, NY

David Chao, NASA Glenn Research Center, Cleveland, OH

Ronald Sicker, NASA Glenn Research Center, Cleveland, OH

Tibor Lorik, ZIN Technologies, Cleveland, OH

Louis Chestney, ZIN Technologies, Cleveland, OH

John Eustace, ZIN Technologies, Cleveland, OH

John Zoldak, ZIN Technologies, Cleveland, OH

Peter C. Wayner Jr., The Isermann Department of Chemical and Biological Engineering, Rensselaer Polytechnic Institute, Troy, NY

The constrained vapor bubble (CVB) experiment is one of the significant investigations into fluid physics in microgravity conditions being undertaken by NASA. Destined to fly on the International Space Station by the end of 2009, the experiment will yield data that will provide new insight into the complex phenomena occurring in systems such as nucleate boiling and heat pipes. It will consist of a transparent heat pipe made from quartz using pentane as working fluid with an extraordinary level of instrumentation allowing us, for the first time, to study every aspect of the heat pipe in great detail. Earth's gravity based data was collected at NASA Glenn using an identical hardware to that from the ISS. This data will be compared with the space based data to better understand how the interfacial forces are affected in the microgravity environment where the Bond number will be very small. Here we present earth's gravity environment data and try to match it with predictions from a theoretical model. The model uses the Young-Laplace equation to calculate the fluid pressure gradient and solves the momentum equation in the liquid and vapor phases. It also takes into account the heat conduction in the solid quartz and calculates the temperature profile along the heat pipe device which can be matched with the experimental data. Predictions will also be made about how the heat pipe device is expected to behave in space.

See more of this Session: Fundamental Research in Transport Processes
See more of this Group/Topical: Engineering Sciences and Fundamentals

Numerical Study of Contact Line Jumping in Drop Dewetting

Moniraj Ghosh, Chemical and Biomolecular Engineering, Johns Hopkins University, Baltimore, MD

Kathleen J. Stebe, Chemical and Biomolecular Engineering, UNIVERSITY OF PENNSYLVANIA, Philadelphia, PA

In this work we demonstrate the boundary conditions applied at the contact line can profoundly affect the resulting dynamics of a drop dewetting from an unfavorable surface.

In a 2-D simulation, we study the dynamics of a drop whose contact line encounters a change in its surface energy from a well wet region to a poorly wet region. The dynamics of contact line pinning, depinning and hopping is studied as the film thins, forming a ridge which propagates toward the bulk drop as the contact line recedes. We simulate the motion of the drop in the lubrication limit using two commonly adopted mechanisms to prevent the singular behavior at the contact line. In one approach, we assume the existence of an infinite, ultrathin, equilibrium liquid film ahead of the contact line that is stabilized by the balance of surface tension and disjoining pressure. In the other, we allow finite slip at the contact line.

When adopted for wetting studies, these two mechanisms predict largely the same wetting dynamics and free surface shapes outside the cutoff region. Simulations using either cut-off mechanism generate the universal (and experimentally observed) Hoffman's law ($\theta^3 \infty Ca$) at low to moderate capillary number. Thus, macroscopic wetting dynamics are insensitive to the local displacement mechanism close to the contact line.

In contrast, we find that the contact line boundary conditions give differing dynamics for drop dewetting. We compare the basic features of the evolution the film as given by the two models. Finally, we study how surfactants affect the flow.

See more of this Session: Fundamental Research in Transport Processes
See more of this Group/Topical: Engineering Sciences and Fundamentals

Differential Kinetics of Polyox Degradation During Drag Reduction in a Smooth Pipe

Preetinder S. Virk, Department of Chemical Engineering, Massachusetts Institute of Technology, Cambridge, MA

The effectiveness of drag-reducing polymer additives invariably decreases with increasing downstream distance, on account of polymer degradation by scission in the turbulent flow. The specific objective of the present work was to differentiate between the degradation of polymer in the entrance region and the degradation that occurs in the downstream test pipeline. The importance of distinguishing between these two types of degradation is that only the latter, "intrinsic", degradation in the test pipeline section is meaningful for scaling up laboratory results to real pipelines. The present experiments employed 0.2 to 20 wppm aqueous solutions of a polyethyleneoxide (PEO) polymer, molecular weight ~5x106, flowing through a smooth 5 mm ID test pipe, with electro-polished bore, of length about 180 diameters, made up of six pressure-tapped segments. The test pipe was part of a single-pass progressive cavity pump-driven flow system fed from two 180 liter tanks that held premixed polymer solutions. Pressure differences between each of six tap pairs were measured by a transducer and logged on a data acquisition computer. Degradation was sensitively detected as a bifurcation between the pressure drops measured between upstream (j) and downstream (k) segments of the test pipe at fixed flowrate. In the polymeric regime of drag reduction, degradation was found to be insignificant from onset to a characteristic "falloff" wall shear stress ƒäw^ and became significant thereafter with the ratio of falloff to onset wall shear stresses ~5 essentially independent of polymer concentration. The severity of degradation, measured by the estimated ratio of (still-active/original) polymer concentrations at a section, was found to be essentially independent of polymer concentration but increased significantly with increasing wall shear stress. At a fixed wall shear stress, for example Tw = 200 Pa, corresponding to solvent Re ~ 50000, the apparent first-order degradation rate constant kapp ~ 2 s-1. An attempt is made to physically interpret these degradation results in terms of the polymeric sublayer model of drag reduction.

See more of this Session: Fundamental Research in Transport Processes
See more of this Group/Topical: Engineering Sciences and Fundamentals

Simulation of Electrohydrodynamic Atomization for Enhanced Particle Collection Efficiency in an Encapsulation Chamber

Abstract ID: 165682
Alireza Rezvanpour and Chi-Hwa Wang

Department of Chemical and Biomolecular Engineering, National University of Singapore, Singapore 117576

Introduction

Electrohydrodynamic atomization (EHDA) is a promising method for the fabrication of micro- and nano-sized particles with narrow size distribution and better morphologies than a few conventional methods of particle fabrication such as spray drying. Further enhancement in monodispersity and morphology of particles can also be achieved using an enclosed shuttle chamber with controlled solvent evaporation rate during the fabrication process. The objective of this study is to simulate the fluid and particle dynamics in the EHDA chamber, thereby providing a means of predicting collection efficiency prior to experimental verification.

EHDA process involves a combination of hydrodynamics, aerodynamics and electrostatics. Several factors influence the process such as inert gas flow rate, solution flow rate, nozzle-to-collecting plate distance, droplet size, charge-to-mass ratio, etc. Computational fluid dynamics (CFD) softwares like FLUENT can be used to model the mechanical section. Although FLUENT does not offer a direct solution for the electrostatic field, its user-defined functions can be used to solve the Poisson field by including it into the general scalar transport equations within FLUENT [1].

Forces governing particle motion

For particles moving in a fluid, buoyancy, gravitation and drag are three mechanical forces that influence their trajectories. Buoyancy occurs as a result of the uneven pressure exerted by the fluid at different regions of a particle's surface. When coupled together with gravitational force, the combined effect is known as buoyant weight, and it is mathematically expressed as:

$$\mathbf{F_B} = \mathbf{g}(\rho_p - \rho_f)v_p \qquad (1)$$

Where $\mathbf{g}$ is the gravitational acceleration, ρ_p and ρ_f are densities of the particle and fluid respectively, and v_p is the volume of a particle. The buoyant weight is an option selectable in the Discrete Phase Model of FLUENT, which is used to simulate particle trajectory. On the other hand, drag force occurs as a result of friction between the particle and the fluid. Drag force and Reynolds equations are expressed as:

$$\mathbf{F}_D = \frac{3\mu C_D \mathrm{Re}\, v_p}{4d_p^2}(\mathbf{u}_f - \mathbf{u}_p) \qquad (2)$$

$$\mathrm{Re} = \frac{\rho_f d_p}{\mu}(u_p - u_f) \qquad (3)$$

Where C_D is the drag coefficient, $\mathbf{u}_p$ and $\mathbf{u}_f$ are velocities of the particle and fluid respectively, μ is the viscosity of the fluid, and d_p is the volume of a particle. The drag coefficient is calculated internally by FLUENT using correlations. Besides mechanical forces, an electrical force also acts on each particle as a result of the electric potential field around it. Two dominant classes of entities contribute to electric potential (Φ) in the EHDA chamber. The first type of objects is stationary surfaces which are charged conductors. These comprise of the nozzle, ring and collection plate. The electric potential due to such entities can be obtained by numerically solving the Laplace equation with the appropriate boundary conditions.

$$\nabla^2.(\varepsilon\Phi) = 0 \qquad (4)$$

where ε is the electrical permittivity of vacuum. The second type of entities is the moving particles themselves, which are charged. With the addition of such entities, the Laplace Equation is no longer valid. Instead, the Poisson equation must be solved numerically for the overall electric potential.

$$-\nabla^2.(\varepsilon\Phi) = \rho \qquad (5)$$

where ρ is the known as space charge, and it is the charge per unit volume present at each point within a system. Note that the Laplace equation is just a simplified form of the Poisson equation with space charge set to zero. Hence the Poisson equation has to be solved when both types of entities are present. The resulting electric potential field will give rise to an electrical force as follow:

$$\mathbf{F}_E = -q\nabla\Phi \qquad (6)$$

where q is the charge of a particle. After combining all the forces acting on a particle and then dividing the overall force equation by the mass of each particle, the acceleration of the particle is obtained. This is the actual equation that FLUENT solves (with the exception of the electrical force term). Note that q_m in the acceleration equation denotes charge-to-mass ratio.

$$\frac{d\mathbf{u}_p}{dt} = \frac{\mathbf{g}(\rho_p - \rho_f)}{\rho_p} + \frac{3\mu C_D \mathrm{Re}}{4\rho_p d_p^2}(\mathbf{u}_f - \mathbf{u}_p) - q_m\nabla\mathbf{\Phi} \qquad (7)$$

Implementation of electrical force in FLUENT

The generic force equation in FLUENT is exactly identical to the acceleration equation derived in the previous section except that the electrical force term is removed (see below).

$$\frac{d\mathbf{u}_p}{dt} = \frac{\mathbf{g}(\rho_p - \rho_f)}{\rho_p} + \frac{3\mu C_D \mathrm{Re}}{4\rho_p d_p^2}(\mathbf{u}_f - \mathbf{u}_p) + \mathbf{B} \qquad (8)$$

In its place, however, is an arbitrary body force term ($\mathbf{B}$) that can be altered using the "User Defined Functions" module in FLUENT to become electrical force [2]. Hence,

$$\mathbf{B} = -q_m\nabla\mathbf{\Phi} \qquad (9)$$

Besides that, another barrier to inclusion of the electrical force term is that FLUENT does not calculate electric potential, which is required to compute electrical force. Fortunately, FLUENT has a generic transport equation that can be for an arbitrary, user-defined scalar [2]. For an arbitrary scalar ϕ, the user-defined scalar transport equation had the following form,

$$\frac{d\rho\varphi}{dt} + \nabla(\rho\mathbf{u}\phi - \Gamma\nabla\varphi) = S_\phi \qquad (10)$$

where Γ and S_ϕ are the diffusion coefficient and source term respectively, while the second term (leftmost in the bracket) represents convective flux of the scalar ϕ. Since steady-state simulations are carried out, the first term on the left-hand side of the transport equation is automatically zero. In order to modify the transport equation into the Laplace equation, the convective flux term must be disabled. In addition, ε will be used as the value for Γ. Lastly, the source term is set to zero. In order to change from the Laplace equation to the Poisson equation, the source term of the transport equation must be given the value of the space charge. The space charge at a spatial location can be computed as the charge-to-mass ratio of a particle multiplied by the mass concentration of the particles (C) at the location. Thus, the source term has the following mathematical form [1]:

$$S_\phi = \rho = q_m C \tag{11}$$

The concentration of particles simulated with Fluent's Discrete Phase Model (DPM) will be calculated by the following equation when simulations are performed with coupled interactions between the fluid and particle phases [4]:

$$C = \frac{1}{\delta_v}\sum \dot{m}\left(t_{out} - t_{in}\right) \tag{12}$$

where δ_v is the volume of a computation cell, $\dot{m}$ is the mass flow rate of a particle stream passing through the cell, and t_{in} and t_{out} denotes the time that the particle stream enters and exit the cell respectively. By replacing $\dot{m}$ with the charge flow rate of a particle stream I_P, the expression for the space charge. Hence,

$$S_\phi = \frac{1}{\delta_v}\sum I_P\left(t_{out} - t_{in}\right) \tag{13}$$

EHDA CHAMBER

The geometry of the chamber used in this simulation study is depicted in Figures 1 and 2. Pure nitrogen flowing through the single inlet is subjected to a sudden expansion before entering the chamber interiors. A small nozzle, charged to a constant voltage, is located near the top centre of the chamber and is aligned with the axis of the chamber protrusions. Surrounding the nozzle is a ring electrode, which is also positively-charged with respect to the earth. Near the exit of the chamber, an earthed collection plate is used for particle deposition. Nitrogen exits the chamber through the sudden contraction and is released to the atmosphere.

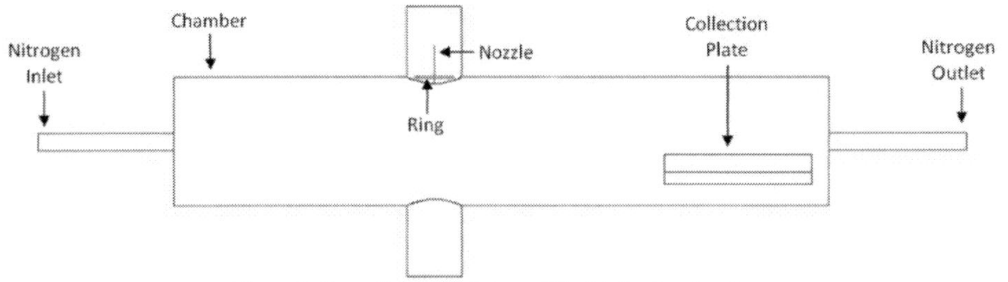

Figure 1: Two-dimensional view of the fabrication chamber.

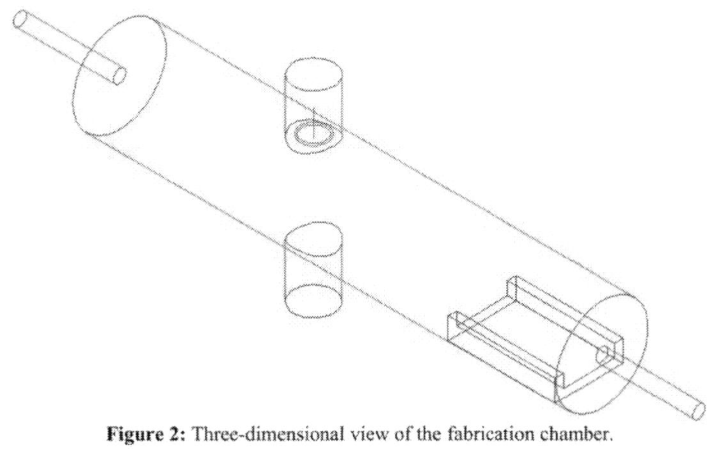

Figure 2: Three-dimensional view of the fabrication chamber.

SIMULATION PARAMETERS

In this study, the inlet nitrogen flow rate, EHDA solution flow rate, and electric potentials of the ring and nozzle are varied to determine their effects on the collection efficiency of the particles.

Nitrogen Flow Field within Chamber

In this simulation, due to the low nitrogen flow rate used (max 25 L/min) and the large chamber inlet and interior diameters (1.5 cm and 10.8 cm respectively), it is assumed that nitrogen flow within the chamber is laminar. Hence the "Laminar Viscous Model" is selected in Fluent. In addition, because the nitrogen velocity profile at the inlet is immaterial to overall accuracy, nitrogen flow is assumed to be at constant velocity normal to the inlet area (i.e. plug flow) using the "Velocity Inlet" boundary condition. The outlet of the chamber is set using the "Pressure Outlet" boundary condition; pressure at the outlet is set to zero gauge pressure. All other flow related conditions, such as no-slip condition at chamber walls, are at their default settings.

Initial Conditions for Particle Trajectory Simulations

Dichloromethane (DCM) solvent with 7% w/v PLGA P2191 (roughly 5 wt% PLGA P2191 was arbitrarily chosen as the EHDA solution for all simulations [5]. Following the works of Ganan-Calvo [6], the initial diameter d_d of the droplet formed by EHDA and the total current I generated by the process to the EHDA solution flowrate Q and solution properties can be empirically correlated using dimensionless quantities as such:

$$\frac{d_d}{d_0} = 0.92\left(\frac{Q}{Q_0}\right)^{0.5} - 0.01 \qquad (14)$$

$$\frac{I}{I_0} = 2.4\left(\frac{Q}{Q_0}\right)^{0.25} + 1.72 \qquad (15)$$

where d_0, I_0 and Q_0 are reference values. These were based on the relationships by Gavan-Calvo and have the following expressions:

$$d_0 = \left(\frac{\gamma \varepsilon_0^2}{\rho K^2}\right)^{\frac{1}{3}} \qquad (16)$$

$$I_0 = \left(\frac{\gamma^2 \varepsilon_0}{\rho} \right)^{0.5} \quad (17)$$

$$Q_0 = \frac{\gamma \varepsilon_0}{\rho K} \quad (18)$$

Where ρ and K are the density and electrical conductivity of the EHDA solution and γ is the gas-liquid surface tension. Values for DCM+PLGA P2191 (7% w/v) were obtained from [1] and tabulated in Table 1. These parameters were used to calculate the initial droplet diameter and total current at the EHDA solution flowrates used, and these results were tabulated in Table 2.

Table 1: Properties used for calculation of droplet diameter and current in Table 3.

Density of DCM+PLGA P2192 (7% w/v)	1376 kg/m^3
Electrical conductivity of DCM+PLGA P2192 (7% w/v)	1.5 x 10^{-8} s/m
Surface tension of pure DCM	4.4 x 10^{-4} kg/m-s
Electrical permittivity of vacuum	8.85 x 10^{-12} C^2/N-m^2

Table 2: Particle parameters used at each flowrate.

EHDA Solution Flowrate (mL/hr)	Droplet Diameter (μm)	Total Current (nA)
0.5	17.0697	5.64832
1.0	24.9367	5.98363
1.5	30.9732	6.20846
2.0	36.0622	6.38238
2.5	40.5457	6.52617

The ratio of the droplet diameter to the jet diameter is around 1.89. Using this relation, the EHDA solution flow rate divided by the cross-sectional area of the jet provides an estimate for the initial particle velocity, which is approximately 2 m/s [3]. To provide diversity in the particle trajectories, the particles were injected into the fluid using a solid-cone injection scheme with 0.1° cone angle and zero radius, as shown in figure 3, with initial velocities of 2 m/s.

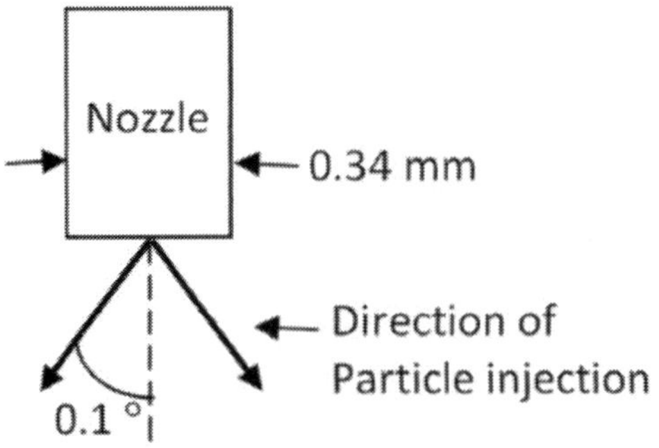

Figure 3: Particle injection pattern at nozzle.

Simulation of Particle Trajectories

For simplicity, the porosity of particles and internal diffusion of DCM and PLGA within the fabricated particles were ignored. Instead, simulations were carried out by selecting particles as multi-component with 5 wt% PLGA and the rest as DCM, and particles were assumed to be totally mixed throughout the simulation; the density of the particle is computed by assuming volume additivity using the densities of pure DCM (1326 kg/m^3) and PLGA (936.71 kg/m^3). The rate of evaporation of DCM vapor from the particle's surface was computed using a Sherwood number correlation [Fluent 6.3 User's Guide, page 22-71] and assuming that the partial pressure of DCM vapor at the particle's surface follows Raoult's law.

To obtain the convective mass transfer coefficient from the Sherwood number, the diffusion coefficient of DCM in nitrogen gas is required as a material property input. Since nitrogen is non-polar and DCM has a small dipole moment, Chapman-Enskog theory (Equation 17) was used to calculate $D_{DCM\text{-}N2}$ as 9.58113 x 10^{-6} m^2/s. This value was assumed to be constant despite the temperature of the particle dropping as DCM vaporizes. Note also that heat and mass transfer was selected to be coupled.

$$D_{AB} = \frac{0.0018858T^{1.5}}{P\sigma_{AB}^2\Omega_D}\left(\frac{1}{M_A}+\frac{1}{M_B}\right)^{0.5} \qquad (19)$$

where T is temperature, M is the molar mass, P is pressure, σ_{AB} is the average collision diameter (the values are tabulated [7]) and Ω is a dimensionless temperature-dependent collision integral (the values are tabulated [7] but usually of order of 1).

Boundary Conditions Common to All Simulations

Gravity is set to 9.807 m/s^2. The boundary conditions of the user-defined scalar that represent electric potential are as follow: (1) All surfaces of the nozzle are set at a constant value stated in Table 1, (2) the ring surface is set at a constant value stated in Table 1, (3) surfaces of the collector plate are set at a constant value of zero and (4) the flux of the user-defined scalar normal to each of the remaining surfaces are set to zero (note that in this manner, the inlet and outlet are treated as though they are insulators). All surfaces were set to be isothermal at 300 K and particles were selected to be trapped only at the collector surface; it is recognized by Fluent as 'escaped' when it touches the ring, nozzle, chamber walls, and chamber inlets and outlets. Note that having a trapped status means that all DCM still present in the particle will automatically vaporize once the particle reaches the collector surface while the escaped status means that the whole particle's mass is lost from the system. This is to allow easy computation of collection efficiency (CE) as

$$CE = \frac{Trapped}{Trapped + Escaped + Incomplete} \times 100 \qquad (20)$$

In another words, Particle trajectories in simulations will end in one of the following three ways: 1) coming into contact with the collection plate and are counted as "Trapped", 2) coming into contact with boundaries other than the collection plate and are counted as "Escaped", and 3) remain circulating within the chamber due mainly to drag and electrical forces and are counted as "Incomplete".

Solution Strategy for Simulations using Poisson Field

For all simulations, the particles are not tracked initially. The flow field and the user-defined scalar are solved until their residuals become constant. For the user-defined scalar's transport equation, the Laplace equation is used. After these two fields had converged, particle injections are switched on. At the same time, the Laplace field is changed to the Poisson field by including the space charge as a source term for the user-defined scalar. Each particle tracking iteration is set to take place once every 15 iterations of the flow field and the user-defined scalar. After 500 iterations (around 32 or 33 DPM iterations), the simulations are stopped. Note that due to the cross-flow nitrogen, slight changes in the trajectories of the particles will result in an entirely changed electric potential field; hence it is computational impossible to obtain convergence for particle trajectory simulations. Instead, the results of the last 10 DPM iterations are recorded for calculation of efficiencies.

RESULTS AND DISCUSSION

Effect of Varying Nitrogen Flow rate on Collection Efficiency

Figure 4 shows the variation of collection efficiency versus nitrogen flow rate, EHDA solution flow rate, ring electrical potential and nozzle electrical potential. From figure 4-a, it could be observed that at an optimal nitrogen flow rate of 17.5 L/min, the highest collection efficiency was obtained. Increasing in nitrogen flow rate below 17.5 lit/min results in enhanced turbulence and eddy of nitrogen flow inside the chamber. This enhancement can indeed increase the corresponding drag force. However, increasing in nitrogen flow rate from 17.5 lit/min can increase this turbulence which can cause the particles to circulate inside the chamber and increase the chance of particle deposition on the chamber's walls before reaching the collecting plate.

Figure 4-b shows the variation of collection efficiency versus solution flow rate. As seen in the figure, collection efficiency generally increases with the EHDA solution flow rate until the flow rate is 1.5 mL/hr. Thereafter, collection efficiency apparently reaches to a constant value around 45%. This is perhaps due to the increased charge flow rate resulting in a greater electric deposition force exerted on the particles.

Figure 4-c shows the variation of collection efficiency versus ring electric potential. As seen in the figure, collection efficiency tends to increase as the ring electric potential increases. This effect is observed because higher ring potential results in greater electrical force exerted on the chamber region between the ring and the collection plate.

Figure 4-d shows the variation of collection efficiency versus nozzle electric potential. From the figure, there was no apparent trend between collection efficiency and the nozzle electric potential. This is perhaps because changes in the nozzle electrical field had minimal effects on the electrical field strength in the region between the ring electrode and the collection plate.

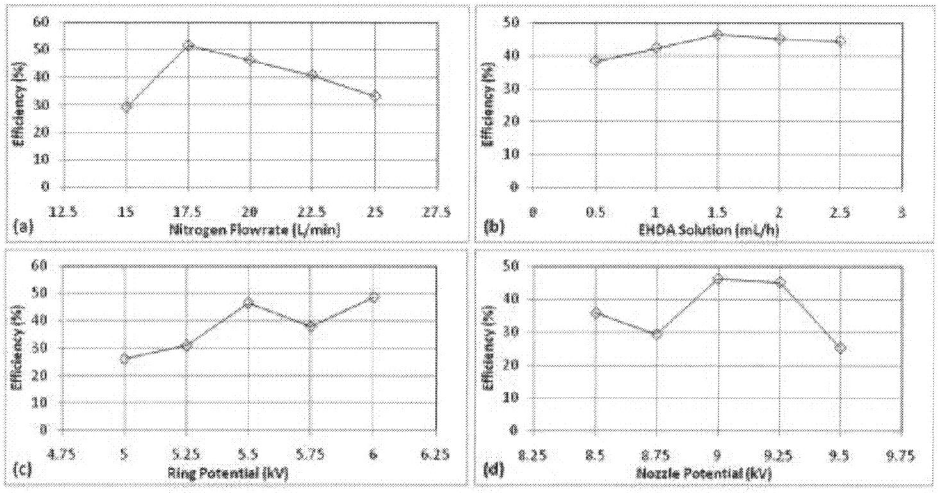

Figure 4: Variation of collection efficiency versus (a) nitrogen flow rate (b) EHDA solution flow rate (c) Ring electrical potential, (d) Nozzle electrical potential.

Figure 5 shows the post-processing results of one of the simulation results. In this trial, nitrogen flow rate is 22.5 L/min, solution flow rate is 1.5 mL/hr, nozzle voltage is 9 kV and ring voltage is 5.5 kV. These figures confirm that the maximum dispersity and turbulence at this nitrogen flow rate can be seen in collecting region which can results in increasing in flight time of the particles. The effect of increasing in flight time can be seen in figure 5-d.

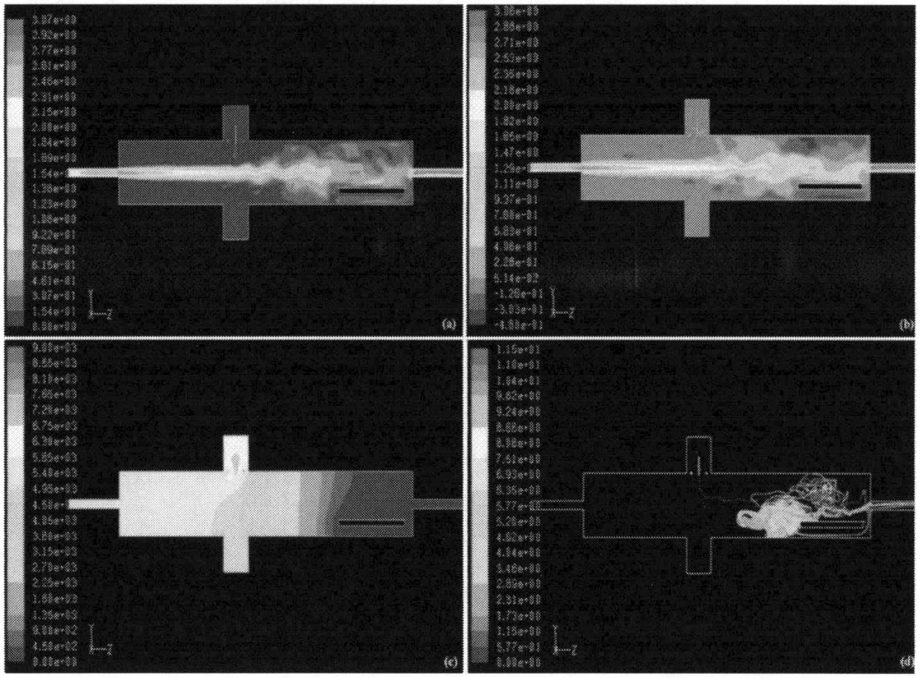

Figure 5: Post-processing results of simulation of EHDA process in encapsulation chamber which shows (a) Flow field (velocity magnitude) (b) Flow field (z-velocity) (c) Electric potential field (d) Particle trajectory.

CONCLUSIONS

Simulating the fluid and particle dynamics in the EHDA chamber was investigated in this study. The results of this study can help to provide a means of predicting collection efficiency prior to experimental verification. The first part of the simulations focused on identifying the effects of the charge-to-mass ratio of particles on their collection efficiency. This was necessary because usage of the chamber restricts measurements of the actual charge-to-mass ratio. In the second part of the study, the Poisson equation was solved for the prediction of electrical force due to charge repulsion in the simulation of particle trajectories. For particles moving in a fluid, buoyancy, gravitation and drag are the three major mechanical forces that influence their trajectories. All the forces acting on a particle were combined to determine its acceleration and trajectory based on the Lagrangian approach.

The results confirm that there is a maximum value for nitrogen flow rate and EHDA solution flow rate in which the maximum collection efficiency can be observed. Moreover, collection efficiency tends to increase as the ring electric potential increases; while, there is no apparent trend between collection efficiency and the nozzle electric potential.

ACKNOWLEDGEMENT

This work is supported by the National University of Singapore under the grant number R279-000-222-112.

REFERENCES:

[1] Zhao SX, Adamiak K, Peter Castle GS. The implementation of poisson field analysis within fluent to model electrostatic liquid spraying. Canadian Conference on Electrical and Computer Engineering 2007;1456-1459.

[2] FLUENT 6.3 User's Guide, Fluent Inc., 2006.

[3] Xie JW, Lim LK, Phua YY, Hua JS, Wang CH. Electrohydrodynamic spraying functioning modes: A critical review. Journal of Aerosol Science 1994;25:1021-1036.

[4] Ye Q, Steigleder T, Scheibe A, Domnick J. Numerical simulation of the electrostatic powder coating process with a corona spray gun. Journal of Electrostatics 2002;54:189-205.

[5] Yao J, Lim LK, Xie JW, Hua JS, Wang CH. Characterization of electrospraying process for polymeric particle fabrication. Journal of Aerosol Science 2008;39:987-1002.

[6] A. M. Ganan-Calvo, J. C. Lasheras, J. Davila and A. Barrero, The electrostatic spray emitted from an electrified conical meniscus, J. Aerosol Sci., Vol. 25, No. 6, pp. 1121-1142, 1994.

[7] J. Hirschfelder, C. F. Curtiss, R. B. Bird: Molecular Theory of Gases and Liquids. Wiley, New York, 1954.

CONDUCTION CONTROLLED FREEZING of PCMs IN CYLINDERS

Subrahmaniyam Susarla, -------------, METHODIST COLLEGE OF ENGINEERING AND TECHNOLOGY, Hyderabad, India

Freezing and melting of materials in cylinders have several applications including solar energy and processing industry.A knowledge of partial and total freezing rates is vital for design of such systems.In renewable energy literature the material undergoing phase change is often referred as PCM. It is known that freezing rate depends on geometry of PCM holder,any initial superheat of PCM,temperature difference between PCM and cooling surface and thermophysical properties of solid and liquid.In addition,the nature of solid/liquid interface during freezing complicates the process.From literature and our preliminary experiments ,significant differences between predictions and data were noted for PCMs n -eicosane and n-octadecane.The practical frozen mass is above theoretical value at any instant for isothermal cooling surface condition. In the present work,we intend to make a detailed study using three PCMs.In addition the process of freezing is carried out in cylinders of different diameters.The general details are given below. PCM: n-octadecane(Melting point 28C) Lauric acid (MP 42C) Wax(MP 58-60C) Volume:30 to 500cc Cylinder(L/d);1 to 5 Temperature drop across PCM:3 to 50C Isothermal water baths are used for freezing of PCM. The experiments are being coducted with negligible superheat.The freezing process is therefore predominantly conduction controlled.The partial frozen mass and time data obtained would cover a broad range of mentioned variables as well as a range of thermophysical properties of PCM.This would enable a critical evaluation of literature models and also the limitations of theoretical models would be discussed.

See more of this Session: Fundamental Research in Transport Processes
See more of this Group/Topical: Engineering Sciences and Fundamentals

Density Functional Theory Study of the Initial Oxidation of the Pt(111) Surface

Jeffery M. Hawkins, Chemical Engineering, University of Florida, Gainesville, FL

Jason F. Weaver, Chemical Engineering, University of Florida, Gainesville, FL

Aravind R. Asthagiri, Department of Chemical Engineering, University of Florida, Gainesville, FL

Expanding the fundamental understanding of catalytic behavior under oxygen-rich conditions on transition metal surfaces, such as Pt(111), has recently garnered interest, since the ability to operate internal combustion engines in fuel-lean, oxygen-rich environments could lead to dramatic improvements in fuel efficiency. However, new catalysts capable of handling the increased NOx emissions would be required. One area of particular importance in the development of such catalysts is to understand the transition of oxygen phases on metal surfaces from chemisorbed oxygen to surface and bulk oxides, since these oxide phases can have dramatically different reactivity. We have used density functional theory (DFT) calculations to examine the initial stages of oxidation of the Pt(111) surface and have determined that subsurface oxygen is not the precursor to oxidation. Instead, we predict a strong preference for the formation and growth of one-dimensional Pt oxide chains within the p(2×1) structure which exhibit large buckling and modification of the charge of the surface Pt atoms. Furthermore, half of the oxygen atoms in the Pt oxide chains reside near hcp sites, contrary to reports that oxygen atoms reside only on fcc sites on Pt(111). Our results agree well with a recent scanning tunneling microscopy study and suggest a precursor mechanism to the oxidation of metal surfaces involving Pt oxide chain formation and growth on terraces at moderate oxygen coverages. We will also briefly discuss the implications of these new structures to current models of CO and NO oxidation on Pt(111).

See more of this Session: Fundamentals of Surface Reactivity
See more of this Group/Topical: Engineering Sciences and Fundamentals

Preparation Characterization and Surface Chemical Reactions of Model Cu-ZnO/Silica Sorbents Upon H2S Adsorption

Divya Repala, Chemical engineering, Auburn University, Auburn, AL

Alexander Samokhvalov, Center for Microfibrous Materials Manufacturing, Department of Chemical Engineering, Auburn University, Auburn, AL

Bruce J. Tatarchuk, Chemical Engineering Department, Auburn University, Auburn, AL

Metal dopants are known to play significant role in improving the performance of ZnO-based supported sorbents used for H2S removal. Lifespan of the supported doped sorbents depends on the conditions of oxidative regeneration of the sulfided ("spent") sorbents that affect changes in oxidation, coordination states, dispersion and ion diffusion off and towards the active surface. Well-characterized model sorbents are often used in combination with surface science techniques to study the above processes. Among model sorbents, those prepared under the conditions resembling industrial ones, i.e. pressure, temperature and choice of precursors are especially useful.

Model supported sorbents have been prepared by depositing mixed Cu-Zn oxides of variable stochiometry from aqueous solutions of nitrate precursors onto oxidized silicon wafers by chemical spray pyrolysis technique under controlled atmosphere conditions ("calcined sorbents"). Further, model "calcined" sorbents were reacted with H2S gas forming "sulfided sorbents". Morphology of model sorbents was studied by SEM-EDX. Oxidation states and atomic ratios of the elements were confirmed by XPS.

Regeneration of "sulfided" sorbents by oxygen was performed in-situ by temperature-dependent XPS. Changes in oxidation states of Sulfur and metal ions present in the sorbents with respect to the temperature and partial pressure of oxygen have been analysed.

See more of this Session: Fundamentals of Surface Reactivity
See more of this Group/Topical: Engineering Sciences and Fundamentals

CO Dissociation Induced by Adsorbed Oxygen and Water On Ir(111)

Ming Pan, Chemical Engineering, The University of Texas at Austin, Austin, TX

Son Hoang, Chemical Engineering, The University of Texas at Austin, Austin, TX

Jinlong Gong, Chemical Engineering, The University of Texas at Austin, Austin, TX

C. Buddie Mullins, Chemical Engineering, The University of Texas at Austin, Austin, TX

It is well-known that CO dissociation is a critical step for Fischer-Tropsch synthesis (FTS) which has been widely investigated theoretically and experimentally. In the FTS process, carbon monoxide reacts with hydrogen to produce hydrocarbon alternatives to petroleum. Due to the fundamental importance of C-O bond cleavage for the FTS process, we have studied dissociation of CO on the Ir(111) single crystal surface under ultra high vacuum (UHV) conditions by employing isotopically labeled reactants and temperature programmed desorption (TPD) measurements. Firstly, our investigations indicated that dissociation of CO is immeasurable by our techniques on clean Ir(111). However, we found that co-adsorbed atomic oxygen induces CO dissociation on the Ir(111) surface [e.g., 13C16O was detected desorbing from 13C18O and 16O co-adsorbed on Ir(111)]. We observed similar phenomenon on CO covered Ir(111) when water alone or both water and oxygen were co-adsorbed, likely indicative of the importance of the formed hydroxyl groups since we have found a ~2% dissociation probability of water and a strong interaction between adsorbed atomic oxygen and water on Ir(111). When H218O and 16Oa were co-adsorbed on Ir(111), the TPD spectra show the production of H216O, 16O18O, and 18O2, as well as the occurrence of a new desorption feature at 235 K indicating dissociation of water in which adsorbed hydroxyl groups (·OH) are likely intermediates. To our best knowledge, this is the first observation of hydroxyl group and oxygen atom induced CO dissociation on metal surfaces. These findings provide deeper insight into reaction mechanisms involving co-adsorbed carbon monoxide and oxygen (e.g., CO oxidation) on metal-based catalysts.

See more of this Session: Fundamentals of Surface Reactivity
See more of this Group/Topical: Engineering Sciences and Fundamentals

Structure Sensitivity in the Reaction of Ethyl Groups On Model Chromia Surfaces: Ethyl Chloride On α-Cr_2O_3 (10$\underline{1}$2) and (0001)

John D. Brooks, Chemical Engineering, Virginia Polytechnic Institute and State University, Blacksburg, VA

David F. Cox, Chemical Engineering, Virginia Polytechnic Institute and State University, Blacksburg, VA

The reactions of ethyl intermediates are studied on both the α-Cr_2O_3 (10$\underline{1}$2) and (0001) surfaces for insight into the chemical details of the catalytic dehydrogenation of ethane over chromia [1]. The α-Cr_2O_3 (10$\underline{1}$2) and (0001) surfaces were chosen because they exhibit the lowest free energies of the low-index surfaces [2], with (10$\underline{1}$2) being the predominantly exposed face on microcrystalline α-Cr_2O_3 powders [3]. The two surfaces differ primarily in terms of the local coordination of the surface Cr^{3+} cations: five coordinate for (10$\underline{1}$2), and three coordinate on (0001).

Ethyl (CH_3CH_2-) intermediates are formed from the dissociative adsorption of ethyl chloride, via C-Cl bond cleavage. In thermal desorption on the (10$\underline{1}$2) surface, the surface ethyl group reaction is initiated via a rate-limiting β-hydride elimination near 500 K in thermal desorption experiments forming CH_2=CH_2 and surface hydrogen. Subsequently, two parallel competing reactions form CH_3-CH_3, via α-hydrogen addition to remaining surface ethyl groups, and H_2, via the combination of two surface hydrogen atoms. On the (0001) surface, two seperate reaction channels are observed for surface ethyl species that appear to occur via the same elementary reaction steps seen on (10$\underline{1}$2), and produce similar products: CH_2=CH_2, CH_3-CH_3 and H_2. The lower-temperature reaction channel produces all three products near 500 K with approximately the same activation barrier to reaction observed on the (10$\underline{1}$2) surface. A higher-temperature reaction channel at 600 K produces CH_2=CH_2 and H_2, but no CH_3-CH_3. The lack of CH_3-CH_3 production from the higher energy channel suggests isolated reaction sites where remaining ethyl group coverage is limited, and the recombination of surface hydrogen from the rate-limiting production of CH_2=CH_2 is the preferred pathway.

The chlorine freed from the dissociative adsorption of ethyl chloride on both surfaces binds at the surface Cr sites and inhibits the surface chemistry by simple site blocking. No surface carbon deposition is observed from reactions of ethyl intermediates on either surface and indicates that ethyl intermediates are not principle coke formers in the dehydrogenation of ethane over (10$\underline{1}$2) or (0001) facets of α-Cr_2O_3.

References

1. B.M. Weckhuysen, R.A. Schoonheydt, Catalysis Today 51 (1999) 223.

2. P.J. Lawrence, S.C. Parker, P.W. Tasker, Journal of the American Ceramic Society 71 (1988) C389.

3. D. Scarano, G. Spoto, S. Bordiga, G. Ricchiardi, A. Zecchina, Journal of Electron Spectroscopy and Related Phenomena 64-5 (1993) 307.

See more of this Session: Fundamentals of Surface Reactivity
See more of this Group/Topical: Engineering Sciences and Fundamentals

Reaction of Methylene On α-Cr_2O_3(1012): A Prototype for Surface Carbene Coupling Reactions On Non-Reducible Oxides

Chad M. Byrd, Chemical Engineering, Virginia Polytechnic Institute and State University, Blacksburg, VA

David F. Cox, Chemical Engineering, Virginia Polytechnic Institute and State University, Blacksburg, VA

CH_2I_2 has been used to examine the reaction of methylene on á-Cr_2O_3(1012). XPS shows that C-I bond breaking occurs at temperatures lower than 250 K, and indicates that both methylene and I adatoms bind at surface Cr cation sites. Thermal desorption experiments indicate multiple reaction routes for surface methylene including dehydrogenation to H_2 and surface carbon, hydrogenation to CH_3 and methane and coupling to ethene in two separate reaction channels with different activation barriers. Since surface Cr cations on Cr_2O_3(1012) expose predominantly one coordination vacancy, the coupling reaction indicates that methylene surface diffusion likely plays an important role in the coupling reaction.

Density functional theory (DFT) has been used to examine the diffusion and coupling of methylene. DFT indicates that methylene binds preferentially to surface Cr cations as seen experimentally with XPS, but also predicts a local minimum at bridge sites between neighboring coordinately-unsaturated surface cations and anions. Two possible diffusion paths have been identified: (1) a lower-barrier route involving direct cation-to-cation diffusion across a surface trough, and (2) a higher-barrier route involving indirect cation-to-cation diffusion through the bridging methylene species. These predicted barriers to diffusion are in good agreement with experimental activation barriers to coupling observed in thermal desorption experiments.

See more of this Session: Fundamentals of Surface Reactivity
See more of this Group/Topical: Engineering Sciences and Fundamentals

Density Functional Theory Studies of the Reactivity of Oxide Phases On Pd(111)

Aravind R. Asthagiri, Department of Chemical Engineering, University of Florida, Gainesville, FL

Jeffery M. Hawkins, Chemical Engineering, University of Florida, Gainesville, FL

Jason F. Weaver, Chemical Engineering, University of Florida, Gainesville, FL

Understanding the reactivity of oxide phases on transition metal surfaces under oxygen-rich conditions is valuable to an array of catalytic processes. Several oxide phases that form on Pd(111), such as PdO(101) and Pd5O4, have recently been identified, and their structures have been determined through low-energy electron diffraction (LEED) and scanning tunneling microscopy (STM). We have performed density functional theory (DFT) calculations on the adsorption and reactivity of these oxide phases to H, H2, H2O, and CO. We will discuss the behavior of the oxide phases in the presence of these adsorbates, as well as compare our results to the those of the pure Pd(111) surface. In general, the two-dimensional surface oxides have very distinct reactive behavior from the three-dimensional oxides that form under higher oxygen coverages. We will also compare our DFT results with recent temperature programmed desorption (TPD) studies.

See more of this Session: Fundamentals of Surface Reactivity
See more of this Group/Topical: Engineering Sciences and Fundamentals

Evaluating Metal Organic Frameworks as Smart Nanoporous Preconcentrators for Explosive Sensing

Ruichang Xiong, Chemical and Biomolecular Engineering, University of Tennessee, Knoxville, TN

Johnathan P. Luna, Chemical and Biomolecular Engineering, University of Tennessee, Knoxville, TN

David J. Keffer, Chemical and Biomolecular Engineering, University of Tennessee, Knoxville, TN

Miguel Fuentes-Cabrera, Center for Nanophase Materials Sciences and Computer Science and Mathematics Division, Joint Institute for Computational Science, University of Tennessee and Oak Ridge National Laboratory, Oak Ridge, TN

Donald M. Nicholson, Computer Science and Mathematics Division, Oak Ridge National Laboratory

Andrea Mcihalkova, Computational Center for Molecular Structure and Interactions, Jackson State University

Tetyana Petrova, Computational Center for Molecular Structure and Interactions, Jackson State University

Jerzy Leszczynski, Computational Center for Molecular Structure and Interactions, Jackson State University

Khorgolkhuu Odbadrakh, Department of Physics, West Virginia University

James P. Lewis, Department of Physics, West Virginia University

In order for explosive molecules to be detected, the amount delivered to a sensor must exceed its detection threshold. Pre-concentrators facilitate this process by extracting explosive molecules from the atmosphere and delivering them to sensors in amounts required for reliable detection. Current pre-concentrators are not very selective and hence deliver material that is predominately contaminant. With the ability to tailor Metal Organic Frameworks (MOFs), there is hope that a pre-concentrator can be designed to selectively adsorb explosive molecules in the presence of contaminants. In order to test the feasibility of selectively pre-concentrating explosive molecules for detection, molecular simulations of RDX within IRMOFs (IRMOF-1, IRMOF-3, IRMOF-10, IRMOF-10 with amine group) were performed with and without contaminants. The simulations give new insight into the competitive nature of the adsorption and the binding of the explosive molecules to the framework. Adsorption isotherms, mobility of the RDX within the framework, the occupancy of each type of cage within the unit cell, and the framework's effect on the configuration of the explosive molecules were investigated by grand canonical Monte Carlo (GCMC) and molecular dynamics (MD).

See more of this Session: Interfacial Aspects in Nanosensors
See more of this Group/Topical: Engineering Sciences and Fundamentals

Optical Sensing at Nanostructured Surfaces: Surface Plasmon Enhanced Transmission through Surfaces with Nanostructured Topology

Andrew C. Hillier, Iowa State University, Ames, IA

Wei-Hsun Yeh, Department of Chemical and Biological Engineering, Iowa State University, Ames, IA

Bipin Singh, Iowa State University, Ames, IA

Sensors based on surface plasmon resonance (SPR) have become increasingly popular as a label-free method for measuring the binding of analytes to functionalized surfaces and in the detection of immobilized biomolecules. One of the key attributes of SPR sensing is that it eliminates the complex labeling/conjugation steps required of competing techniques that utilize fluorescently labeled molecules in detection assays. In addition, the large field-enhancement associated with SPR is the basis for many surface analytical techniques such as surface-enhanced Raman scattering and surface-enhanced fluorescence.

Although SPR is typically achieved with prism-coupled configurations, it has recently been discovered that plasmon-enhanced transmission of light can be achieved at nanostructured surfaces, such as nanometer-sized hole and slit arrays. In this work, we demonstrate the appearance of surface plasmon enhanced optical transmission through nanostructured grating surfaces. Enhanced light transmission is observed over a narrow wavelength range corresponding to conditions that excite surface plasmons at the gold-air interface. We explore the nature of these transmission peaks to identify the origins of the enhanced transmission as well as the impact of changing the surface profile, including the grating pitch and amplitude, in order to fine tune the optical response. A novel chirped grating based upon spontaneous buckling of an elastomeric film is used to modulate the grating structure and demonstrate the impact of surface topology. In addition, the utility of this device as an ex-situ sensor is demonstrated by measuring the thickness of various thin films.A model immunosensing application is also examined involving the formation of immunocomplexes between bovine serum albumin (BSA) and anti-BSA. This work demonstrates the utility and versatility of this grating-based SPR sensing technique.

See more of this Session: Interfacial Aspects in Nanosensors
See more of this Group/Topical: Engineering Sciences and Fundamentals

Doped Metal Oxide Nanowire Sensors for Chemical Detection

Xiaopeng Li, Chemical Engineering, University of Massachusetts Lowell, Lowell, MA

Erica Chin, Chemical Engineering, University of Massachusetts Lowell, Lowell, MA

Zhiyong Gu, Chemical Engineering, University of Massachusetts Lowell, Lowell, MA

Hongwei Sun, Mechanical Engineering, University of Massachusetts Lowell, Lowell, MA

Pradeep U. Kurup, Civil and Environmental Engineering, University of Massachusetts Lowell, Lowell, MA

Metal oxides have been widely used as sensor materials for the detection of various types of chemicals. Metal oxide nanowires for chemical sensors have been shown to have higher sensitivity and faster response time compared to tin film sensors owing to their large specific surface (surface area to volume ratio). Developing nanowire sensors with a wider range of chemical selectivity is still a challenging issue. The incorporation of dopants, such as nickel (Ni), indium (In), platinum (Pt), and antimony (Sb), into metal oxides is an efficient way to significantly increase the chemical selectivity of sensors. In this research, single- or multi-segmented nanowires that include tin (Sn) segments are synthesized by electrodeposition technique using nanoporous templates. Dopants are incorporated into the sensor segment during electrodeposition. Electron microscopy (SEM and TEM) along with EDAX are used to characterize the nanowires fabricated. Either single nanowire sensor or sensor arrays are formed by using conventional microfabrication technique (photolithography and metal deposition) or dielectrophoretic (DEP) assembly technique. The nanowires are thermally annealed to convert tin (Sn) into tin oxide (SnO_2) and integrated with microheaters to form a functional sensor device. Nanowire sensor arrays with different dopants are fabricated and exposed to various chemicals (including explosives). Sensors response (electrical resistance measurements) will be studied to improve and optimize the sensor performance (in terms of sensitivity, selectivity, and response time).

See more of this Session: Interfacial Aspects in Nanosensors
See more of this Group/Topical: Engineering Sciences and Fundamentals

A Liposome Assay for Evaluating the Ocular Toxicity of Surfactants

Brett A. Howell,Chemical Engineering, University of Florida, Gainesville, FL

Yash Kapoor, Chemical Engineering, University of Florida, Gainesville, FL

Anuj Chauhan, Chemical Engineering, University of Florida, Gainesville, FL

The ocular toxicity of various compounds is often determined by the Draize eye test, which has been criticized in the past for its lack of reproducibility and the cruelty associated with harsh testing conditions for animals. We have developed a liposome based in vitro assay for estimating ocular toxicity. The release of calcein dye from liposomes induced by interactions with surfactants was studied and correlated to Draize eye scores. First, the liposome assay was conducted for various surfactants at identical concentrations and correlated to the Draize scores. Next, mechanistic and geometric considerations were used to determine the appropriate surfactant concentration that should be used in the liposome assay. Correlations between percentage dye released and the Draize scores were drastically improved after surfactant concentrations were chosen to be CMC/200, where CMC is the critical micelle concentration. With this choice of surfactant concentration, excellent correlations were obtained with Draize scores from three separate sources and four different ocular surfactant loadings (Pearson = 0.99, 0.82, 0.78, 0.74; Spearman = 0.94, 0.79, 0.79, 0.85). Subsequently, the ocular toxicities of six non-ionic surfactants, Brij 700, Brij 58, Brij 56, Brij 78, Brij 97, and Brij 98 were shown to be minimal based on the proposed correlations. The modified liposome assay developed here could be used in combination with other in vitro assays to obtain initial estimates for ocular toxicities, thereby minimizing the need for the Draize eye test.

See more of this Session: Interfacial Aspects in Nanosensors
See more of this Group/Topical: Engineering Sciences and Fundamentals

Engineering Protein Presenting Patch Surfaces for Biological Study/Application

ZiQiu Tong, Chemical and Biomolecular Engineering, Johns Hopkins University, Baltimore, MD

Moniraj Ghosh, Chemical and Biomolecular Engineering, Johns Hopkins University, Baltimore, MD

Christina Alves, Chemical and Biomolecular Engineering, Johns Hopkins University, Baltimore, MD

Konstantinos Konstantopoulos, Chemical and Biomolecular Engineering, Johns Hopkins University, Baltimore, MD

Kathleen J. Stebe, Chemical and Biomolecular Engineering, UNIVERSITY OF PENNSYLVANIA, Philadelphia, PA

Cell-surface interactions under flow via specific binding depend strongly on surface functionalization and flow regime. To this end, we use micro-contact printing method in conjunction with microfluidics tools to create multifunctional patchy-surfaces with precise control over feature geometry and molecular species for the application of cancer cell detection. We then exploit the patchy surfaces to study interactions between cancer cells and selectin-functionalized surfaces in microfluidic devices. Patchy surfaces are of particular interest for cellular adhesion, as the residence time of a cell over a functionalized patch is defined by the ratio of the patch length in the direction of flow over cell streaming velocity.

To create a multifunctional surface, a micro-contact printing stamp is used to transfer octadecytrichlorosilane onto a glass surface defining active patterned regions for subsequent protein adsorption. The glass substrate is then backfilled with PEG-silane to inactivate the remaining surface. A microfluidic device is then affixed to the substrate to deliver proteins onto the active regions. The number of types of proteins and number of patches can be easily controlled simply by manipulating microfluidics design. We first demonstrated the feasibility of creating 45μm x 45μm patches of two different fluorescently labeled proteins. Using the same approach to functionalize a glass surface with two different antibodies, we were able to isolate diseased cells (colon cancer cells) from normal blood cell (leukocytes) suspension onto prescribed microdomains. Such multifuntional surfaces, we believe, could be useful in biological studies as well as development of biosensors for cancer cell detection.

We further exploit functionalized surfaces affixed to a microfluidic device as an in vitro hemodynamic flow model to study 2D cancer cell binding kinetics. In particular, a glass surface is engineered to present patches of selectin molecules, adhesion proteins involving in cancer cell metastasis. The patches are of uniform width but differing lengths in the direction of flow. Cancer cells are introduced over the patterned glass surface via microfluidic device at several different shear stresses relevant to physiological flow conditions. The cell residence time is proportional to the length of the protein patch. Our preliminary data showed that at a specific shear stress, a critical selectin patch size is required to initiate the binding event of cancer cells to seletins. We are also investigating the extent of cell binding influenced by selectin site density and selectin types (i.e., P- and L- selectins).

See more of this Session: Interfacial Aspects in Nanosensors
See more of this Group/Topical: Engineering Sciences and Fundamentals

Control of Lateral Interactions of Surface-Immobilized Antibodies Via Substrate NANO-Patterning

Chang-Hoon Chai, Rutgers University, New Brunswick, NJ

Paul Takhistov, School of Environmental & Biological Sciences, Rutgers University, New Brunswick, NJ

Lateral interaction between immobilized Ab on the surface of immunosensor is important factor that determines biosensor's performance. Such interactions hider the surface diffusion of Ab in liquid solution onto the host surface and increase the degree of non-specific Ab immobilization onto the host surface. We propose to control lateral interactions via topological modification of the host surface.

To investigate the effect of the topography of nano-patterned surface on Ab immobilization and their lateral interaction, nano-porous substrates were prepared by electrochemical anodization of aluminum. The effect of dimension of nano-porous structure on the amount of immobilized Ab was investigated by measuring the fluorescence intensity of FITC-Ab immobilized immunosensor based on nano-porous substrate. Effect of the nano-patterns on the sensitivity of biosensor was tested for detection of Staphylococcus enterotoxin B in complex matrix. Our data show an improved sensitivity and selectivity of developed nanosensor.

See more of this Session: Interfacial Aspects in Nanosensors
See more of this Group/Topical: Engineering Sciences and Fundamentals

Anomalously Slow Diffusion in Protein-Coated Nanopores

Ian I. Suni, Department of Chemical and Biomolecular Engineering, Clarkson University-, Potsdam, NY

Ruth Baltus, Chemical and Biomolecular Engineering, Clarkson University-, Potsdam, NY

Rajdeep Singh, Chemical and Biomolecular Engineering, Clarkson University-, Potsdam, NY

Peanut protein Ara h 1 is detected by the change in electrical conductivity through an Au-coated nanoporous polycarbonate membrane sensor onto which the peanut antibody is immobilized. Introduction of peanut protein Ara h 1 causes a reduction in the membrane conductivity by binding to antibodies immobilized onto the pore walls, reducing the effective pore diameter. The observed conductivity changes are studied as a function of time and protein concentration, and these results depend strikingly on the nominal pore diameter. For 15 nm diameter pores, the conductivity change saturates at low peanut protein concentration, with no significant time lag in the sensor response. For 30 and 50 nm diameter pores, a time lag of tens of miutes is observed in the sensor response that appears to be consistent with mass transport limitations. However, back-of-the-envelope calculations sugest that transport through these membranes should not be rate-limting. These calculations quantitatively include the effects of hindered transport, as described by the Renkin equation. These results demonstrate that axial protein diffusion is anomalously slow within these protein-coated nanoporous membranes.

See more of this Session: Interfacial Aspects in Nanosensors
See more of this Group/Topical: Engineering Sciences and Fundamentals

Synthesis of Polystyrene-Silica Composite Particles via One-Step Nanoparticle-Stabilized Emulsion Polymerization

*Huan Ma and Lenore L. Dai**

School of Mechanical, Aerospace, Chemical and Materials Engineering

Arizona State University, Tempe, AZ 85287-6006

**Corresponding author. Email address: lenore.dai@asu.edu*

1 Introduction

Polymer composites are ubiquitous in daily lives such as in automotives, aircrafts, electronics, appliances, and sporting goods. Conventional polymer composites are made by post-formulating and mixing polymer resins and inorganic fillers [1]. In conventional polymer composites, the inorganic fillers are randomly dispersed in the polymer matrix, often in the format of aggregation. The inclusion of inorganic fillers significantly increases the physical, such as mechanical, optical, electrical, and thermal properties of the polymer matrixes [2-5], although the randomness and aggregation often result in uncontrolled or limited property enhancement. Here we report the success of preparing core-shell structured polymer composite particles via one-step nanoparticle-stabilized emulsion polymerization. It is worthwhile to note that the composite structure here is opposite from the often reported core-shell structure in which inorganic particles serve as the core and polymer serves as the shell [6-11]; here the polymer serves as the core and the inorganic particles serve as the shell. Such materials provide a new class of building blocks and can "exhibit unusual, possibly unique, properties which cannot be obtained simply by co-mixing polymer and inorganic particles." [12] Surprisingly, there has been remarkably little work in this field. Our approach continues the very recent effort of emulsion or dispersion polymerization to fabricate this class of materials, in addition to the developed post-surface-reaction [13], electrostatic deposition [14], and layer-by-layer self-assembly [7,15-17].

In the past several years, new chemical methods have been utilized to prepare organic-inorganic composite particles, including emulsion [18-21], dispersion [22-27], miniemulsion [28-32], and inversed emulsion polymerization [33]. The stabilizing effect is usually provided by surfactants, co-monomers, initiators, or cooperatively with nanoparticles. Very recently, polymerization solely stabilized by nanoparticles has received increasing attention, due to its simplified composition and preparation steps. However, the polymerization mechanism and the nanoparticle incorporation mechanism are still unclear and currently under active investigation [20,34]. During the progress of this work, Schmid and coworkers [22] reported the dispersion polymerization of styrene using silica nanoparticles as the sole stabilizing agent. Polystyrene-silica core-shell composite particles were successfully prepared by one-step alcoholic dispersion polymerization using cationic initiator AIBA (2,2'-azobix(isobutyramidine) dihydrochloride) and negatively-charged silica nanoparticles. The silica incorporation is likely promoted by the electrostatic attraction between the cationic initiator and the anionic nanoparticles [22]. This may be evidenced by the fact that in the absence of the electrostatic interaction, the work using nonionic initiator AIBN (2,2'-azobisisobutyronitrile) only yields latex particles with surface Si/C ratio of 0.01 [22]; in other words, the latex particles are poorly covered with silica nanoparticles. Here we report the success of preparing polystyrene-silica core-shell structured composite particles by emulsion polymerization using a nonionic initiator, VA-086 (2,2'-azobis(2-methyl-N-(2-hydroxyethyl) propionamide)), and discuss the role of silica nanoparticles as the sole stabilizing agent.

2 Experimental

2.1 Materials

IPA-ST silica solution, obtained from Nissan Chemicals, is 10-15 nm silica nanoparticles dispersed in 2-isopropanol. The silica concentration is 30-31% by weight. Nonionic azo initiator VA-086 (98%, 2,2'-azobis(2-methyl-N-(2-hydroxyethyl)propionamide)), styrene monomer (99.9%), and HPLC grade water were purchased from Wako Chemicals, Fisher Scientific, and Acro Organics, respectively. All materials were used as received.

2.2 Synthesis of polystyrene-silica composite particles

First, water, IPA-ST and styrene were agitated mechanically at 600 rpm for 8 minutes using Arrow 6000 (Arrow Engineering) in an ice bath to emulsify. Second, the emulsion was degassed with nitrogen and kept in nitrogen atmosphere under magnetic stir. When the temperature was raised to 70 °C, the initiator solution was added to start the polymerization. The composite particles were sampled at different time intervals ranging from 3 hours to 24 hours. Before characterization, samples were washed twice by centrifuging-redispersing cycles using an Eppendorf 5810R centrifuge. In each cycle, the sample was centrifuged at 7000 rpm for 5 min, the supernatant was replaced with water and the sediment was redispersed by shaking manually.

2.3 Hydrofluoric (HF) etching

To remove the silica shell, HF etching procedure described by Han and coworkers [35] was adopted. Approximately 1 ml of original composite particle dispersion was added to 50 ml 10% HF aqueous solution. Then the mixture was stirred at room temperature (approximately 23 °C) for 2 hours. After settling for several hours, the supernatant was replaced with water for at least three times and the sediments were obtained.

2.4 Particle size, morphology and composition characterization

Particle size and distribution were measured by a Microtrac Nanotrac Particle Size Analyzer using the dynamic light scattering (DLS) technique. The washed composite particles were further dispersed to proper concentrations with water before measurements. Each sample was tested for at least twice and had consistent results. A scanning electron microscope (SEM, Hitachi S-4300) was used to observe the surface morphology of the composite particles. SEM samples were prepared by drying a drop of washed composite particle dispersion on newly cleaved mica substrate, and applying Au/Pd coating of 10-15 nm thick using a sputter coater. In order to obtain the cross-section image of the composite particles, the particles were firstly embedded in a LX-1122 epoxy resin block and cut into 90-

150 nm sections using a Reichert-Jung UltraCut E microtome equipped with a glass knife at room temperature. Then, the thin sections were collected with copper grids and observed using the transmission electron microscope (TEM, Hitachi H-8100). The chemical composition of the nanoparticles were analyzed using energy dispersive X-ray spectrometry (EDX), built in the S-4300 SEM, and thermogravimetric analysis (TGA, Mettler-Toledo SDTA851e). In EDX, pure carbon substrate was selected. In TGA, Samples were heated to 800 °C at 10 °C/min in air.

3 Results and Discussion

It is worthwhile to note that we carefully selected VA-086 as the initiator. VA-086 is a water-soluble nonionic initiator and no success has been reported in surfactant-free emulsion polymerization of styrene [36]. In order to verify the sole stabilizing effect of silica nanoparticles, emulsifier-free emulsion polymerization using VA-086 as the initiator in the absence of nanoparticles was performed. No polystyrene particle formation was observed in the product, evidenced by SEM experiments. These experiments show that the initiator VA-086 has little effect on stabilizing the system in emulsion polymerization and therefore silica nanoparticles are the only source of stabilizer when present. In addition, VA-086 is neutral in charge thus is expected to minimize any electrostatic interactions with the negatively-charged silica nanoparticle surfaces which may complicate identifying silica nanoparticles as the sole stabilizer in emulsion polymerization.

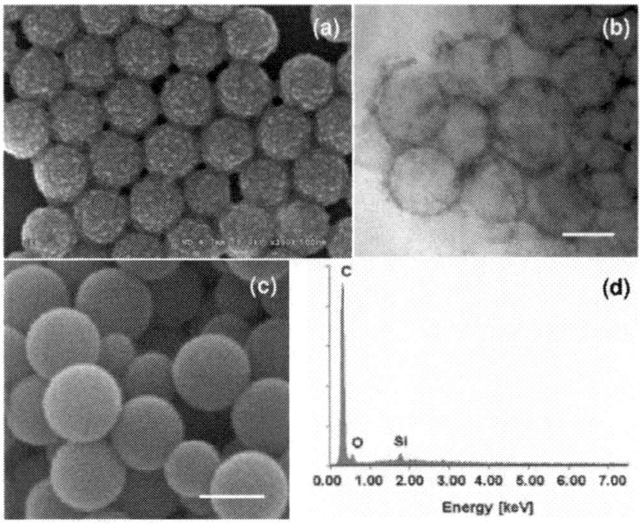

Figure 1. (a) An SEM image of composite particles, (b) a TEM image of cross-sectioned composite particles, (c) an SEM image of composite particles after HF etching, and (d) the EDX spectrum of composite particles. All images are representative and the scale bars in (b) and (c) represent 200 nm.

Unless noted, composite particles were prepared by formulating 8 ml styrene, 52.5 ml water, 20 g IPA-ST, and 0.06 g initiator VA-086. Figure 1(a) is a representative SEM image of composite particles sampled at 5 hour reaction time. The roughness of the composite particle surfaces suggests that the composite particles are covered by silica nanoparticles. The particle sizes, characterized by DLS, are 203.9 ± 51.6 nm in diameter with mono-distribution (the standard deviation here represents the width of the particle size distribution), consistent with SEM observations. The core-shell structure can be clearly observed in the TEM image presented in Figure 1(b). In many regions, the thickness of the shell is close to the size of one silica nanoparticle (10-15 nm), which may suggest a monolayer coverage. The silica shell can be removed by excess amount of hydrofluoric acid (HF) solution. As shown in Figure 1(c), after HF etching treatment, the particle surfaces become smooth and reveal the polystyrene core.

The particle composition was characterized by EDX and TGA. The presence of Si and O elements in the composite particles is indicated by the EDX analysis, as shown by the intensity peaks in Figure 1(d). The silica content is quantitatively determined by TGA. We assume that the major weight loss during heating is associated with the thermo-oxidative degradation of polystyrene and the residue close to 800°C is solely silica. The silica content is approximately 20 wt%, which is significantly higher than the silica content of particles (1.1 wt%) prepared by the dispersion polymerization using nonionic initiator AIBN [22]. The improvement is likely due to the distinct polymerization mechanisms. In contrast to the dispersion polymerization in which the polystyrene monomers are dissolved in alcohols, the emulsion polymerization here contains distinguished liquid-liquid interfaces due to the immiscibility between the monomers and the aqueous continuous phase. Thus the nanoparticles, even in the absence of electrostatic interactions, are thermodynamically favorable to self-assemble and remain at the liquid-liquid interfaces, following the same argument in sold-stabilized emulsions [37-42]. At the initial stage of polymerization, the nanoparticles provide stability to the monomer droplets. During the nucleation stage, silica nanoparticles remain at the interfaces between the monomer and continuous phases. It is worthwhile to note that the role of silica nanoparticles described here is not the same as that in the polymerization involving oppositely charged initiator and nanoparticles [22]. In the latter case, the initiator molecules or residues adsorb onto the silica nanoparticle surfaces after initiation [22] thus the silica nanoparticles function as the surface-active initiator residue in that polymerization. This is our first attempt of applying the concept of solid-stabilized emulsions in polymerization, after studying the fundamentals of solid-stabilized emulsions [40-42], utilizing them as templates to investigate the dynamics of particles [43-45], and developing microrheology at liquid-liquid interfaces [46-47]. Such application may open new avenues of synthesizing advanced materials utilizing the concept of solid-stabilized emulsions. It is also worthwhile to note that the synthesized "core-shell" composite particles have various potential applications in photonic crystals, optical switches, and display devices. For example, recent work [48] demonstrates that thin films made of polystyrene-silica core-shell composite particles remain transparent, even if the silica loading is as high as 39%. The transparency is due to the well-dispersibility of silica (shell) in the polymer matrix, which cannot be achieved simply by co-mixing the polystyrene and silica particles.

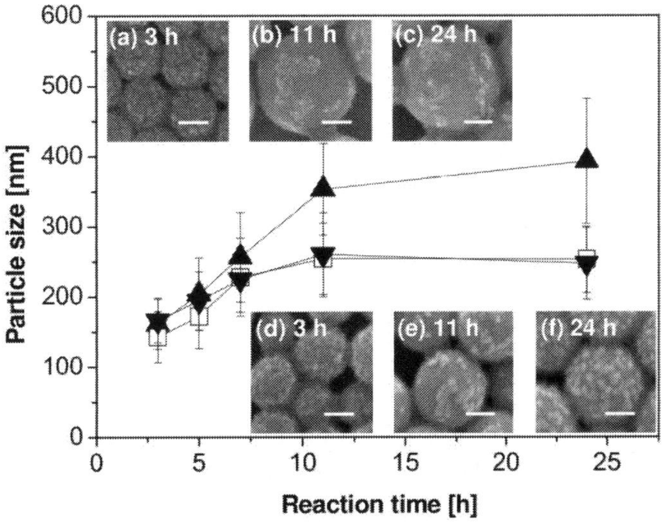

Figure 2. Plot of particle size at 50% of the particle size distribution versus reaction time and representative SEM images with different initiator concentrations: 0.83 wt% (▲, inset images a, b and c), 2.5 wt% (□) and 4.2 wt% (▼, inset images d, e and f). The error bars indicate the width of the particle size distribution and the scale bars represent 100 nm.

In addition, the influence of initiator concentration was studied by increasing the initiator amount from 0.06 g to 0.18 g and 0.30 g (0.83 wt%, 2.5 wt% and 4.2 wt%, respectively, based on monomer amount), maintaining other experimental parameters. Samples were taken at different time intervals between 3 and 24 hours and analyzed by SEM and DLS. Figure 2 is the plot of particle size measured by the dynamic light scattering (DLS) technique versus the reaction time. During the first 3 to 7 hours of reaction, the particles in all three systems experienced comparable particle size growth. The two systems with higher initiator concentrations ceased to grow soon after 7 hours of reaction time, whereas the system with the lowest initiator amount continued to grow significantly and produced the largest composite particles at 24 hour reaction time. The trend is confirmed by SEM experiments (images not shown). Comparing the particle sizes at 24 hour reaction time, a decrease in particle size occurred when the initiator amount was increased from 0.06 g to 0.18 g. The dependence of particle size on initiator concentration herein is in reverse trend with that in solid-stabilized alcoholic dispersion polymerization reported by Schmid and coworkers [22]. The difference probably originates from the distinct particle growth mechanisms [49]. In our emulsion polymerization, a larger number of nuclei likely formed at the initial stage of polymerization, so the particle size decreased with increasing initiator concentration, given a fixed amount of monomer and assuming the same monomer conversion. Further increasing the initiator amount from 0.18 g to 0.30 g did not have significant influence on particle size. One possible explanation would be that the available amount of silica nanoparticles did not provide sufficient stability to an even larger number of nuclei. With the particle size growth, the changes in surface coverage occurred and were observed using SEM. Selected representative images are shown in the insets of Figure 2. In most cases, silica coverage decreased with reaction time after 5 hours. This is likely due to the increase of surface area during particle growth. One exception is the system with the highest initiator concentration (4.2 wt%, based on monomer amount). An increase of silica coverage at 24 hour reaction time was repeatedly observed. The underlying mechanism for the unexpected coverage increase at 24 hour reaction is still unclear. To the best of our knowledge, this is the first study of tracking the change of surface coverage along the reaction time.

4 Conclusions

In summary, polystyrene-silica core-shell structured composite particles were successfully synthesized by nanoparticle-stabilized emulsion polymerization. Silica nanoparticles act as the sole stabilizer during the emulsion polymerization, following the same mechanism in solid-stabilized emulsions. In the polymerization formulated with 8 ml styrene, 52.5 ml water, 20 g IPA-ST, and 0.06 g initiator VA-086, composite particles sampled at 5 hour reaction time exhibit core-shell structure in which the polystyrene core is covered by silica. The silica content, determined by TGA, is approximately 20 wt%. In addition, we further explore the polymerization mechanism by studying the particle growth as a function of initiator concentration and reaction time: when the initiator/monomer ratio is increased from 0.83 wt% to 2.5 wt%, the particle size at 24 hour reaction time decreases for a fixed monomer amount, probably due to a larger number of nuclei at the initial stage of polymerization. The dependence of particle size on initiator concentration herein is in reverse trend with that in solid-stabilized alcoholic dispersion polymerization [22]. Further increasing the initiator/monomer ratio to 4.2 wt% does continually decrease the particle size, which may be limited by the stabilization provided by a fixed concentration of silica nanoparticles. The surface coverage also changes with initiator concentration and reaction time although the underlying mechanism is not fully understood.

Acknowledgements

We would like to acknowledge the Department of Chemical Engineering, the Department of Chemistry, and the Imaging Center at Texas Tech University for instrumental usage and the National Science Foundation for financial support (CBET-0625191 and CBET-0644850).

References

[1] P.C.P. Watts, W.K. Hsu, G.Z. Chen, D.J. Fray, H.W. Kroto, D.R.M. Walton, J. Mater. Chem. 11 (2001) 2482.

[2] C. Sanchez, B. Julian, P. Belleville, M. Popall, J. Mater. Chem. 15 (2005) 3559.

[3] B.R. Saunders, M.L. Turner, Adv. Colloid Interface Sci. 138 (2008) 1.

[4] A. Leszczyńska, J. Njuguna, K. Pielichowski, J.R. Banerjee, Thermochimica Acta 454 (2007) 1.

[5] G. Wirnsberger, B.J. Scott, B.F. Chmelka, G.D. Stucky, Adv. Mater. 12 (2000) 1450.

[6] E. Bourgeat-Lami, J. Lang, J. Colloid Interface Sci. 197 (1998) 293.

[7] F. Caruso, Adv. Mater. 13 (2001) 11.

[8] S. Gu, T. Kondo, M. Konno, J. Colloid Interface Sci. 272 (2004) 314.

[9] Y.K. Guo, M.Y. Wang, H.Q. Zhang, G.D. Liu, L.Q. Zhang, X.W. Qu, J. Appl. Polym. Sci. 107 (2008) 2671.

[10] D. Nagao, M. Yokoyama, S. Saeki, Y. Kobayashi, M. Konno, Colloid Polym. Sci. 286 (2008) 959.

[11] W. Tianbin, K. Yangchuan, European Polym. J. 42 (2006) 274.

[12] C. Barthet, A. J. Hickey, D.B. Cairns, S.P. Armes, Adv. Mater. 11 (1999) 408.

[13] D.E. Lynch, Y. Nawaz, T. Bostrom, Langmuir 21 (2005) 6572.

[14] A. Dokoutchaev, J.T. James, S.C. Koene, S. Pathak, G.K.S. Prakash, M.E. Thompson, Chem. Mater. 11 (1999) 2389.

[15] F. Caruso, R.A. Caruso, H. Mohwald, Science 282 (1998) 1111.

[16] F. Caruso, A.S. Susha, M. Giersig, H. Möhwald, Adv. Mater. 11 (1999) 950.

[17] R.A. Caruso, A. Susha, F. Caruso, Chem. Mater. 13 (2001) 400.

[18] J. Lee, C.K. Hong, S. Choe, S.E. Shim, J. Colloid Interface Sci. 310 (2007) 112.

[19] M.J. Percy, J.I. Amalvy, D.P. Randall, S.P. Armes, S.J. Greaves, J.F. Watts, Langmuir 20 (2004) 2184.

[20] M. Chen, S. Zhou, B. You, L. Wu, Macromolecules 38 (2005) 6411.

[21] X. Cheng, M. Chen, S. Zhou, L. Wu, J. Polym. Sci. A: Polym. Chem. 44 (2006) 3807.

[22] A. Schmid, S. Fujii, S.P. Armes, C.A.P. Leite, F. Galembeck, H. Minami, N. Saito, M. Okubo, Chem. Mater. 19 (2007) 2435.

[23] A. Schmid, S. Fujii, S.P. Armes, Langmuir 22 (2006) 4923.

[24] A. Schmid, S. Fujii, S.P. Armes, Langmuir 21 (2005) 8103.

[25] J. Yang, T. Hasell, W.X. Wang, J. Li, P.D. Brown, M. Poliakoff, E. Lester, S.M. Howdle, J. Mater. Chem. 18 (2008) 998.

[26] J. Hong, H. Han, C. K. Hong, S.E. Shim, J. Polym. Sci. A: Polym. Chem. 46 (2008) 2884.

[27] M. J. Percy, V. Michailidou, S.P. Armes, C. Perruchot, J.F. Watts, S.J. Greaves, Langmuir 19 (2003) 2072.

[28] X. Qiao, M. Chen, J. Zhou, L. Wu, J. Polym. Sci. A: Polym. Chem. 45 (2007) 1028.

[29] F. Tiarks, K. Landfester, M. Antonietti, Langmuir 17 (2001) 5775.

[30] Y. Luo, C. Dai, W. Chiu, J. Polym. Sci. A: Polym. Chem. 46 (2008) 1014.

[31] S.A.F. Bon, P.J. Colver, Langmuir 23 (2007) 8316.

[32] S. Cauvin, P.J. Colver, S.A.F. Bon, Macromolecules 38 (2005) 7887.

[33] D.J. Voorn, W. Ming, A.M. van Herk, Macromolecules 39 (2006) 2137.

[34] G.K. Agarwal, J.J. Titman, M.J. Percy, S.P. Armes, J. Phys. Chem. B 107 (2003) 12497.

[35] Han, S.P. Armes, J. Colloid Interface Sci. 262 (2003) 418.

[36] H. Torii, K. Fujimoto, H. Kawaguchi, J. Polym. Sci. A: Polym. Chem. 34 (1996) 1237.M.G.

[37] A.D. Dinsmore, M.F. Hsu, M.G. Nikolaides, M. Marquez, A.R. Bausch, D.A. Weitz, Science 298 (2002) 1006.

[38] Y. Lin, H. Skaff, T. Emrick, A.D. Dinsmore, T.P. Russell, Science 299 (2003) 226.

[39] H. Xu, S. Melle, K. Golemanov, G.G. Fuller, Langmuir 21 (2005) 10016.

[40] S. Tarimala, L.L. Dai, Langmuir 20 (2004) 3492.

[41] L.L. Dai, R. Sharma, C.-Y. Wu, Langmuir 21 (2005) 2641.

[42] M.X. Luo, O.A. Mazyar, Q. Zhu, M.W. Vaughn, W.L. Hase, L.L. Dai, Langmuir 22 (2006) 6385.

[43] S. Tarimala, S. Ranabothu, J.P. Vernetti, L.L. Dai, Langmuir 20 (2004) 5171.

[44] C.-Y. Wu, S. Tarimala, L.L. Dai, Langmuir 22 (2006) 2112.

[45] S. Tarimala, C.-Y. Wu, L.L. Dai, Langmuir 22 (2006) 7458.

[46] J. Wu, L.L. Dai, Appl. Phys. Lett. 89 (2006) 094107.

[47] J. Wu, L.L. Dai, Langmuir 23 (2007) 4324.

[48] A. Schmid, J. Tonnar, S. P. Armes, Advanced Materials 20 (2008) 3331.

[49] T. Yamamoto, M. Nakayama, Y. Kanda, K. Higashitani, J. Colloid Interface Sci. 297 (2006) 112.

Tunable Core-Shell Nanoarchitectures Containing Encapsulated Beta-Carotene

Kristin Gilida, Department of Chemical and Biochemical Engineering, Rutgers University, Piscataway, NJ

Kenneth Morabito, Division of Engineering and Warren Alpert School of Medicine, Brown University, Providence, RI

Michael Huang, Division of Engineering and Warren Alpert School of Medicine, Brown University, Providence, RI

Anubhav Tripathi, Division of Engineering and Warren Alpert School of Medicine, Brown University, Providence, RI

Nina C. Shapley, Department of Chemical and Biochemical Engineering, Rutgers University, Piscataway, NJ

New hybrid architectures incorporating synthetic polymers and nanoparticles are being developed to achieve protection of natural dyes and other bioderived active species from UV radiation, oxygen and moisture. The research focuses on the design and evaluation of modular core-shell nanoarchitectures, with the potential for new technology in sensing, energy harvesting, pharmaceutical delivery, protection and decontamination. Investigating the encapsulation and protection of beta-carotene is useful for multiple purposes, since beta-carotene is a natural pigment as well as a therapeutic biomolecule.

Uniform and core-shell polymer (PMMA) microspheres are produced by using the solvent evaporation technique, while nanoparticles are synthesized from the phase inversion method before incorporation into the microparticle core or shell. Particle synthesis is undertaken both in bulk systems and on a high precision microfluidic platform, where superior control of the particle size and composition is accomplished. We examine the particle morphology by optical and electron microscopy and evaluate the protective properties through UV/Visible spectrophotometry. The ability to encapsulate beta-carotene in any compartment of the hybrid microparticle can enable systematic study of protection mechanisms at multiple length scales.

See more of this Session: Particle Synthesis and Stabilization
See more of this Group/Topical: Engineering Sciences and Fundamentals

Hybrid Colloid-Nanowire Structure and Assembly

Yingying Sha, Chemical Engineering, University of Massachusetts Lowell, Lowell, MA

Zhiyong Gu, Chemical Engineering, University of Massachusetts Lowell, Lowell, MA

A novel type of composite particles, microsphere-nanowire hybrid colloids, has been fabricated using self-assembly coupled particle formation process. Metallic nanowires are synthesized using electrodeposition in nanoporous templates, and the size of the nanowires ranges from 30-200 nm in diameter and micron-meters in length. Either polystyrene or silica particles are used as colloidal seeds. Polystyrene particles are synthesized by emulsion polymerization, while silica particles by sol-gel process. During colloidal formation, nanowires are added into the emulsions to initiate self-assembly to obtain colloid-nanowire structures. Results have been compared for using ionic vs. nonionic surfactants. Surface coverage of nanowires on colloids and the morphology of hybrid structures can be controlled by changing the relative concentrations between the colloids and nanowires, surfactant type, nanowire addition time, and the length of the nanowires. The network of the assembled structure can be obtained by sintering the hybrid colloids at high temperature or by the utilization of an external field (magnetic or electric).

See more of this Session: Particle Synthesis and Stabilization
See more of this Group/Topical: Engineering Sciences and Fundamentals

Templated Fabrication of Multifunctional Gold Nanoflask Particles for Controlled Drug Delivery

Nicholas Linn, Chemical Engineering, University of Florida, Gainesville, FL

Chih-Hung Sun, Chemical Engineering, University of Florida, Gainesville, FL

Peng Jiang, Chemical Engineering, University of Florida, Gainesville, FL

Hollow particles play an important role in microencapsulation - a process that has been widely exploited for controlled release of drugs. Here we report the fabrication of hollow gold nanoflask particles by a simple templating process. Non-close-packed colloidal monolayers prepared by a simple spin-coating process are used as templates. A thin film of gold can be sputtered on the surface of the particles and the silica templates can be removed later on to form flask-shape hollow nanoparticles. The resulting particles exhibit tunable plasmonic properties which can be easily tuned by changing the template particle size and the shell thickness. The interior of the hollow nanoshells can be filled up with chemicals (e.g., laser dye) and the opening can then be sealed by a thin layer of polymer. We have demonstrated that the encapsulated chemicals can be delivered in a controlled manner when a near-IR laser beam is illuminated on these composite nanoparticles. In addition, we observe the strong in-situ optical adsorption and heating by the gold nanoflask particles, promising for developing hypothermal treatment. Moreover, the templated nanoflask particles exhibit strong surface-enhanced Raman scattering and are promising for in-situ biosensing.

See more of this Session: Particle Synthesis and Stabilization
See more of this Group/Topical: Engineering Sciences and Fundamentals

More Benign Synthesis and Extraction of Ligand-Free Palladium Nanoparticles

Juncheng Liu, Department of Chemical Engineering, Auburn University, Auburn, AL

Nicholas Ruffini, Department of Chemical Engineering, Auburn University, Auburn, AL

Christopher Roberts, Department of Chemical Engineering, Auburn University, Auburn, AL

Pamela Pollet, School of Chemistry and Biochemistry, Georgia Institute of Technology, Atlanta, GA

Veronica Llopis Mestre, School of Chemical & Biomolecular Engineering, Georgia Institute of Technology, Atlanta, GA

Charles Liotta, School of Chemistry and Biochemistry, Georgia Institute of Technology, Atlanta, GA

Charles A. Eckert, School of Chemical & Biomolecular Engineering and Specialty Separations Center, Georgia Institute of Technology, Atlanta, GA

We demonstrate a more benign synthesis method for small, relatively nondisperse, metal nanoparticles for catalysis. By eliminating separate capping agents and using a recyclable solvent, this technique avoids the massive waste steams associated with current processes.

Metal nanoparticles offer size-dependent optical, electronic, magnetic, catalytic and other properties. But their syntheses generally require reduction of metal salts in aqueous or organic solutions with capping ligands. These caps serve to passivate the surface of the formed particles and to suppress the growth of the metal clusters that would otherwise aggregate due to the high surface energy of the system. However, the capping ligands may reduce the surface area of the nanoparticles available for catalytic applications. Synthesis often uses excess capping ligands, which then require downstream separations generating waste streams.

In general, a fast, homogeneous reduction reaction is a prerequisite to produce small metal nanoparticles. And it would be desirable to produce small, uniform metal nanoparticles in a solvent medium without the need for the addition of capping ligands and the resulting purification processes. An ideal solvent medium would not only exhibit favorable solvation of a metal salt (e.g. Na_2PdCl_4) and of a reducing agent (e.g. $NaBH_4$), but would also exhibit sufficient interaction with the surface of the formed metal nanoparticles to passivate and stabilize the nanoparticle dispersion as created within this solvent medium effectively. Dimethyl sulfoxide (DMSO) is a favorable solvent for ionic solids, as well as both polar and polarizable molecules. In addition, it is relatively inexpensive, stable, and exhibits low vapor pressure and a high boiling point of 189 ºC. Here we present the first successful synthesis and stabilization of 3.5 nm Pd nanoparticles (standard deviation of 0.5 nm) within DMSO via a fast, homogeneous reduction of a Pd salt using $NaBH_4$ in the absence of traditional capping ligands. These Pd nanoparticles were found to be extremely stable and did not exhibit precipitation and/or agglomeration within the DMSO solvent after more than 9-months of aging time. Moreover, these Pd nanoparticles were conveniently separated from the DMSO solvent medium via vacuum freeze drying by taking advantage of the high freezing point of DMSO (18.45 ºC). We have also successfully extracted the Pd metal nanoparticles from the DMSO phase into an organic phase, thereby providing a means of recycling the DMSO solvent.

See more of this Session: Particle Synthesis and Stabilization
See more of this Group/Topical: Engineering Sciences and Fundamentals

Stability of DPPC Vesicles in Aqueous Electrolyte Solutions and Comparisons to Predictions From the DLVO Theory

Yoonjee Park, School of Chemical Engineering, Purdue University, West Lafayette, IN

Ruiyi Huang, School of Chemical Engineering, Purdue University, West Lafayette, IN

David S. Corti, School of Chemical Engineering, Purdue University, West Lafayette, IN

Elias I. Franses, School of Chemical Engineering, Purdue University, West Lafayette, IN

Dipalmitoylphosphatidylcholine (DPPC) lipid is the major component of lung surfactant, which stabilizes the lung by reducing the surface tension at the air/liquid interface of the lung alveoli. DPPC is zwitterionic and insoluble in aqueous solutions, forming dispersions of DPPC vesicles only after extensive sonication above the chain melting temperature, 41º C. The vesicle dispersions are thermodynamically unstable, and they can remain colloidally (kinetically) stable for days, depending on the concentration, temperature, and ionic strength. Their colloidal stability is important for the shelf life and for producing low dynamic surface tension minima (DSTM) because the coagulated dispersions have higher DSTM.

The stability of the DPPC vesicle dispersions in aqueous solutions, characterized by the stability ratio W (the inverse of which is a measure of the probability that a collision will lead to coagulation), was determined with a series of measurements of dispersed particle sizes with dynamic light scattering (DLS) at 25º C. Several vesicular dispersions in water, 1 mM, 10 mM, 150 mM NaCl solution, or phosphate buffer saline (PBS) solution were tested to find out how electrostatic forces may affect the stability ratio W. The zeta potentials were -30, -14, 1±6, 2±3, and -20 mV, respectively. These values are not consistent, probably due to their relatively low values. Nonetheless, the results indicate that electrostatic effects may play a role and some ions may adsorb on the surface of the zwitterionic vesicles. Moreover, the values of zeta potential varied with time as some coagulation proceeded.

The stability ratio W was also predicted from the interaction potential Φmax, and the Fuchs-Smoluchowski theory for dilute dispersions. This quantity was determined from a new dimensionless group formulation of the DLVO (Derjaguin-Landau-Verwey-Overbeek) theory, with measured values of vesicle radius, zeta potential, Debye length (from the ionic strength), and published measured values of the Hamaker constant of the lipid and water. The predicted W values were generally smaller than the measured values, indicating that some other forces (hydration or other) play a role, or that the DLVO predictions were inaccurate. The effective values of Φmax were estimated from the data and the DLVO formalism.

See more of this Session: Particle Synthesis and Stabilization
See more of this Group/Topical: Engineering Sciences and Fundamentals

Structure and Colloidal Stability of Nanosized Zeolite Beta Precursors

Nathan D. Hould, Chemical Engineering, University of Delaware, Newark, DE

Raul F. Lobo, Center for Catalytic Science and Technology, Department of Chemical Engineering, Univesity of Delaware, Newark, DE

During the synthesis of zeolite beta in solutions of molar composition 1 SiO2 / 80 H2O / 0.25 TEA2O / 4 CH3CH2OH / 0.06 Na2O / 0.01 Al2O3 three populations of particles are observed by small angle scattering. First, at room temperature, silica self-assembles into primary particles (< 3 nm). Upon heating the least stable primary particles aggregate into secondary particles. Some of the secondary particles are stable as spheroidal monomer particles but others are unstable and aggregate into small clusters. After 4 days of heating most silica comprising secondary particles remains amorphous (by attenuated total reflectance - Fourier transform infrared spectroscopy) but their composition is similar to zeolite beta (by scattering length density matching using small angle neutron scattering) showing that organization of the silica occurs after structure directing agents are occluded into secondary particles. After heating solutions containing Al for 6 days we start to observe Al-secondary particles aggregating into tertiary particles (zeolite beta crystals, < 200 nm, volume fraction 1.1 * 10^-4). The coherence length of the zeolite beta structure in tertiary particles increases sigmoidally over time with most of the changes occurring between 12 days and 18 days of heating when the majority of Al and Al-secondary particles are consumed by the tertiary particles. This shows that as silica reorganizes from amorphous into zeolite beta Al-secondary particles become less stable (in the colloidal sense) and aggregate with tertiary particles. The observed mechanism for zeolite beta nucleation is different from classical nucleation theory where the crystalline order fluctuates simultaneously with the density since we observe that there is a step- wise aggregation of precursor particles and a continuous evolution of their structure towards zeolite beta.

See more of this Session: Particle Synthesis and Stabilization
See more of this Group/Topical: Engineering Sciences and Fundamentals

The Bulk Viscosity of Suspensions

Manuj Swaroop, Chemical Engineering, California Institute of Technology, Pasadena, CA

John F. Brady, California Institute of Technology / Division of Chemistry and Chemical Engineering, Pasadena, CA

Suspension flows can lead to variations in particle volume fraction, thus making the particle phase compressible on a macroscopic scale. The stress in such a flow is characterized by an effective bulk viscosity in addition to the effective shear viscosity of the suspension. The bulk viscosity of a suspension is defined analogous to that for a pure fluid as the constant of proportionality relating the deviation of the trace of the macroscopic stress from its equilibrium value to the average rate of expansion. The equilibrium macroscopic stress is the sum of the fluid pressure and the osmotic pressure of the suspended particles. Variations in particle volume fraction are modeled by having a compressible fluid expand uniformly at a constant rate, causing the particles suspended in it to move apart. The expansion flow drives the suspension microstructure out of equilibrium and the thermal motion of the particles tries to restore the equilibrium. The rigid particles cannot expand, creating a disturbance flow that contributes to the total mechanical pressure in the system, thereby changing the effective bulk viscosity. Explicit formulae have been derived to compute the bulk viscosity for all volume fractions of suspended rigid particles and for all expansion rates. The hydrodynamic forces between particles including the strong lubrication interactions near contact play an important role at high concentrations. The bulk viscosity of concentrated suspensions with full hydrodynamic interactions is determined via direct simulation by adapting the Stokesian Dynamics paradigm to allow for a uniform rate of expansion.

See more of this Session: Particulate and Multiphase Flow I
See more of this Group/Topical: Engineering Sciences and Fundamentals

Probe Mobility in Finite Stokes Number Suspensions: From Viscous to Granular Microrheology

Ashley G. Smart, Division of Chemistry and Chemical Engineering, California Institute of Technology, Pasadena, CA

John F. Brady, California Institute of Technology / Division of Chemistry and Chemical Engineering, Pasadena, CA

Microrheology - whereby viscoelastic properties of a complex fluid are inferred from the motion of a small, immersed probe - has emerged as a useful tool for studying suspensions, particularly those systems that are too small for traditional techniques. In this small limit, particle inertia is typically negligible. Microrheology, however, may be an equally valuable paradigm for larger, inertial suspensions such as granular flows, where micro-scale phenomena underly processes like mixing and segregation. In this talk, we outline a framework for passive and active microrheology in sheared, finite Stokes number suspensions, thus connecting the non-inertial and inertial limits. In the small Stokes regime (non-inertial), probe mobility is characterized by a competition between forcing and shearing, resulting in force-thinning behavior under passive forcing that gives way to a viscous plateau under active forcing. By contrast, in the large Stokes (inertial, or granular) regime - characterized by competition between forcing and collision-induced diffusion - microviscosity is constant for passive forcing, and force thickens as Pe^2 under active forcing. Particle dynamics results for a large range of Stokes numbers, volume fractions, and elastic coefficients collapse predictably as a function of Peclet number and shear rate-normalized forcing in the high and low Stokes regimes, respectively.

See more of this Session: Particulate and Multiphase Flow I
See more of this Group/Topical: Engineering Sciences and Fundamentals

Rheology and Microstructural Development in Oscillating Suspensions of Non-Colloidal Spheres

Hyun-Ok Park, Chemical Engineering, University of Florida, Gainesville, FL

Jason E. Butler, Chemical Engineering, University of Florida, Gainesville, FL

Rheological measurements are used to assess the existence and rate of microstructural changes in oscillating suspensions of non-colloidal spheres. Though slightly non-linear, naively separating the stress response into components in and out-of-phase with the strain reveals that the two components change at different rates. Rapid and large changes in the storage modulus occur over the first 15-20 oscillations for concentrated suspensions in the vicinity of a strain amplitude of one. The loss modulus is much larger than the storage modulus and appears to remain constant over the first twenty oscillations. However, the loss modulus does change slowly, consequently requiring large numbers of oscillations to reach steady state. The results are qualitatively independent of the shear cell geometry, suggesting that shear-induced particle migration is unimportant and that the observed behavior results from changes in the suspension microstructure. Comparison of the experimental results at small total strains to simulation results provides insight into the rapid changes of the storage modulus. Extending the range of volume fractions and total strains beyond previous investigations, as well as a more detailed study of the response at small total strain, has provided a more comprehensive view of the rheology.

See more of this Session: Particulate and Multiphase Flow I
See more of this Group/Topical: Engineering Sciences and Fundamentals

Combined Hydrodynamic and Static Force Interactions of Two Rough Spheres in Nonlinear

Marc S. Ingber, Mechanical Engineering, University of New Mexico, Albuquerque, NM

The phenomenon of particle migration in nonlinear viscous shear flow has been under investigation now for several decades. Although there has been significant progress in the understanding of suspension flows, there still remains several significant open questions. For example, the migration rate appears to scale with slightly less than the cube of the characteristic particle size in Couette flow whereas, for Poiseuille flow, the rate appears to scale with slightly less than the square of the characteristic particle size. The resolution of this discrepancy in the migration rate between Poiseuille and Couette flow is currently an open question. As another example, in periodic flow, it has been shown that there is a critical amplitude below which the flow is reversible, and above which, irreversibilities accumulate rapidly with increasing amplitude. Further, for large amplitude oscillatory Poiseuille flows, particles migrate to the low shear-rate region of the flow field, whereas for small amplitude, particles actually migrate to the high shear-rate region of the flow field. However, current rheological models do not differentiate between small and large amplitude osciallation. Hence, another open question is to determine the physics currently lacking in these rheological models.

In fact, the underpinnings of current rheological models are based almost entirely on multi-body interactions, and hence, the answers to these open questions have been primarily sought by analyzing dense suspension flows. In the current research, a much simpler system is considered of two rough spheres undergoing nonlinear shear flow in which the hydrodynamic interactions are augmented by an adjustible electrostatic force. Numerical simulations are performed using a traction-corrected boundary element formulation. It is shown in this research that the magnitude and even the direction of the particle migration is strongly affected by the static force. Many of the behaviors observed in dense suspensions can be replicated in these two-particle systems. Some conjectures are put forward based on these two-particle simulations that could potentially resolve outstanding issues in suspension modeling.

See more of this Session: Particulate and Multiphase Flow I
See more of this Group/Topical: Engineering Sciences and Fundamentals

The Shear-Induced Diffusivity of a Dilute Suspension in Bounded Systems

Ankit Rohatgi, Department of Chemical and Biomolecular Engineering, University of Notre Dame, Notre Dame, IN

David T. Leighton, Jr., Department of Chemical and Biomolecular Engineering, University of Notre Dame, Notre Dame, IN

In a recent paper, Zurita-Gotor & Blawzdziewicz, 2007 identified the role of wall reflection on fluid and particle trajectories for the flow of bounded dilute suspensions. In unbounded systems particle and fluid trajectories are either open or closed, but in both cases tracers and particles return to their original streamlines. Thus such binary interactions do not yield self- or gradient-diffusion. The presence of a no-slip boundary, however, leads to a class of open "flipping" trajectories which are capable of producing dispersion, and this mechanism has been demonstrated to be the dominant source of shear induced self-diffusivity measured by Zarraga & Leighton, 2002 and Beimfohr et. al.,1993. This was the case even though the experimental particle diameter to gap width ratios were 1:20 or smaller.

In this work, we explore this phenomenon further by examining the influence of wall induced trajectories on both shear-induced self- and gradient-diffusivities. The trajectories are determined using the stresslet reflection from a single wall to obtain an analytical expression for the fluid and particle velocity distributions (e.g., Blake & Chwang, 1974). This is integrated to obtain the resulting diffusivities using the method of da Cunha & Hinch, 1996. With two walls, fluid and particle trajectories are obtained using an infinite sequence of stresslet reflections. The diffusivities calculated from binary interactions between a tracer and a suspension sphere scale linearly with concentration as expected, consistent with experimental observations in the dilute limit. The self-diffusivity arising from one wall is given analytically by $D_S = \frac{1}{4}\dot{\gamma}\phi\frac{a^3}{h}$, which is independent of the tracer size. For large tracers, it is shown that the tracers interact with multiple particles which results in a decrease in the self-diffusivity. For two walls, the self-diffusivity is given analytically and is consistent with both the calculations of Zurita-Gotor & Blawzdziewicz, 2007 and measurements of Zarraga & Leighton, 2002.

Of considerable interest is the gradient diffusivity, which controls the flux due to a gradient in concentration. Due to the symmetry of the interactions, the wall reflection contribution to this diffusivity in both gradient and vorticity directions for equal size tracers is identically zero. For fluid tracers, however, the gradient diffusivity in the velocity gradient direction scales as $\dot{\gamma}\phi\frac{a^3}{h}$, while that in the vorticity direction is again zero. Surprisingly, the symmetry of both fluid-particle and particle-particle interactions in the presence of a wall leads to the possibility of a flux in vorticity direction proportional to the second derivative $\dot{\gamma}\phi a^3\frac{\partial^2 c}{\partial y \partial z}$. This leading order term has the potential to destabilize a concentration polarization layer in the dilute limit.

References

Beimfohr, S., Looby, T., Leighton, D. T., Measurement of the shear-induced coefficient of self-diffusion in dilute suspensions, *in Proceedings of the DOE/NSF Workshop on Flow of Particles and Fluids*, Ithaca, NY, 1993.

Blake, J. R., Chwang, A. T., Fundamental singularities of viscous flow. Part I: The image systems in the vicinity of a stationary no-slip boundary, *J. Eng. Math.*, **8**(1), 23-29, 1974.

da Cunha, F.R., Hinch, E.J., Shear-induced dispersion in a dilute suspension of rough spheres, *J. Fluid. Mech., 309, 211-223, 1996.*

Zarraga, I.E., Leighton, D.T., Measurement of an unexpectedly large shear-induced self-diffusivity in a dilute suspension of spheres, *Phys. Fluids*, **14**(7), 2194-2201, 2002.

Zurita-Gotor, M., Blawzdziewicz, J., Wajnryb, E. Swapping Trajectories: a new wall-induced cross-streamline particle migration mechanism in a dilute suspension of spheres, *J. Fluid. Mech. 592, 447-469, 2007.*

See more of this Session: Particulate and Multiphase Flow I
See more of this Group/Topical: Engineering Sciences and Fundamentals

Modeling of Particle Transport at Low Reynold's Number in Complex Channels Using a Dynamic Boundary-Integral Method

John F. Ashley, Department of Chemical and Biological Engineering, University of Colorado at Boulder, Boulder, CO

Alexander Zinchenko, Department of Chemical and Biological Engineering, University of Colorado at Boulder, Boulder, CO

Robert H. Davis, Department of Chemical and Biological Engineering, University of Colorado at Boulder, Boulder, CO

Christopher N. Bowman, Department of Chemical and Biological Engineering, University of Colorado at Boulder, Boulder, CO

We have developed the first theoretical model and numerical simulation method for the 'pinched-flow fractionation' (PFF) phenomenon in a microfluidic device. Pinched-flow fractionation has been demonstrated as an efficient, practical way to achieve accurate particle sizing and separation using fluid focusing and spreading in a microchannel (Yamada et al., 2004, Anal. Chem. 76:5465-5471). The device consists of two input channels that meet to form a plane-parallel pinch area, followed by an expansion area and outlet. One of the input channels contains a dilute mixture of polydisperse particles made neutrally buoyant in the stream by a buffer solution, while the second input channel contains only a buffer solution. The higher volumetric flow rate from the inlet channel without particles causes the particles to be pushed close to one wall in the pinch area of the device. The particles then effectively separate by the spreading streamlines in the expansion area according to their sizes, as the particle centers can come no closer than one radius from the device wall in the pinch area. A theoretical description of this phenomenon based on calculation of particle trajectories has not been attempted thus far, to our knowledge.

We have developed a novel and suitable 3D boundary-integral method for low-Reynolds- number dynamical simulations of a single spherical particle in such a microchannel, assuming small volume fraction in the inlet to neglect particle-particle interactions. Complexity of channel geometry, where, outside of the pinch area, the overall device size is much larger than the particle size, provides a challenge for a numerical solution. Additionally, the need to resolve lubrication effects between the particle and channel walls when in close proximity presents a computational challenge. To provide a feasible solution, we first calculate the fluid flow in the entire channel without a particle, from the solution of a 2D boundary-integral problem with prescribed inlet and outlet Poiseuille flow profiles having specified flow rates at low Reynolds numbers. The 3D particle is then embedded in a computational cell, roughly 10-20 times larger than the particle size, but still much smaller than the overall device dimension. The cell, dynamically constructed at each time step, is an intersection of a large cube around the particle and the channel domain. Thus, we refer to this approach as a 'dynamic boundary-integral method', as the computational domain changes while the particle moves through the device. The problem is formulated as a 'completed-double-layer' boundary-integral equation for the surface potential, with the fluid velocity on the cell boundary taken from the 2D solution, and solved iteratively at each time step. High particle surface triangulations (up to 15,640 elements) and high-resolution adaptive mesh on the cell boundary allow us to accurately capture particle-wall lubrication for small separations between the particle and channel wall (~1% of particle radius).

This method is systematically used to study the effects of particle radius, the ratio of flow rates, and the channel geometry on particle trajectories and the efficiency of size segregation. Since the centers of the larger particles travel further from the nearby wall in the pinch region due to steric limitations, their trajectories upon leaving the pinch region to the outlet area also further from the device wall. Moreover, the trajectories of the larger particles deviate from the streamlines by a greater amount than that observed for smaller particles. Our calculations may be used to assess when the dynamic boundary-integral method provides a more accurate representation of particle trajectories than approximating them as fluid streamlines. Our approach is quite general for microchannel geometries consisting for an arbitrary number of rectangular panels, and it is not limited by a spherical particle shape. In the future, the model may be adapted to follow the motion of deformable drops through complex complex microfluidic devices.

See more of this Session: Particulate and Multiphase Flow I
See more of this Group/Topical: Engineering Sciences and Fundamentals

Stokesian Interactions Between Two Spheres Inside a Narrow Cylindrical Conduit

Shahin Navardi, Department of Mechanical Engineering, Texas Tech University, Lubbock, TX

Sukalyan Bhattacharya, Department of Mechanical Engineering, Texas Tech University, Lubbock, TX

Hanyan Wu, Mechanical Engineering, Texas Tech University, Lubbock, TX

Creeping motion of multi-particle systems inside a cylindrical channel is an important phenomenon which affects many processes like blood-flow in bio-conduit or transportation in microfluid channels. Such systems involve both interparticle and particle-wall interactions which so far have not been fully analyzed when the size of the confining conduit is comparable to the size of the suspended bodies.

In this talk, we present our semianalytical formulation which addresses this unsolved problem. Our mathematical procedure is based on basis-function expansion of Stokesian solutions leading to an efficient simulation technique. Though the method can be used for any number of interacting particles, here we concentrate on a two-sphere system explaining dynamics of both particles under different flow-situations and geometric configurations.

First, we consider two closely situated identical spheres inside a narrow cylinder, and find their hydrodynamic frictions for motion in quiescent flow as functions of the particle-positions and cylinder-sphere size-ratios. Next, we determine the force and torque on fixed spheres in parabolic flow, as well as the motion of freely suspended system. Then, we repeat similar calculations for cylinder-bound spheres of asymmetric sizes. Finally, we analyze the flow where a large fixed sphere is blocking the conduit and a small suspended sphere is interacting with pressure-driven flow near the constriction. The last results are especially important because they can describe how suspended bodies like blood-cells behave inside a vessel partially blocked by depositions.

See more of this Session: Particulate and Multiphase Flow I
See more of this Group/Topical: Engineering Sciences and Fundamentals

Bimodal Vs. Monomodal Suspension Flows through An Abrupt Contraction-Expansion

Nancy F. Zeng, Department of Chemical and Biochemical Engineering, Rutgers University, Piscataway, NJ

Nina C. Shapley, Department of Chemical and Biochemical Engineering, Rutgers University, Piscataway, NJ

The goal of this study is to understand the correspondence between concentrated bimodal and monomodal suspensions flowing through an abrupt contraction-expansion, which is a model geometry for materials processing operations. Previous studies have compared particle concentration profiles for monomodal and bimodal suspensions in unidirectional flows, emphasizing the differences between monomodal and bimodal suspension behavior. However, similar behavior can be identified clearly in the abrupt contraction-expansion flow, where distinctive particle concentration distributions for each bulk particle volume fraction form downstream of the expansion section.

In this work, suspensions of neutrally buoyant, noncolloidal spheres in viscous, Newtonian liquids undergo pressure-driven flow in an abrupt, axisymmetric 1:4 contraction-expansion. Nuclear magnetic resonance imaging (NMRI) is used to measure the steady-state particle concentration distribution and flow field. By adopting varied particle materials (i.e. rigid polymer vs. hydrogel) we can monitor distributions of small and large particles through the contrast in the spin-spin (T2) relaxation time among rigid solids, gels and suspending fluids. Results to date suggest that the large particle distributions observed in bimodal suspensions resemble those of monomodal suspensions of large particles with the same average steady shear viscosity.

See more of this Session: Particulate and Multiphase Flow I
See more of this Group/Topical: Engineering Sciences and Fundamentals

Inertially-Driven Migration of Spherical Particles in Microchannel Flow

Pandurang M. Kulkarni, Benjamin Levich Institute and Department of Chemical Engineering, The City College of New York, New York, NY

Katherine J. Humphry, Physics Department, Harvard University, Cambridge, MA

Howard A. Stone, Division of Engineering and Applied Sciences, Harvard University, Cambridge, MA

Jeffrey F. Morris, Benjamin Levich Institute and Department of Chemical Engineering, The City College of New York, New York, NY

Suspended particles are known to migrate in pressure-driven flow at finite-inertia conditions. This is the "tubular pinch" phenomenon, successfully described by theory for the point particle limit in both channel and tube geometries. Rather surprisingly, particle migration is observed to occur regularly in microfluidic applications, particularly when the suspending fluid is aqueous and thus of low viscosity. In this work we focus on the migration and equilibrium positions of suspended spheres in rectangular microchannel geometries. The finite size is found to be a critical determinant in the behavior when the ratio of particle size (a) to the narrow channel dimension (H) satisfies $a/H > 1/10$. We first review the theory for point particles and existing results for finite-size particles in inertial flows; these show a bifurcation in the equilibrium position for tube and channel flows at large conduit Reynolds number which is apparently due to the finite particle size. We then present a comprehensive examination by both experiment and lattice-Boltzmann based simulation of the migration behavior of particles in rectangular microchannels, showing the dependence of the behavior on the aspect ratio of the channel (W/H), the dimensionless particle size (or a/H), and the Reynolds number. Scaling of the transient data shows that the rate of approach to the ultimate equilibrium position scales linearly with Re, the shear-rate based Reynolds number at the particle scale.

See more of this Session: Particulate and Multiphase Flow I
See more of this Group/Topical: Engineering Sciences and Fundamentals

Buckling Transition in Wall-Bounded Suspension Flows

Jerzy Blawzdziewicz, Mechanical Engineering, Yale University, New Haven, CT

Eligiusz Wajnryb, Institute of Fundamental Technological Research, Polish Academy of Sciences, 00-04 Warszawa, Poland

In this talk we analyze new structural transformations that occur in periodic square arrays of microspheres. During the array evolution, the initial perturbations observed as density waves and vertical particle scatter slowly dissipate and there occurs a sudden transition from the square to hexagonal order with a large number of mobile defects.

The resulting suspension structure resembles a hexatic phase. We explain this transformation in terms of a buckling instability that produces vertical displacements of particle rows. Our simulations predict that at the early stages of the array evolution random transverse particle displacements stabilize the particle lattice, preventing coherent movement of particle rows. We discuss implications of this novel stabilizing mechanism for systems such as compressed membranes undergoing thermal fluctuations.

See more of this Session: Particulate and Multiphase Flow I
See more of this Group/Topical: Engineering Sciences and Fundamentals

Stabilizing Influence of Topography On Thin Liquid Films Flowing Over Locally Heated Surfaces

Jeffrey M. Davis, Chemical Engineering, University of Massachusetts, Amherst, MA

Naveen Tiwari, Chemical Engineering, University of Massachusetts, Amherst, MA

Thin liquid films flowing over surfaces with localized heating develop a pronounced ridge at the upstream edge of the heater due to the gradient in surface tension that opposes the flow. This ridge becomes unstable to transverse perturbations above a critical Marangoni number and evolves into an array of rivulets, which corresponds to the rivulet instability observed in experiments that can lead to film rupture. Similar fluid ridges form near topographical variations on isothermal surfaces at small capillary numbers, but these ridges are strongly stable to perturbations because the topography induces a net capillary pressure gradient that causes rearrangement of fluid in the flow direction. In this work, a long-wave, lubrication analysis is used to study the influence of topographical features on the linear stability of thin liquid films that flow over a locally-heated surface. The effects of basic step-down and mound features in the flow direction are analyzed. In contrast to liquid films on heated, horizontal surfaces, for which topography is destabilizing, even such non-optimized topography is found to be effective at stabilizing the flowing film with respect to rivulet formation and subsequent rupture. The critical Marangoni number at the instability threshold increases substantially (by over a factor of four) with appropriate topography, even for non-zero Biot numbers.

An energy analysis is used to provide insight into the mechanism by which the topography stabilizes the flow. While the streamwise gradient in the capillary pressure induced by the topographical feature is the expected stabilization mechanism because it is strongly stabilizing for isothermal flow over topography, this term is destabilizing for topographical features on the locally-heated surface. Upon the addition of topography, stabilization is primarily due to the streamwise capillary flow induced by the variation in the curvature of the perturbation in the flow direction. This term provides additional energy damping primarily because of the effect of the topography on the steady base state, through which the various physical effects isolated in the energy analysis are nonlinearly coupled. Because the stabilizing effect of the topographical features is only weakly sensitive to the governing parameters and particular temperature profile, the use of such features could be a simple but effective method of stabilization for sufficiently thin films.

See more of this Session: Stability and Nonlinear Hydrodynamics
See more of this Group/Topical: Engineering Sciences and Fundamentals

Linear Stability Analysis of a Model Planar Free Surface Viscoelastic Displacement Flow

Arash Abedijaberi, Chemical and Biomolecular Engineering, University of Tennessee in Knoxville, Knoxville, TN

Bamin Khomami, Department of Chemical and Biomolecular Engineering, Material Research and Innovation Laboratory (MRAIL), University of Tennessee, Knoxville, TN

The linear stability of a model planar stagnation viscoelastic free surface displacement flow, utilizing the Oldroyd-B and the FENE-P models is studied via generalized eigenvalue analysis. Specifically, we have examined the effect of the ratio of solvent to total solution viscosity (η), the Weissenberg number (Wi) and the Capillary number (Ca) on the stability of the flow by monitoring the leading discrete mode. It is demonstrated that long waves are the most dangerous modes and their growth rate is a function of Wi, η, and Ca. Specifically, the disturbance growth rate is enhanced with increasing Wi and Ca and decreases with increasing η. Eigenfunction analysis shows that the perturbation normal stresses in the flow direction and the perturbation pressure, associated with the most dangerous mode, are localized near the free surface stagnation point. Furthermore, it is shown that elastic destabilization of the flow occurs due to the alteration of the interface normal stress balance at the free surface as a result of formation of an elastic stress boundary layer near the free surface stagnation point.

See more of this Session: Stability and Nonlinear Hydrodynamics
See more of this Group/Topical: Engineering Sciences and Fundamentals

Prediction of Atomization in Two-Phase Turbulent Flow

Eric W. Smith, Chemical and Biomolecular Engineering, University of Notre Dame, Notre Dame, IN

Mark J. McCready, Chemical and Biomolecular Engineering, University of Notre Dame, Notre Dame, IN

Randy M. Roberts, Advanced Production Systems, Chevron Energy Technology Company, Houston, TX

The interfacial behavior of a two phase stratified flow can be classified as smooth, wavy, or rolling. An initial linear instability of the smooth interface leads to periodic waves and a second instability of the wavy interfaces leads to formation of roll waves. It is the existence of roll waves that provides the mechanism for atomization, and so prediction of this second transition is the goal of this study. Ultimately, the work presented here is to aid in the design of future pipelines to reduce failures, reduce maintenance expenses, and ensure platform safety.

Previous experimental work (Bruno and McCready, 1988) had shown that roll waves grow out of a linearly unstable long wave. However, quantitatively correct predictions had not been possible because previous stability analyses had not correctly described the wavy interface basestate nor accounted for the turbulent gas flow.

In this work a k-epsilon turbulence model is used and adapted for gas-liquid flows. Models of this type are commonly used for engineering calculations and are general enough to apply to all conditions found in oil-gas pipelines. Since short waves always exist before long waves, finding the neutral stability of long waves requires the basestate velocity to account for the effects of a wavy interface. This effect, similar to single phase flow with one roughened wall, is a shift of the maximum velocity away from the roughened surface. This effect is well documented in literature (Akai 1981, Krogstad 1992, Lorencez 1997). For gas-liquid flow an alteration was needed to account for the fact that the relaminarization at the interface is not the same as at the wall. By adjusting the effective y^+ at the interface, the base state calculations match the asymmetric gas-liquid profiles found in literature.

The effect of turbulence is not confined to the basestate. The Orr-Sommerfeld equation is augmented with terms that arise from eddy viscosity arguments. To solve the linear stability problem, as is common with a highly non-linear coupled system of differential equations, an iterative method was implemented, which requires an initial guess of the three dependent variables: velocity (u), turbulent kinetic energy (k), and turbulent dissipation (e). Profiles produced by a smooth interface proved to be insufficient as initial guesses. Furthermore, in any given iteration, a value of k and e had to be given as a boundary condition. To remedy the problem, the values of k and e at the interface were set in proportion to the values of k and e at the effective y^+ near the wall. Keeping these values merely proportional to the local turbulence maximums decoupled the interface and the wall regions in the sense that the profile near the interface was allowed to have a different magnitude of turbulence than the wall, which is what is observed in smooth interface calculations and expected because of the difference in velocities.

After ensuring numerical accuracy for gas Reynolds numbers as large as 10^6 and applying multiple numerical and solution methods as a means of verification, the effective y^+ at the interface can be adjusted as the single empirical variable in order to match the atomization data. The calculations are compared in the figure below with data taken from Ivan Mantilla (Tulsa, 2007). It is expected that this single parameter, which is the effective surface roughness, can be predicted from the linear growth rates of the smooth interface.

This talk will describe the implementation of the numerical scheme, the verification, by comparison with experiments, of the correctness of solution to the basestate problem and predictions of atomization onset for a range of conditions again in comparison with experimental results.

Funding provided by the Chevron Energy Technology Company

See more of this Session: Stability and Nonlinear Hydrodynamics
See more of this Group/Topical: Engineering Sciences and Fundamentals

Growth and Nonlinear Resonant Interactions of Interfacial Waves in Stratified Channel Flows

Bryce K. Campbell, Mechanical Engineering, Massachusetts Institute of Technology, Cambridge, MA

Yuming Liu, Mechanical Engineering, Massachusetts Institute of Technology, Cambridge, MA

Randy Roberts, Chevron Energy Technology Company, Houston, TX

This work considers nonlinear wave-wave interactions and interfacial wave resonances in a two-fluid stratified flow through a horizontal channel for the purpose of understanding mechanisms capable of generating slug flow. An efficient high-order spectral method is developed for the simulation of the generation and nonlinear evolution of interfacial waves. The method is based on a potential flow formulation which includes normal viscous stresses and a pressure forcing term at the interface respectively for modeling of the damping effect and surface shear effect by the upper fluid. The method is capable of accounting for the nonlinear interactions of a large number of wave components in a broadband spectrum, and obtains an exponential convergence of the solution with the number of spectral modes and interaction order.

It is known that depending on the relative velocity between the two phases, two instability mechanisms can create initial wave disturbances on the interface: the classic Kelvin-Helmholtz instability or shear instability due to turbulent fluctuations in the upper phase. These instability mechanisms on the initial growth of the disturbances on the interface are well understood; however, the subsequent nonlinear evolution and the effects of wave resonances on the interface remain a subject of interest and will be the focus of this study.

The Kelvin-Helmholtz (KH) model is the traditional model for predicting interfacial instability in stratified flows. This linear instability occurs when the lower phase inertia and the pressure from the upper phase overcome the stabilizing effects of gravity and surface tension. This mechanism is most unstable to short waves and does not allow for modal interactions. When nonlinearity is included, additional second-order sum- and difference-wavenumber modes as well as higher harmonics must be accounted for. Given an initial spectrum of stable modes, sum-wavenumber modes and higher harmonics are generated due to nonlinear wave-wave interactions, which can be unstable to KH. The generated second-order difference-wavenumber modes are stable to KH. For certain combinations of wave modes, we found that there exist simultaneous coupled second-harmonic (overtone) and triad resonances. Due to the overtone resonance, energy from the KH unstable second harmonic is transferred to the (stable) first harmonic. Meanwhile, this first harmonic transfers its energy to other two (stable) wave modes in the resonant triad. These coupled resonant wave-wave interactions result in rapid growth of long wave components on the interface.

For flows which are KH stable; the shear instability can cause the growth of relatively short waves on the interface. As in the case of KH instability, the overtone and triad resonances also occur. The overtone resonance transfers energy from the dominant unstable mode to its subharmonic (creating a period doubling phenomenon). Similarly the triad resonance enables the transfer of energy from unstable mode(s) to stable mode(s). These resonant wave-wave interactions provide a mechanism for the growth of long waves by taking energy from (relatively short) unstable waves.

Direct comparisons between the numerical simulations and existing laboratory experiments are made. Good agreements between them are obtained. The findings in this study improve the understanding of the underlying mechanisms which cause short waves to evolve into large amplitude liquid slugs through nonlinear wave-wave interactions.

See more of this Session: Stability and Nonlinear Hydrodynamics
See more of this Group/Topical: Engineering Sciences and Fundamentals

AC Electrohydrodynamic Instabilities in Thin Liquid Films

Scott A. Roberts, Chemical Engineering and Materials Science, University of Minnesota, Minneapolis, MN

Satish Kumar, Chemical Engineering and Materials Science, University of Minnesota, Minneapolis, MN

When DC electric fields are applied to a thin liquid film, the interface may become unstable and form a series of pillars. In this paper, we apply lubrication theory to examine the possibility of using AC electric fields to exert further control over the size and shape of the pillars. For perfect dielectric films, linear stability analysis shows that the influence of an AC field can be understood by considering an effective DC field. For leaky dielectric films, Floquet theory is applied to carry out the linear stability analysis, and it reveals that high frequencies may be used to inhibit the accumulation of interfacial free charge, leading to a lowering of growth rates and wavenumbers. Nonlinear simulations confirm the results of the linear stability analysis while also uncovering additional mechanisms for tuning overall pillar height and width. The results presented here may be of interest for the controlled creation of surface topographical features in applications such as patterned coatings and microelectronics.

See more of this Session: Stability and Nonlinear Hydrodynamics
See more of this Group/Topical: Engineering Sciences and Fundamentals

An Electrokinetic Vortex Instability at the Interface of a Permselective Membrane: Beyond Limiting Currents

Gilad Yossifon, Department of Chemical and Biomolecular Engineering, University of Notre Dame, Notre Dame, IN

Hsueh - Chia Chang, Department of Chemical and Biomolecular Engineering, University of Notre Dame, Notre Dame, IN

We examine the important limiting-current phenomenon, when a DC current penetrates a permselective nanoporous membrane or nanochannel, and showed that it breaks down beyond a critical voltage to produce an over-limiting current because of a nonlinear hydrodynamic instability. Unlike zero-flux Poisson-Boltzmann equilibrium distributions, a constant current flux through the membrane produces an electroneutral diffusion layer with a non-equilibrium linear concentration distribution that extends from the membrane interface to the electrode. The diffusion-limited current corresponds to the limit when the interfacial ion concentration approaches zero. This is, however, a singular limit as the local chemical potentials of the ions would approach negative infinity in a logarithm manner. Instead, an extended polarized layer occurs between the electroneutral diffusion layer and the thin electric double layer (EDL) at the interface with a thickness that exceeds the Debye layer. Rubinstein et al (Desalination, 69, 101 (1988)) first predicted that over-limiting currents beyond the limiting current occur because of a vortex instability associated within this extended polarized layer. We recently verified this instability experimentally with a nanoslot by rejuvenating the interfacial vortices repeatedly with a low-frequency AC field, such that an ionic fluorescent dye trace the vortices under confocal microscopy before complete mixing occurs (PRL, 101, 254501 (2009)). The diffusion layer is found to be established by a self-similar constant-flux diffusion front whose position advances diffusively as sqrt(Dt). The vortex instability arrests the front growth once it has advanced to a critical position, which is a function of the applied voltage. This position corresponds to the thickness of the bifurcated vortex layer and a new diffusion layer thickness which offers a good prediction of the over-limiting current (PRE, 79, 046305 (2009)). We examine this length scale selection mechanism of the vortex instability here by examining the stability of the diffusion front. The instability is attributed to the interaction between the diffusion front and the polarized layer to produce the classical coupled dynamics of two interfaces. The positive feedback mechanism is due to the enhanced Maxwell pressure, when the separation between the two layers decreases, which drives an electro-osmotic vortex that further compresses the separation. Using a spectral theory for transient base states (Phys. Fluid, 14, 999 (2002)), we find that the dominant wavelength increases as sqrt(Dt), as in miscible fingering, in quantitative agreement with the experiments. The critical time for the destruction of the diffusion front to establish the steady vortex array is found by matching the disturbance concentration to the base-state concentration. It produces a voltage-dependent vortex layer thickness that will be compared to the experimental data.

See more of this Session: Stability and Nonlinear Hydrodynamics
See more of this Group/Topical: Engineering Sciences and Fundamentals

Universal Electrokinetic Mobility of Highly-Charged Particles of Arbitrary Shape, Size and Concentration

Todd M. Squires, Chemical Engineering, University of California at Santa Barbara, Santa Barbara, CA

In the limit of thin double-layers and negligible surface conduction, Morrison showed the electrophoretic mobility of a uniformly-charged particle to depend *only* on its zeta potential ζ -- not on its shape, size, or even concentration. Finite surface conduction causes this argument to fail. We treat the opposite limit, wherein particles are so highly charged that surface conductivity dominates completely (Bikerman-Dukhin number $Du \rightarrow \infty$), and show their electrokinetic mobility to be again independent of shape, size or concentration. Significantly, our result holds for *any* mean-field model of the double-layer and for any `phoretic' forcing, including electrophoresis and diffusiophoresis. The (standard) Guoy-Chapman model gives a $Du \rightarrow \infty$ electrokinetic mobility of $\varepsilon \ln 4\, k_B T/e\eta$, independent even of ζ.

See more of this Session: Stability and Nonlinear Hydrodynamics
See more of this Group/Topical: Engineering Sciences and Fundamentals

Numerical Simulation of Gas-Liquid Flow in a Flat Bubble Column Using the Lattice-Boltzmann Scheme

Radompon Sungkorn, Institute for Process and Particle Engineering, Graz University of Technology, Graz, Austria

Stefan Radl, Institute for Process and Particle Engineering, Graz University of Technology, Graz, Austria

Jos Derksen, Chemical & Materials Engineering, University of Alberta, Edmonton, AB, Canada

Johannes Khinast, Institute for Process and Particle Engineering, Graz University of Technology, Graz, Austria

Bubble columns are widely used in the chemical and pharmaceutical industry to produce a variety of products. It has been known that these devices are characterized by a high degree of unsteadiness and complexity[1]. Thus, a detailed understanding of the liquid-phase flow field and the bubble dispersion pattern is crucial in the design and scale-up of bubble columns.

By using computational fluid dynamics (CFD) modeling it is possible to gain insight into the details of flow and mixing in these multiphase systems. Although significant efforts have been invested in the field of bubble column simulation, challenges still remain. For example, fully resolved 3D simulations of commercial-scale bubble columns are prohibitively expensive for conventional CFD models. Therefore, we focus on developing a reliable computational tool with high efficiency in parallel computation.

In this work we studied gas-liquid flow in a laboratory-scale bubble column (the so-called Becker case[2]) by means of Large Eddy Simulation (LES) combined with Lagrangian particle tracking with two-way coupling. This approach is known as the Eulerian-Lagrangian (EL) method. The same lattice-Boltzmann scheme as employed by Derksen[3] is extended for discretizing the Navier-Stokes equations. To the best of the author's knowledge, this is the first application of the lattice-Boltzmann scheme with Lagrangian particle tracking to simulate bubbly flows. We have chosen this scheme because of its outstanding computational efficiency especially on parallel computer platforms[3]. Our work also includes an efficient scheme for checking the distance between the particles and the cells[4]. Excellent agreement with experimental data[2] for the mean velocity field as well as the transient bubble dispersion patterns is obtained. Also, we show that our code is significantly more efficient in terms of computation time and speed-up compared to conventional CFD codes.

References

(1) Hu G, Celik I. Eulerian-Lagrangian based Large-Eddy Simulation of a Partially Aerated Flat Bubble Column. Chemical Engineering Science 2008;63:253-271.

(2) Becker S, De Bie H, Sweeney J. Dynamic Flow Behaviour in Bubble Columns and Loop Reactors: Part II. Comparison of Detailed Experiments and Flow Simulations. Chemical Engineering Science 1994;49:5747-5762.

(3) Derksen JJ, Van den Akker HEA. Large Eddy Simulations on the Flow Driven by a Rushton Turbine. AIChE Journal 1999;45:209-221.

(4) Radl S, Khinast JG. Multiphase Flow and Mixing in Dilute Bubble Swarms. Submitted to AIChE Journal.

See more of this Session: Stability and Nonlinear Hydrodynamics
See more of this Group/Topical: Engineering Sciences and Fundamentals

STUDY ON THREE-PHASE FLOW BEHAVIOR UNDER ELECTROMAGNETIC FIELD

*Ze Sun , Ping Li, Guimin Lu, Bing Li, Jianguo Yu**

National Engineering Research Center for Integrated Utilization of Salt Lake Resources, East China University of Science and Technology, Shanghai, China

Abstract

With the development of powerful CFD solver and faster and cheaper computers, multiphase flow models have been improved significantly during the last decade, and some complex multiphase flow behaviors also can be predicted by numerical simulation. In this paper, the three-phase flow behaviors under the electromagnetic field will be investigated by CFD simulation and PIV measurement. These complex transport phenomena frequently take place in the metallurgical processes. For example, in the electrolysis process of the molten magnesium salt, magnesium is produced from cathode as droplets, while the by-product chlorine gas is produced from anode as bubbles, both trying to rise the surface of the electrolyte as a result of the density difference with the electrolyte, so a three-phase flow under the stronger electromagnetic field will be formed in the electrolytic cell. In this work, we focus the research on the three-phase flow behavior in the advanced diaphragmless electrolytic cell, where the cell with the dimension 2.91x1.87x1.40m, 8 anodes with dimension 0.95x1.14x0.15m, 9 cathodes with dimension 0.95x1.05x0.05m, and the anode and cathode distance as 0.07m. Firstly, the numerical simulations on the electric field and the magnetic field are done, respectively, where the order coupling method to calculate the coupled field is adopted and the current distribution is regarded as the input of the magnetic field distribution. Ansys 11.0 software is used in the simulation. Furthermore, the numerical simulation on the flow field considering the effect of electromagnetic field is done. The governing equation of the internal flow field in the electrolytic cell is established, the standard k-e turbulence model and the VOF multiphase flow model are selected for a comprehensive description of the flow characteristics of multiphase flow in the electrolytic cell, Lorentz force is added as the momentum source term in order to combine the effects of magnetic field on the flow field. FLUENT6.3 software is used in the simulation.

Keywords: Magnesium electrolyzer; Three-phase flow; Electromagnetic field; CFD;

1 Introduction

* Corresponding author: 86-21-64252826 Fax: 86-21-64252826
E-mail: jgyu@ecust.edu.cn

In magnesium electrolyzer, there exist three-phase flow including liquid magnesium, molten electrolyte and chlorine gas under 700°C. The chlorine gas goes up rapidly because of high difference in density between molten electrolyte and chlorine gas. The electrolyte circulation from electrochemical compartment to service compartment was formed because of chlorine gas going up. When electrolysis begins, strong current, above 100kA or more, flow through the electrolyzer. As a result, Lorentz forces are induced. So, the three-phase flow field is affected by electromagnetic field through Lorintz forces.

During electrolysis process, molten electrolyte circulation is the major reason for decline of current efficiency. Better circulation better current efficiency in electrolysis magnesium industry. Many scientists and engineers pay attention to this field. V.V. Burnakin used H_2O and argon gas to describe the hydrodynamics of a two-phase flow[3]. He controlled gas flow rate corresponding to current density from 0.1–0.4 A/cm^2. Brunakin built equations for two-phase flow characteristics in a magnesium electrolyzer[4]. V.V. Agalakov[9] investigated the flow field of the multiphase flow in the electrolyzer by the computational numerical simulation with the two-dimensional mathematical model under the following assumptions and simplify.

- Only focused on the flow phenomena
- Two-dimensional model
- Steady flow
- Bubbles had uniform radius
- No interaction between bubbles
- Flow field without electromagnetic force

S.A. Shcherbinin[6-8] built a three-dimensional mathematical model to simulate the temperature and electric fields in the magnesium electrolyzer. Author's previous work established mathematical model of flow field and validated the math model using particle image velocimetry (PIV) data[10].
In this paper, behaviors of three-phase flow affected by electromagnetic field in magnesium electrolyzer were investigated by computational fluid dynamics (CFD). Three dimensional model was built to study electromagnetic field by combining commercial software ANSYS and Fluent. Connection between finite element software ANSYS and control volume software Fluent was though user defined function (UDF).

2 Coupled fields calculation process

In the whole calculation process, coupled fields calculation was the main characteristic. Firstly, in order to get the electromagnetic field distribution, coupling calculation of electric field and electromagnetic field were undertaken, where electric field were as electromagnetic field's input. Secondly, considering effects of electromagnetic field during flow field calculation, flow field coupled electromagnetic field through taking Lorentz forces as source term in momentum equation. The coupled calculation process was shown in figure 1.

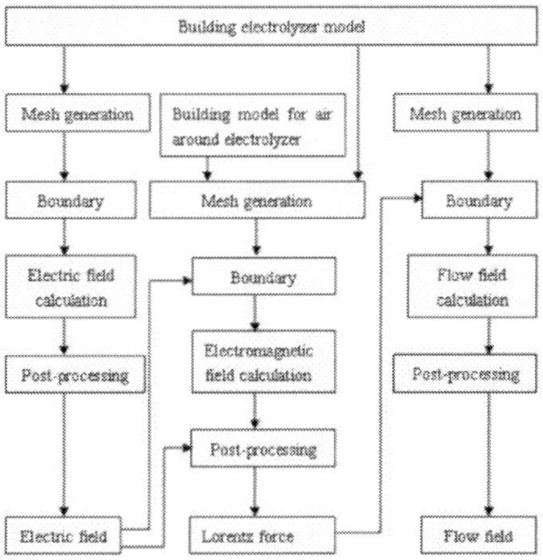

Figure 1 coupled field calculation process

3 Mathematical Model

Numerical simulation of three-phase flow behaviors in magnesium electrolyzer involves several subjects including electromagnetism and fluid dynamics .The model leans on the following basic hypotheses:

1. The electromagnetic fields satisfy the stationary Maxwell's equations, and Ohm's law is moreover supposed to be valid in all the cells.
2. The fluids are immiscible, incompressible and Newtonian.
3. The chlorine gas is as another continuous phase.
4. The electrical conductivity r and magnetic K are constant, respectively.
5. Fluid without chemical reactions.

Electromagnetism is based on Maxwell equations and Ohm's law. The differential form of Maxwell partial differential equations is given as follows:

$$\nabla \times H = J + \frac{\partial D}{\partial t} \tag{1}$$

$$\nabla \times E = -\frac{\partial B}{\partial t} \tag{2}$$

$$\nabla . D = \rho \tag{3}$$

$$\nabla . B = 0 \tag{4}$$

Where H is the magnetic intensity vector, J is the conduction current density, D is the electric displacement vector, E is the electric field intensity vector, B is the magnetic flux density vector and ρ is the free charge volume density.

The electromagnetic constitutive equations of simple isotropic material mediums can be expressed as:

$$D = \varepsilon E \tag{5}$$

$$B = \mu H \tag{6}$$

$$J = \gamma E \tag{7}$$

Where $\varepsilon = \varepsilon_r \varepsilon_0$ is the absolute dielectric constant, $\mu = \mu_r \mu_0$ is the absolute permeability and γ is the electric conductivity.

Fluid dynamics equations:

Fluid dynamics is based Navier–Stokes equations, for multi-phase, volume of fluid (VOF) was an effective method for three-phase flow in magnesium electrolyzer[10]. For the qth phase, the equations have the following forms:

Continuity equation

$$\frac{1}{\rho_q}\left[\frac{\partial}{\partial t}\left(\alpha_q \rho_q\right)+\nabla\cdot\left(\alpha_q \rho_q \vec{v}_q\right)=S_{\alpha q}+\sum_{p=1}^{n}\left(\dot{m}_{pq}-\dot{m}_{qp}\right)\right] \tag{8}$$

Where α_q was the volume fraction of qth phase in discrete volume; ρ_q was the density of qth phase; m_{pq} *and* m_{qp} were mass exchange between q^{th} phase and p^{th} phase, in electrolyzer, the mass exchange was zero; $S_{\alpha q}$ was the source term, in electrolyzer, $S_{\alpha q} = 0$.

The volume fraction equation will not be solved for the primary phase; the primary-phase volume fraction will be computed based on the following constraint:

$$\sum_{q=1}^{n}\alpha_q = 1 \tag{9}$$

Considering the effects of electromagnetic field on molten salt, the Lorentz force was as source term in momentum equation:

$$\frac{\partial}{\partial t}\left(\rho \vec{v}\right)+\nabla\cdot\left(\rho \vec{v}\vec{v}\right)=-\nabla_p+\nabla\cdot\left[\mu\left(\nabla\vec{v}+\nabla\vec{v}^T\right)\right]+\rho\vec{g}+\vec{F} \tag{10}$$

Where F is the Lorentz force, $F = J \times B$.

A single momentum equation is solved throughout the domain, and the resulting velocity field is shared among the phases.

The turbulence model

In order to get the solution of Navier–Stokes equations, turbulence model is essential to close set of equations. Turbulence model divides three kinds: direct numerical simulation (DNS), large eddy simulation (LES) and Reynolds-Averaged Navier-Stokes (RANS) Models.

The best method is DNS. This method does not rely on any model to compute turbulence, but the size of cells appears to be imposed by the smallest turbulence scales, the Kolmogorov scales. There is however a strong limitation to the use of DNS due to memory limitations imposed by today's computers. Indeed, the number of degrees of freedom required to characterize a velocity field u_i that corresponds to a turbulent flow is known to increase as $Re^{9/4}$ in three-dimensional (3D) turbulent systems. DNS of the Navier–Stokes equations are thus limited to moderately small Reynolds number flows.

Large-eddy simulation (LES) is an intermediate technique between the direct simulation and modeling the Reynolds stresses. In LES, the contribution of the large energy-carrying structures is computed exactly. Since the small scales turbulence tends to be more isotropic than the large, the use of isotropic models seems reasonable, and allows to solve the average Navier-Stokes equation on coarser meshes than those used in DNS.

The most used model among the turbulence models is the Reynolds-Averaged Navier-Stokes (RANS) Models. The k-e model is the most popular one. In the case of the k-ε model, two transport equations are considered; one for the computation of the turbulent kinetic energy and one for the dissipation rate.

$$\frac{\partial}{\partial t}(\rho k)+\frac{\partial}{\partial x_j}(\rho u_j k)=\frac{\partial}{\partial x_j}\left[\left(\frac{\mu_t}{\sigma_{k0}}\right)\frac{\partial k}{\partial x_j}\right]+G_k-\rho\varepsilon \tag{11}$$

$$\frac{\partial}{\partial t}(\rho\varepsilon)+\frac{\partial}{\partial x_j}(\rho u_j\varepsilon)=\frac{\partial}{\partial x_j}\left[\left(\frac{\mu_t}{\sigma_{\varepsilon 0}}\right)\frac{\partial \varepsilon}{\partial x_j}\right]+\frac{\varepsilon}{k}(C_{\varepsilon 1}G_k-C_{\varepsilon 2}\rho\varepsilon) \tag{12}$$

Where $G_k=\mu_t\frac{\partial u_i}{\partial x_j}(\frac{\partial u_j}{\partial x_i}+\frac{\partial u_i}{\partial x_j})$, C_μ , $C_{\varepsilon 1}$, $C_{\varepsilon 2}$, $\sigma_{\varepsilon 0}$, σ_{k0} are constants , They are C_μ=0.09 , $C_{\varepsilon 1}$=1.44 , $C_{\varepsilon 2}$=1.92 , $\sigma_{\varepsilon 0}$=1.3 , σ_{k0}=1.0 respectively.

4 Mesh Generation

Table 1 and table 2 were the dimensions of the geometry of real model. After building finite element model, the distribution of magnetic field can be obtained.

Table 1 Dimensions of geometry

	Electrolyzer	Anode	Cathode	Distance between anode and cathode
Y direction/m	2.91	0.15	0.05	0.07
X direction/m	1.87	0.95	0.95	-----
Z direction/m	1.40	1.14	1.05	-----

Table 2 Properties of materials

Material / Properties	electrolyte	Graghite	Cast steel	Refractory material	Air
Magnetic Permeability	1.0	1.0	1.0	1.0	1.0
electrical conductivity (Ω•m)	4.0×10^{-3}	8.5×10^{-6}	1.3×10^{-7}	∞	∞

Grid of electromagnetic field

During calculation of electromagnetic field, finite element method was employed. This method involves dividing the domain of the solution into a finite number of simple sub-domains (the finite elements), and then solving the finite-dimensional one to obtain an approximate solution. Although the results from the FE analysis belong to an approximate field, it is not impossible to enhance the accuracy by defining finer elements and providing more accurate material properties. According to its symmetry, half of electrolyzer could be used to solve this issue. Therefore, the geometric model for calculation given by Figure 2(a) is the right part of the electrolyzer only, and its mesh partitioning is shown as Figure 2 (b).

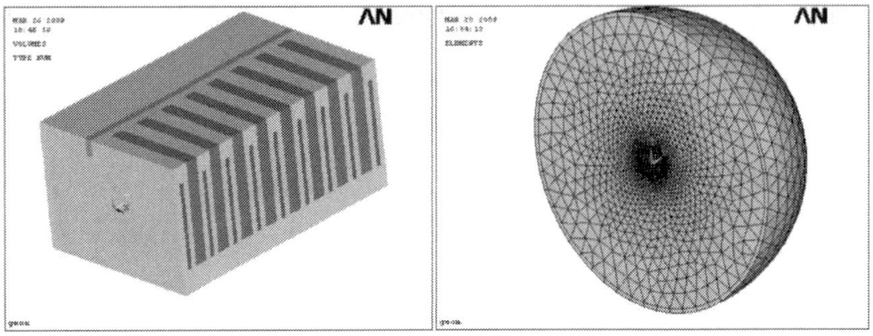

(a) (b)

Figure 2 finite element model for magnesium electrolyzer

Grid of flow field

During calculation of flow field, control volume technique was employed. Control volume technique converted a general scalar transport equation to an algebraic equation that can be solved numerically. This control volume technique consisted of integrating the transport equation about each control volume, yielding a discrete equation that expresses the conservation law on a control-volume basis.

The GAMBIT software release 2.2 was used to generate structure and unstructured mesh. The grid of the electrolyzer showed in figure 3

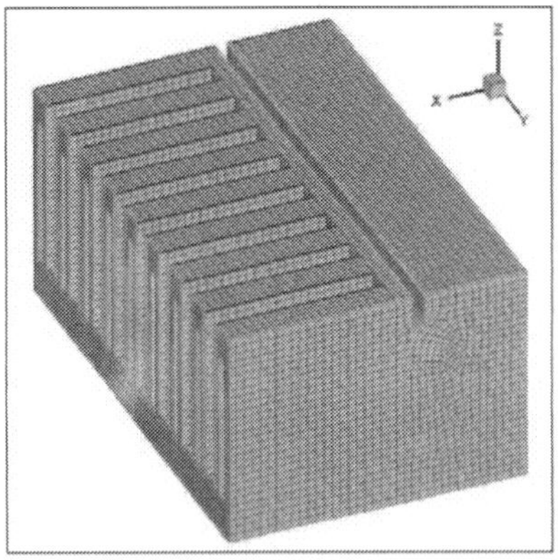

Figure 3 Grid of calculation domain for flow field

5 Boundary Conditions and Solution

The solution can be made by using magnetic vector potential (MVP) method in ANSYS 11.0. The finite elements type SOLID 69 was used to get electric field and SOLID97 to get electromagnetic field. The flow field distribution was obtained by using Fluent 6.3. To solve the governing partial differential equations, Navier–Stokes equations, for the conservation of mass, momentum, and scalars such as turbulence in integral form, a control-volume technique and SIMPLE procedure are used. The momentum source term, Lorentz forces, was read by user defined function (UDF). Convergence was assumed for each time step as all residuals fall below 10^{-3}. All simulations were performed on a personal computer equipped with two processors, Intel core 6300 and 2 GB main memory running under the Windows operating system.

Electric field calculation :

Applied loads:

Voltage was 4.95V which was the value in plant ; the potential of cathode was 0V ; y=0 was the symmetric plane.

Solve:

GUI : Main Menu > Solution > Current SL

Electromagnetic field calculation:

Taking the electric field as input, current distribution was read by GUI:

GUI : Preprocessor > Loads > -Load-Apply > -Magnetic-Excitation > -Curr Density- > From Elec An

Solve:

GUI : Main Menu > Solution > Current SL

6 Results and discussion

The distribution of electric field

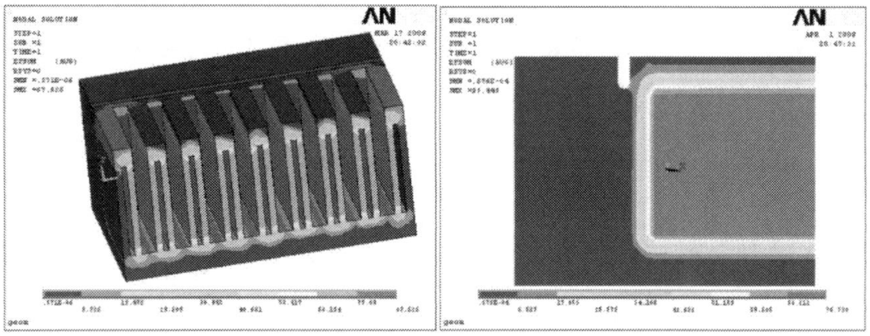

(a) (b)

Figure 4 contour of electric field

Figure 4 (a) was the contour of electric field of electrolyzer. As the figure shown, voltage energy was concentrated in electrochemical compartment, especially in the space between anode and cathode. From anode to the top of cathode, the value of voltage decreased from maximum to zero. So, in service compartment, magnesium did not affected by electric field because of magnesium carrying no charge. This was good for collection of magnesium in service area.

Figure 4 (b) was the contour of voltage on the surface paralleling to work surfaces of electrodes. In the picture, maximum voltage was a constant on the surface of cathode. The value of voltage declined rapidly with leaving the surface of cathode.

The distribution of electric field

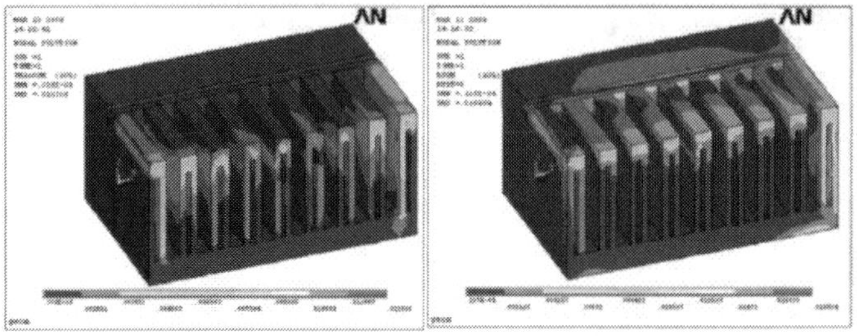

(a) (b)

Figure 5 contours of electromagnetic field(a) and contours of Lorentz forces(b)

The results show that magnetic Field and Lorentz force distributes vertically symmetrical along the electrolysis compartment, decreasing gradually from the end toward the middle, and the maximum Lorentz force reaching 0.0135N at corner.

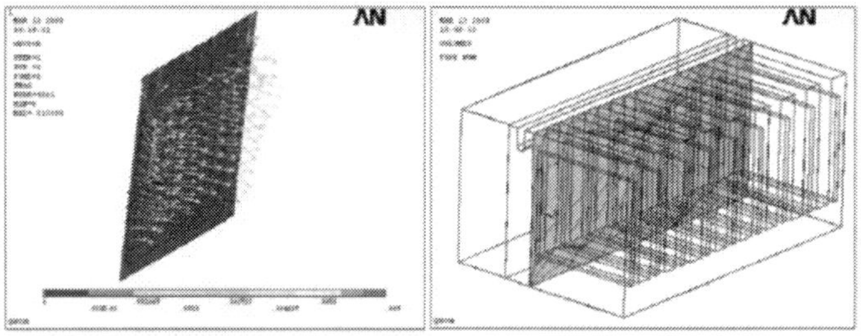

(a) (b)

Figure 6 vectors of Lorentz(a) and location of plane(b)

In order to get more details of Lorentz forces, a plane vertical to work surfaces of electrodes was shown in figure 6. Figure 6(b) was its location. As shown in figure 6, vectors of Lorentz forces pointed service compartment from electrochemical compartment. These directions of Lorentz forces were benefit for electrolyte circulation forming.

In the process of flow field calculation, numerical Lorentz forces must be gotten, table 3 was the numerical Lorentz Forces in magnesium electrolyzer.

Table 3 Distribution of Lorentz force

X /m	Y /m	Z /m	Fx /N	Fy /N	Fz /N
-0.900E-01	0.275E+01	0.934E-01	0.206E-02	0.857E-17	-0.124E-02
-0.400E-01	0.275E+01	0.935E-01	0.167E-02	0.693E-17	-0.111E-02
0.100E-01	0.275E+01	0.936E-01	0.117E-02	0.481E-17	-0.815E-03
0.600E-01	0.275E+01	0.937E-01	0.800E-03	0.330E-17	-0.611E-03
…	…	…	…	…	…
⋮					
…	…	…	…	…	…
-0.302E+00	0.279E+01	0.520E+00	0.219E-05	-0.155E-05	-0.350E-06

-0.302E+00	0.284E+01	0.520E+00	0.262E-05	-0.234E-05	0.437E-06
-0.302E+00	0.289E+01	0.520E+00	0.263E-05	-0.373E-05	0.115E-0.5

Flow field of three-phase flow under electromagnetic field

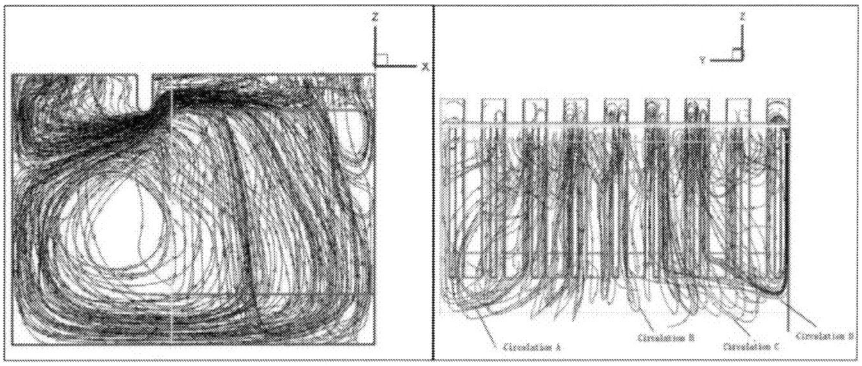

Figure 7 Streamlines of flow field in electrolyzer: (a)Y direction (b)X direction(in front)

Though analyzing the flow field obtained, we can see that the whole circulation distributes symmetrically along the electrolyzer. Electrolyte in the pair of the cathode and anode did parallel movement between the surfaces of the electrodes. Liquid magnesium can be delivered effectively to the collection compartment by electrolyte circulation. Other circulation between the cathode and anode deflect to the side wall instead of parallel to the surface of cathode and anode. The velocity of the electrolyte near the partition wall reached to the maximum while decreasing gradually away from the wall in the electrolysis compartment and collection compartment. The location of the maximum velocity in the electrolysis compartment was on the surface of the partition wall, reached to 164mm/s.

7 Conclusions

In the paper, combining ANSYS and Fluent software, electric field, electromagnetic field and flow field in magnesium eletrolyzer were gotten through post-processing. Coupling the electromagnetic field's effects, electrolyte circulations' types were obtained, and optimum circulation was showed through analyzing.

(1) Electric field concentrated in the electrolysis compartment; magnetic Field and Lorentz force distributes vertically symmetrical along the electrolysis compartment, decreasing gradually from the end toward the middle.

(2) Electrolyte circulations divided two different types: parallel to working surfaces of electrodes and not parallel to working surfaces of electrodes. The circulations paralleling to working surfaces of electrodes can transport liquid magnesium efficiently from electrochemical compartment to service compartment.

(3) The research indicates that the structure optimization of the electrolyzer should be designed to ensure electrolyte circulation parallel to the surfaces of electrodes and removing the adverse effects of the electrolyzer flow field.

8 Acknowledgements

The project is supported by National Nature Science Foundation of China (No. 50874048).

References

1. Horst E. Friedrich, and Barry L. Mordike.(2006), "Magnesium Technology". Germany, Springer
2. Zabelin, I. V., Ventsev S. D. and Kharko A. I., et al(1992), "Flow velocity of electrolyte in magnesium production cell", *Tsvetnye Metally (Moscow, Russian Federation),* May, pp.45-47
3. Burnakin V. V., Shestakov V. M., and Sorokous V. G., et al(1989), "Hydrodynamics of flow in the interelectrode space of magnesium electrolyzers", *Tsvetnaya Metallurgiya,* April, pp. 62-67
4. Burnakin, V. V., Polyakov, P. V., and Shestakov, V. M., et al(1979), "Analytical calculation of characteristics of two-phase flows in magnesium electrolytic cells", *Tsvetnye Metally (Moscow, Russian Federation)* 8, pp. 73-74
5. Korobov M. A(1983), "Mathematical model of a magnesium electrolytic cell", *Tsvetnye Metally (Moscow, Russian Federation)* 5, pp. 53-55
6. Scherbinin S. A., Yakovleva G. A., and Fazylov A. F(1994), "Mathematical model of thermal and electric fields in a magnesium electrolytic cell", *Tsvetnye Metally (Moscow, Russian Federation)*, 4, pp.60-64
7. Shcherbinin S. A., Yakovleva G. A., and Kazylov A. K(1994), "Numerical study of thermal and electric fields of a magnesium electrolytic cell", *Tsvetnye Metally (Moscow, Russian Federation)*, 6, pp. 68-71
8. Shcherbinin S. A., Fazylov A. R., and Yakovleva G. A(1997), "Mathematical simulation of three-dimensional thermal and electric fields of magnesium electrolyzer", *Tsvetnye Metally (Moscow, Russian Federation)*, 5, pp. 79-83
9. Agalakov V. V., Shcherbinin S. A., and Yakovleva G. A(1997), "Mathematical simulation of gas-hydrodynamic processes in a magnesium electrolyzer", *Tsvetnye Metally (Moscow)*, 7, pp.74-77
10. Ze Sun, He-nan Zhang, and Ping Li, et al(2009), "Modeling and simulation of the flow field in the electrolysis of magnesium". *JOM Journal of the Minerals, Metals and Materials Society.* May, pp. 29-33
11. Derksen J.J(2008)., "Scalar mixing by granular particles", *AIChE J,* July, pp. 1741-1747
12. Haroun Mahgerefteh, Garfield Denton, and Yuri Rykov(2008), "A hybrid multiphase flow model", AIChE J, September, pp. 2261-2268
13. Ville Alopaeus, Pasi Moilamen, and Marko Laakkomen(2009), "Analysis of stirred tanks with two-zone models", *AIChE J,* October, pp.2545-2552

Two-Phase Orifice Outflow Model

Haroun Mahgerefteh*, Navid Jalali
Department of Chemical Engineering,
University College London, London, WC1E 7JE
United Kingdom

*Author for correspondence (h.mahgerefteh@ucl.ac.uk)

Abstract

This paper describes the development of a transient outflow model for predicting pressurised two-phase discharge through an orifice. A phase dependent discharge coefficient is used to account for the vapour void fraction of the fluid in the narrowest flow cross section. The extended ω-method on the other hand accounts for the thermodynamic non-equilibrium effects at the beginning of nucleation for low vapour mass fractions. The liquid /gas phase slip is determined using a semi-theoretical correlation employing a two-phase multiplier.

The discharge Model is validated against experimental data for a variety of hydrocarbons containing different upstream liquid mass fractions in the range 0.1 - 0.7 up to a maximum pressure of almost 60 bar. The respective maximum and average errors of 5 % and 1.3 % between model predictions and measured data are observed.

1. Introduction

The study of two-phase flow through restrictions such as nozzles and orifice plates are not only of interest in flow rate measurement but also provide useful guidelines for sizing of pressure relief valves.

The most commonly used approach for sizing relief valves is based on API 520 [1,2] recommendations. Recent studies [3] involving direct comparison with experimental data have shown significant inaccuracies when the percentage of vapour in the upstream fluid mixture becomes appreciable.

Generally, two-phase flow through nozzle and orifice can be bounded between the homogeneous frozen and homogeneous equilibrium flow models In the former case, no momentum, heat and mass transfer between the phases are considered. In the case of homogeneous equilibrium model (HEM) however, it is assumed that the constituent phases travel with the same velocity and are in thermodynamic equilibrium [4].
In most cases the HEM model for two-phase flow, for both frozen and flashing flows, is shown to be adequate in long pipelines when the thermodynamic state and fluid properties are accurately predicted. However, for flows in short pipes and nozzles, such as those in relief valves, the difference between the experimental results and the predicted discharge rate by the HEM can be as much as 300% [5].

In this paper a modification of the Homogenous equilibrium model is presented for predicting outflow through an orifice. Phase slip is accounted for using a semi-empirical correlation. In the case of flashing two-phase flows on the other hand, boiling delay is accounted for using the ω method coupled with a phase dependent discharge coefficient. The model's performance is evaluated by comparison with published experimentally measured discharge data for a range of different hydrocarbon mixtures.

2. Theory

In this paper, the HEM model is only applied to cases where the flow through the orifice is choked. Therefore the fluid passing through the throat is flowing on the constant entropy line. An energy balance across the orifice gives:

$$H_u + \frac{1}{2}u_u^2 = H_o + \frac{1}{2}u_o^2 \quad (1)$$

Where,

H_u and H_o = the upstream and orifice enthalpy per unit mass, respectively

u_u and u_o = the upstream and orifice velocities, respectively

by neglecting the upstream velocity, The speed that the fluid travels through the orifice is the maximum velocity that the fluid can ravel and is defined as:

$$c^2 = \left(\frac{\partial P}{\partial \rho}\bigg|_S \right) \quad (2)$$

Where, c is the speed of sound in the fluid.

Assuming the ratio of the orifice-pipe diameters is small, the velocity upstream can be neglected with respect to the velocity at the orifice [6]. Thus, Eq. (1) can be re-casted in terms of mass discharge rate:

$$G = [2(H_u - H_o)]^{1/2} / v \quad (3)$$

Where,

G= is the mass discharge flux.

υ= is the specific volume of the two phase mixture.

From second law of thermodynamic for an isentropic process, Eq. (3) can be re-written as

$$G = \frac{[2\int_{P_u}^{P_o} - vdP]^{1/2}}{v} \quad (4)$$

For an isentropic two-phase expansion, the parameter ω found to be a unique function of the stagnation pressure for a given fluid and is given by [7].

$$\omega = \frac{\frac{v}{v_u} - 1}{\frac{P_u}{P} - 1} = \frac{x_u v_{v,u}}{v_u} + \frac{C_{pl,u} T_u P_u}{v_u} (\frac{v_{vl,u}}{h_{vl,u}})^2 \quad \textbf{(5)}$$

Where,

x = quality or gas mass fraction.

v_v = vapour specific volume.

v_l = liquid specific volume.

v_{vl} = difference between specific volume in vapour and liquid phases.

v = specific volume of the two-phase mixture.

$C_{pl,u}$ = specific heat capacity of liquid at upstream condition. .

T = temperature.

Grolmes and Leung [8] correlated the P-v relation in Eq. (4) and proved that using ω, defined in Eq. (5), satisfies the critical pressure ratio, $\eta = \frac{P_o}{P_u}$ in the non-linear equation given below:

$$\eta^2 + (\omega^2 - 2\omega)(1-\eta)^2 + 2\omega^2 \ln \eta + 2\omega^2 (1-\eta) = 0 \quad \textbf{(6)}$$

This alternative, approximate solution for two-phase flow known as Omega method is based on the assumption of HEM. It is modified herewith to allow for the thermodynamic non-equilibrium behaviour.

The critical mass velocity for chocked flow is determined from the change in specific volume with respect to change in pressure on the constant entropy line (also known as compressibility of two-phase flow) [9]. Thus, for a homogeneous two-phase flow mixture this derivative can be expanded as:

$$\frac{dv}{dP} = x\frac{dv_v}{dP} + v_{vl}\frac{dx}{dP} \quad \textbf{(7)}$$

The depressurization process of a two-phase fluid is usually accompanied with boiling of the liquid. Henry and Fauske (1971) introduce this non-equilibrium behaviour of the flow as a boiling delay factor N.In analogy; Diener & Schmidt [10] expressed the real evaporation rate, second term of Eq. (7), as:

$$\frac{dx}{dp} = \frac{dx_e}{dp} N \tag{8}$$

Where,

$\frac{dx_e}{dp}$ = evaporation rate at equilibrium

Diener & Schmidt [10] defined the boiling delay experimentally, based on 1300 experimental data, as a power law:

$$N = [x_e(p_{crit})]^a \tag{9}$$

Where,

a = power law exponent (a value of 2/3 is used for short nozzle and orifice).

p_{crit} =critical pressure at the narrowest cross sectional area of the flow.

The boiling delay factor N, the nucleation of bubbles, varies between zero (i.e. frozen flow) and 1 (i.e. equilibrium flow).

Since the discharge of fluid through an orifice is very fast, there is not sufficient time for heat transfer between the two phases thus leading to large non equilibrium effects.

The extended Omega takes account for non-equilibrium behaviour of the depressurization process caused by the boiling delay and is given as [4]:

$$\omega = \omega_{N=1}\left[x_u + \frac{C_{pl,u}T_uP_u}{v_u}\left(\frac{v_{vl,u}}{(h_{vl,u})^2}\right)\ln\left(\frac{1}{\eta}\right)\right] \tag{10}$$

The above requires that the ω parameter to be first calculated from Eq. (5) and then corrected using Eq. (10).

In addition to non equilibrium behaviour added to the HEM model, the effect of phase slip must also be taken in to consideration. Expressing the slip as a density ratio of the two phases [11]:

$$S = \left[\frac{\rho_f}{\rho_g}\right]^y \tag{11}$$

With parameter y ranging from 0 for homogenous flows (no phase slip) to 0.5 where maximum phase slip is observed. Based on the above, Simpson et al. [11] proposed a semi theoretical two-phase multiplier in a general form in:

$$\phi^2 = k\left[1 + x\left(\left[\frac{\rho_f}{\rho_g}\right]^y - 1\right)\right]\left[1 + x\left(\left[\frac{\rho_f}{\rho_g}\right]^{1-y} - 1\right)\right] \tag{12}$$

Where, k is the ratio of two-phase to all liquid flow coefficient which allows for changes in flow pattern and is considered to be equal to 1. The parameter y is given as 1/6 by Simpson et al. [11]. However, this is modified to 1/3 based on the work done by Moody [12] which is widely used in industry.

Eq. (13) is represented in form of an effective specific volume, based on earlier work carried out by Lottes [13], where the mass flow rate considering the slip between the phases can be calculated using a slip correction factor as defined below:

$$\frac{G_{HEM,Slip}}{G_{HEM}} = \varphi = \sqrt{\frac{\upsilon_{TP}}{\upsilon_e}} \tag{13}$$

$$\upsilon_e = \upsilon_l \phi^2 \tag{14}$$

Where,

G_{HEM} =calculated mass flow rate assuming Homogenous Equilibrium Model.

υ_{TP} =specific volume of two-phase at upstream condition.

υ_e =an effective specific volume.

υ_l = Liquid specific volume.

The discharge coefficient, C_d for two-phase flow through an orifice varies as the liquid mass fraction changes at orifice throat. A correlation for C_d as a function of the liquid alone and vapour alone discharge coefficients, weighted by means of the void fraction of the two-phase mixture at the condition in the narrowest cross section of the flow is used in this study.

3. Results and Discussion

Richardson et al. [3] conducted a series of experiments in which natural gas with known composition, temperature and pressure was discharged through sharp edge orifice plates with different sizes. In order to study the effect of premixing of the feed on the discharge rate through the restriction, mixtures were fed into the unit in well-mixed (i.e. when the thermodynamic equilibrium is reached) form as well as poorly-mixed (i.e. the feed is mixed using a mixing valve prior to discharge). Pressure, temperature and flow rate were recorded every 3 seconds.

The performances of four orifice flow models are investigated in this study based on comparison with the experimental data reported by Richardson et al. [3]. model A is the Simpson's two-phase multiplier and Omega method. For, model B, model A is modified by accounting for boiling delay. The calculated discharge rate and the percentage error for models A and B compared to experimental data are presented in Table 1.

Pressur e (bar)	Temperatu re (C)	Measured Discharge Rate(kg/s)	Model A		Model B	
			Discharge Rate (Kg/s)	Error (%)	Discharge Rate (Kg/s)	Error (%)
13.3	236.15	0.495	0.697	-20.2	0.639	-14.4
15.2	245.85	0.655	1.014	-35.9	0.895	-24
17.6	252.55	0.842	1.327	-48.5	1.138	-29.6
19.7	258.55	1.042	1.638	-59.6	1.356	-31.4
20.8	262.55	1.208	1.829	-62.1	1.462	-25.4
22.5	265.85	1.363	1.994	-63.1	1.543	-18
33.9	244.85	1.276	1.794	-51.8	1.638	-36.2
36.7	251.95	1.493	2.118	-62.5	1.883	-39
39.3	255.75	1.66	2.329	-66.9	2.038	-37.8
41	259.45	1.817	2.491	-67.4	2.138	-32.1
42.9	261.95	1.95	2.608	-65.8	2.209	-25.9
43.8	263.65	2.038	2.67	-63.2	2.235	-19.7
57.2	255.05	2.111	2.775	-66.4	2.543	-43.2
58.3	257.45	2.217	2.863	-64.6	2.597	-38
59	259.35	2.309	2.922	-61.3	2.627	-31.8
59.6	261.45	2.391	2.978	-58.7	2.646	-25.5
60.3	262.85	2.464	3.018	-55.4	2.66	-19.6
60.6	264.05	2.528	3.049	-52.1	2.659	-13.1
61.6	265.05	2.595	3.063	-46.8	2.679	-8.4

Table 1. Calculated discharge rate and percentage error from experimental results for Model A: Simpson et al. two-phase multiplier (slip correlation) and the original Omega, Model B: Simpson et al. two-phase multiplier (slip correlation) and the Non-Equilibrium Omega; experimental data is from 15mm sharp edge orifice.

The results obtained from model A and B produce absolute average errors of 56% and 29% respectively. A considerable over prediction is observed in model A. However accounting for boiling delay as incorporated in model B results in significant though unacceptable improvement .In order to improve the prediction of the model the value for parameter y is revised to 1/3, according to findings of Moody [12]. Model C is the modified two-phase multiplier using the revised value for parameter y and the Omega method with consideration of boiling delay factor (Non-Equilibrium behaviour). Model D is also the modified two-phase multiplier (same as model C) and the Omega method. The results obtained using model C and D are presented in Table 2.

Pressure (bar)	Temperature (C)	Measured Discharge Rate(kg/s)	Model C		Model D	
			Discharge Rate (Kg/s)	Error (%)	Discharge Rate (Kg/s)	Error (%)
13.3	236.15	0.495	0.52	-2.5	0.567	-7.2
15.2	245.85	0.655	0.728	-7.3	0.825	-17
17.6	252.55	0.842	0.931	-8.9	1.085	-24.3
19.7	258.55	1.042	1.124	-8.2	1.359	-31.7
20.8	262.55	1.208	1.233	-2.5	1.542	-33.4
22.5	265.85	1.363	1.324	3.9	1.711	-34.82
33.9	244.85	1.276	1.328	-5.2	1.454	-17.89
36.7	251.95	1.493	1.545	-5.2	1.738	-24.54
39.3	255.75	1.66	1.689	-2.9	1.930	-27.068
41	259.45	1.817	1.793	2.4	2.089	-27.2
42.9	261.95	1.95	1.871	7.9	2.209	-25.9
43.8	263.65	2.038	1.907	13.1	2.279	-24.14
57.2	255.05	2.111	2.113	-0.2	2.306	-19.5
58.3	257.45	2.217	2.175	4.2	2.398	-18.16
59	259.35	2.309	2.214	9.5	2.463	-15.45
59.6	261.45	2.391	2.248	14.3	2.530	-13.93
60.3	262.85	2.464	2.272	19.2	2.578	-11.48
60.6	264.05	2.528	2.283	24.5	2.612	-8.49
61.6	265.05	2.595	2.305	29	2.655	-6

Table 2. Calculated discharge rate and percentage error from experimental results (poorly mixed) for model C: Modified Simpson et al. two-phase multiplier (slip correlation) and the Non-Equilibrium Omega, and model D: Modified Simpson et al. two-phase multiplier (slip correlation) and the original Omega; experimental data is from 15mm sharp edge orifice.

The general trend shows a good improvement in the results whereby the absolute average error for pressures up to 59 bar using model C is 5% and 23% when model D is used.

Interestingly, according to the results shown in Table 2, model C starts under predicting the discharge rate from pressures above 50 bar. On the other hand pressures over 50 bar the original Omega is found to show good agreement with the experimental data.

To investigate this further, other experimental data within this pressure range are compared with the calculated discharge rate using Model C and D and presented in Table 3 and 4.

Pressure (bar)	Temperature (C)	Measured Discharge Rate(kg/s)	Model C		Model D	
			Discharge Rate (Kg/s)	Error (%)	Discharge Rate (Kg/s)	Error (%)
51.9	251.55	0.829	0.855	-2.6	0.929	-10
52.8	257.05	0.915	0.908	0.7	1.013	-9.8
55.8	261.75	1.044	0.966	7.8	1.107	-6.3
58.1	265.65	1.144	0.995	14.9	1.177	-3.3
60	268.05	1.229	0.998	23.1	1.211	1.8

Table 3. Calculated discharge rate and percentage error from experimental results (poorly mixed) for model C: Modified Simpson et al. two-phase multiplier (slip correlation) and the Non-Equilibrium Omega, and model D: Modified Simpson et al. two-phase multiplier (slip correlation) and the original Omega; experimental data is from 10mm sharp edge orifice.

Pressure (bar)	Temperature (C)	Measured Discharge Rate(kg/s)	Model C		Model D	
			Discharge Rate (Kg/s)	Error (%)	Discharge Rate (Kg/s)	Error (%)
53.6	256.55	0.884	0.886	-0.2	0.968	-8.4
57.1	262.25	1.019	0.969	5	1.092	-7.3
59.1	265.75	1.112	1.006	10.6	1.162	-5
60.5	268.85	1.197	1.009	18.8	1.2	-0.3

Table 4. Calculated discharge rate and percentage error from experimental results (well mixed) for model C: Modified Simpson et al. two-phase multiplier (slip correlation) and the Non-Equilibrium Omega, and model D: Modified Simpson et al. two-phase multiplier (slip correlation) and the original Omega; experimental data are from 10mm sharp edge orifice.

4. Conclusion

This paper aimed to show the main features of the two-phase discharge from openings such as phase slip and the Non-Equilibrium behaviour.

Using these characteristics the HEM model was modified in four different models. Amongst these models, the model, with the modified Simpson et al. two-phase multiplier which takes into account the non-equilibrium, was found to show a good agreement between the calculated discharge rates and the experimental results for pressures up to 50 bar.

For pressures between 50 bar and 60 bar model D, instead, is in agreement with the experimental data. This model has the same two-phase multiplier with no consideration for boiling delay. The reason is not yet known and further investigation is required in order to find out what aspect of the Model is causing this transition.

References:

American Petroleum Institute, 1993, Recommended Practice 520, Sizing, Selection and Installation of Pressure-Relieving Devices in Refineries: Part 1-Sizing and Selection, 6th edition.

American Petroleum Institute, 2000, Recommended Practice 520, Sizing, Selection and Installation of Pressure-Relieving Devices in Refineries: Part 1—Sizing and Selection, 7th edition.

Richardson, S. M., Saville, G., Fisher, A., Meredith, J., Dix, M. J., Experimental Determination of Two-Phase Flow Rates of Hydrocarbons Through Restrictions, Trans IChemE Process Safety and Environmental Protection, 84(B1): 40–53, 2006.

Diener, R., Schmidt, J., Sizing of Throttling Device for Gas/Liquid Two-Phase Flow Part1: Control Valve, Orifices, and Nozzle, Process Safety Progress, 23, 335-344, 2004.

Sallet, D.W., Evaluation of two-Phase Flow Models for Flashing Flow In Nozzles, Process Safety Progress, 19(1), 32-39,1997.

Vennard, J. K. *Elementary Fluid Mechanics*, 7th. Edition, Wyley, New York (1996).

Leung, J.C., A generalized correlation for one-component homogeneous equilibrium flashing choked flow, AIChE Journal, 32(10), 1743-1746, 1986.

Grolmes, M. A., and J. C. Leung, “Scaling Considerations for Two- Phase Critical Flow,” T. N. Veziroglu and A. E. Bergles, eds., Multi- Phase Flow and Heat Transfer III, A: Fundamentals, 549, Elsevier, Amsterdam, 1984.

Leung, J.C., Similarity between Flashing and Nonflashing Two-Phase Flows, AIChE Journal, 36(5), 797-800, 1990.

Diener, R., Schmidt, J., Sizing of Throttling Device for Gas/Liquid Two-Phase Flow Part2: Control Valve, Orifices, and Nozzle, Process Safety Progress, 24, 29-37, 2005.

Simpson, H.C., Rooney, D.H., and Grattan, E., Two phase flow through gate valves and orifice plates, Int. Conf on the Physical Modelling of Multi-Phase Flow, Coventry, UK, April 19–20, 1983.

Moody, F.I., Maximum flow rate of a single component two phase mixture, J. Heat Transfer 87, 134-142, 1965.

Lottes, P., Expansion losses in two-phase flow, Nuclear Science and Engineering, 9, 26–31, 1961.

Measurement of Critical Locus for Carbon Dioxide + Acetonitrile System with Variable-Volume View Cell

Takeshi Furuya, Nanotechnology Research Institute, National Institute of Advanced Industrial Science and Technology (AIST), Tsukuba, Ibaraki, Japan

Satoshi Yoda, Nanotechnology Research Institute, National Institute of Advanced Industrial Science and Technology (AIST), Tsukuba, Ibaraki, Japan

Yoshihiro Takebayashi, Nanotechnology Research Institute, National Institute of Advanced Industrial Science and Technology (AIST), Tsukuba, Ibaraki, Japan

Kiwamu Sue, Nanotechnology Research Institute, National Institute of Advanced Industrial Science and Technology (AIST), Tsukuba, Ibaraki, Japan

Toshihiko Ohta, Aichi works, Taketoyo plant, NOF corporation, Aichi, Japan

Kazuhiro Mae, Department of Chemical Engineering, Kyoto University, Kyoto, Japan

Phase equilibrium data are most important information for the process development with supercritical fluids. In recent years, a mixture of supercritical carbon dioxide (scCO2) and organic solvent, expanded liquid and co-solvent, are paid much attention to expand the capability and the applicability of scCO2 as a solvent.

The critical locus data determine the boundary conditions, temperature and pressure, for the phase separation of the scCO2 and organic solvent system. Based on the critical locus data of the scCO2 and organic solvent mixture, the phase of the system, homogeneous or phase separation, can be controlled to adjust the temperature and pressure of the system. However, the critical locus data for the scCO2 and organic solvent systems are very scarce and difficult to measure. A rapid and accurate measurement of the critical locus data is desired for an effective design of a solvent with scCO2 and organic solvent mixture.

From this viewpoint, we have developed a variable-volume view cell with data acquisition system for the critical locus data measurements. The cell consists of 70 mm in outer diameter and 15 mm in inner diameter. The maximum temperature and pressure of the cell are 473 K and 70 MPa, respectively. The piston is moved with geared motor and the piston speed is variable in the range of 0.001 to 3.5 mm/minutes. The volume of the cell can be changed from 3.87 to 19.95 mL. The piston position is detected continuously with linear scale and inputted with the inside pressure data of the cell in the control system. The piston movement is strictly and automatically controlled by the sequence using both data for ensuring precise measurements. Data acquisition system enables to record the temperature and pressure inside the view cell continuously. Critical temperatures and critical pressures for the CO2 + acetonitrile system were measured at different compositions with this system. Rapid and accurate critical locus measurement was ensured with this variable-volume view cell system.

See more of this Session: Thermodynamic Properties and Phase Behavior II
See more of this Group/Topical: Engineering Sciences and Fundamentals

Investigation of the Solubility of Carbon Dioxide in Complex Solvents by Experiment and Molecular Simulation

Christian S. Schacht, Engineering Thermodynamics, Delft University of Technology, Delft, Netherlands

Thijs J.H. Vlugt, Engineering Thermodynamics, Delft University of Technology, Delft, Netherlands

Theo W. De Loos, Engineering Thermodynamics, Delft University of Technology, Delft, Netherlands

Joachim Gross, Engineering Thermodynamics, Delft University of Technology, Delft, Netherlands

Abstract

Formerly, the development of solvents for a certain separation task was a tedious issue, including the synthesis of the new compounds as well as their experimental investigation. Improved descriptions of molecular interactions enable us to estimate the properties of process solvents based on their molecular structure. Consequently, molecular simulations can be applied as a tool to rationalize the search for new process solvents.

Dendritic and hyperbranched polymers (HBPs) are currently being considered as solvents or entrainers within the field of thermal separations [1,2]. Hyperbranched polymers exhibit a high thermal stability as well as low melt viscosity. Furthermore, they have a negligible vapor pressure and their physical properties, such as solubility, selectivity and glass transition temperature can be tailored via the large number of functional and core groups [3,4]. Within this work, a structured approach is shown to evaluate the applicability of HBPs as process solvents for the removal of CO2 from gas streams. Canonical ensemble molecular simulations based on the TraPPE force fields [5] combined with the Widom test particle insertion method was used to compute Henry's law constants of CO2 in different model compounds. The method is verified by comparison with literature data and experimentally determined vapor liquid equilibria of hyperbranched polyglycerol + co-solvent + CO2 [6]. Subsequently, this combination of Molecular Dynamics simulations and particle insertion is used to analyze the influence of combinations of ether- and alcohol-groups on the solubility of CO2. Furthermore, the impact of branched versus linear structures as well as the number of functional groups per molecule is shown.

Literature

[1] M. Seiler, Fluid Phase Equilib. 241 (2006) 155-174.

[2] J. Rolker et al. , Ind. Eng. Chem. Res. 46 (2007) 6572-6583.

[3] B. Voit, J. Polym. Sci. A : Polym. Chem. 38 (2000) 2505-2525.

[4] C.R. Yates et al., Eur. Polym. J. 40 (2004) 1257-1281.

[5] J.M. Stubbs et al., J. Phys. Chem. B 108 (2004) 17596-17605.

[6] M. Kozlowska et al., Proceedings of ECCE-6 2 (2007)

See more of this Session: Thermodynamic Properties and Phase Behavior II
See more of this Group/Topical: Engineering Sciences and Fundamentals

Modelling the Phase Behaviour of the CO2+H2O+Amine Mixtures Using Transferable Parameters with SAFT-VR

Niall Mac Dowell, Chemical Engineering, Imperial College London, London, United Kingdom

Amparo Galindo, Centre for Process Systems Engineering, Imperial College London, London, United Kingdom

Claire S. Adjiman, Department of Chemical Engineering, Imperial College London, London, United Kingdom

George Jackson, Centre for Process Systems Engineering, Imperial College London, London, United Kingdom

The reduction in CO_2 emissions from anthropogenic sources has become a topic of widespread interest in recent years. As the power generation sector is by far the largest stationary-point-source of CO_2, being responsible for approximately 35% of total global CO_2 emissions[1] this issue has particular relevance for the energy sector. The current method of choice for large-scale CO_2 capture is amine-based chemisorption; typically in packed columns, with the solvent of choice being a primary alkanolamine: monoethanolamine (MEA). Despite the widespread use if this technique, there is an ongoing debate as to how best to model these systems, owing to the complex reactions and phase equilibrium they exhibit. To capture the interactions that occur in these systems, we use the statistical associating fluid theory (SAFT)[2]. This is a molecular approach, specifically suited to hydrogen-bonding, chain-like fluids. In this contribution we use the SAFT approach for potentials of variable range (SAFT-VR[3]) to calculate the fluid phase behaviour of amine + H_2O + CO_2 mixtures. The molecules are modelled as homonuclear chains of bonded square-well segments of variable range, and a number of short-ranged off-centre attractive square-well sites are used to mediate the anisotropic effects due to association in the fluids.

We model MEA as a pair of tangentially bonded homonuclear spherical segments. Six distinct association sites are required to mediate the hydrogen bonding interactions exhibited by this molecule. In discriminating between potential models of MEA, we consider both symmetric models - where no distinction is made between the functional groups on MEA - and asymmetric models of MEA - where we explicitly consider the multifunctional nature of MEA. We find that despite giving an adequate description of the MEA + H_2O phase behaviour a purely symmetric model of MEA is not suitable for describing either the phase behaviour MEA + CO_2 binary mixture or that of the MEA + CO_2 + H_2O mixture. Further, as the chemistry of this system is well known, we note that it is not possible to preserve the stoichiometry of this mixture with a purely symmetric model of MEA. In order to reduce the number of adjustable parameters required to describe MEA, we exploit the molecular nature of SAFT-VR by transferring parameters from SAFT-VR models of alkanols and alkylamines developed in this work. We select our asymmetric model of MEA, by consideration of its ability to represent vapour-liquid equilibrium (VLE) data as well as enthalpy of vaporisation and interfacial surface tension data. Models for H_2O and CO_2 are taken from previous work[4,5] with two effective "reaction" sites being incorporated in the CO_2 model to mediate the chemical interaction between the MEA and CO_2. In modelling binary mixtures of MEA+H_2O, parameters are transferred from binary mixtures of ethanol+ H_2O and ethylamine+ H_2O. Unlike interaction parameters to describe the MEA+ CO_2 mixture are determined by comparison with ternary experimental data. Our final calculations for the ternary mixture predict the phase behaviour of with an absolute average deviation in the mole fraction of CO_2 in the liquid phase 0.01. The excellent accuracy of the asymmetric MEA model and of the MEA+H2O+CO2 mixture model over a wide range of conditions makes the proposed approach suitable for incorporation in process modelling.

Building on this success, we also investigate alternative solvents. There is a well recognised requirement for an improvement on the MEA based capture methods[6], with emphasis on increasing CO_2 loading capacity and absorption rate. Recent work[7] has shown that alkylamines could be ideal candidates to address both of these requirements. However, before these solvents can be deployed in practice, a detailed knowledge of their phase behaviour in aqueous CO_2 mixtures is required. The unlike interaction parameters required to describe aqueous alkylamine mixtures are obtained by comparison to experimental VLE data. Due to the lack of data for the $RNH_2 + CO_2$ mixture, we transfer binary interaction parameters obtained from NH_3+ CO_2 mixture. Our predictions confirm that ternary mixtures of RNH_2+H_2O+CO_2, for alkyl chains with more than 2 carbon atoms, exhibit liquid-liquid immiscibility, the extent of which increases with alkyl chain length. The consequence of this phenomenon is that the aqueous phase separates into a water-rich and amine-rich region. This phase split results in one phase of essentially pure water, and a concentrated amine-CO_2 phase, containing only a small amount of water. This provides a valuable basis for experimentation and preliminary assessment of alkylamines as solvent alternatives.

1. Steeneveldt, R. et al., ChERD, 84(A9): 739-763, 2006

2.Chapman, W.G. et al., Ind. Eng. Chem. Res., 29, 1709-1721, 1990

3.Gil-Villegas, A. et al., Chem. Phys. 106 (10), 1997

4.Clark, G. N. I. et al., Mol. Phys.,04(22-24):3561-3581, 2006

5.Galindo, A. et al., J. Phys. Chem. B., 106: 4503-4515, 2002

6.Wolsky, A. M., Daniels, E. J. and Jody, B. J., Environ. Prog., 13, 214, 1994

7.Singh, P., Niederer, J. P. M. and Versteeg, G. F. , Int. J. of Greenhouse Gas Control, 1, 2007, 5 - 10

See more of this Session: Thermodynamic Properties and Phase Behavior II
See more of this Group/Topical: Engineering Sciences and Fundamentals

Hydrogen Bonding in Ultra-High Pressure Water: From Open HB Networks to Closed-Packing Coordination and Their Link to Induced Charge Asymmetry

Ariel A. Chialvo, Chemical Science Division, Oak Ridge National Laboratory, Oak Ridge, TN

Juske Horita, Chemical Sciences Division, Oak Ridge National Laboratory, Oak Ridge, TN

Water is usually described as a fluid having a tetrahedral liquid structure based on the number of nearest neighbors in its first coordination shell given by the volume integral over the oxygen-oxygen pair radial distribution function, gOO(r). The tetrahedral distribution of water molecules is the direct manifestation of an underlying hydrogen bonding network, characterized at ambient conditions by the first two peaks of the gOO(r) at ~2.8Å and ~4.5Å, respectively, as well as the first peak of the corresponding gOH(r) at ~1.8Å, and obeying the intermolecular angle O-H...O ~ 104.5 degrees. Because water intrinsic polarizability, the approaching pair of molecules experience an additional attractive force beyond that from purely Coulombic interactions, i.e., a multi-body induced polarization resulting from the surrounding environment. This polarization manifests as a net increase of the total dipole moment of the water molecule from that of the original permanent dipole moment of the isolated molecules, as well as an enhancement of the charge asymmetry between the two hydrogen sites. While much of what we know to date about water's HB behavior has been (and still being) gained by studying non-polarizable models, they are basically unable to account for charge asymmetry. In contrast, polarizable water models are inherently charge-asymmetric, i.e., charge asymmetry arising as the direct response to the inhomogeneous nature of the local electric field around the water's interactions sites, a fluctuating quantity whose average behavior cannot be reproduced by pairwise potential models.

The tetrahedral character of the HB network in water is currently analyzed in terms of the so-called 'tetrahedral parameter' qT, a measure of the tendency of the four nearest water-oxygens to a central water-oxygen to arrange in a tetrahedral network (ERRINGTON and DEBENEDETTI, 2001). While a useful framework, a tetrahedral arrangement of water-oxygen does not necessarily guarantee a tetrahedral HB network, i.e., this scenario suggests that in order to gain better understanding on the water's HB behavior, we must also characterize the corresponding angular H...O-H distributions and the evolution of the second-nearest neighbors population including the inconspicuous participation of interstitial (i.e., so-called fifth) neighbors around the central water molecule.

Moreover, one disturbing fact in this context is the lack of an unambiguous HB definition, a condition that complicates the interpretation of experimental rawdata, because the (temporal and spatial) sensitivity of the probe depends on what particular feature of the HB interaction the technique is targeting. In other words, oftentimes we find that the links between the experimental signal and the microscopic picture are unsupported conjectures, and consequently, inducing controversial claims. Under these circumstances, molecular simulation can provide valuable microscopic information about the HB behavior, its connection with particular features of the interaction model, and its correspondence with experimental probes.

Toward that end, in this presentation: (a) we address explicitly the induced charge asymmetry behavior as predicted by our Gaussian Charge Polarizable water model (CHIALVO and CUMMINGS, 1998; PARICAUD et al., 2005), (b) we discuss its significance and implications in understanding the HB structure under the light of the recent controversial interpretation of XAS data (WERNET et al., 2004), and (c) we characterize the dramatic change of water coordination from the open tetrahedral network to the closed-packed twelve-coordination under extreme aqueous environments such as T>500C and P>5GPa.

REFERENCES:

Chialvo, A. A. and Cummings, P. T., 1998. Simple Transferable Intermolecular Potential for the Molecular Simulation of Water over Wide Ranges of State Conditions. Fluid Phase Equilibria 150-151, 73-81.

Errington, J. R. and Debenedetti, P. G., 2001. Relationship between structural order and the anomalies of liquid water. Nature 409, 318-321.

Paricaud, P., Predota, M., Chialvo, A. A., and Cummings, P. T., 2005. From dimer to condensed phases at extreme conditions: Accurate predictions of the properties of water by a Gaussian charge polarizable model. Journal Of Chemical Physics 122.

Wernet, P., Nordlund, D., Bergmann, U., Cavalleri, M., Odelius, M., Ogasawara, H., Naslund, L. A., Hirsch, T. K., Ojamae, L., Glatzel, P., Pettersson, L. G. M., and Nilsson, A., 2004. The structure of the first coordination shell in liquid water. Science 304, 995-999

See more of this Session: Thermodynamic Properties and Phase Behavior II
See more of this Group/Topical: Engineering Sciences and Fundamentals

Solvent Properties of Water+Methanol Binary Mixture at High Temperatures Via MD Simulation

Hiroshi Inomata, Chemical Engineering, Tohoku University, Sendai, Japan

Takumi Ono, Chemical Engineering, Tohoku University, Sendai, Japan

Shunsuke Kobayashi, Chemical Engineering, Tohoku University, Sendai, Japan

Taka-aki Hoshina, Chemical Engineering, Tohoku University, Sendai, Japan

Yoshiyuki Sato, Chemical Engineering, Tohoku University, Sendai, Japan

Alcohol and water are expected to become very effective solvents at higher temperatures because of their thermodynamically and chemically stability. The properties of these components are very similar at room temperature and completely miscible but become fairly different at supercritical state. This is one reason why supercritical water and supercritical methanol have been applied separately to various polymer-decomposition processes with the aim of inducing different reaction mechanism and product selectivity.

For wider applicability of these solvents at higher temperature, we have tried to use the mixture of alcoholand water at higher temperature. We focused on density and have measured the density of methanol+water mixture with a newly developed vibration tube densimeter in the temperature ranging from 573 to 673 K under pressures. The experimental results indicated that the excess molar volumes VE for methanol+water mixture was positive at 673 K and 25 MPa in the whole composition range and changed from negative to positive with increasing the methanol composition (XM) at 623 K and 20 MPa. This suggested that methanol+water solution may have attractive and repulsive force dependent on the composition at high temperature, which can be attributable to the balance of potential energy and kinetic energy.

We disclose the above behavior by using MD simulation[1][2]. The calculated radial distribution function and potential mean force indicated the possible change in intermolecular force from attractive to repulsive with increasing XM. The calculated diffusion coefficients of water(DW) and methanol(DM) were also evaluated from mean square displacement. The values of DW and DM exhibited similar behavior and increased drastically at XM = 0.2 ~ 0.4, indicated that the molecular mobility was dominated by water at lower XM and by methanol at higher XM and that the transition was induced by the hydrogen bonding network structure change in the presence of the methanol molecules in this region.

[1] T. Honma et al., Fluid Phase Equilibria, 194-197 271. (2002).

[2] T. Honma et al., J. Phys. Chem. A, 107 3960. (2003).

See more of this Session: Thermodynamic Properties and Phase Behavior II
See more of this Group/Topical: Engineering Sciences and Fundamentals

Thermodynamic modeling of water-acid gases-alkanolamine systems

Negar Sadegh[*1], *Kaj Thomsen*[1], *Erling Halfdan Stenby*[1], *Georgios Kontogeorgis*[1]

The share of natural gas in the world energy panorama has been appreciably growing for the last 20 years. This trend is expected to increase in the next few decades with the progressive replacement of fuel oil and coal by this relatively environment-friendly source of energy. However, this development will depend on the progress of gas processing technologies to give access to reserves now not exploitable.

More than 30% of available gas fields are acid, containing large quantities of CO_2 and H_2S and other sulfur compounds. The technologies generally employed to remove these impurities are very often based on their absorption in chemical or/and physical solvents.

In the gas processing industry absorption with chemical solvents has been used commercially for the removal of acid gas impurities from natural gas. The currently preferred chemical solvent technology for carbon capture is the use of amine-based chemical absorbent. Alkanolamines, simple combinations of alcohols and ammonia, are the most commonly used category of amine chemical solvents for CO_2 capture. Internationally many large oil and gas fields have high CO_2 and H_2S content (>50 mol%) and also traces of carbonyl sulfide (COS) and mercaptanes (RSH). It is known that many of the most used thermodynamic models for the simulation of acid gas solubility in alkonolamines (Kent Eisenberg, Deshmuk-Mather, electrolyte-NRTL) may give large errors when extrapolated to high pressures, high wt% amine, mixed CO_2 and H_2S gasses, gasses with traces of COS, RSH. So accurate estimation of thermodynamic equilibrium between high pressure natural gasses (up to 100 bar) with high amounts of CO_2 and H_2S and alkonolamines such as MEA, MDEA and DEA is crucial for obtaining a good design of amine based acid gas removal process. In this work the Extended UNIQUAC model is used for representation of the behavior of acid gas-alkanolamine mixtures over an extensive pressure range, with emphasis on mixed acid gases and presence of carbonyl sulfide and mercaptanes.

[1] Center of Phase Equilibria and Separation Processes (IVC-SEP), Department of Chemical and Biochemical engineering, Denmark Technical University. Email Addresses: Negar Sadegh: nes@kt.dtu.dk; Kaj Thomsen: kth@kt.dtu.dk; Erling Halfdan Stenby: ehs@kt.dtu.dk; Georgios Kontogeorgis: gk@kt.dtu.dk

Model parameters are optimized from the available experimental data. As a foundation for this modeling, a large literature review on experimental works for water-acid gases-alkanolamine (MEA/MDEA) systems has been done.

MDEA has the advantage that no carbamate is formed and the enthalpy of reaction is smaller than that of either MEA or DEA. Hence the cost of regeneration is lower. However, MDEA solutions react slowly with CO_2 and if bulk removal of carbon dioxide is desired, the result is a larger number of trays or an increased height of packing compared with MEA or DEA.

Monoethanolamine (MEA) has been the most widely employed gas treating alkanolamine solvent due to its high reactivity, low cost, ease of regeneration, and low solubility of hydrocarbons. The disadvantage includes a high enthalpy of reaction with carbon dioxide and formation of carbamate. Methyldiethanolamine (MDEA) solution frequently is used for selective removal of H_2S from gas streams which contain both CO_2 and H_2S. However, MDEA is also useful for bulk CO_2 removal because of the low heat of reaction with acid gases, which leads to lower energy requirement for regeneration.

Introduction:

The problem with thermodynamic modeling of acid gas treating plants is that the vapor–liquid equilibrium (VLE) data reported for these systems are not generally very consistent.

Moreover, other indispensable data such as binary VLE of alkanolamine–water system, excess enthalpy data and standard state properties of alkanolamines for both gas and liquid phases are scarce and those available from a handful of sources show discrepancies.

Developing empirical correlations based on the existing data, using excess Gibbs energy based activity coefficient models and application of equations of state which are based on excess Helmholtz energy are three different approaches for thermodynamic modeling of chemical absorption of CO_2.

Correlations could be very precise and calculations are often not cumbersome. Yet, empirical expressions typically fail when being extrapolated to conditions other than what they are based on.

One example is the simple correlation of Kent and Eisenberg [1] for CO_2 solubility in aqueous MEA and diethanolamine (DEA).

Gabrielsen et al. [2] presented another very simple correlation for the calculation of the partial pressure and enthalpy of absorption of CO_2 in MEA, DEA and MDEA.

An equation of state can be easily extended to predict the solubility of more than a single gas in the solution; it also can be used to calculate the properties such as density of both liquid and vapor phases. However, the performance of an equation of state to a great extent depends on the mixing rules chosen and an unsuitable choice can lead to erroneous results. Chunxi and Fürst [3], Solbraa [4] and Huttenhuis et al. [5] have used the Fürst and Renon [6] equation of state to represent CO_2 solubility in aqueous MDEA. Solbraa [4] has also applied the Cubic Plus Association (CPA) model proposed by Kontogeorgis et al. [7] for the MDEA system. Using electrolyte activity coefficient models is the most obvious alternative. Posey and Rochelle [8] applied electrolyte NRTL (e-NRTL) model to predict the CO_2 solubility in MDEA solution. This model had been used earlier by Austgen et al. [9,10] for MEA, DEA and mixtures of MDEA with MEA and DEA.

The purpose of the present work is to apply the extended UNIQUAC model [11] to estimate various thermodynamic properties of the alkanolamine systems required for the design natural gas treating units. The extended UNIQUAC model [K. Thomsen, P. Rasmussen, Chem. Eng. Sci. 54 (1999) 1787–1802] was applied to the thermodynamic representation of carbon dioxide and hydrogen sulfide absorption in aqueous monoethanolamine (MEA), methyldiethanolamine (MDEA) and mixtures of the two alkanolamines (MEA–MDEA). All the essential parameters of the model are simultaneously regressed to a collective set of data on the single MEA and MDEA systems and mixed systems. The model parameters are determined based on a large number of data covering the temperature and pressure of the reversible absorption process (both absorption and desorption) and exceeding far beyond.

Phase and Chemical Equilibria:

At its extraction, natural gas contains carbon dioxide and hydrogen sulfide that must be removed. Today's processes are mainly based on chemical absorption of acid gases by aqueous solutions of alkanolamines. Both acid gases are weak electrolytes and they partially dissociate in the aqueous phase to form a complex mixture of molecular and ionic species. In a closed system at constant temperature and pressure, physical equilibrium governs the distribution of molecular species between the liquid phase and the vapor phase. Simultaneously, in the liquid phase, alkanolamine reacts with CO_2 and H_2S, either through acid-base buffer mechanisms or through a direct reaction of CO_2 with primary or secondary alkanolamine to produce ionic species (Appendix A): physical and chemical equilibrium are highly coupled. Representation of the vapor-liquid equilibrium (VLE) behavior of acid gas-alkanolamine-water systems is made complex by the large number of chemical reactions which occur in the system. It requires that both physical and chemical equilibria be accounted for in a thermodynamically rigorous way.

Chemical Dissociations:

$2\,H_2O \leftrightarrow OH^- + H_3O^+$ (1)

$CO_2 + 2H_2O \leftrightarrow HCO_3{}^- + H_3O^+$ (2)

$HCO_3{}^- + H_2O \leftrightarrow CO_3{}^- + H_3O^+$ (3)

$H_2S + H_2O \leftrightarrow HS^- + H_3O^+$ (4)

$HS^- + H_2O \leftrightarrow S^{2-} + H_3O^+$ (5)

$R_1R_2R_3NH^+ + H_2O \leftrightarrow R_1R_2R_3N + H_3O^+$ (6)

$R_1R_2NCOO^- + H_2O \leftrightarrow R_1R_2NH + HCO_3^-$ (7)

$R_1R_2R_3$N is the alkanolamine chemical formula:

- *Monoethanolamine:* $R_1 = R_2 = H \; and \; R_3 = CH_2 - CH_2 - OH$; primary alkanolamine
- *Diethanolamine:* $R_1 = H \; and \; R_2 = R_3 = CH_2 - CH_2 - OH$; secondary alkanolamine

Methyldiethanolamine: $R_1 = CH_3 \; and \; R_2 = R_3 = CH_2 - CH_2 - OH$; tertiary alkanolamine.

Extended UNIQUAC Model:

The extended UNIQUAC model as presented by Thomsen and_Rasmussen [11] is used for the thermodynamic calculations of this work.

Extended UNIQUAC is the original UNIQUAC equation (Abramsand Prausnitz [12] and Maurer and Prausnitz [13]) combined with the Debye–Hückel term. The latter term is added to account for the electrostatic interactions caused by the presence of the ionic species in the solution.

Therefore, the excess Gibbs energy is expressed as the combination of three terms: the entropic and enthalpic terms of the original UNIQUAC equation to consider the non-electrostatic interactions and, the electrostatic term (Debye–Hückel):

$$\frac{G^E}{RT} = \left(\frac{G^E}{RT}\right)_{UNIQUAC\ Entropic} + \left(\frac{G^E}{RT}\right)_{UNIQUAC\ Enthalpic} + \left(\frac{G^E}{RT}\right)_{Debye-Huckel}$$

Parameter Regression Database:

Table 1 shows the experimental data that are used for parameter estimation of the extended-UNIQUAC model.

Table 1. Experimental data used for parameter estimation

Type of Data	Temperature Range (K)	Pressure Range (MPa)	Number of Data Points	Reference
Gas solubility	298.16-338.34	0.483-3.962	200	14
Gas solubility	308.15-338.15	-	222	15
Gas solubility	288.70- 588.70	0.34-20.27	96	16
Gas solubility	353.15-373.15	-	85	15
Gas solubility	313.15-373.15	-	407	17
Gas solubility	298.15-393.15	-	287	18
Gas solubility	273.58- 413.15	-	149	19
Gas solubility	298.15- 333.15	-	193	20
Gas solubility	275.25- 302.97	-	65	21
Gas solubility	313.15- 373.15	-	54	22
Gas solubility	323.10-373.10	0.0002-20.2	7	23
Gas solubility	323.15-366.90	-	50	24
Gas solubility	339.10	0.215- 0.951	8	25
Gas solubility	283-296	0.69	87	26
Gas solubility	303-323	-	47	27
Gas solubility	323-393	-	11	28
Gas solubility	298.16-338.3	0.483-3.96	52	29
Gas solubility	303.15-318.15	0.101	4	30
Gas solubility	298.15- 313.15	-	10	31
Gas solubility	313.15-393.15	-	23	32

Conclusion:

The modeling results are in good agreement with the experimental data. In this work it has been shown that the extended UNIQUAC model can accurately represent physical and chemical equilibria (VLE, Speciation), over a wide range of temperature and pressure. It can be concluded that Extended UNIQUAC is a good thermodynamic model for the design of natural gas treating units.

References:

[1] R.L. Kent, B. Eisenberg, Hydrocarbon Proc. 55 (1976) 87–90.

[2] J. Gabrielsen, M.L. Michelsen, E.H. Stenby, G.M. Kontogeorgis, Ind. Eng. Chem. Res. 44 (2005) 3348–3354.

[3] L. Chunxi,W. Fürst, Chem. Eng. Sci. 55 (2000) 2975–2988.

[4] E. Solbraa, Ph.D. Thesis, Norwegian University of Science and Technology, 2002.

[5] P.J.G. Huttenhuis, N.J. Agrawal, E. Solbraa, G.F. Versteeg, Fluid Phase Equilib. 264 (2008) 99–112.

[6] W. Fürst, H. Renon, AIChE J. 39 (1993) 335–343.

[7] G.M. Kontogeorgis, E.C. Voutsas, I.V. Yakoumis, D.P. Tassios, Ind. Eng. Chem. Res. 35 (1996) 4310–4318.

[8] M.L. Posey, G.T. Rochelle, Ind. Eng. Chem. Res. 36 (1997) 3944–3953.

[9] D.M. Austgen, G.T. Rochelle, X. Peng, C.-C. Chen, Ind. Eng. Chem. Res. 28 (1989) 1060–1073.

[10] D.M. Austgen, G.T. Rochelle, C.-C. Chen, Ind. Eng. Chem. Res. 30 (1991) 543–555.

[11] K. Thomsen, P. Rasmussen, Chem. Eng. Sci. 54 (1999) 1787–1802.

[12] D.S. Abrams, J.M. Prausnitz, AIChE J. 21 (1975) 116–128.

[13] G. Maurer, J.M. Prausnitz, Fluid Phase Equilib. 2 (1978) 91–99.

[14] A. Chapoy, A. H. Mohammadi, B. Tohidi, A. Valtz, D. Richon, Ind. Eng. Chem. Res. 44 (2005) 7567-7574.

[15] R.MacGREGOR, The Canadian Journal of Chemical Engineering. 69 (1991) 1357-1366.

[16] GPA Research Report. PR-48.

[17] J. Lee, F.D.Otto, A.Mather, Journal of Chemical and Engineering Data. 20 (1975) 161-163.

[18] F. Y. Jou, A. E. Mather, F. D. Otto, Ind. Eng. Chem. Process Des. Dev. 21 (1982) 539-544.

[19] J.H. Jones, H. R. Froning, E.E. Claytor, JR, J.Chem.Eng.Data. 4 (1959) 85-92.

[20] E. Riegger, H.V.Tartar, E.C.Lingafelter, 66 (1944) 2024-2027.

[21] A. A. Doubal, J. P. Riley, 26 A (1979) 259-268.

[22] F. Jou, J. J. Carroll, A. E. Mather, F. D. Otto, The Canadian Journal of Chemical Engineering. 71 (1993) 264-268.

[23] D. Koschel, J. Y. Coxam, L. Rodier, V. Majer. Fluid Phase Equilibria. 247 (2006) 107–120.

[24] A. Barreau, E. Blanchon le Bouhelec, K.N. Habchi Tounsi, P. Mougin, F. Lecomte, Oil & Gas Science and Technology – Rev. IFP. 61 (2006) 345-361.

[25] E. Blanchon Le Bouhelec-Tribouillois, P. Mougin, A. Barreau, I. Brunella, D. Le Roux and R. Solimando, Oil & Gas Science and Technology – Rev. IFP. 63 (2008) 363-372.

[26] P.J.G. Huttenhuis, N.J. Agrawal, J.A. Hogendoorn, G.F. Versteeg, Journal of Petroleum Science and Engineering. 55 (2007) 122–134.

[27] M. Kundu, S. S. Bandyopadhyay. The Canadian Journal of Chemical Engineering. 83 (2005) 344-353.

[28] D. Koschel, J.Y.Coxam, V.Majer. Ind. Eng. Chem. Res. 46 (2007) 1421-1430.

[29] A. Chapoy, A. H. Mohammadi, B. Tohidi, A. Valtz, D. Richon. Ind. Eng. Chem. Res.44 (2005) 7567-7574.

[30] N. Ai, J. Chen, W. Fei, J. Chem. Eng. Data. 50 (2005) 492-496

[31] M. H. Jenab, M. A. Abdi, S. H. Najibi, N. S. Matin. J. Chem. Eng. Data. 50 (2005) 583-586.

[32] M. E. Rebolledo-Libreros, A. Trejo, Fluid Phase Equilibria. 224 (2004) 83–88.

Vapor-Liquid Equilibrium Modeling for Carbon Dioxide + Difluoromethane (R-32) and Carbon Dioxide + Pentafluoroethane (R-125) Mixtures Using Helmholtz Energy Equations of State

Ryo Akasaka, Faculty of Humanities, Kyushu Lutheran College, Kumamoto, Japan

Recently, pure carbon dioxide (CO2) has gotten a lot of attention as natural, nontoxic, and inflammable working fluid for heat pumps. However, this type of heat pump needs to be operated at a very high pressure (usually over 15 MPa). Refrigerant mixtures of CO2 with other refrigerants with lower boiling point can reduce the high operating pressure. On the other hand, difluoromethane (R-32) has high flammability. The global warming potentials (GWPs) of R-32 and pentafluoroethane (R-125) are far from negligible. Therefore, mixing these refrigerants with nonflammable and low-GWP substances such as CO2 is particularly promising.

This work deals with the modeling of the vapor-liquid equilibrium (VLE) of the CO2 + R-32 and CO2 + R-125 mixtures. This is done with a simple multi-fluid model based on Helmholtz energy equations of state. For pure CO2, R-32, and R-125, very accurate and wide-ranging Helmholtz energy equations of state are available. The multi-fluid model is the best approach to combine these pure-fluid equations for representations of mixture properties.

Although typical multi-fluid models developed for industrially important mixtures have many adjustable parameters, the model used here reduces the number of adjustable parameters so that it is applicable to fluids with limited experimental datasets. The model contains only three parameters that have to be optimized. Values for the parameters have been determined by fitting procedure to experimental VLE data of the CO2 + R-32 and CO2 + R-125 mixtures. With the parameters, the model can correlate the bubble-point pressures with the accuracy of 2%. Moreover, the model can predict the vapor densities within 0.8% for the CO2 + R-32 mixture and 0.2% for the CO2 + R-125 mixture. This model will be helpful for assessments of refrigeration systems using the mixtures.

See more of this Session: Thermodynamic Properties and Phase Behavior II
See more of this Group/Topical: Engineering Sciences and Fundamentals

Transport and Reaction Processes in Non-Homogeneous Porous Media for Fuel Cell Applications

Sesha H. Vemuri, Chemical Engineering, Carnegie Mellon University, Pittsburgh, PA

Parag Jain, Chemical Engineering, Carnegie Mellon University, Pittsburgh, PA

Dehee Kim, Chemical Engineering, Carnegie Mellon University, Pittsburgh, PA

Lorenz T. Biegler, Chemical Engineering, Carnegie Mellon University, Pittsburgh, PA

Myung S. Jhon, Chemical Engineering, Carnegie Mellon University, Pittsburgh, PA

The key technological issues in polymer electrolyte fuel cell (PEFC) technology stem from the performance inefficiencies within its membrane electrode assembly (MEA). A typical MEA comprises of the gas diffusion layer (GDL), the catalyst layer (CL), and the polymer electrolyte membrane, which are heterogeneous porous material sub-components, and through which heat & mass transport processes occur with or without electrochemical reaction. For overcoming the performance limitations within MEA, a detailed understanding of these transport processes is required. Traditional non-homogeneous media theories [1-3] estimate the effective molecular transport coefficients (without fluid flow information) through averaged phase hold-up and/or tortuosity parameters, therefore, do not account for the microscopic structural information within the media.

We recently adopted an agglomerate CL model, which captures the essential transport processes and accounts for species, proton, and electron transports within the GDL/CL sub-components as well as reaction/transport processes within the CL, and incorporated it into our in house state-of-the-art interior point optimization algorithm, IPOPT. [4] We examined the effect of various agglomerate shapes (sphere, cylinder, and plate) on simulations and optimizations via introducing agglomerate effectiveness factor and effective transport properties. We also investigated the agglomerate size effect via scaling analysis and obtained up to three-fold enhancement in the overall current density as the agglomerate size decreased by ten times.

We further incorporated mesoscopic structural information for estimating transport properties in non-homogeneous media via lattice Boltzmann methods (LBMs). The advantages of clear physical pictures, geometric flexibility, an inherently transient nature, and fully parallel algorithms, make LBM to be an attractive tool for multi-phase, multi-scale simulation for PEFC devices.[5-7] LBM has been employed previously for simulating individual transport processes (momentum, energy, or mass) however, simulating combination of these phenomena, present in a PEFC MEA requires a clever modification in the LBM scheme via equilibrium distribution function or relaxation time. We first developed the standard LBM with Bhatnagar-Gross-Krook approximation for porous media, and verified with the Maxwell-Eucken formulation, and accounted for local structural information via correlation functions. Prior to the real application to a PEFC MEA, we extend our LBM scheme to systematically estimate the effective transport properties in non-homogeneous media especially including fluid flow.

Our methodology will be used for addressing PEFC sub-component level issues through meso-/micro-scopic modeling, and is planned for integration with entire PEFC system with optimization.

References:

1. J. C. Maxwell, A Treatise on Electricity and Magnetism (Clarendon, Oxford, 1881), Vol. 1, p. 435.

2. D.A.G. Brugemann, Ann. Phys.(5) 24, 637 (1935).

3. Jianfeng Wang, James K. Carson, Mike F. North, Donald J. Cleland, International Journal of Heat and Mass Transfer 49 (2006) 3075-3083

4. P. Jain, L.T. Biegler, and M.S. Jhon, Electrochem Solid-State Lett, 11 (10), B193 (2008).

5. W.T. Kim, S.S. Ghai, Y. Zhou, I. Staroselsky, H.D. Chen, and M.S. Jhon, IEEE Trans. Magn., 41, 3016 (2005).

6. W.T. Kim, M.S. Jhon, Y. Zhou, I. Staroselsky, and H.D. Chen, J. Appl. Phys., 97, 10P304 (2005).

7. Y.S. Xu, Y. Liu, X.Z. Xu, and G.X. Huang, J. Electrochem. Soc. , 153, A607 (2006).

See more of this Session: Fundamental Research in Transport Processes II
See more of this Group/Topical: Engineering Sciences and Fundamentals

A Lagrangian Particle Method as An Alternative to Solute Lattice Boltzmann: Investigation of Mass Transfer in Biological Porous Media

Roman S. Voronov, School of Chemical, Biological, and Materials Engineering, University of Oklahoma, Norman, OK

Samuel VanGordon, Bioengineering, University of Oklahoma, Norman, OK

Bonnie Landy, Bioengineering, University of Oklahoma, Norman, OK

Vassilios I. Sikavitsas, Chemical, Biological, and Materials Engineering, University of Oklahoma, Norman, OK

Dimitrios V. Papavassiliou, Chemical, Biological, and Materials Engineering, University of Oklahoma, Norman, OK

An effective and inherently parallelizable numerical method for the simulation of microfluidic flows in complicated geometries is the Lattice Boltzmann Method (LBM). Although several methods exist for modeling solute transport with LBM (for example in multi-component LBM the second component can mimic a solute, when its non-local interaction with the bulk fluid is grossly reduced), these methods are set in the Eulerian framework. Useful statistical quantities, such as the solute survival distance, survival time, effective diffusion coefficient, collision frequency, etc. cannot be extracted directly from such simulations. Moreover, just as in the case of classical LBM, the range of solute diffusivity that can be modeled is limited by the numerical solute's relaxation time. As an alternative to these methods, our work presents a Lagrangian approach to simulating convective scalar transfer in conjunction with LBM.

A house hybrid MPI/Open MP parallelized scheme is employed in order to take advantage of the inherent LBM parallelizability. Macroscopic mass transfer is modeled using the Lagrangian scalar tracking (LST) method (Papavassiliou, Int. J. Heat Mass Transfer, 45(17), 3571-3583, 2002) in conjunction with the LBM algorithm. In LST, the motion of scalar markers is used to synthesize the scalar profile. The trajectories of these markers are composed by a convection part (obtained using the velocity field from the LBM simulations) and a diffusion part (i.e., Brownian motion obtained from a mesoscopic Monte-Carlo approach). This method is resourceful in terms of computational efficiency, in that it can be used to simulate various Schmidt number fluids with a single flow field resulting from an LBM simulation. In addition, modifications of the LST algorithm allow the simulation of a whole spectrum of solute reaction rates, also using just a single flow field obtained from LBM.

The presentation will include the description of the numerical methodology and the validation of the method for known cases of flow in porous media. The usefulness of LST will be demonstrated for the particular case of flow through biodegradable synthetic porous polymer scaffolds, which are used in flow perfusion bioreactors for the growth of bone tissue as potential replacement therapies for damaged or lost bone tissue. Knowledge of the mechanism of mass transfer of nutrients to osteoblast cells growing within the porous scaffold can be obtained from LST simulations, allowing the optimization of the 3D scaffold structure for enhanced tissue growth. Micro-Computed Tomography scans of the bone tissue growth along with computer simulation results are used to obtain insights into the tissue growth process with the ultimate goal of being able to predict where the tissue will grow a priori - based only on the plain scaffold geometry and the flow rates.

See more of this Session: Fundamental Research in Transport Processes II
See more of this Group/Topical: Engineering Sciences and Fundamentals

Separation of SWCNTs by Type Using Electric Field, Field-Flow Fractionation*

Frederick R. Phelan Jr.† and Barry J. Bauer

Polymers Division
National Institute of Standards and Technology
Gaithersburg, MD, USA

ABSTRACT

Modeling of the separation of single wall carbon nanotubes (SWCNTs) in Electric Field, Field-Flow Fractionation (EF-FFF) is examined using a Brownian dynamics method in which the forces and torques arising from an AC electric field acting in the field direction have been incorporated. Two mechanisms for type separation – *i.e.*, separation of metallic from semiconducting tubes – are elucidated. In a uniform electric field, E-field aligned rods can be induced to exhibit steric inversion, which leads to rapid type separation under steric mode conditions. In a non-uniform electric field, the dielectrophoretic (DEP) force can be used to induce type separation under normal mode conditions. Both methods rely on exploiting polarizability differences between metallic and semiconducting tubes. The relative efficiency of the two methods is compared and discussed.

Keywords: Brownian Dynamics; Dielectrophoresis; Field-Flow Fractionation; Nanotubes; Separations; SWCNTs; Polarizability

1 INTRODUCTION

The manufacturing of single wall carbon nanotubes (SWCNTs) results in mixtures that are polydisperse with respect to length, diameter and chirality, the latter of which strongly affects the tube electronic properties. For nanotubes to achieve their full potential in applications, it is necessary to be able to separate them according to these characteristics. Separations are particularly critical for nanotube applications in electronics, where the chirality determines whether the tubes are metallic or semiconducting in nature.

One possible technique for achieving this is field-flow fractionation (FFF), depicted in Figure 1. In FFF, a mixture to be separated (typically, a dilute solution of particulates) is driven through a channel while a field is applied in a direction perpendicular to the streamwise direction [1-3]. The perpendicular field may be a second flow field, an electric field, or a temperature gradient, amongst a number of possibilities. The perpendicular field is chosen so that the interaction between the field and the streamwise parabolic velocity profile, promotes a separation of components based on their mobility in the field direction.

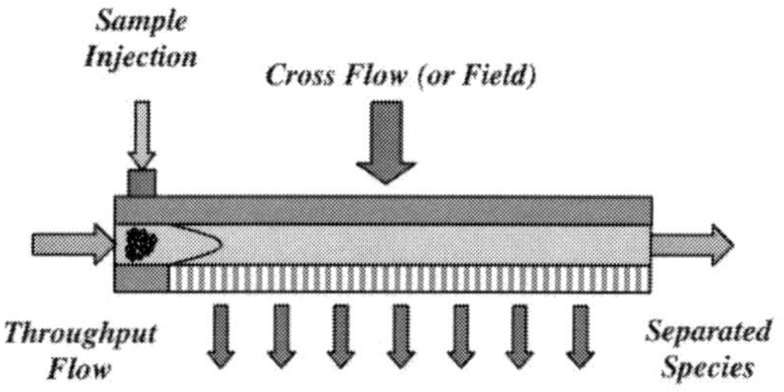

Figure 1: Montage of field-flow fractionation techniques. In the asymmetric variation of classical flow-FFF, the lower wall is porous.

In the family of FFF techniques, a number of variations can be fashioned to promote separations according to the various nanotube properties. Classical flow-FFF, in which the perpendicular field is a cross-flow, generates separations based on the difference in the diffusion coefficients of smaller and larger particles [1-3]. For SWCNTs, flow-FFF has been shown to promote length based separations [4-7]. Electric Field, Field-Flow Fractionation (EF-FFF), in which an electric field is applied in addition to the cross flow, can be expected to promote separations based on the different electronic properties of the particles, in particular, the differing electrophoretic mobility [8,9] and the polarizability [10,11] of metallic and semiconducting tubes. Sedimentation-FFF (Sd-FFF), in which a centripetal force is applied in the field direction [12,13], generates separations based on density differences. For SWCNTs this can be expected to promote diameter separations since SWCNT density varies mainly with diameter [14,15].

In EF-FFF with an AC electric field, the primary variable for controlling separations is the complex

† Corresponding author. Tel.: (301) 975-6761; E-mail address: frederick.phelan@nist.gov.

polarizability. The principal values for the complex polarizability of anisotropic particles are given by

$$\hat{\alpha}_{ii} = \varepsilon_m \frac{\left(\hat{\varepsilon}_p - \hat{\varepsilon}_m\right)}{\hat{\varepsilon}_m + L_{ii}\left(\hat{\varepsilon}_p - \hat{\varepsilon}_m\right)} \quad (1.1)$$

where $\hat{\varepsilon}_p$ and $\hat{\varepsilon}_m$ are the complex permittivities of the particle and medium, given by

$$\hat{\varepsilon}_p = \varepsilon_p - i\frac{\sigma_p}{\omega} \quad (1.2)$$

$$\hat{\varepsilon}_m = \varepsilon_m - i\frac{\sigma_m}{\omega} \quad (1.3)$$

ε_p and ε_m are the permittivity of the particle and medium, σ_p and σ_m are the conductivity of the particle and medium, L_{ii} are shape factors expressed in terms of elliptic integrals [16-18], ω is the angular frequency of the applied field and i is the imaginary number. Analytical expressions for shape factors for various cases of practical interest are shown in Table 1.

Figure 2 shows the real part of anisotropy of the polarizability based on the cylinder (slender body) model for 1 μm SWCNTS in an aqueous solution based on the physical properties for metallic and semiconducting tubes (Table 2). While there is some difference in the values across the entire range of frequencies, what is most pertinent is that the value for the semiconducting tubes drops off towards zero in the Megahertz range, meaning that metallics will be selectively active in this regime. The same hold true for the real part of the axial polarizability as shown in Figure 3. While the exact differences in real systems will depend on the specific solvent/surfactant combinations, in general the differences in polarizability between metallic and semiconducting SWCNTs offers a basis for type separation in both FFF and other liquid based separation schemes [10,11].

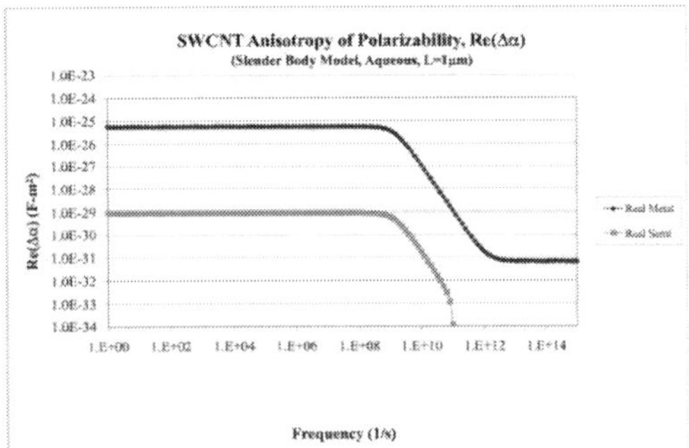

Figure 2: Real part of the anisotropy of the polarizability (F-m^2) as a function of frequency for metallic and semiconducting SWCNTs.

In what follows, two mechanisms for type separation of metallic from semiconducting tubes using EF-FFF are analyzed theoretically and numerically using a Brownian dynamics simulation. First, the case of separation using a uniform AC electric field applied in the gradient direction is examined. It is shown that E-field aligned rods can be induced to exhibit steric inversion, which leads to rapid type separation under steric mode conditions. Next, the case of a non-uniform AC electric field also acting in the gradient direction is examined. For this case it is shown that the dielectrophoretic (DEP) force can be used to induce type separation under normal mode conditions by selectively retaining metallic tubes. Both methods rely on exploiting polarizability differences between metallic and semiconducting tubes. The relative efficiency of the two methods is compared and discussed.

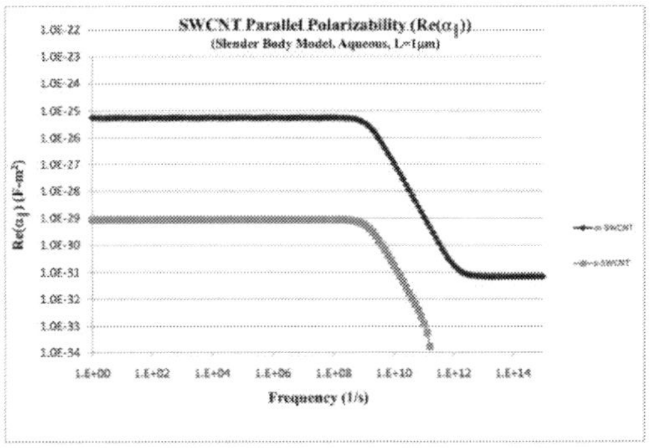

Figure 3: Real part of the axial polarizability (F-m^2) as a function of frequency for metallic and semiconducting SWCNTs.

2 SIMULATION

Full details of the Brownian dynamics simulation used to investigate the separation of rodlike particles in flow-FFF and EF-FFF are found in references [4,5,19-21]. A summary of pertinent details of the numerical method are discussed below.

2.1 EF-FFF Model Equations

The rodlike nanotubes are modeled as prolate ellipsoids, with major axis radius a and minor axis radius b. The governing equations for an ensemble of ellipsoids in EF-FFF, individually denoted by the superscript (i), can be written as [4,5,22-24]

$$\frac{d}{dt}\left(\underline{R}^{(i)}\right) = \underline{v}\left(\underline{R}^{(i)}\right) + \left[\underline{\underline{\zeta}}^{(i)}\right]^{-1} \cdot \left(\underline{F}_B^{(i)} + \underline{F}_{DEP}^{(i)} + \underline{F}_{EP}^{(i)}\right) \quad (1.4)$$

$$\frac{d}{dt}\left(\underline{p}^{(i)}\right) = -\underline{\underline{W}} \cdot \underline{p}^{(i)} + \lambda_p^{(i)}\left(\underline{\underline{D}} \cdot \underline{p}^{(i)} - \underline{\underline{D}} : \underline{p}^{(i)}\underline{p}^{(i)}\underline{p}^{(i)}\right) - \underline{p}^{(i)} \times \left[\underline{\underline{\xi}}^{(i)}\right]^{-1} \cdot \left(\underline{\mathrm{T}}_B^{(i)} + \underline{\mathrm{T}}_E^{(i)}\right) \quad (1.5)$$

where $\underline{R}$ and $\underline{p}$ are the position and orientation vectors of the particles, $\underline{F}_B$ and $\underline{\mathrm{T}}_B$ are the Brownian force and torque, $\underline{F}_{DEP}$ is the dielectrophoretic force, $\underline{F}_{EP}$ is the electrophoretic force, $\underline{\mathrm{T}}_E$ is the torque due to the electric field , $\underline{\underline{\zeta}}$ and $\underline{\underline{\xi}}$ are hydrodynamic resistance tensors, $\underline{\underline{D}}$ and $\underline{\underline{W}}$ are the stretching and vorticity tensors, respectively, and $\lambda_p = \frac{\Re^2 - 1}{\Re^2 + 1}$ with $\Re = \frac{a}{b}$. The electrophoretic force is zero in all AC fields, whether uniform or non-uniform, and thus, is not considered in the present analysis. The dielectrophoretic is zero for the case of a uniform AC electric field.

Eqs. (1.4) and (1.5) are integrated for an ensemble of particles with respect to initial conditions described by FFF flow-focusing operations; analytical kinematics are used for the flow field [4,5]. In order to integrate the time dependant equations for an AC field at the high frequencies used in these types of operations, time-averaged expressions for the DEP force and electric field torque must be used. The time-average DEP force is given by the expression [17,25,26]

$$\left\langle \underline{F}_{DEP} \right\rangle = \frac{1}{2} V \,\mathrm{Re}\left[\left(\underline{\underline{\hat{\alpha}}} \cdot \underline{\hat{E}}\right) \cdot \nabla \underline{\hat{E}}^*\right] \quad (1.6)$$

where $\underline{\hat{E}}$ is the electric field phasor vector, $\underline{\hat{E}}^*$ is the complex conjugate, V is the particle volume, and $\underline{\underline{\hat{\alpha}}}$ is the complex polarizability tensor. The time-average torque in Eq. (1.5) due to an AC electric field is given by

$$\left\langle \underline{\mathrm{T}}_E \right\rangle = \frac{1}{2} V^{(i)} \,\mathrm{Re}\left[\Delta\hat{\alpha}^{(i)}\left(\underline{p}^{(i)}\underline{p}^{(i)} \cdot \underline{\hat{E}}\right) \times \underline{\hat{E}}^* + \hat{\alpha}_\perp^{(i)} \underline{\hat{E}} \times \underline{\hat{E}}^*\right] \quad (1.7)$$

where $\Delta\hat{\alpha} = \hat{\alpha}_\parallel - \hat{\alpha}_\perp$, and $\hat{\alpha}_\parallel$ and $\hat{\alpha}_\perp$ are the axial and perpendicular components of the polarizability.

2.2 Boundary Conditions

An important part of the scheme is the interaction of the particles with the boundaries. The cross flow velocity continually drives the particles towards the accumulation wall and the diffusion step being random may also cause the particles to collide with this boundary. To handle this, a no-penetration boundary condition is used, based on the rod size and 3-D orientation [5,19,20] as show in Figure 4, in which the particle velocity is set to its center of mass value with respect to its distance from the wall in the local velocity field. This distance is given by $\delta^{(i)} = a\left|p_y^{(i)}\right|$, where $p_y^{(i)}$ is the y-component of the orientation vector and the boundary condition for the velocity is $v_x = v_x\left(\delta^{(i)}\right)$.

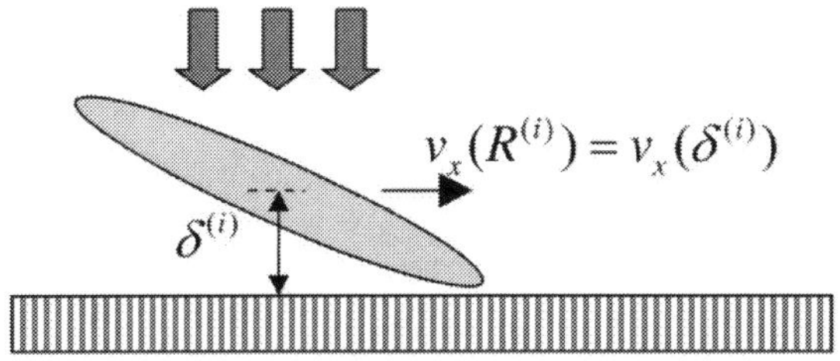

Figure 4: Boundary condition for ellipsoidal particles.

3 EF-FFF WITH UNIFORM AC FIELD

A number of different mechanisms can be exploited to achieve separation in FFF operations. Normal mode separation applies to particles which are small enough to undergo significant Brownian motion, and whose size is small compared to the cross flow gap size. In this case, smaller particles, which are more diffusive, have an average position closer to the centerline and elute faster than larger particles. Steric mode separation occurs when the particle layer in FFF is strongly compressed along the accumulation wall and due to size exclusion, larger particles are more highly entrained by the throughput flow and elute more quickly than smaller ones. This turnaround is called steric inversion.

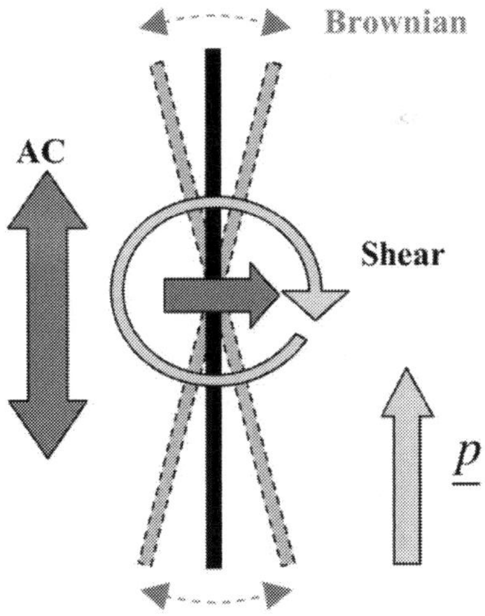

Figure 5: Forces on a rodlike particle subject to both shear flow, translational and rotational Brownian motion, and an AC electric field. Such a situation acts to induce wobbly alignment and modifies the FFF elution characteristics.

The steric inversion of spheres suggests that an even stronger steric effect might be obtained for rodlike particles due to their great length, by orienting them normal to the flow direction. This could be useful in the context of nanotube separation by preferentially orienting either metallic or semiconducting types relative to the other. This would create different elution rates for the two particle types. One way to induce an orientation of nanotube particles is by subjecting the particles to an electric field which acts in the gradient direction. When an anisotropic particle is subject to an electric field, a net torque which acts to align the particle in the field direction is produced [17]. In the present analysis, we consider the case of a uniform AC electric field acting in the gradient direction in EF-FFF. In such a case, the orientation torque produced by the electric field is in competition with both the shear field and Brownian motion as depicted in Figure 5. The degree of alignment depends on the polarizability of the particle, and the electric field strength. In this section, we develop an analytical expression to describe the Retention of oriented tubes, analyze the orientation as a function of electric field strength, shear rate and Brownian motion, and finally, analyze the elution of oriented particles.

3.1 Retention Theory

Particle separation in flow-FFF can be described by a theoretical variable called the retention, R, which represents the ratio of the average residence time of non-retained tracers, t_0, to the average retention time of the particles, t_r, *i.e.*, $t_r = t_0 / R$, where $0 \le R \le 1$. For the case of normal mode separation, the retention is given by

$$R = 6\lambda\left[\coth\left(\frac{1}{2\lambda}\right) - 2\lambda\right] \tag{1.8}$$

where λ is an inverse Peclet Number (Pe) given by

$$\lambda = \frac{D}{|v_c|H} = \frac{1}{Pe} \tag{1.9}$$

v_c is the cross flow velocity, D is the diffusion coefficient of the particle in the cross flow direction, and H is the cross-flow thickness.

A Retention model which takes into account both normal mode diffusion and steric effects for spheres in FFF was derived by Giddings [1]

$$R = 6\left(\alpha - \alpha^2\right) + 6\lambda(1-2\alpha)\left[\coth\left(\frac{1-2\alpha}{2\lambda}\right) - \frac{2\lambda}{1-2\alpha}\right] \tag{1.10}$$

where $\alpha = r/H$ is the ratio of the particle radius to gap thickness. For the case of negligible particle size, Eq. (1.10) reduces to Eq. (1.8). It can be seen that, theoretically, normal mode elution is independent of the particle shape and depends only on the particle diffusion coefficient. Steric mode elution becomes independent of diffusion, and depends mainly on the particle dimensions.

Little is known about how steric mode separations are effected by particle shapes other than spheres. For the case a nanotube oriented perpendicular to a surface, the boundary condition of the particle becomes as is shown in Figure 6, and the particle velocity will be at its maximum possible steric exclusion value, similar to the case for spheres. Under this condition, an appropriate expression for the Retention for the case of perfect orientation can be obtained by the modification of Eq. (1.10) with the substitutions $\alpha \to \alpha' = \frac{1}{2}\frac{L}{H}$, where L is the axial length of the rod, and $\lambda \to \lambda' = \frac{D_{\parallel}}{|v_c|H}$, where $D_{\parallel}$ is the axial diffusion coefficient of the rod (see Appendix), yielding the expression

$$\begin{aligned} R &= 6\left(\alpha' - (\alpha')^2\right) \\ &+ 6\lambda'(1-2\alpha')\left[\coth\left(\frac{1-2\alpha'}{2\lambda'}\right) - \frac{2\lambda'}{1-2\alpha'}\right] \end{aligned} \tag{1.11}$$

We compare this expression with simulation results for highly oriented tubes.

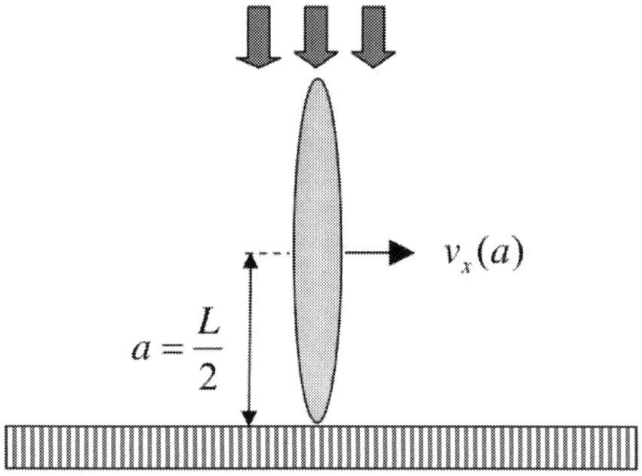

Figure 6: Boundary condition on a rodlike particle oriented normal to the flow direction.

3.2 Electro-Orientation

The orientation behavior of rodlike particles in a steady shear flow subject to a uniform AC electric field and Brownian motion is examined by solving Eq. (1.5) together with Eq. (1.7). The motivation for this calculation is to determine the effect of shear on particle orientation, and to determine range of electric field and polarizability needed to orient particles in EF-FFF. Theoretically, it is expected that the degree of particle orientation will depend on the ratio of the dimensionless groups $N_{DE} = \frac{V\,\mathrm{Re}[\Delta\hat{\alpha}]E_{rms}^2}{kT}$, which is the ratio of the dielectrophoretic potential energy to kT, and $Pe_r = \frac{\eta\dot{\gamma}Y_c^{(i)}}{kT}$ which is the Peclet number for rotational diffusion (where $Y_c^{(i)}$ is rotational friction coefficient, see Appendix). At lower values of Pe_r when the particles are more diffusive, higher values of N_{DE} are needed to orient the particles.

The particle orientation in a uniform electric AC field is characterized by an order parameter given by [27,28]

$$S = \left\langle \frac{3\cos^2\theta - 1}{2} \right\rangle \tag{1.12}$$

where θ is the angle between the particle orientation and the electric field vector, $\langle ... \rangle = \int_0^\pi (...) f(\theta)\sin\theta d\theta$, denotes an average over a potential, $f(\theta)$ is the potential

$$f(\theta) = \frac{e^{-\frac{U(\theta)}{kT}}}{\int_0^\pi e^{-\frac{U(\theta)}{kT}} \sin\theta d\theta} \tag{1.13}$$

and $U(\theta)$ is the interaction energy between the particle and the electric field

$$U(\theta) = -\frac{1}{2} V\,\mathrm{Re}[\Delta\hat{\alpha}]E_{rms}^2 \cos^2\theta \tag{1.14}$$

The value of S is equal to 1 when the particles are perfectly oriented with the field direction, 0 when they are oriented randomly with respect to the field direction, and a value of -½ when they are oriented normally to the field.

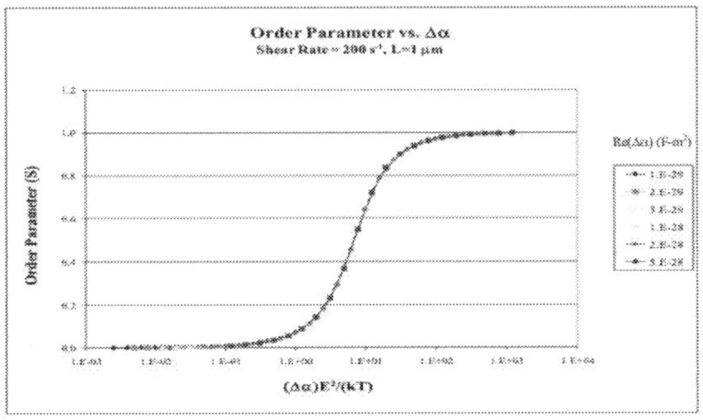

Figure 7: Order parameter at constant shear rate as a function of effective polarizability (F-m^2).

Orientation results for a particle with a 1000 nm length and 1 nm diameter, at a constant shear rate of 200 are shown in Figure 7. The orientation is plotted as a function of the dimensionless group $N_{DE} = \frac{V\,\mathrm{Re}[\Delta\hat{\alpha}]E_{rms}^2}{kT}$, which is the ratio of the dielectrophoretic potential energy to kT. Effective polarizibility values, $\Delta\alpha = V \cdot \mathrm{Re}[\Delta\hat{\alpha}]$, were chosen in a range consistent with values obtained from analytical models and recent experimental investigations [28-31]. Shear rates were chosen in a range consistent with the maximum rates seen in experimental FFF studies [6,19]. The figure shows that the particle orientation data at constant shear rate collapses onto a single master curve.

The particle orientation at constant polarizabilty as a function of shear rate is shown in Figure 8. These results also collapse onto a single curve and indicate that at this size scale, the particle orientation is insensitive to the shear flow, and that the orientation is mainly a competition between the Brownian motion and the torque due to the electric field. Thus we expect that in electro-orientation EF-FFF the average orientation is constant in the gradient direction, and that high shear rates at the lower boundary are not a hinderance to generating a steric mode separation.

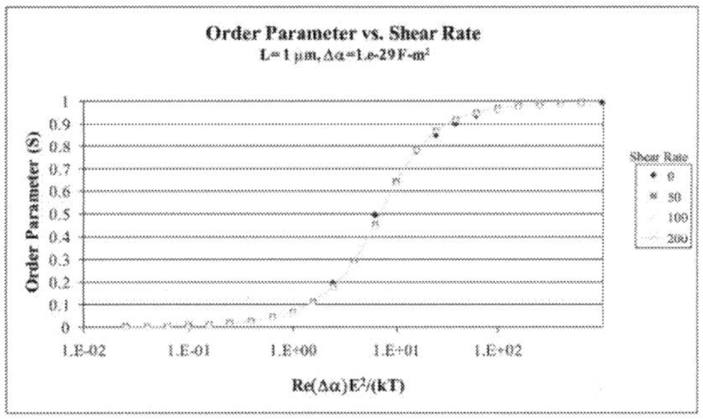

Figure 8: Order parameter at constant polarizability as a function of shear rate.

3.3 EF-FFF with Electro-Orientation

Simulation results are shown in Figure 9 for the processing parameters given in Table 2, and particle lengths ranging from 10 nm to 1000 nm, with a 1 nm diameter. An ensemble size of 1000 particles was used in all cases. Results for the steric inversion expression, Eq. (1.11), are plotted vs. EF-FFF results for $S = [0.1, 0.99, 1]$, and also the case of normal mode flow-FFF, which is for the most part is nearly random orientation or $S = 0$. The results for $S = 1$ were obtained by setting the particle initial condition at an orientation normal to the flow direction, and turning off all rotational motion.

A number of observations can be made from the Figure. First, for the oriented particles, steric inversion is clearly observed. All the EF-FFF curves diverge sharply from the flow-FFF curve (which are governed by normal mode elution) at a length of about 100 nm. For the given conditions, the steric transition occurs when the particle length is approximately 400 nm. The greatest degree of inversion occurs for the modified steric inversion theory of Giddings, *i.e.*, Eq. (1.11). This acts as a lower bound on the elution times while the flow-FFF curve acts as an upper bound. The simulation results for $S=1$ (constrained rotational motion) are in close agreement with Eq. (1.11). The data for $S=0.99$ and $S=0.1$ show that as the average particle orientation decreases, there is still steric inversion, but the elution times begin to slow down. However, mergence with the flow-FFF curve is slow.

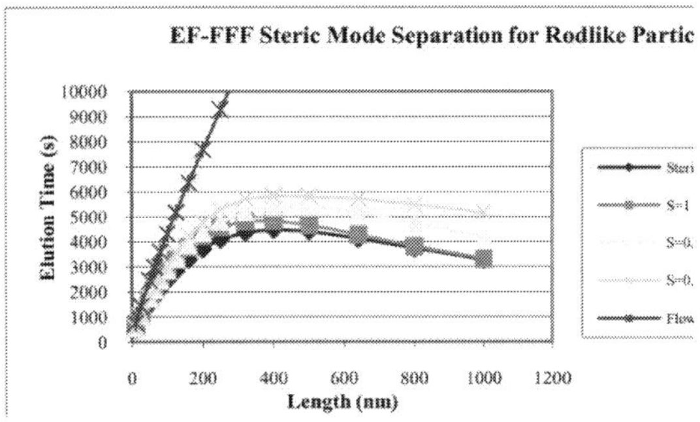

Figure 9: Comparison of elution time for rodlike particles under normal mode conditions, and for rods oriented perpendicular to flow direction.

These results can be used in the context of nanotube type separation by selectively orienting the more active metallic tubes relative to the semis by controlling the relative polarizability of the two different types by means of frequency, *cf.* Figure 2, and appropriate solvent/surfactant combinations. Orienting one type relative to the other will allow the orienting and non-orienting types to be separated into two fractions. The first fraction will contain oriented rods of all sizes, and non-oriented Brownian rods whose size is such that their elution time is less than the elution time corresponding to the steric inversion point for the oriented rods. The second fraction will contain all the longer non-oriented rods. For the case shown in Figure 9, the non-orienting cutoff size in the first fraction is about 150 nm; smaller cutoff sizes can be obtained by optimizing parameter conditions. Both fractions may then be further refined using size separation only.

4 DEP-FFF

Another means to separate particles in electric field, FFF is through the use of a non-uniform electric field, which generates a DEP force on particles. DEP has been used in conjunction with FFF in a number of studies [10,32-37] and has also been used as a means to separate nanotubes in non-FFF studies [11,38-40]. In practice, there are two ways to implement DEP-FFF depending on the size of the particles [35]. In levitation DEP-FFF, the DEP force is used to counterbalance the gravitational forces on particles, and is therefore used to lift particles away from the accumulation wall. In retention DEP-FFF, the DEP force is used in addition to the cross flow to counterbalance the Brownian force, and is therefore used to pull particles towards the retaining wall. In this case we analyze that latter which is the appropriate choice for SCWNTs which are highly Brownian.

4.1 Electrode Configuration

A typical configuration for DEP-FFF consists of an array of interdigitated electrodes along the lower wall [35]. A 2-D cross section depicting the electrodes and the approximate electric field lines generated by this type of arrangement is shown in Figure 10.

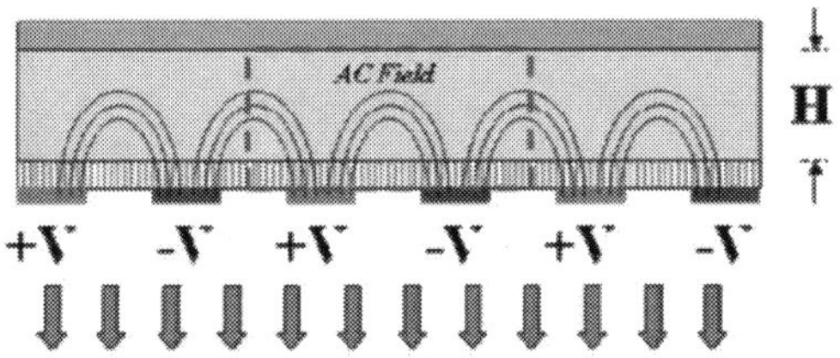

Figure 10: Electrode configuration in a typical retention DEP-FFF setup and approximate sketch of the electric field lines. The repeat unit for the voltage calculation is outlined. Particles which experience "positive" DEP move towards the accumulation wall.

The configuration is characterized by an electrode width, L_1, and an electrode spacing L_2, as depicted in Figure 11. Typical values for the electrode width and spacing are on the order of 50 μm. Since an FFF apparatus is on the order of 25 cm, there are on the order of 600 electrode pairs in a setup.

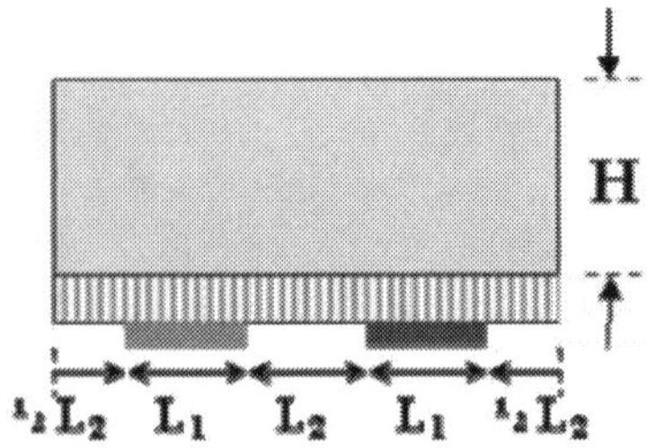

Figure 11: Repeat unit for voltage calculation in DEP-FFF electrode configuration.

4.2 Computation of Average Force

Because of the large number of electrode pairs in an actual FFF device, it is not practical to compute the detailed DEP force through all the electrode pairs over the entire length of the device. Thus, the average DEP forces over the space of a single repeat unit are used in the FFF calculation.

The net motion due to the DEP force in Eq. (1.4) for particles experiencing rotational motion during flow is given by the expression

$$\left[\underline{\underline{\zeta}}^{(i)}\right]^{-1}\cdot\left\langle \underline{F}_{DEP}^{(i)}\right\rangle = \underline{\underline{M}}^{(i)}\cdot\nabla\left|\underline{\hat{E}}\right|^{2} \qquad (1.15)$$

where

$$\underline{\underline{M}}^{(i)} = \frac{1}{2}\frac{V}{\eta}\mathrm{Re}\left[\left(\frac{\hat{\alpha}_{\parallel}}{X_{A}^{(i)}}-\frac{\hat{\alpha}_{\perp}}{Y_{A}^{(i)}}\right)\underline{p}^{(i)}\underline{p}^{(i)}+\frac{\hat{\alpha}_{\perp}}{Y_{A}^{(i)}}\underline{\underline{I}}\right] \qquad (1.16)$$

and $X_A^{(i)}$ and $Y_A^{(i)}$ are the axial and transverse drag coefficients for the particles defined in the Appendix. Since for nanotubes it is expected that the components of $\underline{\underline{\hat{\alpha}}}$ are positive and dominated by $\hat{\alpha}_{\parallel}$, according to Eq. (1.15) SWCNTs should experience "positive" DEP and move in the direction of the gradient $\nabla\left|\underline{\hat{E}}\right|^2$. Due to symmetry, the average force on the particles in the x-direction is zero, but is mainly downward and acts to move the particles towards the wall in the y-direction.

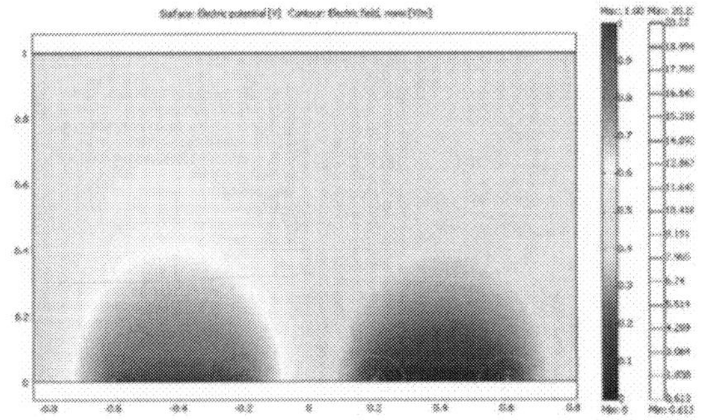

Figure 12: COMSOL solution for voltage showing electric field lines in dimensionless repeat unit.

To solve for the electric field needed to compute the DEP force, the electric potential in the fluid is computed by solving the equation

$$\nabla\cdot\left(\sigma_m \nabla V\right) = 0 \qquad (1.17)$$

where σ_m is the conductivity of the fluid and V is the voltage. The solution is obtained using COMSOL Multiphysics‡ [41]. The components of the electric field vector are then computed from the relation $\underline{E} = \nabla V$, and these values are used to calculated the electric field phasor vector $\underline{\hat{E}}$.

Figure 12 shows the COMSOL solution for the voltage and electric field lines. The problem is scaled to a unit cross-sectional thickness and voltage. The average (dimensionless) value of $\nabla_y\left|\underline{E}\right|^2$ in the gradient direction used to compute the DEP force is shown in Figure 13. The solution for $\nabla_y\left|\underline{E}\right|^2$ is converted to real units using the scale factor V_0^2/H^3, where V_0 is the applied voltage. The gradient is sharp and highly non-linear near the boundary.

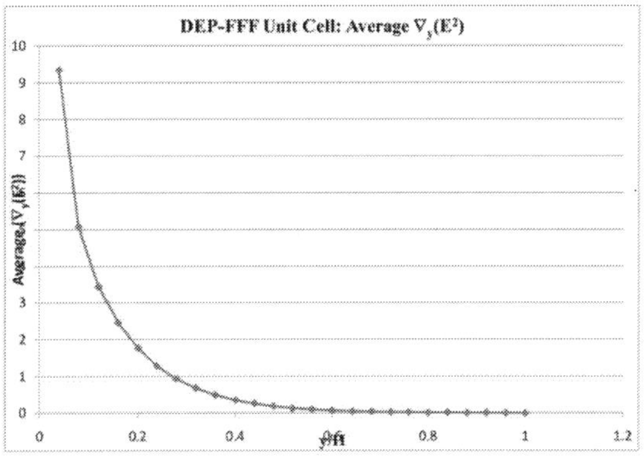

Figure 13: The average value of $\nabla_y\left|\underline{E}\right|^2$ over a repeat unit in the gradient direction.

4.3 Results for DEP-FFF

Simulation results in the form of elution profiles for DEP-FF are shown in Figs. 14-15 for 100 nm particles with a 1 nm diameter, for the processing parameters given in Table 4. An ensemble size of 1000 particles was used in all cases. The electrode width and spacing were set to 50 μm. The polarizability for the different particle types was computed using cylinder model with the property values listed in Table 2.

‡ Identification of a commercial product is made only to facilitate reproducibility and to adequately describe procedure. In no case does it imply endorsement by the National Institute of Standards and Technology (NIST) or imply that it is necessarily the best product for the procedure.

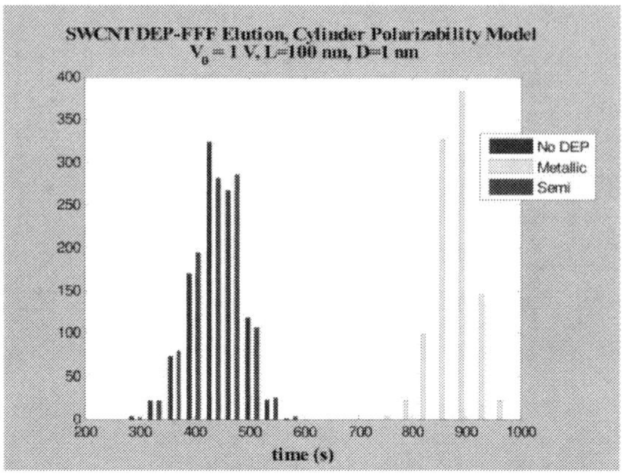

Figure 14: Results for DEP-FFF elution of m-SWCNTs and s-SWCNTs using the cylinder polarizability model.

Figure 14 shows results for DEP-FFF elution of m-SWCNTs and s-SWCNTs for an applied voltage of 1 V and at frequency of 10 Ghz, in comparison with the elution based on flow-FFF (which is not sensitive to type). The s-SWCNTs are slightly more retained, but the overall elution profile is not much different than that for flow-FFF (No DEP). By comparison, the m-SWCNTs are much more highly retained and the average elution time changes by over 400 s.

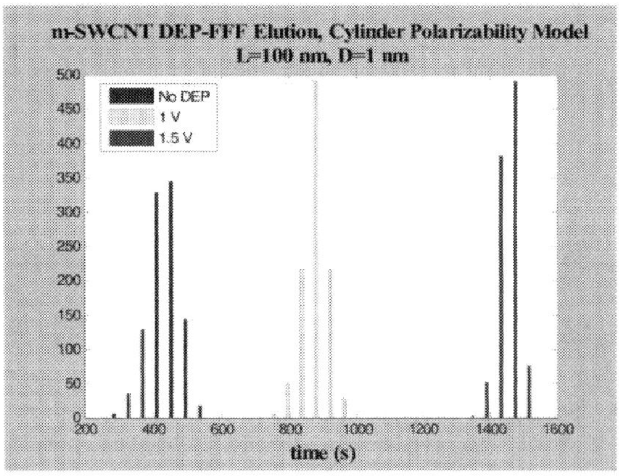

Figure 15: Results for DEP-FFF elution of m-SWCNTs using the cylinder polarizability model for different voltages.

The sensitivity of the elution of the m-SWCNTs is further shown in Figure 15, which compares flow-FFF elution with results for m-SWCNTs at 1 V and 1.5 V, again at 10 GHz. Due to the highly non-linear nature of the gradient $\nabla_y |\underline{E}|^2$ near the boundary, even the small change in voltage produces a large change in the average elution time. This sensitivity suggests a scheme where an initially large voltage is used to effectively trap m-SWCNTs until adequate separation with s-SWCNTs is achieved, which is then tapered off to allow the m-SWCNTs to elute. Current work is focused on this.

5 SUMMARY AND CONCLUSIONS

We have investigated particle separation mechanisms in EF-FFF using a Brownian dynamics method. Two types of separation mechanisms have been investigated based on both uniform and non-uniform electric fields. Both methods exploit polarizability differences between metallic and semiconducting tubes. The application of the present results to real systems will depend on the specific solvent/surfactant combinations which certainly effect the polarizability, and possibly effect the hydrodynamic properties of the particles. A greater understanding of such effects are needed. Based on the best data currently available, the following conclusions can be drawn from the present work:

- In the range shear rates common for FFF, shear flow has little effect on the electo-orientation of rodlike particles in EF-FFF with a uniform electric field, as Brownian motion is the dominant opposition force.
- Rods oriented normal to the flow direction can be expected to exhibit steric inversion. The steric effect grows stronger with increasing particle orientation. Eq. (1.11) provides a useful expression for the Retention of perfectly oriented particles.
- The use of steric inversion could be useful in the context of nanotube separation by preferentially orienting metallic tubes relative to the semiconducting types. The first fraction can be obtained much more rapidly than for normal mode elution and will be mainly metallics. The second fraction will be purely semiconducting.
- Based on the theoretical values of the polarizability for SWCNTs, metallics will be much more highly retained in DEP-FFF than semiconducting tubes. This suggests that a separation scheme using programmable voltages (high to low) would be effective in sorting m-SWCNTs and s-SWCNTs into two fractions.
- The implementation of electro-orientation based EF-FFF with a uniform electric field is potentially simpler to implement and control, but based on available polarizability data, will require somewhat higher voltages than DEP-FFF.
- Properly programmed DEP-FFF potentially offers more control over type separation than electro-orientation EF-FFF, as the sterically separated

phase in the latter, will contain a small amount of the non-oriented phase.

REFERENCES

1. J. C. Giddings, Separation Science and Technology **13,** 241 (1978).
2. J. Janca, in *Field-Flow Fractionation*, (Marcel Dekker, New York, 1987).
3. in *Field-Flow Fractionation Handbook*, (Wiley-Interscience, 2000).
4. F. R. Phelan Jr. and B. J. Bauer, Chemical Engineering Science **62,** 4620 (2007).
5. F. R. Phelan Jr. and B. J. Bauer, Chemical Engineering Science **64,** 1747 (2009).
6. J. Chun, J. A. Fagan, E. K. Hobbie, and B. J. Bauer, Analytical Chemistry **80,** 2514 (2008).
7. M. H. Moon, D. J. Kang, J. H. Jung, and J. M. Kim, Journal of Separation Science **27,** 710 (2004).
8. S. K. Doorn *et al.*, Journal of Physical Chemistry B **107,** 6063 (2003).
9. B. Suarez, B. M. Simonet, S. Cardenas, and M. Valcarcel, Journal of Chromatography A **1128,** 282 (2006).
10. H. Q. Peng *et al.*, Journal of the American Chemical Society **128,** 8396 (2006).
11. M. J. Mendes, H. K. Schmidt, and M. Pasquali, Journal of Chemical Physics B **112,** 7467 (2008).
12. J. C. Giddings, M. N. Myers, M. H. Moon, and B. N. Barman, Acs Symposium Series **472,** 198 (1991).
13. J. C. Giddings, M. H. Moon, P. S. Williams, and M. N. Myers, Analytical Chemistry **63,** 1366 (1991).
14. M. S. Arnold *et al.*, Nature Nanotechnology **1,** 60 (2006).
15. J. A. Fagan, M. L. Becker, J. Chun, and E. K. Hobbie, Advanced Materials **20,** 1609 (2008).
16. H. A. Pohl, in *Dielectrophoresis*, (Cambridge University Press, London, 1978).
17. T. B. Jones, in *Electromechanics of Particles*, (Cambridge University Press, New York, 1995).
18. A. Sihvola, in *Electromagnetic Mixing Formulas and Applications*, (INSPEC Inc., 2000).
19. F. R. Phelan Jr. and B. J. Bauer, Chemical Engineering Communications **in preparation,** (2008).
20. F. R. Phelan Jr., S. Kuruvilla, and B. J. Bauer, AICHE Journal **in preparation,** (2008).
21. F. R. Phelan Jr., S. Kuruvilla, and B. J. Bauer, *Steric Mode Separation of Nanotubes Using Electric Field, Field-Flow Fractionation* , (AICHE, AIChE 2008 Annual Meeting, 2008).
22. A. Satoh, in *Introduction to Molecular-Microsimulation of Colloidal Dispersions*, (Elsevier Science B.V., Amsterdam, The Netherlands, 2003).
23. S. Kim and S. J. Karrila, in *Microhydrodynamics: Principles and selected applications*, (Butterworth-Heinemann, Stoneham, 1991).
24. F. R. Phelan Jr., S. Kuruvilla, and B. J. Bauer, Microfluidics and Nanofluidics **submitted,** (2009).
25. T. B. Jones, IEEE Engineering in Medicine and Biology Magazine **22,** 33 (2003).
26. T. B. Jones, Biophysical Journal **64,** 1588 (1993).
27. C. T. O'Konski, K. Yoshioka, and W. H. Orttung, Journal of Chemical Physics **63,** 1558 (1959).
28. J. A. Fagan, V. Bajpai, B. J. Bauer, and E. K. Hobbie, Applied Physics Letters **91,** 213105-213105-3 (2007).
29. K. Bubke *et al.*, Applied Physics Letters **71,** 1906 (1997).
30. M. S. Brown, J. W. Shan, C. Lin, and F. M. Zimmerman, Applied Physics Letters **90,** 203108 (2007).
31. K. Donovan and K. Scott, Physical Review B **72,** 195432-195432-8 (2005).
32. G. H. Markx, J. Rousselet, and R. Pethig, Journal of Liquid Chromatography & Related Technologies **20,** 2857 (1997).
33. Y. Huang *et al.*, Journal of Hematotherapy & Stem Cell Research **8,** 481 (1999).
34. J. Yang *et al.*, Biophysical Journal **78,** 2680 (2000).
35. X. B. Wang *et al.*, Analytical Chemistry **72,** 832 (2000).
36. J. Vykoukal *et al.*, Lab on A Chip **8,** 1386 (2008).
37. T. S. Leu and C. Y. Weng, Journal of Micro-Nanolithography Mems and Moems **8,** (2009).
38. F. Hennrich, R. Krupke, M. M. Kappes, and H. V. Lohneysen, Journal of Nanoscience and Nanotechnology **5,** 1166 (2005).
39. S. Banerjee, T. Hemraj-Benny, and S. S. Wong, Journal of Nanoscience and Nanotechnology **5,** 841 (2005).
40. Z. Chen *et al.*, Analytical Chemistry **78,** 8069 (2006).
41. in *COMSOL MP 3.5 User's Guide*, (COMSOL AB, Burlington, MA, 2009).

6 TABLES

Case	L_{11}	L_{22}	L_{33}
Sphere	⅓	⅓	⅓
Cylinder	0	½	½
Ellipsoid	$\frac{1-s^2}{2s^3}\left(\ln\frac{1+s}{1-s}-2s\right)$	$\frac{1}{2}(1-L_{11})$	$\frac{1}{2}(1-L_{11})$

Table 1 – Polarizability shape factors.

Component	Relative Dielectric Constant	Conductivity (A/V-m)
m-SWCNT	1.E+04	1.E+08
s-SWCNT	2.5	1.6E+04
Water	80.1	1.0

Table 2 – Properties used for computing complex polarizability in Figures 2-3.

Processing Parameters	
Length, L (m)	0.25
Thickness, H (m)	0.00025
Width, W (m)	0.02
Focus Point, x_f (m)	0.01
T (K)	293
Viscosity (Pa-s)	0.001
Q_x (m^3/s)	2.5×10^{-8}
Q_y (m^3/s)	2.0×10^{-7}
Q_f (m^3/s)	2.0×10^{-8}

Table 3 – Processing parameters used in the EF-FFF with electro-orientation simulations.

Processing Parameters	
Length, L (m)	0.25
Thickness, H (m)	0.00025
Width, W (m)	0.02
Focus Point, x_f (m)	0.01
T (K)	293
Viscosity (Pa-s)	0.001
Q_x (m^3/s)	2.5×10^{-8}
Q_y (m^3/s)	2.0×10^{-8}
Q_f (m^3/s)	2.0×10^{-8}

Table 4 – Processing parameters used in the DEP-FFF simulations.

7 ACKNOWLEDGEMENTS

FRPJr thanks Prof. T.B. Jones of the University of Rochester for his guidance in deriving the appropriate torque expressions for AC fields. Sibu Kuruvilla of the University of Illinois Urbana Champaign did the electro-orientation calculations and the some of the EF-FFF calculations. Milena R. Bobea Rodriguez of the University of Puerto Rico did the DEP calculations.

8 APPENDIX – ELLIPSOID FRICTION COEFFICIENTS

$D_\parallel$, the axial diffusion coefficient of the ellipsoid, is given by

$$D_\parallel = \frac{kT}{\eta X_A} \quad (1.18)$$

where X_A, the axial friction coefficient of an ellipsoidal particle is given by

$$\frac{X_A}{6\pi a} = \frac{8}{3} \cdot \frac{s^3}{-2s + (1+s^2)Z} \quad (1.19)$$

with

$$s = \sqrt{1-\Re^2} \quad (1.20)$$

$$Z = \ln\frac{(1+s)}{(1-s)} \quad (1.21)$$

$D_\perp$, the transverse diffusion coefficient of the ellipsoid, is given by

$$D_\perp = \frac{kT}{\eta Y_A} \quad (1.22)$$

where Y_A, the transverse friction coefficient of an ellipsoidal particle is given by

$$\frac{Y_A}{6\pi a} = \frac{16}{3} \cdot \frac{s^3}{2s + (3s^2-1)Z} \quad (1.23)$$

D_r, the rotational diffusion coefficient of the ellipsoid, is given by

$$D_r = \frac{kT}{\eta Y_C} \quad (1.24)$$

where Y_C, the rotational friction coefficient of an ellipsoidal particle is given by

$$\frac{Y_C}{8\pi a^3} = \frac{4}{3} \cdot \frac{s^3\left(2-s^2\right)}{-2s + (1+s^2)Z} \quad (1.25)$$

EXPERIMENTAL AND NUMERICAL STUDY OF HYDRODYNAMIC DISPERSION AND COLLOIDAL TRANSPORT IN SOLID CELLULAR FOAM

Tyler R. Brosten[1], Einar O. Fridjonsson[1], Joseph D. Seymour[1], Robert S. Maier[2] and Sarah L. Codd[1]

[1]Magnetic Resonance Lab, College of Engineering, Montana State University, Bozeman MT USA
[2]U.S. Army Engineer Research and Development Center, Vicksburg MS USA

Introduction

The transport of a colloidal suspension through a porous material is an important process found within a broad range of disciplines; e.g. geophysics, engineering, and medicine. The importance of colloidal transport (e.g. microbe, radio-nuclide, etc.) in the subsurface has motivated much theoretical and experimental work focused upon understanding the mechanisms of colloid transport and deposition in porous media; e.g. [1,2]. The practical challenges associated with experimentally observing the transport of colloids in opaque media have limited past techniques to classic breakthrough methods [2-4]. This work presents results from nuclear magnetic resonance (NMR) measurement of moderate concentration (volume fraction $\Phi = 0.15$), neutrally charged, Brownian (hard sphere particle radius of 1.25μm), colloidal suspension transport through a high porosity open cell polymer foam. The recent development of core shell oil filled particles [5] has allowed for the separation of liquid and solid-phase dynamics in NMR measurements of colloidal suspensions [6-8]. The proven application of NMR in the study of single phase transport in porous media, [9], coupled with recent developments in NMR active colloidal suspensions have established all the experimental details necessary to directly probe the transport of colloidal suspensions in porous media. The experimental observations of colloid transport in this work are limited to the long time limit of transport; times much greater than the transient so-called "clean-bed removal" filtration phase [3]. In this "stationary" regime the interplay of colloid attachment and flow has equilibrated so that the solute and colloid dynamics are stationary.

Materials and Methods

Polyurethane open-cell foam (Foamex, Inc.) was used as the porous media. The foam has a porosity of 97% and an approximate average pore size of 110PPI (pores per inch). Note that the 110PPI designation given by the manufacturer is only an approximate characterization of the pore structure. The NMR transport measurements employed in this work consisted of standard pulsed gradient spin echo (PGSE) methods. The reader is referred to the text of Callaghan [10] for a detailed discussion of these methods and related NMR fundamentals. The NMR transport experiments began by first flushing the sample with de-ionized water (Millipore, Inc.) for about 12 hours at 20mm/s using a peristaltic pump (Masterflex, Inc.). Transport measurements of pure water flow from a Pharmacia P500 pump were then taken for comparative purposes. After the pure water measurements were finished the sample was pre-equilibrated with the colloidal suspension through 6 hours of continuous up-flow at a mean axial flow velocity of $\langle u \rangle = 1$ mm/s. The flow of colloidal suspension was driven by a programmable syringe pump (KDS Scientific, Inc.) interfaced to a 50 ml glass syringe (Hamilton, Inc.).

The core-shell oil particles were manufactured using the method of Loxley and Vincent [5]. The microspheres are short range repulsive due to short chain PVA molecules which adhere to the surface and exhibit hard sphere behavior [11]. The core of these particles are liquid oil and therefore are "NMR active". The suspension dynamics studied in this work are limited to a colloidal volume fraction of Φ = 0.15 with a particle diameter of d = 2.5±0.9μm.

Results and Discussion

Displacement propagator dynamics

The temporal evolution of the displacement propagator, $P(\mathbf{R},\Delta)$, was measured as a function of observation time for both pure water and the colloidal suspension. The phase separated propagator of the colloidal suspension was obtained by integrating the spectral dimension of the two dimensional Fourier transformed signal over the water and oil peaks. The phase separated axial and transverse displacement propagators of the colloidal suspension are compared in Figure 1 for $\langle u \rangle$ = 1.0 mm/s at several observation times.

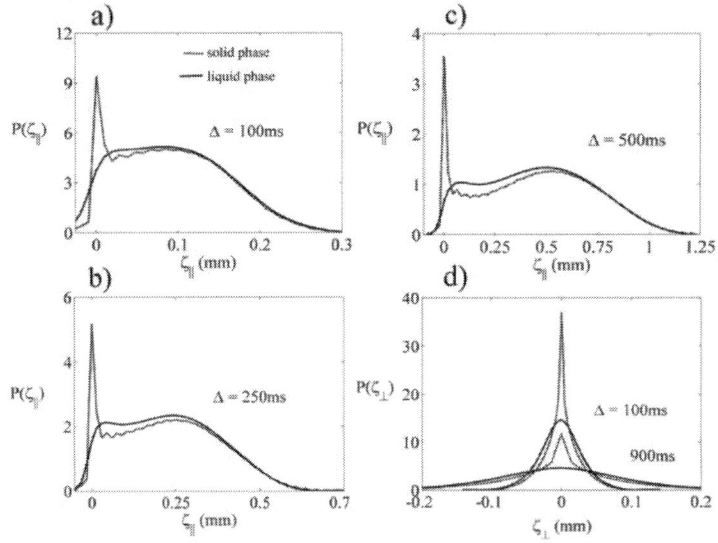

Figure 1: Axial and transverse propagator temporal evolution of the phase separated colloidal suspension at a mean flow velocity of 1 mm/s; a) - c) colloidal suspension axial propagators, d) colloidal suspension transverse propagators

The axial propagator of the colloidal suspension solid-phase, Figs. 1 a) through c) contains a significant portion of the population near axial displacement length $\zeta_{||} = \langle u \rangle \Delta$ 0, indicating hold-up of colloidal particles. The variance or second moment, σ^2, of the displacement propagator can be determined from a low q fit to the PGSE signal amplitude. A time dependent dispersion coefficient is defined as the variance divided by the observation time, $D = \sigma^2/(\Delta)$. The two quantities share the same information content but in different form. The time dependent phase-separated variance and dispersion coefficient are shown in Figure 2 for the transverse and axial direction. 2 a) & c) contain the axial dispersion dynamics as a function of mean displacement and 2 b) & d) contain the transverse dynamics. The data shown in Figures 1 and 2 was obtained following at least 6 hours of continuous flow of suspension.

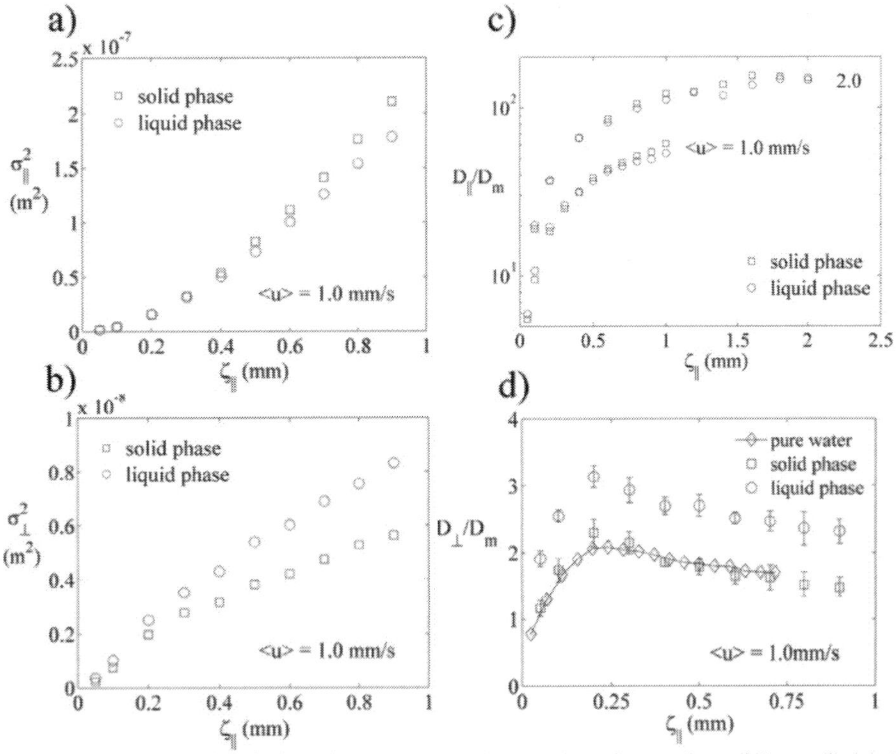

Figure 2: Phase separated axial and transverse dispersion dynamics of the colloidal suspension within the foam pore structure, a) & b) axial and transverse displacement variance as a function of mean displacement for a mean flow velocity of 1 mm/s, c) & d) axial and transverse effective dispersion coefficient as a function of mean displacement, D_m = 2x10^{-9} m^2/s, also shown in d) are the dynamics of pure water at a mean flow velocity of 1mm/s

The linear growth of the axial and transverse variance, Fig. 2 a) & b), at long observation time, $\zeta_{||}$ > 0.6 mm, suggests the existence of a unique asymptotic dispersion coefficient for both phases. To our knowledge these are the first measurements directly demonstrating this behavior of a colloidal suspension in a porous media. At long times the axial variance of the colloidal suspension solid-phase grows faster than the liquid-phase variance. There is a transition point in the axial dynamics before which the liquid-phase axial dispersion is greater than the solid-phase, Fig. 2 c). This short time difference is the result of a number of potential factors including the significantly greater molecular mobility of the liquid-phase and also perhaps a greater scattering of the liquid-phase throughout the convective velocity field. In contrast the liquid-phase transverse dispersion is consistently greater than the solid-phase dispersion. The *backscattering* (increase then decrease) effect seen in the transverse dispersion dynamics of both phases, Fig. 2 d), is the result of oscillatory streamlines within the structure. Numerical lattice Boltzmann simulations of single phase transport in a 3D foam structure show this equivalent behavior. The experimental and numerical data of the backscattering effect and a nonequilibrium statistical mechanics description will be discussed in detail in a separate work. For pure water, at high Peclet number, the maximum of the transverse dispersion curves occurs at a fixed mean displacement of l = 250μm; a number which is on the order of the wavelength of the axial oscillations of streamlines within the pore space. This fixed displacement is a characteristic length of the structure and will be used in the Peclet number definition of the foam structure.

Asymptotic dispersion dynamics

The phase-separated axial and transverse asymptotic dispersion coefficients are plotted in Figure 3 as function of Peclet number. The Peclet number was varied by adjusting the mean flow velocity. In the convection dominated regime, i.e. Pe > 100, classical dispersion theory predicts a power law scaling of the asymptotic axial dispersion coefficient with Pe [12]. Power law fits to the axial asymptotic dispersion coefficients are shown in Fig. 3 a). The asymptotic coefficients shown in Figure 3 were determined from the long time behavior of $D(\Delta)$.

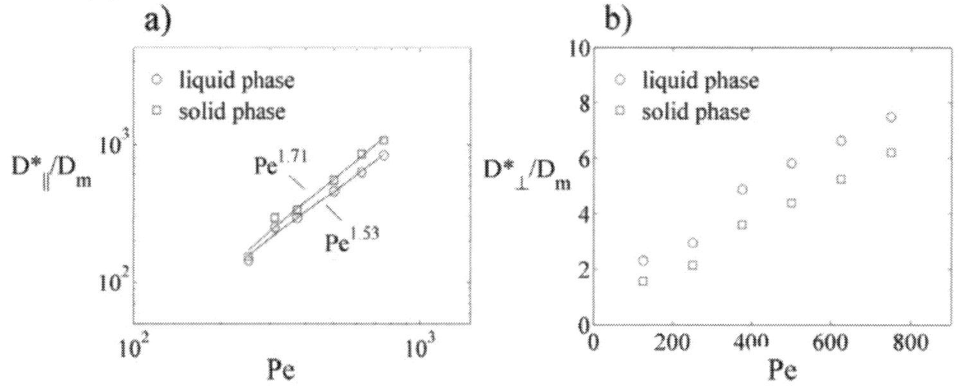

Figure 3: Phase separated a) axial and b) transverse asymptotic dispersion coefficients of the colloidal suspension as a function of Peclet number and non-dimensionalized by the molecular diffusion D_m = 2x10^{-9} m^2/s

The asymptotic axial dispersion coefficient of the colloidal solution solid-phase was found to scale like $Pe^{1.71}$ while the liquid-phase was found to scale like $Pe^{1.53}$; for pure water the coefficient scales like $Pe^{1.81}$ (determined from a linear fit to the long time variance). Therefore, as mentioned in the earlier transverse dispersion discussion, the dynamics of the colloidal solution liquid-phase are more significantly driven by mechanical effects then either the solid-phase or pure water dynamics. The solid-phase dynamics however appear more like Taylor dispersion, where scaling is Pe^2.

Summary

The active transport of a colloidal suspension within cellular foam was probed using NMR. The dynamics of both the liquid and solid-phase of the suspension were presented over a range of observation times and mean flow velocities. The deposition of colloidal particles onto the solid pore structure were shown to cause a transformation of the solid pore structure which in turn alters the macro-scale dynamics of both phases relative to single phase transport. Asymptotic dispersion coefficients of both phases are measured and scaling with Peclet number examined.

References

1. T. R. Ginn, B. D. Wood, K. E. Nelson, T. D. Scheibe, E. M. Murphy, and T. P. Clement (2002), "Processes in microbial transport in the natural subsurface," *Advances in Water Resources,* vol. 25, pp. 1017-1042
2. A. B. Kersting, D. W. Efurd, D. L. Finnegan, D. J. Rokop, D. K. Smith, and J. L. Thompson (1999), "Migration of plutonium in ground water at the Nevada Test Site," *Nature,* vol. 397, pp. 56-59
3. J. N. Ryan and M. Elimelech (1996), "Colloid mobilization and transport in groundwater," *Colloids and Surfaces a-Physicochemical and Engineering Aspects,* vol. 107, pp. 1-56
4. X. Q. Li, T. D. Scheibe, and W. P. Johnson (2004), "Apparent decreases in colloid deposition rate coefficients with distance of transport under unfavorable deposition conditions: A general phenomenon," *Environmental Science & Technology,* vol. 38, pp. 5616-5625
5. A. Loxley and B. Vincent (1998), "Preparation of poly(methylmethacrylate) microcapsules with liquid cores," *Journal of Colloid and Interface Science,* vol. 208, pp. 49-62
6. J. R. Brown, J. D. Seymour, S. L. Codd, E. O. Fridjonsson, G. R. Cokelet, and M. Nyden, "Dynamics of the solid and liquid phases in dilute sheared Brownian suspensions: Irreversibility and particle migration," *Physical Review Letters,* vol. 99, pp. -, Dec 14 2007.
7. P. Stilbs, "Fourier transform pulsed-gradient spin-echo studies of molecular diffusion," *Progress in Nuclear Magnetic Resonance Spectroscopy,* vol. 19, pp. 1-45, 1987.
8. H. Wassenius, M. Nyden, and B. Vincent, "NMR diffusion studies of translational properties of oil inside core-shell latex particles," *Journal of Colloid and Interface Science,* vol. 264, pp. 538-547, Aug 15 2003.
9. J. D. Seymour and P. T. Callaghan, "Generalized approach to NMR analysis of flow and dispersion in porous media," *Aiche Journal,* vol. 43, pp. 2096-2111, Aug 1997.
10. P. T. Callaghan, *Principles of nuclear magnetic resonance microscopy.* Oxford [England] New York: Clarendon Press; Oxford University Press, 1991.
11. H. Wassenius, M. Nyden, and B. Vincent (2003), "NMR diffusion studies of translational properties of oil inside core-shell latex particles," *Journal of Colloid and Interface Science,* vol. 264, pp. 538-547
12. J. Salles, J. F. Thovert, R. Delannay, L. Prevors, J. L. Auriault, and P. M. Adler (1993),

Reaction-Diffusion Based Measurement of Protein-Ligand Affinity

Paul J. Wesson, Chemical and Biological Engineering, Northwestern University, Evanston, IL

Yanhu Wei, Chemical and Biological Engineering, Northwestern University, Evanston, IL

Bartosz A. Grzybowski, Department of Chemical and Biological Engineering and Department of Chemistry, Northwestern University, Evanston, IL

Many methods to measure protein ligand binding/dissociation constants require an intrinsic shift in a "signal" from the protein (i.e. fluorescence, NMR spectra, etc.), thus they require some optimization for each protein measured. However, by using reaction-diffusion (RD), one can accurately measure protein-ligand dissociation constants, Kd, for arbitrary protein-ligand combinations.

In this talk, we show that Kd may be measured using a general RD process. Here, we first "ink" a patterned agarose stamp with a protein, then apply the stamp to a poly(acryl amide) gel which contains immobilized ligand. As the protein diffuses from the stamp into the gel, it intereacts with the ligand: when the protein is bound to the ligand it is stationary, however, when it is not bound to the ligand it freely diffuses. After a prescribed period of time, the stamp is removed from the gel, and the gel is stained with either Coomassie blue or a combined Coomassie blue/silver staining method. The resulting pattern of stained protein may then be fit using the RD equations to determine the dissociation constant for that protein/ligand pair.

See more of this Session: Fundamental Research in Transport Processes II
See more of this Group/Topical: Engineering Sciences and Fundamentals

Measurement of Temperature Distribution Over 2000°C in Premixed Flame by Using Laser Speckle Method

Ken-ichiro Tanoue, Tokiwadai 2-16-1, Department of Mechanical Engineering, Graduate School of Science and Engineering, Yamaguchi University, Ube, Japan

Makoto Takayanagi, Tokiwadai 2-16-1, Department of Mechanical Engineering, Graduate School of Science and Engineering, Yamaguchi University, Ube, Japan

Yuuto Ogura, Tokiwadai 2-16-1, Department of Mechanical Engineering, Graduate School of Science and Engineering, Yamaguchi University, Ube, Japan

Tatsuo Nishimura, Tokiwadai 2-16-1, Department of Mechanical Engineering, Graduate School of Science and Engineering, Yamaguchi University, Ube, Japan

Temperature distribution of premixed flame over 2000°C between methane gas and air has been developed using laser speckle method. The inner diameter of the pipe exit was 13mm. The gas flow rate of methane changed from 0.7 L/min to 2L/min. The flow rate of air was ten times of the methane flow rate. The premixed flame consists of two different kinds of flames. One is the internal flame which the maximum temperature is greater than 2000°C. The other is the diffusion flame around the internal flame. From the movie of the speckle patterns, it was found that the motion of the premixed flame was unsteady and oscillatory. The more methane flow rate made not only the longer the period of the oscillation but also the more the unsteady domain of the diffusion flame. Furthermore, from Fermat's principle and Gradstone-Dale's equation, the temperature distribution of the unsteady and oscillatory premixed flame was predicted.

See more of this Session: Fundamental Research in Transport Processes II
See more of this Group/Topical: Engineering Sciences and Fundamentals

Microfabricated Deflection Tensiometers: A Novel Technique in Interfacial Tensiometry

Zachary A. Zell, Chemical Engineering, University of California, Santa Barbara, Santa Barbara, CA

SiYoung Q. Choi, Chemical Engineering, University of California, Santa Barbara, Santa Barbara, CA

L. Gary Leal, Chemical Engineering, University of California, Santa Barbara, Santa Barbara, CA

Todd M. Squires, Chemical Engineering, University of California at Santa Barbara, Santa Barbara, CA

Monolayers of phospholipids and fatty acids provide simple model systems for fluid-fluid interfaces in biological systems and microemulsions. Although the Wilhelmy plate technique is the most conventional method for measuring surface pressure, limitations exist when dealing with small samples or enclosed systems. Inspired by the idea of Hu et al [Langmuir, 2003, 19, 100], who related the elastic deformation (deflection) of a teflon wire loop to the surface pressure, we have developed a new technique for microtensiometry. Using a simple design equation derived from elastic beam theory, we microfabricate micro-tensiometers and validate their use by comparing measured surface pressure isotherms for fatty acids and phospholipid monolayers at air-water and fluid-fluid interfaces against simultaneous Wilhelmy plate measurements, with excellent quantitative agreement. Our technique offers a high degree of design flexibility due to control of geometry, materials and surface chemistry, is relatively inexpensive and enables measurements in small or enclosed systems. Finally, minor modifications enable our tensiometers to function as Langmuir micro-troughs for insoluble surfactants.

See more of this Session: Novel Experimental Methods for the Study of Interfacial Phenomena
See more of this Group/Topical: Engineering Sciences and Fundamentals

Microtensiometer to Directly Probe Kinetic Adsorption Constants

Nicolas Alvarez, Chemical Engineering, Carnegie Mellon University, Pittsburgh, PA

Lynn M. Walker, Chemical Engineering, Carnegie Mellon University, Pittsburgh, PA

Shelley L. Anna, Departments of Mechanical Engineering and Chemical Engineering, Carnegie Mellon University, Pittsburgh, PA

Prediction of dynamic properties of interfaces requires detailed knowledge of the rates of diffusion and adsorption/desorption for surfactants at these interfaces. The coefficients of diffusion, adsorption, and desorption should be independent of the measurement technique. However, the approaches currently used to obtain these parameters are highly context dependent and can lead to unphysical trends such as a concentration-dependent diffusion coefficient and large mismatches between predicted and observed dynamic behavior. In pendant drop studies, one is restricted to conditions in which diffusion limits mass transport, or cases in which timescales for diffusion and sorption kinetics are comparable. Recently, theoretical and numerical work has demonstrated that a characteristic lengthscale governs the transition from diffusion-limited to kinetic-limited mass transfer for spherical interfaces [1]. This theory predicts that if the spherical interface has a radius much smaller than the intrinsic length scale, the mass transport process is kinetically limited. To experimentally validate this predicted transition to kinetic-limited mass transport, we have developed a micro-tensiometer to measure the dynamic surface tension at 10 to 100 micron diameter spherical interfaces. Using this new technique, we have shown that a transition from diffusion-limited to kinetic-limited dynamics is achieved as the radius decreases for a series of alkyl polyethylene oxide (CiEj) surfactants. We describe a new scaling for the diffusion timescale which better captures the physics of diffusion limited adsorption of surfactants onto spherical interfaces. Finally, using experiments, numerics, and scaling analyses, we demonstrate that the microtensiometer enables the direct measurement of sorption kinetic constants for these surfactants.

[1] Jin, F., Balasubramaniam, R., and Stebe, K.J., J. Adhesion. 80 (2004) 773.

See more of this Session: Novel Experimental Methods for the Study of Interfacial Phenomena
See more of this Group/Topical: Engineering Sciences and Fundamentals

A Test Platform for Systematic Investigation of Surface Phenomena in MEMS

Naveed Ansari, Department of Chemical Engineering, Auburn University, Auburn, AL

W. R. Ashurst, Chemical Engineering, Auburn University, Auburn, AL

Strong adhesion, friction and wear are the major obstacles on the road to commercialization of MEMS devices. Investigating the fundamental mechanisms involved in these surface phenomena is critical to many basic and applied problems in the field of micro- and nano- technology. This paper presents a test platform for investigating the mechanisms involved in surface interactions between two contacting surfaces.The test platform is built on (relatively smooth) SOI wafer in order to reduce the influence of roughness of contacting surfaces on surface properties, to facilitate a better understanding of the fundamental mechanisms involved. To achieve consistent results and eleminate the influence of fabrication process variations on surface properties, the test platform includes several different micromechanisms on the same chip to systematically investigate surface properties. Microinstruments to determine structural properties are also incorporated to provide the capability of direct measurement of necessary device parameters. The test platform facilitates the understanding of interactions and the factors that influence the interactions between two contacting MEMS surfaces. It can aslo be used to study the effect of the strategies that are used to improve the reliability of MEMS devices on the surface properties of contacting MEMS surfaces.

See more of this Session: Novel Experimental Methods for the Study of Interfacial Phenomena
See more of this Group/Topical: Engineering Sciences and Fundamentals

A Magnetic Colloidal Force Probe to Measure Nanometer Spacing Via Optical Imaging

Dichuan Li, Chemical and Biomolecular Engineering, Rice University, Houston, TX

Sibani Lisa Biswal, Chemical and Biomolecular Engineering, Rice University, Houston, TX

Surface force versus distance laws between colloidal particles are of fundamental importance in understanding their physical and chemical properties. Compared with conventional measurements such as surface forces apparatus (SFA) and atomic force microscope (AFM), a relatively simple method, referred to as magnetic chaining technique (MCT), has been developed to measure the repulsive inter-particle force-distance relations.

In the MCT, an ensemble of magnetic colloidal particles forms linear chains of particles due to each particle acquiring an induced dipole in an externally applied uniform magnetic field. Inter-particle forces can be calculated with high precision, as their magnetic dipole interactions have a resolution of 0.2pN. The inter-particle distance is measured by observing the back-scattered Bragg diffraction spectra peak wavelength, which has a resolution of 4nm. This resolution is relatively high considering its easy-to-use nature. However, a large volume of magnetic colloidal particles with a homogeneous particle size distribution are needed to achieve a high resolution.

In this talk, we present a novel approach to overcome these limitations. Specifically, we image a chain of micron-sized magnetic particles via an optical microscope under various magnetic field strengths allowing us to probe the interparticle forces between the particles. In our technique, less than a few micro-liters of sample suspension with low particle concentration (50ppm) is required. Additionally, our technique is not as sensitive to size distribution, as long as all the particles in a chain can be optically resolved. Finally, by proper image processing, our technique brings the distance resolution down to angstrom level. We will demonstrate our extremely simple and versatile force probing technique by measuring electrostatic forces between charged particles and steric repulsions between polymer grafted particle surfaces.

See more of this Session: Novel Experimental Methods for the Study of Interfacial Phenomena
See more of this Group/Topical: Engineering Sciences and Fundamentals

Disparity in Adsorption of Octadecyltrichlorosilane Islands On Rough Silicon Oxide Surfaces

Aimee Rae Poda, Chemical Engineering, Auburn University, Auburn, AL

The details of molecular assembly on topographical asperities, plays an important role in systems such as MEMs where SAM coated asperities must resist wear and adhesion during controlled or accidental contact during device operation. Many discrepancies associated with tribological performance have been identified as SAMs have been applied to surfaces of varying topographical complexity. SAM growth kinetics and interfacial properties as deposited on MEMs devices are often inferred from information learned on smooth surfaces. Questions remain as to whether the topographical morphology of the MEMs device will play an important role in the quality of improvement afforded from a deposited SAM. Research has shown that monolayers need to maintain a specific packing density and molecular orientation as deposited on surfaces of low roughness in order to provide a high tribological response. This research focuses on the interfacial structure of self-assembled coatings as influenced by surface topography and its implications toward tribological performance. Emphasis is placed on OTS islands produced from the (LC) phase due to the fact that this phase has shown the greatest propensity for friction improvement. Densely packed OTS islands resulting from low temperature deposition conditions are deposited onto native silicon oxide surfaces and onto silicon oxide surfaces created via molecular vapor deposition (MVD). The utilization of an MVD silica surface allows for an examination of a reactive silica surface with a variable degree of surface roughness as well as different topographical asperities (i.e. peak sizes, peak shapes). This silicon surface due to its amorphous nature allows for a true investigation of adsorption of OTS islands as influenced by topographical asperities eliminating the complexities associated with a polysilicon surfaces often influenced by dopant levels and water accumulation at crystal grain boundaries. In order to analyze the adsorption of the OTS islands and provide topographic resolution for AFM imaging the samples are exposed to a dilute etchant which removes only the MVD deposited silica that is not 'protected' by densely packed OTS islands. -->

See more of this Session: Novel Experimental Methods for the Study of Interfacial Phenomena
See more of this Group/Topical: Engineering Sciences and Fundamentals

Active Interfacial Microrheology of Protein Solutions

Prajnaparamita Dhar, Chemical Engineering, University of California, Santa Barbara, Goleta, CA

Joesph A. Zasadzinski, Department of Chemical Engineering, University of California, Santa Barbara, CA

Thomas Martin Fischer, Non-linear dynamics, Universitat Bayreuth, Bayreuth, Germany

Yanyan Cao, Chemistry, Pennsylvania State University, University Park, PA

T. Mallouk, Chemistry, Penn State University, University Park, PA

We present measurements of the surface shear viscosity of an albumin solution using a new nano-needle rheometer design, based on active microrheology techniques. The reorientation dynamics of Nickel nanorods(length=3µm, diameter =300 nm) placed at an air/albumin solution interface, in the presence of an externally applied magnetic field, is directly related to the surface shear viscosity of the monolayer. We present the surface viscosity of albumin monolayer both as a function of a range of bulk concentrations (0.02 mg/ml - 2 mg/ml) of albumin as well as a function of time. Our results indicate that the surface viscosity of albumin attains a near saturation above the bulk saturation concentration (0.1 mg/ml). However, it continues to increase by orders of magnitude (10-9 Ns/m-10-6 Ns/m) with time long after attaining the equilibrium surface pressure. This dramatic change in the surface viscosity of albumin with aging is a direct indication of the formation of complex protein networks at the interface which is likely a result of structural evolution of the albumin molecules with time. In addition to measuring the surface viscosity of a simple 2-d viscous system with the same order of precision as passive one and two particle microrheology techniques (10-9 Ns/m) (Prasad et al PRL 2006), this device is also capable of measuring surface viscosities of more complex systems including higher concentrations of albumin forming denser or more complex interfacial networks. This has grave implications in biology. Inhibition of lung surfactants by serum molecules is believed to be a consequence of lung injury resulting in adult respiratory distress syndrome. Surface viscosity measurements of the albumin molecules at the interface may lead to more insights into the mechanisms of inhibition reversal and re-spreading of lung surfactant in the presence of serum proteins.

See more of this Session: Novel Experimental Methods for the Study of Interfacial Phenomena
See more of this Group/Topical: Engineering Sciences and Fundamentals

Nanoscale Mapping of Thin Films and Monolayer Coatings On Solid Interfaces Using Three-Dimensional Atom Probe Tomography

Andrew C. Hillier, Iowa State University, Ames, IA

Yi Zhang, Iowa State University, Ames, IA

Three dimensional atom probe tomography (3DAPT) is a unique tool in the arsenal of nanoscale interfacial characterization methods. It can uniquely provide information about both the structure and composition of materials at near atomic-scale resolution. Although generally used for the characterization of three-dimensional metal and metal alloy phases, recent developments now allow atom probe tomography to be applied to soft materials and interfacial films. The work presented here focuses on the application of 3DAPT to thin film coatings and monolayers at interfaces.

The 3DAPT functions by pulse-mode field evaporation of a sample and projection of the resulting evaporated ions or molecular fragments into a position sensitive time of flight mass spectrometer. The location and identity of the projected fragments can then be resolved and reconstructed into a three-dimensional image of the sample at nanometer-scale or better resolution. In this talk, I will describe several examples of the application of 3DAPT to interfacial systems. Monolayer coatings of adsorbed ions, such as iodide, chloride or bromide anions, and self assembled alkanethiolates have been imaged on gold electrode surfaces. Various electrodeposited polymers, including poly(aniline) and poly(phenylene oxide), have been analyzed as thin films on palladium surfaces. The 3DAPT provides information about the chemical composition of the polymer films through the mass of the evaporated fragments and information about the structure of the polymer layer as well as details about the polymer-metal interface can be resolved. Complex coatings that consist of metal nanoparticles encapsulated within electrodeposited polymer films have also been imaged.

See more of this Session: Novel Experimental Methods for the Study of Interfacial Phenomena
See more of this Group/Topical: Engineering Sciences and Fundamentals

A General Method for Accurate Size Determination From 2-D Images

Md. Iqbal Hossain, School of Chemical and Biomedical Engineering, Nanyang Technological University, Singapore, Singapore

Tao Chen, School of Chemical and Biomedical Engineering, Nanyang Technological University, Singapore, Singapore

Yanhui Yang, School of Chemical and Biomedical Engineering, Nanyang Technological University, Singapore, Singapore

Raymond Lau, School of Chemical and Biomedical Engineering, Nanyang Technological University, Singapore, Singapore

Size distribution of the discrete phase objects such as particles, bubbles, and droplets can enhance the understanding of mixing, heat and mass transfer properties as well as complex phase interactions etc in multiphase flow systems. Numerous techniques including laser Doppler anemometry, tomography, intrusive probe measurement, and direct imaging have already been developed in order to determine the size distributions. Direct imaging is preferred over other techniques to give direct measurement of size and shape.

Discrete phase objects such as particles, bubbles, and droplets can be present at various distances from the imaging device when images are captured. Hence, the horizontal location of the object will need to be known in order to determine the actual size of an individual object. However, it is impossible to determine the horizontal location of the object from a two-dimensional image. A common magnification scale is normally used for all the objects in an image. For imaging devices such as borescope, there is substantial change in magnification scale with horizontal distance. The use of a single calibration scale would yield a deviation from the actual size. In this study, a statistical method is proposed to reconstruct the actual size distribution from the measured size distribution using a single calibration scale. The proposed method has been verified experimentally using particles of a known distribution. The stability of the proposed method is also analyzed theoretically against different viewable volumes, degree of change in magnification with horizontal locations, as well as different types of distributions.

See more of this Session: Novel Numerical Methods in Fluid Mechanics
See more of this Group/Topical: Engineering Sciences and Fundamentals

Coarse-Grain Molecular Fluid Mechanics

Alan Graham, Los Alamos National Lab, Los Alamos, NM

Antonio Redondo, Theoretical Divsion, Los Alamos National Lab, Los Alamos, NM

Shihai Feng, Material Science and Technology, Los Alamos National Lab, Los Alamos, NM

Patrick Reardon, Material Science and Technology, Los Alamos National Lab, Los Alamos, NM

In this investigation we explore the use of coarse-grain molecular simulations to explore fluid mechanics and boundary conditions in liquids. In these numerical experiments, coarse-grain molecular liquids are subjected to simple shear flow, pressure-driven flow, and falling-ball rheometer. Temporal and spatial averaging is used to average both the momentum transport and the boundary conditions. The key to success of a first-principle multiscale approach to fluid mechanics is insuring the application of the appropriate conservation laws in deriving the potentials based on molecular-scale simulations. Very large-scale molecular dynamic simulations will be compared with coarse-grain and continuum level simulations in a variety of flow fields.

See more of this Session: Novel Numerical Methods in Fluid Mechanics
See more of this Group/Topical: Engineering Sciences and Fundamentals

A Cartesian Cut Cell Method for Gas/Solids Flows

Jean-François Dietiker, West Virginia University Research Corporation, Morgantown, WV
Chris Guenther, National Energy Technology Laboratory,
U.S. Department of Energy, Morgantown, WV
Madhava Syamlal, National Energy Technology Laboratory,
U.S. Department of Energy, Morgantown, WV

Abstract

The implementation of a Cartesian cut cell method into the multiphase flow solver MFIX (Multiphase Flow with Interphase Exchange) is presented. The solids phase is treated as an interpenetrating phase, and mass and momentum balance equation are solved for gas and solids phases. Momentum transfer between the solids and gas phase is achieved through an empirical drag law. The cut cell method described by Kirkpatrick et al. to represent curved boundaries for the solution of the Navier-Stokes equations on a staggered three-dimensional Cartesian grid is extended to gas/solids flows.

The background Cartesian grid is intersected with curved or sloping surfaces describing the boundaries, resulting in the generation of cut cells. A special treatment is required to compute fluxes through cut cell faces, and the contribution to the wall shear stress must be included. The same treatment applies for both gas and solids phases.

The accuracy of the method is first tested for simple single phase benchmark cases such as the flow in a channel skewed with respect to the background grid, and the flow past a cylinder. Results show a clear improvement with the cut cell method compared to the traditional staircase step representation of boundaries. The method is subsequently applied to several gas/solids flows.

1. Introduction

The practical utilization of a Computational Fluid Dynamics (CFD) tool relies on its ability to accurately model the physics involved in the targeted application, its numerical efficiency, accuracy, and capability to represent complex geometries. Many computational techniques based on the finite volume method use unstructured grids to represent arbitrary geometries, for which a significant effort must be spent on grid generation, and special care must be employed to avoid highly-skewed cells. The implementation of higher-order schemes is also difficult to program.

On the other hand, the utilization of structured grids aligned with the Cartesian coordinate system leads to fairly simple discretization of the governing equations, and ease in programming the numerical method. However, structured grids have the major disadvantage of not being able to accurately represent curved or sloping boundaries. Typically, a boundary that is not aligned with the grid is approximated by a series of steps, referred to as staircase representation. Accuracy is improved by grid refinement, to the expense of computational resources. Alternative solutions include the use of overlapping grids [1,2], or flexible mesh embedding techniques [3,4], where information must be transferred between grids. The Immersed boundary method (IBM) [5-8] has been successfully implemented to represent complex geometries, where body forces are introduced to represent the presence of the boundaries.

Another method that has gained popularity over the last few years is the Cartesian cut cell technique, where a curved or sloping boundary is intersected with the background Cartesian grid, and boundary cells are truncated to conform to the shape of the boundary. The usual discretization schemes can be applied to interior cells, and the cut cells require special treatment. This method has been used to solve Euler [9] and Navier-Stokes [10] equations. Recently, Kirkpatrick et al. [11] described the implementation of the cut cell technique for staggered three-dimensional Cartesian grid into a Navier-Stokes solver. The boundary geometry is represented by quadric surfaces, and the treatment of boundary cells is easy to implement into an existing flow solver.

The objective of the current research activity is to implement a Cartesian cut cell technique into the multiphase flow solver MFIX (Multiphase Flow with Interphase Exchange), following the description of Kirkpatrick et al. A brief description of the method is provided in the next section. Simple gas phase benchmark cases are used to assess the accuracy of the method. Next, the method is applied to two Gas/Solids flows.

2. Numerical method

MFIX is a multiphase flow solver developed at the National Energy Technology Laboratory (NETL). It has the capability to describe the hydrodynamics, heat transfer, and chemical reactions occurring in a fluid/particle system. Typical applications include, but are not limited to, fluidized beds, bubbling beds, spouted beds, circulating fluidized beds, and risers. MFIX uses a three-dimensional Cartesian or cylindrical coordinate system. In the continuum model, the solids phase is treated as an interpenetrating phase, where mass and momentum balances equation are solved for gas and solids phases. Momentum transfer between the solids and gas phase is achieved through an empirical drag law. Details about the governing equations and numerical techniques of MFIX can be found in References 12 to 14.

The steps involved in the solutions are summarized as follows: 1) Pre-processing: the curved or sloping boundaries are represented by quadric surfaces, and computational cells are truncated to conform to the shape of the boundaries. Velocity nodes are moved to the center of cut cell faces. 2) Flow solution: computation of fluxes through cut cell faces are modified based on new geometry and location of velocity nodes, and the contribution of the wall shear stress due to the cut face is included. 3) Post-processing: the data visualization must incorporate the representation of the cut cells.

2.1 preprocessing

The procedure to identify cut cells is as follows. An intersection search between the boundaries and the background grid is performed along each edge of the cell. Once all intersections points are determined within a cell, the connectivity is established for each cut-cell and its faces, and re-ordered if necessary. Faces are described as convex polygons, and the face areas are computed similar to convex polygon areas. To compute cut-cell volumes, cut-cells are split into pyramids, and volumes of pyramids are computed and added to form the cell volume. Velocity nodes are placed at the center of pressure cell faces. All interpolation correction terms and non-orthogonality correction terms are computed from the new velocity node locations. An example of a computational grid, with the location of velocity nodes is shown in Fig. 1a, where the thick solid line represents the wall boundary. There are now three types of cells: standard (uncut) cells, cut cells that require special treatment to incorporate the presence of the solid wall, and blocked cells that are excluded from computations since they are located outside the active computational domain.

2.2 Cut cell treatment

The treatment of cut cells is based on a no-slip velocity boundary condition at the wall. As a first step in the implementation process, gas and solids phases are treated the same, i.e., no-slip boundary condition is imposed at the wall for both gas and solids phases. When a velocity component is interpolated near the wall, the location of the velocity node and wall velocity are taken into account. Due to the relocation of velocity nodes at the cut cell face centers, the velocity u_e may not be located at the center of the east face (Fig. 1b). Instead, the velocity at the face center u_{ec} is computed assuming zero velocity at the wall using the ratio of normal distances to the wall:

$$u_{ec} = \frac{\Delta h_{ec}}{\Delta h_e} u_e = \alpha_e u_e \tag{1}$$

In the u-momentum cell, other velocity components are interpolated at the face center based on the location of adjacent velocity nodes (Fig. 1c). For example, the v-component of velocity along the north face would be interpolated from

$$v_n = \theta_{ne} v_{ne} + \theta_{nw} v_{nw} \tag{2}$$

where $$\theta_{ne} = \frac{\Delta x_w}{\Delta x_{we}}, \quad \theta_{nw} = \frac{\Delta x_e}{\Delta x_{we}}, \text{ and } \Delta x_{we} = \Delta x_w + \Delta x_e \tag{3}$$

Other velocity components are interpolated in a similar fashion in the v and w-momentum cells. Due to the relocation of velocity nodes at the face centers, vector $\vec{S}$ joining points P and E, may not be perpendicular to the east face (Fig. 1d). Therefore, the derivative $\partial u/\partial x$ is approximated by equation (4) using the fact that velocity is zero at the wall. The vector $\vec{N}$ is perpendicular to the wall and passes through point e.

$$\frac{\partial u}{\partial x} \approx \frac{u_E - u_P}{|S_x|} - \frac{1}{S_x}\left(S_y \frac{\partial u}{\partial y} + S_z \frac{\partial u}{\partial z}\right) \tag{4}$$

where $$\frac{\partial u}{\partial y} \approx \frac{N_y u_e}{\Delta h_e} \quad \text{and} \quad \frac{\partial u}{\partial z} \approx \frac{N_z u_e}{\Delta h_e} \tag{5}$$

therefore $$\frac{\partial u}{\partial x} \approx \frac{u_E - u_P}{|S_x|} - \frac{S_y N_y + S_z N_z}{S_x \Delta h_e} u_e = \frac{u_E - u_P}{|S_x|} - \text{NOC}\, u_e \tag{6}$$

The non-orthogonality correction term $\text{NOC} = \frac{S_y N_y + S_z N_z}{S_x \Delta h_e}$ can be computed during preprocessing.

It is assumed that in the cut cell, the velocity is tangential to the wall. The shear force acting on the fluid due to the wall shear stress is parallel to the velocity vector (Fig. 1e), and can be written as

$$\vec{F} \approx -\mu \Delta A_{cut} \frac{\partial \vec{u}}{\partial n} \approx -\mu \Delta A_{cut}\left(\frac{u}{\Delta h}\vec{i} + \frac{v}{\Delta h}\vec{j} + \frac{w}{\Delta h}\vec{k}\right) = F_x \vec{i} + F_y \vec{j} + F_z \vec{k} \tag{7}$$

where ΔA_{cut} is the surface area of the cut face. Therefore, the contribution of the cut face can be applied implicitly along each direction (e.g., $F_x = -\mu \Delta A_{cut} \frac{u}{\Delta h}$ in the x-direction).

2.3 Problematic cells

A wide variety of cut cells are typically generated, and cut cells vary in shape and size. The generation of small pressure cells may lead to stability issues. Small cells can be removed from the computational domain by slightly altering the geometry, without significant loss of accuracy. When the boundary intersects the background grid near a corner point, the intersection point is moved to the corresponding corner point (i.e., the intersection point is "snapped" onto the grid). A tolerance, specified by the user defines the sensitivity of the snapping procedure. In this study, no stability issue was observed, and small pressure cells were not removed. Due to the staggered grid formulation, velocity cells require pairs of pressure nodes (East/West, North/South and Top/Bottom) to compute pressure gradients. However, when some pressure cells are removed (blocked cells), the corresponding velocity cells will only have one pressure node and the pressure gradient cannot be computed in that direction. When this situation occurs, the momentum equation is not solved for this cell, and a zero-velocity is specified to enforce no-slip boundary condition. The treatment of single-pressure velocity cells effectively limits their size to half the size of a standard cell, in any direction. This criterion automatically avoids problems linked to small velocity cells.

2.4 Cost of cut-cell treatment

The treatment of cut cells involves additional interpolation and correction terms. The geometric parameters are computed during pre-processing and do not impact the overall efficiency of the flow solver. Additional computation is required during flow solution to incorporated modifications involving terms described in the previous section (interpolation correction, non-orthogonality correction, and wall shear stress term). Table 1 shows examples of overhead resulting from the cut-cell treatment. For a two-dimensional computation with 7.25% of cut cells, the resulting overhead was 5.52%, where the overhead is the additional computational time compared with the same grid where the cut cells are replaced by standard cells, and no special treatment other than the usual boundary condition is applied. A three-dimensional computation with 3.70% of cut cells had an overhead of 2.49%. Therefore, with a ratio overhead/cut cell less than one, the cost associated with the cut cell treatment can be considered negligible, especially if it leads to a substantial gain in accuracy.

Table 1. Cost of cut cell treatment.

Geometry	*Grid size*	*Number of cells*	*Standard cells*	*Cut cells*	*Blocked cells*	*Overhead*	*Overhead/ cut cell*
2D	40x80	3200	71.50%	7.25%	21.25%	5.52%	0.75
3D	60x100x30	180000	9.55%	3.70%	86.75%	2.49%	0.67

3. Results

3.1 Channel flow

The first test case is used to asses the accuracy of the implementation of the cut cell technique for a two-dimensional gas flow in a channel. The channel is skewed at angles β ranging from zero to 45 degrees with respect to the vertical direction (Fig. 2a). Two grid resolutions have been investigated, with 7 cells (Grid A) and 15 cells (Grid B) across the channel width. Parabolic velocity profiles are prescribed at the inlet, which assures that the flow is fully developed.

Comparison with the exact solution is performed by extracting velocity components along 20 horizontal lines across the channel. Results for $\beta = 20.0$ degrees are presented in Fig. 2.With the coarse grid, the staircase method provides non-smooth velocity profiles with large errors compared to the analytical solution (Fig. 2b). The misrepresentation of the boundary leads to large velocity fluctuations that propagate to the center of the channel. Although the accuracy is improved with a finer grid (Fig. 2c), velocity gradients near the wall are not well resolved, and as a consequence, the velocity along the channel centerline is under-predicted. On the other hand, the cut cell technique provides smooth and accurate velocity profiles in the entire computational domain, even for the coarse grid (Figs. 2d and 2e).

The x and y velocity component errors are plotted as a function of the skew angle β in Fig. 3, for first and second order spatial discretization. As expected, both methods provide the least error when

β=0.0 degree, i.e., when the channel walls are aligned with the grid. The cut cell technique clearly outperforms the staircase method for all non-zero angles. The error tends to decrease as the angle increases with the staircase method, whereas it slightly increases with the cut cell method. The cut cell technique error is typically one order of magnitude smaller than the staircase error. At β = 45.0 degrees, errors made on the x and y components of the velocity are identical due to the interchangeable role of the x and y coordinates at this particular angle.

3.2 Flow over a cylinder

Next, the flow over a circular cylinder is investigated. For a Reynolds number of Re =40, the flow separates, but remains steady behind the cylinder. This is a common test case found in the literature to validate IBM methods. Grid convergence analysis is performed by solving the flow field on five grid resolutions, ranging from 20 to 160 cells per diameter, near the cylinder and grid sizes are summarized in Table 2. The domain extends 20 diameters upstream, and 40 diameters downstream to limit the effect of inflow and outflow boundary conditions to the solution.

Table 2. Grid systems for the flow over a cylinder.

GRID	*IMAX x JMAX*	*cells/diameter*
A	120x80	20
B	200x140	40
C	320x240	80
D	420x320	120
E	520x420	160

The surface pressure distribution is shown in Fig. 4. Although the general trend is captured by the staircase method, large pressure oscillations are observed, even as the grid is refined. The cut cell technique provides accurate results, with smooth pressure distribution along the entire cylinder surface, and grid convergence is observed. Surface vorticity distribution is shown in Fig. 5, and provides a measure of the accuracy of velocity gradient at the wall. The staircase representation leads to large fluctuations in surface vorticity, which are still present as the grid is refined. The coarse grid tends to widely under-predict the surface vorticity. The cut cell technique lead to a smooth vorticity distribution and accuracy improves as the grid is refined, until grid convergence is achieved.

Grid convergence analysis is illustrated in Fig. 6. Integrated quantities such as drag coefficient converge as the grid is refined with both techniques, with a faster rate of convergence achieved with the cut cell technique. This suggests that while both techniques are accurate for very fine grids, the cut cell technique is able to accurately predict the drag coefficient with a coarse grid, and provides a substantial gain is computational time required to obtain the solution. The effect of grid resolution on the separation angle is shown in Figure 6b, where the separation angle is measured as the location where the vorticity changes sign. The staircase representation largely under-predicts the separation angle with a coarse grid due to its inability to accurately resolve velocity gradients. The cut cell technique is practically insensitive to grid resolution.

Table 3 summarizes results for the steady flow around a cylinder, for the finest grid, and second-order spatial discretization. Drag coefficient and pressure coefficients are accurately computed with the staircase and cut cell technique, as they match experimental results or numerical results found in the literature [15-22]. The separation angle and separation length Lr/D are under-predicted with the staircase method, but are accurately predicted with the cut cell technique. This suggests that the staircase representation leads to large errors for separated flows, and may not be adequate at larger Reynolds number, where vortex shedding occurs. To verify this assertion, the Reynolds number was increased to values ranging from Re=80 to Re=320. Figure 7 shows the vortex shedding frequency as a function of the Reynolds number. The staircase method under-predicts the vortex frequency for all Reynolds numbers investigated in this study. The cut cell technique provides accurate results that compare well with experimental data [23, 24], and are almost identical to results obtained with Immersed Boundary Method (Silva et al., 2003). The inset in Fig. 7 shows instantaneous contours of the vorticity magnitude.

Table 3. Summary of results for the cylinder flow.

Author(s)	*Method*	*Total drag coeff.*	*Pressure drag coeff.*	*Friction drag coeff.*	*Front pressure coeff.*	*Rear pressure coeff.*	*Sep. Angle (Deg.)*	*Lr/D*
Grove et al. (1963)	Exp.		0.935		1.190			
Takami & Keller (1969)	Num.	1.536			1.141		53.55	2.35
Son & Hanratty (1969)	Num.	1.510					53.90	
Dennis & Chang (1970)	Num.	1.522	0.998	0.524	1.144		53.80	2.35
Collins & Dennis (1973)	Num.	1.560					53.60	2.15
Dennis (1976)	Num.	1.494			1.142			
Nieuwstadt & Keller (1973)	Num.	1.550			1.120		53.34	2.18
Fornberg (1980)	Num.	1.498			1.140		55.00	2.24
Kirkpatrick et al. (2003)	Num.	1.535			1.0		53.55	2.26
MFIX, Staircase, GRID E, 2nd	Num.	1.537	1.004	0.533	1.184	-0.484	51.75	2.209
MFIX, Cut Cell, GRID E, 2nd	Num.	1.542	1.010	0.532	1.192	-0.480	53.74	2.267

3.3 Spouted bed

The flow in a spouted bed is investigated in this section. The simulation setup corresponds to the experimental data of Liu et al. [25]. The total bed height is 60 cm, with a width of 15.2 cm and depth of 1.5 cm. The inlet width is 1.9 cm, and the inlet port height is 10 cm. Initial bed height are H_b = 5.0, 7.5, and 10.0 cm. The superficial gas velocity U_g ranges from 1.34 to 1.95 m/s. The solids phases consists of glass beads particles having an average diameter of 2.0 mm and density of 2380 kg/m^3. The grid is fairly coarse with 71 cells in the x-direction and 108 in the y-direction. The flow is assumed to remain two-dimensional since the ratio width/depth is 10. The grid system near the spout is shown in Fig. 8a. An instantaneous view of the void fraction is shown in Fig. 8b. The experiments show that a dominant frequency can be extracted from the time-series signal of either pressure drop or solids vertical velocity using Power Spectral Density distribution. The dominant frequency is plotted versus superficial gas velocity in Fig. 9, for several initial bed heights. The dominant frequency is well predicted for H_b = 10.0 cm and H_b = 7.5 cm. For the lowest value of initial bed height (H_b = 5.0 cm), the flow did not exhibit coherent behavior, and the dominant frequency is under estimated. For a given bed height, the dominant frequency tends to slightly increase as the superficial velocity increases. If the superficial velocity is held constant, an increase in the bed height from H_b = 7.5 cm to 10.0 cm, leads to a smaller dominant frequency, i.e., a higher mass of solids tends to respond slower to the gas flow.

3.4 Ozone Decomposition

Next, ozone decomposition simulation in a cylindrical vessel is conducted with second-order spatial discretization. This type of simulation is typically conducted with cylindrical coordinate systems due to the vessel geometry, but it may lead to inaccuracies as a result of the inability of the flow to cross the centerline. A full three-dimensional simulation is desirable but much more costly in terms of computational resources. The simulation corresponds to the experiments of Fryer and Potter [26]. The cylindrical vessel has a diameter of 22.9 in. The bed is fluidized with a mixture of 90 % dry air, and 10% ozone. The average particle diameter is 117 μm, and its density is 2650 kg/m^3. Other simulation conditions can be found in Ref. 27. A typical simulation requires around 30 seconds of simulation time, and statistics are gathered after 10 seconds to discard the effect of the initial transient flow (initially the bed does not contain any ozone). Figure 10a shows the outlet mass fraction of ozone C_{out}/C_{in} (ozone conversion) versus gas velocity, for an initial bed height of 11.5 cm. The 2D axisymmetric results show that a fine grid is required to obtain accurate results, and grid convergence was achieved for Grid C [27]. However, using the same grid resolution as Grid C for a full three-dimensional computation is still prohibitively expensive, and 3D simulations where conducted with the coarsest grid only (Grid A).

The 3D grid has 56 cells along the height of the cylinder and 36 in the other directions (Grid A). This corresponds to a grid size of 18x56 in the 2D axisymmetric case, where only half of the centerplane is represented. Once the intersection between the cylinder and the background grid is found, 72% of the cell are standard interior cells, 11% are cut cells, and the remaining 17% are excluded from computation (blocked cells), as they are located outside of the cylinder. Results obtained with Grid A are similar to the 2D axisymmetric case and 3D case, for gas velocities ranging from 4 to 12 cm/s (Fig. 10a). However, with the 2D axisymmetric grid, no statistically meaningful results could be obtained at a gas velocity of 2 cm/sec, and the outlet mass fraction of ozone is under predicted at a gas velocity of 14 cm/s. The three-dimensional computation with cut cell technique shows better prediction at these two extreme flow conditions. Isosurfaces of void fraction are shown in Fig. 10b, and illustrate the bubbling bed. Additional simulations with finer grids are under way to assess the effect of grid resolution on accuracy of the cut cell technique for this type of flow.

Conclusions

The Cartesian cut cell technique has been implemented into the multiphase flow solver MFIX. Boundary cells are truncated to conform to the shape of the curved of sloping boundary. The cut cells require additional treatment to represent the boundary as a no-slip wall. Results show that the cut cell technique is efficient and provide a substantial gain in accuracy when compared to the usual staircase representation. Velocity gradients are well predicted near the wall. Pressure and velocity disturbances introduced with a staircase representation are eliminated with the cut cell technique. Acceptable level of accuracy that could be obtained with a very fine grid using a staircase representation can now be obtained with the cut cell technique with a much coarser grid. Currently the no-slip wall boundary condition applies to both gas and solids phase. Future plans include the implementation of free-slip and partial slip boundary conditions.

References

[1] Meakin, R. L., "Composite Overset Structured Grids," Chapter 11, Handbook of Grid Generation, CRC Press, 1999.

[2] Petersson, N. A., "Hole-Cutting for Three-Dimensional Overlapping Grids," SIAM Journal on Scientific Computing, Vol. 21, No. 2, 1999, pp. 646-665.

[3] Albone, C. M., "Embedded Meshes of Controllable Quality Synthesized from Elementary Geometric Features," AIAA Paper 92-0662, 1992.

[4] Albone, C. M., and Joyce, G., "A Flow-Field Solver using Overlying and Embedded Meshes together with a Novel Compact Euler Algorithm," Lecture Notes in Physics, Vol. 323, 1989, pp. 106-110.

[5] Ye, T., Mittal, R., Udaykumar, H.S., and Shyy, W., "An Accurate Cartesian Grid Method for Viscous Incompressible Flows with Complex Immersed Boundaries," Journal of Computational Physics, Vol. 156, 1999, pp. 209-240.

[6] Taira, K., and Colonius, T., "The Immersed Boundary Method: a Projection Approach," Journal of Computational Physics, Vol. 225, 2007, pp. 2128-2137.

[7] Lima E Silva, A. L. F., Silveira-Neto, A., and Damasceno, J. J. R., "Numerical Simulation of Two-Dimensional Flows over a Circular Cylinder using the Immersed Boundary Method," Journal of Computational Physics, Vol. 189, 2003, pp. 351-370.

[8] Tseng, Y.H., and Ferziger, J.H., "A Ghost-Cell Immersed Boundary Method for Flow in Complex Geometry," Journal of Computational Physics, Vol. 192, 2003, pp. 593-623.

[9] Yang, G., Causon, D. M., and Ingram, D. M., "Calculation of Compressible Flows about Complex Moving Geometries using a Three-Dimensional Cartesian Cut Cell Method," International Journal for Numerical Methods in Fluids, Vol. 33, No. 8, 2000, pp. 1121-1151.

[10] Tucker, P.G., and Pan, Z., "A Cartesian Cut Cell Method for Incompressible Viscous Flow," Applied Mathematical Modelling, Vol. 24, 2000, pp. 591–606.

[11] Kirkpatrick, M. P., Armfield, S.W. and Kent, J. H. "A Representation of Curved Boundaries for the Solution of the Navier-Stokes Equations on a Staggered Three-Dimensional Cartesian Grid," Journal of Computational Physics, Vol. 184, 2003, pp. 1-36.

[12] https://mfix.netl.doe.gov/documentation/Readme.pdf

[13] Syamlal, M., Rogers, W., O'Brien, T. J. MFIX Documentation: Theory Guide; DOE/METC-94/1004 (DE94000087), U.S. Department of Energy: Morgantown, WV, 1993. https://mfix.netl.doe.gov/documentation/Theory.pdf.

[14] S. Benyahia, M. Syamlal, T.J. O'Brien, "Summary of MFIX Equations 2005-4," From URL http://www.mfix.org/documentation/MfixEquations2005-4-3.pdf, July 2007.

[15] Grove, A. S., Shair, F. H., Petersen, E.E, and Acrivos, A., "An Experimental Investigation of the Steady Separated Flow Past a Circular Cylinder," Journal of Fluid Mechanics, Volume 19, No.1, 1963, pp. 60-80.

[16] Takami, H., and Keller, H., "Steady Two-Dimensional Viscous Flow of an Incompressible Fluid past a Circular Cylinder," The Physics of Fluids Supplement II, 1969, pp. 51-56.

[17] Son, J. S., and Hanratty, T. S., "Numerical Solution for the Flow around a Cylinder at Reynolds Numbers of 40, 200, and 500," Journal of Fluid Mechanics, Vol. 35, No.2, 1969, pp. 369-386.

[18] Dennis, S.C.R, and Chang, G.-Z., "Numerical Solutions for Steady Flow past a Circular Cylinder at Reynolds numbers up to 100," Journal of Fluid Mechanics, Vol. 42, No. 3, 1970, pp. 471-489.

[19] Collins, W.M., and Dennis, S.C.R., "Flow past an Impulsively Started Circular Cylinder," Journal of Fluid Mechanics, Vol. 60, 1973, pp. 105-127.

[20] Dennis, S.C.R, "Numerical Method for Calculating Steady Flow past a Cylinder," in Proceedings of the Fifth International Conference on Numerical Methods in Fluid Dynamics, Lecture Notes in Physics, Vol. 59, 1976, pp. 165-172.

[21] Nieuwstadt, F., and Keller, H.B., "Viscous Flow past Cylinders," Computers and Fluids, Vol. 1, 1973, pp. 59-71.

[22] Fornberg, B., "A Numerical Study of Steady Viscous Flow past a Circular Cylinder," Journal of Fluid Mechanics, Vol. 98, No. 4, 1980, pp. 819-855.

[23] Williamson, C.H., "Vortex Dynamics in the Cylinder Wake", Annual Review of Fluid Mechanics. Vol. 28, 1996, pp. 477-539.

[24] Rosko, A., "On the Wake and Drag of Bluff Bodies," Journal of the Aeronautical Sciences, Vol. 22, No. 2, 1955, pp. 124-132.

[25] Liu, G.-Q., Li, S.-Q., Zhao, X.-L. and Yao, Q., "Experimental Studies of Particle Flow Dynamics in a Two-Dimensional Spouted Bed," Chemical Engineering Science, Vol. 63, No. 4, 2008, pp. 1131-1141.

[26] Fryer, C., and Potter, O. E., "Experimental Investigation of Models for Fluidized Bed Catalytic Reactors," AIChE Journal, Vol. 22, No. 1, 1976, pp. 38-47.

[27] Syamlal, M., and O'Brien, T. J., "Fluid Dynamic Simulation of O3 Decomposition in a Bubbling Fluidized Bed," AIChE Journal, Vol. 49, No. 11, 2003, pp. 2793-2801.

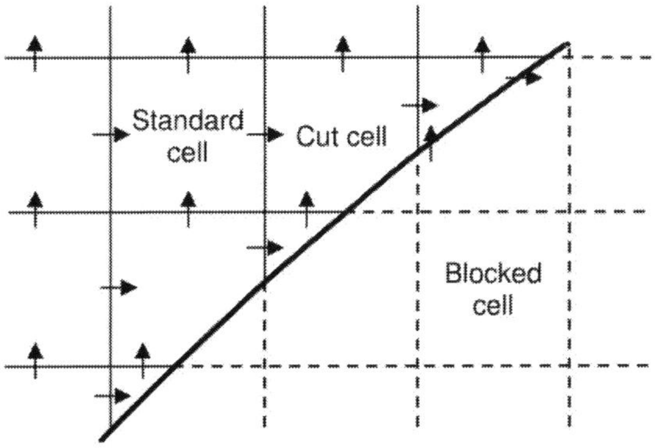

(a) Scalar cut-cells with location of velocity nodes.

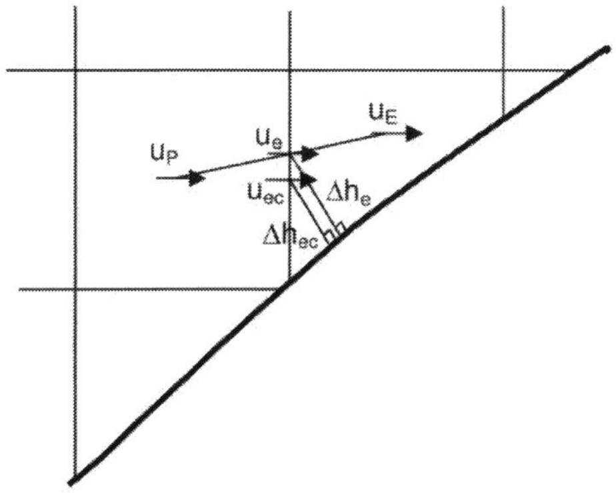

(b) Interpolation correction for u_{ec} at east face of u-velocity cells.

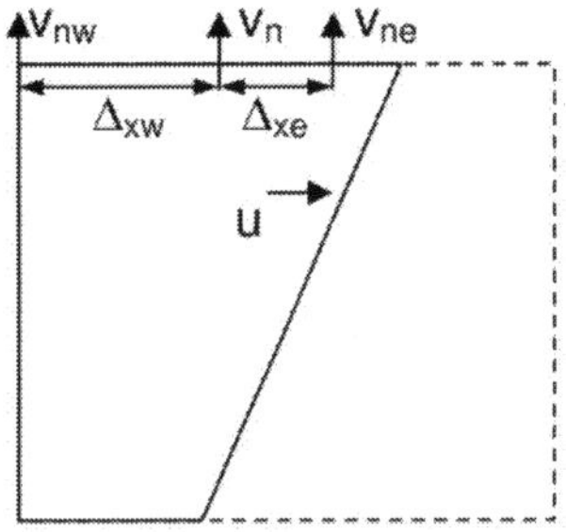

(c) Interpolation of v_n at north face of u-velocity cell.

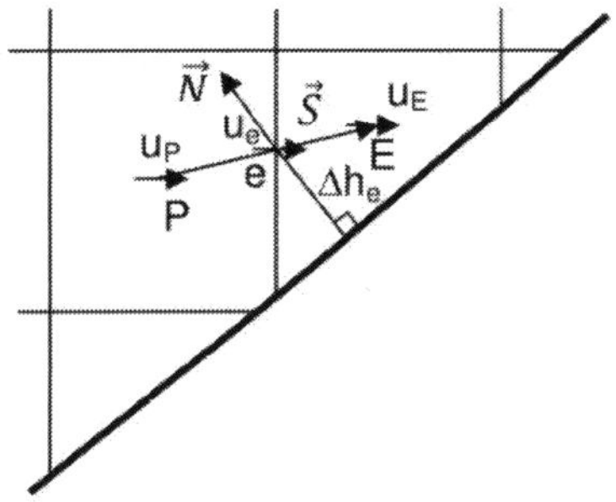

(d) Non-orthogonality correction for the derivative $\partial u/\partial x$ at east face of u-velocity cells.

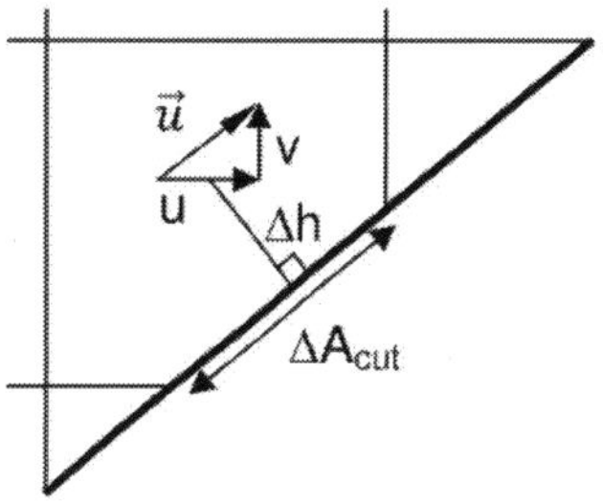

(e) Wall shear stress arising from the cut-face.

Figure 1. Cut cell treatment.

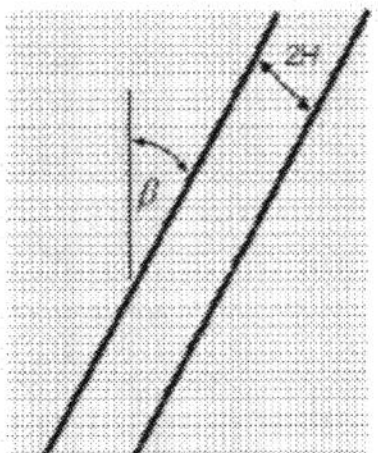

(a)
Schematic

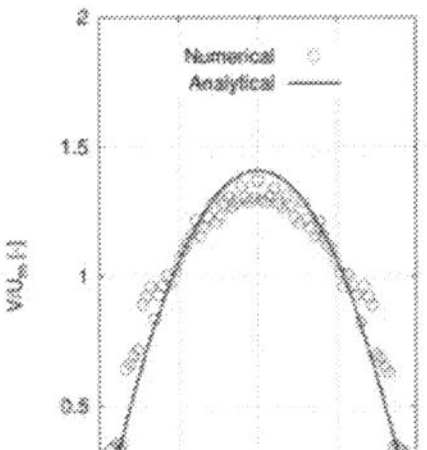

(b)
staircase
coarse grid

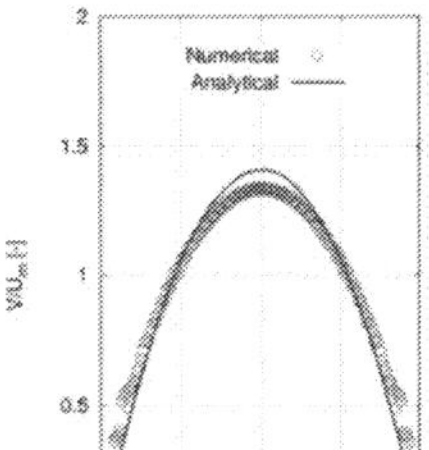

(c)
staircase
fine grid

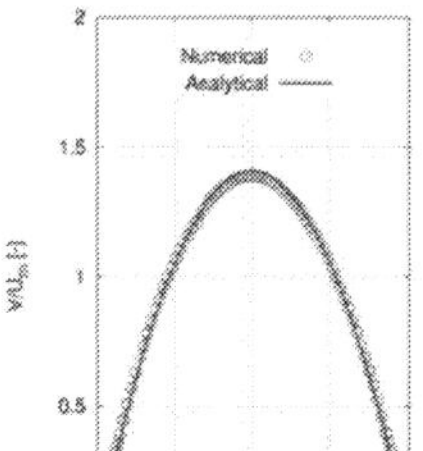

(d)
cut cell
coarse grid

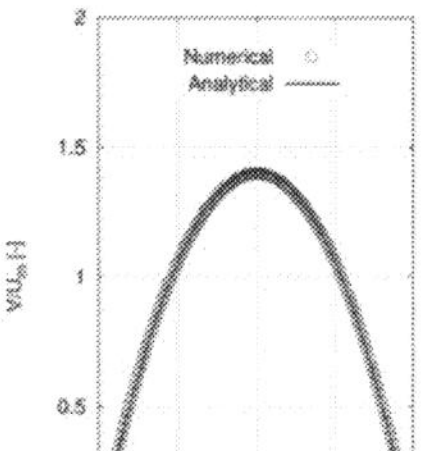

(e)
cut cell
fine grid

Figure 2. Schematic and velocity profiles for the channel flow.

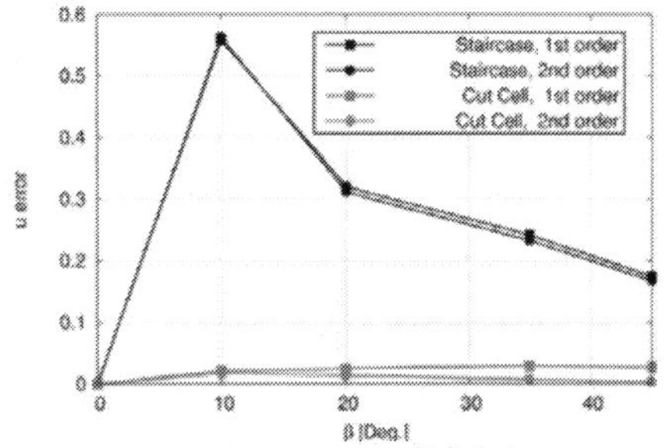

(a) u error, Grid A

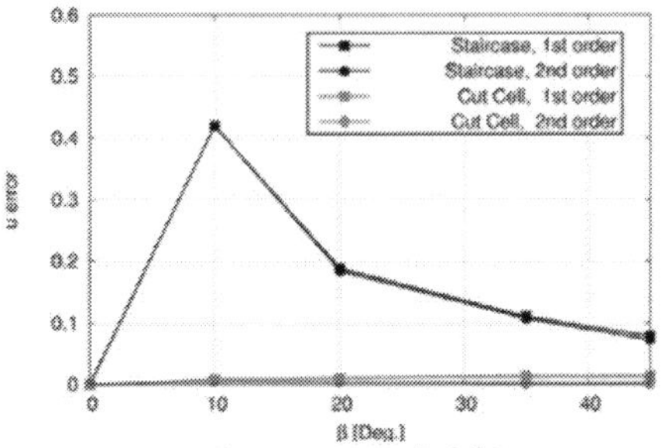

(b) u error, Grid B

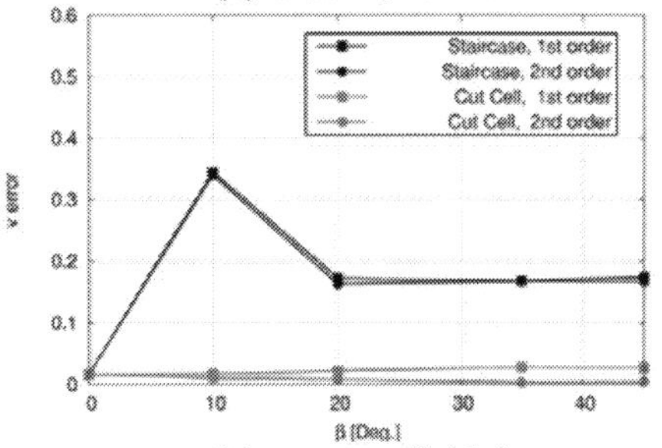

(c) v error, Grid A

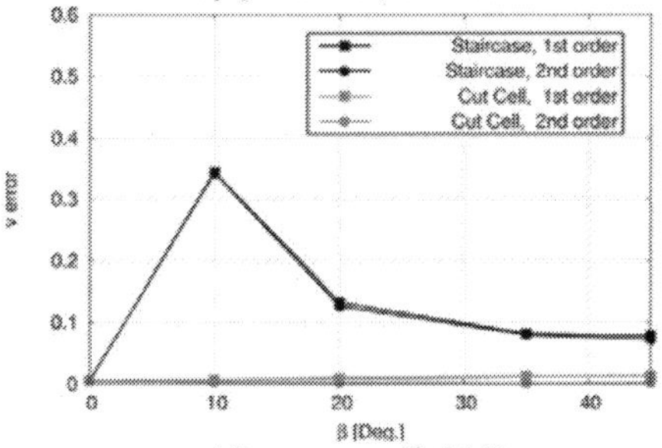

(d) v error, Grid B

Figure 3. Error on velocity components as a function of skew angle.

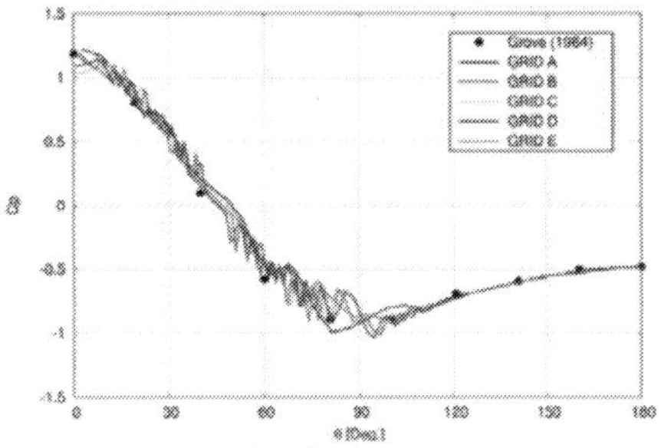

(a) Staircase

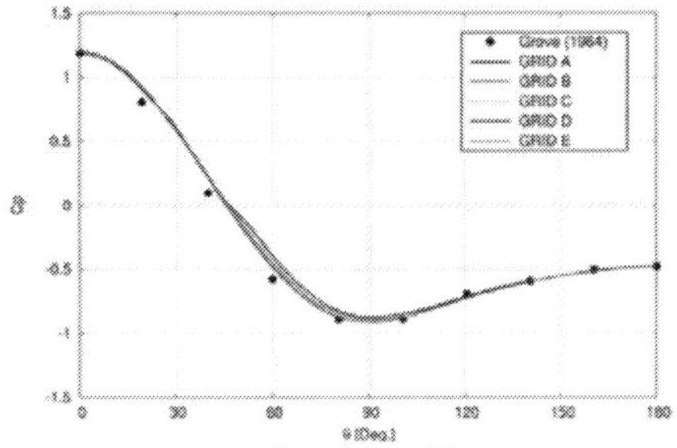

(b) Cut cell

Figure 4. Surface pressure distribution.

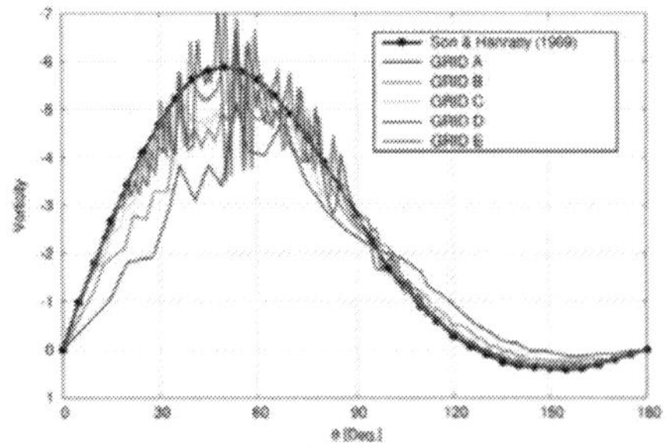

(a) Staircase

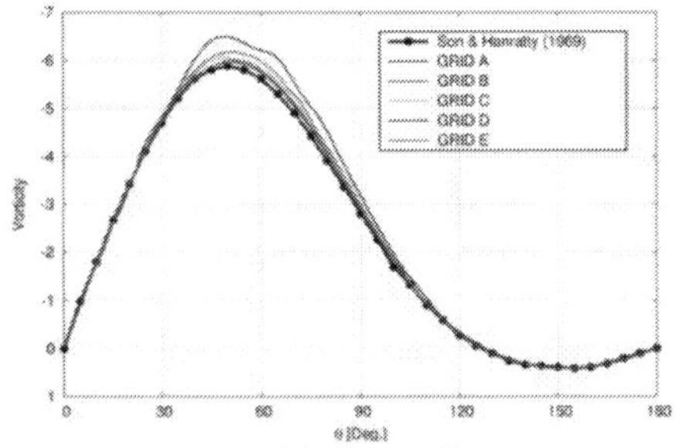

(b) Cut cell

Figure 5. Surface vorticity distribution.

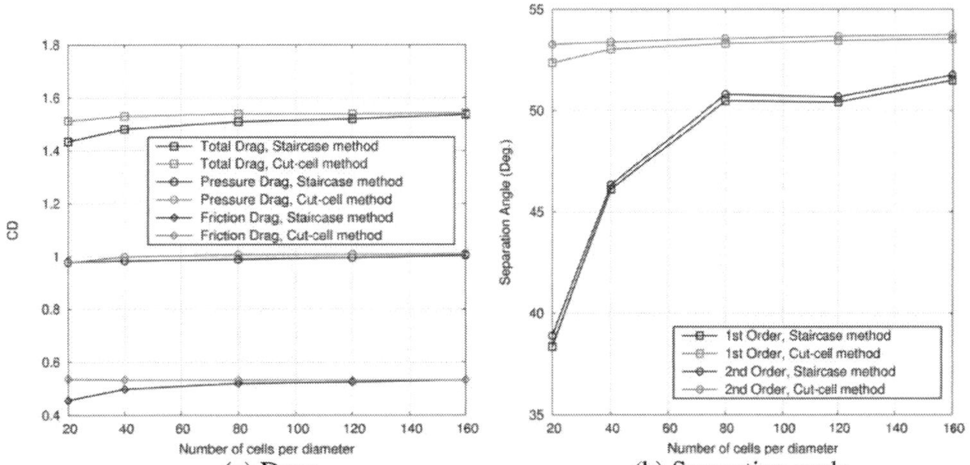

(a) Drag (b) Separation angle

Figure 6. Grid convergence.

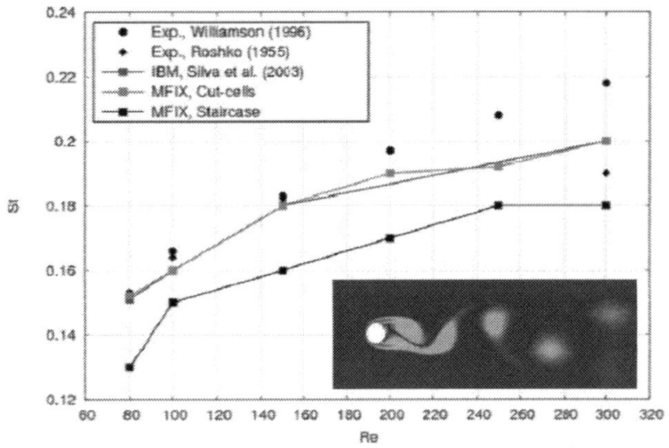

Figure 7. Vortex shedding frequency versus Reynolds number.

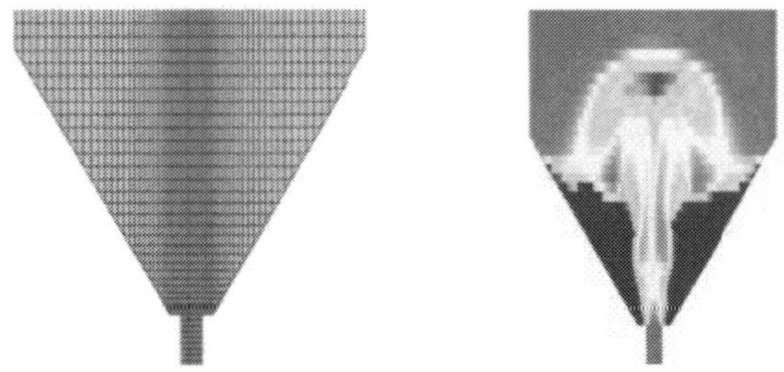

(a) Grid system (b) Contour of void fraction

Figure 8. Spouted bed simulation.

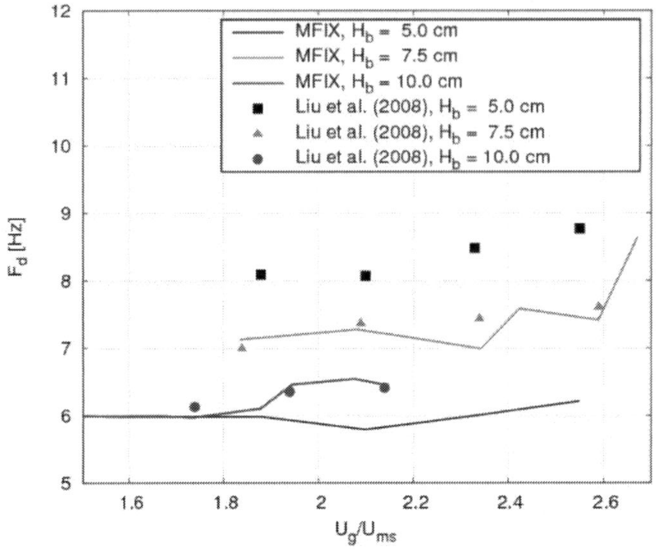

Figure 9. Dominant frequency versus superficial gas velocity.

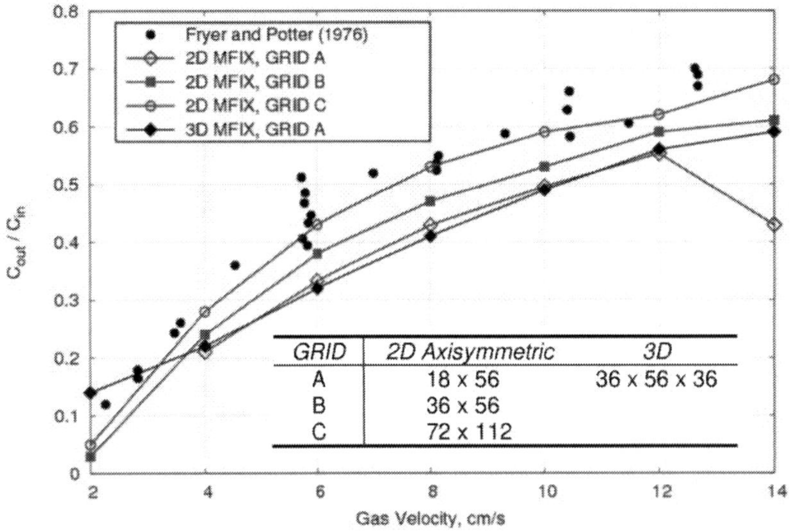

(a) Outlet mass fraction of ozone vs. gas velocity

(b) Isosurfaces of void fraction

Figure 10. Ozone decomposition.

RANS and LES of Single and Multiphase Flows Using the Immersed Surface Technology Combined with Block Mesh Refinement (IST/BMR)

Narayanan Chidambaram, R&D, ASCOMP GmbH, Zurich, Switzerland

Djamel Lakehal, R&D, ASCOMP GmbH, Zurich, Switzerland

Heterogeneous nucleate boiling and departure is approached here using detailed CMFD simulations, namely with the help of interface tracking methods (level set method). The interface is captured and the corresponding interphase heat transfer (jump in energy) is directly simulated by integrating the energy flux-balance around the interface. The objective of this work is to develop a time-coarse-grained model that represents only the later stage of nucleation and smooth out its early stage, which is prohibitively expensive.

See more of this Session: Novel Numerical Methods in Fluid Mechanics
See more of this Group/Topical: Engineering Sciences and Fundamentals

Predicting Turbulent Flows Using the Immersed Surface Technology Combined with Block Mesh Refinement (IST/BMR)

Daniel Caviezel, R&D, ASCOMP GmbH, Zurich, Switzerland

Narayanan Chidambaram, R&D, ASCOMP GmbH, Zurich, Switzerland

Djamel Lakehal, R&D, ASCOMP GmbH, Zurich, Switzerland

This talk will center on using a fast and versatile grid generation technique for the LES of industrial fluid mechanics problems. The approach known as The Immersed Surfaces Technique (IST) dispenses from the use of conventional structured grids and finite-element unstructured type of grids, in that solid bodies are immersed within a Cartesian grid from a CAD file. The idea (which differs from the Immersed Boundaries of Peskin) is to represent solid walls by a Level Set function representing the exact distance to the surface, which is zero at the surface, positive in the fluid(s) and negative in the solid. The fluid(s) and the solid have their own material properties, based on the Level Set function. The technique has the major advantage to solve conjugate heat transfer problems and fluid-structure interactions. In practice, the CAD file of the solid is first immersed into a cubical grid covered by a Cartesian mesh. The Navier-stokes equations are modified to account for the presence of the solid level set function. The treatment of viscous shear at the solid surfaces is handled using information buried in the solid level-set function. The BMR technique (Block Mesh Refinement) was developed to help better solve the boundary layer zone when use is made of the IST technique discussed above. In BMR, more refined sub-blocks are automatically generated around solid surfaces; with dimensions made dependent on the Reynolds number (the sub-block scale should always be set such that it covers the boundary layer thickness). An unlimited number of sub-blocks of different refinement can be generated, with connectivity between the blocks matching up to 1-to-8 cells. The method has been validated for canonical laminar and turbulent flows, and is now under evaluation for its extension to LES. Various examples will be presented; limitations and issues will be discussed as well.

See more of this Session: Novel Numerical Methods in Fluid Mechanics
See more of this Group/Topical: Engineering Sciences and Fundamentals

Numerical Evolution of Model and Parametric Sensitivities in Turbulent Reacting Flow Simulations

Venkat Raman, Aerospace and Engineering Mechanics, University of Texas at Austin, Austin, TX

Yonduck "James" Sung, Department of Mechanical Engineering, University of Texas at Austin, Austin, TX

Computational modeling of turbulent reacting flows has become an invaluable tool in the design and scale-up of chemical reactors. In spite of vast advances in the understanding of basic flow and reaction physics, computational models still lack broad predictive capability. A key source of error is the modeling coefficients used in the description of the different physical processes such as turbulence or chemical reactions. These coefficients are often derived empirically, without accounting for the interactions amongst the processes. Hence, the simulations are highly sensitive to the values used for these coefficients. Developing better models requires that the impact of coefficient variations on simulation results is understood. In order to estimate this sensitivity, we propose a novel sensitivity evolution methodology. We develop a non-intrusive technique, where partial differential equations governing the evolution of the sensitivity of the turbulent flow field to the model coefficients are solved along with the modeled flow equations. This methodology is applied to the large eddy simulation of gas-phase aerosol synthesis process. Here, the uncertainties in the model coefficients for aerosol nucleation, aggregation, and growth are studied. In addition, the propagation of errors due to uncertain boundary conditions is also studied. Interpretation of the sensitivity results and their use in model development will be discussed.

See more of this Session: Novel Numerical Methods in Fluid Mechanics
See more of this Group/Topical: Engineering Sciences and Fundamentals

Numerical Study of Bubble Dynamics with Advanced Lattice Boltzmann Simulation

Zhao Yu, Chemical and Biomolecular Engineering, Ohio State University, Columbus, OH

Liang-Shih Fan, Department of Chemical and Biomolecular Engineering, Ohio State University, Columbus, OH

In this study, the dynamics of bubbles will be simulated using an advanced lattice Boltzmann method (LBM). The new LBM takes advantage of both the adaptive mesh refinement (AMR) technique and the multiple relaxation time (MRT) algorithms, so that bubbles with large deformation and high Reynolds number can be simulated. High grid resolution is applied near the bubble surface as well as the regions with high vorticity in order to improve accuracy. The MRT technique is found to greatly enhance the numerical stability at low viscosities, and consequently enables the study of high-Reynolds number bubble flows that are relevant to many real-world problems. The computation results for flow field around buoyant rising bubbles under various conditions are presented. In particular, the wake structure of the bubbles as a result of their deformation and high Reynolds number is analyzed. Different patterns of bubble motion are observed in the simulations, including both rectilinear and oscillatory rise. The bubble trajectory is obtained from the simulation, and the relation between the transient forces on the bubble and the flow field is discussed. It is found that the wake structure has important effects on the motion of the bubble. Simulations are also carried out to investigate the hydrodynamic interactions between bubbles, which affect the bubble shape and velocity through the change of local flow field, particularly through the bubble wake interaction.

See more of this Session: Novel Numerical Methods in Fluid Mechanics
See more of this Group/Topical: Engineering Sciences and Fundamentals

Rapid and Robust Optimization of Transport Limited Processes in Laminar Flows

Daniel Lester, Materials Science and Engineering, CSIRO, Melbourne, Australia

Guy Metcalfe, Materials Science and Engineering, CSIRO, Melbourne, Australia

Many fluidic processes occur in low Reynolds number (laminar) flows. As such flows are generally non-mixing, transport and dispersion of relevant species are often rate-limiting with respect to the overall process. Examples occur in a wide range of applications involving small length scales (e.g. microfluidics), small velocity scales (e.g. porous media), large viscosities (e.g. polymer, food, and mineral processing), or rheologically sensitive fluids (e.g. biological fluids and pharmaceutical processing). In such flows, species of interest are advected along streamlines of the flow, and often transverse motion is limited to comparatively slow molecular diffusion: species distributions evolve along the streamlines of the flow but global mixing throughout the flow domain is limited.

In low Reynolds number, laminar flows viscous damping dominates, and these flows have velocity fields which are regular in the dynamical systems sense, such that the associated streamlines do not naturally promote good mixing. However, as identified by Aref [1], while laminar flows generate regular velocity fields, laminar flows can give rise to Lagrangian chaos; i.e. particle paths generated by laminar advection can form a chaotic tangle. This chaotic advection or Lagrangian turbulence arises from the notion that the advection equation describing motion of a passive tracer

$$\frac{\partial \mathbf{x}}{\partial t} = \mathbf{v}(\mathbf{x}, t) \quad (1)$$

represents a dynamical system rich enough to generate Lagrangian chaos. As such, "good mixing" corresponds to global chaos throughout the flow domain, ensuring complete mixing to small length scales within finite time. For incompressible fluids, (1) represents a conservative system to which the tools of Hamiltonian mechanics (Lyapunov exponents, Poincaré sections) may be applied in order to help optimize mixing.

Note that the advection equation (1) is purely kinematic; it is universal with respect to fluid rheology, flow geometry, velocity and length scales and so applies to all deforming media. Indeed, the main necessary condition for the attainment of chaos is that (1) contains at least 2.5 degrees of freedom. There is significant scope for simple temporal perturbations to generate chaos in otherwise non-mixing laminar flows.

As all species are also subject to some level of molecular diffusion and chaotic dynamics in (1) generates extremely small length scale striations, diffusion cannot be ignored in any physical system. Hence mixing is correctly cast as scalar dispersion, as described by the advection diffusion equation

$$\frac{\partial \phi}{\partial t} = -\mathbf{v} \cdot \nabla \phi + \frac{1}{Pe} \nabla^2 \phi = L[\phi] \quad (2)$$

where *f* is the concentration of scalar to be mixed, and *Pe* is the Peclét number quantifying the relative timescales of advection and diffusion. As species with high diffusivities naturally disperse quickly, chaotic advection is most effective for high *Pe*, and in the singular limit $Pe \rightarrow \infty$, the Hamiltonian system (1) is recovered. However, at finite *Pe* the system (2) is dissipative, so Hamiltonian mechanics and related tools no longer apply.

Judicious programming of the velocity field **v** in (2) can result in significant enhancement of the dissipation (mixing) rate of *f*. Optimized mixing rates scale as $\sqrt{Pe}$, hence improvements of multiple orders of magnitude are possible for $Pe > 10^4$. However, an open problem is how to optimize the advective velocity for such a goal. Given parameterization of the velocity field by a parameter set *c*, which contains a number of tunable flow control parameters (such as boundary motion or forcing parameters), then scalar transport varies over the multidimensional parameter space $\mathbf{Q}=\{c \times Pe\}$. Clearly, direct numerical solution of (2) over this space represents an unfeasible computational problem, especially given numerical evidence [2] suggesting fractal distributions of mixing enhancement at high *Pe*.

Although there exist rapid methods for simulating chaotic advection with dissipation, most of these methods do not correctly treat the diffusion operator in (2). The so-called matrix mapping method [3], using a discrete map to rapidly project the scalar distribution forward at each time step, however dissipation occurs via numerical errors, which do not correspond to physical diffusion. Operator-splitting methods project the solution sequentially via advection and diffusion operators, which is shown by Giona et al [4] to generate spurious scalar transport which does not capture the complex intertwining of diffusion with chaotic advection. Likewise, with respect to optimization, given the complex and fractal nature of the enhancement rate distribution, rapid optimization techniques [5] are likely to fail over **Q**, and only high resolution global elucidation of the global structure of transport over **Q** can robustly identify optimum stirring protocols.

The dissipative nature of (2) gives rise to a contraction of phase space, such that under normal conditions, solutions to (2) take the form of so-called "strange eigenmodes" [6, 7] which are superposed exponentially decaying spatio-temporal patterns which govern the homogenization of *f*. As the slowest decaying eigenmode is also the most regular, the majority of the initial variance in *f* is mapped to this dominant eigenmode. As such, within a short time, the decay rate of the dominant eigenmode dictates the rate of scalar dispersion in the flow domain, and this metric is exact in the asymptotic limit of complete mixing. Therefore, quantification of the dominant strange eigenmode associated with the advection diffusion operator *L* in (2) for a given point in **Q** corresponds to quantification of the dispersion rate and pattern for the corresponding flow field and Peclét number.

By exploiting both the symmetries inherent to chaotic flows and the dissipative nature of the advection diffusion equation (2), in this paper we present a rapid method which accurately resolves the dominant strange eigenmodes over the parameter space **Q**. As these eigenmodes quantify both the mixing rate and dominant mixing pattern, the rate and nature of mixing over this space can be rapidly and accurately resolved, facilitating laminar flow optimization and exploitation of the potential benefits offered by chaotic advection. Mathematical justification and computational details of this method are described in a separate paper [2], and in this paper we briefly describe the method and focus on applications of the method to various flow scenarios.

To achieve global resolution of **Q**, we assume that the all of the flows over the parameter space *c* can be composed from a small set of so-called steady "base flows" $\mathbf{v}_i$. This is not as restrictive as it may first appear as the linear nature of laminar flows renders symmetries inherent to such flows, and this method acts to exploit these symmetries. By applying the mapping operations of scaling, superposition, reflection and reorientation to the base flows, a much greater set of composite steady flows can be generated. By applying transient amplification to this set of steady flows, the full set of transient flows across *c* can be constructed. The main point is that the full set of transient flows $\mathbf{v}(c)$ can be composed from a small set of steady base flows $\mathbf{v}_i$ via mappings and transient amplification.

By representing the advection operator for each base flow through an appropriate spectral expansion over the physical flow domain, the building blocks of a spectral representation (denoted *L*') of the advection-diffusion operator *L* in (2) can be constructed. Although spectral expansion of the base flows advection operator may be computationally expensive, this only needs to be performed once for each base flow. Subsequent composition of *L'* in spectral space is quite inexpensive, and so the operator can be rapidly composed over the entire parameter space **Q**. The dominant eigenvector and eigenvalue of *L'* correspond respectively to a spectral approximation to the dominant strange eigenmode and its decay rate. As such, construction of *L*' and solution of the dominant eigenproblem over **Q** generates a map of both the mixing rate and dominant mixing pattern over the entire parameter space.

These results allow simple identification of the optimal velocity field $\mathbf{v}(c)$ for a particular value of *Pe* to achieve chaotic advection and hence optimization of transport-limited processes. We show examples of application of the method to several industrially relevant flows, and illustrate how the method may be applied to a wide variety of flow scenarios.

[1] Aref, H., 1984, Stirring by chaotic advection. J. Fluid Mech. **143**:1-21.

[2] Lester, D. R., Metcalfe, G., Rudman, M. and Blackburn, H., 2008, Global parametric solutions of scalar transport, J. Comp. Phys., **227**, 3032-3057.

[3] Galaktionov, O. S., Anderson, P. D., Peters, G. W. M., and Meijer, H. E. H., 2002, Mapping approach for 3D laminar mixing simulations: application to industrial flows, Int. J. Num. Methods. Fluids., **40**, 345-351.

[4] Giona, M., Adrover, A., Cerbelli, S., 2005, On the use of the pulsed-convection approach for modeling advection-diffusion in chaotic flows - A prototypical example and direct numerical simulations, Physica A, **348**, 37-73.

[5] Gibout, S., Le Guer, Y., Schall, E., 2006, Coupling of a mapping method and a genetic algorithm to optimize mixing efficiency in periodic chaotic flows, Comm. Nonlin. Sci. Num. Sim., **11**, 413-423.

[6] Liu, W. and Haller, G., 2004, Strange Eigenmodes and Decay of Variance in the Mixing of Diffusive Tracers, Physica D, **188**, 1-39

[7] Pierrehumbert, R. T., 1994, Tracer microstructure in the large-eddy dominated regime, Chaos, **4**, 6, 1091-1110

See more of this Session: Novel Numerical Methods in Fluid Mechanics
See more of this Group/Topical: Engineering Sciences and Fundamentals

A non-stiff cytoskeleton-based continuum computational modeling for the dynamics of erythrocytes in strong viscous flows

Panagiotis Dimitrakopoulos
Department of Chemical and Biomolecular Engineering,
Bioengineering Graduate Program
University of Maryland
College Park, MD, USA

Walter R. Dodson III
Fischell Department of Bioengineering
University of Maryland
College Park, MD, USA

Introduction

The motion of erythrocytes through vascular vessels has long been recognized as a fundamental problem in physiology and biomechanics; unfortunately, studies on the flow dynamics of erythrocytes still constitute a challenging problem in any type of research owing to the micron-size of the cell and its complex multi-layered membrane. In the area of our interest (i.e. computational investigations), several continuum and molecular-based models have been developed in the recent decades to study erythrocytes. In the continuum models, treating the erythrocyte membrane as a two-dimensional elastic solid with large area-dilatation modulus results in a very stiff computational problem especially for three-dimensional investigations. On the other hand, cytoskeleton-based molecular algorithms were able to model efficiently the global area-incompressibility of the skeleton, but their applicability to flow problems is usually restricted owing to large computational cost.

Results

In this talk, we show that, by combining the current experience on erythrocyte computational algorithms via both continuum and molecular modelings, we develop a cytoskeleton-based continuum algorithm, based on the Skalak *et al.*constitutive law, which accounts for the global area-incompressibility of the skeleton via a non-stiff, and thus efficient, procedure. In addition, we investigate the erythrocyte dynamics in shear flows for moderate and strong capillary numbers and small to moderate viscosity ratios. These conditions correspond to a wide range of surrounding medium viscosities and shear flow rates, and match those used in ektacytometry systems and the tank-treading frequencies in vivo.

Conclusions

Our computational results for the cell deformability and tank-treading frequency are in excellent agreement with experimental findings from ektacytometry and capture very accurately the shear modulus of the erythrocyte membrane. To the best of our knowledge, since the development of Skalak *et al.*constitutive law nearly 35 years ago, our work is the first one to develop a non-stiff continuum-based algorithm for the accurate determination of erythrocyte dynamics.

Turbulent Flow of Dilute Aqueous Suspensions of Single Wall Carbon Nanotubes in a Smooth Pipe

Preetinder S. Virk, Department of Chemical Engineering, Massachusetts Institute of Technology, Cambridge, MA

The object of the present work was to experimentally explore the turbulent flow of dilute aqueous suspensions of single-walled carbon nanotubes in a smooth pipe. The industrial grade single-walled carbon nanotubes, of average diameter and length (d, l) = (1.1 nm, 17500 nm), were suspended at concentrations c from 0.01 to 100 wppm in a solution containing 900 wppm of polyvinylpyrolidone surfactant in deionized water. A seamless stainless steel test pipe with electropolished bore, of internal diameter and bore roughness (D, k) = (4.58 mm, 0.18 microns), was used, comprizing 6 sections, each of L/D = 30 with a pressure tap near its downstream end. The test pipe was part of a single-pass, progressing cavity pump-driven flow system that could traverse Reynolds numbers from 8300 to 67000 and pipe wall shear stresses from 10 to 500 Pa. Friction factors obtained with deionized water showed that fully-developed turbulent flow was established at axial distances x/D > 60 from the pipe entrance at all but the lowest Re; also pressure drops between taps 4&5, across pipe section 5 at 120 < x/D < 150, yielded the relation 1/Åãf = (3.85±0.14) log ReÅãf + (0.48±0.47) that was close to the Prantdl-Karman equation for Newtonian turbulent pipe flow. On account of their scarcity, nanotube suspensions were tested in brief ÅgscanningÅh runs of 30 s duration at each of four Re = (8300, 16700, 33000, 67000). First, two scans at c = 0, that is, of the 900 wppm PVP surfactant solution alone, served to establish a basis, yielding P-K results of 1/Åãf (11.1±0.1, 12.2±0.1, 13.2±0.1, 14.5±0.1) at ReÅãf (750, 1370, 2520, 4580). Thereafter, scans of SWNT suspensions at each of c = (0.01, 0.10, 1.0, 10, 100) wppm provided sets of 1/Åãf (11.2±0.2, 12.2±0.1, 13.2±0.0, 14.5±0.0) at ReÅãf (740, 1375, 2520, 4590) that were virtually indistinguishable from the basis scans at c = 0 wppm. The dilute aqueous solutions of c = 0.01 to 100 wppm SWNTs in 900 wppm PVP surfactant thus exhibited Newtonian turbulent flow behavior to within the experimental error of 0.1 units of 1/Åãf over the ReÅãf range from 750 to 4600.

See more of this Session: Particulate and Multiphase Flow II
See more of this Group/Topical: Engineering Sciences and Fundamentals

Complex Flows of Cellulosic Fiber Suspensions Visualized Using Magnetic Resonance Imaging

Emilio Tozzi, Chemical Engineering and Materials Science, University of Calfiornia Davis, Davis, CA

M. J. McCarthy, Food Science & Technology, University of Calfiornia Davis, Davis, CA

S.P. Shoemaker, Food Science & Technology, University of Calfiornia Davis, Davis, CA

D. Lavenson, Chemical Engineering & Materials Science, University of Calfiornia Davis, Davis, CA

R.L. Powell, Chemical Engineering & Materials Science, University of Calfiornia Davis, Davis, CA

Efficient processing of fibrous biomass requires adequate rheological characterization of fiber suspensions. Most published data on biomass rheology has been obtained using rotational rheometers and interpreted through generalized-Newtonian models. Due to the large size of biomass particles, fast settling, entanglements and migration occur. These phenomena cause difficulties in rheological studies made with rotational rheometers and analyzed via generalized-Newtonian models. Direct imaging of velocity profiles provides a way of characterizing flow in the presence of such non-idealities. We used magnetic resonance imaging to measure velocity profiles of cellulosic fiber suspensions flowing in an horizontal tube. Pressure drop and video images were simultaneously recorded. We observed a strong influence of fiber length, concentrations and flow rates on velocity profiles and pressure drops. At low flow rates and low concentrations vertically asymmetric velocity profiles were observed due to settling. At increased concentrations, entanglement of longer fibers was detected as solid-like regions moving with uniform velocity. Disruption of fiber entanglements at higher flow rates was observed in some cases. Advantages and challenges of using MRI for rheological characterization of biomass suspensions having various particle sizes are discussed.

See more of this Session: Particulate and Multiphase Flow II
See more of this Group/Topical: Engineering Sciences and Fundamentals

Modeling Inclined Plane Flow of Particle-Laden Thin Films

Nebojsa Murisic, Mathematics, University of California, Los Angeles, Los Angeles, CA

Andrea L. Bertozzi, Mathematics, University of California, Los Angeles, Los Angeles, CA

The flows of particle-laden liquid films are essential in numerous applications, such as geophysical (erosion, debris flow, and turbidity currents) and industrial ones (commercial oil-sand and bitumen operations, semiconductor manufacturing, pharmaceutical production, and food processing). In addition, better understanding of the physical mechanisms involved in particle-laden thin film flows will undoubtedly prove useful in dealing with various related problems involving multi-phase flows.

The motivation for this work stems from a series of experiments carried out at the UCLA Applied Mathematics Laboratory and reported in [1]. In these experiments, the focus was on suspensions of polydisperse glass beads flowing down an inclined plane. By varying both the particle concentration and the inclination angle, three distinct flow regimes were observed. At low inclination angles and concentrations, rapid sedimentation occurred, resulting in particles forming a uniform particulate layer behind the advancing front, with clear liquid film flowing over this layer and eventually undergoing the fingering instability, typical for clear liquid thin film flows (see, e.g. [2]). On the other hand, at high concentrations and inclination angles it was found that particle-rich ridge forms at the advancing front, suppressing the fingering instability. Finally, a well-mixed regime was identified for intermediate concentrations and inclination angles, characterized by occurrence of regular fingers containing particles.

A theoretical model based on lubrication approximation was derived in [1]. It incorporated basic physical mechanisms, including Stokesian settling. However, the main focus was on a reduced governing system of conservation laws, which were used to develop a shock-dynamics theory relevant to this problem. While this theory introduced a gas dynamics technique to a contact line problem, it failed to fully capture the experimentally observed flow regimes.

The main goal of this study is to better understand the mechanisms responsible for different flow regimes, by achieving both qualitative and quantitative agreement between the theory and the experimental data. To this end, we derive a novel mathematical model, based on Navier-Stokes equations for an incompressible viscous fluid and the appropriate equation for particle mass concentration. The main building blocks of this model are as follows: (i) the thin film flow is such that lubrication approximation is appropriate; (ii) surface tension and gravity effects are considered; (iii) both the shear-induced particle migration and modified Stokesian settling are included into the model for the evolution of particle mass concentration; (iv) the interaction between the liquid and the solid substrate is modeled using a disjoining pressure approach (including both van der Waals and structural components); (v) the dependence of the dynamic contact angle on the local particle concentration is included; (vi) the possibility for solutal Marangoni effect is taken into account.

The model for shear-induced particle migration is based on a theory developed in [3]. While Brownian diffusive flux is neglected (due to a large Peclet number), both the effect of spatially varying interaction frequency between the particles and the effect of spatially varying viscosity are included in this model. These effects consider pair-wise interactions between particles which lead to corresponding net fluxes, entering the governing equation for particle mass concentration. To the best of our knowledge, this is is the first time shear-induced migration has been included into a lubrication approximation framework, and employed in modeling flows of particle-laden thin films.

The interaction between the liquid and the solid substrate is modeled using a two-component disjoining pressure approach. This model is a combination of van der Waals and so-called "structural" forces, the latter taking into account direct interaction between particles and the solid substrate. Van der Waals forces, which are dominant over scales smaller than particle diameter, are described using a two-term model (see e.g. [4]). This model includes repulsive and attractive intermolecular forces leading to a stable precursor film. On the other hand, the structural component of the disjoining pressure acts on longer scales (longer than particle diameter), it is oscillatory in nature and gives rise to particle layering. The modeling of this component is based on statistical mechanics; we include it into our theoretical framework in a manner similar to the one discussed in [5]. The dependence of the dynamic contact angle on the local particle concentration in the vicinity of the advancing front is based on Cox-Voinov law, with our approach relying on the discussion in [6].

We present the derivation process in some detail and show how it eventually leads to a pair of coupled quasi-2D partial differential equations governing the evolution of liquid thickness and particle mass fraction. We develop a numerical code based on finite differences in order to solve this coupled system. In addition, we present the results of our recent experiments, carried out as an extension of experimental work reported in [1]. We proceed by comparing our numerical results with both the experimental data from [1] and our own. Finally, we outline the plan for future work, including more precise experiments which are to be carried out at the Applied Mathematics Laboratory as part of the REU (Research Experience for Undergraduates) program at UCLA.

References:

[1] Zhou, J., Dupuy, B., Bertozzi, A.L., Hosoi, A.E., 2005, Theory for Shock Dynamics in Particle-Laden Thin Films, Phys. Rev. Lett. 94, 117803.

[2] Kondic L., 2003, Instabilities in Gravity Driven Flow of Thin Fluid Films, SIAM Review 45, 95.

[3] Phillips R.J., Armstrong R.C., Brown R.A., Graham A.L., Abbott J.R., 1992, A Constitutive Equation for Concentrated Suspensions that accounts for Shear-Induced Particle Migration, Phys. Fluids A 4, 30.

[4] Schwartz L.W., Eley R.R., 1998, Simulation of Droplet Motion on Low-Energy and Heterogeneous Surfaces, J. Colloid Interface Sci. 202, 173.

[5] Matar O.K., Craster R.V., Sefiane K., 2007, Dynamic Spreading of Droplets containing Nanoparticles, Phys. Rev. E 76, 056315.

[6] Ralston J., Popescu M., Sedev R., 2008, Dynamics of Wetting from an Experimental Point of View, Rev. Mater. Res. 38, 23.

See more of this Session: Particulate and Multiphase Flow II
See more of this Group/Topical: Engineering Sciences and Fundamentals

Rheology of a Dilute Emulsion of Low-Viscosity Drops at High Shear Rates

Michael Loewenberg, Chemical Engineering, Yale University, New Haven, CT

Pieter Jan Antoon Janssen, Department of Chemical and Biological Engineering, University of Wisconsin Madison, Madison, WI

Patrick D. Anderson, Eindhoven University of technology, Eindhoven, Netherlands

A theoretical analysis is presented for the rheology of a dilute emulsion of low-viscosity drops or bubbles in shear flow at high shear rates. In the high-shear-rate limit, the predicted relative shear viscosity of the emulsion is 1 - 2c, where c is the volume fraction of drops. A first normal stress difference proportional to the square-root of the drop viscosity ratio is also predicted; in a suspension of bubbles, the first normal stress decreases with the 3/4 power of the capillary number. Confining boundaries reduce the normal stress but do not strongly influence the shear viscosity.

See more of this Session: Particulate and Multiphase Flow II
See more of this Group/Topical: Engineering Sciences and Fundamentals

Nonlinear Interactions in Electrophoresis of Ideally Polarizable Particles

David Saintillan, Department of Mechanical Science and Engineering, University of Illinois at Urbana-Champaign, Urbana, IL

Jae Sung Park, Department of Mechanical Science and Engineering, University of Illinois at Urbana-Champaign, Urbana, IL

In the classical analysis of electrophoresis, particle motion is a consequence of the interfacial fluid slip that arises inside the ionic charge cloud or Debye screening layer surrounding the particle surface when an external field is applied. Under the assumptions of thin Debye layers, weak applied fields, and zero polarizability, it can be shown that the electrophoretic velocity of a collection of particles with identical zeta potential is the same as that of an isolated particle, unchanged by interactions. When some of these assumptions are relaxed, nonlinear effects may also arise and result in relative motions. First, the perturbation of the external field around the particles creates field gradients, which may result in nonzero dielectrophoretic forces (DEP) due to Maxwell stresses in the fluid. In addition, if the particles are able to polarize, they can acquire a nonuniform surface charge, and the action of the field on the dipolar charge clouds surrounding them drives disturbance flows in the fluid, causing relative motions by induced-charge electrophoresis (ICEP). These two nonlinear effects are first analyzed in detail in the prototypical case of two equal-sized ideally polarizable spheres carrying no net charge, using accurate boundary-element simulations, along with asymptotic calculations by the method of twin multipole expansions and the method of reflections. Based on this two-particle model, we find that both types of interactions result in significant relative motions and can be either attractive or repulsive depending on the configuration of the spheres. Numerical simulations of a full-scale suspension undergoing DEP and ICEP with periodic boundary conditions are also performed using an efficient smooth particle-mesh Ewald algorithm, and results are reported on the suspension microstructure, velocity statistics and particle hydrodynamic diffusion.

See more of this Session: Particulate and Multiphase Flow II
See more of this Group/Topical: Engineering Sciences and Fundamentals

Self-Consistent Fluctuating Hydrodynamics Simulations of Thermal Transport in Nanoparticle Suspensions

Rui Qiao, Mechanical Engineering, Clemson University, Clemson, SC

Ping He, Mechanical Engineering, Clemson University, Clemson, SC

We present the mesoscopic simulation of heat conduction in nanoparticle suspensions (nanofluids) by using the energy-conserving dissipative particle dynamics (eDPD) method. Through coarse graining, our simulations probe the thermal and momentum transport in nanofluids at a length scale much greater than that in atomistic methods. We show that our simulations model the fluctuating hydrodynamics in nanofluids in a thermodynamically self-consistent manner, which is critical for resolving the current controversies on mechanisms of heat conduction in nanofluids. Simulation results indicate that the Brownian motion of nanoparticles plays a negligible role in determining the thermal conductivity of nanofluids at least within the framework of fluctuating hydrodynamics at mesoscales.

See more of this Session: Particulate and Multiphase Flow II
See more of this Group/Topical: Engineering Sciences and Fundamentals

DRAG FORCE FORMULATION IN MACROSCOPIC PARTICLE MODEL AND ITS VALIDATION

Madhusuden Agrawal[*], *Ansys Inc., Houston, TX 77084, USA*
Shinichi Ookawara, Tokyo Institute of Technology, Tokyo 152-8552, Japan
Kohei Ogawa, Tokyo Institute of Technology, Tokyo 152-8552, Japan

Introduction

Particulate flows in which particle size is comparable to channel dimensions have gained considerable importance with growing interest in biomedical and micro-chemical technologies (Hessel *et al.*, 2005). In most particles tracking methods implemented in commercial CFD codes, particle drag and lift forces are evaluated through specified models based on the particle and flow properties at the point. Most of these models for drag and lift forces are usually derived with the assumption that a single particle moves in infinitely quiescent media or infinitely homogeneous sheared flow. When the particle size is comparable to a channel or particle exists near wall region in a device, the accuracy of these models cannot be assured since these assumptions are not valid, viz., the particle volume cannot be treated as a point compared with the scale of velocity profile. For these scenarios, it is necessary to take into account particle physical volume to model for both hydrodynamic and wall effects in such narrow spaces. Although direct numerical simulation associated with dynamic meshing around a moving particle is expected to be the most accurate method for this purpose, the computing cost is not practical. In our previous study (Ookawara *et al.*, 2005), a novel approach Macro Particle Model (MPM) proposed by Agrawal *et al.* (2004) was validated in terms of a falling velocity of sphere in a quiescent Newtonian liquid in a cylindrical pipe. This model was subsequently applied to simulate particle behavior in microseparator/classifier developed by Ookawara *et al.* (2004), in which wall effects on particle drag and lift would be important factors (Ookawara *et al.*, 2007). The model prediction was found to well correspond to the experimental visualization (Oozeki *et al.*, 2008).

Although there are many different, corrected or extended models for particle drag and lift which take into account the length from wall, shear rate range, particle Reynolds number, volume fraction etc, it is impossible to find a CFD code which is equipped with all models ever proposed. In the absence of any unified model valid for all the conditions and regions, the direct numerical simulation of coupled fluid and particle motion is a potential option to resolve the situation. But unfortunately, these direct numerical simulations can be very computationally intensive, which are impractical to use in most real applications. It is highly expected that the MPM will be a useful tool instead of real direct numerical simulation in situations where validated models for drag and lift do not exist or suitable models cannot be determined.

It is well established that the terminal velocity of a sphere, falling in a quiescent Newtonian liquid in a cylindrical tube, is considerably reduced by the presence of a nearby wall compared with that of a sphere falling in infinite media and that the particle velocity profile depends only on the diameter ratio of sphere to tube in the viscous regime. Therefore, this problem is selected for a quantitative validation of the MPM in this study. In MPM, solid body velocity, which describes the particle motion, is fixed in computational cells that are contained

*Corresponding Author: madhusuden.agrawal@ansys.com

within particle's physical volume which effectively transfer momentum from particle to fluid cells. On the other hand, drag and lift forces as well as torques acting on the particle are evaluated based on local pressure field, flow pattern and shear stress distribution. Since the MPM is capable of predicting the interacting particle motion and fluid flow around it, it can be regarded an approximated or quasi- direct numerical simulation. The falling velocity predicted by the previous MPM was larger compared with the correlation because the drag force in the previous formulation largely corresponded to virtual mass force that was required to accelerate fluid surrounding the particle. In the present study, the pressure and viscous components of drag forces are additionally incorporated into the drag formulation in MPM based on pressure and shear stress distributions around the particle. In this paper, the falling velocity predicted by the improved MPM is compared with the previous predictions as well as with the correlation. In the MPM approach, the mesh density, especially number of fluid cells in the diameter of particle, is an important variable for the accuracy of the drag prediction. A mesh sensitivity analysis has also been performed in this study.

Theoretical Background

In Stokes regime, a terminal velocity, V_∞, and relaxation time, τ_∞, of spherical particle (whose density and diameter are ρ_p and d respectively) falling in an infinite Newtonian fluid (whose density and viscosity are ρ_f and μ) is described as:

$$V_\infty = \frac{gd^2(\rho_p - \rho_f)}{18\mu} \quad \text{and} \quad \tau_\infty = \frac{(\rho_p + \rho_f/2)d^2}{18\mu} \qquad \ldots (1)$$

The terminal velocity (V) of a sphere falling in a quiescent Newtonian liquid is considerably reduced by a wall effect compared with in infinite media which is expressed by a wall factor (f) as: $f = \frac{V}{V_\infty}$ ... (2)

Generally this wall factor (f) is a function of a particle Reynolds number and the sphere to tube diameter ratio ($\lambda = d / D$). However for very low (as well as for very high) Reynolds number flows, the wall factor is independent of Reynolds number and only a function of sphere to tube diameter ratio (λ).

The following theoretical equation by Haberman and Sayre (1958) is widely accepted to express the wall effect in the viscous regime:

$$f = \frac{1 - 2.105\lambda + 2.0865\lambda^3 - 1.7068\lambda^5 + 0.72603\lambda^6}{1 - 0.75857\lambda^5} \qquad \ldots (3)$$

The upper limit of viscous regime is dependent on λ as listed in Table 1 (Chhabra *et al.*, 2003).

Numerical Methods

In the MPM approach, particle is treated in a Lagrangian frame of reference. The particle is assumed to span several computational cells. Cells that contain at least one node within the region occupied by the particle are considered as being "touched" by the particle. At every time step, a solid body velocity that describes the particle motion is fixed in these cells as shown in Figure 1. The particles have six degrees of freedom so both translational and rotational velocity components of particle motion are taken into account, while fixing the velocities in fluid cells. For a touched fluid cell at time t, the fixing of particle velocity is performed as:

$$V_{\text{f, after fixing}}(t) = \alpha_p(t)V_p(t) + \{1-\alpha_p(t)\}V_{\text{f, before fixing}}(t) \qquad \ldots(4)$$

where $\alpha_p(t)$ is particle volume fraction in the cell at time, t. By fixing the rigid body motion of the particle to fluid cells, momentum is effectively transferred from the particle to fluid.

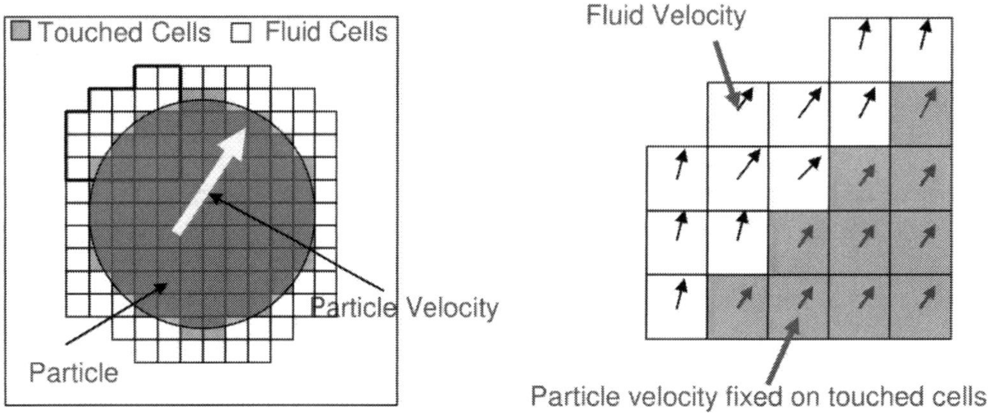

Figure 1. Fluid cells touched by particle and fixing of particle velocity in touched cells.

Drag and lift forces as well as torques on the particle are explicitly calculated based on velocity, pressure and shear stress distribution in fluid cells in and around the particle. Pressure, viscous as well as virtual mass components of drag forces are included in particle forces/torque calculations. New velocities (linear as well angular) and positions of the particles are obtained at each time step based on these forces and torques.

The integral of the momentum change, linear as well as angular, gives the viscous mass component of drag force and torque experienced by each particle, which is expressed as follows (in the *i* direction):

$$R_{\text{m},i} = \left(\sum_{\text{cells}} m_f V_{f,i} - \sum_{\text{cells}} m_f V_{p,i}\right)(1/\Delta t) \qquad \ldots(5)$$

where m_f, $V_{f,i,}$, $V_{p,i}$ and Δt are fluid mass in a cell, velocity of fluid and particle in *i* direction and time step, respectively. Integration in the above equation is performed on all cells touched by particle volume.

Similarly, the *i* component (in Cartesian coordinates) of pressure ($R_{p,i}$) and viscous ($R_{v,i}$) forces acting on particle surface is formulated as:

$$R_{p,i} = \sum_{\text{cells}}\left(Pd^2\,\mathbf{r}\cdot\mathbf{x}_i/|\mathbf{r}|\right) \ , \quad R_{v,i} = \sum_{\text{cells}}\sum_{j}\left\{\boldsymbol{\tau}_{ji}\, d^2\left(-\mathbf{r}\cdot\mathbf{x}_j/|\mathbf{r}\cdot\mathbf{x}_j|\right)\right\} \qquad \ldots(6)$$

In the left equation, P and d^2 is pressure and approximated area of particle surface in a fluid cell that is partially occupied by the particle. The vectors of $\mathbf{r}$ and $\mathbf{x}_i$ are a radius vector from the fluid cell center to particle center and a unit vector for Cartesian coordinates. In the right equation, $\boldsymbol{\tau}_{ji} = \mu\,(\partial u_i/\partial x_j)\,\mathbf{x}_i$ is shear stress in the positive *i* direction on a plane perpendicular to *j* direction and this is the force exerted by the fluid in the region of greater *j* coordinates on the fluid of lesser *j* coordinates. Integration, for pressure and viscous forces, is performed for all cells partially touched by particle volume, which could be good approximation to represent fluid cells close to particle surface.

In addition to correct drag/torque calculations and momentum transfer, MPM is a general purpose module to model for particulate flows in ANSYS/FLUENT. It has built-in algorithm to handle particle-particle and particle-wall collisions including friction dynamics. Table-4 summarizes some of these main features of MPM module. This study involves a single particle motion only so collision and other physics are not included in these MPM simulations.

Problem Setup

The properties of liquid and solid are determined in order to simulate a falling particle in an aqueous solution. The densities of resin sphere ρ_p and liquid ρ_f are kept constant as 1,695 kg/m^3 and 1,020 kg/m^3, respectively. Liquid viscosity is specified in such a way that terminal velocity, V_∞, and relaxation time, τ_∞, are constant regardless of particle diameter, d. This viscosity changes are done as follows:

$$\mu|_{\lambda=\lambda} = (10\lambda)^2 \mu|_{\lambda=0.1} \quad \ldots (7)$$

This viscosity definition will lead to: $\tau_\infty = 1\times10^{-3}$ s and $V_\infty = 3\times10^{-3}$ m/s ...(8)

Diameter D and height H of a tube are set to 0.02 m and 0.06 m, respectively. The given ratio of height to diameter, 3, is regarded as sufficient to achieve the insensitivity of results to the tube height (Higdon and Muldowney, 1995). The particle diameter is varied from 0.002 m to 0.018 m as given in Table 2.

Different mesh density was used to perform mesh sensitivity study. Number of fluid cells in the particle diameter was varied for a fixed tube to sphere diameter ratio (λ=0.5) for the mesh sensitivity study. Figure 2 shows perspective views of the tube geometries with the three mesh densities examined. Figure 3 shows the geometry of the tube and the particle in the middle of the tube.

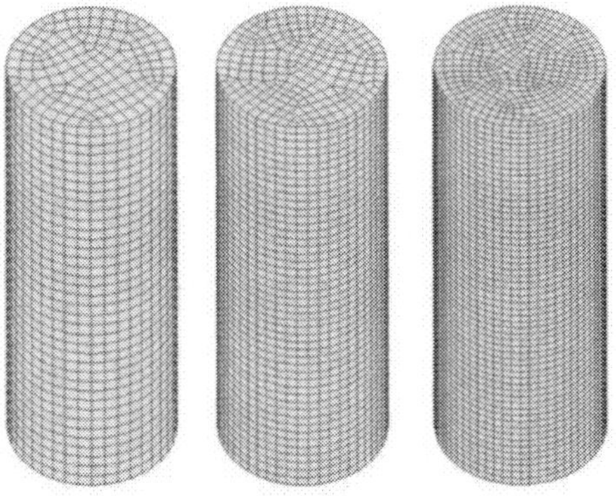

(a) Coarse (b) Medium (c) Fine

Figure 2. Tube geometry with examined mesh densities

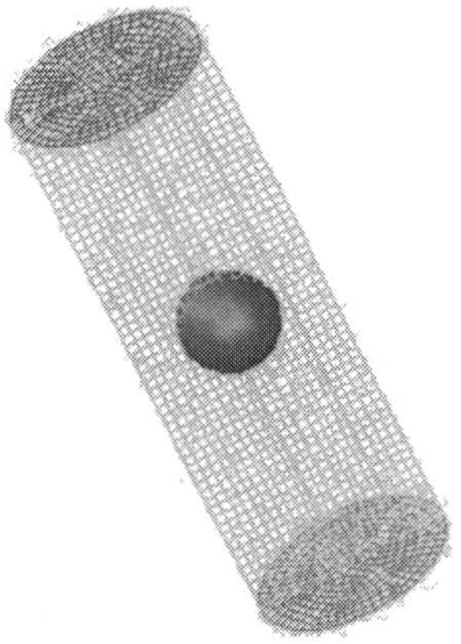

Figure 3. Geometry of the tube and initial particle location

For the mesh size sensitivity study, the number of cells in the particle diameter, total number of cells generated in the tube, and number of cells in the particle volume are shown in Table 3.

The commercial CFD code ANSYS/FLUENT 6.3 which is based on finite volume method, is employed for these simulations. The governing equations are conservation of mass and momentum. Higher order descretization schemes for the pressure and convection terms were used in this study for better accuracy. At time zero, the particle with velocity of zero is released exactly from the tube center, viz., the particle center is located on the tube axis and the distances from the particle center to both top and bottom walls are same initially. The gravity is specified as 9.8 m/s^2 along the tube axis. The no-slip condition is specified at all physical boundaries. The time step size is selected as to be two hundredth of the relaxation time in infinite medium given by Eq. (1). Computational requirements for these simulations are relatively inexpensive; CPU time varies from few minutes to few hours depending on total mesh count.

Results and Discussions

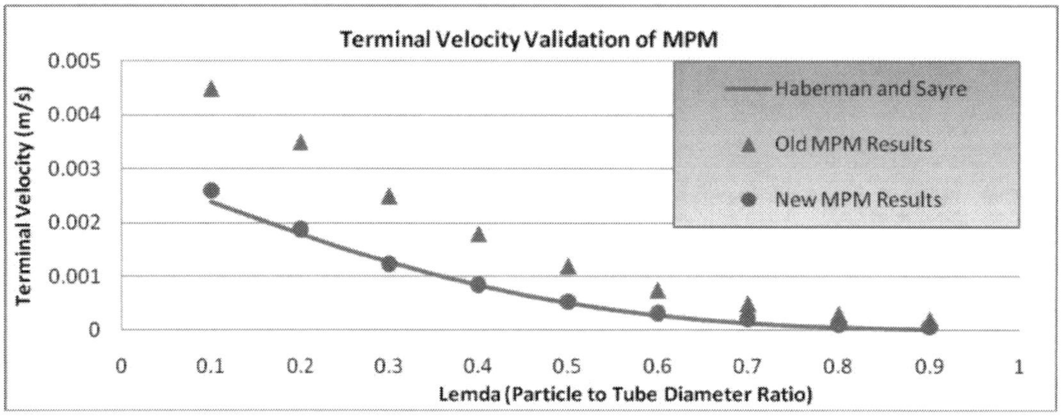

Figure – 4. Comparison of terminal velocity predicted by present and previous MPM with the Haberman and Sayre correlation

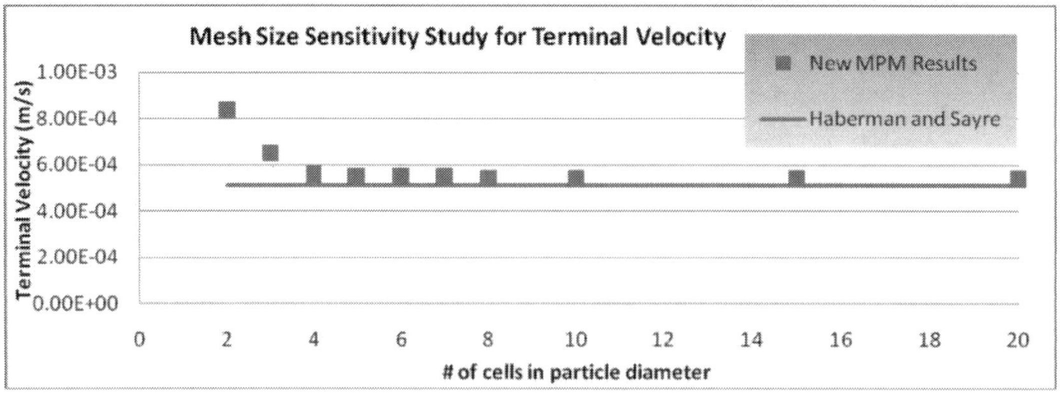

Figure – 5. Mesh size sensitivity study for terminal velocity for λ=0.5

As seen in Figure 4, the present MPM can predict the terminal velocity in the tube that is in better agreement with the correlation compared to the previous MPM for the entire range of particle diameter to tube diameter (λ). With the new drag formulation in MPM, the prediction of terminal velocity for entire range of λ agrees very well with the published correlation of Haberman and Sayre. The comparison suggests that the drag expressed by Eq. (5) largely corresponds to the virtual mass force and hence the drag magnitude is not sufficient especially in the small λ range leading to the larger terminal velocity by the previous MPM drag formulation. By adding the pressure and viscous components of the drag force as expressed by Eq. (6), the total drag magnitude becomes appropriate and hence the terminal velocity is properly predicted in the whole λ range by the present MPM. In this study, the mesh size is adjusted in such a way that there are always 5 cells in particle diameter for all λ values.

It should be noted that in the MPM formulation, the velocity reduced due to wall effects can be predicted without any model or correction in drag evaluation. The particle existence and motion causes the flow field around the particle while the particle motion is determined by the fluid forces caused by the flow field. The fully coupled particle-fluid simulation can be realized with a sufficient accuracy over rather coarse mesh compared to particle size. As seen in Figure 5, terminal velocity prediction is fairly independent of mesh resolution once we have mesh size which has about 4-5 cells in the diameter of the particle.

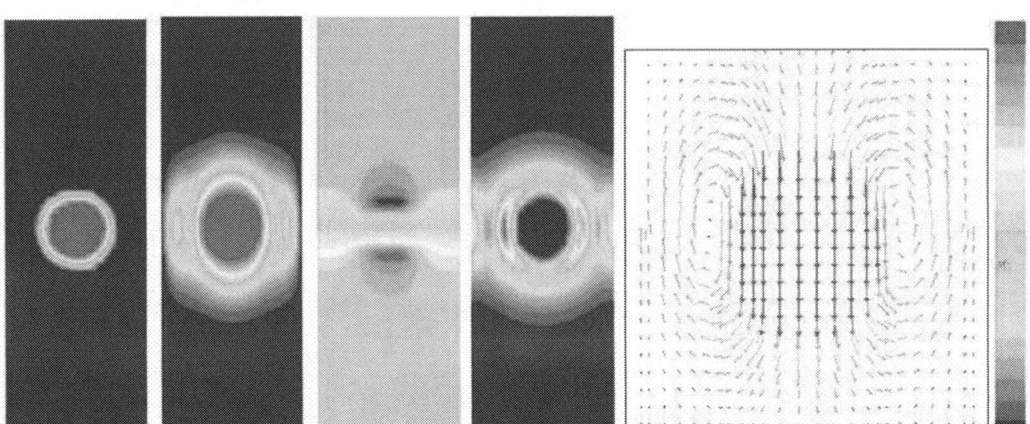

(a) (b) (c) (d) (e) (f)

Figure – 6. Simulation results at for $\lambda = 0.5$ with 10 cells per diameter mesh. (a) Particle volume fraction in range 0 – 1 (b) Velocity magnitude in range 0 – 5.4 e-4 m/s (c) Static Pressure in range -25 – 65 Pa (d) Strain Rate in range 0 – 0.4 /s (e) Velocity vectors in range 0 – 5.4 e-4 m/s (f) color map – increasing in value from blue to red

Figure 6 shows various colored plots of CFD results for λ =0.5 case. The range of each contour map is indicated in the figure caption and it is divided into 20 sub ranges (which increases linearly from blue to red color) whose common color map is shown in figure 6f. Contours of particle volume fraction are shown in figure 6a; region of red indicates instantaneous particle location (or particle volume). Pressure contours plot in figure 6c shows high and low pressure zones are formed adjacent to the lower and upper parts of the particle, respectively. It can be seen in figure 6b that flow field is induced around the particle and the magnitude decays rapidly in the axis direction. Strain rate contours in figure 6d, indicates high shear in partially occupied cells around particle volume, which could be regarded as a good representation of particle surface in MPM. Figure 6e shows the velocity vectors in the vicinity of particle region. This velocity vectors plot clearly shows a region where fluid velocity vectors have the same magnitude and direction as the rigid body motion of the particle. Upward flow seen at both sides of the particle generates circulating flow field around the particle associated with downward flow below and beyond the particle. It can be said that the MPM reasonably predicts flow around the particle induced by the particle motion itself.

Conclusion

It is pointed out that particulate flow in microdevice for particle separation/classification can be characterized by a steep shear rate in the liquid phase when the particle size is comparable to that of the device. The infinite homogeneity around particle in the liquid phase and the moderate flow conditions, which is commonly assumed for derivation of drag and lifts force models, is no longer valid in such microdevices. Since the present study on a falling ball in a tube shows that the MPM is capable of predicting the interacting particle motion and fluid flow around it over a relatively coarse mesh, the MPM can be regarded as implementing a quasi-direct numerical simulation of particulate flow with a reasonable computational cost. Therefore, it is highly expected the MPM will be a useful tool instead of real direct numerical simulation in the situations where validated models for drag and lift do not exist or suitable models cannot be determined. This new MPM model incorporates explicit formulation of viscous and pressure drag components. Validation of this new MPM with an established problem, viz., reduced terminal velocity of a sphere in a cylindrical tube in viscous regime, suggests that it can accurately predict the falling velocity decreasing with the diameter ratio of particle to tube. The results of mesh sensitivity study demonstrate that MPM approach can predict good results in relatively coarse meshes. The macroscopic particle model (MPM) is an innovative approach to model big spherical particles for situations where traditional discrete phase models are not applicable. MPM model is relatively inexpensive computationally compared to direct numerical simulations.

Although this model is implemented in ANSYS/FLUENT 6.3 using user-defined functions and a customized graphical user interface, the approach itself is not limited to FLUENT and is usable in other CFD software also. It will be a further study to examine the model applicability to various situations, to which the ordinary particle tracking method cannot be suited. The particle motion in non-Newtonian fluids, in which the apparent viscosity is much dependent on shear rate and hence it could vary even over the particle surface, is a practical and interesting application.

References

1. Agrawal M., Bakker A., Prinkey M.T., "Tracking Big Particles", *Fluent Newsletter, Fall 2003, page 11.*
2. Agrawal, M., A. Bakker and M.T. Prinkey (2004). "Macroscopic particle model - Tracking big particle in CFD". *The Proceeding of AIChE 2004 Annual Meeting, Austin, USA. 268b.*

3. Chhabra, R. P., S. Agarwal and K. Chaudhary (2003). "A note on wall effect on the terminal falling velocity of a sphere in quiescent Newtonian media in cylindrical tubes". *Powder Tech., 129, 53-58.*
4. Haberman, W. L., R.M. Sayre (1958). "David Taylor Model Basin Report No. 1143". *Dept. of Navy, Washington, DC.*
5. Hessel V., H. Löwe, A. Müller and G. Kolb (2005). "Chemical Micro Process Engineering-Processing and Plants". *Wiley-VCH, Weinheim.*
6. Ookawara, S., R. Higashi, K. Ogawa and D. Street (2004). "Feasibility study on concentration of slurry and classification of contained particles by microchannel". *Chem. Eng. J., 101, 171-178.*
7. Ookawara, S., M. Agrawal, D. Street and K. Ogawa (2005). "Modeling the motion of a sphere falling in a quiescent Newtonian liquid in a cylindrical tube by using the macroscopic particle model". *7th World Congress of Chemical Engineering. Glasgow, Scotland. C39-004.*
8. Ookawara S., M. Agrawal, D. Street and K. Ogawa (2007). "Quasi-direct numerical simulation of lift force-induced particle separation in a curved microchannel by use of a macroscopic particle model". *Chem. Eng. Sci., 62, 2454-2465.*
9. Oozeki, N., S. Ookawara, K. Ogawa, P. Löb and V. Hessel (2008). "Characterization of microseparator/classifier with a simple arc microchannel". *AIChE J., published online: DOI: 10.1002/aic.11650.*
10. Madhusuden Agrawal and Liz Marshall, 2006. "Using a Macroscopic Particle Model for Dilute Solids Suspensions". *AIChE 2006 annual meeting.*
11. Saffman, P. G., 1965, "The lift on a small sphere in a slow shear flow", *J Fluid Mech 22:385-400.*
12. Shelby, J. P., Lim, D.S.W., Kuo, J.S., and Chiu, D.T., 2003, "High radial acceleration in microvortices", *Nature, 425: 38.*
13. Higdon, J. J. L. and Muldowney, G. P., 1995, "Residence functions for spherical particles, droplets and bubbles in cylindrical tubes", *J Fluid Mech, 298:193-210.*

APPENDIX

Table 1. Wall factor f (Haberman and Sayre, 1958) and upper limit Reynolds number of viscous regime (Chhabra *et al.*, 2003).

λ	0.1	0.2	0.3	0.4	0.5	0.6	0.7	0.8	0.9
f	0.79158	0.59534	0.42200	0.27920	0.17036	0.09441	0.04668	0.02040	0.00825
Re_{max}	0.027	0.040	0.050	0.083	0.18	0.52	2.10	8.40	25.17

Table 2. Examined parameters and expected values based on equations 1-4

λ	0.1	0.2	0.3	0.4	0.5	0.6	0.7	0.8	0.9
d [m]	0.002	0.004	0.006	0.008	0.010	0.012	0.014	0.016	0.018
μ [Pa s]	0.49	1.96	4.41	7.84	12.25	17.64	24.01	31.36	39.69
fV_∞ [m/s]	2.375e-3	1.786e-3	1.266e-3	8.376e-4	5.111e-4	2.832e-4	1.400e-4	6.120e-5	2.474e-5
Re	9.886e-3	3.718e-3	1.757e-3	8.718e-4	4.255e-4	1.965e-4	8.328e-5	3.185e-5	1.144e-5
Mesh Count	332500	44775	12750	5814	2670	1600	1050	760	663
#cells/Diameter	5	5	5	5	5	5	5	5	5

Table 3. Mesh properties for mesh sensitivity study

# Cells/Diameter	*2*	*3*	*4*	*5*	*6*	*7*	*8*	*10*	*15*	*20*
Total Mesh Count	300	720	1536	2670	4932	7812	10464	22800	72900	166680
# Cells/particle volume	8.33	20.00	42.67	74.17	137.00	217.00	290.67	633.33	2025.00	4630.00

Table 4. Main features of the Macroscopic Particle Model

Accurate and Fully Coupled Particle-Fluid Interaction	Flow blockage and momentum exchange to fluid due to particles and explicitly calculates particle forces/torques based on local flow pattern, pressure field and shear stresses distribution
Particle rotation and spin	The rotational/spin effect of particle is properly included in drag/torque calculation, collision/friction calculation and linear/angular momentum transfer
Particle-Particle and Particle-Wall collisions	Hard sphere collision algorithm is implemented in MPM for particle-particle and particle-wall collisions where collisions are assumed binary and quasi-instantaneous and particle contact occurs at a point. Motion of wall is taken into account during particle-wall collision calculations
Particle-Particle and Particle-Wall Field forces	In many particle-particle systems, field forces besides gravity are important. For example, particles might agglomerate because of electrostatic or magnetic forces. Such forces are implemented in MPM using a potential force model
Continuous particle injection	Ability to define start time, stop time and interval time for each injection. The MPM will inject new particles at regular intervals from the start time to the stop time.
Particle injection in a cylindrical coordinate	Ability to define initial particle position and velocity in a cylindrical coordinate system.
Random particle injection	Option to inject particles from random locations between specified domain
Moving/rotating injection location	Option to define rotational and/or translational velocity for injection location. MPM calculates a new injection location based on these velocities and releases particles from new positions.
Particle deposition and buildup on defined surfaces	A comprehensive model to capture particle deposit and buildup physics. if any particle collides with predefined surfaces (or with any previously deposited particles) it will deposit there
Particle interaction with continuous phase	Option to have either one way or two ways coupling of particle motion with fluid flow.
Compatibility with Multiphase Models	Options to choose primary phase, secondary phase or mixture phase velocity calculation of drag force on the particle, if the carrying fluid is multiphase in nature
Random Particle Packing with a Size Distribution	MPM can randomly pack particles in a domain (box or cylinder) based on particle size distribution (particle size vs volume fraction data)
Cemented Particles	MPM has the option to freeze selected particles, these particles don't move but affects fluid flow
Brownian Diffusion for sub-micron size particles	MPM has the option to include Brownian diffusion forces, where the components of the Brownian force are modeled as a Gaussian white noise process
Options for other drag forces	For smaller particles, MPM allows use of Morsi-Alekxander based drag calculation.

A Kinetic Model for a Non-Isothermal Granular Gas with Bi-Disperse Particles

Alberto Passalacqua, Department of Chemical and Biological Engineering, Iowa State University, Ames, IA

Prakash Vedula, School of Aerospace and Mechanical Engineering, University of Oklahoma, Norman, OK

Christine M. Hrenya, Department of Chemical and Biological Engineering, University of Colorado, Boulder, CO

R. O. Fox, Department of Chemical & Biological Engineering, Iowa State University, Ames, IA

A seventh-order quadrature-based moment method for the solution of the complete Boltzmann equation (Fox, 2008), with the complete Boltzmann collision integral, is applied to the simulation of a bi-disperse granular mixture.

Two different test cases are considered: an isolated granular mixture of spherical, inelastic spheres, which cools down due to collisions, and a binary granular mixture between two Maxwellian stationary walls at different temperatures.

In the homogeneous case, particles accommodate an homogeneous cooling state, where the temperature ratio of the two species becomes constant, with different granular temperatures for each species. The temperature ratio is studied as a function of mass ratio, size ration, density, and mixture composition. Particles with restitution coefficient equal to 0.95 and 0.8 are considered, and results are compared with molecular dynamics predictions (Dahl et al., 2002).

A mixture characterized by a volume fraction ratio equal to 4, a particle diameter and density ratios of 2, with restitution coefficient equal to 0.9 is considered for the non-homogeneous simulation of a binary mixture between two non-isothermal walls, with a temperature ratio between the walls equal to 10 and a wall distance to particle diameter ratio equal to 15.4. Results for the species granular temperature profiles and for the concentrations are provided and compared with molecular dynamics data of Galvin et al. (2005).

In both applications, the quadrature-based method provides results for the available moments (e.g. number density, mean velocity, stress tensor, energy flux, etc.) in quantitative agreement with the molecular dynamics data.

References

Dahl, S. R., C. M. Hrenya, V. Garzo, and J. W. Dufty. 2002. Kinetic temperatures for a granular mixture. Physical Review E 66, no. 4, 041301.

Fox, R.O. 2008. Higher-order quadrature-based moment methods for kinetic equations. Journal of Computational Physics (submitted).

Galvin, J. E., S. R. Dahl, e C. M. Hrenya. 2005. On the role of non-equipartition in the dynamics of rapidly flowing granular mixtures. Journal of Fluid Mechanics 528, 207-232.

See more of this Session: Particulate and Multiphase Flow II
See more of this Group/Topical: Engineering Sciences and Fundamentals

Relationship between Structure and Transfer Coefficients in Fast Fluidized Beds ---- a Theoretical Analysis

Hongzhong Li*, Baolin Hou

State Key Laboratory of Multi-phase Complex systems
Institute of Process Engineering, Chinese Academy of Sciences

Extended Abstract

It is common knowledge that in the fast fluidized bed there are non-uniformity of radial and axial solids distributions, i.e., axially two regions in an s-shaped voidage profile, and radially a solids concentration varying from dilute at the center to dense near the wall in macroscopic view; meanwhile there are non-uniformity of dispersed phase and cluster phase distributions in microscopic view. In the dispersed phase, the particles are present essentially individually, while in the cluster phase the particles agglomerate with one another and are enmeshed in the dispersed phase. The clusters behave in the way of sometimes aggregating and sometimes dispersing duo to interactions between both fluid-particle and particle-particle, and their shape and size are different and change frequently. Furthermore the distributions of velocities of the gas and particles are also heterogeneous. For particles the velocity is higher in the center of the bed and lower near the wall, and higher in the dispersed phase and lower in the cluser phase; while for gas the velocity is higher in the center of the bed and lower near the wall, and higher in the dispersed phase and lower in the cluster phase. It can conclude that in fast fluidized bed there exist heterogeneous structures in both macro-scale and micro-scale.

Both experimental results and theoretical analysis have shown that there are significant effects of the bed structures on the momentum transfer behavior. For example, under same operating conditions of gas velocity, solids flow rate, and averaged voidage, the drag coefficient of interaction between gas and particle is 18.6 for the homogeneously distributed particles, and while the the drag coefficient of interaction between gas and particle is 2.85 for the heterogeneously distributed particles. It can be imagined that there also are significant effects of the bed structures on the mass and heat transfer behaviors. In the present work a theoretical analysis on the relationships between local bed structure and momentum, mass, and heat transfer coefficients, is given respectively as follows.

(1) Quantitative description of local structure and gas-particles interaction in fast fluidized bed

In general, for describing the local bed structure of the fast fluidized bed quantitatively, 8 parameters are needed. They are voidages in dispersed phase and in cluster phase ε_d, and ε_c respectively, superficial gas velocities in dispersed phase and in cluster phase U_{fd}, and U_{fc} respectively, superficial particle velocities in dispersed phase and in cluster phase U_{pd}, and

* Corresponding author. Tel.: 86-10-62556951; Fax: 86-10-62536108.
E-mail address: hzli@home.ipe.ac.cn (H. Li)

U_{pc} respectively, diameter of cluster d_c, volume friction of cluster phase f. The 8 parameters of local bed structure can be predicted by using the Energy-Minimization Multi-Scale (EMMS) [†]model when the superficial gas velocity U_f, superficial particle velocity U_p, gas density ρ_f, gas viscosity μ_f, particle density ρ_p, and particle diameter d_p are given , which was proposed by Li et al.(1994).

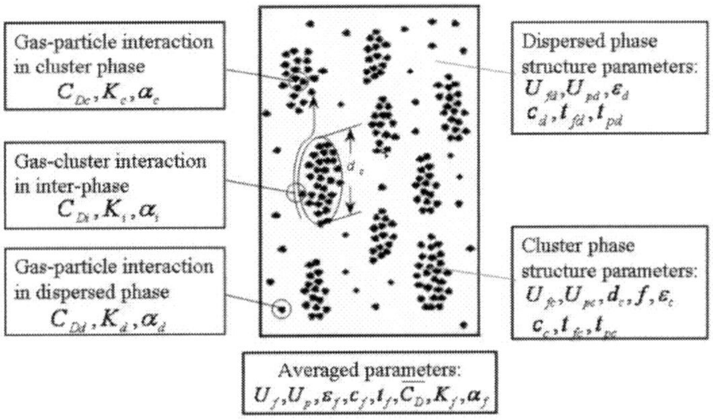

Figure 1. Resolution of structure and gas-particles interaction in fast fluidized bed

In this work, the equation for the drag force of flowing gas in cluster on particles is revised as:

$$\left(1-2\frac{d_p}{d_c}\right)\frac{f(1-\varepsilon_c)}{\frac{\pi}{6}d_p^{\;3}}C_{dc}\frac{1}{2}\rho_f U_{sc}^{\;2}\frac{\pi}{4}d_p^{\;2}=f(\rho_p-\rho_f)g(1-\varepsilon_f)$$

Because the particles on the surface of cluster contact with flowing gas in dispersed phase, these particles are different from the particles inside cluster. In this case, an equivalent number particle, whose surface is equal to the contact area between cluster surface particles and flowing gas in dispersed phase, should be subtracted from the particles of cluster.

In addition, for mass transfer and heat transfer the additional parameters are needed. They are the component concentrations of gas in dispersed phase and in cluster phase c_d, and c_c, the temperatures of gas in dispersed phase and in cluster phase t_{fd}, and t_{fc}, the temperatures of particles in dispersed phase and in cluster phase t_{pd} ,and t_{pc}. They can been obtained by using the mass conservation equation and heat conservation equation for dispersed phase and cluster phase respectively. The parameters for gas-particles interaction i.e., drag coefficient C_D, mass transfer coefficient K_f, and heat transfer coefficient α_f, are the functions of structure parameters. The resolution of structure and gas-particles interaction in fast fluidized bed is shown in Figure 1.

(2) Momentum transfer coefficient---Drag coefficient C_D

For homogeneous local structure or averaged method (Ergun, 1952; Wen and Yu, 1966),

$$C_D = C_{D0}\varepsilon_f^{-4.7} \qquad \text{When } \varepsilon_f > 0.8$$

$$\text{Where } C_{D0} = 0.44 \qquad \text{When } \mathrm{Re}_p > 1000$$

$$C_{D0} = \frac{24}{\mathrm{Re}_p}(1 + 0.15\mathrm{Re}_p^{0.687}) \qquad \text{When } \mathrm{Re}_p < 1000$$

$$C_D = 200\frac{(1-\varepsilon_f)\mu f}{\varepsilon_d^3 \rho_f d_p U_s} + \frac{7}{3\varepsilon_f^3} \qquad \text{When } \varepsilon_f < 0.8$$

For heterogeneous local structure

$$C_D = \frac{f(1-\varepsilon_d)(1-\frac{d_p}{d_{cl}})C_{Dd}U_{sd}^2 + (1-f)(1-\varepsilon_l)C_{Dl}U_{sl}^2 + f(\frac{d_p}{d_{cl}})C_{Di}U_{si}^2}{(1-\varepsilon_f)U_s^2}$$

Where

Drag coefficient for dispersed phase $C_{Dl} = C_{D0}\varepsilon_l^{-4.7}$

Drag coefficient for cluster phase $C_{Dd} = 200\dfrac{(1-\varepsilon_d)\mu f}{\varepsilon_d^3 \rho_f d_p U_{sd}} + \dfrac{7}{3\varepsilon_d^3}$

Drag coefficient for inter-phase $C_{Di} = C_{D0}\varepsilon_l^{-4.7}(1-f)^{-4.7}$ When $\varepsilon_i = \varepsilon_l(1-f) > 0.8$

$$\text{Or } C_{Di} = 200\frac{(1-\varepsilon_i)\mu_f}{\varepsilon_i^3 \rho_f d_{cl} U_{si}} + \frac{7}{3\varepsilon_i^3} \qquad \text{When } \varepsilon_i = \varepsilon_l(1-f) < 0.8$$

(3) Mass transfer coefficient K_f

For homogeneous local structure or averaged method (Jung and La Nauze, 1983),

$$K_f = Sh\frac{D}{d_p} = 2\varepsilon_f\frac{D}{d_p} + 0.69\frac{D}{d_p}\left(\frac{U_s d_p \rho_f}{\varepsilon_f \mu_f}\right)^{\frac{1}{2}}\left(\frac{\mu_f}{\rho_f D}\right)^{\frac{1}{3}}$$

For heterogeneous local structure,

$$K_f =$$

$$\frac{K_l a_p(1-\varepsilon_l)(1-f)(c_s - c_l) + K_d(a_p - a_{cl})(1-\varepsilon_d)f(c_s - c_d) + K_i a_{cl}(1-\varepsilon_d)f(c_s - c_l) + m_a}{a_p(1-\varepsilon_f)(c_s - c_f)}$$

Where

Mass transfer coefficient for dispersed phase

$$K_l = 2\varepsilon_l\frac{D}{d_p} + 0.69\frac{D}{d_p}\left(\frac{U_{sl} d_p \rho_f}{\varepsilon_l \mu_f}\right)^{\frac{1}{2}}\left(\frac{\mu_f}{\rho_f D}\right)^{\frac{1}{3}}$$

Mass transfer coefficient for cluster phase

$$K_d = 2\varepsilon_d\frac{D}{d_p} + 0.69\frac{D}{d_p}\left(\frac{U_{sd} d_p \rho_f}{\varepsilon_d \mu_f}\right)^{\frac{1}{2}}\left(\frac{\mu_f}{\rho_f D}\right)^{\frac{1}{3}}$$

Mass transfer coefficient for inter-phase

$$K_i = 2\varepsilon_l(1-f)\frac{D}{d_{cl}} + 0.69\frac{D}{d_{cl}}\left(\frac{U_{si} d_{cl} \rho_f}{\varepsilon_l(1-f)\mu_f}\right)^{\frac{1}{2}}\left(\frac{\mu_f}{\rho_f D}\right)^{\frac{1}{3}}$$

(4) Heat transfer coefficient α_f

For homogeneous local structure or averaged method (Rowe, Clayton, and Lewis, 1965; Jung and La Nauze, 1983),

$$\alpha_f = 2\varepsilon_f\frac{\lambda}{d_p} + 0.69\frac{\lambda}{d_p}\left(\frac{U_s d_p \rho_f}{\varepsilon_f \mu_f}\right)^{\frac{1}{2}}\left(\frac{c_p \mu_f}{\lambda}\right)^{\frac{1}{3}}$$

For heterogeneous local structure,

$$\alpha_f = \frac{\alpha_l a_p(1-\varepsilon_l)(1-f)(t_{pl} - t_{fl}) + \alpha_d(a_p - a_{cl})(1-\varepsilon_d)f(t_{pd} - t_{fd}) + \alpha_i a_{cl}(1-\varepsilon_d)f(t_{pd} - t_{fl}) + h_{af}}{a_p(1-\varepsilon_f)(t_p - t_f)}$$

Where

Heat transfer coefficient for dispersed phase

$$\alpha_l = 2\varepsilon_l \frac{\lambda}{d_p} + 0.69\frac{\lambda}{d_p}(\frac{U_{sl} d_p \rho_f}{\varepsilon_l \mu_f})^{\frac{1}{2}}(\frac{c_p \mu_f}{\lambda})^{\frac{1}{3}}$$

Heat transfer coefficient for cluster phase

$$\alpha_d = 2\varepsilon_d \frac{\lambda}{d_p} + 0.69\frac{\lambda}{d_p}(\frac{U_{sd} d_p \rho_f}{\varepsilon_d \mu_f})^{\frac{1}{2}}(\frac{c_p \mu_f}{\lambda})^{\frac{1}{3}}$$

Heat transfer coefficient for inter-phase

$$\alpha_i = 2\varepsilon_l (1-f)\frac{\lambda}{d_p} + 0.69\frac{\lambda}{d_{cl}}(\frac{U_{si} d_{cl} \rho_f}{\varepsilon_l (1-f)\mu_f})^{\frac{1}{2}}(\frac{c_p \mu_f}{\lambda})^{\frac{1}{3}}$$

(5) Discussion: Effect of aggregating and dispersing of particles on mass and heat transfer

There are two cases for the cluster aggregating and dispersing. One case is the single particle is striped from the surface of the clusters, or is adhered to the surface of the clusters. During the process of stripping, the slip velocity between gas and single particle decreases due to terminal velocity of the cluster larger than that of single particle, which results in a decrease of mass and heat transfer coefficients. While during the process of adhering, the slip velocity between gas and single particle increases resulting in a increase of mass and heat transfer coefficients. At the stable condition the striped rate equals the adhered rate, and the negative and positive actions are counteracted. So that the effects of the process of stripping and adhering on the mass and heat transfer coefficients can be neglected.

Another case is the big cluster is broken into several small clusters, or several small clusters are aggregated into a big cluster. During the process of breaking, the slip velocity between gas and small cluster decreases due to terminal velocity of the big cluster larger than that of small cluster, which results in a decrease of mass and heat transfer coefficients. While during the process of aggregating, the slip velocity between gas and small cluster increases resulting in a increase of mass and heat transfer coefficients. At the stable condition the broken rate equals the aggregated rate, and the negative and positive actions are counteracted. So that the effects of the process of breaking and aggregating on the mass and heat transfer coefficients can also be neglected.

From the above discussion It can be concluded that maybe the added mass M_a and added heat quantity H_a in Equations, which represent the mass and heat quantities transferred from particles to gas due to aggregating and dispersing of the particles, can be neglected.

(6) Experimental verification

The momentum transfer model developed in this work is a slightly improved Yong's model. Yong's multi-scale structure model has been validated with experimental data. So it is not necessary to verify this improved model again. In order to test and validate the developed mass transfer theory in this work, we employ the experimental data reported by Subbarao(2002). Figures 2 shows comparison of experimental data with the theoretical predictions for the effect of solids circulation flux. Although there are certain errors between the experimental data and theoretical predictions, the errors are acceptable.

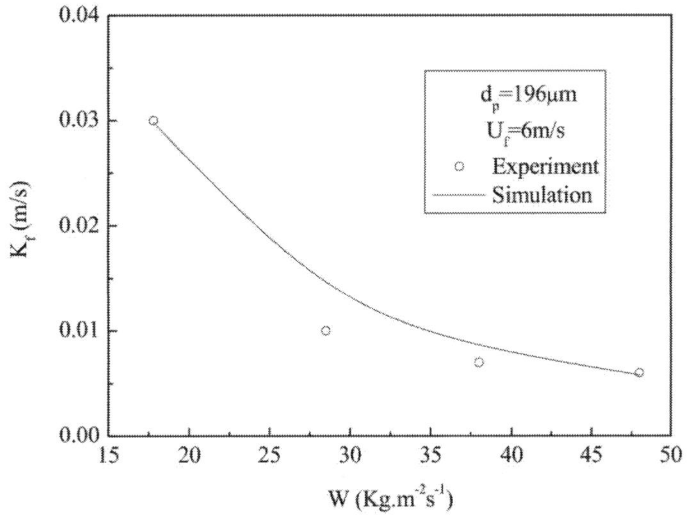

Fig. 2. Comparison of Subbarao's experimental data with theoretical model predictions

In order to test and validate the developed heat transfer theory in this work, we employ the experimental data reported by Watanabe (1991).

Fig.3 shows the temperature profiles in the riser. They are experimental data from Watanabe (1991) and simulation results from this work. From Fig.3 it can be found that with increase of the bed height the gas temperature is quickly increased and reachs to constant, but the solid temperature is quickly decreased and reaches the same temperature with the gas. By comparing the experimental data and the simulation results, we find that the difference between the simulation and the experiment data for the gas temperature is bigger than that for solid temperature. This may be attributed to no considering the thermal conduction between gas and wall in the simulation.

From Fig. 3 we can see that although there are certain errors in theoretical predictions, the errors are acceptable.

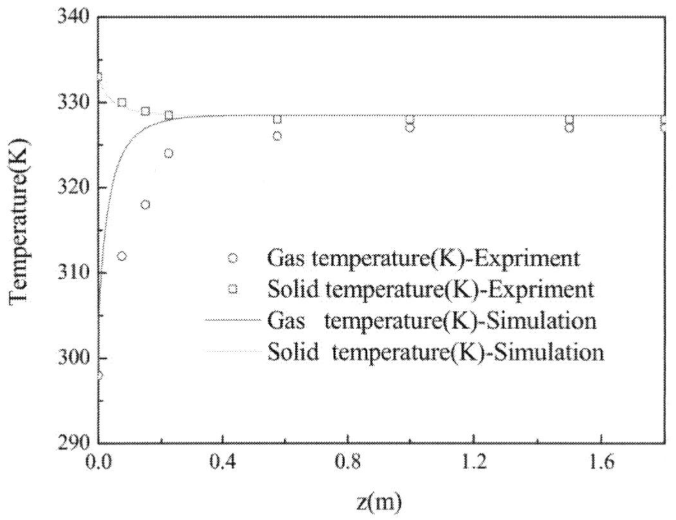

Fig. 3 Comparison of the experimental data and simulation results for the temperature profiles in the riser

In conclusions, there are significant effects of the heterogeneous gas-particles flow structure on the drag coefficient, the mass transfer coefficient and the heat transfer coefficient in fast fluidized beds. The correlation equations from traditional averaged methods for these coefficients are not suitable to the heterogeneous gas-particles flow structure in fast fluidized beds. In the present work a theoretical analysis on the relationships between local bed structure and momentum, mass, and heat transfer coefficients, is given. The effects of aggregating and dispersing of the particles on the mass and heat transfer are discussed. The theoretical prediction was compared with the experiment data in the literature, The theoretical values are in good agreement with the experimental data.

In order to verify the applicability of the present theoretical analysis, the experiments on hydrodynamic flow, mass transfer, and heat transfer are being conducted. It is hoped to publish the full results soon.

References

Ergun, S., 1952, Fluid Flow through Packed Columns. Chem. Eng. Prog. 48, 89-94.

Jung, K., La Nauze, R. D., 1983, Sherwood Numbers for Burning Particles in Fluidized Beds. in: Kunii, D., Cole, S. S. (Eds.), Fluidization IV. Engineering Foundation, New York. pp. 427-434.

Li, J., Kwauk, M., 1994, Particle-fluid Two-phase Flow: The Energy-Minimization Multi-Scale Method. Metallurgical Industry Press, Beijing. pp. 23-40.

Rowe, P. N., Clayton, K. T., Lewis, J. B., 1965, Heat and Mass Transfer from a Single Sphere in an Extensive Flowing Fluid. Trans. Inst. Chem. Engrs. 43, 14-31.

Subbarao, D., Gambhir, S., 2002, Gas to particle mass transfer in risers. In J. R. Grace, J. Zhu, H. de Lasa (Eds.) , Circuliating Fluidized Bed Technology VII, Canadian Society for Chemical Engineering, Ottawa, Canada, pp. 97-104.

Watanabe, T., Chen, Y., Hasatani, M., et al., 1991, Gas-solid heat transfer in fast fluidized bed. In P. Basu, M. Horio, M. Hasatani (Eds.), Circuliating Fluidized Bed Technology III, Pergamon Press, Oxford, pp.283-288.

Wen, C. Y., Yu, Y. H., 1966, Mechanics of Fluidization. Chem. Eng. Prog, Symp. Ser. 62 (62), 100-111.

Particle Adhesion in Microelectronics Processing

Caitlin Kilroy, Purdue University, West Lafayette, IN

Stephen P. Beaudoin, School of Chemical Engineering, Purdue University, West Lafayette, IN

As demand grows in the microelectronics industry for smaller, faster, and cheaper devices, optimization of the adhesion of particles to semiconductor surfaces during process steps such as lithography and chemical mechanical planarization (CMP) becomes increasingly important. Proper modeling of these interactions can result in the reduction of defects caused by abrasive damage and remnant particles. This study includes combining the effects of both van der Waals (vdW) and electrostatic double layer forces (EDL) in order to predict the adhesive forces for particles ranging from the micron-scale to the nano-scale. This requires a comprehensive characterization of the geometry and morphology of the particles and substrates to include roughness, which will then be simulated for a range of different orientations such as would be found in real-life particle-substrate interactions. An "effective" Hamaker constant is obtained by comparing modeled and experimental adhesion force data. In the case of interactions within electrolyte solutions such as CMP slurry, EDL force calculations accounting for the morphologies of rough particles and substrates are included in the determination of the overall force. By using these results, predictions can be made for manipulating the process conditions according to the desired adhesion of various particles throughout the fabrication of microelectronics devices.

See more of this Session: Semiconductor Surface Chemistry
See more of this Group/Topical: Engineering Sciences and Fundamentals

RHEED-TRAXS Development for Real Time, in-Situ Stoichiometry Analysis During Thin Film Deposition

Bing Sun, Chemical Engineering, Northeastern University, Boston, MA

Zhuhua Cai, Chemical Engineering, Northeastern University, Boston, MA

Martina List, Chemical Engineering, Northeastern University, Boston, MA

Katherine S. Ziemer, Chemical Engineering, Northeastern University, Boston, MA

Multifunctional heterostructures of functional oxides, such as ferroelectric barium titanate and piezoelectric lead zirconate titanate, integrated on semiconductor platforms are of interest to the development of next-generation smarter, smaller and more energy efficient devices. As these multi-element oxide materials exhibit measurably different functional properties with structure changes or stoichiometry changes of less than one percent[1], Molecular Beam Epitaxy (MBE) is used to take advantage of this sensitivity of functional properties by controlling the stoichiometry and structure of each layer. In most multi-element oxide material systems, the same crystal structure can represent a range of relative stoichiometries. Thus, an accurate real time stoichiometry control technique is needed. Reflection High Energy Electron Diffraction - Total Reflection Angle X-ray Spectroscopy (RHEED-TRAXS) has the potential to achieve real time stoichiometry control and explain the growth mechanism in the ultra-high vacuum (UHV) environment of MBE. When incident RHEED electrons with energy in the range of 12-20 keV hit the sample surface, characteristic x-rays fluorescence representative of the film surface stoichiometry is emitted. By measuring the x-ray at or close to its total reflection angle RHEED-TRAXS is reported to probe only the top 20Å of group V elements[2]. While the proof of concept for RHEED-TRAXS has been demonstrated through qualitative analysis of grown films [3] and recently in growing films [4], thorough quantification studies of RHEED-TRAXS systems are not seen in literature. It is unclear as to the relative impact of various MBE system parameters, including geometry, on x-ray signal and hence RHEED-TRAXS spectra.

This work demonstrates RHEED-TRAX as a real-time surface chemistry characterization tool in ultra-high and high vacuum processing environments. The impact of various factors in the MBE system and RHEED-TRAXS collection chamber such as geometry of the optical path, aperture size and emission current of the RHEED electron beam has been identified and evaluated. The Mg Kα line and Si Kα line signal intensity changes are followed using RHEED-TRAXS during the deposition process of magnesium oxide (MgO) films on silicon carbide (SiC) substrates. Both the Si signal attenuation and the Mg signal intensity time profile are used in conjunction with X-ray Photoelectron Spectroscopy (XPS) to study the growth process with varied growth conditions. XPS is also used as a reference to quantify the RHEED-TRAXS spectra. RHEED electron beam with 12.5keV kinetic energy and 7.0μA emission current, along with 30 seconds acquisition time are used to collect spectra for quantification during film deposition.

References: 1. Kanno, I., et al., Crystallographic characterization of epitaxial Pb(Zr,Ti)O3 films with different Zr/Ti ratio grown by radio-frequency-magnetron sputtering. Journal of Applied Physics, 2003. 93(7): p. 4091-4096.

2.Braun, W. and K.H. Ploog, Real-time surface composition and roughness analysis in MBE using RHEED-induced X-ray fluorescence. Journal of Crystal Growth, 2003. 251(1-4): p. 68-72.

3 Shigetomi, J., et al., Initial growth stage of indium arsenide/gallium arsenide studied by RHEED-TRAXS method. Journal of Crystal Growth, 1991. 111(1-4): p. 110-14.

4 Vanmil, B., Gallium nitride growth by RF-plasma assisted molecular beam epitaxy: Determination of

surface stoichiometry by RHEED-TRAXS, annealing of gallium nitride:beryllium and the effects of active nitrogen species, surface polarity, and excess gallium-overpressure on high temperature limits. 2005. p. 110 pp.

See more of this Session: Semiconductor Surface Chemistry
See more of this Group/Topical: Engineering Sciences and Fundamentals

Defect Engineering for Ultra-Shallow Junctions Using Solid-Solid Interfaces

Prashun Gorai, University of Illinois, Urbana-Champaign, Urbana, IL

Yevgeniy Kondratenko, University of Illinois, Urbana-Champaign, Urbana, IL

Edmund Seebauer, University of Illinois, Urbana-Champaign, Urbana, IL

The behavior of defects within silicon can be changed significantly by controlling the chemical state at nearby surfaces or solid interfaces. Experiments have shown that certain chemical treatments change the ability of a free surface to act as a "sink" for point defects such as interstitials. When the surface is made chemically active, this ability rises. The surface can then remove Si interstitials selectively over impurity interstitials, leading to less transient enhanced diffusion and better dopant activation during annealing after implantation. End-of-range damage can be removed more efficiently as well. Such behavior is kinetically quantified through an annihilation probability. Although annihilation probabilities for interstitials have been measured under various conditions for free surfaces, very little understanding exists for the corresponding quantity at solid-solid interfaces.

The present work seeks to fill that gap through measurements of annihilation probabilities at interfaces between silicon and several kinds of oxides and nitrides. The method employs implantation of isotopically labeled Si (mass 30) in Si host lattice, annealing, and simulation of the subsequently measured SIMS profiles. Marked differences are observed among the various interfaces.

See more of this Session: Semiconductor Surface Chemistry
See more of this Group/Topical: Engineering Sciences and Fundamentals

Chemical Mechanisms of Interface Degradation Between Barium Hexaferrite Thin Films On Wide Bandgap 6H-SiC

Zhuhua Cai, Chemical Engineering, Northeastern University, Boston, MA

Trevor L. Goodrich, Chemical Engineering, Northeastern University, Boston, MA

Zhaohui Chen, Department of Electrical and Computer Engineering, Northeastern University, Boston, MA

Bing Sun, Chemical Engineering, Northeastern University, Boston, MA

Vince Harris, Department of Electrical and Computer Engineering, Northeastern University, Boston, MA

Katherine S. Ziemer, Chemical Engineering, Northeastern University, Boston, MA

There is a growing interest for the integration of multifunctional ferrimagnetic-ferroelectric oxides on wide bandgap semiconductors, such as silicon carbide (6H-SiC), for next generation, tunable electronic and microwave devices. Ferrimagnetic barium hexaferrite (BaM) is ideal for microwave device applications because of its high permeability (> 100) and strong uniaxial magnetocrystalline anisotropy (17 kOe). Effective integration of device quality BaM on SiC has been demonstrated using a two-step process strategy with a molecular beam epitaxy (MBE) 10 nm MgO (111) single crystal template film and 100 nm BaM seed layer followed by pulsed laser deposition (PLD) 500 micrometer thick films of BaM. In optimizing the magnetic properties of the BaM film, variations in magnetic properties were linked to BaM/SiC interface degradation. X-ray photoelectron spectroscopy (XPS), in both angle resolved and argon etch depth profiling, along with reflection high-energy electron diffraction (RHEED) and x-ray diffraction (XRD), provided detailed chemical and structural characterization of the interface layers after different post growth treatments. Chemical mechanisms of interface degradation and proposed strategies for maintaining interface chemical integrity will be discussed.

See more of this Session: Semiconductor Surface Chemistry
See more of this Group/Topical: Engineering Sciences and Fundamentals

Structure, Thermodynamics, and Solubility in Polyomino Fluids

Brian C. Barnes, Department of Chemistry, Washington University in St. Louis, St. Louis, MO

Daniel W. Siderius, Department of Chemistry, Washington University in St. Louis, St. Louis, MO

Lev D. Gelb, Department of Chemistry, Washington University in St. Louis, St. Louis, MO

The directed self-assembly of small molecules or objects in two dimensions is an ongoing challenge in nanotechnology and surface science [1,2]. As the number and complexity of components increases, the microstructure of the resulting phase, and location of any phase transitions, is difficult to predict. In simple systems such as rods, nematic transitions are readily observed [3,4]. When a variety of sizes [5], or sizes and shapes [6] are present, there can be rich behavior for both interacting and non-interacting objects.

In order to better understand the self-assembly of small molecules and nanoparticles adsorbed at interfaces, we have performed Monte Carlo simulations of hard polyominoes, connected shapes which occupy a certain number of lattice sites. Calculations are performed in the grand canonical ensemble, and are analogous to real systems in which molecules or nanoparticles reversibly adsorb to a surface or interface from a bulk reservoir. The model studied is athermal; objects in these simulations avoid overlap but otherwise do not interact. As a result, all of the behavior observed is entropically driven.

We have extensively studied the family of tetrominoes (polyominoes that occupy four lattice sites, of which there are seven) as both pure and mixed fluids, and identified a wealth of fascinating behavior, including non-ideal mixing, pronounced "microscale phase separation", and short-range geometric ordering[7]. In subsequent analyses, we demonstrate how the observed phenomena may be explained using classical thermodynamic concepts such as virial coefficients, volume of mixing, and Henry's Law. No sharp phase transitions have been observed, but this appears to be due to the small size of tetrominoes.

We then examined larger polyominoes, which have more structure than tetrominoes and, as a result, even more interesting phase behavior. For pure fluids, we recover the isotropic-nematic transition for rod-like species of sufficient length-to-width ratio. Polyominoes of sufficient size are demonstrated to exhibit a first-order phase transition similar to that of hard disks. We also examine more exotic shapes, and identify self-ordering that resembles crystallization, dimerization in mixtures, and immiscibility. These phenomena are again rationalized using classical thermodynamic concepts, from which we identify heuristics that qualitatively predict trends in self-assembly.

[1] Cicoira and Rosei, Surf. Sci., **600**, 1 (2006)

[2] Bowden et al., Science, **276**, 233 (1997)

[3] Bates and Frenkel, J. Chem. Phys., **112**, 10034 (2000)

[4] Ghosh and Dhar, Euro. Phys. Lett., **78**, 20003 (2007)

[5] Panagiotopoulos, J. Chem. Phys., **123**, 104504 (2005)

[6] Davis et al., Ind. Eng. Chem. Res., **45**, 5421 (2006)

[7] Barnes, Siderius, and Gelb, Langmuir, DOI:10.1021/la900196b (2009)

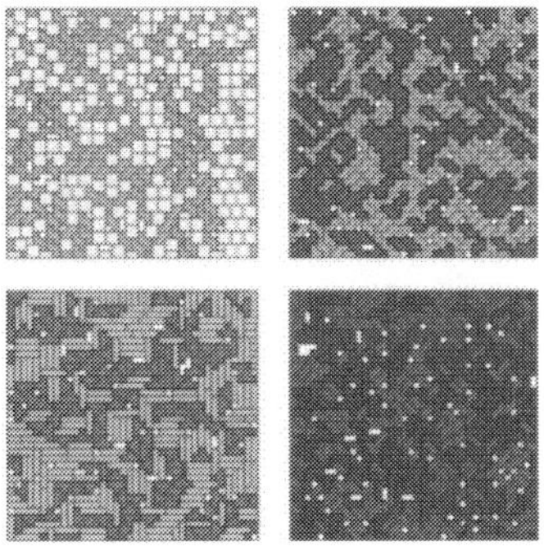

See more of this Session: Thermodynamic Properties and Phase Behavior III

See more of this Group/Topical: Engineering Sciences and Fundamentals

Classical Density Functional Theory of Freezing Based On a Bridge Functional: Applications to Colloidal Systems

Anurag Verma, Chemical Engineeering, University of Massachusetts, Amherst., Amherst, MA

David M. Ford, Chemical Engineering, University of Massachusetts, Amherst, Amherst, MA

Colloidal phase transitions caused by long and short range interparticle interactions are important in various modern fabrication processes with applications in supercomputing, synthesis of proteins and nano/micro devices. We hereby explore classical Density Functional Theory as a tool for making predictions of freezing transitions in colloids. In our approach we show an extension of the traditional perturbative DFT approach that employs a closure relation in form of a bridge functional. The shape of this bridge functional is qualitatively and quantitatively different from other conventional closures like HNC, PY, MS, VM etc., which are known to give good results in inhomogeneous liquid states. For hard sphere potentials, the correct form of the bridge functional produces results as accurate as MC simulations. We further explore the uniqueness of this finding and apply this novel approach to predict phase transitions in colloidal systems with varying interparticle interactions.

See more of this Session: Thermodynamic Properties and Phase Behavior III
See more of this Group/Topical: Engineering Sciences and Fundamentals

Monte Carlo Simulation of Dense Packings of Hard Tetrahedra

Amir Haji-Akbari, Chemical Engineering, University of Michigan- Ann Arbor, Ann Arbor, MI

Michael Engel, Chemical Engineering, University of Michigan- Ann Arbor, Ann Arbor, MI

Aaron S. Keys, Department of Chemical Engineering, University of Michigan, Ann Arbor, MI

Xiaoyu Zheng, Mathematics, Kent State University, Kent, OH

Rolfe Petschek, Physics, Case Western Reserve University, Cleveland, OH

Peter Palffy-Muhoray, Liquid Crystal Institute, Kent State University, Kent, OH

Sharon C. Glotzer, Chemical Engineering, University of Michigan, Ann Arbor, MI

The densest packing of tetrahedra remains an unsolved problem, and there has been much recent debate. We simulate dense packings of mathematically smooth, hard regular tetrahedra using NPT Monte Carlo simulations and determine the density- pressure equation of state. We find disordered and ordered packings with densities that significantly exceed the hard-sphere FCC packing density of 0.740480490. Our findings thus demonstrate that tetrahedra obey Ulam's conjecture that spheres pack with a lower maximum packing density than any other hard convex object, despite recent conjecture to the contrary. We show that the system is able to achieve such high packing densities by the local ordering of tetrahedra into certain favorable motifs, forming larger structures that pack efficiently but are overall jammed.

See more of this Session: Thermodynamic Properties and Phase Behavior III
See more of this Group/Topical: Engineering Sciences and Fundamentals

Phase Behavior of Associating Fluids in Contact with a Solid Surface

Sandip Khan, Department of Chemical Engineering, Indian Institute of Technology Kanpur, Kanpur, India

Sang Kyu Kwak, Division of Chemical and Biomolecular Engineering, Nanyang Technological University, Singapore, Singapore

Jayant K. Singh, Chemical Engineering, Indian Institute of Technology, Kanpur, Kanpur, India

Interaction of fluids with solids in terms of confinement or near the surface is natural phenomena in our daily life. Influence of solid interaction not only change the fluid properties(critical properties, free energies etc) but also induce different phase transitions such as layering, prewetting and capillary condensation. The occurrence of these phases depends on relative strength of fluid-fluid and fluid-solid interaction.

In this work, we present a systematic study of the prewetting behavior of Lennard-Jones based dimer forming associating fluids on a model solid CO_2 surface for various associating strengths using grand-canonical transition matrix Monte Carlo (GC-TMMC) and histogram reweighting techniques. Occurrences of prewetting transition is observed for associating strengths ranging from ε_{af} =2.0, 4.0, 6.0, 8.0 and 10.0. Wetting temperature, T_w, increases with associating energy and attains a maxima at a moderate associating strength. Further increase in associating energy decreases the wetting temperature, which is in contrast to the DFT prediction. Critical prewetting temperature, Tpwc, on the other hand, always increase with association energy. Boundary tension of thin-thick film via GC-TMMC and finite size scaling is found to increase at a constant T/T_{pwc} with increasing association strength up to 8.0. Subsequent increase in associating strength decreases the boundary tension, which is akin to the behavior of the vapor-liquid surface tension of bulk associating fluids. We also report structural, orientation order and monomer fraction of the thin and thick films of associating fluids.

See more of this Session: Thermodynamic Properties and Phase Behavior III
See more of this Group/Topical: Engineering Sciences and Fundamentals

Wetting Behaviors of Water in Confined Graphite and Rutile Slits

Xin Feng, State Key Laboratory of Materials-oriented Chemical Engineering, Nanjing University of Technology, Nanjing, China

Mingjie Wei, State Key Laboratory of Materials-oriented Chemical Engineering, Nanjing University of Technology, Nanjing, China

Qing Shao, State Key Laboratory of Materials-oriented Chemical Engineering, Nanjing University of Technology, Nanjing, China

Linghong Lu, State Key Laboratory of Materials-oriented Chemical Engineering, Nanjing University of Technology, Nanjing, China

Xiaohua Lu, State Key Laboratory of Materials-oriented Chemical Engineering, Nanjing University of Technology, Nanjing, China

The surface science enjoys an increasing interest, and many researchers develop attentions on wetting behaviors of water on hydrophobic and hydrophilic surfaces. Moreover, the wetting behaviors of nano-scale confinements draw attention of researchers who focus on the fields such as membrane separation, biological channel, heterogeneously porous catalysts and so on.

Rutile (110) is well accepted as the representative bulk model oxide and is used to discover the distinct phenomena on metal oxide surfaces which represent the hydrophilic ones. Meanwhile graphite (100) is the exact model for hydrophobic surfaces. Furthermore the surface effect will be enhanced in nano-scale pores and gaps, which are difficult to characterize in experiments. However, molecular simulation offers an alternative method to get insight on the molecular level information about the fluid around the surfaces. In this work, we study the static structure properties of water confined to slab geometry of graphite (100) or rutile (110) with molecular dynamics simulation, and then try to find out the distinction in these two kinds of slits.

We build a model that contains two slabs of graphite(100) or rutile(110) opposite to each other with water molecules between them, and their distance is ranging from 0.4 to 2.0 nm by NVT-ensemble (T=300K) MD simulations. The long-range electrostatic interaction plays an important role on the interactions between atoms of the rutile slabs and water molecules. Significant water layers were observed along the normal direction of theses rutile surface. The Water molecules in the closest layer are absorbed at the five-coordinated Ti and two-coordinated O (or bridging O) atoms on the surfaces. On the other hand, water molecules prefer to keep distance (about 0.2 nm) from the graphite surfaces. Further, the dipole of water molecules near the (graphite or rutile) surfaces was found to have considerable preferential orientation, mainly because of long-rang electrostatic interactions of the surface atoms for rutile surfaces and repulsion force for graphite surfaces. With the decrease of the distance between slabs, water molecules have smaller diffusion coefficient in rutile slits while approximate diffusion coefficients in graphite slits.

See more of this Session: Thermodynamic Properties and Phase Behavior III
See more of this Group/Topical: Engineering Sciences and Fundamentals

Solvation of a Spherical Cavity in Simple Liquids: Interpolating Between the Limits

Jianzhong Wu, Department of Chemical and Environmental Engineering, University of California, Riverside, Riverside, CA

Dissolution of a solute molecule into a solvent necessitates creation of a cavity devoid of solvent molecules. The cavity solvation free energy is exactly known at both very small and large length scales but in between it can only be estimated by various approximations. Guided by simulation results for the solvation of small cavities and by density functional theory, we analyze the size dependence of the solvation free energy, contact density of solvent molecules, and the interfacial tension for a spherical cavity in a Lennard-Jones fluid or in a system of hard spheres. Unlike cavity formation in the hard-sphere system, a quadratic curvature expansion is insufficient to connect smoothly the exact results in the microscopic and macroscopic limits for the cavity surface tension (or equivalently, the contact solvent density) in Lennard-Jones fluids. Considering the sensitivity of solvation to molecular details at small length scales, we conjecture that, for practical purposes, a heuristic approach may be sufficient to link the thermodynamic limit at large length scales and exact results of cavity formation at very small length scales.

See more of this Session: Thermodynamic Properties and Phase Behavior III
See more of this Group/Topical: Engineering Sciences and Fundamentals

Evaluation of Bridge Function for Hard Sphere Fluid Confined in Slit Pore Via Transition Matrix Monte Carlo Mayer Sampling

Wen Wen Chen, Division of Chemical and Biomolecular Engineering, Nanyang Technological University, Singapore, Singapore

Sang Kyu Kwak, Division of Chemical and Biomolecular Engineering, Nanyang Technological University, Singapore, Singapore

Integral equation theory (IET) has been around for many years to study properties of fluids and most its related-studies were devoted to predict the properties of the bulk fluids via determining appropriate bridge functions for IET closures. In this study, we redirect the IET approach towards the prediction of thermodynamic properties of slit-pore confined fluids, of which roles find very much significance in industrial applications. As it is hard to find a direct evaluation of the confined fluid bridge function, we adopt and modify Mayer-sampling technique, which is applied to predict the properties of hard-sphere confined in hard planar walls. The simpler one-particle Ornstein-Zernike (OZ) equation with a hypernetted chain (HNC) closure is used and we have constructed Transition Matrix Monte Carlo Mayer-sampling technique to evaluate the bridge coefficients (clusters) expanded in density for this study. The calculated bridge function containing b2(r), b3(r) and b4(r) coefficients is included in iterative calculation of the closure to obtain the density profiles, which are in turn used to provide the equation of state of the target fluids.

See more of this Session: Thermodynamic Properties and Phase Behavior III
See more of this Group/Topical: Engineering Sciences and Fundamentals

The Shape of a Sessile Drop

Bharadwaj Prabhala, Mechanical Engineering, Tennessee Technological University, Cookeville, TN

Mahesh Panchagnula, Mechanical Engineering, Tennessee Technological University, Cookeville, TN

Venkat Subramanian, Chemical Engineering, Tennessee Technological University, Cookeville, TN

Srikanth Vedantam, Mechanical Engineering, National University of Singapore, Singapore, Singapore

Contact angle is an important parameter which characterizes the interaction between the liquid and solid surface. Based upon the contact angle solid substrates are primarily classified as hydrophilic, hydrophobic and super hydrophobic substrates. In this study, we investigate the dependence of the local contact angles on the configuration of the contact line shape of a sessile liquid drop on a solid substrate. We use the numerical algorithm 'Surface Evolver' which is an interactive program for the study of surfaces shaped by surface tension, gravitational and other energies. The algorithm calculates the velocity at each vertex and the vertex is moved in the direction such that the surface evolves towards a minimum energy. The shapes and the energies of the drop are computed using the Surface Evolver. An analytical solution based on perturbation expansion is developed to predict the shape of the sessile drop for a given contact line description. The shape of the contact line is also varied in the same manner and the exact drop shape is computed in Surface Evolver. The Root Mean Square error is calculated by comparing the radius at all the vertices between the analytical and numerical approaches for varying contact angles and amplitudes of undulations. We show that the applicability of the analytical solution is quite widespread.

See more of this Session: Poster Session: Interfacial Phenomena
See more of this Group/Topical: Engineering Sciences and Fundamentals

Constructing Free Energy Landscapes in Interfacial Colloidal Systems Using a Fokker-Planck Formalism

Ray M. Sehgal, Chemical Engineering, University of Massachusetts, Amherst, Amherst, MA

Daniel J. Beltran-Villegas, Chemical and Biomolecular Engineering, Johns Hopkins University, Baltimore, MD

Michael A. Bevan, Chemical and Biomolecular Engineering, Johns Hopkins University, Baltimore, MD

David M. Ford, Chemical Engineering, University of Massachusetts, Amherst, Amherst, MA

The self- and directed- assembly of colloidal (nano- to micro- scale) particles into structures within materials and devices is an emerging paradigm with wide-ranging technological impact. However, the ability to create a target structure with an acceptably small level of defects is still lacking; systems too easily become dynamically arrested in undesired disordered, or defect-rich, states. We have been exploring a synergistic combination of recent advances in digital microscopic imaging techniques with free energy calculation methods to create free energy landscapes that could be used in monitoring and controlling self-assembly. Following recent computational work by Y.G. Kevrekidis and co-workers, we analyze short-time trajectories of an assembling colloidal system (as obtained by a microscopy experiment) and extract the coefficients of the Fokker-Planck equation that describes the underlying probability distribution function. We show examples for two types of problems. In the first, the stochastic variables are simply the real-space positions of a single colloidal particle associating with a structural surface feature. In the second, a small set of order parameters are used as the stochastic variables in a many-particle system undergoing directed assembly into a crystalline object.

See more of this Session: Poster Session: Interfacial Phenomena
See more of this Group/Topical: Engineering Sciences and Fundamentals

Effective Core and Corona Block Sizes for Approximating Micelle Formation Pressures and Micelle Aggregation/Precipitation Pressures in Polystyrene-Block-Polybutadiene and Polystyrene-Block-Polyisoprene Solutions in near Critical Propane From SAFT

Winoto Winoto, Chemical & Petroleum Engineering, University of Wyoming, Laramie, WY

Maciej Radosz, University of Wyoming, Laramie, WY

The statistical associating fluid theory version referred to as SAFT1 is demonstrated to represent the bulk and micellar phase transitions of polystyrene-block-polybutadiene and polystyrene-block-polyisoprene copolymer solution in propane when reduced segment approximation (RSA) and secluded segment approximation (SSA) are applied to calculate effective corona (diene) and core (styrene) block sizes. The RSA considers a reduced segment number for the corona-forming block to compensate the supposedly soluble block upon micellization transition, hence RSA can fit SAFT1 to micellization pressure of polystyrene-block-polydiene solution in propane. The SSA considers the micelle core excluded from interaction with the solvent during precipitation of micelle from the solution. The SSA can fit SAFT1 to micellar cloud pressure of polystyrene-block-polydiene solution in propane.

See more of this Session: Poster Session: Interfacial Phenomena
See more of this Group/Topical: Engineering Sciences and Fundamentals

Surface-Induced Phase Separation of O/W Nano-Emulsion

Phong Huynh, Rutgers University, New Brunswick, NJ

Paul Takhistov, School of Environmental & Biological Sciences, Rutgers University, New Brunswick, NJ

The hydrophobic properties of substrates are known to affect wetting behavior of the deposited multiphase droplets. In this study, decomposition of O/W nano-emulsion droplets placed onto various wettable surfaces with controlled hydrophobicity is investigated. It was found that thin film spreading of the oil phase over the substrate surfaces is one of the primary reasons for emulsion decomposition. Our data shows that decomposition occurs first near the droplet pole and not at the contact line. Various techniques (phase contrast optical microscopy, gravimetric measurements and Raman spectroscopy) have been used to reveal the mechanism of the emulsion destabilization. It was found that process of nano-emulsion decomposition on highly hydrophobic surfaces can be described as two-stage process with critical point at which the value of dynamic contact angle is suddenly increased. We propose a simple mass balance model to explain observed phenomena.

See more of this Session: Poster Session: Interfacial Phenomena
See more of this Group/Topical: Engineering Sciences and Fundamentals

Raleigh-Bernard Instability IN Evaporating Zein-Alcohol Mixtures and ITS EFFECT ON PROTEIN Self Assembly

Chang-Hoon Chai, Rutgers University, New Brunswick, NJ

Paul Takhistov, School of Environmental & Biological Sciences, Rutgers University, New Brunswick, NJ

Zein, a prolamine protein is a promissing material to develop a control release system of poorly soluble drugs. Zein can encapsulate such compounds and its aggregation can be controlled from nanosacle particles to macro-sized continuous film structures. However the use of zein as building block in control release system requires understanding of factors controlling its self-assembly. In our study zein is dissolved in ethanol-water mixture of various concentrations. By the evaporation of ethanol, zein forms various types of self-assembling structures: from nano-aggregates to continuous zein films. It was found experimentally that Raleigh-Bernard instability is controlling factor that determines type of self assembling structures. We propose simple scaling model which allow to determine critical regimes and processing parameters to obtain desired structures protein assemblies. Additional DLVO-based analysis of zein aggregative stability is in good agreement with experimental data.

See more of this Session: Poster Session: Interfacial Phenomena
See more of this Group/Topical: Engineering Sciences and Fundamentals

Corrosion Mechanisms Occurring On AM60 Mg Exposed to a 3.5% NaCl Aqueous Solution: Experimental Effects

Holly J. Martin, Center for Advanced Vehicular Systems (CAVS), Mississippi State University, Mississippi State, MS

M.F. Horstemeyer, Center for Advanced Vehicular Systems (CAVS), Mississippi State University, Mississippi State, MS

P.T. Wang, Center for Advanced Vehicular Systems (CAVS), Mississippi State University, Mississippi State, MS

The use of lightweight metallic alloys, such as cast magnesium, has recently gained traction in the automobile industry [1, 2]. The low density (1.74 g/cm3), high mechanical stiffness, excellent castability, and easy machinability make cast magnesium a strong candidate for structural components that are intricate and lightweight [2]. However, these benefits are greatly affected by a high corrosion rate when compared to aluminum or steel. In addition to a relatively high corrosion rate, magnesium has a high electrochemical potential, causing magnesium to corrode easily in the presence of seawater [3]. The high electrochemical potential of magnesium allows for coupon dissolution when magnesium is placed in a NaCl aqueous solution.

The high corrosion rate has relegated the structural use of magnesium to areas that are not exposed to the environment, including car seats and internal electronic boxes [4, 5]. Therefore, the use of cast magnesium as a structural material requires a detailed understanding of the metal's response to corrosion conditions. Corrosion mechanisms in cast magnesium have been linked to the microstructure, suggesting that the performance of magnesium in corrosive environments could be significantly improved [6-8]. In order to improve corrosion resistance, additional elements have been introduced in an effort to develop new magnesium alloys [4]. These new magnesium alloys must be rigorously evaluated to determine the change in galvanic corrosion within the alloy due to the introduction of additional elements [4].

In the Center for Advanced Vehicular Systems (CAVS) at Mississippi State University, the electrochemistry between a conductor, AM60 magnesium (Mg), and an electrolyte, sodium chloride, is being investigated. Three different mechanisms are responsible for the corrosion of magnesium, including pitting, intergranular corrosion, and general corrosion. The influence of aluminum (A: 5.5-6.5%) and manganese (M: 0.25% min) on the corrosion mechanisms is also being studied. Previous work using polished AM60 Mg demonstrated that pitting is initially the dominant corrosion mechanism [5, 9]. Following the initial exposure to the NaCl aqueous solution, changes in the surface resulted in changes to pit characteristics, including pit depth, pit area, and pit volume, likely the result of the formation of a protective Mg(OH)2 film [9]. This film was eventually degraded as the dominant corrosion mechanism changed from pitting to general [9]. Intergranular corrosion also contributed to changes in the pit characteristics. While the corrosion mechanisms of AM60 Mg have been previously studied, these results were gathered using highly polished AM60 Mg coupons. The corrosion effects of 3.5% NaCl on cast AM60 have not previously been studied. Two different methods of salt corrosion were investigated, including constant immersion and salt-spraying. The research presented will cover the effects of immersion and salt-spray on the corrosion mechanisms of AM60 Mg.

[1] Jambor, A.; Beyer, M. Mater. Des. 18 (1997) 203-209.

[2] Froes, F.H.; Eliezer, D.; Aghion, E.L. JOM. 50 (1998) 30-34.

[3] Shaw, B.A. Corrosion Resistance of Magnesium Alloys, in: L.J. Korb, ASM (Eds.), ASM Handbook, Vol.

13A: Corrosion, Ninth Ed., ASM International Handbook Committee, Metals Park, 2003, 692.

[4] Makar, G.L.; Kruger, J. Int. Mater. Rev. 38 (3) (1993) 138-153.

[5] Song, G.; Atrens, A. Adv. Eng. Mater. 5 (12) (2003) 837-858.

[6] Lunder, O.; Aune, T.K.; Nisancioglu, K. NACE Corros. 43 (1987) 291-295.

[7] Song, G.; Atrens, A.; Dargusch, M. Corros. Sci. 41 (1999) 249-273.

[8] Ambat, R.; Aung, N.N.; Zhou, W. Corros. Sci. 42 (2000) 1433-1455.

[9] Martin, H.J.; Alvarez, R.B.; Horstemeyer, M.F.; Chandler, M.Q.; Williams, N.; Wang, P.T.; Ruiz, A. Metall. Trans. A. In Review.

See more of this Session: Poster Session: Interfacial Phenomena
See more of this Group/Topical: Engineering Sciences and Fundamentals

Effects of Hydrocarbons on Viscosity of Water Containing Rheology-Modifiers

Kyung C. Kwon, Chemical Engineering, Tuskegee University, Tuskegee, AL

Valerie M. Gordon, Chemical Engineering, Tuskegee University, Tuskegee, AL

Tertiary crude oil recovery techniques are employed to increase production of crude oils extracted from an oil field. Using tertiary crude oil recovery techniques, 30-60% of the reservoir's original crude oil can be extracted compared with 20-40 % using primary and secondary recovery. Crude oil's viscosity is reduced to increase its production in the tertiary oil recovery method by heating oils with steam to extract easily the crude oil, burning a portion of the crude oil to heat the surrounding oil in the in-situ burning, using occasionally detergents, and flooding carbon dioxide.

On the contrary to the conventional tertiary crude oil recovery techniques, a new tertiary oil recovery technique will be developed to displace crude oil with injected water containing rheology modifiers without reducing viscosity of a body of crude olis. Upon contacting the rheology modifier in injected water with crude oil, viscosity of the injected water containing rheology modifier will increase at the water-crude oli interface. Roles of various rheology modifiers in water will be investigated for the viscosity increase at the interface of injected water containing rheology modifier and crude oils with our viscometer and its novel viscosity equations.

Selected viscosity modifiers include alcohols such as methanol, ethanol, isopropanol, and n-butanol, ketones such as acetone and methyl isobutyl ketone, tetrahydrofuran, 2-methyl tetrahydrofuran, sodium dodecyl sulfate, and addition compounds such as urea, thiourea, hydroquinone, cyclodextrins, picric acid, and sodium chloride.

See more of this Session: Poster Session: Interfacial Phenomena
See more of this Group/Topical: Engineering Sciences and Fundamentals

Extraction of Phenolic Compounds From Aqueous Solution Using a Slightly Hydrophobic Nonionic Surfactant

Rico Tedyono, Department of Chemical Engineering, National Cheng Kung University, Tainan, Taiwan

Bing-Hung Chen, Department of Chemical Engineering, National Cheng Kung University, Tainan, Taiwan

A simple and rapid L_3-phase extraction technique was developed to recover mixtures of phenolic compounds, including 8 priority phenolic pollutants, from aqueous solutions using nonionic surfactant Tergitol 15-S-5 at ambient temperature. The readily biodegradable Tergitol 15-S-5 surfactant is a mixture of species with secondary ethoxylated alcohols, and exhibits as a dispersion of the lamellar phase in water $(L_\alpha + W)$ at ambient temperature (24°C). Hence, salting-out electrolytes were in need to induce the phase separation along the boundary of $L_3 + W$ phases at ambient temperature. Effects of different electrolytes, such as monovalent LiCl, NaCl and KCl, as well as bivalent $MgCl_2$ and $CaCl_2$, to the L_3-based extraction process were studied. The strategy to achieve a high preconcentration factor was successfully demonstrated with increasing concentration of salting-out additives and/or decreasing the surfactant concentration, which could be attributable to increase the volume of water phase. The optimum conditions for the extraction and preconcentration of these selected phenolic compounds have been established as follows, 0.2% wt of Tergitol 15-S-5 + 5.6% wt of NaCl. Moreover, the logarithms of distribution coefficients of these phenolic compounds between coexisting L_3 phase and water phase was found to be linearly proportional to the logarithms of *n*-octanol/water distribution coefficients of these phenolic ompounds.

See more of this Session: Poster Session: Interfacial Phenomena
See more of this Group/Topical: Engineering Sciences and Fundamentals

Synthesis and Characterization of Novel Nanomaterials: Gold Nanoshells with An Organic-Inorganic Hybrid Core

Alisha D. Peterson, Chemical & Biomedical Engineering, University of South Florida, Tampa, FL

Vinay K. Gupta, Chemical & Biomedical Engineering, University of South Florida, Tampa, FL

Gold nanoshells, which are materials containing a thin shell of gold around a core microparticle of silica, are of great interest due to their unique optical properties. Potential applications for these novel nanomaterials include sensing, imaging, drug delivery, and solar energy conversion. Our interest is on nanoshells that use a core formed from an organic-inorganic hybrid material. By utilizing co-polymerization of organic and inorganic monomeric materials, we are developing gold nanoshells that incorporate either titania or silica. Tuning the organic framework and the cross-linking in the polymer core can provide a greater degree of versatility to change optical properties of the resulting nanomaterials. In this presentation, we will detail the preparation and characterization of these novel nanomaterials. Results from transmission electron microscopy (TEM), dynamic light scattering, and UV-Vis spectroscopy will be used to compare the gold nanoshells based on novel hybrid core particle to the traditional silica cores.

See more of this Session: Poster Session: Interfacial Phenomena
See more of this Group/Topical: Engineering Sciences and Fundamentals

Enhanced Kinetics of CO2 Enclathration with a Mixed Organic Layer

Junshe Zhang, Chemical Eng, Department of Chemical Engineering, The City College of New York, New York, NY

Jae W. Lee, Department of Chemical Engineering, The City College of New York, New York, NY

Trapping CO2 in hydrates is one of the new technologies for CO2 capture and storage. This work presents the rapid CO2 with a small amount of cyclopentane (CP). The enclathration was carried out in a 474 cm3 non-stirred batch reactor with 100 and 200 cm3 water. At volume ratios of CP to water of 0.01 to 0.1, the maximum enclathration rate is 0.32 mol h-1 at CO2 pressures ranging from 1.9 to 3.4 MPa and at 274 K. CO2 hydrates (sI) and CO2 + CP (sII) binary hydrates coexist at the end of enclathration. The mole ratio of CO2 entrapped in the sI hydrates to that in the sII binary hydrates is higher with 100 cm3 water than with 200 cm3 water. The same trend is also observed for the total amount of CO2 encalthrated. The enclathration rate depends not only on the water volume but also on the pressure. The enclathration rate and the water conversion reach a maximum at 3.06 MPa and then decrease as the pressure increases from 1.9 to 3.4 MPa with 100 cm3 water and 5 cm3 CP. The water conversion to clathrate hydrates reaches 52 % within 2 h. This accelerated enclathration provides one stepping-stone to develop the hydrate-based CO2 capture and storage technique

See more of this Session: Poster Session: Interfacial Phenomena
See more of this Group/Topical: Engineering Sciences and Fundamentals

Methyl Glucamine and Glycoside Surfactant Headgroup Synergism Comparison with Alkyl Sulfates

Kendall Werts, Dr. Brian P. Grady

1. Introduction

Biosurfactants are a very important class of surfactant. Biosurfactants can be defined as surfactants that come from renewable resources. They can also be defined as surfactants synthesized by biological organisms or surfactants that are degradable and/or nontoxic in nature [1]. Developing better biosurfactants is an excellent step in making many industries more environmentally conscience. But biosurfactants often do not show as promising properties as more commonly used surfactants, and developing new biosurfactants with better properties is difficult and expensive. A different approach is finding biosurfactants that show synergism, which is fairly common, with more commonly used surfactants since most practical applications of surfactants are already mixtures of surfactants [2]. These synergistic interactions between different types of surfactants have long been studied to determine the reason for such interaction and the best way to determine which surfactants will work better together [3-5]. It has been found that mixed micelle formation is greatly influenced by differing headgroups [6]. Cationic and anionic surfactants show synergism with nonionic surfactants, because nonionic headgroups act as spacers between the rest of the positively or negatively charged molecules at interfaces so that they can stay closer together with less effect from repulsion [7].

Regular Solution Theory. In order to accurately predict the minimum CMC as well as synergism in surface tension efficiency for a given surfactant mixture, nonideal solution theory has been used successfully [5]. The basic equations for nonideal solution theory for mixed micelles are:

$$\alpha C_{12} = X C_1 \exp\left[\beta(1-X)^2\right] \quad (1)$$

$$(1-\alpha)C_{12} = (1-X)C_2 \exp\left[\beta X^2\right] \quad (2)$$

where α is the mole fraction of surfactant 1 in solution, X is the mole fraction of surfactant 1 in the mixed micelle, C_{12}, C_1, and C_2 are the critical micelle concentrations of the mixed surfactant, pure surfactant 1 and pure surfactant 2 respectively, and β is the interaction parameter. A negative deviation from ideality gives a negative β parameter and indicates synergy. If these equations are solved for synergistic conditions, we arrive at the requirements for synergy to exist, which are [3]:

1. β must be negative
2. $\left|\ln\left(\frac{C_1}{C_2}\right)\right| < |\beta|$

Synergism is dependent on the pure component CMCs and the β parameter, which demonstrates the amount of interaction between the surfactant molecules. In order to determine which surfactant headgroups interact the most usefully, surfactants can be matched according to their CMCs. If the individual CMCs of the two surfactants in solution are quite similar, observed synergy should be mostly due to the interaction of the surfactant molecules.

Surfactants in Study. Linear alkyl sulfates are commonly used surfactants in dish soap, shampoo, and detergent. These surfactants are very common because they show excellent properties for multiple uses, so a few linear alkyl sulfates were chosen to use for this study. N-Methyl glucamines (MEGA) are interesting surfactants that show promising properties and good interaction with sodium dodecyl sulfate [8-11]. They are rather inexpensive and easy to find in a range of alkyl chain lengths, so a few of this type of surfactant were used in this study. MEGA surfactants are formed through an enzymatic amidification of an N-methyl glucamide and a fatty acid [12]. Alkyl polyglycosides (APG) are an interesting and widely studied biosurfactant. They are also used in detergents and cleaning products and show high biodegradability and low toxicity [13]. The alkyl glycosides show slightly stronger interactions with other surfactants than alkyl maltosides, and they are also easy to find inexpensively in a range of alkyl chain lengths and show good interaction with alkyl sulfates, so a few of these alkyl glycosides were used in this study [14-17].

2. Materials and Methods

S12S was matched by CMCs from literature with MEGA10 and 9GPN, while S9S was matched with MEGA8 and 7GPN. S10S was matched with MEGA9 and 8GPN, but these experiments have not been performed yet. The CMCs found in literature are all listed in table 1.

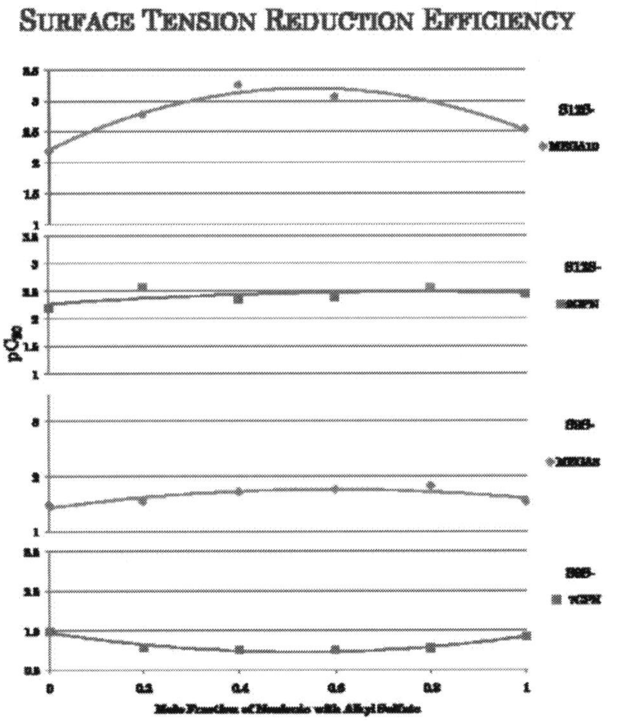

Figure 1. Surface tension reduction efficiency for each mole fraction in each system.

Solubility experiments were performed by adding 2.5 mL of a surfactant solution to 2.5 mL of decane and mixing with a vortexer for about 2 seconds once a day for 3 days. The vials were allowed to sit for 2 weeks after they were filled and the volume of the different phases was measured. Critical micelle concentration experiments were performed with a CAHN 312 DCA measuring surface tension of different concentrations of surfactant. The CAHN uses a Wilhelmy Plate method with a silica plate probe.

3. Results and Discussion

Surface tension reduction efficiency is the negative log of the change in concentration required to drop the surface tension 20 dyne/cm. Calculated efficiencies as they vary with mole fraction are shown in figure 1. Efficiency is seen in the S12S-MEGA10 and S12S-9GPN systems. Figure 1 shows an increased efficiency in the S12S-MEGA10 mixture. For the S12S-9GPN system, the

CMC RESULTS

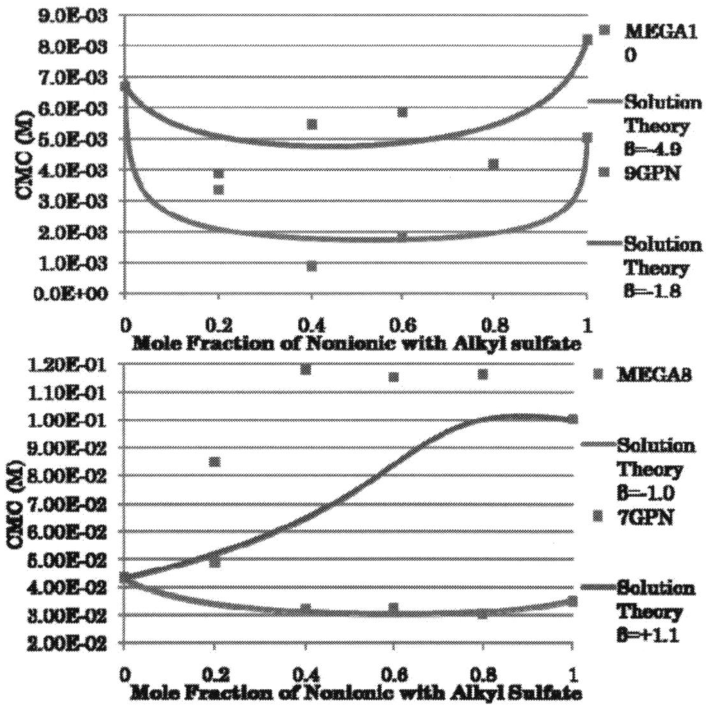

Figure 2. Regular Solution Theory Predictions for CMC and measured CMCs for S9S-MEGA10 and S9S-9GPN.

efficiency is only slightly increased with increased mole fraction of 9GPN. The S9S-MEGA8 system again shows synergism in surface tension reduction efficiency in the mixture as opposed to the pure components. The S9S-7GPN system, however, actually shows a decrease in efficiency in the mixture and a higher efficiency on the pure components.

Figure 2 shows regular solution theory predictions for the CMCs at different mole fractions for each system. Regular solution theory fits the S12S-MEGA10 data with a β of -4.9, which shows a strong negative deviation from ideality and thus strong synergism. The S12S-9GPN data is fit by a β of -1.8 which still shows synergism, but considerably less than in the S12S-MEGA10 system. Figure 2 also shows the CMC and Regular Solution Theory plots for S9S-MEGA8 and S9S-7GPN systems. S9S and MEGA again show synergism with a β of -1.0; however the S9S-7GPN predictions with a β of +1.1 show antagonism which was unexpected.

The ability for these surfactant solutions to emulsify decane was studied and these results are demonstrated in figure 3. The emulsification volume as weight percent increased was plotted for each mole fraction studied. The S12S-MEGA10 system emulsifies more decane at lower weight percents for 0.8 mole fraction MEGA10, but 0.4 and 0.6 mole fraction show similar results. Even the 0.2 mole fraction mixture shows better results than the pure components. For the S12S-9GPN system, synergism is seen for mole fractions above 0.2 9GPN. The plot of emulsification volume vs. weight percent for neither S9S-MEGA8 nor S9S-7GPN shows much synergism. This is not surprising since the S9S does not appear to emulsify the decane at all.

EMULSION RESULTS

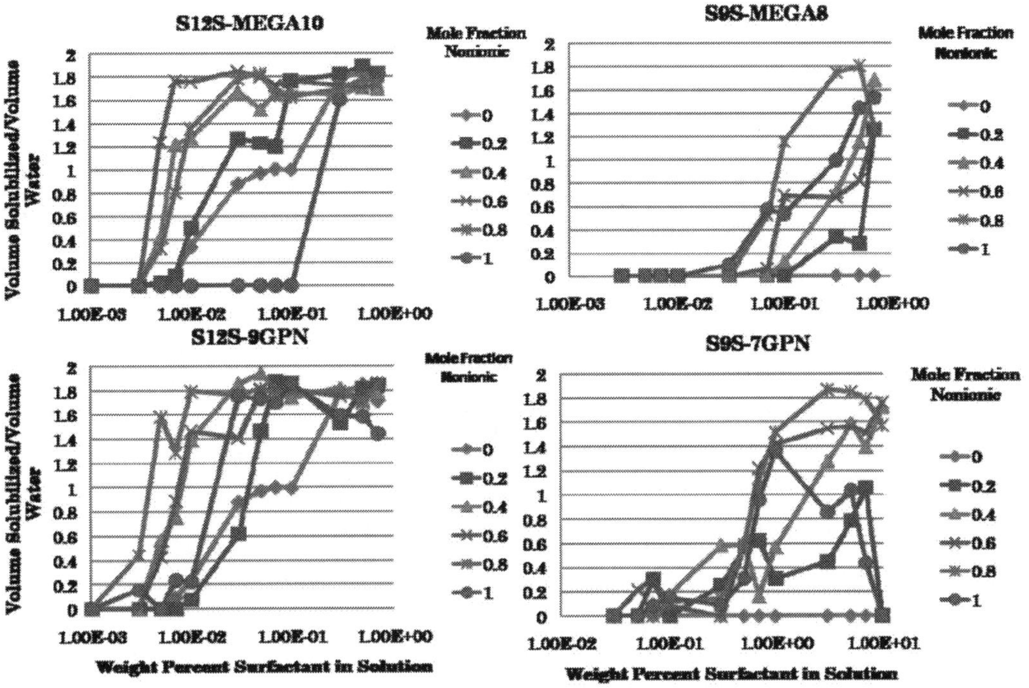

Figure 3. Solubilization of decane in different weight percent solutions of surfactant for mole fractions of each system

4. Future Work

Two more useful pairing can be made for these surfactant systems. They are S10S-MEGA9 and S10S-8GPN. Surface tension, CMC and solubility experiments will be performed for these last 2 pairs of surfactants, hopefully with the results similar to current findings. In addition to this, a few more tests can be run to make sure current data is accurate, including surface tension experiments at higher concentrations to get better fits to CMC lines.

5. Conclusions

This study demonstrated that synergism between alkyl sulfates and methyl glucamines is greater than that of alkyl sulfates and glycosides. Synergism in solubilization power of decane was seen in both pairs in all cases. Synergy in CMC reduction, surface tension reduction, and surface tension reduction efficiency was seen in the MEGA cases as well as the larger tail-group GPN case. In all cases, the alkyl sulfate-MEGA pairing out-performed the alkyl sulfate-GPN pairing, sometimes quite dramatically. Since similar individual CMCs cause synergy to be reliant mostly on the interaction parameter β, it can be assumed that MEGA and alkyl sulfates interact more usefully than GPN and alkyl sulfates. Since a greater amount of interaction should stem from the polarity of the headgroups, it can also be assumed that the sulfate and glucamine headgroup interaction is the main driving force behind the greater synergy between these two surfactants. The future work should also confirm this result.

References

1. Holmberg, K., *Natural surfactants.* Current Opinion in Colloid & Interface Science, 2001. **6**(2): p. 148-159.
2. Sierra, M.L. and M. Svensson, *Mixed micelles containing alkylglycosides: Effect of the chain length and the polar head group.* Langmuir, 1999. **15**(7): p. 2301-2306.
3. Rosen, M.J., *SELECTION OF SURFACTANT PAIRS FOR OPTIMIZATION OF INTERFACIAL PROPERTIES.* Journal of the American Oil Chemists Society, 1989. **66**(12): p. 1840-1843.
4. Rosen, M., *Molecular interactions and the quantitative prediction of synergism in mixtures of surfactants*, in *Horizons 2000 – aspects of colloid and interface science at the turn of the millenium*. 1998. p. 35-41.
5. Hua, X.Y. and M.J. Rosen, *SYNERGISM IN BINARY-MIXTURES OF SURFACTANTS .1. THEORETICAL-ANALYSIS.* Journal of Colloid and Interface Science, 1982. **90**(1): p. 212-219.
6. Bakshi, M.S., et al., *Mixed-micelle formation by strongly interacting surfactant binary mixtures: effect of head-group modification.* Colloid and Polymer Science, 2002. **280**(11): p. 990-1000.
7. Zhou, Q. and M.J. Rosen, *Molecular interactions of surfactants in mixed monolayers at the air/aqueous solution interface and in mixed micelles in aqueous media: The regular solution approach.* Langmuir, 2003. **19**(11): p. 4555-4562.
8. Zhu, Y.P., et al., *Surface properties of N-alkanoyl-N-methyl glucamines and related materials.* Journal of Surfactants and Detergents, 1999. **2**(3): p. 357-362.
9. Hierrezuelo, J.M., J. Aguiar, and C.C. Ruiz, *Stability, interaction, size, and microenvironmental properties of mixed micelles of decanoyl-N-methylglucamide and sodium dodecyl sulfate.* Langmuir, 2004. **20**(24): p. 10419-10426.
10. Hierrezuelo, J.M., J. Aguiar, and C.C. Ruiz, *Role of the head group on the mixed micellization process in binary systems containing a sugar-based surfactant: decanoyl-N-methylglucamide.* Molecular Physics, 2005. **103**(24): p. 3299-3308.
11. Ko, J.-S., et al., *Adsorption and Micelle Formation of Mixed Surfactant Systems in Water. IV. Three Combinations of SDS with MEGA-8, -9, and -10.* Journal of Oleo Science, 2004. **53**(3): p. 109-126.
12. Maugard, T., et al. *Enzymatic amidification for the synthesis of biodegradable surfactants: Synthesis of N-acylated hydroxylated amines.* 1998: Elsevier Science Bv.
13. von Rybinski, W. and K. Hill, *Alkyl polyglycosides - Properties and applications of a new class of surfactants.* Angewandte Chemie-International Edition, 1998. **37**(10): p. 1328-1345.
14. Rosen, M.J., *Surfactants and Interfacial Phenomena.* 3 ed. 2004, Hoboken, New Jersey: John Wiley & Sons, Inc.
15. *Anatrace 2009-2010 Catalogue*, A. (Affymetrix), Editor. 2009.
16. Ranganathan, R., L. Tran, and B.L. Bales, *Surfactant- and salt-induced growth of normal sodium alkyl sulfate micelles well above their critical micelle concentrations.* Journal of Physical Chemistry B, 2000. **104**(10): p. 2260-2264.
17. Patist, A., et al., *On the measurement of critical micelle concentrations of pure and technical-grade nonionic surfactants.* Journal of Surfactants and Detergents, 2000. **3**(1): p. 53-58.

The Discharge of Thin Film Li Electrode under Transport Control

Chuang-Yaw Lin, Guang-Yau Wu and Shi-Chern Yen

Introduction

Li-ion battery research will help to understand the mathematical simulation of Li-ion battery charge and discharge behavior. Li-ion in positive and negative diffusion behavior of particles dominated the charge and discharge characteristics of lithium batteries. The literature for solid-state diffusion model of lithium batteries was based on Fick's law [1-3]. G. Ning.and Popov [4] simplified the solid-state diffusion model with the volume-averaged solid phase concentration and interface concentration. V. Srinivasan and J. Newman [5] considered the diffusion behavior with moving boundary model. Q. Zhang and R. E. White [6] expressed with moving boundary model and shrinking core model to describe the Li-ion transport during phase transition.

In this study, simplified counter diffusion model and Fick's law diffusion model are considered with the boundary there is a maximum value of one case and compare the distribution of Li-ion concentration and relative potential with Li^+/Li in MCMB electrode.

Electrochemical modeling

In this section, Fick's law and simplified counter diffusion models are considered.

Fick's Law

The common model of secondary lithium batteries is the Fick's 2nd law, using this simple law we can solve the diffusion equation in a spherical particle

$$\frac{\partial \theta}{\partial t} = D_s\left[\frac{1}{r^2}\frac{\partial}{\partial r}\left(r^2\frac{\partial \theta}{\partial r}\right)\right] \quad (1)$$

where θ is the dimensionless concentration of lithium inside a spherical electrode particle. The diffusion coefficient Ds is assumed to be constant. The boundary conditions are

$$\frac{\partial \theta}{\partial r}(0,t)=0 \quad (2)$$

and

$$\theta(R,t)=1 \quad (3)$$

where R is the radius of the particle. The battery is usually fully charged initially

$$\theta(r,0)=1 \quad (4)$$

Simplified counter diffusion model (modified Fick's law)

In this case we consider the interaction about Li^+ and solid particles are counter diffusion. First we define a system with two components, component 1 is Li^+ and component 2 is the solid particles. Then following the Fick's 2nd law

$$\begin{cases}\dfrac{\partial \theta_1}{\partial t} = \nabla \cdot (D_1 \nabla \theta_1 - \upsilon \theta_1) \\ \dfrac{\partial \theta_2}{\partial t} = \nabla \cdot (D_2 \nabla \theta_2 - \upsilon \theta_2)\end{cases} \tag{5}$$

where θ_1 is the dimensionless concentration of lithium. θ_2 is the dimensionless concentration of solid particle. υ is the velocity relative to r axis. And the diffusion coefficient D_1 and D_2 are assumed to be constants.

The dimensionless total concentration (θ) is given by the equation $\theta = \theta_1 + \theta_2 = 1$. Equations (1-5) for two components then lead to

$$\frac{\partial \theta}{\partial t} = \frac{\partial \theta_1}{\partial t} + \frac{\partial \theta_2}{\partial t} = \nabla \cdot (D_1 \nabla \theta_1 + D_2 \nabla \theta_2 - \upsilon \theta) \tag{6}$$

If it is assumed that the molar volume of the alloy is independent of composition, the total concentration (C) is a constant, and $\partial \theta / \partial t = 0$. Thus the relative velocity will be given by

$$\upsilon = D_1 \nabla \theta_1 + D_2 \nabla \theta_2 \tag{7}$$

The simplified counter diffusion model can be obtained by substituting equation (7) in one of the equations (5) and assumed $\partial \theta_1 / \partial r = -(\partial \theta_2 / \partial r)$. This gives for the spherical coordinate

$$\frac{\partial \theta_1}{\partial t} = \frac{1}{r^2} \frac{\partial}{\partial r}\left[r^2 (\theta_1 D_2 + \theta_2 D_1) \frac{\partial \theta_1}{\partial r} \right] \tag{8}$$

Last we consider a characteristic situation. The particle diffusivity is such very small that we can ignore the $\theta_1 D_2$ term. Therefore equation (8) can be simplify

$$\frac{\partial \theta_2}{\partial t} = \frac{D_1}{2} \frac{1}{r^2} \frac{\partial}{\partial r}\left(r^2 \frac{\partial \theta_2^2}{\partial r} \right) \tag{9}$$

The boundary conditions are

$$\frac{\partial \theta_2}{\partial r}(0, t) = 0 \tag{10}$$

and

$$\theta_2 (R, t) = 1 \tag{11}$$

The initial condition is

$$\theta_2 (r, 0) = 1 \tag{12}$$

Dimensionless

Although we did the governing equation before, we need an equation easy to program and compare. For this reason, we can rewrite the equation by some dimensionless group (show in Table 1).

Table 1. Dimensionless equations of Fick's Law and simplified counter diffusion model

System	Dimensionless Group	Dimensionless Equation
Fick's Law	$\eta = \frac{r}{R}, \tau = \frac{t}{T}, T = \frac{R^2}{D_S}$	$\frac{\partial \theta}{\partial \tau} = \frac{1}{\eta^2} \frac{\partial}{\partial \eta}\left(\eta^2 \frac{\partial \theta}{\partial \eta} \right)$

Simplified counter diffusion model		$\frac{\partial\theta_2}{\partial\tau}=\frac{1}{2}\frac{1}{\eta^2}\frac{\partial}{\partial\eta}\left(\eta^2\frac{\partial\theta_2^2}{\partial\eta}\right)$

Mathematical Analysis

Usually, complicated models must be solved using numerical techniques. Numerical techniques are very general, and valid for most models. The most common numerical techniques for modeling electrochemical systems are: the finite difference method (FDM), the control volume formulation (CVF), and the finite element method (FEM)[6-8].

Here, we select the finite difference method to predict our model. In mathematics, the finite-difference methods are numerical methods for approximating the solutions to differential equations using finite difference equations to approximate derivatives.

Linearization Method

In the original case, the Fick's law can be linearized easily. By using the Taylor's expansion of θ, equation (1) will be changed to

$$\frac{\theta_{\eta,\tau+1}-\theta_{\eta,\tau}}{\Delta\tau}=\left[\frac{1}{\eta\Delta\eta}+\frac{1}{(\Delta\eta)^2}\right]\theta_{\eta+1,\tau+1}-\frac{2}{(\Delta\eta)^2}\theta_{\eta,\tau+1}+\left[\frac{1}{(\Delta\eta)^2}-\frac{1}{\eta\Delta\eta}\right]\theta_{\eta-1,\tau+1} \quad (13)$$

where $\tau = t\,D_S/R^2$ is the dimensionless time and $\eta = r/R$ is the dimensionless position in radius. The value of latter concentration is θ with lower index τ+1. Oppositely, the value of earlier concentration is θ with lower index τ-1. The meaning of θ with lower index η+1 and η-1 is the concentration near the position η.

This linearization method is useless to linearize equation (9) directly, because this it is non-linear partial differential equation (PDE). In other words, we need a new linearization method that is Richtmyer's Linearization Method.

Richtmyer's Linearization Method

Richtmyer's et al. [9] present a linearization method of non-linear PDE

$$\frac{\partial U}{\partial t}=\frac{\partial^2 U^m}{\partial x^2} \quad (14)$$

where m is a positive integral that is bigger than 2. And they approximate by the implicit weighted average difference. At last, using Taylor's expansion can linearize the differential equation.

$$\begin{aligned}&-2(m_1+m_2)\theta_{\eta+1,\tau}\theta_{\eta+1,\tau+1}+(1+4m_2\theta_{\eta,\tau})\theta_{\eta,\tau+1}-2(m_2-m_1)\theta_{\eta-1,\tau}\theta_{\eta-1,\tau+1}\\&=-(m_1+m_2)\theta^2_{\eta+1,\tau}+2m_2\theta^2_{\eta,\tau}-(m_2-m_1)\theta^2_{\eta-1,\tau}+\theta_{\eta,\tau}\end{aligned} \quad (15)$$

Where m_1 and m_2 are the coordinate parameter, they will be changed in different coordinate. If we consider the system in a spherical coordinate, m_1 and m_2 will be $\frac{\Delta\tau}{2\eta\Delta\eta}$ and $\frac{\Delta\tau}{2(\Delta\eta)^2}$.

Newton-Raphson method

In numerical analysis, Newton's method is perhaps the best-known method for finding successively better approximations to the zeroes (or roots) of a real-valued function [10-12]. Given a function $f(x)$ and its derivative $f'(x)$, we begin with a first guess x_0. A better approximation x_1 is

$$x_1 = x_0 - \frac{f(x_0)}{f'(x_0)} \tag{16}$$

According to this method, we defined the difference function

$$\begin{aligned} f_i(\theta_x) = & -2(m_1+m_2)\theta_{\eta+1,\tau}\theta_{\eta+1,\tau+1} + (1+4m_2\theta_{\eta,\tau})\theta_{\eta,\tau+1} - 2(m_2-m_1)\theta_{\eta\text{-}1,\tau}\theta_{\eta\text{-}1,\tau+1} \\ & + (m_1+m_2)\theta^2_{\eta+1,\tau} - 2m_2\theta^2_{\eta,\tau} + (m_2-m_1)\theta^2_{\eta\text{-}1,\tau} + \theta_{\eta,\tau} \end{aligned} \tag{17}$$

and determining the number of nodes, to arrange the matrix them. Then we find the approximate solution by C++ program.

Result and Discussion

Consider the mass transfer of lithium ion in battery discharge situation. The lithium ion diffusion in the anode particles is show in Fig.1. This result is due to assume without ion in the electrode in the beginning, then the lithium ion will diffuse into the anode particles (MCMB) gradually. Here, the concentration diagram of Li-ion is show that the diffusion process only has the transfer resistance of anode material. Therefore, the MCMB particle is saturated with lithium ion in a very short time. In the other hand, we have to consider the effect of the formerly entered ion that is the Modified Fick's Law. The result is also show in Fig.1. The curve is lower than the Fick's law, because the saturation time is delayed. In addition to, the slope is small than the Fick's law before $\tau = 0.5$ can show the effect of the formerly entered ion. The lithium ion inside the particle could obstruct the mass transfer, so the flux of the Modified Fick's Law will be small than the Fick's law. Moreover, the concentration of lithium ion increases along with the time such that the resistance also increases, so we can find the slope is changed slowly.

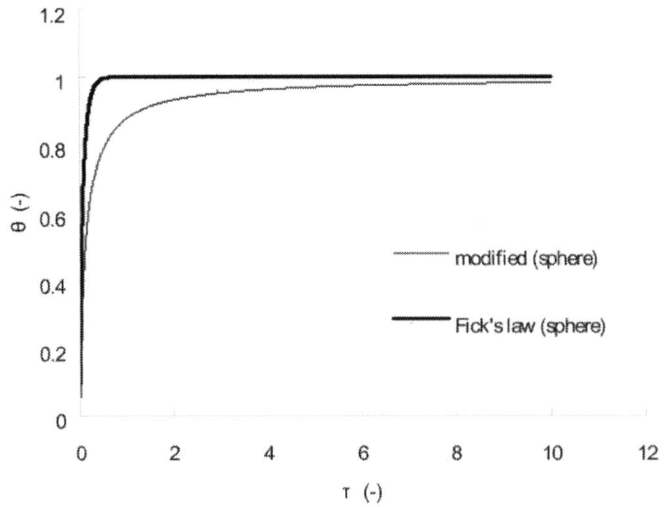

Fig.1 The Li-ion concentration diagram with Fick's law and modified Fick's law model

The main point of this simulation is the variation of potential along with time, and we could calculate the potential by the concentration distribution. First, we do the volume integral over the particle and find the average concentration of particles. And then we can obtain the corresponding potential by experimental data (show in Fig. 2) where potential is base on Li^+/Li. The trend is show in Fig. 3. The potential is decrease along with time, because the particles fill with lithium ion. Therefore, the potential of MCMB will be close to the potential of lithium. In our simulation, we can find the potential is equal to 0.02 V after $\tau = 6$.

Moreover, the Fick's Law and Modified Fick's Law have some difference in the potential profile. This result is due to the flux of Fick's Law is bigger than the Modified Fick's Law, so the potential of Fick's Law will reach the saturation potential quickly.

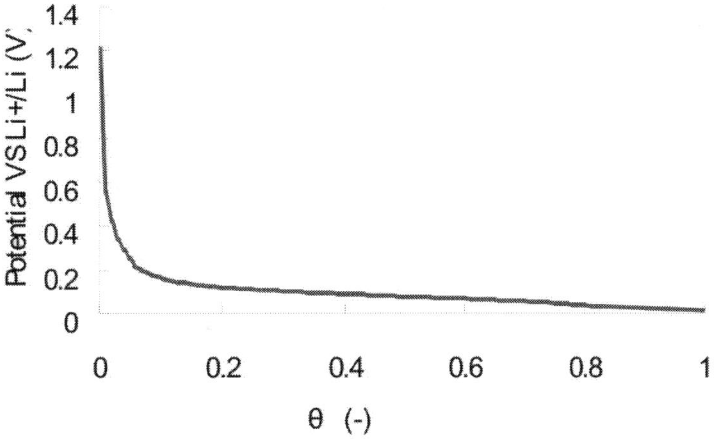

Fig.2 The open circuit potential (OCP) of MCMB T-cell

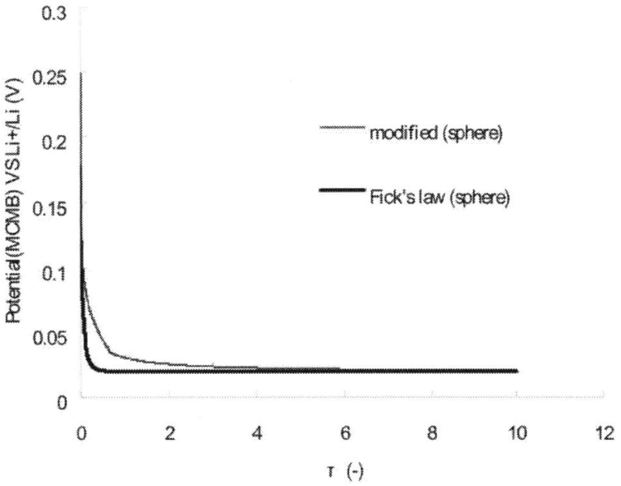

Fig.3 The potential distribution of MCMB particle with Fick's law and modified Fick's law model

Conclusion

The Li-ion transport phenomenon in solid-state is powerful to influence the potential of electrode. In this study, the simplified counter diffusion and Fick's law models are considered and find that the rate of Li-ion diffusion in simplified counter diffusion slows than Fick's law model. Due to the vacancy effect, the diffusion rate of Li-ion is not as Fick's law model so quickly and results in an average potential of MCMB particle is higher than Fick's law model.

Notation

D_S	Mass transfer coefficient of lithium ion, $[L^2/t]$
D_1	Mass transfer coefficient of lithium ion, $[L^2/t]$
D_2	Mass transfer coefficient of solid particle, $[L^2/t]$
m_1	Coordinate parameter, $[t/L^2]$
m_2	Coordinate parameter, $[t/L^2]$
r	Position in radius, $[L]$
R	Radius of solid particle, $[L]$
t	Time, $[t]$
T	Reference time, $[t]$
η	Dimensionless position in radius, [-]
θ	Dimensionless concentration of lithium ion in the Fick's law, [-]
θ_1	Dimensionless concentration of lithium ion in modified Fick's law, [-]
θ_2	Dimensionless concentration of solid particles in modified Fick's law, [-]
$\theta_{i,j}$	Dimensionless concentration at position i at time j, [-]
ν	Velocity relative to r axis, $[L/t]$
τ	Dimensionless time, [-]

References

1. M. Doyle, T.F. Fuller, J. Newman, *J. Electrochem. Soc.* 140 (1993) 1526.
2. Parthasarathy M. Gomadam[a], John W. Weidner[a,*], Roger A. Dougal[b], Ralph E. White, *J. of Power Sources* 110 (2002) 267–284.
3. Thomas F. Fuller, Marc Doyle, and John Newman, *J. Electrochem. Soc.*, 141, 1994.
4. Gang Ning, Ralph E. White, Branko N. Popov, *Electrochimica Acta* 51 (2006) 2012–2022.
5. V. Srinivasan and J. Newman, *J. Electrochem. Soc.*, **151**, A1517 (2004)
6. Oliver Rübenkönig, *The Finite Difference Method (FDM) - An introduction*, (2006) Albert

Ludwigs University of Freiburg
7. K.W. Morton and D.F. Mayers, *Numerical Solution of Partial Differential Equations*, An Introduction. Cambridge University Press, 2005.
8. Autar Kaw and E. Eric Kalu, *Numerical Methods with Applications*, (2008)
9. Richtmyer, R. D., and Morton, K. W. (1967) *Difference Method for Initial-Value Problems*. Interscience Publishers.
10. Tjalling J. Ypma, Historical development of the Newton-Raphson method, SIAM Review 37 (4), 531–551, 1995. doi:10.1137/1037125.
11. P. Deuflhard, Newton Methods for Nonlinear Problems. Affine Invariance and Adaptive Algorithms. Springer Series in Computational Mathematics, Vol. 35. Springer, Berlin, 2004. ISBN 3-540-21099-7.
12. C. T. Kelley, Solving Nonlinear Equations with Newton's Method, no 1 in Fundamentals of Algorithms, SIAM, 2003. ISBN 0-89871-546-6.

Metal-Dispersed Mesoporous Silica by in-Situ and Post Impregnation

Ki Woong An, Chemical Engineering & Material Science, Chung-Ang University, Seoul, South Korea

Min Jung Kim, Chemical Engineering & Material Science, Chung-Ang University, Seoul, South Korea

Sun-Geon Kim, Chemical Engineering & Material Science, Chung-Ang University, Seoul, South Korea

Metal-dispersed mesoporous silica was prepared by in-situ synthesis and post impregnation, respectively. In-situ synthesis of the metal-dispersed mesoporous silica was performed by the addition of metal precursor into the precursor solution in the aerosol reactor. Silver grew in the mesopores surrounded by the solidified silica wall. The incorporation of the metals catalyzed the solidification of the silica phase, considerably improving the regularity of the mesophases. The post impregnation with various precursor solutions containing silver was carried out for the mesoporous silica. The dispersion of metal nanoparticles was significantly improved with 3-aminoisopropyltetraethoxysilane (APS) and mercaptoisopropyltetraethoxysilane (MPTS). The use of ethylene glycol as a solvent also improved the metal dispersion.

See more of this Session: Poster Session: Interfacial Phenomena
See more of this Group/Topical: Engineering Sciences and Fundamentals

Derivation of Modified Langmuir Isotherm and Application to Antifreeze Protein Multimers

Özge Can, Department of Chemical Engineering, University of Delaware, Newark, DE

Ali Ghoorchian, Chemical and Biomedical Engineering, Cleveland State University, Cleveland, OH

Nolan Holland, Chemical and Biomedical Engineering, Cleveland State University, Cleveland, OH

Antifreeze proteins (AFPs) are ice binding proteins found in various organisms allowing them to survive at subzero temperatures by inhibiting ice crystal growth. Upon lowering the solution temperature, AFPs cause thermal hysteresis, which is the difference between the freezing and melting temperatures and it is the common measure of their activity that can be experimentally determined. This interesting property has made them suitable candidates for several applications such as subzero preservation of cells, tissues, organs, frozen foods, and heat transfer fluids in refrigeration systems. These real life applications, however require a better understanding of their mechanism of action on the ice surface that would help to design superior molecules that are more suitable for economical use.

This study aims to better understand the equilibrium behavior of the AFP/ice system by the application of adsorption isotherms. An adsorption equilibrium model is derived based on the classical Langmuir isotherm in order to determine the surface concentration of the AFP molecules on the ice surface, which is experimentally impossible to determine. The link between thermal hysteresis and surface coverage was shown for the one-, two- and three-domain type III fish AFPs that were designed and expressed in E.Coli expression system. The equilibrium binding constant and hence the adsorption free energy for the type III AFP domain is also presented.

See more of this Session: Poster Session: Interfacial Phenomena
See more of this Group/Topical: Engineering Sciences and Fundamentals

Microbial Inhibition of Corrosion-the Role of Biofilms

Prithiviraj Chellamuthu, Molecular and Computational Biology, University of Southern California, Los Angeles, CA

Kenneth Nealson, Department of Earth Sciences, University of Southern California, Los Angeles, CA

In the United States alone, the cost of corrosion (both money spent to stop it, and money spent to repair it) is estimated to be as much as $364 billion per year. To put this in perspective, the amount spent each year is roughly equivalent to the cost of the damage done by Hurricane Katrina. The problem is widespread, affecting systems ranging from water sprinklers to oil well rigs, airplanes, and fuel storage tanks. It is commonly believed microorganisms increase the corrosion rate, but recent work has shown that microorganisms can inhibit corrosion in various metals, including iron, copper, and aluminum. Recent investigations in microbiologically influenced corrosion inhibition (MICI) have produced models: bacteria neutralizing the corrosive reactants and bacteria forming a protective film on the metal surface.

The mechanism through which the bacteria alter the corrosion kinetics in decreasing the corrosion rate is not well understood. The thermodynamics of MICI is determined by the microenvironments on the metal surfaces, biofilm dynamics, and media interfaces. That is, because of microbial activity changing pH, Eh, and other important variables, the microbes can greatly alter the kinetics of the system. Understanding how each variable is microbially affected requires a clear knowledge of the metabolic and electron transfer activities of relevant microbes. Hence, in this study a well characterized model organism Shewanella oneidensis MR-1 was used to understand relevant mechanisms, especially the role of biofilm formation on the metal surface.

A galvanic cell was set up with an aluminum rod as the anode and a copper rod as the cathode. An Ag/AgCl was used as the reference electrode was used for all the experiments. The electrolytes used were PIPES buffered (50mM, pH 7.0) Luria-Bertani media or PIPES buffer (50mM, pH 7.0) with lactate as carbon source. The cells were maintained under aerobic or anaerobic conditions with air or nitrogen being flushed to maintain conditions. To understand the importance of biofilm in corrosion inhibition, MR-1 a good biofilm forming bacteria and a poor biofilm forming mutant variant of MR-1, ΔpilD, were used. Non-destructive electrochemical techniques such as potentiodynamic polarization and electrochemical impedance spectroscopy (EIS) were employed to study corrosion reaction kinetics. Planktonic cell counts were performed at the beginning and at the end of the experiments. The exposed metal surfaces were analyzed using scanning electron microscopy (SEM) and deep ultra-violet microscopy, a non destructive method eliminating the need for sample processing or fixing.

The power output of the cell increased significantly under aerobic conditions when compared to the anaerobic condition. Under aerobic conditions, the polarization resistance (Rp) i.e. corrosion resistance of the MR-1 cell increased. Unexpectedly running the cell under anaerobic conditions had the opposite effect on Rp.

Bacterial-conditioned media of MR-1 (spent media, bacteria removed by centrifugation) also exhibited some corrosion inhibition properties compared to sterile media indicating that compounds synthesized by the bacteria present in the media might offer protection to the metal electrodes, even in the absence of biofilms.

Also in order to identify the compounds that potentially offer protection to the metals, experiments are being carried out with PIPES minimal media, instead of a rich PIPES buffered LB media. Experiments with the ΔpilD (biofilm formation mutants) are also being carried out and these experiments should help us understand the importance of biofilms in protecting the metal from corrosion.

See more of this Session: Poster Session: Interfacial Phenomena
See more of this Group/Topical: Engineering Sciences and Fundamentals

Absorption of Carbon Dioxide Into Aqueous Surfactant Solutions

Matthew E. Schroeder, Chemical Engineering, Kansas State University, Manhattan, KS

Lance D. Williamson, Chemical Engineering, Kansas State University, Manhattan, KS

Larry Glasgow, Kansas State University, Manhattan, KS

The absorption of carbon dioxide into aqueous surfactant solutions at elevated pressures is of interest in certain military applications. We have conducted a study of the absorption process on a laboratory scale (using beverage cylinders) to better understand the approach to saturation. Since the solubility is pressure-dependent, the interfacial equilibrium composition (XA0) changes over the course of the absorption. We began by studying the diffusion-limited process in which the head space above the quiescent surfactant solution was pressurized with carbon dioxide (with a typical initial pressure of about 55 bar). Experimental data were obtained by recording the gas pressure over time; the decreasing pressure was converted to absorption rate. A Fickian diffusion model was compared with the data.

The rate of absorption under quiescent conditions is impractically slow, so we devised a roller apparatus driven by a variable-speed DC motor. In this manner the pressurized cylinders could be rotated slowly in a horizontal position, providing greater interfacial area and continuous renewal of the liquid at the interface. A model of the resulting liquid flow has been constructed and tested against experimental data. As expected, this arrangement significantly accelerates the approach to saturation, while minimizing the temperature increase resulting from viscous dissipation.

See more of this Session: Poster Session: Interfacial Phenomena
See more of this Group/Topical: Engineering Sciences and Fundamentals

Characterization of Commercial Antifoams for Use in Immunoassay Diluents

Daniel L. Obrzut, Diagnostics R&D, Abbott Laboratories, Abbott Park, IL

Joan E. Peart, Diagnostics R&D, Abbott Laboratories, Abbott Park, IL

Alfredo R. Narváez, Diagnostics R&D, Abbott Laboratories, Abbott Park, IL

Immunoassays are a key tool for clinical and research laboratories. Immunoassays are often solution-based kits made up of an active ingredient(s) in a diluent(s) and executed in multi-step automated systems. The diluents of the solutions are formulated to prevent any of the assay components from aggregating or dropping out of solution while ensuring accurate, optimal, and repeatable results. These diluents can include buffers, salts, chelators, proteins, surfactants, and antifoams. Antifoam is used in these formulations to deter foam formation due to the presence of foaming agents (e.g. surfactants and proteins). Antifoam is added at the minimum effective concentration so foam is eliminated without interfering with the assay. Selection of antifoam and antifoam concentration is very assay dependent since the presence of foaming agents varies by the diluent formulation.

In this study, several commercial, polydimethylsiloxane-based antifoams were tested to evaluate their potential for use in immunoassays. Shake testing was performed to determine the minimum antifoam concentration needed to defoam solutions of common diluent formulations. The shake test evaluated how much foam initially forms, how fast foam is eliminated, and how many times foam can be eliminated rapidly. The effect antifoams had on the rheological properties of diluents was evaluated using an oscillatory rheometer. Since the antifoams of interest are hydrophobic in nature, their effect on the air-liquid interface was analyzed using Brewster Angle Microscopy to address their impact on automatic pipetting liquid detection. These three methods assist with the characterization of antifoams for use in immunoassays.

See more of this Session: Poster Session: Interfacial Phenomena
See more of this Group/Topical: Engineering Sciences and Fundamentals

Adsorption Isotherm and Structure of Water of Methyl, Carboxylic Acid, and Hydroxyl Terminated Self-Assembled Monolayers

Aimee Tu, Chemical Engineering, Pennsylvania State University, University Park, PA

Anna Barnette, Chemical Engineering, Pennsylvania State University, University Park, PA

Seong H. Kim, Chemical Engineering, Pennsylvania State University, University Park, PA

Recently a lot of research has been done in order to understand water adsorption properties on self-assembled monolayers (SAMs). This is important because water is commonly found on surfaces due to atmospheric humidity. The advances made in understanding interfacial water have contributed to various nanotechnology and biomaterials applications. This study examines the effects of surface chemistry on the adsorption isotherm and structure of water layers on three thiol-based self-assembled monolayers (SAMs) on gold with controlled terminal groups - methyl, carboxylic acid, and hydroxyl groups. The adsorption isotherms are measured with polarization modulation reflection absorption infrared spectroscopy (PM-RAIRS), ellipsometry, and quartz crystal microbalance (QCM), and the structure of the water layer is deduced from the OH stretching vibration peak positions in PM-RAIRS. The CH3-terminated SAM shows no uptake of water due to the hydrophobic nature of the methyl group. The adsorption isotherm of water on the COOH-terminated SAM shows a strong humidity dependence. The water layer on the COOH surface seems to have a broad distribution of liquid and solid-like structures. The OH-terminated SAM keeps a large amount of water molecules on it even at dry conditions and the adsorption isotherm exhibits little humidity dependence.

See more of this Session: Poster Session: Interfacial Phenomena
See more of this Group/Topical: Engineering Sciences and Fundamentals

Interactions Between Cationic Bilayers in the Presence of Polyelectrolytes

Travers H. Anderson, Chemical Engineering, University of California, Santa Barbara (UCSB), Santa Barbara, CA

Hongbo Zeng, Chemical and Materials Engineering, University of Alberta, Edmonton, AB, Canada

Jacob N. Israelachvili, Chemical Engineering, University of California, Santa Barbara (UCSB), Santa Barbara, CA

Colloidal vesicle dispersions are commonly used in both consumer products and medical applications. Often additives such as polyelectrolytes are added to the dispersions to stabilize the dispersions or give desired consumer affects to the product. The colloidal properties of these dispersions are determined by interactions between the vesicles and polyelectrolytes making up the dispersion. We have used a Surface Forces Apparatus (SFA) to measure the interactions between positively charged di (tallowethyl-ester) dimethyl ammonium chloride bilayers in the presence and absence of positively charged poly (diallyldimethylammonium chloride) (polyDADMAC) in calcium chloride solutions. The effects of salt concentration, electrolyte concentration and molecular weight have been studied for this system. In colloidal vesicle dispersions of this kind we have found that the addition of polyelectrolyte has been shown to destabilize the suspension. This is due to a depletion attraction interaction between the vesicles in the dispersion as well as a decrease in the electrostatic repulsion between the vesicles due to the presence of the polyelectrolyte. This study has implications for anticipating the behavior of vesicle dispersions in complex mixtures.

See more of this Session: Poster Session: Interfacial Phenomena
See more of this Group/Topical: Engineering Sciences and Fundamentals

Formation of High Aspect-Ratio Solid Silicon Wires within Optical Fibers Using High Pressure Chemical Fluidic Deposition

Banafsheh Keshavarzi, Chemical Engineering, The Pennsylvania State University, University Park, PA

Neil Baril, Department of Chemistry, The Pennsylvania State University, University Park, PA

John V. Badding, Department of Chemistry, The Pennsylvania State University, University Park, PA

Ali Borhan, Chemical Engineering, The Pennsylvania State University, University Park, PA

High-pressure chemical fluidic deposition is used for the formation of semiconductor micro and nanowires within the capillary spaces of highly-ordered microstructured optical fibers. We consider deposition of silicon under high pressure flow of a silane/helium mixture within extreme aspect-ratio microcapillaries. Although a complete filling of the microcapillary is ideally desired, experimental observations show that the thickness of the deposited silicon layer varies with axial position. It is shown that, in the case of flowing deposition, the temperature and concentration profiles along the capillary length during deposition lead to the formation of two separate nodes in the deposition profile. The formation of the initial node is shown to be due to an instability resulting from the coupling between flow and reaction. A complete filling of the region between these nodes, several centimeters in length, was achieved by optimizing the deposition conditions. Deposition within sealed capillaries also demonstrated a filling profile possessing a completely filled region centimeters in length.

See more of this Session: Poster Session: Interfacial Phenomena
See more of this Group/Topical: Engineering Sciences and Fundamentals

An Experimental Study of the Radial Penetration of Non-Newtonian Fluids in Thin Porous Substrates

Shirin Sharifi Khobdeh, Chemical Engineering, The Pennsylvania State University, University Park, PA

Ali Borhan, Chemical Engineering, Pennsylvania State University, University Park, PA

We present the results of an experimental study of the radial capillary penetration of non-Newtonian fluids in thin porous substrates. The porous substrates included paper and thin packed beds of spherical glass beads. The porous substrates were sufficiently thin that the liquid completely penetrated across the porous substrate in all cases, leading to the formation of identical nearly circular liquid stains on both sides of the substrate. Measurements of the time-evolution of stain area were used to characterize the spreading rate. For fixed porous substrate, the spreading rates for non-Newtonian fluids were compared to those for Newtonian fluids. The effect of the size of the liquid reservoir on the rate of the radial spreading was also examined by considering both infinite liquid reservoirs and finite drops.

See more of this Session: Poster Session: Interfacial Phenomena
See more of this Group/Topical: Engineering Sciences and Fundamentals

Micellization of Alkyl-Propoxy-Ethoxylate Surfactants in Aqueous Solutions: Thermodynamics and Structure

Biswajit Sarkar and Paschalis Alexandridis

Department of Chemical and Biological Engineering
University at Buffalo, The State University of New York (SUNY)
Buffalo, NY 14260-4200, USA

Abstract

The self-assembly of an alkyl-propoxy-ethoxylate surfactant in aqueous solution has been investigated with a focus on (i) the thermodynamics of micellization and (ii) the structure of the micelles. The results are compared to those for alkyl-ethoxylate amphiphiles with the aim to elucidate the role of the middle, propoxy, block in this novel alkyl-propoxy-ethoxylate surfactant, which exhibits a "graded" hydrophobic-hydrophilic character.

Introduction

Alkyl-ethoxylates (C_iEO_j) are nonionic surfactants with low critical micelle concentrations (cmc) in water, compatibility with other surfactants, and mild nature. These surfactants are widely used in personal care products, detergency, paint formulations, controlled drug delivery, etc.[1, 2] Considering its broad technical application, the micellization of these surfactants in aqueous solutions has been extensively investigated.[3-5] The following trends have emerged: the log(cmc) decreases linearly with the alkyl chain length[3]; the cmc decreases with a temperature increase[4]; the cmc increases as the poly(ethylene oxide) (PEO) block length increases.[5]

Alkyl-propoxy-ethoxylates ($C_iPO_nEO_j$) are a novel class of nonionic surfactants that have been used successfully in nanomaterial synthesis.[6] The middle poly(propylene oxide) (PPO) block possesses a hydrophobicity/hydrophilicity intermediate to that of alkyl and poly(ethylene oxide). Recently, our group investigated the aqueous phase behavior and structure of $C_iPO_nEO_j$ surfactants using a mean field lattice model.[7, 8] We found the thermotropic phase behavior to dominate over the lyotropic behavior, and several ordered phases to form at lower temperature. The domination of thermotropic phase transitions over lyotropic was explained on the basis of the decrease in solvent quality for both PEO and PPO blocks, and the decrease in incompatibility between alkyl and PEO blocks at elevated temperature. Here we turn our attention to the dilute solution behavior of the $C_{13}PO_{12.2}EO_{17}$ surfactant in water.

Results and Discussion

Micelle Formation

The cmc of the $C_{13}PO_{12.2}EO_{17}$ surfactant in aqueous solution was determined using DPH UV-vis absorption spectroscopy[9] and pyrene fluorescence spectroscopy[10]. The DPH absorption intensity is a function of the concentration of solubilized DPH. At surfactant concentrations below the cmc, DPH was not solubilized in a hydrophobic environment and the UV-vis absorption intensity due to DPH was very low. As the surfactant concentration increased, micelles formed and DPH was solubilized in the hydrophobic micelle core, resulting in a characteristic spectrum with the main peak at 356 nm.[9] UV-vis absorption spectra of DPH/surfactant/water for varying surfactant concentration are presented in Figure 1. The absorption intensity at 356 nm has been plotted vs. the logarithm of surfactant concentration, and the cmc was determined from the first inflection point of this curve (as indicated in Figure 1).

The fluorescence emission intensity of pyrene-containing aqueous surfactant solutions (at an excitation wavelength of λ = 335 nm) includes five characteristics vibronic peaks. The ratio of the first to the third vibronic peak (I_1/I_3) of the pyrene emission spectrum decreases with decreasing polarity of the pyrene microenvironment[11] (emission spectra of varying concentration and the variation in I_1/I_3 ratio at different surfactant concentrations are presented in Figure 1). As the surfactant concentration increased toward the cmc, pyrene started to partition to hydrophobic sites and, as a result, the I_1/I_3 ratio started to decrease. At concentrations above the cmc, pyrene has accumulated into hydrophobic moieties and the I_1/I_3 ratio value has stabilized. The I_1/I_3 ratios for aqueous surfactant solutions have been plotted against the logarithm of surfactant concentration, and the cmc was determined from the second inflection of this curve, as indicated in Figure 1.[10]

The cmc of the surfactant at 25°C is 0.02 wt%. With an increase in temperature from 25°C to 40°C, the cmc decreased to 0.01 wt%. This cmc decrease was more pronounced for $C_{13}PO_{12.2}EO_{17}$ than for C_iEO_j ($C_{12}EO_6$) surfactants.[12] This can be explained on the basis of higher dehydration of PPO over PEO at higher temperature.[9]

Thermodynamics of micellization

The micellization of nonionic surfactants can be described with the closed association model which assumes an equilibrium between individually dissolved surfactant molecules and micelles.[9] According to this model, the change in the standard free energy (ΔG^o) for transfer of one mole of surfactant from an aqueous solution to the micellar phase is given by:[9]

$$\Delta G^o = R\,T\ln(X_{cmc}) \qquad (1)$$

where R is the ideal gas constant, T the absolute temperature, and X_{cmc} the surfactant mole fraction at the critical micellization concentration.

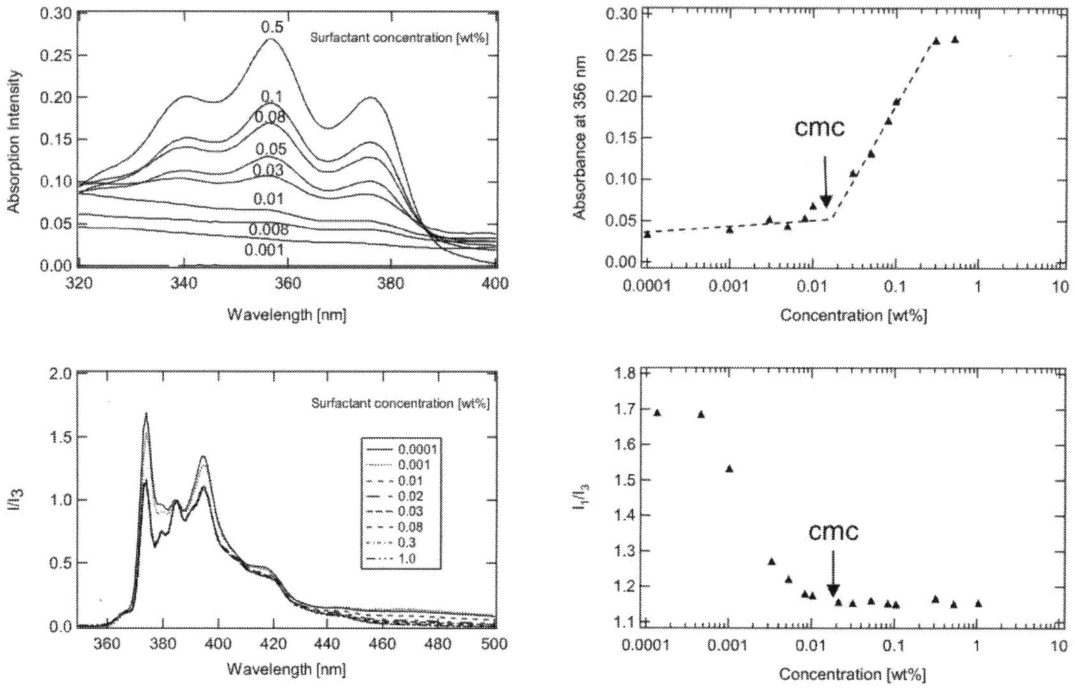

Figure 1. (top, left) DPH UV-vis spectra at various surfactant concentrations in aqueous solution at 25°C. (top, right) Absorbance due to DPH at 356 nm plotted vs. surfactant concentration at 25°C. (bottom, left) Variation in the pyrene fluorescence emission spectra (normalized with respect to the I_3 intensity) with increasing surfactant concentration in aqueous solution at 25°C. (bottom, right) I_1/I_3 plotted vs. surfactant concentration at 25°C. The cmc is determined from the first inflection point of absorbance at 356 nm vs. surfactant concentration curve and from the second inflection point of I_1/I_3 ratio vs. surfactant concentration curve, as indicated by the arrows.

The standard enthalpy of micellization (ΔH^o) (assumed independent of temperature for the temperature range considered) can be obtained through the following equation:

$$\Delta H^o_{mic} = -RT^2\left(\frac{\partial \ln X_{cmc}}{\partial T}\right) \qquad (2)$$

The standard entropy of micellization (ΔS^o) can be calculated from the free energy and enthalpy:

$$\Delta S^o_{mic} = (\Delta H^o_{mic} - \Delta G^o_{mic})/T \quad (3)$$

The ΔG^o, ΔH^o, and ΔS^o values calculated from the experimental data according to the above equations are -33.2 kJ/mol, 34.1 kJ/mol, and 0.22 kJ/mol/K, respectively. The positive enthalpy of micellization indicates that the transfer of a surfactant molecule from solution to a micelle is an enthalpically unfavorable, endothermic process. Such a positive ΔH^o has been ascribed to an energy penalty associated with hydrogen bond breaking during micellization.[9]

The unfavorable enthalpy is overwhelmed by a favorable (positive) entropy. The positive values of ΔS^o can be viewed in the context of the very low solubility of hydrocarbons in water. Introduction of hydrocarbon molecules into water results in a substantial decrease of water entropy. Micellization is associated with the transfer of alkyl chains from aqueous solution to oil-like micelle cores, restoring the water structure and hence decreasing the water entropy. The gain in the water entropy dominates over the loss in entropy due to the localization of hydrocarbon chains in the micelle core.[9]

We compare now the standard enthalpy and entropy of micellization of $C_{13}PO_{12.2}EO_{17}$ and C_iEO_j surfactants. Both ΔH and ΔS are positive indicating that the micellization of C_iEO_j surfactants is endothermic and entropy driven similarly to the micellization of $C_iPO_nEO_j$ surfactants. In order to decouple the influence of the hydrophilic head group from the overall chemical composition, we normalized ΔH and ΔS with N_{EO} for both types of amphiphiles. A higher value of $\Delta S/N_{EO}$ is observed for $C_{13}PO_{12.2}EO_{17}$ compared to C_iEO_j. This may be attributed to the presence of PPO in $C_{13}PO_{12.2}EO_{17}$. Higher values of $\Delta H/N_{EO}$ are also observed for $C_{13}PO_{12.2}EO_{17}$ compared to C_iEO_j.

Micelle Structure

We utilized Small angle X-ray scattering (SAXS) to study the structure of the $C_{13}PO_{12.2}EO_{17}$ surfactant micelles in aqueous solution. The angular distribution of the scattered electrons was recorded in a two dimensional detector. The scattering intensity was obtained by averaging the intensity of all points in the 2-D detector space for a scattering vector value, q. The reduction of data to absolute scale was done following Ref[13]. The SAXS scattering intensity originating from a system of particles can be written as[14]

$$I(q) = n_p P(q) S(q) \quad (4)$$

where n_p is the particle number density, P(q) the particle form factor and S(q) the particle structure factor. The form factor captures the structure of the particle (micelle) while the structure factor accounts for inter-particle interactions. We found the ellipsoid core shell form factor[15] the most appropriate to describe the micelle internal structure. The core containing alkyl chains is assumed to be solvent-free, whereas the shell is highly solvated. We assumed the minor core radius of the micelles to be equal to the extended alkyl chain length. The scattering results from the differences in electron density between core and shell, and between the shell and bulk solvent. Inter-particle interactions are accounted through the hard sphere structure factor S(q).[16]

Scattering data together with the fit (solid line) are presented in Figure 2. The major core radius, minor and shell radius, and hard sphere radius obtained from this fit are 38 Å, 43 Å, 70 Å, and 80 Å, respectively. The micelle association number is 259.

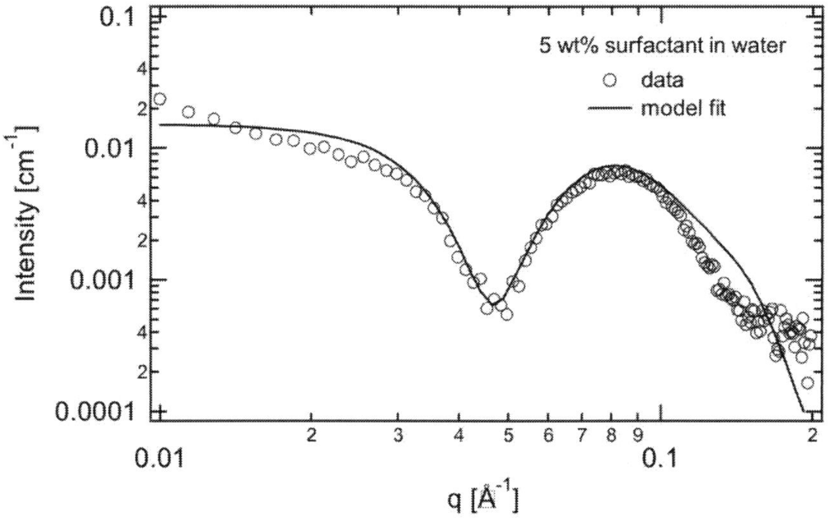

Figure 2. Reduced SAXS intensity data for 5 wt% aqueous solutions of $C_{13}PO_{12.2}EO_{17}$ surfactant at 25°C. The solid line represents a fit to the core shell oblate ellipsoid.

It is interesting to compare the structure of the $C_{13}PO_{12.2}EO_{17}$ micelles with that of the well studied C_iEO_j micelles. Preu *et al.* used an ellipsoidal core shell model to describe the structure of $C_{12}EO_{23}$ (Brij 35) micelles.[17] The minor core radius obtained from their study was found to vary within 15-17 Å which is comparable with our results. However, they reported a smaller major core radius compared to our result. The Generalized Indirect Fourier Transform (GIFT) method was used to obtain the $C_{12}EO_{23}$ micelle structure in aqueous solution.[18] The micelle core radius was found to vary within the fully stretched chain length of the alkyl chain. At higher surfactant concentration (above 10 wt%) an ellipsoidal (decaying rod-like) shape was found.[18]

References

1. Alexandridis, P.; Lindman, B., *Amphiphilic block copolymers: Self-assembly and applications*. Elsevier: Amsterdam ; New York, 2000.
2. Rhein, L. D.; Schlossman, M.; O'Lenick, A.; Somasundaran, P., *Surfactants in personal care products and decorative cosmetics*. Third ed.; CRC Press: 2006; Surfactant Science Series. Vol. 135.
3. Meguro, K.; Takasawa, Y.; Kawahashi, N.; Tabata, Y.; Ueno, M., Micellar properties of a series of octaethyleneglycol-n-alkyl ethers with homogeneous ethylene-oxide chain and their temperature-dependence. *Journal of Colloid and Interface Science* **1981**, 83, (1), 50-56.
4. Rusdi, M.; Moroi, Y.; Hlaing, T.; Matsuoka, K., Micelle formation and surface adsorption of octaethylene glycol monoalkyl ether (CnE8). *Bulletin of the Chemical Society of Japan* **2005,** 78, (4), 604-610.
5. Barry, B. W.; Eleini, D. I. D., Surface properties and micelle formation of long-chain polyoxyethylene nonionic surfactants. *Journal of Colloid and Interface Science* **1976,** 54, (3), 339-347.
6. Bollmann, L.; Urade, V. N.; Hillhouse, H. W., Controlling interfacial curvature in nanoporous silica films formed by evaporation-induced self-assembly from nonionic surfactants. I. Evolution of nanoscale structures in coating solutions. *Langmuir* **2007,** 23, (8), 4257-4267.
7. Shusharina, N. P.; Alexandridis, P.; Linse, P.; Balijepalli, S.; Gruenbauer, H. J. M., Phase behavior and structure of an ABC triblock copolymer dissolved in selective solvent. *European Physical Journal E* **2003,** 10, (1), 45-54.
8. Shusharina, N. P.; Balijepalli, S.; Gruenbauer, H. J. M.; Alexandridis, P., Mean-field theory prediction of the phase behavior and phase structure of alkyl-propoxy-ethoxylate "graded" surfactants in water: Temperature and electrolyte effects. *Langmuir* **2003,** 19, (10), 4483-4492.
9. Alexandridis, P.; Holzwarth, J. F.; Hatton, T. A., Micellization of poly(ethylene oxide)-poly(proplene oxide)-poly(ethlene oxide) triblock copolymers in aqueous-solutions - thermodynamics of copolymer association. *Macromolecules* **1994,** 27, (9), 2414-2425.
10. Lin, Y.; Alexandridis, P., Cosolvent effects on the micellization of an amphiphilic siloxane graft copolymer in aqueous solutions. *Langmuir* **2002,** 18, (11), 4220-4231.
11. Nivaggioli, T.; Alexandridis, P.; Hatton, T. A.; Yekta, A.; Winnik, M. A., Fluorescence probe studies of Pluronic copolymer solutions as a function of temperature. *Langmuir* **1995,** 11, (3), 730-737.
12. Goto, A.; Takemoto, M.; Endo, F., A thermodynamic study on micellization of nonionic surfactant in water and in water-ethanol mixture by gel-filtration. *Bulletin of the Chemical Society of Japan* **1985,** 58, (1), 247-251.
13. Pedersen, J. S., A flux- and background-optimized version of the NanoSTAR small-angle X-ray scattering camera for solution scattering. *Journal of Applied Crystallography* **2004,** 37, 369-380.
14. Lin, Y.; Alexandridis, P., Small-angle neutron scattering characterization of micelles formed by poly(dimethylsiloxane)-graft-polyether copolymers in mixed polar solvents. *Journal of Physical Chemistry B* **2002,** 106, (47), 12124-12132.

15. Berr, S. S., Solvent isotope effects on alkyltrimethylammonium bromide micelles as a function of alkyl chain-length. *Journal of Physical Chemistry* **1987,** 91, (18), 4760-4765.
16. Alexandridis, P.; Yang, L., SANS investigation of polyether block copolymer micelle structure in mixed solvents of water and formamide, ethanol, or glycerol. *Macromolecules* **2000,** 33, (15), 5574-5587.
17. Preu, H.; Zradba, A.; Rast, S.; Kunz, W.; Hardy, E. H.; Zeidler, M. D., Small angle neutron scattering of D2O-Brij 35 and D2O-alcohol-Brij 35 solutions and their modelling using the Percus-Yevick integral equation. *Physical Chemistry Chemical Physics* **1999,** 1, (14), 3321-3329.
18. Tomsic, M.; Bester-Rogac, M.; Jamnik, A.; Kunz, W.; Touraud, D.; Bergmann, A.; Glatter, O., Nonionic surfactant Brij 35 in water and in various simple alcohols: Structural investigations by small-angle X-ray scattering and dynamic light scattering. *Journal of Physical Chemistry B* **2004,** 108, (22), 7021-7032.

Mechanisms of Anti-Fouling Properties of Polypeptoid Polymers Adsorbed On Various Surfaces: A Study by Molecular Dynamics Simulations

Sung Hyun Park, Department of Biomedical Engineering, Northwestern University, Evanston, IL

Igal Szleifer, Department of Boimedical Engineering, Northwestern University, Evanston, IL

Protein adsorption and cellular adhesion to surfaces is one of the main reasons for damages and malfunctioning of medical devices such as cardiovascular implants, stents, and catheters. Development of anti-fouling surfaces has been major topic of research in recent years, where coating surfaces with poly(ethylene glycol) (PEG)-based polymers and other alternatives being one of the most commonly explored methods. Recently, poly-N-substituted glycine oligomers or polypeptoids grafted onto surfaces through a peptide with 3,4-dihydroxy-L-phenylalanine(DOPA)-lysine(LYS) sequence have been shown to exhibit excellent anti-fouling properties on various surfaces such as teflon and titanium oxide. In this study, we investigated the detailed mechanisms of surface adsorption and anti-fouling properties of polypeptoid polymers on a variety of surfaces with an atomic resolution by performing classical molecular dynamics simulations. The effects of different variables, such as the polymer chain length, variations in monomer structure, polymer thickness, and surface density and packing, on the antifouling properties have been examined. The adsorption mechanisms of DOPA-LYS peptide units in association with robust attachment of the polypeptoid polymers to the surfaces have also been studied.

See more of this Session: Poster Session: Interfacial Phenomena
See more of this Group/Topical: Engineering Sciences and Fundamentals

Phase Discontinuity in Porous Media with Different Wetting Properties

Trung V. Nguyen, University of Kansas, Lawrence, KS

Clint D Frye, University of Kansas, Lawrence, KS

When two porous media with different wetting properties are placed in contact the force balance requires the capillary pressure at the interface of the two media to be continuous. Since porous media with different wetting properties have different capillary pressure curves this leads to a phase discontinuity at the interface. Even though this boundary or interfacial condition is well established and has been used in many mathematical models, including fuel cell models, it has not been experimental demonstrated. This work shows two experiments that were conducted to visually demonstrate this interfacial phenomenon, one with small beads and another with fibers of different wetting properties.

See more of this Session: Poster Session: Interfacial Phenomena
See more of this Group/Topical: Engineering Sciences and Fundamentals

A Molecular Dynamics Study of C12E6 and SDS Surfactants at the Silica-Water and Air-Water Interfaces

Liu Shi, School of Chemical Biological and Materials Engineering, The University of Oklahoma, Norman, OK

Naga Rajesh Tummala, School of Chemical Biological and Materials Engineering, The University of Oklahoma, Norman, OK

Alberto Striolo, School of Chemical Biological and Materials Engineering, The University of Oklahoma, Norman, OK

NVT molecular dynamics simulations have been carried out to investigate the adsorptions of hexaethylene glycol monododecyl ether (C12E6) and sodium dodecyl sulfate (SDS) surfactants at the silicon dioxide-water and water-air interfaces. The simulations were performed at room temperature and at different surfactant coverages, from an infinite-diluted coverage to full monolayer. We will discuss structural and dynamic properties of the molecular adsorption of surfactants by presenting results for density profiles, end-to-end distances, tilt angles, radial distribution functions, and mean square displacement of both water and surfactants. The results show an entangled structure of surfactant chains at both silica-water and air water interfaces and the interface properties are dependent of the degree of coverage. The simulations will be compared to experimental data obtained as a function of the density of surfactant molecules per unit interfacial surface area. The results are expected to secure advances in novel technologies such as admicellar polymerization.

See more of this Session: Poster Session: Interfacial Phenomena
See more of this Group/Topical: Engineering Sciences and Fundamentals

The Influence of Bilayer Composition On the Gel to Liquid Crystalline Transition

K. G. Ayappa, Department of Chemical Engineering, Indian Institute of Science, Bangalore, India

Ananya Debnath, Chemical Engineering, Indian Institute of Science, Bangalore, India

V Kumaran, Chemical Engineering, Indian Institute of Science, Bangalore, India

Prabal K. Maiti, Center for Condensed Matter Theory, Department of Physics, Indian Institute of Science, Bangalore, India

We report molecular dynamics simulations of bilayers using a united atom model with explicit solvent molecules. The bilayer consists of single tail cationic surfactant, behenyl trimethyl ammonium chloride (BTMAC) with stearyl alcohol (SA) as the co-surfactant. We study the gel to liquid crystalline transitions in the bilayer by varying the amount of water at fixed BTMAC to SA ratio as well as by varying the BTMAC to SA ratio at fixed water content. The bilayer is found to exist in the tilted, $L_{\beta'}$ phase at low temperatures and for the compositions investigated in this study, the $L_{\beta'}$ to L_{α} melting transition occurred in the temperature range of 330-338 K. The larger area per head group associated with the quarternary amine surfactant BTMAC decreases the transition temperature with an increase in the BTMAC composition. For the highest BTMAC to SA composition (2:3 molar ratio) we observe an increase in the d-spacing prior to the melting transition. This increase in d-spacing is accompanied by a sharpening in the water density variation across the head group region of the bilayer. Signatures of this swelling effect which can be observed in the alkane density distributions, area per head group and membrane thickness, are attributed to the hydrophobic effect. At a fixed bilayer composition, the transition temperature (>338 K) from the $L_{\beta'}$ to L transition obtained for a

the high water content bilayer (80 wt %) is similar to that

obtained with low water content (54.3 wt %), confirming that the

melting transition at these water contents, is dominated by chain melting.

See more of this Session: Poster Session: Interfacial Phenomena
See more of this Group/Topical: Engineering Sciences and Fundamentals

Characterizing Structure in Assembled Systems Using Shape-Matching

Aaron S. Keys, Chemical Engineering, University of Michigan, Ann Arbor, MI

Christopher R. Iacovella, Chemical Engineering, University of Michigan, Ann Arbor, MI

Sharon C. Glotzer, Chemical Engineering, University of Michigan, Ann Arbor, MI

In many computer science applications, a technique known as "shape matching" is used to identify unknown structures. Common applications include retrieving fingerprints from a criminal database or verifying signatures electronically. In the context of materials research, we can use shape matching methods to identify structures formed by assembled particle systems. Our scheme involves associating "shape descriptors" with structural patterns of interest in our particle systems and then characterizing the patterns by comparing shape descriptors with reference structures. We show that shape matching techniques can be applied to a variety of structural characterization problems such as local and global identification and classification, automated phase diagram mapping, and constructing spatial and temporal structure correlation functions. Our techniques are applicable over a wide range of systems, both simulated and experimental, provided particle positions are known or can be detected with high accuracy. To aid in the development and dissemination of these techniques, we have implemented a C++/python library to perform shape matching analysis, which will be made available at matforge.org.

See more of this Session: Poster Session: Interfacial Phenomena
See more of this Group/Topical: Engineering Sciences and Fundamentals

Prediction of Physical Properties of An Emerging Class of Compounds

Paul Blowers, Chemical and Environmental Engineering, The University of Arizona, Tucson, AZ

Hydrofluoroethers (HFEs) are the primary candidate materials that will replace HFCs in industrial practice if the United States government signs legislation to reduce global climate change. The US EPA recently announced on April 24, 2009 that six compounds or classes of compounds that contribute to global climate change are human health hazards. While carbon dioxide is the largest gas of concern, hydrofluorocarbons (HFCs) are included due to their long term impacts through extended atmospheric lifetimes.

While about 50 HFEs have some physical property data available in the literature, many of these possible HFC alternatives have none. This paper uses computational chemistry to predict properties relevant for screening a larger set of these compounds for industrial use. While vapor pressures have been predicted in the past, this work is the first to estimate boiling points using quantum chemical means. Heat capacities, important for solvent re-use and for determining work requirements in compressors are also explored.

Potential environmental impacts and decision trade-offs are explored. Global warming potentials, kinetic degradation pathways, and likely end-points for removal mechanisms are detailed in the context of the pending HFC regulations.

See more of this Session: Poster Session: Thermodynamics and Transport Properties
See more of this Group/Topical: Engineering Sciences and Fundamentals

Prediction of Vapour-Liquid Equilibria and Second-Derivative Properties with the SAFT-Gamma Group Contribution Approach

Thomas Lafitte, Centre for Process Systems Engineering, Imperial College London, London, United Kingdom

Vassilis Papaioannou, Centre for Process Systems Engineering, Imperial College London, London, United Kingdom

Claire Adjiman, Centre for Process Systems Engineering, Imperial College London, London, United Kingdom

Amparo Galindo, Centre for Process Systems Engineering, Imperial College London, London, United Kingdom

George Jackson, Centre for Process Systems Engineering, Imperial College London, London, United Kingdom

The SAFT-&gamma EOS[1] has recently been developed as group contribution methods derived entirely within an equation of state to describe fluids of heteronuclear molecules which are formed from fused square-well segments of different types. The different groups are then defined in terms of one or multiple[2] spherical segments characterized by size and attractive energy (well-depth and range) parameters, and shape factor parameter which describes the contribution that each segment makes to the overall molecular properties. For associating groups a number of bonding sites are also included on the segment introducing two additional energy and range parameters. The formulation of SAFT-&gamma as a continuum fluid theory offers the key advantage that the binary interaction parameters between groups can be obtained directly from an examination of pure component properties alone. This is particularly useful in predicting the properties of fluid mixture (vapour-liquid and liquid-liquid equilibria, VLE and LLE) when no experimental data on the mixture are available. In this work we will present the extension of the theory to deal with heteronuclear molecules formed from Mie (Lennard-Jonesium interactions of varying repulsive and attractive range) segments of different type. Using this more realistic inter-segment potential as an elementary building block of a homonuclear SAFT-like EOS[3] has proven previously to be highly accurate in predicting second-order derivative properties (e.g., isobaric heat capacity, isobaric thermal expansivity, speed of sound) of complex fluids. First, in order to assess the accuracy of the SAFT-&gamma Mie EOS obtained, the thermodynamics properties are compared with computer simulation results for chains of fused heteronuclear Mie segments. Regarding the good agreement obtained in each cases, we then show that the use of generalized Lennard-Jones potential of this type in the context of the SAFT-&gamma group contribution approach allows for the simultaneous description of both VLE and second-derivative properties of pure fluids (alkanes, alkanols, n-alkylbenzenes, mono-and diunsaturated hydrocarbons) and their mixtures.

[1] A. Lymperiadis, C. S. Adjiman, A. Galindo and G. Jackson, J. Chem. Phys. 127, 234903 (2007)

[2] A. Lymperiadis, C. S. Adjiman, G. Jackson and A. Galindo , Fluid Phase Equilib.. 274, 85(2008)

[3] T. Lafitte, D. Bessieres, M. M. Piñeiro and J.-L. Daridon, J. Chem. Phys., 124, 024509 (2006).

See more of this Session: Poster Session: Thermodynamics and Transport Properties
See more of this Group/Topical: Engineering Sciences and Fundamentals

PREDICTION OF SATURATED VAPOR PRESSURE WITH SIMPLE STRUCTURE PROPERTY- PROPERTY RELATIONSHIPS

Neima Brauner, School of Engineering, Tel-Aviv University, Tel-Aviv, Israel
Mordechai Shacham, Chem. Eng. Dept., Ben-Gurion University, Beer-Sheva, Israel

Introduction

Pure component vapor pressure data are essential for process and product design, in assessing the environmental impact of a chemical compound and in modeling some types of toxicity (Dearden, 2003). At present, vapor pressure data are available only for a small fraction of the compounds of interest to the chemical industry. Even if the data are available they may not cover the full temperature range of interest. In product design vapor pressure values may be required for substances that have not been synthesized yet. Thus, prediction of saturated vapor pressure data is often essential.

Current methods used to predict temperature-dependent properties can be classified into "group contribution" methods, methods based on the "corresponding-states principle" (for reviews of these methods see, for example, Poling *et al*, 2001 and Godavarty et al.,2006), and "asymptotic behavior" correlations (see, for example, Marano and Holder,1997). These methods rely on several other property values, such as normal boiling temperature: T_b, critical temperature: T_c critical pressure, P_c and acentric factor: ω. However, such data for properties may not be available for a target compound for which the vapor pressure has to be predicted. Moreover, experimental values for P_c and calculated values for ω are usually of low precision causing deterioration of the accuracy of the predicted vapor pressure values. Furthermore, these methods contain adjustable parameters that were fitted to a training set, which may not represent well enough the target compound. A detailed discussion of these issues can be found, for example in Vetere (2006).

In recent years, there has been increasing interest in using molecular descriptors integrated into Quantitative Structure Property Relationships (QSPR) for prediction of vapor pressure. However, the great majority of the currently available QSPR models are limited to prediction at a single temperature of 298.15 K. The exceptions are the methods of Godavarty et al. (2006), which combine the use of the corresponding-state principle with a QSPR (thus using certain properties and adjustable parameters) and Yaffe and Cohen (2001) who's QSPR is neural-network which may not be readily applicable for potential users.

Shacham et al., (2009) suggested QSPR-based methods for predicting temperature-dependent variation of vapor pressure of chemical compounds. These methods are based on the identification of potential predictive compounds, which are structurally similar to the target compound (a similarity group) and for which data for a vapor pressure related properties (e.g., normal boiling temperature, T_b or acentric factor, ω) are available. For this step, the algorithm of the Targeted QSPR method of Brauner et al. (2006) is employed. If the target compound belongs to a homologous series additional members of these series can serve as part or the whole similarity group. In the following step, stepwise regression is used to identify molecular descriptors which are collinear with T_b and ω for the members of the similarity group. These descriptors are used to develop simple structure-structure relationships to predict the saturation temperature of the target compound: T_t^s at a specified pressure (*Ts-QSPR* method) or the reduced vapor pressure of the target compound $P^s_{R,t}$ at a specified reduced temperature (*Prs-QSPR* method). In the following sections the two methods will be described in some detail, their use will be demonstrated for predicting T^s_t and $P^s_{R,t}$ for members of the aldehyde (alkanal) homologous series, and their relative merits will be discussed.

The Ts-QSPR and Prs-QSPR methods

The Ts-QSPR method is based on the short-cut QS2PR method of Cholakov et al. (2007) and Shacham et al., (2009). Using two predictive compounds (No. 1 and 2), which are similar to the target compound (i.e. belong to the same homologous series), an unknown property of the target compound: y^p_t can be predicted using the known property values of the two predictive compounds: y^p_1 and y^p_2, through the following relationship

$$y^p_t = \frac{\zeta^j_2 - \zeta^j_t}{\zeta^j_2 - \zeta^j_1} y^p_1 + \frac{\zeta^j_t - \zeta^j_1}{\zeta^j_2 - \zeta^j_1} y^p_2 \quad \textbf{(1)}$$

where ζ^j_1, ζ^j_2 and ζ^j_t are descriptor values of the 1st and 2nd predictive compounds and the target compound, respectively. The method applicable only if the descriptor ζ^j is collinear with the property y^p for the similarity group. In case of T^s prediction, the descriptor used must be collinear with the normal boiling temperature (the value of T^s at atmospheric pressure). If T_b is known (or can be predicted) for the target compound, T_b values can replace the descriptors in Eq. 1.

For calculation of the T^s values for the predictive compounds, equations that provide the vapor pressure as function of temperature are needed. We have been using the Riedel equation:

$$\ln(P^s_i) = A_i + \frac{B_i}{T} + C_i \ln T + D_i T^{E_i} \quad \textbf{(2)}$$

where A_i, B_i, C_i, D_i, and E_i are coefficients characteristic to a specific compound. The coefficients are determined by regression of vapor pressure data. Coefficients for a large number of compounds are available in the DIPPR database (Rowley, 2006). For calculation of $T^s_i(P)$ the equation has to be rewritten in an implicit form:

$$f(T^s_i) = \ln(P) - \left(A_i + \frac{B_i}{T} + C_i \ln T + D_i T^{E_i} \right) \quad \textbf{(3)}$$

For each pressure value this equation has to be solved iteratively to yield T^s_i. The T^s_i values are then introduced into the property-property correlation (Eq. 1) to obtain the respective T^s_t value.

To predict T_t^s for a wide range of pressure values it is recommended to convert the problem into a differential-algebraic system, and to solve this system using is the "controlled integration" technique of Shacham and Brauner (1997). Using this method Eq. 3 is rewritten:

$$\varepsilon = \ln(P) - \left(A_i + \frac{B_i}{T} + C_i \ln T + D_i T^{E_i} \right) \quad \textbf{(3a)}$$

and a system of differential equations is solved:

$$\frac{dT_2^s}{dt} = K_c \varepsilon_2 \quad \frac{dP}{dt} = 1 \quad \frac{dT_1^s}{dt} = K_c \varepsilon_1 \quad \textbf{(4)}$$

Here two "proportional controllers" are used to change the saturation temperature of the predictive compounds according to the pressure change. The proportional gain, K_c, is selected large enough so as to keep ε below a pre-specified tolerance. The integration starts from atmospheric pressure where $T_i^s = T_b$, thus $\varepsilon_1 = \varepsilon_2 = 0$. The integration can proceed in the positive direction (high pressure range), or the negative direction (low pressure range). At each integration point, T_t^s is calculated using Eq. 1.

The Prs-QSPR method is based on the "two reference fluid" estimation method, which is discussed in some detail, for example, by Poling et al. (2001). Using "two reference fluid" method the reduced vapor pressure of the target compound ($P_{R,t}^s$), at a particular T_R value, is estimated using the following equation:

$$\ln(P_{R,t}^s) = \ln(P_{R,1}^s) + \left[\ln(P_{R,2}^s) - \ln(P_{R,1}^s)\right] \frac{\omega_t - \omega_1}{\omega_2 - \omega_1} \quad \textbf{(5)}$$

where ω_1 and ω_2 are acentric factor values of the predictive compounds and ω_t is the acentric factor of the target compound.

The enhancements introduced in the PRs-QSPR method are related to the way the predictive compounds are selected (high level of similarity with the target compound) and the replacement of the acentric factor with a descriptor that is collinear with the acentric factors of the similarity group members. It is important to mention that the $P_{R,t}^s(T_R)$ value obtained from eq. 5 has to be usually converted to a $P_t^s(T)$ value, and T_C and P_C of the target compound are needed for this conversion. If these data are not available, they can be predicted the same way as the T_b and ω values are predicted.

To compare the prediction uncertainty involved in these two methods, namely, the error δT^s when T^s is predicted using eq. 1, and the error $\delta\left[\ln(P_{R,t}^s)\right]$ when eq. 4 is used, they must be brought into a common basis. The error propagation formula is used to convert the uncertainty δT^s to uncertainty in $\delta\left[\ln(P_{R,t}^s)\right]$. For the Riedel equation the following error conversion formula applies:

$$\delta[\ln(P_R^s)] = \frac{d[\ln(P_{R,1}^s)]}{dT}\delta T = \left(-\frac{B_i}{T^2} + \frac{C_i}{T} + E_i D_i T^{E_i - 1}\right)\delta T \qquad \textbf{(6)}$$

In the following sections the use of the proposed methods will be demonstrated.

Prediction of T^s for the Aldehyde Series Using the Ts-QSPR method

Some of the data required for predicting vapor pressure for members of the alkanal homologous series are shown in Table 1. Ten members of these series; from propanal (C_3H_6O) through tridecanal ($C_{13}H_{26}O$), are included. The properties T_b, ω, T_C, P_C are shown in the Table. These properties are taken from the DIPPR database (Rowley, 2006). The values of the molecular descriptors, which are collinear with the four properties, are also shown in Table 1. These descriptors were calculated using the Dragon program (Todeschini et al. 2006). The descriptor with the highest level of co-linearity with a particular property was selected from a database containing ~1500 descriptors for each compound by means of the JMP (http://www.jmp.com/) statistical software.

In Figure 1 T_b is plotted versus the descriptor *ESpm04d* (a 2-D edge adjacency index: "spectral moment 04 from edge adjacency matrix weighted by dipole moments") for the alkanal homologous series. Observe that there is a high level of co-linearity between T_b and this descriptor. This relationship can be represented by the equation T_b = -1308.9+376.59 *ESpm04d*, with correlation coefficient R^2 = 0.9997.

To demonstrate prediction of saturation temperature using the Ts-QSPR method, pentanal (C5H10O) and decanal (C10H20O) are selected as predictive compounds No. 1 and No. 2, respectively, and heptanal (C7H14O) is designated as the target compound. In Table 2 the Polymath^{TM} (a product of Polymath Software, http://www.polymath-software.com) program for predicting T^s of heptanal in the high pressure range is shown.

In the 1st line of Table 2 the differential equation which changes the pressure in the positive direction is shown. In the following two lines, the differential equations which change the temperature of the predictive compounds to follow the saturation curve (using the controlled integration technique (Eq. 4)) are displayed. The value of K_c = 100000 was selected in order to keep the value of $\varepsilon < 0.01$ (Eq. 3a). The T_b values of the predictive compounds (lines 5 and 6) are used only for setting the initial temperature values for integration. The *ESpm04d* descriptor values for the predictive and the target compounds are shown in lines 7 through 9. It is important to point out that for the target compound the only data provided is the descriptor value, which can be calculated based on its known molecular structure.

Lines 10 through 13 present the Riedel equations for the predictive compounds (as provided in the DIPPR database). The ranges of the applicability of the equations, in terms of temperature and pressure are also provided (as comments). The boundaries of the applicability intervals are important for determination of the extent of the integration. The integration bounds are set so that the equations are used only within their valid range of applicability. In lines 14 and 15, the ε values are calculated for the two predictive compounds (Eq. 3a). The saturation temperature is predicted for the target compound in line 16, using Eq. 1. In this case descriptor values are used for the prediction. However, if T_b for the target compound is available (or can be predicted) all the descriptors in Eq. 1 can be replaced by the respective T_b values.

The initial values of the variables and the final value of t are shown in lines 17 through 21. The initial pressure set as atmospheric pressure and the initial saturation temperatures are set as the boiling point temperatures of the predictive compounds. The final integration range (time) is set so as not to exceed the critical pressure of decanal (2.6 MPa).

The plot of the saturation temperatures of the predictive and target compounds versus pressure is shown in Figure 2. For the target compound, heptanal, two curves of T_s are displayed. One is the value calculated by Riedel's equation (which is available in the DIPPR database) the other one is based on the predicted values computed by the program of Table 2. Observe that the two curves are hardly distinguishable.

Figure 3 shows the prediction uncertainty: $err = T^s_{Riedel} - T^s_{Pred.}$ versus T_S for heptanal. Two cases are considered. In one case the descriptor: *ESpm04d* is used for the prediction with Eq. 1, in the second case T_b is used. Observe that the two curves are completely parallel and the error increases toward the higher temperature region. The maximal error is larger when the descriptor is used for the prediction, however it does not exceed 1%. This is the maximal experimental error level of T_b for most of the compounds included in this study. Thus, we may conclude that the prediction error is within the bounds of the experimental uncertainty.

The calculations for the low pressure range can be carried out using the program shown in Table 2. The differential equation describing the variation of P must be changed to: $dP/dt = -1$ and the final value of t must be set at a value <0.10135, to ensure that the pressure does not become negative. The prediction errors in the low pressure range are even smaller than in the high pressure range.

Prediction of P^s_R for the Aldehyde Series Using the Prs-QSPR method

When using the Prs-QSPR ω *should be replaced by a descriptor in Eq. 5.* In Figure 4 ω is plotted versus the descriptor *pilD* (a 2-D "walk and path counts" descriptor: "conventional bond order ID number") for the alkanal homologous series. The co-linearity between ω and the *pilD* descriptor can be represented by the equation ω = 0.1378+0.0142 pilD, with correlation coefficient R^2 = 0.9967. Observe that the spread of the ω values around the straight line is larger than in the case of T_b (see Figure 1). This can be explained by the higher accuracy of the measured T_b values. To calculate acentric factor the vapor pressure at T_R = 0.7 is used, and the uncertainty of this value is usually higher than the uncertainty of the T_b values.

For this study the same set of predictive compounds (pentanal (C5H10O) and decanal (C10H20O)) and target compound (heptanal (C7H14O)) are used. In Table 3 the Polymath program for predicting $P^s_{R,t}$ of heptanal from T_R = 0.4 up to T_R = 0.999 is shown. In the 1st line of Table 3 the differential equation which changes T_R in the positive direction is shown. The P_C and T_C data of the predictive compounds are included in lines 2 through 5. The *pilD* descriptor values for the predictive and the target compounds are introduced in lines 6 through 8. As in the case of the Ts-QSPR method only the descriptor value should be provided for the target compound in order to predict $P^s_{R,t}$.

The reduced temperatures are converted to temperature values in lines 9 and 10 and in lines 11 through 14 the Riedel equations for the predictive compounds with their ranges of applicability are shown. The vapor pressure values are converted to $\ln P^s_{R,j}$ values, required by Eq. 5, in lines 15 and 16. The $\ln P^s_{R,t}$ value is predicted in line 17 using Eq. 5. For the target compound only the descriptor value is needed in order to obtain the prediction. However, if ω

for the target compound is available (or can be predicted) all the descriptor values in Eq. 5 can be replaced by the respective ω values.

The initial values of t and T_R and the final value of t are in lines 18 through 20. The integration starts from T_R = 0.4 (the lower temperature limit of the Riedel equation for decanal) and stops at the close vicinity of the critical point (T_R = 0.999).

Figure 5 shows the prediction uncertainty: $err = \ln P^s_{R,Riedel} - \ln P^s_{R,Pred}$ versus $\ln P^s_{R,t}$ for heptanal. Two cases are considered. In one case the descriptor: *pilD* is used for the prediction with Eq. 5, in the second case ω is used. The error is larger when the descriptor is used for prediction and it increases toward the low pressure range. The relative error seems to be much larger than in the case of T^s. However, using Eq. 6 to bring the prediction errors of the two methods to the same basis the differences between the two methods are not very significant. Obviously, the Prs-QSPR method requires more data for the target compound (ω, T_C, P_C or the respective co-linear descriptors) than the Ts-QSPR method (T_b or a collinear descriptor). More data means additional possible sources of errors, especially for high carbon number compounds, which may disintegrate below the critical point.

Conclusions

The applications of two new methods for predicting the vapor pressure curve for a target compound were demonstrated, using the aldehyde homologous series as an example. The advantages of the proposed methods over the traditional methods are the following.

- Only structural information (no measured property values) are needed for the target compound, however if measured values are available they can simplify the prediction procedure and often yield more accurate predictions;
- Predictive compounds similar to the target are selected in a systematic manner ;
- The temperature - vapor pressure relationships of the predictive compounds are used only in their valid range of applicability;
- No the need to use generalized correlations with adjustable coefficients;
- It is possible to predict either saturation temperature, or vapor pressure, giving more flexibility regarding the range and uncertainty of the predictions .

The methods were applied also for prediction of vapor pressure in wide temperature ranges for additional homologous series (*n*-Alkane, 1-alkene, 1-Alcanol, *n*-Alcanoic Acid and Alkyl-Benzene series). The results show that the methods enable the prediction of vapor pressure within experimental uncertainty, depending on the level of similarity between the predictive compounds and the target compound. In cases of longer range interpolation or extrapolation, some degradation of the prediction is noticed; however, even in those cases the proposed methods are superior to the traditional prediction techniques.

References

1. Brauner, N.; Stateva, R. P.; Cholakov, G. St.; Shacham, M. A. Structurally "Targeted" QSPR Method for Property Prediction. *Ind. Eng. Chem. Res.* **45**, 8430–8437 (2006)
2. Dearden, J. C. "Quantitative Structure–Property Relationships for Prediction of Boiling Point, Vapor Pressure, and Melting Point", *Environmental Toxicology and Chemistry*, **22**(8), 1696–1709 (2003).
3. Godavarthy, S.S.; Robinson, R.L.; Gasem, K.A.M. "SVRC-QSPR model for predicting saturated vapor pressures of pure fluids", *Fluid Phase Equilibria* **246**, 39-51, 2006.
4. Marano, J.J., Holder, G.D., "General Equations for Correlating the Thermo-physical Properties of n-Paraffins, n-Olefins and other Homologous Series. 2. Asymptotic Behavior Correlations for PVT Properties", *Ind. Eng. Chem. Res.*, **36**, 1887-1894 (1997).
5. Poling, B.E., Prausnitz, J. M., O'Connel, J. P., Properties of Gases and Liquids, 5th Ed., McGraw-Hill, New York (2001).
6. Rowley, R. L.; Wilding, W. V.; Oscarson, J. L.; Yang, Y.; Zundel, N. A. DIPPR Data Compilation of Pure Chemical Properties Design Institute for Physical Properties; Brigham Young University Provo Utah, 2006. http//dippr.byu.edu (accessed Sept. 2009).
7. Shacham, M., Cholakov, G. St., Stateva R. P., Brauner, N., "Quantitative Structure – Structure – Relationships for Prediction of Phase Equilibrium Related Properties", *Ind. Eng. Chem. Res.*, Submitted for publication, 2009
8. Todeschini, R.; Consonni, V.; Mauri, A.; Pavan, M. DRAGON user manual, Talete SRL:Milano, Italy, 2006, (http://www.talete.mi.it)
9. Vetere, A., "Again the Riedel Equation", *Fluid Phase Equilibria* **240**, 155-160(2006)
10. Yaffe D and Cohen Y, "Neural network based temperature-dependent quantitative structure property relations (QSPRs) for predicting vapor pressure of hydrocarbons", Journal *of Chemical Information and Computer Sciences* **41**, 463-477(2001)

Table 1. Physical Property and Molecular Descriptor Data for the Aldehyde Series

No.	Compound	Normal Boiling Temp. (K) Value	Descriptor ESpm04d	Acentric Factor Value	Descriptor piID	Critical Pressure (Pa) Value	Descriptor ESpm12x	Critical Temperature (K) Value	Descriptor ESpm02d
1	PROPANAL (C3H6O)	321.12	4.326	0.255909	8.094	5260000	12.364	504.4	2.418
2	BUTANAL (C4H8O)	347.95	4.403	0.277417	10.886	4320000	12.703	537.2	2.582
3	PENTANAL (C5H10O)	375.7	4.473	0.347159	13.832	3970000	12.913	566.8	2.723
4	HEPTANAL (C7H14O)	425.95	4.602	0.42794	20.109	3160000	13.224	616.8	2.956
5	n-OCTANAL (C8H16O)	443.65	4.66	0.463565	23.411	2960000	13.349	638.1	3.055
6	NONANAL (C9H18O)	468.15	4.715	0.511744	26.809	2730000	13.46	658	3.145
7	DECANAL (C10H20O)	486	4.768	0.582008	30.294	2600000	13.56	674.2	3.228
8	N-UNDECANAL (C11H22O)	506.15	4.818	0.62049	33.859	2000000	13.651	692	3.304
9	DODECANAL (C12H24O)	523.15	4.865	0.674965	37.498	1860000	13.735	708	3.375
10	TRIDECANAL (C13H26O)	540.15	4.91	0.716147	41.206	1740000	13.812	724	3.441

Table 2. Polymath Program for predicting Saturation Temperature of Heptanal in the High Pressure Region Using the Ts-QSPR Method

No.	Command / # Comment*
1	d(P)/d(t)=1 #Mpa
2	d(T1)/d(t)=Kc*eps1 #Predictive Comp. 1, K
3	d(T2)/d(t)=Kc*eps2 #Predictive Comp. 2, K
4	Kc=100000
5	NBP_C5H10O = 375.7 # K
6	NBP_C10H20O = 486 # K
7	D1=4.473
8	D2=4.768
9	Dt=4.602
10	# Min_T=182, Max_T=566.1, Min_Val=0.052282, Max_Val=3968500, Error:< 3%
11	VP_C5H10O=(exp(149.58 - 8890 / T1 - 20.697 * ln(T1) + 0.022101 * T1))*1e-6 #MPa
12	# Min_T=267.15, Max_T=674.2, Min_Val=0.48648, Max_Val=2598900, Error:< 3%
13	VP_C10H20O=(exp(201.64 - 15133 / T2 - 26.264 * ln(T2) + 1.4625E-05 * T2 ^ 2))*1e-6 #MPa
14	eps1=P-VP_C5H10O
15	eps2=P-VP_C10H20O
16	Tt=(D2-Dt)*T1/(D2-D1)+(Dt-D1)*T2/(D2-D1) # Target Compound Saturation Temp., K
17	t(0) = 0
18	t(f) = 2.498650
19	P(0)=0.10135 #MPa
20	T1(0)=375.7 # K
21	T2(0)=486 # K

* Text that starts with the # sign is considered as a "comment" up to the end of the line. Line numbers were added.

Table 3. Polymath Program for predicting log(P_R) of Heptanal in Using the Prs-QSPR Method

No.	Command / # Comment
1	d(Tr)/d(t)=1
2	PC_C5H10O = 3970000 # Pa
3	PC_C10H20O = 2600000 # Pa
4	TC_C5H10O = 566.8 # K
5	TC_C10H20O = 674.2 # K
6	D1=13.832
7	D2=30.294
8	Dt=20.109
9	T1=Tr*TC_C5H10O
10	T2=Tr*TC_C10H20O
11	# Min_T=182, Max_T=566.1, Min_Val=0.052282, Max_Val=3968500, Error:< 3%
12	VP_C5H10O=exp(149.58 - 8890 / T1 - 20.697 * ln(T1) + 0.022101 * T1) #Pa
13	# Min_T=267.15, Max_T=674.2, Min_Val=0.48648, Max_Val=2598900, Error:< 3%
14	VP_C10H20O=exp(201.64 - 15133 / T2 - 26.264 * ln(T2) + 1.4625E-05 * T2 ^ 2) #Pa
15	lnPr1=ln(VP_C5H10O/PC_C5H10O)
16	lnPr2=ln(VP_C10H20O/PC_C10H20O)
17	lnPrt=lnPr1+(lnPr2-lnPr1)*(Dt-D1)/(D2-D1)
18	t(0) = 0
19	t(f) = 0.599
20	Tr(0)=0.4

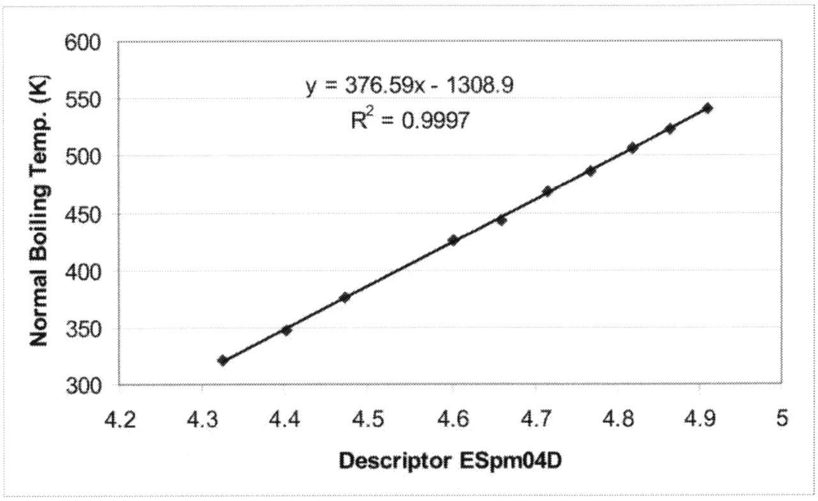

Figure 1. Plot of T_b Versus the Molecular Descriptor *ESpm04D* for the Aldehyde Series

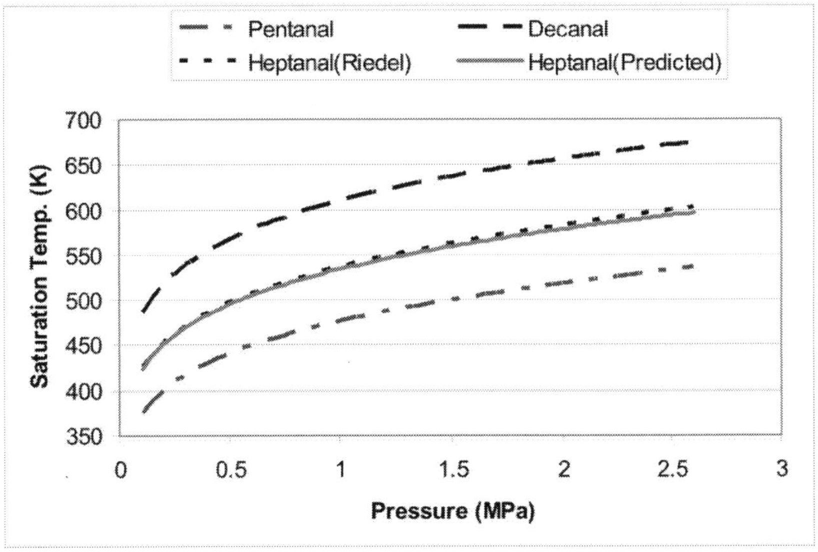

Figure 2. Calculated (Riedel) and Predicted Saturation Temperatures for Heptanal

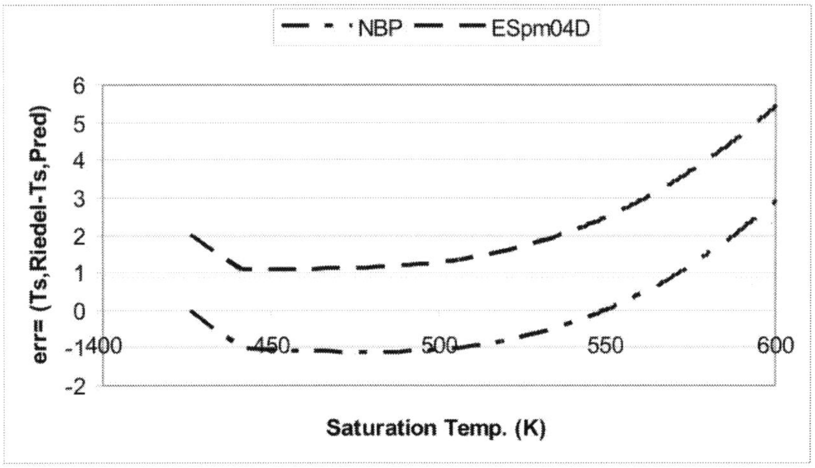

Figure 3. Prediction Uncertainty for Heptanal when T_b and Descriptor ESpm04D are used for prediction with the Ts-QSPR method

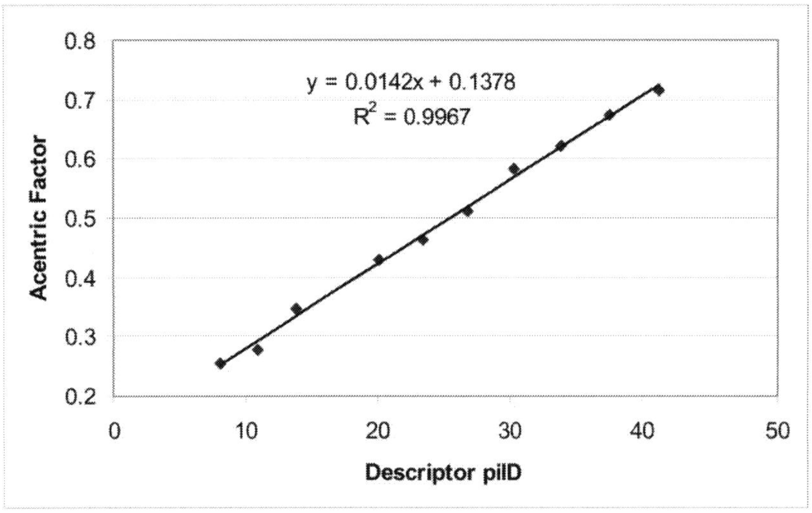

Figure 4. Plot of ω Versus the Molecular Descriptor *piID* for the Aldehyde Series

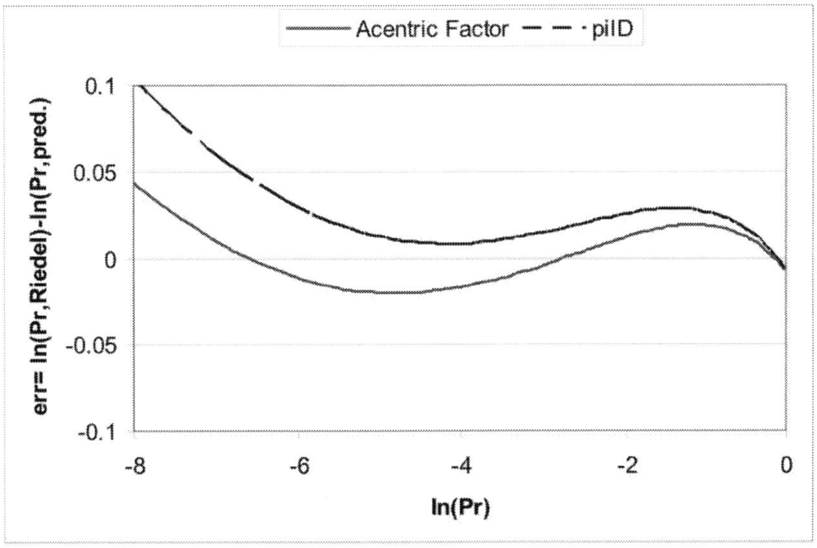

Figure 5. Prediction Uncertainty for Heptanal when ω and Descriptor pilD are used for prediction with the Prs-QSPR method

Direct Calculation of PC-SAFT Parameters From Experimental LVE Data and From a GC-Plus Approach

Romain Privat, Department of Chemical & Biochemical Engineering, CAPEC, DK-2800 Lyngby, Denmark

Rafiqul Gani, Department of Chemical & Biochemical Engineering, Technical University of Denmark, Lyngby, Denmark

Jean-Noel Jaubert Jaubert, University of Nancy, Nancy, France

Since several years, SAFT equations of state (EoS) are drawing the attention of many researchers in thermodynamics. The PC-SAFT (that is, the perturbed chain SAFT EoS) is one of the simplest and widely used among this family of models. Stemming from molecular theory and thus more complex than cubic EoS, PC-SAFT equations offer not yet explored possibilities in modelling fluid phase equilibria. When one wants to use them, at least three pure component parameters have to be considered. These are: the number m_i of segments per chain in molecule i, the segment diameter σ_i in molecule i and ε_i, the depth of pair potential.

In order to make the PC-SAFT equations more widely used and able to represent a wide range of organic systems, the group contribution (GC) concept can be employed. Such GC-methods are nevertheless restricted to the number of groups defined by the method. Only very common groups such as methyl, ethyl, hydroxyl, carboxyl and so on, are in general defined by the GC-methods, thus excluding systems with specific atom groups which are essential for industrial applications. To avoid such limitations, an alternative consists in creating the missing groups through the connectivity index concept.

In this work, a robust and widely applicable fitting procedure has been developed for the direct calculation of the PC-SAFT parameters from experimental LVE data. Using these fitted parameters, two methods aimed at predicting the set of parameters (m_i, σ_i and ε_i) for a pure component were developed. The first one is a classical GC-method with many distinct functional groups and thus applicable to a large set of pure compounds. This method covers a wider range of chemicals and leads generally to smaller deviations on the estimation of LVE properties than other published methods.

The second one relies on the atom connectivity approach and allows to create missing groups when a component cannot be described by the first GC-method.

Our predictive model has been tested on hundreds of pure component property data from organic chemicals as well as on multicomponent mixture data including polymers, with very promising results.

See more of this Session: Poster Session: Thermodynamics and Transport Properties
See more of this Group/Topical: Engineering Sciences and Fundamentals

Isobaric Vapor-Liquid Equilibrium of 1-Propanol and Benzene System

Dong-Yuk Kang, Department of Advanced Chemicals and Engineering, Chonnam National University, Gwangju, South Korea

Hoi-Gu Jang, Department of Advanced Chemicals and Engineering, Chonnam National University, Gwangju, South Korea

Chang-Nam Han, Department of Advanced Chemicals and Engineering, Chonnam National University, Gwangju, South Korea

Dong Lyun Cho, Department of Advanced Chemicals and Engineering, Chonnam National University, Gwangju, South Korea

Choon-Hyoung Kang, Applied Chemical Engineering Department and the Research Institute for Catalysis, Chonnam National University, Kwangju, South Korea

In this study, isobaric vapor-liquid equilibria of a binary mixture of 1-propanol and benzene, which shows severe nonideal behavior with azeotropic compositions, were measured by using a recirculating equilibrium cell from Illudest at various pressures ranging from 20kPa to 70kPa. The measured VLE data were satisfactorily correlated with UNIQUAC and NRTL models along with the thermodynamic consistency test. The excess molar volume of the mixture was also measured using a vibrating densitometer.

See more of this Session: Poster Session: Thermodynamics and Transport Properties
See more of this Group/Topical: Engineering Sciences and Fundamentals

Thermal Cracking and Effects of Oxygen On Oxidation of Tricyclodecane

Joong Yeon Kim, Department of Chemical & Biological Engineering, Korea University, Seoul, South Korea

Sun Hee Park, Department of Chemical & Biological Engineering, Korea University, Seoul, South Korea

Byung-Hee Chun, Department of Chemical & Biological Engineering, Korea University, Seoul, South Korea

Jeong Sik Han, Agency for Defense Development, Daejon, South Korea

Byung Hun Jeong, Agency for Defense Development, Daejon, South Korea

Sung Hyun Kim, Department of Chemical & Biological Engineering, Korea University, Seoul, South Korea

A test was implemented to determine thermal cracking kinetics of tricyclodecane(TCD) without catalysts such as metal and to find out effects of oxygen on oxidative reaction of TCD. TCD can be used as a coolant as well as a fuel for spacecraft or aircraft. As technologies for aircraft are developing fast, aircrafts encounter with more heat load and the temperature change of the fuel is larger. If fuels are exposed to high temperatures, formation of insoluble products takes place and build-up of these insoluble products cause fuel system failure.

Thermal cracking of TCD were carried out in a batch-type reactor whose inner wall was covered with quartz. Flow rate can affect features of TCD's thermal cracking in a flowing condition. Experiments were implemented to find out thermal cracking kinetics of TCD at a static condition without flow characteristics. We also investigated effects of oxygen for TCD's oxidation.

First, the concentrations were measured with respect to time and conversions were figured out. A reaction rate was estimated as first order reaction from conversion measurements with respect to time. Activation energy of thermal cracking of TCD was calculated as 261.1 kJ/mol from Arrhenius' law. In order to compare thermal cracking to catalytic cracking of TCD, titanium was put on the bottom of the reactor covered with quartz in the same conditions above. The conversions of TCD with titanium were higher than those of TCD without titanium and activation energy was 200.5 kJ/mol. Comparing the results between two types of test, metals are detrimental to thermal stability because of reduction of activation energy for thermal cracking of TCD.

Another test was also implemented to investigate effects of oxygen on TCD's oxidation because air flows into system continuously in the industrial air-breathing systems. The reactor wall was covered with quartz to prevent effect of metal as catalyst on oxidation of TCD. Three types of TCDs were prepared to determine effect of oxygen; TCD without oxygen by bubbling He gas through the TCD, TCD with excess of oxygen by bubbling oxygen through the TCD, and air-saturated TCD. TCD with excess of oxygen and air-saturated TCD had same tendency of oxidation rate in the early stage of test, while oxidation of TCD without oxygen showed slower oxidation rate than those of former two cases. However, oxidation rate of air-saturated TCD was getting slower than that of TCD with excess of oxygen gradually. The rate of oxygen consumption was independent of the oxygen concentration under the sufficient oxygen concentration. However, at a low enough oxygen concentration, the rate-determining step for TCD's oxidation is related to the reaction in which alkyl radical of TCDs react with oxygen. The consumption of oxygen(i.e. oxidation rate) is the first order in oxygen. Thus, air-saturated TCD showed that its oxidation rate decreased gradually.

The thermal stability of TCD decreased as that the conversion of TCD increased, because catalysts reduced the activation energy of its thermal cracking. However, it's difficult to avoid large conversion due to metal catalysts because aircraft systems mostly consist of metals. Thus, further researches are necessary to find metals which can withstand high temperatures and reduce the conversion of TCD related to thermal cracking and to study additives which are able to reduce effects of metals. Oxygen concentration didn't affect the TCD's oxidation under the sufficient oxygen concentration. However, the oxidation rate of TCD was dependent on oxygen concentration under the insufficient oxygen concentration. Thus, additional experiments will be done to find out additives which can reduce oxidation rate of TCD under the industrial air-breathing systems.

See more of this Session: Poster Session: Thermodynamics and Transport Properties
See more of this Group/Topical: Engineering Sciences and Fundamentals

Molecular Dynamics of Pseudomonas Cepacia Lipase in Different Buffers and Correlations with Its Biocatalytic Activity

Samrish Variyath, Chemical Engineering, University of Cincinnati, Cincinnati, OH

JungSeung Kim, Chemical Engineering, University of Cincinnati, Cincinnati, OH

Stephen W. Thiel, Chemical and Materials Engineering, University of Cincinnati, Cincinnati, OH

Neville G. Pinto, Chemical & Materials Engineering, University of Cincinnati, Cincinnati, OH

Vadim Guliants, Department of Chemical and Materials Engineering, University of Cincinnati, Cincinnati, OH

Lipase from Pseudomonas cepacia (EC 3.1.1.3) (PcL) is one of the most versatile biocatalysts used in various industries, such as food, oleochemicals and pharmaceuticals. For these practical applications, the optimization of its biocatalytic behavior is essential. Providing a favorable micro-environment can enhance biocatalytic activity of this lipase. Consequently, the suitable selection of buffer components is highly important for the practical applications of this biocatalyst. In particular, the specific ion effect has not been elucidated fully for the enzyme activity because of its complicated nature.

In this study, the specific cation effect on the hydrolytic activity of PcL has been investigated for the NaCl, KCl, CaCl2, NaH2PO4 and K2HPO4 buffer solutions under ambient conditions employing molecular dynamics (NAMD) and experimental biocatalytic activity assays. The PME (Particle Mesh Algorithm) was used to describe the long range interactions of the cations with the enzyme employing the CHARMM force field with explicit water (TIP3P) solvent. The motions of each residue in the catalytic triad (Ser87, Asp264 and His286) and lid domains of the enzyme have been investigated by these computational methods. Experimentally, the specific activities of the enzyme were measured by a colorimetric assay of p-nitrophenyl acetate hydrolysis. The specific ion effects for PcL activity under different solution pH and ionic strength have been demonstrated by the molecular dynamics simulations.

See more of this Session: Poster Session: Thermodynamics and Transport Properties
See more of this Group/Topical: Engineering Sciences and Fundamentals

Dynamic Modeling of the Modifed Claus Process

Richard Turton, Chemical Engineering, West Virginia University, Morgantown, WV

Debangsu Bhattacharyya, Department of Chemical Engg., West Virginia University, Morgantown, WV

Stephen Zitney, Collaboratory for Process and Dynamic Systems Research, Office of Research & Development, National Energy Technology Laboratory, U.S. Department of Energy, Morgantown, WV

Dustin D. Jones, Chemical Engineering, West Virginia University, Morgantown, WV

For consideration in Environmental Remediation and Waste Management

The purpose of the current study is to investigate the transient behavior of the Claus process used for sulfur removal from coal-fired IGCC plants. To accomplish this, chemical kinetics for reactions taking place within the reaction furnace and the waste heat boiler were identified and used as the model basis rather than the more commonly used methods of restricted equilibrium or correlations. To model the system, a dynamic, pressure driven simulation of the process was carried out within the Aspen Plus Dynamics® process simulator. It was desired to determine the effects of feed composition, temperature and pressure on the performance of the Claus process and how effects of variation in feed conditions will propagate in the system. Verification of the developed steady state model was accomplished using available published plant data for the output of the reaction furnace and waste heat boiler as well as a comparison of the results of this model with predicted products from TSWEET®. In determining the dimensions of the reaction furnace, typical configurations and sizes were found within the literature. To determine heat transfer coefficients of the waste heat boiler along the length of the waste heat boiler, Aspen HTFS+® was used and the variations in the heat transfer coefficient were modeled across the length of the waste heat boiler, typically beginning near 80 W m-2 K-1 and dropping to 20 W m-2 K-1 towards the end.

The configuration of interest for this study was a split flow furnace with a 95% oxygen feed, two catalytic reactors and sulfur condensers. The feed to the process is high concentration NH3 and CO2 stream from a sour water stripper and a high concentration CO2 and H2O acid gas, with furnace residence times of 0.5 to 2.0 seconds. Waste heat boiler configurations were made to ensure a mass flux of 6 - 7 kg m-2 s-1 through the tubes.

Results of the steady state and dynamic simulations will be discussed and the implications of changes in IGCC feeds and operating variables on the composition of the tail gas from the Claus unit will be discussed.

See more of this Session: Poster Session: Thermodynamics and Transport Properties
See more of this Group/Topical: Engineering Sciences and Fundamentals

The Use of a Parallel Fermentation System for Optimization of Saccharomyces Cerevisiae Ethanol Production in Batch Culture

Aubrey Tiernan, Chemical Engineering, University of Massachusetts, Amherst, MA

Timothy Hanly, Chemical Engineering, University of Massachusetts, Amherst, MA

Michael A. Henson, Chemical Engineering, University of Massachusetts, Amherst, MA

Batch cell culture remains an important technology for the large-scale production of many biochemical products, including commodity biochemicals such as ethanol. We have used dynamic extensions of metabolic flux balance models to show that the optimal operating policy for Saccharomyces cerevisiae ethanol production in batch culture consists of a fully aerobic phase that maximizes biomass production followed by an anaerobic phase that maximizes ethanol synthesis. Our computational results suggest that ethanol productivity is sensitive to the switching time between the aerobic and anaerobic growth phases. Validation of these predictions requires time consuming fermentation experiments in which the ethanol productivity is determined for different aerobic-anaerobic switching times to establish the optimal point as well as the curvature near the optima.

In this contribution, we demonstrate the use of a 4-bioreactor parallel system to establish and validate optimal fermentation operating policies using S. cerevisiae ethanol production from glucose media as a model system. Each bioreactor has a 250 mL working volume and independent gas pumps, dissolved oxygen, temperature and pH probes, and stirring system for the execution of parallel batch experiments in precisely controlled growth environments. The four bioreactors are shown to yield reproducible biomass, glucose and ethanol concentration measurements when inoculated from the same shake flask and grown under the same aerobic conditions. The parallel fermentation data are further validated with experiments performed on a standard laboratory bioreactor with 1.25 L working volume and by comparison to predictions from a dynamic flux balance model. Having established that the parallel system produces reliable data, our computational results for the optimal aerobic-anaerobic switching time are evaluated by performing parallel experiments in which each bioreactor is subjected to a different switching time. With the continuing miniaturization and parallelization of bioreactor technology, our results indicate that parallel systems hold great promise for fermentation process optimization.

See more of this Session: Poster Session: Thermodynamics and Transport Properties
See more of this Group/Topical: Engineering Sciences and Fundamentals

Methodology for Toxic and Flammable Gas Sensors Positioning: Multi Grid Module

Giordhanne Bruno Carpaneda Gimenes[1]
Antonio Tavernard Pereira Neto[1]
Igor Ribeiro de Souza[1]
Caio Veiga Penna Delgaudio[1]
Valter Cesar de Souza[1]
Flavio Waltz[1]

[1] *Chemtech Serviços de Engenharia e Software Ltda.*
Rio de Janeiro – Brasil.

Resume

Computational fluid dynamics (CFD) and mechanical systems optimization concepts are currently quite important tools for engineering problems analysis. A typical application of such tools is gas dispersion studies, a theme of great importance for the oil and gas industry. An important safety item for oil and gas production installations (especially oil rigs) is the use of flammable and/or toxic gas sensors to detect gas leaks. The importance of these sensors relies on potential costs compensations and the reduction of environmental impacts, life loss, and profit loss. The methodology for positioning gas sensors uses CFD tools. Associating simulation results with coherent mathematical treatment, this methodology presents improvements in relation to other studies of gas dispersion and sensor positioning. The methodology output is a probability profile that shows the most probable areas to find the gas in detectable concentrations, which allows one to define the best location for the sensors. This methodology was already presented during the 2008 AIChE Spring Meeting, which took place in New Orleans, LA (USA).This work presents an upgrade in the methodology: the possibility of multi grid treatment, allowing one to create the probability profiles using simulations with different grids as inputs, which was not possible before. The strategy was implemented using interpolation tools and concerns the migration of every different grid to a standard one to be interpreted by the methodology mechanisms. This module makes the methodology more flexible while allows the engineer to adopt grids which fits better in each of his cases.

1. Introduction

CHEMTECH's methodology for toxic and flammable gas sensors positioning is based on the main concepts of safety engineering, defined in the available literature and regulation procedures. To allocate the sensors, several factors must be considered:

a) The relative density among the air and the gas and flammable vapors dispersed;
b) Location of the most probable leaking sources;
c) Streamlines from the natural or mechanical ventilation;
d) Zones where vapors can accumulate;
e) Presence of inhibition agents.

To reduce the occurrence of false alarms, the use of sensors triads is recommended. The alarm will sound only if two of the three sensors detect the leaked gas.

The type and levels of the system alarm depend on standards and technical recommendations, the gas' toxicity and its lower limit of flammability, the area occupation, and the actions to be taken after the alarm.

The sensors must be allocated close to the possible leaking points and in the confined rooms where the presence of the gas is possible. The paths of the gas clouds and its dispersion's characteristics depend on the gas density and on the ventilation patterns of the area. The density can be used to determine the height related to the emission source in which the sensors should be allocated.

The positioning of the sensors should be done with sensors covering zones (instead of specific points), in a way that each sensor could alert about the presence of gas from different source, bringing the system's efficiency and the utility of the equipment.

In the employed methodology statistical analysis is add to the points described above in order to relate the results achieved by the simulation with the wind's probability for each direction and the leakage probability. This approach generates a probability profile, when the chance of existence of gas in certain computational volume is represented.

The methodology can be described as follows: in a certain scenario, n simulations are done, one for each wind direction. To the simulation results, corresponding to the values of volumetric fraction of the gas in the cells of that scenarios, names as S1, S2, ..., Sn are attributed. The sensors sensibility occurs in a range of concentrations that is related to the Lower Flammable Limit (LFL) or to the Toxicity Limit for that substance. To deal with this, a filter is needed, in order to select the volumetric fractions that are into the sensor's range. The filter is applied to the simulations results S1, S2, ..., Sn, generating the filtered files S1', S2', S3', ..., Sn'. The gas will only reach certain places of the domain if a preferential wind flows in that direction. This event probability is, then, calculated from the leakage probability and from the wind directions probabilities, applying Bayes Theorem[1]. Considering the possibility of wind for n directions, the result obtained using Bayes Theorem is used to ponder the new result files (Sn'), generating the files S1'', S2'', ..., Sn''. Following this step, these files are compared, condensed, and normalized in the final file C1*, which represents the probability profile to find the gas in a certain range of concentrations inside the domain. This allows the engineer to allocate the sensors in a optimized way. Figure 1 resumes the methodology.

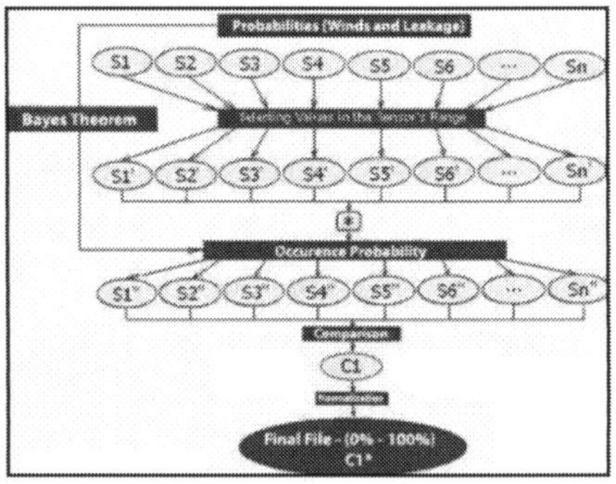

Figure 1 Methodology representation

[1] This methodology was published in TN Petróleo, 2007's July/August edition.

The use of several types of grids is necessary in many occasions, when the distance among the leakage points does not allow the adoption of only one mesh configuration. The use of different grids, with different refining degrees, for each leaking scenario produces more detailed and accurate results concerning the dangerous gas concentration. The adoption of a unique highly refined grid does not add great improvements to the simulation results and, moreover, demands highly computational efforts, when compared to the cases where the grid is refined specifically to one scenario.

Figure 2 shows a situation where the simulated leakage points are located distant one to the other and the adoption of one specific grid for each point becomes necessary.

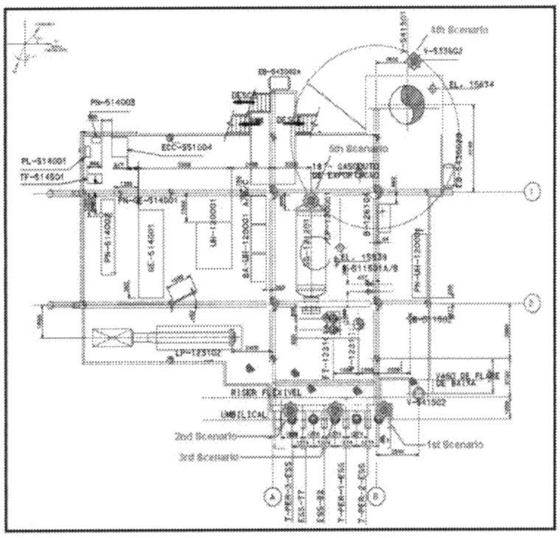

Figure 2 Possible scenarios' positioning in a sensors' allocation study

Although simulated with different grids, the results are related to the same study's context, to the same geometry and domain. However, the application of the Methodology for Sensors Allocation is only possible if the results were represented in equivalent grids. So, a Grid's transition step becomes necessary before the application of the methodology. The algorithm employed to allow those changes is the focus of this paper. Its functionality can be observed in Figure 3.

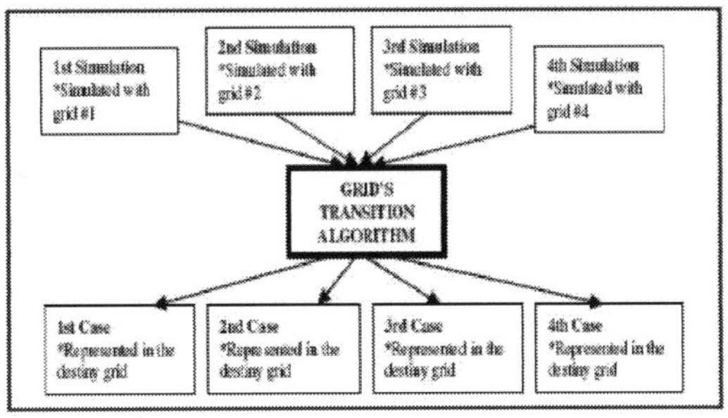

Figure 3 Grid's Transition functionality

2. Methodology

The calculations that were done by the algorithm are based in comparisons between the dimensions of each cell in the destination grid and the cells in the original simulation. The value of the variable in each cell of the destination grid is calculated depending on the relation between the destination grid and the original one, in that region. There are three possible cases. They are presented below, and so is their calculation approach.

First Case: If the new cell is positioned totally into a cell in the original grid, the variable value in the new grid's cell will be the same as the value of the cell that it is inside;

Second Case: If the new cell intersects the original grid in N parts in such a way that each Xn% of its volume coincides with a cell Gn in the original grid, the value of the variable will be calculated from the weighted average considering the intersection volume percentage. Figure 4 represents what was described above;

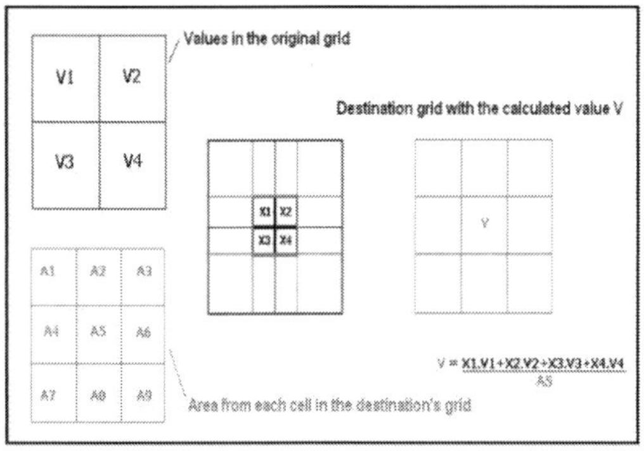

Figure 4 Second Case Representation for a 2D Grid (the destination grid's cells are in red)

Third Case: If the cell in the destination grid is bigger than the corresponding one in the original grid with N cells (or fractions of it) inside, as above the value is calculated trough the weighted average considering the intersection volume percentage. Figure 5 represents what was said.

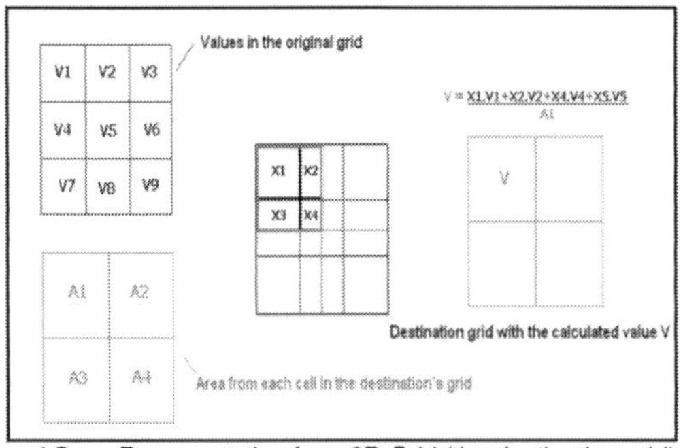

Figure 5 Second Case Representation for a 2D Grid (the destination grid's cells are in red)

3. Case Study

To apply the methodology, a case study was developed considering simulations with the same standard applied by the Special Projects Division (DPE in Portuguese) from CHEMTECH in the Sensors Allocation Projects. A hypothetic oil platform's geometry was used (showed in Figure 6) with dimensions and properties close to some platforms already studied by the authors of this article.

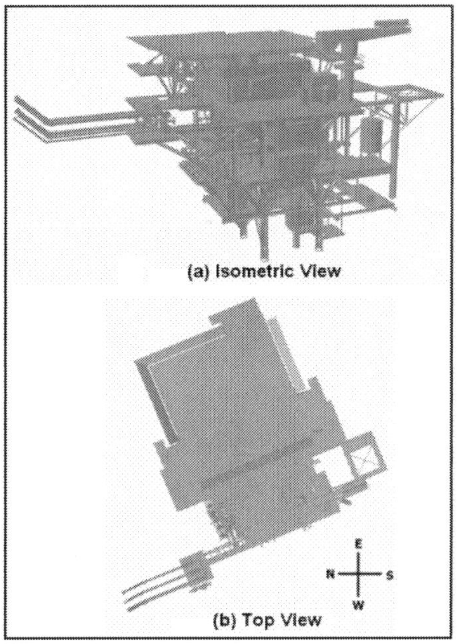

Figure 6 Platform's geometry

In order to explore the algorithm nine simulations were done, corresponding to the scenarios 2 and 5 in the Figure 2. The wind direction and the probabilities used in each case are presented in Table 1.

Table 1 Simulation's Scenarios

Scenario	Wind	Wind Occurrence Probability (%)	Leakage Probability (%)
2	Calm (NE)	27,45	15
2	N	22,98	15
2	NE	27,45	15
2	NW	3,07	15
2	W	2,44	15
5	Calm (NE)	27,45	10
5	E	18,99	10
5	NE	27,45	10
5	SE	11,07	10

For each scenario a grid was used. They are presented in Figure 7.

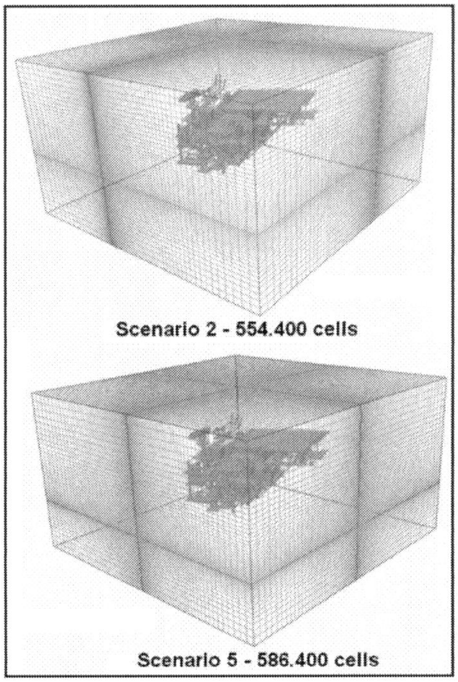

Figure 7 Grids used in the simulations

The destination grid used to represent and compare the results achieved is showed in Figure 8.

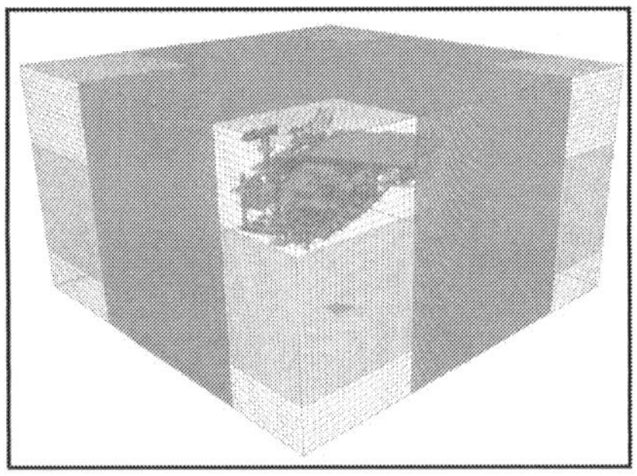

Figure 8 Destination grid - 7.840.000 cells

4. Results

From the described methodology for the grid's transition nine result files were obtained through the simulations and all of them were represented in the destination grid. The methodology to generate the probability profile was applied to these nine files, generating the profile represented in the destination grid. The probability profiles obtained are showed below, in Figure 9 and Figure 10.

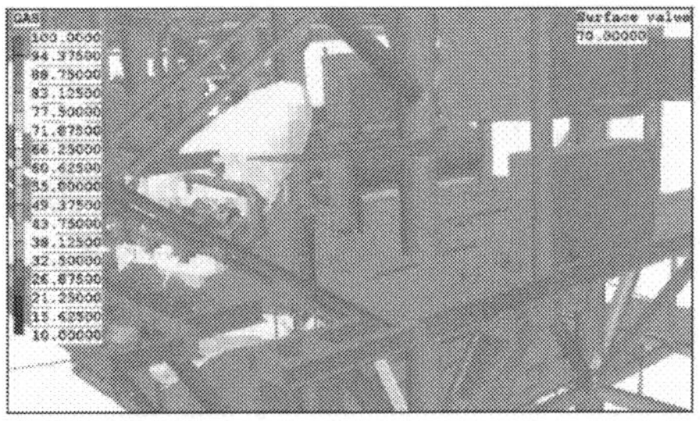

Figure 9 Surface representing the cells with a probability of 70% to find the gas

Figure 10 Surface representing the cells with a probability of 90% to find the gas

It is important to compare the results achieved in the original grid and in the destination grid, in order to validate the change. After the transition for the new grid, the result for the scenario must be as the one in the original grid. Figure 11 shows the concentration profiles for the scenario 2 and confirms that the destination grid reproduces with fidelity the results simulated in the original one.

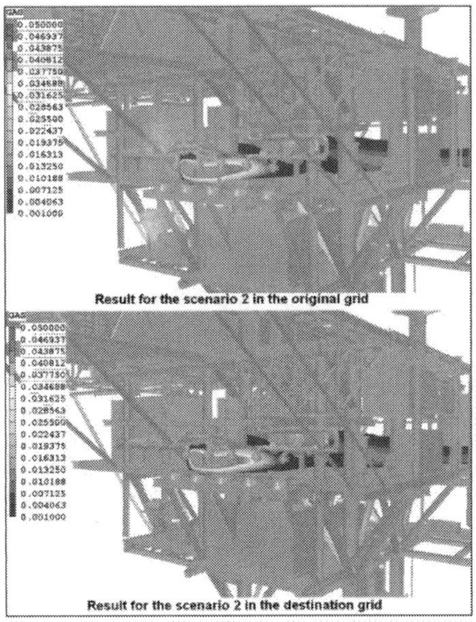

Figure 11 Comparison between the result in the old and in the new grid

The probability gradient can be generated as well. Figure 12 shows two possible views for these profiles. They relate the results achieved in both scenarios trough the grid's transition algorithm. This kind of representation is the more useful for the sensor's positioning, they allow the engineer to choose the best locations for the sensors.

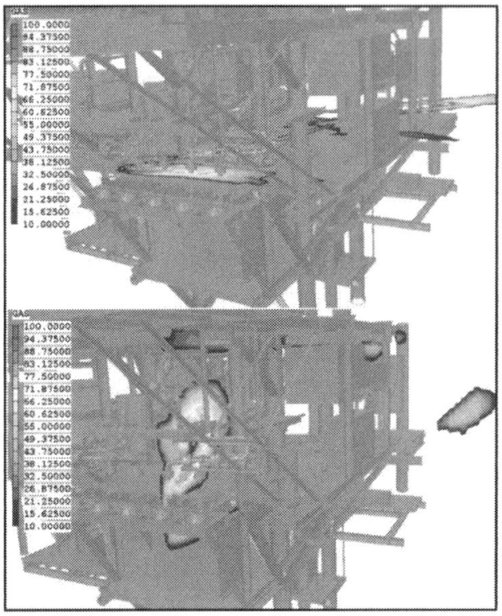

Figure 12 Planes representing the probability gradient

Figure 12 shows that the region in which the probabilities are higher correspond to the proximities of the leakage scenario 2, which happens with more frequency.

5. Conclusion

The grid's transition methodology allows the generation of a probability profile that summarizes the results from several simulations, regarding many leakage points simulated in different grids. This represents an improvement to the sensors allocation methodology concerning gas leak. The final probability gradient profile provides a global view of all the leaking scenarios, allowing the engineer to choose the best positioning for the sensors in the detection system.

The grid's transition methodology makes the methodology for sensors positioning more flexible since it provides to the engineer total freedom to adopt specific grids that better fits to each leakage scenario.

References

[1] CARPANEDA GIMENES, G. B. et all, Alocação de Sensores em Plataformas, TN Petróleo, Ano X, jul/ago de 2007, Número 55;

[2] CARPANEDA GIMENES, G. B. et all, Development of a Methodology for Toxic and Flammable Gases Sensors Positioning in Oil Platforms, AIChE Spring Meeting, New Orleans, LA, USA, April 2008;

[3] HSE-UK. Health and Safety Executive in Great Britain. The selection and use of flammable gas detectors. Novembro, 2004. Disponível em http://www.hse.gov.uk, acessado em novembro/2006

Experimental Densities for the [BMIM][BF4] + Water + 1-Propanol Mixture From 313 to 343 K and Pressures up to 20 Mpa

Luis E. Camacho-Camacho, Insituto Politecnico Nacional-ESIQIE, Mexico, Mexico

Luis A. Galicia-Luna, INSTITUTO POLITECNICO NACIONAL-ESIQIE, Mexico, Mexico

(P, rho, T, x) properties of [BMIM][BF4] + water + 1-propanol mixtures were measured at (313 to 343) K and pressures from 1 to 20 MPa. This work follows other similar ones to investigate the influence of water and alcohols in the (P, rho, T) properties of ionic liquids at high pressures.

Densities of ternary mixtures were measured by means of a Vibrating Tube Densimeter which is thermorregulated by using a liquid bath. Data obtained with this apparatus have shown to be very accurate when is compared with those obtained by other authors. An open literature searching let us know that there is no information about high-pressure data for this ternary mixture. Uncertainties of values were estimated as 0.2 kg m-3 for density, 0.03 K for temperature, 0.008 MPa for pressure.

[1] L. E. Camacho-Camacho, L. A. Galicia-Luna. 2008 AIChE Annual Meeting. Philadelphia, PA, USA. November 16-21, 2008.

[2] L. E. Camacho-Camacho, L. A. Galicia-Luna. 17th Symposium on Thermophysical Properties. Boulder, CO, USA. June 21-26, 2009.

[3] L. E. Camacho-Camacho, L. A. Galicia-Luna. 24th European Symposium on Applied Thermodynamics ESAT 2009. Santiago de Compostela (SPAIN). June 27 - July 1, 2009.

[4] L. Z. Zhang, J. Z. Han, Rong. J. Wang, Xing. Y. Qiu, and Jian. B. Ji. J. Chem. Eng. Data 2007, 52, 1401-1407

[5] L. Z. Zhang, D. -S. Deng, J. -Z. Han, D. -X. Ji, and J. -B. Ji. J. Chem. Eng. Data 2007, 52, 199-205

See more of this Session: Poster Session: Thermodynamics and Transport Properties
See more of this Group/Topical: Engineering Sciences and Fundamentals

A Network-Based Study On the Progression of Astrocytoma in Human Beings

Chunjing Wang, Chemical and Biomolecular Engineering, University of Illinois, Urbana-Champaign, Urbana, IL

James A. Eddy, Bioengineering, University of Illinois at Urbana Champaign, Urbana, IL

Nathan D. Price, Department of Chemical and Biomolecular Engineering, Institute for Genomic Biology, University of Illinois Urbana Champaign, Urbana, IL

Many diseases, including cancer, are known to emerge from cumulative interactions between multiple genes and other molecular agents within biological networks. Examining the disease-perturbed networks specific to a certain malignant phenotype helps us to gain insights into the biology and understand the mechanisms of the disease. Astrocytomas account for roughly 75% of neuroepithelial tumors and are usually classified into four grades. While the lower (I and II) grades astrocytomas have lower occurrence rate, higher grades (III and IV) occur much more frequently and represent higher mortality rates. A lot of effort has been made to identify possible classifiers (both as subnetworks or as individual genes) to accurately classify the correct stage, so that clinicians could choose the optimal therapy for the patients. A network-based binary classification method-differential ranking conservation (DIRAC) has become available recently and promises to identify accurate pathway biomarkers for different cancer phenotypes. In this project, we applied DIRAC to identify the most differentially expressed sub-networks which appeared central to different stages of astrocytoma. We investigated how these pathways evolved over different stages and how they interact to bring about the known disease phenotype. From the hypothetical process network map we constructed, we conclude that in low grades astrocytomas, the most perturbed subnetworks are the signaling pathways or related to angiogenesis. As the disease progresses to higher grades, more and more cell cycle and cell apoptosis pathways become involved. This shift of central biological events may help us to infer the underlying mechanism of astrocytomas. We also applied the well-known gene set enrichment analysis (GSEA) to the same patients and compared the resulting perturbed networks. Ultimately we hope to design and develop novel strategies for therapeutic intervention.

See more of this Session: Poster Session: Thermodynamics and Transport Properties
See more of this Group/Topical: Engineering Sciences and Fundamentals

Rapid Acid/Base Reactions That Produce Energetic Explosive Intermediates

Clint, J. Brown, Chemical and Materials Engineering, University of Alabama in Huntsville, Huntsville, AL

Paul, M. Frisby, Chemical and Materials Engineering, University of Alabama in Huntsville, Huntsville, AL

James, E. Smith Jr., Chemical and Materials Engineering, University of Alabama in Huntsville, Huntsville, AL

Rapid Acid/Base Reactions that Produce Energetic Explosive Intermediates

C. Brown, P.M. Frisby, and J. E. Smith, Jr.†

Department of Chemical and Materials Engineering

University of Alabama in Huntsville

Huntsville, Alabama 35899

The recent development and extension of a thermally controlled advanced laser diagnostic system was used to measure the chemical delay time (CDT) for hypergolic combustion with respect to temperature, including some intriguing results associated with cryogenic combustion. The fuels concentrated on in this study were Anhydrous Hydrazine (AH), Monomethylhydrazine (MMH), and Unsymmetrical Dimethyl Hydrazine (UDMH), as shown in Figure1.

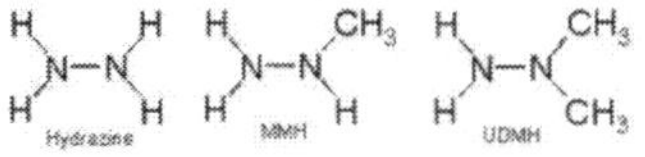

Figure1: Structures of Hydrazine, MMH, and UDMH

These fuels were reacted with red fuming nitric acid (RFNA) under a blanket of Argon gas to prevent reactions with oxygen and to minimize the intrusion of water vapor, particularly at cryogenic temperatures. Advanced thermal stages have been added to the existing CDT laser diagnostic system, which provides kinetic results for the hypergolic fuels. The temperature range for Hydrazine was varied from below its freezing point to 202.7°F (360K) due to evaporation, while MMH was studied between 160°F (344K) and -50°F (288K) and the UDMH was studied at temperatures from above 140 to below -60°F Since this system has a high acid demand, and the composition of the acid can change rapidly during oxidizer lead, the implication of studying the O/F ratio at cryogenic temperatures minimizes fuming in RFNA system permits higher O/F ratios while maintaining the integrity of the acid. A system was designed utilizing liquid nitrogen and a heavy stainless steel reaction vessel to achieve and maintain low temperatures (Brown, et. al. 2009a). For temperatures above ambient, a ceramic system described in Dasarathy, et. al 2005 was employed, and the data indicated that the CDT of UDMH increased when heated and was attributed to the increase in oxidizer demand at higher temperatures. To study the effect of higher O/F ratios in the UDMH system, oxidizer lead was required. Measurements were made at cryogenic temperatures approaching the freezing point of UDMH to determine the effect on CDT for UDMH. At temperatures above 20°F, the fuming from the RFNA disrupted the CDT measurements producing high relative error due to the NO_2 vapors interfering with the laser diagnostic system. The newly developed cryogenic technique inhibited the release of NO_2 allowing for study of the effect of higher O/F ratios. Oxidizer to fuel ratios were varied for UDMA/RFNA from 2-4 at a temperature of -30°F (272 K). (Brown, et al., 2009b)

The CDT for Hydrazine was approximately 620 ms just above room temperature but decreased significantly to an average of 45.4 ms as it approached its boiling point, which is a 92 percent decrease in CDT above room temperature. Below room temperature the CDT increased to about 2000 ms at the AH reported freezing point of 34.5°F (274.5K). (Brown, et al., 2009a) The CDT of MMH at room temperature is around 200 microseconds and around 5 milliseconds at -35°F (236K). The CDT data at temperatures below ambient are not only slower than room temperature but also becomes more unpredictable due to the increased thermal energy barrier for reactive decomposition(Frisby, et al., 2009). Eventually, the CDT data enter a mass transfer limited region in which the combustion severely slowed. The results indicate that CDT of UDMH at a higher oxidizer to fuel ratio (4/1) decrease by 100% when compared to the results obtained at 2/1, supporting results obtained by Farmer, 1997. (Brown, et al., 2009b). Previous studies showed that kinetic modeling of the CDT versus the absolute temperature followed an Arrhenius like behavior (Dasarathy, et al., 2005). Using similar modeling as the previous studies, AH/RFNA below room temperature was an extension of this behavior. We further examined the bipropellant system below the freezing point of AH, at -5°F. As the acid droplet hit the frozen AH, a surface reaction began to melt the fuel, forming a slush-like phase, which continued to melt until combustion was achieved at a much slower overall CDT of 6275 ms, since this time now includes the mass transfer associated with the melt (Brown, et al., 2009a). Results of the kinetic modeling and photographic results of the slush like phase will be presented.

Experiments were designed to determine the liquid temperature at the point to ignition by placing a small open junction Type-E thermocouple into the combustor. The temperature results were used to determine the presence of explosive intermediates by comparing them to the explosive boiling points of known species. Figure 2 displays the temperature and CDT results for MMH. The temperature at the point of ignition for this compound is 152°F (340K) (Frisby, et al., 2009). This temperature corresponds to the boiling point of methyl nitrate (CH_3NO_3) at 150°F (338K), which undergoes explosive decomposition at this temperature (Sax, et al., 1987).

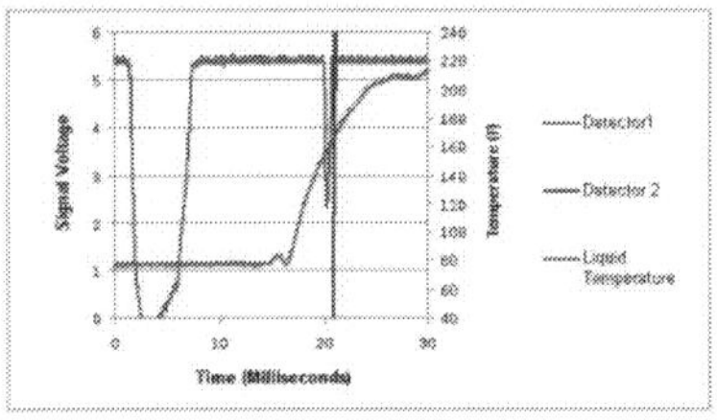

Figure 2. Temperature of MMH combustion along with DSO signal reading.

Figure 3 shows similar results for UDMH. The CDT is defined as the time between Point C, which is the initial evolution of gas and Point D, which is the appearancce of flame. The standard temperature used to represent the ignition is halfway in between Point C and Point D. As seen in Figure 3, this temperature is at 150°F (338K) indicating that the explosive intermediate of UDMH combustion is also methyl nitrate (Brown, et al., 2009b).

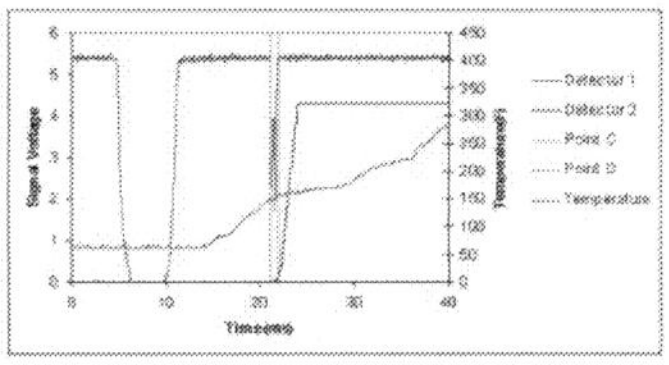

Figure 3. Temperature of UDMH combustion along with DSO signal reading.

Figure 4 displays the temperature reading for AH (Brown, et al., 2009a). The results show that the compound without the methyl groups has a different reaction mechanism. The temperature of 315°F (430K) indicates that the intermediate species in this reaction is ammonium azide (NH_4N_3) which has a boiling point of 320°F (433K) at which point it also explodes (Sax, et al., 1987). This is due to ammonium azide decomposing into the explosive hydrazoic acid (HN_3) at this temperature. Following the ignition of hydrazine there are plumes of smoke shown in the Detector 1 as the temperature continues to rise, this reading indicate that ammonium nitrate (NH_4NO_3), which has an explosive boiling point of 410°F (483K) and other explosive species may be present in competing reactions with ammonium azide.

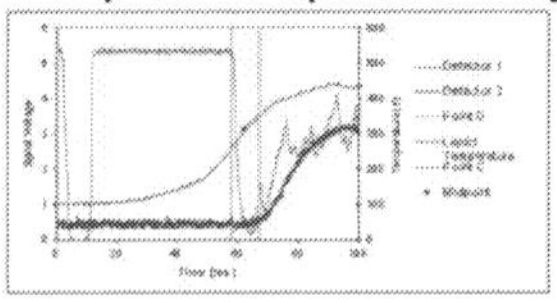

Figure 4: Temperature of AH combustion with DSO signal readings.

The results indicate a similar ignition temperature for MMH and UDMH with methyl nitrate being the explosive intermediate, and an ignition temperature of AH indicating ammonium azide as the intermediate. As the liquid temperature rises and some of the fuel evaporates, this salt decomposes, igniting the mixture in the vapor phase thus explaining why the chemical delay time region exists. As it appears the intermediate explosive compounds of the fuel/RFNA reaction made within the reaction just in time, future work will be focused on attempting to isolate the intermediates of the reactions to verify the findings of the temperature study.

References:

Brown, C, Frisby, P.M., and Smith, Jr, J.E. "Kinetic Modeling and Mass Transfer Effects below the Freezing Point for the Hypergolic Combustion of Anhydrous Hydrazine Reacted with Red Fuming Nitric Acid,". Paper # 523, Proceedings of 56th JANNAF Propulsion Meeting, 35th Propellant and Explosives Development and Characterization Subcommittee, Las Vegas, Nevada April 14-17, 2009

Brown, C, Frisby, P.M., and Smith, Jr, J.E. "The Use of Cryogenic Temperatures to Study High Oxidizer to Fuel Ratios in the UDMH/RFNA System," Paper # 521, Proceedings of 56th JANNAF Propulsion Meeting, 35th Propellant and Explosives Development and Characterization Subcommittee, Las Vegas, Nevada April 14-17, 2009

Dasarathy R.B., Hampton C.S., and Smith, Jr. J.E., "The Relationship Between the Chemical Delay Time and Temperature for Hypergolic Bipropellants, "Proceedings of 2nd JANNAF Liquid Propulsion Subcommittee Meeting, December 2005.

Farmer, M. J., "A Study of the Reaction Rates of Hypergolic Propellants," Masters Thesis, Department of Chemical and Materials Engineering, University of Alabama in Huntsville, May 1997.

Frisby, P.M., Brown, C, and Smith, Jr, J.E. "The Thermal Performance and Kinetic Behavior of Monomethylhydrazine Reacted with Red Fuming Nitric Acid," Paper # 522, Proceedings of 56th JANNAF Propulsion Meeting, 35th Propellant and Explosives Development and Characterization Subcommittee, Las Vegas, Nevada April 14-17, 2009

Sax, N. I. and Lewis, R. J., Hazardous Chemicals Desk Reference, Van Nostrand Reinhold Company Inc.: New York, 1987.

See more of this Session: Poster Session: Thermodynamics and Transport Properties
See more of this Group/Topical: Engineering Sciences and Fundamentals

Solubility of CO_2 in Aqueous Solutions of Several Classes of Amines

Ahmed Nouacer, Université Houari Boumediène, Laboratoire Thermodynamque des Composés Organiques et Modélisation, Algeria

Ahmed Hajjaji, Analitycal science laboratory, Thermodynamic and online analysis Team, Claude Bernard University Lyon - France, Lyon, France

Nadia Baskali, Analitycal science laboratory, Thermodynamic and online analysis Team, Claude Bernard University Lyon - France, Lyon, France

Ilham Mokbel, Analitycal science laboratory, Thermodynamic and online analysis Team, Claude Bernard University Lyon - France, Lyon, France

Farid Belaribi, Université Houari Boumediène, Laboratoire Thermodynamque des Composés Organiques et Modélisation, Algeria

Ramy Abou Naccoul, Analitycal science laboratory, Thermodynamic and online analysis Team, Claude Bernard University Lyon - France, Lyon, France

Jacques Jose, Analitycal science laboratory, Thermodynamic and online analysis Team, Claude Bernard University Lyon - France, Lyon, France

Aqueous solutions of amine are used for removing CO_2 and H_2S from natural gas streams and refinery process streams. They are also used to remove CO_2 from combustion gases and may receive wide application for abatement of greenhouse gases due to their basic character.

The objective of this work is to study the absorption capacity of CO_2 in aqueous amines from several classes selected by a Quantitave Structure Activity Relationship (QSAR) method in order to find new chemical solvents. Among the twenty retained molecules are morpholine, 4-methyl-morpholine, tris(hydromethyl)aminoethane, 2-2-dimethylaminoethoxy-ethanol, 1,2-bis-dimethylaminoethane, 1,2-bis-2-aminoethoxy-ethane.

In this aim we realized a static apparatus based on the drop of the pressure of CO_2 contained in the reservoir. The solubilities of the previously quoted amines were determined at 2 different temperatures corresponding to the process operation: 40°C (CO_2 absorption temperature) and 80°C (regeneration temperature).

The present study was carried out in the frame work of the ANR CapCO2 (Agence Nationale de la Recherche which launched a CO_2 program in 2005).

See more of this Session: Poster Session: Thermodynamics and Transport Properties
See more of this Group/Topical: Engineering Sciences and Fundamentals

Removal of Sulfur Compounds From Solvent Extracts of Low Rank Coal

Kwang B. Yi, Greenhous Gas Research Center, Korea Institute of Energy Research, Daejeon, South Korea

Chang Hyun Ko, Korea Institute of Energy Research, Daejeon, South Korea

Jong-Ho Park, Korea Institute of Energy Research, Daejeon, South Korea

Jong-Nam Kim, Korea Institute of Energy Research, Daejeon, South Korea

Recently, Korea and Japan are investigating low-rank coal upgrading technology using solvent extraction. The solvent extraction is the technology to separate only combustible matters from low-rank coal and has been known as environmenatlly sound and cost effective. Furthermore, the upgraded coal can be directly used in a gas turbine since it does contain trace-amount of ash. However, there is still small amount of sulfur (100 to 1000 ppmw) exist in the upgraded coal which might cause corrosion in blades and boiler component. Therefore, the remained sulfur need to be removed for improving coal quality. Korea Institute of Energy Research (KIER) has been recently developed desulfurization sorbent and the process that can be directly attached to the solvent extraction process. Because the desulfurization condition is identical with the extraction process, no further energy consumption is required to obtain low sulfur coal. The performance of the developed sorbent in a lab-scale experimental apparatus at various condtions will be presented.

See more of this Session: Poster Session: Thermodynamics and Transport Properties
See more of this Group/Topical: Engineering Sciences and Fundamentals

Nitriding of Metal Powder

Stephanie Egger, Metallurgy, University of Leoben, Leoben, Austria

Harald Raupenstrauch, Chair of Thermal Processing Technology, University of Leoben, Leoben, Austria

Nitriding of metal is an interesting part of surface treatment for increasing physical properties like corrosion resistance, hardness and chemical resistance.

Atomisation of liquid metal with nitrogen leads to an instable condition. Because of this fact the reaction with a nitriding gas shows other behaviour than not atomised metal powder.

Ammonia is an important nitriding gas because of the lower dissociation temperature than nitrogen. The sintering temperature of metal powder is lower than the dissociation temperature of molecular nitrogen.

A common nitriding temperature of atomised metal powder is between 350°C to 600°C, wherefore ammonia is used for the favour nitriding gas.

For analysing the kinetic nitriding process measurements are made with a thermo gravimetric analyser (TGA).

Therefore measurements with several gas additions (CO, N2, O2, air), gassing rates, pre-treatments and different temperatures are investigated. Thereby the increasing of the mass is relevant for the kinetic analyses.

Further the content of alloying elements shows the influences to the nitriding process.

Aims of this work are the optimal nitriding conditions to decrease the nitriding duration and increase the product quality and reproducibility of nitration.

See more of this Session: Poster Session: Thermodynamics and Transport Properties
See more of this Group/Topical: Engineering Sciences and Fundamentals

Development of Advanced Coal Gasification Kinetics Models for CFD (and process simulation) Codes

Richard Turton, Chemical Engineering, West Virginia University, Morgantown, WV

Chris Guenther, National Energy Technology Laboratory, U.S. Department of Energy, Morgantown, WV

Kiran Chaudhari, Chemical Engineering, West Virginia University, Morgantown, WV

The objective of this research is to implement detailed kinetic expressions for the gasification of a wide variety of coals in existing CFD codes such as MFIX. These expressions describe the fundamental steps taking place in the gasification of coal, namely, coal devolatilization, tar-gas chemistry, soot formation, the heterogeneous and homogeneous gasification reactions along with combustion reactions. For this purpose, the data generated by PC Coal Lab licensed by Niksa Energy Associates LLC will be used to simulate the gasification of various coals in CFD codes. The CFD code of most interest is MFIX, which is used to describe the hydrodynamics, heat transfer and chemical reactions in the reacting fluid-solids systems comprising the contents of a gasifier. The implementation of coal gasification kinetics in MFIX is currently done through a set of subroutines making up the Carbonaceous Chemistry for Continuum Modeling (C3M) code. C3M has default gasification kinetics for only a few coals. While on the other hand, PC Coal lab can predict the gasification kinetics for over 2000 coal species. Therefore this project focuses on the development of a seamless connection between PC Coal Lab and the C3M .The interface will be designed to allow MFIX to transfer information through C3M to PC Coal Lab, run PC Coal Lab with the input data from MFIX, send kinetic information back to C3M in a form that allows C3M to update and continue the MFIX simulation using updated parameters from PC Coal Lab.

See more of this Session: Poster Session: Thermodynamics and Transport Properties
See more of this Group/Topical: Engineering Sciences and Fundamentals

Molecular Dynamics Study On the Melting Point of Ionic Liquid [Emim]Br

Haijun Feng, School of Chemistry and Chemical Engineering, South China University of Technology, Guangzhou, China

Jian Zhou, School of Chemistry and Chemical Engineering, South China University of Technology, Guangzhou, China

How to get the melting points of ionic liquids below the room temperature is the major problem in the applications of ionic liquids. High costs in experiments, lack of precise descriptions of microstructures and complicated effects, all these make the experimental determination of melting points of ionic liquids difficult, however molecular dynamics simulations can complement to experimental investigation. In this work, molecular dynamics simulations are carried out to study the melting point of ionic liquids [emim]Br by the direct heating method, void-introduced method, sandwich method and NVE ensemble method, to find the most suitable method to predict the melting point by simulation. By analyzing the variations of the nonbonded energy, density, radial distribution function, diffusion coefficient and translational order parameter of [emim]Br with temperature, the results of the four methods are 548±8 K,370±6 K,373±4 K and 335±3 K, respectively, which is approximately 55.7%,5.1%,6.0% and -4.8% higher than the experimental value of 352 K. The advantages and disadvantages of each method and the mechanism of melting are also discussed.

See more of this Session: Poster Session: Thermodynamics and Transport Properties
See more of this Group/Topical: Engineering Sciences and Fundamentals

Improved Implementation of Kirkwood-Buff Solution Theory in Periodic Molecular Simulations

Dean R. Wheeler, Chemical Engineering, Brigham Young University, Provo, UT

Joseph W. Nichols, Chemical Engineering, Brigham Young University, Provo, UT

Stan G. Moore, Chemical Engineering, Brigham Young University, Provo, UT

Kirkwood-Buff (KB) solution theory is a means to obtain certain thermodynamic derivatives from knowledge of molecular distributions. These thermodynamic properties include isothermal compressibility, partial molar volumes, and chemical potential derivatives. Traditional KB theory requires that one make integrals over radial distribution functions; these integrals suffer inaccuracies due to finite-distance truncation effects and their use in closed systems. In this work we discuss how best to minimize these inaccuracies under traditional KB theory. In addition we implement a new method for calculating KB quantities in molecular simulations with periodic boundary conditions and particularly within the canonical ensemble. The new method is based on a finite-Fourier-series expansion of molecular concentration fluctuations and leads to more reliable results for a given computational effort. The procedure is validated and compared to the original method for a nonideal liquid mixture of Lennard-Jones particles intended to imitate a real system, carbon tetrafluoride and methane.

See more of this Session: Poster Session: Thermodynamics and Transport Properties
See more of this Group/Topical: Engineering Sciences and Fundamentals

REACTION PATHWAY ANALYSIS FOR ELECTROLYSIS OF ALCOHOLS AND SUGARS IN SUB-CRITICAL WATER

Mitsuru SASAKI[1,], Takanari OSHIKAWA[1], Hiromichi KOGA[1], Asli YUKSEL[1], Wahyudiono[2], and Motonobu GOTO[2]*

[1] *Graduate School of Science and Technology, Kumamoto University, Kumamoto, Japan*

[2] *Bioelectric Research Center, Kumamoto University, Kumamoto, Japan*

* Corresponding author, TEL: +81-96-342-3664, FAX: +81-96-342-3665, e-mail: msasaki@kumamoto-u.ac.jp

Introduction

Nowadays, biodiesel fuels have garnered much attention because they possess various advantages as alternatives to petroleum-based fuel, *e.g.*, they are renewable, possess a favorable energy balance, have lower harmful emissions and are non-toxic. However, this rise in the production of biodiesel has created the problem of disposing of the glycerol that is generated as the major by-product. Presently, in many biodiesel plants with low capacity, this crude glycerol is discharged in the wastewater as main by-product together with some methanol, alkali and low concentration of fatty acids. As a result of this, both environmental pollution and economical losses are generated. Glycerol is used in medical, pharmaceutical, and personal care preparations, mainly as a means of improving smoothness, providing lubrication and as a humectant. It is found in cough syrups, elixirs and expectorants, toothpaste, mouthwashes, skin care products, shaving cream, hair care products, and soaps. Additionally, in foods and beverages, glycerol serves as a humectant, solvent and sweetener, and may help preserve foods. It is also used as filler in commercially prepared low-fat foods (*e.g.*, cookies), and as a thickening agent in liqueurs. Glycerol and water also serve as preservatives for certain types of leaves.

Recently, the conversion of glycerol to various compounds has been investigated. Catalytic dehydration of glycerol in sub- and supercritical water and its reaction mechanism have been studied. It was found that the OH group on various catalyst can catalyze the dehydration of glycerol to 3-hydroxypropionaldehyde, 1,3-propanediol, acrolein and other degradation products through subsequent hydrogenation[1-3]. In another study, ionic reactions and pyrolysis of glycerol as competing reaction pathways in near- and supercritical water have been investigated. It was shown that the products of the glycerol degradation are methanol, acetaldehyde, propionaldehyde, acrolein, allyl alcohol, ethanol, formaldehyde, carbon monoxide, carbon dioxide, and hydrogen[4, 5]. Moreover, pyrolysis of glycerol has been studied in steam with the major products being acrolein, formaldehyde and acetaldehyde[6, 7]. At higher temperatures, other products such as carbon dioxide, molecular hydrogen, ethylene and methane are observed, indicative of more complex chemistry.

One method of particular interest is the conversion of glycerol to lactic acid. Kishida *et al.*[8] showed that glycerol can be converted into lactic acid by hydrothermal decomposition experiments of glycerol in sub-critical water under alkaline conditions. Studies on the conversion pathway have suggested that glycerol is first decomposed to pyruvaldehyde with elimination of hydrogen by H^- shift to the adjacent hydrogen. The pyruvaldehyde is then converted into lactic acid ion by the benzilic acid rearrangement. In this study, the main product was lactic acid. In addition, small amounts of pyruvaldehyde, acetic acid and formic acid were detected. In another study, the conversion mechanism of cellulosic biomass to lactic acid in subcritical water without the addition of any alkaline catalyst was discussed. The results showed that this conversion mechanism appeared to follow the same pathways elucidated in the conversion of sugar to lactic acid in alkaline solution[9].

The main purpose of this work is to convert glycerol inside the model biodiesel wastewater into more value-added chemicals, especially lactic acid which plays an important role in several biochemical processes by using both batch and flow reactors. We additionally aimed to clarify the gaseous and liquid products, effects of electrolysis current and alkali concentration on the decomposition of glycerol. In this study, sub-critical water was selected as a reaction medium because it is cheap, non-toxic, non-flammable, non-explosive and offers essential advantages compared to other substances, particularly in the field of "green chemistry"[10]. Recent studies have demonstrated that heating water above its boiling point under enough pressure to maintain its liquid state causes some physicochemical changes that are useful in reducing the rate problems as well as increasing the solubility of organics in water. For example, the dielectric constant is much lower, and the number of hydrogen bonds is diminished. Additionally, the ion product or dissociation constant (K_w) for water as it approaches the critical point is about three orders of magnitude higher than it is for ambient liquid water. As a result, water becomes an excellent solvent for organic compounds. It can also boast a higher H^+ and OH^- ion concentration than liquid water under certain conditions[11, 12]. In the case of economic feasibility, sub-critical water minimizes the amount of energy consumed for electrolysis. Because the "current efficiency" as a function of the reaction progress increases with increasing the temperature of water toward its critical point, which also increases the economical feasibility of the electrochemical reactions[13]. In addition, reactions of glucose and fructose have been carried out by using the both reactors in order to understand reaction behavior and mechanism for alcohols and sugars and evaluate possibility of the development of chemicals recovery from waste biomass resources with the hydrothermal electrolysis technique.

Experimental and analysis

The experiments were done first by using batch reactor and then continued with flow type reactor. For the first set of experiments, aqueous electrolyte solutions were prepared by dissolving an alkali (NaOH) in deionized water and then a given amount of glycerol was charged in the solutions. Then series of experiments were done by adding methanol to the feed solution in order to understand the effects of methanol concentration on the decomposition of glycerol inside the model biodiesel wastewater. The batch and flow reactors were shown in Figure 1 (a) and (b), respectively.

(a)

(b)

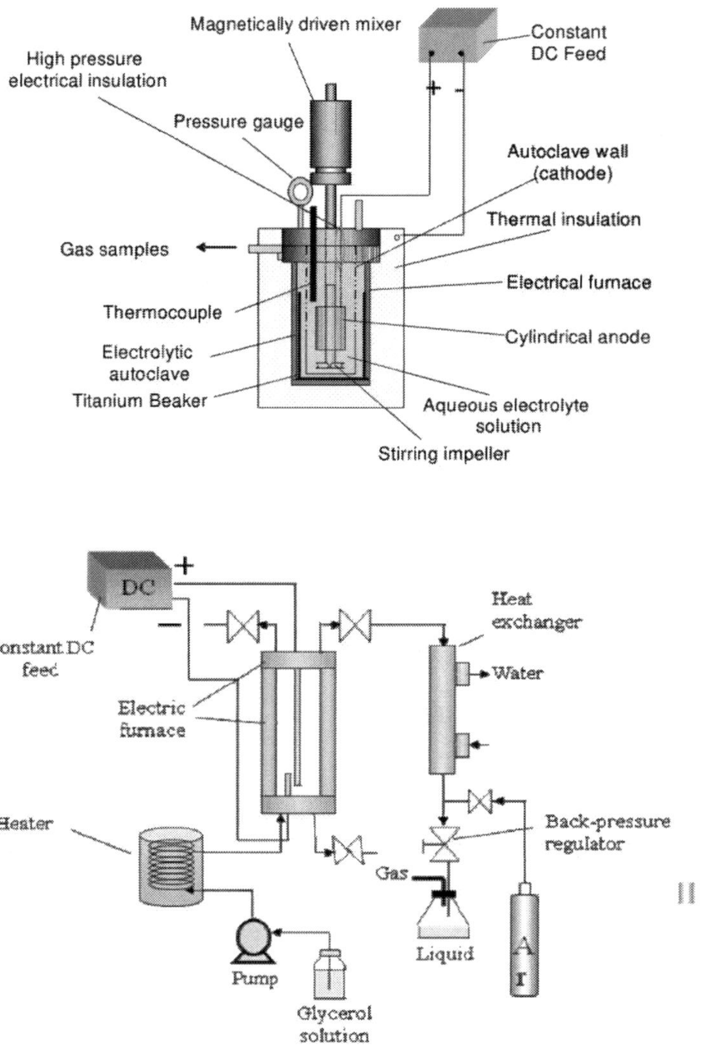

Figure 1: **(a)** Batch reactor; **(b)** Continuous flow reactor for hydrothermal electrolysis experiments.

In the batch system, a cylindrical iridium plate electrode (23 mm in diameter, 30 mm in length), coaxial to the titanium beaker (74 mm in diameter, 100 mm in length), was used as an anode and the titanium beaker was employed as a cathode to promote cathodic protection of the autoclave from corrosion. In flow system, titanium beaker acts as both anode and cathode. Before starting each experiment with autoclave, system was purged with inert gas (argon) and then initially charged with 3 MPa argon gas. The inner pressure of the autoclave increased as a function of temperature, and the typical reaction pressure was 10 MPa. For flow experiments, the reaction pressure was kept constant at 10 MPa. In both systems, reactors were heated from room temperature to the desired temperature (280

°C). The temperature was measured using a thermocouple inserted in the reactors during the experiment. as the temperature of the autoclave reached 280 °C. Once the systems had been heated to the desired temperature, a constant electric current was passed between the electrodes for hydrothermal electrolysis experiments. No current was passed for hydrothermal degradation runs. In the case of sugars, glucose (99% purity, Wako), fructose (99%, Wako) and xylose (99%, Wako) were employed as starting materials and acetic acid was used as an electrolyte. They were reacted under various operating conditions as with the case of glycerol.

At the end of reaction, both gas and liquid samples were collected for analysis. Solid samples were not specifically analyzed, because no measurable solid products were observed. Gas formed during the reaction was collected in sampling plastic bags and analyzed by gas chromatography with a thermal conductivity detector (GC-TCD). Liquid products were identified by high performance liquid chromatography (HPLC) and quantified by gas chromatography with a flame ionization detector (GC-FID). Total organic carbon in the liquid product solution was monitored by using TOC analyzer.

Results and discussion

After the hydrothermal electrolysis of glycerol with alkali lactic acid (L- and D-) and hydrogen gas were formed as main organic acid and a gaseous product, respectively. Additionally, small amounts of formic acid, glycolic acid and aldehydes were produced. Figure 2 shows the glycerol conversion under various operating conditions by batch reactor.

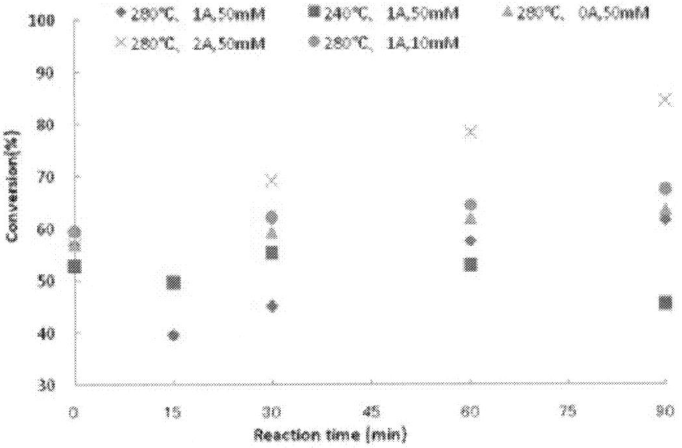

Figure 2: Conversion of glycerol under various operating conditions by batch reactor

Conversion of glycerol under alkaline conditions increased with increasing electrolysis current. When 2 A current was applied during 90 min, it reached its maximum value as 84 % at 280 °C. As for NaOH effect, by increasing the concentration of alkali from 10 to 50 mM only 6 % difference in the conversion value was observed at the end of 90 min. Figure 3 shows the optimum conditions for the production of lactic acid from the hydrothermal electrolysis of glycerol under alkaline conditions by using autoclave. As seen below, at 280 °C, with 50 mM NaOH by applying 1 A current, maximum lactic acid yield was raised to 35 % after 90 min electrolysis.

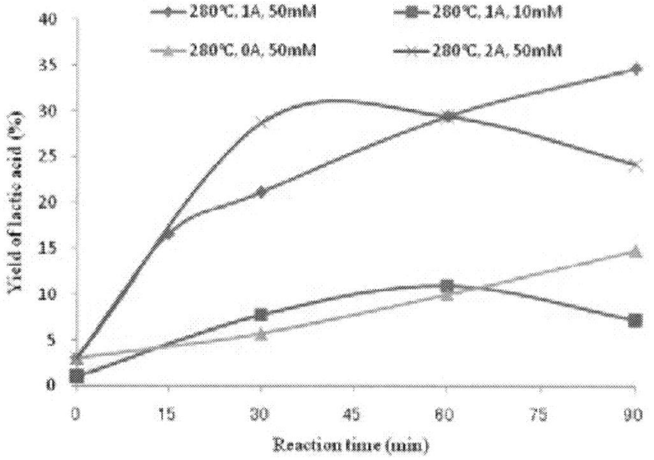

Figure 3: Optimum conditions for lactic acid production from glycerol by batch electrolysis in sub-critical water.

In the second part of this study, glycerol was treated under alkaline conditions by the addition of methanol by using flow type reactor. Here as gas product, CO_2 was also generated together with H_2. Figure 4 shows the volume percentage of gas products generated by flow type reactor.

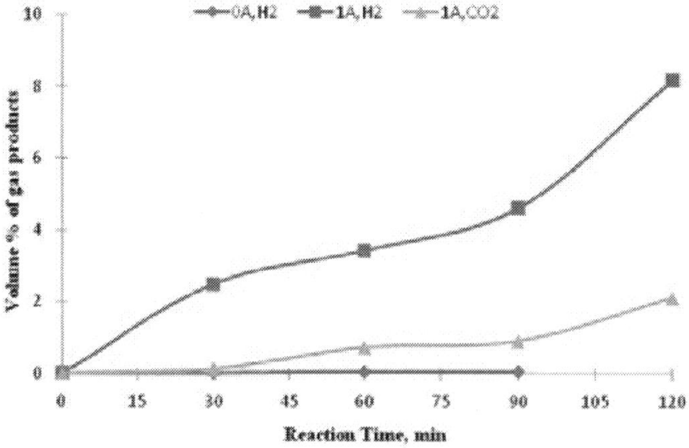

Figure 4: Volume percentage of gas products at 280 °C with 0.1 M glycerol, 50 Mm NaOH and 0.05 M methanol

By the addition of methanol, glycerol conversion to CO_2 was promoted. However, in this case, the yield of organic acids was much lower than in the first part around 5-6 %. In addition to this, glycerol conversion was 60-70 % within 90 min and around 85 % within 90-120 min under the influence of 1 A current. Here, we tried to consider a possible reaction pathway for glycerol on the basis of the experimental results obtained in this study. As described before, the main products are lactic acid and hydrogen, but some products could be obtained on this study. Figure 5 shows typical results obtained by the batch experiments at 280 °C. At the reaction (electrolytic) time was zero, glycerol conversion has already reached around 60 % and formic acid was mainly formed. With current loadings (especially 50 mM), yields of D- and L-lactic acids increased with the increase in electrolytic time. In all the experimental conditions, the existence of glyceraldehyde could be confirmed by HPLC analysis of the product aqueous solutions. This suggests that glycerol could convert at first into glyceraldehyde and subsequently to lactic acids through the electrolysis reaction as shown in Figure 6. At the present, however, only qualitative analysis has been performed and could not clarify the reaction pathway and mechanism more in detail. Quantitative analysis of glyceraldehyde will help us to elucidate the reaction pathway for glycerol and also explore optimum condition where lactic acids or glycolic acid could be obtained efficiently with the hydrothermal electrolysis.

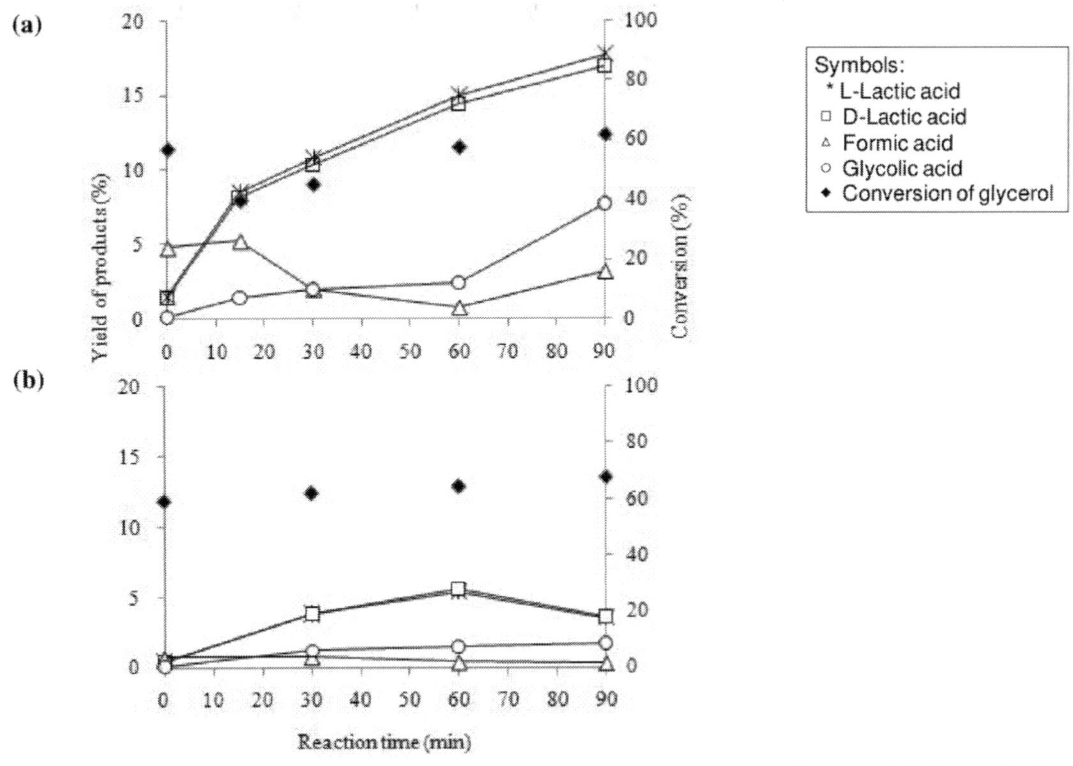

Figure 5: Time course of glycerol conversion and product yields (a) 50 mM-NaOH and (b) 10 mM-NaOH. Experimental conditions were 280 °C, 0.1 M glycerol, 0.05 M methanol.

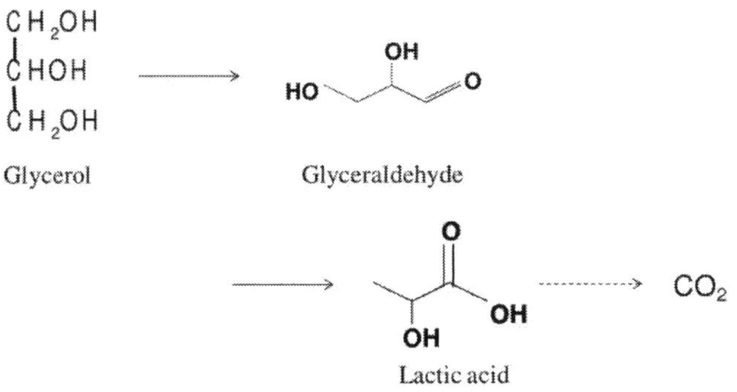

Figure 6: Possible reaction route on the hydrothermal electrolysis of glycerol. In the present, glyceraldehyde could be qualitatively detected in all the experimental conditions. Of course, side reactions of glycerol to formic acid or glycolic acid take place at this atmosphere.

Conclusions

The development of a specially designed batch autoclave and flow type reactor for the hydrothermal electrolysis of glycerol with an alkali and analysis techniques to investigate the conversion of glycerol to organic acids, especially lactic acid, in sub-critical water has led to several conclusions. After the hydrothermal electrolysis of glycerol with NaOH, hydrogen gas was generated as a gaseous product, whereas lactic acid was produced as the main liquid product. Additionally, small amounts of formic acid, glycolic acid and glyceraldehyde were generated as minor products. Increasing the concentration of NaOH led to a great increase in the yield of lactic acid with a yield of 34.7% at 280 °C with a NaOH concentration of 50 mM at the end of 90 min reaction time. It should be noted that a very small amount of lactic acid was formed even after 90 min electrolysis time at 10 mM NaOH. Therefore, as these results show, the concentration of NaOH greatly enhanced the conversion of glycerol into lactic acid. When methanol was added as a reactant in the experiments with flow type reactor, carbon dioxide generation was promoted in addition to the hydrogen gas. However, the yields of organic acids were much lower than it was at batch electrolysis in sub-critical water. In addition, possible reaction pathway for glycerol was considered based on the experimental results. Reactions of sugars (glucose and fructose) were now under investigation and will be reported in the near future.

Acknowledgements

This work was supported by the Kumamoto University Global COE Program "Global Initiative Center for Pulsed Power Engineering".

References

1. L. Ott, M. Bicker and H. Vogel (2006), *Green Chem.*, 8, 214.
2. V. Lehr, M. Sarlea, L. Ott and H. Vogel (2007) *Catalysis Today*, 121, 121.
3. M. Watanabe, T. Lida, Y. Aizawa, T. M. Aida and H. Inomata (2007) *Bioresource Technology*, 98, 1285.
4. W. Buhler, E. Dinjus, H. J. Ederer, A. Kruse and C. Mas (2002) *Journal of Supercritical Fluids*, 22, 37.
5. T. Valliyappan, N. N. Bakhsi and A. K. Dalai, *Bioresource Technology*, 99, **2008**, 4476
6. M. J. Antal, W. S. L. Mok, J. C. Roy, A. T. Raissi and D. G. M. Anderson (1985) *J. Anal. Appl. Pyrolysis*, 8, 291.
7. Y. S. Stein, M. J. Antal and M. J. Jones (1983) *J. Anal. Appl. Pyrolysis*, 4, 283.
8. H. Kishida, F. Jin, Z. Zhou, T. Moriya and H. Enomoto (2005) *Chemistry Letters*, 34, 11.
9. F. Jin, Z. Zhou, H. Enomoto, T. Moriya and H. Higashijima (2004) *Chemistry Letters,* 33, 2.
10. T. Rogalinski, K. Liu, T. Albrecht and G. Brunner (2008) *J. of Supercritical Fluids*, 46, 335.
11. T. Clifford, *Fundamentals of Supercritical Fluids* (New York: Oxford University Press) (1998) p. 23.
12. P. E. Savage (1999) *Chem. Rev.*, 99, 603.
13. F. S. Asghari and H. Yoshida (2008) *J. Phys. Chem. A.*, 112, 7402.

Interactions Between Lipid Bilayers and Carbon-Based Nanoparticles

Young-Min Ban, Department of Chemical Engineering, University of Florida, Gainesville, FL

Dmitry I. Kopelevich, Department of Chemical Engineering, University of Florida, Gainesville, FL

In this talk, we present results of coarse-grained molecular dynamics (MD) simulations of interactions between dipalmitoylphosphatidylcholine (DPPC) lipid membranes and carbon-based nanoparticles (carbon nanotubes and fullerenes). Understanding these interactions is important for development of biomimetic devices and assessment of potentially harmful effects of nanomaterials on living cells. The considered nanoparticles are hydrophobic and therefore they tend to reside in the interior of the lipid bilayer. It is observed that these nanoparticles may affect several physical properties of the membrane, including its bending modulus, spontaneous curvature, and the lateral pressure profile. Fluctuations of tilt of the lipid molecules are also affected in the direct neighborhood of the nanoparticles. In addition, we observe that the dynamics of the thermal fluctuations of the membrane is significantly affected by the nanoparticles, especially when the latter approach the border between the hydrophilic head-groups and the hydrophobic tail-groups of the lipids.

The information obtained from MD simulations is used to construct a meso-scale model for lipid vesicles with nanoparticles embedded in their membranes. This model is used to assess possibility of the nanoparticle-induced membrane instability. The talk is concluded by a discussion of possible effect of the embedded nanoparticles on biophysical processes in cellular membranes.

See more of this Session: Poster Session: Thermodynamics and Transport Properties
See more of this Group/Topical: Engineering Sciences and Fundamentals

Equations for the thermophysical properties of dimethyl carbonate

Yong Zhou, Jiangtao Wu*

State Key Laboratory of Multiphase Flow in Power Engineering, Xi'an Jiaotong University, Xi'an Shaanxi 710049, People's Republic of China

ABTRACT: Dimethyl carbonate (DMC) is a nonirritating and nontoxic chemical, further more; it can be used as a fuel additive. In this work, the thermophysical properties formulations for dimethyl carbonate has been developed based upon a selection of available experimental thermophysical property data. In determining the coefficients of the equation of state, multi-property fitting methods were used; including single-phase pressure-volume-temperature (*pVT*), heat capacity, vapor pressure, and saturated liquid density data. The equation of state has been developed to conform to the Maxwell criterion for two-phase liquid-vapor equilibrium states and is valid for temperatures from 280 K to 400 K and pressure to 60 MPa, and the extrapolation behavior of the equation of state at high temperatures and high pressures is reasonable. The deviation of the density $(100\times(\rho_{cal}-\rho_{exp})/\rho_{exp})$ is generally less than 0.1% and does not exceed 0.2% except for very few data points in liquid region. The uncertainty of the vapor pressure of the equation of state is generally less than 2.0%. The deviation of viscosity equation developed is generally less than 2.0%, and the absolute average deviation (AAD) between experimental and calculated values is 1.03%. In additional, the extended corresponding states (ECS) model with R134a as the reference fluid was used to represent viscosities and thermal conductivities, and the absolute average deviations are 1.34% and 1.47%, respectively. Detail comparisons between experimental and calculated data have been performed in the work.

Key words: dimethyl carbonate; equation of state; thermophysical properties; ECS.

1. Introduction

In recent years, researchers has found that dimethyl carbonate can be used as fuel additive because in that case the vapor pressure of fuels could be reduced and in turn reduce their emissions to the atmosphere by evaporation[1], thus, carbon dioxide and smoke decrease while the carbonate addition increases; on the other hand, NO_x does not increase if dimethyl carbonate increases[2]. Dimethyl carbonate is a strong contender to help the refining industry meet the Clean Air Act, because dimethyl carbonate has about three times the oxygen content as methyl *tert*-butyl ether. Dimethyl carbonate has a good blending octane, it does not phase separate in a water stream like some alcohols do, and it is both low toxicity and quickly biodegradable[3]. Therefore, the thermal properties of dimethyl carbonate are very important. In recent years, a lot of experimental data of dimethyl carbonate, such as *pVT* property, viscosity, thermal conductivity, etc. have been measured by the author's group.

This work has developed the equation of state and the viscosity equation for dimethyl carbonate on the basis of an extensive collection of experimental data. As the viscosity and thermal conductivity experimental data are very few, the extended corresponding states (ECS) model with R134a as the reference fluid is used to represent the viscosities and thermal conductivities. This work has fitted the coefficients for dimethyl carbonate to improve the representation of the viscosities and thermal conductivities.

*Corresponding Author

Email: jtwu@mail.xjtu.edu.cn, Fax: 86-29-82668789

2. Experimental Data

A literature research has been launched on dimethyl carbonate, and all the data found are listed in Table 1. Although there are many researchers have carried out research on the thermophysical properties of dimethyl carbonate, dimethyl carbonate's experimental data are few and scattered, even worse their precisions are not high compared to the reference fluid and some refrigerants. Literature survey found that there are as many as thirty-seven articles containing dimethyl carbonate's pVT experimental data, but unfortunately, there are only four articles[4-7] containing lots of experimental data, and the other articles only contain the experimental data under the normal pressure. Even worse, the purpose of obtaining the experimental data is just to verify the purity of sample in many articles. Therefore, there are many duplicate experimental data points, and there are some deviations between them.

There are only four articles[8-11] containing the experimental vapor pressure data over the wide range. Just like the pVT property, there are many documents measuring the normal boiling point just to test the purity of the sample. The experimental ranges of the four articles almost overlap, but the experimental data sets deviate from each other largely. There are two data sets of dimethyl carbonate's saturated liquid density published[8,12].

A good equation of state should have the ability to represent the energy property perfectly. But there are only three sets (eighteen data points) of dimethyl carbonate's isobaric heat capacity found, and there exist an obvious deviation of experimental values at the same state[13-15]. Based on our analysis, we think the experimental data of Comelli F. et al.[15]'s are more reliable than the experimental data of Pardo J. M. et al.[13,14]'s so the weights of Comelli F. et al.[15]'s experimental data we set are bigger than the weights of Pardo J. M. et al.[13,14]'s during fitting the equation of state.

The equation of state developed in this work is formulated in terms of the Helmholtz energy. The Helmholtz energy is split up into two parts. One part describes the property of the ideal gas at the given temperature and density and the second part describe the residual property of the real fluid. Correlations for the ideal gas part of the Helmholtz energy can be derived from the isobaric heat capacity which are based either on experimental data in the gas phase or on spectroscopic data. If sufficiently data for isobaric heat capacity of ideal gas are available, the development of a correlation for ideal gas Helmholtz energy is easy. But the author has not found any experimental data for dimethyl carbonate's isobaric heat capacity of ideal gas, this work obtained the property by group contribution method, and the result is perfect.

There are twelve articles found containing the viscosity experimental data of dimethyl carbonate, the range of which is form 283 to 353 K, and pressures up to 100 MPa in liquid region; temperatures form 282 to 383 K on the saturated line. There are no dilute gas viscosity experimental data found, so collision integral model was used. There are two articles containing dimethyl carbonate's thermal conductivity data, and the range is form 280 to 390 K, pressures up to 30 MPa in liquid region. The viscosity and thermal conductivity experimental data are very limit, so the extended corresponding states model was used to represent the properties.

Table 1 Experimental data for dimethyl carbonate

Year	Author	No. of points	Temperature range (K)	Pressure range (MPa)	AAD$^+$ (%)
pVT					
1988	Isaias Garcia et al.[16]	1	298	0.101	0.01
1991	Juan A. Gonzalez et al.[17]	1	298	0.101	0.01
1992	Isaias Garcia et al.[18]	1	298	0.101	0.02

1995	F. Romolo et al.[19]	1	298	0.101	0.02
1996	F. Romolo et al.[20]	3	298-313	0.101	0.00
1997	F. Comelli and F. Romolo[21]	2	298-313	0.101	0.00
1998	A. Pal et al.[22]	3	298-318	0.101	0.06
1998	J. Garcia et al.[23]	1	298	0.101	0.03
1999	J. M. Pardo et al.[14]	4	278-308	0.101	0.01
1999	A. Pal et al.[24]	1	298	0.101	0.01
1999	A. Rodrguez et al.[25]	1	298	0.101	0.02
2000	E. R. Lopez et al.[26]	6	278-323	0.101	0.01
2001	L. Lugo et al.[4]	81	278-353	0.1-25	0.02
2001	M. J. P. Comunas et al.[5]	97	283-353	0.1-60	0.03
2002	Jose M. Pardo et al.[13]	4	288-308	0.101	0.01
2002	L. Mosteiro et al.[27]	2	288-298	0.101	0.00
2003	Esther Romano et al.[28]	5	288-313	0.101	0.02
2004	Jacobo Troncoso et al.[6]	90	283-328	0.1-40	0.04
2004	Tojo Jose et al.[29]	5	293-318	0.101	0.02
2004	Ana B. Pereiro et al.[30]	5	293-313	0.101	0.00
2004	Prasad T. E. Vittal et al.[31]	1	293	0.101	0.02
2004	A. Rodriguez et al.[32]	1	298	0.101	0.02
2004	X. Ma et al.[33]	1	290	0.101	0.82
2004	M.A. Rivas et al.[34]	7	288-328	0.101	0.01
2004	C. Yang et al.[35]	7	293-343	0.101	0.12
2004	Jose Tojo et al.[29]	5	293-318	0.101	0.02
2005	J.M. Pardo et al.[36]	4	288-308	0.101	0.01
2005	P. Ma and N. Li[37]	7	293-343	0.101	0.02
2005	Eva M. Terrado et al.[38]	3	278-298	0.101	0.06
2006	A. Rodriguez et al.[39]	4	293-313	0.101	0.00
2006	F. Comelli et al.[15]	2	298-308	0.101	0.00
2006	F. Wang et al.[40]	2	298	0.101	0.02
2007	M.A. Iglesias-Otero et al.[41]	7	293-323	0.101	0.01
2008	Jaime Wisniak et al.[42]	5	293-313	0.101	0.00
2008	M.A. Rivas et al.[43]	5	288-328	0.101	0.01
2009	X. Meng et al.[7]	40	283-353	0.1-20	0.12
2009	Laura Mosteiro et al.[44]	5	288-308	0.101	0.01
Vapor pressure					
1987	B. Jiang and J. Zhang[9]	38	288-370	0.0042-0.1269	1.18
1997	W. V. Steele et al.[8]	18	310-397	0.0133-0.27	0.56
1997	F. Comelli and R. Francesconi[21]	2	313-363	0.015-0.101	0.70
1999	Y. Shi et al.[45]	1	363	0.101	0.50
2001	X. Ma et al.[46]	1	363	0.101	0.18
2002	A. Rodriguez et al.[47]	131	326-411	0.0267-0.3839	1.17
2003	L. Zhang et al.[48]	1	363	0.101	0.20
2003	A. Sporzynski et al.[49]	1	363	0.101	1.31
2004	T. E. Vittal Prasad et al.[31]	1	361	0.0958	0.70
2004	J. Tojo et al.[29]	1	363	0.101	0.79
2004	Svetlana A. Kozlova et al.[11]	16	274-304	0.0019-0.01	1.36
2006	F. Comelli et al.[15]	1	363	0.101	0.19
2009	Y. Xing et al.[50]	4	358-373	0.0864-0.1382	0.39
Saturated liquid density					
1997	W. V. Steele et al.[8]	5	308-393		0.07
2003	S. Bi et al.[12]	21	282-383		0.03
Isobaric heat capacity					
1999	J. M. Pardo et al.[14]	3	288-308	0.101	4.31
2002	J. M. Pardo et al.[13]	3	288-308	0.101	4.50
2006	F. Comelli et al.[15]	10	288-333	0.101	0.59
Viscosity					
2003	S. Bi et al.[12]	21	282-383	Saturated liquid	0.70/1.38*
1998	A. Pal et al.[22]	3	298-318	0.101	1.34/0.78*

1999	A. Pal et al.[24]	1	298	0.101	1.54/1.18*
2001	M. J. P. Comunas et al.[5]	40	293-353	0.1-100	1.13/1.63*
2003	E. Romano et al.[28]	5	288-313	0.101	0.59/0.58*
2003	A. Rodrguez et al.[51]	4	298-313	0.101	1.39/1.19*
2004	A. Rodriguez et al.[32]	4	293-313	0.101	1.39/1.19*
2004	Ana B. Pereiro et al.[30]	5	293-313	0.101	0.62/0.09*
2005	P. Ma and N. Li[37]	7	293-343	0.101	0.76/1.64*
2006	A. Rodriguez et al.[39]	4	293-313	0.101	0.60/0.31*
2006	C. Yang et al.[35]	7	293-343	0.101	1.12/1.98*
2009	X. Meng et al.[7]	40	283-353	0.1-20	1.17/1.13*
Thermal conductivity					
2004	X. Jin et al.[52]	19	280-361	0.101	3.27*
2008	L. Song et al.[53]	109	290-390	0.1-30	1.14*

+ $\text{AAD}=\frac{100}{n}\sum_{i=1}^{n}\left|\frac{\eta_{i\exp}-\eta_{ical}}{\eta_{i\exp}}\right|$

* AAD of ECS model between calculated and experimental values

3. Equation of State

3.1 The Formulation of the Equation of State

The equation of state developed in this work for dimethyl carbonate was formulated with the Helmholtz energy as the fundamental property with independent variables of temperature and density

$$f(T,\rho)=f^{\circ}(T,\rho)+f^{r}(T,\rho) \quad \text{Eq. 1}$$

The functional form could be explicit in the dimensionless Helmholtz energy using independent variables of dimensionless density and temperature. The form of the equation of state becomes

$$\phi(\tau,\delta)=\phi^{\circ}(\tau,\delta)+\phi^{r}(\tau,\delta) \quad \text{Eq. 2}$$

where $\tau=T_c/T$, $\delta=\rho/\rho_c$, the critical temperature is 557.0 K, the critical density is 3.974 mol/L. All thermodynamic properties could be calculated through the derivatives of the dimensionless Helmholtz energy.

The ideal gas Helmholtz energy is given by

$$f^{\circ}(T,\rho)=h^{\circ}(T)-RT-Ts^{\circ}(T,\rho) \quad \text{Eq. 3}$$

The enthalpy of ideal gas, h^{o} is only a function of temperature, T entropy s^{o} is a function of temperature T and density ρ. These properties could be presented by the ideal gas isobaric heat capacity formula.

$$f^{\circ}(T,\rho)=\left(\int_{T_0}^{T}c_p^{\circ}dT+h_0^{\circ}\right)-RT-T\left[\int_{T_0}^{T}\frac{c_p^{\circ}-R}{T}dT-R\ln\left(\frac{\rho}{\rho_0^{\circ}}\right)+s_0^{\circ}\right] \quad \text{Eq. 4}$$

where R=8.314472 J/(mol·K) is universal constant, ρ_0°, h_0° and s_0° are properties at reference state (T_0, p_0). So the dimensionless ideal gas Helmholtz energy is given by

$$\phi^{\circ}(\tau,\delta)=\frac{h_0^{\circ}\tau}{RT_c}-\frac{s_0^{\circ}}{R}-1+\ln\frac{\delta\tau_0}{\delta_0\tau}-\frac{\tau}{R}\int_{\tau_0}^{\tau}\frac{c_p^{\circ}}{\tau^2}d\tau+\frac{1}{R}\int_{\tau_0}^{\tau}\frac{c_p^{\circ}}{\tau}d\tau \quad \text{Eq. 5}$$

As shown in above formulation, the dimensionless ideal gas Helmholtz energy could be derived by ideal gas isobaric heat capacity with the reference state given. In this work, the reference state is the normal boiling point with the enthalpy and entropy of the saturated liquid equal zero. But there are no ideal gas heat capacity experimental data for dimethyl carbonate found, thus, this work use the group contribution method to obtain the ideal gas heat capacity formula which is given by

$$c_p^{\mathrm{o}} = n_0^{\mathrm{o}} + \sum_{i=1}^{3} n_i^{\mathrm{o}} T^i \qquad \text{Eq. 6}$$

where the coefficients are given in Table 2.

Table 2 Coefficients of the ideal gas isobaric heat capacity formula

i	n_i^{o}
0	51.07
1	0.17084
2	6.62E-5
3	-8.8E-8

Nonlinear fitting techniques have been used to fit the coefficients and exponents of the Helmholtz energy. Many constraints have been added to the fit to ensure the perfect extrapolation behavior of equation of state. The equation of the residual Helmholtz energy obtained in this work is

$$\phi^{\mathrm{r}}(\tau,\delta) = \sum_{i=1}^{I_{\mathrm{pol}}} n_i \delta^{d_i} \tau^{t_i} + \sum_{i=I_{\mathrm{pol}}+1}^{I_{\mathrm{pol}}+I_{\mathrm{exp}}} n_i \delta^{d_i} \tau^{t_i} e^{-\delta^{p_i}} \qquad \text{Eq. 7}$$

where the coefficients and exponents obtained in this work are given in Table 3. The equation of state is valid from 280 K to 400 K, pressures up to 60 MPa. The uncertainty in density of the equation of state is 0.1% and does not exceed 0.2% except for very few data points in liquid region; the uncertainty in heat capacity is 2.0%; the uncertainty in vapor pressure is 2.0%; and the uncertainty in saturated liquid density is 0.05%. The extrapolation behavior of the equation of state outside the above range is reasonable and detail analysis is performed in the following.

Table 3 Coefficients and exponents of the Eq. 7

i	n_i	t_i	d_i	p_i
1	0.2492612946365D+01	0.46	1	-
2	-0.4077704794363D+01	1.00	1	-
3	0.1283556275792D+00	0.46	3	-
4	0.3098163467220D-04	1.00	8	-
5	0.1161327624207D+01	1.76	2	1
6	0.7375983136613D+00	2.27	1	1
7	-0.4819345992168D-01	1.40	5	1
8	-0.5836535783693D+00	5.00	1	2
9	-0.1228066747506D+00	4.72	4	2
10	-0.5458572751320D-01	7.16	3	2

3.2 The representation behavior of the equation of state

There are considerable deviations between the four data sets[8-11] for vapor pressures of dimethyl carbonate. Comparisons of experimental vapor pressures with values calculated by the equation of state are presented in Fig. 1 as $100\times(p_{\mathrm{exp}}-p_{\mathrm{cal}})/p_{\mathrm{exp}}$. As shown in Fig. 1, most of the deviations of the data are within 2.0%. There are two reasons why the deviations are so large: first, the experimental data sets deviate from each other largely; second, the vapor pressures of dimethyl carbonate at low temperature are very small. The equation of state represents the saturated liquid density experimental data within 0.05% except for one data point of S. Bi et al.[12] at 282 K as shown in Fig. 2.

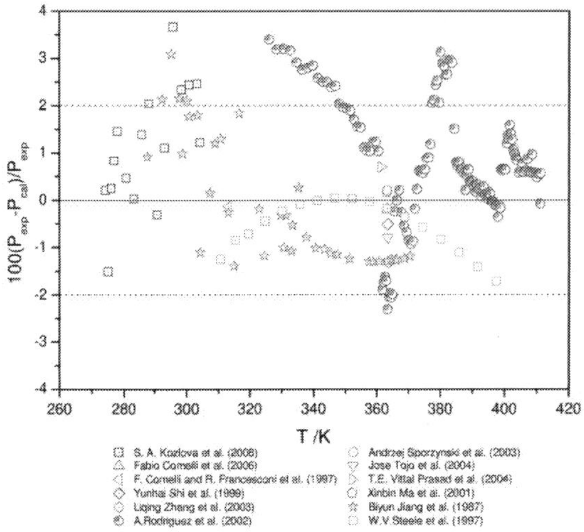

Fig. 1 Comparisons of vapor pressures calculated with equation of state to the experimental data

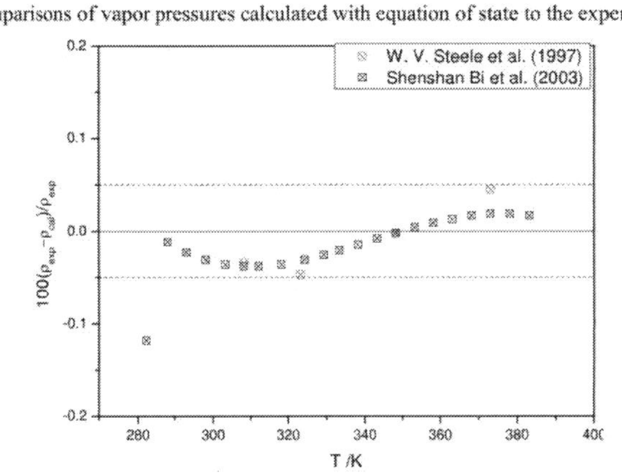

Fig. 2 Comparisons of saturated liquid density calculated with equation of state to the equation of state

In the liquid region, deviations between the experimental *pVT* data and the values calculated from the equation of state are within 0.1%, except for the data of X. Meng et al.[7] which are obtained by vibrating string viscosimeter, the deviations are within 0.2%. Comparisons of for experimental *pVT* data sets[4-7] with values calculated from the equation of state are presented in Fig. 3 as $100\times\left(\rho_{\text{exp}}-\rho_{\text{cal}}\right)/\rho_{\text{exp}}$, and Comparisons of experimental *pVT* data sets at normal atmosphere pressure with values calculated from the equation of state are presented in Fig. 4. Comparisons of experimental isobaric heat capacity data with values calculated from the equation of state are present in Fig. 5 as $100\times\left(c_{p\text{exp}}-c_{p\text{cal}}\right)/c_{p\text{exp}}$.

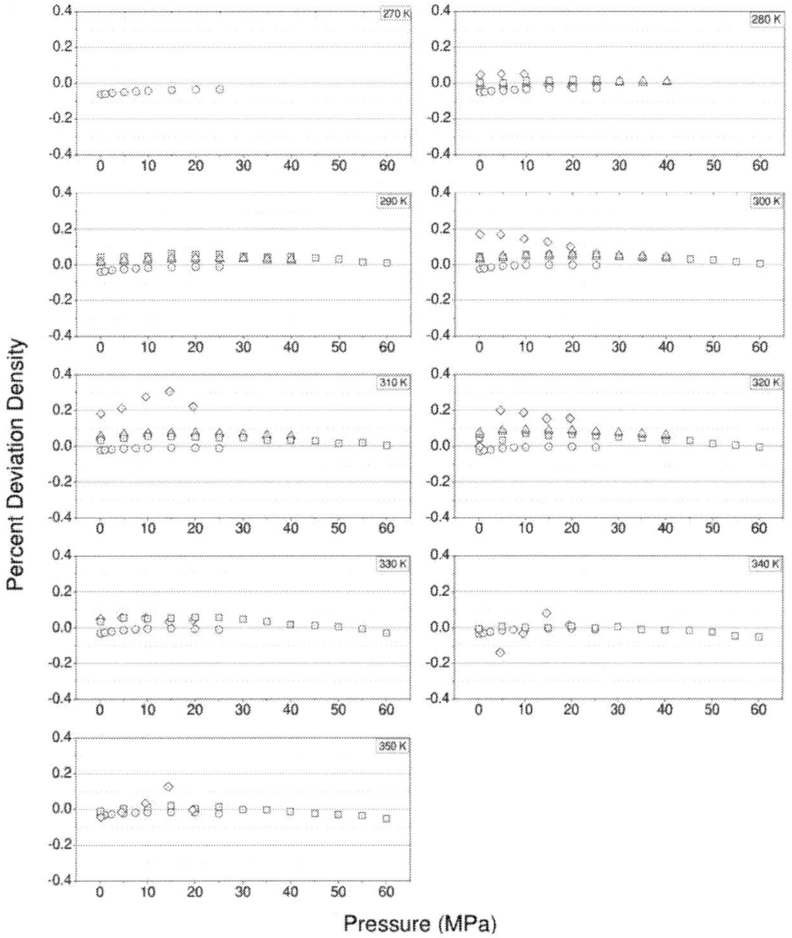

○ L. Lugo et al.（2001） □ M. J. P. Comunas et al.（2001）

△ Jacobo Troncoso et al.（2004） ◇ Meng et al.（2009）

Fig. 3 Comparisons of densities calculated with the equation of state to experimental data

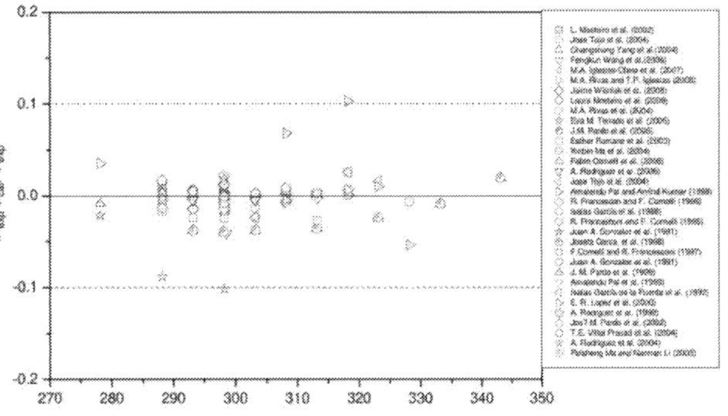

Fig. 4 Comparisons of densities calculated with the equation of state to experimental data at normal atmosphere pressure

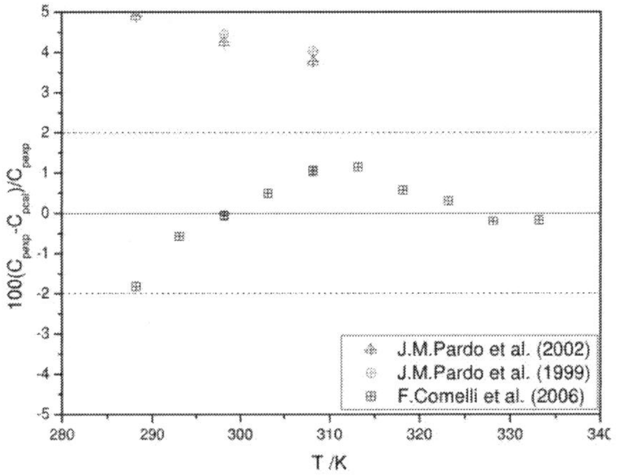

Fig. 5 Comparisons of isobaric heat capacities calculated with the equation of state to experimental data

3.3 The extrapolation behavior of the equation of state

The equation of state should have reasonable extrapolation behavior, and plot of constants property line on various thermodynamic coordinates are useful in assessing the behavior of the equation of state. The equation of state developed in this work was used plot the plots of temperature versus isochoric heat capacity (Fig. 6), isobaric heat capacity (Fig. 7) and speed of sound (Fig. 8), and the plots of density versus gruneisen value (Fig. 9) and temperature (Fig. 10). These plots indicate that the equation of state behavior reasonable within the valid range and that the extrapolation behavior is reasonable at higher temperatures and pressures.

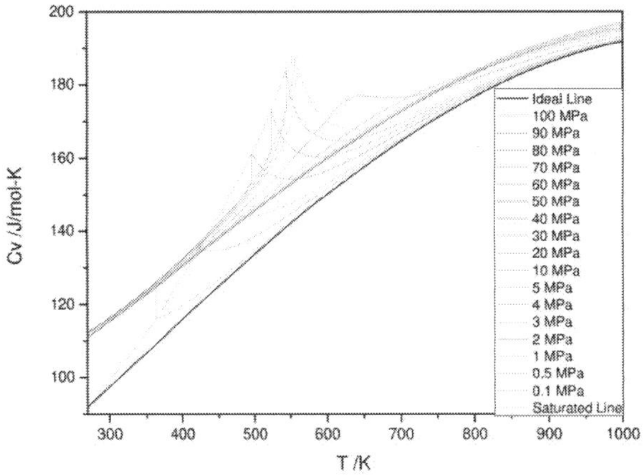

Fig. 6 Isochoric heat capacity versus temperature diagram

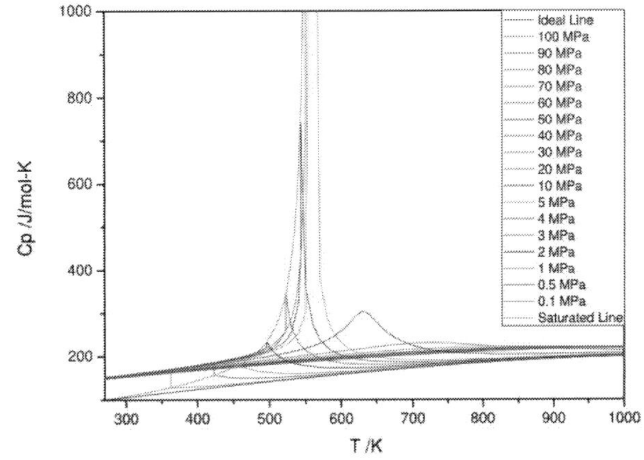

Fig. 7 Isobaric heat capacity versus temperature diagram

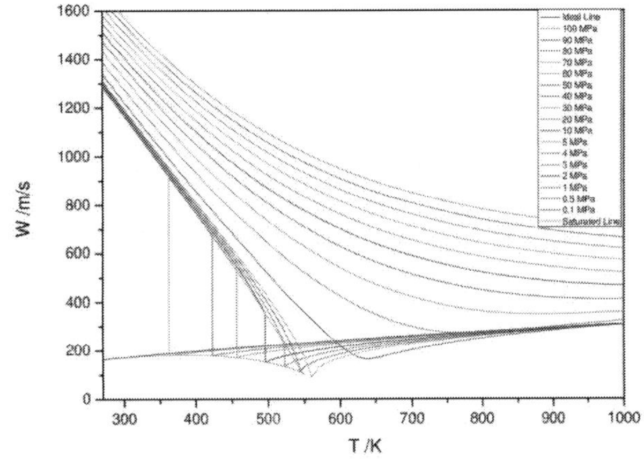

Fig. 8 Speed of sound versus temperature diagram

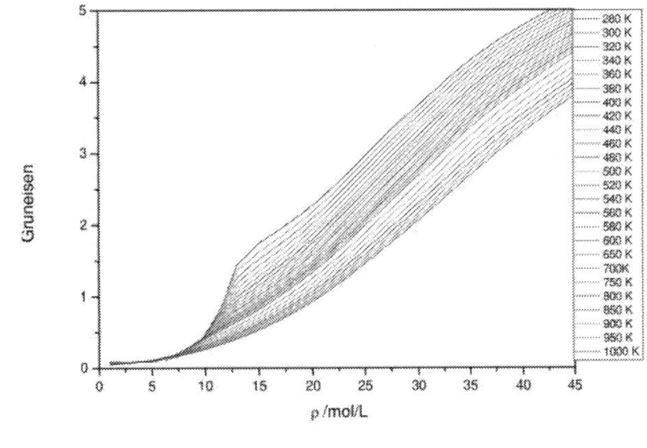

Fig. 9 Gruneisen value versus density diagram

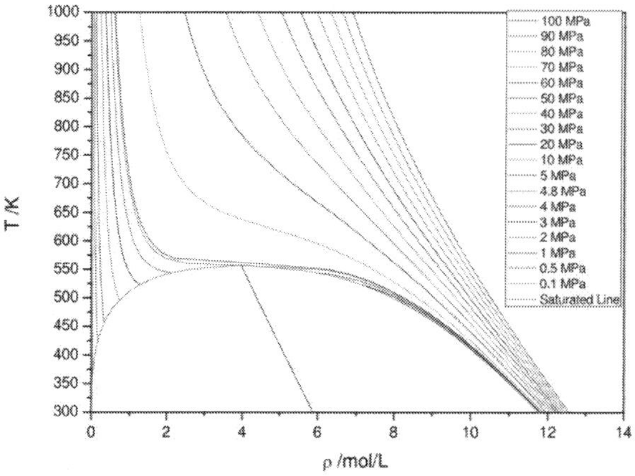

Fig. 10 Temperature versus density diagram

4 Viscosity Equation

4.1 The formulation of the viscosity equation

The viscosity, η, of a pure fluid is considered as a sum of three contributions[54]

$$\eta(T,\rho)=\eta^{\circ}(T)[1+B_{\eta}(T)\rho]+\Delta\eta_{h}(T,\rho)+\Delta\eta_{c}(T,\rho) \qquad \text{Eq. 8}$$

where, the term, $\eta^{\circ}(T)$, represents the viscosity in the limit of zero density, $B_{\eta}(T)$ is the second virial coefficient for viscosity based on the Rainwater-Friend theory[55], $\Delta\eta_{h}(T,\rho)$ is the residual contribution of the real fluid, and the term, $\Delta\eta_{c}(T,\rho)$, represents the critical enhancement of the viscosity. In this work, the term $B_{\eta}(T)$ and $\Delta\eta_{c}(T,\rho)$ ware set to zero. The viscosity equation is expressed in terms of temperature and density, but the experimental data are always published versus temperatures and pressures, thus, densities have to be calculated. Even though, the equation of state of dimethyl carbonate developed in this work which is claimed is valid only up to 60 MPa, the equation of state has reasonable extrapolation as analyzed above, thus, the equation of state developed in this work has been employed to calculate all densities. But the data exceed these limits and the results of these extrapolations should be viewed with caution.

$\eta^{\circ}(T)$ can not be measured directly, the data are extrapolated from the available experimental data at low density to zero density, which can entail significant errors[56]. But there was no this type data found, the empirical function was used. According to the rigorous kinetic theory of gases[57], the viscosity, $\eta^{\circ}(T)$, for spherical particles is given by

$$\eta^{\circ}(T)=5\sqrt{mkT/\pi}\Big/\left(16\sigma_{0}^{2}\Omega^{(2,2)^{*}}\right) \qquad \text{Eq. 9}$$

where $k=1.3806505\times10^{-23}\,\mathrm{K}^{-1}$ is the Boltzmann constant, $m=M/N_{\mathrm{A}}$ is the molecular mass with M being the molar mass, and $N_{\mathrm{A}}=6.0221415\times10^{23}\,\mathrm{mol}^{-1}$ is the Avogadro's constant. The collision diameter σ_{0} is defined as the smallest separation distance where the intermolecular potential function is equal to zero, and $\Omega^{(2,2)^{*}}$ is a collision integral that depends upon the potential function. Neufeld et al.[58] has developed empirical correlations for the temperature dependencies of the transport collision integrals of the Lennard-Jones 12-6 potential, which represent $\Omega^{(2,2)^{*}}$ in terms of the reduced temperature $T^* = k_B T / \varepsilon_0$ which is given by

$$\Omega_{LJ}^{(2,2)^*} = 1.16145 T^{*-0.14874} + 0.52487 e^{-0.77320 T^*} + 2.16178 e^{-2.43787 T^*} \quad \text{Eq. 10}$$

As there are very few data for the viscosity of dimethyl carbonate in the dilute gas region, the values of σ_0 and ε / k_B is determined using the method developed by Chung et al.[59], that is

$$\sigma_0 = 0.0809 V_c^{1/3} \quad \text{Eq. 11}$$

$$\frac{\varepsilon}{k_B} = \frac{T_c}{1.2593} \quad \text{Eq. 12}$$

The term, $\Delta\eta_h(T,\rho)$, fitted in this work is given by

$$\Delta\eta_h(T,\rho) = \sum_i n_i \tau^{t_i} \delta^{d_i} + \sum_i n_i \tau^{t_i} \delta^{d_i} e^{-\delta^{p_i}} \quad \text{Eq. 13}$$

where the coefficients and exponents obtained in this work are list in Table 4. As mentioned above, the viscosity experimental data for dimethyl carbonate are only located in liquid region temperatures from 283 to 353 K and pressures up to 100 MPa, and saturated liquid line temperatures from 282 K to 383 K. Thus, this is the valid region for the viscosity equation of dimethyl carbonate developed in this work. But the extrapolation behavior of the equation is reasonable in other region.

Table 4 Coefficients and exponents of Eq. 13

i	n_i	t_i	d_i	p_i
1	0.507808723533D+01	-0.1000	4	-
2	-0.567340207974D-01	-3.0968	10	1
3	0.832177458368D-02	-2.8945	12	1
4	0.354598383878D+02	0.0731	2	2
5	0.513527991450D-01	-3.9871	0	3

4.2 The representation and extrapolation behavior of the viscosity equation

The comparison has been carried out in this section. As shown in Fig. 11, the saturated viscosity data of Bi et al.[12] are represented generally in 1.0%. The uncertainty in liquid region is generally less than 2.0% which are shown in Fig. 12 and Fig. 13. There are totally fifty-one experimental data points at atmospheric pressure, temperatures from 283 to 353 K, but the data deviate from each other largely, thus, the larger deviations appear in this region as shown in Fig. 13.

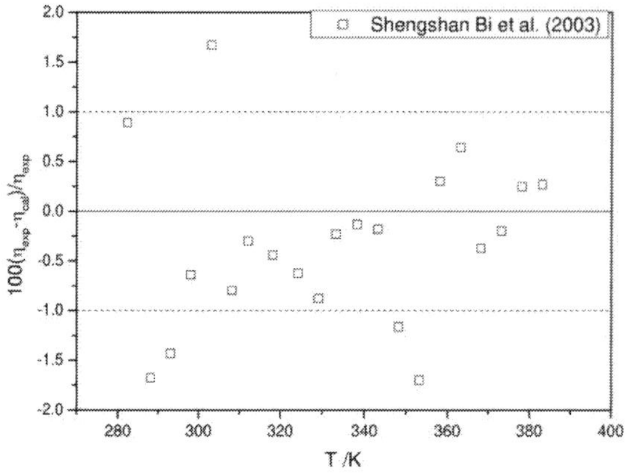

Fig. 11 Comparison of saturated liquid viscosity with the viscosity equation to the experimental value

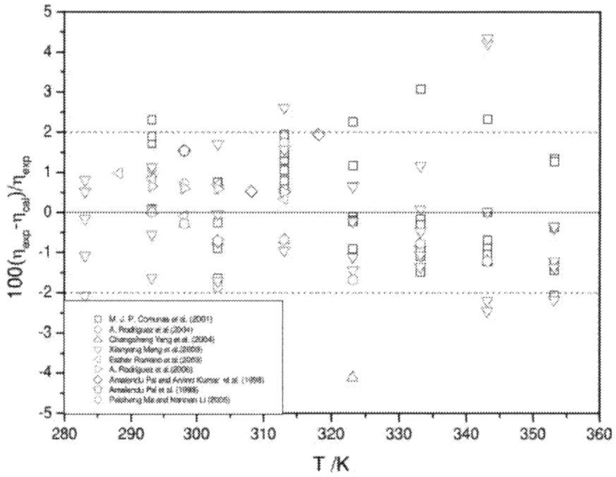

Fig. 12 Comparison of viscosity with the viscosity equation to the experimental value

The plot of viscosity versus temperature has been plot to verify the extrapolation behavior of the viscosity equation developed in this work. The plot show that the viscosity equation developed in this work is reasonable as shown in Fig. 14.

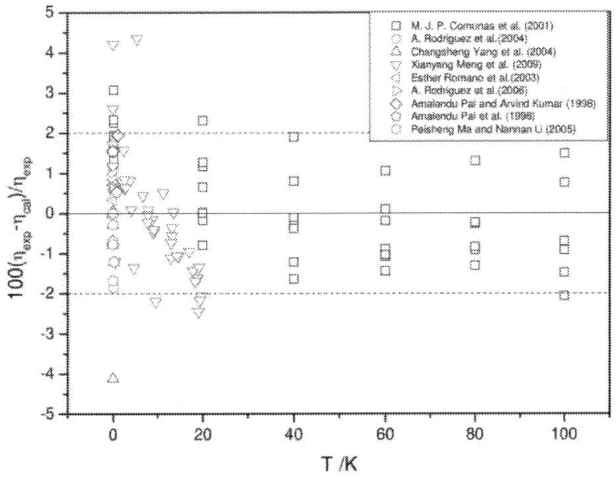

Fig. 13 Comparison of viscosity with the viscosity equation to the experimental value

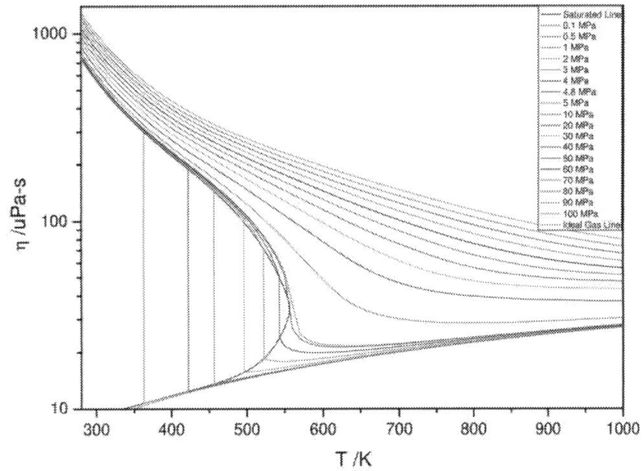

Fig. 14 Viscosity versus temperature diagram

5 Extended Corresponding States Model for Viscosity and Thermal Conductivity of dimethyl carbonate

In 1981, Ely and Hanley published extended corresponding states (ECS) model for viscosity[60], and in 1983, the model was applied to thermal conductivity[61], and Since then, there have been several modifications. In 2003, Huber et al.[62] applied the ECS viscosity model developed by Klein et al.[63] and modified ECS thermal conductivity model developed by McLinden et al.[64] to pure refrigerants. This work present the coefficients for dimethyl carbonate without changing the model form Klein et al.[63] and Huber et al.[62].

5.1 Viscosity model

The viscosity of a pure fluid could be represented as a sum of a dilute gas and a residual contribution, then apply the ECS model to the residual contrbution[60]

$$\eta(T,\rho)=\eta^{\circ}(T)+\Delta\eta(T,\rho)=$$

$$\eta^{\circ}(T)+\Delta\eta_{\mathrm{ref}}(T_{\mathrm{ref}},\rho_{\mathrm{ref}})F_{\eta}(T,\rho) \qquad \text{Eq. 14}$$

where the superscript $^{\circ}$ denotes a dilute-gas value which is given by Eq. 9 and the subscript $_{\mathrm{ref}}$ denotes a reference fluid value which is evaluated at a conformal temperature and density. T_{ref} and ρ_{ref} are given by

$$T_{\mathrm{ref}}=T/f \qquad \text{Eq. 15}$$

$$\rho_{\mathrm{ref}}=\rho h \qquad \text{Eq. 16}$$

where equivalent substance reducing ratios f and h are given by

$$f=\frac{T_{\mathrm{c}}}{T_{\mathrm{c\,ref}}}\theta \qquad \text{Eq. 17}$$

$$h=\frac{\rho_{\mathrm{c\,ref}}}{\rho_{\mathrm{c}}}\phi \qquad \text{Eq. 18}$$

where θ and ϕ are shape factors which are accurately determined by the equation of sate developed in this work using the method presented in Klein et al.[63].

$F_{\eta}(T,\rho)$ in Eq. 14 is given by

$$F_{\eta}=f^{1/2}h^{-2/3}\left(\frac{M}{M_{\mathrm{ref}}}\right)^{1/2} \qquad \text{Eq. 19}$$

To improve the representation of the viscosity model, an empirical correction factor, $\psi(\rho_{\mathrm{r}})$, has been used in Eq. 14$\eta^{\circ}(T)+\Delta\eta_{\mathrm{ref}}(T_{\mathrm{ref}},\rho_{\mathrm{ref}})F_{\eta}(T,\rho)$ Eq. 14 to instead ρ_{ref}, that is

$$\rho_{\mathrm{ref}}'=\rho_{\mathrm{ref}}\psi(\rho_{\mathrm{r}}) \qquad \text{Eq. 20}$$

where $\rho_{\mathrm{r}}=\rho/\rho_{\mathrm{c}}$, and $\psi(\rho_{\mathrm{r}})$ is given by

$$\psi(\rho_{\mathrm{r}})=\sum_{i=0}^{n}a_i\rho_{\mathrm{r}}^{i} \qquad \text{Eq. 21}$$

where the coefficients a_i are found by fitting the experimental viscosity data, as listed in Table 5. The comparisons of thermal conductivities calculated by the ECS model to the experimental data are shown in Fig. 15. Over all, the AAD between the experimental and the calculated values is 1.34%. Table 1 gives the AAD of each data set.

Table 5 Coefficients in Eq. 21 and Eq. 33

i	a_i	b_i
0	0.922569354	1.114854816
1	0.009859550	-0.025217961

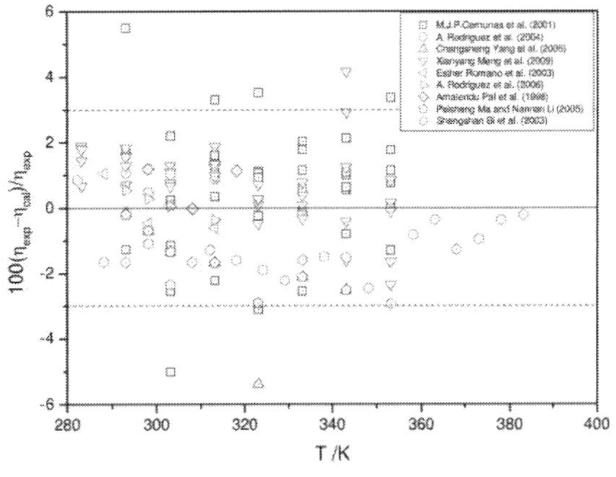

Fig. 15 Comparisons of viscosities calculated by the ECS model to the experimental data

5.2 Thermal conductivity model

The thermal conductivity of a pure fluid could be represented as a sum of translational and internal modes of energy transfer[61]

$$\lambda(T,\rho)=\lambda^{\text{int}}(T)+\lambda^{\text{trans}}(T,\rho) \qquad \text{Eq. 22}$$

An Eucken correlation for the internal contribution was used to represent the internal contribution

$$\lambda^{\text{int}}(T)=\frac{f_{\text{int}}\eta^{\text{o}}}{M}\left(C_{\text{p}}^{\text{o}}-\frac{5}{2}R\right) \qquad \text{Eq. 23}$$

where η^{o} is the dilute gas viscosity as given by Eq. 9, C_{p}^{o} is given by Eq. 6, and f_{int} is set to 0.00132 in this work as there are no dilute gas thermal conductivity experimental data available. Similar to the expression of viscosity, the translational contribution can be represented as a dilute gas contribution, a residual contribution, and a critical enhancement

$$\lambda^{\text{trans}}(T,\rho)=\lambda^{\text{o}}(T)+\lambda^{\text{r}}(T,\rho)+\lambda^{\text{crit}}(T,\rho) \qquad \text{Eq. 24}$$

The dilute gas thermal conductivity, $\lambda^{\text{o}}(T)$, is given by[61]

$$\lambda^{\text{o}}(T)=\frac{15R\eta^{\text{o}}}{4\text{M}} \qquad \text{Eq. 25}$$

where dilute gas viscosity, η^{o}, is given by Eq. 9. The residual contribution is represented using ECS

$$\lambda^{\text{r}}(T,\rho)=\lambda_{\text{ref}}^{\text{o}}(T_{\text{ref}},\rho_{\text{ref}})F_{\lambda} \qquad \text{Eq. 26}$$

where F_{λ} is given by

$$F_{\lambda}=f^{1/2}h^{-2/3}\left(\frac{M_{\text{ref}}}{M}\right)^{1/2} \qquad \text{Eq. 27}$$

The critical enhancement is represented using the model developed by Olchowy and Sengers[65]

$$\lambda^{\text{crit}}(T,\rho)=\frac{\rho C_P R_0 k_B T}{6\pi\eta\xi}(\Omega-\Omega_0) \quad \text{Eq. 28}$$

where Ω, Ω_0, and ξ are given by $\Omega=\frac{2}{\pi}\left[\left(\frac{C_P\text{-}C_V}{C_P}\right)\arctan(q_d\xi)+\frac{C_V}{C_P}(q_d\xi)\right]$ Eq. 29Eq.29, Eq. 30,and Eq.31, respectively.

$$\Omega=\frac{2}{\pi}\left[\left(\frac{C_P\text{-}C_V}{C_P}\right)\arctan(q_d\xi)+\frac{C_V}{C_P}(q_d\xi)\right] \quad \text{Eq. 29}$$

$$\Omega_0=\frac{2}{\pi}\left\{1-\exp\left[\frac{-1}{(q_d\xi)^{-1}+\frac{1}{3}\left(\frac{(q_d\xi)\rho_C}{\rho}\right)^2}\right]\right\} \quad \text{Eq. 30}$$

$$\xi=\xi_0\left(\frac{P_C\rho}{\Gamma\rho_C^2}\right)^{\nu/\gamma}\left[\left.\frac{\partial\rho(T,\rho)}{\partial P}\right|_T-\frac{T_R}{T}\left.\frac{\partial\rho(T_R,\rho)}{\partial P}\right|_T\right]^{\nu/\gamma} \quad \text{Eq. 31}$$

where R_0= 1.03, γ= 1.239 and ν= 0.63 are universal constants[66]. For the reference temperature, $T_R=1.5T_C$ is selected where the critical enhancement is assumed to be negligible. It is reasonable to fix the values of Γ and ξ_0 as 0.0496, and 1.94×10^{-10}m, respectively. As there are no data for dimethyl carbonate in the critical region, the cutoff parameter is fixed as 5×10^{-10}m[62]. The isobaric heat capacity, isochoric heat capacity, and partial derivative $\partial\rho/\partial P|_T$ are calculated by the equation of state developed in this work. The viscosity is calculated using the viscosity equation obtained in this work.

To improve the representation of the thermal conductivity model, an empirical correction factor, $\chi(\rho_r)$, has been used in Eq. 26 to instead ρ_{ref}, that is

$$\rho'_{\text{ref}}=\rho_{\text{ref}}\chi(\rho_r) \quad \text{Eq. 32}$$

where $\rho_r=\rho/\rho_c$, and $\chi(\rho_r)$ is given by

$$\chi(\rho_r)=\sum_{i=0}^{n}b_i\rho_r^i \quad \text{Eq. 33}$$

where the coefficients b_i are found by fitting the experimental thermal conductivity data, as listed in Table 5. The comparisons of thermal conductivities calculated by the ECS model to the experimental data are shown in Fig. 16. Over all, the AAD between the experimental and the calculated values is 1.47%. Table 1 gives the AAD of each data set.

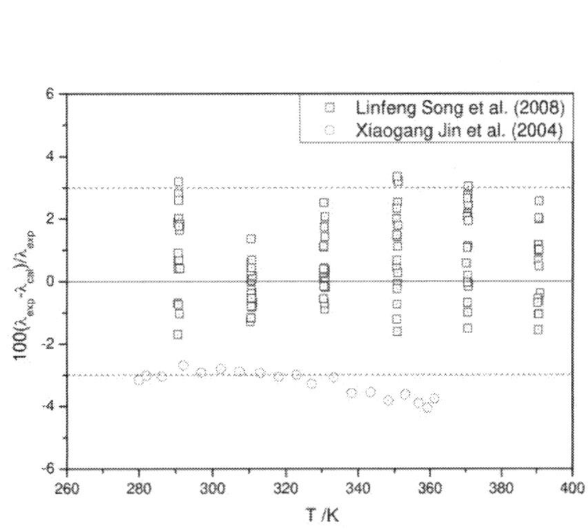

Fig. 16 Comparisons of thermal conductivities calculated by the ECS model to the experimental data

6 Conclusions

The equation of state for dimethyl carbonate, valid from 280 to 400 K, pressures up to 60 MPa, was fitted using the Helmholtz energy as the fundamental prosperity and temperature and density as the independent variables. The equation of state is based on the experimental pVT data, isobaric heat capacities, vapor pressures and saturated liquid densities. The uncertainty in density of the equation of state is 0.1%; the uncertainty in heat capacity is 2.0%; the uncertainty in vapor pressure is 2.0% at temperatures from 280 to 400 K; and the uncertainty in saturated liquid density is 0.05% at temperatures from 280 to 400 K. In the critical region, there are no experimental data found, so the behavior of the equation of state could not be verified. Although the equation of state developed in this work is the first available equation for the properties of dimethyl carbonate.

The viscosity equation of dimethyl carbonate developed in this work is valid from 253 to 353 K with pressures up to 100 MPa in liquid region. The uncertainty of the viscosity equation is generally less than 2.0% and the absolute average deviation (AAD) between the experimental and the calculated values is 1.03%. As the viscosity and thermal conductivity data are few, the ECS model was tried to represent the properties. The AADs between calculated and experimental viscosity and thermal conductivity data are 1.34%, and 1.47%, respectively.

7 Acknowledgments

The author would like to thank Eric W. Lemmon, National Institute of Standards and Technology, U.S., for his kind help and discussion. This research is supported by the Foundation for the Author of National Excellent Doctoral Dissertation (Grant 200540) and Fok Ying Tung Education Foundation (Project Number 111060).

Reference

[1] Katrib Y, Deiber G, Mirabel P, et al. Atmospheric Loss Processes of Dimethyl and Diethyl Carbonate [J]. Journal of Atmospheric Chemistry, 2002, 43 (3): 24.
[2] Wang F, Wu J, Liu Z. Surface tension of dimethyl carbonate (C3H6O3)[J]. Fluid Phase Equilibria, 2004, 220 (1).
[3] Pacheco MA, Marshall CL. Review of Dimethyl Carbonate (DMC) Manufacture and Its Characteristics as a Fuel Additive[J]. Energy & Fuels, 1997, 11 (1): 28.
[4] Lugo L., Comunas, P. M. J., et al. (p, Vm, T, x) measurements of dimethyl carbonate + octane binary mixtures I.

Experimental results, isothermal compressibilities, isobaric expansivities and internal pressures[J]. Fluid Phase Equilibria, 2001, 186 (1-2): 21.
[5] Comunas MJP, Baylaucq A, Boned C, et al. High-pressure measurements of the viscosity and density of two polyethers and two dialkyl carbonates[J]. International Journal of Thermophysics, 2001, 22 (3): 20.
[6] Troncoso J, Bessieres D, Cerdeirina CA, et al. p.rho.Tx Data for the Dimethyl Carbonate - Decane System[J]. Journal of Chemical & Engineering Data, 2004, 49 (4): 5.
[7] Meng X, Zheng P, Wu J, et al. Viscosity and sensity measurements of dimethyl carbonate[J]. Journal of Engineering Thermophysics, 2009, 30 (1): 5.
[8] Steele WV, Chirico RD, Knipmeyer SE, et al. Vapor Pressure, Heat Capacity, and Density along the Saturation Line.. Measurements for Dimethyl Isophthalate, Dimethyl Carbonate, 1,3,5-Triethylbenzene, Pentafluorophenol, 4-tert-Butylcatechol, .alpha.-Methylstyrene, and N,N'-Bis(2-hydroxyethyl)ethylenediamine[J]. Journal of Chemical & Engineering Data, 1997, 42 (6): 13.
[9] Jiang B, Zhang J. Determination and correlation of vapor pressures of dimethyl carbonate and 2,3-dichloropropene[J]. Journal of Tianjin University, 1987, (3): 9.
[10] Rodr´ıguez A, Canosa J, Dom´ınguez A, et al. Isobaric vapour–liquid equilibria of dimethyl carbonate with alkanes and cyclohexane at 101.3 kPa[J]. Fluid Phase Equilibria, 2002, 198: 15.
[11] Kozlova SA, yanenko VNE, Georgieva M, et al. Vapour pressure and enthalpy of vaporization of aliphatic dialkyl carbonates[J]. Journal of Chemical Thermodynamics, 2008, 40: 5.
[12] Bi S, Wu J, Liu Z, et al. Experimental investigation of the viscosity of saturated liquid dimethyl carbonate[J]. Journal of Xi'an Jiaotong University, 2003, 37 (9): 3.
[13] Pardo JM, Gonzalez-Salgado D, Tovar CA, et al. Comparative study of the thermodynamic behaviour of the binary mixtures dimethyl carbonate + (benzene, n-heptane, cyclohexane, or toluene)[J]. Canadian Journal of Chemistry, 2002, 80 (4): 9.
[14] Pardo JM, Tovar CA, Cerdeirina CA, et al. Excess molar volumes and excess molar heat capacities of (dimethyl carbonate, or diethyl carbonate + n-heptane) at several temperatures[J]. Journal of Chemical Thermodynamics, 1999, 31 (6): 10.
[15] Comelli F, Francesconi R, Bigi A, et al. Excess Molar Enthalpies, Molar Heat Capacities, Densities, Viscosities, and Refractive Indices of Dimethyl Sulfoxide + Esters of Carbonic Acid at 308.15 K and Atmospheric Pressure[J]. Journal of Chemical & Engineering Data, 2006, 51 (2): 6.
[16] Garcia I, Cobos JC, Gonzalez JA, et al. Thermodynamics of binary mixtures containing organic carbonates.1. Excess enthalpies of dimethyl carbonate + hydrocarbons or + tetrachloromethane[J]. Journal of Chemical & Engineering Data, 1988, 33 (4): 4.
[17] Gonzalez JA, Garcia I, Cobos JC, et al. Thermodynamics of binary mixtures containing organic carbonates. 4. Liquid-liquid equilibria of dimethyl carbonate + selected n-alkanes[J]. Journal of Chemical & Engineering Data, 1991, 36 (2): 3.
[18] Fuente IGdl, Gonzalez JA, Cobos JC, et al. Excess molar volumes for dimethyl carbonate + heptane, decane, 2,2,4-trimethylpentane, cyclohexane, benzene, toluene, or tetrachloromethane[J]. Journal of Chemical & Engineering Data, 1992, 37 (4): 3.
[19] Romolo F, Fabio C. Excess Molar Enthalpies, Densities, and Excess Molar Volumes of Binary Mixtures Containing Esters of Carbonic Acid at 298.15 and 313.15 K[J]. Journal of Chemical & Engineering Data, 1995, 40 (4): 14.
[20] Francescon R, Comelli F. Vapor-Liquid Equilibria, Excess Molar Enthalpies, and Excess Molar Volumes of Dimethyl Carbonate + 1,2-Epoxybutane at 288.15, 298.15, or 313.15 K[J]. Journal of Chemical & Engineering Data, 1996, 41 (4).
[21] Comelli F, Francesconi R. Isothermal Vapor-Liquid Equilibria Measurements, Excess Molar Enthalpies, and Excess Molar Volumes of Dimethyl Carbonate + Methanol, + Ethanol, and + Propan-1-ol at 313.15 K[J]. Journal of Chemical & Engineering Data, 1997, 42 (4).
[22] Pal A, Kumar A. Excess Molar Volumes, Viscosities, and Refractive Indices of Diethylene Glycol Dimethyl Ether with Dimethyl Carbonate, Diethyl Carbonate, and Propylene Carbonate at (298.15, 308.15, and 318.15) K[J]. Journal of Chemical & Engineering Data, 1998, 43 (2): 5.
[23] Garcia J, Lugo L, Comunas MJ, et al. Experimental excess volumes of organic carbonate + alkane systems. Estimation of the parameters of the Nitta-Chao model for this kind of binary mixture[J]. J Chem Soc, Faraday T rans, 1998, 94 (12): 6.
[24] Pal A, Dass G, Kumar A. Excess Molar Volumes, Viscosities, and Refractive Indices of Tetraethylene Glycol Dimethyl Ether + Dimethyl Carbonate, + Diethyl Carbonate, and + Propylene Carbonate at 298.15 K[J]. Journal of Chemical & Engineering Data, 1999, 44 (1): 4.
[25] Rodrguez A, Canosa J, Tojo J. Physical Properties of the Ternary Mixture Dimethyl Carbonate + Methanol + Benzene and Its Corresponding Binaries at 298.15 K[J]. Journal of Chemical & Engineering Data, 1999, 44 (6): 6.
[26] Lopez ER, Lugo L, Comunas MJP, et al. Temperature dependence of the excess molar volume of (dimethyl

carbonate, or diethyl carbonate + toluene) from T = 278.15 K to 323.15 K[J]. Journal of Chemical Thermodynamics, 2000, 32 (6): 12.

[27] Mosteiro L, Mascato E, González-Salgado D, et al. Excess molar volume of dimethyl carbonate and p-xylene and n-decane at 288.15 and 298.15 K[J]. Journal of Thermal Analysis and Calorimetry, 2002, 70 (1): 7.

[28] Romano E, Trenzado JL, Gonzalez E, et al. Thermophysical properties of four binary dimethyl carbonate + 1-alcohol systems at 288.15–313.15 K[J]. Fluid Phase Equilibria, 2003, 211 (2): 12.

[29] Tojo J, Canosa J, Rodrguez A, et al. Densities and Excess Molar Properties of Dimethyl Carbonate with Alkanes (C6 to C10) and VLE of Dimethyl Carbonate with Alkanes (C9 to C10) at 101.3 kPa[J]. Journal of Chemical & Engineering Data, 2004, 49 (1): 8.

[30] Pereiro AB, Rodriguez A, Canosa J, et al. Density, Viscosity, and Speed of Sound of Dialkyl Carbonates with Cyclopentane and Methyl Cyclohexane at Several Temperatures[J]. Journal of Chemical & Engineering Data, 2004, 49 (5): 8.

[31] Prasad TEV, Banerjee S, Babu JK, et al. Boiling temperature measurements on the binary mixtures formed by dimethyl carbonate with some chloroethanes and chloroethylenes at 95.8 kPa[J]. Fluid Phase Equilibria, 2004, 218 (2): 3.

[32] Rodriguez A, Canosa J, Dominguez A, et al. Viscosities of dimethyl carbonate with alcohols at several temperatures UNIFAC-VISCO interaction parameters (-OCOO-/alcohol)[J]. Fluid Phase Equilibria, 2004, 216 (1): 8.

[33] Ma X, Liu X, Li Z, et al. Vapor-liquid equilibria for the ternary system methanol + dimethyl carbonate + dimethyl oxalate and constituent binary systems at different temperatures[J]. Fluid Phase Equilibria, 2004, 221 (1-2): 6.

[34] Rivas MA, Pereira SM, Banerji N, et al. Permittivity and density of binary systems of {dimethyl or diethyl carbonate} + n-dodecane from T = (288.15 to 328.15) K[J]. Journal of Chemical Thermodynamics, 2004, 36 (3): 9.

[35] Yang C, Xu W, Ma P. Excess molar volumes and viscosities of binary mixtures of dimethyl carbonate with chlorobenzene, hexane, and heptane from (293.15 to 353.15) K and at atmospheric pressure[J]. Journal of Chemical & Engineering Data, 2004, 49 (6): 7.

[36] Pardo JM, Tovar CA, Troncoso J, et al. Thermodynamic behaviour of the binary systems dimethyl carbonate + n-octane or n-nonane[J]. Thermochimica Acta, 2005, 433: 6.

[37] Ma P, Li N. Measurement and correlation of densities and viscosities for correlative systems of dimethyl carbonate[J]. Journal of Tianjin University, 2005, 38 (7): 9.

[38] Terrado EM, Pardo JI, Urieta J, et al. Solubilities of Nonpolar Gases in Dimethyl Carbonate and Diethyl Carbonate[J]. Journal of Chemical & Engineering Data, 2005, 50 (2): 5.

[39] Rodriguez A, Pereiro AB, Canosa J, et al. Dynamic viscosities of the ternary liquid mixtures (dimethyl carbonate + methanol + ethanol) and (dimethyl carbonate + methanol + hexane) at several temperatures[J]. Journal of Chemical Thermodynamics, 2006, 38 (5): 15.

[40] Wang F, Wu J, Liu Z. Surface Tensions of Mixtures of Diesel Oil or Gasoline and Dimethoxymethane, Dimethyl Carbonate, or Ethanol[J]. Energy & Fuels, 2006, 20 (6).

[41] Iglesias-Otero MA, Troncoso J, Carballo E, et al. Density and refractive index for binary systems of the ionic liquid [Bmim][BF4] with methanol, 1,3-dichloropropane, and dimethyl carbonate[J]. Journal of Solution Chemistry, 2007, 36 (10): 12.

[42] Wisniak J, Cortez G, Peralta RD, et al. Density, excess volume, and excess coefficient of thermal expansion of the binary systems of dimethyl carbonate with butyl methacrylate, allyl methacrylate, styrene, and vinyl acetate at T =(293.15, 303.15, and 313.15)K[J]. Journal of Chemical Thermodynamics, 2008, 40 (12): 13.

[43] Rivas MA, Iglesias TP. On permittivity and density of the systems {triglyme + (dimethyl or diethyl carbonate)} and formulation of .epsilon. in terms of volume or mole fraction[J]. Journal of Chemical Thermodynamics, 2008, 40 (7): 11.

[44] Mosteiro L, Mariano AB, Casas LM, et al. Analysis of Surface Tension, Density, and Speed of Sound for the Ternary Mixture Dimethyl Carbonate + p-Xylene + n-Octane[J]. Journal of Chemical & Engineering Data, 2009, 54 (3): 7.

[45] Shi Y, Li W, Tu J. Vapor-liquid equilibria for binary systems of methanol-isoamyl acetate, dimethyl carbonate-isoamyl acetate, and methanol-1,2-propanediol at 101.325kPa. [J]. Journal of Chemical Engineering of Chinese Universities, 1999, 13 (2): 5.

[46] Ma X, Li Z, Wang B, et al. Vapor-Liquid Equilibria for the Binary System of Dimethyl Carbonate(1)-Dimethyl Oxalate(2)[J]. Journal of Chemical Engineering of Chinese Universities, 2001, 15 (3): 4.

[47] A.Rodriguez, J.Canosa, A.Dominguez, et al. Isobaric vapour–liquid equilibria of dimethyl carbonate with alkanes and cyclohexane at 101.3 kPa[J]. Fluid Phase Equilibria, 2002, 198: 15.

[48] Zhang L, Chen J, Ding J. Predication of atmospheric VLE for the ternary system dimethyl carbonate-methanol-n-hexane[J]. Speciality Petrochemicals, 2003, (3): 4.

[49] Sporzynski A, Szurgocinska M, Domanska U, et al. Thermodynamics of Mixtures Containing Organic Carbonates. 14. Excess Molar Gibbs Energies for 1-Hexanol + Dimethyl or Diethyl Carbonate Systems at 353.15 and 363.15 K.

Comparison with ERAS Results[J]. Ind Eng Chem Res, 2003, 42 (19): 7.
[50] Xing Y, Shao D, Fang W, et al. Vapor pressures and flash points for binary mixtures of tricyclo [5.2.1.0^{2,6}] decane and dimethyl carbonate[J]. Fluid Phase Equilibria, 2009, 284: 5.
[51] Rodrguez A, Canosa J, Domnguez A, et al. Viscosities of Dimethyl Carbonate or Diethyl Carbonate with Alkanes at Four Temperatures. New UNIFAC-VISCO Parameters[J]. Journal of Chemical & Engineering Data, 2003, 48 (1): 6.
[52] Jin X, Wu J, Liu Z, et al. The thermal conductivity of dimethyl carbonate in the liquid phase[J]. Fluid Phase Equilibria, 2004, 220 (1): 4.
[53] Song L, Wu J, Liu Z. Thermal conductivity of dimethyl carbonate in liquid phase[J]. Journal of Chemical Industry and Engineering (China), 2008, 59 (3): 6.
[54] Vogel E, Küchenmeister C, Bich E, et al. Reference Correlation of the Viscosity of Propane[J]. Journal of Chemical & Engineering Data, 1998, 27 (5): 24.
[55] Rainwater JC, Friend DG. Second viscosity and thermal-conductivity virial coefficients of gases: Extension to low reduced temperature[J]. Phys Rev A, 1987, 36 (8): 5.
[56] Huber ML, Laesecke A. Correlation for the Viscosity of Pentafluoroethane (R125) from the Triple Point to 500 K at Pressures up to 60 MPa[J]. Ind Eng Chem Res, 2006, 45: 7.
[57] Hirschfelder JO, Curtiss CF, Bird RB. Molecular Theory of Gases and Liquids[M]. New York: Wiley, 1964.
[58] Neufeld PD, Janzen AR, Aziz RA. Empirical Equations to Calculate 16 of the Transport Collision Integrals (l,s)* for the Lennard-Jones (12–6) Potential[J]. J Chem Phys, 1972, 57: 3.
[59] Chung TH, Ajlan M, Lee LL, et al. Generalized multiparameter correlation for nonpolar and polar fluid transport properties[J]. Ind Eng Chem Res, 1988, 27: 9.
[60] Ely JF, Hanley HJM. Prediction of Transport Properties. 1. Viscosity of Fluids and Mixtures[J]. Ind Eng Chem Fundam, 1981, 20: 10.
[61] Ely JF, Hanley HJM. Prediction of transport properties. 2. Thermal conductivity of pure fluids and mixtures[J]. Ind Eng Chem Fundam, 1983, 22: 8.
[62] Huber ML, Laesecke A, Perkins RA. Model for the Viscosity and Thermal Conductivity of Refrigerants, Including a New Correlation for the Viscosity of R134a[J]. Ind Eng Chem Res, 2003, 42 (13): 16.
[63] Klein SA, McLinden MO, Laesecke A. An improved extended corresponding states method for estimation of viscosity of pure refrigerants and mixtures[J]. International Journal of Refrigeration, 1997, 20 (3): 10.
[64] McLinden MO, Klein SA, Perkins RA. An extended corresponding states model for the thermal conductivity of refrigerants and refrigerant mixtures[J]. International Journal of Refrigeration, 2000, 23 (1): 43.
[65] Olchowy GA, Sengers JV. A simplified representation for the thermal conductivity of fluids in the critical region[J]. International Journal of Thermophysics, 2005, 10 (2): 10.
[66] Krauss R, Weiss VC, Edison TA, et al. Transport Properties of 1,1-Difluoroethane (R152a)[J]. International Journal of Thermophysics, 1996, 17 (4): 27.

Analysis of Thermodynamic Properties of Coarse Grained Surfactant Models

Samantha A. Sanders, Chemical Engineering, Princeton University, Princeton, NJ

Athanassios Z. Panagiotopoulos, Chemical Engineering, Princeton University, Princeton, NJ

Surfactants are used for a wide range of industrial and biological applications. Composed of a hydrophilic head group and a hydrophobic tail group, they self-assemble into various structures in aqueous solutions. Atomistic models cannot reach the necessary time and length scales for self-assembly in dilute solutions. Therefore, various coarse grained models have been developed. A recent group of coarse grained models provides semi-quantitative information. However, these models have not been tested on their ability to reproduce experimentally measured critical micelle concentrations (cmc's). In this work, we analyze the MARTINI model (Marrink et al., J. Phys. Chem. B, 2004, 108, 750-760; Marrink et al., J. Phys. Chem. B, 2007, 111, 7812-7824) and the model developed by Shinoda and coworkers. (Shinoda et al., Molecular Simulation, 2007, 33, 27-36), for their ability to reproduce the cmc's of various zwitterionic and non-ionic surfactants. We obtain long (several ns) molecular dynamics trajectories at multiple temperatures allowing the determination of cmc's and aggregate size distributions and their trends. Our results suggest that that the models underpredict the critical micelle concentrations by 1-2 orders of magnitude for zwitterionic surfactants, with smaller but still significant deviations for nonionic ones. While the monomer surfactant concentrations reach equilibrium, the aggregate size distributions do not reach equilibrium. Overall, while the results from these models are in the correct regions, the underprediction indicates the need of further refinement in order to create quantitative, transferrable models for surfactant self-assembly.

See more of this Session: Poster Session: Thermodynamics and Transport Properties
See more of this Group/Topical: Engineering Sciences and Fundamentals

Computer-Aided Molecular Design of Ionic Liquids for the Pretreatment of Biomass Using the Signature Molecular Descriptor

Donald P. Visco, Chemical Engineering, Tennessee Technological University, Cookeville, TN

Derick C. Weis, Chemical Engineering, Tennessee Technological University, Cookeville, TN

Cellulose is an abundant, renewable resource available from plants that can be turned into ethanol and help replace fossil fuels. Potential feed stocks for cellulose include waste products like corn stover and sugar cane bagasse, or planting a specific crop of switchgrass for harvest. Currently, the cost of cellulosic ethanol is too high for practical consideration. Ionic liquids have been identified for a pretreatment step in the production of cellulosic ethanol as a way to reduce the cost. Dissolving cellulose in an ionic liquid disrupts the ordered crystalline structure to a more amorphous form.[1] The dissolved cellulose can be regenerated by adding water, and then more easily broken down into fermentable sugars by enzymatic hydrolysis. In addition, ionic liquids are preferred over traditional pretreatment solvents because of environmental and safety considerations.

Some important properties for developing more effective ionic liquids for cellulose pretreatment include melting point, viscosity, and cellulose solubility. One of the current challenges associated with designing innovative ionic liquids for cellulose pretreatment is accurately estimating physical properties to save time and resources. In fact, there are an estimated 1018 possible combinations of cation/anion pairs available to create ionic liquids.[2] Ideally, one would like to design new ionic liquids on a computer which are predicted to possess the desired properties before actually performing experiments in the lab. Applying this approach is much quicker, less expensive and allows for a more comprehensive search for the most suitable potential candidates. In this work, we use a computer-aided molecular design technique known as inverse design.[3] Quantitative structure-property relationships (QSPRs) using the Signature molecular descriptor were created to predict melting point, viscosity, and cellulose solubility. The inverse-QSPR algorithm was employed to develop new ionic liquids with predicted properties suitable for cellulose pretreatment. A focused database of ionic liquids was created to provide a starting point for the evaluation and testing of the new and/or non-intuitive pre-treatment candidates. Results on the synthesis and evaluation will be reported in the poster presentation.

1. R. P. Swatloski, S. K. Spear, J. D. Holbrey and R. D. Rogers, J. Am. Chem. Soc., 2002, 124, 4974-4975.

2. A. R. Katritzky, R. Jain, A. Lomaka, R. Petrukhin, M. Karelson, A. E. Visser and R. D. Rogers, J. Chem. Inf. Comput. Sci., 2002, 42, 225-231.

3. D. P. Visco, Jr., R. S. Pophale, M. D. Rintoul and J. L. Faulon, J. Mol. Graph. Model., 2002, 20, 429-438.

See more of this Session: Poster Session: Thermodynamics and Transport Properties
See more of this Group/Topical: Engineering Sciences and Fundamentals

Determination of Pitzer Parameters for Pb(II) and Fe(II) Solubilities in Concentrated Brines in the Presence of Environmental Ligands

Ahmed E. Ismail, Performance Assessment and Decisions Analysis, Sandia National Laboratories, Carlsbad, NM

Laurence H. Brush, Repository Performance, Sandia National Laboratories, Carlsbad, NM

Je-Hun Jang, Repository Performance, Sandia National Laboratories, Carlsbad, NM

Shelly R. Johnsen, Repository Performance, Sandia National Laboratories, Carlsbad, NM

Martin B. Nemer, Performance Assessment and Decisions Analysis, Sandia National Laboratories, Carlsbad, NM

Yongliang Xiong, Repository Performance, Sandia National Laboratories, Carlsbad, NM

Assessing the performance of the U. S. Department of Energy's Waste Isolation Pilot Plant (WIPP), a deep geologic repository for defense-related transuranic waste in southeastern New Mexico, requires a detailed understanding of the response of concentrated brines to both minerals found in the surrounding rock formation and man-made materials that are emplaced in the repository. The geochemistry model developed for the WIPP is based on the activity coefficient model of Kenneth Pitzer. Parameter sets are available for many of the chemical species in the repository, including both common brine constituents such as Na^+, Mg^{2+}, HCO_3^-, SO_4^{2-} and Cl^-, as well as radionuclides such as Am(III) and Th(IV). In some scenarios iron and lead could be available in significant quantities in the repository through their use in waste packaging materials. There is limited information available in the literature for Pb(II) and Fe(II) in the complex brines observed at the WIPP. In addition, because of the basic pH and the strongly reducing conditions expected in the repository, both Pb(II) and Fe(II) are expected to form complexes with inorganic ligands such as hydroxide and chloride, and organic ligands such as EDTA and oxalate.

Consequently, we have undertaken an experimental and modeling effort to determine Pitzer parameters for the interactions of Pb(II) and Fe(II) complexes with other major species in the brine (Na^+, Mg^{2+}, Cl^-, HCO_3^-, and SO_4^{2-}). Experimental measurement of Pb(II) and Fe(II) solubilities in relatively simple solutions are combined with a free-energy minimization scheme to determine the new Pitzer parameters for each pair of species studied. Subsequent validation testing will assess the influence of the new parameters for Pb(II) and Fe(II) complexes on the WIPP geochemical model.

Sandia is a multi-program laboratory operated by Sandia Corporation, a Lockheed Martin Company, for the United States Department of Energy's National Nuclear Security Administration under Contract DE-AC04-94AL85000. This research is funded by WIPP programs administered by the Office of Environmental Management of the U.S Department of Energy.

See more of this Session: Poster Session: Thermodynamics and Transport Properties
See more of this Group/Topical: Engineering Sciences and Fundamentals

Viscosity of Binary Mixtures of Soybean + Canola, + Coconut, and + Corn Biodiesel at Different Temperatures

Filipe Xavier Feitosa, Chemical Engineering Department, Universidade Federal do Ceará, Fortaleza, Brazil

Tamires Barreto Cidade, Chemical Engineering Department, Universidade Federal do Ceará, Fortaleza, Brazil

Carolina B. Veloso, Chemical Engineering Department, Universidade Federal do Ceará, Fortaleza, Brazil

Célio L. Cavalcante Jr., Chemical Engineering, Universidade Federal do Ceará, Fortaleza, Brazil

Monica C. G. Albuquerque, Chemical Engineering Department, Universidade Federal do Ceará, Fortaleza, Brazil

Hosiberto Batista de Sant Ana, Chemical Engineering Department, Universidade Federal do Ceará, Fortaleza, Brazil

Renewable energy has gained an important attention on the last decades due to the decreasing oil supply and increasing environmental consciousness. The production of biodiesel from vegetable oils is one of the alternatives on the world strategy to expand renewable energy sources. Vegetable oils are biodegradable, non-toxic, and can significantly contribute to reduce pollution, when transesterified to biodiesel.

Soybean, canola, coconut and corn oils are different kinds of vegetable oil that can be used to produce biodiesel. Biodiesel is a group of monoalkyl esters of long chain fatty acids derived from renewable feedstock, which has been increasingly used as substitute of conventional diesel fuel. It may be directly used in diesel vehicles with little or no modification. These biodiesels were produced by transesterification method using basic homogeneous catalyst.

The knowledge of thermodynamic properties of biodiesel (and their blends) play an important role on the understanding of the nature of intermolecular interactions between the different molecules of biodiesel produced by vegetable oils used on its production, which affect their thermodynamic properties. We believe that there is still a lack of systematic studies on data of thermodynamics property of biodiesel blends and biodiesel/diesel over the whole composition range at different operational conditions of pressure and temperature

In the present paper, we present values of viscosity and density of binary mixtures of soybean + canola, + coconut, + corn biodiesel oils over the whole composition range at temperatures from 273.15 to 373.15 K and at atmospheric pressure. Densities and viscosities data were correlated with R-square adjustment and Andrade equation, respectively, with good agreement. The deviations viscosities have been calculated from the experimental data. The deviations have been correlated using the Redlich-Kister polynomial equation.

See more of this Session: Poster Session: Thermodynamics and Transport Properties
See more of this Group/Topical: Engineering Sciences and Fundamentals

Modeling the Phase Behavior, Excess Enthalpies and Henry's Constants of the H2O + H2S Binary Mixture Using the SAFT-VR+D Approach

M. Carolina Dos Ramos, Department of Chemical and Biomolecular Engineering, Vanderbilt University, Nashville, TN

Clare McCabe, Department of Chemical and Biomolecular Engineering, Vanderbilt University, Nashville, TN

Due to quality regulations and environmental restrictions, sulphur compounds must be monitored and removed from the final products in a range of industrial operations, including waste treatment plants, in pulp and paper operations, and in natural gas and petroleum reservoirs. Therefore, an accurate knowledge of the phase behavior of mixtures of water and hydrogen sulphide is crucial to plant design and operation within the chemical industry [1]. Additionally, both hydrogen sulphide and water are very interesting molecules to study due to the presence of their permanent dipole moment and hydrogen bonding interactions. The SAFT-VR+D approach [2] is based on a version of the statistical associating fluid theory that was developed to model associating dipolar fluids by explicitly accounting for dipolar interactions and their effect on the thermodynamics and structure of a fluid. In the SAFT-VR+D approach this is achieved through the use of the generalized mean spherical approximation (GSMA) to describe a reference fluid of dipolar associating square well segments. In the present work, we modify the SAFT-VR+D approach to study mixtures of dipolar associating fluids of arbitrary size, based on the work of Cummings and Blum [3], and to explicitly account for the dipolar interactions of each polar compound and examine the high-pressure phase diagram and other thermodynamic properties of the binary systems of water + hydrogen sulphide. Both water and hydrogen sulphide are modeled as associating spherical molecules with four off-centre sites to mimic hydrogen bonding and an embedded dipole moment (μ) to describe their polarity. Using simple Lorentz-Berthelot combining rules, the theory is able to predict the phase behavior of the H_2O + H_2S system from pure component parameters: type *III* phase behavior according to the classification of Scott and van Konynenburg [4] with a three-phase line at low temperatures and its corresponding upper critical end point, and two separate critical lines, one at high temperatures that runs continuously from a gas-liquid critical line to a liquid-liquid critical line at high pressures and a second gas-liquid critical line located at low temperatures. The phase behavior is predicted by the SAFT-VR+D approach in excellent agreement with the experimental data, without requiring binary interaction parameters. In addition, the theory is able to predict in a qualitative manner the excess molar enthalpies and Henry's constants as a function of temperature.

[1] F. Al-Awadhy, I. Kocabas, J. H. Abou-Kassem, and M. R. Islam, *Energy Sources* **27** (1-2), 3 (2005); A. Chapoy, A. H. Mohammadi, B. Tohidi, A. Valtz, and D. Richon, *Industrial & Engineering Chemistry Research* **44** (19), 7567 (2005).

[2] H. Zhao and C. McCabe, *Journal of Chemical Physics* **125**, 104504 (2006); H. Zhao, Y. Ding, and C. McCabe, *Journal of Chemical Physics* **127**, 084514 (2007).

[3] P. T. Cummings and L. Blum, *Journal of Chemical Physics* **84**, 1833 (1986).

[4] P. H. van Konynenburg and R. L. Scott, *Philosophical Transactions of the Royal Society of London*, Series A **A298**, 495 (1980); R. L. Scott and P. H. van Konynenburg, *Discussions of the Faraday Society* **49**, 87 (1970).

See more of this Session: Poster Session: Thermodynamics and Transport Properties
See more of this Group/Topical: Engineering Sciences and Fundamentals

Low Vapor Pressure (down to 10-6 Pa) Determination of 12 Anthropogenic Polycondensed Hydrocarbons Using Dynamic Saturation Apparatus with Online Detection

Ramy Abou Naccoul, Analitycal science laboratory, Thermodynamic and online analysis Team, Claude Bernard University Lyon - France, Lyon, France

Ilham Mokbel, Analitycal science laboratory, Thermodynamic and online analysis Team, Claude Bernard University Lyon - France, Lyon, France

Antonio Carlo Razzouk, Analitycal science laboratory, Thermodynamic and online analysis Team, Claude Bernard University Lyon - France, Lyon, France

Evelyne Rauzy, Chemistry, Université de la Méditerranée, Marseille

Joseph Saab, Chemistry department, Holy Spirit University Kaslik, Jounieh, Lebanon

Naim Ouaini, Chemistry department, Holy Spirit University Kaslik, Jounieh, Lebanon

Jacques Jose, Analitycal science laboratory, Thermodynamic and online analysis Team, Claude Bernard University Lyon - France, Lyon, France

Knowledge of data describing phase equilibria of key compounds contained in petroleum and coal liquids is of major interest for modeling the behavior of reservoir and coal fluids over a wide range of temperatures and pressures. While abundant data of reasonable accuracy are accessible for components of low and medium molar mass, less reliable information is available for characterizing high boiling fractions containing considerable amount of polycondensed hydrocarbons. Verified data for compounds of high molar mass are obviously needed for better characterization of reservoir fluids.

The aim of our work is to present the vapor and sublimation pressures of a series of Polycyclic Aromatic Hydrocarbons, PAH, (1,2,3,4, Dibenzanthracene, benzo(k) fluoranthene (a,c), Chrysene, 1,2 Benzanthracene, Coronene, 1,2,5,6, Dibenzanthracene (a,h), 1,2,7,8 Dibenzanthracene (a,j), Benzo(b) fluoranthene, Ovalene, etc). The measurements were carried out by a gas saturation method. The used apparatus has been designed and developed in our laboratory. The principle is based on relative method using a reference compound along with the sample. Experimental measurements were made for 12 PAH having 3 to 9 arenes. The pressure ranged from 100 Pa to 5.10-6 Pa corresponding to an interval of temperatures between 295 K to 513 K. The validation of the apparatus and the procedure in the quoted pressure range were carried out by studying normal alkanes (hexadecane and eicosane) and phenanthrene (aromatic compound with 3 arenes). The experimental results showed an excellent agreement with the available literature values: standard deviation of 5.5 % overall the studied temperature range. In a second step, the collected data were used to perform a cubic type equation of state based on a group contribution method.

See more of this Session: Poster Session: Thermodynamics and Transport Properties
See more of this Group/Topical: Engineering Sciences and Fundamentals

Phase Behavior of Arbutin in a Mixture of Ethanol and Supercritical Carbon Dioxide

Chang-Nam Han, Department of Advanced Chemicals and Engineering, Chonnam National University, Gwangju, South Korea

Dong-Yuk Kang, Department of Advanced Chemicals and Engineering, Chonnam National University, Gwangju, South Korea

Dong Lyun Cho, Department of Advanced Chemicals and Engineering, Chonnam National University, Gwangju, South Korea

Jae-Soon Shin, Department of Applied Chemical Engineering and The Research Institute for Catalysis, Chonnam National University, Gwangju, South Korea

Choon-Hyoung Kang, Department of Applied Chemical Engineering and The Research Institute for Catalysis, Chonnam National University, Gwangju, South Korea

Phase behavior of the ternary systems of arbutin, which is well known to be effective material for skin lightening agent, in a solvent mixture of ethanol and supercritical carbon dioxide was investigated. The solubilities of arbutin in the mixture of ethanol and carbon dioxide were determined as functions of temperature, pressure and solvent composition by measuring the cloud points of the ternary mixtures at various conditions using a high-pressure phase equilibrium apparatus equipped with a variable-volume view cell. On the basis of ethanol, arbutin loading was kept at 1.5wt% throughout this study and the pressure and temperature were varied up to 20MPa and 333.15K, respectively. For a range of CO_2 loading less than 33wt% on the basis of ethanol, the cloud point was not observed. On the contrary, solid remained undissolved when CO_2 loading exceeded 54wt%. Between these loadings, steep and thus very sensitive to pressure variation solubility curves, which extends downward to a vaporization boundary, were found.

See more of this Session: Poster Session: Thermodynamics and Transport Properties
See more of this Group/Topical: Engineering Sciences and Fundamentals

Transformation of Calcium Sulphate Dihydrate to Anhydrous Calcium Sulphate in Aqueous Solutions

Ghazal Azimi, Chemical Engineering and Applied Chemistry, University of Toronto, Toronto, ON, Canada

Vladimiros G. Papangelakis, Chemical Engineering and Applied Chemistry, University of Toronto, Toronto, ON, Canada

The mechanism of transformation of calcium sulphate dihydrate ($CaSO_4{\cdot}2H_2O_{(s)}$) to anhydrous calcium sulphate ($CaSO_{4(s)}$) in aqueous solutions was investigated. The kinetics of the transformation was followed by monitoring the changes both in the concentration of dissolved calcium in the solution measured by ICP-OES and in the mole fraction of dihydrate/anhydrite in the solid phase obtained from X-ray diffraction and thermo gravimetric analysis (TGA). It was found that the transformation occurs in such a way that dihydrate, which is the metastable phase under the conditions studied, dissolves and anhydrite, the stable phase, precipitates from the solutions. During the dehydration of calcium sulphate dihydrate at temperatures above 90°C, metastable hemihydrate ($CaSO_4{\cdot}1/2H_2O_{(s)}$) may form as an intermediate stage. Addition of an anhydrite seed to the solution diminishes the induction period and accelerates the transformation process.

See more of this Session: Poster Session: Thermodynamics and Transport Properties
See more of this Group/Topical: Engineering Sciences and Fundamentals

Measurement of Bubble Points of CO2 in Ionic Liquids, [BMP][Tfo]and [P14,6,6,6][Tf2N] Using Variable-Volume View Cell

Ha Na Song, Department of Chemical and Biomolecular Engineering, Sogang University, Seoul, South Korea

Jun Hyuk Yim, Department of Chemical and Biomolecular Engineering, Sogang University, Seoul, South Korea

Byung-Chul Lee, Department of Chemical Engineering and Nano-Bio Technology, Hannam University, Daejeon, South Korea

Jong Sung Lim, Department of Chemical and Biomolecular Engineering, Sogang University, Seoul, South Korea

The bubble point pressure of carbon dioxide (CO2) and ionic liquids mixture were measured by using a high-pressure experimental apparatus equipped with a variable-volume view cell, at various CO2 compositions in the temperature range from 303.15 to 373.15 K and the pressure range up to about 70 MPa. We selected ionic liquids, 1-butyl-1-methylpyrrolidinium trifluoromethanesulfonate ([BMP][TfO]) and trihexyltetradecyl- phosphonium bis(tri-fluoromethylsulfonyl)imide ([P14,6,6,6] [Tf2N]) for experimental measurement in this study. The modified Lydersen-Joback-Reid method and Peng-Robinson Equation of state have been applied to correlate the experimental data. The experimentally measured bubble point data gave relatively good agreement with those calculated data.

See more of this Session: Poster Session: Thermodynamics and Transport Properties
See more of this Group/Topical: Engineering Sciences and Fundamentals

Vapor-Liquid Equilibrium Measurement for the Binary System of Carbon Dioxide (CO2) + 1,1,1-Trifluoroethane (R-143a) at High Pressures

Seol A. Kim, Department of Chemical and Biomolecular Engineering, Sogang University, Seoul, South Korea

Ah Ram Kim, Department of Chemical and Biomolecular Engineering, Sogang University, Seoul, South Korea

Ki-Pung Yoo, Department of Chemical and Biomolecular Engineering, Sogang University, Seoul, South Korea

Jong Sung Lim, Department of Chemical and Biomolecular Engineering, Sogang University, Seoul, South Korea

PTxy measurements for the binary system of carbon dioxide (CO2) + 1,1,1-trifluoroethane (HFC-143a) were measured at seven equally spaced temperatures of (273.15, 283.15, 293.15, 303.15, 313.15, 323.15 and 333.15) K, by using a circulation-type equilibrium apparatus in which both vapor and liquid phases were recirculated. The experimental data were correlated with the Peng-Robinson equation of state (PR-EoS) using the Wong-Sandler mixing rule combines with the NRTL excess Gibbs free energy model and the Carnahan-Starling-De Santis equation of state (CSD EoS). Almost all the calculated values with these models show good agreement with the experimental data.

See more of this Session: Poster Session: Thermodynamics and Transport Properties
See more of this Group/Topical: Engineering Sciences and Fundamentals

Consideration of the Entropy in the Prediction of Stable Crystalline Polymorphs

Tai Boon Tan, Chemical and Biological Engineering, University at Buffalo, The State University of New York, Buffalo, NY

Nancy Cribbin, Chemical and Biological Engineering, University at Buffalo, The State University of New York, Buffalo, NY

Andrew Schultz, Chemical and Biological Engineering, University at Buffalo, The State University of New York, Buffalo, NY

David A. Kofke, Chemical and Biological Engineering, University at Buffalo, The State University of New York, Buffalo, NY

In order to predict the stable crystalline polymorph, we can compare the free energy of each polymorph and select the one with the lowest free energy. Most of the current prediction techniques only take into account the energetic contribution to the free energy and have ignored the entropic contribution. This is mainly because determining the entropic contribution is another degree of complication. In order to improve the accuracy of these approximation methods, we have included the neglected entropic term into our free-energy calculation by using methods that build on the lattice-dynamics approach. This idea is motivated by many previous studies, which have reported that in many cases, the difference between the energies of the most stable polymorphs is relatively small and the stability is believed in some cases to be influenced by entropy.

Calculation of true free energy is performed by computing the difference with respect to a known reference. In this work, we use a harmonic reference system with spring constants given to match the configurational correlations measured in the target system. We consider various perturbation techniques that compute the free energy difference between the target and reference systems while avoiding lengthy thermodynamic integration procedures. The basic methods we look at are free-energy perturbation approaches involving a single intermediate stage, which include overlap sampling and umbrella sampling. Such methods require only one or two simulations (of the target and/or reference system) to yield a result, and for small enough systems we show that these methods can be effective. In larger or more difficult systems we consider a process of switching on various harmonic modes in groups, and evaluate the free-energy change for each before summing to determine the total difference. We consider whether and how this process may be abbreviated in a way that does not require transformation of all harmonic modes. Different prototype systems are studied and discussed in the context of these methods.

See more of this Session: Poster Session: Thermodynamics and Transport Properties
See more of this Group/Topical: Engineering Sciences and Fundamentals

Modelling the Liquid Phase Behaviour of Aqueous Non-Ionic Amphiphilic Mixtures in the Presence of Oil and Salt Modelled with SAFT-VRE

Jens M. A. Schreckenberg, Department of Chemical Engineering, Imperial College London, London, United Kingdom

Claire S. Adjiman, Chemical Engineering, Imperial College London, London, United Kingdom

George Jackson, Centre for Process Systems Engineering, Imperial College London, London, United Kingdom

Amparo Galindo, Centre for Process Systems Engineering, Imperial College London, London, United Kingdom

The liquid phase behaviour of aqueous alykylpolyoxyethylene (CiEj) nonionic amphiphilic mixtures is of great interest in many industrial processes. We propose a model for these mixtures with the SAFT-VR [1] framework and study the coexistence properties of these systems in the presence of oil and salt. The surfactant is modelled with short-range attractive interaction sites that allow the explicit treatment of hydrogen bonding between the water and surfactant molecules. A transferable model for CiEj amphiphiles of varying size is presented, which reproduces the region of liquid-liquid immiscibility of the solutions. We introduce an n-alkane model and study the complex three component (water + amphiphile + oil) phase behaviour; under certain conditions these mixtures exhibit three coexisting liquid phases. We describe the liquid phase behaviour of these systems at ambient pressure and different temperatures comparing with experimental data and provide predictions the high pressure phase behaviour.

The influence of strong electrolytes on the phase behaviour of these systems is studied using the extension of SAFT-VR for electrolytes [2], in which additional terms for the long-range Coulombic interaction and the solvation energies of the ions are incorporated. The Coulombic forces are modelled with the mean spherical approximation in the primitive model. We find the dielectric constant of these systems, which appears in the Coulombic term, has a crucial effect on the phase behaviour of the systems. We study the phase behaviour of water + amphiphile + salt mixtures and predict salting out in agreement with experimental data.

[1] A. Gil-Villegas, A. Galindo, P. J. Whitehead, S. J. Mills, G. Jackson, and A. N. Burgess, "Statistical associating fluid theory for chain molecules with attractive potentials of variable range (SAFT-VR)", Journal of Chemical Physics, 106, 4168-4186 (1997).

[2] A. Galindo, A. Gil-Villegas, G. Jackson, and A. N. Burgess, "SAFT-VRE: Phase behaviour of electrolyte solutions using the statistical associating fluid theory for potentials of variable range", Journal of Physical Chemistry B, 103, 10272-10281, (1999).

See more of this Session: Poster Session: Thermodynamics and Transport Properties
See more of this Group/Topical: Engineering Sciences and Fundamentals

Modelling Real Fluids and Their Mixtures with An Improved SAFT-VR Equation of State for Segments Interacting through Mie Potentials (SAFT-VR Mie)

Thomas Lafitte, Centre for Process Systems Engineering, Imperial College London, London, United Kingdom

Anastasia Apostolakou, Centre for Process Systems Engineering, Imperial College London, London, United Kingdom

Amparo Galindo, Centre for Process Systems Engineering, Imperial College London, London, United Kingdom

Claire Adjiman, Centre for Process Systems Engineering, Imperial College London, London, United Kingdom

George Jackson, Centre for Process Systems Engineering, Imperial College London, London, United Kingdom

The concept of using the more realistic Mie potential (Lennard-Jonesium interactions of the general form $u(r)=e\ C(l_1,l_2)\{(s/r)^{l_1}-(s/r)^{l_2}\}$) as the elementary building block of homonuclear chains of Mie segments within a SAFT-like EOS has proven to be very fruitful in order to model simultaneously the fluid phase behaviour and second-derivative properties of real fluids[1]. Recently, an improved version of the SAFT-VR Mie EOS has been presented. This new theory which is based on a more accurate expression of both the Helmholtz free energy and contact radial distribution function of a fluid of chain molecules interacting through the Mie potential has been shown to reproduce the Monte Carlo simulation data for a broad interval of repulsive and attractive ranges. In this work, we have applied this new SAFT-VR Mie equation of state to real substances (alkanes, alkanols, amines, carboxylic acids, water) and selected binary mixtures. The heat capacity in the compressed liquid region, generally presented as a severe consistency test for EOS, was estimated with percent average absolute deviations (AAD%) close to 2% in each cases. The speed of sound is also accurately captured with deviations below 4%. In addition, we will show that the new EOS also provides a more accurate description of the VLE of the pure substances studied. In order to emphasize the relevance of using a variable repulsive range through the Mie potentials, a comparison with other SAFT-like EOS will be presented (SAFT-VR SW, soft-SAFT[2], PC-SAFT[3]).

[1] T. Lafitte, D. Bessieres, M. M. Piñeiro and J.-L. Daridon, J. Chem. Phys., 124, 024509 (2006).

[2] F. J. Blas and L. F. Vega, Mol. Phys., 92, 135 (1997).

[3] J. Gross and G. Sadowski,Ind. Eng. Chem. Res., 40, 1244 (2001).

See more of this Session: Poster Session: Thermodynamics and Transport Properties
See more of this Group/Topical: Engineering Sciences and Fundamentals

Position Group Contribution Method for the Prediction of standard enthalpies of vaporization of Organic Compounds

Qiang WANG [a,b]*, *Qingzhu JIA* [a,b], *Shaokao XIE* [b], *Jingwen CHAI* [b], *Peisheng MA* [b]

a. School of Material Science and Chemical Engineering, Tianjin University of Science and Technology, 13St. TEDA, Tianjin, 300457, People's Republic of China

b. School of Chemical Engineering and Technology, Tianjin University, Tianjin 300072, People's Republic of China

* To whom correspondence should be addressed. E-mail: *wang_q@tust.edu.cn*

Abstract

Recently, our Laboratory proposed a model for the prediction of critical properties of organic compounds. The higher prediction accuracy of the proposed method shown in our previous works suggests that it is possible to use a similar framework to predict the three critical properties, T_c, P_c, V_c of organic compounds containing various functionalities.

In this work our model, a combined approach of group contribution and position distribution function, is extended for the prediction of standard enthalpies of vaporization of organic compounds ranging form C2 to C18. The results indicate that our model provides very satisfactory results. The overall average absolute difference for standard enthalpies of vaporization predictions of 223 organic compounds is 2.78 % means relative derivation at boiling point.

Keywords: Standard enthalpies of vaporization; Prediction; Position distribution function

1. Introduction

Enthalpy of vaporization is an important parameter that is utilized for the design and operation of vapor–liquid equilibrium-based processes, where enthalpy of vaporization must be known for at least one temperature. It is not always possible, however, to find experimental values of properties for the compounds of interest in the literature. Since, it is not practical to measure them as the need arises, estimation methods are generally employed in this and other similar situations.

There exist many empirical correlations that allow one to calculate the enthalpy of vaporization of pure fluids (Carruth and Kobayashi (1972), Svoboda and Smolova, 1994; Liley, 2003; Meyra et al., 2004).[1-4] Certainly, the property may also be calculated by means of group contribution models (Basarova and Svoboda, 1995; Tu and Liu, 1996; Li et al., 1997; Fau ´ndez et al., 2000; Cuadros et al., 2003).[5-10] However, most group contribution methods have a serious problem that they cannot distinguish among structural isomers. To overcome these limitations, esveral attemptes have been reported in the literature. Constantinou and Gani,[11] Marrero and Gani,[12] and Dalmazzone et al.[13]developed methods which perform estimations at different levels. Recently, Depends on three parameters related to the shape and size of the molecule and to the intermolecular interactions, Mulero et al.[14] have proposed a model for the prediction of the enthalpy of vaporization at 273.15K of organic compounds.

Recently, our Laboratory proposed a model for the prediction of critical properties of organic compounds. The higher prediction accuracy of the proposed method shown in our previous works suggests that it is possible to use a similar framework to predict the three critical properties, T_c, P_c, V_c of organic compounds containing various functionalities.[15-19] Therefore, the objective of this work is the application our new model in the prediction of the enthalpy of vaporization of organic compounds at the normal boiling point and 298.15K using a database that covers a variety of chemical classes.

2. Method proposed in this work

The standard enthalpies of vaporization function is constructed by all groups' contribution as well as the position correlation factor. The position correlation factor were used to take into account longer distance interactions, which could distinguish most isomer include *cis-* and *trans-* or *Z-* and *E-* structure of organic compounds for their thermodynamics properties.

Here, the position distribution function for the estimation of standard enthalpies of vaporization is expressed as:

$$H = H_0 + \sum_i A_i N_i + \sum_j A_j \tanh(N_j / N) + \sum_k A_k P_k + a_1 \exp(1/M) + a_2 \exp(1/N) \quad (1)$$

$$N = \sum_i N_i + \sum_j N_j$$

Parameter A_i or A_j stands for i or j group contributions, N_i for the number of each group that carbon element forms the centre of the group in the molecular formula, N_j for the number of each group that non-carbon element forms the centre, N for total number of groups, P_k for the position correlation factor, and a_1, a_2 for parameters of the model. The set of contributions that allowed to minimize the residual estimation difference was then computed by regression. V_0 is constant and M is molecular weight. According to the IUPAC nominating method, we draw the structures and assign the P_k values for the relevant positional correction. Table 1 reports the values computed for the group contributions A_i. And our method developed is applicable only to comparatively low-molar-mass compounds involving carbon chain from C2 to C18.

Conclusion

Based on position group contribution method, a new method recently proposed for the estimation of the critical properties and boiling point as well as melting point , T_C, P_C, V_C, T_b, T_f is extended to the prediction of standard enthalpies of vaporization. Contributions for compounds containing carbon, hydrogen, oxygen, nitrogen, chlorine and sulphur were reported, and that position distribution function has been developed which could distinguish between the thermodynamic properties of most isomers of organic compounds including *cis-* and *trans-* or *Z-* and *E-* structures. The results indicate that our model provides highly significances. The overall means absolute difference for enthalpies of vaporization predictions of 223 organic compounds is 2.78%. The higher prediction accuracy of the proposed method shown in our previous works and this work suggests that it is possible to use a totally same framework to predict the critical properties and the thermodynamics properties, of organic compounds containing various functionalities.

Literature Cited

1. Carruth, G. F. and Kobayashi, R. (1972). Ind. Eng. Chem. Fundam.,11, 509

2. Svoboda, V. and Smolova, H. (1994). Fluid Phase Equilib.,97, 1

3. Liley,P.E.(2003). Ind.Eng.Chem.Res., 42, 6250.

4. Meyra, A.G., Kuz, V. A., and Zarragoicoechea,G. J.(2004). FluidPhaseEquilib., 218,205.

5. Basarova, P. and Svoboda, V. (1995).Fluid Phase Equilib.,105, 27

6. Tu,C.H. and Liu,C.P.(1996). FluidPhaseEquilib., 121,45.

7. Li et al., 1994; (4) Li, P.; Ma, P. S.; Yi, S. Z,; Zhao, Z. G.; Cong, L. Z. (1994). Fluid Phase Equi. 101, 101–119

8. Li, P.; Liang, Y.H., Ma, P. S.; Zhu, C., (1997). Fluid Phase Equi. 101, 101–119

9. Faundez, C. A., Mulero, A., and Cuadros, F.(2000). J. Phase Equilib., 21,364

10. Cuadros, F., Mulero, A., Okrasinski, W., Faundez, C. A., and Parra, M.I.(2003). Rev.Chem.Eng.,19,387

11. Constantinou and Gani,5 (5) Constantinou, L.; Gani, R. New group contribution method for estimating properties of pure compounds. AIChE J. 1994, 40, 1697-1710

12. Marrero and Gani,6 (6) Marrero, J.; Gani R. Group-contribution based estimation of pure component properties. Fluid Phase Equilibria 2001, 183–184, 183–208

13. Dalmazzone,D., Salmon, A., and Guella, S., 2006. Fluid Phase Equilibria, 242, 29-42

14. Mulero, A., CACHADINA, I., CUADROS, F.,2006,Chem.Eng.Comm.,193:192-205

15. Wang, Q.; Ma, P. Sh; Jia, Q. Zh.; Xia, Sh. Q. Position Group Contribution Method for the Prediction of Critical Temperatures of Organic Compounds. J. Chem. Eng. Data 2008, 53, 1103-1109

16. Wang, Q.; Jia, Q. Zh.; Ma, P. Sh. Position Group Contribution Method for the Prediction of Critical Pressure of Organic Compounds. J. Chem. Eng. Data 2008,

17. Jia, Q. Zh.; Wang, Q.; Ma, P. Sh. Position Group Contribution Method for the Prediction of Critical Volume of Organic Compounds. J. Chem. Eng. Data 2008, 53 (11), 2606-2612

18. Wang, Q.; Ma, P.; Wang, C.; Xia, S. Position Group Contribution Method for the Prediction of Boiling Point of Organic Compounds. Chin. J. Chem. Eng. 2009, 17 (2), 254-258

19. Wang, Q.; Ma, P.; Nai, S. Position Group Contribution Method for the Prediction of Melting Point of Organic Compounds. Chin. J. Chem. Eng. 2009, 17 (3), 468-472

Prediction of Vapor-Liquid and Liquid-Liquid Equilibria of Mixtures From Peng-Robinson+COSMOSAC Equation of State

Chieh-Ming Hsieh, Department of Chemical Engineering, National Taiwan University, Taipei, Taiwan

Shiang-Tai Lin, Department of Chemical Engineering, National Taiwan University, Taipei, Taiwan

In this work we demonstrate successful vapor-liquid equilibrium (VLE) and liquid-liquid equilibrium (LLE) predictions for mixture fluids from a novel approach which combines cubic equations of state and first principle calculations. In particular, the energy parameter a(T,x) and volume parameter b(x) in Peng-Robinson equation of state (PR EOS) are respectively obtained from the result of first principle solvation calculations. As a consequence, the phase equilibrium of mixtures can be determined without using any experimental data. In our previous work, we have used this method to predict the critical properties, vapor pressure (as a function of temperature), the liquid phase density, and the complete phase diagram for pure substances. The same method has also been tested for vapor-liquid equilibrium of mixture fluids without using any binary interaction parameters.

In this work, we examined this method in describing LLE. In particular, the same set of parameters is used in describing VLE and LLE of pure and mixture fluids. Therefore, this method is capable to describe the phase diagram over a large temperature/pressure range, and the vapor-liquid-liquid equilibrium. The overall average deviation in LLE composition for 57 binary systems is found to be 6.16 %, similar to those from the modified UNIFAC and UNIFAC-LLE models.

See more of this Session: Poster Session: Thermodynamics and Transport Properties
See more of this Group/Topical: Engineering Sciences and Fundamentals

Phase Behavior of the System Hyperbranched Polyglycerol - Methanol - CO2

Christian S. Schacht, Engineering Thermodynamics, Delft University of Technology, Delft, Netherlands

Theo W. De Loos, Engineering Thermodynamics, Delft University of Technology, Delft, Netherlands

Joachim Gross, Laboratory for Process Equipment, Delft University of Technology, Delft, Netherlands

Abstract

Dendritic and Hyperbranched Polymers (HBPs) are currently being considered as solvents or entrainers within the field of thermal separations [1,2]. To determine the applicability of HBPs as process solvents for CO_2 removal from gas streams, the vapor-liquid equilibria of different HBP - Co-solvent - CO_2 (Hyperbranched polyglycerol-methanol-CO_2) systems are measured (compare Figure 1) and the influence of the molecular mass on the phase equilibrium is analyzed. Methanol is used as a co-solvent, in order to keep the viscosity sufficiently low. Unlike usual polymer phase equilibria, these systems show a distinct dependence of the bubble point pressure on the molar mass. These molecular structures still require adequate models to describe, extrapolate and predict their thermodynamic behavior. An attempt to model these systems was done by using the PC-SAFT EOS for branched molecules, which has been presented earlier [4]. Furthermore, the phase envelopes of the ternary system linear Polyglycerol (Mw=2800)-methanol-CO_2 is determined to evaluate the influence of the branched structure on the phase behavior.

References

[1] M. Seiler, Fluid Phase Equilib. 241 (2006) 155-174.

[2] J. Rolker et al., Ind. Eng. Chem. Res. 46 (2007) 6572-6583.

[3] J. Gross, G. Sadowski, Ind. Eng. Chem. Res. 40 (2001) 1244-1260.

[4] M.K. Kozlowska et al., J. Phys. Chem. B, 113 (2009) 1022-1029.

See more of this Session: Poster Session: Thermodynamics and Transport Properties
See more of this Group/Topical: Engineering Sciences and Fundamentals

Molecular Modeling of High-Energy Materials

Nandhini Sokkalingam, Department of Chemical Engineering, Detroit, MI

Jeffery J. Potoff, Chemical Enginering, Wayne State University, Detroit, MI

In this work force fields (molecular models) have been developed for six energetic compounds, 2,4-dinitroanisole (DNAN), N-methyl-p-nitroaniline (MNA), dinitropyrazole (DNP), nitrotriazolone (NTO), methyl-trinitroimidazole (MTNI) and TATB (trinitrotriaminobenzene) categorized by the army as potential insensitive munitions (IM) compounds. IM compounds exhibit low shock sensitivity and high thermal stability over traditional explosive compounds, such as TNT. The force fields are used to predict pharmacokinetic properties such as octanol-water partition coefficients and Henry's law constants along with other thermophysical properties such as vapor-liquid equilibria, vapor pressure, critical parameters and normal boiling points. The validation of the developed force fields is accomplished with the scarce experimental data available. The octanol-water partition coefficients and Henry's law constant predicted for DNAN and MNA are in good agreement with the experiment. For DNP, NTO, MTNI and TATB, materials for which limited experimental data are available, the predicted partition coefficients agree well with values calculated with group contribution methods.

See more of this Session: Poster Session: Thermodynamics and Transport Properties
See more of this Group/Topical: Engineering Sciences and Fundamentals

Screening Ionic Liquids for Process Applications Utilizing COSMO Calculations

Donna L. Gaya, Department of Chemical and Biomolecular Engineering, University of South Alabama, Mobile, AL

Kevin N. West, Department of Chemical and Biomolecular Engineering, University of South Alabama, Mobile, AL

James H. Davis Jr., Department of Chemistry, University of South Alabama, Mobile, AL

Ionic liquids are promising process solvents for a variety of industrial and commercial applications. Their vanishingly-low vapor pressures and the ability to tune their properties by altering the molecular structure make them highly attractive candidates to replace volatile organic solvents. Additionally, the ability to contact a gas or volatile liquid stream with an ionic liquid with no liquid loss makes them well suited for absorption and separations processes. However, determining the optimal ionic liquid for a particular application can be cumbersome, as the potential number of anion/cation combinations is exceedingly high, making experimental screening impractical.

Recently, the use of COnductor-like Screening MOdels (COSMO), such as COSMO-RS and COSMO-SAC, has been shown as an effective tool for screening solvents for particular applications. These models utilize quantum chemical calculations to predict interaction energies between surface segments on molecules, which are then used in combination with statistical mechanics to predict phase behavior. We have used COSMO models to screen a series of ionic liquids for high solubility for a variety of compounds relevant to medical applications and present the results of theses calculations, along with comparison to experimental data. Additionally, we compare the results from different implementations of the COSMO model in various software packages.

See more of this Session: Poster Session: Thermodynamics and Transport Properties
See more of this Group/Topical: Engineering Sciences and Fundamentals

Electrical Conductivity of Carbon Filled Liquid Crystal Polymer Composites for Fuel Cell Bipolar Plate Applications

Jason M. Keith, Department of Chemical Engineering, Michigan Technological University, Houghton, MI

Julia A. King, Department of Chemical Engineering, Michigan Technological University, Houghton, MI

Rodwick L. Barton Carter, Department of Chemical Engineering, Michigan Technological University, Houghton, MI

Electrically conductive resins are needed for use as fuel cell bipolar plates. Often, bipolar plates are manufactured with a single type of graphite powder in a thermosetting resin. In this research, varying amounts of three different electrically conductive carbon fillers (carbon black, synthetic graphite, and carbon fiber) were added to Vectra A950RX liquid crystal polymer in various concentrations. The maximum filler volume fraction was 12% for carbon black, 65% for synthetic graphite, and 45% for carbon fiber. The resulting single filler composites were tested for electrical resistivity (1/electrical conductivity).

In addition, the effects of single fillers and combinations of different carbon fillers were studied via a factorial design. The results indicated that for the composites containing only single fillers, synthetic graphite, followed by carbon fiber, cause a statistically significant decrease in composite electrical resistivity.

Furthermore, three models were developed for the single filler composites: Mamunya, Additive, and Global Effective Media (GEM) models. The model results suggest that the Mamunya and Additive models yield better correlations with the single filler experimental data. However, the GEM model can also be applied to predict the electrical conductivity of composites with multiple fillers. This was done for carbon fiber and synthetic graphite fillers with excellent agreement between model predictions and the experimental data.

See more of this Session: Poster Session: Thermodynamics and Transport Properties
See more of this Group/Topical: Engineering Sciences and Fundamentals

Liquefaction of Solid Southern Pine Wood with Water, Hydrogen, and Catalysts at High Temperature and Pressure

Kyung C. Kwon, Chemical Engineering, Tuskegee University, Tuskegee, AL

Valerie M. Gordon, Chemical Engineering, Tuskegee University, Tuskegee, AL

Nader Vahdat, Chemical Engineering, Tuskegee University, Tuskegee, AL

Tamara Floyd, Chemical Engineering, Tuskegee University, Tuskegee, AL

Howard Mayfield, Afrl/rxql, Tyndall Air Force Base, Panama, FL

Solid biomass will be liquefied with solvents, hydrogen-nitrogen mixture, and catalysts, using a batch reactor at high temperature and pressure. Liquefied biomass will be upgraded with hydrogen -nitrogen mixture and formulated catalysts or commercial catalysts readily available to petroleum refineries. The research will involve investigating effectiveness of solvents, catalysts, and hydrogen-nitrogen mixture for the liquefaction of solid biomass. Efficiency of various solid biomass feedstocks for liquefied biomass production will be evaluated under various experimental liquefaction conditions. Water will be used as a main solvent, whereas hydrogen donor solvents such as 1-methyl-naphthaline, tetralin and isopropanol in small amounts, homogeneous catalysts, and other heterogeneous catalysts will be added to water-biomass slurries. Water-insoluble liquefied biomass products will be analyzed by measuring their oxygen contents and viscosity values. The liquefied biomass obtained from the liquefaction process will be deoxygenated, isomerized, and hydro-cracked with hydrogen-nitrogen mixture and catalysts such as in-house formulated catalysts and commercial catalysts to reduce their oxygen contents, lower their viscosity values, and decrease average molecular weights.

See more of this Session: Biomass and Biorenewables Processing Under Pressure
See more of this Group/Topical: Engineering Sciences and Fundamentals

Effects of Varying Water-to-Glycerin Molar Ratio for Supercritical Water Reformation of Glycerin

Michael S. Stever, Jason W. Picou, Jared S. Bouquet, and Sunggyu Lee,
Missouri University of Science and Technology, Rolla, MO 65409
and
Tae-Hoon Lim and Byung Gwon Lee
Korea Institute of Science and Technology, Seoul, KOREA

Abstract

The production of biodiesel has increased in recent years in response to rising demands for renewable alternative energy and the use of green-fuel technologies. A common byproduct generated in the industrial manufacture of biodiesel is glycerin. While glycerin has beneficial properties for a variety of end uses, its over-abundance on the market place has rendered a new challenge for finding other economically viable and sizable end-uses for glycerin. This paper addresses a novel use of glycerin as a starting material for hydrogen generation via non-catalytic reformation using supercritical water. Additionally, effects on the water-gas shift reaction are considered. An experimental feasibility study of the novel conversion process was conducted on a 0.4 L tubular reactor constructed of Haynes® Alloy 230. The examined variables for the novel reaction process involved a water-to-glycerin molar ratio ranging from 3:1 to 24:1, a reactor temperature between 868 to 973 K, and space time ranging from 98 to 157 seconds. The effects of the reaction variables, in particular the water-to-glycerin molar feed ratio, on the gaseous hydrogen production, as well as on the overall process chemistry, were mechanistically and kinetically elucidated.

Background

In reformation of glycerin, as is the case with all hydrocarbons, numerous chemical reactions other than the reformation reaction take place, both simultaneously and consecutively, including dehydrogenation, pyrolytic decomposition, methanation, polymerization, and the water-gas shift reaction.[1, 2]

Reformation: $C_3H_8O_3 \rightarrow 3CO + 4H_2$

Dehydrogenation: $C_aH_b \rightarrow C_aH_{b-2} + H_2$

Pyrolytic Decomposition: $C_aH_b \rightarrow C_xH_y + C_zH_w$

Polymerization/Cyclization: $C_2H_4 \rightarrow$ coke

Methanation: $CO + 3H_2 \rightarrow CH_4 + H_2O$

Water-Gas Shift: $CO + H_2O \rightarrow CO_2 + H_2$

The forward water gas shift reaction is one of great significance for environmental and energy technology applications, since the water-gas shift reaction further converts water and carbon monoxide into highly desirable product gas, hydrogen. The forward reformation and dehydrogenation reactions are also desirable for hydrogen production, while methanation is not. Motivation for this research is largely driven by a need to produce hydrogen from abundantly available renewable resources. Hydrogen produced from reforming of commercial waste or byproduct allows an undesirable or less desirable product to be converted into a clean alternative fuel that has a variety of current and futuristic applications in industry.

Experimental Section

Apparatus

The supercritical water reformation experiments were performed using a custom-designed experimental process system, a schematic process flow diagram of which is shown in Figure 1. The system is comprised of five main components: reactant feed system, Haynes® Alloy 230 reactor, radiant reactor heaters, integrated heat exchanger, product sample collection system, and data acquisition and control system. The reactant feed system consists of a high-pressure reciprocating micro-metering pump and a balance for the purpose of detailed mass balances. A four-zone heater was used to provide heat to the reactor for isothermal conditions. In addition, sampling ports on the system allow for collection of liquid and gaseous products at ambient conditions.[3, 4, 5]

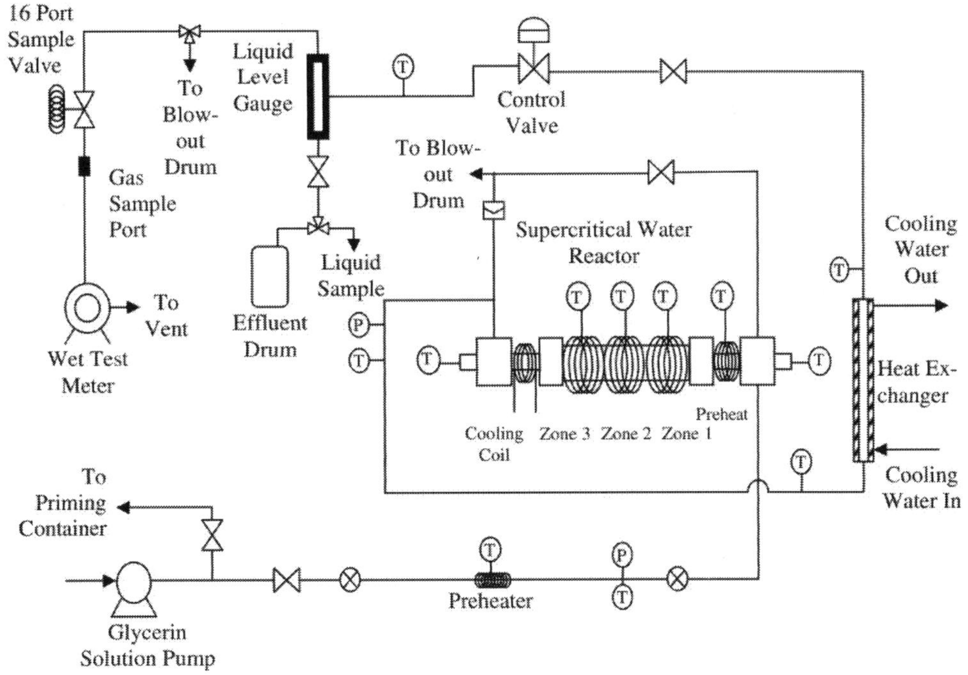

Figure 1: *Process flow diagram of supercritical water reformation unit at MS&T.*

Procedure

The reformation experiments were conducted by the following procedure: system start-up to operating conditions, sample and data collection, and system shut-down. System start-up involved bringing the internal temperature and pressure of the supercritical water reactor (SWR) to desired operating conditions. Once the operating temperature was reached water was fed into the system at the desired flow rate and the temperature and pressure were allowed to stabilize before glycerin solutions were fed into the reactor. The experimental run was started after the temperature, pressure, and reactant flow rates were stabilized. The data acquisition and control system based on National Instruments' Labview® system recorded inlet and outlet pressures, and temperatures at various locations within the system. As experiments progressed, liquid and gaseous samples were taken for analysis. In addition, the mass flow-rate of premixed glycerin solution fed into the system was recorded. When an experimental run concluded, the reactor was cleaned by feeding air. The air reacted with residuals in the reactor to produce syngas. Air was fed until gas samples confirmed product gas was no longer being formed. Then, operating conditions were changed for the next run and the system allowed to stabilize. When experiments were finished the system was shut-down and the reactor was depressurized and allowed to cool.[4, 5]

Results and Discussion

A total of seven experiments were conducted to determine the effects of varying the water-to-glycerin (W/G) molar ratio on product gas yields of hydrogen, carbon monoxide and carbon dioxide. For this work, product gas yields were defined as moles of gas produced per mole of glycerin fed into the reactor. Temperatures were maintained between 868 and 874 K for one set of experiments and between 971 and 973 K for the other set of experiments. Space times were kept between 98 and 101 seconds and pressure was approximately 24.1 MPa. The results are summarized in Table 1.

Table 1: Effects of varying water-to-glycerin (W/G) molar ratio on product gas yields for temperatures at 873 K and 973 K, space time at 100 s, and pressure at 24.1 MPa

Run #	T/K	space time/s	P/MPa	W/G molar ratio	Gas Yield (mol gas/mol glycerin fed)		
					H_2	CO	CO_2
WG-29	868	101	24.1	3.0	0.37	1.18	0.15
WG-20	874	100	24.1	13.5	2.82	0.32	1.58
WG-19	874	98	24.0	24.0	3.43	0.24	1.82
WG-27	971	102	24.1	3.0	1.17	0.77	0.77
WG-22	973	100	24.1	13.5	2.11	0.24	1.50
WG-36	971	100	24.1	13.5	2.06	0.21	1.54
WG-17	971	102	24.0	24.0	3.23	0.24	1.77

Figure 2 is gaseous product yield as a function of water-to-glycerin ratio and shows that the hydrogen yield increases with increasing water-to-glycerin ratio. This indicates that more dilute solutions will produce more moles of hydrogen per mole of glycerin fed. A greater increase in the hydrogen yield was observed at 873 K experiments in comparison to 973 K experiments. In addition, hydrogen yield was greater for the 873 K experiments versus the 973 K experiments at similar water-to-glycerin feed ratios except the lowest ratio of 3. This finding is quite different from that for other nonoxygenated aliphatic hydrocarbons, for which higher temperatures always resulted in higher yields of hydrogen for the temperature ranges investigated. Also, an increase in hydrogen and carbon dioxide gas yields, along with a decrease in carbon monoxide yields suggests that the water-gas shift reaction is more favored at dilute water-to-glycerin feed ratios.

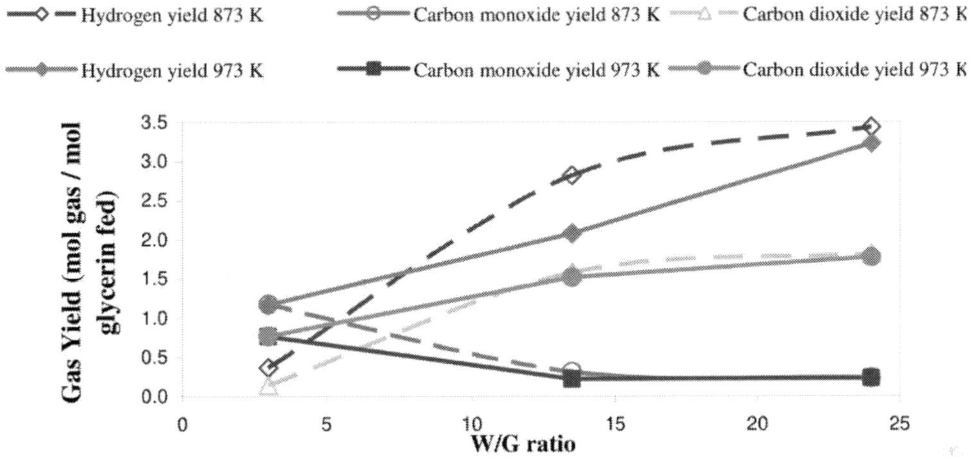

Figure 2: *Product gas yields with varying water-to-glycerin molar ratios at space time 100 seconds.*

A total of 8 different experiments were conducted to determine the effects of varying the water-to-glycerin molar ratio and temperature at a longer space time upon product gas yields. Temperatures were held between 870 and 873 K for one set of experiments and 965 and 969 K for another set. Space time was maintained between 143 and 157 seconds, and pressure was approximately 24.1 MPa. The results are summarized in Table 2.

Table 2: Effects of varying water-to-glycerin molar ratio on product gas yields for temperatures at 873 K and 973 K, space time at 150 s, and pressure at 24.1 MPa

Run #	T/K	space time/s	P/MPa	W/G molar ratio	Gas Yield (mol gas/mol glycerin fed) H_2	CO	CO_2
WG-28	870	157	24.2	3.0	0.44	1.21	0.22
WG-21	873	143	24.0	13.5	2.58	0.26	1.57
WG-18	871	145	24.1	24.0	3.10	0.17	1.75
WG-26	968	151	24.0	3.0	1.14	0.80	0.80
WG-23	969	150	24.0	13.5	1.97	0.21	1.49
WG-35	968	147	24.1	13.5	2.07	0.19	1.57
WG-16	969	147	23.9	24.0	3.28	0.25	1.74
WG-37	965	145	24.2	24.0	2.95	0.19	1.76

Figure 3 exhibits similar trends in gas yields as may be seen in Figure 2. However, the results were not as pronounced as in Figure 2. The CO and CO_2 gas yields were similar, whereas the hydrogen yield at 24 water-to-glycerin molar ratio for the 873 K experiments was almost 10 percent higher for the experiments conducted at 100 seconds than those at 150 seconds. This suggests that longer reactor space times beyond the optimal level do not enhance the overall hydrogen yield from the reformation process, via a number of possible reactions involving hydrogen as a reactant.

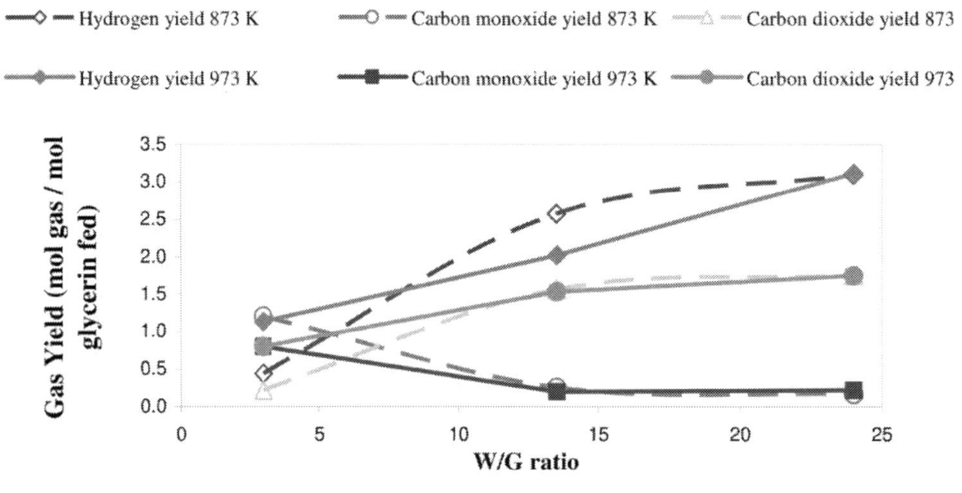

Figure 3: *Product gas yield with varying water-to-glycerin molar ratios at space time 150 seconds.*

The conversion of carbon and hydrogen fed into the reactor as glycerin into gaseous products collected at ambient conditions for all fifteen experiments for this work are summarized in Table 3.

Table 3: Effects of varying water-to-glycerin molar ratio, space time, and temperature on carbon and hydrogen conversion at 24.1 MPa

Run #	T/K	space time/s	P/MPa	W/G molar ratio	Carbon conversion	Hydrogen conversion
WG-29	868	101	24.1	3.0	69.8%	5.3%
WG-20	874	100	24.1	13.5	97.5%	40.6%
WG-19	874	98	24.0	24.0	100.0%	49.4%
WG-27	971	102	24.1	3.0	88.7%	16.9%
WG-22	973	100	24.1	13.5	97.4%	30.4%
WG-36	971	100	24.1	13.5	99.8%	29.6%
WG-17	971	102	24.0	24.0	99.9%	46.4%
WG-28	870	157	24.2	3.0	76.8%	6.4%
WG-21	873	143	24.0	13.5	95.4%	37.1%
WG-18	871	145	24.1	24.0	98.2%	44.6%
WG-26	968	151	24.0	3.0	94.3%	16.3%
WG-23	969	150	24.0	13.5	96.4%	28.3%
WG-35	968	147	24.1	13.5	101.5%	29.8%
WG-16	969	147	23.9	24.0	98.9%	47.2%
WG-37	965	145	24.2	24.0	100.3%	42.5%

As shown in Figure 4, higher water-to-glycerin ratios at higher temperature achieves higher carbon gasification percentage in comparison to lower temperature glycerin-rich runs; however, at approximately 15 water-to-glycerin molar ratio the increase in temperature does not have a significant impact on carbon gasification. In addition, the longer space times do not appear to enhance carbon gasification.

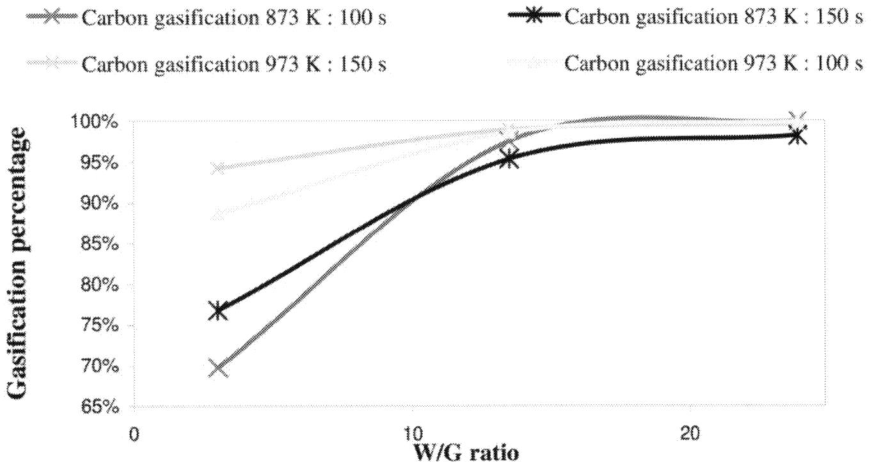

Figure 4: *Carbon gasification percentages with varying water-to-glycerin molar ratios while varying temperature and space time.*

Figure 5 shows the variation in hydrogen gasification with varying water-to-glycerin molar ratios. As with the carbon gasification, higher temperatures at longer space times do not appear to enhance the overall hydrogen gasification process. At approximately 8 water-to-glycerin molar the lower temperatures resulted in higher hydrogen gasification yields. At 24 water-to-glycerin molar ratios, the lower temperature and lower space time experiments achieved the highest amount of hydrogen gasification. The longer space times in the reactor may favor hydrogen consuming reactions such as methanation and hydrogenation of unsaturated hydrocarbons such as ethylene. This finding warrants further in-depth mechanistic study.

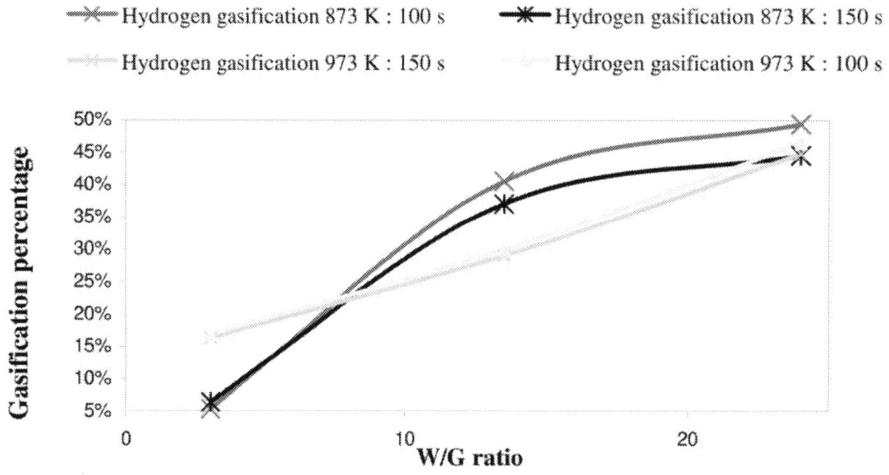

Figure 5: *Hydrogen gasification percentages with varying water-to-glycerin molar ratios while varying temperature and space time.*

Conclusions

Seven experiments were conducted at temperatures of 873 and 973 K and space time of 100 seconds and eight experiments were conducted at 873 and 973 K and space time of 150 seconds to determine the effects of varying water-to-glycerin molar ratios on hydrogen, carbon monoxide, and carbon dioxide product gas yields. It was found that increasing the water-to-glycerin molar ratio resulted in an increase in hydrogen and carbon dioxide formation. This is indicative that dilute solutions to a certain extent will produce a greater amount of hydrogen at the evaluated temperatures and space times. It is possible that the reformation reaction is promoted in a large excess condition of water and methanation is unfavored with more dilute solutions resulting in higher hydrogen yields. In addition, it is evident that the forward-water gas shift reaction is favored at higher water-to-glycerin molar ratios and is actively taking place at the conditions studied.

All fifteen experiments were also analyzed to determine the effects of varying water-to-glycerin molar ratios on carbon and hydrogen conversion. It was found that increasing water-to-glycerin molar ratio resulted in higher carbon and hydrogen conversions. This suggests that more glycerin reacted at higher water-to-glycerin molar ratios to form greater amounts of carbon and hydrogen product gases. It was also found that temperatures greater than 873 K and space times greater than 100 seconds did not increase carbon or hydrogen conversion at larger water-to-glycerin molar ratios.

References

1. Byrd, A. J.; Pant, K. K.; Gupta, R. B. Hydrogen production from glycerol by reforming in supercritical water over Ru/Al_2O_3 catalyst. *Fuel* **2008**, 87, 2956-2960.
2. Kruse, A. Hydrothermal biomass gasification. *J. Supercrit. Fluids* **2009**, 47, 391-399.
3. Lee, S.; Lanterman, H.B.; Wenzel, J.E.; Edwards, N.; Adams, A.; Wootton, J.; Garcia, A. Noncatalytic direct reformation of JP-8 fuel in supercritical water. Preprints-American Chemical Society, Division of Petroleum Chemistry 2006, 51, 521-523.
4. Wenzel, J.E.; Niemoeller, A.; Stever, M.S. Kinetics of reforming ethanol into hydrogen in a supercritical water medium. Presented at 2007 AICHE Annual Meeting, Salt Lake City, 2007.
5. Wenzel, J.E.; Picou, J.; Lee, S. Control of ethanol dehydration in the supercritical water reforming of ethanol into hydrogen. Presented at 2008 AICHE Annual Meeting, New Orleans, 2008.

Roles of Subcritical Water in the Conversion of Switchgrass

*Leming Cheng, Shirley Liu, and X. Philip Ye**
Department of Biosystems Engineering and Soil Science, The University of Tennessee
2506 E. J. Chapman Dr., Knoxville, TN 37996-4514
** Corresponding author: (865)974-7129, xye2@utk.edu*

1. Introduction

Biomass feedstock is a promising alternative resource to fossil resources, which can provide clean fuel and, as a bonus, value-added chemical products with nearly neutral CO_2 emission; this is especially true for agricultural waste biomass or energy crops that can be produced economically and efficiently. However, up to now, biomass has not been utilized as efficiently as it could be. One of the main obstacles is that the water content of most waste biomass is much higher than that of fossil fuels. For example in a corn-ethanol energy consumption analysis, it is reported that [1] approximately 50% of the total energy requirements are consumed by distillation and drying steps. These two inefficient steps are primarily a result of the energy requirements for volatilizing water. In other traditional thermochemical processes, such as pyrolysis, gasification, or combustion conversion routes, water is typically separated thermally, by vaporization or similar means going through phase change from liquid to gas.

Recently, water in its sub- and supercritical states has increasingly attracted attention as a reaction medium for the efficient production of clean fuels and valuable chemicals from such diverse raw materials as fossil fuels, solid waste, and biomass [2-3]. Subcritical water defines a state in the temperature range of 100-374°C and pressure high enough to ensure that water is in liquid state. Because the reaction medium is water, the wet biomass can be directly used as feedstock and the water contained in biomass can be removed in its liquid phase by a non-evaporative dewatering process under higher pressures [4]. In this way, about 40% of latent heat can be saved by removing the inert moisture as liquid water [5]. Furthermore, water in either the supercritical or subcritical state possesses unique solvent and transport properties compared to those of water in the liquid or steam state under ambient pressures, having liquid-like densities and gas-like diffusivities [6]. These properties make subcritical water a very promising reaction medium for the conversion of biomass. Thus, lignocellulosic materials will be readily decomposed into chemicals by hydrolysis and further reactions. The produced products can then be dissolved and dispersed in the water phase, and char-forming reactions can be suppressed.

However, the majority of previous studies have focused on model compounds of heterogeneous biomass. Little has been done thus far on the detailed reaction characteristics of real lignocellulosic biomass because of the very complex nature of these materials. The overall objective of this study is to gain a better understanding of the characteristics of the rapid decomposition of switchgrass as well as an understanding of the formation of value-added chemical products when the reaction medium is mild subcritical water rapidly rising to required temperatures. Because fast pyrolysis is also regarded as a promising way of converting biomass into bio-crude oil, an effort is also made to better understand the different reaction pathways of switchgrass converted in subcritical water vs. fast pyrolysis by comparing products obtained from same switchgrass sample in our lab. The roles of subcritical water in switchgrass conversion are discussed in the context of the comparison.

2. Experimental

2.1 Biomass Samples

In this study, one of the dominant tall-grass species in Central North America and dedicated energy crops, switchgrass *(P. virgatum)*, was selected as a raw material because of its hardiness in poor soil and climate conditions, low fertilization and herbicide requirements, and rapid growth cycle. The chemical composition analysis of the switchgrass can be found in a previous publication [7].

2.2 Experimental Apparatus & Product Separation and Analysis

Conversion of biomass in subcritical water was conducted in a ~11-ml batch stainless steel reactor (OD 16.0 mm, ID 11.4 mm, Length 105 mm, Swagelok Caps). In each run, a pre-calculated amount of water was fed with 1g of dry biomass sample into a reaction vessel made of stainless steel (SS316). Detailed information about the reaction process and product separation can be found in [7].

Analysis of the gas products, including CO_2, CO, H_2 and CH_4, was performed on a gas chromatography-thermal conductivity detection (GC-TCD) with argon as the carrier gas. The Water Soluble (WS) components were measured by HPLC (Waters Delta 4000 LC system) equipped with a BIO-RAD column (Aminex HPX-87P) and a refractive index (RI) detector. Calibration was achieved using cellobiose, glucose, xylose, galactose, arabinose, mannose, furfural and 5-HMF standards (all from ACROS company) to obtain the concentration of each product. The Methanol Soluble (MS) analysis was conducted on a GC-MS (Shimadzu QP2010S) with a Restek Rtx-1701 capillary column (60m*0.25mm*0.25μm). Infrared spectra were measured for the raw material and Methanol Insoluble (MI) samples to obtain information about the main organic components by an Attenuated Total Reflection Fourier Transform Infrared (ATR-FTIR) spectrometer (Varian 3100 FT-IR with Varian 400 UMA) equipped with a liquid nitrogen-cooled mercury cadmium telluride (MCT) detector.

3. Results and Discussion

3.1 Decomposition Characteristics

Table 1 shows the yields of gas, WS, MS, and MI as a function of reaction time at different conversion temperatures. The conversion of switchgrass at 1s is higher than 40 wt% at 250 °C, implying that the decomposition reaction rate of switchgrass in subcritical water is very rapid; the yield of WS products is higher than 37 wt% at the reaction time of 1s, which decreases slightly with further prolongation of the reaction time. The formation of MS and gas products is favored with an increase of reaction time at 250 °C. Significant increases in both the conversion and the gas yield at 300 °C are observed, compared with those at 250 °C; the MS yield decreases from 30.0 wt% to 20.5 wt% when reaction time increases from 1s to 15s, then levels off with further increase of reaction time, indicating the secondary decomposition of WS products into MS fractions. These results indicate that the non-catalytic decomposition process in subcritical water can be divided into the WS and MS formation, consequent secondary decomposition, and further fragmentation of those products. The WS and MS products can be further decomposed into gas products, especially at higher temperatures.

Table 1. Conversion of switchgrass in subcritical water vs. fast pyrolysis

Temperature (°C)	Residence time (s)	Product distribution (wt% dry biomass basis)					
		Residue	MS	WS	MS+WS	Gas and loss[b]	Conversion
250	1	58.16	9.16	37.23	46.39	ND[c]	41.84
	15	54.84	8.74	34.31	43.04	2.11	45.16
	30	47.37	10.84	34.73	45.58	7.05	52.63
	45	44.16	13.05	29.59	42.65	13.20	55.84
	60	44.21	13.68	24.95	38.63	17.16	55.79
300	1	35.37	30.00	30.06	60.06	4.57	64.63
	15	29.37	20.53	26.64	47.16	23.47	70.63
	30	23.72	20.21	26.80	47.01	29.27	76.28
	45	21.11	22.84	24.60	47.44	31.46	78.89
	60	20.63	20.11	23.66	43.77	35.60	79.37
350	1	17.16	22.74	26.90	49.64	33.20	82.84
	15	12.74	29.47	18.90	48.37	38.89	87.26
	30	10.53	29.47	18.62	48.10	41.38	89.47
	45	10.74	28.84	18.90	47.75	41.52	89.26
	60	10.05	24.00	17.36	41.36	48.59	89.95
Fast pyrolysis [a]	2	21.2~41.1	-	-	12.9~43.6	24.0~47.2	58.9~78.8

a: The fast pyrolysis data were obtained using a fluidized reactor from the same batch of switchgrass sample at 450~550°C [8].

b: The gaseous product yields plus possible loss were obtained via mass balance.

c: ND: Not detected.

The biomass conversion reaches ~90 wt% when the reaction time is longer than 15s at 350 °C, demonstrating that most of the lignocellulosic structures of biomass have been broken down at a reaction temperature significantly lower than that which occurs in the fast pyrolysis process (where switchgrass conversion achieves only 58.9~78.8 wt% at temperature range of 450~550°C) [8], as shown in Table 1. The liquid yield of SCW (WS+MS) varies in the range of 38.63~60.06 wt%, which is higher than that of fast pyrolysis reaction. It is also found that at each subcritical conversion temperature, the liquid yield decreased with the increase of reaction time because of fragmentation of liquid products. The fact that the conversion rate of biomass in subcritical water was very rapid at a relatively lower temperature compared with that observed in the fast pyrolysis implies that the subcritical water plays a very important role in promoting switchgrass conversion by hydrolysis reaction, because 1) the change of cellulose structure from crystalline to amorphous in subcritical water might leads rapid decomposition rate of cellulose [1]; 2) high dielectric constant of subcritical water might act as acid or base catalysts for hydrolysis reaction through ether or ester linkages in lignocellulosic materials; and 3) char-forming reactions are suppressed since the products can then be dissolved and dispersed in the subcritical water phase.

3.2 WS and MS Products

In SCW conversion, main products in the WS fraction detected by HPLC are oligosaccharides, cellobiose, glucose, xylose, 5-HMF and furfural, as shown in Table 2. Oligosaccharides can still be significantly detected at shorter reaction times; with the increase of reaction time, the content of oligosaccharides tend to decrease while those of furfural and 5-HMF increase. The formation of typical hydrolysis products and their dehydration derivatives indicates that hydrolysis of cellulose and hemicellulose is the major decomposition pathway for cellulose and hemicellulose. With the increase of reaction temperature, the swelling or dissolving effect of subcritical water plays an important role in breaking down cellulose structures through hydrogen bonds and ether bonds. However, the rate of secondary decomposition of glucose, xylose, and oligosaccharides also significantly increases with the increase of temperature. Thus, chemical products by means of decarboxylation, decarbonylation, and dehydration reactions will be formed by the further decomposition of monosaccharides and oligosaccharides.

Table 2. Common liquid products detected from SCW reaction vs. fast pyrolysis

Category	**Origin**	**SCW liquid product (WS+MS)**	**Fast pyrolysis crude oil** [8] [a]
Saccharides and dehydration derivatives	Cellulose and hemicellulose	Oligosaccharides, cellobiose, glucose, xylose, 5-HMF furfural	Levoglucosan[b]; Pentopyranose; Furfural [b] 2-(Hydroxymethyl)furan Hydroxyacetaldehyde [b] Acetol
Phenol related derivatives	Lignin	3-ethylphenol, Ethylguaiacol; Guaiacol; 4-Ethyl-phenol; 2,4-Dimethoxyphenol; 3-Methyl-2-methoxyphenol 4-Methyl-2-methoxyphenol 4-Ethyl-2-methoxyphenol	Phenol [b]; Catechol [b] 2-Methyl-1,2-benzenediol 3-Methyl-phenol 4-(Hydroxymethyl)phenol 4-Ethyl-phenol 2-Methoxy-4-methyl-phenol 2,4-Dimethoxyphenol 2-Methoxy-4-vinylphenol 4-Ethyl-2-methoxy-Phenol 2-Methoxy-3-(2-propenyl)-phenol
Other main products	Fragmentation of Saccharides	Acetic acid, 1-Hydroxy-2-propanone	Acetic acid [b], Propanoic acid [b] 1-Hydroxy-2-propanone [b]

[a] The fast pyrolysis data were obtained using a fluidized reactor from the same batch of switchgrass sample at 450~550°C.

[b] Main products

Important chemicals identified in fast pyrolysis oil that are believed from cellulose and hemicellulose are also shown in Table 2. Those of considerable proportions are hydroxyacetaldehyde, furfural, levoglucosan, and their decomposition fragments such as acetic acid and acetol [9]. There are more small molecular products obtained in fast pyrolysis oil than those in subcritical water, indicating that the cleavage of C-C bonds plays a more prominent role in decomposition of cellulose structure in fast pyrolysis.

Figure 1 shows a typical GC-MS spectra of the MS fraction of switchgrass conversion under 350 °C, 20MPa, and 60s. The main compounds found in the MS fraction by GC-MS fall mainly in the following categories: sugar-derived compounds such as furfural and furans, lignin-derived compounds such as phenol-related compounds, and some fatty acid compounds. The latter is possibly produced from the hydrolysis of waxy ester polymers, e.g., cutin on the surface of switchgrass plant. The fact that some sugar decomposition compounds detected in the MS fraction corresponds to the decreasing yield of WS under this reaction condition implies that part of the WS products can be decomposed into MS at higher temperatures. It is noteworthy to point out that many lignin-derived compounds, which are typical lignin fragments via hydrolysis breakdown, such as 3-ethylphenol, ethylguaiacol, and guaiacol etc., are obviously identified using the NIST Mass Spectra database, implying that these lignin decomposition products can be well recovered in MS. Lignin has many phenolic structures as well as functional groups such as hydroxyl, carbonyl, and methoxy groups. The most important ether bond in lignin is that between one propyl side chain of a hydroxyphenylpropane unit and the hydroxyl group attached to a benzene ring of another hydroxyphenylpropane unit [10].

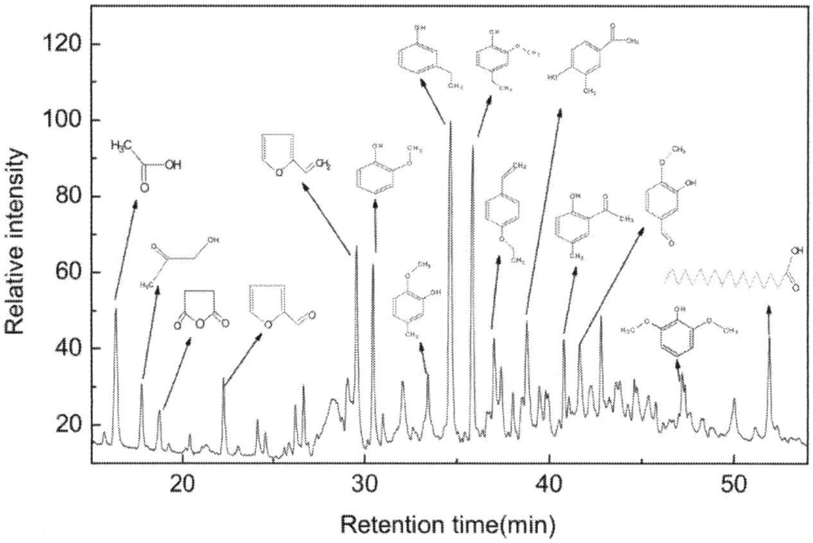

Figure 1. Typical GC-MS spectra of MS fraction of switchgrass conversion (350 °C, 20MPa, 60s)

Evidently in Figure 1, most of the phenolic chemicals have intact functional groups, such as 3-ethylphenol, ethylguaiacol, and guaiacol etc., while no significant phenol and catechol were detected even at 350 °C and 60s. On the contrary, for the phenol derivatives produced in fast pyrolysis (Table 2), relatively higher concentration of phenol, catechol, and trace amount of many phenol derivatives with different functional groups were detected. The chemistry behind this difference is that pyrolysis reaction can take place via the cleavage of ethers, the C-C or similar bond energy linkages, while hydrolysis reaction in SCW can only take place at ethers and ester linkages. Another possible reason is that fragmentation reaction of these phenolic chemicals is enhanced at higher reaction temperatures. This result indicates that lignin hydrolysis in subcritical water, rather than fragmentation reactions, played an important role during the decomposition of lignin structure in this rapid SCW conversion. Instead of forming complex liquid products and char due to the significant fragmentation and condensation reactions under high reaction temperatures and longer reaction time, this selective breakdown of the bonds in lignin structure in subcritical water under relatively mild conditions as compared with fast pyrolysis, will result in many useful, functional products of high values.

3.3 Gas Products

Table 3 shows the gas yields and compositions under SCW reaction (20MPa, 60s) vs. fast pyrolysis. The gas yield increases from 17.2 wt% at 250°C to 48.6 wt% at 350°C, indicating that higher temperatures favor the formation of gas products. The main content of the gas product is CO_2 and CO, and trace amounts of H_2 and CH_4 are also detected in subcritical water process. A slightly increasing trend of concentrations of H_2 and CH_4 is observed with the increase of temperature, indicating that the fragmentation of WS and MS is enhanced to some extent at the higher temperature. The contents of CO, H_2, and CH_4 are significantly lower than those generated in fast pyrolysis. The very low contents of H_2 and CH_4 produced in SCW compared with those obtained in fast pyrolysis might be the result of relatively lower extent of fragmentation, such as the intermediate products being broken down into simpler molecules or knocking off branches of the functional groups attached to aromatic rings. The higher CO_2 concentration and lower CO concentration in SCW conversion compared with those in fast pyrolysis might be the result of enhanced water-gas reaction in subcritical water. The CO_2-rich gas obtained under subcritical water conditions can be easily collected with minimal CO_2 separation cost in the future biorefinery, further reducing CO_2 emissions.

Table 3. Gas yields and compositions under SCW reaction (20MPa, 60s) vs. fast pyrolysis

Temperature (°C)	**Gas yield[a] (wt% dry biomass basis)**	**Gas composition (mol %) N_2 free basis**			
		H_2	CO	CH_4	CO_2
250	17.16	0.17	4.27	0.02	95.53
300	35.60	0.49	7.46	0.13	91.91
350	48.59	0.95	6.62	0.32	92.11
450[b]	24.0~47.2%	5.1	57.6	7.8	29.5

a: The loss is included in gas yield.
b: Typical gas composition in switchgrass fast pyrolysis [9].

3.4 Solid Product

No black char formation was observed in any of the MI residues in SCW conversion, even at the temperature of 350°C. In fast pyrolysis, char with a black color and high content of carbon will be produced. This difference may be attributed to the high rising rate of temperature, the rapid decomposition of the biomass, and the good solvating property of subcritical water. To further understand the properties of MI fraction produced in SCW conversion, ATR-FTIR analysis of the MI samples is conducted. The results of MI samples, in comparison with standard organosolv lignin (Sigma-Aldrich) and raw switchgrass material are presented in Figure 2. Figure 2A shows the overlaid spectra of raw switchgrass and MI products obtained at 250°C and 30s. The bands at 1,033cm^{-1}

and 1,056 cm^{-1} represent the C-O stretch for the $O-CH_3$ and C-OH of cellulose and hemicellulose structures in biomass, which are overlapped and combined in the raw switchgrass samples but can be clearly identified in the MI products. Typical hemicellulose and cellulose structure such as CH bending (1,373cm^{-1}), C-O-C asymmetry stretching (1,160cm^{-1}), C-O stretching (1,033 and 1,056cm^{-1}) and anomeric carbon vibration (900cm^{-1}) can also be detected. These results indicate that the compact matrix of lignocellulosic material is significantly decomposed at 250°C, obviously exposing the cellulose structure.

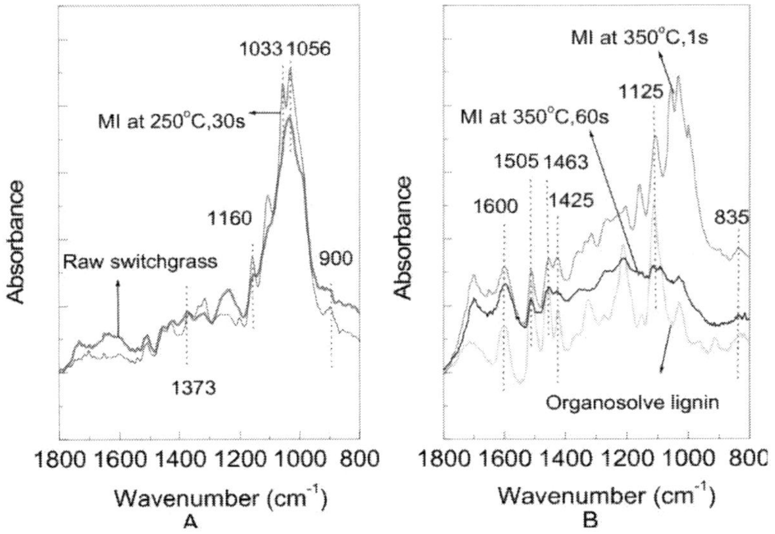

Figure 2. ATR-FTIR spectra of MI fractions, organosolv lignin, and raw switchgrass materials

Figure 2B compares the spectra of standard organosolv lignin and two MI products obtained at 350°C for 1s and 60s respectively. It can be clearly seen that aromatic rings were well represented in two MI samples and the standard lignin as peaks at 1,600 cm^{-1}, 1,505 cm^{-1}, 1,463 cm^{-1} and 1,425cm^{-1}, as well as peaks detected at 1,125 cm^{-1} and 835cm^{-1} (C-H in plane deformation and bending). All these peaks are of typical lignin structure [11-12], suggesting that the unconverted lignin fraction can be well presented by FTIR. It is observed that the main peaks in the spectrum of MI obtained at 350°C and 60s are similar to those of standard organosolv lignin, indicating that the hemicellulose and cellulose structure is nearly completely decomposed and the remaining solid product is mainly of lignin structure. This result confirms the unique solvating and transport properties of subcritical water, potentially opening a new way to efficiently utilize the different constituents in lignocellulosic biomass using subcritical water technology.

4. The roles of subcritical water in biomass conversion (a brief summary)

From the above comparisons of product yields and properties of different products produced in SCW vs. those in fast pyrolysis, a better understanding of the different chemical pathways of biomass conversion can be concluded, as outlined in Figure 3. Generally, high conversion and high liquid product yield can be obtained at a temperature lower than that used in fast pyrolysis of switchgrass.

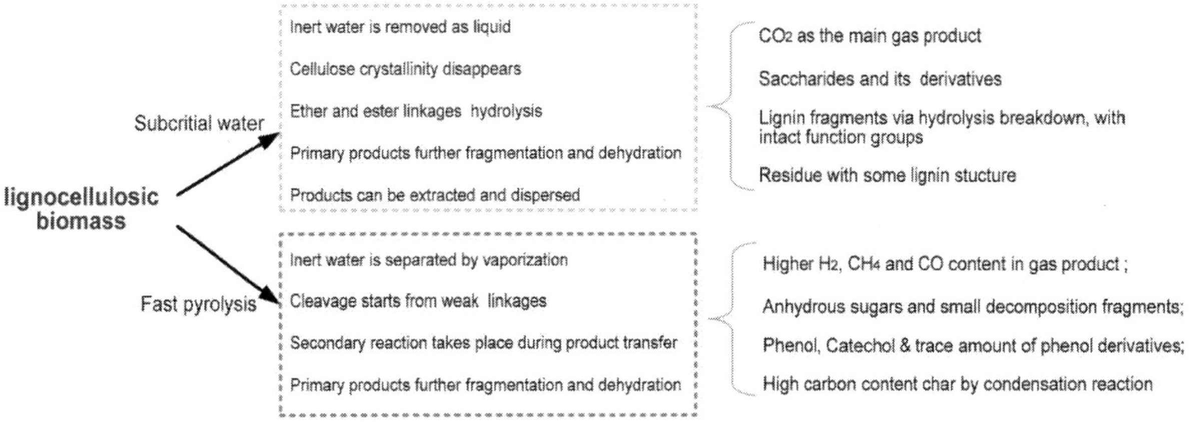

Figure 3. Decomposition profile of switchgrass in subcritical water vs. that in fast pyrolysis

The subcritical water plays important roles in promoting switchgrass conversion, which can be summarized as follows: 1) efficient removal of inert moisture in wet biomass as liquid water; 2) facilitation of the conversion of crystalline cellulose to amorphous form by reducing hydrogen bonding and the hydrolysis of β(1→4)-glycosidic bonds in cellulose [1]; 3) subcritical water itself can act as acid and basic catalyst for the hydrolysis reaction because of lower dielectric constant, leading to the formation of sugars and its dehydration derivatives; 4) char-forming reactions are suppressed because the products can then be extracted and dispersed in the subcritical water phase, indicating that high conversion of lignocellulosic biomass can be achieved; 5) water-gas shift reaction taking place in SCW converts CO to H_2 and CO_2; 6) the lignin hydrolysis, rather than fragmentation reactions, plays a prominent role during the decomposition of lignin structure, leading to the formation of many chemicals with intact phenolic functional groups.

ACKNOWLEDGMENTS

The project was financially supported by the Tennessee Agricultural Experiment Station (HATCH project No. TEN00325).

References

[1] Peterson AA, Vogel F, Lachance RP, Fröling MJ and Antal JM (2008), Thermochemical biofuel production in hydrothermal media: A review of sub- and supercritical water technologies. Energy & Environmental Science, 1, pp33-65.
[2] Cheng L., Zhang R. and Bi J (2004), Pyrolysis of a low rank coal in sub- and supercritical water, Fuel Process Technol., 85, pp921-932.
[3] Watanabe M., Hirakoso H., Sawamoto S., Adschiri T. and Arai K. (1998), Polyethylene conversion in supercritical water, J. Supercritical Fluids, 13, pp 247–252.
[4] Nakagawa H., Namba A., Böhlmann M. and Miura K. (2004), Hydrothermal dewatering of brown coal and

catalytic hydrothermal gasification of the organic compounds dissolving in the water using a novel Ni/carbon catalyst, Fuel 83, pp719-725.
[5] Cheng L and Ye X P (2009), Recent progress in converting biomass to biofuels and renewable chemicals in sub- or supercritical water. Biofuels (Accepted)
[6] Akiya N. and Savage P.E. (2002), Roles of water for chemical reactions in high-temperature water, Chem. Rev., 102, pp2725-2750.
[7] Cheng L, Ye XP, He R and Liu S (2009), Investigation of rapid conversion of switchgrass in subcritical water. Fuel Processing Technology ,90(2), pp 301-311.
[8] He R., Ye X.P., English B.C. and Satrio J. A. (2009), Influence of pyrolysis conditions on switchgrass bio-oil yield and physicochemical properties, Bioresource Technology ,100, pp 5305–5311
[9] Boateng A.A., Daugaard D. E., Goldberg N. M. and Hicks K. B. (2007), Bench-Scale fluidized-bed pyrolysis of switchgrass for bio-oil production, Ind. Eng. Chem. Res., 46, pp1891-1897.
[10] Saisu M., Sato T., Watanabe M., Adschiri T. and Arai K. (2003), Conversion of lignin with supercritical water-phenol mixtures, Energy Fuels, 17, pp 922-928.
[11] Qian Y., Zuo C., Tan H. and He J. (2007), Structural analysis of bio-oils from sub-and supercritical water liquefaction of woody biomass, Energy, 32, pp 196-202.
[12] Ehara K., Takada D. and Saka S. (2005), GC-MS and IR spectroscopic analyses of the lignin-derived products from softwood and hardwood treated in supercritical water, J. of Wood Science, 51, pp 256-261.

Measurement of Diffusion in Dimethyl Ether - Water Systems for Extraction of Aqueous Biomass

Steve J. Tallon, Industrial Research Ltd, Lower Hutt, New Zealand

O. J. Catchpole, Industrial Research Ltd, Lower Hutt, New Zealand

K. Fenton, Industrial Research Ltd, Lower Hutt, New Zealand

This paper describes processes for extraction of lipids from aqueous feed streams using the near critical solvent dimethyl ether (DME). DME has a moderate mutual solubility with water which enables lipids to be extracted directly from aqueous streams. Both neutral and polar lipids can be extracted in good yield from a range of feedstocks including marine, algal, plant, egg, dairy, and fermented feedstocks. Results are presented on measurements of the rate of diffusion of water in DME using the Taylor-Aris method. These measurements, combined with thermodynamic and equilibria data for the DME-water system enable modelling of the contacting dynamics in the process.

See more of this Session: Biomass and Biorenewables Processing Under Pressure
See more of this Group/Topical: Engineering Sciences and Fundamentals

Development of a Dynamic Method for Measuring Bio-Solute Solubilities in Subcritical Water

Keerthi Srinivas, Chemical Engineering, University of Arkansas, Fayettville, AR

Dongfang Zhang, Chemical Engineering, University of Arkansas, Fayettville, AR

Fernando Montanes, Instituto de Fermentaciones Industriales, CSIC, Madrid, Spain

Jerry W. King, Chemical Engineering, University of Arkansas, Fayettville, AR

The knowledge of the aqueous solubilities of bioactive compounds such as sugars and flavonoids as a function of temperature is significant for a wide range of applications in food, pharmaceutical and biofuels technology. A few studies have been conducted in the past to measure the solubility of sugars and flavonoid compounds in water. However, such studies were primarily conducted in a batch vessel at temperatures below the boiling point of water. We have attempted to measure the solubility of some sugars and flavonoids in water at subcritical conditions using a dynamic flow system. The apparatus, based on a modification of the system used by Miller and Hawthorne [1], allows water to flow through a saturation cell comprised of the selected solute and a dispersant, at a constant flow rate and at a particular temperature. The concentration of sugars in the outlet sampling vial is analyzed using techniques such as HPLC and HPSEC.

Using the above-mentioned apparatus, the aqueous solubilities of sugars were measured between temperatures 25 and 200C, and that of flavonoids were measured between 25 and 140C. The effect of temperature, pressure and flow rates on the aqueous solubilities of these bio-solutes was studied using this technique. It was found that the solubilities showed an exponential trend with an increase in temperature. There was close to a four-fold increase in the solubilities of the sugars in water above its boiling point while the flavonoids showed a two-fold increase. Typical sugar solubilities between 20 - 120C ranged from 470 - 2380 g/L for glucose and from 500 - 1337 g/L for maltose monohydrate. However, the flavonoids were comparatively less soluble in water with typical solubilities at the same temperatures ranging between 29 - 773 g/L for protocatechuic acid and 2*10^(-3) - 0.25 g/L for quercetin. Assuming the water activity to be unity, basic solution thermodynamic theory can be used to calculate the heat, enthropy and Gibbs free energy of solutions for these compounds in water. Such data can be used to examine the effect of water as a solvent for extraction and hydrolysis of such solutes from biomass substrates.

Pressure plays a minor role in the solubility of such solutes in subcritical water and was used as a mode to study and maintain an undisrupted flow of water through the flow solubility apparatus. Higher flow rates were maintained especially at higher temperatures to prevent the thermal degradation of the solutes in water. Due to the limited availability of solubility data in the literature and no data for some solutes, especially flavonoids, theoretical models such as UNIFAC and Hansen three-dimensional solubility parameter concept were used to examine the accuracy of such measurements. Such experimentally-measured solubility data along with its theoretical correlations can be used in optimizing the design parameters for extraction and/or reaction of such solutes from natural products.

[1] David J. Miller and Steven B. Hawthorne, Solubility of liquid organics of environmental interest in subcritical (hot/liquid) water from 298 K to 473 K, J.Chem.Eng.Data, 45 (2000), pp. 78-81.

See more of this Session: Biomass and Biorenewables Processing Under Pressure
See more of this Group/Topical: Engineering Sciences and Fundamentals

Processing of Unsaturated Fatty Acids in Supercritical Carbon Dioxide

Darrell L. Sparks, Chemical Engineering, Mississippi State University, Mississippi State University, MS

L. Antonio Estevez, Department of Chemical Engineering, University of Puerto Rico, Mayaguez, PR

Rafael Hernandez, Chemical Engineering, Mississippi State University, Mississippi State University, MS

Todd French, Chemical Engineering, Mississippi State University, Mississippi State University, MS

Currently, lipids derived from plants, animals, and microorganisms are being considered as promising renewable feedstocks to synthesize a wide range of fuels and chemicals. For example, many lipid sources contain unsaturated fatty acids, which can be oxidized to form a variety of products such as diacids and epoxides. These chemicals are valuable intermediates for the formulation of pharmaceuticals, herbicides, detergents, plasticizers, lubricants, paints, and other useful products. The use of supercritical fluids, such as carbon dioxide (SC-CO_2), in processing lipid components is being heavily researched; however, experimental solubility data primarily exists for lipids composed of higher molecular weight fatty acids such as oleic acid. In this research, the solubility of small-chain fatty acids in SC-CO_2 was determined experimentally and the applicability of common solubility models was evaluated. Additionally, this paper will show how supercritical carbon dioxide can be used to enhance fatty-acid oxidation with potassium permanganate.

See more of this Session: Biomass and Biorenewables Processing Under Pressure
See more of this Group/Topical: Engineering Sciences and Fundamentals

Lipid Processing Using Critical Fluids

Fred Joseph Eller III, New Crops & Processing Technology, NCAUR-ARS-USDA, Peoria, IL

A lipase (Novozyme 435) interesterification of soybean oil triacylglycerides with ethyl ferulate to produce feruloylated acylglycerols has been previously been described and patented. These feruloylated acylglycerols, termed SoyScreen™, are believed to have potential as natural sunscreens and as antioxidants for the cosmetic and food industry.

Critical fluid technology has been used in several ways to improve this lipid processing method. Liquid carbon dioxide (L-CO_2) has been used to purify the SoyScreen™ product from the reaction mixture. Factors such as fractionation column length, solvent to feed ratio (S:FR) and feed rate (FR) were studied. Both by-product fatty acid ethyl esters (FAMEs) and unreacted ethyl ferulate are efficiently separated from the crude reaction mixture using a L-CO_2 counter-current fractionation method to yield an essentially pure product. In addition, to purify the ethyl ferulate used in this reaction, a method utilizing supercritical carbon dioxide in conjunction with a silica column was also developed to provide nearly snow-white ethyl ferulate used in the interesterification reaction.

Partially deacylated high oleic sunflower oil (PDHOSO) has been determined to offer some advantages over soybean oil as a starting material to produce the feruloylated acylglycerols. The PDHOSO is produced by enzymatic propanolysis and the crude product contains a mixture of mono-, di-, and triacylglcerides as well as by-product fatty acid propyl esters (FAPEs). However, before the PDHOSO can subsequently be used to produce the final feruloylated acylglycerol product, the FAPEs must be removed from the crude reaction mixture. Subsequently, a continuous counter-current fractionation L-CO_2 method was developed to remove the by-product FAPEs from the propanolysis reaction mixture. Parameters such as S:FR and FR were examined to optimize the separation. This process yielded a glyceride product purity of over 98% as well as an extract containing over 97% FAPEs.

Critical fluid technology is being studied as a potential means to enhance the synthesis of SoyScreen™ by cycling CO_2 through a bed of Novozyme 435 containing triacylglycerides and ethyl ferulate. In this process, CO_2 removes FAMEs from the reaction mixture and precipitates them in a reduced pressure receiver before recycling the CO_2. In this manner, the by-product FAMEs are removed from the reaction mixture allowing the reaction to proceed further to the right than otherwise possible.

See more of this Session: Biomass and Biorenewables Processing Under Pressure
See more of this Group/Topical: Engineering Sciences and Fundamentals

Optimization of Hot Pressurized Solvent Extraction Systems for the Recovery of Bio-Active Solutes

Jerry W. King, Chemical Engineering, University of Arkansas, Fayettville, AR

Keerthi Srinivas, Chemical Engineering, University of Arkansas, Fayettville, AR

Dongfang Zhang, Chemical Engineering, University of Arkansas, Fayettville, AR

Jeana K. Monrad, Department of Food Science, University of Arkansas, Fayetteville, AR

Luke Howard, Department of Food Science, University of Arkansas, Fayetteville, AR

The use of hot pressurized solvents above their boiling point is well documented in the literature and applied for processing of agricultural and food substrates, conversion of biorenewable fuels, and material science. Studies in our research group have focused on the extraction of bio-active solutes from such substrates as grape marc, fruit berries, and tree bark; the primary target solutes being flavonoids exhibiting antioxidant properties. Pressurized hot solvent extraction above the boiling point of the solvent results in very rapid solute fluxes from the substrate due to enhanced solubility and accelerated mass transfer rates. To optimize the extraction process both modeling and experimental measurements have been made on flavonoid recoveries from grape marc using a Hansen three dimensional solubility parameter approach or semi-empirical equations for solute solubility in subcritical water. These approaches allow prediction of both the solute miscibility with the pressurized fluid as well as a semi-quantitative estimate of the magnitude of solubility enhancement as a function of temperature.

Studies have been performed using batch, semi-continuous batch, as well as accelerated solvent extraction (ASE) methods to study the recovery of anthocyanins from freeze-dried grape marc. These have included the environmentally-benign agents: supercritical carbon dioxide, subcritical water, and ethanol as a cosolvent. ASE experiments have been run using an increasing ethanol concentration at temperatures between 100- 140C by increasing the ethanol from 10-90 vol. %. Typically, as little as 10 vol. % ethanol, can recovery anthocyanins at approximately the 500 ìg/ 100 g dry weight level, thereby minimizing ethanol use for anthocyanin recovery. Separate ASE experiments have shown that acidified ethanol (pH = 2.3) can recover the target anthocyanins at a 95% level, while neat ethanol and water yield recoveries of 87 and 86%, respectively. Mineral acid - acidified water yielded only 45% anthocyanin recovery.

In an attempt to replace the mineral acid - based acidification for aqueous extractions, our group have conducted batch extractions using high pressure SC-CO2 acidification of water - which will acidify water to a pH under 3.0. Using SC-CO2 pressures up to 60 MPa we have achieved 52% recovery of the anthocyanins without resorting to mineral acid addition using acidified methanol as a baseline (total anthocyanin) extraction solvent. Since higher anthocyanin yields result from the addition of ethanol aside from its prophylactic effect on retarding degradation of the target solutes (1st order extraction kinetics), a combination of a low percentage ethanol combined with CO2-acidified water may be an optimal solvent medium for the extraction of this flavonoid class from grape marc using very short extraction times (1 minute) at 120C.

See more of this Session: Biomass and Biorenewables Processing Under Pressure
See more of this Group/Topical: Engineering Sciences and Fundamentals

Film Formation From Aqueous Suspensions of Polymer-Silica Nanocomposites

Milan J. Patel, Department of Chemical Engineering & Biotechnology, University of Cambridge, Cambridge, United Kingdom

Venkata R Gundabala, Department of Chemical Engineering and Biotechnology, Pembroke Street, Cambridge, UK, University of Cambridge, Cambridge, United Kingdom

Alexander F. Routh, Department of Chemical Engineering and Biotechnology, Pembroke Street, Cambridge, UK, University of Cambridge, Cambridge, United Kingdom

Polymer films find wide application in the coatings, adhesives, and cosmetic industries. These are often formed from a suspension of colloidal polymer particles that come into contact and coalesce as the solvent (either organic or aqueous) evaporates. After evaporation and coalescence, the film resembles a compressed packed bed of zero voidage although the particle surfaces are still identifiable. Polymer molecules subsequently diffuse across the once-distinct interparticle boundaries giving a continuous polymer film as a final product.

Higher performance polymer films may be obtained by using nanocomposite particles as building blocks for the films. These are either polymer-polymer or polymer-inorganic nanocomposites in type. Polymer-polymer nanocomposites are useful in that they combine the properties of widely different polymers to provide multi-purpose industrial materials. Polymer-inorganic nanocomposites often include a silica or clay particle within a polymer particle. The latter class shows good potential in providing tough, optically transparent, scratch-resistant coatings. And last but not least, the solvent employed may be aqueous or organic, with the use of the former demonstrating a clear environmental advantage.

In the current paper, our main focus is on polymer (latex)-inorganic (silica) nanocomposites. Almost all applications of these nanocomposites demand that the suspensions form a continuous, crack-free film under suitable conditions, and the contribution of this work is a process model predicting the stable environmental conditions (minimum temperature) for which a given suspension of latex-silica nanoparticles will form a crack-free film. This parameter is termed the Minimum Film Formation Temperature (MFFT). The most conveniently controlled process parameter when manufacturing the feedstock suspension is the volume fraction (i.e. loading) of silica within each nanocomposite particle. The approach of the process model is as follows.

Film formation has been studied from a modelling standpoint for over fifty years. The process is generally agreed to comprise three stages, namely excess solvent evaporation, particle coalescence and finally polymer interdiffusion to form a continuous film. The focus of this work is on the intermediate stage. Existing models relate to simple single polymer particles. Here we extend the modelling to include a hard component (silica) within the particles.

A driving force is required to induce particle coalescence and there are three potential candidates. The first is surface tension between the polymer and the aqueous solvent. When particles come into contact they coalesce relatively rapidly at the contact point and are resisted by the inherent stiffness of the particles. The mechanical constitutive behaviour of the polymer (latex) may be elastic, viscous, or some combination. In this model, the latex is assumed to be linear viscoelastic and incompressible.

When the characteristic solvent evaporation time is large, or the latex viscosity and/or shear modulus is small, the nanoparticles will coalesce into larger and larger groups prior to full evaporation of the solvent. An apparently continuous film is formed upon the substrate although at this stage the particle boundaries are still identifiable within the film. As this process is complete before the solvent evaporates, a rarefied layer of solvent remains on top of the film. This mode of film formation is termed wet sintering.

If the polymer is slightly more resistant to deformation, solvent will evaporate until the underlying particles are compressed into a contacting network or particulate skeleton, although the porosity is still well above zero at this time. At this point, a second potential driving force arises, namely capillary pressure at the air-polymer-water interface at the top of the particulate skeleton. This stress acts perpendicularly to the film surface and tends to compress the skeleton, driving coalescence. This mode of film formation is termed capillary deformation (CD).

The use of even stiffer polymers or a smaller evaporation time means evaporation is essentially complete prior to the onset of significant coalescence within the skeleton. However, the dry skeleton still experiences air-polymer surface tension and will slowly coalescence under its influence. Due to the obvious parallels with wet sintering, this mode of film formation is termed dry sintering (DS).

In cases intermediate between CD and DS, a mixed mode of film formation arises where the solvent front recedes in the general direction of the substrate surface with DS above (behind) the front and CD below (ahead) of the front. This mode is termed the receding water front (RWF) mode.

This work extends a previous process model of film formation (Routh and Russel 1999; Routh and Russel 2001) to include a hard component and each of the above formation mechanisms is predicted as a function of the environmental conditions, in particular the ambient temperature and humidity, and the silica loading.

The polymer viscosity decays exponentially with increasing temperature (allowing more rapid particle coalescence) whereas the evaporation rate changes much more slowly, particularly if the air is saturated with vaporised solvent and/or the rate of mass transfer of vapour away from the film is limited. If one assumes the influence of humidity is small and the mass transfer rate is relatively independent of temperature (at least when compared to the change in viscosity), one may perform a temperature-time superposition such that for a given suspension and evaporation rate, the temperature is all that is required to predict which mode of film formation occurs and thus the time required for complete formation. CD is the most industrially relevant mode and the existing model provides the minimum temperature at which a given film will deform by this mode (MFFT mentioned previously).

The current work extends the existing model to films incorporating nanocomposites featuring silica inclusions; this particular problem has received little attention in the modelling literature. The model is mostly analytical with the final ODEs solved numerically. The inclusions are assumed to be rigid relative to the polymer which is assumed to feature linear viscoelastic, incompressible rheology. The particles are assumed to be of a single size and each is assumed to be loaded with a single cylindrical inclusion of identical dimensions that is initially centrally located within the particle. The range of inclusion-particle volume fractions studied was 0-10% which corresponds to the range most easily manufactured in parallel experimental work by the authors' collaborators.

The major terms arising in the extended model are the interfacial normal and shear stresses that exist between the particle and inclusion. These correspond to surface tension and hydrodynamic drag stress acting to oppose coalescence, respectively.

Interestingly, the results obtained predict that surface tension has no effect on the time required or alternatively the difficulty of film formation, and thus has no influence on the MFFT. However, the validity of this result is restricted to linear viscoelastic polymer particles. If a drag stress is included, the results predicted are unaffected by the magnitude of the surface tension stress. The model produces a dimensionless group that incorporates the effects of interfacial shear stress (divided by capillary pressure), inclusion-particle volume fraction and the shape of the inclusions (aspect ratio). The parameters are strongly coupled. Thus, while the film formation time may be processed into a MFFT via a temperature time superposition, the drag dimensionless group does not permit simple manipulation into an inclusion volume fraction (an industrially relevant parameter).

The model was applied to the four film formation modes described previously (WS, CD, RWF and DMS). As expected the MFFT rose with increasing values of the drag dimensionless group, indicating a higher inclusion loading and/or the use of more aspherical and more 'sticky' or rougher inclusions retards film formation.

References

Routh, A. F. and W. B. Russel (1999). "A Process Model for Latex Film Formation: Limiting Regimes for Individual Driving Forces." Langmuir 15(22): 7762-7773.

Routh, A. F. and W. B. Russel (2001). "A Process Model for Latex Film Formation: Limiting Regimes for Individual Driving Forces." Langmuir 17(23): 7446-7447.

See more of this Session: Colloidal Dispersions I
See more of this Group/Topical: Engineering Sciences and Fundamentals

The Microstructural Origins of Gel Rheology in Dense Mixtures of Colloids and Adsorbing Polymer

Katie L. Pickrahn, Chemical Engineering and Materials Science, University of California, Irvine, CA

Bharath Rajaram, Chemical Engineering and Materials Science, University of California, Irvine, CA

Ali Mohraz, Chemical Engineering and Materials Science, University of California, Irvine, CA

We investigate the link between the microstructure, dynamics, and rheological properties in dense mixtures of charge-stabilized silica colloids and oppositely-charged poly(ethylene imine) (PEI) polymer in a refractive index matched solution of dimethylsulfoxide and water. Such systems form self-supporting gels over a finite range of polymer concentrations, due to polymer-induced interparticle bridging. Dynamic rheological measurements show a reentrant behavior with increasing polymer concentration, in which both the storage modulus and yield stress of the gel initially grow, but go through a maximum and subsequently drop with rising polymer concentrations. Analysis of the spatial configuration and dynamics of the particles by quantitative confocal microscopy reveals the microstructural origins of this behavior. In the first regime where gel stiffness increases with polymer concentration, the particle dynamics are remarkably heterogeneous, due to the coexistence of an arrested gel phase with freely diffusing particles and small clusters. We segregate the particle populations based on their mobility and quantify the microstructure of the arrested phase. Both the overall particle dynamics and gel stiffness are found to correlate with the effective volume fraction of the arrested phase in this regime. However, in the range of polymer concentrations studied, the reentrant rheology is not captured by the particle dynamics. Instead, it is accompanied by subtle changes in the microstructure of the arrested phase and possibly in the strength of interparticle bonds. The implications of this behavior for complex fluids engineering of colloid/polymer mixtures are discussed.

See more of this Session: Colloidal Dispersions I
See more of this Group/Topical: Engineering Sciences and Fundamentals

Consequences of Shape Anisotropy On Colloidal Glasses

Ryan C. Kramb, Department of Chemical and Biomolecular Engineering, University of Illinois at Urbana-Champaign, Urbana, IL

Charles F. Zukoski, Department of Chemical and Biomolecular Engineering, University of Illinois at Urbana-Champaign, Urbana, IL

Rui Zhang, Department of Materials Science and Engineering, University of Illinois at Urbana-Champaign, Urbana, IL

Kenneth S. Schweizer, Department of Materials Science and Engineering, University of Illinois at Urbana-Champaign, Urbana, IL

Theoretical predictions on glasses formed from hard particles show that a moderate deviation from spherical shape decreases the elastic modulus in the glass state at a specified volume fraction. The particles used in these calculations consist of two fused spheres (homodicolloid shape) with a variable long direction, L, and a fixed short direction, D. The aspect ratio, L/D is varied from 1.0 to 2.0 and a minimum in the elastic modulus is found at L/D=1.4. It is also found that making the modulus nondimensional by multiplying G' by L^2D/kT, and rescaling the absolute volume fraction on the glass point (φ/φ_g-1), forces all data to collapse onto a single curve. Here we present recent experimental work that confirms many of the predictions of the model.

Using seeded emulsion polymerization (with or without divinyl benzene crosslinker), polystyrene particles are synthesized as either spheres or homodicolloids with and aspect ratio of 1.3. The particles are stabilized with the addition of a nonionic surfactant monolayer to the surface. The ionic strength of the suspension is tuned such that the particles experience nearly hard interaction potentials. The particles are concentrated to above the glass transition volume fraction and the elastic modulus is measured. The results show that the dicolloids do, in fact have a lower modulus than the spheres at the same volume fraction. In addition, by adjusting the particle diameter to account for soft Coulomb repulsions and scaling on the theoretical predictions of the glass transition volume fraction, we also see collapse of the two shapes as predicted by the model. The slope of the modulus data, however, is higher than predicted. This is likely due to the fact that the experiments were done at a frequency of 1Hz while the theory predications are for zero frequency. The confirmation of two (G' spheres > G' dicolloids, collapse of shapes onto single curve) out of three predictions of the model by these experiments represents a powerful confirmation of the physical basis of the theory and the inherent assumptions that it embodies.

See more of this Session: Colloidal Dispersions I
See more of this Group/Topical: Engineering Sciences and Fundamentals

Microstructure of State Behavior of Polymer Suspension

So Youn Kim, Chemical and biomolecular engineering, University of Illinois at Urbana-Champaign, Urbana, IL

Charles F. Zukoski, Chemical and biomolecular engineering, University of Illinois at Urbana-Champaign, Urbana, IL

When colloids are suspended in a solvent and polymer is added, particles experience a net attraction. This depletion attraction leads to liquid/liquid phase separation or gelation depending on the polymer to particles size ratios and the polymer concentration. Most studies of depletion forces have focused on polymer-colloid mixtures where the polymer does not adsorb to the particle surface and have investigated the dilute polymer solution behavior. Our goal is to explore depletion forces as a function of polymer concentration and strength of attraction between particles and polymer segments in concentrated polymer solutions.

This work is important in developing an understanding of the state of dispersion of particles in polymer nanocomposites. Application of these materials depends on the state of aggregation of particles in the polymer melt which is related with polymer induced particle interactions. Our studies will investigate the role of polymer adsorption in mitigating depletion forces in polymer solutions and in the melt state. Our studies will enable a deeper understanding of how to tune particle-polymer interactions to ensure desired properties in nanocomposites.

In previous work, we confirmed that in polyethylene oxide, PEO, melts polymer adsorb to silica particles such that suspensions are stable and experience essentially hard sphere repulsions. Here we look at the role of adsorption on the polymer induced particle interactions in a tertiary system consisting of particles, polymer and low molecular weight solvent. In these studies, 44 nm diameter particles are suspended in ethanol and PEO is added. From Small Angle X-ray Scattering (SAXS), we verified that at a low polymer concentration, the particles aggregate. As the polymer concentration increases, the particles restabilize depending on the polymer to particle size ratios and interaction between polymer and particle. Our studies confirm that the second virial coefficient moves from positive to negative and back to positive as the polymer concentration increases. Depletion restabilization is a poorly understood phenomena of significance in both the production of nanocompistes and in the formulation of many suspensions.

Here, we focus on strength of attraction between particles and polymer segments in concentrated polymer solutions by investigating structure factors at high polymer concentrations which is believed where particles are stable.

Depletion forces change with polymer molecular weight and the strength of the polymer segment interactions with the particles surface. The polymer segment interactions can be modified by the substitution of polymer such as PTHF, which is known adsorb less strongly to the particles surface. Our goal in these studies is to demonstrate the impact of changes in segment /surface interaction strength on particle stability and suspension microstructure.

Characterizing this specific system provides invaluable information about particle packing and particle interaction, ultimately, this work will lead to greater insights in how to control particle aggregation in polymer nanocomposites.

See more of this Session: Colloidal Dispersions I
See more of this Group/Topical: Engineering Sciences and Fundamentals

Contributions of Dynamical Heterogeneities to Non-Linear Rheology of Confined Supercooled Colloidal Liquids Under Oscillatory Shear

Yingxi Elaine Zhu, Department of Chemical and Biomolecular Engineering, University of Notre Dame, Notre Dame, IN

Prasad S. Sarangapani, University of Notre Dame, Notre Dame, IN

We have recently investigated the structural dynamics of confined amorphous colloidal thin films under oscillatory shear using a home-designed micron-gap rheometer interfaced with a confocal microscope. We focus on model "hard-sphere" poly-(methyl methacrylate) (PMMA) particles of 1.2 um in diameter suspended in the density and refractive index matched nonpolar solvent at particle volume fraction of 0.43-0.57. We simultaneously visualize the dynamical response of confined PMMA particles between two solid surfaces at narrow gap spacing of 5-80 μm to applied shear deformation and measure the viscoelasticity of PMMA colloidal thin films. For confined PMMA colloids under shear where an applied deformation is sufficiently large to induce non-linear rheological responses, we find significant non-universality in their plastic behavior. Non-affine motion, determined by subtracting the globally uniform strain from the bare particle coordinates, reveals that particles move as cooperatively rearranging groups that exhibit strong anisotropy and coexist with long lived immobile regions at narrow gaps in contrast to the isotropy observed in the bulk; the length scales of these groups typically approach the order of spatial confinement dimension. We also examine the role of the film-thickness dependent anisotropy on shear thickening observed in the limit of large strain amplitudes.

See more of this Session: Colloidal Dispersions I
See more of this Group/Topical: Engineering Sciences and Fundamentals

Nanostructure and Properties of Nanoparticle Dispersions with Tunable Interaction Potentials

Aaron P. R. Eberle, Chemical Engineering, University of Delaware, Newark, DE

Norman J. Wagner, Chemical Engineering, University of Delaware, Newark, DE

Colloidal systems ranging from proteins to intermediates in the manufacturing of particulate suspensions such as paints, and inks can exist in distinct thermodynamic equilibrium states similar to molecular systems. For instance, isolated, condensed, and ordered particles are analogous to gas, liquid, and crystalline states. Additionally, aggregation, gelation, stable and metastable phases, and arrested states give rise to a rich phase diagram that is dependent on the delicate interplay between interaction potentials and physical barriers with respect to the thermal energy. Such complex systems are of fundamental scientific interest and pose challenges to industrial formulation and application because of the hierarchy of structures that connect particle properties to bulk material properties. Understanding the mechanisms leading to aggregation with the ultimate goal of designed control is of central importance to achieving a desired particulate microstructure and related bulk mechanical properties.

In this work we explore the relationship between colloidal interaction and structure relating to thermodynamic states and phases in both dynamic and static environments. Specifically, we explore the mechanisms of leading to aggregation, structure, and mechanical properties of silica nanoparticle gels and attractive driven nanoparticle glasses formed by thermally quenching a model dispersion. The model thermoreversible system is composed of nano-sized silica coated with a relatively thin oligomeric surface layer (end-grafted 1-octadecanol) that provides steric stability in a good solvent. In a poor solvent, short range attraction drives aggregation, flocculation, gelation, phase separation, and crystallization depending on the particle volume fraction and quench path.

Single particle properties are studied using fiber optic quasi-elastic light scattering (FOQELS), small-angle neutron scattering (SANS), and Zeta potential measurements. At the particle surface the molecular phase behavior of the brush is studied using neutron reflectivity and is directly connected to the change in interparticle potential with contrast match variation SANS experiments. The nanostructure is connected to the colloidal phase behavior as a function of particle volume fraction and temperature using static-SANS. Finally, rheological measurements allow for the connection between interparticle potential, structure and bulk mechanical properties at corresponding concentrations and temperatures. The goals of this study are to identify the hierarchical structures that quantitatively relate the state of the gel or glass, interparticle potential, and particle properties to the bulk rheology.

See more of this Session: Colloidal Dispersions I
See more of this Group/Topical: Engineering Sciences and Fundamentals

Experimental Investigation of Breakage of DLCA Colloidal Aggregates in Shear and Elongational Flows

Yogesh M. Harshe, Chemistry and applied biosciences, Institute for chemical and bioengineering, ETH Zurich, Zurich, Switzerland

Miroslav Soos, Department of Chemistry and Applied Biosciences, Institute for Chemical and Bioengineering, ETH Zurich, Zurich, Switzerland

Marco Lattuada, Institute for Chemical and Bioengineering, Department of Chemistry and Applied Biosciences, ETH Zurich, Zurich, 8093, Switzerland

Massimo Morbidelli, Department of Chemistry and Applied Biosciences, Institute for Chemical and Bioengineering, Zurich, Switzerland

Abstract:

Breakage and/or reconstructing of colloidal aggregates are encountered in the processing of colloidal suspensions. The final average aggregate size and aggregate size distribution is decided by the initial aggregates' size, imposed flow field, and very importantly the initial structure of the aggregates. Except few systematic experimental studies scarcely some literature has been devoted in this study. The dynamic evolution of such aggregate properties could be used to have an insight into the kinetics of the breakage mechanism. The understanding of this process enables to estimate some lumped parameters which can then be used in Population Balance Equations as breakage kernels.

In the present attempt we have focused our attention to pure breakage process of Diffusion Limited Cluster Aggregates (DLCA) in shear and elongational flows. Initially, DLCA aggregates are produced by carrying diffusion limited cluster aggregation starting with different primary particle sizes to produce aggregates of a specific size. The aggregates were then exposed to various shear rates for both shear and elongational flows. The time evolution of fractal dimension (D¬f) and radius of gyration (Rg) was tracked by light scattering measurements and image analysis technique, wherever possible. The obtained data was compared with the previous literature and new scaling laws were defined. The results were qualitatively compared with the Stokesian dynamic modeling.

Keywords:

Breakage, DLCA, aggregates, Stokesian dynamics

See more of this Session: Colloidal Dispersions I
See more of this Group/Topical: Engineering Sciences and Fundamentals

Alcohols Adsorbed On SiO2 Under Ambient Conditions - From Fundamentals of Adsorption Isotherm and Molecular Orientation to Applications in Tribology

Anna Barnette, Chemical Engineering, Pennsylvania State University, University Park, PA

David B. Asay, Pennsylvania State University, University Park, PA

Seong H. Kim, Chemical Engineering, Pennsylvania State University, University Park, PA

The adsorption isotherm and orientation of small alcohol molecules on clean, hydrophilic silicon oxide surfaces under ambient conditions are studied and their effects on adhesion, friction, and wear of silicon oxide surfaces are explained. The attenuated total reflection infrared spectroscopy (ATR-IR) study reveals that as the isotherm thickness and the molecular orientation of the adsorbed alcohol layers are closely related. Even though the thickness of the adsorbed alcohol layers is only an order on 1 nm, its effects on adhesion, friction, and wear of the silicon oxide surfaces is quite remarkable. The nano-asperity capillary force due to the adsorbed alcohol layer is experimentally measured with an atomic force microscopy (AFM) and theoretically explained. The influence of chain length on friction forces is investigated for ethanol, 1-butanol, and 1-pentanol vapor-phase lubricants on silicon oxide surfaces with AFM. The experimental friciton-load curves are analyzed using contact mechanics to reveal the effects of the adsorbed alcohol layers on friction. Finally, the wear prevention effects of alcohols are experimentally and theoretically investigated.

See more of this Session: Fundamentals of Interfacial Phenomena I - Wetting and Interfacial Forces
See more of this Group/Topical: Engineering Sciences and Fundamentals

Liquid-to-Solid Phase Transition: Drainage of a Simple Nonpolar Fluid Under Nanometer Confinement

Yajie Lei, Department of Mechanical and Aerospace Engineering, George Washington University, Washington, DC

Yongsheng Leng, Department of Mechanical and Aerospace Engineering, George Washington University, Washington, DC

Ultrathin liquid films are ubiquitous in many technological problems, including wetting, adhesion, nanofluidics, and boundary lubrication of friction contacts. In surface force measurements, there are long-standing controversial debates concerning the thermodynamic properties of nanometer confined fluids. In this presentation, we report the dynamics thinning (squeeze-out) of a liquid argon between two solid boundaries. Using a liquid-vapor (LV) molecular ensemble and a steered molecular dynamics approach, we studied the mechanical properties of liquid argon during approach and retraction processes. Our simulation results show that when the gap distance is less than 2.5 nm (about 7 molecular diameters of argon), the solidification to a crystalline structure of the confined fluid is a spontaneous process and does not depend on whether the contact between the fluid molecules and the solid boundary is commensurate or not. The solvation force is calculated during the approach of the upper solid surface. The periodic disordered-liquid-to-ordered-solid phase transitions of the confined films are further reflected by the oscillations of solvation forces. The HCP compact structure of nonpolar argon in crystallized state was observed, accompanied by significant vacancy diffusion at larger distances. The non-symmetric force profiles during approach and retraction indicate that the compression strength is much larger than tensile strength of this HCP crystal. This study reveals a fundamental difference in the properties between aqueous and nonpolar fluids under nanometer confinement.

See more of this Session: Fundamentals of Interfacial Phenomena I - Wetting and Interfacial Forces
See more of this Group/Topical: Engineering Sciences and Fundamentals

Electrowetting in a Nanopore Via Capillary Condensation

Rohini Gupta, Chemical and Biomolecular Engineering, Johns Hopkins University, Baltimore, MD

Joelle Frechette, Chemical and Biomolecular Engineering, Johns Hopkins University, Baltimore, MD

Electrowetting is an effective means to move liquid droplets inside a micro/nanofluidic device. In this talk we will show how we can implement and characterize nanoscale electrowetting in the Surface Forces Apparatus (SFA). Using the SFA, we measured capillary adhesion between hydrophobic surfaces in the presence of condensable vapors. We characterized how adhesion depends on surface chemistry and partial pressure and compared our results with that predicted from the Kelvin equation. We have also modified the SFA to allow for external potential control and have investigated and characterized how we can favor the condensation of undersaturated water vapors on hydrophobic surfaces by charging the hydrophobic surface. We compare our measurements with predictions from the Young-Lippmann-Kelvin equation. Our measurements are complemented by dynamic contact angle and macroscopic electrowetting measurements.

See more of this Session: Fundamentals of Interfacial Phenomena I - Wetting and Interfacial Forces
See more of this Group/Topical: Engineering Sciences and Fundamentals

Wetting Behavior of a Nanodroplet On a Heterogeneous Solid Surface

Hongfei Wu, Chemical Engineering, Pennsylvania State University, University Park, PA

Ali Borhan, Chemical Engineering, Pennsylvania State University, University Park, PA

Kristen A. Fichthorn, Chemical Engineering, Pennsylvania State University, University Park, PA

There is ample evidence in nature and laboratory experiments that modifying the topography of a solid surface can change its wettability. In this study, the wetting behavior of a sessile Lennard-Jones liquid droplet on a heterogeneous solid surface is investigated using molecular dynamics (MD) simulations. The solid surface is physically patterned with a periodic array of grooves embedded in an otherwise flat surface. Simulations of droplet spreading on the patterned surface show that the relative magnitudes of the length scales characterizing the surface heterogeneity compared to the drop size play an important role in determining the hydrophobic characteristics of the surface. The computed surface wettability phase diagram exhibits transitions between the Wenzel and Cassie modes of wetting as the step height, width and separation distance are varied. The transition between wetting modes is discussed through comparison of a simple model with the simulated results, as well as in terms of the interplay between the bulk liquid chemical potential, the liquid-solid interfacial tension, and the topology of the liquid-solid potential-energy surface induced by surface roughness.

See more of this Session: Fundamentals of Interfacial Phenomena I - Wetting and Interfacial Forces
See more of this Group/Topical: Engineering Sciences and Fundamentals

Labyrinthine Instability of Thin Liquid Films

Parthasakha Neogi, Chemical and Biological Engineering, Missouri University of Science and Technology, Rolla, MO

When thin liquid films on solid surfaces breakup due to instability, one of the configurations which emerges at large times out of nucleation of holes and dewetting, is well known and has been studied in detail. Another large time behavior is also well know and is called wavy instability but this configuration has not been studied. We show here that this form of instability is called labyrinthine instability in magnetic fluids and describes quite well the wavy instability in thin films here. It is known that these originate from spinodal decomposition, and the geometry of two coexistent phases have been modeled here following similar treatment in the magnetic systems. The dimensions of the cells have been determined by minimizing the total energy and the results are compared with experiments.

See more of this Session: Fundamentals of Interfacial Phenomena I - Wetting and Interfacial Forces
See more of this Group/Topical: Engineering Sciences and Fundamentals

Capillary Assembly in Surfactant Monolayers

Marcello Cavallaro Jr., Chemical & Biomolecular Engineering, University of Pennsylvania, Philadelphia, PA

Eric P. Lewandowski, Chemical & Biomolecular Engineering, University of Pennsylvania, Philadelphia, PA

Kathleen J. Stebe, Chemical & Biomolecular Engineering, University of Pennsylvania, Philadelphia, PA

In this work we study capillary assembly in the presence of an insoluble surfactant monolayer by analyzing the rate of approach of particles attracted to each other by capillary interactions. The method simply requires video imaging of particles as they approach one another at an interface. In our case, cylindrically shaped particles in the zero Bond number limit are used so that gravity effects are neglected and the surface distortion depends solely on particle geometry and wetting properties. Far-field assembly is driven by the overlap of quadrupolar modes of the distortions created by the particles. For surfactants that form Newtonian surface layers on Newtonian subphases, the distance between the particles versus time to contact obeys a power law and yields a prefactor proportional to the surface viscosity. Surface viscosity measurements for pentadecanoic acid as a function of surface coverage are reported.

See more of this Session: Fundamentals of Interfacial Phenomena I - Wetting and Interfacial Forces
See more of this Group/Topical: Engineering Sciences and Fundamentals

Modifying Properties of Polymer Nanostructure through Liquid Phase Reactive Treatments

Richard A. Lawson, School of Chemical & Biomolecular Engineering, Georgia Institute of Technology, Atlanta, GA

David E. Noga, School of Chemical & Biomolecular Engineering, Georgia Institute of Technology, Atlanta, GA

Clifford L. Henderson, School of Chemical & Biomolecular Engineering, Georgia Institute of Technology, Atlanta, GA

As pattern sizes shrink below 50 nm in semiconductor devices, the resist film thicknesses and resist feature sizes used in the lithography steps for patterning these devices are also rapidly decreasing in size. This reduction in film thickness was originally moivated by limitations in the depth of focus of the optical projection exposue tools used to pattern such resist films. However, it has also become clear that such film thickness reductions are important in mitigating pattern collapse problems that occur due to capillary force induced collapse of such fine polymeric resist nanostructures during wet processing and drying of the resist pattern. This reduction in film thickness is a concern as film thicknesses approach and shrink below 100 nm. The sub-100 nm film thickness regime is where ultra-thin film effects have already been shown to become a major issue due to the confinement of the material at the interfaces and the large surface to volume ratio in such thin films; i.e. the interfacial layer between the substrate and the resist and the interfacial layer at the free surface of the resist film become significant relative to the total film thickness. These ultra-thin film effects have shown to result in dramatic changes in the properties of the materials such as raising or lowering the glass transition temperature, changing the mobility of the material, and orders of magnitude reduction in the diffusivity of small molecules such as acid and water. Among the most important changes for resist and lithography applications may be the potential change in the mechanical properties of these films as the film thickness shrinks and/or the patterned feature size shrinks. We have carefully studied the effect of polymer film thickness and patterned polymer feature size using carefully designed pattern collapse test structures that are used to determine the effective resist polymer modulus and critical stress at the point of pattern collapse for varying film thicknesses, pattern sizes, and types of photoresists. High resolution e-beam lithography was used to generate 2-D line pattern collapse tet structures that provide the ability to calculate the capillary forces generated during drying the patterned features from their as-developed wet state. These experiments have clearly shown that effective polymer modulus decreases with both decreasing film thickness and polymer nanostructure width. What is needed is a way to mitigate this reduction in mechanical properties that tend to exacerbate pattern collapse problems in a resist material as both feature size and film thickness are reduced.

To that end, we have developed a new liquid phase reaction chemistry based on carbodiimide chemistry that covalently cross-links the solid phase resist nanostructures on its exposed surface while still in the wet processing state. The goal in this process is to use the high surface to volume ratios in such thin films and nanostructures to allow for significant modification and imporvement of the effective modulus of the nanostructure through only this surface cross-linking. Chemistry has been developed that will cross-link either pendant alcohols or carboxylic acids, meaning that this reactive rinse can be applied to a majority of all commercial photoresists. The liquid phase reaction is carried out by simply rinsing a patterned photoresist with a water solution with the reagents dissolved into it. Due to the very high surface area of the patterned surface, the reaction is carried out very quickly (i.e. less than one minute) at room temperature. Using the pattern collapse test structures mentioned earlier, we have shown that the use of this reactive rinse provides a significant improvement in the pattern collapse behavior in photoresists and a large measurable increase in the effective modulus of the polymer nanostructure. Since collapse occurs due to capillary forces on the resist features during drying, the reactive rinse can be applied at the end of the normal wet resist processing sequence to prevent pattern collapse but yet not affect any of the previous processing of the resist and also not adding any significant complexity to the resist process. The influence of this process on resist performance will also be discussed in detail including: (1) spectroscopic studies have been carried out to examine the extent of cross-linking during the reactive rinse process and to determine its location in the polymer nanostructure, (2) the effect of the rinse on polymer swelling and distortion, and (3) the effect of the rinse of nanostructure roughness.

See more of this Session: Fundamentals of Interfacial Phenomena I - Wetting and Interfacial Forces
See more of this Group/Topical: Engineering Sciences and Fundamentals

A Brownian Dynamics Study of the Effect of Solvent Quality On the Coil-Stretch Transition

Eric S. G. Shaqfeh, Departments of Chemical and of Mechanical Engineering, Stanford University, Stanford, CA

Shikha Somani, Chemical Engineering, Stanford University, Stanford, CA

Ravi Prakash Jagadeeshan, Monash University, Melbourne, VIC, Australia

When long chain, linear polymers in dilute solutions are subjected to purely elongation flows or elongation-dominated mixed flows, the solution properties show a sharp variation near a critical flow rate where the conformation of the polymer changes from a coiled to a stretched state or vice versa. Since the solution properties change dramatically at this so-called coil-stretch transition, it is important in many applications. This transition is primarily characterized by the critical flow rate for a given polymer molecular weight and solvent quality. While the effect of solvent quality on equilibrium properties of polymer solutions has been widely studied, there have been relatively few attempts at examining its effect on properties far from equilibrium. One such study is by Cifre and Torre1, where they examined the effect of solvent quality on the critical elongation rate for coil-stretch transition by Brownian dynamics simulations and found that there was no effect on the transition properties. However, certain important aspects of the newly developed method of successive fine graining (SFG2, 3) and new theories about coil-stretch hysteresis4-7 were not considered in their work. We revisit the specific problem of steady, homogeneous elongational flow, and attempt to accurately determine the dependence of the critical elongational rate on solvent quality and chain length,.We have rigorously included, effects such as fluctuating hydrodynamic interactions (HI) and excluded volume (EV) in our calculations. The solvent effect is captured systematically using the EV parameter 'z' which is the appropriate parameter in perturbation expansions for determining the effect of solvent quality about the 'theta' state. Exact results are obtained numerically using Brownian dynamics simulations, while the Successive Fine Graining (SFG) technique has been used to eliminate parameter dependence in the results of the bead-spring chain simulations. Contrary to earlier findings, it is observed that with improvement in solvent quality, even though the critical strain rate decreases, critical Deborah number (based on either longest relaxation time(λ1) or zero shear rate viscosity(λη,0)) increases and this is solely due to an increase in relaxation time. The approach is then further extended to study the effect of solvent quality on the phenomena associated with polymer conformational hysteresis, including the nature of transition states and the activation energy for the transition.

References:

1. Cifre; J. G. H. and Torre; J. G. D. L. J.Rheol. 1999, 43, 2, 339-358

2. Sunthar P. and Prakash J. R. Macromolecules 2005, 38, 617-640

3. Sunthar P.; Nguyen D. A.; Dubbelboer R.; Prakash J. R. and Sridhar T. Macromolecules 2005, 38, 10200-10209

4. Schroeder C. M.; Babcock H. P.; Shaqfeh E. S. G. and Chu S. Science 2003, 301, 5639, 1515-1519

5. Schroeder C. M.; Shaqfeh E. S. G. and Chu S. Macromolecules 2004, 37, 24, 9242-9256

6. Hsieh C. C. and Larson R. G. J. Rheol. 2005, 49, 5, 1081-1089

7. Beck, V.A. and E.S.G. Shaqfeh, J. Rheol., 2007 51(3), 561-574

See more of this Session: Non-Newtonian Flows
See more of this Group/Topical: Engineering Sciences and Fundamentals

Understanding the Dynamics of Viscoelastic Turbulent Flows and Polymer Drag Reduction in Minimal Flow Units

Li Xi, Department of Chemical and Biological Engineering, University of Wisconsin-Madison, Madison, WI

Michael D. Graham, Department of Chemical and Biological Engineering, University of Wisconsin-Madison, Madison, WI

Adding a small amount of flexible polymers into a turbulent flow can cause substantial reduction in its friction drag by influencing the fundamental dynamics of the self-sustaining turbulent structures. In this study, we isolate the essential components of the self-sustaining process by conducting direct numerical simulations (DNS) in minimal flow units (MFU) of viscoelastic turbulent flows. These MFU solutions obtained at various polymer parameters recover all key transitions in viscoelastic turbulent flows reported previously in experiments at much higher Re, including the onset of drag reduction, low degree of drag reduction (LDR), high degree of drag reduction (HDR) and maximum drag reduction (MDR). At the MDR regime, the mean velocity profile is insensitive to changing polymer parameters. The LDR-HDR transition is characterized by a sudden increase in the minimal box size of sustaining turbulence, which may correspond to a qualitative change in the self-sustaining mechanism. With these solutions, we study the effects of polymer on turbulent coherent structures and the dynamics of the self-sustaining process, based on which the mechanism behind the multistage transition scheme in viscoelastic turbulent flows can be better understood.

See more of this Session: Non-Newtonian Flows
See more of this Group/Topical: Engineering Sciences and Fundamentals

Shear Banding and Non-Equilibrium Phase Behavior of Concentrated Wormlike Micellar Solutions

Matthew E. Helgeson, Chemical Engineering, MIT, Cambirdge, MA

Matthew D. Reichert, Chemical Engineering, University of Delaware, Newark, DE

Eric W. Kaler, Stony Brook University, Stony Brook, NY

Norman J. Wagner, Chemical Engineering, University of Delaware, Newark, DE

Shear banding is a flow instability encountered in a variety of complex fluids, including highly entangled polymers, colloidal suspensions, and, most prominently, wormlike micelles (WLMs). Although shear banding has been identified in a number of WLM-forming surfactants, its underlying mechanisms are still largely unknown. Here, we present the rheological behavior of concentrated cationic WLMs near an equilibrium isotropic-nematic (I-N) transition as a function of composition and temperature to determine the relationship between shear banding, fluid microstructure and underlying surfactant phase behavior. A novel combination of conventional rheometry, velocimetry, and spatially-resolved flow-small angle neutron scattering reveals that shear banding is due to a first-order, shear-induced transition to a paranematic state. The shear rheology of isotropic WLM solutions are well-described by the Giesekus constitutive model, which provides a quantitative discrimination between banding and non-banding WLMs through the drag anisotropy coupling parameter. This anisotropy parameter is shown to correlate with the order parameter describing the relative distance to the equilibrium I-N transition, as well as the critical flow-alignment required for the onset of shear banding. Combining this information with measurements of the critical shear rates for banding allows the construction of a non-equilibrium state diagram near the I-N transition in terms of the Weissenberg number and the compositional order parameter. This non-equilibrium phase diagram explains a number of experimental observations in shear banding fluids, including spatiotemporal fluctuations and rheo-chaos during shear banding.

See more of this Session: Non-Newtonian Flows
See more of this Group/Topical: Engineering Sciences and Fundamentals

Microscale Shear Flow of Focal Conic Defects in Layered Liquids

Sourav Chatterjee, Dept. of Mechanical Engineering, Carnegie Mellon University, Pittsburgh, PA

Shelley L. Anna, Departments of Mechanical Engineering and Chemical Engineering, Carnegie Mellon University, Pittsburgh, PA

Intermolecular interactions in liquid crystals and concentrated surfactant solutions lead to unique microstructures including lamellae, in which parallel layers are incompressible but bend easily. In such systems, planar layers are easily destabilized via external fields and nearby surfaces to produce topological defects of the order of tens of microns in size. These microscopic defects play a leading role in the flow behavior of such materials, and therefore impact numerous industrial applications including optoelectronic devices and displays and the processing of coatings, adhesives, and biomaterials to encapsulate drugs. To examine the interaction between such microscale defects and flow, we have developed a shear cell to impose a linear Couette flow in a microscale thin gap, while allowing for real time microscopic visualization. We use the shear cell to visualize the dynamics of defect formation in initially defect-free samples of a common small-molecule thermotropic liquid crystal, 8CB. We observe that the formation of focal conic defects, a specific topological defect typically found in thermotropic smectic liquid crystals, is triggered by edge effects and occurs in a series of phases, marked by distinct changes in the birefringence intensity. The defects are seen to annihilate partially or completely on reverse shear. The effect of shear rate and strain amplitude on defect formation and annihilation is studied.

See more of this Session: Non-Newtonian Flows
See more of this Group/Topical: Engineering Sciences and Fundamentals

Structures in Extensional Flow of Visco-Elastic Fluids

Dennis A. Siginer, Department of Mechanical Engineering & Department of Mathematics, Petroleum Institute, Abu Dhabi, United Arab Emirates

F. Talay Akyildiz, Department of Mathematics, Petroleum Institute, Abu Dhabi, United Arab Emirates

Exact solutions of partial differential equations describing the evolution of viscoelastic stresses are presented in the extensional flow of fluids whose constitutive response may be described by Oldroyd and Giesekus models. Singular solutions for a range of values of the Weissenberg number are presented and the results of Renardy (J. Non-Newtonian Fluid Mech. 138 (2006) 204-205), and Becherer et al. (J. Non-Newtonian Fluid Mech. 153 (2008) 183-190) are deduced for the case of upper convected Maxwell (UCM) fluid and one independent spatial variable. The effect of variable boundary data at the inflow can be studied as viscoelastic stresses over two spatial variables are considered. As such, our results are applicable to a wider range of flow problems.

See more of this Session: Non-Newtonian Flows
See more of this Group/Topical: Engineering Sciences and Fundamentals

The Extrusion of a Model Yield Stress Fluid Imaged by MRI Velocimetry

Brooks D. Rabideau, Université Paris-Est - Institut Navier - LMSGC, Champs sur Marne, France

Nicolas Roussel, Laboratoire Central des Ponts et Chaussées, Paris, France

Christophe Lanos, LGCGM équipe Matériaux et Thermique de l'Habitat, IUT de Rennes, Rennes, France

Philippe Coussot, Université Paris-Est - Institut Navier - LMSGC, Champs sur Marne, France

Extrusion tests were performed by forcing a well-characterized model yield stress fluid from a cylindrical cartridge through various cylindrical extrusion dies using a variety of different piston velocities. For our tests the Bingham number in the die ranged from 0.1 to 10. MRI techniques allowed for the non-invasive determination of the local velocity within the extruded material in the range. The velocity profile within a very long die was determined by MRI and agreed very well with the Herschel-Bulkley theory of flow in a conduit using the parameters determined from independent rheometrical tests, validating both the rheological approach and the accuracy of the MRI techniques. Although the velocity was determined by MRI in the upper and lower zones separately, the intersection of these zones showed great agreement, providing an entire view of the extrusion process. In our range of Bingham number the velocity field for a given contraction ratio appeared similar, with an heart of acceleration above the die and lateral dead zones varying negligibly with the piston velocity. Finite element simulations employing two different models were used in separate simulations: a biviscous model and an elastic solids model. It was found that to accurately reproduce the flow characteristics of extrusion this elastic stress model should be used. The MRI results show that for our range of values the piston velocity has little effect on the size of the weak velocity regions, and with an appropriate scaling the central, longitudinal velocity follows a master curve. A decreasing contraction ratio, on the other hand, appears to increase the size of the weak velocity region, in strong contrast with the simulation results.

See more of this Session: Non-Newtonian Flows
See more of this Group/Topical: Engineering Sciences and Fundamentals

TERMINAL VELOCITY AND DRAG CURVE OF NON – SPHERICAL PARITICLES IN LIQUIDS

Subrahmaniyam Susarla, Jeevan Garmilla, Raghu Kattekola, Santhosh Kumar Gujje, Swamynath Nomula*

University College of Technology, Osmania University, Hyderabad, 500007

*Methodist College of Engineering and Technology, Abids, Hyderabad, 500001

Email: ssubrahmaniyam@yahoo.co.in

INTRODUCTION:

Settling of particles in liquids has wide range applications in chemical industries. The phenomena of spherical particles terminal velocity and calculation of drag co-efficient are well researched. However, the settling of non – spherical particles in Newtonian/Non – Newtonian liquids is still under development. Several attempts to correlate data are seen in literature [1 – 12]. In addition, development of a common correlation of drag coefficient or Newtonian and Power law liquids with modified Reynolds number is also found in literature. For Newtonian liquids, Yow et – al [12] gave correlations of drag co-efficient in terms of Reynolds number and Sphericity using a wide base of literature data. The basis of their correlation is Kaskas equation. The non-spherical particles with well defined shapes are cubes, cones, cylinders, octahedrons, tetrahedrons and planes. Correlations for drag co-efficient covering a wide range of Reynolds numbers for cones settling in Newtonian and Power – law fluids are available in literature [3]. The wall effects on terminal velocities are given for spherical particles in Newtonian liquids [2].

Experimental work and Results:

Experimental work was carried out to generate terminal velocity data covering non-spherical shape of particles of different shapes as well as to study the wall effect in different diameter glass columns. The experimental procedure is a conventional one i.e. the distance travelled by the particle and time of travel were noted. Several Newtonian liquids such as glycerin, glycerin – water mixture, caster oil, vegetable oil and honey were used. Non – spherical particles of materials glass, steel, Al, Br, Cu and Teflon were used. Partial data are shown in figures 1 – 9.

The experiments indicate a Reynolds Number variation of 0.1 to 25 and Fig.1 shows the variation of terminal velocity with equivalent diameter (sphere of equal volume as particle) in the column of 55mm. The terminal velocity increase with equivalent diameter is as expected.

Increasing density of the particle, terminal velocity is seeing to increase at any equivalent diameter as for the case of a sphere.

Fig: 2a shows variation of terminal velocity with equivalent diameter in a column of 22mm diameter with glycerin as liquid. A significant wall effect is seen for all the particles. The terminal velocity increases with increasing diameter initially and after a certain value a increase in diameter further leads to low terminal velocity. In this case a sphere shows a lowest terminal velocity. The effect of wall on terminal velocities is also seen from Fig: 2b using Caster oil. However, the spheres show a different trend. In general wall effect results are similar to those reported in literature [2].

As mentioned earlier several correlations of C_d vs. N_{Re} are available in literature for Newtonian liquids. We have chosen the equations of Yow et-al (2005). Figures 3, 4 and 5 show data of cylinders, cones and cubes along with predictions from Yow et-al. In these graphs, sphericity is the parameter. A satisfactory agreement of data and correlation is seen for glycerin. Similar agreement with Caster oil and a few compositions of glycerin was recorded by authors.

In order to obtain a general correlation data of drag co-efficient vs. Re based on equivalent diameter was plotted for all shapes of particles studied. These are approximately 110 data points. The following correlation was obtained.

$C_d = 30.5\ N_{Re}^{-0.794}$ (correlation co-efficient 0.935)

(This is shown in Fig: 9)

This correlation does not include a sphericity factor and is very approximate one for calculation of drag coefficients. Further analysis of data being carried out to develop a correlation for drag co-efficient applicable for small diameter tubes.

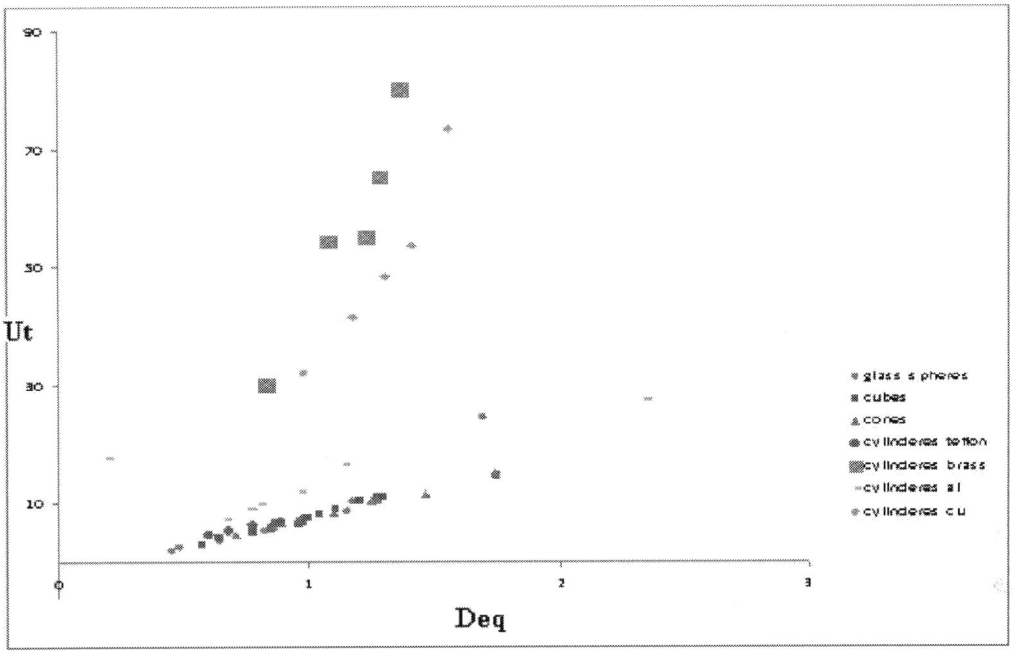

Fig: 1 Variation of terminal velocity with equivalent diameter in caster oil (parameter = density of the particle)

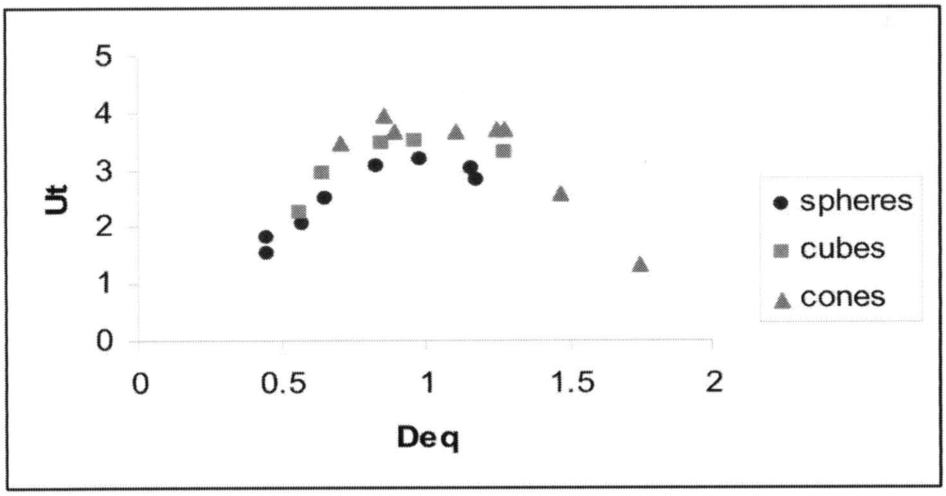

Fig: 2a Variation of terminal velocity with particle equivalent diameter in glycerin

(Column diameter 22mm)

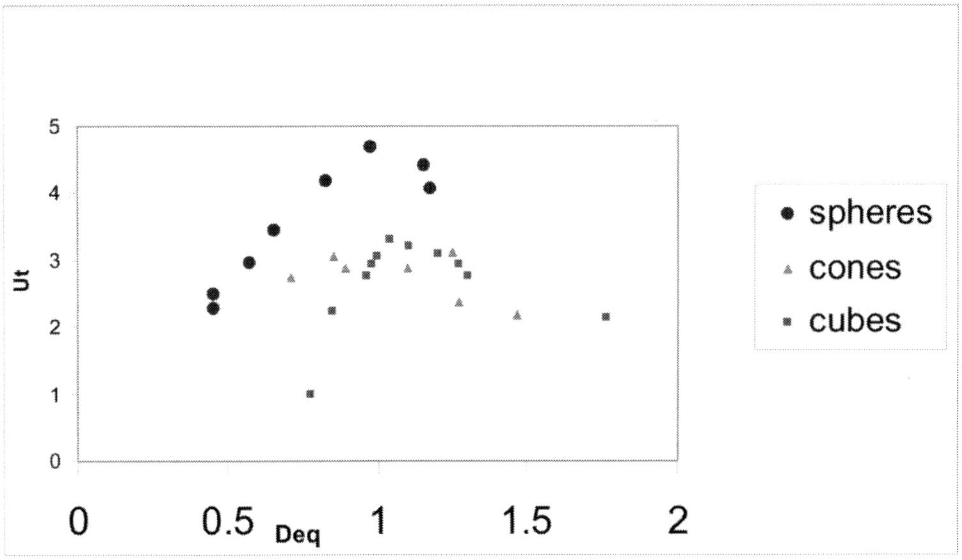

Fig: 2b Variation of terminal velocity with particle equivalent diameter in caster oil

(Column 22mm)

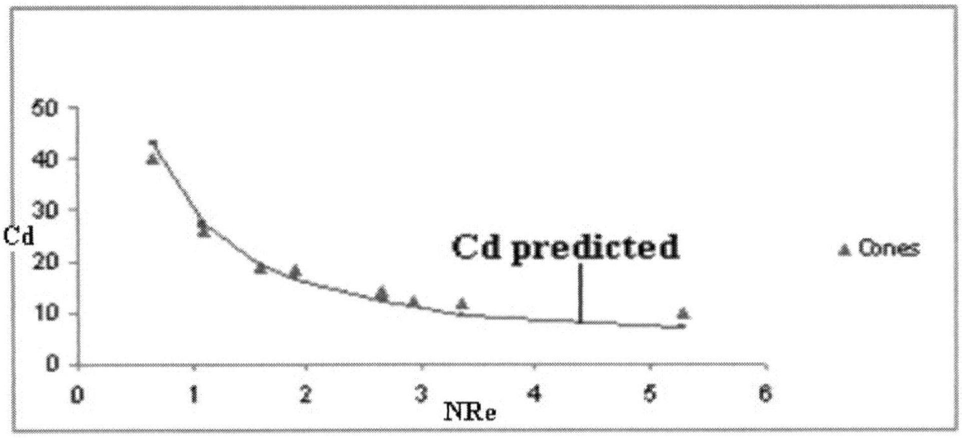

Fig: 3 Variation of drag co-efficient with Reynolds number for cones in glycerin, based on Deq comparison with Yow et-al

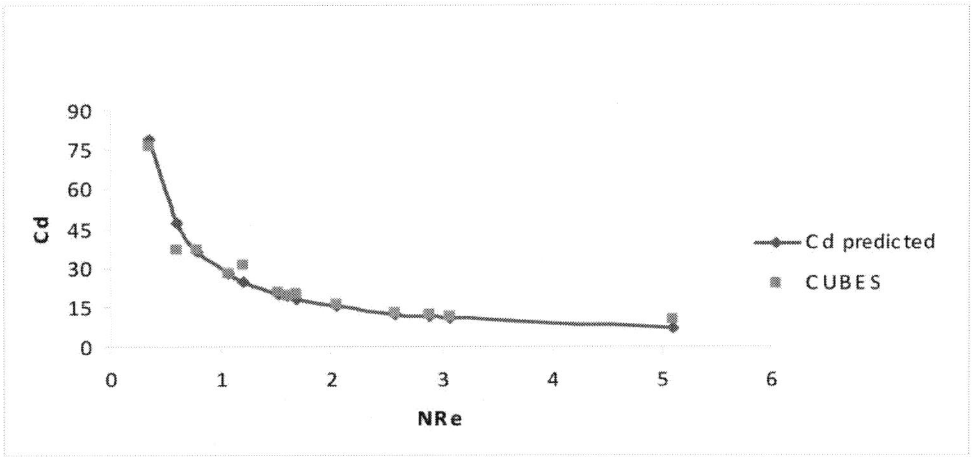

Fig: 4 Variation of drag co-efficient with Reynolds number for cubes in glycerin, based on Deq comparison with Yow et-al

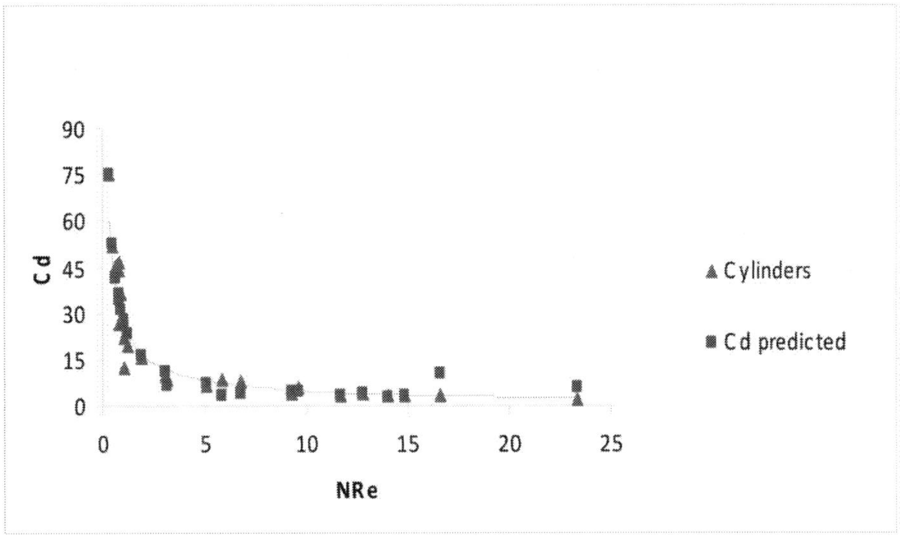

Fig: 5 Variation of drag co-efficient with Reynolds number for cylinders in glycerin, based on Deq comparison with Yow et-al

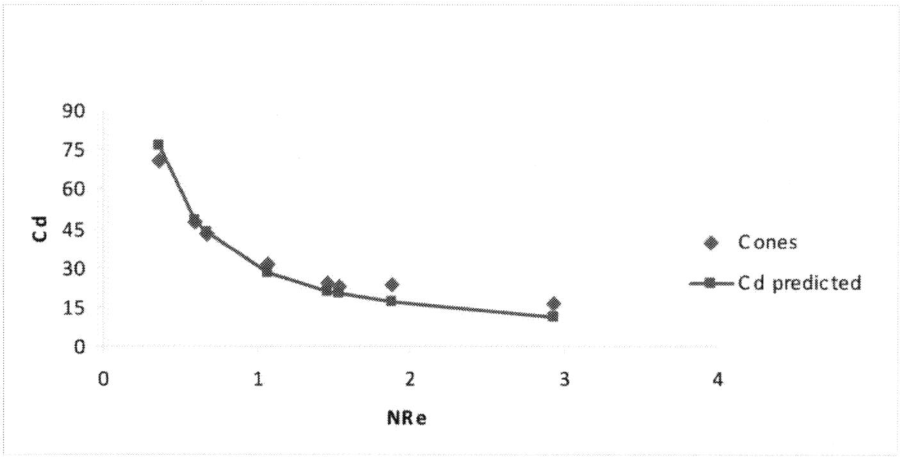

Fig: 6 Variation of drag co-efficient with Reynolds number for cones in Caster oil, based on Deq comparison with Yow et-al

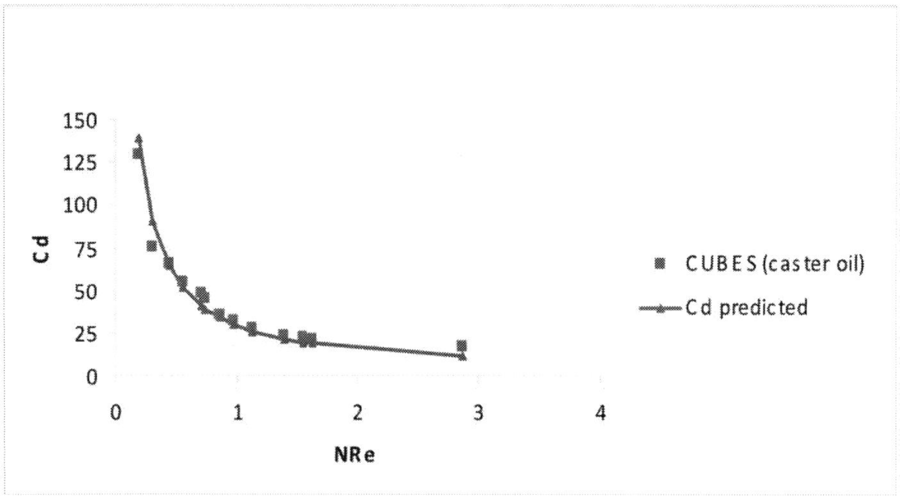

Fig: 7 Variation of drag co-efficient with Reynolds number for cubes in Caster oil, based on Deq comparison with Yow et-al

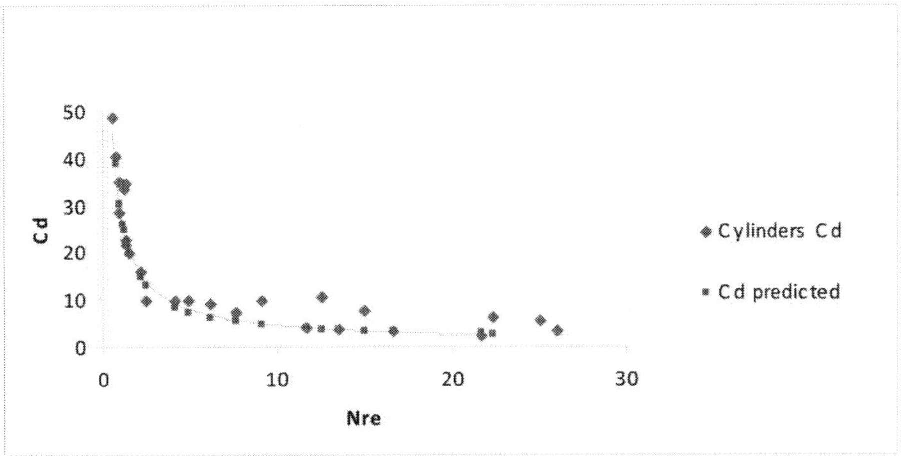

Fig:8 Variation of drag co-efficient with Reynolds number for cylinders in Caster oil, based on Deq comparison with Yow et-al

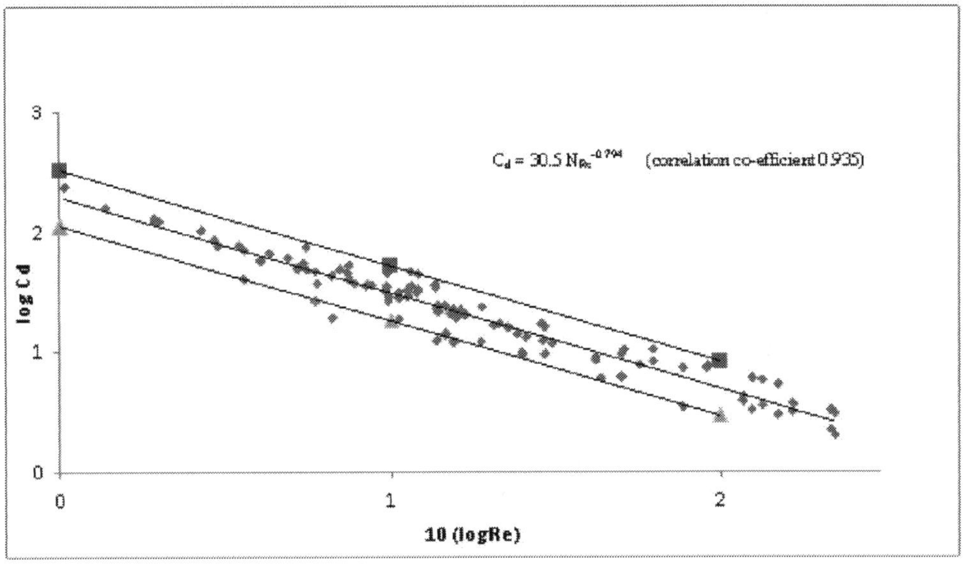

Fig: 9 Correlation data of drag coefficient vs Reynolds based on the equivalent diameter for all the shapes of particles (caster oil and glycerin in 55mm diameter column)

References:

1. Agarwal N., Chhabra R. P., Settling Velocity of Cubes in Newtonian and Power Law Liquids, Powder Technology, Vol: 178, 2007, PP: 17 – 21

2. Ataide C. H., Pereira F. A. R., Borrozo M. A. S., Wall effects on the terminal velocity of spherical particles in Newtonian and Non – Newtonian fluids, Brazilian Journal of Chemical Engineering, Vol: 16, Issue: 4, 1999, PP: 387 – 394.

3. Borah A., and Chhabra R. P., Drag on freely falling cones in Newtonian and Power Law Fluids, The Canadian Journal of Chemical Engineering, Vol: 83, 2005, PP: 559 – 565.

4. Cheng N. S., Comparison of formulas for drag coefficient and settling velocity of spherical particles, Power Technology.

5. Dewsbury K. H., Karamanev D. G., Margaritis A., Dynamic behavior of free raising Buoyant solid spheres in non – Newtonian liquids, AIChE Journal, Vol: 46, Jan 2000, No. 1, PP: 46 – 51

6. Dolecek P., Bendova H., Siska B., Machac I, Fall of spherical particles through a carreau fluid, International Conference of the Slovak Society of Chemical Engineering, Tatranske Matliare, May 2004, PP: 397 – 403

7. Felice R. D., and Kehlenbeck, Sedimentation velocity of solids in finite size vessels, Chemical Engineering Technology, Vol: 23, No.12, 2000

8. Landman K. A. White L. R., Determination of the Hindered Settling Factor for Flocculated Suspensions, AIChE Journal, Vol: 38, No. 2, Feb 1992.

9. Saito F., Kamiwano M., Aoki R., Free Settling of non – spherical particles in a stagnant liquid, Particulate Science and Technology 2, 1984, PP: 247 – 258

10. Shah S. N., Fadili Y. E. and Chhabra R. P., New model for single spherical particle settling velocity in Power Law (visco – inelastic) fluids, International Journal of Multiphase flow, Vol: 33, 2007, PP: 51 – 56

11. Valentik L., Whitemore R. L., The terminal velocity of spheres in Bingham Plastics, British Journal of Applied Physics, Vol: 16, 1965

12. Yow H. N., Pitt M. J., and Salman A. D., Drag correlations for particles of regular shapes, Advanced Power Technology, Vol: 16, 2005, PP: 363 – 372.

Effect of Shear On Nanoparticle Dispersion in Polymer Melts: A Coarse-Grained Molecular Dynamics Study

Vibha Kalra, School of Chemical and Biomolecular Engineering, Cornell University, Ithaca, NY

Fernando Escobedo, School of Chemical and Biomolecular Engineering, Cornell University, Ithaca, NY

Yong L. Joo, School of Chemical and Biomolecular Engineering, Cornell University, Ithaca, NY

Coarse-grained molecular dynamics simulations have been conducted to study the effect of shear flow on polymer nanocomposite systems. In particular, the interactions between different components have been tuned such that the nanoparticle-nanoparticle attraction is stronger than nanoparticle-polymer interaction, and therefore, the final equilibrium state for such systems is one with clustered nanoparticles. In the current study, we focus on how shear flow affects the kinetics of particle aggregation in systems with polymers of different chain lengths. The particle volume fraction and size are kept fixed at 0.1 and 1.7 MD units, respectively. Through this work, shear has been shown to significantly slow down nanoparticle aggregation, an effect that was found to be a strong function of both polymer chain length and shear rate. To understand our findings, a systematic study on effect of shear on particle diffusion and an analysis of relative time scales of different mechanisms causing particle aggregation have been conducted. The aggregation rate obtained from the time scale analysis is in good agreement with that determined from the aggregation time derived from the pair correlation function monitored during simulations.

See more of this Session: Non-Newtonian Flows
See more of this Group/Topical: Engineering Sciences and Fundamentals

Stabilization and Control of Rheological Properties for Concentrated Colloidal Suspensions Under High Ionic Strength

Jaehun Chun, Pacific Northwest National Laboratory, Richland, WA

Adam P. Poloski, Defense Nuclear Facilities Safety Board, Washington, DC

Control of the stability and rheological properties of concentrated colloidal suspensions has been an important but challenging issue for various applications such as ceramic processing and nuclear waste treatment, one that fundamental science has yet to address fully. Rheological modifiers can be categorized into two types in terms of colloidal interaction implemented for enhancement of stability and rheological properties: electrostatic and steric interactions. Binding agents such as weak acids strengthen an electrostatic repulsion, whereas nonionic surfactants introduce a steric repulsion. We investigated various rheological modifiers on an iron oxide-rich nuclear waste simulant characterized by high ionic strength and high pH. From rheological analysis, it was found that citric acid is one of the best rheological modifiers for this simulant. Further analysis also supports that better stability is achieved by the addition of citric acid. This work is important to the waste treatment processing at Hanford and Savannah River sites, since controlling rheological properties enables us not only to increase the solids concentration in the process and reduce operational cost accordingly but also to ensure sustainable operations.

See more of this Session: Non-Newtonian Flows
See more of this Group/Topical: Engineering Sciences and Fundamentals

Viscoelasticity During Scale-up in Extrusion of Polymer Blends

Suresh Ahuja

ABSTRACT

In processing of polymers especially blends and composites, continuous mixers are used such as co-rotating twin-screw extruders [ZSK). Scale-up rules are commonly based on the principle of similarity. In the case of twin- screw extruders these rules have to be adjusted because of geometrical and thermodynamic restrictions Geometrical restriction is dependent on screw diameter and center- line distance. The other restriction is the heat transfer from the polymer melt during compounding process. Scale-up process using the dimensionless temperature and dimensionless power consumption are more useful than the scaling rule of mass flow rate to screw diameter. Materials used in our study were blends of composites containing cross-linked polymers. In these materials elastic modulus was greatly higher than viscous modulus. Polymers with different levels of cross-linking with and without fillers were dynamically analyzed in a rotational rheometer. Equilibrium plateau modulus was computed using time temperature superposition. Equilibrium plateau modulus characteristic of polymer network showed power law dependence on melt flow index in a small twin-screw extruder. On increasing the size of the extruder power law index decreased which was associated with lower dimensionless temperature(ratio of set temperature to die temperature) and lower dimensionless power parameters.

INTRODUCTION

Properties of material components and the conditions under which they are processed are significant variables in determining the properties of their output. Polymer blends, polymer composites, polymerization or cross-linking reaction and devolatilization are produced in continuous mixers like extruders. For reasons of economy the development of these compounds is generally carried out on lab scale extruders. After modifying a raw plastic into a compound with the required properties the process has to be transferred on a production extruder. By using scale–up rules, operating and design data of known machines, can be applied to larger and smaller extruders. These scale-up rules are commonly based on the principle of similarity. Most scale-up rules consist of simple power-type relations that were derived from analytical expressions describing individual steps in extrusion sequence. In the case of twin- screw extruders these rules have to be adjusted because of geometrical and thermodynamic restrictions Geometrical restriction is dependent on screw diameter and center- line distance. The other restriction is the heat transfer from the polymer melt during compounding process. Extrusion process can be analyzed by calculating the energy, specific energy, and temperature rise during the process with respect not only to the viscosity of the melt, but also to the screw geometry, (location and number of transport elements, kneading sections and blisters, pitch, positive or negative, screw clearance, and flight width) and screw speed. In geometrical scaling, mixing is kept constant. Throughput, energy, and torque all scale with the third power of the diameter. Geometrical scaling is, however, allowed only if both extruders are operating under fully adiabatic conditions. As soon as heat exchange with the barrel wall is more important in the process, throughput can only scale with the second power of the diameter [1]. If the temperature in the small extruder is different in the large extruder, viscosities are expected to be different. Mixing rules does not provide the same laws as scaling according to the rules of equal thermal development

Flow behavior can be simulated by using finite element method with linear interpolation for pressure and a linear interpolation enriched with a bubble node at the center for velocity and using the static condensation technique. Shear and elongational viscosity models were the truncated power-law models [2].

$$\eta_s = Ae_{II}^{n-1} \quad \text{for} \quad e_{II} \geq e_0, \qquad \eta_s = \eta_0 \quad \text{for} \quad e_{II} \leq e_0$$

$$\eta_a = Be_{II}^{m_a-1} \quad \text{for} \quad e_{II} \geq e_0, \qquad \eta_a = 3\eta_0 \quad \text{for} \quad e_{II} \leq e_0$$

and

$$\eta_p = Ce_{II}^{m_p-1} \quad \text{for} \quad e_{II} \geq e_0, \qquad \eta_p = 4\eta_0 \quad \text{for} \quad e_{II} \leq e_0$$

----------------------(1)

where ηs is the shear viscosity, ηa and ηp are respectively the axisymmetric and planar elongational viscosities, η0 is the zero-shear viscosity, n, ma and mp are the power-law indices, and A, B and C are the consistency coefficients

During scale-up, several similarities should exist. Geometric similarity exists if the ratio between any two length parameters in the large scale equipment is the same as the ratio between the corresponding lengths in the small scale model. It can be ensured by keeping the same rotational speed, through put proportional to the cube of the diameter and by designing the die such that pressure ratio equals the ratio between end viscosities. For hydrodynamic similarity two requirements should be fulfilled: The dimensionless flow profiles should be equal and for twin screw extruders, both extruders should have the same (dimensionless) filled length. Equal dimensionless flow profiles lead to equal shear rates in corresponding locations, but not to equal velocities. Similarity in residence times means equal residence times in the small scale and large scale equipment. Equal residence time can be achieved if the volume divided by the throughput remains Constant. Two types of thermal similarities can be used: equal average end temperatures and similar temperature profiles. The concept of equal average end temperatures can be applied if the extruder operates adiabatically. From the dimensionless energy equation it follows that thermal similarity can be attained if the dimensionless numbers of Graez and Brinkmann are the same for both sizes of machines. The Graez number and the Brinkmann number are given below [3].

$$G_Z = \frac{UH^2}{aL} \text{ and } B_r = \frac{\mu U^2}{\lambda \Delta T} \quad \text{--(2)}$$

where λ is the thermal conductivity, T is the temperature, a is the throttle coefficient, L is the screw length, H is the channel depth, μ is the viscosity and U=πND with N is the screw speed and D is the screw diameter. By keeping the screw diameter constant the influence of throughput on the system energy balance is specified for several polymers.

A theoretical approach for partly filled, geometrically similar systems is validated by means of experimental data extracted from a Coperion W&P ZSK-30. The results provide a basis to transfer operating points for co-rotating twin screw extruder [4].

EXPERIMENTAL

Polyesters with different crosslink density were subjected to twin- screw extrusion under different through puts with different die temperatures. The extrudate materials were analyzed for melt index and Equilibrium Elastic Modulus. The Equilibrium Elastic Modulus was analyzed by subjecting the materials to dynamic sinusoidal deformation at linear visco-elastic strains, using time-temperature superposition and using polynomial fits to get plateau modulus.

RESULTS AND DISCUSSION

Crosslinked polymers with different levels of cross-linking with and without fillers were analyzed and equilibrium plateau modulus was obtained using time temperature superposition. Dependence of melt flow index of materials, MFI on Ge showed interesting features. Power law exponent varied with the starting crosslinked polyester and screw speed. Ge(1) and Ge(2) runs used 30% and 20% gel, respectively and gave lower power law exponent. The Ge(3) runs used polyesters which varied in gels from 5% to 30%, whereas, the Ge(4) runs used polyesters that varied in gels from 10% to 20%.

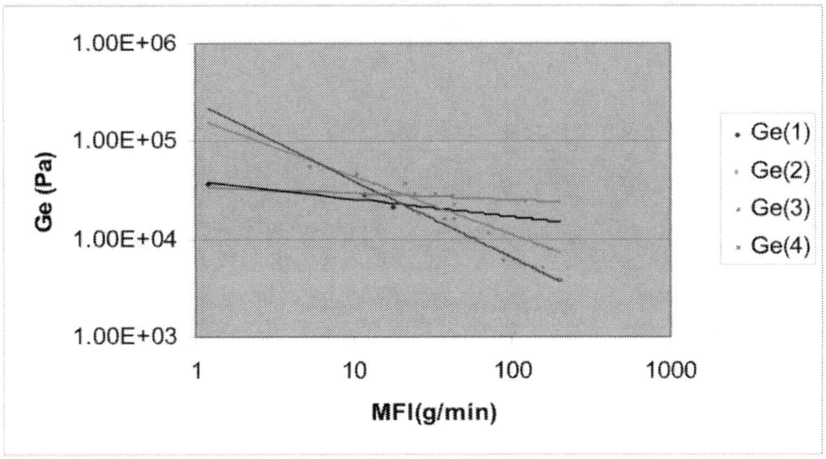

Figure 1 shows power law fits between Ge and MFI for polyesters with gel (cross-link density).

As shown in Figure 2, the affect of scale-up for cross-linked polyesters is to change power law exponent of Ge dependence on MFI. This effect is obvious at the lower MFI in ZSK2, a scale-up extruder where shear stresses are higher with higher die temperature 140C compared to ZSK1 where die temperature was 120C at the same screw speed of 300RPM.

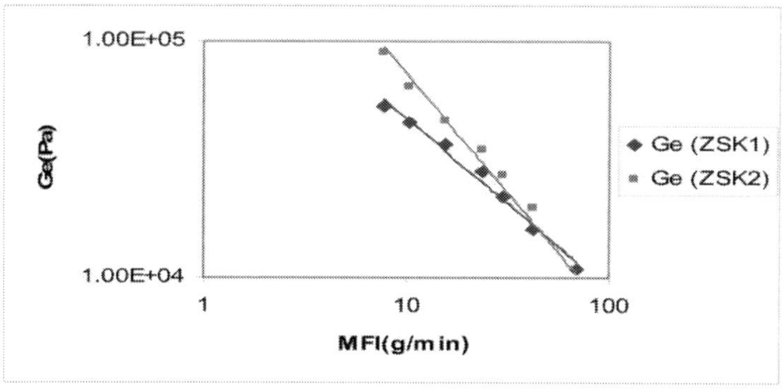

Figure 2 shows the dependence of Ge on MFI

CONCLUSION

In reviewing scale up models power law viscous models are used and visco-elasticity is not considered and temperature variation from extruder to extruder can vary from frictional heating. In cross-linked polymers and composites, elasticity is significantly higher. A simple empirical power law model of correlation between Equilibrium Elastic modulus and MFI is presented. Future work would be to use a viscoelastic constitutive equation in connection with temperature development in laminar flow.

ACKNOWLWDGEMENT

Support of Xerox Corporation and Thermal Analysis was critical in this work.

REFERENCES

1. H. E. H Meijers and. P. H M Elemens (1998), Polymer Engineering and Science, 28(5), pp 275-290
2. A. Shah and M. Gupta (2003), ANTEC, pp*322-326*
3. L P.B.M Janssen, and L. Mościcki (2006) TEKA Kom. Mot. Energ. Roln., 2006, 6A, pp82–91
4. H. Potente; K Kretschmer, , T.C Pohl, (2002), PPS-18

Application of the GC-SAFT-VR Equation to Predict Fluid Phase Behavior of Associating Systems

M. Carolina Dos Ramos, Department of Chemical and Biomolecular Engineering, Vanderbilt University, Nashville, TN

Nurul A. Aziz, Department of Chemical and Biomolecular Engineering, Vanderbilt University, Nashville, TN

Joel R. Westwood, Department of Chemical and Biomolecular Engineering, Vanderbilt University, Nashville, TN

Jessica D. Haley, Department of Chemical and Biomolecular Engineering, Vanderbilt University, Nashville, TN

Clare McCabe, Department of Chemical and Biomolecular Engineering, Vanderbilt University, Nashville, TN

The statistical associating fluid theory for potentials of variable range (SAFT-VR) [1] is a widely used molecular-based equation of state that has been successfully applied to study a wide range of fluid systems. It provides a framework in which the effects of molecular shape and interactions on the thermodynamics and phase behavior of fluids can be separated and quantified. In the original approach, molecules were modeled as chains composed of identical segments, where the heterogeneity of molecules in terms of structure and functional groups was described implicitly through effective parameters, leading to the use of binary interaction parameters in order to describe mixture phase behavior in non-ideal systems. To overcome this limitation, in recent work the theory was extended to model chains composed of segments of different size and/or energy of interaction, thus enabling the development of a group-contribution SAFT-VR approach (GC-SAFT-VR) [2]. In previous work, parameters for a range of different key functional groups (CH_3, CH_2, CH, CH_2=CH, C=O, C_6H_5, CH_3O and CH_2O esters groups) have been determined by fitting to experimental vapor pressure and saturated liquid density data for a number of small molecules containing the functional groups of interest. Transferability of the parameters was tested by comparing the theoretical predictions with experimental data for pure fluids and binary mixtures not included in the fitting process, as well as by studying the VLE and LLE of small molecules in polymer systems [3]. In this work, we further test the GC-SAFT-VR approach through the study of phase behavior in associating systems such as alcohols, amines, aldehydes, and carboxylic acids, and their mixtures. New functional groups (OH, NH_2, CH=O, COOH) are defined and their molecular parameters characterized by again fitting to experimental vapor pressure and saturated liquid density data for selected small molecules. Transferability of the parameters is tested by comparing the theoretical predictions with experimental data for pure fluids and binary mixtures not included in the fitting process. The GC-SAFT-VR approach is found to predict the phase behavior of the systems studied in excellent agreement with experimental data without adjusting the group parameters to binary mixture data; thus achieving the goal of developing a predictive tool for studying a wide range of fluid phase equilibria.

[1] A. Gil-Villegas, A. Galindo, P. J. Whitehead, S. J. Mills, G. Jackson, and A. N. Burgess, *Journal of Chemical Physics* **106** (10), 4168 (1997); A. Galindo, L. A. Davies, A. Gil-Villegas, and G. Jackson, *Molecular Physics* **93** (2), 241 (1998).

[2] Y. Peng, K. D. Goff, M. C. dos Ramos, and C. McCabe, *Fluid Phase Equilibria* **277** (2), 131 (2009).

[3] Y. Peng, K. D. Goff, M. C. dos Ramos, and C. McCabe, *Submitted to Fluid Phase Equilibria* (2009).

See more of this Session: Thermodynamic Properties and Phase Behavior IV
See more of this Group/Topical: Engineering Sciences and Fundamentals

COSMO-RS for the Estimation of Pka in Non-Aqueous Solvents

Frank Eckert, COSMOlogic GmbH&CoKG, Leverkusen, Germany

Andreas Klamt, COSMOlogic GmbH&CoKG, Leverkusen, Germany

Michael Diedenhofen, COSMOlogic GmbH&CoKG, Leverkusen, Germany

Experimental measurement of non-aqueous phase pKa values is still a demanding task. The rapid development of efficient quantum chemical (QC) methods in the last years has opened new perspectives for the rigorous prediction of liquid phase pKa. The Conductor-like Screening Model for Real Solvents (COSMO-RS) [1,2], is a combination of QC with a statistical thermodynamics method for local surface interactions, which takes into account local deviations from dielectric behaviour as well as hydrogen bonding. In this approach all information about solutes and solvents is extracted from initial continuum solvation QC calculations. COSMO-RS is widely validated in the chemical engineering thermodynamics for the prediction of activity coefficients, solubility and vapour pressures of demanding compounds in pure and mixed solvents. This presentation reports some novel extensions and applications of the COSMO-RS methodology towards the prediction of acidity and basicity pKa of solutes in non-aqueous solvents such as acetonitrile, dimethylsulfoxide (DMSO), 1,2-dimethoxyethane (DME) and heptane. Correlating the quantum chemical dissociation free energy ΔGdiss of a molecule solvated in a solvent, with its experimental pKa via a linear free energy relationship (LFER) a theoretical a priori prediction method for pKa in non-aqueous solvents has been achieved, which has the regression constant and the slope as only adjusted parameters. The method is not restricted to certain compound classes. For all solvents considered in the study a standard deviation below 1.5 pKa units was found in the correlations and independent test sets. In concordance with findings for aqueous pKa predictions [3] for all non-aqueous solvents the slope of experimental pKa vs. ΔGdiss was found to be smaller than the value of 1/RTln(10) that is expected from electrostatic theory.

[1] F. Eckert, A. Klamt, AIChE Journal, 2002, 48, 369.

[2] A. Klamt, F. Eckert, Fluid Phase Equilibria, 2000, 172, 43.

[3] A. Klamt, F. Eckert, M. Diedenhofen, M. E. Beck, J. Chem. Phys. A., 2003, 107, 9380.

See more of this Session: Thermodynamic Properties and Phase Behavior IV
See more of this Group/Topical: Engineering Sciences and Fundamentals

A Thermodynamic Model for Electrolyte Solutions with Partial Ionization

Ricardo Macías-Salinas, ESIQIE-Departamento de Ingeniería Química, Instituto Politécnico Nacional, México, D.F., Mexico

A Helmholtz-energy-based model has been developed to describe the thermodynamic properties of single strong electrolytes (of the alkali-halide type) in water within a wide range of temperatures from 25 °C to near the critical point of the solvent. The new model contains three major contributions: (1) a discrete-solvent term to account for short-range interactions between uncharged particles using the Peng-Robinson equation of state, (2) a stepwise hydration equilibrium between cation and water, and (3) a charge-charge interaction term given by the explicit mean-spherical-approximation (MSA) expression. The final model contains only three adjustable parameters: ionization constant, solvated diameter of the cation and the Peng-Robinson short-range interaction parameter between cation and water. Only the dissociation constant was considered to change with temperature. The proposed model was applied to the representation of mean ionic activity coefficients, osmotic coefficients and volumetric properties of various 1-1, 2-1, 1-2 and 2-2 electrolytes in water over wide range of temperatures, pressures and salt molalities. The results indicated a good agreement between the experimental data and those calculated using the present model.

See more of this Session: Thermodynamic Properties and Phase Behavior IV
See more of this Group/Topical: Engineering Sciences and Fundamentals

Modeling of Aqueous Electrolyte Solutions - From Simple Ions to Polyelectrolytes -

Christoph Held, Department of Biochemical and Chemical Engineering, Technische Universität Dortmund, Dortmund, Germany

Gabriele Sadowski, Department of Biochemical and Chemical Engineering, Technische Universität Dortmund, Dortmund, Germany

Shahbaz Naeem, Department of Biochemical and Chemical Engineering, Technische Universität Dortmund, Dortmund, Germany

Electrolyte solutions play an important role in biotechnology and chemical engineering. To describe systems containing water and ions, an Equation of State (the PC-SAFT EoS) has been combined with a Coulomb approach (Debye-Hückel) [1] and applied to numerous aqueous inorganic salt solutions [2]. ePC-SAFT was found to be an excellent model for salt solutions containing of nearly spherical ions. In order to model systems with extensive ion-pair formation (e.g. solutions containing cadmium halides), a chemical-reaction approach accounting for dissociation/association equilibrium within ePC-SAFT [1] led to a successful description of experimental data [3].

Based on the results of those inorganic-electrolyte systems, phase equilibria of more complex, polyelectrolyte solutions are investigated. As the Debye-Hückel contribution alone is not able to describe the electrostatic interactions of polyelectrolyte chains, the ePC-SAFT model had to be extended to model polyelectrolyte systems. For that purpose, the reference term "hard-chain" was replaced by a "charged-hard-chain" contribution (as proposed in [4]). Moreover, the counter-ion condensation present in polyelectrolyte solutions was explicitly accounted for.

This approach is successfully applied to model the phase behavior of aqueous poly(acrylic acid) solutions as function of the degree of neutralization (different amount of charged monomer units in the polymer backbone). Moreover, the influence of added salt on the phase behavior of polyelectrolyte solutions can be described in good agreement with experimental data.

References

[1] L.F. Cameretti, G. Sadowski, J.M. Mollerup, Ind. Eng. Chem. Res. 44 (2005) 3355-3362; ibid., 8944.

[2] C. Held, L.F. Cameretti, G. Sadowski, Fluid Phase Equlilib. 270 (2008) 87-96.

[3] C. Held, G. Sadowski, Fluid Phase Equlilib. (2009), 279 (2009) 141-148.

[4] J. Jiang, J. Feng, H. Liu, Y. Hu, J. Chem. Phys. 124 (2006) 144908(1-6).

See more of this Session: Thermodynamic Properties and Phase Behavior IV
See more of this Group/Topical: Engineering Sciences and Fundamentals

Thermodynamic Properties of Polar Fluids

Richard J. Sadus, Centre for Molecular Simulation, Swinburne University of Technology, Hawthorn, Victoria 3122, Australia

The thermodynamic properties of fluids are ultimately determined by the nature and type of intermolecular interactions. Simple fluids such as, the inert gases can often be adequately described by intermolecular potentials that incorporate the contributions for dispersion forces and repulsive interactions. In contrast, fluids of practical interest commonly involve polar forces arising from interactions between permanent dipoles and/or quadrupoles. Indeed, there are many examples of fluids that have both dipolar and quadrupole interactions. It is difficult to isolate the affect of such competing non-polar interactions on experimentally observed properties. It is in this context that molecular simulation has a useful role. In this work, we report molecular dynamics simulation of the effect of dipole and quadrupole interactions on the thermodynamics and phase behaviour of both pure fluids and fluid mixtures. In particular, mixtures of different types of polarity are studied allowing us to observe the influence of dissimilar polar interactions.

See more of this Session: Thermodynamic Properties and Phase Behavior IV
See more of this Group/Topical: Engineering Sciences and Fundamentals

Thermodynamic Modeling of the Sulfuric Acid - Water - Sulfur Trioxide System with Symmetric Electrolyte NRTL Model

Chau-Chyun Chen, R&D, Aspen Technology, Inc., Burlington, MA

Huiling Que, R&D, AspenTech Limited, Shanghai, China

Shu Wang, R&D, Aspen Technology, Inc., Burlington, MA

Yuhua Song, R&D, Aspen Technology, Inc., Burlington, MA

Properties and phase Equilibria of the sulfuric acid - water - sulfur trioxide system is of great interest to the industry. Two recent application examples are dew point calculations of flue gas in oxyfuel combustion and process design for Sulfur-Iodine Cycle thermal chemical decomposition of water for hydrogen production. Recently developed symmetric electrolyte NRTL activity coefficient model is utilized to investigate the thermodynamics of the sulfuric acid - water - sulfur trioxide system. Partial dissociations of sulfuric acid and bisulfate ion in aqueous phase are considered. Hydrations of proton from the partial dissociations of sulfuric acid are assumed in the form of mono-hydrate and di-hydrate. Decomposition of sulfuric acid to water and sulfur trioxide is also accounted for. Excellent match between model predictions and available literature data are achieved for osmotic coefficient, liquid phase speciation, heat of solution, liquid phase heat capacity, and vapor-liquid equilibrium of the sulfuric acid - water - sulfur trioxide system over the whole concentration range.

See more of this Session: Thermodynamic Properties and Phase Behavior IV
See more of this Group/Topical: Engineering Sciences and Fundamentals

Developing Speciation Reactions for Properties and Phase Equilibria of Complex Systems

John P. O'Connell, Chemical Engineering, University of Virginia, Charlottesville, VA

James E. Murphy, University of Virginia, Charlottesville, VA

Paul M. Mathias, Fluor Corporation, Aliso Viejo, CA

Fluid systems of current processing interest often involve aqueous ionic solutions with complex energy and phase behavior. Developing property and phase equilibrium models for these systems is often challenging.

In lieu of fully quantum-mechanically based molecular simulations, descriptions of the commonly found strong association and solvation effects use a chemical theory approach where reactions form new species which show nonideality. Efficient development of successful models requires a priori determination of the likely species with confirmation from both theory and experiment.

The talk will suggest strategies for developing speciation reactions and describe their application in processes including thermochemical decomposition of water to make hydrogen, post-combustion carbon capture, and chemicals manufacture.

See more of this Session: Thermodynamic Properties and Phase Behavior IV
See more of this Group/Topical: Engineering Sciences and Fundamentals

Droplet motion in a microfluidic channel with a constriction

Moon Soo Lee
Department of Chemical and Biomolecular Engineering,
University of Maryland
College Park, MD, USA

Panagiotis Dimitrakopoulos
Department of Chemical and Biomolecular Engineering,
Bioengineering Graduate Program
University of Maryland
College Park, MD, USA

Introduction

The study of the droplet behavior through a constriction in a rectangular channel is a problem encountered in a broad range of applications in microfluidic devices and enhance oil recovery.

Results

Utilizing our three-dimensional spectral boundary element algorithm, we investigate the motion, deformation and critical blocking conditions of a viscous droplet in a square channel at low-Reynolds-number flows.

Conclusions

In the talk we will present the effects of several parameters affecting the dynamics of the droplet as it is squeezed through the constriction including droplet size, size of constriction and non-symmetrical constriction shapes.

The Emptying of Liquid From Gravure Cells

Shawn Dodds, Chemical Engineering and Materials Science, University of Minnesota, Minneapolis, MN

Satish Kumar, Chemical Engineering and Materials Science, University of Minnesota, Minneapolis, MN

The transfer of liquid from one surface to another is important in a wide variety of both natural and industrial settings, including printing processes such as gravure, lithography, and microcontacting. In these processes it is desirable to maximize the amount of liquid transferred from the printing image to the final substrate in order to minimize fouling of the image surface, as well as to avoid waste of potentially valuable and/or toxic compounds. To accomplish this requires that the printing liquid dewets the image surface, leaving behind as little residual liquid as possible. Therefore, it is important to understand how the surface properties of both the image and the final substrate affect the dewetting process, and thus the liquid transfer.

As a first step in understanding liquid transfer in printing processes, we use the Galerkin/finite element method to study the stretching of a finite volume of Newtonian liquid confined between a stationary trapezoidal cavity, which represents a model gravure cell, and a moving flat plate. The contact lines are allowed to slip along both surfaces, and we evaluate the effect of gravure cell geometry, contact angle, and capillary number on the fraction of liquid transferred to the moving plate. As the gravure cell is made wider at a fixed capillary number, the fraction of liquid transferred is increased. However, liquid transfer is often limited by contact line pinning, which prevents the gravure cell from being completely emptied.

To complement this computational study, we have also used flow visualization to monitor the stretching of a liquid drop both from beneath and beside the drop. The side view is used to track the breakup of the drop during stretching, allowing us to quantify liquid transfer using image analysis. Viewing from beneath the drop allows us to track the motion of the contact line, making it possible to observe both contact line pinning events and non-axisymmetric modes of breakup. By performing both sets of visualizations, we can better understand how surface properties affect the motion of the contact line, as well as the liquid transfer.

See more of this Session: Poster Session in Fluid Mechanics
See more of this Group/Topical: Engineering Sciences and Fundamentals

Using the Level Set Method to Model Droplet Motion through An Ambient Miscible Fluid

Ileana Carpen, Chemical Engineering, Tennessee Technological University, Cookeville, TN

Manohar Gottapu, Chemical Engineering, Tennessee Technological University, Cookeville, TN

Droplet systems (liquid/liquid, liquid/gas) have widespread applications in the medical, industrial, and research fields, examples of which are aerosolized drugs, spray drying, printing, and liquid-liquid extraction. Most applications use immiscible fluid pairs, which are also the most studied and best understood cases. The classical 1911 Hadamard-Rybczynski solution broke new ground when the two scientists presented the creeping flow solution for the flow around and in an immiscible drop. Miscible pairs have been less tractable for analytical/modeling studies, and comparably little work has been done in this area. Arce and coworkers have presented low Reynolds number results for the buoyancy-driven motion of a droplet in a partially miscible high viscosity fluid which are qualitatively different from the classical motion of an immiscible droplet. Their experimental studies have shown that a puddle region develops on the back side of the drop due to convective-diffusive flows. In this paper, we present our modeling results for a single droplet partially miscible with an infinite fluid phase and compare the predictions of this study with the experimental results of Arce and coworkers.

See more of this Session: Poster Session in Fluid Mechanics
See more of this Group/Topical: Engineering Sciences and Fundamentals

Hydrophobicity Effect On Gas-Liquid Mass Transfer in Microchannels

Gengzhi Yu, Mechanical Engineering, University of Washington, Seattle, WA

Amy Shen, Mechanical Engineering, University of Washington, Seattle, WA

Anthony Lawrenz, Materials Science and Engineering, University of Washington, Seattle, WA

Experiments for mass transfer of liquid (water) and gas (air) inside microchannels have been conducted under atmospheric pressure while varying the channel hydrophobicity to investigate the effect of surface wettability on the mass transfer rates. The results show that both the flow patterns and the mass transfer rates are dependent on the surface hydrophobicity of the channel. Both chemical and physical methods are used for the mass transfer measurements.

See more of this Session: Poster Session in Fluid Mechanics
See more of this Group/Topical: Engineering Sciences and Fundamentals

Gas Hold-up Profiles Investigation in Ionic Liquid Column

Haifeng Dong, State Key Lab of Multi-phase Complex Systems, Institute of Process Engineering, Chinese Academy of Sciences, Beijing, China

Lidong Zhao, State Key Lab of Multi-phase Complex Systems, Institute of Process Engineering, Chinese Academy of Sciences, Beijing, China

Lu Liu, State Key Lab of Multi-phase Complex Systems, Institute of Process Engineering, Chinese Academy of Sciences, Beijing, China

Xiaoling Wang, State Key Lab of Multi-phase Complex Systems, Institute of Process Engineering, Chinese Academy of Sciences, Beijing, China

Xiangping Zhang, State Key Lab of Multi-phase Complex Systems, Institute of Process Engineering, Chinese Academy of Sciences, Beijing, China

Suojiang Zhang, State Key Lab of Multi-phase Complex Systems, Institute of Process Engineering, Chinese Academy of Sciences, Beijing, China

Ionic liquids (ILs) are melts of salts which have much lower melting points than normal salts, usually less than 373K. The compounds contain many ion-pairs, not like molecular compounds, such as water and many other organic solvents. In contrast with salt solutions, the ILs consist solely of ions, without the presence of any solvent. They have a non-measurable vapour pressure, high thermal stability, ionic conductivity and a large electrochemical window and are able to solvate a variety of organic and inorganic species. These intrinsically useful characteristics have led to ionic liquids being regarded as greener alternatives to volatile organic solvents, namely as reaction media for synthesis, catalysis and biocatalysis In present work, the gas phase was air and the liquid phase was BMImBF4. The electrical resistance tomography (ERT) has been for the first time extended to measure void fraction wave characteristics in a bubble column (Fig.1, 0.203m diameter, 2.5m height) under high temperature (up to 473K) and high pressure (up to 3MPa). The effects of IL¡¯s temperatures, column pressure, axial liquid height (HD/D) and superficial gas velocities for radial variation of gas hold up and local holdup have been investigated. The holdup derived from ERT was also compared with the hold-up values measured by pressure transmitters. The hold-up profiles were found to depend strongly on superficial gas velocities and IL¡¯s temperature (Fig.2), while column pressure and axial liquid height (HD/D) influenced slightly(Fig.3). Key words: Ionic liquid; Bubble column; Hold up; Electrical resistance tomography. Acknowledgement: This work was supported by the National Natural Science Funds for Distinguished Young Scholar (No. 20625618) and the Instrument Developing Project of the Chinese Academy of Sciences(No. Y2005013)

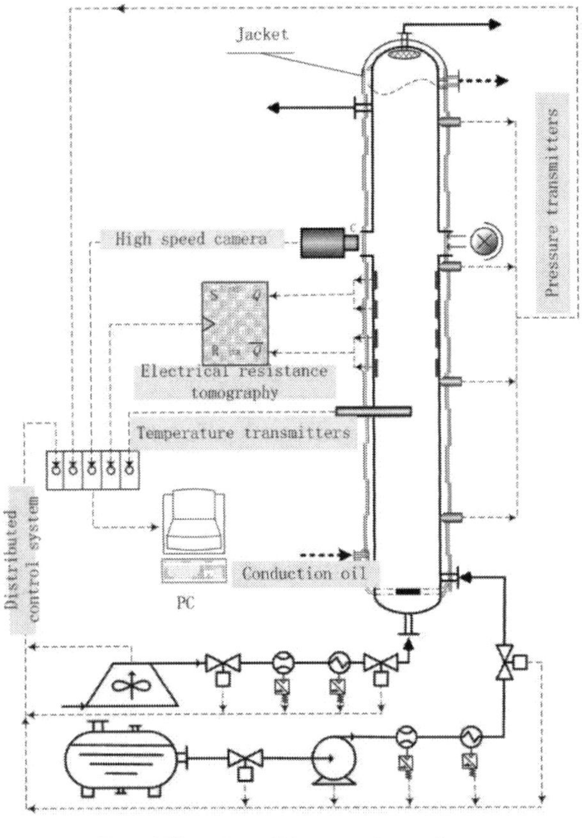

Fig.1 Schematics of the experimental setup.

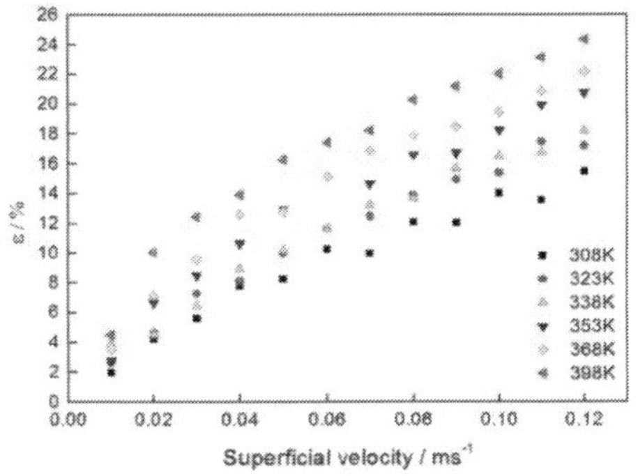

Fig.2 Holdup in the column at different superficial gas velocity and temperature.

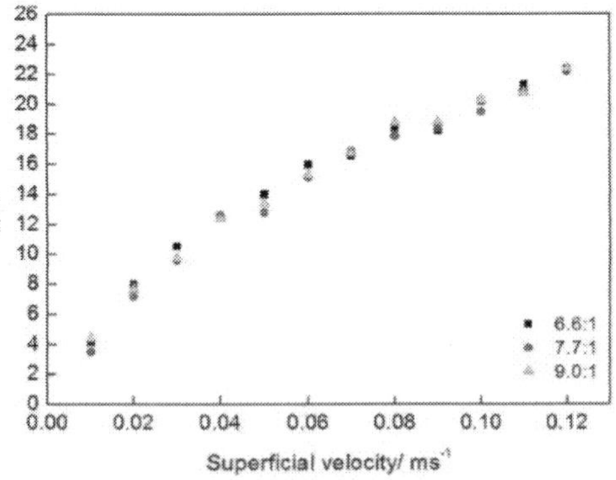

Fig.3 Gas holdup versus H_D/D at 368K.

See more of this Session: Poster Session in Fluid Mechanics
See more of this Group/Topical: Engineering Sciences and Fundamentals

Effect of Confinement-Geometry On the Suspension-Rheology and the Pressure-Drop Inside a Conduit

Shahin Navardi, Department of Mechanical Engineering, Texas Tech University, Lubbock, TX

Sukalyan Bhattacharya, Department of Mechanical Engineering, Texas Tech University, Lubbock, TX

Hydrodynamic interactions between a confining boundary and a suspended or deposited particle affect the dynamics of a number of colloidal systems. For example, a fixed particle deposited on the wall of a cylindrical conduit enhances the flow-resistance in the channel by inducing additional pressure-drop. Similarly, required pressure-gradient to drive a flow through a duct increases in presence of suspended particles leading to an increase in effective viscosity which is greater than the nominal viscosity of the solvent. In this talk, we present our results which provide a unified description of these phenomena.

In our analysis, we consider a sphere in a cylinder under creeping motion assumption. The key step of the mathematical formulation is a semianalytical approach where the Stokesian field is expanded in terms of two sets of basis functions corresponding to the particle and the conduit geometry, respectively. The obtained flow-solution is used in determination of far-field pressure governing either the pressure-drop across a fixed blockage or the rheology of the suspension of freely moving particles. Furthermore, we use a novel lubrication theory to independently verify these analytical results.

Our results include the following components. First, we consider a spherical obstruction inside a conduit and determine the pressure-drop across it in presence of a parabolic flow quantifying the increase in resistivity of the channel. Then, we account for the pressure generated by an axially translating as well as azimuthally rotating sphere in static fluid inside the cylinder. Finally, we combine these findings to compute the pressure-field across a freely translating and rotating particle in parabolic flow. We relate the last result to the effective viscosity of a dilute suspension, and describe the effect of the confinement on the rheology. Such geometric effect is absent in an unbounded domain where the viscosity of the solution is expressed by Einstein's relation. We have checked that under the limit of very large conduit-diameter our calculation approaches to the Einstein's result.

See more of this Session: Poster Session in Fluid Mechanics
See more of this Group/Topical: Engineering Sciences and Fundamentals

Particle Migration and Torque Evolution in Oscillatory Torsional Flows of Concentrated Suspensions

Kapil V. Deshpande, Department of Chemical and Biochemical Engineering, Rutgers University, Piscataway, NJ

Nina C. Shapley, Department of Chemical and Biochemical Engineering, Rutgers University, Piscataway, NJ

In this study, we observe particle motion and rheology in oscillatory torsional flow of a monomodal 0.4 particle volume fraction suspension over a duration of 4800 oscillations. The oscillation amplitude covers the transition range below the minimum strain for the formation of steady-state microstructure and the oscillation frequency is also varied. We detect slow, radially outward drift of tracer particles, which is mainly a function of the oscillatory strain amplitude with slight dependence on the frequency of oscillation. We compare measured particle drift velocities to predicted values estimated from the suspension balance model. Simultaneously, we capture the torque evolution. At the largest strain amplitudes and lowest frequencies, the torque increases as the flow evolves, as was previously observed in steady flow, and a local minimum is detected at intermediate strain values, in agreement with other studies of oscillatory flow. In summary, the results suggest the interplay at intermediate oscillation amplitudes between radial particle migration driven by the overall particle stress balance and microstructural rearrangement driven by local self-assembly of the particles into layers.

See more of this Session: Poster Session in Fluid Mechanics
See more of this Group/Topical: Engineering Sciences and Fundamentals

Gel Formation and Aging in Concentrated Nanoparticle Suspensions

Subramanian Ramakrishnan, Department of Chemical and Biomedical Engineering, Florida A&M University - Florida State University, Tallahassee, FL

Robert L. Leheny, Johns Hopkins University, Baltimore, MD

Hongyu Guo, Johns Hopkins University, Baltimore, MD

James Harden, Physics, University of Ottawa, Ottawa, ON, Canada

We report a combined x-ray photon correlation spectroscopy (XPCS) and rheometry study of the evolution of concentrated suspensions of nanometer-scale colloids undergoing gelation and aging. The suspensions are comprised of silica colloids, 45 nm in diameter, coated with octadecyl-hydrocarbon chains in decalin at colloidal volume fractions ranging from 0.20 to 0.43. At high temperatures, decalin is a good solvent for the chains, which form a solvated brush that stabilizes the colloids. At low temperature, the solvent quality is poor, leading to a weak, temperature-dependent, short-range attraction between the colloids that drives a reversible gel transition. Following a quench through this transition, the shear modulus grows at a rate that depends strongly on temperature. The intermediate scattering function measured with XPCS displays two features, a plateau value that provides information about constrained local dynamics in the gel and a terminal relaxation time that provides information about relaxation of residual stress. From the wave-vector dependence of the plateau value, a localization length can be extracted. At lower colloidal volume fraction, the relationship between the localization length and the shear modulus agrees quantitatively with the prediction of Chen and Schweitzer based on a simplified mode coupling theory [1], but deviations from the prediction are observed at early times for higher volume fraction.

[1] Y.-L. Chen and K. S. Schweitzer, J. Chem. Phys. 120, 7212 (2004).

See more of this Session: Poster Session in Fluid Mechanics
See more of this Group/Topical: Engineering Sciences and Fundamentals

Analysis of Structures Formed by Application of Step Strain to Dense Colloidal Suspensions

Laura T. Shereda, Chemical Engineering, University of Michigan, Ann Arbor, Ann Arbor, MI

Ronald G Larson, Chemical Engineering, University of Michigan, Ann Arbor, Ann Arbor, MI

Michael J. Solomon, Chemical Engineering, University of Michigan, Ann Arbor, Ann Arbor, MI

Under shear flow dense colloidal suspensions have demonstrated shear thinning and shear thickening rheological behavior and have been studied extensively due to both fundamental interest of the complex system and crystal formation under shear flow (Krishnamurthy, 2005, J. Rheology. Behavior of colloidal suspensions has been studied in both hard sphere and charged stabilized systems in both simple and oscillatory shear by means of dynamic light scattering (Ackerson et al, 1988, PRL). Ackerson et al (Ackerson 1991, Physica A) established stress and strain criteria for colloidal crystal formation under shear flow, however, in such methods information on the disorder to order transition is difficult to acquire. (Cerda 2008). In order to understand the formation of colloidal crystals under shear flow we studied structural changes in dense colloidal suspensions by direct visualization using confocal microscopy. A parallel plate assembly with a stationary bottom plate and a moveable top plate is loaded with suspensions of charge stabilized, refractive index matched poly (12-hydroxystearic acid) stabilized poly(methyl methacrylate) colloids of size about one micron suspended in the viscous solvent dioctyl phthalate (colloid volume fraction ~ 0.35) with small quantities of photopolymer. Samples were imaged, a step strain was applied, and samples were immediately quenched by exposure to UV light and imaged again. Strain amplitudes were of order of amplitude 1. Particle centroids were located in 3D by means of quantitative image processing. Local crystallinity was quantified by application of local bond order parameter criteria developed by ten Wolde et al. (ten Wolde et al., 1996, Journal of Chemical Physics). We determined the effect of an application of a precise amount of step strain at controlled shear rate in a parallel plate device and compare the results to previous shear studies as well as our previous work with spin coating (Shereda 2008, PRL).

See more of this Session: Poster Session in Fluid Mechanics
See more of this Group/Topical: Engineering Sciences and Fundamentals

Orientational Ordering in Colloidal Rods by Sedimentation and Centrifugation

Deshpremy Mukhija, Chemical Engineering, University of Michigan, Ann Arbor, MI

Aayush A. Shah, Macromolecular Science and Engineering, University of Michigan, Ann Arbor, MI

Michael J. Solomon, Chemical Engineering, University of Michigan, Ann Arbor, MI

Sedimentation and the effective gravitational force of centrifugation were applied to produce 3D ordered assemblies of monodisperse colloidal rods that vary from 2 to 5 microns in the length and are of aspect ratio from 3.6 to 8.0. By means of confocal laser scanning microscopy and image processing, we quantified the extent of ordering, identified variables that controlled the assembly, and evaluated possible mechanisms to explain the orientational ordering. The quality of orientational order was quantified by direct calculation of the nematic order parameter (S). The dimensionless magnitude of the applied field was assessed by means of the Peclet number, a parameter that compares the time scale for rotational diffusion to the time scale for sedimentation. Experimental data showed that the effect of field strength on order parameter is a function that displays a complex dependence on aspect ratio. The scaling of the results for all aspect ratio with the equilibrium isotropic-nematic volume fraction suggested that the achievement of orientational order in the sediment is principally determined by its local volume fraction. Non- equilibrium effects, however, perturbed quantitative details of the dependence of S on the local sediment volume fraction from the known behavior for equilibrium rod suspensions.

See more of this Session: Poster Session in Fluid Mechanics
See more of this Group/Topical: Engineering Sciences and Fundamentals

Migration Mechanisms for Rigid Polymers and Nanorods Sheared at High Peclet Number

Joontaek Park, Chemical Engineering, University of Florida, Gainesville, FL

Jason E. Butler, Chemical Engineering, University of Florida, Gainesville, FL

Mechanisms leading to the net migration for rigid, Brownian rods sheared at high rates are explored. At these large, but finite, values of the Peclet number, the competition between the randomization caused by Brownian rotation and tendency of the shearing flow to align the rods breaks the symmetry about the flow direction of the orientation distribution. The asymmetric distribution of the rod's orientation can then couple with inhomogeneous shear flows to generate a net migration. Examples include the migration away from bounding walls in simple shear flow and Poiseuille flow, where the inhomogeneous field is produced by hydrodynamic interactions with the walls, as well as the net migration of rods toward the axis of rotation in torsional flows. These effects are not inertial, as the studies are performed under the conditions of Stokes flow and each mechanism is analogous with similar phenomena for flexible polymers, though the rigid rods require a much higher value of the Peclet number before the migration becomes significant. Rheological impacts of the migration mechanism are slight for the case of simple shear flow, but can be significant for torsional flows.

See more of this Session: Poster Session in Fluid Mechanics
See more of this Group/Topical: Engineering Sciences and Fundamentals

Using Rheo-SALS to Study Shear Induced Phase Separation (SIPS) in Aqueous Solutions of Cationic Surfactant and Salt

Prachi Thareja, Chemical Engineeirng, University of Delaware, Newark, DE

Ingo Hoffmann, Institut für Chemie Technische Universität Berlin, Berlin, Germany

Matthew E. Helgeson, Chemical Engineering, University of Delaware, Newark, DE

Matthew W. Liberatore, Department of Chemical Engineering, Colorado School of Mines, Golden, CO

Michael Gradzielski, Institut für Chemie Technische Universität Berlin, Berlin, Germany

Norman J. Wagner, Chemical Engineering, University of Delaware, Newark, DE

Cationic surfactants in the presence of salt can self - assemble into long, flexible chains, known as "thread" or "worm like micelles" (WLM). In semi - dilute concentration range, the entanglement of these worm like micelles give rise to transient networks, forming viscoelastic solutions. These solutions are shown to exhibit very well defined "Maxwellian" linear rheology with a single relaxation time, but an unusual flow behavior under shear. Specifically, there have been recent studies elucidating the formation of shear banding, i.e., the splitting of a viscometric flow into bands of different local shear rates. In the present study, we investigate further the nonlinear shear rheology of cationic surfactant solutions in conjunction with small angle light scattering (SALS): specifically near the miscibility gap of 40mM Erucyl bis(hydroxyethyl) methylammonium chloride (EHAC) and sodium salicylate (NaSal), by using a rheo-SALS instrument (TA instruments) developed in our group at the University of Delaware.

Isotropic solutions of a high NaSal concentration (200-800mM), show shear-banding that manifests rheologically as a stress plateau between 25 to 45°C. Interestingly, this stress plateau is observed to be very weakly dependent on the temperature, but the critical shear rates vary systematically with temperature. The SALS pattern in flow-vorticity plane show no scattering below the critical shear rate and a "butterfly" pattern with enhanced scattering in the flow direction. This is indicative of a shear induced phase separation (SIPS) with large structures aligned along the vorticity direction. The isotropic solutions with low NaSal (<20mM) do not show any evidence of shear-banding or SIPS. We observe a maximum in zero shear viscosity for which the NaSal concentration is independent of the temperature. This behavior is unlike that of cationic WLM surfactant solutions that exhibit shear banding in the vicinity of an isotropic-nematic phase transition, pointing to the possibility of different underlying mechanisms driving shear banding in highly branched WLM solutions.

See more of this Session: Poster Session in Fluid Mechanics
See more of this Group/Topical: Engineering Sciences and Fundamentals

Direct Comparison of the Static and Dynamic Properties of a Semi-Flexible Polymer Calculated Using Lattice-Boltzmann and Brownian Dynamics Simulations

A.J.C Ladd, Chemical Engineering, University of Florida, Gainesville, FL

Rahul Kekre, Department of Chemical Engineering, University of Florida, Gainesville, FL

Jason E. Butler, Chemical Engineering, University of Florida, Gainesville, FL

We compare the results of lattice-Boltzmann and Brownian dynamics simulations of the static and dynamic properties of linear chain molecules. Results of published studies using either of the methods show similar results, but different choices of polymer model preclude a quantitative comparison. Here, simulations of an identical model of a linear polymer chain enable the quantitative conclusion that results from the lattice-Boltzmann method agree with those from Brownian dynamics within one percent. Presented results include a systematic study of parameters that may affect the accuracy of the of the lattice-Boltzmann simulations, including the temperature, mass of the polymer, viscosity of the fluid, and the lattice-Boltzmann grid spacing with respect to the polymer size.

See more of this Session: Poster Session in Fluid Mechanics
See more of this Group/Topical: Engineering Sciences and Fundamentals

Hemodynamic forces exerted on moving and adherent leukocytes in post-capillary blood vessels

Shugi Kuriakose
Department of Chemical and Biomolecular Engineering,
University of Maryland
College Park, MD, USA

Panagiotis Dimitrakopoulos
Department of Chemical and Biomolecular Engineering,
Bioengineering Graduate Program
University of Maryland
College Park, MD, USA

Introduction

In this talk we investigate the hemodynamic forces on a moving leukocyte in the presence of an adherent leukocyte or a protuberance on the blood vessel wall. In addition, we investigate the hemodynamic forces on an adherent leukocyte in the presence of a moving leukocyte. This study extends the previous work of Wang and Dimitrakopoulos on the nature of hemodynamic forces on the vascular endothelium and adherent leukocytes [*Phys. Rev. Lett.*, **96**, 028106 (2006); *Phys. Fluids*, **18**, 087107 (2006)].

Results

Leukocytes are blood cells which defend the body against infections. These cells move towards the site of infection and thus their margination is governed by hemodynamic forces. The leukocytes then adhere to the vessel wall at the site of infection. Hemodynamic forces exerted on leukocytes adherent to the surface of blood vessels play a pivotal role in the cell adhesion process since they need to be balanced by the receptor-ligand binding forces on the surface of the cells.

Conclusions

In this talk we investigate the effects of flow strength, size of blood vessel, adhesion spreading, and distance between the moving and adherent leukocytes to the hemodynamic forces on both cells.

Effects of Velocity Structure On Turbulent Heat Transfer From the Wall

Anjani Kalyan Karna, Chemical, Biological and Materials Engineering, The University of Oklahoma, Norman, OK

Dimitrios V. Papavassiliou, Chemical, Biological, and Materials Engineering, University of Oklahoma, Norman, OK

This work involves numerical experiments that were conducted in a turbulent channel flow using Direct Numerical Simulation (DNS) in conjunction with the tracking of thermal walkers released from the wall of the channel in this flow field. This is a Lagrangian approach to the dispersion process. Insights about the mechanism of turbulent transport can be obtained by observing the behavior of heat markers that are released from a single instantaneous line heat source.

The goal is to investigate the correlation between the velocity and the temperature field in wall turbulence. The simulations were conducted in a computational box of dimensions 3800x600x1900 in viscous wall units, and for different Prandtl number fluids (Pr = 0.1, 0.7, 6, 20, and 50). The half channel height in viscous wall units was h = 300. Each computation was repeated for 5 different initial velocities of the flow field, in order to release heat in independent velocity field. The total number of the heat markers was 260,100, and they were released from a single line source at the wall. The trajectories of the heat markers were monitored in space and time as they moved through the channel. The behavior of these heat markers was observed in 12 successive locations downstream form the source ranging from h to 12h. In each of these locations the flow structures that carried thermal markers towards the channel center (i.e., those that transfer heat away from the wall) and flow structures that carried thermal markers towards the channel walls (i.e., those that transfer heat towards the wall) were studied as a function of the Prandtl number. The characteristic length scales of the flow that are contributing most to heat transfer were determined based on these structures. The characteristic time scales were obtained based on the calculation of Lagrangian correlation coefficients for the markers that are captured in each of these downstream locations.

See more of this Session: Poster Session in Fluid Mechanics
See more of this Group/Topical: Engineering Sciences and Fundamentals

Two Different Heat Transfer Asymptotes and Interplay of Elasticity and Inertia in Heat Transfer Enhancement in Laminar Flow in Straight Non-Circular Tubes

Dennis A. Siginer, Department of Mechanical Engineering & Department of Mathematics, Petroleum Institute, Abu Dhabi, United Arab Emirates

Mario F. Letelier, Department of Mechanical Engineering, Universidad de Santiago de Chile, Santiago, Chile

The fully developed thermal field in constant pressure gradient driven laminar flow of a class of nonlinear viscoelastic fluids with instantaneous elasticity in straight pipes of arbitrary contour with constant wall flux is investigated. The nonlinear fluids considered are constitutively represented by a class of single mode, non-affine constitutive equations. The driving forces can be large. Asymptotic series in terms of the Weissenberg number Wi are employed to expand the field variables. A continuous one-to-one mapping is used to obtain arbitrary tube contours from a base tube contour. The analytical method presented is capable of predicting the velocity and temperature fields in tubes with arbitrary cross-section. Heat transfer enhancement due to shear-thinning is identified together with the enhancement due to the inherent elasticity of the fluid. The latter is to a very large extent the result of secondary flows in the cross-section but there is a component due to first normal stress differences as well. Increasingly large enhancements are computed with increasing elasticity of the fluid as compared to its Newtonian counterpart. Order of magnitude larger enhancements are possible even with slightly viscoelastic fluids. The coupling between inertial and viscoelastic nonlinearities is crucial to enhancement. The asymptotic independence of Nu from elasticity with increasing Wi is shown analytically for the first time, that is functional dependance Nu = f(Pe,Wi) becomes Nu= f(Pe)with increasing Wi. Isotherms for the temperature field are discussed for non-circular contours such as the ellipse and the equilateral triangle together with the behavior of the average Nusselt number Nu, a function of the Reynolds Re, the Prandtl Pr and the Weissenberg Wi numbers. The change of type of the vorticity equation governs the trends in the behavior of Nu with increasing Wi and Re. The implications on the heat transfer enhancement is discussed in particular for slight deviations from Newtonian behavior where a rapid rise in enhancement seems to occur as opposed to the behavior for larger values of the Weissenberg number where the rate of increase is much slower. The asymptotic independence of Nu from elasticity with increasing Wi is related to the extent of the supercritical region controlled by the interaction of the viscoelastic Mach number M and the Elasticity number E, which mitigates and ultimately cancels the effect of the increasingly strong secondary flows with increasing Wi to level off the enhancement. The physics of the interaction of the effects of the Elasticity E, Viscoelastic Mach M, Reynolds Re and Weissenberg Wi numbers on generating the heat transfer enhancement is discussed. The existence of a heat transfer asymptote in laminar flow of viscoelastic fluids is shown for the first time. In addition similar to the heat transfer asymptote in turbulent pipe flows the counterpart in laminar flows in non-circular tubes delineates the region where enhancement is a function only of inertia. A different asymptote corresponds to different cross-sectional shapes in straight tubes.

See more of this Session: Poster Session in Fluid Mechanics
See more of this Group/Topical: Engineering Sciences and Fundamentals

A Model of HfO2 Chemical Vapor Deposition Chemistry and Fluid Mechanics On H-Si Surfaces

Kejing Li, Chemical and Biological Engineering, University of Alabama, Tuscaloosa, AL

Elizabeth Moore, National Institute of Standards and Technology, Gaithersburg, MD

Lin Zhang, Centre for Composite Materials and Structures, Harbin Institute of Technology, Harbin, China

William Kimes, National Institute of Standards and Technology, Gaithersburg, MD

Tonya Klein, Chemical and Biological Engineering, University of Alabama, Tuscaloosa, AL

We present a model describing the analysis of adsorption of hafnium tert-butoxide (HTB) and tetrakis dimethylamino hafnium (TDMAH) onto hydrogen terminated Si at reaction rate limited regime between 60-250ºC, during the steady state chemical vapor deposition (CVD) of high dielectric material HfO2 in a tubular flow reactor. A set of 30 elementary gas phase and surface reactions is used to simulate the adsorption process. Transport species from gas phase to surface is accounted using multicomponent transport properties. The reaction mechanisms are deduced by ab initio theoretical calculation estimates. Experimental measurement of film growth rate is performed by in situ ATR-FTIR and interferometer. Velocity, temperature, and concentration profiles along the Si substrate are determined by finite element analysis calculations.

See more of this Session: Poster Session in Fluid Mechanics
See more of this Group/Topical: Engineering Sciences and Fundamentals

Modeling of Aggregation and Breakup of Colloidal Particles in a Turbulent Stirred Tank through a Multizonal PBE Model

Matthäus U. Bäbler, Department of Mechanical and Process Engineering, Institute of Process Engineering, ETH Zurich, Zurich, Switzerland

Marco Mazzotti, Department of Mechanical and Process Engineering, Institute of Process Engineering, ETH Zurich, Zurich, Switzerland

Miroslav Soos, Department of Chemistry and Applied Biosciences, Institute for Chemical and Bioengineering, ETH Zurich, Zurich, Switzerland

Massimo Morbidelli, Department of Chemistry and Applied Biosciences, Institute for Chemical and Bioengineering, ETH Zurich, Zurich, Switzerland

Processing of particulate systems often involves aggregation and breakup phenomena driven by turbulent fluid motions. These phenomena may either be desired, e.g., flocculation to facilitate solid-liquid separation, dispersion of a powder, etc., or they may be an unwanted side effect, e.g., coagulation in emulsion polymerization, agglomeration in solution crystallization, etc. In this work we address the aggregation and breakup of small colloidal particle clusters in a diluted turbulent stirred tank reactor for which we develop a multizonal population balance equation (PBE) model.

The PBE model combines a detailed aggregation rate model [1] with a comprehensive breakage rate model [2]. The aggregation rate model [1] accounts for colloidal and hydrodynamic interactions among the colliding clusters. For the former, the van der Waals force between the primary particles in the two approaching clusters that are the nearest is considered. To model the hydrodynamic interactions, the porous clusters are described as uniformly permeable spheres where Brinkman's equation is adopted to describe intra-aggregate flow. For the breakage rate model [2], it is assumed that there exists a critical hydrodynamic stress above which a cluster of given mass breaks up. Furthermore, it is assumed that the breakup of an individual cluster is instantaneous once the stress exceeds the critical stress. In this framework, the breakage kinetics is governed by turbulent fluctuations that lead to a fluctuating hydrodynamic stress acting on the clusters. To describe the critical hydrodynamic stress, a recently developed scaling model for dense fractal clusters is used [3] whereas turbulent fluctuations on a local scale are described through a multifractal model [2,4].

Finally, to account for the spatial variation of the flow properties in the stirred tank reactor, i.e., the large scale heterogeneity, a multizonal approach is used. Thereby, the flow domain is divided into a small number of zones where in each of them the flow is assumed homogenous. Regarding the complexity of both the aggregation and breakage model, this multizonal approach presents a fair compromise between numerical load and accuracy. Two implementations of the multizone model are investigated. In the first one, the zones are predefined [5] and CFD is used to determine the zone properties and interzonal fluxes. In the second one, the zone properties are determined empirically and the interzonal fluxes are obtained by fitting to experiments. The model is compared to the aggregation of a fully destablized polystyrene latex in a stirred tank reactor [6,7]. Good agreement for the cluster mass distribution at steady state at various stirring speeds and various solid volume fractions is found in the case of the empirical zone model whereas the CFD model is insufficient in describing the variation in the solid volume fraction.

[1] Bäbler M.U., AIChE J. 54, 1748 (2008)

[2] Bäbler M.U., Morbidelli M., Baldyga J., J. Fluid Mech. 612, 261 (2008)

[3] Zaccone A. et al., Phys. Rev. E, submitted (2009)

[4] Meneveau C., Sreenivasan K. R., J. Fluid Mech. 224, 429 (1991)

[5] Bourne J.R., Yu S., Ind. Eng. Chem. Res. 33, 41 (1994)

[6] Soos M., Moussa A.S., Ehrl L., Sefcik J., Wu H., Morbidelli M., J. Colloid Interface Sci. 319, 577 (2008)

[7] Moussa A.S., Soos M. Sefcik J., Morbidelli M., Langmuir 23, 1664 (2007)

See more of this Session: Poster Session in Fluid Mechanics
See more of this Group/Topical: Engineering Sciences and Fundamentals

Deep Electrochemical Etching of Silicon

Joel Nino G. Bugayong, Purnima Narayanan, John C. Flake

Deep structures in silicon substrates are typically created by anisotropic etching in aqueous KOH solutions or in plasma processes such as Reactive Ion Etching (RIE). Electrochemical etching and photoelectrochemical etching are well known techniques and are commonly used to form porous silicon but have not been used for deep anisotropic etching. In this work we evaluate the potential for deep electrochemical etching of silicon without formation of porous silicon.

In this process, silicon wafers are patterned using a mask as shown in Figure 1. The wafer may be chemically etched using KOH to form an initiator pit, or inverted pyramid, revealing (111) surfaces. This initiator pit helps provide anisotropy in the subsequent electrochemical etch.

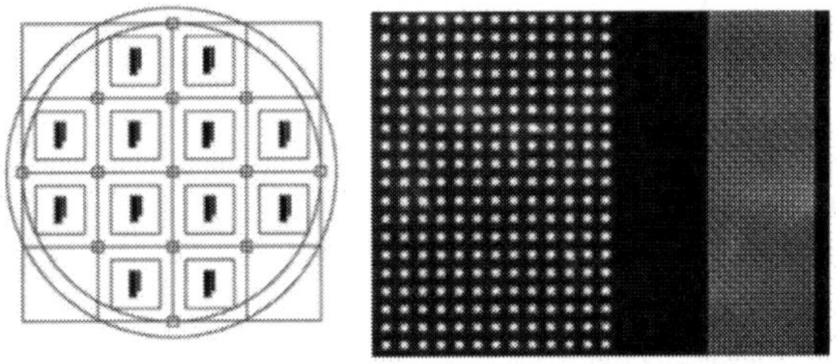

Fig.1. (left) AutoCAD drawing of a reticle (right) microscope image of a processed photomask.

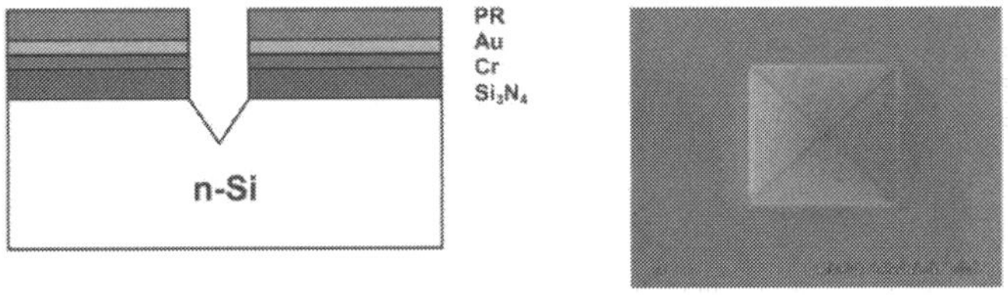

Fig.2. (left) Illustration of n-Si after UV lithography, RIE, and KOH etching (right) SEM of a KOH pit.

Silicon is electrochemically etched in the presence of hydrofluoric acid (HF) to form a soluble silicon hexafluoride species.

$$2h^+ + 6HF + Si \rightarrow SiF_6^{2-} + H_2 + 4H^+$$

The etch mechanism is believed to include electrochemical and chemical components. In the electrochemical step, holes (or electrons missing from bonding orbital) at the surface initiate the attack of fluorine or oxygen species in the solution. For example, when n-type silicon is anodized in HF, the electrode is under depletion and acts like a Schottky diode under forward bias. As shown in voltammogram in Figure 3 of a silicon sample under illumination in a HF-containing electrolyte, photogenerated holes migrate to the surface and initiate etching with an associated photocurrent. The first step(s) are typically photoelectrochemical in nature and subsequent steps are typically chemical reactions. The light-initiated reaction is localized in regions where photogenerated holes appear at the silicon-electrolyte interface and is affected by drift and diffusion. Meanwhile, p-type wafers may be anodically etched without light in a similar process; however, the localization must be achieved with masking methods. In the chemical step, the breaking of Si backbonds may be favored in areas with weakly bonded sites such as "hanging bonds" at the interface of (111) planes. The high electronegativity of F aids in the breaking of Si-Si backbonds. In patterns with wide pitched structures, macropores with porous sidewall branching is typically observed. This may be attributed to excessive hole diffusion in p-type wafers or over aggressive chemical etching. In patterns with tight pitches, there is no sidewall etching or pore branching, as shown in Figure 4. This is due to the field exclusion, since the distance between pore tips are smaller than twice the space charge region (SCR) width and results in the exclusion of a drift component to bring holes to the surface.

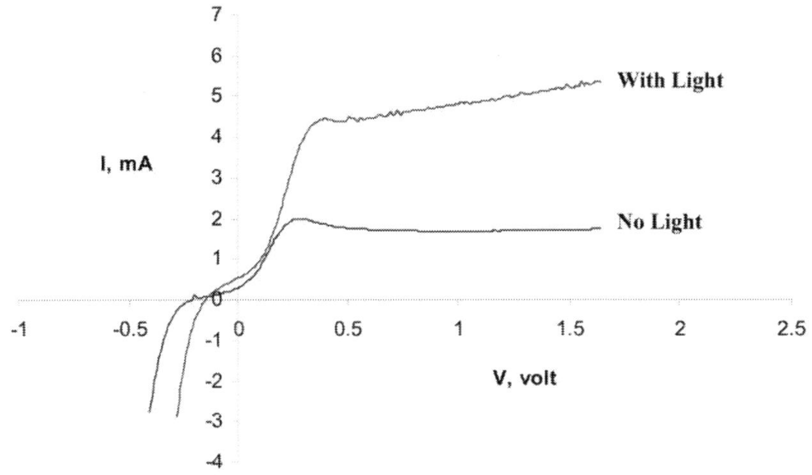

Fig.3. Cyclic voltammogram of blank n-Si 10-20 Ω-cm using 633nm HeNe laser, at an electrolyte composition of 23:27:50 HF:H_2O:EtOH (200 mV/s sweep rate).

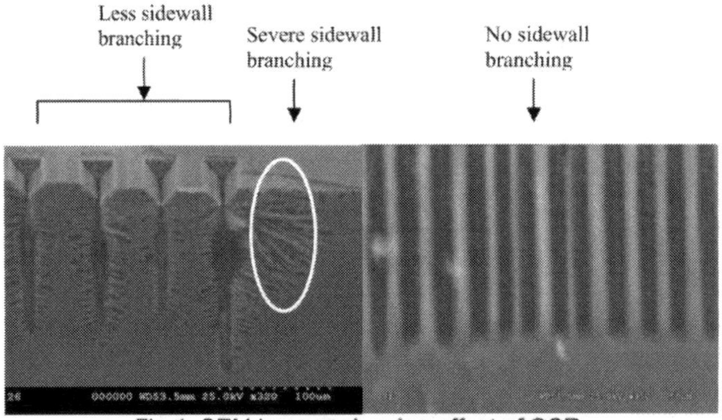

Fig.4. SEM images showing effect of SCR.

In this case (large pitches), sidewall etching may be prevented by tailoring the chemical reaction or interface such that there is a strong etch rate dependence on crystal plane.

Chemical selectivity may be achieved using the same principals associated with conventional KOH etching of silicon. Etch rates and mechanisms are dependent on crystalline plane and surface termination. For example, a difluoride terminated intermediate (e.g. a (100) plane) may support a faster etch rate relative to a monofluoride (111) or (110) plane. Furthermore, molecular size or transport restrictions of reactants or products at the (110)-like sidewall structure of <100> pores may favor attack of the backbonds by one fluoride or hydroxide.

Modulating Layer-by-Layer Assembly of Oppositely Charged Nanoparticles Using Amphiphilic Molecules

Daeyeon Lee, Chemical and Biomolecular Engineering, University of Pennsylvania, Philadelphia, PA

Jeanne Ho, Chemical and Biomolecular Engineering, University of Pennsylvania, Philadelphia, PA

Xinyu Cynthia Lau, Chemical and Biomolecular Engineering, University of Pennsylvania, Philadelphia, PA

Aqueous-based layer-by-Layer (LbL) assembly of charged species provides a versatile method to create conformal thin films on a variety of substrates with a precise control over the physical and chemical properties of thin films. While recent studies have shown that LbL assembly of oppositely charged nanoparticles can lead to generation of multifunctional thin films, these multifunctional thin films could only be assembled in a very narrow pH range. Such a limitation significantly hampers the widespread application of the LbL method to generate all-nanoparticle thin films with different functionalities. In this study, we demonstrate that the LbL assembly of oppositely charged nanoparticles outside the narrow pH ranges can be successfully achieved by adding low molecular weight amphiphilic molecules to nanoparticle suspensions. In this talk, the effect of concentration of these amphiphilic molecules on the composition and growth behavior of LbL thin films comprising oppositely charged nanoparticles will be described. The effect of calcination on the structural properties of the LbL thin films is also investigated.

See more of this Session: Colloidal Dispersions II
See more of this Group/Topical: Engineering Sciences and Fundamentals

Structural Transitions of Nanoparticle Assemblies Mediated by Polymer Interactions at An Air-Water Interface

Kung-Po Chao, Chemical and Biomolecular Engineering, Rice University, Houston, TX

Sibani Lisa Biswal, Chemical and Biomolecular Engineering, Rice University, Houston, TX

There is currently a large effort being invested in synthesizing and characterizing various types of nanoparticle assemblies, including quantum dots, plasmonic metal nanostructures, and varying aspect ratio particles such as tetrapods. These particles are useful when arranged into more complex structures and devices but finding strategies to order these structures has been difficult. One of the deterrents in following the ordering has been the inability to follow the details of the assembly process in real time. We will be presenting a method that track nanoparticle ordering at an interface with high spatial and temporal resolution in order understand the forces that control nanoparticle assembly. In particular, we are investigating the phase behavior of a polymer- nanoparticle assembly on an air-water interface. The system is a mixture of poly(L-lactic acid) (PLLA)and nanoparticles which are mixed together and placed on a Langmuir Trough. Various particle/polymer ratios are studied at different surface pressures. We find that phase separation and ordering of the particles exist for various surface pressures. For example, when surface pressure is around 20 mN/m, we observe a particle structure transition from nanoparticle islands to lassos, as observed via transmission electron microscopy (TEM). By tuning the concentration ratios of polymer and nanoparticles at constant surface pressure, we could achieve the same effects as well. Finally, we use Brewster Angle Microscopy (BAM) to monitor the structural ordering of the nanoparticles at the air-water interface in real-time.

See more of this Session: Colloidal Dispersions II
See more of this Group/Topical: Engineering Sciences and Fundamentals

Effect On Nonadsorbed Nanoparticles On near-Contact Dynamic Interactions Between Surfaces

John Y. Walz,Chemial Engineering, Virginia Tech, Blacksburg, VA

Clayton T. McKee, Chemical Engineering, Virginia Tech, Blacksburg, VA

Colloidal probe atomic force microscopy was used to measure the effect of scan rate on the interaction between a micron-sized colloidal particle (test particle) and a flat plate in aqueous solutions containing dispersed spherical nanoparticles. The nanoparticle concentrations investigated were sufficiently large that long-range oscillations in the force profile were clearly observed, suggesting ordering of the nanoparticles in the confined region between the test particle and plate. The measurements allowed investigating (1) the effect of scan rate (rate of approach or retraction between the test particle and plate) on the nanoparticle ordering, and (2) the effect of the nanoparticles on the hydrodynamic drag force exerted on the test particle. To understand the effect of scan rate on nanoparticle ordering, a dimensionless Peclet number for the gap region was developed and evaluated. For the effect of the nanoparticles on the test particle drag force, it was found that even at nanoparticle volume fractions of several percent by volume, the drag force exerted on the test particle was well described using equations developed for simple Newtonian fluids.

See more of this Session: Colloidal Dispersions II
See more of this Group/Topical: Engineering Sciences and Fundamentals

Investigation Into the Role of Particle Size Ratio and Material On Nanoparticle Haloing

Xiaoting Hong, Chemical Engineering, University of Louisville, Louisville, KY

Gerold A. Willing, Chemical Engineering, University of Louisville, Louisville, KY

Nanoparticle Haloing is a colloidal stabilization method that relies on a small volume fraction of highly charged nanoparticles to stabilize a small volume fraction of negligibly charged colloidal scale particles. While volume fraction plays a significant role in the stabilization of the colloidal particles, it is expected that the size ratio between the two particles will also have a strong influence on the stabilization mechanism as the size ratio will not only change the nanoparticle density within the gap between colloidal particles but it will also change the charge distribution within the gap as well. In this talk, we will investigate the role of the size ratio between the two particles in the stability of a Nanoparticle Haloing system. Additionally, we will touch on the role of nanoparticle charge on the stabilization mechanism by changing the metal oxide used for the nanoparticle. These investigations are vital for the further development and generalization of Nanoparticle Haloing as a colloidal stabilization technique.

See more of this Session: Colloidal Dispersions II
See more of this Group/Topical: Engineering Sciences and Fundamentals

Electric Field-Induced Optical Anisotropy of Model Rod-Like Colloids, Carbon Nanotubes

Ji Yeon Huh, Polymers Division, National Institute of Standards and Technology, Gaithersburg, MD

Angela R. H. Walker, Physics Lab, National Institute of Standards and Technology, Gaithersburg, MD

Erik K. Hobbie, Polymers Division, National Institute of Standards and Technology, Gaithersburg, MD

Jeffrey A. Fagan, Polymers Division, National Institute of Standards and Technology, Gaithersburg, MD

The high aspect ratio of carbon nanotubes exhibits a fundamental structural anisotropy that plays a critical role in various applications including nanoscale transistors, electrochemical electrodes, and flexible conductive films. In this work, we investigate the orientation of length/ chirality-sorted carbon nanotubes in aqueous suspensions by measuring the linear dichroism in response to high frequency electric fields. Based on the strong field-induced alignment with the nematic order parameter, the anisotropic polarizability of carbon nanotubes is determined and further compared to that of other high-aspect-ratio rod-like colloids.

See more of this Session: Colloidal Dispersions II
See more of this Group/Topical: Engineering Sciences and Fundamentals

Collisions and Coalescence of Micro-Bubbles Measured Using Atomic Force Microscopy

Ivan U. Vakarelski, PFPC and the Department of Chemical and Biomolecular Engineering, The University of Melbourne, Parkville, Australia

Rogerio Manica, Institute of High Performance Computing, Singapore, Singapore

Raymond R. Dagastine, PFPC and the Department of Chemical and Biomolecular Engineering, The University of Melbourne, Parkville, Australia

G. W. Stevens, PFPC and the Department of Chemical and Biomolecular Engineering, The University of Melbourne, Parkville, Australia

Franz Grieser, PFPC and the School of Chemistry, The University of Melbourne, Parkville, Australia

D.Y.C. Chan, PFPC and the Department of Mathematics and Statistics, The University of Melbourne, Parkville, Australia

The dynamic forces between micro-bubbles are crucial for understanding applications as wide ranging as foams in mineral processing to ultrasound contract agents. We employ gas bubbles in water, the simplest and arguably the most shape sensitive soft matter systems, to quantitatively link the dynamic coupling of bubble shape with external forces that control their behavior. We have extended the experimental and theoretical methods developed to study the dynamic interactions between droplets using Atomic Force Microscopy (AFM) or the interactions between a micro-bubble and a flat surface in aqueous solution to study the collisions and coalescence behavior of two micro-bubbles in aqueous solution. These measurements show that bubble collisions, even at speeds comparable to Brownian motion, are dependent on a combination of equilibrium surface forces, hydrodynamic drainage forces and interfacial deformation. Quantitative modeling, which shows excellent agreement between experiments and calculations, demonstrates that the variations in the dynamic forces arise from changes in both the equilibrium surface forces and hydrodynamic drainage effects. These results have implications in applications as wide ranging as froth floatation to micro-fluidics.

See more of this Session: Colloidal Dispersions II
See more of this Group/Topical: Engineering Sciences and Fundamentals

Effect of Melt Molecular Weight and Temperature On the Rheology of Dewetted Polymer-Grafted Nanoparticles in Polymer Melts

Maura McEwan, Chemical Engineering, University of Virginia, Charlottesville, VA

David L. Green, Department of Chemical Engineering, University of Virginia, Charlottesville, VA

The effects of melt molecular weight and temperature on the rheology of polymer-grafted nanoparticles have been investigated in chemically-identical homopolymer melts. The suspensions were formulated such that de-wetting took place between the graft polymer and the melt polymer. Dewetting at the graft polymer/melt polymer interface, controlled by the molecular weights of the graft and melt polymer, N and P, respectively, lead to particle attractions, which lead to a thixotropic rheological response and viscosity-stress curves that can be scaled by temperature and swelling ratio (P/N). The onset of elasticity is denoted by the storage modulus (G') overtaking the loss modus (G"), which occurs at a particle core volume fraction of $\varphi = 0.15$, which is essentially independent of the swelling ratio for particles of core radius R = 100 nm. The strength of attraction was modeled using a square well potential based on the swelling ratio, where the interaction potential was used to scale the rheological data. The implications of the power-law dependence of the storage modulus on particle volume fraction are discussed.

See more of this Session: Colloidal Dispersions II
See more of this Group/Topical: Engineering Sciences and Fundamentals

Slip-Link Simulations of Entangled DNA with Comparison to Experiments

Eric S. G. Shaqfeh, Departments of Chemical and of Mechanical Engineering, Stanford University, Stanford, CA

Ajey Dambal, Chemical Engineering, Stanford University, Stanford, CA

Amit Kushwaha, Chemical Engineering, Stanford University, Stanford, CA

The nonequilibrium dynamics of entangled polymers is a particularly complex process because of the interplay between the many known chain mechanisms including reptation, tube-length fluctuations, constraint renewal, and chain stretch. Recently, slip-link simulations, pioneered by Hua and Schieber (J. Chem. Phys. 1998), have been developed with the purpose of describing all mechanisms active in entangled polymer dynamics without ad hoc approximations. These simulations, though still a developing area of research, have turned out to be a powerful tool since their higher level of coarse-graining, relative to molecular dynamics, allows for simulations at much longer time scales. We have implemented such a sliplink-based model, originally proposed by Masubuchi et al. (J. Chem. Phys. 2001). The original version of model is further refined to account for more accurate description of chain end dynamics, finite extensibility effects and variable subchain friction for nonequilibrium dynamics (Masubuchi et. al., J. Chem. Phys. 2008). The model, along with the "individual" mechanisms of reptation and tube length fluctuation, also incorporates the collective contributions arising from the 3D network structure of the entangled chains, including constraint release. With the new code, we have simulated both the static and dynamic responses of monodisperse linear worm-like chains and compared the simulations results with the experimental data reported by Jary et al. 1999 and Teixeira et al. 2007 for entangled DNA in shear flow. The simulations were run for monodisperse linear chains including up to 30 entanglements. Additionally, we also have the ability to perform implicit simulations. In the present work, we demonstrate (i) the effect of different constraint renewal (CR) procedures on the dynamics of the entangled chain network, such as the decrease in the number of entanglements and the effect of increased chain stretch on decreasing CR frequency (ii) in shear flow, the dynamic mechanisms that create the plateau in the steady shear stress and (iii) using modified Kraynik-Reinelt boundary conditions, we examine the dynamic chain mechanisms in extensional and mixed planar flows. The simulation model is able to capture the rheological behavior of entangled systems both in the linear and nonlinear range, though we demonstrate sensitivity to the procedure for creating re-entanglement.

See more of this Session: Complex-Fluid and Bio-Fluid Dynamics I
See more of this Group/Topical: Engineering Sciences and Fundamentals

DNA Dynamics in Nanoslits and Nanochannels : Roles of Semi-Flexibility and Hydrodynamic Screening

Yeng-Long Chen, Institute of Physics and Research Center for Applied Science, Academia Sinica, Taipei, Taiwan

Jen-Fang Chang, Institute of Physics, Academia Sinica, Taipei, Taiwan

Classical polymer physics has been shown to be extremely successful in predicting flexible polymer conformation and dynamics in confined environments. Less is understood about the dynamics of confined semi-flexible chains. In sub-100nm channels, nano-fabrication has facilitated studies of DNA conformation and dynamics near and below the DNA persistence length ($\sim$50nm), which provides a good model for studying semi-flexible polymer dynamics. We investigate semi-flexible chain dynamics in nanochannels and nanoslits with Brownian dynamics simulations, Monte Carlo simulations, and optical microscopy. Our study explores how the bending energy, the conformational entropy, and the DNA-surface interactions affect the macromolecular conformation. The importance of intra-chain hydrodynamic interactions and how hydrodynamic screening affect chain dynamics is examined. Our study finds that the competition between chain semi-flexibility and confinement can lead to damping of low frequency chain relaxation modes.

See more of this Session: Complex-Fluid and Bio-Fluid Dynamics I
See more of this Group/Topical: Engineering Sciences and Fundamentals

Collision of a Large DNA Molecule in An Array of Obstacles

Jaeseol Cho, Department of Chemical Engineering and Materials Science, University of Minnesota, Minneapolis, MN

Kevin D. Dorfman, Department of Chemical Engineering and Materials Science, University of Minnesota, Minneapolis, MN

We have investigated the electrophoretic collision of T4 DNA (169 kbp) in an array of obstacles using a recently developed Boundary Element Method/Brownian dynamics simulation routine. Starting from a relaxed configuration in the gap between the posts, the molecule is driven into the obstacle under the influence of a relatively strong electric field. The magnitude of the field and the array geometry (1 μm diameter posts arranged in a hexagonal array with a 3 μm pitch) are similar to those used in experiments. We found that, in contrast to a relatively short molecule such as λ-DNA (48.5 kbp), the ensemble of collisions for the longer T4 DNA is considerably more complex. In addition to the single post collision dynamics observed by Randall and Doyle, we observed two array-based collision modes that we refer to as "A" and "M" collisions. The latter modes require that the unraveling chain possess sufficient stored length so that one arm of the chain can collide with a second obstacle. We will discuss the role of these additional collision modes on modeling the electrophoretic mobility and effective diffusivity (dispersivity) of the DNA in an array.

See more of this Session: Complex-Fluid and Bio-Fluid Dynamics I
See more of this Group/Topical: Engineering Sciences and Fundamentals

Selecting Optimal Flow Regimes for Rapid Microscale PCR Using Natural Convection

Radha Muddu, Department of Mechanical Engineering, Texas A&M University, College Station, TX

Yassin A. Hassan, Departments of Nuclear and Mechanical Engineering, Texas A&M University, College Station, TX

Victor M. Ugaz, Artie McFerrin Department of Chemical Engineering, Texas A&M University, College Station, TX

The lack of rapid, affordable, and easy to use medical diagnostic technologies is one of the most critical issues confronting global public health, particularly in resource-limited settings where dedicated laboratory facilities do not exist. Many assays depend in some way on the polymerase chain reaction (PCR) to amplify an initially dilute target DNA sample to a detectable concentration level. But a major challenge to the development of instrumentation to meet these needs is the highly inefficient design of conventional PCR thermocycling hardware that is slow, expensive, and consumes considerable electrical power to repeatedly heat and cool the reagent mixture. Natural convection has emerged as a promising alternative thermocycling approach that has the potential to address these needs using an inherently simple design (similar in principle to a lava lamp) that consumes minimal power and is well-suited for use in portable applications.

One of the challenges facing design of convective thermocyclers, particularly at the micro-scale, is the need to precisely control the spatial velocity and temperature distributions within the reactor to ensure that the reagents sequentially occupy the correct temperature zones for a sufficient period of time. Here we describe results of a new effort to probe the full 3-D velocity and temperature distributions in miniaturized convective thermocyclers. This analysis is challenging because the thermal and geometric conditions of convective PCR can produce flow profiles exhibiting characteristics that can range from laminar, to transitional, to turbulent. Here, we highlight two reactor geometries that exhibit very different flow characteristics ranging from stable closed circulatory loops extending between the two temperature extremes to more complex paths that continually evolve over time and provide increased exposure to intermediate temperatures.

A surprising and unexpected result of this analysis is the discovery of a subset of complex flow trajectories that appear to be highly favorable for PCR due to a synergistic combination of (1) continuous exchange among flow paths that provides an enhanced opportunity for reagents to sample the full range of optimal temperature profiles, and (2) increased time spent within the extension temperature zone--the rate limiting step of PCR. This hypothesis is confirmed by experiments that show PCR amplification is achievable in under 10 minutes in reactors designed to generate these flow conditions. Our results suggest that by selecting the optimal range of flow trajectories, it is possible to achieve orders of magnitude enhancements in cycling time (~ 20 s per cycle), a capability that is an inherent feature of the underlying physics of convective flow.

See more of this Session: Complex-Fluid and Bio-Fluid Dynamics I
See more of this Group/Topical: Engineering Sciences and Fundamentals

Dynamics of Suspensions of Elastic Capsules Flowing in Confined Geometries

Pratik Pranay, Department of Chemical and Biological Engineering, University of Wisconsin Madison, Madison, WI

Pieter Jan Antoon Janssen, Department of Chemical and Biological Engineering, University of Wisconsin Madison, Madison, WI

Michael D. Graham, Department of Chemical and Biological Engineering, University of Wisconsin Madison, Madison, WI

Modeling the behavior of fluid-filled capsules (which can be considered as a simple representation of red blood cells and vesicles), is not only important to understand biological processes, such as blood flow in the microcirculation, but also to help design and improve microfluidic devices for characterizing or separating such particles.

The present work describes simulations of large numbers of deformable capsules with various properties in confined geometries. Our algorithm incorporates a General-Geometry-Ewald-Like method (GGEM) for efficiently calculating hydrodynamic interactions (O(N)) in an immersed-boundary method. This allows for a detailed description of the particle interface combined with a large degree of freedom to model the confining domain (grooved channels, cylindrical and spherical obstructions).

With our algorithm, we have addressed several issues. The ability to quickly simulate large number of particles enables examinations not only of the competition between shear-induced diffusion and wall-induced hydrodynamic migration of single particles, but also exploration of concentration effects and segregation by size, shape and/or deformability. The flow through a slit containing an array of pillars, which is a model representation of the interalveolar sheet structure of the capillaries in the lungs, has been examined as well. Combined with the simulation of grooved channels, we propose a methodology to separate these cells depending on their deformability and size.

Finally, the effect of addition of long-chained polymer molecules in blood flow, known to lower blood pressure, is investigated.

See more of this Session: Complex-Fluid and Bio-Fluid Dynamics I
See more of this Group/Topical: Engineering Sciences and Fundamentals

Complex Dynamics of Vesicles and Red Blood Cells in Viscous Flows

Petia M. Vlahovska, Thayer School of Engineering, Dartmouth College, Hanover, NH

Red blood cells in shear flow can tank-tread, tumble, or "swing" (tank-treading accompanied by oscillations in the inclination angle). I will present a model that quantitatively describes this behavior. The analysis accounts for the fact that the membrane is deformable, incompressible, and resistant to bending and shearing. In shear flows, the theory shows that a closed lipid membrane (vesicle or red cell) deforms into a prolate ellipsoid, which tumbles at low shear rates, and swings at higher shear rates. The amplitude of the oscillations decreases with shear rate. In quadratic flows, the theory predicts that red blood cells always migrate towards the flow centerline unlike drops, whose direction of migration depends on the viscosity ratio. These results improve understanding of blood rheology in the microcirculation.

See more of this Session: Complex-Fluid and Bio-Fluid Dynamics I
See more of this Group/Topical: Engineering Sciences and Fundamentals

The Dilute Rheology of Swimming Suspensions: A Simple Kinetic Model

David Saintillan, Department of Mechanical Science and Engineering, University of Illinois at Urbana-Champaign, Urbana, IL

A simple kinetic model is presented for the shear rheology of a dilute suspension of particles swimming at low Reynolds number. If interparticle hydrodynamic interactions are neglected, the configuration of the suspension is characterized by the particle orientation distribution, which satisfies a Fokker-Planck equation including the effects of the external shear flow, rotary diffusion, and particle tumbling. The orientation distribution then determines the leading-order term in the particle extra stress in the suspension, which can be evaluated based on the classic theory of Hinch and Leal [J. Fluid. Mech. 52(4):683-712 (1972)], and involves an additional contribution arising from the permanent force dipole exerted by the particles as they propel themselves through the fluid. Numerical solutions of the steady-state Fokker-Planck equation were obtained using a spectral method, and results are reported for the shear viscosity and normal stress differences in terms of flow strength, rotary diffusivity, and correlation time for tumbling. It is found that the rheology is characterized by much stronger normal stress differences than for passive suspensions, and that tail-actuated swimmers result in a strong decrease in the effective shear viscosity of the fluid.

See more of this Session: Complex-Fluid and Bio-Fluid Dynamics I
See more of this Group/Topical: Engineering Sciences and Fundamentals

Velocity and Stress Correlations in Suspensions of Swimming Microorganisms: Theory and Simulation

Patrick T. Underhill, The Isermann Department of Chemical and Biological Engineering, Rensselaer Polytechnic Institute, Troy, NY

Michael Graham, University of Wisconsin, Madison, WI

Large collections of swimming microorganisms are receiving much recent attention because they are able to produce collective motions on a scale much larger than the scale of a single organism. In particular, the collective behavior leads to velocities larger than that of an isolated organism, fluid structures larger than the size of an organism, enhanced transport in the fluid, and enhanced stress fluctuations which produce altered rheological properties. It is important to understand the mechanism for these collective behavior because of the possible importance in swarming, biofilm formation, and bacterial infections. A better understanding would also allow for design of artificial systems consisting of interacting swimming agents. We have developed a new theory which shows how some of these phenomena arise from the hydrodynamic interactions between the organisms and compare the predictions with the results from experiments and computer simulations. In this way we can understand how the behavior scales with concentration, the importance of the method of swimming used, the influence of run-and-tumble like motions of the organisms, and how the interactions can lead to large-scale fluid structures. For example, in periodic geometries, the orientational correlations of the microorganisms lead to large-scale fluid structures that depend logarithmically on the simulation box size.

See more of this Session: Complex-Fluid and Bio-Fluid Dynamics I
See more of this Group/Topical: Engineering Sciences and Fundamentals

Reaction-Driven Propulsion of Bacteria

Misael Diaz, Department of Chemical Engineering, University of Puerto Rico at Mayagüez, Mayagüez, PR

Ubaldo M. Cordova-Figueroa, Department of Chemical Engineering, University of Puerto Rico at Mayagüez, Mayagüez, PR

Recent studies have shown that the 'non-motile' bacteria Syntrophus aciditrophicus has the ability to sense its chemical environment-an essential step for chemotaxis typically observed in 'motile' microorganisms (e.g., Escherichia coli). In light of these findings, new questions arise. Could self-propulsion take place naturally in non-motile metabolically active bacteria without the aid of external gradients or forces? What conditions are necessary for this to occur? These questions motivate us to propose a simple model for self-propulsion of a 'non-motile' metabolically active bacterium in the presence of an uniform reactant concentration. It has been observed that a biological cell permeable to fluid molecules but not to solutes-a semipermeable membrane-moves away from concentrated regions. This phenomenon results from an imbalance in the osmotic pressure between the two fluids separated by a cell membrane, leading to a fluid flow from inside the cell to high concentration regions in the outer fluid and from low concentration regions in the outer fluid to inside the cell. Such motion is known as osmophoresis. In the case of metabolically active non-motile bacteria it is unclear how osmophoresis will be affected by a chemical reaction at the surface of the cell that can alter the concentration field and, therefore, the osmotic pressure gradient 'sensed' by the bacterium. In this study, we consider an asymmetric surface chemical reaction at the bacterium surface (e.g., degrading transmembrane enzymes) causing a local osmotic pressure gradient autonomously and thus fluid flow across the membrane. We demonstrate that this fluid flow propels the bacterium toward less concentrated regions and that its velocity depends on the reaction speed, its size, and membrane and solvent properties.

See more of this Session: Complex-Fluid and Bio-Fluid Dynamics I
See more of this Group/Topical: Engineering Sciences and Fundamentals

Dynamic Strategies for Target-Site Localization by Gene Regulatory Proteins

Mario Diaz de la Rosa, Chemical Engineering, Stanford University, Stanford, CA

Elena Koslover, Biophysics Program, Stanford University, Stanford, CA

Andrew Spakowitz, Chemical Engineering, Stanford University, Stanford, CA

Gene expression is orchestrated by a host of regulatory proteins that coordinate the transcription of DNA to RNA. Regulatory proteins function by locating specific binding sequences of DNA and binding to these sequences to form the transcription initiation complex. In many instances, these regulatory proteins only have several hundred copies that must efficiently locate target sequences on the genome-length DNA strand. The non-specific binding of regulatory proteins to random sequences of DNA is believed to permit the protein to slide along the DNA in a stochastic manner. Periodically, a thermal kick or an interaction with another bound protein will disengage the regulatory protein from the DNA surface, leading to three-dimensional diffusion. Eventually, the protein will reattach to the DNA at some new location that is dictated by both the diffusivity of the protein and the DNA configuration. Cycling through these random events constitutes a search strategy for the target site. We build a reaction-diffusion theory of this search process in order to predict the optimal strategy for target site localization. The statistical behavior of the DNA strand acts as a necessary input into the theory, and we consider several governing behaviors for the DNA strand. We explore the impact of DNA configuration on target site localization in order to predict how protein expression will vary under different experimental conditions.

See more of this Session: Complex-Fluid and Bio-Fluid Dynamics I
See more of this Group/Topical: Engineering Sciences and Fundamentals

Surfactant Adsorption On Structure II and Semi-Clathrate Hydrates

Chi Lo, Chem Eng Dept, The City College of New York, New York, NY

Junshe Zhang, Chemical Eng, Department of Chemical Engineering, The City College of New York, New York, NY

Alexander Couzis, Chemical Engineering, City College of the City University of New York, New York, NY

Ponisseril Somasundaran, Columbia University, New York, NY

Jae W. Lee, Department of Chemical Engineering, The City College of New York, New York, NY

Clathrate hydrates are supramolecular crystalline compounds, in which small molecules are encaged in a polyhedral framework composed of hydrogen bonded water molecules. Three major structures of clathrate hydrates have been reported: sI, sII, and sH hydrates. The structure of clathrate hydrates is predominantly determined by the size of hydrate formers, e.g. methane forms sI, cyclopentane forms sII, and a mixture of methane and neohexane forms sH. One volume of clathrate hydrates can encage about 170 volumes of gas at STP conditions if all the cavities are singly occupied. Therefore, clathrate hydrates are being considered as a promising storage medium for natural gas and hydrogen. Despite of this potential, one big obstacle to industrial application of the hydrate-based gas storage is unfavorable formation kinetics. Surfactants have been known to accelerate enclathration kinetics. Thus, the interaction between surfactants and hydrates provides insight into the role of surfactants in promoting hydrate formation. This work present the adsorption behavior of sodium dodecyl sulfate (SDS) on cyclopentane (CP) hydrates and its derivative surfactant on tetrabutylammonium bromide (TBAB) hydrates. Cyclopentane (CP) is a hydrophobic former while tetrabutylammonium bromide (TBAB) is a salt that forms semi-clathrate hydrates. The adsorption on these two hydrates was studied by z-potential and pyrene fluorescence measurements. CP hydrates have a negative surface charge in the absence of SDS and it decreases to a minimum as the SDS concentration increases from 0 to 0.17 mM. Then, it increases with further increased SDS concentrations. The adsorption density of DS- on CP hydrates reaches a saturated value at 1.73 mM SDS. The micropolarity parameter of the TBAB hydrate/water interface starts to increase rapidly at 0.17 mM SDS and levels off at 1.73 mM SDS. The presence of Br- in TBAB hydrate suspensions could compete with TBADS (from association of DS- and TBA+) and DS- for the adsorption on the hydrate surface but they have a much stronger affinity for the hydrates than Br-. From the fluorescence measurements, it was found that the micropolarity of the hydrate/water interface is mainly dependent on the polarity of hydrate formers.

See more of this Session: Fundamentals of Interfacial Phenomena II - Surfactants and Liquid/Solid Interfaces
See more of this Group/Topical: Engineering Sciences and Fundamentals

Composition-Dependent Adhesion Forces Between Cyclopentane Hydrate Particles

Zachary M. Aman, Chemical Engineering, Colorado School of Mines, Golden, CO

Laura E. Dieker, Chemical Engineering, Colorado School of Mines, Golden, CO

Guro Aspenes, Chemistry, University of Bergen, Bergen, Norway

Carolyn A. Koh, Chemical Engineering, Colorado School of Mines, Golden, CO

Amadeu K. Sum, Chemical Engineering, Colorado School of Mines, Golden, CO

E. Dendy Sloan, Chemical Engineering, Colorado School of Mines, Golden, CO

Both anti-agglomerants and naphthenic acids are key elements of flow assurance in oil pipelines and understanding their role in hydrate agglomeration is critical to hydrate plug prediction and prevention. Adhesion forces between cyclopentane hydrates were measured at atmospheric pressure and 3.2 °C using an improved micromechanical apparatus. Due to the complexity of crude oil systems, a series of model oils were also prepared by adding surface-active components to 200 cP mineral oil, as analogs to the crude oil systems. After hydrate stabilization in pure cyclopentane, a variable weight percent of model oil was added to the hydrate system in order to study the isolated effects of individual naphthenic acid constituents or model anti-agglomerants in mineral oil. Addition of 1 wt% sorbitan monooleate (Span 80, a model anti-agglomerants) to mineral oil reduced particle-particle adhesion force by 37% compared to pure mineral oil. Addition of a 0.6 wt% commercial naphthenic acid mixture to mineral oil reduced particle-particle adhesion force by 79% compared to pure mineral oil. The observed reduction in adhesion force was constant between 1 and 30 wt% addition of model oil. These results indicate that both model anti-agglomerants and commercial naphthenic acid mixtures are active on the surface of hydrate particles in the organic phase, drastically reducing the agglomeration tendency between such particles.

See more of this Session: Fundamentals of Interfacial Phenomena II - Surfactants and Liquid/Solid Interfaces
See more of this Group/Topical: Engineering Sciences and Fundamentals

The Formation of a Micelle-Free Zone in the Transport of Surfactant From An Aqueous Micellar Solution to An Oil Phase

Nikhil S Bhole, Fenfen Huang and Charles Maldarelli
Levich Institute and Department of Chemical Engineering,
The City College and The Graduate Center of the City University of New York,
New York, New York, United States

1 Background of surfactant transport process

Rapid surface tension relaxation of a newly formed air-water or oil-water interface is important in many industrial applications (1; 2), examples include formation of a highly textured foam or fine emulsions (3; 4), wetting of solid surfaces by an aqueous drop in the textile or agrochemical industries (5; 6; 7). This rapid surface tension relaxation is often achieved by addition of amphiphilic molecules or surface active agents (surfactants), which preferentially adsorb at the interfaces reducing surface or interfacial tensions. To achieve this rapid relaxation of surface tension very high bulk concentrations of surfactants are used. It is well known fact that as the surfactant bulk concentration in the aqueous phase (C_{bulk}) increases beyond a certain critical concentration, surfactant molecules begin to self-assemble into structures that shield the hydrophobic tails from the aqueous phase and expose polar groups to the water (8; 9; 10). These aggregates are called micelles and the critical concentration above which these are seen is the critical micelle concentration (CMC, C_{CMC}). There are many different shapes of aggregates that surfactants can assemble into, like spherical, cylindrical, lamellar etc. These micelles are not stable but continuously exchange monomers with the surrounding. The high concentrations used in the industrial applications are often above the CMC and hence models which describe transport from micellar solutions are required. The purpose of this paper is to develop and experimentally verify a class of models for this transport.

To begin with, we consider an surfactant transport process from an aqueous solution to an air-water interface. Consider an aqueous solution of surfactant monomers with a bulk concentration below the CMC. Initially surfactant monomers are uniformly distributed throughout the bulk. When a clean air-water interface is suddenly created, surfactant monomers from the sublayer region just underneath the interface, kinetically adsorb onto the air-water interface. This adsorption causes depletion in sublayer monomer concentration, which generates a concentration gradient driving a monomer diffusion flux from the bulk phase to the interface. This surfactant transport process involving only monomers is well understood and has been reviewed thoroughly c.f. (11; 12; 13; 14; 15; 16; 17; 18). Diffusion in the bulk can be modeled by Fick's law while adsorption at the interface can be modeled by Langmuir or Frumkin rate law expressions. Rate expression for Frumkin kinetics is as follows,

$$\frac{d\Gamma}{dt} = \beta C_s(\Gamma_\infty - \Gamma) - \alpha \Gamma e^{K\Gamma/\Gamma_\infty} \tag{1}$$

where Γ is the surface concentration or surface density, C_s is the monomer sublayer concentration, α and β are the desorption and adsorption kinetic rate constant, Γ_∞ is the maximum packing density and K is the interaction parameter in the adsorbed layer. $K = 0$ gives us the Langmuir kinetic rate law.

When the bulk concentration increases above the CMC, micelles begin to form. Many experimental studies have shown that as the bulk concentration increases beyond the C_{CMC}, the rate of adsorption and hence the surface tension relaxation continues to accelerate (19; 20; 21). Role of micelles in accelerating the tension relaxation is not yet thoroughly understood. Micelle break-up and reassembly kinetics tries to maintain the monomer concentration at the CMC. Noting this, the transport picture is fairly clear: Monomers from the sublayer adsorb onto the interface causing depletion in monomer concentration. This is continuously replenished by kinetic break-up of micelles, thereby increasing the monomer concentration in the sublayer generating large kinetic rates of adsorption. This micelle break-up also creates a micellar diffusive flux from the bulk towards the interface, which limits the availability of micelles near the interface.

Modeling the micelle break-up contribution to the flux of monomer to the interface is not straight-forward as micelle break-up kinetics is difficult to describe. Rigorously it is described by a step-wise disassociation model by Aniansson and Wall (22; 23; 24; 25). In this model an aggregated species S_i (i is number of monomers in aggregate) disassociates releasing one monomer at a time:

$$\mathrm{S}_i \underset{k_{-i}}{\overset{k_i}{\rightleftharpoons}} \mathrm{S}_{i-1} + \mathrm{S}; \quad i = 2, 3 \ldots$$

where k_i and k_{i-1} are the rate constants for forward and backward reaction step respectively and are dependent upon i. This step-wise disassociation dictates that the micellar solution is polydisperse, consisting of a large population of monomers and oligomers, a small population of rare aggregates (medium aggregation number) and a large population of larger aggregates narrowly distributed about a preferred aggregation number σ. Two time scales are associated with this disassembly, a fast time-scale (τ_F) is related to loss of single monomer from an aggregate and a slow time scale (τ_{SL}) is related to time required for complete disassociation of a preferred aggregate through the step-wise disassociation pathway.

A simplified approach to model diffusion and disassembly kinetics is to avoid the details of the step-wise process and assume that micelles break-down into monomers in one single step (26; 27; 28; 29);

$$\mathrm{S}_\sigma \underset{k_{-\sigma}}{\overset{k_\sigma}{\rightleftharpoons}} \sigma \mathrm{S}$$

In this scheme, micellar solution is assumed to be composed of monodispersed micelles with aggregation number σ and monomers. This simplified model has only one time scale, τ_{SL}. Several authors have studied micelle break-up kinetics with step-wise disassociation, when the micellar solution is subjected to small perturbations from equilibrium (21; 30; 31; 32; 33; 34; 35). In these models, time scales of micelle break up are comparable to diffusional time scale, τ_D.

The above models are restrictive because of the limitation that the deviations from the equilibrium are small. However, when a clean interface is created, the conditions go far away from the equilibrium. In such case, other time scales involved, such as diffusion time scale, can become larger than the slowest micellar kinetic time scale ($\tau_D >> \tau_{SL}$). In this limit of infinitely fast micelle break-up, micelle-monomer equilibrium is always maintained wherever micelles are present (quasi-equilibrium). In other words, when total bulk concentration is above C_{CMC} monomer concentration is always maintained at CMC. In this limit, surfactant transport process in presence of micelles

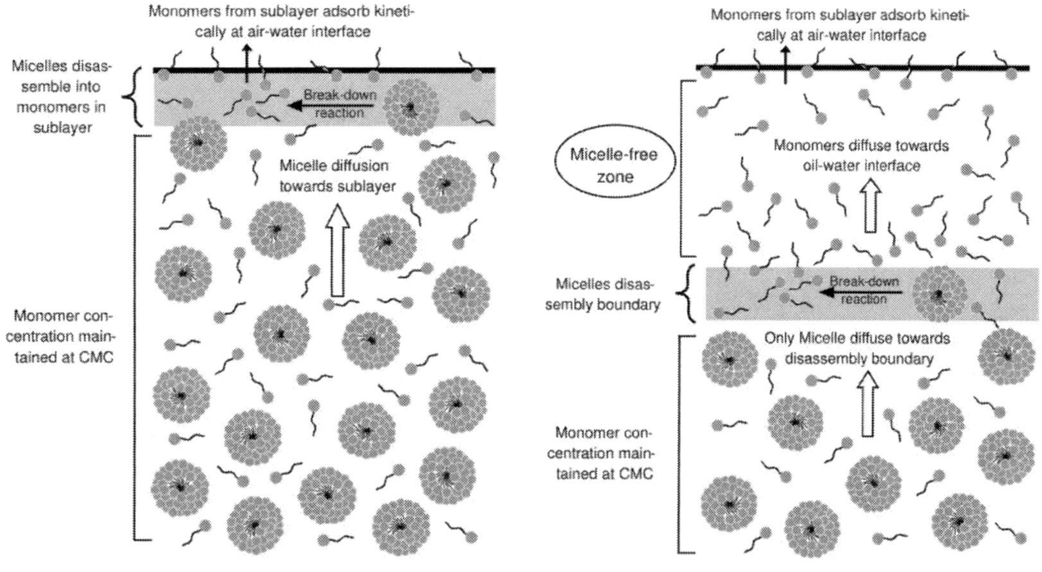

(a) High micelle concentration

(b) Low micelle concentration

Figure 1: Pictorial representation of surfactant transport process from an aqueous micellar solution to an initially clean air-water interface.

naturally gets divided into two regimes; in high micelle concentration regime (Fig. 1a), monomer depletion caused by the adsorption at the interface is replenished by the micelle break-up in the sublayer as the larger diffusive flux of micelles assures presence of micelles all the way up to the sublayer. In this regime, only micelles diffuse and monomer flux is zero, as they are maintained at fixed concentration C_{CMC}. On the other hand at lower micelle concentration regime (Fig. 1b), diffusion flux of micelles cannot keep up with the adsorption flux at the interface and micelle break-down take place away from the interface. This reaction boundary separates the two distinct zones, micelle zone and a micelle-free zone; maintaining monomer concentration in micelle zone always equal to CMC. Thus, in the micelle zone, only micelles diffuse till the micelle disassembly boundary, where break-down reaction supplies monomers to the micelle-free zone and these monomers then diffuse to and subsequently adsorb onto the interface.

In this paper we will be focusing onto the infinitely fast kinetic regime of transport and in particular where micelle-free zone is observed. This zone represents a kind of a barrier to the surfactant transport process preventing micelles to reach up to the interface. Theoretical model in the limit of fast micelle break-up was first used previously by Song et al (36) for a surfactant transport from an aqueous micellar solution to an air bubble interface. The predicted surface tension relaxation agreed with experimental measurements as obtained from pendant bubble tensiometry for micelle concentrations giving rise to micelle-free zone. Our objective is to provide an experimental evidence for the formation and evolution of the micelle-free zone and also to compare the theoretical solutions using the fast kinetics model to the experiments.

To simplify our study of micelle-free zone, instead of allowing the clean air-water interface to drive the process we study the transport from an aqueous micellar solution to an initially clean oil phase. The reason for using the oil is two fold: first the air-water interface has a limited capacity

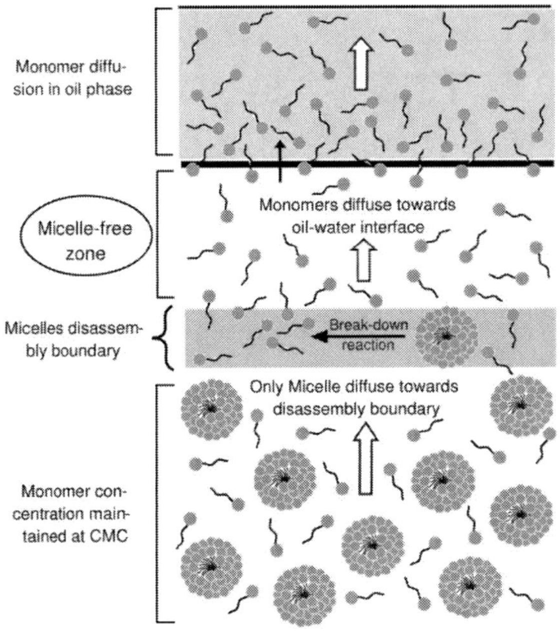

Figure 2: Representation of overall process of surfactant transport from an aqueous micellar solution to an initially clean oil phase.

to adsorb surfactant, consequently micelle-free zones are smaller and more difficult to observe and secondly, also it is difficult to create a clean (surfactant-free) air-water interface. An oil phase can have a much larger capacity to pick up surfactant, depending upon the partition coefficient of the surfactant and the volume of oil phase used, this leads to large micelle-free zones. So the overall process when the micelle-free zone is present is as shown in Fig. 2; micelles diffuse in the bulk aqueous phase up to the micelle disassembly boundary where they break down into monomers, surfactant monomers then diffuse from the boundary to the sublayer where they kinetically adsorb at the oil-water interface. Adsorbed monomers then desorb into the oil phase and diffuse further into that phase.

In this paper we focus our attention to the transport of non-ionic polyethoxylated surfactant $C_{14}E_6$ ($CH_3(CH_2)_{13}(OCH_2CH_2)_6OH$), which forms globular aggregates above its critical micelle concentration of 0.009 mol/m^3. There are several reasons for choosing this surfactant. Firstly its monomer adsorption kinetics is well characterized (36). Second, being a nonionic surfactant it has a large partition coefficient in the oil phase. And third, it has correct time-scales of diffusion and micelle break-up kinetics; slow time scale for micelle break-up (τ_{SL}) has been measured experimentally for $C_{12}E_6$ and $C_{12}E_8$ are 10 and 4 s respectively (37; 2), and we expect that for $C_{14}E_6$ it should be of the same order. The diffusion time scale for the transfer of surfactant from water to oil can be estimated as, $\tau_D = l^2/D$. For aqueous phase of $l = O(1\text{mm})$ and $D = O(10^{-10}\text{m}^2/\text{s})$, diffusion time scale, $\tau_D = O(10^4)$. As seen before when $\tau_D >> \tau_{SL}$, micelles break-up infinitely faster as compared to diffusion and micelle-free zone is formed.

To visualize the micelle transport experimentally, we use a small hydrophobic dye molecule,

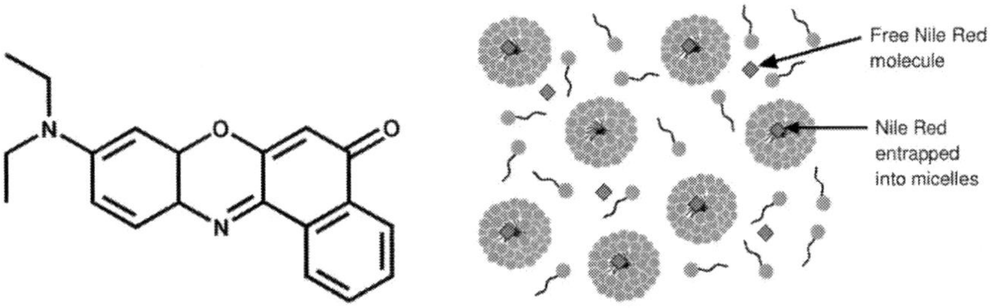

(a) Chemical structure (b) Nile Red in surfactant solution

Figure 3: Nile Red molecule; chemical structure and being incorporated into micelles.

Nile Red (9-diethylamino-5H-benzo[α]phenoxazine-5-one) which is sensitive to the hydrophobicity/hydrophilicity of its environment. Nile Red dye, Fig. 3a, has been well characterized and its emission is dependent upon the environment (38; 39; 40). In hydrophobic environment it exhibits a strong fluorescence, while in an aqueous environment its fluorescence is almost quenched. In a hydrophobic environment, it shows a maximum excitation for a wavelength near 540 nm and a fluorescence emission maximum near 640 nm. However depending upon the polarity of the surrounding medium Nile Red shows large bathochromic or solvatochromic shift (39). Being a low molecular weight, small, hydrophobic molecule it strongly partitions into the micellar core, where it sees a hydrophobic environment and thus exhibiting a strong fluorescence. Upon micelle disassembly the dye is released into the aqueous medium, and its fluorescence is quenched. This property of the Nile Red can be exploited to visualize the micelle and micelle-free zones by detecting its emission in horizontal planes as a function of distance from the oil-water interface, using confocal fluorescence microscopy. In following section (section 2), the experimental measurements of the micelle-free zone are presented. Theoretical solution of the zone dynamics are presented in section 3, along with a comparison of the simulations with the data. The paper concludes with the summary in section 4.

2 Experimental

2.1 Materials and Methods

Surfactant used, poly (ethylene glycol) n-alkyl ether $C_{14}E_6$ was obtained from Nikko Chemicals, Japan and its aqueous solutions were prepared from deionized water purified by a Milli-Q filter system equipped with an Organex-Q column to remove trace organic contaminants (Millipore, MA, USA) . The resistivity of the deionized water was about 18 MΩcm. Nile Red (Molecular Probes, Inc., Eugene, OR, USA) was first dissolved in ethanol (Reagent grade, Sigma-Aldrich, St. Louis, MO, USA) before adding to the aqueous surfactant solution. Nile Red concentration in the surfactant solutions was maintained at 0.2 μg/mL, within its water solubility limit (0.3 μg/mL). Oil used, Hexadecane >99% purity was obtained from Acros Organics, NJ, USA. All chemicals were used as received without further purification.

An inverted Confocal Scanning Laser Microscope (CSLM; Leica TCS SP2 AOBS, employing an acoustic-optical beam splitter and a prism spectrophotometer detector; 63x objective, NA = 1.4), was used for detection of fluorescence from the Nile Red. Nile Red was excited with two laser beams, 514 and 543 nm and its emission was collected between 580–750nm. Chambers for

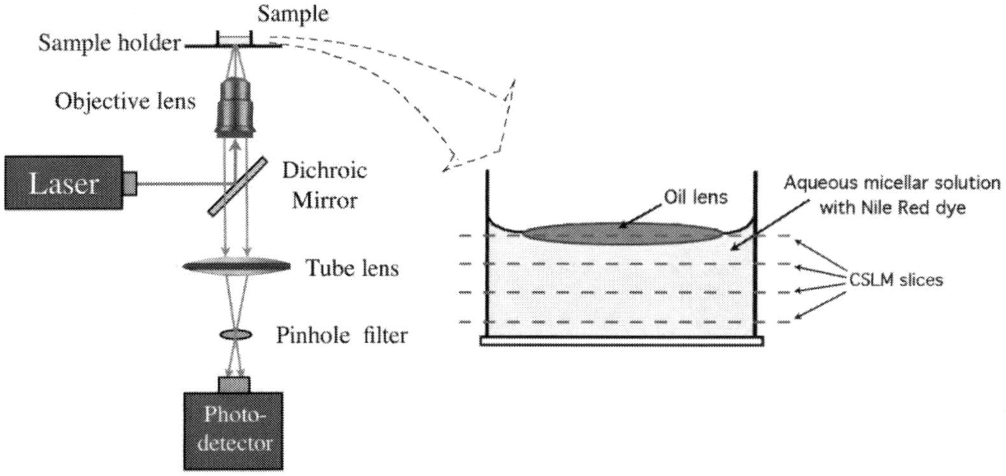

Figure 4: Experimental arrangement showing schematics of Confocal Scanning Laser Microscope and a sample, oil lens on the top of aqueous solution.

the confocal experiments were obtained from Electron Microscopy Sciences, Hatfield, PA, (Lab-Tek chamber slides with 2 chambers, each with 24 x 20 mm in size), and were made up of plastic walls and soda lime glass (~160μm thick) at the bottom. Experimental arrangement is as shown in Fig. 4, 1.5mL of an aqueous micellar solution with a Nile Red was taken in a chamber slide. After sufficient time, a fresh oil measuring 40μL was gently placed on the top of the aqueous phase. For such small volume of oil, it does not spreads till the side walls of the chamber but forms a lens of about 15–17 mm in size. For this arrangement, water thickness was 1.9 mm and oil thickness was about 300 μm. As the water thickness is very small compared to the oil lens diameter, the transport of surfactant essentially take place only in one dimension. Fluorescence signal obtained from Nile Red was measured by CSLM. To obtain fluorescence signal as a function of a distance, the sample was optically sliced by physically moving the sample stage vertically so that the focal plane of an objective is moved through the sample and average fluorescence intensity was obtained from each individual slices. Thus, presence or absence of a fluorescence signal at a certain position indicates presence or absence of micelles.

2.2 Reference Fluorescence Profile

Fig. 5 shows the CSLM generated fluorescence signal intensity as a function of distance for below CMC concentration of $C_{14}E_6$. Surfactant Concentration is $0.75 \times C_{\text{CMC}}$ and that of a Nile Red is 0.2μg/ml. No oil was used in this experiment. Fig. 5 shows that for below CMC concentration of $C_{14}E_6$, a very weak fluorescence signal was obtained from the bulk aqueous phase as the Nile Red molecules sees an aqueous environment. Also, the fluorescence signal intensity decays as a function of a distance as we move further away from glass-water interface. As the laser source is below the glass chamber, more and more light is absorbed by the dye molecules as it travels through the solution. And thus available laser intensity decreases depending upon the extinction coefficient of the dye molecule, which in turn causes decaying fluorescence signal with distance. We also observe strong fluorescence peaks at the air-water and the water-glass interface, which

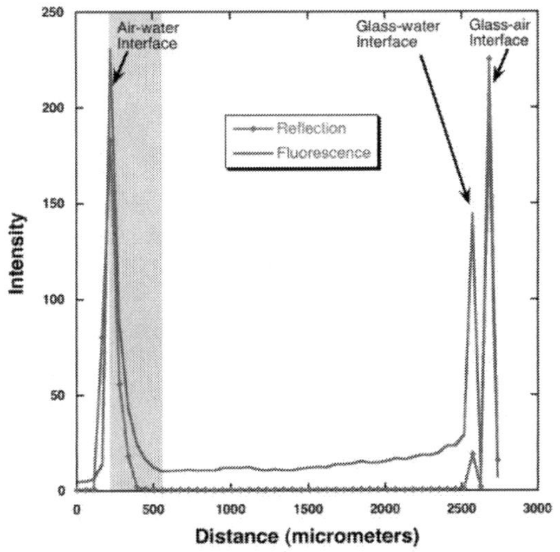

Figure 5: Fluorescence signal as a function of distance as obtained from CSLM for $0.75\times C_{\mathrm{CMC}}$ $C_{14}E_6$ concentration with Nile Red concentration at 0.2 μg/ml

is attributed to the presence of an adsorbed monolayer of surfactants at these interfaces traps the Nile Red molecules in their hydrophobic chains. As the Nile Red trapped in the monolayers sees a hydrophobic environment, it exhibits a strong fluorescence signal. Positions of all the three interfaces were also confirmed by reflection microscopy in which excitation wavelength is included in the collecting window and the peaks from both fluorescence and reflection microscopy were found to overlap on each other.

We would also like to emphasize that the shaded region in Fig. 5, which we call it as a shadow, emerges from the air-water interface and extends into the bulk aqueous phase for as long as 300μm. This shadow masks the underlying actual fluorescence from the bulk phase. We think these high intensities in the shadow region could be due to the strong out-of-focus light from the interface or due to the reflection from the exciting laser which CSLM is not able to filter out. This often causes a problem since we lose the resolution in locating zero fluorescence region or micelle free zone for the first 300μm or so.

Plotted in the Fig. 6 is the fluorescence signal intensity profile for surfactant concentrations above CMC, i.e. in the presence of micelles. Nile Red molecules in presence of micelles partitions themselves into the hydrophobic core and hence a very strong fluorescence signal was observed from the bulk aqueous phase. Here we clearly see the exponentially decaying fluorescence signal as we move away from the laser source. Again we observe the peaks at the interfaces due to trapped dye in a monolayer at the air-water interface and may be in a bilayer on the glass-water interface. The shadow was still present at the interface and masks the actual fluorescence underneath.

2.3 Visualization of micelle-free zone

When an fresh oil lens is placed gently on the surface of the aqueous phase, equilibrium near the oil-water interface is disturbed as the adsorbed surfactant molecules at the interface now start to

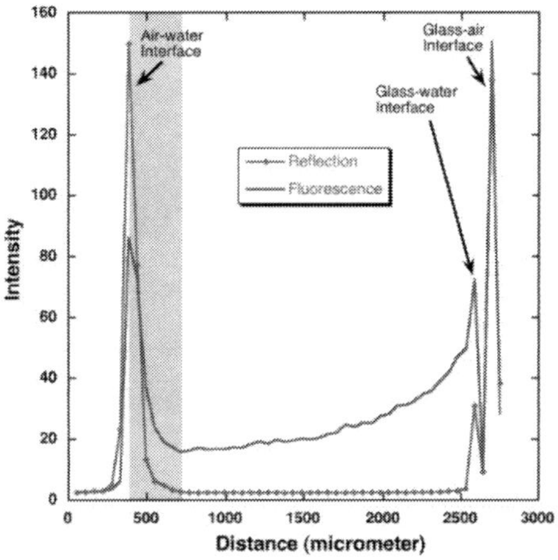

(a) 1.5×C_{CMC}

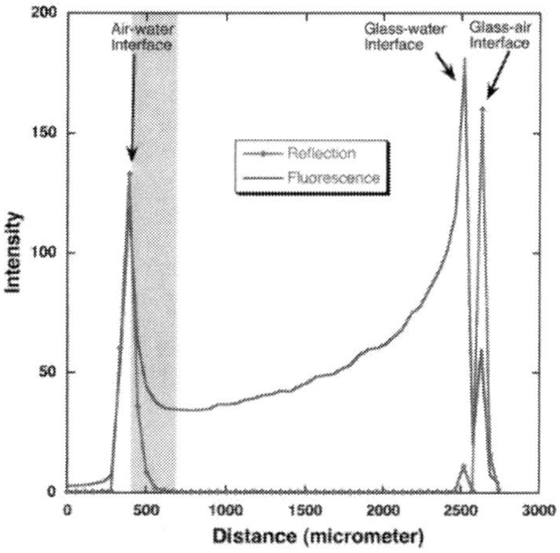

(b) 2×C_{CMC}

Figure 6: Fluorescence signal as a function of distance obtained from CSLM for $C_{14}E_6$ concentration above CMC with Nile Red concentration at 0.2 μg/ml

desorb and dissolve into the oil phase. As a result a micelle-free zone is developed which is shown in Fig. 7 at different time stages. Raw data obtained from the CSLM was processed to remove the weak fluorescence from free Nile Red molecules in aqueous phase by subtracting the reference data for 0.75×C_{CMC} obtained at the same exact settings. Reference data of 0.75×C_{CMC} was first smoothened by fitting a polynomial before subtracting from the raw data to avoid any irregularities.

Fig. 7a shows fluorescence profile after 18 sec of placing an oil lens on the top of 2×C_{CMC} aqueous surfactant solution. At this time, even if micelle-free zone develop it was not distinguishable as it was masked by the shadow. In order for us to locate the micelle-free zone it needs to evolve out of shadow. For some later time, say 320 sec (Fig. 7b), we clearly see the distinction between fluorescence and no fluorescence region and micelle-free zone boundary could be easily located at 330μm from the oil-water interface. As oil starts to pick-up the hydrophobic Nile Red as well, fluorescence intensity in the oil phase also climbs up. However as the time progresses, for example at 1280 sec in Fig 7c, it becomes more and more difficult to demarcate the micelle-free zone boundary. As the Nile Red diffuses out into the oil phase, corrected fluorescence signal intensity goes negative near the oil-water interface because we are subtracting fluorescence signal of 0.2μg/mL free Nile Red from less concentrated aqueous solution. In this case, micelle-free zone boundary was located at the point where fluorescence intensity changes sign from negative to positive.

With time, micelle-free zone first develops at the oil-water interface and penetrates into the bulk aqueous phase. After placing the oil lens on the top, the sample was continuously scanned for fluorescence profile with CSLM and plotted in Fig. 8 is the data of fluorescence intensity as a function of distance at different time stages. Fig. 8 clearly indicates the growing micelle-free zone with time. Also dye signal in oil phase gives us incite into the Nile Red diffusion process in the oil phase. Initially a peak was seen at the oil-water interface for time t=101 sec. As the time progress bulk oil phase away from the interface start fluorescing as Nile Red diffuse in.

Number of such experiments were performed at different initial concentrations and data was analyzed, as mentioned before, to get the micelle-free zone boundary position as a function of time,

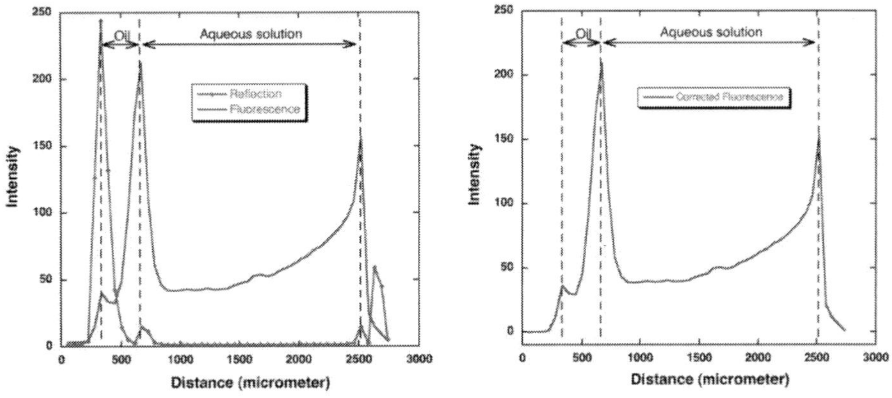

(a) t=18 sec

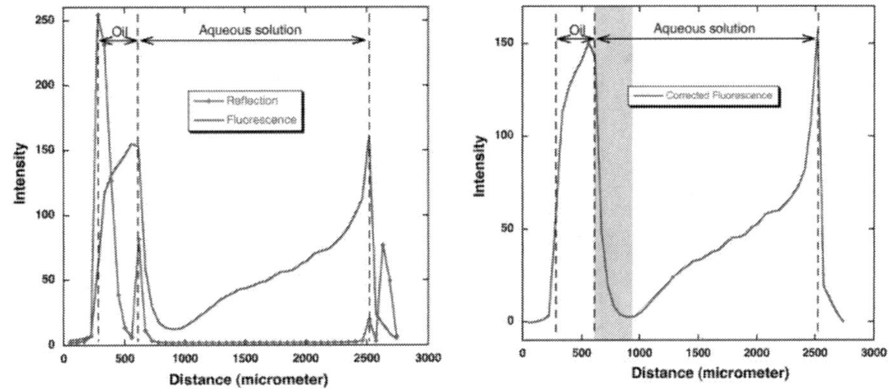

(b) t=320sec

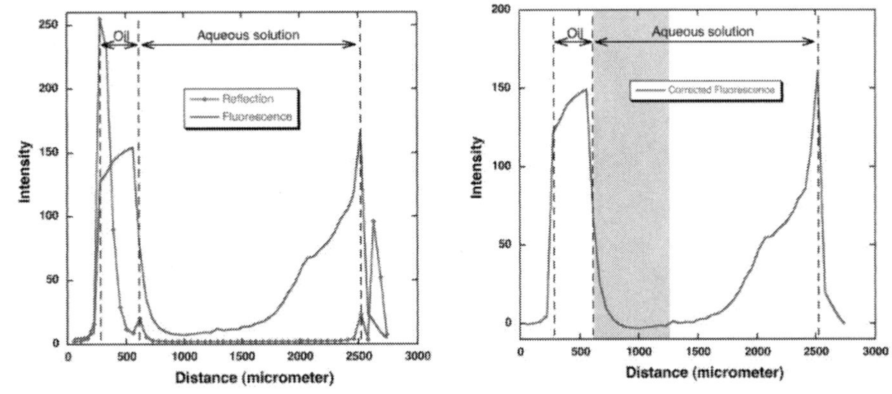

(c) t=1280sec

Figure 7: Visualization of micelle-free zone at different time stages for $2\times C_{\mathrm{CMC}}$ initial concentration. Fig on the left shows raw data obtained from the CSLM, while Fig on the right shows the baseline-corrected fluorescence profile obtained after subtracting the reference $0.75\times C_{\mathrm{CMC}}$. Shaded region is the no fluorescence region or micelle-free zone.

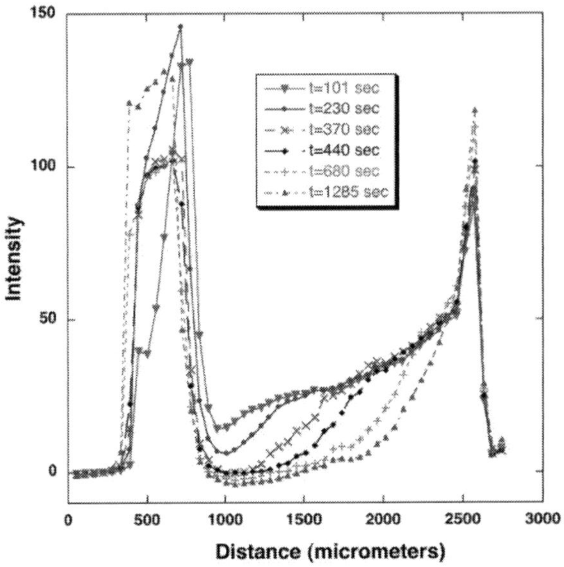

Figure 8: Corrected fluorescence Intensity vs distance obtained from CSLM at different time steps for initial bulk concentration of $2\times C_{\mathrm{CMC}}$.

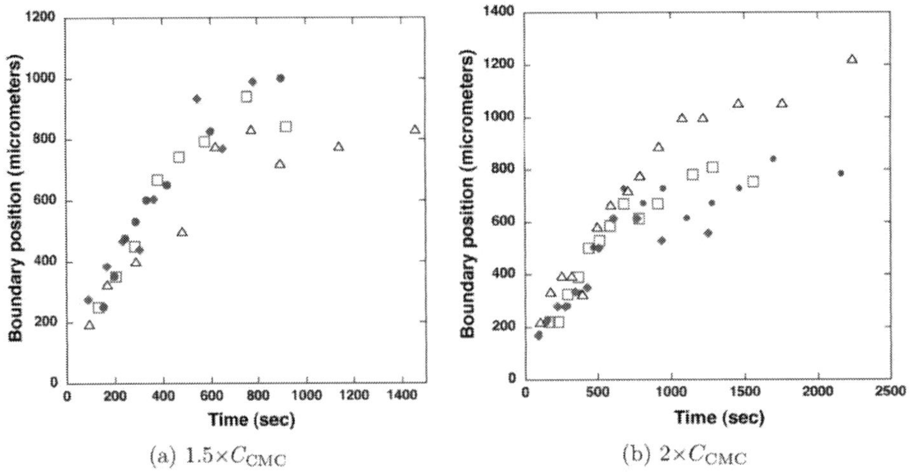

(a) $1.5\times C_{\mathrm{CMC}}$ (b) $2\times C_{\mathrm{CMC}}$

Figure 9: Micelle-free zone boundary movement with time as obtained from CSLM experiments for $1.5\times C_{\mathrm{CMC}}$ and $2\times C_{\mathrm{CMC}}$ initial bulk concentration. Each different type of symbol represent different set of experiment.

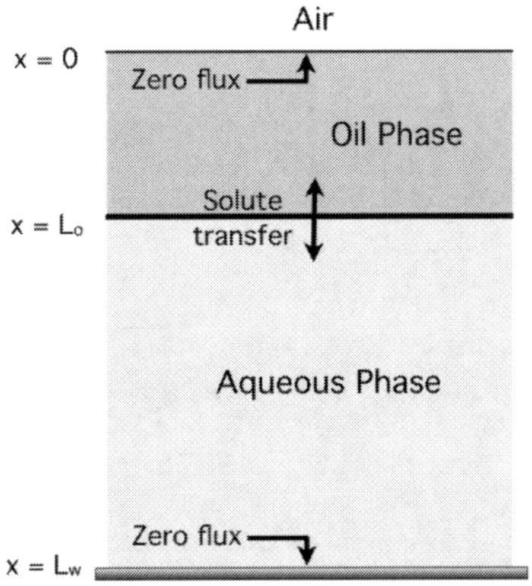

Figure 10: Simulation arrangement of oil and water layer for development of 1-D transport model.

which is plotted in Fig. 9. Different types of symbols in the graphs represents different sets of experiment conducted at identical conditions and the moving boundary position with time was observed to be consistent and reproducible. Note here that, initially zone boundary travels faster and on later time boundary movement slows down which is evident from change in a slope in Fig. 9a and 9b. Reason for this slowing down of boundary movement is that initially oil phase is clean and adsorption flux at the oil-water interface is very high as oil continuously withdraws surfactant from the interface. As a result, depletion in the aqueous phase also take place at faster rate. But with time, oil phase begin to saturate with surfactants and thus adsorption flux is reduced, slowing down the depletion in the water. Another important factor that could be causing change in the slope is secondary diffusion; as we have an oil lens and not an oil layer, diffusion flux from the side walls to the center counters the depletion process.

3 Theoretical transport model

3.1 Model formulation

In this section, 1-D diffusional transport model from a finite length aqueous micellar solution to an initially clean oil phase (Fig. 10) is presented in order to replicate the experimental arrangement so that later on comparison could be made between the two. In formulating the model assumption is made that time scale for the kinetic break down of the micelles is infinitely faster than the diffusional time scale. This assumption implies that wherever the bulk concentration in aqueous phase is above CMC, only micelles diffuse and monomer concentration is constant; while when bulk concentration is below CMC micelles are absent and only monomers diffuse. This gives rise to two different zones in the aqueous phase (micellar and micelle-free). Diffusion process in the oil phase and two aqueous zones is governed by Fick's law

$$\begin{aligned}
\frac{\partial C_o}{\partial t} &= D_o \frac{\partial^2 C_o}{\partial x^2} && (0 < x < L_o) \\
\frac{\partial C_w^m}{\partial t} &= \frac{\partial}{\partial x}\left(D_w^m \frac{\partial C_w^m}{\partial x}\right) && (L_o < x < L_{zone}) \\
\frac{\partial C_w^{mic}}{\partial t} &= \frac{\partial}{\partial x}\left(D_w^{mic} \frac{\partial C_w^{mic}}{\partial x}\right) && (L_{zone} < x < L_o + L_w)
\end{aligned} \tag{2}$$

where subscript o, w indicates oil and water phases respectively, subscript $zone$ indicates the micelle-free zone, superscripts m and mic indicate quantities related to monomers and micelles respectively. Coordinate x is measured from the oil-air interface into the bulk and L is the length of a respective phase, t is time, $C(x,t)$ is the bulk surfactant concentration, and D is the bulk diffusivity.

To solve these equations using a front trapping method a new variable C_w is introduced, which is total bulk concentration in the aqueous phase defined as,

$$C_w = C_w^m + \sigma \cdot C_w^{mic} \tag{3}$$

where σ is aggregation number of a micelle, i.e number of monomers present per micelle ($\sigma \approx 100$ for $C_{14}E_6$). The governing equation for aqueous phase, valid over entire domain, in terms of this total concentration becomes,

$$\frac{\partial C_w}{\partial t} = \frac{\partial}{\partial x}\left(D_w \frac{\partial C_w}{\partial x}\right) \qquad (L_o < x < L_o + L_w) \tag{4}$$

where D_w is a piecewise continuous step function, which is equal to the monomer diffusivity when $C_w < C_{\mathrm{CMC}}$ and equal to the micelle diffusion coefficient for $C_w > C_{\mathrm{CMC}}$. At the micelle-free zone boundary, i.e. $C_w = C_{\mathrm{CMC}}$, micelle diffusion flux to the boundary is matched with the monomer diffusion flux out of the boundary according to the conservation equation.

At the beginning of the process, surfactant is present only in the aqueous phase,

$$\begin{aligned}
C_o(x, t=0) &= 0 && (0 < x < L_o) \\
C_w(x, t=0) &= C_{bulk} && (L_o < x < L_o + L_w)
\end{aligned} \tag{5}$$

At the two boundaries, namely oil-air and glass-water interfaces, zero flux boundary conditions are imposed,

$$\begin{aligned}
\left[\frac{\partial C_o}{\partial x}\right]_{x=0} &= 0 \\
\left[\frac{\partial C_w}{\partial x}\right]_{x=L_o+L_w} &= 0
\end{aligned} \tag{6}$$

At the oil-water interface, surfactants kinetically adsorb from aqueous phase and further desorb into the oil phase. Accumulation of surfactants onto the oil-water interface could be obtained from the mass balance equation,

$$\frac{\partial \Gamma}{\partial t} = D_w \left[\frac{\partial C_w}{\partial x}\right]_{x=L_o^+} - D_o \left[\frac{\partial C_o}{\partial x}\right]_{x=L_o^-} \tag{7}$$

where $\Gamma(t)$ is surface density. In order to close the problem we need two more conditions at this oil-water interface. We assume a local equilibrium between the two phases and also assuming that the interface too is in local equilibrium with both phases,

$$C_o(L_o, t) = K_p \cdot C_w^m(L_o, t) \tag{8}$$

$$\Gamma(t) = \Gamma_\infty \cdot \frac{\beta C_w^m(L_o, t)}{\beta C_w^m(L_o, t) + \alpha} \tag{9}$$

where, K_p $(= C_o/C_w^m)$ is the partition coefficient of surfactant between oil and water, α and β are the kinetic rate constants for desorption and adsorption in Langmuir isotherm, Γ_∞ is the maximum packing density. Initial condition for the surface density is also given by the same relation, i.e by assuming that the interface was in equilibrium with the aqueous phase at $t = 0$.

3.2 Numerical formulation and code validation

The system of equations was solved numerically with finite volume method, for which the space domain was discretized into N equal sized control volumes over which mass balance must satisfy. Algebraic equations for the concentrations at each node were obtained from the differential equations by employing a fully implicit finite volume method. Algebraic equations when written in a matrix form, generates a tri-diagonal matrix, which could be easily inverted to obtain the concentrations in each phases and surface density at the oil-water interface. From the total concentration in aqueous phase, micelle and monomer concentrations can be calculated, which in turn gives the micelle and micelle-free zones. A FORTRAN code has been developed to generate and solve the tri-diagonal matrix for the concentrations.

For the code validation purpose consider an aqueous phase, with uniformly distributed surfactants at below CMC concentration, exposed to an initially clean air-water interface. There was no oil phase involved, but the clean air-water interface drives the transport process. Adsorption at the air-water interface has been modeled with non-equilibrium Henry's isotherm. Thus in this simplified case, the governing equations and the boundary conditions are as follows,

$$\frac{\partial C_w}{\partial t} = D_w \frac{\partial^2 C_w}{\partial x^2} \qquad (0 < x < L) \tag{10}$$

$$\frac{d\Gamma}{dt} = \beta C_w(x=0)\Gamma_\infty - \alpha\Gamma = D_w \left[\frac{\partial C_w}{\partial x}\right]_{x=0} \tag{11}$$

and

$$D_w \left[\frac{\partial C_w}{\partial x}\right]_{x=L} = 0 \tag{12}$$

Initial conditions are $\Gamma(t=0) = 0$ and $C_w(x, t=0) = C_{bulk}$. An exact solution for this problem is obtained by Laplace transform which in dimensionless form is,

$$\tilde{\Gamma}(\tau) = \frac{k}{N_l k + 1} - \sum_{n=1}^{\infty} \frac{2k^2 N_l e^{-\beta_n^2 \tau}}{[(N_l k + 1) - (2N_d - N_l^2 k^2 - N_d N_l k) \cdot \beta_n^2 + N_d^2 \beta_n^4]} \tag{13}$$

where $\tilde{\Gamma} = \Gamma/\Gamma_\infty$, $\tau = D_w t/L^2$, $N_d = D_w/(L^2\alpha)$, $N_l = \Gamma_\infty/(C_{bulk}L)$, $k = \beta C_{bulk}/\alpha$ and β_n's are the roots of equation,

$$\frac{N_l k \beta_n}{(N_d \beta_n^2 - 1)} = \tan \beta_n \tag{14}$$

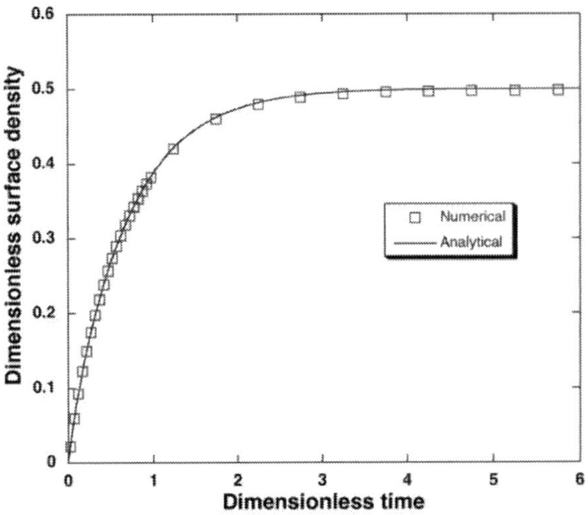

Figure 11: Comparison between $\tilde{\Gamma}(\tau)$ obtained from analytical solution, eq. 13, and that obtained numerically with parameters $K = N_d = N_l = 1$

$\tilde{\Gamma}(\tau)$ obtained from this relation compared well with that obtained from the finite volume method code incorporating non-equilibrium Henry's isotherm, Fig. 11.

3.3 Theoretical results and comparison with experiments

Numerical results for the 1-D transport of $C_{14}E_6$ surfactant from micellar aqueous solution to an initially clean hexadecane phase are reported. Partition coefficient K_p for the $C_{14}E_6$ between oil and water has been measured by surface tension measurements using pendant bubble tensiometry and a very large partition coefficient, of about 350, was observed in these experiments. Adsorption-desorption kinetic parameters at oil-water interface of $C_{14}E_6$ used were same as those of air-water interface. Reason being, for such a large partition coefficient, oil phase maintains water sublayer concentration practically at zero and the kinetics at the oil-water interface no longer matters. We have tried to alter the kinetic parameters by couple of order of magnitudes in our simulations but the micelle-free zone boundary movement was unaltered. Thus, following Song et al. (36), we have used $\alpha/\beta = 8.35 \times 10^{-}6$ mol/m^3, $\Gamma_\infty = 2.4 \times 10^{-}6$ mol/m^2 and monomer diffusivity in water $D_w^m = 4 \times 10^{-}10$ m^2/s, $C_{CMC} = 0.009$ mol/m^3. From the studies of Tiberg et al.(41; 42), the micelle diffusion coefficient of $(1/4)D_w^m$ at dilute concentrations (1-20$\times C_{\rm CMC}$) is used. D_o was scaled with hexadecane viscosity relative to water, i.e. viscosity of hexadecane is 3 cP and that of water is 1cP, thus D_o was taken to be $(1/3)D_w$. Other parameters were used based on experimental arrangement, such as $L_o = 300\mu$m, $L_w = 1.9$ mm and respective initial bulk concentration in the aqueous phase.

Fig. 12 shows numerical results obtained from our simulations for 2$\times C_{\rm CMC}$ initial concentration with entire space domain discretized into $N = 400$ equal sized control volumes. From the total concentration obtained, micelle and monomer concentration could be calculated which is plotted in Fig. 12a and 12b, where the concentrations are nondimensionalized by $C_{\rm CMC}$. Micelle distribution shows that for certain region near the oil-water interface shows absence of micelles, micelle-free zone, and micelle concentration then sharply increases in the micelle zone owing to the smaller micelle diffusion coefficient. In other words, to maintain the certain diffusion flux towards the oil-water interface, due to smaller diffusion coefficient micelle concentration gradient needs to be large. Also note that because of very large partition coefficient of surfactant, oil phase keeps the sublayer monomer concentration very close to zero and monomer concentration gradient is almost constant owing to larger monomer diffusion coefficient, especially in the earlier time steps. Fig. 12c shows dimensionless total concentration in both oil and water phases. Discontinuity in the total concentration gradient in aqueous phase when $C_w = C_{\mathrm{CMC}}$ is natural, even though the diffusion flux is continuous, as D_w is assumed to be a step function. Surfactant concentration in the oil phase also starts to climb up with time as oil starts picking up the surfactant monomers from the oil-water interface. And finally in Fig. 12d micelle-free zone boundary position is plotted against time, where we see that the micelle-free zone originates at the oil-water interface, and penetrates into the bulk aqueous phase depleting entire sample out of micelles for given parameters. Effect of initial concentration on the micelle-free zone boundary movement is also seen in Fig. 12d, as the concentration increases diffusion flux is increased and thus it takes longer for micelle-free zone to travel certain distance. Effect of other adjustable parameters such as length of oil and water phases was observed to be negligible.

This micelle-free zone boundary position as a function of time can be compared to the experiments. However experimental arrangement that we have, has a large area to volume ratio. Aqueous phase dimensions in the chamber slide are 2.4 cm × 2 cm × 0.19 cm, and surfactant molecules from the aqueous phase have tendency to adsorb on the plastic side walls, bottom glass slide as well as on the air-water interface when the solution was poured into the chamber, thus there is always a loss of surfactant molecules from the bulk resulting in reduced bulk phase concentration. To minimize this loss due to adsorption we rinse the chamber couple of times with the sample solution so as to saturate the walls and bottom glass slide of the chamber. However, loss due to adsorption at the air-water interface is always going to be present. Therefore we need to take into account this loss of surfactant monomers from the aqueous bulk phase before we compare theory and experiments. Based on the head group area of 50 $\mathring{A}^2$ for $C_{14}E_6$ molecule, we have computed the loss of surfactant molecules to an air-water interface and experiments were then compared with corrected concentration. For given geometry of the chamber $0.13 \times C_{\mathrm{CMC}}$ concentration is depleted from the 1.5 mL of the sample solution, thus we compare experiments with 1.5 and $2 \times C_{\mathrm{CMC}}$ solution and simulations with initial concentration of 1.37 and $1.87 \times C_{\mathrm{CMC}}$ respectively in Fig. 13. Figure shows very good agreement between experiments and theory, especially at the earlier time. During later time, agreement is poor because of several reasons; most important of which is secondary transport. As we have an oil lens and not a layer of oil, surfactant transport is not perfectly unidirectional, especially when oil begins to saturate with surfactant. In other words, surfactants not only diffuse in the vertical z-direction but they also diffuse radially in from the chamber walls towards center. Effect of this secondary transport is that it slows down micelle-free zone boundary movement as it has an additional diffusive flux of surfactant to the depleted region. Second reason for poor agreement in later time is attributed to the loss of Nile Red dye molecules to the oil phase. As Nile Red diffuse into the oil, overall signal in the bulk aqueous phase is weakened; leading to increased uncertainty in locating the micelle-free zone or no fluorescence region.

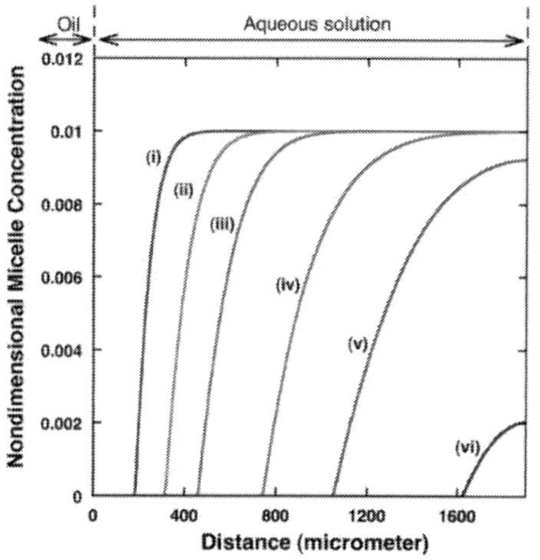

(a) Nondimensional micelle concentration $C_w^{mic}/C_{\mathrm{CMC}}$ distribution with distance (in micrometers) from the oil-water interface.

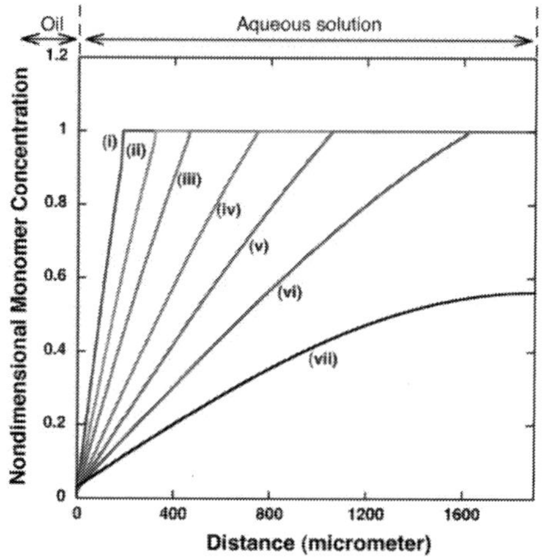

(b) Nondimensional monomer concentration C_w^{m}/C_{CMC} distribution with distance (in micrometers) from the oil-water interface.

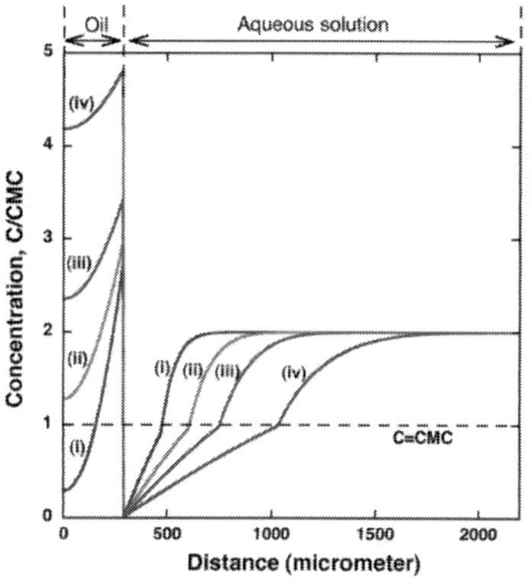

(c) Nondimensional total concentration C/C_{CMC} distribution with distance (in micrometers) from the air-oil interface.

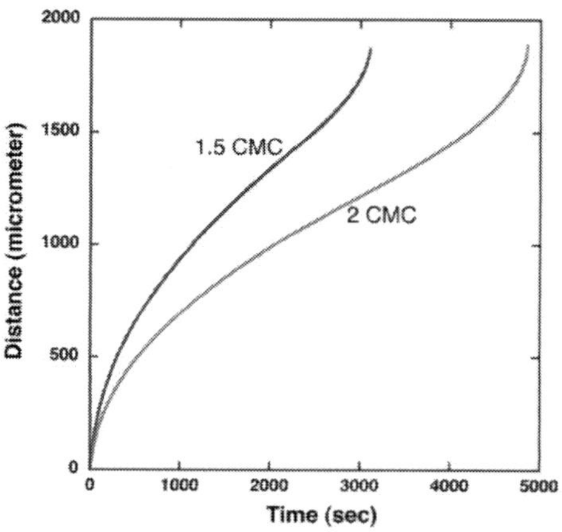

(d) Micelle-free zone boundary position (in micrometers) from the oil-water interface plotted against time (in seconds) for different initial concentrations.

Figure 12: Simulation results for $C_{14}E_6$ surfactant at different time intervals (i) 90sec, (ii) 226sec, (iii) 451sec, (iv) 1128sec, (v) 2256 sec, (vi) 4512 sec and (vii) 6769 sec. Initial aqueous phase concentration in Fig. (a), (b) and (c) is 2CMC.

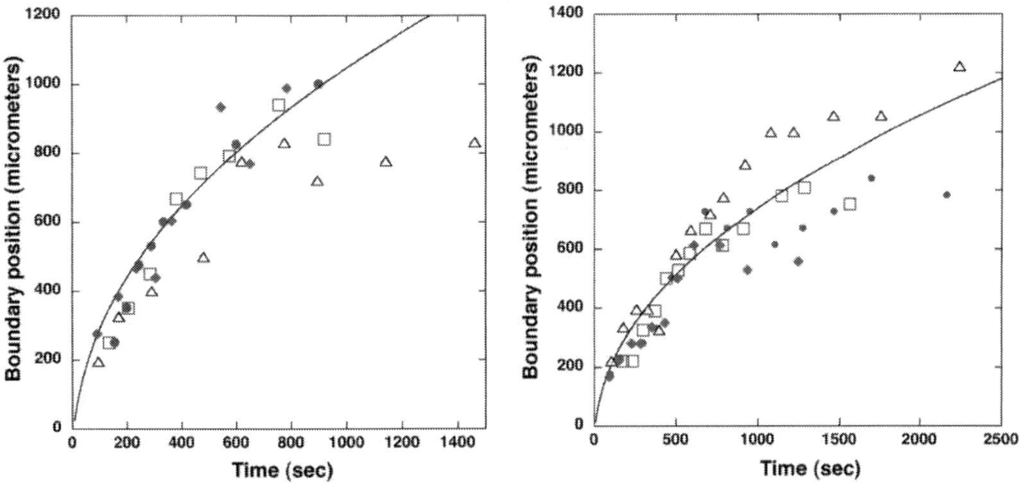

(a) 1.5×C_{CMC} (b) 2×C_{CMC}

Figure 13: Comparison between theory and experiments for the micelle-free zone boundary position vs time. Symbols represent experiments and solid line is theoretical prediction for the corrected concentration.

4 Summary

To summarize, surfactant transport process from an aqueous micellar solution to an initially clean oil phase has been studied experimentally as well as theoretically. Many experimental studies have confirmed that presence of micelles accelerates the surfactant transport process. When the surfactant is soluble in both phases, over all transport process involves transport of solute across the interface. This process is governed by diffusion in both phases and kinetic adsorption/desorption at the oil-water interface. When initially clean oil phase comes in contact with the aqueous surfactant solution at equilibrium, oil continuously withdraw surfactant from the oil-water interface which in turn drives adsorption of the monomers from the aqueous sublayer onto the interface. Depletion of the monomers from the sublayer, disturbs the micelle-monomer equilibrium causing micelles to break-up to replenish the monomers. This micelle break-up drives the diffusion of micelles from bulk to the interface. When diffusion flux of micelles is very small compared to adsorption flux of monomers at the interface, micelles completely disappear from a region near the interface and micelle-free zone is formed which is studied here in detail. To visualize the micelle-free zone, an environment sensitive dye molecule (Nile Red) was added into the micelles. It has shown to exhibit a strong fluorescence when the dye is incorporated into the micelles but fluorescence was almost quenched for Nile Red in the free state. Confocal Scanning Laser Microscopy was used to detect the fluorescence from the sample, showing two distinct regions; fluorescent and non-fluorescent, indicating micelle zone and micelle-free zone respectively. Micelle-free zone boundary position is tracked with time to compare with the theory.

A theoretical model has been developed for the surfactant transport from micellar solution to initially clean oil phase, in the limit of fast micelle break-up, where micellization kinetics is avoided as the separation of time scales is valid for chosen surfactant $C_{14}E_6$. In this limit of quasi-equilibrium, wherever micelles are present monomer concentration is always maintained at C_{CMC}. Thus, when the micelle-free zone formed only micelles diffuse to the micelle disassociation boundary supplying monomers to the micelle-free zone, which then diffuse further towards sublayer and adsorbs onto the interface. Micelle-free zone boundary position is located where total bulk concentration in aqueous phase is equal to C_{CMC}. This micelle-free zone boundary evolves from the sublayer and travels into the bulk. Boundary position movement is obtained theoretically and it is compared well with the theory.

References

[1] A. Patist, J. Kanicky, P. Shukla, and D. Shah, "Importance of micellar kinetics in relation to technological processes", *Journal of Colloid and Interface Science* **245**, pp. 1–15 (2002).

[2] A. Patist, S.G. Oh, R. Leung, and D.O. Shah, "Kinetics of micellization: Its significance to technological processes", *Colloids and Surfaces A: Physicochemical and Engineering Aspects* **176**, pp. 3–16 (2001).

[3] S. Oh, S. Klein, and D. Shah, "Effect of micellar lifetime on the bubble dynamics in sodium dodecyl sulfate solutions", *AICHE Journal* **38**, pp. 149–152 (1992).

[4] S. Oh and D. Shah, "Relationship between micellar lifetime and foamability of sodium dodecyl sulfate and sodium dodecyl sulfate/1-hexanol mixtures", *Langmuir* **7**, pp. 1316–1318 (1991).

[5] R. Daniel and J. Berg, "A simplified method for predicting the dynamic surface tension of concentrated surfactant solutions", *Journal of Colloid and Interface Science* **260**, pp. 244–249 (2003).

[6] M. Rosen and Z. Zhu, "Enhancement of wetting properties of water insoluble surfactants via solubilization", *Journal of the American Oil Chemists Society* **70**, pp. 65–68 (1993).

[7] S. Oh and D. Shah, "Effect of micellar lifetime on the wetting time of cotton in sodium dodecyl sulfate solutions", *Langmuir* **8**, pp. 1232–1233 (1992).

[8] Milton Rosen, *Surfactants and Interfacial Phenomena*, Wiley New York third edition (2004).

[9] D. Evans and H. Wennerstrom, *The Colloid Domain Where Physics, Chemistry, Biology and Technology Meet*, Wiley-VCH New York 2nd edition (1999).

[10] C. Tanford, *The Hydrophobic Effect*, Wiley New York 2nd edition (1980).

[11] A. Prosser and E. Franses, "Adsorption and surface tension of ionic surfactants at the air-water interface: review and evaluation of equilibrium models", *Colloids and Surfaces A: Physicochemical and Engineering Aspects* **178**, pp. 1–40 (2001).

[12] Chien-Hsiang Chang and E Franses, "Adsorption dynamics of surfactants at the air/water interface: A critical review of mathematical models, data and mechanisms", *Colloids and Surfaces A: Physicochemical and Engineering Aspects* **100**, pp. 1–45 (1995).

[13] G. Kretzschmar and R. Miller, "Dynamic properties of adsorption layers of amphiphilic substances at fluid interfaces", *Advances in Colloid and Interface Science* **36**, pp. 65–124 (1991).

[14] R. Miller, P. Joos, and V. Fainerman, "Dynamic surface and interfacial tensions of surfactant and polymer solutions", *Advances in Colloid and Interface Science* **49**, pp. 249–302 (1994).

[15] R. Miller and G. Kretzschmar, "Adsorption kinetics of surfactants at fluid interfaces", *Advances in Colloid and Interface Science* **37**, pp. 97–121 (1991).

[16] R. Miller, G. Loglio, U. Tesei, and K. Schano, "Surface relaxations as a tool for studying dynamic interfacial behavior", *Advances in Colloid and Interface Science* **37**, pp. 73–96 (1991).

[17] Paul Joos and Valentin B. Fainerman, *Dynamic Surface Phenomena*, VSP Amsterdam (1999).

[18] S. S. Dukhin, G. Kretzschmar, and R. Miller, *Dynamics of Adsorption at Liquid Interfaces: Theory, Experiment and Applications* volume 1, Elsevier Amsterdam (1995).

[19] T.H. Illiev and C.D. Dushkin, "Dynamic surface tension of micellar solutions studied by the maximum bubble pressure method 1. experiment", *Colloid and Polymer Science* **272**, pp. 1157–1165 (1994).

[20] J.V. Mitrancheva, C.D. Dushkin, and P. Joos, "Kinetics of the surface tension of micellar solutions: Comparison of different experimental techniques", *Journal of Colloid and Polymer Science* **274**, pp. 12–35 (1996).

[21] B.A. Noskov, "Kinetics of adsorption from micellar solutions", *Advances in Colloid and Interface Science* **95**, pp. 237–293 (2002).

[22] S. N. Wall and E.A.G. Aniansson, "Numerical calculations on the kineticsof stepwise micelle association", *J. Phys. Chem.* **84**, pp. 727–736 (1980).

[23] E.A.G. Aniansson and S.N. Wall, "On the kinetics of step-wise micelle association", *The Journal of Physical Chemistry* **78**(10), pp. 1024–1030 (1974).

[24] E.A.G. Aniansson and S.N. Wall, "A correction and improvement of " on the kinetics of step-wise micelle association" by e.a.g. aniansson and s.n. wall", *The Journal of Physical Chemistry* **79**(8), pp. 857 (1975).

[25] E.A.G. Aniansson, S.N.Wall Wall, M. Almgren, H. Hoffmann, I. Kielmann, W. Ulbricht, R. Zana, J. Lang, and C. Tondre, "Theory of the kinetics of micellar equilibrium and quantitative interpretation of chemical relaxation studies of micellar solutions of ionic surfactants", *The Journal of Physical Chemistry* **80**(9), pp. 905–922 (1976).

[26] K. B. Danov, P.M. Vlahosvska, T. Horozov, C.D. Dushkin, P.A. Kralchevsky, A. Mehreteab, and G. Broze, "Adsorption from micellar solutions: Nonlinear theory and experiment", *Journal of Colloid and Interface Science* **183**, pp. 223–235 (1996).

[27] R. Miller, "Adsorption kinetics of surfactants from micellar solutions", *Colloid and Polymer Science* **259**, pp. 1124–1128 (1981).

[28] J. Lucassen, "Adsorption kinetics in micellar systems", *Faraday Discussions of the Chemical Society* **59**, pp. 76–86 (1976).

[29] Gordon C. Kresheck, Eugene. Hamori, Gary Davenport, and Harold A. Scherage, "Determination of the dissociation rate of dodecylpyridinium iodide micelles by a temperature-jump technique", *The Journal of American Chemical Society* **88**, pp. 246–253 (1966).

[30] A. Johner and J. Joanny, "Block copolymer adsorption in a selective solvent: A kinetic study", *Macromolecules* **23**, pp. 5299–5311 (1990).

[31] B. Noskov, "Diffusion of micellar surfactants in liquids. the influence of the fast relaxation process.", *Kolloidn. Zh.* **52**, pp. 509–514 (1990).

[32] C.D. Dushkin and I.B. Ivanov, "Effect of the polydispersity of diffusing micelles on the surface elasticity", *Colloids and Surfaces* **60**, pp. 213–233 (1991).

[33] C.D. Dushkin, I.B. Ivanov, and P.A. Kralchevsky, "The kinetics of the surface tension of micellar surfactant solutions", *Colloids and Surfaces* **60**, pp. 235–261 (1991).

[34] C.D. Dushkin and I.B. Ivanov, "An asymptotic approach to the kinetics of formation of polydisperse micelles", *Journal of Surface Science and Technology* **6**, pp. 269 (1990).

[35] V.B. Fainerman, "Kinetics of surfactant adsorption from micellar solutions (theory)", *Kolloidn. Zh.* **43**, pp. 94–100 (1981).

[36] Q. Song, A. Couzis, P. Somasundaran, and C. Maldarelli, "A transport model for the adsorption of surfactant from micelle solutions onto a clean air/water interface in the limit of rapid aggregate disassembly relative to diffusion and supporting dynamic tension experiments", *Colloids and Surfaces A: Physicochemical Engineering Aspects* **282-283**, pp. 162–182 (2006).

[37] Julian Eastoe, James Dalton, Philippe Rogueda, Esther Crooks, Alan Pitt, and Elizabeth Simister, "Dynamic surface tensions of nonionic surfactant solutions", *Journal of Colloid and INterface Science* **188**, pp. 423–430 (1997).

[38] Guillermo R. Castro, Bridget K. Larson, Bruce Panilaitis, and David L. Kaplan, "Emulsan quantitation by nile red quenching fluorescence assay", *Applied Microbiology and Biotechnology* **67**(6), pp. 767–770 (2005).

[39] M.C.A. Stuart, J.C. van de pas, and J.B.F.N. Engberts, "The use of nile red to monitor the aggregation behavior the use of nile red to monitor the aggregation behavior in ternary surfactant–water–organic solvent systems", *Journal of Physical Organic Chemistry* **18**, pp. 929–934 (2005).

[40] P Greenspan and SD Fowler, "Spectrofluorometric studies of the lipid probe, nile red", *J. Lipid Res.* **26**(7), pp. 781–789 (1985).

[41] J. Brinck, B. Jonsson, and F. Tiberg, "Kinetics of nonionic surfactant adsorption and desorption at the silica-water interface: One component", *Langmuir* **14**, pp. 1058–1071 (1998).

[42] F. Tiberg, B. Jonsson, and B. Lindman, "Ellipsometry studies of the self assembly of nonionic surfactants at the silica-water interface: Kinetic aspects", *Langmuir* **10**(3714-3722) (1994).

Structure and Transport in Monolayers of Charged Colloid Particles at Oil-Water Interfaces: The Role of Oil Soluble Surfactant

Virendra Singh, School of Chemical & Biomolecular Engineering, Georgia Institute of Technology, Atlanta, GA

Qiong Guo, School of Chemical & Biomolecular Engineering, Georgia Institute of Technology, Atlanta, GA

Sven H. Behrens, School of Chemical & Biomolecular Engineering, Georgia Institute of Technology, Atlanta, GA

The behavior of colloidal particles in liquid-liquid interfaces is relevant to numerous commercial products and industrial processes, such as food processing, pharmaceutical formulation, flotation, oil extraction and materials fabrication. Over the past century, colloid and surface forces have been well understood in bulk phase, but the presence of an "oil-water interface" introduces significant complexity to the colloidal interactions. We have studied the absorption, structure and interaction of charged polystyrene particles at the oil (decane) - water interface by optical microscopy. The electrostatic interaction between charged particles at a water-low dielectric medium interface leads to the formation of a typical crystalline 2-D monolayer at the interface. Such particle monolayers are remarkably insensitive to the addition of salt in the aqueous phase, but small amounts of surfactant in either liquid phase disturb the crystalline order and thus promote liquid-like structures. Two-dimensional aggregation of particles was observed when increasing the surfactant concentration in the oil phase. We discuss the role of interfacial and electrostatic energy for the observed behavior and propose a multi-faceted effect of the added surfactant.

See more of this Session: Fundamentals of Interfacial Phenomena II - Surfactants and Liquid/Solid Interfaces
See more of this Group/Topical: Engineering Sciences and Fundamentals

Characterization of the Effect of Produced Water Composition On Residence Time of Oil-Water Dispersions in Batch Separation Cell*

N. Urdaneta, The University of Tulsa, Tulsa, OK

L. Gomez, The University of Tulsa, Tulsa, OK

R. Mohan, The University of Tulsa, Tulsa, OK

O. Shoham, The University of Tulsa, Tulsa, OK

S. Simpson, Chevron Energy Technology Company, Houston, TX

D. Gallup, Chevron Energy Technology Company, Houston, TX

C. Avila, Chevron Energy Technology Company, Houston, TX

G. Kouba, Chevron Energy Technology Company, Houston, TX

During the oil production process, the oil/water mixture passes through restrictions, expansions and shearing devices that affect the flow characteristics of dispersions. The oil industry is looking at different ways to characterize oil-water dispersions in order to properly select, size, and predict the performance of downstream equipment such as separators. One of the parameters that surface facilities engineers need for designing separators is the residence time necessary to completely separate oil/water dispersions. In the literature, there is little work that show and explain the effect of produced water salinity on the oil-water dispersions separation profiles (residence time). This study attempts to determine how the residence time changes as function of the water composition, utilizing the Dispersion Characterization Rig (DCR) at the University of Tulsa. The DCR is a state-of-the-art laboratory equipment that can be used to characterize the flow and separation behavior (residence times) of complex oil-water dispersions under controlled conditions. The facility operates under the same principle as the batch separator.

The objectives of this study are twofold: (1) Evaluation of model fluids for produced water, which were prepared using tap water and Sodium Chloride (NaCl), and (2) Evaluation of produced water samples from three different fields. The experimental matrix includes the evaluation of 4 model fluids with concentrations from 10000 to 150000 ppm NaCl in tap water and 3 produced waters (A, B and C) with different chemical composition. These waters were evaluated for 3 water cuts (25%, 50% and 75%) and 3 pressures at the choke valve (0, 50, 100 psi), using constant values of temperature, cell volume and mixture velocity. Preliminary results show that the concentration of salt in the water affects the residence time to completely separate oil/water dispersions. These experiments indicate that the separation behavior taking place is the result of the physicochemical properties of the oil-water interface. The results can be used to validate model predictions of separation of oil/water dispersions.

* These activities are supported by Chevron through the TU-CoRE project

See more of this Session: Fundamentals of Interfacial Phenomena II - Surfactants and Liquid/Solid Interfaces
See more of this Group/Topical: Engineering Sciences and Fundamentals

Adsorption Kinetics of Alkyl Polyethylene Oxide Surfactants at the Water-Oil Interface

Wingki Lee, Carnegie Mellon University, Pittsburgh, PA

Lynn M. Walker, Chemical Engineering, Carnegie Mellon University, Pittsburgh, PA

Shelley L. Anna, Chemical Engineering and Mechanical Engineering, Carnegie Mellon University, Pittsburgh, PA

The adsorption kinetics of alkyl polyethylene oxide (CiEj) nonionic surfactants onto an initially clean air-water interface has been well studied over the past two decades. In principle, the adsorption of surfactant at an air-liquid interface should be very similar to that at a liquid-liquid interface. However, we have found that the diffusion, adsorption, and desorption coefficients for the air/water/surfactant system do not provide the correct scaling to describe a surfactant-mediated tipstreaming phenomenon for producing micron-scale droplets in a microfluidic device, even though the phenomenon is believed to be controlled in particular by the desorption rate of surfactant. As a result, we have undertaken a systematic study to explore the differences between the surfactant dynamics at air-water and oil-water interfaces. In this work, we use the pendant drop method to obtain the dynamic surface tension for a homologous series of three nonionic C_iE_8 (i = 10, 11, and 12) surfactants adsorbed onto an initially clean water-silicone oil interface. Since density differences are small in oil-water systems, the experiments are performed at low Bond number, and we use a recently published fitting routine to extract the instantaneous interfacial tension. We compare our experimental results with the published sorption constants for the air/water system obtained using mixed diffusion-kinetic models, and find that the two systems have different rate constants. The impact of these results on complex interfacial flows will be discussed.

See more of this Session: Fundamentals of Interfacial Phenomena II - Surfactants and Liquid/Solid Interfaces
See more of this Group/Topical: Engineering Sciences and Fundamentals

Effects of Added Surfactant On Particle Stabilized Emulsion Droplets

Sven H. Behrens, School of Chemical & Biomolecular Engineering, Georgia Institute of Technology, Atlanta, GA

Qiong Quo, Chemical and Biomolecular Engineering, Georgia Institute of Technology, Atlanta, GA

Virendra Singh, Chemical and Biomolecular Engineering, Georgia Institute of Technology, Atlanta, GA

It is well known that colloidal particles can strongly adsorb to an oil-water interface and stabilize emulsion droplets. The properties of any particle stabilized liquid-liquid interface can be further tuned with surfactant additives, and the effects of added surfactant can be surprisingly varied and complex. Surfactant molecules can adsorb to the liquid interface, thereby lowering the interfacial tension and the adsorption strength of the particles. Alternatively, the surfactant can adsorb to the particle surface and enhance or lower the particles' stabilizing effect by altering their wetting behavior. Depending on its composition and architecture, the surfactant can also induce and electric surface charge at the particle-oil interface; and reversed surfactant micelles in the bulk oil phase can provide electrostatic screening of such particle surface charge. We report experimental evidence for each of these scenarios, and discuss the delicate interplay of competing effects when surfactants meet a particle-stabilized emulsion droplet.

See more of this Session: Fundamentals of Interfacial Phenomena II - Surfactants and Liquid/Solid Interfaces
See more of this Group/Topical: Engineering Sciences and Fundamentals

Implantation of Ketoprofen Into a Biodegradable Suture Using Liquid and Supercritical Carbon Dioxide

Randy D. Weinstein, Department of Chemical Engineering, Villanova University, Villanova, PA

Kenneth Muske, Chemical Enigneering, Villanova University, Villanova, PA

Sherrie-Ann Martin, Chemical Engineering, Villanova University, Villanova, PA

Liquid and supercritical carbon dioxide have the ability to swell and plasticize many polymers. Furthermore, the properties of carbon dioxide such as its high diffusivity, low viscosity, environmental friendliness, inertness, and ease of recycling make it an ideal solvent for food and pharmaceutical processing. In this study we explore the use of liquid and supercritical carbon dioxide for the creation of a sustained release drug delivery device. An anti-inflammatory (ketoprofen) is dissolved into carbon dioxide at various temperatures (0 - 55 oC) and pressures (65 - 300 bar) and then exposed to biodegradable suture made of poly(lactide-co-glycolide) copolymers. The effect of temperature, pressure, exposure time and processing conditions are explored on the ability of the substrates to absorb ketoprofen. Uniformity of the absorption is examined by dissolution studies whereby the substrates are exposed to various acid and basic conditions to either slowly or quickly dissolve. The mass transfer of the pharmaceutical out of the substrate is then measured and modeled. Additionally modeling is made of the anti-inflammatory solubility in carbon dioxide as well as the diffusion and mass transfer of the drug into the suture.

See more of this Session: Materials Synthesis and Processing with Compressed or Supercritical Fluids I: General
See more of this Group/Topical: Engineering Sciences and Fundamentals

Measurement of Relative Recalcitrance of Biomass Substrates Using Hot Water/ CO2 Reactive Pretreatment

Dongfang Zhang, Department of Chemical Engineering, University of Arkansas, Fayettville, AR

Jerry W. King, Chemical Engineering, University of Arkansas, Fayettville, AR

Ya-Jane Wang, Department of Food Science, University of Arkansas, Fayetteville, AR

Biomass extracts, particularly sugars from biomass hydrolysis are important products for further conversion in the production of bioethanol. This research is concerned with investigating a more environmentally-benign method for conducting biomass hydrolysis with the aid of subcritical water and supercritical carbon dioxide (SC-CO2). Current methods of degrading biomass utilize acidic or basic medium which needs to be neutralized after the conclusion of the reactive pretreatment step. Carbon dioxide at elevated pressures can acidify the aqueous medium allowing for the depolymerization of the constituent carbohydrate polymers (cellulose and hemicellulose) to monomeric sugars, glucose and xylose from the biomass substrate.

An orthogonal experimental design L9(34) was used to determine the most effective reaction conditions for select biomass substrates, such as corn cob, corn stover and switch grass. The typical range of parameters investigated consisted of temperatures 160-180C, CO2 pressures from 15-45 MPa, 30-90 minutes reaction time, and various sieved particles of each biomass substrate. Small samples of each target biomass were placed in a tubular reaction vessel filled with neat water and rapidly brought to one of the above temperatures, to serve as a baseline for assessing the effect of CO2 on the depolymerization process. Likewise, similar experiments were performed with the aid of an Applied Separations Spe-ed unit to pump SC-CO2 into the tubular reaction vessel contained the biomass substrate and water. Using identical conditions as those employed for depolymerization in neat water, carbonated hot water hydrolysis was performed on each biomass substrate in triplicate.

The sugar yields were calculated based on the size exclusion chromatographic (SEC) determination of the monomeric sugar content of the resultant hydrolyzate. SEC was also used to access the relative recalcitrance of the above biomass substrates to depolymerization under hot carbonated water reactive pretreatment. Calibration of the biomass depolymerization profiles obtained during SEC was done relative to pullulan and sugar standards. Substantial shifts were observed for each type of biomass, including pine tree residues, various rice substrates, and assorted corn cob products in addition to the previously mentioned substrates. Some of the above biomass substrates showed a greater resistance to depolymerization under identical conditions relative to the other substrates which were highly converted into lower oligomeric and monomeric sugar -containing hydrolyzates. Results compared with traditional dilute sulfuric acid - based hydrolysis will be reported as well as in terms of the recalcitrance of the targeted biomass to hydrolysis using subcritical water - SC-CO2 mixtures.

See more of this Session: Materials Synthesis and Processing with Compressed or Supercritical Fluids I: General
See more of this Group/Topical: Engineering Sciences and Fundamentals

AUTHOR INDEX

AUTHOR INDEX

AUTHOR INDEX

AUTHOR INDEX

AUTHOR INDEX

AUTHOR INDEX

AUTHOR INDEX

AUTHOR INDEX

AUTHOR INDEX

AUTHOR INDEX

AUTHOR INDEX

AUTHOR INDEX

AUTHOR INDEX

AUTHOR INDEX

Engineering Sciences and Fundamentals

Presentations at the 2009 AIChE Annual Meeting

Nashville, Tennessee, USA
8-13 November 2009

Volume 2 of 2

ISBN: 978-1-61567-925-6

Printed from e-media with permission by:

Curran Associates, Inc.
57 Morehouse Lane
Red Hook, NY 12571

Printed by Curran Associates, Inc. (2010)

For permission requests, please contact AIChE
at the address below.

AIChE
3 Park Avenue
New York, NY 10016-5991

Phone: (203) 702-7660
Fax: (203) 775-5177

www.aiche.org

Additional copies of this publication are available from:

Curran Associates, Inc.
57 Morehouse Lane
Red Hook, NY 12571 USA
Phone: 845-758-0400
Fax: 845-758-2634
Email: curran@proceedings.com
Web: www.proceedings.com

TABLE OF CONTENTS

VOLUME 1

VOLUME 2

Quantitative Analysis and Structural Characterization of Carbonaceous Pitches

Sourabh U. Kulkarni, Department of Chemical & Biomolecular Engineering, Clemson University, Clemson, SC

Mark C. Thies, Department of Chemical and Biomolecular Engineering, Clemson University, Clemson, SC

Petroleum pitches can serve as precursors for wide range of carbon products, such as high thermal conductivity carbon fibers, carbon electrodes, carbon fibers for fuel cell substrates, and carbon/carbon composites. Research increasingly suggests that the molecular composition of the pitch and the structures present in the pitch have a significant impact on the properties of the final carbon products, but to date this understanding is primarily qualitative because of an inability to isolate the pitches into fractions with narrow mol wt distribution that can be characterized easily. We have recently developed a technique called Dense Gas Extraction (DGE) for the fractionation of petroleum pitches, with the goal of obtaining fractions with relatively narrow mol wt distribution. Such fractions can serve as standards for structural characterization and quantitative analysis.

In this work, we report separation of petroleum pitches into fractions with narrow mol wt distribution using both continuous and semi-batch DGE processes and application of these fractions as standards for structural characterization and quantitative analysis. Matrix-Assisted, Laser Desorption/Ionization time-of-flight mass spectrometry (MALDI) was used for the quantitative analysis of the petroleum pitches. Pitch fractions containing 98% monomer and 97% dimer were isolated by DGE, and were used to study the effect of oligomeric composition on MALDI response. Also, both new and conventional techniques were applied to identify the specific structures that exist in petroleum pitches. Pitch fractions containing 98% monomer and 97% dimer were characterized using MALDI, MALDI-Post Source Decay, and Field Desorption mass spectrometry; and 1H-NMR, UV-Vis, and FT-IR spectroscopy and specific molecular structures that are significant constituents of petroleum pitches were proposed.

See more of this Session: Materials Synthesis and Processing with Compressed or Supercritical Fluids I: General
See more of this Group/Topical: Engineering Sciences and Fundamentals

Measurement of Isobaric Vapor-Liquid Equilibria Necessary for Distillation Process Design Using Automatic VLE Apparatus

Katsumi Tochigi, Department of Materials and Applied Chemistry, Nihon University, Tokyo, Japan

Naoki Kamihama, Department of Materials and Applied Chemistry, Nihon University, Tokyo, Japan

Hiroyuki Matsuda, Department of Materials and Applied Chemistry, Nihon University, Tokyo, Japan

Kiyofumi Kurihara, Materials and Applied Chemistry, Nihon University, Tokyo, Japan

Katsumi Yokoyama, Package Development, Omega Simulation Co. Ltd., Tokyo, Japan

Vapor-liquid equilibrium is a key thermophysical property necessary for distillation process design. Tetrahydrofuran (THF) is produced by the dehydrocyclization of 1,4-butanediol. The key step in purification is breaking of the THF + water azeotrope. The n-paraffines like n-pentane, n-hexane and n-heptane are considered as the azeotropic solvents. The vapor-liquid equilibria for the systems containing THF are important for discussing the azetropic distillation process.

This paper deals with the measurement of bubble points for four binary systems, THF + water, + n-pentane, + n-hexane, + n-heptane at reduced pressure using automatic VLE apparatus. It is confirmed that THF + water and THF + n-hexane systems are azeotropic mixtures, though THF + n-pentane and THF + n-heptane are non-azeotropic mixtures. The NRTL parameters have been then determined using experimental bubble points. The NRTL parameters will be useful for discussing distillation process.

See more of this Session: Thermodynamic Properties and Phase Behavior V
See more of this Group/Topical: Engineering Sciences and Fundamentals

Amorphous Polystyrene-Block-Polybutadiene and Crystallizable Polystyrene-Block-(hydrogenated polybutadiene) Solutions in Compressible near Critical Propane and Propylene - Hydrogenation Effects

Winoto Winoto, Chemical & Petroleum Engineering, University of Wyoming, Laramie, WY

Maciej Radosz, University of Wyoming, Laramie, WY

Kunlun Hong, Center for Nanophase Materials Science, Oak Ridge National Laboratory, Oak Ridge, TN

Jimmy Mays, Center for Nanophase Materials Sciences, Oak Ridge National Laboratory, Oak Ridge, TN

Polystyrene, polybutadiene, hydrogenated polybutadiene, and styrene diblock copolymers of these homopolymers can form homogenous solutions in compressible solvents, such as propane and propylene, which separate into two bulk phases upon reducing pressure. The cloud and micellization pressures for homopolymer and diblock copolymers are generally found to be higher in propane than in propylene, except for hydrogenated polybutadiene and polystyrene-block-(hydrogenated polybutadiene). Hydrogenated polybutadiene homopolymers and copolymers exhibit relatively pressure-independent crystallization and melting observed in both propane and propylene solutions.

See more of this Session: Thermodynamic Properties and Phase Behavior V
See more of this Group/Topical: Engineering Sciences and Fundamentals

BWR EQUATION OF STATE FOR 1-PENTYNE AND VLE CALCULATION CONCERNING DIMETHYL ETHER

Tomoya Tsuji[1], *Tomonori Nakazato*[1], *Takeshi Kondo*[2],
Reppei Uematsu[2], *Toshihiko Hiak*[1]*i*, *Mitsuo Namba*[2]
[1]*Nihon University, Narashino, Chiba, Japan*
[2]*The High Pressure Gas Safety Institute of Japan, Machida, Tokyo, Japan*

Introduction

Dimethyl ether (DME) has been mainly used as a propellant for splay cans. A large scale DME plant, with 80000 ability a year, has been already operated in Japan. So, DME is expected to be an alternative clean energy for liquefied petroleum gas (LPG), because the physical properties, such as molecular weight, vapor pressure, and so on, are similar to those of LPG. In addition, SOx is not produced at the combustion. However, now in Japan, *tert*-Butyl mercaptan (TBM) has been widely used as an odorant for LPG. Then, the sulfur free odorant should be required for DME. A national project has been promoted by Ministry, Economy, Trade and Industry, Japan, and the High Pressure Gas Safety Institute of Japan proposed some alternative odorants [1]. Some alkines are thought to be a new odorant for LPG, but there is no experimental data concerning physical properties. Then, in this study, saturated vapor pressure and P-V-T relationship were measured for 1-pentyne, and bubble point pressure for a binary, dimethyl ether (DME) + 1-pentyne. Using experimental data of vapor pressure and P-V-T relationship, Benedict-Webb-Rubin (BWR) equation of state [2] was successfully developed for 1-pentyne. Finally, vapor-liquid equilibrium (VLE) were evaluated by use of the BWR eq.

Experimental Section

Materials

1-Pentyne was supplied from Sigma-Aldrich Inc. St. Louise, U. S., and the stated purity was no less than 99%. Pentane was a special grade reagent from Wako Pure Chemical Industries, Ltd., Osaka, Japan, and the purity no less than 98.0 %. To remove dissolved gases, these chemicals were used after boiling under the atmospheric pressure. DME was purchased from Mitsubishi Gas Chemical Co. Ltd., Tokyo, and the purity was no less than 99.9 %. DME was used as received without further purifications.

Equipment and Procedure

Three apparatus were employed in this study. **Figures 1 to 3** showed the schematic diagram of experimental apparatus. All apparatus were based on a static method, and basically had the same equipments except for the sample cell, the temperature controller and the pressure sensor. In case of the vapor pressure measurement up to 263.11 K, the apparatus, shown in Fig.1, was employed. The cell was made of Pyrex Glass, and its inner volume was about 37 cm^3. The maximum working pressure was thought to be 7.50 MPa [3]. The cell was set in a low temperature thermostatic water bath TRL-080S, Thomas Kagaku Co. Ltd., Tokyo, filled with ethanol. A CCD camera equipped with the bath to observe the inside of the cell. The temperature and the pressure were measured by a precision thermometer F250 MKII, Automatic Sensor Laboratories, Milton Keynes, U. K., and a highly reliable pressure transducers PHS-2KA, Kyowa Electronic Instrument Co. Ltd., Tokyo,

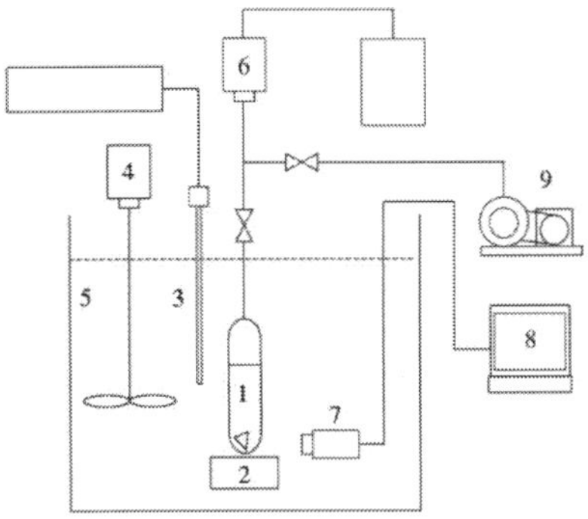

1: View Cell 2: Magnetic Stirrer 3: Thermistor Thermometer 4: Agitator 5: Ethanol Bath 6: Absolute Pressure Sensor 7: CCD Camera 8: Monitor 9: Vacuum Pump

Figure 1 A Static type Apparatus for Saturated Vapor Pressure Measurement

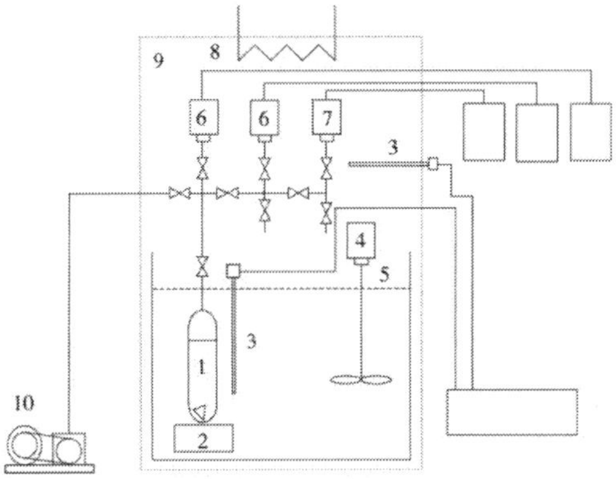

1: View Cell 2: Magnetic Stirrer 3: Thermistor Thermometer 4: Agitator 5: Water-Ethylene glycol Bath 6: Pressure Sensor 7: Absolute Pressure Sensor 8: Heater 9: Air Chamber 10: Vacuum Pump

Figure 2 A Static type Apparatus for Saturated Vapor Pressure and Bubble Point Pressure Measurement

respectively. The resolution of the thermometer was 0.01 K. The capacity of the pressure transducer was 200 kPa, and the resolution 0.01 kPa.

In case of the vapor pressure measurement from 273.15 to 310.15 K and the bubble point pressure at 303.15 K, the apparatus, shown in Fig.2, was employed. The cell and the pressure transducer were the same as that up to 263.11 K. The cell was set in a low temperature thermostatic water bath TRL-101FEZ, Thomas Kagaku Co. Ltd., Tokyo, filled with aqueous ethylene glycol solution. The temperature were measured by a thermistor thermometer D641, Takara Thermistor .Co., Ltd. Yokohama, Japan. The precision of the thermometer was 0.01 K. In the bubble point pressure measurement, the composition was determined by weighing mass of each component in the cell. The mass was measured with a direct reading balance AV 1581, EXACT Co., Kamagaya, Japan, with a maximum capacity of 1 kg, the precision 0.1 mg.

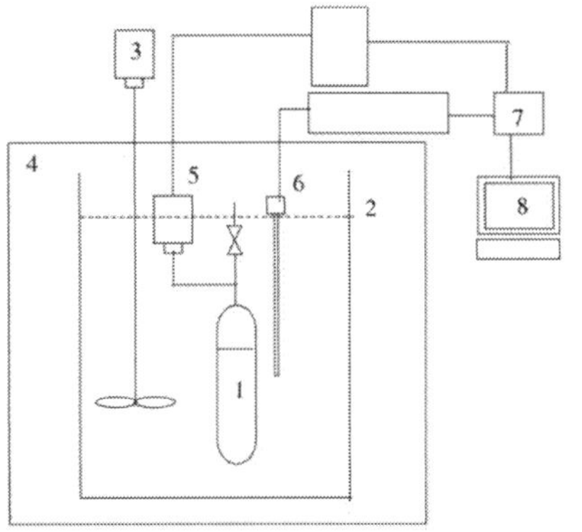

1: High Pressure Cell 2: Oil Bath 3: Agitator 4: Air Chamber 5: Absolute Pressure Sensor 6: Thermistor Thermometer 7: Data Logger 8: Personal Computer

Figure 3 A Static type Apparatus for Saturated Vapor Pressure and P-V-T relationship Measurement

In case of the vapor pressure measurement from 333.67 to 463.04 K and the P-V-T relationship at 485.00 K, the apparatus, shown in Fig.3, was employed. The cell, used for vapor pressure measurement, was a double-ended DOT-compliant sample cylinder 304L-HDF-2-40, Swagelok Co., Solon, U. S., and that for P-V-T relationship was a single-ended miniature sample cylinder

SS-4CS-TW-50, Swagelok Co. These cells were made of stainless steal. The maximum working pressure was up to 12.4 MPa for 304L-HDF-2-40, and 6.89 MPa for SS-4CS-TW-50, and the inner volume about 40 cm^3 and 50cm^3, respectively. Prior to the P-V-T measurements, the precise volume of SS-4CS-TW-50 was calibrated up to 0.001 cm^3 by the experimental data of P-V-T relationship and the standard data for pentane at 485.00 K. In the measurement, the cell was set in a thermostatic oil bath T-305, Thomas Kagaku Co. Ltd., Tokyo. The temperature was measured by a thermistor thermometer D641, from 333.67 to 383.82 K, or a precision thermometer, F250 MKII from 392.84 to 485.00 K. The pressure was measured by a highly reliable pressure transducers PHS-2KA, Kyowa Electronic Instrument Co. Ltd., Tokyo with the maximum capacity of 5 MPa and the resolution 0.1 kPa. In the measurement of P-V-T relationship, the mass of the gas was determined by the direct-reading balance.

.

Results and Discussion

Experimental Data

Figure 4 showed the experimental result of saturated vapor pressure for 1-pentyne. The reliability of the data has been already ensured by measuring for pentane in the same temperature range. As shown in the figure, the saturated vapor pressure of 1-pentyne was similar to that of pentane. The experimental data were correlated with the following Antione equation:

$$\log_{10} P^{\circ}[kPa] = A - \frac{B}{T[K] - C} \tag{1}$$

where three constants, *A*, *B*, *C*, were optimized with the experimental data. The calculation results were shown in Fig.4. Though the upper limit of temperature, that is the critical point, were not known, Antoine eq. showed a good reproducibility.

Figure 5 showed the experimental result of P-V-T relationship of 1-pentyne at 485.00 K. The reliability was also confirmed by measuring P-V-T relation of pentane at the same temperature. In this

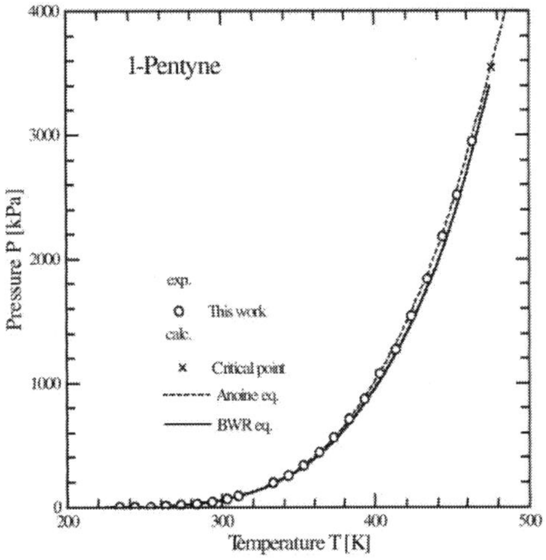

Figure 4 Saturated Vapor Pressure for 1-pentyne

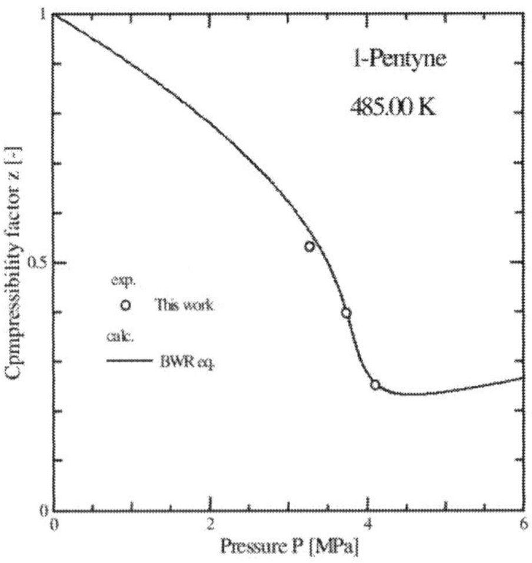

Figure 5 P-V-T Relationship of 1-pentyne (485.00 K)

study, just three data can be obtained. All data was obtained within one day after setting in the thermostat bath. When two days had passed after setting, the pressure seemed to be decreased gradually. Then, the phenomena were due to the polymerization reaction of 1-pentyne [4]. Considering the pressure dependence of the compressibility factor, the experimental temperature, 485.00 K, was thought to be slightly higher than the critical temperature.

Figure 6 showed the experimental result of the bubble point pressure for DME + 1-pentyne at 303.15 K. The reliability of the bubble point pressure was also checked for DME + pentane at the same temperature. As shown in the figure, the bubble point pressure was almost followed with Raoult's law.

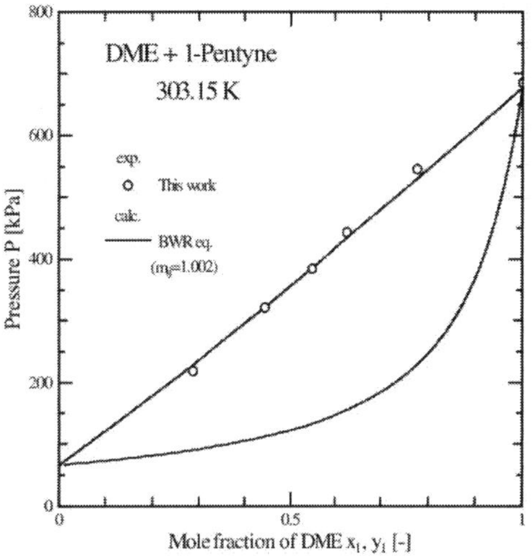

Figure 6 VLE for DME + 1-Pentyne at 303.15 K.

BWR eq. for 1-pentyne

BWR equation of state is virial expansion type and it was widely used in the evaluation of physical properties for light hydrocarbons. BWR eq. is given by:

$$P = RT\rho + (B_0RT - A_0 - \frac{C_0}{T^2})\rho^2 + (bRT - a)\rho^3 + a\alpha\rho^6 + \frac{c\rho^3}{T^2}(1+\gamma\rho^2)\exp(-\gamma\rho^2) \quad (2)$$

where the eight constants, A_0, B_0, C_0, a, b, c, α, γ are available the literature. For example, the constants for propane, butane and pentane, were reported by Cooper and Goldfrank [5]. Recently, those for DME was also reported by Tsuji et al. [6]. However, those for 1-pentne have not been determined. Then, the constants were optimized by applying corresponding state theory developed in our previous study [6]. The procedure was described below:

Applying Joffe's method [7], BWR eq. can be converted to the following non-dimensional form:

$$P_r = \frac{T_r}{v_r} + (B_{0r}T_r - A_{0r} - \frac{C_{0r}}{T_r^2})\frac{1}{v_r^2} + (b_rT_r - a_r)\frac{1}{v_r^3} + \frac{a_r\alpha_r}{v_r^6} + \frac{c_r}{T_r^2 v_r^3}(1+\frac{\gamma_r}{v_r^2})\exp(-\frac{\gamma_r}{v_r^2}) \quad (3)$$

where P_r, T_r and v_r are reduced properties using the critical temperature and pressure.

$$P_r = \frac{P}{P_c} \quad (4) \qquad T_r = \frac{T}{T_c} \quad (5) \qquad v_r = \frac{P_c}{\rho RT_c} \quad (6)$$

Compared with eq. (2), the non-dimensional constants in eq. (3) are given by:

$$A_{0r} = A_0\frac{P_c}{R^2T_c^2} \quad (7) \qquad B_{0r} = B_0\frac{P_c}{T_c} \quad (8) \qquad C_{0r} = C_0\frac{P_c}{R^2T_c^4} \quad (9)$$

$$a_r = a\frac{P_c^2}{R^3T_c^3} \quad (10) \qquad b_r = b\frac{P_c^2}{R^2T_c^2} \quad (11) \qquad c_r = c\frac{P_c^2}{R^3T_c^5} \quad (12)$$

$$\alpha_r = \alpha\frac{P_c^3}{R^3T_c^3} \quad (13) \qquad \gamma_r = \gamma\frac{P_c^3}{R^2T_c^2} \quad (14)$$

If the corresponding state theory is perfect, the same value can be obtained for non-dimensional constants for any substance. However, the values were slightly different each other. So, a linear relation, similar to Lee-Kesler eq., was assumed for non-dimensional constants:

$$Q_r = \frac{Q_{r,R1} - Q_{r,R2}}{\omega_{R1} - \omega_{R2}}(\omega - \omega_{R2}) + Q_{r,R2} \quad (Q_r : A_{0r}, B_{0r}, C_{0r}, a_r, b_r, c_r, \alpha_r, \gamma_r) \tag{15}$$

where ω is acentric factor given by the following definition:

$$\omega = -\log_{10} \left. \frac{P^s}{P_c} \right|_{T_r=0.7} - 1.000 \tag{16}$$

ω_{R1} and ω_{R2} is that for reference substance. In this study, butane and pentane were employed as the references, However, the critical temperature and pressure were not available for 1-pentyne. Then, we assumed that the critical temperature and pressure for 1-pentyne should be located at the extrapolation of the saturated vapor pressure curve. In the optimization, at first, the critical temperature was assumed. After evaluating the critical pressure by eq. (1), the acentric factor was obtained by (16). Using the values of T_c, P_c, ω, the non-dimension constants, Q_r, were determined by eq. (15). Applying eqs.(4)-(14), the P-V-T relationship was calculated. These procedures were repeated until the best reproducibility will be observed in the P-V-T relationship for 1-pentye. In this study, the acentric factor should be ω=0.267. The value was slightly lager than that of pentane, ω_{R2}=0.251 [8]. The determined critical point was shown in Fig. 4, and the calculation results for the saturated vapor pressure and the P-V-T relationship was shown in Figs. 4 and 5. BWR eq. showed a good reproducibility up to the vicinity of the critical point.

For calculation for VLE, the following mixing rule, Stotler-Benedict eq. [9], was adopted.

$$A_0 = \sum_i \sum_j x_i x_j m_{ij} (A_{0i} A_{0j})^{1/2} \tag{17}$$

$$B_0 = \sum_i \sum_j x_i x_j \frac{B_{0i} + B_{0j}}{2} \tag{18}$$

$$C_0 = \sum_i \sum_j x_i x_j (C_{0i} C_{0j})^{1/2} \tag{19}$$

$$a = \left(\sum_i x_i a_i^{1/3} \right)^3 \tag{20}$$

$$b = \left(\sum_i x_i b_i^{1/3} \right)^3 \tag{21}$$

$$c = \left(\sum_i x_i c_i^{1/3} \right)^3 \tag{22}$$

$$\alpha = \left(\sum_i x_i \alpha_i^{1/3} \right)^3 \tag{23}$$

$$\gamma = \left(\sum_i x_i \gamma_i^{1/3} \right)^3 \tag{24}$$

where m_{ij} is a binary interaction parameter. In this study, the parameters were optimized by the experimental data of bubble point pressure for DME + 1-pentyne. In the calculation, optimized value of the binary interaction parameter, m_{ij}=1.002, was obtained. The calculation result was shown in Fig. 6. As shown in the figure, the BWR eq. should be predicted not only for the saturated vapor pressure and the P-V-T relationship but for VLE.

Conclusion

Using reliable experimental data of saturated vapor pressure and P-V-T relationship, BWR constants were successfully determined. The BWR eq. showed a good reproducibility not only for saturated vapor pressure and P-V-T relationship but for VLE for DME + 1-pentyne. Considering the calculation results for the binary, the equation will be useful at the evaluation of other thermophysical properties, for example, drawing Mollier chart.

References

1. http://www.khk.or.jp/english/activities.html.
2. M. Benedict, G. B. Webb, L. C. Rubin (1940), J. Chem. Phys., 8, pp.334.
3. T. Tsuji and T. Hiaki (2004), Research Report in Industrial Technology, Nihon University A, 37, pp. 27.
4. M. Tamura, private communication.
5. H. W. Cooper and J. C. Goldfrank (1967), Hydrocarbon Processing, 46, pp.141.
6. T. Tsuji, T. Hiaki, M. Namba, AIChE 2006 Annual Meeting, San Francisco, CA.
7. J. Joffe (1949), Chem. Eng. Prog., 45, 160.
8. R. C. Reid, J. M. Prausnitz, B. E. Poling (1986), in "The Properties of Gases & Liquids 4th Ed.", McGraw-Hill.
9. H. H. Stotler and M. Benedict (1953), Chem. Eng. Progr. Symp. Ser. No. 6, 49, pp. 25.

A Note On Stability and Uniqueness of Solutions to Multi Component Flash Under Equilibrium

Mohit Aggarwal, Chemical Engineering, Carnegie Mellon University, Pittsburgh, PA

B. Erik Ydstie, Chemical Engineering, Carnegie Mellon University, Pittsburgh, PA

Abstract

The stability and uniqueness of a multi-component flash without reactions has been studied by many authors but not yet proved for a general system [6, 4, 1, 7]. In this work, we present a proof of the global stability and uniqueness of the solution to a multi-component flash system with two phases and arbitrary number of independent reactions. The only assumption made is that the system is under thermodynamic equilibrium with respect to vapor-liquid equilibria and chemical reactions. Our method is based on the concept of convergent systems [5] and Lyapunov's second theorem on stability. The basic argument is that if all the solutions of a system "forget" their initial conditions and converge to each other then the system converges to a unique solution only. We use the second law of thermodynamics to define a Lyapunov function W and prove that the variation or distance between two nearby trajectories of the process goes to zero with time. W is closely related to the Availability function which, as defined by Keenan [3], is the maximum amount of work done by a system in equilibrating against another system. W represents the maximum work that can be extracted if two states of the process converge to each other. It can be interpreted as a metric in the thermodynamic space. W is equivalent to the second order variation of entropy. In this regard it is similar to the Lyapunov function used by Glansdorff and Prigogine [2]. They analyzed local stability around an equilibrium point while we are able to prove global stability and uniqueness. Furthermore, our analysis addresses the semi-definite nature of the Lyapunov function due to the degeneracy in the entropy surface. This allows us to study stability with respect to convective fluxes of systems with multiple phases. We use the entropy balance to derive the balance for the Lyapunov function and show that its rate of change with respect to time is negative and less than or equal to a factor of its own value. This implies exponential convergence and hence proves Lyapunov stability. The Lyapunov function is only positive semi-definite in the extensive variable space therefore we need to apply inventory control for the convergence of extensive variables. Using Gibbs phase rule, it is shown that for the system to converge to a steady state we need to control as many independent extensive variables as there are phases present. Finally we present some 1 numerical experiment results to support the arguments in the paper.

The aim of this paper is to introduce a novel method to analyze stability of chemical processes. The method is not limited to the multi-component flash and can be applied to a more general class of processes like heterogeneous systems with multiple phases. It can be easily used to analyze the stability of process networks like a distillation column which is of much practical significance. We propose the extension to kinetic systems operating away from chemical equilibrium.

References

[1] M. F. Doherty and J. D. Perkins. On the dynamics of distillation processes--iv uniqueness and stability of the steady state in homogeneous continuous distillations. Chemical Engineering Science, 37(3):381 -- 392, 1982.

[2] P. Glansdorff and I. Prigogine. Thermodynamic Theory of Structure, Stability and Fluctuations. Wiley-Interscience, New York, NY, 1971. [3] J. H. Keenan. Thermodynamics. John Wiley & Sons, Inc., 1941.

[4] A. Lucia. Uniqueness of solutions to single-stage isobaric flash processes involving homogeneous mixtures. AIChE Journal, 32(11):1761 -- 1770, 1986.

[5] A. Pavlov, A. Pogromsky, N. van de Wouw, and H. Nijmeijer. Convergent dynamics, atribute to boris pavlovich demidovich. Systems & Controls Letters, 52:257--261, 2004.

[6] H. H. A. Rosenbrock. Theorem of dynamic conversion for distillation. Trans. Inst. Chem. Eng., 38:279--287, 1960.

[7] P. Rouchon and Y. Creff. Geometry of the flash dynamics. Chemical Engineering Science, 48(18):3141 --3147, 1993. 2

See more of this Session: Thermodynamic Properties and Phase Behavior V
See more of this Group/Topical: Engineering Sciences and Fundamentals

Evaluating Rosenfeld's Conjecture On the Relation Between Entropy and Diffusivity

J. Richard Elliott, Chemical and Biomolecular Engineering, University of Akron, Akron, OH

Rosenfeld has suggested a relationship between the self-diffusivity and the entropy departure function

$\Delta s = (S - Sid)/R$.[1] The suggested correlation is:

$\rho^{1/3} D / T^{1/2} \sim B\exp(A^* \Delta s)$

This relationship is largely empirical, but it has been investigated by Mittal et al.[2] They found that the correlation was quite accurate for hard spheres and Lennard-Jones spheres, but less so for square well spheres. If successful, such a correlation would provide a valuable link between thermodynamic properties and transport properties. Whereas perturbation theories of thermodynamic properties have been quite successful, no comparable theory exists for transport properties.

We have recently developed a correlation for diffusivity on n-alkane chains ranging from ethane to C60. The correlation is based on molecular simulations of the fused-hard sphere chains as reference fluids combined with an empirical correction for the effects of attractive forces. It is accurate to 8.6% average absolute deviation (AAD), despite the diffusivity being very sensitive to variations in temperature and density. Similarly, Elliott and Gray have published relations to characterize the terms of thermodynamic perturbation theory (TPT) for n-alkane chains.[3] Combined, these results provide a basis for evaluating Rosenfeld's conjecture. In particular, the entropy departure for hard chain fluids (given by A0) can be compared to their simulated self-diffusivities. The entropy departure for the complete atomistic potential can be computed from second order TPT using the terms for A1, A2. The trends of the self-diffusivity vs. entropy can thus be evaluated for both the reference fluids and their full potentials. We report the results of this comparison and its implications for the relation between thermodynamics and transport properties.

Reference:

1. Rosenfeld, Y., Phys. Rev. A 1977, 15, 2545.

2. Mittal, J.; Errington, J. R.; Truskett, T. M., Relationships between Self-Diffusivity, Packing Fraction, and Excess Entropy in Simple

Bulk and Confined Fluids. J. Phys. Chem. B 2007, 111, 10054-10063.

3. Elliott, J. R.; Gray, N. H., Asymptotic Trends in Thermodynamic Perturbation Theory. J. Chem. Phys. 2005, 123, 184902.

See more of this Session: Thermodynamic Properties and Phase Behavior V
See more of this Group/Topical: Engineering Sciences and Fundamentals

Quantitative Structure-Property Relationship Model for Prediction of Octanol-Water Partition Coefficient

K. M. Yerramsetty, School of Chemical Engineering, Oklahoma State University, Stillwater, OK

B. J. Neely, School of Chemical Engineering, Oklahoma State University, Stillwater, OK

K. A. M. Gasem, School of Chemical Engineering, Oklahoma State University, Stillwater, OK

The octanol-water partition coefficient (K_{ow}) is a thermophysical property describing the hydrophobicity of a molecule as expressed by the ratio of the organic and aqueous solubilities. This property has been found useful in inferring transport mechanisms and molecular distribution in environmental areas and bio-applications. For example, K_{ow} is a significant property in describing the permeation of a compound through the skin in transdermal drug delivery.

In this study, we present a new quantitative structure property relationship (QSPR) model for predicting K_{ow} utilizing a wide range of molecular species data from the PHYSPROP physical property database. Using the random diversity sampling technique, which insures a diverse sampling of the molecular species in the database, 5598 and 4144 molecules were assigned to training and validation sets, respectively. Non-linear QSPR models involving robust neural network algorithms with non-linear descriptor reduction techniques based on differential evolution were developed.

We improve on other literature models in several aspects, including (a) use of multiple QSPR software to provide descriptors assuring both model superiority and stability, (b) automated optimization of molecular structure conformations, (c) a wholly nonlinear descriptor reduction strategy, and (d) use of robust non-linear neural networks with multiple initializations to ensure network stability. Employing these state-of-the-art nonlinear algorithms, we have modeled accurately log K_{ow} with a root mean square error of 0.24 and 0.39 in the training and validation sets, respectively. Further, our models compare favorably to those available in literature using a standardized, literature test set of molecules.

See more of this Session: Thermodynamic Properties and Phase Behavior V
See more of this Group/Topical: Engineering Sciences and Fundamentals

Inertial Particle Acceleration Statistics in near-Wall Turbulence

Valentina Lavezzo, Department of Mechanical Engineering, University of Udine, Udine, Italy

Alfredo Soldati, Department of Mechanical Engineering, University of Udine, Udine, Italy

Sergiy Gerashchenko, Department of Mechanical & Aerospace Engineering, Cornell University, Ithaca, NY

Zellman Warhaft, Department of Mechanical & Aerospace Engineering, Cornell University, Ithaca, NY

Lance R. Collins, Department of Mechanical & Aerospace Engineering, Cornell University, Ithaca, NY

The motion of water droplets over a flat plate mounted in a wind tunnel with high levels of free stream turbulence is measured using high-speed particle tracking. The accelerations are calculated by convolution of the measured particle trajectories with a Gaussian smoothing and differentiating filter. The longitudinal acceleration mean and variance are found to increase with increasing particle inertia, contrary to what had been found for inertial particles in homogeneous isotropic turbulence. To further investigate this trend, we use direct numerical simulations of channel flow laden with particles. By comparing results with and without gravity, we see that the puzzling reversal in the trend is due to the complex interaction of the particles with turbulence, the mean shear and gravity.

See more of this Session: Turbulent Flows
See more of this Group/Topical: Engineering Sciences and Fundamentals

Momentum and Scalar Transport at the Turbulent/Non-Turbulent Interface of a Jet

Jerry Westerweel, Laboratory for Aero & Hydrodynamics, Delft University of Technology, Delft, Netherlands

A longstanding question about unconfined, but localised, turbulent flows is to describe and quantify the characteristic features of the inhomogeneous transition zone between fully developed turbulence and the decaying fluctuations outside the turbulent region. Recent findings indicate the existence of a finite jump in the tangential velocity at the interface. This is associated with small-scale eddying motion at the outward propagating interface (nibbling) by which irrotational fluid becomes turbulent.

In order to investigate the dynamics and transport processes at the continuous, and well defined bounding interface between the turbulent and non-turbulent regions of flow, conditionally sampled measurements with particle

image velocimetry (PIV) of a turbulent round submerged liquid jet have been taken at Re = 2000. The jet carries a fluorescent dye measured with planar laser-induced fluorescence (LIF), and the surface discontinuity in the scalar concentration is identified as the fluctuating turbulent jet interface. Thence the mean outward 'boundary entrainment' velocity is derived and shown to be a constant fraction (about 0.07) of the the mean jet velocity on the centerline. Profiles of the conditional mean velocity, mean scalar, and momentum flux show that at the interface there are clear discontinuities in the mean axial velocity and mean scalar, and a tendency towards a singularity in mean

vorticity. These actual or asymptotic discontinuities are consistent with the conditional mean momentum and scalar transport equations integrated across the interface. Measurements of the fluxes of turbulent kinetic energy and enstrophy are consistent with computations in showing that for a jet flow the entrainment process is dominated by small scale eddying at the highly sheared interface ('nibbling'), with large scale engulfing making a small (less than 10%) contribution consistent with concentration measurements showing that the interior of the jet is well mixed. To explain the difference between velocity and scalar profiles, their conditional mean gradients are defined in terms of a local eddy viscosity and eddy diffusivity and the momentum and scalar fluxes inside the interface. Since the eddy diffusivity is larger than

the eddy viscosity, the scalar profile is flatter inside the interface so that the scalar discontinuity is relatively greater than the mean velocity discontinuity.

Theoretical arguments are proposed for how the vortex sheet

develops and how the internal structure of the interface layer relates to the inhomogeneous rotational and irrotational motions on each side, and why the dominant entrainment process of jets and wakes differs from that of free shear layers.

See more of this Session: Turbulent Flows
See more of this Group/Topical: Engineering Sciences and Fundamentals

Microscale Fluid Flow Visualization in CIJR and MIVM and CFD Model Validation Using Confocal μ-LIF

Yanxiang Shi, Department of Chemical and Biological Engineering, Iowa State University, Ames, IA

Michael G. Olsen, Department of Mechanical Engineering, Iowa State University, Ames, IA

Rodney O. Fox, Department of Chemical and Biological Engineering, Iowa State University, Ames, IA

The process of producing nanoparticles via precipitation requires supersaturation due to rapid mixing effect at the smallest scale. This effect can be achieved in microscale reactors such as confined impinging jet reactor (CIJR) and multi-inlet vortex mixer (MIVM) at high Reynolds numbers where strong turbulence can be observed. With the help of computational fluid dynamics (CFD), "experimentless" design of such reactors can be possibly realized. Previously, the CFD model, DQMOM-IEM, developed by the Fox group at Iowa State University has been compared not only with the conversion of 2,2-dimethoxypropane (DMP) that was obtained experimentally by the Prud'homme group at Princeton University [1,2], but also with the microscale particle image velocimetry (μ-PIV) results [3,4]. The model predicted both the conversion of DMP and the velocity field accurately.

First, in order to extend the validation and to get an intuitive understanding of the mixing, the pH indicator phenolphthalein is employed to visualize the impinging jets. One stream of CIJR contains NaOH and phenolphthalein and the other HCl and phenolphthalein (two streams for each solution in MIVM). The fluid turns fuchsia once these two streams come into contact, which illustrates the molecular scale mixing. Since previous experiments were done for $Re_j = 600$ and 1000 (defined as the inlet Reynolds number), where the flow is turbulent, both of them are also chosen in this work for comparison. Qualitatively, these experimental results give satisfactory agreement with simulations.

Next, the concentration field is measured using the microscale laser induced fluorescence (μ-LIF). Up till this point, only passive scalar mixing is accomplished with confocal technology since it has the advantage of optical sectioning which makes the LIF measurement more accurate. The fluorescent dye rhodamine 6G is used as the passive scalar and the flow is turbulent. The mixture-fraction mean and variance obtained from the micromixing model are compared to that from μ-LIF. Correspondingly, the large-scale segregation (LSS) and small-scale segregation (SSS) are analyzed.

[1] Y. Liu and R. O. Fox. CFD Predictions for Chemical Processing in a Confined Impinging-Jets Reactor. *AIChE Journal.* 2006(52):731-744.

[2] Y. Liu, C. Cheng, Y. Liu, R. K. Prud'homme and R. O. Fox. Mixing in a Multi-Inlet Vortex Mixer (MIVM) for Flash Nano-precipitation. *Chemical Engineering Science.* 2008(63):2829-2842.

[3] Y. Liu, M. G. Olsen and R. O. Fox. Turbulence in a Microscale Planar Confined Impinging-Jets Reactor. *Lab on a Chip.* 2009(9):1110-1118.

[4] J. C. Cheng, M. G. Olsen and R. O. Fox. Investigation of Fluid Dynamics in a Multi-Inlet Vortex Reactor by Computational Fluid Dynamics & Microscopic Particle Image Velocimetry. *Applied Physics Letter.* (In Press)

See more of this Session: Turbulent Flows
See more of this Group/Topical: Engineering Sciences and Fundamentals

Open-Channel Flow Turbulence and Air-Water Gas Transfer

Damon E. Turney, Bren School of Environmental Science & Management, UCSB, Santa Barbara, CA

Sanjoy Banerjee, Chemical Engineering, City College of New York, New York, NY

Transfers of mass across air-water interfaces have gained renewed attention due to efforts to track pollutants and nutrient chemicals in environmental systems such as rivers, lakes and oceans. Despite many decades of investigation of this subject, predictions of the transfer rate still carry uncertainty factors of two or greater, mostly because the fluid motions very near the air-water interface have eluded measurement and characterization. This talk presents results from new particle image velocimetry (PIV) experimental techniques that measure and characterize the water motions within one millimeter of the air-water interface. Laboratory open-channel flows of varying bulk velocity and depth were used for generating the turbulence studied here. The air-water transfer rate is measured by stripping the oxygen from the upstream water and measuring the rate of uptake of the gas at downstream locations. Bulk turbulence characteristics are also measured with PIV. Several theories for predicting the interfacial transfer rate are tested with the experimental data. The surface divergence theory and surface renewal theories are seen to agree with measurements in open-channel flows, but both require poorly understood corrective coefficients. A better understood theory of interfacial transfer is formulated and tested, with the goal of eliminating corrective coefficients. Surfactants are seen to have a large effect on interfacial motion despite significant efforts to clean the interface. Future directions for research of air-water gas transfer are proposed.

See more of this Session: Turbulent Flows
See more of this Group/Topical: Engineering Sciences and Fundamentals

On the Use of V-LES as An Alternative to LES for Industrial Turbulent Flows

Mathieu Labois, ASCOMP GmbH, Zurich, Switzerland

Narayanan Chidambaram, R&D, ASCOMP GmbH, Zurich, Switzerland

Djamel Lakehal, R&D, ASCOMP GmbH, Zurich, Switzerland

The limits of the RANS approach for turbulent single phase and multi-phase flow have been reached. The competition between industrial key players in automotive and land-based and aerospace technologies in particular is promoting the development of new modeling strategies that put flow unsteadiness at the forefront of their predictive capabilities. This added to the constant growth of computer resources in terms of speed and mass storage helps replace the LES concept at the front stage of applied CFD. Be it as it may, the LES strategy is still too expensive for various types of flows, in particular those in which steady state conditions are difficult to achieve (e.g. strong non-homogeneity of turbulence). This is the reason why alternative approaches based on scale separation are indispensable; including the V-LES (Very Large Eddy Simulation), the filter-based technique and the so-called SAS variant. The idea is to combine RANS and LES for a specific flow, where the knowledge of the size of the most important scales is key. Here the fow is decomposed into resolved and subscale part, the latter being independent from the grid (in contrast to the SGS in LES), but is dependent on the flow (and thus the obstacles). Larger scales than this specific scale are directly resolved, meanig actually no model is included, but the sub-scale part is modelled, but with a more refined statistical turbulence model than the 0-equation Smagorinsky one, i.e. a k-epsilon or even a Reynolds stress model. The approaches cited above differ in the way the transtion between resolved and sub-scale turbulence is achieved. We discuss our latest results, in which we have conducted a systematic sensitivity analysis of the involved parameters. The V-LES is shown to be a viable alternative to LES for very high Re flow conditions.

See more of this Session: Turbulent Flows
See more of this Group/Topical: Engineering Sciences and Fundamentals

Extended Abstract

The URAPS-Closure for the Normalized Reynolds Stress

K. S. Koppula[‡], A. Bénard[*], and C. A. Petty[‡]
[‡]Department of Chemical Engineering and Materials Science
[*]Department of Mechanical Engineering
Michigan State University
East Lansing, MI 48824

The unclosed RANS-equation for a constant property Newtonian fluid is an exact equation that relates the mean velocity field to the mean pressure field and the Reynolds stress. An algebraic turbulence model for the Reynolds stress and the continuity equation for constant density fluids provide a low-order closure for the RANS-equation.

Koppula (2009) has developed a new algebraic closure model for the normalized Reynolds stress that is realizable and universal for all turbulent flows. Universality stems directly from an analysis of the equation-of-motion in a non-inertial frame of reference. The resulting closure can be applied in either inertial or non-inertial frames regardless of the class of benchmark flows used to determine the phenomenological closure parameters. The closure model for the normalized Reynolds (NR-) stress is formulated as a non-negative mapping of the NR-stress into itself and is thereby realizable for all turbulent flows. The theory depends on the relative importance of four local time scales: a viscous time scale; a turbulent time scale; a time scale related to the mean field velocity gradient; and, a time scale associated with the non-inertial frame of reference.

The preclosure equation shifts the turbulence closure problem from the NR-stress to the normalized prestress. The prestress is caused by pressure fluctuations and fluctuations in the instantaneous Reynolds stress. A self-consistent hypothesis, similar to the classical Rotta-conjecture for the pressure/strain rate correlation, is used to relate the prestress to the NR-stress. The resulting anisotropic prestress (APS-) closure, which generalizes the isotropic prestress (IPS-) closure of Parks et al. (1998) and Weispfennig et al. (1999), is combined with the preclosure equation for the NR-stress to produce a universal realizable anisotropic prestress (URAPS-) closure. The URAPS-closure provides a resolution to one of the key questions in turbulence modeling: Can a low-order closure model for the NR-stress be formulated that is realizable and universal for all turbulent flows independent of the specific benchmark flows used for calibration?

The NR-stress predicted by the URAPS-closure for spanwise rotation of asymptotic homogeneous shear (see Koppula et al., 2009) and for spanwise rotation of fully-developed channel flows (see Figures 2-5 below) are compared with available DNS results. The agreement between the URAPS-closure predictions and the DNS results justifies the use of the new closure as a low-order closure of the RANS-equation.

ACKNOWLEDGEMENTS

This research was partly funded by the National Science Foundation Industry/University Cooperative Research Center for Multiphase Transport Phenomena (NSF/ECC 0331977). The authors gratefully acknowledge the members of the Center for their support. KSK, AB, and CAP sincerely thank Satish Mathu for the production of Figures 2-5.

REFERENCES

Iwamoto, K., 2002, "Database of Fully Developed Channel Flow", THTLAB Internal Report No. ILR-0201., THTLAB, The Univ. of Tokyo

Koppula, K. S., 2009, "Universal Realizable Anisotropic Prestress Closure for the Normalized Reynolds Stress", Ph.D. Dissertation, Michigan State University.

Koppula K. S., A. Bénard, C. A. Petty, 2009, "Realizable algebraic Reynolds stress closure", Chemical Engineering Science, doi:10.1016/j.ces.2009.04.040 (in press).

Parks, S. M., K. Weispfennig, and C.A. Petty, 1998, "An Algebraic Preclosure Theory for the Reynolds Stress", Phys. Fluids, **10** (3), 645-653.

Weispfennig, K., S. M. Parks, and C. A. Petty, 1999, "Isotropic Prestress for Fully Developed Channel Flows", Phys. Fluids, **11**(5), 1262-1271.

Wu, H. and N. Kasagi, 2002, "Effects of Combined Streamwise and Spanwise Rotations on Turbulent Channel Flow", The Fifth JSME-KSME Fluids Engineering Conference, Nov. 17-21, Nagoya, Japan.

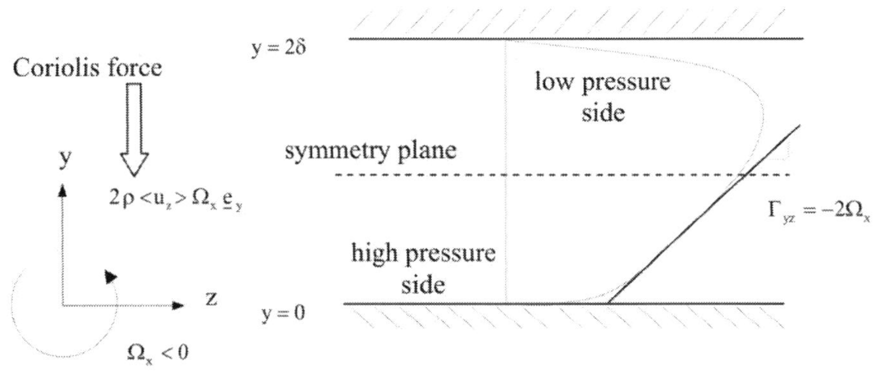

Figure 1. Spanwise rotation of fully-developed channel flows.

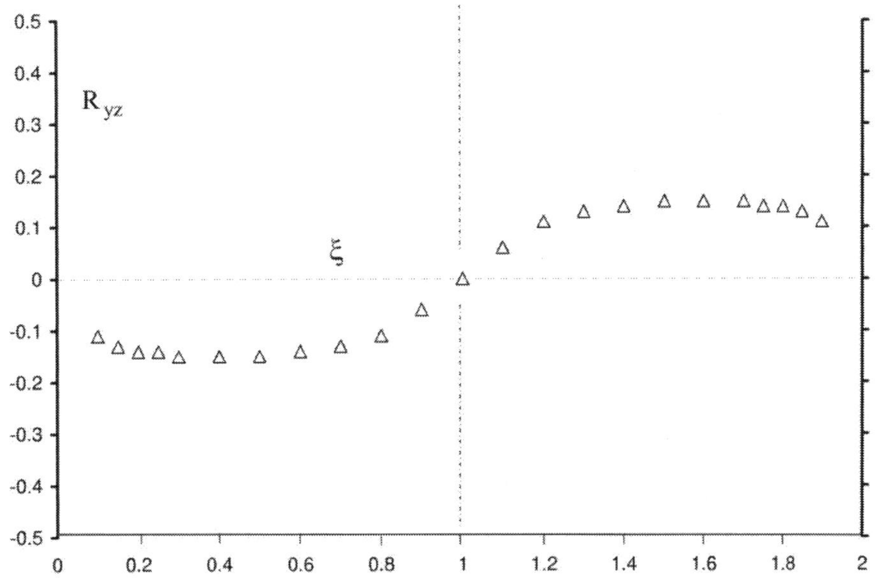

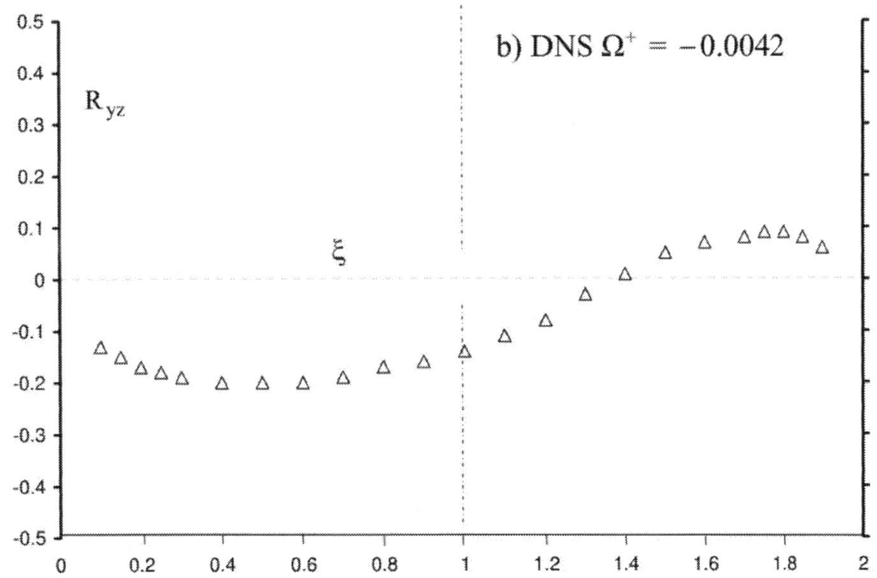

Figure 2. Influence of spanwise rotation on the shear component of the normalized Reynolds stress for fully-developed channel flow predicted by DNS (Iwamoto, K., 2002; Wu, H. and N. Kasagi, 2002)

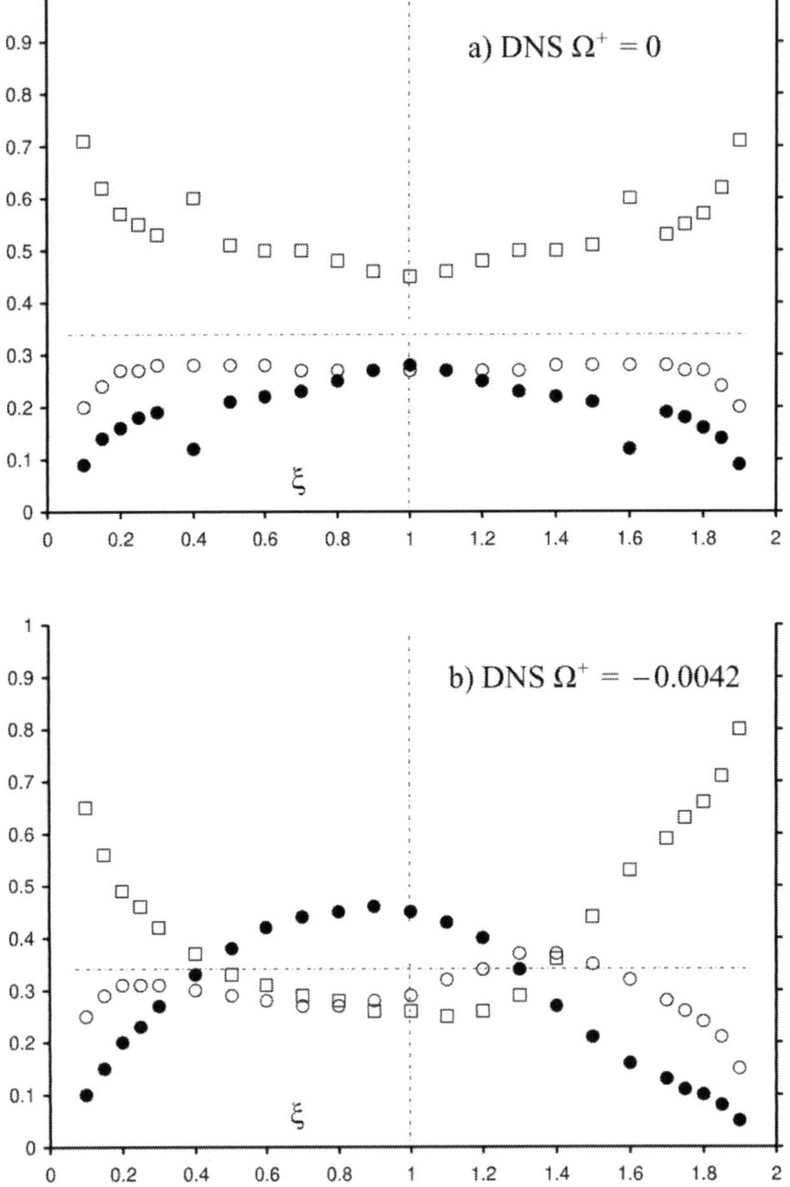

Figure 3. Influence of spanwise rotation on the redistribution of energy for fully-developed channel flow: R_{xx}, O; R_{yy}, •; R_{zz}, □ (Iwamoto, K., 2002; Wu, H. and N. Kasagi, 2002)

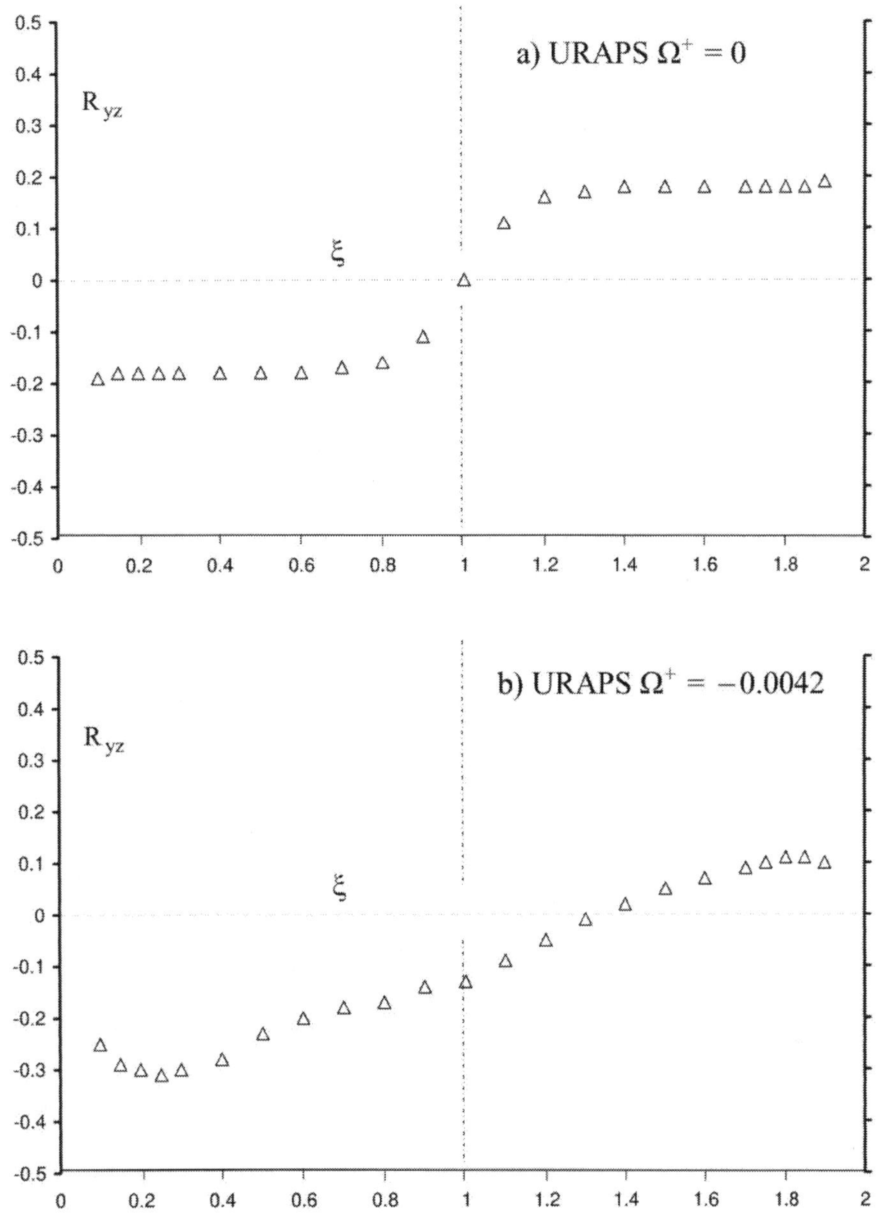

Figure 4. Influence of spanwise rotation on the shear component of the normalized Reynolds stress for fully-developed channel flow (Koppula, 2009).

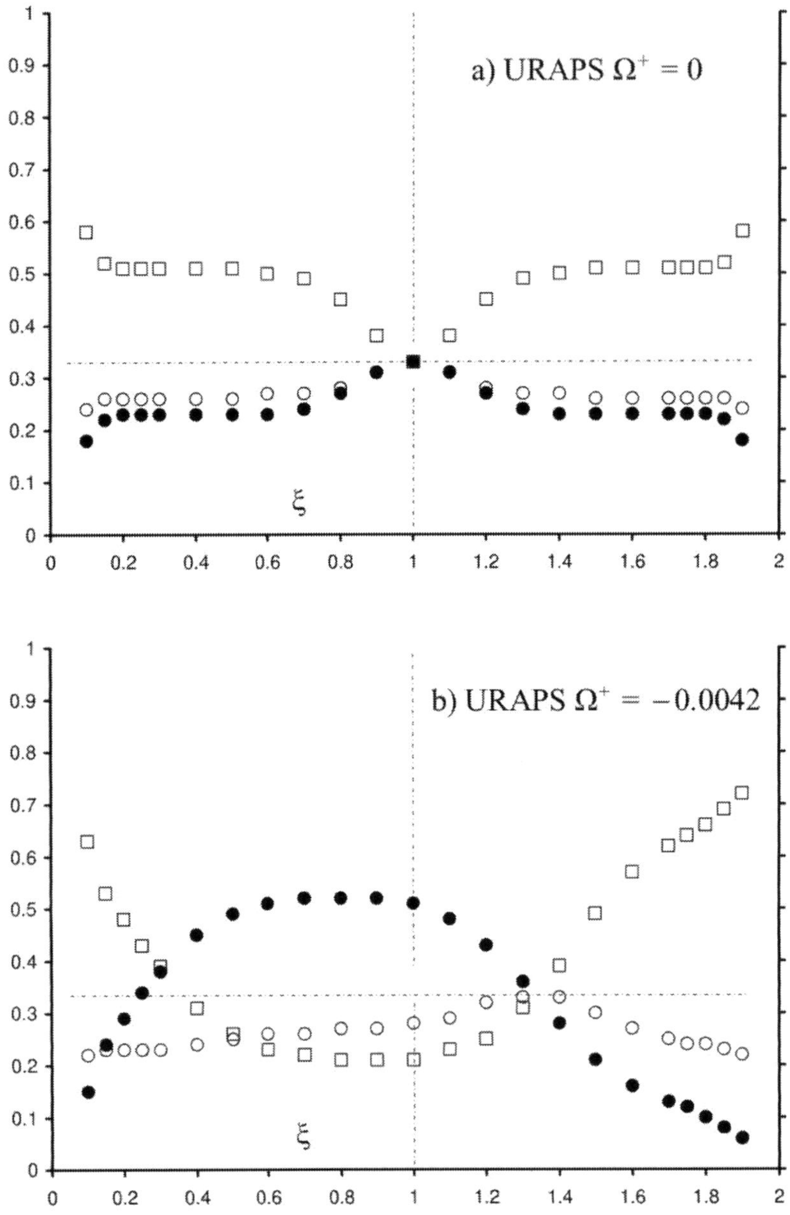

Figure 5. Influence of spanwise rotation on the redistribution of energy for fully-developed channel flow: R_{xx}, O; R_{yy}, •; R_{zz}, □ (this research)

RANS and LES Studies of Turbulent Flow and Heat Transfer for a Single Sphere, with Applications to High-Re Flow in Packed Tubes

Anthony G. Dixon, Worcester Polytechnic Institute, Worcester, MA

M. Ertan Taskin, Worcester Polytechnic Institute, Worcester, MA

Michiel Nijemeisland, Johnson Matthey, Cleveland, United Kingdom

Hugh Stitt, Johnson Matthey, Cleveland, United Kingdom

Reynolds-Averaged Navier-Stokes (RANS) modeling and large eddy simulation (LES) of flow around a sphere at subcritical turbulent Re (400 - 20,000) is described. Comparisons are made to standard correlations for the drag coefficient (C_D), separation angle, reattachment length and local heat transfer Nusselt number (Nu_{loc}), with good agreement found at the lower Re for both approaches. At higher Re the RANS methods cannot reproduce the trends in C_D as eddy-shedding and wake dynamics become more important, but good agreement is obtained using LES, at the cost of a very fine mesh and integrating over a long enough time period for steady averages to be collected. Reasonable results are obtained from the steady k-ω SST method, provided a fine enough mesh is used to resolve the boundary layer, for quantities that are not strongly linked to form drag, such as Nu.

The fine meshes required for the best quality results prohibit extension to assemblies of particles, such as fixed beds, at this level of detail. Results are reported here of mesh coarsening studies, to determine the meshing strategy to retain the desired features of the flow, while reducing mesh size to a practical level. For example, mesh refinement to resolve eddies in the wake of the sphere, which is essential for good prediction of C_D for the single particle study, is not necessary for particle beds in which wake formation is suppressed.

See more of this Session: Turbulent Flows
See more of this Group/Topical: Engineering Sciences and Fundamentals

DNS of the Velocity and Temperature Fields in a Small Room

John B. McLaughlin, Clarkson University, Potsdam, NY

Xinli Jia, Clarkson University, Potsdam, NY

Goodarz Ahmadi, Clarkson University, Potsdam, NY

Jos Derksen, Department of Chemical & Materials Engineering, University of Alberta, Edmonton, AB, Canada

This talk presents the results of a numerical study of the velocity and temperature fields in a model of a small room containing a seated mannequin. Results are also presented for the trajectories and ultimate fate of small particles that are introduced through the air inlet. The study was motivated by an experimental study carried out by Professor M. Glauser's group at Syracuse University. In the experimental study, air entered the room through a floor vent and exited through a ceiling vent on the other side of the room. A table was placed in front of the floor vent, and a mannequin was seated at the table. The mannequin could be electrically heated so that its surface temperature was 37C. Velocity and temperature measurements were carried out at several points and flow visualization experiments were conducted. It was noted that, if present at all, turbulent fluctuations were small.

The objective of the simulations was to obtain a more detailed understanding of the flow in the room. Of specific interest were the effects of the table, chair, and mannequin on the ultimate fates of small particles that were introduced into the room with the air in the inlet. The importance of the thermal plume around the mannequin was of particular interest since the thermal plume plays a role in transporting particles from near the floor to the breathing zone and may also prevent particles from reaching the breathing zone from above.

The simulations were performed with a single phase version of a lattice Boltzmann method (LBM) that was originally developed for two-phase flows by Inamuro et al. (J. Comp. Phys. 198, 628-644, 2004). The method has the advantages of simplicity comparable to the widely used BGK LBM and stability at much higher Reynolds numbers than may be achieved with the BGK LBM. Particle trajectories were computed by integrating a particle equation of motion that included aerodynamic drag and gravity. The one-way coupling assumption was invoked since the particles ranged in size from 2 to 16 micrometers and their concentration was dilute.

See more of this Session: Turbulent Flows
See more of this Group/Topical: Engineering Sciences and Fundamentals

Behavior of Forwards and Backwards Dispersion in Turbulent Channel Flows

Chiranth Srinivasan, Chemical Engineering, University of Oklahoma, Norman, OK

Dimitrios Papavassiliou, Chemical Engineering, University of Oklahoma, Norman, OK

ABSTRACT

This study utilizes a direct numerical simulation of turbulent flow in an infinite channel in conjunction with the tracking of scalar markers to investigate turbulent dispersion. The channel essentially comprises of a computational block, which is 4πh × 2h × 2πh in the x, y, z dimensions (where h is the half channel height and h =150). The four important regions of wall bounded turbulent flows, namely the viscous sub-layer, the transition region, the logarithmic layer, and the outer region of the flow at the center of the channel are considered. Scalar markers captured in these four turbulent flow regions are correlated both forwards and backwards in time. It is found that backwards and forwards dispersion proceed at different rates at different regions of the flow. A change in the rates of backwards and forwards dispersion is also seen with a change in Prandtl number. However, for Prandtl number six and higher, there are negligible differences in the behavior of backwards and forwards dispersion. The observed differences are associated with the form of the probability density function of the velocity fluctuations, and more specifically with the skewness of the probability density function. The velocity autocorrelation coefficients are used to determine the principal direction of heat flow in the channel with time. The gradient of heat flux and the direction of the transfer of heat show a difference for the backwards and forwards dispersion. The Prandtl number also plays a significant role in deciding the amount and the principal direction of heat transfer.

See more of this Session: Turbulent Flows
See more of this Group/Topical: Engineering Sciences and Fundamentals

Experimental Studies for the Effect of Pressure Drop Across Horizontal Pipe and Control Valve in Series: Comparison with ASPEN

Ahmad.J Rusumdar, Department of Modelling, Leibniz Institute for Tropospheric Research, Leipzig, Germany

Suresh Tavva, Department of Chemical Engineering, National Institute of Technology, Tiruchirappalli, India

Arivazhagan M, Department of Chemical Engineering, National Institute of Technology, Tiruchirappalli, India

The contact of two or more immiscible fluids is encountered widely in the chemical and petroleum industries; among this two-phase gas- liquid flow through pipes and valves is a complex process. One of the problems that are hardly understood is how the two phases are distributed among the pipe lines and control valves. Studies on operating characteristics of control valves with two phase flow have not been given much attention in the literature despite its industrial importance during the design and selection as well as plant operations. However, a considerable work has been performed to study the pressure drop effects in two phase flow through pipes and different geometrical shapes of flow ducts. In this current work an experimental facility has been set up to study the pressure drop characteristics in a horizontal pipe and control valve in series under two phase flow conditions. By using the commercial ASPEN PLUS 11.1 package were employed to compare the experimental results. To calculate the simulated the pressure drop in pipe and control valve, for the typical fluid system of water, kerosene, diesel, palm oil separately (liquid phase) with the air (gas phase). The pressure drop characteristics are studied at constant valve position over a range of liquid flow rates from 50 LPH to 150 LPH and the gas flow rate varied from 25LPH to100 LPH. The above procedure were repeated for different valve openings (25% to 100%) The numerical results were validated against experimental data. The prediction of the pressure drop characteristics in pipe and valve were within an average error of about ±8 %. Comparisons of experimental and simulated profiles were found to be in good agreement.

See more of this Session: Turbulent Flows
See more of this Group/Topical: Engineering Sciences and Fundamentals

Dynamic Surface Tension and Dispersion Stability of Aqueous Lipid/Protein Dispersions

Yoonjee Park, School of Chemical Engineering, Purdue University, West Lafayette, IN

Elias I. Franses, School of Chemical Engineering, Purdue University, West Lafayette, IN

A mixture of lipid and protein compounds called "lung surfactant" concentrates, or adsorbs, at the interface of an aqueous layer which lines the lung alveoli. Dipalmitoylphosphatidylcholine (DPPC) lipid is zwitterionic, insoluble, and the major component of natural lung surfactant and of some commercial synthetic lung surfactant replacement formulations. Sonicated dispersions of DPPC vesicles at 0.1 wt% (1000 ppm) in aqueous isotonic buffer solutions produced previously (S. H. Kim, Y. Park, S. Matalon, and E.I. Franses, 2008, Colloids and Surfaces B, 67: 253-260) are colloidally stable for hours and days, and they have very low (< 10 mN/m) dynamic surface tension minima (DSTM) under pulsating area condition at 37 and 25º C. The vesicles are stabilized in part electrostatically due to negative charge from adsorbed ions (zeta potential = -20±3 mV), and in part by other forces.

The DPPC dispersions may also have to be colloidally stable and produce low DSTM in the presence of blood serum proteins such as albumin or fibrinogen, because such proteins can be released from the blood stream to the alveolar lining layer when the lung alveolar tissue is injured. When a stable solution of 1000 ppm BSA (bovine serum albumin), which is negatively charged, is mixed with a DPPC vesicle dispersion, it becomes colloidally unstable, aggregating within minutes and hours. This implies some heterocoagulation. The DSTM's become much higher because the protein may adsorb more than the lipid at the surface and may block in part the adsorption of the lipid and the lowering of the DST.

DPPC lipid vesicles have been modified to reduce the possibility of aggregation with the protein by the addition of a small weight fraction of a neutral lipid with PEG (polyethyleneglycol) groups. The mixed lipid vesicles were found to be more stable than DPPC, remaining stable for months. Moreover, the mixed vesicles show no tendency to aggregate with the albumin molecules for days, probably because of a steric repulsive mechanism. The mixed vesicles are, therefore, more "biocompatible." Finally, the mixed lipid dispersions have low DSTM without or with the albumin.

See more of this Session: Fundamentals of Interfacial Phenomena III - Biological and Organic Molecules at Interfaces
See more of this Group/Topical: Engineering Sciences and Fundamentals

Synergistic Interactions Among Biosurfactants and Other Surfactants

Jun Wu, Earth and Environmental Engieering, Columbia University, New York, NY

Ponisseril Somasundaran, Columbia University, New York, NY

Kevin Jarrell, Modular Genetics, Inc, Cambridge, MA

Due to the increasing demand for environmental friendly reagents, greener surfactants are recently receiving the warranted attention. The unique surface activity and biodegradability make this group of surfactants leading candidates as the next generation reagents. One of the unique properties of greener surfactants is the synergistic interactions between each other as well as conventional surfactants. In this work, we selected several greener surfactants, sugar based alkyl glycosides and amino acid based acyl-glutamate for exploring their potential synergism in mixtures

under different solution conditions. It is noted that synergy under various conditions of temperature, pH and salinity depends very much upon the chemical structure of the components as well as mixing ratio and even the sequence of mixing. The results are analyzed to obtain relationship between packing of surfactant species at various interfaces, solubility and resultant synergism/antagonism.

See more of this Session: Fundamentals of Interfacial Phenomena III - Biological and Organic Molecules at Interfaces
See more of this Group/Topical: Engineering Sciences and Fundamentals

Separation of Protein Isoforms Based On Differential Adsorption

Anuj Chauhan, Chemical Engineering, University of Florida, Gainesville, FL

Madhukar S. Patel, Engineering, Brown University, Providence, RI

Lea Mouallem, Engineering, Brown University, Providence, RI

Glareh Azadi, Engineering, Brown University, Providence, RI

Anubhav Tripathi, Division of Engineering and Warren Alpert School of Medicine, Brown University, Providence, RI

It has been found that an excess or insufficient amount of a specific isoform has the tendency to cause a variety of diseases. Creating a technique to separately identify isoforms will therefore, allow for the prevention of diseases such as various cancers and Alzheimer's. This work tests the difference in adsorption between two monomeric fluorescent protein isoforms. The potential difference in adsorption between the two isoforms will allow one to develop an apparatus to separately identify protein isoforms, based on this difference in adsorption. The fraction adsorbed was used to calculate the percent protein adsorbed for completely folded and partially unfolded protein isoforms. The difference in adsorption ranged from 11 to 53 percent was found in partially denatured proteins. The data has also shown only a slight difference in adsorption when the proteins are in their native states. The type of bead used affected the magnitude and pattern of the adsorption difference between the partially denatured isoforms. The effect of bead surface and protein conformation on adsorption will be discussed.

See more of this Session: Fundamentals of Interfacial Phenomena III - Biological and Organic Molecules at Interfaces
See more of this Group/Topical: Engineering Sciences and Fundamentals

Characterization and Applications of Hydrocarbon/Fluorinated and Fully Fluorinated Catanionic Vesicles

Vivian A Ojogun[1], Hans-Joachim Lehmler[2] and Barbara L Knutson[1,*]

[1]Department of Chemical and Materials Engineering, University of Kentucky, Lexington, KY 40506-0046

[2]Department of Occupational and Environmental Health, University of Iowa, Iowa City, IA 52242-5000

Spontaneously formed vesicles in dilute aqueous mixtures of cationic and anionic surfactants serve as potential substitutes to conventional meta-stable liposomes in template synthesis, molecular encapsulation and transport applications. Although electrostatics govern intra- and inter-aggregate interactions, the bilayer matrix of the catanionic surfactant pair exerts strong influence on phase behavior and vesicle characteristics (size and stabilization mechanism).

We compare the impact of mixed hydrocarbon/fluorocarbon bilayer in cetylpyridinium bromide (CPB)/sodium perfluorooctanoate (SPFO) to the fully fluorinated bilayer system of 1,1,2,2-tetrahydroperfluorododecylpyridinium chloride (HFDPC)/(SPFO) on aggregate phase behavior and catanionic vesicle formation. Previously, formation of stable unilamellar vesicles was established in the anionic-rich phase region at lower surfactant concentrations for HFDPC/SPFO (< 1% wt/wt) than in CPB/SPFO (> 1% wt/wt) with visual inspection, dynamic light scattering and transmission electron microscopy images. Subsequent templating of silica hollow spheres revealed greater colloidal stability in HFDPC/SPFO than in CPB/SPFO. The same techniques are applied in development of partial phase diagrams for both CPB/SPFO and HFDPC/SPFO with assignment of the various aggregate phase boundaries.

Further characterization of vesicle formation and bilayer matrix is provided with fluorescence spectroscopic measurements of pyrene and pyrene derivatized carboxylic acid probes. The selection of pyrene derivatives; 1-pyrenebutanoic acid (PBA), 1-pyrenedecanoic acid (PDA) and 1-pyrenehexadecanoic acid (PHA) with increasing hydro-alkyl chains lengths (C3 – C15) facilitates access to various depths of the bilayer and inform on differences in bilayer chain packing organization.

Relative encapsulation and retention of a model cationic dye, rhodamine 6G (R6G) was determined for CPB/SPFO and HFDPC/SPFO vesicles using a combination of size exclusion chromatography, DLS and UV techniques. Results suggests that due to increased hydrophobicity and lipophobicity, the fully fluorinated bilayer in HFDPC/SPFO, provides a greater barrier effect to permeation than in the mixed CPB/SPFO vesicles. This demonstrates effective tuning of catanionic surfactant properties to specific applications.

Interfacial Tension of Pure Substances and Binary Mixtures by Density Functional Theory Combined with Peng-Robinson and Cubic-Plus-Association Equations of State

Zhidong Li, Reservoir Engineering Research Institute (RERI), Palo Alto, CA

Abbas Firoozabadi, Reservoir Engineering Research Institute (RERI), Palo Alto, CA

We develop a semi-empirical density functional theory (DFT) and investigate the interfacial tension of several pure substances (N2, CO2, H2S, normal alkanes from C1 to nC10, and H2O) and binary mixtures (C1/C3, C1/nC5, C1/nC7, C1/nC10, CO2/nC4, N2/nC5, N2/nC6, N2/nC8, N2/nC10, nC6/nC7, nC6/nC8, nC6/nC10, and CO2/H2O). The theory is combined with the Peng-Robinson (PR) and cubic-plus-association (CPA) equations of state. The weighted density approximation (WDA) is adopted to extend the bulk excess Helmholtz free energy to the inhomogeneous interface. Besides, a supplementary term, quadratic density expansion (QDE), is introduced to account for the long-range characteristic of intermolecular dispersion attractions, which cannot be accurately described by the WDA. In the bulk limit, the QDE vanishes and the theory is reduced to the PR or CPA. For pure substances, the potential expansion energy is the only adjustable parameter in the QDE and determined by using a single measured interfacial tension at the lowest temperature examined. We faithfully predict the interfacial tensions of pure substances and mixtures over a wide range of temperatures and pressures.

See more of this Session: Fundamentals of Interfacial Phenomena III - Biological and Organic Molecules at Interfaces
See more of this Group/Topical: Engineering Sciences and Fundamentals

Understanding the Impact of Chemical Patterns On the Interactions and Self-Assembly of Organic Macromolecules at Interfaces

Claribel Acevedo-Velez, Department of Chemical and Biological Engineering, University of Wisconsin-Madison, Madison, WI

William C. Pomerantz, Department of Chemistry, University of Wisconsin-Madison, Madison, WI

Juan J. de Pablo, Department of Chemical and Biological Engineering, University of Wisconsin-Madison, Madison, WI

Samuel H. Gellman, Department of Chemistry, University of Wisconsin-Madison, Madison, WI

Nicholas L. Abbott, Department of Chemical and Biological Engineering, University of Wisconsin-Madison, Madison, WI

Although it is generally appreciated that the interactions of organic macromolecules depend strongly on the patterns of chemical functional groups presented at their surfaces, surprisingly few methods allow for precise manipulation of these chemical patterns. This presentation will describe an investigation of the interfacial interactions of organic molecules formed from non-natural ß-amino acids (ß-peptides). This class of molecule allows for precise manipulation of surface chemical patterns via sequence control during step-wise organic synthesis. By using single-molecule force spectroscopy, the impact of specific chemical patterns on forces generated between ß-peptide-based molecules presented at interfaces has been investigated. These measurements and others involving helical ß-peptides will be reported in this talk. Overall, the measurements provide fundamental insights into the mechanisms through which subtle changes in the patterns of chemical functional groups presented by organic macromolecules can be amplified into remarkably dramatic changes in their self-assembly.

See more of this Session: Fundamentals of Interfacial Phenomena III - Biological and Organic Molecules at Interfaces
See more of this Group/Topical: Engineering Sciences and Fundamentals

Tribometric Comparison of n-Alkanethiolate and n-Alkyl Trichlorosilane Monolayers

Brandon D. Booth, Department of Chemical and Biomolecular Engineering, Vanderbilt University, Nashville, TN

Steven G. Vilt, Department of Chemical and Biomolecular Engineering, Vanderbilt University, Nashville, TN

Clare McCabe, Department of Chemical and Biomolecular Engineering, Vanderbilt University, Nashville, TN

G. Kane Jennings, Department of Chemical and Biomolecular Engineering, Vanderbilt University, Nashville, TN

In this presentation we offer a quantitative comparison of the frictional performance for monolayers derived from n-alkanethiolates on gold and n-alkyl trichlorosilanes on silicon. Pin-on-disk microtribometry provided frictional analysis at applied normal loads from 10 to 1,000 mN at a speed of 0.1 mm/s. For monolayers prepared from both n-alkanethiols on gold and n-alkyl trichlorosilanes on silicon, a critical chain length of at least 8 carbons is required for beneficial tribological performance at an applied load of 9.8 mN. The direct comparison between the tribological stability of alkanethiolate and silane monolayers shows that monolayers prepared from n-octadecyl dimethylchlorosilane and n-octadecyl trichlorosilane withstood normal loads at least 30 times larger than those that damaged octadecanethiolate SAMs. Collectively, our results show that the tribological properties of monolayer films are dependent on their internal stabilities, which are influenced by cohesive chain interactions (van der Waals) and the adsorbate-substrate bond.

See more of this Session: Fundamentals of Interfacial Phenomena III - Biological and Organic Molecules at Interfaces
See more of this Group/Topical: Engineering Sciences and Fundamentals

Particle Charging and Stabilization in Non-Polar Liquids by Reversed Surfactant Micelles

Qiong Guo, School of Chemical and Biomolecular Engineering, Georgia Institute of Technology, Atlanta, GA

Virendra Singh, Georgia Institute of Technology, Atlanta, GA

Sven H. Behrens, School of Chemical & Biomolecular Engineering, Georgia Institute of Technology, Atlanta, GA

It is well known that ionic surfactants such as AOT are able to induce strong charging effects in nonpolar liquids. They avoid the large energetic barrier associated with the introduction of small ions in low dielectric media by forming charged micellar aggregates. Here we report evidence that nonionic surfactants have a very similar effect on nonpolar oils. The content of water solubilized in the micelle core, the overall micelle size and the fraction of charged micelles are extracted from Karl Fischer titration, dynamic light scattering, and precision conductometry data. Electrokinetic studies of added colloidal particles further reveal the micelles' ability to create and screen electric charges on surface of suspended solid particles. We interpret the observed behavior using a statistical mechanics approach originally proposed for ionic surfactant based microemulsions. Our findings suggest that nonionic surfactants can play a key role in the industrially important task of controlling charges in nonpolar solvents.

See more of this Session: Colloidal Dispersions III
See more of this Group/Topical: Engineering Sciences and Fundamentals

Screened Pair Interaction of Charged Colloidal Particles in Liquid-Liquid Interfaces

Carlos Espinosa, Chemical and Biomolecular Engineering, Georgia Institute of Technology, Atlanta, GA

Virendra Singh, School of Chemical and Biomolecular Engineering, Georgia Institute of Technology, Atlanta, GA

Sven H. Behrens, School of Chemical & Biomolecular Engineering, Georgia Institute of Technology, Atlanta, GA

The interaction of charged colloidal particles adsorbed to oil-water interfaces is still a subject of intensive research and scientific debate. The relative stability of such particles against aggregation compared to particles in aqueous bulk solution or in air-water interfaces has led to the controversial proposition that particles can sustain electric surface charges on the surface portion in contact with the oil phase, even if the latter has a very low electric permittivity. We show that this notion is supported by liquid-structure analysis of interfacial particle ensembles observed with digital video microscopy. Interaction measurements in the presence of surfactant micelles in the oil phase in fact suggest the surprising possibility of screening electrostatic interaction through the oil with the help of non-ionic surfactants.

See more of this Session: Colloidal Dispersions III
See more of this Group/Topical: Engineering Sciences and Fundamentals

Using Magnetic Nanoparticles to Probe the Stability of Polymer-Coated Colloids

Adriana P. Herrera, Chemical Engineering, University of Puerto Rico, Mayaguez Campus, Mayaguez, PR

Carola Barrera, Chemical Engineering, University of Puerto Rico, Mayaguez Campus, Mayaguez, PR

Yashira Zayas, Chemical Engineering, University of Puerto Rico, Mayaguez Campus, Mayaguez, PR

Carlos Rinaldi, Chemical Engineering, University of Puerto Rico, Mayaguez Campus, Mayaguez, PR

Cobalt ferrite nanoparticles with narrow size distribution were prepared and functionalized with covalently bonded homo-polymers of N-isopropyl-acrylamide such as pNIPAM and pNIPMAM. The temperature induced transitions of colloidal suspensions of these nanoparticles were studied through a combination of techniques such as differential scanning calorimetry (DSC), small angle neutron scattering (SANS), dynamic light scattering (DLS), and AC susceptibility measurements. In the pNIPAM coated nanoparticles excellent agreement was found for a transition temperature of ~30ºC by all four methods, although the AC susceptibility measurements indicated aggregation which was not evident from the DLS results. For the pNIPMAM coated nanoparticles DLS and AC susceptibility measurements indicated aggregation at a temperature of ~33-35ºC, much lower than the transition temperature peak at 40 ºC observed by DSC. These experiments and observations demonstrate the possibility of using AC susceptibility measurements to probe transitions in colloidal suspensions induced by external stimuli. Because magnetic measurements do not require optical transparency, these methods could be applied even in concentrated or opaque systems, in which light scattering techniques encounter technical problems.

See more of this Session: Colloidal Dispersions III
See more of this Group/Topical: Engineering Sciences and Fundamentals

Impact of Solvent Quality On Nanoparticle Dispersion in Semidilute and Concentrated Polymer Solutions

Nupur Dutta, Chemical Engineering, University of Virginia, Charlottesville, VA

David L. Green, Department of Chemical Engineering, University of Virginia, Charlottesville, VA

Our objective is to quantify the impact of solvent quality on the dispersion of polymer-grafted nanoparticles in semidilute and concentrated polymer solutions. These systems are ubiquitous to industry (e.g., paints, coatings, cosmetics) and medicine (e.g., blood, proteins, pharmaceuticals); however, the factors that control nanoparticle dispersion, such as interfacial wetting, solution temperature and solvent quality, have not been well elucidated. To this end, we use photon correlation spectroscopy to capture the dynamics of polydimethylsiloxane-grafted nanoparticles (PDMS-g-silica) in simple bromocyclohexane solutions as well as complex semidilute and concentrated PDMS-bromocyclohexane solutions. Bromocyclohexane is a theta solvent for PDMS and we vary the temperature to quantify the effect of good, theta, and poor solvent conditions on wetting of the graft polymer and nanoparticle dispersion . Moreover, we use static light scattering to quantify the interactions between the PDMS-g-silica nanoparticles through their second virial coefficient, which we seek to compare to those derived from theory. By relating experiment and theory our ultimate goal is to develop general results for predicting nanoparticle stability in semidilute and concentrated polymer solutions under conditions of variable solvency.

See more of this Session: Colloidal Dispersions III
See more of this Group/Topical: Engineering Sciences and Fundamentals

Quantitative Analysis of Synthetic Platelet Exfoliation in Water and Platelet Aspect Ratio Distribution

Harry J. Ploehn, Chemical Engineering, University of South Carolina, Columbia, SC

Hongsheng Gao, Chemical Engineering, University of South Carolina, Columbia, SC

Peter A. Barber, Chemistry & Biochemistry, University of South Carolina, Columbia, SC

Hans-Conrad zur Loye, Chemistry & Biochemistry, University of South Carolina, Columbia, SC

We use atomic force microscopy (AFM) and dynamic light scattering (DLS) for quantitative analysis of exfoliation and aspect ratio distribution of suspended, exfoliated inorganic platelet materials in water, including natural Na+-montmorillonite (MMT) clay as well as synthetic calcium niobate perovskite ($KCa_2Nb_3O_{10}$). Tapping-mode AFM with appropriate image analysis allows us to unambiguously identify exfoliated platelets and to quantify the distributions of platelet thickness and aspect ratio. MMT disperses in water (as is well known) but only partially exfoliates. We use dry weight analysis, DLS, and AFM to assess the impact of sample preparation procedures for maximizing exfoliation and obtaining high quality AFM images. Unexfoliated MMT and other contaminant particles can be removed by low speed centrifugation, but large exfoliated platelets are increasingly removed as centrifugal acceleration increases. Dilution of MMT suspensions with acetone prior to deposition onto mica leads to high quality AFM images with many isolated platelets. Quantitative analysis shows that the distribution of MMT platelet thickness is narrow and centered near 1.0 nm, and that the platelet aspect ratio distribution is log-normal, which has not been previously reported. We find that the average aspect ratio measured by AFM correlates linearly with the effective spherical particle diameter measured by DLS. The perovskite does not exfoliate in water despite having a higher cation exchange capacity, but cation exchange with tetrabutyl ammonium does promote exfoliation. DLS and AFM results help us optimize perovskite solution preparation and show that the perovskite's aspect ratio also has a log-normal distribution.

See more of this Session: Colloidal Dispersions III
See more of this Group/Topical: Engineering Sciences and Fundamentals

Evidence for the Autophobic Dewetting of Compatibilized Droplets in Immiscible Melts

Jeremy Fowler, Chemical Engineering, University of Virginia, Charlottesville, VA

Renlong Gao, Department of Chemistry, Virginia Tech, Blacksburg, VA

Tomonori Saito, Chemistry, Virginia Tech, Blacksburg, VA

Timothy E. Long, Department of Chemistry, Virginia Tech, Blacksburg, VA

David L. Green, Department of Chemical Engineering, University of Virginia, Charlottesville, VA

Our goal is to quantify how the wetting of diblock copolymers governs the dynamic behavior of polymer droplets in immiscible homopolymer blends. This system in which the copolymer migrates to the interface between the homopolymers and emulsifies the dispersed phase is analogous to a system of nanoparticles grafted with and dispersed in the same homopolymer. Our previous work with nanoparticles shows that a wetting diagram of the graft and matrix homopolymers predicts that particles disperse in the complete wetting region and aggregate in the incomplete wetting region. Our hypothesis is that droplets with diblock copolymers at their surfaces will behave similarly to the nanoparticles. To this end, polyisoprene (PI) matrices and diblock copolymers of poly(isoprene)-b-poly(dimethylsiloxane) (PI-b-PDMS) were synthesized by anionic polymerization. PDMS droplets were homogenized with the matrix and copolymers to formulate model systems. Using optical microscopy with a shear stage, flow was used to manipulate droplet sizes and to probe dynamic behavior. At high copolymer coverages and low shear, we observed that droplets that were once emulsified and dispersed, aggregated and formed clusters. The point at which the droplets aggregate depends on the matrix molecular weight; in shorter matrices, more copolymer must be added to induce droplet aggregation. This is consistent with autophobic dewetting, which has been recently observed for highly grafted nanoparticles in homopolymer melts. Moreover, the surface coverages at which droplets aggregate correspond to the predicted behavior for autophobic dewetting, indicating that a wetting diagram can also be used to predict the regions over which diblock copolymers emulsify polymer droplets in homopolymer matrices. The occurrence and behavior of these aggregates can be explained and quantified by analyzing dynamic response to flow conditions.

See more of this Session: Colloidal Dispersions III
See more of this Group/Topical: Engineering Sciences and Fundamentals

Particle-Size and Chain-Length Effects On Forces Between Functionalized Silica Nanoparticles

Ahmed E. Ismail, Performance Assessment and Decisions Analysis Department, Sandia National Laboratories, Carlsbad, NM

J. Matthew D. Lane, Surface and Interface Sciences, Sandia National Laboratories, Albuquerque, NM

Chris Lorenz, Division of Engineering, King's College, London, London, United Kingdom

Gary S. Grest, Surface and Interface Sciences, Sandia National Laboratories, Albuquerque, NM

To prevent the flocculation and phase separation of nanoparticles in solution during industrial processing, nanoparticles are often functionalized with short chain surfactants. Here we present fully-atomistic molecular dynamics simulations which characterize how these functional coatings affect the interactions between nanoparticles and with the surrounding solvent. In particular, we examine the behavior of silica nanoparticles coated with poly(ethylene oxide) oligomers in water.

We have shown in previous work that the hydrodynamic behavior of the functionalized nanoparticles is different than uncoated particles [Lane et al., Phys. Rev. E (2009)]: the force oscillation observed for uncoated particles is not observed for functionalized nanoparticles. However, in equilibrium systems, the thermal energy of the particles is insufficient to permit significant depletion of water molecules between the nanoparticles. In the present work, we extend the results of our previous study to determine the effects of particle sizes (5-20 nm) and chain lengths (6-100 repeat units) on the equilibrium interactions between nanoparticles. While particle size plays a dominant role in controlling the long-range interaction between particles, the effects of the functionalized groups are observed primarily in the short-range interactions.

This work is supported by the Laboratory Directed Research and Development program at Sandia National Laboratories. Sandia is a multiprogram laboratory operated by Sandia Corporation, a Lockheed Martin Company, for the United States Department of Energy under contract DE-AC04-94AL85000.

See more of this Session: Colloidal Dispersions III
See more of this Group/Topical: Engineering Sciences and Fundamentals

Simulation of Particles with Anisotropic Rigidity in Shear Flow Using a Minimal Bead-Spring Model

Nobuhiko Watari, Macromolecular Science and Engineering Center, University of Michigan, Ann Arbor, MI

Ronald G. Larson, Chemical Engineering, University of Michigan, Ann Arbor, MI

We show that an achiral particle with anisotropic rigidity can migrate in the vorticity direction in shear flow due to shear-induced chirality. A minimal "tetrumbbell" model of such a particle is constructed from four beads and six springs to make a tetrahedral structure. A combination of two different spring constants corresponding to "hard" and "soft" springs yields ten distinguishable tetrumbbells, which when simulated in shear flow with hydrodynamic interactions between beads but no Brownian motion at zero Reynolds number, produces five different types of behavior in which seven out of ten tetrumbbell structures migrate in the vorticity direction due to shear-induced chirality. Some of the structures migrate in the same direction along the vorticity direction even when the shear flow is reversed, which is impossible for permanently chiral objects.

We also show that this approach with simple bead-spring model to the dynamics of an object with a complex structure can be applied to simulations of swimming of flagellar bacteria.

See more of this Session: Complex-Fluid and Bio-Fluid Dynamics II
See more of this Group/Topical: Engineering Sciences and Fundamentals

An Active Microrheological Technique to Determine Normal Stress Differences of Complex Fluids

Aditya S. Khair, Chemical Engineering, University of California at Santa Barbara, Santa Barbara, CA

Todd M. Squires, Chemical Engineering, University of California at Santa Barbara, Santa Barbara, CA

Passive microrheology is a popular technique for measuring the linear-response rheological properties of complex fluids, through the fluctuating motion of embedded colloidal probes. Active microrheology, in contrast, aims to infer non-linear rheological information using actively driven probes. Presently, however, the focus of active microrheology has been limited to attempting to extract the material viscosity as a function of driving speed. In this talk, we propose a new active microrheological technique to determine both the first (N_1) and second (N_2) normal stress differences of a complex fluid, a task which is notoriously difficult in macroscopic mechanical rheometry. Specifically, we consider a two-point active microheology "experiment" in which a pair of colloidal probes are driven with constant velocity through a second order fluid --- the prototypical model for a large class of complex fluids in the limit of gentle deformations. We calculate the relative forces acting on the probes as they are pulled parallel and perpendicular to their line of centres, from which N_1 and N_2 are determined. Finally, we discuss the experimental feasibility of the proposed technique.

See more of this Session: Complex-Fluid and Bio-Fluid Dynamics II
See more of this Group/Topical: Engineering Sciences and Fundamentals

Vibrational Effects On Particle Sedimentation in Complex Fluids

Patrick T. Spicer, Procter & Gamble CETL, West Chester, OH

Marco Caggioni, Microstructured Fluids Group, Procter and Gamble Co., West Chester, OH

Particle sedimentation in complex fluid products is rarely desirable and rheological modification is frequently used to guard against such instabilities. However, a fluid rheology that prevents sedimentation under static conditions may not be useful during transport when a wide range of vibrational frequencies is encountered. This work studies the vibration-induced sedimentation of particles suspended in yield-stress fluids using video observations. Comparisons are made with theoretical predictions and national transport guidelines, while the effects of thixotropy are also examined.

See more of this Session: Complex-Fluid and Bio-Fluid Dynamics II
See more of this Group/Topical: Engineering Sciences and Fundamentals

Scaling the Long-Term Shear Stability of Aqueous Pigment Dispersions

Evagelos K. Athanassiou, Department of Chemistry and Applied Biosciences, ETH Zurich, 8093 Zurich, Switzerland

Hanspeter Gradischnig, BMW AG, Munich, Germany

Peter Siemsen, BMW AG, Munich, Germany

Wendelin J. Stark, Department of Chemistry and Applied Biosciences, ETH Zurich, 8093 Zurich, Switzerland

Industrial application of paints today involves the handling of millions of tons of pigment dispersions. Shear and mechanical stress during pumping, storage or processing significantly affect color and shelf life. At present, mainly empirical test series and pilot scale process models are used to assess the stability of such dispersions and provide a marginal guidance when developing or testing large scale paint applications. Scaling the stability against shear stress is particularly important for water based metal pigment dispersions where damage to the pigment flakes causes their deterioration. The present work investigates the scaling of shear stress from production line (>1000 t/y) to laboratory scale (100 g scale). We analytically show why the integrated shear stress on a specific dispersion can be deconvoluted into a series of individual shear stress contributions. Based on the fluid mechanics (laminar flow) during the transport process these individual contributions can be accurately reproduced in the laboratory using a Couette-type shearing setup similar to a classical rheometer. As a representative example, a stable and a shear-sensitive paint from an automotive production line where sampled over 4 month production. The evolving deterioration, color changes and dispersion instability could be accurately reproduced in a single week using the here outlined scaling method. The combination of fluid mechanic computation and corresponding analytical methods offers a time and cost effective lab-scale alternative for the development of automotive metallic aqueous paints with high shear stress stability.

Fig. 1: Scanning electron microscopy images of mica flakes (left). Upon the application of a critical stress the water-based paints become unstable resulting in the destruction of the flakes (right) and the corresponding inhomogeneity of the coloration.

See more of this Session: Complex-Fluid and Bio-Fluid Dynamics II
See more of this Group/Topical: Engineering Sciences and Fundamentals

Colloidal Sorting by Effective Density

Darrell Velegol, Department of Chemical Engineering, The Pennsylvania State University, University Park, PA

Joseph McDermott, Chemical Engineering, The Pennsylvania State University, University Park, PA

Cesar Gonzalez Serrano, Chemical Engineering, The Pennsylvania State University, University Park, PA

Separating one type of molecule from another - whether those molecules are petroleum-based or pharmaceutically-based for instance - is a classic chemical engineering problem. We might design distillation columns or membranes as the separating technology. But suppose now that we have a mixtures of colloidal particles in suspension, say particles ranging from 0.1 microns to 10 microns in diameter. How do we sort those particles from one another by size, shape, or density? Sedimentation separation seems like an obvious choice, since larger particles should settle more quickly than smaller particles. Shouldn't they? In this talk we examine how simple intuition has led our lab group astray several times, as well as how a Rayleigh-Benard instability limits the usual sedimentations - even density gradient sedimentations. We will also examine how we can change the effective density of particles by altering the thickness of the electrical double layer, leading to a simple but powerful colloidal particle sorting technique that takes advantage of the collective nature of the complex fluid.

See more of this Session: Complex-Fluid and Bio-Fluid Dynamics II
See more of this Group/Topical: Engineering Sciences and Fundamentals

Noninvasive Measurement of Yeast Cell Interactions Using Laser Tweezers Optical Trapping

Anne M. Grillet, Microscale Science and Technology, Sandia National Laboratories, Albuquerque, NM

Matthew D. Reichert, Microscale Science and Technology, Sandia National Laboratories, Albuquerque, NM

Timothy P. Koehler, Microscale Science and Technology, Sandia National Laboratories, Albuquerque, NM

Chistopher M. Brotherton, Microscale Science and Technology, Sandia National Laboratories, Albuquerque, NM

Interactions between cells are of great importance for many reasons including biofilm formation, transmission of infectious diseases and biofuel production. Optical tweezers have proven to be a powerful tool for the noninvasive measurement of electrostatic interaction forces between colloidal particles, and a great body of literature exists for interactions of various particles quantified through several analysis methods. Yeast cells serve as a living corollary to spherical colloids, given their relative shape and size and ability to maintain a surface charge. In addition, yeast flocculation is a hugely important and well studied phenomenon that has obvious impact for beer production. In this study, optical tweezers are used to trap two living yeast cells, bring them into contact, and observe the force required to separate the cells all the while simultaneously monitoring the interaction force between the cells. Using the direct force measurement technique, attractive forces of ~500 femtoNewtons were measured between two approaching cells. Eventually the cells touched and stuck to one another. The force required to separate the two cells was also measured.

Acknowledgements:

This work was supported by the Laboratory Directed Research and Development program at Sandia National Laboratories. Sandia is a multiprogram laboratory operated by Sandia Corporation, a Lockheed Martin Company for the United States Department of Energy's National Nuclear Security Administration under contract DE-AC04-94AL85000.

See more of this Session: Complex-Fluid and Bio-Fluid Dynamics II
See more of this Group/Topical: Engineering Sciences and Fundamentals

Diffuse-Interface Modeling of Phase Segregation in Amphiphilic Liquid Mixtures

Andrea G. Lamorgese, Chemical Engineering, City College of New York, New York, NY

Sanjoy Banerjee, Chemical Engineering, City College of New York, New York, NY

Adding a surfactant to a binary mixture of oil and water can produce many different complex structures on the mesoscale. Depending on the temperature and concentration of each component, the amphiphile can self-assemble and force the oil/water mixture into a number of different equilibrium structures. This process is controlled by the competing attractive/repulsive interaction among the species and can result in, for example, lamellar, gyroid or hexagonal arrangements of the oil and water domains. Although the equilibrium behavior of such amphiphilic systems is well understood, their rheology and the dynamics of self-assembly have been investigated to a lesser extent. These properties are important not only from a fundamental standpoint, but also because of their widespread applications. In this study we are motivated by an optimization of surfactant/de-emulsifier manipulation for improving the separation behavior during extraction and refining of heavy oil mixtures. To date, several models have been proposed for amphiphilic systems, classified in terms of type of approach: molecular dynamics, time-dependent Ginzburg-Landau (diffuse-interface) approaches, and membrane approaches. While molecular dynamics simulations can be useful for investigating structural properties of adsorbed amphiphiles, these simulations are unable to access the meso- and micro-scales of the self-assembly process. We also focus on diffuse-interface models since most theories on phase-separation dynamics are constructed using a mesoscale continuum theory which relates short-scale modification of hydrodynamics to molecular interactions. Although most mesoscale simulations of amphiphilic fluids have been perfomed using lattice-gas or lattice-Boltzmann schemes, these numerical models are limited to low accuracy and, in some cases, modeling uncertainties are introduced as a result of bottom-up approaches. Such simplifications make the resulting predictions, at best, qualitative and the models unpredictive. In this study, we extend our previous work on mixing and phase separation of binary mixtures to the modeling of ternary (or quaternary) liquid mixtures as regular solutions with a Flory-Huggins and Cahn-Hilliard representation of the excess and non-local components of the free energy of mixing. In this diffuse-interface model convection and diffusion are coupled via a non-equilibrium Korteweg force expressing the tendency of the ternary (or quaternary) system to minimize its free energy. Using a hybrid compact/pseudo-spectral discretization, the governing equations are integrated in a two- or three-dimensional periodic box and in a channel for simulating gravity-driven segregation of an initially phase-separated mixture at equilibrium. Subsequently we will present a comprehensive assessment of the model by comparing the numerical predictions with experimental results.4.192.17 on 5-11-2009-->

See more of this Session: Complex-Fluid and Bio-Fluid Dynamics II
See more of this Group/Topical: Engineering Sciences and Fundamentals

Flow Injection of Polymers Into Nanopores

O. Berk Usta, Chemical and Petroleum Engineering, University of Pittsburgh, Pittsburgh, PA

Anton Markejestein, Laboratory for Aero & Hydrodynamics, Delft University of Technology, Delft, Netherlands

Issam Ali, Department of Physics, Sultan Qaboos University, Al-Khodh

Anna C. Balazs, Department of Chemical and Petroleum Engineering, University of Pittsburgh, Pittsburgh, PA

Julia Yeomans, Theoretical Physics, University of Oxford, Oxford, United Kingdom

We use a mesoscale simulation to measure the strength of the velocity flux needed to push a polymer into a narrow channel. We find excellent agreement with the prediction by Sakaue et al [1], based on a de-Gennes blob model of the polymer, that the critical flux for translocation depends linearly on the temperature, but is independent of the length of the polymer chain or the width of the channel.

[1] T. Sakaue, E. Rapha¨el, P.G. de Gennes and F. Brochard-Wyart, Europhys. Lett. 72 83 (2005)

See more of this Session: Complex-Fluid and Bio-Fluid Dynamics II
See more of this Group/Topical: Engineering Sciences and Fundamentals

Nanoscale Ionic Materials (NIMs) : Slow Glassy Dynamics

Praveen Agarwal, Chemical and Biomolecular Engineering, Cornell University, Ithaca, NY

Haibo Qi, Chemical and Biomolecular Engineering, Cornell University, Ithaca, NY

Lynden Archer, Cornell University, Ithaca, NY

Nanoscale ionic materials (NIMs) are a recently discovered class of organic-inorganic hybrid materials which are able to relax to equilibrium in absence of any solvent. Linear rheology of these materials manifest classical traits of soft glasses, including a yield stress, slow dynamics and divergence of the viscosity. The frequency response of NIMs in the nonlinear shear regime reveals several heretofore unexplored features of soft glasses. In particular, we report that the dynamic response of NIMs at multiple, discrete strains can be superimposed to produce universal master curves spanning fifteen or more decades in time. This universal behavior, termed 'time strain superposition' (TSS), is analogous to time temperature superposition in many regards, including the fact that the shift factors obey a WLF-like relation. We discuss these findings using recent models for soft glasses, as well as older phenomenological models based on the free-volume concept.

See more of this Session: Complex-Fluid and Bio-Fluid Dynamics II
See more of this Group/Topical: Engineering Sciences and Fundamentals

Time Strain Superposition in Nanoparticle Organic Hybrids

Haibo Qi, Chemical and Biomolecular Engineering, Cornell University, Ithaca, NY

Praveen Agarwal, Chemical and Biomolecular Engineering, Cornell University, Ithaca, NY

Lynden Archer, Cornell University, Ithaca, NY

We report frequency-dependent shear rheology of dense suspensions of hairy nanoparticles. These materials mimic soft glass materials in many ways including a yield stress, slow dynamics and divergence of the viscosity. We find that measurements performed over a limited range of frequencies and varying shear strains can be superposed onto a master curve that spans more than 22 decades in frequency. SGR model predicts consistent result with experimental data only at strains lower than yield strain. For larger strains, the strain-dependent horizontal shift factor is Surprisingly fit well by the empirical WLF equation for amorphous polymers, with temperature replaced by strain.

See more of this Session: Complex-Fluid and Bio-Fluid Dynamics II
See more of this Group/Topical: Engineering Sciences and Fundamentals

Measurement and Prediction of the Mean Droplet Size in Water-in-Oil Emulsions

John Boxall, Chemical Engineering, Colorado School of Mines, Golden, CO

Jon Monserud, Chemical Engineering, Colorado School of Mines, Golden, CO

Carolyn A. Koh, Chemical Engineering, Colorado School of Mines, Golden, CO

E. Dendy Sloan, Chemical Engineering, Colorado School of Mines, Golden, CO

Amadeu K. Sum, Chemical Engineering, Colorado School of Mines, Golden, CO

David T. Wu, Chemical Engineering and Chemistry Departments, Colorado School Of Mines, Golden, CO

Droplet sizes were measured using a Particle Video Microscope (PVM) probe and a Focused Beam Reflectance Method (FBRM) probe for a variety of oils and shear rates. Two regimes in turbulent mixing flow were identified: i) the inertial sub-range where droplets are on the size scale of the smallest eddies in the turbulent mixing, and ii) the viscous sub-range in which droplets are smaller than the smallest eddies. A universal droplet size curve was generated using dimensionless numbers that included the Reynolds number, Weber number, and a non-dimensional droplet diameter. The curve showed good agreement with the PVM data and the transition from the inertial to viscous sub-range. The droplet size distribution for various oils matched a log-normal distribution using parameters (mean and standard deviation) predicted from correlations using the shear in the system, density, viscosity, and interfacial tension of the oil. The average difference between the surface area calculated from the PVM droplet size distribution and from the predicted log-normal distribution was 11.7%.

See more of this Session: Emulsions and Foams
See more of this Group/Topical: Engineering Sciences and Fundamentals

Electric Fields in Coflowing Liquids as a Route to Fine Emulsions

Venkata R. Gundabala, School of Physics, Georgia Institute of Technology, Atlanta, GA

Alberto Fernandez-Nieves, School of Physics, Georgia Institute of Technology, Atlanta, GA

Liquid-liquid emulsion droplets in the micron and sub-micron size range find applications in several fields such as cosmetics, drug delivery, and food industry. Microfluidics offers a promising route to produce these droplets by decoupling the droplet size from its composition. Here we propose a novel method for producing emulsion droplets in the sub-micron size range in a microfluidic device that couples electric and hydrodynamic forces. Electric fields are applied to a conducting liquid coflowing with an outer insulating liquid in a glass-based coflow geometry. We investigate the drop formation mechanism as a function of operating parameters such as the applied electric field strength, flow rates of the inner and outer liquids, and device geometry. Two modes of operation for the stable generation of emulsion droplets are observed: Cone-Jet mode and Whipping mode. Both the modes operate in a steady-state manner over long periods of time, allowing continuous generation of monodisperse emulsion droplets. For a given conductivity of the inner liquid, the smallest possible droplet diameter is obtained for small inner flow rates, high outer flow rates and high applied voltages. Measured currents from the experiments indicate a strong dependence on applied voltage and a weak dependence on inner flow rate indicating the charge transport mechanism to be predominantly conductive. We envisage this observation to be a consequence of the small size ratio of capillary tip to the jet diameter which prevents the establishment of convective regime. This result is in stark contrast with what is expected in an electrospray where the outer medium is air rather than a coflowing insulating liquid.

See more of this Session: Emulsions and Foams
See more of this Group/Topical: Engineering Sciences and Fundamentals

Stability and Interfacial Structure of Pickering Emulsions Stabilized by Nanoparticles

Danilo Pozzo, Chemical Engineering, University of Washington, Seattle, WA

Kjersta Larson, Chemical Engineering, University of Washington, Seattle, WA

Particle-stabilized emulsions (Pickering Emulsions) are formed when small particles adsorb at the interface between the oil and water phases. The focus of this study is in the interfacial structure and stability of oil in water emulsions that are stabilized by silica nanoparticles. These Pickering emulsions are examined using a combination of small-angle x-ray scattering, dynamic light scattering and microscopy. Small-angle x-ray scattering provides the ability to examine the particle arrangement at the oil-water interface in situ. The particle organization at the interface can be correlated to the stability and drop coalescence of the emulsions. Dynamic light scattering and microscopy are used to quantitatively relate the oil droplet size and macroscopic stability of the emulsion to changes in the nano-scale structure. Pickering emulsions of micron sized particles are known to be sterically stabilized by close-packed microparticles that form an elastic shell. In contrast, nanoparticles are not close-packed at the oil-water interface and the emulsion is found to be electrostatically stabilized.

See more of this Session: Emulsions and Foams
See more of this Group/Topical: Engineering Sciences and Fundamentals

Hetero-Aggregates as Particulate Emulsifiers and the Role of Surface Roughness

Adriana San Miguel, Chemical and Biomolecular Engineering, Georgia Institute of Technology, Atlanta, GA

Sven H. Behrens, Chemical and Biomolecular Engineering, Georgia Institute of Technology, Atlanta, GA

Particles adsorbed to the surface of emulsion droplets act as emulsifiers and stabilizers in so-called Pickering emulsions. Surface roughness has been proposed as one of the key parameters determining the particles' emulsifying properties, but some confusion persists about its true impact: although surface roughness reportedly lessens a particle's ability to stabilize emulsion droplets, weakly flocculated particles with their intrinsic aggregate roughness have been hailed as ideal Pickering emulsifiers. Here we report systematic studies on particles with tunable roughness. These particles consist of a micron-sized core decorated with oppositely charged polymer nanoparticles that can be dissolved by pH adjustment and adsorbed to the same core in the form of a smooth macromolecular film. Progressive dissolution of the adsorbed pH-responsive nanoparticles leads to continuous degrees of surface roughness while maintaining the same surface chemistry. We discuss the observed effect of varying particle roughness on wetting and emulsification properties.

See more of this Session: Emulsions and Foams
See more of this Group/Topical: Engineering Sciences and Fundamentals

Cyclohexane-in-Water Emulsion as a Sacrificial Template for the Synthesis of Multilayered Polyelectrolyte Microcapsules

Sachin Khapli, Department of Chemical and Biological Engineering, Polytechnic Institute of New York University, Brooklyn, NY

Jin Ryoun Kim, Department of Chemical and Biological Engineering, Polytechnic Institute of New York University, Brooklyn, NY

Jin Kim Montclare, Department of Chemical and Biological Sciences, Polytechnic Institute of New York University, Brooklyn, NY

Rastislav Levicky, Department of Chemical and Biological Engineering, Polytechnic Institute of New York University, Brooklyn, NY

Maurizio Porfiri, Department of Mechanical and Aerospace Engineering, Polytechnic Institute of New York University, Brooklyn, NY

Stavroula Sofou, Department of Chemical and Biological Engineering, Polytechnic Institute of New York University, Brooklyn, NY

We report the application of cyclohexane-in-water emulsions as novel sacrificial templates for the fabrication of hollow microcapsules through layer by layer assembly of polyelectrolytes, polystyrene sulfonate (PSS) and polyallylamine hydrochloride (PAH). We demonstrate that aqueous-dispersed cyclohexane droplets can be frozen at manageable temperature by taking advantage of the unusually high melting point of cyclohexane (6°C) and used as solid templates for the layer by layer assembly of polyelectrolytes. Ethanol-assisted removal of cyclohexane from emulsions coated with eleven polyelectrolyte layers leads to the formation of hollow microcapsules preserving the spherical geometry. Majority of the microcapsules (>90%) are prepared with intact polyelectrolyte membranes as measured by the osmotic pressure induced deformation of microcapsules. This work provides a new route for the synthesis of multilayered microcapsules under mild operating conditions and would enable the use of sensitive biomolecules as building blocks.

See more of this Session: Emulsions and Foams
See more of this Group/Topical: Engineering Sciences and Fundamentals

Superstabilization and Functionalization of Foams and Emulsions by Particles with Engineered Structure and Properties

Orlin D. Velev, Department of Chemical and Biomolecular Engineering, North Carolina State University, Raleigh, NC

Sejong Kim, Department of Chemical and Biomolecular Engineering, North Carolina State University, Raleigh, NC

Qixin Zhong, Department of Food Science and Technology, University of Tennessee, Knoxville, TN

Harry Barraza, Unilever Research & Development, Unilever, Bebington, United Kingdom

We will discuss how particles with anisotropic shape and controlled properties can be synthesized and used in the formation of Pickering foams and emulsions with new functionality. We have designed processes for the large-scale formation of rod-like polymer particles. These rod-like particles act as "superstabilizers" of foams and emulsions by forming rigid adsorption shells around the bubbles or droplets. The superstabilization effect is made possible by rod entanglement, formation of hairy shells, and sustaining of thick water films. Similar long-term stabilization was achieved with irregular particles from hydrophobically modified cellulose (HMC), formed by a water-only pH-jump precipitation process. Such particles can also be used to enrich the Pickering foams and emulsions with additional features such as color, specific rheological response, stability against drying and controlled release of components. We will discuss how the addition of pH-sensitive dyes during the process of formation of HMC particles leads to co-precipitation and strong adsorption of the dyes on the cellulose particle surfaces. These strongly colored HMC particles act not only as strong stabilizers of the foams, but also allow their intense and selective coloration without any coloring of the solution medium. The results illustrate how particle stabilizers can be used to make dispersion systems of unusual properties.

See more of this Session: Emulsions and Foams
See more of this Group/Topical: Engineering Sciences and Fundamentals

CO2-Soluble Surfactants for Enhanced Oil Recovery Mobility Control Via Thickening or in-Situ Foam Generation

Yee Soong, National Energy Technology Laboratory, U.S. Department of Energy, Pittsburgh, PA

Dazun Xing, Chemical and Petroleum Engineering Department, University of Pittsburgh, Pittsburgh, PA

Bing Wei, Chemical and Petroleum Engineering Department, University of Pittsburgh, Pittsburgh, PA

Robert M. Enick, Chemical and Petroleum Engineering Department, University of Pittsburgh, Pittsburgh, PA

Julian Eastoe, School of Chemistry, University of Bristol, Bristol, United Kingdom

Azmi Mohamed, School of Chemistry, University of Bristol, Bristol, United Kingdom

Kieran Trickett, School of Chemistry, University of Bristol, Bristol, United Kingdom

The very low viscosity of scCO2 diminishes the effectiveness of CO2 flooding of oil reservoirs. There are two strategies for using CO2-soluble compounds to decrease the mobility of supercritical carbon dioxide. The first involves the "direct thickening" of CO2, which is accomplished by dissolving an associative thickener in the scCO2 that forms viscosity-enhancing macromolecules in solution. The second strategy is to inject a CO2 surfactant solution into the porous media (which contains both brine ands oil) that will generate a low mobility system of CO2 droplets separated by surfactant-stabilized brine lamellae that bridge pore throats.

Direct thickening was accomplished with surfactants that formed cylindrical, rather than spherical, micelles in scCO2. The surfactants employed divalent cations (Ni, Co) rather than a monovalent cation (Na). Therefore, each surfactant had two tails (rather than one). Further, each tail was a double-tail or triple-tail that was tailored to be CO2-philic, consisting of either highly fluorinated alkanes or highly branched hydrocarbon groups. High pressure SANS was employed to establish whether the micelles were cylindrical or spherical. Further, the dimensions of the micelles were determined. Cloud point pressures of surfactant solutions (1-10wt% surfactant) were determined for the dry and wet (W = 0 - 15) systems using a non-sampling technique, and viscosity was determined using a falling cylinder technique. The CO2 viscosity was doubled using several weight percent of a fluorinated surfactant in the presence of water (viscosity results for the non-fluorous surfactants will also be presented). Direct thickening was also attempted with hydroxyaluminum disoaps. Although hydroxyaluminum di(2-ethyl hexanoate) is a prolific thickener for light alkanes at very dilute concentration, it is CO2-insoluble. Therefore the (2-ethyl hexyl) tails were replaced with highly branched, oxygenated hydrocarbon tails that have been shown to be CO2-philic when incorporated into other types of small compounds. Unfortunately, the aluminum disoaps tested to date have been CO2-insoluble, even in the presence of high concentrations of organic co-solvents.

With regard to our second technique, we have established the identity of several non-ionic, hydrocarbon-based, commercially available, inexpensive surfactants that can dissolve in CO2 at typical EOR reservoirs conditions to a high enough concentration (~0.01-0.1wt%) to form relatively stable emulsions/foams of liquid or supercritical CO2 droplets separated by films of brine. Although these surfactants are CO2-soluble, they are even more water-soluble, therefore they tend to form emulsions in which the brine is the low-volume (~15%) continuous phase. When mixed with dense CO2 and water, these surfactants can form a white emulsion of CO2 droplets that slowly collapses over a period of time encompassing tens of hours. The emulsions are less stable as the dissolved solids content of the brine (e.g. the concentration of NaCl) increases over the range of values associated with EOR projects. The performance of the best surfactants identified in this study was contrasted with that of the DOW Tergitol TMN surfactants previously identified by Keith Johnston of the University of Texas at Austin for this EOR application. The surfactants identified in this study appear to form more stable emulsions than Tergitol TMN, especially when tested with highly saline brine.

At this point in time, the use of CO_2-soluble nonionic surfactants to form in-situ foams (as opposed to the direct thickening) appears to be the more promising and commercially viable mobility control technology.

See more of this Session: Emulsions and Foams
See more of this Group/Topical: Engineering Sciences and Fundamentals

The Effect of Interfacial Slip On the Coalescence of Two Equal-Sized Drops in a Head-On Collision

Kostas Tsiglifis, Chemical engineering, University of California at Santa Barbara, Santa Barbara, CA

Arun Ramachandran, Chemical Engineering, University of California at Santa Barbara, Santa Barbara, CA

Anshuman Roy, Chemical engineering, University of California at Santa Barbara, Santa Barbara, CA

Gary Leal, University of Calif. at Santa Barbara, Santa Barbara, CA

A fundamental question that arises in the area of coalescence of two drops in a flow is the dependence on various parameters of the drainage time t_d, which is the time from the instant the drops collide to film rupture and coalescence. In a recent experimental investigation from our group [Hsu *et al.* (2008)], the drainage time t_d for two equal-sized drops undergoing a head-on collision was found to scale with capillary number Ca as $t_d \sim Ca^{m}$ for $10^{-4} < Ca < 10^{-2}$, where the scaling exponent *m* was always found to lie between 1 for low capillary numbers to 4/3 for high capillary numbers. Hsu *et al.* (2008) were successful in reasonably describing both limits of the exponent by scaling theories for a fixed viscosity ratio *l* between the droplet and suspending fluids. What they were unable to explain, however, was the curious dependence of the exponent *m* on *l*. For drop radii smaller than 27 microns, the exponent was observed to increase in experiments as *l* was raised from 0.19 to 6.8, a change which is not captured by the scaling theories. Coupling this dependence with an earlier observation (Park *et al.*, 2004) that flow-induced coalescence between drops is facilitated by increasing the molecular weight of the polymeric suspending fluid, we speculate that these changes occur because of the phenomenon of interfacial slip between the droplet and suspending fluids. Numerical simulations from our group have indicated that the droplets may come as close as a few nanometers before undergoing coalescence. Such thin film dimensions are comparable to the thickness of the diffuse interface between the polymeric drop and suspending fluids where variations in concentration and entanglement densities occur. Obviously, this invalidates our assumptions of zero-interfacial thickness and no-slip applied in both scaling theory and simulations. In this work, we examine the effect of interfacial slip on the dependence of the drainage time with capillary number for different viscosity ratios via boundary integral simulations. Interfacial slip is modeled via the Navier-slip condition, and the slip parameter employed in the simulations is predicted using the work of Goveas and Frederickson (1998). We show the boundary integral equations with the slip boundary condition can be converted to two Fredholm integral equations of the *second* kind, which can be solved iteratively to yield the stress and velocity fields on the interface at every instant of time. For the droplet coalescence problem, our results indicate that drainage time does decrease with the introduction of slip as expected. Also, the exponent *m* relating the dependence of drainage time with capillary number is greater in the presence of slip. These results are verified against the scaling analysis presented by Hsu *et al.* (2008). However, with the current parameters we have employed to describe the slip phenomenon, the changes are overpredicted, i.e. drainage times are shorter and the exponent *m* is greater than experimentally observed values. Possible reasons are explored.

References

Goveas, J. L. and G. H. Fredrickson, "Apparent slip at a polymer-polymer interface," *Eur. Phys. J. B* **2**, 79-92 (1998).

Hsu A. S., A. Roy and L. G. Leal, "Drop-size effects on coalescence of two equal-sized drops in a head-on collision," *J. Rheol.* **52** (6), 1291-1310 (2008).

Park, C. C., F. Baldessari, and L. G. Leal, "Study of molecular weight effects on coalescence: Interface slip layer," *J. Rheol.* **47**, 911-942 (2003).

See more of this Session: Interfacial Flows
See more of this Group/Topical: Engineering Sciences and Fundamentals

Drop Breakage in Sinusoidal Constrictions: Analysis, CFD Simulation and Experimental Validation

Wen Deng, Geological and Atmospheric Sciences, Iowa State University, Ames, IA

Igor Beresnev, Geological and Atmospheric Sciences, Iowa State University, Ames, IA

William Gaul, Chemical & Biological Engineering, Iowa State University, Ames, IA

R. Dennis Vigil, Chemical & Biological Engineering, Iowa State University, Ames, IA

Robert Ewing, Agronomy, Iowa State University, Ames, IA

Surface tension driven flow analysis in a core-annular geometry for arbitrary fluid viscosities leads to an evolution equation describing the temporal dynamics of the fluid-fluid interface. Its useful applications occur, for example, in environmental remediation and petroleum recovery. For sinusoidally constricted pores, a purely geometric criterion exists that enables or prohibits core fluid breakup in the neck of the constrictions. The geometrically favoring condition sets up capillary pressure gradients that ensure a continuous outflow of the core fluid from the neck into the "crests" of the profile. Such behavior is indeed observed in the numerical solutions of the evolution equation, in CFD simulations, and in optical experiments carried out in a sinusoidal flow cell. The drop breakup is typically achieved through the formation and growth of "wavy" disturbances from the initial interface profile, which touch the axis of the capillary in several places producing satellite drops. The final shapes of the profile just before the snap-off depend significantly on the slope of the initial configuration. Smaller slopes and thinner initial annuli slow down the breakup process. Finally, if the geometry prohibits the snap-off altogether, initial interface configurations decay into constant-radius surfaces. The main inferences of the analysis validate well (quantitatively) against CFD simulations and experiments.

See more of this Session: Interfacial Flows
See more of this Group/Topical: Engineering Sciences and Fundamentals

Deformation and Rotation of a Drop in a Uniform Electric Field

Paul Salipante, Thayer School of Engineering, Dartmouth College, Hanover, NH

James Hanna, Thayer School of Engineering, Dartmouth College, Hanover, NH

Petia M. Vlahovska, Thayer School of Engineering, Dartmouth College, Hanover, NH

Dynamics of drops in electric fields is a challenging problem because drop shape is not given a priori but is governed by the balance of interfacial, viscous and electric stresses. We present a theoretical and experimental study of drop behavior in a uniform electric field. We develop a theoretical analysis based on the Taylor's leaky dielectric model and small drop deformations that quantitatively describes the behavior of drops in DC fields, e.g. deformation and rotation. The theory is compared to experiments. Effects of surfactant are discussed.

See more of this Session: Interfacial Flows
See more of this Group/Topical: Engineering Sciences and Fundamentals

Folding Films, Inverse Coarsening, and the Bubble-Bursting Cascade

James C. Bird, School of Engineering and Applied Sciences, Harvard University, Cambridge, MA

Rielle de Ruiter, Wageningen University

Laurent Courbin, Institut de Physique de Rennes - UMR CNRS 6251

Howard A. Stone, Harvard University, Cambridge, MA

Bubble rupture is one of the principle mechanisms in which foams are assumed to coarsen, creating a smaller population of larger bubbles over time. Here, however, we demonstrate that for a large range of situations bubbles that rupture on an interface `inverse coarsen', leading to a larger population of smaller bubbles. We present high speed images and numerical simulations to demonstrate that when a bubble bursts the retracting film can fold and entrap air. The folding leads to a torus of entrapped air which breaks up into a ring of daughter bubbles. These results have broad implications for any process involving bubbles on an interface, including air-water transfer and interfacial foam coarsening.

See more of this Session: Interfacial Flows
See more of this Group/Topical: Engineering Sciences and Fundamentals

The Effect of Surfactants On the Breakup of An Axisymmetric Laminar Jet

Justin Walker, Chemical & Biomolecular Engineering, University of Maryland, College Park, MD

Richard V. Calabrese, Chemical & Biomolecular Engineering, University of Maryland, College Park, MD

The breakup of a laminar axisymmetric jet is a well studied fluid dynamics phenomenon. However, the impact of surface active agents on jet breakup has received limited attention. Many multiphase contactors function by forcing an immiscible fluid through an orifice or nozzle, or by otherwise destabilizing deformed droplets. Many of these processes can be mimicked on a fundamental level by the discharge of an immiscible axisymmetric laminar jet. Once exposed to the surrounding (matrix) phase, waves form on the surface of the jet causing it to destabilize. At a certain distance from the capillary tip (the breakup length), the jet disintegrates into a series of discrete droplets. Often, the jet breaks up into large primary drops and smaller satellite droplets, resulting in a multi-modal droplet size distribution. In this study, the Expectation-Maximization algorithm is used to decouple these modes and determine separate droplet size distributions for each mode.

Clean and surfactant-laden water jets were injected into otherwise still air; and silicone oils of various viscosity grades were injected into clean water and aqueous surfactant solutions. The capillaries used to form the jets have diameters from 200 to 800 microns and are long enough to assure fully developed flow at the capillary tip. Several non-ionic surfactants, all insoluble in silicone oils, were added in turn to the aqueous phase. It should be noted that our two series of experiments are quite different. In the first series, surfactant must diffuse from within the jet to the water-air interface. In the latter, surfactant must diffuse from within the surrounding matrix phase to the oil jet interface. However, surfactant transport to the interface always occurs within the aqueous phase.

Two distinct CCD cameras were used to image droplets. A low frame rate (30 fps), high resolution camera was used to measure breakup length and to quantify droplet population statistics. A high frame rate (5,000 fps), moderate resolution camera was employed to observe detailed breakup phenomena. An automated image processing algorithm was used to acquire droplet geometry and population statistics. To aid in data interpretation, interfacial tension and other relevant interfacial properties were acquired/estimated using both static and dynamic Pendant Drop techniques.

Since our last report a parametric study has been performed to explore a wide range of flow rates, capillary diameters, surfactant types, and surfactant concentrations in both the air-water and water-oil systems. We will report the results of this study, and demonstrate the effect of surfactant concentration, jet discharge flow rate, oil phase viscosity and capillary diameter on jet breakup length; and on primary and satellite droplet size distributions and relative population size. The data will be interpreted with respect to the relevant physicochemical phenomena.

See more of this Session: Interfacial Flows
See more of this Group/Topical: Engineering Sciences and Fundamentals

Stretching and Slipping of Liquid Bridges near Plates and Cavities

Shawn Dodds, Chemical Engineering and Materials Science, University of Minnesota, Minneapolis, MN

Marcio S. Carvalho, Mechanical Engineering, Pontificia Universidade Catolica do Rio de Janeiro, Rio de Janeiro, Brazil

Satish Kumar, Chemical Engineering and Materials Science, University of Minnesota, Minneapolis, MN

The dynamics of liquid bridges are relevant to a wide variety of applications including high-speed printing, extensional rheometry, and floating-zone crystallization. Although many studies assume that the contact lines of a bridge are pinned, this is not the case for printing processes such as gravure, lithography, and microcontacting. To address this issue, we use the Galerkin/finite element method to study the stretching of a finite volume of Newtonian liquid confined between two flat plates, one of which is stationary and the other moving. The contact lines are allowed to slip, and we evaluate the effect of the capillary number and contact angle on the amount of liquid transferred to the moving plate. At fixed capillary number, liquid transfer to the moving plate is found to increase as the contact angle on the stationary plate increases relative to that on the moving plate. When the contact angle is fixed and the capillary number is increased, the liquid transfer improves if the stationary plate is wetting, but worsens if it is non-wetting. The presence of a cavity on the stationary plate significantly affects the contact line motion, often causing pinning along the cavity wall. In these cases, liquid transfer is controlled primarily by the cavity shape, suggesting that the effects of surface topography dominate over those of surface wettability. At low capillary numbers, bridge break-up can be understood in terms of the Rayleigh-Plateau stability limit, regardless of the combination of contact angles or the plate geometry. At a high capillary numbers, the bridge is able to stretch beyond this limit, though the deviation from this limit appears to depend on contact line pinning, and not directly on the combination of contact angles or the plate geometry.

See more of this Session: Interfacial Flows
See more of this Group/Topical: Engineering Sciences and Fundamentals

Fluid Mechanics of Rinsing Flows

Tienyi Theresa Hsu, Chemical Engineering, Stanford University, Stanford, CA

Gerald G. Fuller, Chemical Engineering, Stanford University, Stanford, CA

Curtis W. Frank, Chemical Engineering, Stanford University, Stanford, CA

The fluid mechanics of "rinsing flows", describing the use of a jet of one liquid or air to push away a second, coating liquid on a rigid flat surface, has been investigated. This phenomenon is common to everyday life but despite this, few studies have considered the rinsing flow process in detail. One experimental challenge is that the timescale of the process is short and does not reach a steady state. The present study has employed high speed imaging devices allowing the examination of the initial contact between the two fluids.

It is found that these flows present interesting variations to the classic hydraulic jump with a variety of new phenomena due to the presence of the second, coating layer. For example, a systematic study varying the viscosity ratio between the two fluids reveals a Taylor-Saffman instability at the moving interface between fluid contacts when the viscosity ratio between the coating fluid and the jet fluid is higher than ~25. The hydraulic jump circle size is also a strong function of the viscosity ratio and exhibits a power-law relationship.

In addition, the effect of the viscoelasticity of the coating fluid has been investigated in response to either water or air jets. Three classes of fluids with the same zero-shear viscosity have been investigated: glycerin-water Newtonian liquids, polymeric solutions, and worm-like micelle surfactant solutions. Analysis of the high speed videos reveal that the evolution of the hydraulic jump circle sizes have very different characteristics. For example, for water jet experiments, Newtonian liquids show an inverse exponential-like growth in circle size while polymeric solutions show a linear growth and for surfactant solutions, the circle size first increases then recoils due to the high elasticity of the fluid. In addition, it is observed that the Non-Newtonian liquids are able to suppress surface roughness in the hydraulic jump region due to a coupling of the extensional viscosities of these fluids and stagnation point flows in the vicinity of the hydraulic jump. Both the degree of roughness suppression and hydraulic jump circle size increase with the extensional viscosity of the solution.

The effect of altering the hydrophobicity of the flat surfaces on the efficiency of the rinsing process has also been investigated using scanning electron micrographs and Auger spectroscopy. Under the same conditions, the efficiency is affected by the wettability of the rinsing fluid on the surface and is a function of the radial position from the jet.

See more of this Session: Interfacial Flows
See more of this Group/Topical: Engineering Sciences and Fundamentals

Adhesive Dynamics of Lubricated Membranes

Fong Yew Leong, Large-Scale Complex Systems, Institute of High Performance Computing, Singapore, Singapore

Membrane waves have been observed near the leading edge of a motile cell. Such phenomenon is the result of the interplay between hydrodynamics and adhesive dynamics. In this study, we model membrane behavior on a thin fluid gap supported by adhesive bonds, using lubrication theory and adhesive dynamics. The model yields traveling wave solutions to four adhesion scenarios, namely no adhesion, uniform adhesion, clustered adhesion and focal adhesion. The resulting traveling waves are observed to depend strongly on adhesion and they could be classified as either tension or adhesion waves. In contrast to tension waves, adhesion waves show characteristically amplified peaks and are highly persistent. The transition of tension to adhesion waves is found to depend on a necessary but insufficient criterion. Further analysis shows that strong adhesion leads to dispersion in tension waves but amplification in adhesion waves, resulting in sharp tension to adhesion wave transition points. The present work could explain the strong persistence of the waves observed in adhered cells using differential inference contrast (DIC) microscopy and the observation that the wavelengths decrease shortly after leading edge retraction.

See more of this Session: Interfacial Flows
See more of this Group/Topical: Engineering Sciences and Fundamentals

Acoustically-Driven Programmable Liquid Motion Using Resonance Cavities

Sean M. Langelier, Chemical Engineering, University of Michigan, Ann Arbor, MI

Dustin S. Chang, Chemical Engineering, University of Michigan, Ann Arbor, MI

Ramsey Ibrahim Zeitoun, Chemical Engineering, University of Michigan, Ann Arbor, MI

Mark A. Burns, Chemical Engineering, University of Michigan, Ann Arbor, MI

In the field of microfluidics, device performance and utility are often overshadowed by difficulties and costs associated with operation and control. As a step toward the development of a more suitable platform for microfluidic control, we present a novel distributed pressure control scheme whereby many independently tunable pressure sources can be controlled using the energy present in acoustic waves. We demonstrate how this scheme can be utilized to perform precise droplet positioning, as well as merging, splitting, and sorting within open microfluidic networks. We further show how this scheme can be implemented for control of continuous flow systems, specifically for generation of acoustically tunable liquid gradients. Device operation hinges on a resonance-decoding and rectification mechanism by which the frequency content in a composite acoustic input is transduced into multiple independently buffered output pressures. The device consists of a bank of four uniquely tuned resonance cavities (404 Hz, 484 Hz, 532 Hz and 654 Hz respectively) each being responsible for the actuation of a single droplet, four identical flow-rectification structures, and an acoustic source. Subject to a computer generated acoustic input, cavities selectively amplify resonant tones from the input signal resulting in highly elevated local cavity pressures. Fluidic rectification structures then serve to convert the elevated oscillating cavity pressures into unidirectional flows. The resulting pressure gradients, which are used to manipulate fluids in a microdevice, are tunable over a range of approximately 0-200 Pa with a control resolution of 10 Pa.

See more of this Session: Interfacial Flows
See more of this Group/Topical: Engineering Sciences and Fundamentals

Probing the Dynamics of Microscale Bubble/Droplet Flows with On-Chip Pressure Sensors

Siavash Aslanbeigi, Division of Engineering Science, University of Toronto, Toronto, ON, Canada

Lian Leng, Department of Mechanical and Industrial Engineering, University of Toronto, Toronto, ON, Canada

Axel Günther, Department of Mechanical and Industrial Engineering, Institute of Biomaterials and Biomedical Engineering, University of Toronto, Toronto, ON, Canada

Segmented microflows have found a wide range of applications and allowed to translate chemical and biological processes into continuous-flow formats while minimizing axial dispersion. Most previous studies have relied on optical micrographs of the fluid phase distribution in describing the flow characteristics.

Instead, our contribution focuses on the pressure that is expected to display a characteristic, Capillary-number dependent, spatiotemporal behavior. The characteristic behavior of the pressure signal is important during bubble/droplet break-up and during steady flow along the channel. Estimates of how the pressure varies in these two cases were so far based on analytical or numerical approaches and involve several assumptions: smooth walls, constant interfacial tension and complete wetting. Accurately measuring the dynamic behavior displayed by the pressure in bubble/droplet microchannel flow, therefore, provides for a very sensitive tool to discriminate between flow regimes, determine bubble/droplet velocities and - in multichannel arrangements - achieve flow synchronization.

We integrated piezoresistive pressure transducers in soft lithographically patterned microfluidic devices. On-chip integration significantly reduced the available dead volume and removed any unwanted bubbles in the sensing channel. The integrated pressure sensors were calibrated and electrically amplified. Dynamic pressure measurements at millisecond time resolution and a sensitivity exceeding the capillary pressure by at least one order of magnitude were obtained. The presented results were obtained for segmented gas-liquid and liquid flows at different Capillary numbers (0.001-0.05) and wetting conditions.

See more of this Session: Interfacial Flows
See more of this Group/Topical: Engineering Sciences and Fundamentals

Supercritical Fluid Processing of Nafion® Membranes: Methanol Permeability and Proton Conductivity

Juan C. Pulido, Chemical Engineering, University of Puerto Rico - Mayaguez, Mayagüez, PR

David Suleiman, Chemical Engineering, University of Puerto Rico - Mayaguez, Mayagüez, PR

Nafion® membranes commonly used in direct methanol fuel cells (DMFC), are typically limited by high methanol permeability (also known as the cross-over limitation). These membranes have phase segregated sulfonated ionic domains in a teflon backbone, which make processing challenging and limited by phase equilibria considerations. This study used supercritical fluids (SCFs) as a processing alternative, since the gas-like mass-transport properties of SCFs allow for better penetration into the membranes and the use of polar co-solvents could also influence their morphology, fine-tuning the physical and transport properties. This investigation focused on the effect of SCF processing using SCF CO2 with and without different co-solvents. A variety of co-solvents were used including: acetic acid, acetone, acetonitrile, cyclohexanone, ethanol, isopropanol, methanol, methylene chloride, and tetrahydrofuran. They were selected based on their size, polarity, and chemical nature (e.g., protic or aprotic). Measurements of methanol permeabilities were made using an FT-IR technique, monitoring the methanol concentration with time through the membrane, while proton conductivity was measured using impedance spectroscopy. SCF processing reduced the methanol permeability of Nafion® as much as two orders of magnitude with some co-solvents, while proton conductivity was maintained. In addition, measurements were sensitive to the processing orientation (e.g., normal vs parallel flow). Dynamic mechanical analysis (DMA) revealed an increase in the relaxation around room temperature, which could be an indication of the increased phase segregation after the SCF processing.

See more of this Session: Materials Synthesis and Processing with Compressed or Supercritical Fluids II: Polymers
See more of this Group/Topical: Engineering Sciences and Fundamentals

Gelation, Crytallization and Polymorphic Transformations of Syndiotactic Polystyrene in Acetophenone + Carbon Dioxide Mixtures at High Pressures

Jian Fang, Chemical Engineering, Virginia Tech, Blacksburg, VA

Erdogan Kiran, Virginia Tech, Blacksburg, VA

Syndiotactic polystyrene (s-PS) is known to have five polymorphic states, namely the alpha, beta, gamma, delta, and epsilon forms. The alpha- and beta-forms are typically obtained from melt. They are in planar zigzag chain conformations. While the gamma, delta and epsilon forms are normally formed in the presence of solvents. They are in the 21 helical chain conformations.

How the polymorphic states can be modulated in dense fluids is of current interest. In this presentation, we will report on the gelation and crystallization processes and the polymorphic states that evolve in solutions of s-PS in acetophenone + carbon dioxide fluid mixtures at high pressures.

In pure acetophenone only a gelation phase boundary is observed. Solutions containing 25 wt % CO2 display a sol-gel and a liquid-liquid phase boundary. Solutions containing 40 wt % CO2 display a crystallization and a liquid-liquid phase boundary. It is shown that when the gelation boundary is crossed first, the resulting crystal structure is the ä form. When the L-L phase boundary is crossed first, polymer-rich phase leads to a mixture of "delta + beta" crystal forms, while only the delta-form is found in the polymer-lean phase from solutions with 25 wt % CO2. In solutions with 40 wt % CO2, crossing the crystallization boundary first, leads to the beta form with no presence of the delta crystal form. However, when the liquid-liquid phase boundary is crossed first, the polymer-rich phase forms only the beta crystal form and the polymer-lean phase leads to a mixed crystal structure containing the beta and delta forms.

Formations of the polymorphs were verified by XRD, FTIR and SEM. These observations were then used to propose a phase diagram which provides a rational picture of the consequences of phase separation paths.

See more of this Session: Materials Synthesis and Processing with Compressed or Supercritical Fluids II: Polymers
See more of this Group/Topical: Engineering Sciences and Fundamentals

Specific Interactions in CO_2 + Polymer Systems

Amyn S. Teja, Georgia Institute of Technology, Atlanta, GA

Yanhui Yuan, Chemical and Biomolecular Engineering, Georgia Institute of Technology, Atlanta, GA

It is well known that polymer solubility in CO_2 can be enhanced by the incorporation of carbonyl, silicone, or fluorine groups that exhibit specific interactions with CO_2. More recently, nanoporous polymers for the storage of H_2 and CO_2 have been prepared by CO_2 foaming in localized nanocells within CO_2-philic domains. Knowledge of specific interactions between CO_2 and CO_2-philic groups is essential in the design of these nanoporous polymers. We have developed a compressible lattice model for representation of phase equilibria in CO_2 + polymer systems in which specific interactions play an important role in determining phase equilibria. The model explicitly accounts for weak complex formation in these systems, using two parameters that can be obtained from Fourier transform infrared spectroscopy measurements or by correlation of experimental phase equilibrium data. We have compared our quantified specific interactions with quantum calculations. Furthermore, we show that our model can be used in both low-pressure and high-pressure systems, and is also able to predict glass transition temperatures in polymers swollen by CO_2. This ability has allowed us to fabricate nanoporous polymeric materials using supercritical CO_2 as a blowing agent.

See more of this Session: Materials Synthesis and Processing with Compressed or Supercritical Fluids II: Polymers
See more of this Group/Topical: Engineering Sciences and Fundamentals

Initial Stage of Foaming of PMMA/TMOS/CO_2 Systems

Katsuto Otake, Faculty of Engineering, Department of Chemical Engineering, Tokyo University of Science, Tokyo, Japan

Masaaki Tanaka, Tokyo University of Science, Tokyo, Japan

Takashi Kasahara, Tokyo University of Science, Tokyo, Japan

Tomoki Takahashi, Tokyo University of Science, Tokyo, Japan

Atsushi Shono, Tokyo University of Science, Tokyo, Japan

Takahide Kameda, C. I. Kasei CO., LTD., Tokyo, Japan

Hirobumi Matsumoto, C. I. Kasei CO., LTD., Tokyo, Japan

Satoshi Yoda, Nanotechnology Research Institute, National Institute of Advanced Industrial Science and Technology (AIST), Tsukuba, Ibaraki, Japan

Takeshi Furuya, National Institute of Advanced Industrial Science and Technology (AIST), Tsukuba, Ibaraki, Japan

In recent years, for the effective energy usage, development of high performance heat insulator is the subject of research for engineers. A poly(methyl methacrylate) (PMMA) foam that inside of the foam filled with silica aerogel is regarded as one of the promising candidate for high performance heat insulator. This composite material will be commercially produced with a supercritical carbon dioxide (CO_2) assisted extruding of the mixture of the PMMA and tetramethoxysilane (TMOS). In this work, for the basic understanding of the foaming of the PMMA/TMOS/CO_2 mixtures, in situ observation of the initial stage of the foaming was observed with new type of view cell that was developed in our laboratory.

It was found that with the addition of the TMOS in the PMMA/CO_2 system, the pressure drop which is necessary for the foam formation became monotonously larger with the increase in the amount of TMOS. On the other hand, the mean bubble diameter once decreased with the addition of the TMOS, and increased with the further increase in the amount of TMOS. The dependence of foam density of the amount of TMOS was vice versa. These facts were presumably due to the change of interfacial tension of PMMA/TMOS/CO_2 mixture and CO_2, and of the viscosity of the mixtures. Further experiments should be conducted for the full understandings of the foaming mechanism of this system.

See more of this Session: Materials Synthesis and Processing with Compressed or Supercritical Fluids II: Polymers
See more of this Group/Topical: Engineering Sciences and Fundamentals

Characterization of Linear Low-Density Polyethylene (LLDPE) and Maleated Linear Low-Density Polyethylene (LLDPE-g-MA) Infused with Nanoclay in Supercritical Carbon Dioxide

Matthew Factor and Sunggyu Lee

Department of Chemical & Biological Engineering
Missouri University of Science & Technology
Rolla, MO 65409-1230

Abstract

Nanoclay fillers have the ability to enhance the thermo-mechanical, barrier and flame resistant properties of Linear Low-Density Polyethylene (LLDPE). Although the difficulties with the use of nanoclay fillers lie in the completeness of exfoliation and also in the extent of dispersion into a polymer matrix in order to achieve a maximum level of property enhancement. One method employed to successfully disperse powders into polymers that is both inexpensive and deemed environmentally friendly is supercritical carbon dioxide processing. With supercritical carbon dioxide processing, Cloisite 93A was infused into LLDPE and Maleated LLDPE (LLDPE-g-MA). Upon a visual inspection, clay infusion was achieved near the surface of the polymer pellet, but no clay found its way to the interior of the pellet. And the combination of FTIR and XRD analysis suggests that a high degree of exfoliation was achieved for select runs. This work describes a method to determine the amount of nanoclay infused into a polymer via FTIR analysis. Using this method, the percentage of Cloisite 93A infused into LLDPE and LLDPE-g-MA was determined. Along with dispersion, supercritical carbon dioxide has been used to exfoliate nanoclays, lending to the possibility of the combination of exfoliation and dispersion into a single step.

Introduction

Clay-nanocomposites have received much attention in recent years due to the low clay loadings required to increase thermo-mechanical, barrier and flame resistant properties of polymers[1-5]. In addition, clays have the advantage of being environmentally benign and inexpensive fillers[4,6]. When clays are exfoliated, their property enhancement of polymers far exceeds that of agglomerated clays[1,2,6]. And even greater property enhancement can be achieved if the exfoliated clays are uniformly distributed throughout the polymer matrix. But herein lies the problem as difficulties with clay fillers exist in both their exfoliation and uniform distribution[1,7].

The processing of nanoclays in supercritical carbon dioxide ($scCO_2$) is one method that has been successfully used to exfoliate nanoclays[1]. Exfoliation is accomplished when the molecules of carbon dioxide under supercritical conditions find their way between the clay platelets and upon depressurization, break up the clay stacks into individual platelets[1]. This helps to increase property transfer from the increased surface area interaction between filler and host polymer[1,2,6]. In addition to exfoliating nanoclays, carbon dioxide is an environmentally friendly solvent that has the ability to swell polymers to allow for infusion and dispersion of powders[1,8] such as powdered clay.

Another possible solution to exfoliation and dispersion is to use modified clays or polymers. The modifications make them more compatible with each other, aiding in the exfoliation and distribution of clay[1,3,7]. A higher level of compatibility also increases the efficiency of property transfer from the clay to the polymer. But even with the benefits of modified clays and polymers, the maximum theoretical property enhancement for polymer/clay nanocomposites has eluded researchers.

The experiments outlined in this paper employ both the advantages of supercritical carbon dioxide processing and modified clays and polymers to aid in the exfoliation, infusion and dispersion of clay throughout a polymer matrix. Along with this, a technique is proposed and employed to determine the weight percentage of clay infused into a given polymer. First the infusion of nanoclay Cloisite 93A into linear low-density polyethylene (LLDPE) and linear low-density polyethylene grafted with maleic anhydride (LLDPE-g-MA) is confirmed and quantified in this current work.

Experimental

Materials

Cloisite 93A was obtained from Southern Clay Products, Inc. A quaternary ammonium salt (methyl, dehydrogenated tallow ammonium) is used to modify natural montmorillonite to create Cloisite 93A. The modifier concentration is 90 meq/100g of clay. Ninety percent of the particles by volume are less than 13μm, 50% are less than 6μm, and 10% are less than 2μm. The density of the clay is 1.88 g/cm^3. All clay is dried at a temperature around 80°C in an oven for at least 24 hours before processing. Cloisite 93A was chosen due to the fact that it has a high affinity for carbon dioxide as compared to other modified clays[1].

The polymers used in this experiment were linear low-density polyethylene (LLDPE, Dowlex 2517) and linear low-density polyethylene grafted with maleic anhydride (LLDPE-g-MA, Polybond® 3109). The LLDPE is in pellet form and has a melting point of 124°C, a specific gravity of 0.917 and a melt flow index of 25 g/10 min. The LLDPE-g-MA is also in the form of pellets with a density of 0.926 g/cm^3 and a melt flow index of 30 g/10 min. The graft level of maleic anhydride is 1 wt-% in LLDPE-g-MA and the melting point is 123°C. LLDPE-g-MA was chosen for its increased compatibility with clay over pure LLDPE[3,7].

Processing Procedure

The polymer/clay mixture is prepared by placing 10 grams of polymer and 2.5 g of Cloisite 93A in a stainless steel thimble and mechanically mixing until the polymer is thoroughly coated with the clay. The thimble is then put inside a 300-cc stainless steel autoclave reactor with a magnetically driven impeller. To aid in maintaining a constant temperature, cooling water is sent through the head of the reactor. During heating, to help maintain a uniform reactor temperature, the impeller is driven at around 300 rpm. An experiment is considered to reach its starting time once the desired temperature and pressure are attained. All experiments have a run time of 3 hours and are carried out in a batch mode. After an experimental run is complete, the reactor is depressurized and the samples are removed. The processing conditions are listed in Table 1.

Table 1. Processing Conditions in $scCO_2$ for Polymer-Clay Nanocomposites

Run	T (°C)	P (MPa)	Wt. Ratio (Clay:Polymer)
1	60.0	10.3	0.25 : 1
2	98.9	10.3	0.25 : 1
3	60.0	17.2	0.25 : 1
4	98.9	17.2	0.25 : 1

Characterization

Fourier Transform Infrared Analysis (FTIR) was performed on the samples to determine the percentage of clay infused into the polymer. The samples were analyzed with a Nicolet Nexus 470 FTIR E.S.P. Spectrometer with a resolution of 4 cm^{-1}, an average number of scans of 4, a scanning wavenumber ranging from 400 cm^{-1} to 4,000 cm^{-1}, and an absorption path length of 1 cm. The computer program used to collect the data was Omnic E.S.P. 5.1. Spacing between data was 1.928 cm^{-1}.

In order to determine the weight percent of Cloisite 93A in the samples, calibration standards for FTIR analysis were made by using a batch mixer to create specimens that contained 0, 0.05, 0.10, 0.15, 0.20, and 0.25 wt-% of Cloisite 93A in LLDPE. Standards containing 0, 1, 5, and 13 wt% Cloisite 93A were created using an extruder and a pelletizer. The extruded standards were used initially to approximate the amount of Cloisite 93A in the samples. Whereas the standards created in the batch mixer were employed to precisely quantify minute amounts of clay in the samples. To use the FTIR to analyze the infused polymer, samples were pressed into films that ranged between 50 and 90 μm in thickness for LLDPE/Cloisite 93A and 70 and 180 μm in thickness for LLDPE-g-MA/Cloisite 93A. Films were prepared by pressing polymer pellets between two brass plates at a pressure of 10,000 psi and a temperature of 185°C for 40 to 55 s.

Sample spectra obtained from FTIR analysis in conjunction with the calibration curve were used to determine the amount of Cloisite 93A infused into the polymer. To accomplish this, the following technique adapted from Clark et al.[9] was employed for calculating weight percentage from the absorbance spectra. Absorbance for small concentrations of IR absorbing species is approximated by the Beer-Lambert law:

$$A = \log\left(\frac{I_O}{I_T}\right) = \sum_i \left(K_i \delta C_i\right) \qquad (1)$$

where A, I_O, I_T, K_i, δ, and C_i are the absorbance, intensity of the incident IR beam, intensity of the transmitted IR beam, absorbance coefficient, absorbing layer thickness, and absorbing species concentration, respectively. In this technique, an absorbance equation with contributions from the clay and polymer is combined with an absorbance equation with only a polymer contribution. The Beer-Lambert law used to convey the absorbance equation for a nanocomposite of polymer (LLDPE or LLDPE-g-MA) and Cloisite 93A is

$$A_1 = K_1 \delta C_{93A} + K_2 \delta C_{Polymer} \tag{2}$$

where C_{93A} and $C_{Polymer}$ are the concentrations of Cloisite 93A and the polymer (either LLDPE or LLDPE-g-MA), respectively. The absorption of only the polymer is represented as

$$A_2 = K_3 \delta C_{Polymer} \tag{3}$$

In order to remove the absorbing layer thickness from the calculation and separate the polymer absorbance contribution from the clay absorbance contribution, a peak that had an absorbance contribution from both LLDPE and Cloisite 93A was divided by a peak that only had an absorbance due to LLDPE. This resulted in the linear equation

$$\frac{A_1}{A_2} = \left(\frac{K_1}{K_3}\right)\frac{C_{93A}}{C_{Polymer}} + \left(\frac{K_2}{K_3}\right) \tag{4}$$

The absorption peak attributed to clay within the polymer was found at 522 cm^{-1} on the spectra[10] and is shown in Figure 1. The absorption peak solely dedicated to the polymer (either LLDPE or LLDPE-g-MA) was found at 2019 cm^{-1} on all spectra, shown in Figure 2. The absorbance for each peak was determined by integrating the area under the peak. Using the linear absorbance equation above and the spectra obtained from the polymer/clay standards (Figure 1), a calibration curve was regressed which is shown along with a corresponding square of the Pearson autocorrelation coefficient in Figure 3. Upon analysis, the absorbance ratios from the sample spectra were used with the calibration curve to calculate the weight percentage of Cloisite 93A infused into the polymer. Two calibration curves were used, one for samples with clay percentages ranging from 0 to 0.25 wt-% and another for samples with clay percentages above 0.25 wt-%.

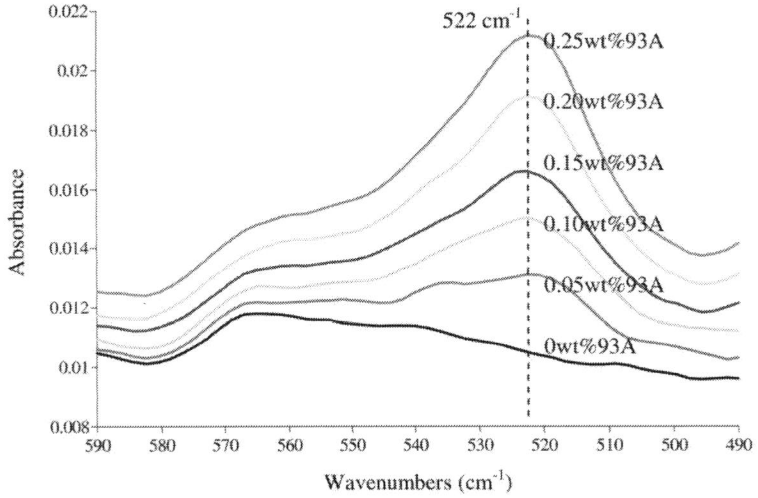

Figure 1. LLDPE/93A Standard Spectra with Absorbance Peak 522 cm^{-1}

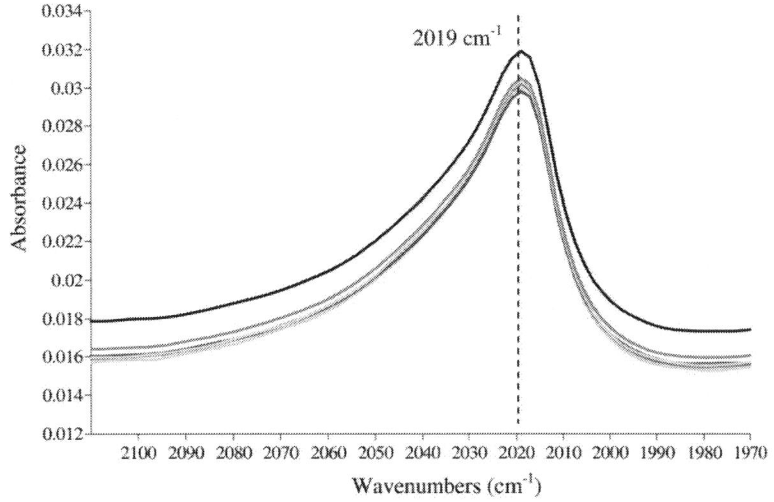

Figure 2. LLDPE/93A Standard Spectra with Absorbance Peak 2019 cm^{-1}

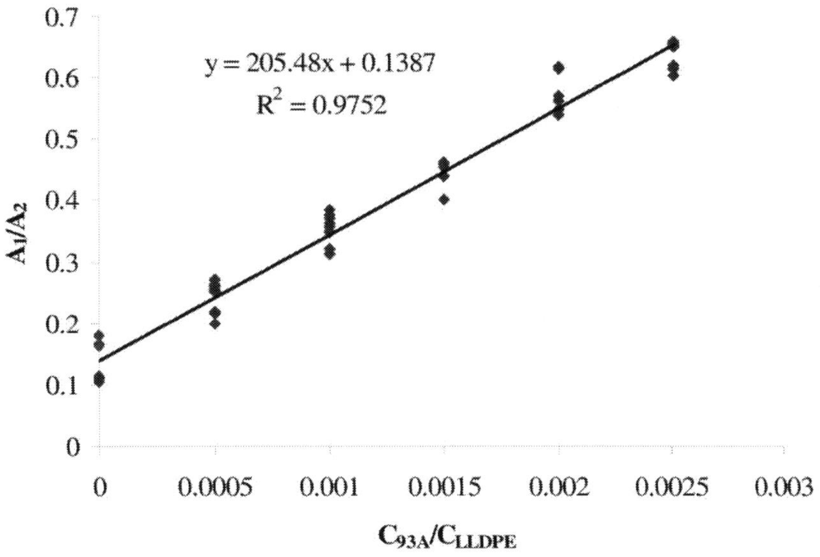

Figure 3. Calibration Curve from LLDPE/Clay Nanocomposites with Various Clay Loadings

Results and Discussion

Nanocomposite samples of LLDPE/Cloisite 93A and LLDPE-g-MA/Cloisite 93A were analyzed using Fourier Transform Infrared (FTIR) Analysis. First, pellets were cleaned of residual clay on their surface to ensure the only clay detected was infused clay.

The clay infused from the outer polymer surface inwards. No clay was observed in the center of the pellet and the majority of the clay remained near the polymer surface. Since clay infusion was not uniform throughout the pellets, the polymer/clay samples were melted, mixed and then pressed into films for FTIR analysis. The mixing of the melted nanocomposites is meant to more uniformly distribute the clay throughout the polymer so an accurate bulk clay percentage in the polymer may be obtained.

Nearly one hundred separate spectra were analyzed to determine the maximum and minimum weight percentages of Cloisite 93A infused into LLDPE and LLDPE-g-MA. The analysis revealed that infusion of Cloisite 93A was accomplished for LLDPE ranging from 0 to 0.16 wt-% Cloisite 93A. Clay was also infused into LLDPE-g-MA in the range of 0.04 to 1.27 wt-% Cloisite 93A. These results show that LLDPE-g-MA has a higher potential than LLDPE for infusing a larger amount of Cloisite 93A at the conditions employed. The grafted maleic anhydride enables a greater infusion of clay since it increases the compatibility between the polymer and clay[3,7].

As seen in Table 2, the low pressure and low temperature experimental run appears to have the highest average infusion of clay into LLDPE. However, due to the high standard deviation in comparison to the average clay infusion, a correlation cannot be confidently drawn between the effects of temperature and pressure. LLDPE-g-MA had its largest clay infusion using a high temperature and high pressure processing environment, listed in Table 3. As suggested by the overlapping clay infusion percentages of both high temperature runs, the high temperature runs for LLDPE-g-MA could produce nanocomposites with equal amounts of clay. The processing conditions with the least ability to infuse clay into LLDPE-g-MA and LLDPE employed the combination of high pressure and low temperature. It may be that the high pressure decreases the mobility of the clay particles and thus the ability of the clay to infuse into the polymer. The high pressure hindrance seems to be overcome in a high temperature environment possibly because the increased heat lowers the fluid density and further softens the polymer, facilitating infusion. In tables 2 and 3 are the average weight percentage and standard deviation of Cloisite 93A infused into LLDPE and LLDPE-g-MA, respectively.

Table 2. Average Weight Percentage and Standard Deviation of Cloisite 93A in LLDPE

T (°C)	P (MPa)	Avg (wt%)	Std Dev (wt%)
60	10.3	0.08	0.04
98.9	10.3	0.05	0.04
60	17.2	0.04	0.06
98.9	17.2	0.07	0.03

Table 3. Average Weight Percentage and Standard Deviation of Cloisite 93A in LLDPE-g-MA

T (°C)	P (MPa)	Avg (wt%)	Std Dev (wt%)
60	10.3	0.48	0.04
98.9	10.3	0.73	0.20
60	17.2	0.07	0.02
98.9	17.2	0.83	0.24

Though the aim of mixing is uniform clay dispersion, it was not achieved in most cases as evidenced visually by white specs in the film and analytically by the large standard deviation present throughout a single sample. This may be due to polymer pellets within the same processing environment being infused with different amounts of clay depending on their location in the reactor. Further work with extended processing times may rectify this. To some extent, the large standard deviations in clay percentages are the cause of defects in the film samples which resulted in oscillations or slight shifts in the spectra. The film defects appear to come from scratches and divots in the brass plates that are transferred to the film during its preparation. In the future, this may be avoided by pressing the polymer samples into films with metal plates free of scratches and divots.

Conclusion

The polymers LLDPE and LLDPE-g-MA were successfully infused with Cloisite 93A in supercritical carbon dioxide and a method was proposed to determine the weight percentage of clay within the nanocomposite. The amount of clay infused in LLDPE is similar for all four of the processing conditions investigated. This result is likely due to the difficulties of infusing clay into LLDPE and also since the infusion amount was small, errors in the spectra and calibration curve make it difficult to differentiate effects of processing conditions. Maleated LLDPE was determined to have a higher capability for clay infusion compared to LLDPE as evidenced by the greater amount of clay present within the sample. When using clay fillers, the maleic anhydride modification also makes maleated LLDPE superior to LLDPE since it has a higher compatibility with the clay[3,7], allowing for a more efficient property transfer from clay to polymer. High pressure combined with low temperature seemed to be the least desirable environment for infusing Cloisite 93A into either polymer. In the future, other methods for preparing samples for FTIR analysis will be investigated in an attempt to create films with a more uniform clay distribution to get a more accurate measure of clay infusion. Furthermore, the infusion process has yet to be optimized for the purpose of increasing the amount of clay infused. Once optimization is achieved, the combination of the steps of exfoliation and infusion can be investigated.

References

1. Horsch, S.; Serhatkulu, G.; Gulari, E.; Kannan, R. M. Supercritical CO_2 Dispersion of Nano-Clays and Clay/Polymer Nanocomposites. *Polymer* **2006**, 47, 7485-7496.
2. Morgan, A. B.; Gilman, J. W. Characterization of Polymer-Layered Silicate (Clay) Nanocomposites by Transmission Electron Microscopy and X-Ray Diffraction: A Comparative Study. *Journal of Applied Polymer Science* **2003**, 87, 1329-1338.
3. Wang, K. H.; Choi, M. H.; Koo, C. M.; Choi, Y. S.; Chung, I. J. Synthesis and Characterization of Maleated Polyethylene/Clay Nanocomposites. *Polymer* **2001**, 42, 9819-9826.
4. Ma, X.; Yu, J.; Wang, N. Production of Thermoplastic Starch/MMT-Sorbitol Nanocomposites by Dual-Melt Extrusion Processing. *Macromolecular Materials and Engineering* **2007**, 292, 723-728.
5. Luyt, A. S.; Geethamma, V. G. Effect of Oxidized Paraffin Wax on the Thermal and Mechanical Properties of Linear Low-Density Polyethylene-Layered Silicate Nanocomposites. *Polymer Testing* **2007**, 26, 461-470.
6. Chen, B. Polymer-Clay Nanocomposites: An Overview with Emphasis on Interaction Mechanisms. *British Ceramic Transactions* **2004**, 103, 6, 241-249.
7. Ryu, S. H.; Chang, Y. W. Factors Affecting the Dispersion of Montmorillonite in LLDPE Nanocomposite. *Polymer Bulletin* **2005**, 55, 385-392.
8. Leavitt, L. The Development of VCI-Infused Biodegradable Polymeric Resin Using Supercritical Carbon Dioxide. Personal Communication, University of Missouri – Columbia, Columbia, MO, 2006.
9. Clark, K. Synthesis of Maleated Poly(Vinylidene Fluoride) in Supercritical Carbon Dioxide Medium. Ph. D. Dissertation, University of Missouri – Columbia, Columbia, MO, 2004.
10. Cervantes-Uc, J. M.; Cauich-Rodriguez, J. V.; Vazquez-Torres, H.; Garfias-Mesias, L. F.; Paul, D. R. Thermal Degradation of Commercially Available Organoclays Studied by TGA-FTIR. *Thermochimica Acta* **2007**, 457, 92-102.
11. Carey, F. A. Infrared Spectroscopy. *Organic Chemistry* **2000**, 4, 519.
12. Park, C. I.; Park, O. O.; Lim, J. G.; Kim, H. J. The Fabrication of Syndiotactic Polystyrene/Organophilic Clay Nanocomposites and Their Properties. *Polymer* **2001**, 42, 7465-7475.
13. Sclavons, M.; Laurent, M.; Devaux, J.; Carlier, V. Maleic Anhydride-Grafted Polypropylene: FTIR Study of a Model Polymer Grafted by Ene-Reaction. *Polymer* **2005**, 46, 8062-8067.
14. Shieh, Y. T.; Su, J. H.; Manivannan, G.; Lee, P. H. C.; Sawan, S. P.; Spall, W. D. Interaction of Supercritical Carbon Dioxide with Polymers. I. Crystalline Polymers. *Journal of Applied Polymer Science* **1996**, 59,695-705.
15. Xu, Q.; Pang, M.; Li, J.; Jiang, Y. Application of Supercritical Carbon Dioxide in the Preparation of Biodegradable Polylactide Membranes. *Journal of Applied Polymer Science* **2004**, 94, 2158-2163.

Viscosity of Biodegradable Polymers Saturated with Carbon Dioxide

Robert D. Bleecher, Chemical Engineering, University of Delaware, Newark, DE

Mark Whitaker, Chemical Engineering, University of Nottingham, Nottingham, United Kingdom

Steven M. Howdle, Chemistry, University of Nottingham, Nottingham, United Kingdom

Annette D. Shine, Chemical Engineering, University of Delaware, Newark, DE

The viscosities of biodegradable polymers saturated with high pressure carbon dioxide were measured experimentally, at temperatures of 40 - 100 degrees C and pressures of 0 - 20 MPa (gage). The viscosities of low molecular weight polycaprolactone, poly(ethylene glycol) and poly(lactide-co-glycolide) were determined using a high pressure rheometer with a rotating, partially submerged cylindrical bob that had been calibrated with Newtonian fluids. Up to an order of magnitude decrease in polymer viscosity occurred due to the presence of carbon dioxide. Data from different temperatures and saturation pressures were correlated as a function of the reduced density predicted by the Sanchez-Lacombe equation of state, suggesting that free volume theory could be used to describe the transport properties. Although each studied polymer followed the Doolittle free volume equation for the zero shear rate viscosity, the universal constants of the WLF equation did not apply. The viscosity behavior is believed to be related to the propensity of each polymer to form drug delivery particles after depressurization from the saturated state.

See more of this Session: Materials Synthesis and Processing with Compressed or Supercritical Fluids II: Polymers
See more of this Group/Topical: Engineering Sciences and Fundamentals

Ion Pair Association in Ultra Supercritical Aqueous Environments: Experimental Difficulties, Molecular Simulation Insights, and Modeling Challenges

Ariel A. Chialvo, Chemical Science Division, Oak Ridge National Laboratory, Oak Ridge, TN

Miroslaw S. Gruszkiewicz, Chemical Sciences Division, Oak Ridge National Laboratory, Oak Ridge, TN

J. Michael Simonson, Center for Nanophase Materials Sciences Division, Oak Ridge National Laboratory, Oak Ridge, TN

Donald A. Palmer, Oak Ridge National Laboratory, Oak Ridge, TN

David R. Cole, Chemical Sciences Division, Oak Ridge National Laboratory, Oak Ridge, TN

Ultra-supercritical (USC) aqueous solutions are extreme environments exhibiting relatively small dielectric permittivity that give rise to complex processes whose mechanisms are poorly understood, mainly because system properties at these conditions are exceptionally difficult to access and measure. Understanding the behavior of aqueous solutions at USC is crucial for the design of more efficient electrical power plants, conditions at which trace-levels of ionic species might induce solid deposition, and consequently, affecting the global cycle efficiency. Moreover, interactions of these extreme environments with cladding and metallic surfaces are mostly unknown because their chemistry at these extreme conditions is poorly understood, a direct consequence of our lack of molecular-based understanding of the speciation behavior of electrolytes in steam environments.

Currently used engineering correlations for the thermophysical properties of high temperature/pressure aqueous electrolyte solutions suffer from an inherent inability to capture the unavoidable tendency of ions toward speciation. This scenario points at two relevant implications: (a) the need for reassessing the way we address the modeling of dissociable solutes in USC aqueous solutions, i.e., more realistic, speciation-based thermodynamic descriptions in terms of ion speciation for all solutes; (b) the need for ways to test the reliability of any macroscopic modeling approach, including the validity of proposed hypotheses, as well as to guide the development of alternatives.

In this context, molecular simulation becomes a versatile tool to gain insight into the microscopic behavior of aqueous electrolyte solutions at USC conditions, to interpret the species solvation behavior, and ultimately, to provide a quantitative assessment of the speciation process (CHIALVO et al., 2002). In fact, molecular simulation provides a unique route to quantifying the dielectric behavior of aqueous electrolyte solutions, the complete characterization of the system microstructure, as well as the interpretation and measurement of the ion-pair formation that will ultimately guide the development of successful macroscopic correlations.

Among experimental approaches, the most promising tool currently available to constraint the speciation characterization of USC steam environments is the flow-through electric conductance apparatus (GRUSZKIEWICZ and WOOD, 1997), a technique that allows reliable measurements at ion concentrations as low as 10-6 molal. Electric conductance measurements obtained at these extremely low ionic strengths can be directly used to determine ionic mobilities without the need for questionable extrapolation of the results obtained at higher concentrations.

In this work we study the Na+Cl- pair association process in USC steam solutions over a wide range of densities, including the zero solvent-density limit. Our main goal is to discuss a new integrated approach comprising molecular simulation and ultra-sensitive electric conductance experiments, to provide new crucially needed data for the association constant in USC environments, as well as to address some relevant issues regarding the modeling of ion association in extreme steam environments (CHIALVO et al., 2009).

REFS:

Chialvo A. A. and Simonson J. M. (2003) "Aqueous Na+Cl- Pair Association from Liquid-like to Steam-like Densities along Near-critical Isotherms". Journal of Chemical Physics 118, 7921-7929

Chialvo A. A., Gruszkiewicz M. S., Simonson J. M., Palmer D. A., and Cole D. R. (2009) "Ion Pair Association in Extreme Aqueous Environments: Molecular-based and Electric Conductance Approaches". Journal of Solution Chemistry (In press)

Gruszkiewicz M. S. and Wood R. H. (1997) Conductance of LiCl, NaCl, NaBr, and CsBr Solutions in Supercritical Water Using a Flow Conductance Cell. Journal of Physical Chemistry B 101, 6549-6559

See more of this Session: Thermodynamics of Energy Systems
See more of this Group/Topical: Engineering Sciences and Fundamentals

Deliquescence and Reaction Kinetics in the Steam Hydrolysis of Sodium Borohydride

Amy M. Beaird, Chemical Engineering, University of South Carolina, Columbia, SC

Lin Yu, Chemical Engineering, University of South Carolina, Columbia, SC

Michael A. Matthews, Chemical Engineering, University of South Carolina, Columbia, SC

The hydrolysis of chemical hydrides, such as sodium borohydride ($NaBH_4$), has been widely studied as a method to store and release hydrogen at the point of use ($NaBH_4 + (2+x)\ H_2O \rightarrow 4H_2 + NaBO_2 \cdot xH_2O +$ heat). Traditionally, efforts have focused on aqueous phase hydrolysis wherein the hydride is stabilized in a caustic solution and passed over a catalyst when hydrogen release is desired. Although significant advances on catalytic materials have been made in this area, the approach requires that the reactants and products remain in solution, thereby demanding copious amounts of water. The aqueous hydrolysis approach inevitably results in an excessively heavy, bulky system. This is one of the major reasons it was given a "no-go" decision by the Department of Energy for automotive hydrogen storage applications.

However, a new approach has been investigated wherein solid $NaBH_4$ is contacted with steam or water vapor just above the boiling point of water, resulting in high hydrogen yields without aid of a catalyst. The purpose of this study was to elucidate the mechanism and physical phenomena associated with steam hydrolysis of sodium borohydride as an alternative pathway for hydrogen storage. Investigation of the effect of temperature on the kinetics of hydrogen release by steam hydrolysis revealed that the reaction is inhibited at higher temperatures. Based on visual observation with a borescope camera during reaction, it is now evident that the crucial first step in the reaction sequence is for hygroscopic sodium borohydride to absorb water and deliquesce, forming a highly concentrated viscous solution which then reacts to form hydrogen. The occurrence of deliquescence depends primarily on the relative humidity and we have determined a threshold exists below which deliquescence cannot occur.

Based on the observation that a concentrated aqueous phase exists in the reaction sequence, we have developed a technique using in-situ ^{11}B-NMR at elevated temperatures for determining the kinetics of hydrogen release decoupled from the deliquescence kinetics. Stable intermediates found in dilute aqueous hydrolysis at low temperatures are not present in the high-temperature concentrated spectra. Fundamental differences in the reaction mechanism of the traditional approach and the steam/water vapor approach are apparent. Utilizing the latter technique may result in a more suitable hydrogen storage methodology.

See more of this Session: Thermodynamics of Energy Systems
See more of this Group/Topical: Engineering Sciences and Fundamentals

Hydrogen-Carbon Dioxide Separation by Electrochemcial Pumping

Jay B. Benziger, Department of Chemical Engineering, Princeton University, Princeton, NJ

Ahmed Abdulla, Princeton University, Princeton, NJ

Steam reforming of hydrocarbons or coal followed by the water gas shift reaction produces a mixture of H2, CO2 and H20. The most common approaches to purify the hydrogen are by alkaline scrubbing of the gas mixture to remove the CO2 or pressure swing adsorption. We report here an alternative method of electrochemical pumping of hydrogen from mixtures of H2, CO2 and H20 using a polymer electrolyte membrane. This process produces high purity hydrogen (< 100 ppm CO2) at >98% recovery and modest power requirements. The equipment is essentially a PEM fuel cell run with an applied voltage to pump the hydrogen. Data is presented that shows the efficiency and recovery of this process. Process designs and an energy comparison to both alkaline scrubbing and PSA shows the electrochemical pumping is more energy efficient and readily adaptable for many a wide range of process scales.

See more of this Session: Thermodynamics of Energy Systems
See more of this Group/Topical: Engineering Sciences and Fundamentals

High Hydrogen Yield From Ammonia Borane Hydrothermolysis for Fuel Cell Based Vehicles

Moiz Diwan, School of Chemical Engineering, Purdue University, West Lafayette, IN

Hyun Hwang, School of Chemical Engineering, Purdue University, West Lafayette, IN

Ahmad Al-Kukhun, School of Chemical Engineering, Purdue University, West Lafayette, IN

Arvind Varma, School of Chemical Engineering, Purdue University, West Lafayette, IN

The widespread use of PEM fuel cells is hindered by the absence of a practical, high density hydrogen fuel source. Among various alternatives, chemical methods of hydrogen storage provide high specific energy at relatively easy storage conditions. Ammonia borane (AB) is a promising hydrogen storage material as it contains 19.6 wt% hydrogen. To release hydrogen from AB, thermolysis and catalytic hydrolysis are usually employed, but these methods require either high temperatures or expensive catalysts. Hydrothermolysis of AB [1], investigated in the present work, allows for hydrogen generation (from both AB and water) at moderate temperatures and without the use of catalyst. This method involves external heating of aqueous AB solutions to temperatures 70-85 °C, under argon pressure (~ 10 bar). To clarify the reaction mechanism, D_2O was used instead of H_2O in certain experiments. It was shown that with increasing AB concentration, the H_2 yield increased, while HD yield decreased. Thus the role of thermolysis (H_2 yield), as compared to hydrolysis (HD yield), increases with AB concentration. This approach provides maximum hydrogen storage capacity of 11 wt%. To our knowledge, the AB hydrothermolysis process is the first one to provide such high hydrogen yield values at T<85 °C. The results indicate that the proposed method is promising for hydrogen storage, and could be used in fuel cell based vehicles.

References:

[1] Diwan, M., Diakov, V., Shafirovich, E., Varma, A., "Noncatalytic Hydrothermolysis of Ammonia Borane", International Journal of Hydrogen Energy, Vol. 33, 2008, pp 1135-1141.

See more of this Session: Thermodynamics of Energy Systems
See more of this Group/Topical: Engineering Sciences and Fundamentals

Novel High Capacity Hydrogen Storage Materials Based On Tetrahydrogenmethane

Maaike C. Kroon, Faculty of Mechanical, Maritime and Materials Engineering, Delft University of Technology, Delft, Netherlands

Hugh Docherty, Department of Chemical and Biomolecular Engineering, Vanderbilt University, Nashville, TN

Peter T. Cummings, Department of Chemical and Biomolecular Engineering, Vanderbilt University, Nashville, TN

Cor J. Peters, Process and Energy, Delft University of Technology, Faculty of Mechanical, Maritime and Materials Engineering, Delft, Netherlands

Geert-Jan Witkamp, Section Process Equipment, Process & Energy Department, Faculty of Mechanical, Maritime and Materials Engineering, Delft University of Technology, Delft, Netherlands

Wendy L. Mao, Geological and Environmental Sciences, Stanford University, Stanford, CA

Hydrogen is considered to be a promising alternative to fossil fuels. Unfortunately, due to the difficulty in finding a practical storage medium, its use as a fuel, for instance for application in the automotive sector, is currently limited. In order to be considered practical, a hydrogen storage material should satisfy a number of requirements, including a high storage capacity (high hydrogen content by mass and volume), ambient storage temperature and pressure, a fast uptake of hydrogen, and reversible hydrogen release. So far, a material that meets these requirements is does not exist.

In this work, we develop novel hydrogen storage materials that fulfills the above-mentioned requirements. The idea is to store hydrogen in the hydrogen-rich crystalline molecular compound tetrahydrogenmethane (CH4(H2)4), which contains as much as 33.4 wt% molecular hydrogen. Not much is known about how this material stores such a remarkable amount of molecular hydrogen: Even something as basic as its structure has not yet been identified. Motivated by this, we investigate the structure of tetrahydrogenmethane, as well as the interaction of hydrogen with this material, using a combined experimental - molecular simulation approach. We also identify promoters that may be added to this material to decrease the necessary storage pressure. Results will be presented at the conference.

See more of this Session: Thermodynamics of Energy Systems
See more of this Group/Topical: Engineering Sciences and Fundamentals

Simulation Studies of Tetrabutyl-Ammonium Bromide Semi-Clathrate Hydrates

Pradeep Venkataraman, Department of Chemical and Biomolecular Engineering, Tulane University, New Orleans, LA

Hank Ashbaugh, Department of Chemical and Biomolecular Engineering, Tulane University, New Orleans, LA

Semi-clathrate hydrates are a class of inclusion compounds where organic ionic co-hosts (e.g., quaternary ammonium salts) stabilize the hydrate cages at temperatures and pressures approaching ambient conditions. These co-hosts are integrated into the clathrate hydrogen-bonding framework, taking the place of one or more water molecules. The broader stability range of these mixed clathrates opens up the possibility of storing hydrogen at lower pressures and higher temperatures for use in fuel cell vehicles. To date, however, there has been little molecular level studies of gas storage in these inclusion compounds. We investigate the stability of semi-clathrate hydrates of tetrabutyl-ammonium bromide (TBAB) and water with methane or hydrogen guests using molecular dynamics simulations. Simulations were performed at varying temperatures and pressures for guest-TBAB ratios of 0, 0.5, 1, 1.5 and 2. Our studies show that while the unit cell volume decreases with the decreasing guest occupancy, the cage shape is preserved. Free energy calculations of guest deletion confirm the preferential stability at higher guest/TBAB ratios and suggest that the conventional van der Waals-Platteeuw model can be used to explain the phase behavior of these inclusion compounds.

See more of this Session: Thermodynamics of Energy Systems
See more of this Group/Topical: Engineering Sciences and Fundamentals

Predictions of Mixed Hydrate Phase Equilibria and the Swapping of CH_4 Hydrate with CO_2 and CO_2+N_2 Mixtures.

Nagasree Garapati[1,2] *and Brian J. Anderson*[1,2]
[1]*National Energy Technology Laboratory, Collins Ferry Road, Morgantown, WV, USA*
[2]*Department of Chemical Engineering, West Virginia University, Morgantown, WV, USA*

Introduction

Natural gas hydrates are likely to contain more carbon than in all other fossil fuels reserves combined worldwide[1]. They can be formed either by bacterial activity at shallow depths which contains mostly CH_4 with trace amounts of C_2H_6 or by thermal pyrolysis of fossil organic matter which contains CH_4 and significant amount of higher hydrocarbons (C_2-C_5) and non-hydrocarbon gases like CO_2[1]. Thus, if CH_4 stored in the natural gas hydrates can be recovered, the hydrates could potentially become energy resource for the future[2]. Different techniques that are currently being proposed for the production of CH_4 from gas hydrates include depressurization, thermal stimulation, and injection of inhibitors, all of which involve dissociation of gas hydrate and the release of significant amount of water causing geomechanical stress on the reservoir leading to subsidence. A newly-proposed method is to replace CH_4 in the hydrate by pure CO_2 and CO_2+N_2 mixture, which serves dual purpose of long-term storage of a greenhouse gas (CO_2) and the production of natural gas (CH_4)[3,4] and this process will enable the ocean floor to remain stable because the hydrate maintain same crystalline structure even after replacement. To understand the production of CH_4 gas from the hydrate reservoirs it is necessary to have knowledge of the phase equilibria of the mixed hydrates like CH_4-C_2H_6, CH_4-CO_2, CH_4-N_2 etc. The phase equilibrium conditions of a pure hydrate changes significantly and the structure of the hydrate formed may also change due to the presence of other hydrocarbons. The structure of the mixed gas hydrate depends on the relative gas composition.

In spite of the large experimental data base of clathrate phase behavior[1], the theory of clathrate phase behavior still depends heavily on the ad hoc fitting of the experimental data. The fitting procedures cannot produce accurate results outside the range of fitting. The reference parameters used in these methods while fitting the intermolecular potential parameters to the experimental data[1,5]obtained by numerically fitting procedures differs from the values determined experimentally[6] or computationally[7] and they are also needed to be adjusted[5] to predict the phase equilibrium and structural transitions of mixed hydrates.

Recently, Bazant and Trout[8] proposed that the spherically averaged intermolecular potential can be determined by the temperature dependent Langmuir constant. The cell potential

[1] Sloan, E. D.; Koh, C. A. *Clathrate hydrates of natural gases*, 3rd Ed, **2007**.
[2] Holder, G. D.; John, V. T.; Yen, S. SPE/DOE symposium on unconventional gas recovery, **1980**.
[3] Ohgaki, K.; Takano, K.; Sangawa, H.; Matsubara, T.; and Nakano, S. *J. Chem. Eng. Japan*. **1996**, 29, 478.
[4] Park, Y.; Kim, D. -Y.; Lee, J. -W.; Huh, D. –G.; Park, K. –P.; Lee, J.; and Lee, H. *PNAS*, **2006**, 103(34), 12690.
[5] Ballard, A. L.; Sloan, E. D. *Gas Hydrates: Challenges for the Future*, **2000**, 912, 702.
[6] Handa, Y. P.; Tse, J. S. *J. Phys Chem.*, **1986**, 90, 5917.
[7] Anderson, B. J.; Tester, J. W.; Trout, B. L. *J. Phys Chem. B*, **2004**, 108, 18705.
[8] Bazant, M. Z.; Trout, B. L. *Physica A*. **2001**, 300, 139

parameters are determined directly from the experimental equilibrium data by analytically solving the integral equation for the Lennard-Jones and Devonshire approximation to the van der Waals and Platteeuw[9] statistical thermodynamics model. The Langmuir constant values are computed from experimental data (for hydrates that occupy only large cage) and *ab initio* potential method (for hydrates that occupy both cages). The resulting potentials are more meaningful and simpler than the numerically-fit Kihara potential. When used with conjunction with variable reference parameters obtained from the empirical correlation developed between the parameters and the size of the guest molecule, the phase equilibrium conditions for mixed hydrates are calculated without adjusting any parameters. The reference parameters for mixed hydrates are calculated from the pure component parameters by using mixing rules and an entropy correction term is added to the mixture reference chemical potential difference, which accounts for the difference in the reference chemical potential values of the guest components.

In this work, the cell potential code developed by Anderson et al.[7] is modified to include a variable reference chemical potential difference and reference enthalpy difference. The model is validated by predicting the structural transitions that are known to occur, such as for example cyclopropane. Along with the equilibrium pressure predictions, the ability of this method to predict the occupancies accurately is another rigid test for the reliability of the method. The cell potential code developed has validated its effectiveness and applicability to mixed hydrate systems by predicting the phase equilibria and structural transitions of the binary hydrates like CH_4-C_2H_6, CH_4-CO_2, CH_4-N_2 and N_2-CO_2 without any adjustable parameters. Additionally, phase equilibria of ternary hydrate CH_4-N_2-CO_2 is predicted and await experimental confirmation.

Theoretical Background

The pressure and temperature conditions for the formation or dissociation of a hydrate are governed by equilibrium thermodynamics. According to standard thermodynamic phase equilibrium criteria, the chemical potential of each component must be the same in every phase.

$$\mu_w^H(T,P) = \mu_w^{L,\alpha}(T,P) \tag{1}$$

$$\Delta\mu_w^{\beta-H}(T,P) = \mu_w^{\beta}(T,P) - \mu_w^H(T,P) = \mu_w^{\beta}(T,P) - \mu_w^{L,\alpha}(T,P) = \Delta\mu_w^{\beta-L,\alpha}(T,P) \tag{2}$$

where $\mu_w^H(T,P)$ is the chemical potential of water in hydrate phase, H, $\mu_w^{L,\alpha}(T,P)$ is the chemical potential of water in liquid aqueous phase, L, or ice phase, α, depending on whether the temperature, T, is above or below 273.15 K, the freezing point of water and $\mu_w^{\beta}(T,P)$ is chemical potential of water in hypothetical empty hydrate lattice, β.

van der Waals and Platteeuw[9], by assuming single guest occupancy, negligible distortions in empty hydrate lattice, and neglecting guest-guest interactions, proposed an equation for the difference in chemical potential between hydrate and empty hydrate lattice as

$$\Delta\mu_w^{\beta-H}(T,P) = -RT\sum_i v_i \ln\left(1 - \sum_j y_{ji}\right) \tag{3}$$

where v_i is the number of type i cavities per water molecule and y_{ji} is the cage occupancy defined as the probability of finding the guest component, J in the cavity of type i.

[9] van der Waals, J. H.; Platteeuw, J. C. *Adv. Chem. Phys.* **1959**, 2, 1.

Mathematically, the cage occupancy y_{Ji} follows Langmuir adsorption isotherm and is related to Langmuir constant as in Equation (4)

$$y_{Ji} = \frac{C_{Ji} f_J}{1 + \sum_J C_{Ji} f_J} \quad (4)$$

where f_J is the fugacity of guest molecule J calculated using Peng-Robinson[10] equation of state and C_{Ji} is the Langmuir constant calculated from van't Hoff temperature dependence equation defined as:

$$C(\beta) = C_0 e^{m\beta} \quad (5)$$

where C_0 and m are specific for each guest component J and cavity type i.

Holder et al.[11] proposed an equation for the chemical potential difference between water in empty hydrate lattice and water in liquid aqueous phase or ice phase as:

$$\frac{\Delta\mu_w^{\beta-L,\alpha}(T,P)}{kT} = \frac{\Delta\mu_w^{\beta-L,\alpha}(T_0,0)}{kT} - \int_{T_0}^{T}\left[\frac{\Delta h_w^{\beta-L,\alpha}(T)}{kT^2}\right]_P dT + \int_{o}^{P}\left[\frac{\Delta V_w^{\beta-L,\alpha}}{kT}\right]_T dP - \ln \gamma_w x_w \quad (6)$$

The first term is the chemical potential difference between the empty hydrate and liquid water at its reference state (273.15 K, 0 bar), and the second term accounts for the change in chemical potential difference due to temperature, at zero pressure.

$$\Delta h_w^{\beta-L,\alpha}(T) = \Delta h_w^{\beta-L,\alpha}(T_0) + \int_{T_0}^{T} \Delta C_P^{\beta-L,\alpha}(T)\, dT \quad (7)$$

$$\Delta C_P^{\beta-L,\alpha}(T) = \Delta C_P^{\beta-L,\alpha}(T_0) + b^{\beta-L,\alpha}(T - T_0) \quad (8)$$

The third term on the right hand side of Equation (6) accounts for the change in chemical potential difference due to pressure, and the last term accounts for the nonidealities of the solution.

The reference parameters are obtained from the correlations developed with respect to the guest molecule size. The mixed hydrate reference parameters are calculated using mixing rules given as:

$$Ref\ parameter_{mix} = \sum_{i=1}^{N} Ref\ parameter_i \times y_i \quad (9)$$

where y_i is the gas phase mole fraction of component i.

Results

Phase Equilibrium Predictions of Simple Gas Hydrates

The phase equilibrium data is predicted using cell potential code for CH_4, C_2H_6, CO_2, C_3H_8, N_2, and C-C_3H_6 and the results are compared with that of the predictions by CSMGEM, a computer program for hydrate thermodynamic calculations. The phase equilibrium data of cyclopropane (C-C_3H_6) are not included in CSMGEM. Figure 1 illustrates the comparison of

---

[10] Peng, D. –Y.; Robinson, D. B. *Ind. Eng. Chem. Fundam.* **1976**, 1, 59.

[11] Holder, G. D.; Corbin, G.; Papadopoulos, K. D. *Ind. Eng. Chem. Fund.* **1980,** 19, 282.

average absolute deviations (AAD) from experimental data using present model and CSMGEM and AAD is calculated using the Equation (10). Table 1 represents the temperature and pressure ranges of the experimental data points of the different guest molecules.

$$AAD\% = 100 \times \frac{1}{N}\sum_{1}^{N}\frac{|P_{exp}-P_{pred}|}{P_{exp}} \qquad (10)$$

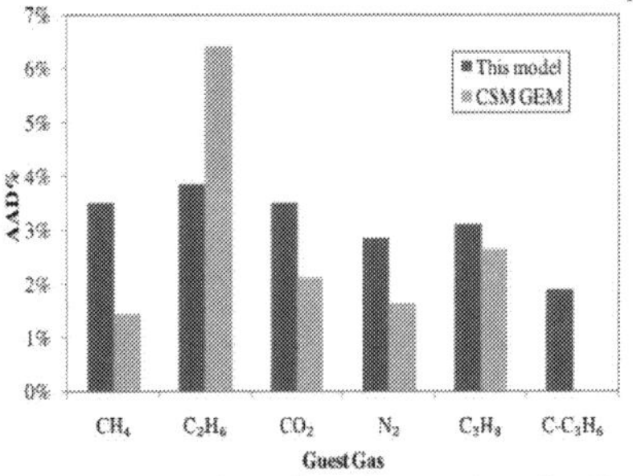

Figure 1 Comparison of Average Absolute Deviations from Experimental Data for Simple Hydrates Predicted by This Model and CSMGEM Software.

Table 1 Temperature and Pressure Range of Experimental Data Points of Various Guest Gas Molecules.

Guest Gas	No. of data Points (Reference)	Temp Range (K)	Pressure Range (bar)
CH_4	98([1-3])	148.8-320.1	0.053-3970
C_2H_6	74([1,2,4-8])	200.8-288.2	0.083-33.6
CO_2	127([1,9-16])	151.5-282.0	0.005-38.4
N_2	46([17])	272.0-286.7	144.8-637.1
C_3H_8	58([1,12,18-23])	273.2-278.4	1.65-5.62
$C\text{-}C_3H_6$	37([24])	237.2-289.4	0.086-5.66

The average absolute difference is more for this model when compared to that of CSMGEM[1]. Much of this difference can be attributed to the equation of state used to calculate the fugacity. The Peng-Robinson[10] equation results in large deviations in the fugacities for hydrate-forming gases at high pressures. However, the occupancies predicted by this model is closer to the experimental values of Ripmeester and Ratcliffe[12] when compared to that of CSMGEM[1]. The predictive ability of the cell potential method can be verified against experimental structural transitions that are known to occur. For example, cyclopropane undergoes a structure transition as a function of temperature; it forms structure II hydrate at temperatures between 257.1 and 274.6 K[13]. In this work the transitions are predicted to occur at 256.4 and 274.6 K respectively.

Phase Equilibrium Predictions of Binary Gas Hydrates

CH_4-C_2H_6 Mixed Hydrate

[12] Ripmeester, J. A.; Ratcliffe, C. I. *Energy Fuels* **1998**, *12*, 197.

[13] Hafemann, D. R.; Miller, S. L. *J. Phys. Chem.* **1969**, 73,1392

The phase equilibria of the binary hydrate CH_4-C_2H_6 are predicted using this method. Interestingly, both CH_4, and C_2H_6 form a structure I hydrate when the gas phase contains only one component but the CH_4-C_2H_6 mixed hydrate forms structure II at various gas mixture compositions[14,15]. The mixed gas hydrate undergoes a transition from structure I to structure II at a y_{CH4}= 0.72-0.75 and the upper transition from structure II to structure I occurs at y_{CH4}=0.992(±0.005)-0.994(±0.007) at 274.2 K[14,15]. Figure 2 represents the structural transitions predicted in this work which occur at y_{CH4} =0.72 from structure I to structure II and at y_{CH4} =0.98 from structure II to structure I at 274.2 K.

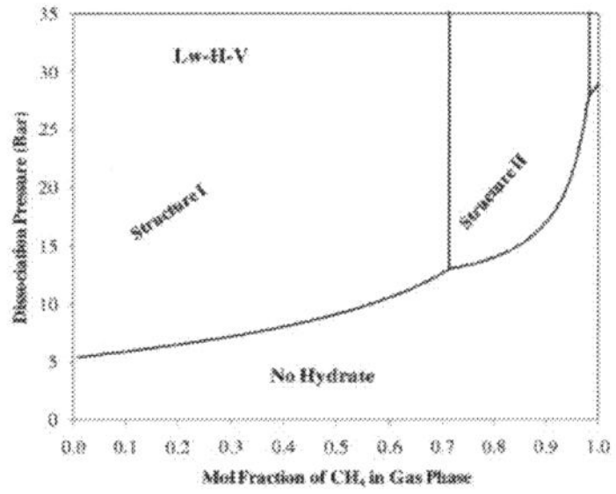

Figure 2 Methane-Ethane (CH_4-C_2H_6) Structural Transition Predictions at 274.2 K. Vertical Lines Indicate the Structural Transition Boundaries.

CH_4-CO_2 Mixed Hydrate

A newly-proposed method of production of CH_4 from the natural gas hydrate is by injection of CO_2 into reservoir. The phase equilibria of CH_4-CO_2 hydrate are predicted in order to understand the replacement process. Figure 3 represents the dissociation pressure error vs. mole fraction of CH_4 in vapor phase.

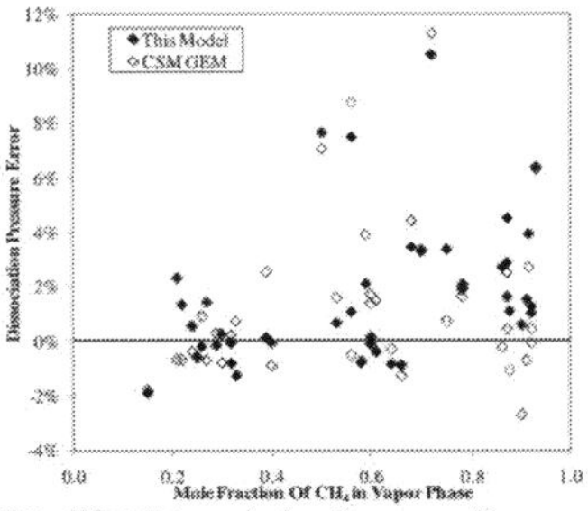

Figure 3 Methane-Carbon dioxide (CH_4-CO_2) Dissociation Pressure Error vs. Mole Fraction of CH_4 in Gas Phase.

CH_4-N_2 Mixed Hydrate

CH_4 as a simple hydrate forms structure I hydrate while N_2 form structure II hydrate, therefore the structure of mixed gas hydrate of CH_4 and N_2 depends on their relative gas composition. The NMR and XRD studies on the mixed hydrate by Lee et al[16] found that the

---

[14] Subramanian, S.; Kini, R. A.; Dec, S. F.; Sloan, E. D. *Chem. Eng. Sci.* **2000**, 55, 1981.
[15] Subramanian, S.; Ballard, A. L.; Kini, R. A.; Dec, S. F.; Sloan, E. D. *Chem. Eng. Sci.* **2000**, 55, 5763.
[16] Lee, J.-W.; Kim, D.-Y.; Lee, H. *Korean J. Chem. Eng.* **2006**, 23, 299.

structural change occurs in the range between 0.2524 and 0.2851 mole fraction of CH_4 in gas phase at various temperatures (in the range from 273.3 to 285 K. In this work the phase equilibrium data and the structural transitions are predicted. It is found that the structural transitions occurs in the range between 0.24 and 0.3 mole fraction of CH_4 in gas phase at temperature range from 273 to 282 K which are shown in Figure 4.

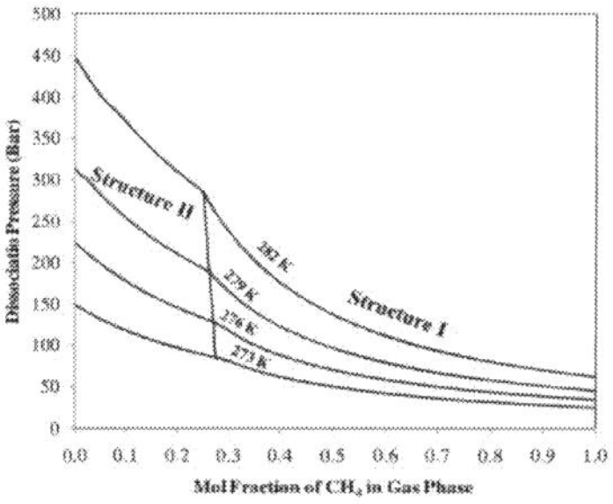

Figure 4 Methane-Nitrogen (CH_4-N_2) Structural Transitions Predictions. The Curves Represent the Phase Equilibria Predictions at Various Temperatures and Vertical Line Indicate the Structural Transition Boundary.

N_2-CO_2Mixed Hydrate

As simple gas hydrates CO_2 and N_2 form structure I and structure II respectively. The structure of mixed hydrate can be either structure I or structure II depending on the relative ratio of guest molecules occupied in the small and large cages. Seo and Lee[17] found that the mixed N_2-CO_2 forms structure I at all compositions above 0.03 mole fraction of CO_2 in the vapor phase with large cages being occupied by CO_2 and small cages by N_2. Figure 5 illustrates structural changes are predicted for CO_2 and N_2 mixed hydrate in the temperature range of 276-282 K. The structural transitions are predicted to occur at low concentration of CO_2 in gas phase, which implies that even at low concentration of CO_2 in gas phase the hydrate phase is enriched by CO_2 resulting in formation of structure I hydrate as shown in Figure 6.

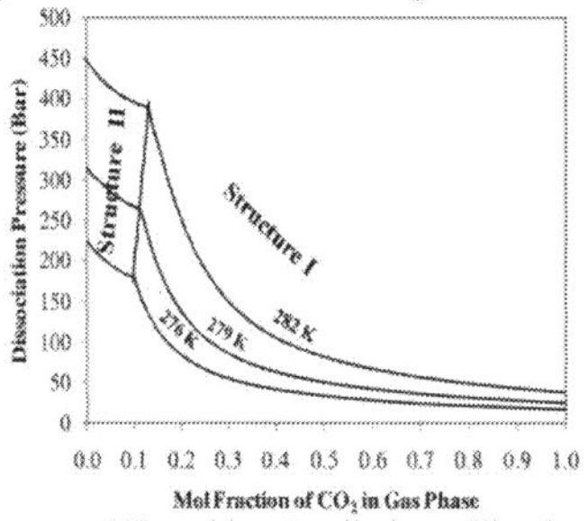

Figure 5 Nitrogen-carbon dioxide (N_2-CO_2) Structural Transition Predictions. The Curve Represent the Phase Equilibria at Various Temperatures and Vertical Line Represent the Structural Transition Boundary.

---

[17] Seo, Y.-T.; Lee, H. *J. Phys. Chem. B.* **2004**, 108, 530.

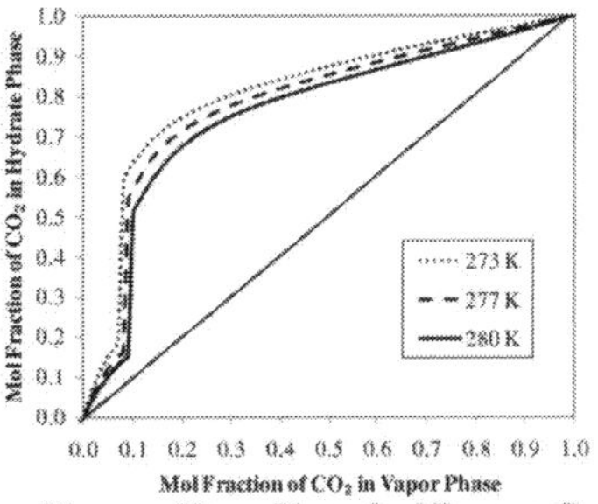

Figure 6 Mole fraction of CO_2 in Hydrate Phase vs Vapor Phase for Nitrogen-Carbon dioxide (N_2-CO_2) Mixed Hydrate.

The average absolute deviations (AAD) of the various mixed hydrates predicted by this model are compared to that of predictions by CSMGEM for the binary hydrates in Figure 7.

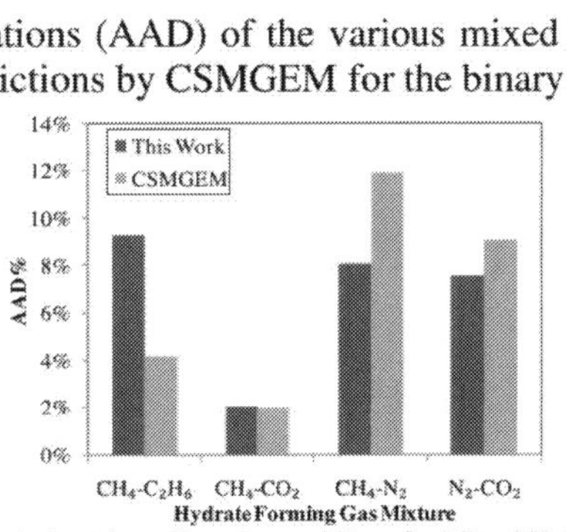

Figure 7 Comparison of Average Absolute Deviations from Experimental Data for Mixed Hydrates Predicted by This Model and CSMGEM Software.

CH_4-N_2-CO_2 Mixed Hydrate

Recent experimental studies by Park et al.[4] have shown that the fraction of recovered methane can be improved to 85% by using a mixture of N_2 and CO_2 when compared with the 64% obtained by pure CO_2. The phase equilibria of the ternary hydrate are predicted and the predictions await experimental confirmation. Figure 8 and Figure 9 represents the ternary plots of equilibrium gas mixture compositions both in gas and hydrate phase are presented at constant temperature, 274.15 K for varying pressures and at constant pressure, 35 bar for varying temperatures respectively.

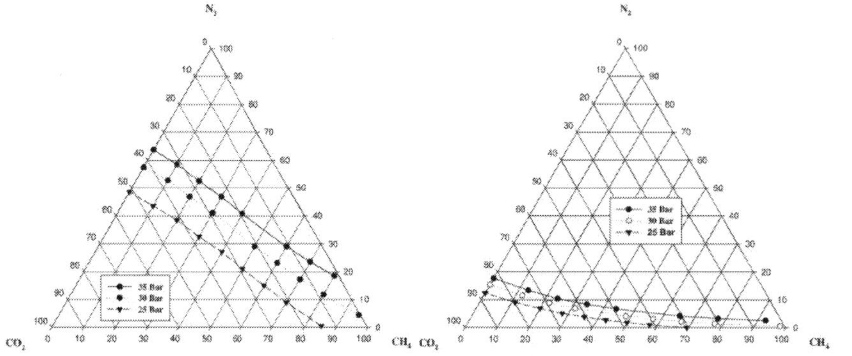

Figure 8 Ternary Plot of Methane-Nitrogen-Carbon Dioxide (CH_4-N_2-CO_2) Compositions at 274.15 K, a) Gas Phase b) Hydrate Phase

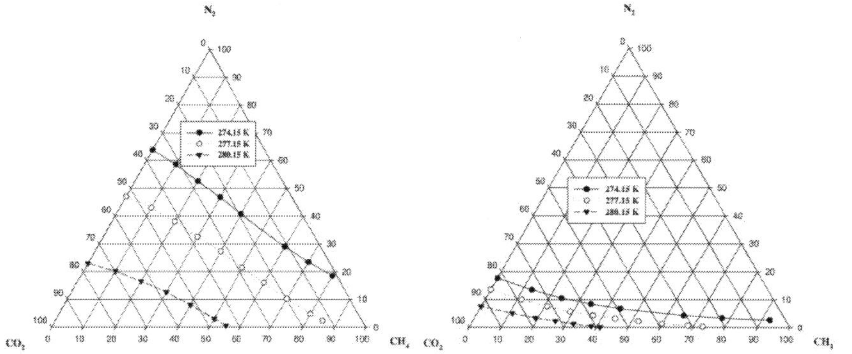

Figure 9 Ternary Plot of Methane-Nitrogen-Carbon Dioxide (CH_4-N_2-CO_2) Compositions at 35 Bar, a) Gas Phase b) Hydrate Phase

Conclusions

- The predictive ability of the cell potential method was verified against experimental structural transitions that are known to occur.
- The cell potential method developed has validated its ability to successfully predict mixed hydrate systems without any adjustable parameters and fitting to the experimental data.
- The phase equilibria of ternary mixed hydrate (CH_4-N_2-CO_2) are predicted at 274.15 K and the predictions await experimental confirmation. Ternary plots of equilibrium gas compositions are presented at constant temperature for varying pressures and at constant pressure for varying temperatures.
- This work will enable accurate simulations and predictions of potential replacement of CH_4 by CO_2 and N_2-CO_2 in natural gas hydrate systems.

Acknowledgements

This technical effort was performed in support of the National Energy Technology Laboratory's on-going research in methane hydrates under the RDS contract DE-AC26-04NT41817.

References

1. Deaton, W. M.; Frost, E. M., Jr.; *U.S. Bureau of Mines Monograph.* ***1946***, 8,101.
2. McLeod, H. O.; Campbell, J. M. *J. Petl. Tech.* **1961**, 13, 590.
3. Holder, G. D.; Grigoriu, G.C. *J. Chem. Thermodyn.* **1980**, 12, 1093.
4. Roberts, O. L.; Brownscombe, E. R.; Howe, L. S. *Oil Gas J.* **1940**, 39, 37.
5. Reamer, H.H.; Selleck, F.T.; Sage, B. H. *J. Petrol Tech.* **1952b,** 4, 197.
6. Falabella, B. J.; Vanpee, M. *I&EC Fundamentals.* **1974,** 13, 228.
7. Holder, G. D.; Hand, J. H. *AIChE J.* **1982,** 28, 44.
8. Avolnitis, D. *Multiphase Equilibria in Oil-Water Hydrate Forming Systems,* M.Sc. Thesis, Heriot-Watt University, Edinburgh, Scotland. **1988.**
9. Miller, S.L.; Smythe, W.D. *Science*, **1970**, *170*, 531.
10. Falabella, B.J. A Study of natural Gas Hydrates, Ph.D. Thesis, University of Massachusetts, University Microfilims, Ann Arbor, 1975
11. Larson, S.D. Phase Studies of the Two-Component Carbon Dioxide-Water system, Involving the Carbon Dioxide Hydrate, University of Illinios, Urbane, IL, 1955.
12. Robinson,D.B.; Mehta, B.R. *J.Can.Pet.Tech.* **1971**, *10*, 33.
13. Ng, H. –J.; Robinson, D. B. *Fluid Phase Equilib.* **1985**, *21*, 145.
14. Unruh, C.H., Katz, D.L. *Trans. AIME* **1949**, *186*, 83.
15. Adisasmito, S; Frank, R.J.; Sloan, E. D. *J. Chem. Eng. Data*, **1991**, *36*, 68
16. Ohgaki, K; Makihara, Y; Takano, K. *J. Chem. Eng. Japan.* **1993**, *26*, 558.
17. van Cleeff. A.; Diepen, G.A.M. *Rec. Trav. Chim.* **1960,** 79, 582.
18. Miller, B.; Strong, E. R. *Am. Gas. Assoc. Monthly,* **1946**, 28, 63.
19. Reamer, H.H.; Selleck, F.T.; Sage, B. H. *Pet. Trans. AIME.* **1952,** 195, 197.
20. Verma, V.K. *Gas Hydrates from Liquid Hydrocarbon-Water Systems,* Ph.D. Thesis, University of Michigan, University Microfilms No. 75-10,324, Ann Arbor, MI. **1974.**
21. Kubota, H.; Shimizu, K.; Tanaka, Y.; Makita, T. *J. Chem. Eng. Japan.* **1984**, 17, 423.
22. Patil, S. L. *Measurements of Multiphase Gas Hydrates Phase Equilibria: Effect of Inhibitors and Heavier Hydrocarbon Components.* M.S. Thesis, University of Alaska, Anchorage, AL. **1987.**
23. Mooijer-van den Heuvel, M. M.; Peters, C. J.; de Swaan Arons, J. *Fluid Phase Equilib.* **2002**, 193, 245.
24. Hafemann, D. R.; Miller, S. L. *J. Phys. Chem.* **1969**, 73, 1392.

Influence of Surface Polarity and Confinement Length Scale On the Dynamics of a Nanoscopic Water Film

Santiago Romero-Vargas Castrillón, Chemical Engineering, Princeton University, Princeton, NJ

Nicolás Giovambattista, Physics Department, Brooklyn College of the City University of New York, Brooklyn, NY

Ilhan A. Aksay, Chemical Engineering, Princeton University, Princeton, NJ

Pablo G. Debenedetti, Chemical Engineering, Princeton University, Princeton, NJ

We use Molecular Dynamics simulations to study the influence of surface polarity and confinement on the dynamics of a nanoscopic water film at T = 300 K and ρ = 1.0 g cm^{-3}. We consider two infinite, atomically detailed silica surfaces reproducing the chemistry of β-cristobalite. By means of a dimensionless parameter, k, we scale the surface Coulombic charges and model systems ranging from hydrophobic apolar (k = 0) to hydrophilic (k = 1.0). We find that both rotational and translational dynamics exhibit a non-monotonic dependence on k characterized by a maximum in the in-plane diffusion coefficient, $D_{||}$, at values between 0.6 and 0.8, and a minimum in the rotational relaxation time, τ_R, at k = 0.6. The slow dynamics observed in proximity of the hydrophobic apolar surface are a consequence of β-cristobalite templating an ice-like water layer. The fully hydrophilic surfaces (k = 1.0), on the other hand, result in slow interfacial dynamics due to the presence of dense but disordered water that forms strong hydrogen bonds with surface silanol groups. In order to study the effect of confinement length scale, we consider a water film confined by hydrophilic (k = 1.0) silica surfaces separated by distances between 0.6 and 5.0 nm. We find that the width of the region characterized by surface-dominated slowing down of water rotational dynamics is ~0.5 nm, while the corresponding width for translational dynamics is ~1.0 nm. The different extent of perturbation undergone by the in-plane dynamic properties is evidence of rotational-translational decoupling. The local in-plane rotational relaxation time and translational diffusion coefficient collapse onto confinement-independent ²master² profiles as long as d ³ 1.0 nm. Long-time tails in the perpendicular component of the dipole moment autocorrelation function are indicative of anisotropic behavior in the rotational relaxation.

See more of this Session: Transport at Interfaces
See more of this Group/Topical: Engineering Sciences and Fundamentals

Transcorneal Penetration of Rhodamine B Across the Rabbit Cornea Mounted in Vitro

Chhavi Gupta, Department of Chemical Engineering, University of Florida, Gainesville, FL

Sangly P. Srinivas, Department of Optometry, Indiana Universtity, Bloomington, IN

Anuj Chauhan, Chemical Engineering, University of Florida, Gainesville, FL

A cornea is three-layer membrane comprising of the outermost epithelium, the middle water-like stroma and then a single layer of endothelium. The mechanisms of transport of drugs across the corneal membrane play an important role in determining the bioavailability of the ophthalmic drugs. This talk will focus on the mechanisms of transport across the corneal membrane of Rhodamine B (RhB), a fluorescent lipophilic molecule.

To experimentally measure the transport of RhB through the rabbit cornea, rabbit corneas were mounted in vitro and perfused with HCO3- Ringers at the anterior and posterior surfaces. About 30 min after mounting, the epithelial surface was exposed to RhB (10 μM) dissolved in the Ringers. Corneal thickness and concentration profile of RhB across the cornea were obtained periodically (every 2-3 min) using a custom-built scanning microfluorometer at a high depth resolution (at least 8 μm using a 40x water immersion objective with 0.75 NA). A multiscale cornea transport model was developed and the transient concentration profiles are fitted to the model to obtain all the parameters. Sensitivity analysis was conducted to determine the reliability of the fitted parameters.

After topical administration, RhB distribution across the corneal membrane showed discontinuities at the cellular boundaries. Specifically, RhB fluorescence was elevated in the lipophilic cellular layers relative to the hydrophilic stromal layer. A mathematical model consisting of three partial differential equations governing diffusion of RhB across the cornea was developed. Additionally slow transport of RhB from the bilayers to hydrophobic sites in the interior of the cells in the epithelium and endothelium was included. Also a transport barrier was included at the interface between the epithelium and the stroma. The experimental data was fitted to the model with reliable parameters except the endothelium parameters, which show less sensitivity due to a limited amount of data from endothelium because of its small thickness.

The mathematical model developed here accurately characterizes the transient solute transport through the corneal membrane. The conventional approaches using a lumped permeability cannot adequately describe the mechanistic of transcorneal penetration of lipophilic drugs particular for delivery by eye drops because the residence time of drugs in eyes is much small than the time needed to establish a pseudo-steady profile, and thus the lumped permeability of the membrane changes with time.

See more of this Session: Transport at Interfaces
See more of this Group/Topical: Engineering Sciences and Fundamentals

Visualization and Interfacial Microrhoelogy of Phospholipid Monolayers with and without Cholesterol

SiYoung Q. Choi, Chemical Engineering, University of California, Santa Barbara, Santa Barbara, CA

Andrew J. Pascall, Chemical Engineering, University of California, Santa Barbara, Santa Barbara, CA

Todd M. Squires, Chemical Engineering, University of California at Santa Barbara, Santa Barbara, CA

Phospholipid monolayers (with and without cholesterol) at the air-water interface serve as model systems for various biological interfaces, e.g. lung surfactant layers and outer leaflets of cell membranes. Although the dynamical (viscoelastic) properties of these interfaces may play a key role in stability, dynamics and function, the relatively weak rheological properties of most such monolayers have rendered their study difficult or impossible. Using a novel technique we have recently developed, we measure the viscoelastic properties of phospholipid monolayers, both with and without cholesterol, at the air-water interface. In particular, our measurements suggest remarkably long relaxation times (~0.1 sec.) Notably, our apparatus enable direct interfacial visualization during the measurement, which enables us to relate the observed interfacial structure and dynamics to the measured viscoelastic behavior. The influence of cholesterol upon these dynamic structures -- and thus rheology -- is visualized, measured and modelled.

See more of this Session: Transport at Interfaces
See more of this Group/Topical: Engineering Sciences and Fundamentals

Investigation of pH-Dependent Formation of Lipid Heterogeneities On Giant Unilamellar Vesicles

Amey Bandekar, Othmer-Jacobs Department of Chemical and Biological Engineering, Polytechnic Institute of NYU, Brooklyn, NY

Stavroula Sofou, Othmer-Jacobs Department of Chemical and Biological Engineering, Polytechnic Institute of NYU, Brooklyn, NY

There is growing evidence that nature uses lipid membranes as a universal wrap around cells to control critical functions by dynamically reorganizing lipids into rafts. These are nanometer- to micron-sized lipid domains of laterally phase separated lipids occurring at the sites of action during cell-to-cell communication. Understanding the biophysical forces among lipids that play a role in these phenomena, and characterizing the potential changes of collective physical properties of membranes caused by these phenomena may potentially impact the control of related cell functions and associated diseases.

We study the kinetics of lipid phase separation, and domain formation and growth in phospholipid bilayers in the form of giant unilamellar vesicles (GUV's) using fluorescence microscopy. The bilayers are composed of: a first lipid type with a zwitterionic phosphatidylcholine (PC) headgroup, a second lipid type with a titratable headgroup such as phosphatidic acid (PA) or phosphatidyl serine (PS), and cholesterol. Phase separation vs. pH is evaluated in these studies. In all preparations GUV's were labeled with C20:0-DiI and (16:0,Bodipy)-PC, both at 0.1 mol %, and observed at 25 °C.

For GUV's composed of DOPC:DSPA:Chol (at 1:1:1 mole ratio) at pH 7.0, the membrane appears homogenously mixed. At the lower pH of 5.0, phase separated domains are observed with complex shapes that could be the result of aggregation and ripening of smaller circular domains. When DSPA is replaced by DSPS, small circular phase separated domains are formed at pH 7.0. At the lower pH values of 6.0 and 5.0, larger phase separated domains with complex shapes are also observed. Phase separated domains should be rich in protonated PA or PS lipid. Domain formation is attributed to decrease in electrostatic repulsion among charged lipids and formation of hydrogen bonding between protonated lipids. We are currently investigating the extent to which heterogeneous short-range lipid order within the domains and diffusion-limited domain ripening affect the formation of the observed complex shapes.

To understand the effect of lipid heterogeneities on altering long range membrane structure, optical observations are compared to membrane characterization using differential scanning calorimetry.

Abbreviations: dioleoyl phosphocholine (DOPC), distearoyl phosphatidic acid (DSPA), distearoyl phosphatidyl serine (DSPS), cholesterol (Chol), 1,1′-dieicosanyl-3,3,3′,3′-tetramethylindocarbocyanine perchlorate (C20:0-DiI), 1-hexadecanoyl-2-(4,4-difluoro-5,7-dimethyl-4-bora-3a,4a-diaza-s-indacene-3-pentanoyl)-sn-glycero-3-phosphocholine (16:0,Bodipy-PC).

See more of this Session: Transport at Interfaces
See more of this Group/Topical: Engineering Sciences and Fundamentals

Dielectrophoresis and AC-Field Induced Encapsulate Release of Micelles in Aqueous Suspensions

Yingxi Elaine Zhu, Department of Chemical and Biomolecular Engineering, University of Notre Dame, Notre Dame, IN

Victoria Froude, University of Notre Dame, Notre Dame, IN

Biocolloids, such as micelles and liposomes, have been of increasing interest as drug delivery systems for controlled release and specific cell targeting via advanced electrokinetic and microfluidic techniques, such as dielectrophoresis (DEP), where colloid particles can be effectively and rapidly manipulated and assembled in suspensions by the use of an AC-electric field. In this work, we examine the dynamic responses of micelles, of 3-5 nm in diameter, to applied AC-electric fields by using fluorescence correlation spectroscopy (FCS) at a single-molecule resolution. We focus on the AC-field induced transport of sodium dodecyl sulfate (SDS) micelles tagged with various fluorescent probes and molecular encapsulates in aqueous media. Micelle assembly concentration and DEP mobility of SDS micelles are accurately examined over a broad range of AC-field frequency (1 KHz -10 MHz) and amplitude (100mV -20V). We observe a strong AC-frequency dependence of micelle concentration between two microelectrodes, from which the DEP crossover frequency switching between the positive and negative DEP response for nanocolloids of less than 10 nm is determined for the first time. Surprisingly, we also observe the AC-field induced dissociation of the micelle structure and the resultant release of fluorescent encapsulates at a characteristic low AC-field frequency of ~1-10 kHz; the dissociation frequency can be tuned by encapsulated molecules with a strong dependence of their surface conductivity, which could have a potential application for controlled drug release by AC-electric fields.

See more of this Session: Transport at Interfaces
See more of this Group/Topical: Engineering Sciences and Fundamentals

Understanding Nanoparticle Diffusion and Exploring Interfacial Nanorheology using Molecular Dynamics Simulations

*Yanmei Song, Mingxiang Luo, and Lenore L. Dai**

Department of Chemical Engineering, Arizona State University, Tempe, AZ 85287

Since Einstein's famous publication [1] on Brownian motion, it is well-known that for a spherical particle larger than the solvent molecules, the diffusion of the particle in a low-viscosity medium can be predicted by the Stokes-Einstein (SE) equation:

$$D = \frac{k_B T}{6\pi\eta a} \tag{1}$$

where D is the diffusion constant, k_B is Boltzmann's constant, T is the absolute temperature, η is the viscosity of the fluid, and a is the particle's radius. However, the motion of the particle suspended in a complex fluid can be very different due to the viscoelastic property of the complex fluid. Another important remaining unanswered question is whether or not the SE equation is applicable to the diffusion of nanoparticles. For example, Tuteja *et al* [2] reported that cadmium selenide nanoparticles (hydrodynamic radius of 4.7 nm) diffuse approximately 200 times faster in a polystyrene melt than prediction from the SE equation by using X-ray photon correlation spectroscopy (XPCS). Understanding nanoparticle diffusion is important due to the tremendous applications of nanoparticles and "any study targeting the spatial assembly of nanostructures requires an understanding of the nanoparticle dynamics to delineate the kinetics" [2].

Furthermore, a relatively new technique based on microparticle motion called microrheology, pioneered by Mason and Weitz,[3,4] has received increasing attention. In microrheology, the particle motion is directly related to the shear modulus of the medium by the generalized Stokes-Einstein equation (GSE) [3,4]:

$$\tilde{G}(s) = \frac{k_B T}{\pi a s \langle \Delta \tilde{r}(s) \rangle} \tag{2}$$

where k_B is the Boltzmann's constant, T is the absolute temperature, a is the particle's radius, and $\langle \Delta \tilde{r}(s) \rangle$ is the unilateral Laplace transform of the mean square displacement $\langle \Delta \tilde{r}(\tau) \rangle$ of the particle. Accordingly, the rheological properties of the material can be extracted by tracking the movements of individual Brownian microparticles embedded in the material. Microrheology has received increasing attention since it not only overcomes the volume limitations of the biomaterial samples (for example, the living cells) but also allows us to obtain extended high frequency responses [3-11]. More importantly, it allows investigation of the local inhomogeneity of material. As a result of all these strengths, microrheology has lead to advancements in many research fields such as living cells [12-16], hydrogels [17], polymer solutions [18-20], membranes [21,22], and various soft biomaterials. Most of the microrheology experiments involve using microparticles as tracers, often due to the limited experimental particle tracking techniques on ultra-small particles. One intuitive question to ask is whether or not nanorheology can be developed based on the diffusion of nanoparticles since nanoscale materials or structures become increasingly important.

Recently, we have pioneered the work of developing interfacial microrheology [23,24], which is important in understanding many phenomena such as stabilizing and processing of foams and emulsions [25,26], proteins films [27,28], biomembranes, and Langmuir-Blodgett films. The single-particle tracking technique has been successfully validated at oil-water interfaces and the complex, loss, and storage moduli of oil-water interfaces as a function of frequency measured from microrheology are compared with those of bulk oils measured from a conventional rhoemeter and developed bulk microrheology [23,24]. However, we found that the nature of the tracer particles plays an important role in the single-particle apparent microrheology at the oil-water interfaces, especially when the oil is viscoelastic. One challenge in the experimental work is the difficulty to quantify the exact immersion depths of the different tracer particles at oil-water interfaces. In this report, we perform MD simulations to investigate the diffusion of nanoparticles and explore the potential of developing single-particle nanorheology based on the motion of the nanoparticles. MD simulation has been a powerful tool for obtaining molecularly detailed information and the underlying physics of various systems. Here we use MD simulation as a tool since it offers unique opportunities to explore the dynamics of nanoparticles with particle size beyond experimental capability and more importantly, well-defined oil-water interfaces may reveal precise location of the nanoparticles at oil-water interfaces.

All MD simulations have been performed using GROMACS 3.3.3 package [29,30]. After the initial configurations were obtained, energy was minimized using the steepest descent method. The leap-frog algorithm was used for integrating Newton's equation of motion with a time step of 0.004 ps. All simulations were carried out under NPT (constant number of molecules, constant pressure, and constant temperature) ensemble using the Berendsen-thermostat to a temperature and pressure bath at 300 K and 1 bar. Periodic boundary conditions (PBC) were applied to all three directions of the simulated boxes. The initial atomic velocities were generated with a Maxwellian distribution at the given absolute temperature. The Particle-Mesh Ewald (PME) method was used for the long range electrostatic interactions. The cut-off for Lennard–Jones forces and Coulomb forces were both set as $r < 1.2$ nm. After the simulation, the physical properties were characterized using the GROMACS analysis tools, and the structures were visualized by Visual Molecular Dynamics (VMD) [31].

The water model used in this work was single point charge (SPC). The spherical modified hydrocarbon nanoparticle (HCP, mean diameter of 1.2 nm) was truncated from a diamond-like lattice made of carbon atoms that bonded in nonplanar hexagonal structure, and to increase the simulation efficiency, saturated with united CH, CH_2 and CH_3 atoms. The force field parameters for PDMS developed by Sok and co-workers [32] were adapted in this work. Each PDMS molecule consists of 10 or 28 repeating units, with the CH_3 groups being taken as the end groups. To increase the computational speed, the united-atom (UA) model was used by treating each CH_3 group in PDMS molecule as a single atom, which reduces the number of atoms per monomer from ten to four.

We have simulated two types of systems: liquid bulk systems with one nanoparticle adding to the center of the simulation box and PDMS (10 repeating units) and water with one nanoparticle at the interface. The detailed information of simulated systems is listed in Table 1. The averaged mass density of simulated water, PDMS with 10 repeating units, and PDMS with 28 repeating units are 971 kg/m^3, 1,017 kg/m^3, and 1,031 kg/m^3, respectively, which are comparable to the experimental values. The viscosities of bulk liquids were calculated using

Table 1. Composition of the Simulation Systems

Systems	PDMS	Water	Nanoparticle	Size of simulation box (nm^3)	Parallel runs
water	0	33163	1^a	10.07×10.07×10.07	8
PDMS (10 repeating units)	513	0	1^a	8.59×8.59×8.59	8
PDMS (28 repeating units)	160	0	1^a	8.23×8.23×8.23	8
PDMS (10 repeating units)-water interface	537	20366	1^b	8.94×9.09×15.92	6

[a] The nanoparticle is non-charged. [b] The nanoparticle is either non-charged or with different negative surface charges.

the periodic perturbation method [33] by performing nonequilibrium molecular dynamics (NEMD) simulations. Four 5 nm simulated boxes were stacked in z direction in order to produce better statistics. The viscosity estimated for SPC water is 0.48 cP. Smith and Van Gunsteren [34] obtained the viscosity of 0.54 cP for a much smaller system of 512 SPC water molecules with a cut-off of 0.9 nm by employing the pressure fluctuations method which is based on equilibrium MD simulations. However because of the large fluctuations of pressure in a simulation box, it is usually difficult to make converged viscosity estimations by using pressure fluctuations [33]. By applying the same NEMD method, we obtained the viscosities of PDMS with 10 repeating units and 28 repeating units of 7.9 cP and 29.8 cP, respectively. For the simulations of liquid-liquid interfaces, the particular importance in determining the adequacy of the potentials used in the simulations is the calculation of the surface/interfacial tension [35]. The interfacial tension of simulated PDMS (10 repeating units)-water interface here is 42.2±0.4 mN/m, which agrees with the experimental value 42.7±0.1 mN/m.

One of our important objectives here is studying the diffusion of tracer nanoparticles and testing whether or not the SE equation is applicable to nanoparticles because the micro/nanorheology is based on the Brownian diffusion of the tracers. Fig. 1a plots the mean square displacements of a single nanoparticle in bulk water and PDMS oils, respectively. As expected, the nanoparticle moves slower with increasing fluid viscosity. Here a local power law of $\langle \Delta r^2(\tau) \rangle$ is assumed as a function of lag time [36], $\langle \Delta r^2(\tau) \rangle = \tau^{\alpha}$, for thermally driven particles moving in fluid. The slope α shall fall between zero and one, with zero corresponding to pure elastic confinement and one corresponding to pure viscous diffusion. The $\langle \Delta r^2(\tau) \rangle$ in xyz directions are shown on a log-log scale in Fig. 1b, with the linear fitted slopes and R^2 noted. The $\langle \Delta r^2(\tau) \rangle$ of the nanoparticles in water and PDMS (10 repeating units) changes linearly with time, suggesting that the motion of the nanoparticle in both water and low viscosity PDMS oil is pure diffusive Brownian motion. It is noticeable that the nanoparticle starts to show sub-diffusive behavior (the slope is slightly less than 1) in PDMS (28 repeating units); however, we still calculated the mean diffusion constant for a rough comparison. The diffusion constants of the nanoparticles were calculated from the mean square displacements obtained from one particle trajectories. The result is averaged over eight parallel runs. In parallel, we estimated the diffusion constants of the nanoparticles by Eq. (1). It is worthwhile to note that we used the simulated viscosities of bulk liquids to estimate the diffusion constants here since the nanoparticles are immersed in such media. The diffusion constants of the nanoparticle in water

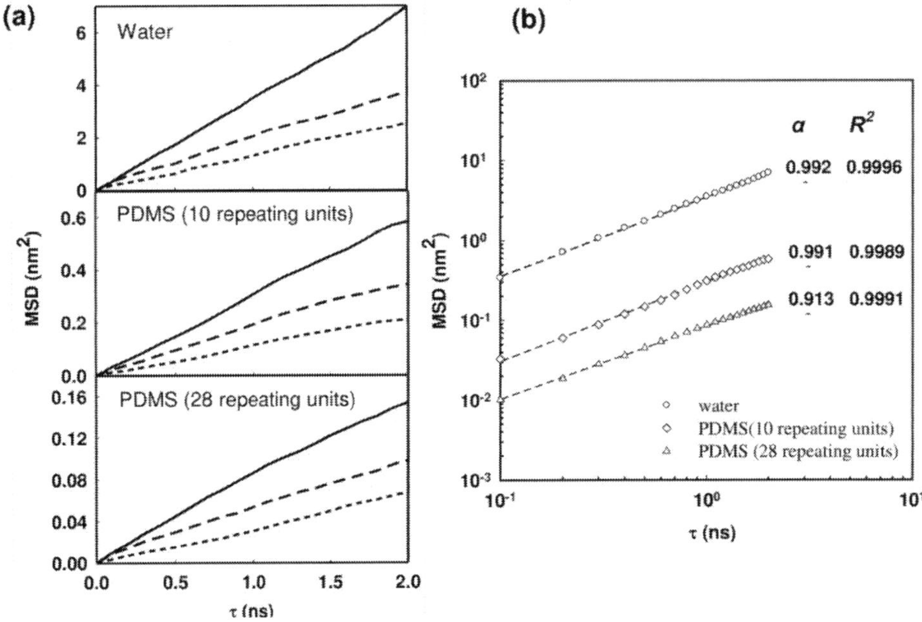

Fig. 1 (a) Mean square displacements of single nanoparticles in bulk liquids. The displacements in xyz, xy, and z directions are represented in sold, long-dash, and short-dash lines, respectively. The xy and z directions are included in order to compare with the mean square displacements of nanoparticles at interfaces. (b) The corresponding logarithm plots of mean square displacements in xyz directions in (a). The dashed lines represent the linear regression of the data with the slope α and R^2 noted next to each line.

Table 2. Diffusion Constants of Nanoparticles in Bulk Water and PDMS Oils

Systems	Viscosity (cP)	D_{MD} ($\times 10^{-5}$ cm^2/s)	D_{SE} ($\times 10^{-5}$ cm^2/s)
SPC Water	0.48	0.635±0.059	0.763
PDMS (10 repeating units)	7.9	0.042±0.008	0.047
PDMS (28 repeating units)	29.8	0.010±0.002	0.012

and PDMS oils are summarized in Table 2. Results clearly show that the simulated diffusion constants for the nanoparticle in all three bulk liquids are reasonably close to that predicted by the SE equation.

So far, we have suggested that the motion of the nanoparticle in bulk water and low viscosity PDMS oils seems to be described by the SE equation. Next, we shift our focus to explore the potential of developing nanorheology using nanoparticles as tracers. In microrheology, the mean square displacement of the particle can be converted into shear modulus by Eq. (2), which can also be written in Fourier domain as [36]:

$$G^*(\omega) = \frac{k_B T}{i\pi a \omega F_u\left[\left\langle \Delta r^2(\tau)\right\rangle\right]} \quad (3)$$

where $F_u\left[\left\langle \Delta r^2(\tau)\right\rangle\right]$ is the Fourier transform of $\left\langle \Delta r^2(\tau)\right\rangle$. Based on Mason's method [35], the shear modulus can be expressed as:

$$\left|G^*(\omega)\right| \approx \frac{k_B T}{\pi a \left\langle \Delta r^2(1/\omega)\right\rangle \Gamma\left[1+\alpha(\omega)\right]} \quad (4)$$

where $\alpha(\omega)$ is the local first logarithmic derivative of $\left\langle \Delta r^2(\tau)\right\rangle$, given by $\left|d\ln(\Delta r^2(\tau)/d\ln(t)\right|_{\tau=1/\omega}$, corresponding to the local power law relation of $\left\langle \Delta r^2(\tau)\right\rangle = \tau^\alpha$. The storage modulus $G'(\omega)$ and loss modulus $G''(\omega)$, defined by $G^*(\omega) = G'(\omega) + iG''(\omega)$, are given as:

$$G'(\omega) = \left|G^*(\omega)\right|\cos(\pi\alpha(\omega)/2) \quad (5)$$

$$G''(\omega) = \left|G^*(\omega)\right|\sin(\pi\alpha(\omega)/2) \quad (6)$$

In this work, Eqs (4)-(6) were used to calculate the shear modulus from the simulated mean square displacement (MSD) data. We calculated the shear moduli of bulk water and PDMS (10 repeating units) oil from simulated 3D mean square displacements. Fig. 2 shows that the loss modulus G'' dominates and no storage modulus G' can be detected over the entire "frequency" range. The "frequency" here is converted from simulation time scale. The results are reasonable since they reflect the pure viscous properties of the pure water and the low viscosity PDMS oil (Fig. 1b). We also noted that for pure viscous fluid, the loss modulus G'' theoretically obeys $G''(\omega) = \eta\omega$ [36], based on the theory of GSE. This indicates that the slopes of the plots in Fig. 2 should give us the viscosities of simulated SPC water and PDMS

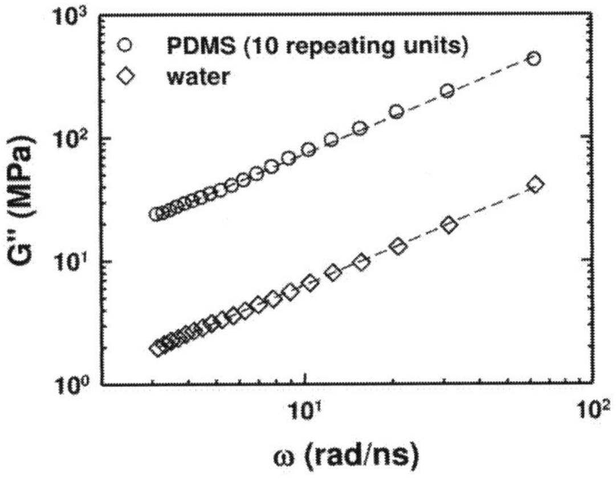

Fig. 2 Simulated shear moduli of pure water and PDMS (10 repeating units) as a function of "frequency". The dashed lines represent the linear regression of the data.

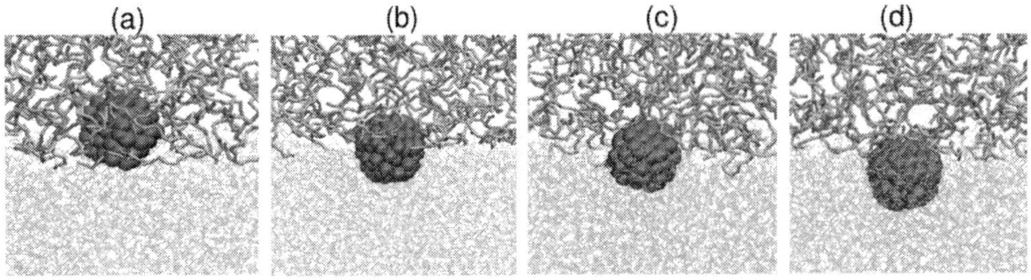

Fig. 3 The immersion of the nanoparticles with different surface charges at the PDMS(10 repeating units) oil-water interfaces: (a) non-charged; (b) 4 negative charges; (c) 6 negative charges; (d) 8 negative charges. The nanoparticle is in red sphere, the water phase is in blue, and the PDMS phase is in green.

with 10 repeating units. The slopes, which are essentially the Newtonian viscosity, are 0.6 cP and 6.9 cP, which reasonably agree with simulated viscosities of 0.48 cP and 7.9 cP, respectively. This is an important step because it confirms the potential for extending the validity of GSE, and in other words, the microrheology method, to nano-sized tracer particles.

We started our study by using the non-charged nanoparticle but have found that the surface charge may be used as one efficient parameter to control the location of the nanoparticles at the oil-water interfaces, likely due to the affinity of the surface charge to the water molecules. Each negative surface charge was added to the nanoparticle by removing one hydrogen atom on the CH_3 united-atoms. Fig. 3 shows the significant change of the immersion depth of the nanoparticle at PDMS oil-water interfaces with different nanoparticle surface charges. The non-charged nanoparticle almost completely immerses in the PDMS oil phase, as shown in Fig. 3a, which is likely due to the hydrophobic nature of the hydrocarbon nanoparticle. It is clear that the nanoparticle immerses more deeply into the water phase with increasing the surface charges, as shown in Fig. 3b and 3c. This is ascribed to the polarities of the nanoparticle and the two liquid phases forming the interface. The PDMS oil is a totally nonpolar substance, which makes it not sustainable to any charges. Eventually, the nanoparticle nearly immerses in the water phase only, as shown in Fig. 3d. The observation here explained by the theory proposed by Nikolaides and Weitz [37], suggesting that the energy of the electrostatic field generated by the charged particles at the liquid interface could be reduced by a deeper immersion of the particles into the water phase, where ions screen the electrostatic field.

Surprisingly, we found that the $\langle \Delta r^2(\tau) \rangle$ of the nanoparticles in xy plane is independent of the location of the nanoparticles (Fig. 4a), although the viscosity ratio of the oil to water is approximately nine. The inset illustrates a typical trajectory of a single nanoparticle (non-charged) in the xy plane at a PDMS oil (10 repeating units)-water interface. The corresponding calculated interfacial loss modulus is shown in Fig. 4b; the interfacial viscosity, calculated from the slops, is 5.7±0.2 cP, which lies between the viscosities of pure water and PDMS oil (10 repeating units). This is an important simulation result because it may suggest that the apparent interfacial nanorheology results of low viscosity oil-water interface by single-nanoparticle tracking are independent of the location of the tracer nanoparticles. Further

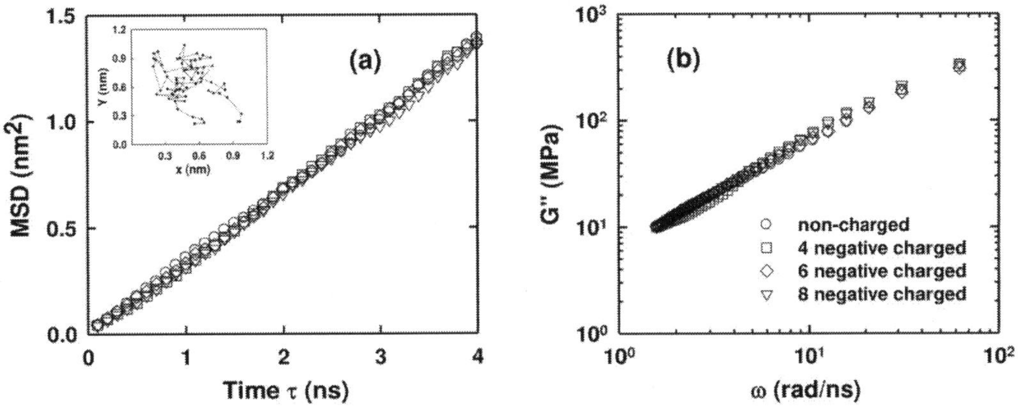

Fig. 4 (a) Effect of surface charges (tracer particle location) on the mean square displacements of nanoparticle at PDMS (10 repeating units) oil-water interface. The inset is the 2D trajectory of the nanoparticle at the interface. (b) Dependence of apparent interfacial nanorheology of PDMS (10 repeating units) oil-water interfaces on the location of the tracer nanoparticle. The non-charged, 4 negatively-charged, 6 negatively-charged, and 8 negatively-charged nanoparticles are represented in circles, squares, diamonds and triangles, respectively.

Table 3. Diffusion Constants of Nanoparticles at PDMS Oil-Water Interfaces

Surface charges of nanoparticle	D_{xyz} ($\times 10^{-5}$ cm^2/s)	D_{xy} ($\times 10^{-5}$ cm^2/s)	D_z ($\times 10^{-5}$ cm^2/s)
no charge	0.077±0.004	0.082±0.009	0.043±0.004
4 negative charges	0.059±0.006	0.082±0.007	0.003±0.001
6 negative charges	0.057±0.007	0.081±0.008	0.017±0.002
8 negative charges	0.067±0.004	0.084±0.005	0.019±0.002

analyses show that the immersion depth doesn't affect the lateral diffusion constants (xy plane) of the nanoparticles at the oil-water interfaces but the motion of the nanoparticle in z direction. The comparison of diffusion constants of the nanoparticles with different surface charges at PDMS oil-water interfaces are summarized in Table 3.

In summary, we have studied the dynamics of single nanoparticles using MD simulations. The diffusive behavior of nanoparticles, calculated from the mean square displacements, in pure water and low viscosity PDMS oil bulk are found to be consistent with the prediction from the Stokes-Einstein equation. In addition, we have calculated the shear moduli and viscosities of the three-dimensional (3D) PDMS oil and water and the two-dimensional (2D) PDMS oil-water interface from the trajectory of the single nanoparticles, which demonstrates the validity of nanorheology. Surprisingly, we found that the lateral diffusion of nanoparticles as well as apparent interfacial nanorheology at the PDMS oil (low viscosity)-water interface are independent of the position of the nanoparticle at the interface.

We thank the Arizona State University Fulton High Performance Computing Initiative (HPCI) for computational resources. We are also grateful to the financial support from the National Science Foundation (CBET-0644850).

[1] A. Einstein, Ann Phys. 17, 549 (1905).
[2] A. Tuteja, M. E. Mackay, S. Narayanan, S. Asokan, and M. S. Wong, Nano Lett. 7, 1276 (2007).
[3] T. G. Mason, and D. A. Weitz, Phys. Rev. Lett. 74, 1250 (1995).
[4] Mason, T. G.; Ganesan, K.; VanZanten, J. H.; Wirtz, D.; Kuo, S. C. Phys. Rev. Lett. 79, 3282 (1997).
[5] F. C. MacKintosh, and C. F. Schmidt, Curr. Opin. Colloid Interf. Sci. 4, 300 (1999).
[6] T. Gisler, and D. A. Weitz, Curr. Opin. Colloid Interf. Sci. 3, 586 (1998).
[7] A. J. Levine, and T. C. Lubensky, Phys. Rev. Lett. 85, 1774 (2000).
[8] F. Gittes, B. Schnurr, P. D. Olmsted, F. C. MacKintosh, and C. F. Schmidt, Phys. Rev. Lett. 79, 3286 (1997).
[9] J. C.Crocker, M. T. Valentine, E. R. Weeks, T. Gisler, P. D. Kaplan, A. G. Yodh, and D. A. Weitz, Phys. Rev. Lett. 85, 888 (2000).
[10] H. Qian, Biophys. J. 79, 137 (2000).
[11] E. K. Hobbie, S. L. Gibson, and S. Kumar, Phys. Rev. Lett. 100, 076001 (2008).
[12] B. Fabry, G. N. Maksym, J. P. Butler, M. Glogauer, D. Navajas, and J. J. Fredberg, Phys. Rev. Lett. 87, 148102 (2001).
[13] M. L. Gardel, F. Nakamura, J. H. Hartwig, J. C. Crocker, T. P. Stossel, and D. A. Weitz, Proc. Nat. Acad. Sci. U.S.A. 103, 1762 (2006).
[14] M. L. Gardel, M. T. Valentine, J. C. Crocker, A. R. Bausch, and D. A. Weitz, Phys. Rev. Lett. 91, 158302 (2003).
[15] A. Palmer, T. G. Mason, J. Y. Xu, S. C. Kuo, and D. A. Weitz, Biophys. J. 76, 1063 (1999).
[16] Y. Tseng, T. P. Kole, and D. A. Weitz, Biophys. J. 83, 3162 (2002).
[17] R. E. Mahaffy, C. K. Shih, F. C. MacKintosh, and J. Kas, Phys. Rev. Lett. 85, 880 (2000).
[18] B. R. Dasgupta, S. Y. Tee, J. C. Crocker, B. J. Frisken, and D. A. Weitz, Phys. Rev. E. 65, 051505 (2002).
[19] B. R. Dasgupta, and D. A. Weitz, Phys. Rev. E. 71, 021504 (2005).
[20] J. J. Elmendorp, Polym. Eng. Sci. 26, 418 (1986).
[21] M. J. Saxton, and K. Jacobson, Annu. Rev. Biophys. Biomol. Struct. 26, 373 (1997).
[22] E. Helfer, S. Harlepp, L. Bourdieu, J. Robert, F. C. MacKintosh, and D. Chatenay, Phys. Rev. Lett. 85, 457 (2000).
[23] J. Wu, and L. L. Dai, Appl. Phys. Lett. 89, 094107 (2006).
[24] J. Wu, and L. L. Dai, *Langmuir.* 23, 4324 (2007).
[25] D. Langevin, Adv. Colloid Interface Sci. 88, 209 (2000).
[26] C. Dicharry, D. Arla, A. Sinquin, A. Graciaa, and P. Bouriat, J. Colloid Interface Sci. 297, 785 (2006).
[27] E. Dickinson, Colloids Surf. *B.* 20, 197 (2001).
[28] B. S. Murray, Curr. Opin. Colloid Interf. Sci. 7, 426 (2002).
[29] E. Lindahl, B. Hess, and D. Van der Spoel, J. Mol. Model. 7, 306 (2001).
[30] D. Van der Spoel, E. Lindahl, B. Hess, G. Groenhof, A. E. Mark, and H. J. C. Berendsen, J. Comput. Chem. 26, 1701 (2005).
[31] W. Humphrey, A. Dalke, and K. Schulten, J. Mol. Graphics. 14, 33 (1996).
[32] R. M. Sok, H. J. C. Berendsen, and W. F. Vangunsteren, J. Chem. Phys. 96, 4699 (1992).
[33] B. Hess, J. Chem. Phys. 116, 199 (2002).

[34] P. E. Smith, and W. F. Van Gunsteren, Chem. Phys. Lett. 215, 315 (1993).
[35] I. Benjamin, Annu. Rev. Phys. Chem. 48, 407 (1997).
[36] T. G. Mason, Rheol. Acta. 39, 371 (2000).
[37] M. G. Nikolaides, A. R. Bausch, M. F. Hsu, A. D. Dinsmore, M. P. Brenner, D. A. Weitz, and C. Gay, Nature. 420, 299 (2002).

Sorbate Diffusion On Different Length Scales in SBA-15 Porous Solids by Pulsed Field Gradient NMR and Zero Length Column Techniques

Amrish R. Menjoge, Department of Chemical Engineering, University of Florida, Gainesville, FL

Qinglin Huang, Department of Chemical Engineering, University of New Brunswick, Fredericton, NB, Canada

Mladen Eic, Department of Chemical Engineering, University of New Brunswick, Fredericton, NB, Canada

Sergey Vasenkov, Department of Chemical Engineering, University of Florida, Gainesville, FL

Silicates SBA-15 containing mesopores and micropores are promising materials for a variety of applications including catalysis, sensor development and drug delivery. Transport of sorbate molecules in the pore networks of SBA-15 particles is expected to play a notable role in these applications. Despite a general consensus on the importance of sorbate transport in SBA-15 materials, properties of this transport have not been investigated as thoroughly as those of zeolites. Here we report results of diffusion studies of guest molecules in selected SBA-15 samples for different length scales of molecular displacements. The diffusion data were obtained by pulsed field gradient (PFG) NMR method under conditions of application of large (up to 30 T/m) gradients as well as by zero length column (ZLC) technique. Comparison of the data obtained by these two techniques for the same sorbate/SBA-15 systems can provide some new insights on the diffusion behavior. In addition, analysis of the dependence of effective diffusivities on the length scale of molecular displacements measured by PFG NMR has the potential to reveal important information on details of molecular transport in SBA-15 materials.

See more of this Session: Transport at Interfaces
See more of this Group/Topical: Engineering Sciences and Fundamentals

Magnetic Enhancement of Phototaxing Catalytic Motors

Neetu Chaturvedi, Chemical Engineering, The Pennsylvania State University, University Park, PA

Darrell Velegol, Department of Chemical Engineering, The Pennsylvania State University, University Park, PA

Yiying Hong, Department of Chemistry, The Pennsylvania State University, University Park, PA

Ayusman Sen, Department of Chemistry, The Pennsylvania State University, University Park, PA

Asymmetric colloidal doublets have been shown to undergo phototaxis. Micron-size colloidal heterodoublets of silver-polystyrene latex microspheres and silver-silica microspheres were fabricated using the "salting out-quenching" technique. The asymmetric colloidal doublets behave as catalytic motors and show self-propulsion and phototaxis under ultraviolet light in the presence of hydrogen peroxide by the mechanism of diffusiophoresis. Magnetic enhancement of phototaxis has been observed with magnetic doublets. Magnetic heterodoublets of silver-magnetic polystyrene latex microspheres were fabricated using the "pH controlled assembly" technique. Catalytic magnetic doublets align themselves in response to an external magnetic field and form exclusion regions around them under ultraviolet light in the presence of hydrogen peroxide. This talk will demonstrate a simple, quick and inexpensive approach for the fabrication of synthetic motors capable of performing directed motion and forming exclusion zones and patterns. The fabrication of small scale synthetic motors is an important step in developing practical nanomachines, which can be used for several applications including roving sensors, drug delivery vectors, and self-assembly of superstructures.

See more of this Session: Colloidal Assembly and Fabrication I
See more of this Group/Topical: Engineering Sciences and Fundamentals

Engineering Anisotropic Colloids for Magnetic Assembly

Jung Hun Song, Department of Chemical Engineering, The City College of City University of New York, New York, NY

Ilona Kretzschmar, Department of Chemical Engineering, The City College of City University of New York, New York, NY

Directed assembly of surface-anisotropic (sa) colloids has gained immense interest in recent years as a potential pathway in controlling and achieving hierarchical colloidal structures. Structures obtained from the assembly of surface-anisotropic particles in external fields are interesting for photonic applications.

Sulfate-terminated polystyrene (PS) colloids are surface-modified using iron deposition in high vacuum and oxygen-rich environments to obtain magnetic Janus colloids. Low concentrations of iron and iron-oxide Janus colloids are subjected to a permanent magnetic field where they are magnetized depending on their cap composition. The assembly behavior of the Janus colloids is studied as a function of (i) degree of oxidation, (ii) cap size, (iii) cap thickness, and (iv) solution medium composition. Three types of assembly behavior are observed varying from no assembly via staggered chains to double chains. Subsequently, staggered and double chains are isolated and characterized using Scanning Electron Microscopy, Transmission Electron Microscopy, and X-ray Energy Dispersive Spectroscopy.

See more of this Session: Colloidal Assembly and Fabrication I
See more of this Group/Topical: Engineering Sciences and Fundamentals

Integrated Self & Directed Assembly of Interfacial Colloidal Crystals Via Electric Field and Depletion Mediated Attraction

Michael A. Bevan, Chemical and Biomolecular Engineering, Johns Hopkins University, Baltimore, MD

The autonomous and reversible assembly of nano- and micro- scale components is an essential process in numerous emerging technologies. As a result, there is great interest in understanding how thermal motion, particle and surface interactions, and external fields can be optimally coupled in assembly processes to elicit desired material and device responses. We approach this problem by directly and quantitatively connecting colloidal interactions, dynamics, and microstructures using optical microscopy (i.e. evanescent wave, video, confocal) and simulation (i.e. Monte Carlo, Stokesian Dynamics) methods. Here we investigate how depletion attraction mediated self-assembly and electric field mediated directed-assembly can be integrated to assemble defect-free colloidal crystals and avoid undesirable states (e.g. amorphous, polycrystalline, or arrested configurations). General strategies investigated include: (1) initiating depletion mediated self-assembly processes and then applying electric field mediated directed-assembly to overcome kinetic bottlenecks, (2) using electric-field directed-assembly to generate non-equilibrium configurations as transition states before initiating depletion mediated self-assembly, and (3) simultaneously using depletion and electric field mediated colloidal crystal assembly. Our findings indicate how integration of self- and directed- assembly can lead to thermodynamically stable equilibrium colloidal crystals via non-equilibrium kinetic pathways.

See more of this Session: Colloidal Assembly and Fabrication I
See more of this Group/Topical: Engineering Sciences and Fundamentals

Directed Self-Assembly of Colloidal Crystals with Electric Fields

Jason M. McMullan, Chemical Engineering, University of Delaware, Newark, DE

Norman J. Wagner, Chemical Engineering, University of Delaware, Newark, DE

Directed Self Assembly (DSA) methods can create useful structures and devices using particles ranging in size from nanoparticles to micron sized particles. In this work, we explore the fundamental mechanics of electrical field driven (dielectrophoretic) DSA with the goal of enabling the engineering of methods for generating two and three-dimensional ordered structures of high fidelity and controlled architecture.

Highly ordered three-dimensional colloidal crystals are assembled from a dilute colloidal suspension by AC electrical fields with frequencies on the order of a kilohertz. This technique exploits dielectrophoretic forces induced in the double layers surrounding the particles by the applied field. Using the standard electrokinetic model, the induced DEP force is predicted to depend on the particle size, charge and electrolyte concentration. Studies on a model colloidal dispersion are compared to theoretical predictions (Hill et al., J Colloid Interface Sci., 258, 56 (2003)) through a predictive model that incorporates double layer polarization. Measurement of fundamental colloidal properties, such as electrophoretic mobility, validates the standard electrokinetic model. Modeling identifies appropriate ranges of solution conditions favoring particle ordering.

These crystalline structures are observed and the extent of order quantified both optically and with scattering techniques. In situ Small Angle Light Scattering (SALS) characterizes the degree of ordering and crystal structure for micron-sized particles. A new Small Angle Neutron Scattering (SANS) sample environment is developed for studying the ordering and crystal structure of nanoparticle assemblies. The kinetics of ordering are measured and compared to predictions based on the forces derived from the standard electrokinetic model. Dielectrophoretic ordering is demonstrated to form both 2D and 3D structures with micron lengthscale features where ordering conditions are chosen to dictate final crystalline structure and strength.

See more of this Session: Colloidal Assembly and Fabrication I
See more of this Group/Topical: Engineering Sciences and Fundamentals

Bio-Inspired Assembly of Nanoplatelets by Electrical Field

Wei-Han Huang, Cheimcal Engineering, University of Florida, Gainesville, FL

Tzung-Hua Lin, Cheimcal Engineering, University of Florida, Gainesville, FL

Inkook Jun, Material Science and Engineering, University of Florida, Gainesville, FL

Peng Jiang, Chemical Engineering, University of Florida, Gainesville, FL

Bottom-up self-assembly of nonspherical colloidal building blocks is of great interest for the development of new materials with potential applications in optoelectronics, photonics, magnetics, catalysis, and mechanics. Oriented assembly of platelet-like particles is particularly interesting as it enables the production of nanocomposites with greatly improved mechanical properties. This is inspired by the ordered assemblies of aragonite (a mineral form of CaCO3) platelets in the nacreous layer of mollusk shells. The intricate layered structures consisting of brittle minerals and soft proteins in these natural composites lead to the unique combination of stiffness, strength, and toughness. Here we report a simple electrophoretic deposition technology that enables rapid production of large-area polymer nanocomposites with layered structures mimicking natural nacres. Uniform, electrostatically stabilized gibbsite nanoplatelets with high aspect ratio are preferentially oriented parallel to the electrode surface when an external direct current electric field is applied. Homogeneous, optically transparent nanocomposites are obtained when the interstitials between the aligned nanosheets are infiltrated with polymer. The resulting ceramic-polymer nanocomposites exhibit excellent mechanical properties.

See more of this Session: Colloidal Assembly and Fabrication I
See more of this Group/Topical: Engineering Sciences and Fundamentals

Dielectrophoretic Assembly of Janus and Patchy Particles in AC Electric Fields

Sumit Gangwal, Department of Chemical and Biomolecular Engineering, North Carolina State University, Raleigh, NC

Amar B. Pawar, Department of Chemical Engineering, The City College of the City University of New York, New York, NY

Ilona Kretzschmar, Department of Chemical Engineering, The City College of the City University of New York, New York, NY

Orlin D. Velev, Department of Chemical and Biomolecular Engineering, North Carolina State University, Raleigh, NC

The application of external fields to Janus particles (whose halves are physically or chemically different) is a rapidly expanding research field, as it can be an efficient means of programmed assembly of novel materials. Janus particles were prepared by coating dielectric microspheres with a conductive metal layer on one hemisphere. The phase space for AC electric field intensity and field frequency was explored for these particles on a glass surface between two electrodes. A rich variety of structures and dynamics were uncovered, which are different than the ones in the directed dielectrophoretic assembly of plain dielectric or plain conductive particles. Even more unusual types of new structures formed when high frequency (> 200 kHz) AC electric fields are applied to suspensions of "patchy" particles. The patchy particles, produced by glancing angle metal deposition, have either a single metallic patch that is ~ 10% of the total latex particle surface or two metallic patches (each covering 25% of the total surface) on opposite poles of the particle. These patchy particles assemble perpendicular to or at an angle to the applied field direction, pre-programmed by the size and orientation of the patches. The patchy particles formed metallodielectric crystals of unusual symmetry at high enough concentrations of particles. The assembly of both Janus and patchy particles in the electric field was further investigated by using electrostatic simulations to calculate the electric energy of the system to determine the most favorable particle orientation and chain configuration. The simulation results indicate that the patchy particles interact as quadruples, which explains why they assemble into multi-directional chains. These experimental and modeling techniques could be used to engineer the processes of directed assembly of other types of anisotropic particles.

See more of this Session: Colloidal Assembly and Fabrication I
See more of this Group/Topical: Engineering Sciences and Fundamentals

Design and Fabrication of Micron-Scale Magnetic, Amphiphilic, Janus Structures

SiYoung Q. Choi, Chemical Engineering, University of California, Santa Barbara, Santa Barbara, CA

Andrew J. Pascall, Chemical Engineering, University of California, Santa Barbara, Santa Barbara, CA

Todd M. Squires, Chemical Engineering, University of California, Santa Barbara, Santa Barbara, CA

We present techniques we have developed for the design and fabrication of micron-scale structures with ferromagnetic or paramagnetic properties (as desired), whose surface chemistry can be controlled to impart solubility or interfacial activity. Using photolithography, we microfabricate structures with specified geometry, incorporating layers of magnetic metals (e.g. Nickel, Cobalt, Permalloy) to provide different magnetic properties as well as a gold layer which enables us to tune surface chemistry using versatile gold-thiol chemistry. Particle shape, size, and thickness can easily be tuned in addition to surface and magnetic properties. We describe the mechanisms that give rise to ferro- or para-magnetic structures, giving basic design rules for these novel structures. While our main purpose for synthesizing these structures is for use as probes for interfacial rheology, we describe more generally some novel feaures and potential applications involving these structures, including interfaces whose properties can be externally tuned, as well as the static and dynamic configurations adopted at fluid/fluid interfaces under applied magnetic fields.

See more of this Session: Colloidal Assembly and Fabrication I
See more of this Group/Topical: Engineering Sciences and Fundamentals

Block Copolymer Electrolytes for Battery Applications

Scott Allen Mullin, Chemical Engineering, University of California, Berkeley, Berkeley, CA

Ashoutosh Panday, Chemical Engineering, University of California, Berkeley, Berkeley, CA

Nitash Balsara, Chemical Engineering, University of California, Berkeley, Berkeley, CA

Energy density is the greatest limiting factor in batteries for electric vehicle applications. Lithium metal batteries address this limitation because lithium has the highest theoretical energy density in terms of both mass (Wh/kg) and volume (Wh/L). Dry solid polymer electrolytes (SPEs), consisting of a lithium salt solvated in a polymer host, are the leading candidate materials for these batteries because they can interface stably with lithium metal. The primary failure mechanism in SPE batteries is the growth of lithium metal dendrites at the anode during recharging. Theoretical work has suggested that improving the mechanical strength of the polymer may prevent dendrite growth - thus forming the basis for our present work. We are investigating the block copolymer poly(styrene)-block-poly(ethylene oxide) (PS-PEO). The polymer microphase separates - forming microdomains of PS and of PEO - and is designed so that the PS phase suppresses lithium dendrite growth, while the PEO phase solvates lithium bis(trifluoromethane)sulfonimide (LiTFSI) salt to form conducting pathways. Complete electrochemical characterization of PS-PEO/LiTFSI mixtures requires measurement of conductivity, salt diffusion coefficient, and lithium ion transference number. The present study covers PS-PEO copolymers that exhibit lamellar and cylindrical morphologies in the absence of salt. The addition of salt affects morphology but the relationships between morphology and electrochemical characteristics have not yet been clarified. Some aspects of these relationships will be presented.

See more of this Session: Interfacial Phenomena in Energy Systems
See more of this Group/Topical: Engineering Sciences and Fundamentals

Microstructure and Capacitance of the Electrical Double Layers at the Interface of Ionic Liquids and Planar Electrodes

Rui Qiao, Mechanical Engineering, Clemson University, Clemson, SC

Guang Feng, Mechanical Engineering, Clemson University, Clemson, SC

We present molecular dynamics simulations of the electrical double layers (EDLs) at the interface of ionic liquids [BMIM][NO3] and planar electrodes. Simulations confirm that a Helmholtz-like interfacial counterion layer exists when the electrode charge density is negative or strongly positive, but the counterion layer is not well-defined when the electrode charge density is weakly positive. The thickness of the EDL, as inferred from how deep the charge separation and orientational ordering of the ions penetrate into the bulk ILs, is about 1.1 nm. The liquid nature of the IL and the short-range ion-electrode and ion-ion interactions are found to significantly affect the structure of the EDL, particularly at low electrode charge densities. Charge delocalization of the ions is found to affect the mean force experienced by the ions and, thus, can play an important role in shaping the EDL structure. The differential capacitance of the EDLs is found to be nearly constant under negative electrode polarization but increases dramatically with the potential under positive electrode polarization. We show that the differential capacitance is a quantitative measure of the response of the EDL structure to a change in electrode charge density. It is found that the [NO3]- ion dominates the response of EDL structure to the change in electrode charge under both positive and negative electrode polarization, which is qualitatively different from that in aqueous electrolytes. Detailed analysis shows that the cation-anion correlations and the strong adsorption of [BMIM]+ ions on the electrode are responsible for the capacitance-potential correlation observed here.

G. Feng, J.S. Zhang and R. Qiao, J. Phys. Chem. C 2009, 113, 4549-4559

See more of this Session: Interfacial Phenomena in Energy Systems
See more of this Group/Topical: Engineering Sciences and Fundamentals

Neutron Imaging of An Enzyme Fuel Cell

Doug S. Aaron, Civil and Environmental Engineering, Georgia Institute of Technology, Atlanta, GA

Costas Tsouris, Oak Ridge National Laboratory, Oak Ridge, TN

Abhijeet P. Borole, Oak Ridge National Laboratory, Oak Ridge, TN

Daniel S. Hussey, National Institute of Standards and Technology, Gaithersburg, MD

David L. Jacobson, National Institute of Standards and Technology, Gaithersburg, MD

Eli Baltic, National Institute of Standards and Technology, Gaithersburg, MD

Platinum is an expensive catalyst that increases the cost of polymer-electrolyte-membrane fuel cells (PEMFCs). In enzyme fuel cells (EFCs), enzymes instead of platinum are used to catalyze the electrochemical reactions. As is the case for PEMFCs, water management affects the performance of EFCs. Water is needed to be present for the enzymes to be active. In previous studies in our laboratory with a three-dimensional cathode EFC, it has been observed that the power output of the EFC was gradually lost. This behavior was attributed to either (i) loss of water in the enzymatic cathode or (ii) loss of enzymatic activity. An attempt was made to measure the water content of the cathode, but these measurements were intrusive to the EFC and interrupted its operation. Here, we employed neutron imaging, which is an ideal technique to in-situ study the water content of an operating fuel cell. Initial experiments verified the technique by introducing a known volume of enzyme solution into the cathode and then determining this volume using neutron imaging. Then, the effects of operating conditions such as feed humidity, feed flow rate, and enzyme solution composition were studied and compared with power output of the EFC. These experiments allowed the separation of contributions from mediator/enzyme decay and dehydration to output loss over time. Electrochemical impedance spectroscopy (EIS) studies were also performed on the EFC to provide a better understanding of how various operating parameters affected internal resistances.

See more of this Session: Interfacial Phenomena in Energy Systems
See more of this Group/Topical: Engineering Sciences and Fundamentals

Interfacial Mass Transport in the Ionic Liquid/1-Octene System

Azita Ahosseini, Chemical & Petroleum Engineering, University of Kansas, Lawrence, KS

Jerzy Petera, Process and Environmental Engineering, The Technical University of Lodz, 90-929 Lodz, Poland

Laurence R. Weatherley, Chemical & Petroleum Engineering, University of Kansas, Lawrence, KS

Aaron M. Scurto, Chemical & Petroleum Engineering, University of Kansas, Lawrence, KS

Biphasic systems with ionic liquids and various solvents including supercritical and compressed gases are used as an efficient platform for reactions and extractions. However, little mass transfer data exists in the literature for any system involving ionic liquids, which impedes development and scale-up. This presentation will illustrate interfacial mass transport in a model biphasic system with 1-hexyl-3-methyl-imidazolium bis[trifluoromethylsulfonyl)amide ([HMIm][Tf_2N]) and 1-octene. The physical, thermodynamic, and mass transport properties are highly concentration dependent and have been measured. From phase equilibrium studies, the IL has very little solubility in the 1-octene phase, but 1-octene has moderate solubility in the IL phase. This simplifies the analysis as mass transfer occurs in only 1 direction. Interfacial mass transfer studies have been conducted for falling droplets of ionic liquid in a continuum of 1-octene in 1 and 2 meter columns. In addition, detailed studies using a high-speed camera have been used to investigate the droplet geometry and flight times. The robust mass transfer model of Weatherley and Petera has been utilized to correlate and predict the mass transfer rates for any geometry and any hydrodynamic regime based upon limited physical property and thermodynamic data.

See more of this Session: Interfacial Phenomena in Energy Systems
See more of this Group/Topical: Engineering Sciences and Fundamentals

Ionic Polymer Lubricants for Boundary Lubrication with Self-Healing Capability

Erik Hsiao, Chemical Engineering, The Pennsylvania State University, University Park, PA

Seong H. Kim, Chemical Engineering, Pennsylvania State University, University Park, PA

As the size of mechanical devices now varies from the macroscale to the nanoscale, more innovative methods are needed to minimize wear caused by adhesion and friction on all scales. Among various coatings applied for this purpose, organic boundary lubrication films is one of the most widely studied and promising approach. However, one big challenge in this case is the lack of self-healing capability of the organic bound films, which limits the durability of the film. In order to overcome this challenge, it is required to synthesize a molecule that can be bound to the surface but still mobile on the surface. For this purpose, we synthesized ionic polymer lubricants (IPLs) by attaching ionic side chains to low molecular weight polydimethylsiloxane (PDMS) backbones. The basic idea is to combine the chemical reactivity of ionic liquids and lubricity of PDMS into one molecule. Thus, IPLs have ionic interactions with the substrates which will provide moderate binding needed for formation of the boundary lubrication layer. They also have a very flexible PDMS backbone that will allow lateral mobility and thus self-healing. Thus they can heal the bare surface exposed by removal of the boundary lubrication molecules via lateral diffusion from surrounding regions. Three cationic IPLs with varying amount of quaternary amine functional groups were synthesized and tested for nano-scale as well as macro-scale lubrication effects of silicon oxide surfaces using atomic force microscopy, scanning polarization force microscopy, pin-on-disc tribometer, and other spectroscopic analysis techniques. The effects of ionic groups in boundary lubrication properties will be discussed.

See more of this Session: Interfacial Phenomena in Energy Systems
See more of this Group/Topical: Engineering Sciences and Fundamentals

Inhibition Effect of Various Surfactants On CO2 Hydrate Formation

Junshe Zhang, Chemical Eng, Department of Chemical Engineering, The City College of New York, New York, NY

Jae W. Lee, Department of Chemical Engineering, The City College of New York, New York, NY

The effect of an ionic (sodium dodecyl sulfate, SDS), cationic (cetyl trimethylammonium bromide, CTAB), and non-ionic (fatty alcohol EO/PO derivative, LS54) surfactant on unsteady-state CO2 absorption in the metastable region of CO2 hydrates and CO2 enclathration with cyclopentane (CP) was investigated in an unstirred batch reactor. These surfactants have no significant effect on the CO2 absorption rate; however, they retard CO2 enclathration and this inhibition becomes significant at 200 ppm SDS and 100 ppm LS54 or CTAB. A layer of clathrate hydrates is visually detected if charging CO2 and CP sequentially. Even though this layer cannot be macroscopically observed when swapping the order of CO2 and CP charge, the existence of this layer is inferred from the unchanged pressure profile. This layer prevents CO2 from contacting with water and thus it delays CO2 enclathration. The habit of clathrate hydrates growing along the reactor wall without agitation is discussed based on the magnitude of temperature spikes at the very early stage of the enclathration.

See more of this Session: Interfacial Phenomena in Energy Systems
See more of this Group/Topical: Engineering Sciences and Fundamentals

Formation and Dissociation of Gas Hydrates in a Seafloor Process Simulator

Costas Tsouris, Oak Ridge National Laboratory, Oak Ridge, TN

J.C. Santamarina, School of Civil and Environmental Engineering, Georgia Institute of Technology, Atlanta, GA

J. Jang, School of Civil and Environmental Engineering, Georgia Institute of Technology, Atlanta, GA

J. Jung, School of Civil and Environmental Engineering, Georgia Institute of Technology, Atlanta, GA

C.J. Rawn, Oak Ridge National Laboratory, Oak Ridge, TN

J.E. Alford, Oak Ridge National Laboratory, Oak Ridge, TN

T.J. Phelps, Oak Ridge National Laboratory, Oak Ridge, TN

Clathrate hydrates are solid crystalline structures in which water hydrogen-bonded cages are stabilized by the presence of gas molecules inside the cages. The conditions for the formation and stability of gas hydrates involve pressures much higher than atmospheric and temperatures a few degrees above the freezing point of water. Natural gas hydrates can be found in coastal sediments or the permafrost, and their exploitation constitutes an attractive future source of energy. CO2 hydrate has been considered as a medium for ocean carbon storage. In this presentation, we will examine interfacial effects on hydrate formation and dissociation. Hydrate is formed either in sediment or in the water column in a 72-L vessel, the ORNL Seafloor Process Simulator (SPS). SPS is made of hastelloy and allows pressures up to 3,000 psi. It has several windows for monitoring devices including thermocouples, pressure and sound velocity transducers, and video cameras. The conditions of hydrate formation and dissociation with respect to the thermodynamic stability diagrams are examined for different types of gas hydrates and for different media.

See more of this Session: Interfacial Phenomena in Energy Systems
See more of this Group/Topical: Engineering Sciences and Fundamentals

Effects of Fluctuating Hydrodynamic Interactions On the Dynamics of Confined Polymer Solutions: Grooved Channels and Pore-Translocation Rates

Juan P. Hernandez-Ortiz, Departamento de Materiales, Universidad Nacional de Colombia, Sede Medellin, Medellin, Colombia

Michael D. Graham, Department of Chemical and Biological Engineering, University of Wisconsin-Madison, Madison, WI

Juan J. De Pablo, Department of Chemical and Biological Engineering, University of Wisconsin, Madison, WI

We present results of simulations of solutions of flexible polymers molecules during flow. A self-consistent coarse-grained Langevin description of the polymer dynamics is adopted in which hydrodynamic effects are included through a Green's function formalism. The *O(N)* General Geometry Ewald-like Method is used to calculate the confined Stokeslet and concentrations in the range from ultra-dilute to semi-dilute (near the overlap concentration). For flowing polymer solutions in a grooved channel, the groove is almost completely depleted of polymer chains at high Weissenberg number in the dilute limit, but at finite concentration this depletion effect is dramatically reduced. Only upon inclusion of hydrodynamic interactions can these phenomena be properly captured. The rate constant for flow-induced translocation of polymer molecules through a square channel has been calculated using the Forward Flux Sampling method. The rate constant is affected by the walls and the confinement provided that hydrodynamic interactions are taken into account. Such interactions lead to a decrease of the rate constant due to the decrease in particle mobility near rigid walls, and an increase due to the increase of the diffusion coefficient of hydrodynamically interacting polymer molecules.

See more of this Session: Microfluidics and Small-Scale Flows I: Fundamentals
See more of this Group/Topical: Engineering Sciences and Fundamentals

Partition Induced Vector Separation in Microfluidic Devices

Jorge A. Bernate, Chemical and Biomolecular Engineering Department, Johns Hopkins University, Baltimore, MD

Carson Riche, Chemical and Biomolecular Engineering Department, Johns Hopkins University, Baltimore, MD

German Drazer, Chemical and Biomolecular Engineering Department, Johns Hopkins University, Baltimore, MD

We study vector separation of particles in microfluidic devices patterned physically and/or chemically. In particular, we consider the case in which the pattern consists of rectangular stripes of bare glass and gold regions. In the purely diffusive case, the particles partition according to their potential energy in the different regions of the pattern. We investigate the chromatographic angle at which different species move animated either by an external force or by a flow field. Partitioning leads to a trajectory angle that differs, in general, from the forcing angle and differences in the partition coefficient for different species lead to vector separation. We solve analytically the particle transport equations under the Fick-Jacobs approximation for an arbitrary two dimensional potential. For the particular case of a periodic array of stripes we arrive at expressions for the trajectory angle and the dispersivity in terms of the forcing orientation angle, the partition coefficient, ratios of Peclet numbers and mobilities in the two stripes, and the relative width of each stripe. Many properties that induce differences in partitioning such us mass, electrical charge, and strength of van der Waals forces, are size dependent. Thus, size-dependent separation is possible. Furthermore, partition based vector chromatography allows separating particles of the same size which present another property that induces differences in partition. As a proof of concept, we separate same size silica and polystyrene particles in a microfluidic device. One promising advantage of this approach over existing alternatives such as field-flow fractionation is that no external field or cross flow is required to achieve separative transport.

See more of this Session: Microfluidics and Small-Scale Flows I: Fundamentals
See more of this Group/Topical: Engineering Sciences and Fundamentals

Particle Interactions with Patchy Surfaces in Confined Microscale Flow: Insights From Modeling

Jeffrey M. Davis, Chemical Engineering, University of Massachusetts, Amherst, MA

Maria M. Santore, Polymer Science and Engineering, University of Massachusetts, Amherst, MA

Ranojoy D. Duffadar, Chemical Engineering, University of Massachusetts, Amherst, MA

Patchy surfaces with nanoscale features can be used to detect, manipulate, and control the motion signatures of colloidal- and micrometer-scale objects in flow. For example, colloidal silica particles in flowing suspensions exhibit skipping, rolling, and firm adhesion on net-repulsive silica surfaces with randomly-distributed, 10-nm cationic patches. Although governed by non-specific colloidal interactions that vary spatially with fluctuations in the local density of adsorbed patches, this system displays selective adhesion, a flow-sensitive adhesion threshold, and biomimetic behavior. This talk is focused on insights provided from a computational model that includes hydrodynamic, colloidal, Brownian, and frictional interactions. The model predictions are in strong agreement with experiment with no adjustable parameters. The simulations provide information not accessible experimentally, including detailed particle trajectories, the evolving particle-surface separation distance, the local densities of adhesive elements beneath arrested and rolling particles, and the binding strength per element. Furthermore, the computations explain the deviations from a DLVO treatment based on the "average" charge of the surfaces. Attention in this talk is focused on the influence of the shear rate on the adhesion threshold (a nano-feature density below which particle adhesion does not occur) and the deposition rates of the particles. The model quantitatively predicts the observed 1/3 power law dependence of the adhesion threshold and the non-monotonic dependence of the deposition rate on the shear rate. This behavior is linked to the RMS variation in the particle-surface separation distance as the particles fluctuate about their secondary minima due to the spatially varying colloidal forces as the particles are transported along the surface by the shear flow. The variation in the RMS separation distance also has a power-law dependence on the shear rate. The coupling of the statistical distribution of the nano-features to the increased size of the locally-attractive region required to adhere a particle as the shear rate increases is used to explain the experimental and computational results.

See more of this Session: Microfluidics and Small-Scale Flows I: Fundamentals
See more of this Group/Topical: Engineering Sciences and Fundamentals

Induced Charge Electrokinetics Over 'controllably Contaminated' Surfaces: The Effects of Dielectric Thin Films and Surface Chemistry

Andrew J. Pascall, Chemical Engineering, University of California, Santa Barbara, Santa Barbara, CA

Todd M. Squires, Chemical Engineering, University of California at Santa Barbara, Santa Barbara, CA

Microfluidic devices have sparked a renewed interest in electrokinetic phenomena for transporting fluids on small scales, and subjected classical electrokinetic theories and understanding to new challenges. For example, AC electric potentials applied to and across metallic surfaces have been shown to drive a time-averaged flow, which could provide an on-chip means to drive high pressures with low voltage. The central physical effect is Induced-charge electro-osmosis, which is a non-linear electrokinetic effect in which an externally applied electric field both induces and drives a layer of charged fluid near an electrically conductive surface[1]. Experimental data on ICEO and related phenomena have shown that the standard theory consistently overpredicts observed slip velocities by up to three orders of magnitude under a variety of conditions[2], suggesting an as yet unknown mechanism for electrokinetic flow supression. Here we presents experiments and theory in which we deliberately and controllably `contaminate' the metallic surface with a thin dielectric film, and derive a theory for ICEO that incorporates both dielectric effects and surface chemistry, both of which act to decrease the slip velocity relative to a `clean' metal. Experimental data for over a thousand combinations of electric field strength and frequency, electrolyte compositions, dielectric thicknesses and surface chemistries show essentially unprecedented quantitative agreement with our theory.

[1]Squires and Bazant. Journal of Fluid Mechanics 2004.

[2]Bazant, et al. arXiv. 0903.4790

See more of this Session: Microfluidics and Small-Scale Flows I: Fundamentals
See more of this Group/Topical: Engineering Sciences and Fundamentals

Concentration Gradient Generation Using Induced Charge Electro-Osmosis

Mranal Jain, Chemical & Materials Engineering, University of Alberta, Edmonton, AB, Canada

Anthony Yeung, Chemical & Materials Engineering, University of Alberta, Edmonton, AB, Canada

Krishnaswamy Nandakumar, Chemical Engineering Program, The Petroleum Institute, Abu Dhabi, United Arab Emirates

Bio-molecule gradients play an important role in the understanding of various biological processes, e.g. in wound healing, cancer metastasis etc. Typically, biological cells are exposed to linear and non-linear bio-molecule concentration gradients and their response is studied for understanding cell growth, cell migration and cell differentiation mechanisms. Traditionally concentration gradients are generated using hydrogels, micropipettes, Boyden chamber, etc. However, these techniques have limitations in terms of gradient stability, reproducibility, predictability and controllability. Although recent studies have demonstrated the use of microfluidic devices for precise, reproducible and stable concentration gradient generation, most of the reported devices are geometrically complex and lack dynamic controllability, which is the ability to alter the gradient characteristics during experimentation.

In this work, a novel microfluidic gradient generator is presented which utilizes induced charge electro-osmosis (ICEO) by introducing conducting obstacles in the microchannel. In comparison with the other reported designs, the proposed method is geometrically simpler and has the unique feature of inducing transverse flows which can be controlled by the magnitude of the applied field. By employing a range of conducting protrusions as obstacles (with different shapes and sizes), transverse convection of varying magnitude can be introduced at different spatial locations in the channel, which can in turn be used to create non-linear as well as asymmetric gradients. The characteristics of the developed concentration gradients are dependent upon the interplay between fixed charge electro-osmotic (FCEO) and ICEO flows. It is shown that the proposed device can switch between linear and non-linear gradients by simply altering the applied electric field, demonstrating the superior controllability of the proposed design. For analysis, both full PNP (Poisson-Nernst-Planck) model and correction method based on slip-velocity approach are used to model the steady ICEO flow. Finally, the formation of user-defined concentration profiles (linear, convex and concave) is demonstrated by varying the conducting obstacle size.

See more of this Session: Microfluidics and Small-Scale Flows I: Fundamentals
See more of this Group/Topical: Engineering Sciences and Fundamentals

Anisotropic Electro-Osmotic Flow Over Super-Hydrophobic Surfaces

Martin Z. Bazant, Chemical Engineering and Mathematics, Massachusetts Institute of Technology, Cambridge, MA

Supreet Bahga, Mechanical Engineering, Stanford University, Stanford, CA

Olga I. Vinogradova, A. N. Frumkin Institute of Physical Chemistry and Electrochemistry, Russian Academy of Sciences, Moscow, Russia

Textured super-hydrophobic (SH) surfaces with trapped gas bubbles are promising for use in microfluidic devices, due to their low viscous resistance, or large effective slip length, in pressure-driven flow. Recently, the possibility of using SH surfaces to amplify electro-osmotic flow has also attracted attention, e.g. in the context of nanofluidic energy conversion (by harvesting streaming current). For textured surfaces, both of these properties of SH surfaces are generally anisotropic, and the slip length and electro-osmotic mobility must be replaced by appropriate tensorial generalizations. Here, we discuss tensorial hydrodynamic slip and calculate the electro-osmotic mobility tensor for a model SH surface consisting of flat stripes of alternating charge and slip length, for arbitrary Debye length in the linear regime. Large enhancements of electro-osmotic flow are predicted only if charge can be maintained on the gas-liquid interfaces.

See more of this Session: Microfluidics and Small-Scale Flows I: Fundamentals
See more of this Group/Topical: Engineering Sciences and Fundamentals

Drag Force in Vibrating Nanowires Using Theory and Simulations

Shriram Ramanathan, Chemical Engineering, Cornell University, Ithaca, NY

Donald L. Koch, Chemical and Biomolecular Engineering, Cornell University, Ithaca, NY

Rustom Bhiladvala, Department of Mechanical Engineering, University of Victoria, Victoria, BC, Canada

Nanoelectromechanical oscillators are very attractive as sensing devices because of their low power requirements as well as the high resolution they promise. Recent attempts to characterize the performance of these devices in fluid have led to disparate results and a standard calculation approach is lacking. In this article, we use our newly developed Bhatnagar-Gross-Krook based Low Mach number Direct Simulation Monte Carlo (BGK-LM-DSMC) method to study the non-continuum drag force acting on a cylinder oscillating normal to a wall. We explore both quasisteady flows in which $\omega\tau_f << 1$ where ω is the oscillation frequency and τ_f is the characteristic fluid relaxation time, as well as unsteady flows in which $\tau_f >> 1$. The drag force is studied as a function of Knudsen number, defined in terms of the mean free path λ and the radius of the cylinder R as $Kn = \lambda/R$. For quasi-steady flows, we also present theoretical calculations for the slip regime, valid for the $Kn << 1$, and a free molecular flow theory which is valid for the $Kn >> 1$ regime. The quasi-steady simulations are in excellent agreement with our theories over a wide range of Kn and also approach continuum predictions as $Kn \rightarrow 0$. Experiments indicate drag forces that are smaller and also show a much higher value of Kn at which the high Kn power-law scaling crosses over to the low Kn constant drag force behavior. Some possible reasons for this discrepancy are discussed. In the unsteady regime, we also explore the effect of varying the oscillation frequency on the drag force. The drag force attains constant values at low and high ω with crossover behavior at intermediate ω.

See more of this Session: Microfluidics and Small-Scale Flows I: Fundamentals
See more of this Group/Topical: Engineering Sciences and Fundamentals

Passive Breakup of Viscoelastic Droplets and Filament Self-Thinning at a Microfluidic T-Junction

Gordon F. Christopher, Dept. of Mechanical Engineering, Carnegie Mellon University, Pittsburgh, PA

Shelley L. Anna, Departments of Mechanical Engineering and Chemical Engineering, Carnegie Mellon University, Pittsburgh, PA

In this talk, we examine a simple experiment in which pre-formed droplets containing a dilute polymer solution enter a T-shaped microfluidic junction and stretch as they pass through the stagnation point. Depending on the initial aspect ratio and speed, droplets may break into two segments. We characterize the breaking-nonbreaking behavior of these droplets and find that viscoelastic droplets are less stable than Newtonian droplets of comparable shear viscosity. When droplets break, we observe that viscoelastic droplet segments are connected by persistent filaments that first undergo stretching due to drag of the outer viscous liquid on the large connected segments. Later, filaments undergo iterated stretching and develop a series of beads along their length. Secondary filaments between beads no longer stretch with the outer flow, but rather they exhibit an exponentially decreasing diameter consistent with elasto-capillary breakup. The relaxation time obtained from the filament diameter profiles is close to the estimated Zimm relaxation time for the polymer solution, but depends on the global flow parameters, including the initial size and speed of the droplet prior to entering the T-junction. Based on the microscale diameters of the filaments, we suggest that splitting of droplets at microfluidic T-junctions may be a useful way to characterize the extensional rheology of low viscosity elastic liquids.

See more of this Session: Microfluidics and Small-Scale Flows I: Fundamentals
See more of this Group/Topical: Engineering Sciences and Fundamentals

Characterization of Bubble Breakup in a Rectangular Microchannel

Gautham C. Unni, Chemical and Biological Engineering, Missouri University of Science and Technology, Rolla, MO

Kimberly H. Henthorn, Missouri University of Science and Technology, Rolla, MO

The production and manipulation of calibrant bubbles is crucial to the operation of some types of self-calibrating biosensors. Interactions between the bubble, liquid, and channel surface plays a critical role in the behavior of the bubbles during manipulation. Channel roughness is directly related to the frictional resistance imparted on the bubbles by the channel surface. This affects the mobility and can lead to undesired bubble breakup. In addition, the degree of hydrophobicity of the channel surface impacts the meniscus shape and movement of the bubble through the channel, which can also contribute to bubble breakup.

In the present study, we characterized the surface roughness of several rectangular PDMS microchannels to determine the relationship between channel roughness and bubble breakup patterns. Profilometry was used to measure the roughness of the bottom surface of the channel, and Scanning Electron Microscopy (SEM) was used to view the side walls and the sharpness of the corners, as they contribute to the overall roughness. Subsequently, the pressure drop across the channel was experimentally measured over a range of fluid velocities and compared to the theoretical pressure drop across the channel. These results have been used to map various bubble breakup patterns and predict the fluid velocities and surface roughnesses that lead to bubble breakup.

See more of this Session: Microfluidics and Small-Scale Flows I: Fundamentals
See more of this Group/Topical: Engineering Sciences and Fundamentals

The effects of hydrophilic surfactant concentration and flow ratio on dynamic wetting in a T-junction microfluidic device

C.Tostado, J.H XU, G.S LUO*

State Key Laboratory of Chemical Engineering, Department of Chemical Engineering, Tsinghua University, Beijing 100084, China

Abstract: An initial study was conducted on the influence of the hydrophilic surfactant concentration and two-phase flow-rate ratio on the aqueous-phase wetting ability (dynamic contact angle) of a water-in-oil (W/O) emulsion in a T-junction poly(methyl methacrylate) microfluidic device. Cetyltrimethyl ammonium bromide (CTAB) was used as the hydrophilic surfactant for the aqueous phase while the organic phase was pure corn oil. Static contact angles and interfacial tensions were measured for emulsions with surfactant concentration ranging from 0 to 0.5% (w/w). Contacts angles for an oil drop on PMMA in water transition from <90° to >90° at CTAB concentrations near the transition concentration of 1 $x10^{-2}$ % (w/w), suggesting a change in the surface energy and a transition from a slightly hydrophobic to hydrophilic PMMA surface. Microchannel flow experiments were conducted to identify an analogous dynamic transition in wetting character, and special attention was paid to the CTAB concentration range between 3.0 $x10^{-3}$ 3 x 10^{-2} to focus on the wetting transition observed in the static case. At concentrations below 1 x 10^{-2} %, the microchannel walls are hydrophobic and stable W:O emulsions are formed. As CTAB increases, "sticky" type flow ensues and both phases wet the channel walls. At the transition concentration, a spontaneous phase-inversion was observed, producing stable O:W plugs. Above the transition concentration, the flow-type returns to W:O plugs, and advancing and receding dynamic contact angles for the aqueous plugs are observed. Dynamic contact angles and wetting responded more to changes in the aqueous phase flow rate than to changes in the organic (continuous) phase flow rate and receding contact angles of the plugs transition from <90° to >90° at a W:O flow ratio near unity. Finally, an equation of state approach was used to estimate the interfacial energies of the emulsions and plug shapes at higher CTAB concentrations. The results indicate that emulsions in which both phases exhibit partial wetting are lowest in energy and changes to either or both of the liquid-liquid and liquid-solid interfacial energies can significantly affect the emulsion type and wetting character.

Predicting the Micellization Behavior of Perfluorinated and Partially-Fluorinated Surfactants in Dilute Aqueous Solution Using Computer Simulations and Molecular-Thermodynamic Theory

Jonathan D. Mendenhall, Chemical Engineering, Massachusetts Institute of Technology, Cambridge, MA

Shangchao Lin, Chemical Engineering, Massachusetts Institute of Technology, Cambridge, MA

Daniel Blankschtein, Chemical Engineering, Massachusetts Institute of Technology, Cambridge, MA

Fluorinated surfactants exhibit superior performance in many practical surface and bulk solution applications relative to their hydrogenated counterparts. An important example is the large reduction in surface tension achieved at a markedly lower surfactant concentration in aqueous solution. One main reason for the improved performance of this class of surfactants is the increased hydrophobicity of fluorinated alkyl groups relative to hydrogenated alkyl groups. In addition to the numerous applications resulting from the unique properties of fluorinated surfactants at surfaces (e.g., use in hydrophobic and oleophobic coatings, use in fire-fighting foams, and use in waxes, polishes, paints, and cleaners), fluorinated surfactants are of interest also for their useful bulk solution behavior. For example, fluorinated surfactants are used in emulsion polymerization applications and have also been investigated as potential artificial blood substitutes. Although the use of fluorinated surfactants is advantageous in many situations, concern has recently been raised regarding the safety of certain fluorinated surfactants, specifically in terms of their environmental toxicity and bioaccumulation. For example, in the case of perfluorooctanoic acid (PFOA), the EPA has created the 2010/15 PFOA Stewardship Program in order to encourage the reduction of industrial emission of PFOA, with the goal of 95% reduction by 2010 and elimination by 2015. In light of this effort, it would be desirable to identify replacement surfactants that have similar performance properties to perfluorinated surfactants, yet do not exhibit the accompanying negative characteristics. Such surfactants could possess architectural differences (e.g., branching or multiple tails) or chemical differences (e.g., partial hydrogenation or inclusion of biodegradable moieties such as esters) relative to their perfluorinated counterparts. Given the enormous investment required to synthesize proposed alternatives, including characterizing them experimentally and evaluating their performance behavior, it would be very advantageous to be able to predict the bulk solution properties of potential surfactant replacements molecularly, thus enabling the formulator to narrow his/her focus to those few candidates that exhibit the desired bulk solution properties. The recently developed combined Computer-Simulation/Molecular-Thermodynamic (CS-MT) theory of Stephenson, Beers, and Blankschtein [J Phys Chem B, 111, 1063-1075, 2007] has been demonstrated to accurately model the process of surfactant micellization in dilute aqueous solution for a variety of linear, fully-hydrogenated surfactants. Here, we demonstrate application of the CS-MT model to perfluorinated and partially-fluorinated surfactants. We first present a novel method that we have developed for general surfactant head/tail identification from Molecular Dynamics simulations. We then discuss general improvements that we have made to the computation of the change in hydration state of atomic groups within a surfactant molecule experienced upon incorporation into a surfactant micelle. Based on these and other molecular inputs, we compute crtitical micelle concentrations (CMC's) and micelle shapes and sizes for a variety of ionic and nonionic (ethoxylated) perfluorinated surfactants, and demonstrate very good agreement with the experimental values, when available. Finally, we present a sample design study illustrating the effect of subtle chemical differences on the bulk surfactant solution properties. In particular, it has long been known that fluorination of the ω-carbon of an otherwise fully-hydrogenated linear alkyl surfactant yields an anomalous 2-fold increase in the CMC [Muller and Birkhahn, J Phys Chem, 71, 952-962, 1967], where a decrease in CMC might be expected based on hydrophobicity arguments. We demonstrate that these types of effects can be captured using the CS-MT model with the addition of an enthalpic-mixing contribution to the free energy of micellization. We present predicted CMC's and micelle shapes and sizes for a series of partially-fluorinated surfactants with increasing fluorination ratio, and compare the predicted values to those available experimentally. In addition, we discuss briefly the application of the CS-MT model to other perturbations of the perfluorinated structure, including challenges posed by other chemical moieties, such as esters, as well as by chain branching.

See more of this Session: Self-Assembly in Solution I
See more of this Group/Topical: Engineering Sciences and Fundamentals

Cyclodextrin Inclusion Compounds for Novel Polymer Self-Assembly

Robert K. Prud'homme, Chemical Engineering, Princeton University, Princeton, NJ

Stephen Lincoln, Chemistry, University of Adelaide, Adelaide, Australia

Xuhong Guo, College of Chemical Engineering, East China University of Science and Technology, Shanghai, China

The existence of tunable interactions in polymer solutions gives the ability to control the rheological properties which plays a key role in the applications. For example, hydrophobically modified water soluble polymers (HMP) have been received considerable interests and numerous applications as rheology modifiers in paints, cosmetics, pharmaceuticals, foods, enhanced oil recovery, water treatment and controlled release of bioactive materials because the hydrophobic interactions and thus the solution viscosity are easily controllable by several ways, such as the polymer structure, molecular weight, concentration, the type of hydrophobic groups and the additions of salt, surfactant or cyclodextrin.

Similar to hydrophobic association, the inclusion or host-guest interaction between cyclodextrins and hydrophobic groups is another reversible physical interaction which can be used to modulate polymer solutions. The internal cavity of CD is hydrophobic and can accommodate suitable hydrophobic groups driven by a non-covalent interaction to form inclusion complex. If CDs are grafted in polymer chains, novel polymer networks can be generated by this kind of host-guest interactions between polymers with hydrophobe and CD side-groups.

In contrast to purely hydrophobically modified polymer association the complexes have a distinct stoichiometry and an unusual temperature dependence.

See more of this Session: Self-Assembly in Solution I
See more of this Group/Topical: Engineering Sciences and Fundamentals

Structure, Rheology, and Network Formation in Wormlike Micelle-Nanoparticle Mixtures

Matthew E. Helgeson, Chemical Engineering, MIT, Cambridge, MA

Florian Nettesheim, Chemical Engineering, University of Delaware, Newark, DE

Kerstin Weiss, RWTH Aachen University, Aachen, Germany

Eric W. Kaler, Stony Brook University, Stony Brook, NY

Norman J. Wagner, Chemical Engineering, University of Delaware, Newark, DE

Wormlike micelles (WLMs) have become widely used in a number of industrial and consumer products and processes where they come in contact with colloidal species, yet relatively little is understood regarding the interactions between WLMs and colloids and resulting changes in macroscopic properties. Here, we present the first systematic study of structure and rheology of ionic WLMs in the presence of model electrostatically stabilized nanoparticles. Structural and thermodynamic measurements in dilute surfactant-nanoparticle mixtures reveal that nanoparticle-induced modification of WLMs is critically linked to interactions at the solution-nanoparticle interface, whereby the adsorption of micelles to an adsorbed surfactant layer leads to the formation of micelle-nanoparticle junctions. These micelle-nanoparticle junctions act as physical cross-links between micelles, and give rise to significant rheological modification of WLMs upon nanoparticle addition. In semi-dilute solutions, this results in a decrease of the entanglement concentration and concomitant increase in viscosity by as much as an order of magnitude for particle concentrations as little as 1 vol%. Additionally, nanoparticle addition imparts viscoelasticity in otherwise Newtonian WLMs, and enhances the viscoelasticity of entangled WLMs. Thus, nanoparticle addition gives rise to a so-called "double network", where the viscoelasticity can be tuned through two energetic scales, the micellar end cap energy and micelle-nanoparticle junction energy, which are readily manipulated by adjusting solution conditions. In concentrated WLMs, nanoparticle addition leads to suppression of the shear banding instability in WLMs. Combined results of rheometry, velocimetry, and structural measurements under shear reveal that suppression of shear banding results from hindered orientational mobility of micelles due to micelle-nanoparticle junctions. The fidelity of these effects over a range of solution conditions and different nanoparticles demonstrates that nanoparticle addition provides robust and unique control over the rheology of wormlike micellar solutions.

See more of this Session: Self-Assembly in Solution I
See more of this Group/Topical: Engineering Sciences and Fundamentals

Structural Studies of Mixtures of Colloidal Semi-Flexible Rods (polyelectrolyte-surfactant aggregates) and Spheres

Viet Lam, Chemical Engineering, Carnegie Mellon University, Pittsburgh, PA

Lynn M. Walker, Chemical Engineering, Carnegie Mellon University, Pittsburgh, PA

The interaction of a polyelectrolyte-surfactant aggregate with semi-flexible rod morphology and silica spheres of different sizes in aqueous environment had been studied using small angle neutron scattering (SANS). The rod-like aggregates have dimensions of 4nm by140nm while the negatively-charged silica spheres range from 2nm to 170nm. This range of relative sizes allows for three possible structures in mixed systems: 1) non-interacting spheres and rods, 2) aggregate covered by silica spheres - taking on a core-shell cylinder morphology, and 3) silica covered by a layer of aggregates - taking on core-shell sphere morphology. We report on a series of contrast matched neutron scattering studies that elucidate the different morphologies as a function of increasing silica size. Solution stability is verified and shown to increase as the size of the silica increases. Preliminary analysis indicates that there is association between aggregates and silica as predicted, and the rod wraps around the silica spheres, even for silica comparable to the radius of the rods. The strong attraction between the rod and the spheres, together with the intrinsic flexibility of the aggregate, is enough to overcome the bending energy required for this morphology to exist. The impact of the morphology on phase stability and macroscopic behaviors will be discussed in the context of the structural characterization.

See more of this Session: Self-Assembly in Solution I
See more of this Group/Topical: Engineering Sciences and Fundamentals

Effects of Sugar Surfactant Architecture On Mixed Micelle Formation with Cationic Surfactants

Suvid Joshi, Chemical and Materials Engineering, University of Kentucky, Lexington, KY

Stephen E. Rankin, Chemical and Materials Engineering, University of Kentucky, Lexington, KY

Sugar surfactants find many applications in industry as nonionic detergents due to their highly hydrophilic polar headgroups and biodegradability. However, pure sugar surfactants sometimes display specific interactions between headgroups that make them difficult to assemble into micelles or lyotropic liquid crystals. Mixed surfactant systems can be used to overcome these difficulties, and are often found to exhibit better surface properties than pure surfactants. Effects of variation in chain length of sugar surfactants mixed with other other non-ionic, cationic and anionic surfactants have been studied in the past. Prior studies relied on critical micellar concentration (CMC) as the means to gain insight into the thermodynamics of such mixed surfactant systems. More direct routes of measuring heats of mixing, solvation and assembly, in particular isothermal titration calorimetry (ITC), provide a desirable alternative because they do not require variable temperature experiments and they can be accomplished with a small amount of surfactant.

This study focuses on the behavior of sugar surfactants with varying structural characteristics (head group size and stereochemistry, tail chain length, and fluorination of the tail) in presence of a common cationic surfactant, cetyltrimethylammonium bromide. ITC is used to directly measure enthalpies of micellization and surfactant mixing and to observe the effects of changing each surfactant characteristic independently. Such a study of mixed surfactant system using ITC will establish how mixed micelles form and how surgar-based surfactants partition into cationic micelles with only small amounts of the sugar-based surfactant. The results of these experiments will help to establish the characteristics of sugar-based surfactants required for effective mixing with CTAB and for the development of synergy in the properties of the mixed surfactant system.

See more of this Session: Self-Assembly in Solution I
See more of this Group/Topical: Engineering Sciences and Fundamentals

High-Pressure Micellar Solutions of Polystyrene-Block-Polybutadiene and Polystyrene-Block-Polyisoprene in Propane Exhibit Cloud-Pressure Reduction and Distinct Micellization End Points

Winoto Winoto, Chemical & Petroleum Engineering, University of Wyoming, Laramie, WY

Sugata P. Tan, Department of Chemical and Petroleum Engineering, University of Wyoming, Laramie, WY

Youqing Shen, University of Wyoming, Laramie, WY

Maciej Radosz, University of Wyoming, Laramie, WY

Kunlun Hong, Center for Nanophase Materials Science, Oak Ridge National Laboratory, Oak Ridge, TN

Jimmy Mays, Center for Nanophase Materials Sciences, Oak Ridge National Laboratory, Oak Ridge, TN

Micellar solutions of polystyrene-block-polybutadiene and polystyrene-block-polyisoprene in propane are found to exhibit significantly lower cloud pressures than the corresponding hypothetical nonmicellar solutions. Such a cloud-pressure reduction indicates the extent to which micelle formation enhances the apparent diblock solubility in near-critical and hence compressible propane. Concentration-dependent pressure-temperature points beyond which no micelles can be formed, referred to as the micellization end points, are found to depend on the block type, size, and ratio. The cloud-pressure reduction and the micellization end point measured for styrene-diene diblocks in propane should be characteristic of all amphiphilic diblock copolymer solutions that form micelles in compressible solvents.

See more of this Session: Self-Assembly in Solution I
See more of this Group/Topical: Engineering Sciences and Fundamentals

Confinement Effects On the Self-Assembly of Organogels

Wanyu Chen, Chemistry, Huazhong University of Science and Technology, Wuhan, China

Yajiang Yang, Chemistry, Huazhong University of Science and Technology, Wuhan, China

Amy Shen, Mechanical Engineering, University of Washington, Seattle, WA

Chang Lee, Mechanical Engineering, Washingtonn University in St. Louis, St. Louis, MO

1,3:2,4-di-p-methylbenzylidene sorbitol (MDBS) is a small organic molecule that is capable of inducing self-assembly in a wide variety of organic solvents and of forming organogels. In this paper, we present a novel approach to tune the network architectures of organogels by utilizing geometric confinement while varying the gelator concentration. Self-assembly of MDBS in propylene carbonate (PC) is investigated in a series of microchannels with widths varying from 20 um to 80 um and the gelator concentration varying from 2%wt to 7%wt.

We demonstrate by optical microscopy, and scanning electron microscopy (SEM) that a transition from fibrillar structure to sheaflike spherulite structure occurs when (a) the channel width is increased for fixed gelator concentrations; (b) gelator concentration is increased for fixed channel widths. The thermal properties of the organogel is measured by differential scanning calorimetry (DSC) to verify the structural difference obtained under confined and unconfined conditions and the structure stability. Our results provide a novel strategy to control the topological structure of self-assembled systems and to modify their thermal properties via geometric confinement.

See more of this Session: Self-Assembly in Solution I
See more of this Group/Topical: Engineering Sciences and Fundamentals

Solubilities of CO and H_2 in Neat and CO_2-Expanded Hydroformylation Reaction Mixtures Containing 1-Octene and Nonanal up to 353.15 K and 9 Mpa

Zhuanzhuan Xie, Chemical & Petroleum Engineering, University of Kansas, Lawrence, KS

William Kirk Snavely, Center for Environmentally Beneficial Catalysis, University of Kansas, Lawrence, KS

Aaron M. Scurto, Chemical & Petroleum Engineering, University of Kansas, Lawrence, KS

Bala Subramaniam, Chemical & Petroleum Engineering, University of Kansas, Lawrence, KS

Accurate knowledge of the phase equilibria of CO_2-expanded hydroformylation reaction mixtures is essential to rational process design and development. Vapor liquid equilibria of the following systems were measured in a variable volume view cell at temperatures ranging from 313.15 to 353.15 K and pressures up to 9 MPa: CO + 1-octene, CO_2 + 1-octene, CO + 1-octene + CO_2, CO + nonanal, CO_2 + nonanal, CO + nonanal + CO_2, H_2 + 1-octene, H_2 + 1-octene + CO_2, H_2 + nonanal and H_2 + nonanal + CO_2. The measured solubilities of CO and H_2 in the liquid phases were consistent with literature values. The presence of CO_2 was found to enhance the solubilities of both CO and H_2 in the liquid phase. The enhancement factor is up to 1.54 for carbon monoxide and 1.82 for hydrogen. The enhancement of CO and H_2 solubilities is slightly higher in nonanal than in 1-octene. The Peng-Robinson equation of state (PR EoS) with van der Waals mixing rules and binary interaction parameters modeled the VLE data adequately, with much better fits for the 1-octene systems compared to the more polar nonanal systems.

See more of this Session: Thermodynamic and Transport Properties Under Pressure
See more of this Group/Topical: Engineering Sciences and Fundamentals

Development of a Molecular Based Computational Approach for Compressed and Supercritical Fluids

Hye Min Kim, Chemical and Biological Engineering, University at Buffalo, The State University of New York, Buffalo, NY

Andrew J. Schultz, Chemical and Biological Engineering, University at Buffalo, The State University of New York, Buffalo, NY

David A. Kofke, Chemical and Biological Engineering, University at Buffalo, The State University of New York, Buffalo, NY

When applied to nonpolar fluids, the virial equation of state has been shown to be describe accurately the PVT behavior of compressed fluids at conditions near the critical point, as well has high-density supercritical fluids. In a polar molecular system, hydrogen bonding can cause a breakdown of the virial series at subcritical densities, in part because the series does not explicitly consider fluid association via hydrogen bonding. Moreover, calculation of the higher order virial coefficients such as B6 for water molecules and other polar systems requires considerable CPU time. Therefore, it would be helpful to find an alternative thermodynamic method which maintains molecular-level detail, but incorporates fluid association. We have used the 2-density series expansion prescribed by Wertheim's association fluid theory for this purpose. We have tested the approach with a single attraction site model comprising a Lennard-Jones interaction and a short-ranged square-well site-site interaction. The cluster diagrams from the theory are evaluated using the Mayer Sampling Monte Carlo algorithm. The cluster diagram expansion of the association theory describes the P-V-T behavior of the model better than the conventional virial expansion. Extensions to other site interaction models and more realistic molecular models are discussed.

See more of this Session: Thermodynamic and Transport Properties Under Pressure
See more of this Group/Topical: Engineering Sciences and Fundamentals

POTENTIOMETRIC PH MEASUREMENTS IN AQUEOUS SOLUTIONS OF CARBOXYLIC ACIDS UNDER HIGH-TEMPERATURE AND HIGH-PRESSURE CONDITIONS

Kiwamu Sue, Yoshihiro Takebayashi, Satoshi Yoda, Takeshi Furuya
Nanotechnology Research Institute, National Institute of Advanced Industrial Science and Technology, Tsukuba, Japan,

Toshihiko Hiaki
College of Industrial Technology, Nihon University, Narashino, Japan

Introduction

Dissociation constants of carboxylic acids at hydrothermal and supercritical conditions have long been a basis for understanding not only fundamental physical and chemical phenomena but also organic and inorganic reaction mechanisms in several applications. Many studies have been carried out on dissociation of aqueous carboxylic acids up to 323 K by several techniques such as potentiometry, calorimetry, and electrical conductivity. However, no measurements have been carried out beyond 323 K except for acetic and β-naphthoic acids [1,2].

In this work, we focus on pH measurements of benzoic acids aqueous solutions by a flow-through potentiometric apparatus [3] at temperatures up to 674 K and pressure of 30 MPa and understanding the relationship between temperature, dielectric constant, and species of carboxylic acids and the determined molal dissociation constants.

Experimental

Two different stock solutions were prepared. The first solution consisted of 10^{-3} mol kg^{-1} HCl + 10^{-1} mol kg-1 NaCl as a reference solution. The second solution consisted of 10^{-2} mol kg^{-1} benzoic acid + 10^{-1} mol kg^{-1} NaCl aqueous solution as a test solution. These solutions were continuously purged with H_2 gas to remove dissolved O_2 and to maintain constant H_2 concentration around the electrodes.

The flow-through potentiometric apparatus and general procedure are described in our previous publication [3]. The cell has Y type passage and pressure balanced hydrogen platinum reference and test electrodes are placed on the left and right side, respectively. These electrodes were sealed by a packing type high-pressure CONAX fitting with Teflon sealant. Each electrode responds to the half cell reaction $2H^+ + 2e^- = H_2$. The H_2 activity can be assumed to be constant over both solutions for keeping same pressure in cell and adding same amount of NaCl into both solutions. Note that we use a standard state of unit activity of H_2 in a hypothetical 1 mol kg^{-1} solution referenced to infinite dilution at any temperature and pressure rather than the customary unit fugacity. The measured potential difference between the electrodes, ΔE, is given by the Nernst equation

$$\Delta E = \frac{RT}{F} \ln \frac{m_{H^+,test}\gamma_{\pm,test}}{m_{H^+,ref}\gamma_{\pm,ref}} + E_{LJ} \qquad (1)$$

where R and F are the gas and Faraday constants, T is the absolute temperature, and E_{LJ} represents the liquid junction potential. In this work, E_{LJ} was assumed to be negligibly small compared to ΔE because concentrations of H^+, OH^-, and benzonate anion were at least two orders of magnitude lower than that of Na^+ and Cl^- at a given conditions[1]. m is (relative) molality, $m/(m_0)$ where $m_0 = 1$ mol kg^{-1}. Subscripts, test and ref, refer to the test and reference solutions, respectively. The molal activity coefficients of ions, γ, in the reference and test compartments are assumed to be almost equal at the given conditions.

Results and Discussion

The experimental ΔE values were measured for test solutions at temperatures ranging from 323 to 674 K and pressure of 30 MPa. The values of m_{ref} at each experimental condition were calculated from dissociation constants data of HCl, NaCl, NaOH, and H_2O and formulation for ionic activity coefficient with electroneutrality and mass balance constraints [3]. Then m_{test} was calculated on the basis of eq (1) with the calculated m_{ref} and the experimentally measured ΔE at given conditions. Molal dissociation constants of benzoic acid, K_a, as a thermodynamic activity product at given temperatures was determined using the values of m_{test}, dissociation constant data of HCl, NaCl, NaOH, and H_2O, and formulation of ionic activity coefficient with mass and charge balance constraints. The present results were compared with the literature data from electrical conductivity [4] as shown in Figure 1. Temperature dependence of the experimental values showed a similar trend up to 523 K.

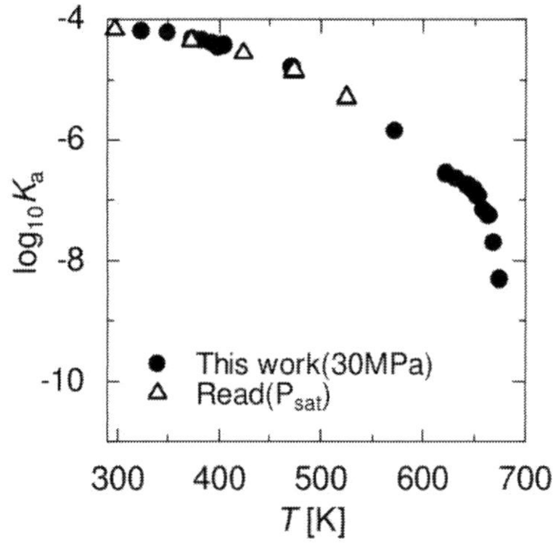

Figure 1 Dissociation constants of benzoic acid at temperatures up to 674 K and pressure of 30 MPa.

To understand the temperature dependence of the dissociation constants at elevated temperatures up to supercritical conditions, Born model [5] was applied. The model was proposed by Xiang and Johnston [5] for describing temperature and pressure dependence of isocoulombic reactions. The model is written by the following relationship

$$\log_{10} K_{ICR}(T,\rho) = \log_{10} K_{ICR}(T_r,\rho_r) + \left(\frac{\partial \log_{10} K_{ICR}}{\partial(1/T)}\right)_{\rho_r}\left(\frac{1}{T}-\frac{1}{T_r}\right) + \frac{36278}{T}\left(\frac{1}{R^*_{OH-}}-\frac{1}{R^*_{A-}}\right)\left(\frac{1}{\varepsilon(T,\rho)}-\frac{1}{\varepsilon(T,\rho_r)}\right) \tag{2}$$

where K_{ICR} is the equilibrium constant for the isocoulombic reaction ($HA + OH^- = A^- + H_2O$, HA:Carboxylic acid) and is expressed as K_a/K_W. T, ρ, and ε are absolute temperature, density and dielectric constant. R^* denotes an ionic radius. Figure 2 shows the K_{ICR} values of benzoic acid reactions converted from the experimentally determined K_a with K_W. The K_{ICR} values of acetic acid calculated from literature [1] are also shown in Figure 2. The relationship between $\log_{10}K_{ICR}$ and $1000/T$ shows a linear trend for both benzoic and acetic acids up to 588 K and drastic increase above 588 K due to the drastic increase in the values of $[1/\varepsilon(T,\rho)-1/\varepsilon(T,\rho_r)]/T$.

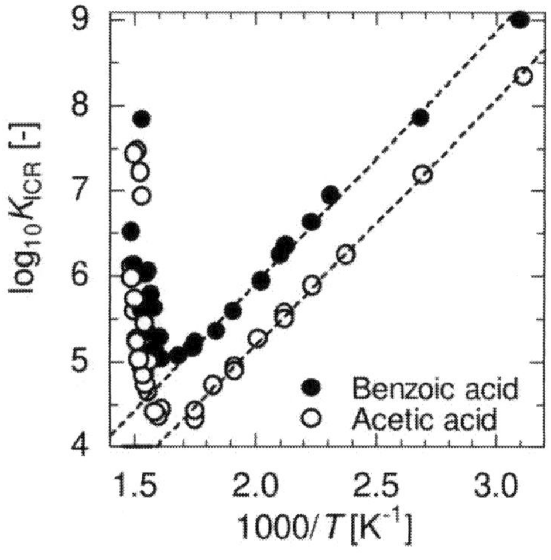

Figure 2 Equilibrium constants of isocoulombic reactions for benzoic and acetic acids at temperatures up to 674 K and pressure of 30 MPa.

For further analysis of the differences between the liner relations described by the first and second terms in eq (2) and the experimental $\log_{10}K_{ICR}$ values at 30 MPa above 588 K, the third term (solvation contribution of ions by changing density and dielectric constant), $\Delta\log_{10}K_{ICR}$, was calculated by subtracting the first and second terms from the experimental $\log_{10}K_{ICR}$ values on each experimental data. $\log_{10}K_{ICR}(T_r, \rho_r)$ of the first term and $(\partial \log_{10}K_{ICR}/\partial(1/T))_{\rho r}$ in the second term were determined through the correlation with the experimental $\log_{10}K_{ICR}$ values up to 588 K. Relationship between $\Delta\log_{10}K_{ICR}$ and $36278[1/\varepsilon(T,\rho)-1/\varepsilon(T,\rho_r)]/T$ are shown in Figure 3. The slopes represent the values of $1/R^*_{OH-}-1/R^*_{A-}$. The increase of $\Delta\log_{10}K_{ICR}$ with increasing $36278[1/\varepsilon(T,\rho)-1/\varepsilon(T,\rho_r)]/T$ means the change of the ionic radii of OH^- and A^-. In addition, the values of $\Delta\log_{10}K_{ICR}$ for benzoic acid became higher than that for acetic acid due to the different radii of benzonate and acetate anions.

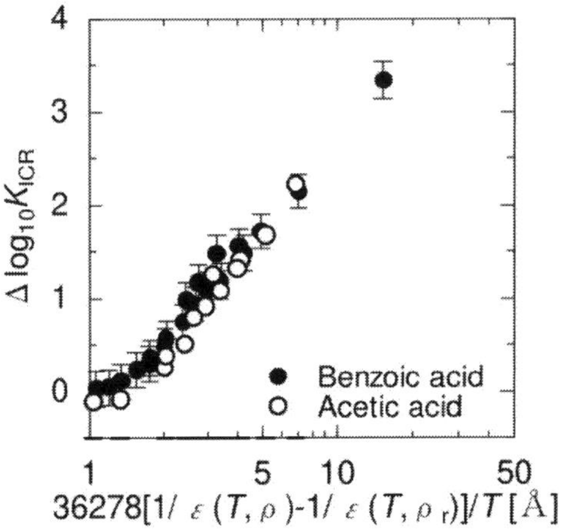

Figure 3 Relationship between the solvation contribution, $\Delta\log_{10}K_{ICR}$, and the value of $36278[1/\varepsilon(T,\rho)-1/\varepsilon(T,\rho_r)]/T$ in the Born model at temperatures up to 674 K and pressure of 30 MPa.

Conclusion

Potentiometric pH measurements of benzoic acid aqueous solutions were carried out at temperatures from 323 K to 674 K and pressure of 30 MPa by a flow-through electrochemical cell. New data of the molal ionization constants for benzoic acid was reported. Born model was successively applied for understanding temperature, dielectric constant, and acid species dependencies of the determined dissociation constants.

Acknowledgement

This research was partially supported by the Ministry of Education, Science, Sports and Culture, Grant-in-Aid for Scientific Research (Young Scientists (A)), 20686053, 2009.

References

1. Sue, K., T. Usami and K. Arai (2003), "Determination of acetic acid dissociation constants to 400 °C and 32 MPa by potentiometric pH measurements," *J. Chem. Eng. Data*, 48, pp. 1081-1084.
2. Xiang, T. and K. P. Johnston (1997), "Acid-base behavior in supercritical water: β-naphthoic acid - ammonia equilibrium," *J. Soln. Chem.*, 26, pp. 13-30.
3. Sue. K., M. Uchida, T. Usami, T. Adschiri, and K. Arai (2004), "Apparatus for direct pH measurement of supercritical aqueous solutions," *J. Supercrit. Fluids*, 28, pp. 287-296.
4. Read, A. J. (1981), "Ionization constants of benzoic acid from 25 to 250 °C and to 2000 bar," *J. Solution Chem.*, 10, pp. 437-450.

5. Xiang, T. and K. P. Johnston (1994), "Acid-base behavior of organic compounds in supercritical water," *J. Phys. Chem.*, 986, pp. 7915-7922.

Simulation of Hypercompressible Flow of Carbon Dioxide through Porous Media

Simon L. Marshall
CSIRO Mathematical and Information Sciences
Gate 5, Normanby Road
Clayton, Victoria, 3168
Australia

Abstract

The dynamics of fluid flow through a porous medium with pressure-independent properties are determined by the equation of state of the fluid, through the isothermal compressibility and the viscosity. These effects are manifested in the hydraulic diffusivity, which, for a real fluid, can vary markedly with pressure in the vicinity of the critical point. Such a strong dependence of the hydraulic diffusivity on pressure produces dynamic behavior characterized by a wide range of time constants. As a result, solution schemes involving partial or full discretization of the flow equation can encounter serious stability problems. Introduction of the pseudopressure, which is defined by a Kirchhoff-type integral transformation, produces a quasilinear diffusion equation, the nonlinearity of which can be removed by introduction of a transformed time coordinate. The latter transformation establishes a nonlinear mapping between the solutions of the linear equation, and the original nonlinear equation. Solutions for the transient pressure fields associated with flow of slightly - supercritical carbon dioxide in porous media of various simple geometries can thus be obtained. This approach permits rigorous analysis of hypercompressibility effects in porous-medium flows of carbon dioxide relevant to geosequestration and supercritical fluid extraction.

1 Introduction

The possible use of subsurface geological formations for disposal of waste carbon dioxide is currently attracting much interest. Simulations of the fluid flow through the porous media can be expected to play a crucial role in assessing the feasibility of such disposal schemes. The predicted flow behavior will depend in an obvious way on the porosity and permeability of the reservoir. But the properties of the fluid are possibly even more significant, considering that the isothermal compressibility can assume extremely large values in the pressure and temperature range of interest. It is well known that accurate prediction of volumetric properties in the critical region is an exacting test of any equation of state; as we will see, the flow of such a fluid through a porous medium poses a correspondingly challenging problem in fluid mechanics.

Mathematical models of flows through porous media are usually based on Darcy's Law,

$$\mathbf{v} = -\frac{K}{\eta}\nabla p, \tag{1-1}$$

where η is the fluid velocity and K is the permeability. The velocity $\mathbf{v}$ is related to the pore fluid density ρ by the continuity equation,

$$\frac{\partial(\rho\phi)}{\partial t} = -\nabla \cdot (\rho\mathbf{v}), \tag{1-2}$$

where ϕ is the porosity of the medium. Use of Darcy's Law for $\mathbf{v}$ results in the 'diffusivity equation'

$$\frac{\partial(\rho\phi)}{\partial t} = \nabla \cdot \left(\rho\frac{K}{\eta}\nabla p\right). \tag{1-3}$$

Solution of this nonlinear equation requires specification of the equation of state relating the pressure and density, and further assumptions concerning the dependence of η, K, and ϕ on p. Functional relationships of the form $\phi = \phi_0 e^{\beta_\phi(p-p_0)}$ and $K = K_0 e^{\beta_K(p-p_0)}$ are often assumed; a recent paper by Marshall [2009] showed that analytical solutions of the nonlinear diffusivity equation are possible if the porosity and permeability moduli β_ϕ and β_K and the isothermal fluid compressibility β are constant, and $\beta_\phi \approx \beta_K$. The concern of this paper is with the application of this equation to flow of carbon dioxide in porous media in another limiting case, where β_ϕ and β_K are insignificant compared with β.

2 Theoretical Analysis

Since the diffusivity equation contains two dependent variables p and ρ, its solution requires specification of an additional relation between these quantities. If the assumption of local equilibrium is justified, this additional relation is provided by the equation of state. The mathematical consequence is that the derivatives of the density can be replaced by the corresponding derivatives of the pressure, by making use of the isothermal compressibility

$$\beta = \frac{1}{\rho}\left(\frac{\partial \rho}{\partial p}\right)_T \tag{2-1}$$

to give

$$\frac{\partial \rho}{\partial t} = \frac{\partial \rho}{\partial p}\frac{\partial p}{\partial t} = \beta\rho\frac{\partial p}{\partial t}; \qquad \nabla\rho = \beta\rho\nabla p. \tag{2-2}$$

Transformations of this type are less easy to use when the compressibility is not constant, as is the case for fluids with equations of state in the form

$$z = \frac{p}{\rho RT} \tag{2-3}$$

where z the compressibility factor. With the additional assumption that the porosity and permeability are effectively constant, the diffusivity equation becomes

$$\nabla \cdot \left[\frac{\rho}{\eta}\nabla p\right] = \frac{\phi}{K}\frac{\partial \rho}{\partial t} \tag{2-4}$$

Now in the left hand side, putting $\rho = p/(zRT)$ leads to

$$\nabla \cdot \left[\frac{\rho}{\eta}\nabla p\right] = \nabla \cdot \left[\frac{p}{\eta RTz}\nabla p\right] = \frac{1}{RT}\nabla \cdot \left[\frac{p}{\eta z}\nabla p\right], \tag{2-5}$$

and the compressibility $\beta = (1/\rho)(\partial\rho/\partial p)_T$ can be used to transform the time derivative of the density into a time derivative of the pressure:

$$\frac{\phi}{K}\frac{\partial \rho}{\partial t} = \frac{\beta\phi}{K}\rho\frac{\partial p}{\partial t} = \frac{\beta\phi p}{KRTz}\frac{\partial p}{\partial t}. \tag{2-6}$$

As first shown by Al-Hussainy et al. [1966], the nonlinear diffusion equation for p can be put into a simpler form by introduction of the pseudopressure ψ defined by

$$\psi = 2\int_{p_0}^{p}\frac{p}{\eta z}dp, \tag{2-7}$$

where p_0 is some arbitrary reference pressure.[1] In these terms, it follows that

$$\frac{\partial\psi}{\partial t} = 2\frac{p}{\eta z}\frac{\partial p}{\partial t}, \qquad \nabla\psi = 2\frac{p}{\eta z}\nabla p. \tag{2-8}$$

The two sides of the transformed diffusivity equation are therefore

$$\frac{\beta\phi}{KRT}\frac{p}{z}\frac{\partial p}{\partial t} = \frac{\beta\phi}{KRT}\frac{\eta}{2}\frac{\partial\psi}{\partial t} \quad \text{and} \quad \frac{1}{RT}\nabla\cdot\left[\frac{p}{\eta z}\nabla p\right] = \frac{1}{2RT}\nabla^2\psi. \tag{2-9}$$

Equating the right hand members of these expressions leads to

$$\frac{\beta\phi\eta}{K}\frac{\partial\psi}{\partial t} = \nabla^2\psi, \quad \text{or} \quad \frac{\partial\psi}{\partial t} = D(\psi)\nabla^2\psi, \tag{2-10}$$

where $D = K/(\phi\beta\eta)$ is the hydraulic diffusivity, understood to be expressed as a function of ψ. If the pseudopressure dependence of the diffusivity is represented as $D(\psi) = D_0 f(\psi)$, equation 2-10 can be put into more convenient form by introduction of the dimensionless coordinates $X_i = x_i/L$ and $\tau = D_0 t/L^2$, where L is a reference length and D_0 is a reference diffusivity. Thus

$$\frac{\partial\psi}{\partial\tau} = f(\psi)\nabla^2\psi, \tag{2-11}$$

where the Laplacian is taken with respect to the X_i. As shown by Wrobel and Brebbia [1987], the remaining nonlinearity in equation 2-11 can be removed by introduction of a transformed time coordinate defined by

$$\frac{d\upsilon}{d\tau} = f(\psi(\mathbf{X},\tau)) \qquad \text{or} \qquad \upsilon = \int_0^\tau f(\psi(\mathbf{X},\tau'))d\tau', \tag{2-12}$$

in terms of which

$$\frac{\partial\psi}{\partial\upsilon} = \nabla^2\psi. \tag{2-13}$$

Since the value of upsilon thus defined is a function of position, this formulation requires iterative solution of a system of nonlinear equations for the values of τ. A novel alternative procedure involves application of the same transformation in reverse, by defining

$$\frac{d\tau}{d\upsilon} = \frac{1}{f(\psi(\mathbf{X},\tau))} \qquad \text{or} \qquad \tau = \int_0^\upsilon \frac{du}{f(\psi(\mathbf{X},u))}, \tag{2-14}$$

which gives a single nonlinear equation for υ, to be solved at each point in space.

---

[1] The factor of 2 is included so that in the low-pressure limit, where z is 1 and η is constant, the pseudopressure becomes equal to p^2/η.

3 Calculations

3.1 Pseudopressure

The first step in the implementation of the analysis presented in the above paragraphs is the calculation of the pseudopressure. From its definition in equation 2-7, this process requires, on the one hand, a suitable equation of state and an expression for the viscosity of the gas viscosity as a function of pressure, and on the other, a method for performing the indicated integration. It is also to be observed that the equation of state is required not only for the determination of the compressibility factor z, but is also involved in the determination of viscosity at high pressure. But since the analytical expressions developed for both purposes are quite complicated in their own right, the integrand $p/(\eta z)$ will in general not be amenable to analytical integration. The errors that could result from numerical integration, table look-up, or curve-fit motivated Klins and Biterge [1987] to develop a semi-analytical approach to the calculation of pseudopressure, while Al-Jawad [1997] advocates an entirely analytical approach based on an equation of state of the Lee-Kesler type. But since neither of these works contains an adequate quantification of the errors of prediction inherent in the equations of state or viscosity correlation, they fail to present a compelling justification for the superiority of analytical formulae in relation to the more numerical alternatives. For this reason, the calculations presented here were performed by a straightforward extension of the cubic spline interpolation subroutines given in *Numerical Recipes* [Press et al., 1992] that also returns interpolated estimates of the derivative and integral of the spline function.

3.1.1 Equation of State

The volumetric properties of carbon dioxide have been extensively studied, possibly as a result of the closeness of its critical temperature (about 31°C) to ambient temperatures. The large body of published experimental data was critically evaluated by Span and Wagner [1996], who developed a reference-quality equation of state that was parameterized to represent the best available measurements of multiple properties to within experimental error. Since this equation is based on an empirical representation of the excess Helmholtz energy as a function of reduced temperature and density, iterative calculation of the density at specified temperature and pressure is always required.

Table 1: Cubic spline calculation of pseudopressure for carbon dioxide at 300 K

p/MPa	ρ/g cm^{-3}	$z = p/\rho RT$	$10^7\eta$/Pa s	$f = 2p/z\mu$	$\psi = \int_{0.1}^{p} f dp$
0.100	0.0018	0.9951	152.10	0.001321	0.000000
0.500	0.0090	0.9753	153.23	0.006692	0.001600
1.000	0.0186	0.9496	154.80	0.013605	0.006664
1.500	0.0287	0.9230	156.58	0.020759	0.015244
2.000	0.0394	0.8952	158.58	0.028177	0.027466
2.500	0.0509	0.8660	160.87	0.035890	0.043470
3.000	0.0634	0.8352	163.50	0.043938	0.063412
3.500	0.0769	0.8026	166.54	0.052373	0.087473
4.000	0.0920	0.7675	170.13	0.061266	0.115861
4.500	0.1088	0.7295	174.44	0.070724	0.148832
5.000	0.1283	0.6875	179.76	0.080908	0.186708
5.500	0.1517	0.6399	186.66	0.092097	0.229898
6.000	0.1816	0.5830	196.30	0.104864	0.279093

3.1.2 Viscosity

Calculation of the viscosity of carbon dioxide at the elevated pressures of interest involved determination of the additive correction to be applied to the low-pressure limiting value. The collision integral required for calculation of this low-pressure limiting value was obtained by use of the empirical approximant developed by Neufeld et al. [1972], as a function of reduced temperatures determined from tabulated Lennard-Jones parameters. Following the recommendations of Reid et al. [1977], the calculation of the additive correction was performed by the method proposed by Dean and Stiel [1965], which involves use of a corresponding-states principle.

3.1.3 Numerical Quadrature

To illustrate the application of cubic spline integration, consider the calculation of the pseudopressure of carbon dioxide at 300 K, at pressures of up to 6 MPa. The pressure, density, viscosity, the integrand and the integral are tabulated in Table 1. Values of the pressure corresponding to specified pseudopressures can be determined by a second call to the spline routine, in which the data abscissae and ordinates are taken from the last and first columns of Table 1. Results are shown in Table 2.

Table 2: Pressure corresponding to given pseudopressures, by reverse cubic spline interpolation.

ψ	p/MPa
0.020000	1.729435
0.040000	2.398477
0.060000	2.922025
0.080000	3.353347
0.100000	3.730683
0.120000	4.066838
0.140000	4.372905
0.160000	4.654534
0.180000	4.916090
0.200000	5.161271
0.220000	5.391542
0.240000	5.607065
0.260000	5.810890

3.2 Hydraulic Diffusivity

To illustrate the calculation of the hydraulic diffusivity, consider isothermal flow of carbon dioxide at 305 K (about 1 degree above its critical temperature) through a porous medium with porosity ϕ=0.3 and a permeability K=1×10^{-15} m^2, or roughly 1 millidarcy. At this pressure and temperature, the Span-Wagner equation of state gives a compressibility 1.0047×10^{-5} Pa^{-1}. Selecting a reference pressure of 0.1 MPa, the hydraulic diffusivity is

$$D_0 = \frac{(1 \times 10^{-15}\mathrm{m}^2)}{(1.5210 \times 10^{-5}\mathrm{Pa\,s})(1.0047 \times 10^{-5}\mathrm{Pa}^{-1})(0.3)} = 2.1488 \times 10^{-5}\mathrm{m}^2\mathrm{s}^{-1},$$

and this value can be used to define the dimensionless time coordinate τ. The hypercompressibility effects become evident when the hydraulic diffusivity and pseudopressure are calculated for pressures between 0.1 and 10 MPa. The reduced hydraulic diffusivity D/D_0 reaches a maximum of about 28 at 4.7 MPa, and passes through a sharp minimum of about 0.5 at 7.52 MPa. An abbreviated table of numerical data for this isotherm is given in Table 3. The second and third columns of this table can be used to set up a cubic spline to calculate the hydraulic diffusivity as a function of the pseudopressure. Any desired accuracy can be achieved, by suitably close spacing of pressures.

Table 3: Hydraulic Diffusivity and Pseudopressure for Carbon Dioxide at 305 K

p/MPa	ψ	D/D_0
0.10	0.000000	1.000000
0.50	0.001573	4.869026
1.00	0.006553	9.392476
1.50	0.014979	13.537829
2.00	0.026964	17.268370
2.50	0.042632	20.541967
3.00	0.062116	23.309495
3.50	0.085570	25.512557
4.00	0.113166	27.080012
4.50	0.145105	27.922343
5.00	0.181627	27.921933
5.50	0.223033	26.914922
6.00	0.269719	24.653626
6.50	0.322255	20.715277
7.00	0.381604	14.202140
7.10	0.394444	12.413169
7.20	0.407666	10.374875
7.30	0.421320	8.002699
7.40	0.435484	5.131624
7.45	0.442806	3.398662
7.50	0.450347	1.332714
7.51	0.451892	0.875117
7.52	0.453454	0.500045
7.53	0.455031	0.501984
7.54	0.456609	0.883259
7.55	0.458182	1.464762
7.60	0.465974	4.126590
8.00	0.526291	19.172360
8.50	0.599059	33.660766
9.00	0.669999	46.035014
9.50	0.739561	57.079003
10.00	0.807987	67.205876

3.3 Pseudotime Transformation

A cubic spline calculated from the second and third columns of Table 3 can be used to generate accurate estimates of the hydraulic diffusivity corresponding to any pseudopressure in the range from 0 to 0.807987, provided that more closely-spaced data points (pressures) are used in the vicinity of the critical pressure. (These have been omitted from the table in the interests of brevity.) If, in particular, a sequence of pseudopressures ψ_k are generated from an analytical or numerical solution of the linear diffusion equation 2-13 at a sequence of pseudotimes v_k, the corresponding values of $f(\psi_k) = D(\psi_k)/D_0$ can be determined and used to construct a table of values of

$$\int_0^{v_k} \frac{1}{f(\psi(\mathbf{X}, u))} du,$$

from which the solution of equation 2-14 at each value of the dimensionless time can be estimated by inverse interpolation.

Consider, for example, radial flow into a cylindrical porous medium, with the pseudopressure held at some constant value ψ_1 on the boundary. (This configuration might be of some interest in supercritical fluid extraction.) Then the appropriate linear diffusion equation for the pseudopressure as a function of pseudotime, for $R \int [0, 1]$, is

$$\frac{\partial \psi}{\partial v} = \frac{1}{R} \frac{\partial}{\partial R} \left[R \frac{\partial \psi}{\partial R} \right] : \qquad \psi(1, v) = \psi_1, \quad \psi(R, 0) = 0. \tag{3-1}$$

From the well known monograph by Carslaw and Jaeger [1959, page 199, equation 8], the solution is

$$\psi(R, v) = \psi_1 \left[1 - 2 \sum_{n=1}^{\infty} e^{-\alpha_n^2 v} \frac{J_0(\alpha_n R)}{\alpha_n J_1(\alpha_n)} \right], \tag{3-2}$$

where J_0 and J_1 are Bessel functions of the first kind, of order zero and one, respectively, and α_n is the nth positive root of J_0. The sequence of pseudopressures would be generated by evaluating this analytical solution at a sequence of pseudotimes.

It is also possible to apply this approach to nonlinear diffusion equations that have simple flux boundary conditions, or boundary conditions of the lumped-capacity type. The approach becomes much more complicated, however, if the boundary conditions are of the Dirichlet type, but depending explicitly on time. In these cases, rewriting the boundary conditions in terms of the pseudotime usually results in expressions of unmanageable complexity.

4 Conclusions

This paper has described a highly efficient numerical scheme by which accurate solutions of the diffusivity equation can be obtained for near-supercritical carbon dioxide in the vicinity of the critical pressure. The three essential elements of this approach are the pseudopressure, defined by a form of Kirchhoff transformation; an analytical or numerical solution of the corresponding boundary value problem for the linear diffusion equation; and a nonlinear transformation of the time coordinate, involving numerical quadrature and iterative solution of a single nonlinear algebraic equation. The salient advantages of this approach are that

- Limitations on accuracy resulting from finite-difference approximations are completely avoided;
- The calculation of the hydraulic diffusivity as a function of pseudopressure can be combined with any compatible solution of the linear diffusion equation;
- The nonlinearities that can give rise to rapid changes in the hydraulic diffusivity with pseudopressure can be dealt with much more accurately in the nonlinear transformation of the time coordinate, than is possible in finite difference methods that require solution of systems of nonlinear equations.

The mapping between the solutions of the linear and nonlinear problems has the effect of greatly extending the applicability of linear analysis (such as Laplace Transformation) to nonlinear phenomena.

Acknowledgments

Financial support of the CSIRO Future Manufacturing National Research Flagship is gratefully acknowledged.

References

R. Al-Hussainy, H.J. Ramey, and P.B. Crawford. The flow of real gases through porous media. *Journal of Petroleum Technology*, 18(5):624–636, 1966.

Mohammed S. Al-Jawad. An efficient method for real gas pseudopressure calculation. *Polymer-Plastics Technology and Engineering*, 36(3):345–362, 1997.

David E. Dean and Leonard I. Stiel. The viscosity of nonpolar gas mixtures at moderate and high pressures. *AIChE Journal*, 11(3):526–532, 1965.

M.A. Klins and M.B. Biterge. A semianalytical approach to pseudopressure calculations. *Journal of Petroleum Technology*, pages 468–471, 1987.

Simon L. Marshall. Nonlinear pressure diffusion in flow of compressible liquids through porous media. *Transport in Porous Media*, 77(3):431 – 446, 2009.

Philip D. Neufeld, A.R. Janzen, and R.A. Aziz. Empirical equations to calculate 16 of the transport collision integrals $\Omega^{(l,s)*}$ for the Lennard-Jones (12-6) potential. *Journal of Chemical Physics*, 57(3):1100–1102, 1972.

William H. Press, Saul A. Teukolsky, William T. Vetterling, and Brian P. Flannery. *Numerical Recipes in FORTRAN 77: The Art of Scientific Computing*. Cambridge University Press, Cambridge, 2 edition, 1992.

R.C. Reid, J.M. Prausnitz, and T.K. Sherwood. *The Properties of Gases and Liquids*. McGraw-Hill, New York, 1977.

R. Span and W. Wagner. A new equation of state for carbon dioxide covering the fluid region from the triple-point temperature to 1100 K at pressures up to 800 MPa. *Journal of Physical and Chemical Reference Data*, 25: 1509–1596, 1996.

L.C. Wrobel and C.A. Brebbia. The dual reciprocity boundary element formulation for nonlinear diffusion problems. *Computer Methods in Applied Mechanics and Engineering*, 65:147–164, 1987.

From Cradle to Burial: High Pressure Phase Equilibrium Behaviour of CO2 During CCS

Haroun Mahgerefteh, Chemical Engineering, University College London, London, United Kingdom

Solomon Brown, Chemical Engineering, University College London, London, United Kingdom

The safe and efficient operation of the entire Carbon Capture and Sequestration chain requires a good understanding of the thermodynamics phase behaviour of CO2. Such information impacts anything from the efficiency of the capture technologies employed, the mechanical integrity of the high pressure pipeline system used for transporting CO2 to the likelihood of the escape of CO2 from geological sequestration sites.

Although significant progress has been made in the development of equations of state for predicting the phase behaviour, all of this work has so far been confined to either pure CO2 or hydrocarbon mixtures containing small amounts of CO2.

Little work on the efficacy of these equations of state for CO2 containing different types of impurities has been conducted. In practice, during CCS, depending on the capture technology, the presence of different impurities such as N2, NOx and SOx within the CO2 stream will be inevitable.

This paper reports data based on a comparison of the performance of the Peng Robinson against the Modified Peng Robinson equations of state in predicting the phase equilibrium behaviour of CO2 in the presence of impurities, especially around the critical region. This is the followed by an examination of the depressurisation thermodynamics trajectory of the CO2 mixture following accidental escape during its transportation to capture using a hypothetical 100 km long , 0.6 m i.d pipeline operating at 110 bara and 32 oC .

See more of this Session: Thermodynamic and Transport Properties Under Pressure
See more of this Group/Topical: Engineering Sciences and Fundamentals

When Does a Vessel Become a Pipe: A Simplified Approach to Pressurised Outflow Simulation

Haroun Mahgerefteh, Chemical Engineering, University College London, London, United Kingdom

Navid Jalali, Chemical Engineering, University College London, London, United Kingdom

Maria Fernandez, Chemical Engineering, University College London, London, United Kingdom

Pressurised pipelines are by far the most widely used method for transporting oil and gas across the globe. In the US alone, almost 40% of its energy demand is supplied through such means. In some other countries this figure approaches 100%.

Given that such pipelines can be several hundreds of kilometres long conveying thousands of tonnes of inventory at pressures as high as 150 bara, their accidental rupture can lead to catastrophic consequences. Indeed there have been numerous incidents involving pipeline ruptures which have resulted in significant number of casualties, property damage and environmental contamination. It is now a statuary requirement to assess all the major hazards associated with pressurised pipelines prior to their commissioning. Pivotal to this exercise is the prediction of mass outflow and its variation with time in the event of pipeline rupture. Such data dictates all the consequences associated with pipeline failure including fire, explosion and toxic release. It lays the foundation for pipeline operators and safety authorities to determine minimum safe distances to populated areas and formulate emergency response planning.

Given the highly unsteady state flows involved during pipeline rupture, the modelling of outflow has been confined to numerical techniques. Despite their success in accurately predicting outflow, a major drawback has been the long computational run times; typically a few days for the simulation of the complete depressurisation of a long (e.g 150 km, 0.7 m i.d) pressurised (e.g 100 bara) pipeline following its puncture.

This study describes the development of a simplified analytical model based on approximating the punctured pipeline as a vessel discharging through an orifice designed to address the above problem. The efficacy of the vessel model in terms of computational run time and accuracy is determined by comparison against an established but computationally demanding rigorous numerical technique based on the solution of the conservation equations using the Method of Characteristics [1-3].

The discharge of different pressurised inventories such as permanent gases (e.g methane, ethane and their mixtures) and two-phase mixtures (e.g. mixture of methane and pentane) are simulated to identify the range of applicability of the vessel model as a function various design and operating parameters These include the ratio of the puncture diameter to the pipeline diameter (0.1 - 0.5), initial line pressure (21.6 bara and 100 bara) and pipeline length (100 m - 5 km). In the case of the permanent gases, the results indicate that the accuracy of the vessel model in terms of the degree of agreement with the numerical model increases with decreasing diameter ratio, line pressure, and pipeline length. The finite differences (±11%) are shown to be primarily due to the isothermal gas expansion assumption incorporated in the vessel model development.

Much the same as the permanent gas inventories, for two-phase mixtures, the vessel model accuracy increases with increasing diameter ratio. However no obvious trends in the degree of agreement are obtained with changes in the line pressure or the pipeline length. This is believed to be due to the complicated relationship between the speed of sound with pressure for the two-phase mixtures in turn controlling the expansion wave velocity within the pipeline and hence the pipeline depressurisation rate. Such phenomena are not considered in the simplified analytical vessel model. Despite this, surprisingly the maximum error in the prediction of the vessel model in the case of the two-phase mixtures is only ±5% which is significantly better (c.f ±11%) than that for the permanent gas predictions.

These acceptable errors coupled with the negligible computational run time indicate the clear advantages of using the vessel model presented in this work for simulating the puncture of long pipelines containing pressurised hydrocarbons for the ranges tested.

References

1.Mahgerefteh, H., & Denton, G., & Rykov, Y., A hybrid multiphase flow model (2008), AIChE Journal, 54(9), 2261-2268.

2.Mahgerefteh, H., & Abbasi, M., Modeling blowdown of pipelines under fire attack (2007), AIChE Journal, 53(9), 2443-2450.

3.Mahgerefteh, H., & Oke, A., 'An efficient numerical simulation for highly transient flows' (2006), Chem.Eng.Sci., 61(15), 5049-5056.

See more of this Session: Thermodynamic and Transport Properties Under Pressure
See more of this Group/Topical: Engineering Sciences and Fundamentals

High-Pressure Transport Properties and Phase Equilibria of Imidazolium Ionic Liquids with CO_2 and R-134a

Azita Ahosseini, Chemical & Petroleum Engineering, University of Kansas, Lawrence, KS

Wei Ren, Chemical & Petroleum Engineering, University of Kansas, Lawrence, KS

Aaron M. Scurto, Chemical & Petroleum Engineering, University of Kansas, Lawrence, KS

Ionic liquids (ILs) and compressed gases are found in a number of proposed processes in extractions, reactions, and engineering applications. This investigation presents experimental and modeling results of the phase behavior, phase equilibria, viscosity and diffusivity of imidazolium ionic liquids with compressed carbon dioxide and 1,1,1,2-tetrafluoroethane (R-134a). Both gases are very soluble in ionic liquids. However with R-134a, all of the ionic liquids studied have regions in temperature and pressure of miscibility and criticality exist at certain temperatures and pressures; Type V systems from the classification scheme of Scott-van Konynenburg. CO_2 has no known regions of miscibility with ionic liquids (possibly Type III). This has important ramifications on potential applications. The presence of the dissolved gases significantly increases the mass transport properties of ionic liquids. While ILs often have higher ambient viscosity than common organic solvents, mixtures of ILs with the gases can decrease the viscosity by 80+%. This decrease in viscosity increases the self-diffusivity of the gases. Effects of the structure of the cation and anion were measured and compared for both thermodynamic and transport properties.

See more of this Session: Thermodynamic and Transport Properties Under Pressure
See more of this Group/Topical: Engineering Sciences and Fundamentals

Interfacial Aqueous Systems: Ion-Ion Correlations and the Electric Double Layer

Dimitrios Argyris, School of Chemical Biological and Materials Engineering, The University of Oklahoma, Norman, OK

Alberto Striolo, School of Chemical Biological and Materials Engineering, The University of Oklahoma, Norman, OK

Aqueous electrolyte solutions at interfaces have generated, and continue to generate, significant research interest due to their importance in numerous phenomena for a wide range of applications. Molecular dynamics simulations have been employed to study structural and dynamic properties of aqueous electrolyte solutions near free standing silica-based surfaces. This study focuses on the properties of various monovalent (Cl-, Na+, Cs+) and divalent ions (S2-, Ca2+, Sr2+) within a few molecular layers from solid-liquid interface. A wide range of ionic strengths was considered in our simulations to investigate the effect on the structure, dynamic properties and accumulation of ions at the solid interface. Two crystallographic faces of the cristobalite crystal with different degrees of hydroxylation were employed to assess the effect of surface roughness and charge density on the properties of interest. The SPC/E model was used to model water-water interactions. Density profiles, radial distribution functions, in-plane density distribution, residence time, and diffusion coefficients were calculated to study the effect of surface properties on the characteristics of confined water and on the formation of the electric double layer.

See more of this Session: Thermodynamics at the Nanoscale I
See more of this Group/Topical: Engineering Sciences and Fundamentals

Influence of Hydrophilic Surface Specificity On the Structural Properties of Confined Water

Ateeque Malani, Department of Chemical Engineering, Indian Institute of Science, Bangalore, India

K. G. Ayappa, Department of Chemical Engineering, Indian Institute of Science, Bangalore, India

Sohail Murad, Chemical Engineering, University of Illinois at Chicago, Chicago, IL

The influence of chemical specificity of hydrophilic surfaces on the structure of confined water in the sub-nanometer regime is investigated using grand ensemble Monte Carlo simulations. The structural variations for water confined between hydroxylated silica surfaces are contrasted with water confined between mica surfaces. Although both surfaces are hydrophilic, our study shows that hydration of potassium ions on the mica surface has a strong influence on the water structure and solvation force response of confined water. In contrast to the disrupted hydrogen bond network observed for water confined between mica surfaces, water between silica surfaces retain their hydrogen bond network displaying bulk-like structural features down to surface separations as small as 0.45 nanometer (nm). Hydrogen bonding of an invariant contact water layer with the surface silanol groups aids in maintaining a constant number of hydrogen bonds per water molecule for the silica surfaces. As a consequence, water depletion and rearrangement upon decreasing confinement is a strong function of the hydrophilic surface specificity, particularly at smaller separations. An oscillatory solvation force response is only observed for water confined between silica surfaces and bulk-like features are observed for both surfaces above a surface separation of about 1.2 nm. We evaluate and contrast the water density, dipole moment distributions, pair correlation functions and the solvation forces as a function of the surface separation.

See more of this Session: Thermodynamics at the Nanoscale I
See more of this Group/Topical: Engineering Sciences and Fundamentals

Protein-Mediated Membrane Phase Behavior: Calculation of Free Energies Associated with Membrane Deformations

Neeraj J. Agrawal, Chemical and Biomolecular Engineering, University of Pennsylvania, Philadelphia, PA

Ravi Radhakrishnan, Department of Bioengineering, University of Pennsylvania, Philadelphia, PA

We present a computational methodology for incorporating thermal effects and calculation of free energies for elastic fluid membranes subject to intrinsic curvature fields using the method of thermodynamic integration. Based on a simple model for the intrinsic curvature imposed by the clathrin coat on the membrane, we employ thermodynamic integration to calcuate free energy change as a function increasing strength of the intrinsic curvature field and a thermodynamic cycle to compute free energy changes for different sizes of the clathrin coat. Through analysis of a simplified one-dimensional analog of the fluid-membrane, we show that the membrane stiffness increases with increasing intrinsic membrane curvature field, thereby renormalizing the membrane bending rigidity. In particular, the stiffness of the membrane for non-zero intrinsic curvature is greater than that for the membrane with zero intrinsic curvature, leading to a reduction in the quasiharmonic entropy. We have confirmed these trends by explicitly computing the free energy changes using thermodynamic integration and by quantifying the loss of entropy accompanied with increasing membrane deformation. One of the main conclusions of this work is that the entropy of a membrane decreases with increasing size of the clathrin coat. However, the contribution of the enrtopic effects to the overall change in the bending free energy is small (5%). We also show that the Fourier modes are not the natural basis when the membrane is subject to an intrinsic curvature field, and the degree of mode-mixing depends on the strength of the intrinsic curvature function. Our results help quantify the free energy change when a planar membrane deforms under the influence of curvature inducing proteins at a finite temperature, and the role of membrane entropy in mediating membrane-induced interactions between curvature-inducing proteins. The thermodynamic integration approach offers a promising avenue for quantifying thermal effects in protein-mediated membrane processes, which are ubiquitous in intracellular signaling and trafficking mechanisms.

See more of this Session: Thermodynamics at the Nanoscale I
See more of this Group/Topical: Engineering Sciences and Fundamentals

Polymer Single Chain Coil-Globule Transitions: A Scaled Particle Theory

David S. Simmons, Chemical Engineering, University of Texas, Austin, Austin, TX

Isaac C. Sanchez, Chemical Engineering, University of Texas, Austin, Austin, TX

The single-chain polymer coil-globule transition (CGT) is of great interest due to its applicability both in biological macromolecules and in synthetic stimuli responsive polymers. The present authors have published a lattice-fluid model for the lower critical solution temperature (LCST)-related CGT and for a theoretical high temperature upper critical solution temperature (UCST)-related CGT. However, more recently an updated model for these transitions has been developed based on a scaled particle theory approach. This model, without the use of adjustable solution parameters, provides a semi-quantitative match with extant experimental measures of CGT behavior. The predicted CGT also reproduces important aspects of a recently proposed qualitative master curve for the phase behavior of polymer solutions. In particular, the model predicts an LCST-related CGT temperature which increases with pressure until it merges at a pressure P_{max} with a UCST-related CGT.

See more of this Session: Thermodynamics at the Nanoscale I
See more of this Group/Topical: Engineering Sciences and Fundamentals

Computational Investigation of Glassy Water in Nano-Scale Confinement

Thomas G. Lombardo, Chemical Engineering, Princeton University, Princeton, NJ

Nicolás Giovambattista, Physics Department, Brooklyn College of the City University of New York, Brooklyn, NY

Pablo G. Debenedetti, Chemical Engineering, Princeton University, Princeton, NJ

We use molecular dynamics simulations to study the structure, thermodynamics, and mechanical properties of glassy water under nanoscale confinement between surfaces with tunable hydrophobicity and hydrophilicity [1]. Surface interactions are modified by scaling the dipole moments of surface charge-groups, giving a homogeneous surface, or by removing charge-groups from the surface, creating a heterogeneous surface with random or ordered hydrophilicty. We study both quenched and mechanically stable configurations (inherent structures) [2] over broad ranges of density and quench rates. The relation between the lateral components of the stress tensor and density, that is to say the equation of state of the energy landscape for this confined system [3], exhibits complex dependence on the surface chemistry and degree of confinement. Under hydrophobic confinement, multiple minima in the landscape equation of state appear which are indicative of a first-order transition between two liquid phases. Isochoric quenches at sufficiently low density result in cohesive or adhesive failure [4], depending on the hydrophobic or hydrophilic nature of the confining surfaces.

[1] T.G. Lombardo, F.H. Stillinger and P.G. Debenedetti, Thermodynamic Mechanism for Solution Phase Chiral Amplification via a Lattice Model. *Proc. Nat'l. Acad. Sci. USA*, DOI: 10.1073/pnas.0812867106, 2009.

[2] Stillinger, F.H., and Weber, T.A., *Phys. Rev. A*, 25, 978, 1982.

[3] Debenedetti, P.G., Stillinger, F.H., Truskett, T.M., and Roberts, C.J., *J. Phys. Chem. B*, 103, 7308, 1999.

[4] Shah, P., and Truskett, T.M., *Mechanics of Mater.*, 38, 924, 2006.

See more of this Session: Thermodynamics at the Nanoscale I
See more of this Group/Topical: Engineering Sciences and Fundamentals

A Monte-Carlo, Single-Chain Mean-Field Theory for Computation of the Free Energy of Packing End-Tethered Poly(Ethylene Glycol) Brushes in Water: Application to Colloidal Steric Stabilization and Alkyl Ethoxylate (CiEj) Nonionic Surfactant Micellization

Jonathan D. Mendenhall, Chemical Engineering, Massachusetts Institute of Technology, Cambridge, MA

Daniel Blankschtein, Chemical Engineering, Massachusetts Institute of Technology, Cambridge, MA

We have recently developed a single-chain mean-field (SCMF) theory based on the work of Carignano and Szleifer [J Chem Phys, 98, 5006-5018, 1993] which allows for implicit solution of the variation in end-tethered poly(ethylene glycol) (PEG) and water densities needed to satisfy incompressibility constraints subject to minimization of the system free energy. The system free energy, in turn, includes inhomogeneous entropic and enthalpic mixing terms based on Flory-Huggins theory, in addition to the usual configurational entropy terms present for chain molecules. We make use of Monte Carlo techniques to compute the change in free energy of these PEG-water systems relative to reference states appropriate to the application. We present two such applications of this packing theory. The first application is the polymer-mediated steric stabilization of colloids (e.g., ink pigments). We consider two approaching plates of equal surface coverage of end-tethered PEG chains as a model for two approaching colloids of large radius and, therefore, low surface curvature. We present profiles of force versus plate separation, computed relative to the reference state of two isolated surfaces using the Overlapping Distribution Method (ODM), for a variety of surface coverages and degrees of polymerization. We make comparison to scaling theories and comment briefly on the effect of simple branching on these profiles. The second application is the micellization of CiEj nonionic surfactants in aqueous solution. The molecular-thermodynamic theory of Puvvada and Blankschtein [J Chem Phys 92, 3710-3724, 1990] has been applied successfully to predict the free energy of micellization of a variety of CiEj nonionic surfactants. From the free energy of micellization, one can readily compute bulk surfactant properties such as the critical micelle concentration (CMC) and the micelle shape and size at thermodynamic equilibrium. However, the theory is practically limited to short ethoxylated chains (i.e., $j<8$) due to a hard-disk approximation invoked to simplify the modeling of the steric free energy-contribution to the free energy of micellization. This free energy penalty, arising from excluded-volume interactions between ethoxylated surfactant heads in the head region of the micelle (i.e., outside of the micelle core-water interface), becomes too severe for large j, as evidenced by an increase in the predicted CMC's relative to those measured experimentally for $j>8$. We demonstrate the use of the Monte Carlo, SCMF theory as a replacement for the hard-disk approximation, with the natural reference state being that of infinite dilution of each ethoxylated head. Since the SCMF theory explicitly accounts for the flexible nature of the ethoxylated heads, the drawbacks of the hard-disk model are avoided.

See more of this Session: Thermodynamics at the Nanoscale I
See more of this Group/Topical: Engineering Sciences and Fundamentals

Alcohol Effects On Trans Nuclear Membrane Transport

Hank Ashbaugh, Department of Chemical and Biomolecular Engineering, Tulane University, New Orleans, LA

Ashish V. Sangwai, Department of Chemical and Biomolecular Engineering, Tulane University, New Orleans, LA

Luigi P. Verdoni, Department of Chemical and Biomolecular Engineering, Tulane University, New Orleans, LA

The nuclear pore complex (NPC) provides a size-selective transport barrier that regulates traffic between the nucleus and cytoplasm. Intrinsically disordered phenylalanine-glycine nucleoporins (FG-Nups) serve as the barrier to transport through the central pore of the NPC. Experimentally it has been observed that alcohols of increasing hydrophobic reversibly diminish the permeability barrier, suggesting the hydrophobic side chains of the FG-Nups microphase separate to form a mesh-like gel that sieves large species. Using molecular simulations, we investigate the solvent-mediated interactions between hydrophobic FG-Nup side chains and the effect of alcohols. Alcohols of increasing hydrophobicity are found to induce side-chain aggregation, in difference to the interpretation that alcohols disperse these amino acids. Our simulations suggest that the decreased selectivity is a result of holes opening within the FG-Nup mesh, providing pathways for trans nuclear membrane transport.

See more of this Session: Thermodynamics at the Nanoscale I
See more of this Group/Topical: Engineering Sciences and Fundamentals

Critical Properties of Fluids in Nanopores: Crossover From 3D to 2D

Jayant K. Singh, Chemical Engineering, Indian Institute of Technology, Kanpur, Kanpur, India

Sudhir K Singh, Department of Chemical Engineering, Indian Institute of Technology, Kanpur, Kanpur, India

We study the effect of surface attraction on simple SW-fluid in slit pores of varying slit width from quasi 3D to 2D regime, by means of grand-canonical transition-matrix Monte Carlo (GC-TMMC) simulation. Our investigations indicate that various equilibrium and critical properties, under confinement of attractive slit pores, show very rich and interesting behaviors. Critical properties under confinement are found to vary rather interestingly as slit width is reduced from 40 molecular diameters (quasi-3D) to 1 molecular diameter (pure-2D). Critical temperatures are found to vary monotonically and approaches the bulk value as slit width becomes infinitely large. Moreover, around five different linear regimes are found for shift in the critical temperature with inverse of the slit width, for different attractive slit pores. On the other hand shift in the critical density and the critical pressure does not show any specific trend. Nevertheless, critical density and pressure show the sign of approaching towards 3D bulk value for infinitely large slit width in some interesting manner. Critical properties, below the slit width of 2 molecular diameters have insignificant effect on the surface characteristics and it behaves as quasi2D system. In this investigation, it has been observed that vapor pressure (spreading pressure) under confinement shows positive and negative deviation around bulk saturation pressure, with change in the slit pore width. Vapor-liquid surface tensions of the fluid for different attractive planar slit-pore of variable slit-widths are also compared. The role of different attractive surfaces on the vapor-liquid surface tension becomes insignificant for slit width less than 3 molecular diameters. But at a larger slit width, surface with lower attraction to the fluid molecules has shown comparatively larger surface tension value. Vapor-liquid Interfacial width is found to increase with decreasing slit width at a given temperature and wall-fluid interaction, which is contrary to what is being observed for fluids confined under hard cylindrical pores. We also present a similar study on cylindrical pores.

See more of this Session: Thermodynamics at the Nanoscale I
See more of this Group/Topical: Engineering Sciences and Fundamentals

Polyamine-Guided Synthesis of Anisotropic, Multicompartment Microparticles

Vinit Murthy, Procter and Gamble, West Chester, OH

Shyam Kadali, Department of Chemical and Biomolecular Engineering, Rice University, Houston, TX

Michael Wong, Chemical and Biomolecular Engineering and Chemistry, Rice University, Houston, TX

Colloidal particles that have nonuniform bulk or surface compositions are of emerging interest because of their potential applications involving advanced chemical storage and delivery and the self-assembly of novel functional materials. Experimental realization of anisotropic particles is much more difficult than that for particles with uniform bulk and surface composition, however. A new wet-chemical synthesis method to anisotropic microparticles is presented. This approach makes convenient use of the unusual observation of a salt-triggered separation of two water-solubilized polyamines into colloidal aggregates with nonuniform polymer composition. The anisotropic structure of these ionically cross-linked aggregates is explained by the difference in surface tensions of the contained single-polymer domains. Contacting the polymer aggregates with silicic acid or 13-nm silica nanoparticles leads to the charge-driven formation of solid or hollow microspheres, respectively. Depending on the poly(lysine)/poly(allylamine) ratio, the nonuniformity of the polymer aggregates translates to surface patches or internal compartments found in the resultant silica/polymer microparticles. Such hybrid materials with their unique structure could serve as a new basis for targeted chemical delivery and controlled release for potential applications in medicine, food, and cosmetics.

See more of this Session: Colloidal Assembly and Fabrication II
See more of this Group/Topical: Engineering Sciences and Fundamentals

Microfluidics-Based Lithography for Fabricating Complex Microparticles and Their Directed Assembly

Priyadarshi Panda, Chemical Engineering, MIT, Cambridge, MA

Ki Wan Bong, Chemical Engineering, MIT, Cambridge, MA

Kai P. Yuet, Chemical Engineering, MIT, Cambridge, MA

T. Alan Hatton, Chemical Engineering, MIT, Cambridge, MA

Patrick S. Doyle, Chemical Engineering, MIT, Cambridge, MA

Microfluidic devices offer the ability to finely control physical and chemical conditions which is advantageous for materials synthesis. Several groups have used multi-phase microflows to produce microparticles and capsules. Surface tension limits these particles to be spheroids. The Doyle group recently showed that more complex shapes and also patterned chemistries within a particle can be attained by coupling microfluidics to photolithography [1]. Continuous-flow lithography (CFL) [1] and the more refined stop-flow lithography (SFL) [2, 3] can generate 2-D extruded particles of any desired shape. To date, SFL has only been demonstrated with limited chemistries and thus the resultant microparticles have limited functionality. In particular, the particles generated using SFL are 2D and any feature added in the third dimension would enhance the practical use of these particles. Further, up until now, SFL has been used to generate polymeric particles. The ability to generate charged or magnetic particles would not only demonstrate the versatility of the process, but also provide the option for the directed assembly of these particles under an applied field. In this work, we demonstrate the ability to generate curved particles, having curvature in the plane orthogonal to the plain of projection of light. We also present the synthesis of anisotropic magnetic microparticles and their assembly into predictable macrostructures under an applied magnetic field.

Polymeric particles of complex architecture are widely used for applications such as photonic materials [4], MEMS [5], biomaterials [6] and self-assembly [7]. Introducing complexity in particle design is important since particle shape can significantly influence particle function [8]. An important facet of shape complexity is the introduction of curvature. It has been shown by molecular simulations that the curvature of particles can be tuned to regulate assembly as demonstrated by viral assembly [9]. Experimental evidence has also been provided for the preferential standing positions of concavo-convex particles with clear analogy to brachiopod or pelecypod shell orientation in moderately turbulent water [10]. Techniques for generating truly 3D non-convex (concave) particles with precise control over particle curvature will therefore by interesting. The Thomas group recently introduced concave curvature in polymeric particles using holographic interference lithography (HIL) which to the best of our knowledge is the only work of its kind. However, at present, their technique is only capable of creating single chemistry particles [11]. The literature is flooded with the generation of spherical magnetic particles. Any attempt at synthesizing non-spherical magnetic particles has been confined to the generation of modifications of spheres i.e., disks and plugs [12]. Magnetic micro-particles of anisotropic shapes have however, not been synthesized by hitherto known techniques. The ability to generate anisotropic magnetic particles and assemble them into predictable macrostructures would find application in a wide range of areas, from photonics [13] to tissue engineering [14].

Microfluidic devices for SFL are made out of PDMS using standard techniques [1]. SFL consists of three steps stop (stopping the flow of polymer solution), polymerize (UV flash through a mask) and flow (flushing out generated microparticles using a pressure pulse). For generating curved microparticles, photocurable solution (polymerization fluid) and tuning fluid are co-flowed and then stopped. Depending on the relative surface energies, curvature develops at the interface between the immiscible fluids. Ultraviolet light is then projected through a mask, solidifying polymeric particles having the shape on the plane of projection of light determined by the mask and the shape in the plane orthogonal to the plane of projection of light determined by the equilibrium curvature. The section of the tuning fluid exposed to UV light does not get polymerized. The chemical programmability of this technique is demonstrated by synthesizing Janus, patched and capped polymeric particles. Finally, the particles generated are grouped into different classes based on 5 different anisotropic axis depending on the chemistries used, curvature and patchiness. Anisotropic magnetic particles were generated using our SFL system by passing a suspension of magnetic beads in poly (ethylene glycol) diacrylate (PEG-DA) through the microfluidic device and flashing UV light through a transparency mask. The magnetic particles are essentially 2D extruded with a shape in the plane of projection of light given by the shape on the mask and straight walls along the plane of projection of light. These particles assemble into different macrostructures on the application of a magnetic field based on the initial shape and aspect ratio of the particles.

This facile approach offers a new method for generating curved microparticles for myriad applications, including photonics, tissue engineering, scattering functions of aerosols, and cosmetics. It also demonstrates the ability to synthesize magnetic microparticles of different shapes and helps envision their assembly for a bottom synthesis of desired macrostructures with potential applications in a variety of fields ranging from bioengineering to photonics and advanced material synthesis.

References

[1] D. Dendukuri, D. C. Pregibon, J. Collins, T. A. Hatton and P. S. Doyle, Nat. Mater. 2006, 5, 365.

[2] D. Dendukuri, S. S. Gu, D. C. Pregibon, T. A. Hatton and P. S. Doyle, Lab Chip 2007, 7, 818.

[3] D. C. Pregibon, M. Toner and P. S. Doyle, Science 2007 315, 1394.

[4] Y. Lu, Y. Yin and Y. Xia, Adv. Mater. 2001, 13, 415.

[5] D. J. Beebe, J. S. Moore, J. M. Bauer, Q. Yu, R. H. Liu, C. Devadoss and B-H Jo, Nature 2000, 404, 588.

[6] R. Langer and D. A. Tirell, Nature 2004, 428, 487.

[7] S. C. Glotzer, Science 2004, 306, 419.

[8] J. A. Champion, Y. K. Katare and S. Mitragotri, Proc. Natl. Acad. Sci. 2007, 104, 11901.

[9] T. Chen, Z. Zhang and S. C. Glotzer, Proc. Natl. Acad. Sci. 2007, 104, 717.

[10] G. V. Middleton, Journal of Sediment. Petrol. 1967, 327, 229.

[11] J.-H. Jang, C. K. Ullal, S. E. Kooi, C. Y. Koh and E. L. Thomas, Nano. Lett. 2007, 7, 647.

[12] D. K. Hwang, D. Dendukuri and P. S. Doyle, Lab Chip 2008, 8, 1640.

[13] Y. Xia, B. Gates, Y. Yin and Y. Lu, Adv. Mater. 2000, 12, 693.

[14] Y. Du, E. Lo, A. Samsher and A. Khademhosseini, PNAS 2008, 105, 9522.

See more of this Session: Colloidal Assembly and Fabrication II
See more of this Group/Topical: Engineering Sciences and Fundamentals

Directed Assembly of Non-Equilibrium Structures Utilizing Optical Trapping and Surface-Modified Colloids

Christopher M. Brotherton, Contraband Detection, Sandia National Laboratories, Albuquerque, NM, USA
Timothy P. Koehler, Microscale Science and Technology, Sandia National Laboratories, Albuquerque, NM, USA
Nelson S. Bell, Electronic and Nanostructured Materials, Sandia National Laboratories, Albuquerque, NM, USA
Anne M. Grillet, Microscale Science and Technology, Sandia National Laboratories, Albuquerque, NM, USA

Colloidal particles modified with photo-sensitive polymeric brushes exhibit reversible phase transformation properties allowing for the control of particle aggregation and suspension rheological response. By manipulating photo-sensitive particles using optical trapping, it is possible to assemble two and three dimensional microstructures. The system demonstrated consists of 2 micron borosilicate particles, coated with polymethyl methacrylate / spirobenzopyran copolymer brushes, suspended in toluene. Particle structures are assembled in the solvent using time-shared optical traps and then exposed to ultraviolet light. The light induces a polarity change in the polymer brushes causing particle-particle adhesion. The resulting two dimensional structure is then attached to a substrate. The process is repeated with additional building blocks to assemble three dimensional microstructures with multiple particle layers. Research activities to develop this assembly capability are presented related to the material requirements needed in optical trapping systems, their application in this photo-sensitive system, and developments in particle assembly.

Introduction

Since the discovery of optical trapping by Arthur Ashkin in 1970 [1], there have been numerous applications in the areas of soft matter, physics and biology [2]. Optical tweezers have become particularly useful in the study of colloidal systems where optical trapping enables the experimental investigation of the microrheology of suspensions [3-5], microstructural mechanics of colloidal aggregates [6, 7] and interactions of colloidal particles [8-10]. However structures assembled using optical traps are only temporary – once the laser is turned off, the particle diffuse away. If the particles are attractive, the microparticles suspensions are not stable and particles aggregate into uncontrolled structures. The assembly of microparticles requires controllable particle-particle interactions in order to fabricate robust persistent engineered microstructures.

In this work, we demonstrate a technique using optical trapping to construct robust three dimensional microstructures using optically switchable functionalized microparticles. Borosilicate microparticles are coated with optically active polymethyl methacrylate / spirobenzopyran copolymer brushes. Spirobenzopyran is a nonpolar molecule that experiences a ring-opening isomerization, when exposed to ultraviolet light (UV - λ = 366nm), resulting in a zwitterionic merocyanine with a higher dipole moment, as shown in figure 1. The reverse reaction is caused by exposure to heat or visible light (λ = 533nm), but is typically much slower than the forward reaction [11]. Using atom-transfer radical polymerization (ATRP), copolymer brushes of polymethyl methacrylate/spirobenzopyran can be attached to silica particles [12-15]. In a nonpolar solvent, such as toluene, the copolymer forms a stable solution. In the zwitterionic form, the copolymer solution phase separates due to the polarity difference between the polymer and the solvent. Likewise, micro particles coated with spirobenzopyran copolymer are stable in nonpolar solvents. When these suspensions are exposed to UV, the polarity change of the spirobenzopyran will cause the particles to develop a strong interparticle attraction and aggregate. In earlier work, this polarity change was used to moderate rheological properties of the suspension [16]. By controlling UV exposure, the particle-particle interactions can be manipulated and used for building microstructures [11, 15].

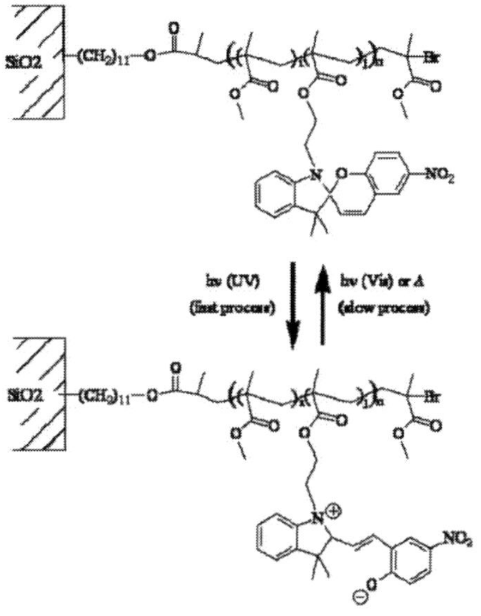

Figure 1. Spirobenzopyran/Zwitterionic merocyanine

In this study, we describe a microfabrication method that exploits the controlled interactions of particles coated with spirobenzopyran copolymer and the precise spatial manipulation of particles using optical trapping. For this method, layers of particles will be oriented using optical trapping and the isomerization of the copolymer brushes will cause the particles to adhere to one another, locking in the structure. Complex three dimensional structures can be constructed by combining multiple particle layers as described below.

Materials and Methods

Particle Preparation

Polymethyl methacrylate / spirobenzopyran copolymer brushes were grafted to 2 micron spherical borosilicate particles (Duke Scientific) using ATRP. Borosilicate was chosen since it has a refractive index (n = 1.52) higher than that of toluene (n = 1.50). The molar concentration of the spirobenzopryan in the copolymer brushes was kept constant at 20%. Following the synthesis, the modified particles were washed four times in toluene (Sigma-Aldrich) to remove any remaining catalysts or reactants. Finally, the particles were placed into toluene for the experiment.

Structure Assembly

Optical trapping was employed to spatially manipulate microparticles using the apparatus detailed in figure 2. A 1064 nm beam (A) emitted from a Ventus Nd:YAG laser is attenuated by a neutral density filter (B) and expanded before being passed to the two-dimensional acousto-optic deflector (AOD). The AOD (C), A.A. Opto-Electronics DTSXY-400-1064, is powered by an Agilent E3640A DC supply and deflects the incident beam at an angle proportional to an input frequency supplied from a control program written in LabVIEW 7.2. The beam exits the AOD and is again expanded before entering a Nikon TE2000-S inverted microscope and passing through filtered beam splitters (E) to overfill the Nikon Plan Fluor 100x/1.3 N.A. oil immersion objective (F). A custom flow chamber (G), consisting of a 2 mm x 3 mm x 50 mm straight channel with a single inlet and outlet, was used to contain the surface modified borosilicate particles dispersed in toluene. The top and bottom of the channel are glass coverslips affixed to the aluminum by two part epoxy to provide optical access to the sample.

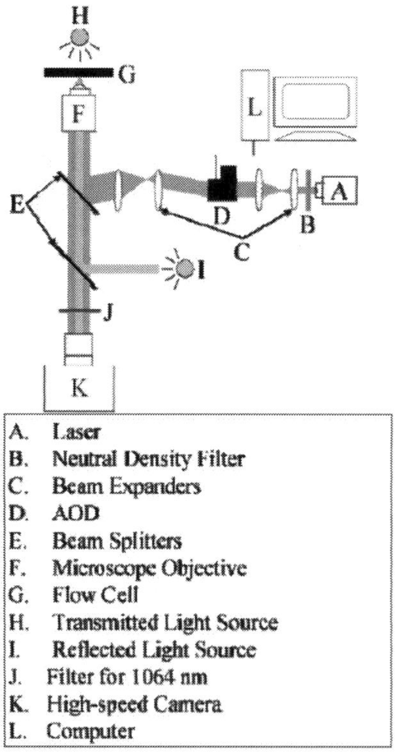

Figure 2. Laser tweezers optical trapping experimental setup

Figures 3 (a-c) illustrates the layer-by-layer assembly procedure. Multiple optical traps are used to arrange a single plane of particles perpendicular to the optical axis of the microscope (figure 3 (a)). The polarity of the particle surface coating is manipulated by exposing the trapped particles to UV using a 120W EXFO X-Cite Series 120 Hg-Arc lamp (figure 2 (I)). After increasing the polarity, the particles adhere to one another and form a single particle layer. Figure 3 (b) illustrates two particle groups. The first has been previous constructed and placed on the substrate. The second particle group remains suspended in the liquid and is moved into position with the optical tweezers. Once in position the second particle group is lowered onto the immobilized particle layer. Figure 3 (c) shows the completed three-dimensional structure. The process can be repeated to include additional particle layers.

Results

Figure 3 (d-f) illustrates the assembly of a 2-layer structure made up of 6 borosilicate particles. In image (d), two groups of particles are evident. The first group is immobilized on the substrate. The second group is controlled using three time-shared optical traps. The field of view is then illuminated using UV light causing a polarity change in the copolymer brushes. After exposure, the particles in the second group adhere to one another forming a particle layer. In the image (e), the second particle group is moved into position directly over the immobilized group. Image (f) shows the completed structure. The second particle group is in direct contact with the immobilized group. Since both layers contain particles with copolymer brushes and have been exposed to UV, adhesion between layers will occur. The resulting structure has dimensions of 4 micron cube.

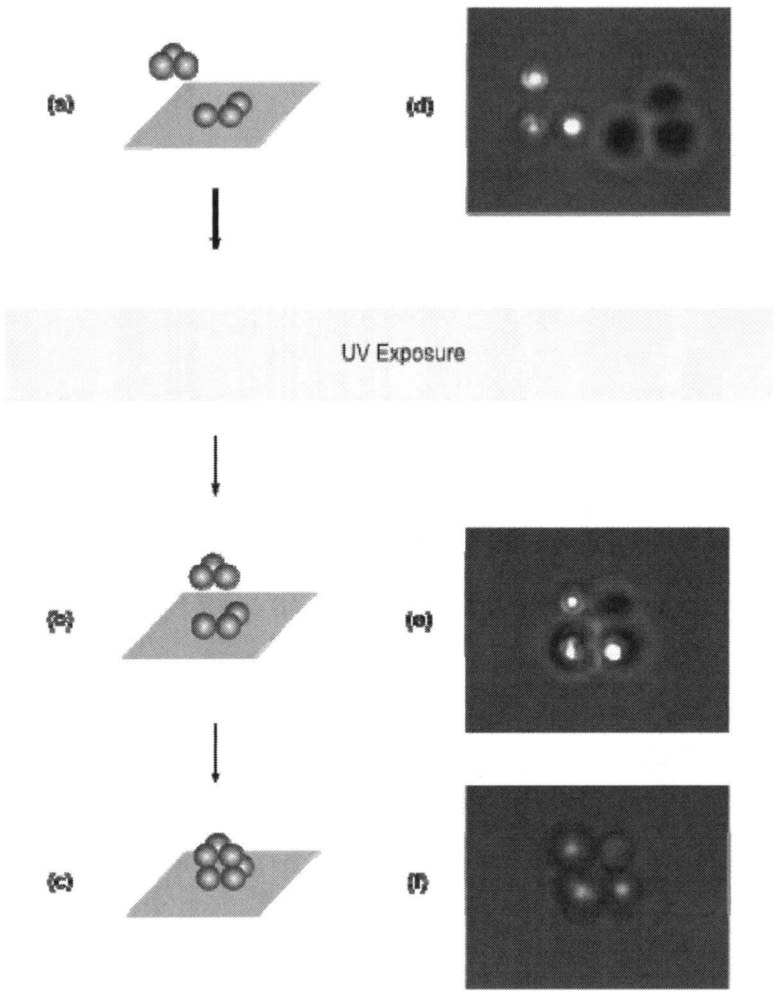

Figure 3. Directed assembly of 3D microstructures using optical trapping

Conclusions and Discussion

A new microstructure fabrication method has been developed that is capable of constructing three-dimensional structures layer by layer using borosilicate particles in toluene. Potential applications for this technique are photonic crystal assembly or assembly of unique crystalline forms that have interesting macroscopic optical, thermal, or mechanical properties. The challenges of this technique are scalability and the required use of a hazardous solvent system. Scalability issues can be addressed by using holographic optical tweezers that allow the fabrication of complex three-dimensional structures directly instead of layer-by-layer. Alternatively, large crystal structures could be assembled via other techniques (*e.g.* convective or self assembly) and then optical tweezers could be used to create strategic defects within the large crystal. The large crystal, with engineered defects, could then be locked in using the photosensitive spirobenzopyran surface chemistry. Finally, the hazardous toluene system was chosen based on polarity and further investigation of other nonpolar solvents may allow for a less hazardous solvent.

Acknowledgements

This work was supported by the Laboratory Directed Research and Development program at Sandia National Laboratories. Sandia is a multiprogram laboratory operated by Sandia Corporation, a Lockheed Martin Company for the United States Department of Energy's National Nuclear Security Administration under contract DE-AC04-94AL85000.

References

1. Ashkin, A., *Acceleration and Trapping of Particles By Radiation Pressure.* Physical Review Letters, 1970. **24**: p. 156-159.
2. Ashkin, A., *History of Optical Trapping and Manipulation of Small-neutral Particle, Atoms and Molecules.* IEEE Journal of Selected Topics in Quantum Electronics, 2000. **6**: p. 841-856.
3. Brau, R.R., et al., *Passive and Active Microrheology with Optical Tweezers.* Journal of Optics A: Pure and Applied Optics, 2007. **9**: p. S103-S112.
4. Crocker, J.C., et al., *Two-point Microrheology of Inhomogeneous Soft Materials.* Physical Review Letters, 2000. **85**: p. 888-891.
5. Meyer, A., et al., *Laser Tweezer Microrheology of a Colloidal Suspension.* Journal of Rheology, 2006. **50**: p. 77-92.
6. Furst, E.M. and A.P. Gast, *Micromechanics of Magnetorheological Suspensions.* Physical Review E, 2000. **61**: p. 6732-6739.
7. Pantina, J.P. and E.M. Furst, *Directed Assembly of Colloidal Aggregates to Probe Near-contact Interactions.* Langmuir, 2004. **20**: p. 3940-3946.
8. Furst, E.M., *Interactions, Structure, and Microscopic Response: Complex Fluid Rheology Using Laser Tweezers.* Soft Materials, 2003. **1**(2): p. 167-185.
9. Sainis, S.K., et al., *Electrostatic Interactions of Collodial Particles in Nonpolar Solvents: Role of Surface Chemistry and Charge Control Agents.* Langmuir, 2008. **24**: p. 1160-1164.
10. Sugimoto, T., et al., *Direct Measurement of Interparticle Forces by the Optical Trapping Technique.* Langmuir, 1997. **13**: p. 5528-5530.
11. Piech, M., et al., *Patterned Colloid Assembly by Grafted Photochromic Polymer Layers.* Langmuir, 2006. **22**: p. 1379-1382.
12. Abbasian, M., et al., *Graft Copolymerization of Styrene or Methyl Methacrylate onto Syndiotactic Polystyrene by ATRP Technique.* Journal of Applied Polymer Science, 2007. **104**: p. 611-619.
13. Kamigaito, M., T. Ando, and M. Sawamoto, *Metal-Catalyzed Living Radical Polymerization.* Chemical Reviews, 2001. **101**: p. 3689-3745.
14. Matyjaszewski, M., et al., *Polymers at Interfaces: Using Atom Transfer Radical Polymerization in the Controlled Growth of Homopolymers and Block Copolymers from Silicon Surfaces in the Absence of Untethered Sacrificial Initiator.* Macromolecules, 1999. **32**: p. 8716-8724.
15. Piech, M. and N.S. Bell, *Controlled Synthesis of Photochromic Polymer Brushes by Atom Transfer Radical Polymerization.* Macromolecules, 2006. **39**: p. 915-922.

16. Bell, N.S. and M. Piech, *Photophysical Effects between Spirobenzopyran-Methyl methacrylate-Functionalized Collodial Particles.* Langmuir, 2006. **22**: p. 1420-1427.

Protected Peptide Nanoparticles: Experiments and Brownian Dynamics Simulations of the Energetics of Assembly

Ting Chen, Chemical Engineering, Princeton University, Princeton, NJ

Suzanne D'Addio, Chemical Engineering, Princeton Unversity, Princeton, NJ

Michael T. Kennedy, Amgen Inc., Thousand Oaks, CA

Aleksander Swietlow, Amgen Inc., Thousand Oaks, CA

Ioannis G. Kevrekidis, Department of Chemical Engineering, Princeton University, Princeton, NJ

Athanassios Z. Panagiotopoulos, Chemical Engineering, Princeton University, Princeton, NJ

Robert K. Prud'homme, Chemical Engineering, Princeton University, Princeton, NJ

Soluble peptides, susceptible to degradation and clearance in therapeutic applications, have been formulated into protected nanoparticles for the first time through the process of kinetically-controlled, block copolymer directed rapid precipitation using Flash NanoPrecipitation. Complimentary Brownian dynamics simulations qualitatively model the nanoparticle formation process. The simulations corroborate the hypothesis that the size of nanoparticles decreases due to increasing supersaturation. Additionally, the influence of the polymer-peptide interaction energy on the efficiency of nanoparticle protection by polymer surface coverage is elucidated in both experiments and simulations.

See more of this Session: Colloidal Assembly and Fabrication II
See more of this Group/Topical: Engineering Sciences and Fundamentals

Aggregation of Two-Dimensional Microparticle Structures in Ferrofluids

Julie E. Byrom, Chemical & Biomolecular Engineering, Rice University, Houston, TX

Sibani Lisa Biswal, Chemical and Biomolecular Engineering, Rice University, Houston, TX

Under the influence of an external magnetic field, magnetic microparticles can be directed to assemble into linear chains. We have shown that once these linear chains are formed, these chains can be cross-linked to form permanent structures that can then be manipulated through external magnetic fields. Conversely, chains of nonmagnetic particles can be formed when they are placed in a ferrofluid medium and subjected to a magnetic field. In a magnetized ferrofluid, these nonmagnetic particles form "magnetic holes" that aggregate similarly to magnetic particles in nonmagnetic mediums.

In ferrofluids, the magnetic dipole of any particle is proportional to the difference of the susceptibility of the particle and the susceptibility of the fluid. Thus, particles with susceptibilities smaller than that of the medium acquire dipoles of direction opposite that of the field while particles with susceptibilities larger than that of the fluid have dipoles in the same direction as the field. Two particles with parallel dipoles align in a head-to-tail fashion along the direction of the field and particles with dipoles that are antiparallel will align in the direction perpendicular to the magnetic field.

By creating systems with particles of different susceptibilities we are able to form structures in two-dimensions, which offers a step forward in creating complex systems. In this talk, we will describe the mechanisms by which these structures are created and the different structures that are formed by mixing magnetic and nonmagnetic particles together in a ferrofluid.

See more of this Session: Colloidal Assembly and Fabrication II
See more of this Group/Topical: Engineering Sciences and Fundamentals

Lock Release Lithography (LRL)

Ki Wan Bong, Chemical Engineering, MIT, Cambridge, MA

Daniel C. Pregibon, Chemical Engineering, MIT, Cambridge, MA

Patrick S. Doyle, Chemical Engineering, MIT, Cambridge, MA

3D functional particles hold distinct advantages as sorting media, smart materials and building blocks for self-assembled, dynamic structures with complex functionality. Despite enormous their potentials, 3D and composite particles are not widely used because current methodologies have limitations in large scale, facile production and suffer from constraints of functionality and morphology. Multiphoton fabrication is a well-known method for synthesizing micro/nano 3D structures as it provides unparalleled control of morphology in all dimensions. In spite of the advantage, this direct drawing technique is prohibitively time-consuming. This method also has not been used to generate multifunctional structures. For higher throughput, 3D particles can be generated using a layer-by-layer process with photo resistst. Unfortunately, these materials are not ideal for many applications, the chemistries are extremely restricted, and the chemical patterning is limited to layered motifs. Three-dimensional particles can alternatively be generated using the PRINT method, where particles are shaped using a 3D mold. While Janus PRINT particles can be made by sequentially filling the mold with different materials, this approach will only yield simple "striped" material patterns on a particle and, to the best of our knowledge, particles with more than 2 stripes have not been synthesized. Here, we introduce lock release lithography (LRL), built off of continuous-flow and stop-flow lithography (SFL). LRL utilizes a combination of channel topography, mask design, and pressure-induced channel deformation to form and release particles in a cycled fashion. This technique provides a means for the high-throughput production of particles with complex 3D morphologies and composite particles with configurable chemistries. The process consists of (1) stopping the flow of a UV-sensitive monomer stream through a microfliudic channel, (2) lithographically printing structures that are "locked" into regions with multi-level channel topography, and (3) inducing channel deformation via high pressure to release structures for harvesting. In the example shown in Figure 1, we use a channel with positive relief features in the topography (ie. post structures protruding from the channel ceiling) to lock in an array of particles that are formed by 75ms of UV exposure through a transparency mask using a standard fluorescence microscope. Particle morphology is defined by a combination mask feature shape and channel topography. Locked into the three-dimensional relief, particles remain immobilized until a relatively high pressure (~ 5 psi) is applied to the poly(dimethylsiloxane) (PDMS) channel to initiate flow and deflect the channel beyond the point of particle release. Using an automated valving system, the flow is then stopped via pressure release, and the process is repeated, thus allowing the formation of 3D particles in an automated, semi-continuous manner. We synthesize a variety of 3D particles such as dishes, cups and table-like structures with unique mask shapes and channel topographies. LRL can be more complicated depending on the mold used to generate the channel topographies, and the mask used to polymerize the particles. Molds generated using standard lithography can be multi-tiered, rounded, or slanted, while virtually any topography can be achieved using multiphoton fabrication. The transparency masks used to generate particles can have virtually any two-dimensional shape, can be greyscale to provide variability in height along particles, and can be used in conjugation with interference masks to give finely tuned microporous structures. However, LRL is not suitable for the preparation of interlocking (like chain links) features or particles with internal hollow structures. Perhaps the most attractive feature of lock release lithography is that the release time is controllable. Because particle release occurs at a critical pressure (related to deformation), lower pressures can be used to exchange monomer without unlocking particles - this allows subsequent addition of new chemistries. As such, LRL can be used efficiently to generate composite particles with multiple precisely engineered chemistries. The process is shown schematically in Figure 2a. First, the multi-inlet channel is filled with chemistry #1 and locked 3D structures are polymerized. Then, by adjusting the pressures of the inlet streams (but keeping them below ~ 1 psi), a second chemistry replaces the first without displacing the locked particle structure. A unique mask can be used with this chemistry to polymerize distinct particle features that are covalently linked to the locked particles via overlap. Finally, a high pressure flow (~ 10 psi) is used to release the composite particles (Fig. 2b-d). This approach can be applied to any number of unique chemistries. Using the process, we generate composite particles with "autumn trees" in a frame (Fig. 2e-g). We also demonstrate a diverse set of functional particles including those displaying heterogeneous swelling characteristics and containing functional entities such as nucleic acids, proteins and beads. The length scales in LRL are ideally suited to generating tissue engineering mesoconstructs each containing multiple cell lines which are precisely positioned within the particle. In addition, the juxtaposition of swelling and stiff materials can be exploited to create particles that can undergo dramatic shape changes. Degradable polymers can be used to create microparticles which controllably evolve or fragment over time. We have believed that this technology will provide a simple but powerful means to mass-produce functional units for microfluidic operations, filtration systems, and tissue engineering constructs.

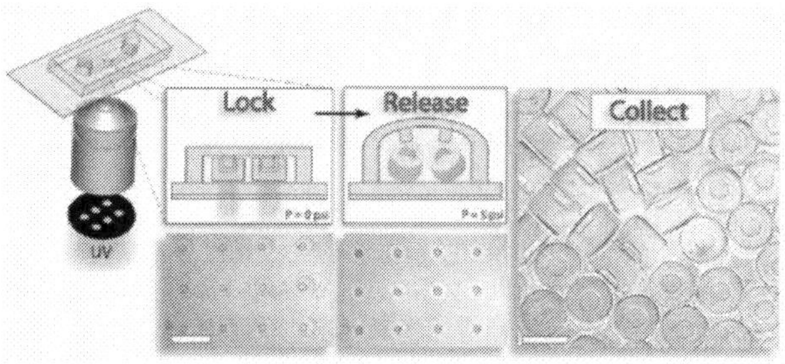

Figure 1. Process of lock release lithography
Scale bars are 100 um ("Lock") and 50 um ("Collect").

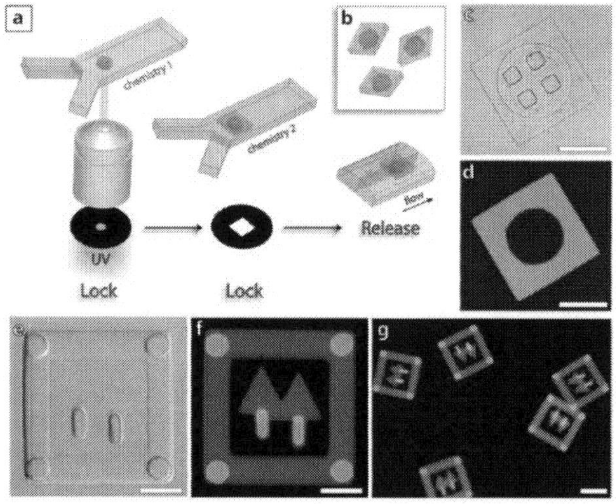

Figure 2. Synthesis of composite particles
Scale bars are 100 um (c,d,g) and 50 um (e,f).

See more of this Session: Colloidal Assembly and Fabrication II
See more of this Group/Topical: Engineering Sciences and Fundamentals

Determination of Viscoelastic Properties of Complex Fluids Using Active Nanorheology Simulations

Swapnil C. Kohale, Department of Chemical Engineering, Texas Tech University, Lubbock, TX

Rajesh Khare, Department of Chemical Engineering, Texas Tech University, Lubbock, TX

Traditionally viscoelastic properties of complex fluids are experimentally measured using a rheometer. In recent years, microrheology has been proposed as an alternative method for determining these rheological properties by using microscale probes. Such measurements require only a small quantity of the sample and furthermore, provide the ability to determine the rheological properties at the microscale. In this work, we extend the approach for the purposes of determination of viscoelastic properties of the complex materials at the nanoscale. The molecular scale structure in the system as well as the specific chemical interactions significantly influence the rheological properties at these length scales. Molecular simulations provide the unique ability to systematically account for these effects.

In previous work, we used molecular dynamics (MD) simulations to study the translational motion of a nanoparticle in simple fluids with an emphasis on elucidating the effects of the channel surface, velocity slip at the nanoparticle surface and the interparticle hydrodynamic interactions on the nanoparticle motion. In this work, we present an active nanorheology MD simulation technique involving oscillatory motion of a nanoparticle in a complex fluid. The specific system that is studied consists of a nanoparticle embedded in a matrix material consisting of either a polymer solution or a polymer melt. The diffusion and hydrodynamic interactions in the system are completely determined by the intermolecular interactions. The response of the complex material to the nanoparticle motion is characterized in terms of the local viscoelastic properties of the matrix. Simulation results are used to elucidate the effects of molecular structure and interactions on the observed rheological behavior.

See more of this Session: Colloidal Hydrodynamics I
See more of this Group/Topical: Engineering Sciences and Fundamentals

Non-Equilibrium Forces in Two-Point Active Microrheology

Eric M. Furst, University of Delaware, Newark, DE

Indira Sriram, University of Delaware, Newark, DE

Using optical tweezers, the forces acting on two micrometer-diameter probe particles are measured as they are driven at a fixed velocity in a direction orthogonal to their line of centers through a quiescent suspension ("bath") of colloidal particles. Direct visualization of the suspension microstructure is achieved using concurrent high-speed confocal microscopy. Far from equilibrium (at Peclet number, Pe >> 1), and in the limit in which the bath particles are nearly equal in diameter to the probe particles, a characteristic non-equilibrium bath microstructure develops around each probe, consisting of a boundary layer upstream of the probe, and a downstream wake depleted of bath particles. As the non-equilibrium microstructures interact at decreasing probe separations, an effective attraction is induced between the probes. This force is a function of probe separation and Peclet number. We will discuss the relationship between the attraction and the combined non-equilibrium bath microstructure for two probe particles in close proximity.

See more of this Session: Colloidal Hydrodynamics I
See more of this Group/Topical: Engineering Sciences and Fundamentals

Nonlinear Microrheology of Brownian Ellipsoids

Ryan J. DePuit, Chemical Engineering, University of California, Santa Barbara, Santa Barbara, CA

Aditya S. Khair, Chemical Engineering, University of California, Santa Barbara, Santa Barbara, CA

Todd M. Squires, Chemical Engineering, University of California, Santa Barbara, Santa Barbara, CA

Microrheology uses colloidal probes to measure bulk viscoelastic properties of soft materials. 'Passive' microrheology exploits the fluctuation-dissipation theorem to quantitatively recover linear response properties. However, this theoretical foundation is limited to near-equilibrium behavior, and does hold true for the investigation of nonlinear behavior. Squires (Langmuir, 2008) discussed a variety of theoretical issues inherent to nonlinear microrheology, including Lagrangian unsteadiness, non-viscometric flows, and inhomogeneous stresses.

To investigate these issues, we have developed a computational system to examine simplest model system that exhibits non-trivial rheology: a spherical colloidal probe translating through a dilute suspension of rigid ellipsoids or rods. We explicitly compute the (Lagrangian) transient, spatially-inhomogeneous microstructure in the bulk around the probe, the associated microstructural stresses upon the fluid, and the consequent retarding force upon the probe. Finally, we examine the effect that these issues have upon the quantitative comparisons between the velocity-dependent "micro-viscosity" and the macroscopically-measured rate-dependent shear viscosity of the same material.

See more of this Session: Colloidal Hydrodynamics I
See more of this Group/Topical: Engineering Sciences and Fundamentals

Nanoscale Viscosity Measurements Using Magnetic Nanoparticles

Carola Barrera, Chemical Engineering, University of Puerto Rico, Mayagüez Campus, Mayagüez, PR

Victoria L. Calero-DdelC, Chemical Engineering, University of Puerto Rico, Mayagüez Campus, Mayagüez, PR

Carlos Rinaldi, Chemical Engineering, University of Puerto Rico, Mayagüez Campus, Mayagüez, PR

Magnetic nanoparticles with diameters of up to 20 nm typically consist of single crystalline domains with an embedded magnetic moment. So-called thermally blocked magnetic nanoparticles behave as colloidal nanoscale permanent magnets which respond to imposed magnetic fields by aligning their magnetic moment through rigid body rotation of the particle. Such response is retarded by the viscous drag of the surrounding medium, thus it is possible that nano-scale rheological information may be obtained through the response of magnetic nanoparticles to magnetic fields. We will illustrate this possibility through a combination of the underlying theory and experimental demonstrations of: (1) comparisons of gelation temperatures obtained through nano- and macro-scale viscosity measurements; (2) quantitative comparisons of the temperature dependent nano- and macro-scale viscosity of mineral oils; and (3) quantitative comparisons between the nano- and macro-scale viscosity for linear poly(ethylene glycol) as a function of molecular weight.

See more of this Session: Colloidal Hydrodynamics I
See more of this Group/Topical: Engineering Sciences and Fundamentals

Steric Effects On Electrophoresis of a Colloidal Particle

Aditya S. Khair, Chemical Engineering, University of California at Santa Barbara, Santa Barbara, CA

Todd M. Squires, Chemical Engineering, University of California at Santa Barbara, Santa Barbara, CA

For over a century, the Poisson-Nernst-Planck (PNP) equations have served as the de facto theoretical model of electrokinetic phenomena. However, the non-interacting, point-sized ions assumed by PNP can lead to impossibly large ion concentrations near even moderately charged surfaces. Here, we revist the classic problem of electrophoresis of a spherical colloidal particle, using modified PNP equations that account for steric repulsion between finite sized ions, through Bikerman's model [1]. Steric effects are controlled by the bulk volume fraction of ions v, and for v= 0 the standard PNP equations are recovered. An asymptotic analysis in the thin-double-layer limit reveals at small zeta potentials (ζ <<i> $k_BT/e \approx 25$ mV) the electrophoretic mobility M_e to increase linearly with ζ for all v, as expected from the Helmholtz-Smoluchowski (HS) formula. For larger ζ, however, it is well known that surface conduction of ions within the double layer reduces M_e below the HS result. Crucially, in the PNP equations surface conduction becomes significant *precisely* because of the aphysically large and unbounded counter-ion densities predicted at large ζ. In contrast, steric effects impose a limit on the counter-ion density, thereby mitigating surface conduction. Hence, M_e does not fall as far below HS for finite sized ions ($v \neq 0$). Indeed, at sufficiently large v, steric effects are so dramatic that a maximum in M_e is not observed for physically reasonable values of ζ (< 10 k_BT/e), in stark contrast to the PNP-based calculations of O'Brien and White [2]. Finally, by calculating a Dukhin-Bikerman number characterising the relative importance of surface conduction, we collapse mobility versus zeta data for different volume fractions onto a single master curve.

[1] J. J. Bikerman, Philos. Mag. 33, 384 (1942)

[2] R. W. O'Brien and L. R. White, J. Chem. Soc. Faraday Trans. II 74, 1607 (1978).

See more of this Session: Colloidal Hydrodynamics I
See more of this Group/Topical: Engineering Sciences and Fundamentals

The Effect of Tracer Properties On Interfacial Particle-Image Velocimetry Results in Microscale Flows

Yutaka Kazoe, Georgia Institute of Technology, Atlanta, GA

Minami Yoda, G. W. Woodruff School of Mechanical Engineering, Georgia Institute of Technology, Atlanta, GA

Surface, or interfacial, effects can be significant over the small spatial scales typical of micro- and nanoscale transport. To date, there are few experimental techniques that can quantify interfacial transport with a resolution fine enough to resolve flows at these spatial scales. This talk describes a particle-image velocimetry (PIV) technique, multilayer nano-PIV (MnPIV), which exploits the exponentially decaying intensity of evanescent-wave illumination to obtain velocities at different distances from the fluid-solid interface within 400 nm of the wall.

Previous experimental studies have demonstrated that velocities obtained with 100 nm diameter fluorescent polystyrene tracers in steady and creeping Poiseuille flow give velocity gradients, after correcting for the nonuniform distribution of tracers, that are within 6% on average of analytical predictions. Nevertheless, the velocities from this PIV technique and for all tracer-based velocimetry techniques, for that matter, are, strictly speaking, tracer velocities, which are then assumed to be identical to the fluid velocities.

Experiments were performed with smaller (40 nm diameter) tracers, to both reduce electrostatic effects and improve the spatial resolution of these data. Results are presented on the effect of tracer size on near-wall particle and velocity distributions in Poiseuille flow, as well as in electroosmotic flow where the effect of particle electrophoresis is non-negligible, though microchannels with a minimum cross-sectional dimension of about 40 um.

See more of this Session: Colloidal Hydrodynamics I
See more of this Group/Topical: Engineering Sciences and Fundamentals

Clogging of Microchannels by Colloidal Particles: From Hydrodynamics to Surface Chemistry

Emilie M. Dressaire, School of Engineering and Applied Sciences, Harvard University, Cambridge, MA

Adeline Perro, Department of Physics, Harvard University, Cambridge, MA

Helene Fay, ENSCPB, Pessac, France

Vinothan N. Manoharan, Department of Physics, Harvard University, Cambridge, MA

Howard A. Stone, School of Engineering and Applied Sciences, Harvard University, Cambridge, MA

Filtration technologies such as membranes rely on the retention of colloidal particles in porous structures. The accumulation of particles contributes to the fouling of the membrane and eventually leads to the clogging of the pores. In order to better understand the clogging process, model porous systems have recently been introduced \footnote{H. M. Wyss, D. L. Blair, J. F. Morris, H. A. Stone, D. A. Weitz, Phys. Rev. E \textbf{74}, 061402 (2006)}. It was shown that the number of particles that pass through a microfluidic channel before it is clogged depends on the ratio of pore to particle size, and is independent of the flow rate and the particle volume fraction.

Here, we report on the clogging of microfabricated channels by solid and soft colloidal particles. We demonstrate that the clogging probability can be strongly modified by a wide array of physicochemical parameters. By varying the features of the colloidal particles (size, density, surface properties, and softness) and the characteristics of the liquid phase (viscosity, ionic strength, and fluid velocity), we show that clogging strongly depends on (i) the ratio of the particle size to channel radius, (ii) the interactions between colloids at short distance compared to the particle size, and (iii) the Reynolds number of the flow. We rationalize our results with a simple statistical model. These results have broad implications for practical applications where the interparticle interactions and/or the flow properties can be altered to favor (or prevent) particles retention or clogging such as filtration membranes (or microfluidic devices).

See more of this Session: Colloidal Hydrodynamics I
See more of this Group/Topical: Engineering Sciences and Fundamentals

Hindered Diffusion of Soft Spheres Under Geometrical Confinement

Jae Kyu Cho, Chemical and Biomolecular Engineering, Georgia Institute of Technology, Atlanta, GA

Victor Breedveld, Chemical and Biomolecular Engineering, Georgia Institute of Technology, Atlanta, GA

We have investigated the Brownian diffusion of fluorescent colloidal particles under geometric confinement via particle tracking video microscopy. The presence of confinement hinders the Brownian diffusion due to hydrodynamic coupling between particles and the confining walls. Hard spheres (polystyrene particle) and soft spheres (polystyrene/hydrogel core/shell particle) were used to compare the effect of softness on hindered diffusion of colloidal particles between parallel walls for a wide range of wall spacings. While the hindered diffusion for hard spheres shows excellent agreement with available model predictions, the hindrance experienced by soft spheres is much less than can be explained with current models: soft spheres in our study exhibit significant diffusion even when the spacing between walls is smaller than the unperturbed diameter of these particles. This anomalous diffusion is most likely due to the deformability and compressibility of the hydrogel particles.

See more of this Session: Colloidal Hydrodynamics I
See more of this Group/Topical: Engineering Sciences and Fundamentals

Computational Study of Lubrication Forces Between Colloidal Particles and Planar Surfaces

Bryce D. Sturtevant, Chemical Engineering, Purdue University, West Lafayette, IN

David S. Corti, Chemical Engineering, Purdue University, West Lafayette, IN

Lubrication forces arise whenever two surfaces, separated by a fluid, move relative to one another in close proximity. This hydrodynamic interaction plays an important role in colloidal aggregation and deposition processes. Lubrication forces are typically analyzed within the framework of continuum hydrodynamics, which dictates that the contact of the two surfaces cannot be achieved due to the infinite force required to squeeze out the interstitial fluid. Yet, there is some uncertainty as to the limiting behavior of lubrication forces as the separation between surfaces approaches microscopic dimensions, where a truly continuum description of the interstitial fluid is precluded. For example, molecular dynamics (MD) simulations of Challa and van Swol (2006, Phys. Rev. E 73, 016306) indicate that lubrication forces remain finite as the surfaces come into contact. These simulations are limited, however, to colloid-solvent size ratios of no more than 15 to 1, whereas some widely used colloidal entities of mesoscopic size (~1μm) have colloid-solvent size ratios of ~1000 to 1. Whether lubrication forces deviate from the continuum predictions at these larger ratios remains unclear. A further concern within the lubrication force regime includes the nature of the interaction of fluid particles with surfaces, specifically with regard to the development of suitable boundary conditions. Typical hydrodynamics calculations rely on either slip or no-slip boundary conditions; it is indefinite as to what extent, in reality, either of these types of boundary condition is applicable for length scales small enough to preclude a continuum representation of the interstitial fluid. To investigate in detail the behavior of lubrication forces between larger colloidal particles and surfaces, we utilize a novel MD simulation that includes a portion of the colloidal surface along with the interstitial fluid coupled to a fluid reservoir. This simulation enables us to probe hydrodynamic effects on length and time scales more relevant to colloids nearing 1μm in diameter. Furthermore, to evaluate the pertinency of slip and no-slip boundary conditions we variously characterize the surface with which the partial colloidal particle interacts as a hard, structureless surface; a purely attractive surface; or a surface composed of Lennard-Jones particles. Simulation results are compared with typical hydrodynamics calculations under both slip and no-slip boundary conditions. Additionally, there is some question as to how well lubrication forces are replicated in coarse-grained simulations of colloidal systems where the interstitial fluid is only represented in an averaged sense. As these techniques are invaluable in determining the behavior of colloidal particles interacting with surfaces, it is imperative that the proper lubrication results be included. To this end, MD simulation results from lubrication studies are correlated with mesoscopic simulations of colloidal systems to ensure that the correct lubrication effects are replicated for colloidal particles in the lubrication regime.

See more of this Session: Colloidal Hydrodynamics I
See more of this Group/Topical: Engineering Sciences and Fundamentals

Particle-Wall Hydronamic Coupling in Multi-Particle Ensembles

Pushkar Lele, Chemical Engineering, University of Delaware, Newark, DE

Eric M. Furst, University of Delaware, Newark, DE

Hydrodynamic coupling of colloidal particles to bounding surfaces significantly influences particle motion and the resulting properties of suspensions near surfaces. Such interactions are ubiquitous in many systems, including complex fluids, proteins and polymer suspensions. This work investigates the multi-body interactions of small ensembles of colloidal particles near a surface, at varying distances from the wall and interparticle separations. We use holographic blinking optical tweezers to trap and position n particles, where $n>2$, in a variety of configurations. The two-dimensional experimental diffusivity tensors of the ensembles are calculated from particle positions, determined by standard particle tracking algorithms. The particle-particle interactions in the bulk away from surfaces can be well approximated by using the Oseen tensor to derive diffusivities along normal modes. Close to the wall ($2 < h/a < 75$), the cluster behavior can be represented by considering two-particle interactions close to surface, following Blake's method of images to satisfy the no-slip boundary condition [1]. For ensembles very close to the surface ($h/a < 1.2$), the two-body solutions are insufficient to describe system dynamics. A full n-body solution using method of images is required to successfully capture the hydrodynamic effects introduced by a bounding surface.

[1] Blake, J. R. Prog. Colloid Polymer Sci., 70 (1971) 303.

See more of this Session: Colloidal Hydrodynamics I
See more of this Group/Topical: Engineering Sciences and Fundamentals

Microfluidic Electrochemistry: A Versatile Platform to Study Reactions at Electrodes

Shahab Shojaei-Zadeh, Benjamin Levich Institute, New York, NY

Joshua Gallaway, Energy Institute, City College of New York, New York, NY

Dan Steingart, Chemical Engineering, Energy Institute, City College of New York, New York, NY

Microfluidics is the science of manipulation and process of small volumes of fluids in micron size channels. Fundamental electrochemical mechanisms such as electrocatalysis, corrosion, and metallization are based on surface reactions occurring at the electrode-electrolyte interface. The high surface area-to-volume ratio in microchannels increases the importance of surface interactions within them. Both "top-down" and "bottom-up" fabrication methods have been developed to manufacture microelectrodes and to assemble them within microfluidic devices. The application of such devices enhances convective transport rates while retaining laminar flow condition required for analytical modeling. Other capabilities such as the ability to use very small quantities of electrolyte, to perform in short time with high sensitivity, to control concentrations of molecules in space and time, and low cost make microfluidic devices a suitable platform to study different electrochemical mechanisms.

The development of surface roughness during electrodeposition of metals has important application for secondary batteries as well as for metallization. During successive battery charging cycles, when metal is re-deposited at the "anode," the evolution of dendritic structures leads to shorting and failure, in some cases catastrophic. The phenomenon of dendrite growth has been a major impediment to the use of large-scale secondary zinc alkaline batteries as well as lithium metal batteries. The microfluidic electrochemical platform is uniquely suited to analysis of dendrite formation, which in most cases is closely tied to mass transport conditions. This is because mass transport and diffusion layers are well-defined at the working electrode surface, electrolyte compositions can be rapidly changed in situ, and visual imaging of the electrode surfaces is straightforward.

To construct a microfluidic electrochemical cell, poly(dimethylsiloxane) (PDMS) microchannels are fabricated using soft lithography techniques. PDMS elastomer (Sylgard 184, Dow Corning) is prepared by mixing and degassing a 10:1 ratio of liquid silicone base and a curing agent. The mixture is then poured either over a microchannel mold and cured in an oven, or cured as a sheet, with the channel geometry then fabricated by laser-cutting. The microelectrodes are fabricated separately via one of a number of techniques. Low profile microelectrodes are printed either directly as an ink jet, or thermally following lithography. Electrodes that are substantial in three dimensions are either patterned as a printed circuit board or embedded in epoxy and cross-sectioned. The combination of these techniques enables precise control of electrode geometries on the micron scale, resulting in reliable knowledge of current densities and therefore quantitative electrochemical experiments. Flow of electrolyte is established in the microchannels with a syringe pump and the region of interest is observed using a microscope. We analyze in situ dynamic observations using a high speed camera. Electrolyte flow rate is used to control mass transfer of metal ions to the working electrode surface, while secondary variables such as electrolyte composition, electrode geometry, and current waveform are varied while surface evolution is monitored.

In this work, we study the alkaline zinc electrode to determine conditions maximizing cycle life, as well as to optimize the experimental system. The goal of finding conditions under which a zinc electrode may be cycled repeatedly will require deposition of dendrite-free deposits. We use silver electrodes as a test system for both dendrite formation as well as phase change studies. By flowing a ZnO KOH solution over two electrodes with a fixed potential difference, we notice silver oxidation and corrosion on the positive electrode and zinc deposition on the negative electrode. Studying this system with microfluidic electrochemistry allows us to understand qualitatively and quantitively the current efficiencies at each electrode, the regimes of electrode transformation as a function of cell potential, electrical current and electrolyte velocity, and the overall reversibility of the system.

See more of this Session: Interfacial Aspects of Electrochemical Systems
See more of this Group/Topical: Engineering Sciences and Fundamentals

Experimental Study On Zinc-Nickel Hydroxide Flow Battery

Yasumasa Ito, Chemical Engineering, City College of New York, New York, NY

Sanjoy Banerjee, Chemical Engineering, City College of New York, New York, NY

Zinc anode rechargeable batteries are amongst the most alternative systems for electrochemical energy storage because of the low cost of zinc and the high energy and power densities obtainable. The hydraulic ion required for the redox reaction with zinc can be supplied by an air cathode or by nickel hydroxide incorporated in a nickel matrix electrode. The primary problem with the use of zinc anodes in rechargeable batteries relates to dendrite formation upon charging. This non-uniformity in electrodeposition occurs due to the system being strongly non-linear and far from equilibrium, the processes being controlled by Poisson-type of equations for the diffusion and electrical fields. The method to change the control mechanism is mass transfer from diffusion control to convection control by making use of flowing electrolyte. This also allows reconditioning of the battery to as new by stripping the deposited zinc by deep discharge when needed.

The purpose of this study is to experimentally investigate the morphology of zinc on charging zinc-nickel hydroxide batteries with flowing electrolyte and to evaluate battery performance to obtain near optimum conditions. Galvanostatic cycling experiments were carried out. The results show that the batteries operate quite stably at the Coulomb efficiency of more than 90% for 300 cycles with reconditioning at every tens of cycles. The electrodeposition of zinc was fairly flat and dendritic structure was hardly observed. The flow velocity of the electrode was set to ~ 10 cm/s and this is slower than those in the past studies. This means that the energy consumption by the pump can be significantly reduced. Although further investigations is required for clarifying the more detailed features and to obtain better performance, it is proved that zinc-nickel hydroxide battery with flowing electrolyte is a reasonable, innovative, and promising battery with high efficiency and long cycle life.

See more of this Session: Interfacial Aspects of Electrochemical Systems
See more of this Group/Topical: Engineering Sciences and Fundamentals

Influence of Thiol Coverage and Functionality On the Potential-Dependent Permeability of Monolayers

Gloria Olivier, Chemical and Biomolecular Engineering, Johns Hopkins University, Baltimore, MD

Joelle Frechette, Chemical and Biomolecular Engineering, Johns Hopkins University, Baltimore, MD

The dynamic response of loosely-packed self-assembled monolayers (SAMs) on gold has been investigated with electrochemical impedance spectroscopy (EIS) under non-faradaic conditions over a large range of applied potentials. The electrochemical response of loosely-packed 16-mercaptohexadecanoic acid (MHA) SAMs is compared with that of densely-packed MHA, hexadecanethiol (HDT), and tetraethylammonium-MHA ion-pair monolayers. We find that the films composed of ion-pairs and loosely-packed MHA have a higher capacitance and are less electrochemically stable than fully-packed SAMs. In sharp contrast to the other SAMs investigated, the loosely-packed MHA SAM exhibits an order-of-magnitude decrease in monolayer resistance (from 107 Ω•cm2 to 106 Ω•cm2) and 15% increase in capacitance as the voltage is scanned in the positive direction from -0.1V to +0.3V (vs SCE). These changes in monolayer resistance and capacitance are reversible, upon scanning the applied potential in the opposite direction, and independent of the electrolyte concentration (0.01 - 1M KCl). Therefore, the change in insulating properties of the loosely-packed MHA film is caused by a voltage-induced structural change (molecular re-organization) within the film itself. Non-faradaic impedance measurements conducted with the loosely-packed MHA film in the de-protonated state (pH 8.5) and protonated state (pH 3) display the same dynamic resistance and capacitance behavior, which suggests that electrostatic attraction between a negatively-charged thiol tailgroup and oppositely-charged gold electrode is not required for conformational change.

See more of this Session: Interfacial Aspects of Electrochemical Systems
See more of this Group/Topical: Engineering Sciences and Fundamentals

Surfactant-Mediated Synthesis of Anisotropic Particles at An Electrode Interface

Noshir Pesika, Chemical & Biomolecular Engineering, Tulane University, New Orleans, LA

Rajesh Venkatasubramanian, Chemical & Biomolecular Engineering, Tulane University, New Orleans, LA

The synthesis and characterization of micro- and nanoparticles, either in suspension or localized on a surface, are current areas of intense scientific interest due to potential applications in a wide range of technologies ranging from data storage, spintronics, quantum computing, drug delivery, and sensors. In this work, the electrochemical deposition of silver at varying overpotentials and in the presence of surfactants is studied. Cyclic voltammetry is used to identify and characterize the different regimes for silver deposition in the presence of CTAB. CTAB promotes the growth of anisotropic (nanowires) silver particles. This technique can potentially be used to grow complex-shaped particles by varying the surfactant(s) and the electrochemical deposition potential.

See more of this Session: Interfacial Aspects of Electrochemical Systems
See more of this Group/Topical: Engineering Sciences and Fundamentals

Colloidal Dispersion Stability of CuPc Aqueous Dispersions: Effects of Electrostatic and Other Forces

Jiannan Dong, School of Chemical Engineering, Purdue University, West Lafayette, IN

David S. Corti, School of Chemical Engineering, Purdue University, West Lafayette, IN

Elias I. Franses, School of Chemical Engineering, Purdue University, West Lafayette, IN

Yan Zhao, Hewlett-Packard Laboratories, Hewlett-Packard Co., Palo Alto, CA

Eric Hanson, Hewlett-Packard Laboratories, Hewlett-Packard Co., Palo Alto, CA

Hou T. Ng, Hewlett-Packard Laboratories, Hewlett-Packard Co., Palo Alto, CA

Dispersions of copper phthalocyanine (CuPc) pigments are used as major colorants of commercial printing inks. The CuPc particles are crystalline, rod-like or cube-like, with sizes in the range from 50 to 100 nm. Their colloidal stability is essential for their shelf life and for their behavior in various printing processes. The particles are functionalized with sulfonate groups, and they are negatively charged.

To test the role of electrostatic forces in their stability, a series of dynamic measurements of dispersed particle sizes was done with dynamic light scattering (DLS) and spectroturbidimetry (ST) at various concentrations of $NaNO_3$ and $CaCl_2$ at 25 °C. The effects of counterion valence and concentration are significant, indicating strong electrostatic repulsions.

The results were used to determine the initial stability ratio W (the inverse of which is a measure of the probability that a collision will lead to coagulation). The DLVO (Derjaguin-Landau-Verwey-Overbeek) theory was applied to make predictions of W for a simple model of monodisperse spheres in dilute dispersion. The Hamaker constant was calculated with a novel combination of an ab initio and an empirical methodology. The DLVO theory was used with two new models for spheres and cubes to predict the potential energy barrier Φmax, which is directly related to W. Electrostatic forces are inferred to play major roles in stabilizing the dispersions. Nonetheless, as the interparticle distance at Φmax becomes smaller than about 2 nm, the DLVO theory predictions for spheres are inaccurate, and inadequate for predicting W. Hence, other forces (e.g. hydration or other) may play important roles, and increase the values W compared to the DLVO-based predictions.

See more of this Session: Interfacial Aspects of Electrochemical Systems
See more of this Group/Topical: Engineering Sciences and Fundamentals

Corrosion Mechanisms Occurring On AE44 Mg Exposed to a 3.5% NaCl Aqueous Solution: Experimental Effects

Holly J. Martin, Center for Advanced Vehicular Systems (CAVS), Mississippi State University, Mississippi State, MS

M.F. Horstemeyer, Center for Advanced Vehicular Systems (CAVS), Mississippi State University, Mississippi State, MS

P.T. Wang, Center for Advanced Vehicular Systems (CAVS), Mississippi State University, Mississippi State, MS

Among the lightweight metallic alloys, those of magnesium present a great interest in the aerospace and automobile industry due to their low density (1.74 g/cm3) and high mechanical stiffness. These benefits are contrasted by a high relative corrosion rate when compared to aluminum or steel. As such, corrosion has limited magnesium's use as a structural component in industry, restricting it to nonexposed environmental locations, such as seats of cars and internal electronic boxes [1, 2]. Furthermore, magnesium alloys have been recognized as metals that corrode easily in the presence of seawater, due to their electrochemical potential as illustrated in the galvanic series [3]. The high electrochemical potential of magnesium allows for coupon dissolution when magnesium is placed in a NaCl aqueous solution. Hence, new developments of Mg alloys need stringent evaluation of the local galvanic corrosion from elements introduced into the alloys with a focus on enhancing the corrosion resistance to pitting [1]. The effects of adding rare earth elements to magnesium die casting were reported more than ten years ago. However, the search for lightweight solutions in the automobile industry has created a renewed interest [4, 5].

In the Center for Advanced Vehicular Systems (CAVS) at Mississippi State University, the electrochemistry between a conductor, AE44 magnesium (Mg), and an electrolyte, sodium chloride, is being investigated. Three different mechanisms are responsible for the corrosion of magnesium, including pitting, intergranular corrosion, and general corrosion. The influence of aluminum (A: 4.7%) and rare earth elements (E: 2.56%) on the corrosion mechanisms is also being studied. Previous work using semi-polished and super-polished AE44 Mg demonstrated that pitting is initially the dominant corrosion mechanism [5, 6]. Following the initial exposure to the NaCl aqueous solution, changes in the surface resulted in changes to pit characteristics, as the dominant corrosion mechanism changed [6]. Initially, the dominant form of corrosion was pitting, followed by a change to intergranular corrosion [6]. General corrosion eventually became the dominant corrosion mechanism [6]. While the corrosion mechanisms of AE44 Mg have been previously studied, these results were gathered using polished AE44 Mg coupons with two different roughness. The corrosion effects of 3.5% NaCl on cast AE44 have not previously been studied. Two different methods of salt corrosion were investigated, including constant immersion and salt-spraying. The research presented will cover the effects of immersion and salt-spray on the corrosion mechanisms of AE44 Mg.

[1] Makar, G.L.; Kruger, J. Int. Mater. Rev. 38 (3) (1993) 138-153.

[2] Song, G.; Atrens, A. Adv. Eng. Mater. 5 (12) (2003) 837-858.

[3] Shaw, B.A. Corrosion Resistance of Magnesium Alloys, in: L.J. Korb, ASM (Eds.), ASM Handbook, Vol. 13A: Corrosion, Ninth Ed., ASM International Handbook Committee, Metals Park, 2003, 692.

[4] Majumdar, J.D.; Galun, R.; Mordike, B.; Manna, I. Mater. Sci. Eng. A., A361 (2003) 119-129.

[5] Blawert, C.; Morales, E.; Dietzel, W.; Kainer, K.U. Mater. Sci. Forum 488-489 (2005) 697-700.

[6] Alvarez, R.B.; Horstemeyer, M.F.; Chandler, M.Q.; Williams, N.; Wang, P.T.; Ruiz, A.; Martin, H.J. Metall. Trans. A. In Review.

See more of this Session: Interfacial Aspects of Electrochemical Systems
See more of this Group/Topical: Engineering Sciences and Fundamentals

Pit Nucleation, Growth, and Coalescence: Modeling the Corrosive Effects of Mg Exposed to a 3.5% NaCl Aqueous Solution

Holly J. Martin, Center for Advanced Vehicular Systems (CAVS), Mississippi State University, Mississippi State, MS

M.F. Horstemeyer, Center for Advanced Vehicular Systems (CAVS), Mississippi State University, Mississippi State, MS

P.T. Wang, Center for Advanced Vehicular Systems (CAVS), Mississippi State University, Mississippi State, MS

The use of lightweight metallic alloys, such as cast magnesium, has recently gained traction in the automobile industry [1, 2]. The low density (1.74 g/cm3), high mechanical stiffness, excellent castability, and easy machinability make cast magnesium a strong candidate for structural components that are intricate and lightweight [2]. However, these benefits are greatly affected by a high corrosion rate when compared to aluminum or steel. In addition to a relatively high corrosion rate, magnesium has a high electrochemical potential, causing magnesium to corrode easily in the presence of seawater [3]. The high electrochemical potential of magnesium allows for coupon dissolution when magnesium is placed in a NaCl aqueous solution. The high corrosion rate has relegated the structural use of magnesium to areas that are not exposed to the environment, including car seats and internal electronic boxes [4, 5]. Therefore, the use of cast magnesium as a structural material requires a detailed understanding of the metal's response to corrosion conditions. Modeling of the pits present on the magnesium (Mg) surface due to corrosion should develop the desired understanding.

In the Center for Advanced Vehicular Systems (CAVS) at Mississippi State University, the electrochemistry between a conductor, Mg, and an electrolyte, sodium chloride, is being investigated. The total corrosion of two different types of Mg coupons, AE44 and AM60, was determined after exposure to a 3.5% NaCl aqueous solution. The total corrosion is determined based on the following model [6]:

$$\Phi_{Total} = \Phi_{GC} + \Phi_{IC}$$

where

Φ_{Total} is the total damage from corrosion arising from any type of corrosion mechanism

Φ_{GC} is the damage from general corrosion (loss of thickness)

Φ_{IC} is the damage from pitting

To calculate the corrosion due to pitting, the following model was developed [6]:

$$\Phi_{IC} = h_p n_p c$$

where

h_p is the pit number density related to nucleation of pits (number per unit area or volume)

n_p is the area of pit growth related to growth of the pits and c is a function of the nearest neighbor diameter related to the coalescence of the pits

Following the initial exposure to the NaCl aqueous solution, changes in the surface resulted in changes to the nucleation, growth, and coalescence properties, as the dominant corrosion mechanism changed [7, 8]. Initially, the dominant form of corrosion was pitting, demonstrated by changes in the pit number density [7, 8]. During an interim period, intergranular corrosion and pitting corrosion were equal mechanisms, demonstrated by changes in both the pit number density and the growth of the pits [7, 8]. Finally, towards the end of the experiments, general corrosion became dominant, as demonstrated by changes in the growth of the pits and the nearest neighbor radius [7, 8]. While the nucleation, growth, and coalescence of was studies using AE44 and AM60, both surfaces were polished, while the corrosion effects of 3.5% NaCl on cast AE44 Mg and AM60 Mg have not previously been studied. Two different methods of salt corrosion were investigated, including constant immersion and salt-spraying. The research presented will cover the effects of immersion and salt-spray on the validity of the model developed to detail the corrosion of magnesium.

[1] Jambor, A.; Beyer, M. Mater. Des. 18 (1997) 203-209.

[2] Froes, F.H.; Eliezer, D.; Aghion, E.L. JOM. 50 (1998) 30-34.

[3] Shaw, B.A. Corrosion Resistance of Magnesium Alloys, in: L.J. Korb, ASM (Eds.), ASM Handbook, Vol. 13A: Corrosion, Ninth Ed., ASM International Handbook Committee, Metals Park, 2003, 692.

[4] Makar, G.L.; Kruger, J. Int. Mater. Rev. 38 (3) (1993) 138-153.

[5] Song, G.; Atrens, A. Adv. Eng. Mater. 5 (12) (2003) 837-858.

[6] Horstemeyer, M.F.; Lathrop, J.; Gokhale, A.M.; Dighe, M. Theor. Appl. Fract. Mech. 33 (2000) 31-47.

[7] Martin, H.J.; Alvarez, R.B.; Horstemeyer, M.F.; Chandler, M.Q.; Williams, N.; Wang, P.T.; Ruiz, A. Metall. Trans. A. In Review.

[8] Alvarez, R.B.; Horstemeyer, M.F.; Chandler, M.Q.; Williams, N.; Wang, P.T.; Ruiz, A.; Martin, H.J. Metall. Trans. A. In Review.

See more of this Session: Interfacial Aspects of Electrochemical Systems
See more of this Group/Topical: Engineering Sciences and Fundamentals

Separation of Plasma From Blood by Crossflow Filtration through Nanoporous Filters From a Microchannel

Michael I. Hill, Chemical Engineering, Columbia University, New York, NY

Edward F. Leonard, Chemical Engineering, Columbia University, New York, NY

Current theories of crossflow filtration assume the existence of a bulk flow of uniform concentration from which particles are filtered through a thin boundary layer at the wall. Crossflow filtration from a microchannel strains this assumption, as the boundary layer will fill the entire microchannel. This is further complicated if one wishes to use layered flows of differing composition in the microchannel; then a single bulk concentration does not exist. The use of nanoporous filters with low flow resistance adds to the complexity, as the pressure drop down the microchannel needs to be balanced against the pressure drop across the filter to avoid backflow of filtrate.

A model of steady crossflow filtration from a microchannel through nanoporous filters at low Reynolds numbers was developed, showing the effect of inlet concentration distribution, dimensionless filter length and Peclet number. This model is currently proving useful in the design of a microfluidic membraneless artificial kidney which must be able to separate plasma from blood at high flows.

See more of this Session: Microfluidics and Small-Scale Flows II: Biological Applications
See more of this Group/Topical: Engineering Sciences and Fundamentals

A Microfluidic Model System for Investigating Transport Phenomena in a Focal Hemostatic Injury Model

Sean F. Maloney, Department of Chemical and Biomolecular Engineering, Institute for Medicine and Engineering, University of Pennsylvania, Philadelphia, PA

Scott L. Diamond, Department of Chemical and Biomolecular Engineering, Institute for Medicine and Engineering, University of Pennsylvania, Philadelphia, PA

Background: Small-scale flows are of critical importance for the effective transport of dissolved gasses, ions, and biomolecules throughout living tissues. In high-pressure closed circulatory systems, a compromised vessel wall induces a series of surface-catalyzed deposition and enzymatic reactions that stop blood loss while maintaining the bulk fluidity. The subsequent reaction-convection-diffusion system coupled to a small-scale flow provides a rich field for the investigation of transport processes of significant fundamental and clinical importance.

Methods: By coupling micropatterning techniques with a microfluidic vessel analog, it is possible to create focal injury analogs with precise spatial control in both the axial and transverse flow directions. The initial surface and inlet bulk conditions can be precisely manipulated, providing for a high degree of control while maintaining the physiologically relevant small-scale flow that provides both convective transport and shear stresses which are crucial to hemostatic processes. Real-time and *post-hoc* analysis of platelet adhesion, secondary signaling, and fibrin formation provides a suite of pertinent output signals that can be monitored under the wide range of flow conditions observed throughout the microvasculature in healthy and diseased states.

Results: The size and composition of molecules exposed at the site of vascular injury have implications on the reactive potential and subsequent hemostatic response under the small-flow conditions found throughout the vascular system. The complex behavior of aggregation and fragmentation of platelets and their subsequent signaling on a reactive surface are dependent on both the flow conditions and the surface composition. The competing phenomena of increased surface flux with increased displacing force and decreased interaction time causes and apparent maximal shear rate for platelet adhesion to a collagen surface of finite length. The polymerization of fibrin fibers from soluble precursors via a series of surface-catalyzed enzymatic reactions under a wide range of flows shows both the complexity and robustness of the system, with significant formation at wall shear rates up to 500sec^{-1}. At higher velocities, the local residence time of the active enzyme thrombin is diminished, limiting its role to that of a soluble secondary platelet activator until the accumulated mass is large enough to form small deformations in the flow field, allowing for localized polymerization in spite of unfavorable bulk flow conditions.

See more of this Session: Microfluidics and Small-Scale Flows II: Biological Applications
See more of this Group/Topical: Engineering Sciences and Fundamentals

Deterministic Lateral Displacement as a Means to Enrich Large Cells for Tissue Engineering

James V. Green, Chemical Engineering, Northeastern University, Boston, MA

Milica Radisic, IBBME/Chemical Engineering and Applied Chemistry, University of Toronto, Toronto, ON, Canada

Shashi K. Murthy, Department of Chemical Engineering, Northeastern University, Boston, MA

Microfluidic separation techniques are attractive for applications in tissue engineering because of their low cost and ability handle small sample volumes. In certain tissue engineering applications, such as cardiac tissue engineering, the enrichment of a functional cell type is carried out from a suspension of digested donor tissue prior to seeding into scaffolds to create functional tissue. If the target cell type is of considerably different size from other cell types present in the digestate, deterministic lateral displacement is a highly attractive mode of separation compared to conventional approaches that rely on sieve-based techniques or pre-plating (where adhesive, non-target cell types are removed by preferential adhesion to cell culture dish surfaces). This presentation will describe the design and operation of a microfluidic device that is designed to enrich cardiac muscle cells by deterministic lateral displacement. This device consists of obstacles places in a staggered arrangement that allows target cells above a specified critical diameter to become progressively displaced laterally while the smaller, non-target cells remain in their streamlines. The target cell purity obtained is >98%.

See more of this Session: Microfluidics and Small-Scale Flows II: Biological Applications
See more of this Group/Topical: Engineering Sciences and Fundamentals

Rapid Arraying of Particles and Cells in Three-Dimensional Microfluidic Devices

Ramsey Ibrahim Zeitoun, Chemical Engineering, University of Michigan, Ann Arbor, MI

Joanna Mirecki-Mullunchick, Materials Science and Engineering, University of Michigan, Ann Arbor, MI

Michael J. Solomon, Chemical Engineering, University of Michigan, Ann Arbor, MI

Mark A. Burns, Chemical Engineering, University of Michigan, Ann Arbor, MI

Direct positioning of particles and cells is used for a variety of applications including molecular detection, self-assembly and tissue or cell engineering. Although there has been a great deal of interest into developing more complex patterning techniques, most suffer from issues in precision, scalability, response times and flexibility. We have developed an arrayed, three-dimensional surface patterned microfluidic device which uses laminar convective flow to position micro-objects on to vacuum actuated holes (drains) into any desired, two-dimensional, computer-programmed pattern. In addition, the device was expanded to independently actuate two independent sets of drains, allowing us to selectively create patterns that have both controlled position and composition. We have used this device to quickly generate anisotropic particles and culture cells from a desired initial pattern and spacing. With rapid, precise and adaptable operation, three-dimensional microfluidic devices can assist novel research and methods by changing an anonymous distribution of particles or cells to an individualized selection.

See more of this Session: Microfluidics and Small-Scale Flows II: Biological Applications
See more of this Group/Topical: Engineering Sciences and Fundamentals

Cell and Particle Behavior in Microfluidic Mixers

Alison M. Hirsch, School of Chemical and Biomolecular Engineering, Georgia Institute of Technology, Atlanta, GA

Boyang Zhang, School of Chemical and Biomolecular Engineering, Georgia Institute of Technology, Atlanta, GA

Hang Lu, School of Chemical and Biomolecular Engineering, Georgia Institute of Technology, Atlanta, GA

Fluid mixing is common in large scale chemical processing. However, many processes, especially applications in bioengineering, are now carried out in microfluidic systems. At the micro-scale, mixing of solutes is predominantly a diffusion process due to the laminar nature of fluid. Many different mixing strategies have been employed to effectively decrease the characteristic length for diffusion. For example, the Staggered Herringbone Array [1] uses carefully designed grooves on the surface of the channel to laterally fold and stretch the fluid, and an even mixing of solutes/solvents can be achieved relatively quickly. However, particle (or cell) mixing behavior in suspension is still not well understood.

To assess the critical factors behind neutrally buoyant particle mixing or focusing, we implemented a series of experiments with varying flow conditions, fluid and particle properties, and geometries. We established an experimental method providing information about particle distribution in three dimensions without using confocal microscopy. Studying the particle distribution at varying Reynolds numbers allows us to discern the relative effects of inertial and viscous forces. Similar experiments have been performed previously by other groups in square channels describing the phenomena at relatively high Reynolds numbers or in channels with relatively large grooves only. Thus, in these situations only one major parameter controls the particle distribution. In contrast, in our experiments where more than one factor is important and no simplifications can be made, our analysis integrates many parameters through dimensional analysis and scaling.

To further understand the physical forces on each particle, Computational Fluid Dynamics models were also used to predict the particle flow characteristics and to calculate relevant forces. The modeling results are able to explain the equilibrium distribution behavior observed in the experiments as well as the paths the particles follow in the process.

By studying particle mixing of various particles under different flow conditions, we hope to provide a more comprehensive view of the particle-fluid behavior in laminar flow. With this knowledge, efficient unit operations in multiphase systems, including mixing and separation, can be designed, particularly in microfluidic technologies for many biological and medical applications that handle cells and beads.

[1]. A. D. Stroock, S. K. W. Dertinger, A. Ajdari, I. Mezic, H. A. Stone and G. M. Whitesides, Science, 2002, 295, 647-651

See more of this Session: Microfluidics and Small-Scale Flows II: Biological Applications
See more of this Group/Topical: Engineering Sciences and Fundamentals

High-Throughput Flow Alignment and Scanning of Barcoded Hydrogel Microparticles

Stephen C. Chapin, Chemical Engineering, Massachusetts Institute of Technology, Cambridge, MA

Daniel C. Pregibon, Chemical Engineering, Massachusetts Institute of Technology, Cambridge, MA

David Appleyard, Chemical Engineering, Massachusetts Institute of Technology, Cambridge, MA

Patrick S. Doyle, Chemical Engineering, Massachusetts Institute of Technology, Cambridge, MA

The ability to accurately detect and quantify biological molecules in a complex mixture is crucial in both basic research and clinical settings. Advancements in the fields of genomics and proteomics require robust technologies that can obtain high-density information from biological samples in a rapid and cost-effective manner. Suspension (particle-based) arrays offer several advantages over conventional planar arrays in the multiplexed detection and quantification of biomolecules, including the use of smaller sample volumes, more favorable probe-target binding kinetics, and rapid probe-set modification. We have previously described a microfluidic-based process (stop-flow lithography, SFL) for rapidly generating geometrically and chemically complex hydrogel microparticles (10,000 per hour) that can be used for the sensitive, multiplexed detection of DNA and RNA oligomers.

We now present a microfluidic system for the rapid flow alignment of these multifunctional hydrogel microparticles, which are designed to bear one or several biomolecule probe regions, as well as a graphical code to identify the embedded probes. Using high-speed imaging, we have developed and optimized a flow-through system that allows for a high particle throughput, ensures proper particle alignment for decoding and target quantification, and can be reliably operated continuously without clogging. A tapered channel flanked by side focusing streams is used to orient the flexible, tablet-shaped particles into a single-file, well-ordered flow in the center of the channel. The effects of channel geometry, particle geometry, particle composition, particle loading density, and barcode design are explored to determine the best combination for eventual use in biological assays. Particles in the optimized system move at velocities of ~50 cm/s and with throughputs of ~40 particles/s. Simple physical models and COMSOL Multiphysics simulations have been used to investigate flow behavior in the device.

The ability to align particles in the channel with high precision enables photomultiplier tube (PMT)-based detection with a simple one-dimensional line scan that integrates particle fluorescence intensity along a thin laser excitation beam established perpendicular to the flow direction. The resulting high-resolution profile is then used to determine the identity of a passing particle, ascertain the probe(s) it carries, and measure the extent of the binding events on the various probe and negative control regions. We present our latest advances in the development of this microfluidic scanning system, with particular emphasis on the instrumentation employed and additional efforts to couple the flow-focusing channels to a liquid-handling system designed to automatically cycle through sample volumes stored in standard formats such as 96-well microtiter plates.

See more of this Session: Microfluidics and Small-Scale Flows II: Biological Applications
See more of this Group/Topical: Engineering Sciences and Fundamentals

Inertial Microfluidics for Multifunctional Microgel Particle Fabrication

John Oakey, Center for Engineering In Medicine, Massachusetts General Hospital, Charlestown, MA

Mehmet Toner, Center for Engineering In Medicine, Massachusetts General Hospital, Charlestown, MA

Microscale hydrogel (microgel) particles have broad utility in applications ranging from biosensing to drug delivery and regenerative medicine. To date, however, the fabrication of microgel particles has been somewhat crude and limiting to potential applications. Microfluidic lithography and microchannel emulsification have recently been shown to be quite powerful techniques for the fabrication of microgel particles with custom shape, composition and behavior. These traits make microfluidic fabrication an ideal strategy for multifunctional microgel formation via particle and cell encapsulation. These techniques suffer only from a lack of precision control over particles and cells within microfluidic flows. Inertial microfluidic focusing has recently been introduced as a high-throughput approach to particle and cell focusing. Inertial focusing exploits hydrodynamic forces arising from interactions between particles and surfaces and other particles. Lift forces, shear gradient forces and Dean drag forces are precisely controlled in microfluidic channels by geometric design to produce well-defined, predictable behavior that may be manipulated through microchannel geometry. We demonstrate the application of inertial focusing to microgel manipulation in microchannels as well as cell and particle focusing for controlled encapsulation within microgels. Focusing within viscous aqueous macromer solutions imposes additional kinetic constraints upon focusing, which must be balanced with the emulsification conditions. Applications for these unique particulate architectures will also be presented.

See more of this Session: Microfluidics and Small-Scale Flows II: Biological Applications
See more of this Group/Topical: Engineering Sciences and Fundamentals

Hydrodynamic Trap for Single Cells and Particles

Melikhan Tanyeri, Chemical and Biomolecular Engineering, University of Illinois at Urbana-Champaign, Urbana, IL

Charles M. Schroeder, Chemical and Biomolecular Engineering, University of Illinois at Urbana-Champaign, Urbana, IL

The ability to trap individual particles, cells and macromolecules has revolutionized many fields of science during the last two decades. Several methods of particle trapping and micromanipulation have been developed based on optical, magnetic and electric fields. In this work, we describe an alternative trapping method, the hydrodynamic trap, based on the sole action of hydrodynamic forces in a microfluidic device. A microfluidic cross slot device is fabricated consisting of two perpendicular microchannels where opposing laminar flow streams converge. In this device, a purely extensional flow field is created at the microchannel junction, thereby resulting in a semi-stable potential well at the stagnation point which enables particle trapping. We implement an automated feedback-control mechanism to adjust the location of the stagnation point which facilitates active particle trapping. Using the hydrodynamic trap, we successfully demonstrate trapping and manipulation of single particles and cells for arbitrarily long observation times. This technique offers a new venue for observation of biological materials without surface immobilization, eliminates potentially perturbative optical, magnetic and electric fields, and provides the capability to change the surrounding medium conditions of the trapped object.

See more of this Session: Microfluidics and Small-Scale Flows II: Biological Applications
See more of this Group/Topical: Engineering Sciences and Fundamentals

Enhanced Gene Delivery by Transverse Cell Migration during Flow-through Electroporation

Jun Wang[1], Yihong Zhan[1], Victor M. Ugaz[4] and Chang Lu[1,2,3]

[1]*Department of Agricultural and Biological Engineering,* [2]*Weldon School of Biomedical Engineering,* [3]*School of Chemical Engineering, Purdue University, West Lafayette, Indiana 47907.*

[4]*Artie McFerrin Department of Chemical Engineering, Texas A&M University, College Station, Texas 77843.*

I. INTRODUCTION

Delivery of exogenous DNA into cells, or gene transfer, is critical for a wide range of applications including basic molecular and cell biology studies, tissue engineering, and new clinical procedures such as cell therapies. Gene transfer using viral vectors is generally considered the most efficient method[1]. However, complex cloning strategies and host inflammatory/immune responses have hindered its applications[2]. Chemical methods based on cationic polymers, lipids or calcium phosphates are used to form complexes with DNA in order to facilitate gene transfer[3; 4; 5]. But chemical transfection results can be variable and heavily influenced by transfection time, construct type and cell type.

Electroporation is the most widely used physical method for gene transfer based on the application of electrical pulses to cells *in vitro* and *in vivo*[6; 7; 8; 9]. During electroporation, an electric field is applied that opens transient pores in the cell membrane that allow impermeant genes to pass down a concentration gradient into the interior of the cell. Electroporation offers a number of advantages over chemical methods, including high survival rates and transfection efficiencies, greatly reduced dependency on cell type, and ability to deliver genes into primary cells that are hard to transfect using chemical methods[10]. Electroporation-based transfection is also amenable to miniaturization for high-throughput operation[11; 12; 13; 14; 15; 16].

Regardless of which transfection technique is employed, the ability to generate a high concentration of DNA on the cell surface is critical to achieving efficient delivery [17; 18]. Unfortunately, a limitation of electroporation methods is the fact that only a small portion of the cell membrane is permeabilized to an extent that permits DNA entry into the cells. This is because electroporation can occur only when the transmembrane potential exceeds a critical threshold value (~0.25-1 V). Assuming a spherical cell shape, the transmembrane potential ($\Delta\psi_E$) is determined by $\Delta\psi_E = 0.75\ g(\lambda)aE\cos\theta$, where $g(\lambda)$ is a complex function of the membrane and buffer conductivities, a is the diameter of the cell, E is the field strength and θ is the angle between the normal to the membrane surface and the field direction[19]. Consequently, application of an electric pulse with optimized intensity and duration only permeabilizes cells in a localized zone near the poles where the surface normal is most closely aligned with the field direction[20; 21; 22]. Moreover, there is only a limited range over which higher electric fields can be applied to extend the size of this zone before the cells become damaged and experience dramatically decreased viability. Previous studies have attempted to address this limitation by applying periodic electric pulses of different polarities and directions[21; 23]. But

[a)]Author to whom correspondence should be addressed; email: changlu@purdue.edu the instrument complexity required to employ these approaches is relatively high, with limited feasibility for scale-up to processing large numbers of cells.

Here we introduce a new approach to overcome these limitations by harnessing inertial Dean flow effects that arise in curved microchannels to expose the cells to a combination of transverse migration and rotation during flow-through electroporation. The resulting complex motion overcomes the major drawback of static electroporation methods by enabling the entire cell surface to be permeabilized and accessed by exogenous DNA. We show that electroporation in this curved spiral-shaped channel design yields a two-fold increase in transfection frequency compared to a straight microchannel. By allowing DNA delivery to occur over the entire cell surface, our method offers a unique and practical approach for performing high-throughput cell transfection with greatly enhanced efficiency.

II. EXPERIMENT

1. Materials. Fluorescent polystyrene beads with diameter 9.9 ± 0.05 □m (s.d.) were purchased from Duke Scientific. The density of beads is 1.05 g/ml according to the protocol. Before experiment, the beads were diluted in phosphate buffered saline (PBS) at a concentration of 1.9×10^6 /ml. The 4.7 Kb plasmid vector pEGFP-C1 (Clontech) coding green fluorescent protein (GFP) was propagated in Escherichia coli and extracted in Tris-EDTA buffer with QIAfilter Plasmid Giga kit (Qiagen). The plasmid was stained with green fluorescent DNA dye YOYO-1(Invitrogen) with a molecular ratio of 1 dye per 5 bp when fluorescent labeling was necessary.

Chinese hamster ovary (CHO-K1) cells were grown at 37 ºC with 5% CO_2 in Dulbecco's modified Eagle's medium (DMEM, Mediatech, Inc.) supplemented with 10% (v/v) fetal bovine serum (FBS, Sigma), penicillin (100 µg/ml, Sigma). The cells were trypsinized and diluted at a ratio of 1:5-1:8 every 2 days to maintain in the exponential growth phase. DT 40 chicken B cells were cultured in RPMI 1640 media supplemented with 10% heat-inactivated fetal calf serum, 1% chicken serum, 50 µM 2-mercaptoethanol, 1 mM sodium pyruvate, 100 IU/ml penicillin G, and 100 µg/ml streptomycin. The cells were subcultured every two days with 1:20 ratio. To observe flowing cells in microfluidic channels, both of CHO cells and DT40 cells were stained by 5 µM nucleic acid dye SYTO 16 (Invitrogen) for 5 min followed by washing.

2. Microchip Fabrication. Microfluidic devices were fabricated on polydimethylsilane (PDMS, General Electric Silicones RTV 615) using standard soft lithography. The microscale patterns were first designed using AutoCAD and then produced on a chrome plate that was employed as a photomask during photolithography of a negative photoresist (SU-8 2025). The fabricated photoresist/silicon wafer structure served as the master for molding PDMS microfluidic chips. The channel depth was ~75 µm (measured by a Sloan Dektak3 ST profilometer). After punching access holes in the PDMS chip, both surfaces of PDMS chip and a pre-cleaned glass slide were treated in plasma cleaner. The PDMS chip was then immediately brought into contact against the slide after treatment to form closed channels.

3. Fluorescence Microscopy. The microfluidic device was mounted on an inverted fluorescence microscope (IX-71, Olympus, Melville, NY). The epifluorescence excitation was provided by a 100W mercury lamp, together with brightfield illumination. The excitation and emission from YOYO-1 stained plasmid or SYTO 16 stained cells were filtered by a fluorescence filter cube (exciter HQ480/40, emitter HQ535/50, and beamsplitter Q505lp, Chroma technology Corp.). To show flow pattern of cells and particles in microfluidic channels, cells are labeled with SYTO 16 and 100~200 consecutive fluorescence images were taken with 40 ms exposure time for each of them and superimposed for image analysis.

The delivery of YOYO-1 labeled pEGFP-C1 into cells was observed using a confocal fluorescence microscope (LSM 510, Carl Zeiss) with a 63X, 1.4 NA oil-immersion objective. In order to observe the delivery of the plasmid under confocal microscope, 500 µg/ml fluorescent plasmid was mixed with CHO cells in the electroporation experiment. These processed cells were moved to a centrifuge tube immediately after electroporation and washed twice to remove nonspecifically adsorbed DNA molecules from the cell surface before imaging.

4. Microchip Operation. The microfluidic EFC device was mounted on an inverted fluorescence microscope (IX-71, Olympus). The inlet of the channel was connected to a syringe pump (PHD infusion pump, Harvard Apparatus) by plastic tubing through which cells or beads were pumped into the microfluidic devices at a constant flow rate. A high voltage power supply (PS350, Stanford Research Systems) was used to generate a constant direct current (DC) voltage between the sample and the outlet reservoirs.

5. Transfection. Cells were centrifuged at 260 g for 5 min and resuspended at 2×10^6 cells/mL in the electroporation buffer (8 mM Na_2HPO_4, 2 mM KH_2PO_4, and 250 mM sucrose, pH=7.2) containing 10 µg/ml plasmid. After electroporation under certain electric parameters in the microfluidic devices, the cells were cultured in DMEM medium for 2 days before quantification of transfection frequency and viability. Transfection frequency is the percentage of EGFP expressing cells among living cells. The cell viability after electroporation was assessed by staining cells with propidium iodide (PI, Invitrogen) and calculated by dividing the number of viable cells by the total number of cells.

III. RESULTS

We first examined the effects of hydrodynamics on the migration of cells and particles (including Chinese hamster ovary (CHO) cells with D~14.6±2.2 µm (s.d.), DT-40 chicken B cells with D~10.8±1.3 µm (s.d.), and polystyrene beads with D~9.9±0.05 µm (s.d.)) in microfluidic channels incorporating a spiral narrow section at the center. The migration of cells or beads in the spiral channel is affected by a balance between (1) transverse lift forces that act stabilize them at distinct equilibrium positions within the microchannel cross section, and (2) a centrifugal Dean force that acts to entrain them in a counter-rotating transverse vortex flow pattern [24]. The interplay between these effects can be seen in **Fig. 1a**, where a suspension of CHO cells is "loosely" focused at the channel centerline throughout the entire spiral section when the flow rate is low (3.75 µl/min). As the flow rate increases above 75 µl/min the Dean drag begins to dominate and the focused stream of CHO cells becomes displaced, crossing the centerline at multiple locations along the spiral section (**Fig. 1a**). This transverse motion indicates that the CHO cells are entrained in the Dean vortex flow pattern at high flow rates. The focused streams cross the centerline more frequently as the flow rate is increased further to 150 µl/min, consistent with the stronger transverse Dean flow effects under these conditions. **Fig. 1b** shows how the transverse position of the cell stream is sensitive to the flow rate at a specific location in the channel.

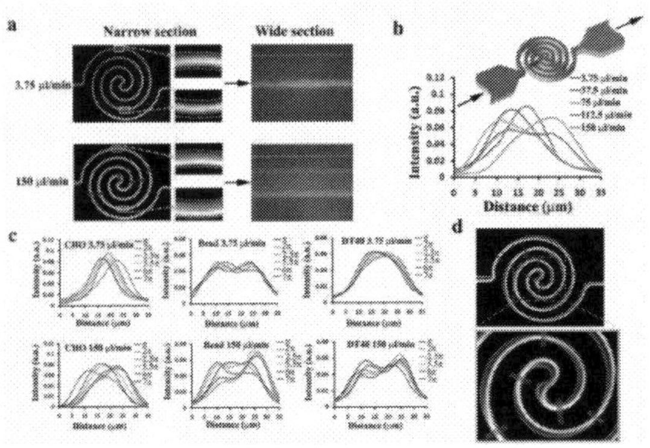

Fig. 1 Particle migration in a flow-through electroporation device with a spiral electroporation section. (a) Overlay fluorescence images of SYTO 16 stained CHO cells flowing in the device at a flow rate of 3.75 µl/min and 150 µl/min. (b) Fluorescence intensity distribution along the width of the narrow section generated by CHO cells under different flow rates. The location where the measurement is taken is labeled by a bar in the figure. (c) Fluorescence intensity distribution along the width of the narrow section generated by CHO cells, polystyrene beads and DT 40 cells at different positions along the channel length (π to 7π with 0 being the entrance and 8π being the exit) and under the flow rates of 3.75 and 150 µl/min. (d) Overlay fluorescence image of polystyrene beads flowing in the spiral section at 150 µl/min. The beads enter the spiral section in two streams that are close to the vertical walls. The blue bars mark the locations where the bead stream shifts across the centerline of the channel.

Somewhat different behavior is observed in a DT40 cell suspension where the cells are focused in a single stream at 3.75 µl/min but split into two streams at 150 µl/min, presumably owing to their smaller size compared to the CHO cells (**Fig. 1c**). The polystyrene beads are even smaller in diameter, and begin to show two focused streams even at 3.75 µl/min. The polystyrene bead and DT40 cell streams also experience significant transverse migration driven by Dean vortices, as evident by numerous centerline crossings throughout the spiral section (**Fig. 1d**). This observation that smaller particles tend to be more strongly affected by the Dean drag is consistent with the physical scaling of the inertial lift force F_z with a^4 while Dean drag F_d scales with a [24; 25].

As a comparison and control, we also studied migration of the cells and polystyrene beads in a device where the spiral channel is replaced with a straight section. In this geometry, the CHO cells gradually focus to the centerline as they travel through the narrow section at the low flow rate of 3.75 µL/min (**Fig. 2a**). In the absence of Dean effects, both the microchannel dimensions and the size of the cells (or particles) determine the position of the focused stream. In

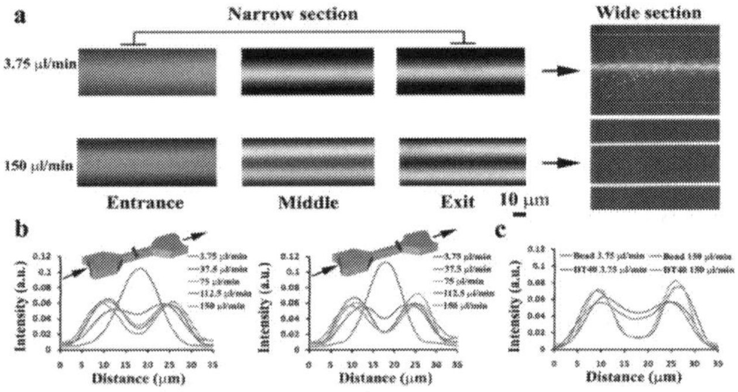

Fig. 2 Particle migration in a flow-through electroporation device with a straight electroporation section. (a) Overlay fluorescence images of SYTO 16 stained CHO cells flowing in the device at a flow rate of 3.75 µl/min and 150 µl/min. Each fluorescent image is generated by superimposing 100-200 consecutive fluorescent images taken in a period of ~15 s while CHO cells flow through. (b) Fluorescence intensity distribution along the width of the narrow section that is generated by labeled CHO cells under different flow rates at the middle (left) and the exit (right) of the straight narrow section. The fluorescence intensity is measured from the overlay fluorescence images. (c) Fluorescence intensity distribution along the width of the narrow section generated by labeled DT40 cells and fluorescent polystyrene beads under flow rates 3.75 and 150 µl/min.

our microchannel design the narrow section incorporates a relatively small cross-section and high aspect ratio (*h*~75μm, *w*~35 μm), while the CHO cells have an average *a/w* of 0.42 where *a* is the cell diameter. Under these conditions, our observations are consistent with previous experiments involving deformable red blood cells, where centerline focusing occurred when *a/w* was larger than 0.4 [26]. At higher flow rates (> 37.5 μL/min) the CHO cells become focused into two streams symmetrically positioned about the centerline (**Fig. 2a,b**), also consistent with previous observations [26; 27]. Here, the high aspect ratio of the narrow section promotes preferential equilibration along the vertical channel walls due to the higher shear rate along the narrowest dimension of the channel[27]. These equilibrium focusing positions move further outward toward the vertical channel walls as the flow rate increases (**Fig. 2b**).

In contrast, the DT40 cells and polystyrene beads do not focus to the center of the straight channel at any of the flow rates tested (3.75-150 μl/min) due to their smaller sizes and lower values for *a/w* (0.31 and 0.28, respectively) **(Fig. 2c)**. Instead, their trajectories are always characterized by two streams symmetrically positioned along both sides of the centerline. Cell deformability does not appear to significantly influence the observed hydrodynamics under these conditions based on the observation that focusing of the rigid polystyrene beads is nearly identical to that of similar-sized DT-40 cells.

We next investigated flow-through electroporation of CHO cells using both straight and spiral shaped channels. In both cases, the wide sections are a factor of 14.3 larger than the narrow section, rendering the electric field intensity higher by the same factor in the narrow section than that in the wide sections when a constant voltage is applied across the channel[28]. When a mixture of cells and plasmid DNA is pumped through the microchannel network, the cells are exclusively electroporated in the narrow section because the field intensity in the wide sections (28-56 V/cm) is substantially lower than the electroporation threshold for most mammalian cells (>300-400 V/cm).

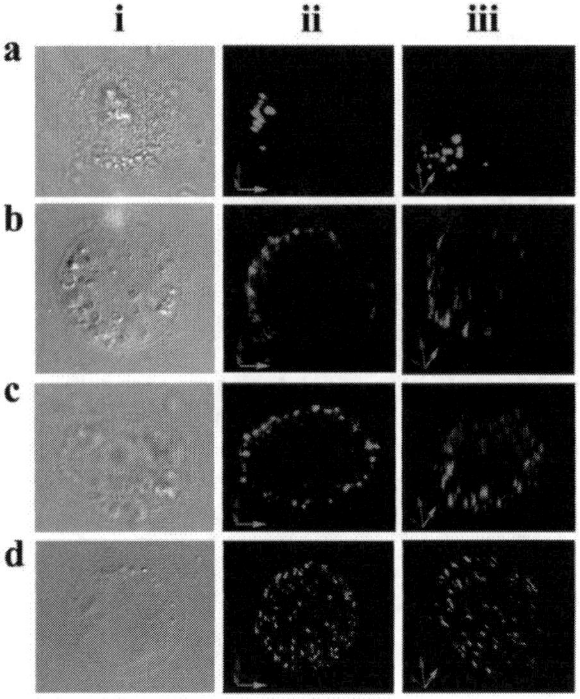

Fig. 3 Phase contrast (i) and 3D reconstruction images generated by confocal fluorescence imaging (ii and iii are from different angles) of CHO cells processed by flow-through electroporation devices under different conditions. The cells are delivered with YOYO-1 labeled plasmid DNA. (a) Cells are processed in a flow-through electroporation with a straight narrow section with an electroporation field intensity of 900 V/cm and a flow rate of 6 μl/min (that yields a field duration of 0.5 ms). The narrow section has a length of 50 μm and a width of 40 μm. (b) and (c) Cells are processed in a flow-through electroporation device with a straight electroporation section. The flow rate is 150 μl/min and the electroporation field intensity and duration are 700 V/cm and 5 ms, respectively. (d) Cells are processed in a flow-through electroporation device with a spiral electroporation section. The flow rate is 150 μl/min and the electroporation field intensity and duration are 700 V/cm and 5 ms, respectively.

We observed the delivery of exogenous DNA molecules (fluorescently labeled with YOYO-1) into the CHO cells using epi- and confocal fluorescence microscopy. Because the DNA molecules can only penetrate the cell membrane within the electroporated zone, and since they are uniformly distributed throughout the microchannel, the fluorescence on or inside the cell membrane essentially "maps" the area that has been electroporated to the extent that allows DNA entry. **Fig. 3a** shows typical 3D reconstruction images generated by confocal fluorescence imaging of a cell after electroporative delivery of labeled plasmid DNA molecules using a short electroporation duration (0.5 ms). These images reveal that delivery is permitted

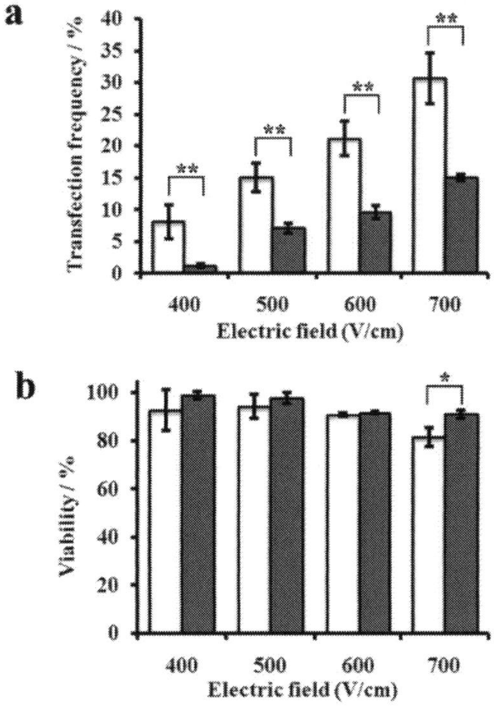

Fig. 4 The delivery and expression of a plasmid DNA vector (pEGFP-C1) in CHO cells by flow-through electroporation devices with different geometries. In both the spiral (white bars) and straight (dark bars) electroporation devices, the electroporation field intensity is varied from 500 to 800 V/cm and the flow rate is at 150 μl/min that yields an electroporation duration of 5 ms. All data points are generated by experiments in triplicate and ~5,000 cells are examined in each trial. Two-tailed student's t test was applied to evaluate the data. (*) indicates significant difference at $P<0.05$ and (**) indicates significant difference at $P<0.01$. (a) Transfection frequency of cells (the number of cell expressing EGFP divided by the number of living cells) examined two days after electroporation. (b) The viability of the cells (the number of living cells divided by the total cell number) examined two days after electroporation.

only within a narrow zone in the vicinity of one pole of the cell, similar to previous observations of static cells suspended in a solution subjected to short pulse electroporation [21]. This result suggests that the cells primarily undergo translational migration through the electroporation section during the short duration. We also tested a longer electroporation duration (5 ms) in the straight channel and found that these conditions produced DNA delivery within a circular stripe around the cell equator (the fluorescent spots formed either a partial circle (**Fig. 3b**) or a full circle (**Fig. 3c**).

When the same flow rate and electric parameters are applied in the spiral channel device, however, the DNA delivery pattern changes dramatically. Instead of being localized within a narrow zone the DNA molecules are now uniformly distributed over the entire cell surface (**Fig. 3d**). This result clearly highlights the influence of the hydrodynamic effects at play in the spiral microchannel that act to permit a large fraction of the cell membrane to experience favorable alignment with the electric field direction as they flow through. Additional epi-fluorescence data confirm that the images in **Fig. 3a-d** are typical of the cell population.

The enhanced DNA delivery based on flow-through electroporation in curvilinear channels has important implications. In **Fig. 4a**, we show the transfection of CHO cells by pEGFP-C1 vector using both straight and spiral shaped microchannels. Under the same flow rate and electric field conditions, the transfection efficiency in the spiral geometry is about two-fold higher than that achieved in the straight channel devices. This level of enhancement is significant given the extremely high manufacturing costs that continue to hinder widespread application of cell-based therapeutics. Moreover, this improvement is even more dramatic considering that the straight microchannel device design (with an electroporation duration >5ms) already provides substantially enhanced DNA uptake relative to conventional methods that deliver DNA only at the poles of a cell (**Fig. 3**). Cell viability after electroporation is similar in both microchannel geometries (**Fig. 4b**).

IV. DISCUSSION

In order to better understand how these hydrodynamic effects act to produce enhanced transfection efficiency, we consider the physics of microscale particle migration in curvilinear channels. These phenomena incorporate a surprising level of complexity, but the key features of their underlying physics can provide a useful basis to interpret our observations [24; 25; 29; 30; 31; 32]. When suspended particles are convected within the laminar flow field inside a curved microfluidic channel, they experience a combination of inertial lift and drag force due to the transverse Dean flow. When the magnitude of these forces is similar, the convected particles migrate to occupy lift-induced equilibrium positions that minimize the Dean flow effects. On the other hand, when the Dean drag is substantially larger than the inertial lift force, the particles become displaced from these equilibrium positions and begin to follow the Dean vortex pattern[25; 32]. In comparison, particles that flow in a straight channel are focused into multiple equilibrium positions due to inertial self-ordering [24; 33]. The strong dependence of these effects on particle size has led to interest in harnessing Dean flow effects for particle sorting and separation applications[24; 25; 32; 34]. However, these effects have not been explored in connection with manipulation of particle orientation or gene delivery into cells.

There are two important Reynolds numbers that characterize the flow of particles in a channel[24; 35]. The channel Reynolds number (R_c) and the particle Reynolds number (R_p) are defined by the below equations:

$$R_p = \frac{U_m a^2}{\nu D_h} = \frac{U_m D_h}{\nu} \cdot \frac{a^2}{D_h^2} = R_c\left(\frac{a}{D_h}\right)^2 \quad [1]$$

where U_m is the maximum velocity in the cross section, and a is the diameter of the particle, υ is kinematic viscosity and D_h is the hydraulic diameter which is defined as $2wh/(w+h)$ (with w and h being the width and depth of the channel). The Dean number De is used to characterize the curvature effects in a curvilinear channel and is defined as:

$$De = \frac{\bar{U} D_h}{\upsilon}\left(\frac{D_h}{2R}\right)^{\frac{1}{2}} = Re\left(\frac{D_h}{2R}\right)^{\frac{1}{2}} \quad [2]$$

where R is the radius of curvature of the channel, $\bar{U}$ is the mean axial velocity, and Re is the Reynolds number ($Re = \frac{\bar{U} D_h}{\upsilon}$).

In our experiments with spiral-shaped channels, we observe that the particles are entrained in Dean vortices when Rc =50-100 and the average De= 13-26 (De varies in the same spiral-shaped channel by a factor of 3.3 due to the various radii). We notice that a smaller particle and a larger Rc favor stronger Dean flow effects which are evident by a larger number of centerline crossings. This trend is consistent with previous work by Di Carlo et al. [24]. However, it needs to be noted that our R_p is substantially larger than their values (2.0-8.5 in this work vs. <1 in Di Carlo et al.) and this may have led to larger values of a/D_h (0.21-0.31 in this work vs. <0.07 in Di Carlo et al.) for the Dean drag dominant regime [24].

In general, we observe that the particle displacements across the centerline occur more frequently per unit contour length of the channel when they are close to the center of the spiral

(**Fig. 1a,d**). This is characteristic of the transverse migration along a Dean vortex. The velocity in the rotational flow due to the Dean drag U_D scales as $U_D \sim De^2 v / D_h$ with v being the kinematic viscosity[24]. Assuming the particle will cover a circular path L_d in the transverse direction with a length of $\frac{3}{2} D_h$ [36], the channel length L_c covered during the period that the particle travels through a complete circular path along a Dean vortex will therefore be

$$L_c = \bar{U} \frac{L_d}{U_d} \sim \frac{3\bar{U} D_h^2}{2 v De^2} . \quad [3]$$

Equation 3 shows that L_c becomes short (i.e., more frequent centerline crossings) close to the spiral center and at high *De*, consistent with our observations.

Based on the pattern of DNA delivery shown in **Fig. 3**, we may hypothesize the types of cell migrations in different devices during flow-through electroporation. Cells experience rotation about an axis that is parallel to the vertical walls of the channel as they travel through the electroporation section, and that these effects become important when the electroporation duration is long enough (e.g. ~5 ms) (**Fig. 5a**). Previous literature indicates that such free rotation of particles occurs in response to imposed shear in a poiseuille flow[37], and that the cell-wall friction due to hydrodynamic interactions can also affect the rotation rate[38]. In a straight channel, the rotation rate Ω_p can be calculated using

$$\Omega_p = 2U_m \frac{| w - (a + 2l) |}{w^2} \quad [4]$$

where l is the distance to the wall[38]. Considering the average CHO cell diameter of 14.6 µm and using the average distance to channel walls shown in **Fig. 2a** yields estimated rotation rates ranging from 2.0×10^3 Hz at a flow rate of 6 µl/min to 5.0×10^4 Hz at 150 µl/min in the straight electroporation section. Such rotation during flow-through electroporation would allow DNA delivery into a circular zone on the cell membrane, as shown in **Fig. 3b** and **3c**. The situation becomes more complicated when the flow-through electroporation occurs in a curved microchannel. In this case the DNA delivery over the entire cell surface (shown in **Fig. 3d**) is a result of the induced cell motion in the spiral channel segment, which involves a combination of rotation and transverse migration following the Dean vortex pattern. **Fig. 5b** illustrates how this process would allow cells to be electroporated over their entire surface. When a cell simultaneously experiences rotation due to the near-wall shear field coupled with transverse motion due to the Dean vortex flow, different parts of the cell surface are

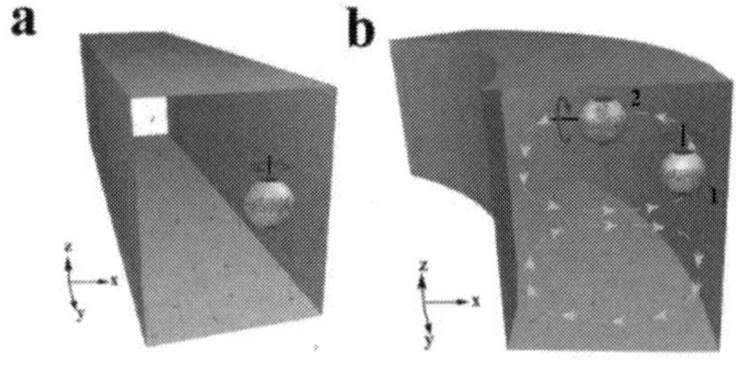

Fig. 5 Proposed mechanisms for cell migration during flow-through electroporation in microchannels of different geometries. (a) The cell migration in a straight channel. The schematic shows that free rotation of a cell in response to the shear field allows it to have electroporation/DNA delivery in a circular stripe around the cell equator. The electric field is along y axis. (b) The cell migration in a curved channel. The schematic shows one possible scenario that allows a cell to be electroporated on its entire surface. The electric field is along y axis. The cell is undergoing both migration entrained in the transverse Dean flow and free rotation when moving from 1 to 2. The cell has electroporation/DNA delivery around its equator at position 1 due to its rotation and electroporation/DNA delivery on a different surface area at position 2, given that the orientations of the cell (relative to the channel) at both positions are identical.

continuously exposed to a sufficiently high transmembrane potential to permit electroporation.

Finally, we note that the electric field has a negligible effect on the migration of cells under the high flow rates imposed in the transfection experiment. At an electric field of 700 V/cm, the electrophoretic mobility of a cell contributes ~0.7 mm/s to the velocity of the cell, while the electroosmotic mobility produces ~1.26 mm/s[39]. Both of these effects are about three orders of magnitude lower than the average velocity of 0.943 m/s generated by the pressure-driven flow when the volumetric flow rate is at 150 μl/min. Electroosmotic flow effects of this magnitude also do not appreciably alter the parabolic velocity profile associated with the pressure-driven flow[40]. We also find that the plasmid DNA molecules (pEGFP-C1, 4.7 kb) used for transfection remain uniformly distributed throughout the channel cross-section in both the straight and spiral geometries (lower fluorescence intensities close to the vertical channel walls reflect a slightly smaller depth due to deformation under pressure generated by the flow[41]).

To summarize, in order to take full advantage of the Dean flow effects for gene delivery enhancement, the balance between inertial lift forces and Dean drag will need to be maintained and both Dean flow induced transverse migration and shear induced rotation will need to be present. In general, as evident from our results and previous work[24], low a/D_h and high R_c will be required to ensure the dominance of Dean drag, with the possibility of using one to compensate the other. A high velocity U_m will be generally favored in order to generate a high enough shear-field induced rotation rate (>10^3 Hz) that allows a cell to finish a number of rotations within the typical millisecond electroporation duration. Finally, as described by equation 3, the contour length of the curved segment L_c that is required for finishing a circle of transverse migration puts a practical limit to how high the velocity and R_c can be.

Flow-through electroporation is advantageous because it allows continuous production of genetically modified cells, as opposed to the batch mode of operation characteristic of conventional electroporators. By harnessing novel features of the underlying physics associated with particle migration in transverse Dean flows, it is possible to achieve considerable enhancements in transfection efficiency with a relatively simple modification to the microchannel geometry. These device designs are extremely versatile, amenable to both scaling up and down for processing cell samples spanning a wide range of volumes, including small volume samples from scarce sources. It is likely that transfection efficiencies can be further improved by optimizing the device design (i.e. the number of narrow/wide sections and the lengths and widths of various sections), the flow rate, and the applied voltage [15]. This study identifies a novel application that links the interesting phenomena associated with particle-laden flows to the important field of gene delivery. Fundamental understanding of the complex interaction between inertial flows and particles will yield further insight into the design of these devices.

ACKNOWLEDGEMENT

The authors thank Life Science Fluorescence Imaging Facility at Purdue University for access of the confocal fluorescence microscope. This work is supported by Wallace H. Coulter Foundation, NSF grant CBET-0747105, a cooperative agreement with the ARS-USDA, project number 1935 42000-035, through the Center for Food Safety Engineering at Purdue University and Bilsland Fellowship (J.W.).

REFERENCES

1. Cone, R. & Mulligan, R. (1984). High-efficiency gene transfer into mammallian cells: generation of helper-free recombinant retrovirus with broad mammalian host range. *Proc. Natl. Acad. Sci. U. S. A.* **81**, 6349-6353.
2. Liu, Q. & Muruve, D. (2003). Molecular basis of the inflammatory response to adenovirus vectors. *GeneTher.* **10**, 935-940.
3. Boussif, O., Lezoualch, F., Zanta, M., Mergny, M. S., D, Demeneix, B. & Behr, J. (1995). A versatile vector for gene and oligonucleotide transfer into cells in culture and in vivo: polyethylenimine. *Proc. Natl. Acad. Sci. U. S. A.* **92**, 7297-7301.
4. Felgner, P., Gadek, T., Holm, M., Roman, R., Chan, H., Wenz, M., Northrop, J., Ringold, G. & Danielsen, M. (1987). Lipofection: a highly efficient, lipid-mediated DNA-transfection procedure. *Proc. Natl. Acad. Sci. U. S. A.* **84**, 7413-7417.
5. Wigler, M., Silverstein, S., Lee, L., Pellicer, A., Cheng, Y. & Axel, R. (1977). Transfer of purified herpes virus thymidine kinase gene to cultured mouse cells. *CELL* **11**, 223-232.
6. Neumann, E., Schaefer-Ridder, M., Wand, Y. & Hofschneider, P. (1982). Gene transfer into mouse lyoma cells by electroporation in high electric fields. *EMBO J.* **1**, 841-845.
7. Prausnitz, M. R., Bose, V. G., Langer, R. & Weaver, J. C. (1993). Electroporation of mammalian skin: a mechanism to enhance transdermal drug delivery. *Proc. Natl. Acad. Sci. U. S. A.* **90**, 10504-8.
8. Rols, M. P., Delteil, C., Golzio, M., Dumond, P., Cros, S. & Teissie, J. (1998). In vivo electrically mediated protein and gene transfer in murine melanoma. *Nat. Biotechnol.* **16**, 168-71.
9. Aihara, H. & Miyazaki, J. (1998). Gene transfer into muscle by electroporation in vivo. *Nat. Biotechnol.* **16**, 867-70.
10. Hamm, A., Krott, N., Breibach, I., Blindt, R. & Bosserhoff, A. (2002). Efficient transfection method for primary cells. *Tissure Eng.* **8**, 235-245.
11. Ryttsen, F., Farre, C., Brennan, C., Weber, S. G., Nolkrantz, K., Jardemark, K., Chiu, D. T. & Orwar, O. (2000). Characterization of single-cell electroporation by using patch-clamp and fluorescence microscopy. *Biophys. J.* **79**, 1993-2001.
12. Huang, Y. & Rubinsky, B. (2003). Flow-through micro-electroporation chip for high efficiency single-cell genetic manipulation. *Sens. Actuator A-Phys.* **104**, 205-212.
13. Fei, Z., Wang, S., Xie, Y., Henslee, B. E., Koh, C. G. & Lee, L. J. (2007). Gene transfection of mammalian cells using membrane sandwich electroporation. *Anal. Chem.* **79**, 5719-22.
14. Valero, A., Post, J. N., van Nieuwkasteele, J. W., ter Braak, P. M., Kruijer, W. & van den Berg, A. (2008). Gene transfer and protein dynamics in stem cells using single cell electroporation in a microfluidic device. *Lab Chip* **8**, 62-67.
15. Wang, H. Y. & Lu, C. (2008). Microfluidic electroporation for delivery of small molecules and genes into cells using a common DC power supply. *Biotechnol and Bioeng.* **100**, 579-586.
16. Skelley, A. M., Kirak, O., Suh, H., Jaenisch, R. & Voldman, J. (2009). Microfluidic control of cell pairing and fusion. *Nat. Methods* **6**, 147-52.
17. Luo, D. & Saltzman, W. M. (2000). Enhancement of transfection by physical concentration of DNA at the cell surface. *Nat. Biotechnol.* **18**, 893-5.
18. Segura, T. & Shea, L. D. (2002). Surface-tethered DNA complexes for enhanced gene delivery. *Bioconjug. Chem.* **13**, 621-9.
19. Weaver, J. C. & Chizmadzhev, Y. A. (1996). Theory of electroporation: A review. *Bioelectrochemistry and Bioenergetics* **41**, 135-160.

20. Tekle, E., Astumian, R. D. & Chock, P. B. (1994). Selective and asymmetric molecular transport across electroporated cell membranes. *Proc. Natl. Acad. Sci. U. S. A.* **91**, 11512-6.
21. Golzio, M., Teissie, J. & Rols, M. P. (2002). Direct visualization at the single-cell level of electrically mediated gene delivery. *Proc. Natl. Acad. Sci. U. S. A.* **99**, 1292-1297.
22. Wang, J., Stine, M. J. & Lu, C. (2007). Microfluidic cell electroporation using a mechanical valve. *Anal. Chem.* **79**, 9584-7.
23. Faurie, U., Phez, E., Golzio, M., Vossen, C., Lesbordes, J. C., Delteil, C., Teissie, J. & Rols, M. P. (2004). Effect of electric field vectoriality on electrically mediated gene delivery in mammalian cells. *Biochim. Biophys. Acta-Biomembr.* **1665**, 92-100.
24. Di Carlo, D., Irimia, D., Tompkins, R. G. & Toner, M. (2007). Continuous inertial focusing, ordering, and separation of particles in microchannels. *Proc. Natl. Acad. Sci. U. S. A* **104**, 18892-18897.
25. Bhagat, A. A. S., Kuntaegowdanahalli, S. S. & Papautsky, I. (2008). Continuous particle separation in spiral microchannels using dean flows and differential migration. *Lab chip*, 1906-1914.
26. Abkarian, M., Faivre, M., Horton, R., Smistrup, K., Best-Popescu, C. A. & Stone, H. A. (2008). Cellular-scale hydrodynamics. *Biomed. Mater.* **3**, -.
27. Bhagat, A., Kuntaegowdanahalli, S. & Papautsky, I. (2008). Enhanced particle filtration in straight microchannels using shear-modulated inertial migration. *Phys. Fluids* **20**, 101702.
28. Wang, H. Y. & Lu, C. (2006). Electroporation of mammalian cells in a microfluidic channel with geometric variation. *Anal. Chem.* **78**, 5158-5164.
29. Dean, W. R. (1927). Note on the notion of fluid in a curved pipe. *Philos. Mag.* **4**, 208-223.
30. Dean, W. R. (1928). The stream-line motion of fluid in a curved pipe. (Second paper.). *Philos. Mag.* **5**, 673-695.
31. Sudarsan, A. P. & Ugaz, V. M. (2006). Multivortex micromixing. *Proc. Natl. Acad. Sci. U. S. A.* **103**, 7228-7233.
32. Seo, J., Lean, M. H. & Kole, A. (2007). Membrane-free microfiltration by asymmetric inertial migration. *Appl. Phys. Lett.* **91**, 033901.
33. Edd, J. F., Di Carlo, D., Humphry, K. J., Koster, S., Irimia, D., Weitz, D. A. & Toner, M. (2008). Controlled encapsulation of single-cells into monodisperse picolitre drops. *Lab Chip* **8**, 1262-1264.
34. Di Carlo, D., Edd, J. F., Irimia, D., Tompkins, R. G. & Toner, M. (2008). Equilibrium separation and filtration of particles using differential inertial focusing. *Anal Chem.* **80**, 2204-2211.
35. Asmolov, E. S. (1999). The inertial lift on a spherical particle in a plane Poiseuille flow at large channel Reynolds number. *J. Fluid Mech.* **381**, 63-87.
36. Ookawara, S., Higashi, R., Street, D. & Ogawa, K. (2004). Feasibility study on concentration of slurry and classification of contained particles by microchannel. *Chem.Eng. J.***101**, 171-178.
37. Zeng, L. Y., Balachandar, S. & Fischer, P. (2005). Wall-induced forces on a rigid sphere at finite Reynolds number. *J. Fluid Mech.* **536**, 1-25.
38. Staben, M. E., Zinchenko, A. Z. & Davis, R. H. (2003). Motion of a particle between two parallel plane walls in low-Reynolds-number Poiseuille flow. *Phys. Fluids* **15**, 1711-1733.

39. Ren, X. Q., Bachman, M., Sims, C., Li, G. P. & Allbritton, N. (2001). Electroosmotic properties of microfluidic channels composed of poly(dimethylsiloxane). *J. Chromatogr. B* **762**, 117-125.
40. Osuga, T. (1999). Comparison of growth of laminar flows in capillary: Poiseuille flow, electroosmotic flow and electroosmotic circulation. *Jpn. J. Appl. Phys.* **38**, 6564-6567.
41. Hardy, B. S., Uechi, K., Zhen, J. & Kavehpour, H. P. (2009). The deformation of flexible PDMS microchannels under a pressure driven flow. *Lab Chip* **9**, 935-938.

Formation of Polymerizable Phospholipid Vesicles Using Microfluidics

Won Wook Jeong, Research Center, LG household & Health Care Ltd., Daejeon, South Korea

Chanjoong Kim, Liquid Crystal Institute, Kent State University, Kent, OH

Liposomes consist of phospholipid bilayer membranes surrounding an aqueous core. They are promising delivery vehicles for drugs, enzymes and bioreactors for biomedical applications. Since phospholipids are an integral component of biological membranes, phospholipid vesicles provide an ideal platform for a study on the physical properties of biomembranes. However, liposomes get easily ruptured even by small environment stresses: shear flow around membrane, pH change, ionic concentration change, or advent of another membrane. Conventional vesicle formation techniques such as hydration and electroformation rely on the self-assembly of phospholipids in an aqueous environment under a shear or an electric field. Due to the random nature of the bilayer folding, these methods typically lead to the formation of vesicles that are nonuniform in both size and shape. Moreover, encapsulation efficiency of those conventional methods is quite low, generally less than 35%.

We enhance stability and monodispersity of liposomes, and achieve high encapsulation efficiency by generating polymerized phospholipid liposomes with a recent technique, microfluidics. We utilize UV-polymerizable phospholipids that have reactive hydrophobic tails. The microfluidic device can generate monodisperse water-in-oil-in-water (W/O/W) core-shell structure in a single step, allowing precise control of the outer and inner drop sizes as well as the number of droplets encapsulated in each larger drop. The phospholipid solution forms the shell of W/O/W double emulsions that are used as templates for the formation of phospholipid membranes. By removing the solvent in oil phase through evaporation and UV-polymerization, we obtain polymerized vesicles.

In conclusion, our technique can be used to create phospholipid vesicles not only having enhanced stability but also maintaining high size uniformity and encapsulation efficiency.

See more of this Session: Microfluidics and Small-Scale Flows II: Biological Applications
See more of this Group/Topical: Engineering Sciences and Fundamentals

Computer Simulation of Protein Aggregation Kinetics Using An Intermediate Resolution Model

Erin M. Phelps, Department of Chemical and Biomolecular Engineering, North Carolina State University, Raleigh, NC

Carol K. Hall, Chemical and Biomolecular Engineering, North Carolina State University, Raleigh, NC

The aggregation of proteins into fibrillar structures is a symptom of over forty known human neurodegenerative diseases including Alzheimer's, Huntington's, and the prion diseases. Computer simulations allow us to study aggregation behavior on a molecular level. We combine an intermediate resolution protein model, PRIME, which captures the essential physical features and interactions of proteins, including realistic protein geometry, hydrogen bonding and hydrophobic interactions, with discontinuous molecular dynamics, a fast alternative to traditional molecular dynamics. Our objective is to identify and quantify the early stages of protein aggregation since association of small groups of peptides into oligomers is believed to be a key event in the onset of Alzheimer's disease. Current work in this area suggests that the peptides follow a nucleated reaction mechanism, which starts with free monomers that form small intermediate structures that are unstable and energetically unfavorable. These intermediates eventually reach a critical size, called the nucleus, which then grows through energetically favorable steps into larger structures known as fibrils. We chose polyalanine (A) as our model system since it is a simple peptide that is easy to model and it forms fibrillar structures in vitro at moderate peptide concentrations. Simulations of a system containing 192 $KA_{14}K$ peptide chains were performed at peptide concentrations of 5 and 10 mM and reduced temperatures of $T^* = 0.125$, 0.13, 0.135, and 0.14. Population data on the various types of small intermediate structures formed, including free monomers, small beta sheets consisting of hydrogen-bonded peptides, and fibrils consisting of hydrophobically stacked beta sheets, were recorded as a function of time. A mathematical model based on a proposed set of reversible association interactions was constructed. The population data gathered from the simulations is used to determine kinetic rate constants in the mathematical model. This optimized mathematical model provides information about the formation of the fibrillization nucleus and the growth of the fibril structure.

See more of this Session: Self-Assembly in Solution II
See more of this Group/Topical: Engineering Sciences and Fundamentals

Theoretical Modeling of the Dynamics of Clathrin Self-Assembly Into Nanoscale Assemblies

Shafigh Mehraeen, Mechanical Engineering, Stanford University, Stanford, CA

Nick Cordella, Chemical Engineering, Stanford University, Stanford, CA

Andrew Spakowitz, Chemical Engineering, Stanford University, Stanford, CA

Clathrin is a cytoplasmic protein that plays a critical role in endocytosis by forming the basket-like cages on the cell membrane that seed the formation of the endosome pits that transport cargo into the cell. A single clathrin complex adopts a pinwheel configuration. The aggregation of many clathrin pinwheels on a membrane or interface leads to lattice-like assemblies with a mixture of 5-, 6-, and 7-member rings. In vitro assembly of clathrin within a solution results in closed, nanoscale assemblies with various shapes and sizes. Our goal in this research is to develop a fundamental theoretical model for the thermodynamics and kinetics of clathrin assembly in 2 and 3 dimensions in order to guide experiments toward the assembly of targeted nanoscale assemblies. Towards this goal, we have developed a theoretical model for clathrin assembly that can address the assembly dynamics in 2 and 3 dimensions. The clathrin are modeled as effective pinwheels that form leg-leg associations and resist elastic bending deformation; thus, the pinwheels are capable of forming the range of ring structures that are observed experimentally. Our theoretical model combines Brownian dynamics simulations with dynamic mean field theory in order to track the motion of hundreds of clathrin pinwheels at sufficiently long time scales to achieve complete assembly. With this theoretical model, we predict the phase diagram for clathrin assembly and perform dynamic simulations for a range of quenches into the phase diagram. The resulting dynamics exhibit the hallmark behavior of spinodal decomposition with subsequent coarsening of ordered domains. The defect structures in the final configurations result in large-scale elastic stresses in the lattice network. We will then proceed to discuss the assembly of specific nanoscale assemblies for application in advanced materials applications.

See more of this Session: Self-Assembly in Solution II
See more of this Group/Topical: Engineering Sciences and Fundamentals

Encapsulation of Vesicles and Particles by Open Bilayers - Statics and Dynamics

Arun Ramachandran, Chemical Engineering, University of California at Santa Barbara, Santa Barbara, CA

L. Gary Leal, Chemical Engineering, University of California at Santa Barbara, Santa Barbara, CA

Jacob N. Israelachvili, Chemical Engineering, University of California, Santa Barbara (UCSB), Santa Barbara, CA

Engineers have exploited the natural compartmentalizing functionality of vesicles to achieve specific objectives such as imaging, separations and drug delivery. Some of these implementations require particles or vesicles to be captured within bigger unilamellar vesicles, which can be achieved by triggering the closure of flat sheets of bilayers [Kisak *et al.* (2002)]. Application of hypotonic osmotic stresses, electric pulses and freezing thawing on vesicle systems also create open bilayers which subsequently may encapsulate other vesicles. However, there has been limited theoretical work in the literature to understand encapsulation by bilayers. The present work represents an effort towards remedying that deficiency.

In this paper, we address two aspects of the encapsulation of entities by open bilayers. First, we look at the steady state configurations for a vesicle being engulfed by a bilayer patch, where both the vesicle and bilayer are composed of charged lipids. We also consider electrolyte concentrations such that vesicle and bilayer surfaces are purely repelling at all separations. Any steady configuration of the bilayer around the vesicle, which is characterized in our simplified model by a curling angle θ, is a result of the balance between the line tension of the bilayer patch attempting to curl around the vesicle and the repulsion between the two attempting to keep them apart. Using a simple description of the energy of the bilayer, it is demonstrated that the bilayer either encapsulates the vesicle completely, or assumes a steady curling angle θ_s around the vesicle with an open line. These steady state calculations can be used to predict the maximum size of a stable hole in a bilamellar (or multilamellar) vesicle, which may be induced by osmosis or electric pulses.

The second problem we consider is the dynamics of encapsulation of a particle by an open bilayer. As the bilayer curls up around the particle, it squeezes fluid out, thus pushing the particle further away. It can be shown that under certain conditions, the capture of a particle depends solely on its initial position, irrespective of the rate at which the bilayer curls up. We identify the parameters necessary for particle capture by simulating the bilayer and particle motions using a boundary integral method.

References

1. Kisak E. T., B. Coldren and J. A. Zasadzinksi, "Nanocompartments Enclosing Vesicles, Colloids, and Macromolecules via Interdigitated Lipid Bilayers", Langmuir **18**, 284-288 (2002).

See more of this Session: Self-Assembly in Solution II
See more of this Group/Topical: Engineering Sciences and Fundamentals

pH-Dependent Lipid Phase-Separation Forming Defective Interfaces: Effects On Lipid Bilayer Fusion

Manali Bhagat, Chemical and Bilogical Sciences and Engineering Department, Polytechnic Intsitute of NYU, Brooklyn, NY

Stavroula Sofou, Othmer-Jacobs Department of Chemical and Biological Engineering, Polytechnic Intsitute of NYU, Brooklyn, NY

Lipid membrane fusion involves the mixing of lipids from separate bilayers. Fusion has been widely studied using model membranes since it is an essential process for cell survival. Fusion takes place in diverse cellular activities such as exocytosis, vesicle transport among cell organelles, release of neurotransmitters at synapses, fertilization, viral infection, and intracellular membrane trafficking. In therapeutic applications using lipid bilayer vesicles as the delivery carrier, in order to deliver the therapeutic agents intracellularly, fusion between the cellular membrane and the lipid vehicle is critical. Fusion can be triggered by different events: one being formation of defects on the approaching bilayers. In this work, we present a mechanism of introducing lipid packing defects on bilayers resulting in complete fusion. Our findings on vesicle aggregation, total lipid mixing assays, and inner leaflet mixing studies support this statement. Fusion is studied using lipid bilayers in the form of unilamellar vesicles. We are studying phospholipid membranes composed of lipid pairs with phosphatidylcholine and phosphatidic acid headgroups that form phase separated domains with lowering pH. We have demonstrated that lipid pairs in the gel state with non-matching acyl-chain lengths form pH-dependent phase separated domains that, in addition, have leaky interfacial boundaries. Our current studies indicate correlation between extensive vesicle fusogenicity and the presence of leaky interfacial boundaries attributed to lipid packing discontinuities within the interfaces of phase separated lipid domains. We hypothesize that in these systems intervesicular fusion is driven by the tendency of the system (vesicle suspension) to minimize the total length of defective interfaces of the phase-separated lipid domains that are characterized by high line tensions.

Increasing aggregation in vesicle suspensions with decrease in pH is observed, and aggregation rates are strongly dependent on the lipid vesicle concentration. Our findings demonstrate complete vesicle fusion with increasing formation of defective interfaces when bilayers are composed of lipid pairs with non-matching acyl-tail lengths. Fusion rates are strongly dependent on the introduced pH-gradient and the lipid vesicle concentration. For bilayers containing lipid pairs with matching acyl-tail lengths, inner monolayer lipid mixing studies demonstrate hemifusion, and suggest an important role for the overall stiffness of the bilayer membrane. These observations support the hypothesis that pronounced hydrophobic mismatch within the interfaces of phase-separated lipid domains is strongly correlated to the cause that drives fusion in these lipid vesicles.

See more of this Session: Self-Assembly in Solution II
See more of this Group/Topical: Engineering Sciences and Fundamentals

Guided Lipid Tubule Formation Via Self-Assembly of Vesicles

Amy Shen, Mechanical Engineering, University of Washington, Seattle, WA

Mike Tan, Department of Medicine, Washingtonn University in St. Louis, St. Louis, MO

Liang Ma, Department of Medicine, Washingtonn University in St. Louis, St. Louis, MO

Elliot Elson, Department of Medicine, Washingtonn University in St. Louis, St. Louis, MO

Vesicle self-assembly offers means to construct micro structures from nano-sized building blocks. However, current vesicle self-assembly processes produce non-oriented structures that vary in shape and size. We show here that, when confined between two glass surfaces, lipid vesicles rapidly assemble into oriented tubules that can grow to be long as 13~mm under ambient conditions without externally supplied flow, temperature control, or catalyzing agents. Tubule membrane and its internal aqueous content can be manipulated by controlling the vesicle's lipid composition and aqueous entrapment. This self-assembly process opens up new pathways for generating complicated and flexible architectures for use in bio-compatible molecular and supramolecular engineering. We demonstrate these possibilities by generating encapsulated siRNA tubules, inter-digitated branching tubules, and polymerized solid tubules in a single-throughput self-assembly process.

See more of this Session: Self-Assembly in Solution II
See more of this Group/Topical: Engineering Sciences and Fundamentals

DNA Surfactants as Probes of Monomer Exchange Dynamics in Self-Assembled Systems

James W. Schneider, Chemical Engineering, Carnegie Mellon University, Pittsburgh, PA

Stephen Istivan, Chemical Engineering, Carnegie Mellon University, Pittsburgh, PA

Angela Holmen, Chemical Engineering, Carnegie Mellon University, Pittsburgh, PA

The dynamics of self-assembly processes have been studied by stopped-flow and phosphorescence methods, to name a few. Knowledge of these time scales is critical when predicting size distributions and solubilization capacity in surfactant microstructures. These data vary significantly with hydrophile-lipophile balance, chain length, and the presence of specific interactions, to name a few.

We present a novel electrophoretic method to characterize the kinetics of monomer exchange using surfactant tails conjugated to DNA oligomers of a precise molecular weight. Here, DNA surfactants are introduced into a capillary and electrophoretically eluted in the presence of nonionic surfactant micelles. Nonionic micelles transiently interact with the DNA surfactant, retarding its electrophoretic mobility. During the run, the highly monodisperse DNA surfactant exchanges among many distinct nonionic micelles, each with widely varying size. The elution time for the DNA surfactant can be mapped to the average size of the micelles and their partition coefficient. The broadness of the peak can be traced to several different mechanisms, including the polydispersity of the micelle population and the number of exchanges that occur. These two effects compete: a large polydispersity is overcome by excessive exchanges, and vice versa. As such, an independent measurement of micelle polydispersity by light scattering, combined with peak broadness from capillary electrophoresis, gives the monomer exchange rate.

We demonstrate that the presence of DNA does not impact the free energy of transfer for n-alkyl surfactant tails with lengths in the range of 12-18 carbons. A detailed model, accounting for effects of diffusion, injection, and detection on peak broadening will be presented along with exchange kinetics for various spherical and rod-like micelles. These are obtained from mixtures of the poly (oxyethylene) surfactants Triton X-100 and C16E6. We also present adsorption-desorption kinetics for DNA surfactants interacting with oil-in-water nanoemulsions and carbon nanotubes obtained using the CE method.

See more of this Session: Self-Assembly in Solution II
See more of this Group/Topical: Engineering Sciences and Fundamentals

Single- and Double-Stranded DNA Simulations Using An Intermediate-Resolution Model

Johnny R. Maury-Evertsz, Chemical and Biomolecular Engineering, North Carolina State University, Raleigh, NC

Carol K. Hall, Chemical and Biomolecular Engineering, North Carolina State University, Raleigh, NC

DNA microarray technology is a powerful tool for biological research, allowing scientists to track the expression patterns of thousands of genes simultaneously. To fully exploit the potential of this emerging technology, a better understanding of the physics underlying the process at the mesoscopic scale is still needed. A new intermediate resolution model for DNA molecules designed for use with discontinuous molecular dynamics (DMD) simulations is presented. The model was developed using a multiscale modeling approach in which the geometric and energetic parameters are obtained by collecting data from CHARMM-based atomistic simulations of single- and double-stranded DNA molecules with explicit solvent and counterions. The sugar, phosphate, and base in the model are each represented by single spheres connected via bonds with appropriate lengths and angles. Hydrogen bonding is modeled using an angle-dependent square-well scheme and base stacking is modeled using explicit intra-strand interactions between nearest neighbor bases. DMD simulations were performed on model single- and double-stranded DNA molecules of 12 different base sequences and several chain lengths at a variety of reduced temperatures. The model reproduces qualitatively: (1) the conformational properties of single-stranded DNA sequences as a function of temperature, (2) the temperature-dependent melting behavior of double-stranded DNA, and (3) the hybridization process in solution.

See more of this Session: Self-Assembly in Solution II
See more of this Group/Topical: Engineering Sciences and Fundamentals

On the Use of Entropy-Scaling to Describe Dynamic Properties of a Dumbbell Model Under Confinement

Ravi Chopra, Chemical and Biological Engineering, University at Buffalo, Buffalo, NY

Thomas M. Truskett, Department of Chemical Engineering and Institute for Theoretical Chemistry, The University of Texas at Austin, Austin, TX

Jeffrey R. Errington, Chemical and Biological Engineering, University at Buffalo, Buffalo, NY

Understanding transport coefficients such as the diffusivity, viscosity, and thermal conductivity of a fluid under confinement is valuable for both scientific research and engineering design. Empirical scaling relationships have emerged as a promising approach for predicting such quantities. In fact, recent studies have demonstrated that entropy- scaling relations provide a robust means to describe the dynamics of bulk and confined atomistic fluids. In this work we examine the extent to which these ideas can be used to characterize molecular fluids under confinement. Through the use of molecular dynamics and transition-matrix Monte Carlo simulations we study how confinement between atomic structured walls modifies the relationships between transport (translational and rotational diffusivity, characteristic relaxation times) and thermodynamic (excess entropy) properties of a dumbbell model.

See more of this Session: Thermodynamics at the Nanoscale - II
See more of this Group/Topical: Engineering Sciences and Fundamentals

Toward Optimal Design of Compound Semiconductor Nanocrystals: First-Principles Analysis, Monte Carlo Simulations, and Experimental Synthesis and Characterization

Sumeet C. Pandey, Department of Chemical Engineering, University of Massachusetts, Amherst, MA

Tejinder Singh, Department of Chemical Engineering, University of Massachusetts, Amherst, MA

T. J. Mountziaris, Department of Chemical Engineering, University of Massachusetts, Amherst, MA

Dimitrios Maroudas, Department of Chemical Engineering, University of Massachusetts, Amherst, MA

Crystalline semiconductor nanocrystals with dimensions over the range from 2 to 10 nm (quantum dots) exhibit size-dependent luminescence due to quantum confinement of excitons. This leads to unprecedented tunability in band gap, which can be controlled by varying the composition, morphology, and size of the nanocrystals. Based on optimization of band gap by such controlled variables, nanocrystals can be designed for pertinent photovoltaic applications, light-emitting devices, and as luminescent biological labels. Synthesis of core/shell quantum-dot structures produces materials that are more stable against photo-oxidation and have improved photoluminescence (PL) efficiencies. Various colloidal synthesis routes to core/shell structures exist, including the coating of a wider-band-gap semiconductor core with a shell of a narrower-band-gap material in a two-step process. However, one-step processes for synthesizing such core/shell structures remain elusive.

In this presentation, we report theoretical and experimental results toward understanding the underlying physics that governs such core/shell structure formation and can lead to one-step synthesis strategies. Specifically, we address the problem of equilibrium surface segregation in ternary II-VI and III-V semiconductor nanocrystals using Monte Carlo (MC) and conjugate gradient (CG) methods based on classical force fields in conjunction with first-principles density functional theory (DFT) calculations. We have conducted MC/CG simulations for structural and compositional relaxation of $In_xGa_{1-x}As$ and $ZnSe_{1-x}S_x$ slab supercells exposing two free surfaces with either [001] or [110] crystallographic orientation. For the simulations, we have used properly modified/extended parameterizations of the valence-force-field (VFF) description of directional bond stretching and bending. The parameterizations were validated by comparisons with DFT calculations of energy differences for various configurations of $In_xGa_{1-x}As$ and $ZnSe_{1-x}S_x$ unreconstructed slabs at different values of the compositional parameter x. The VFF models were then used to analyze surface segregation phenomena and determine the resulting equilibrium concentration profiles in ternary crystalline nanoparticles with well-defined facets and diameters $d > 1.7$ nm. Our relaxation method consists of a multi-stage sequence that includes MC sweeps employing exchanges between In and Ga atoms in $In_xGa_{1-x}As$ structures and between Se and S atoms in $ZnSe_{1-x}S_x$ structures, followed by many continuous-space MC sweeps over all atoms for structural relaxation with an MC step for strain/volume relaxation of the slabs/particles after each such sweep; in all of the above stages, trials are accepted or rejected according to the Metropolis criterion. The MC simulation is preceded and followed by energy minimization according to a CG scheme to account for local structural relaxation. Our DFT calculations were carried out within the generalized gradient approximation (GGA) and the local density approximation (LDA) and employed the projector-augmented-wave method and slab supercell models. We calculated surface energies, model surfactant binding energies, and segregation tendencies for the passivated and unpassivated $In_xGa_{1-x}As$ and $ZnSe_{1-x}S_x$ slabs with[001] and [110] crystallographic orientations. In our experiments, we synthesized nanocrystals using injection of organometallic precursors into hot coordinating solvents. The nanocrystal synthesis was carried out under vacuum and in the presence of argon to avoid reaction with oxygen. The synthesized nanocrystals were then dispersed in butanol.

We present simulation results for the equilibrium concentration distributions in slabs of $In_xGa_{1-x}As$ and $ZnSe_{1-x}S_x$ as a function of composition, x, slab thickness, and slab surface crystallographic orientation, as well as in $In_xGa_{1-x}As$ and $ZnSe_{1-x}S_x$ nanocrystals with well-defined [001] and [110] surface facets as a function of x and nanocrystal size. The results identify the parameter range for the assembly of core/shell structures, such as an In-deficient core surrounded by an In-rich shell in $In_xGa_{1-x}As$ nanocrystals. We also report PL spectra, transmission-electron-microscopy (TEM) images, and X-ray photoelectron spectra (XPS) of our experimentally synthesized nanocrystals. The underlying surface segregation is analyzed in particle-size and composition space and its impact on the experimental one-step synthesis of core/shell nanocrystal structures is discussed.

See more of this Session: Thermodynamics at the Nanoscale - II
See more of this Group/Topical: Engineering Sciences and Fundamentals

Density Functional Theory for the Structure and Solvent Capacity of Nanoparticle-Organic Hybrid Materials

Hsiu-Yu Yu, School of Chemical and Biomolecular Engineering, Cornell University, Ithaca, NY

Donald Koch, School of Chemical and Biomolecular Engineering, Cornell University, Ithaca, NY

Nanoparticle-organic hybrid materials (NOHMs) are a promising new class of materials with unique physicochemical and transport properties. Unlike most nanoparticle systems which aggregate irreversibly due to strong van der Waals attraction, these surface functionalized nanostructures can relax to an equilibrium state and show liquid-like behavior in the absence of solvent.

In a pure NOHM system, the nanoparticle cores are distributed in a homogeneous disordered structure governed by the requirement that the tethered oligomers must fill the interparticle space. We formulate a density functional theory for a system of hard cores with attached bead-spring oligomers in the absence of other solvent molecules. The oligomers are assumed to relax rapidly compared with the cores. The equilibrium concentration field of the oligomers around a core is obtained by minimizing the oligomer's free energy subject to the constraints that the field surrounding a core is normalized and the fluid number density (consisting of oligomers from the test particle and its neighbors) is independent of position outside of the excluded volume of the cores. This constraint leads to an oligomer-configurational entropy penalty for large spaces between the core particles. The static structure factor of the cores is shown to approach zero as the wave number goes to zero. This indicates a deficit of one neighboring core around a test core resulting from the fact that a core carries its own share of fluid.

Nanoparticle ionic materials (a subclass of NOHMs) with amines in the oligomeric chains are being tested experimentally as solvents for CO_2 capture. The low vapor pressure and good thermal stability of these fluids make them promising candidates for this application. We have developed a coarse-grained theory for the salvation properties of a NOHMs fluid, in which a Flory-Huggins interaction parameter describes the chemical affinity of the oligomers for the solute and the space between the cores is uniformly filled by a mixture of oligomers and solute.

See more of this Session: Thermodynamics at the Nanoscale - II
See more of this Group/Topical: Engineering Sciences and Fundamentals

Dynamic Light Scattering in An Aqueous Solution of 3-Methylpyridine

Dimitry A. Ivanov, Department of Chemical and Biomolecular Engineering, University of Maryland, College Park, MD

Mikhail A. Anisimov, University of Maryland, College Park, MD

Jan V. Sengers, University of Maryland, College Park, MD

Andrei F. Kostko, University of Maryland, College Park, MD

We report dynamic light-scattering measurements obtained in an aqueous solution of 3-methylpyridine. The dynamic correlation function of this solution appears to reveal two distinct physical relaxation modes: one associated with a normal molecular diffusion process and another one associated with mesoscopic network relaxation. To obtain further insight into the nature of the apparent mesoscopic structures in the solution we have made systematic studies of the nature of the observed time and wave-number dependence of the dynamics of this mesoscopic relaxation mode as a function of temperature and concentration.

1Research supported by the Petroleum research Fund of ACS.

See more of this Session: Thermodynamics at the Nanoscale - II
See more of this Group/Topical: Engineering Sciences and Fundamentals

Reentrant Surfactant Driven Isotropic-Nematic Transitions in Thin Films

Mark Uline, Department of Biomedical Engineering, Northwestern University, Evanston, IL

Igal Szleifer, Department of Biomedical Engineering, Northwestern University, Evanston, IL

The control of the orientation of nematogens at surfaces and interfaces is of primary importance in the development of liquid crystalline devices as well as many applications in biological systems. One of the ways that the interfacial orientation is controlled is through using surfactant and lipid molecules [1]. Experimental observations have shown that the anchoring ability of surfactants is a non-monotonic function of their surface density [2]. The interplay between the conformational entropy of the chain molecules, the rotational entropy of the nematogens, the intermolecular and molecule-surface interactions, and the packing at the interfaces results in an inhomogeneous molecular distribution close to the interface with complex effective interactions that have large implications for the phase behavior of the nematogen and surfactant molecules.

We present theoretical predictions for the phase behavior of nematogen thin films in the presence of surfactants. The results shed light on the interplay between the conformational entropy of the surfactants, the penetration of the nematogen into the surfactant region, the propagation of the interfacial orientation to the phase behavior of the nematogenic film and the effects of nematogen orientation on the phase transitions of the surfactants themselves. We show that at low surface density of surfactants the nematogens prefer to orient parallel to the surface due to excluded volume interactions. At intermediate surface densities the surfactant induces a homotropic (perpendicular) orientation, where the nematogens penetrate slightly into the surfactant layer. When the surfactant surface is large enough so that the surfactants are highly stretched (since the surfactant chains are self avoiding), the orientation becomes parallel again since the coating is almost equivalent to a hard wall. These orientational effects have important consequences in the isotropic-nematic transition, where we predict a reentrant isotropic-nematic transition that depends on film thickness and surfactant surface density. We also present results on the effect of the presence of nematogens on the phase behavior of the surfactant molecules when there is an additional interaction included that favors the chains aligning perpendicular to the surface [3]. This interaction gives rise to a gel/liquid crystalline phase transition of the surfactant that is shifted due to the presence and concentration of the nematogens.

References

[1] J. M. Brake, M. K. Daschner, Y.-Y. Luk, and N. L. Abbott, Science 302, 2094 (2003).

[2] U. Kühnau, A. Petrov, G. Klose, and H. Schmiedel, Phys. Rev. E 59, 578 (1999).

[3] R. Elliott, K. Katsov, M. Schick, and I. Szleifer, J. Chem. Phys. 122, 044904 (2005).

See more of this Session: Thermodynamics at the Nanoscale - II
See more of this Group/Topical: Engineering Sciences and Fundamentals

Which Factors Control the Onset of the Fluid-Solid Transition of a Nanoconfined Fluid?

Hugh Docherty, Department of Chemical and Biomolecular Engineering, Vanderbilt University, Nashville, TN

Peter T. Cummings, Department of Chemical and Biomolecular Engineering, Vanderbilt University, Nashville, TN

It is well known that the behavior of fluids confined to the order of a few nanometers may differ greatly from that of the corresponding bulk fluid.[1] For example, in the 1980s and early 1990s a large number of studies were reported on a variety of ultrathin liquid films confined between mica surfaces.[2] Although the fluids investigated ranged from linear alkanes[3] to cyclical[4] and globular5 molecules, they exhibited one common feature: When their confinement reached the order of several molecular diameters, a rapid many-orders-of-magnitude increase in the viscosity of the confined fluid was observed, together with behavior typical of the stick-slip response of a crystalline solid structure.[6] Following this work, attention shifted to the nature of the transition from fluid-like to solid-like behavior. Unfortunately, in this case, a consensus has not been achieved and, for over a decade, there has been intense debate as to whether it is a first-order (crystallization) or second-order (vitrification) order phase transition.

Previously, we presented initial results of a computer simulation study of fluids (dodecane and cyclohexane) confined between mica sheets, making use of the latest atomistically-detailed fully-flexible molecular models.[7] These results provided compelling evidence that the transition is consistent with a first-order phase transition. Here we build on that work by applying the same highly-detailed molecular models to an investigation of factors that may explain the experimental discrepancies underlying the intense debate surrounding the effects of nanoconfinement. In particular, we consider the effects of surface shear, contamination of the confined fluid and the relative alignment of the confining surfaces.

[1] Van Alsten, J. and Granick, S., *Langmuir* **6**, 876 (1990).

[2] Alba-Simionesco, C. et al., *Journal of Physics-Condensed Matter* **18**, R15 (2006).

[3] Christenson, H. K., Gruen, D. W. R., Horn, R. G., and Israelachvili, J. N., *Journal of Chemical Physics* **87**, 1834 (1987).

[4] Gee, M. L., Mcguiggan, P. M., Israelachvili, J. N., and Homola, A. M., *Journal of Chemical Physics* **93**, 1895 (1990).

[5] Israelachvili, J. et al., *Journal of Physics-Condensed Matter* **2**, Sa89 (1990).

[6] Klein, J. and Kumacheva, E., *Journal of Chemical Physics* **108**, 6996 (1998).

[7] Docherty, H and Cummings P.T., "*When Nano-Confined Between Mica Sheets, Does Dodecane Undergo a Phase Transition?*" AIChE Annual Meeting 2008, Philadelphia, PA.

See more of this Session: Thermodynamics at the Nanoscale - II
See more of this Group/Topical: Engineering Sciences and Fundamentals

Dielectric saturation of liquid propylene carbonate in electrical energy storage applications

Lu Yang* and Lawrence R. Pratt†

Department of Chemical and Biomolecular Engineering,

Tulane University, New Orleans, LA 70118

(Dated: September 1, 2009)

Abstract

Dielectric characteristics of a molecular model of liquid propylene carbonate are evaluated for utilization in molecular scale simulation of electrochemical capacitors based on nanotube forests. The linear-response dielectric constant of the bulk liquid, and its temperature dependence, is in good agreement with experiment. Dielectric saturation is studied by simulations with static uniform electric fields as large as 4 V/nm. The observed polarization is well-described by the Langevin equation with the low-field/high-field cross-over parameter of 0.07 V/nm. Simulation of liquid propylene carbonate confined between charged parallel graphite electrodes yields a capacitance that depends on the electric potential difference across those thin films. An effective dielectric constant inferred from the capacitance is significantly less than the uniform liquid dielectric constant, but is consistent with the non-linear dielectric response at the strong fields applied to the electrode film. Those saturation effects reduce the weak-field capacitance.

*Current address: Applied Materials, Santa Clara CA 95052; Electronic address: `luckyang@gmail.com`

†Electronic address: `lpratt@tulane.edu`

Carbon nanotube forests have been proposed as electrodes for electrochemical double layer capacitors.[1] The confined electrolyte solution would be an integral part of these systems, and propylene carbonate is a popular solvent choice. Validation of the dielectric characteristics of liquid propylene carbonate without salt would help in considering recent and anticipated simulations.[2] A previous simulation study of uniform liquid propylene carbonate analyzed X-ray scattering results,[3] but not dielectric characteristics. We therefore focus this report on the dielectric characteristics which have not been treated before now. Understanding the dielectric response of those solvents under substantial applied electric fields would also assist the selection of solvents for electrochemical-double-layer capacitors.

I. METHODS

To study the dielectric characteristics of propylene carbonate (PC), we performed all-atom MD simulations of the uniform bulk liquid, with and without a static uniform electric field, and in thin films confined between oppositely charged parallel graphite electrodes. The intermolecular interactions between PC molecules and between PC and graphitic carbon atoms are taken to be the same as in the preceding work on nanotube forests.[2] The molecular dipole moment of the optimized PC molecule is 6.47 D, compared with the experimental values are 4.33 D,[4] 4.98 D,[5] 5.0 D,[6] and 5.36 D.[7] This is consistent with the usual simulation situation where the molecular dipole moment for a *nonpolarizable* molecular model is larger than the experimental dipole moment of the isolated molecule, the difference reflecting a generic medium effect that would derive from molecular polarizabilities if they were to be included. The partial charges of graphitic carbon atoms were set by hand to establish the desired charge state of the graphite electrodes. Periodic boundary conditions were applied for all of the simulations and electrostatic interactions were calculated using the particle mesh Ewald method with a grid spacing of 1 Å. These simulations used the AMBER9 package,[9] except for the calculations that treated the bluk liquid under a uniform external field. In the latter case the NAMD package[11] was used.

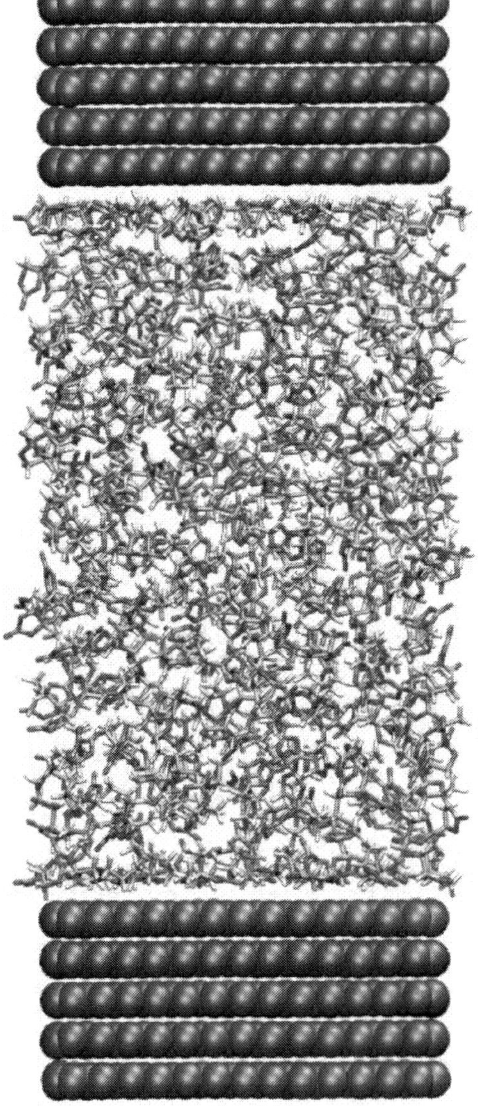

FIG. 1: Simulation geometry and configuration of propylene carbonate sampled from the simulation of the discharged electrodes.

II. RESULTS

Radial distribution functions for the uniform liquid are shown in Fig. 2 which are comparable to previous work.[3] Evaluation of the dielectric constant of the bulk liquid

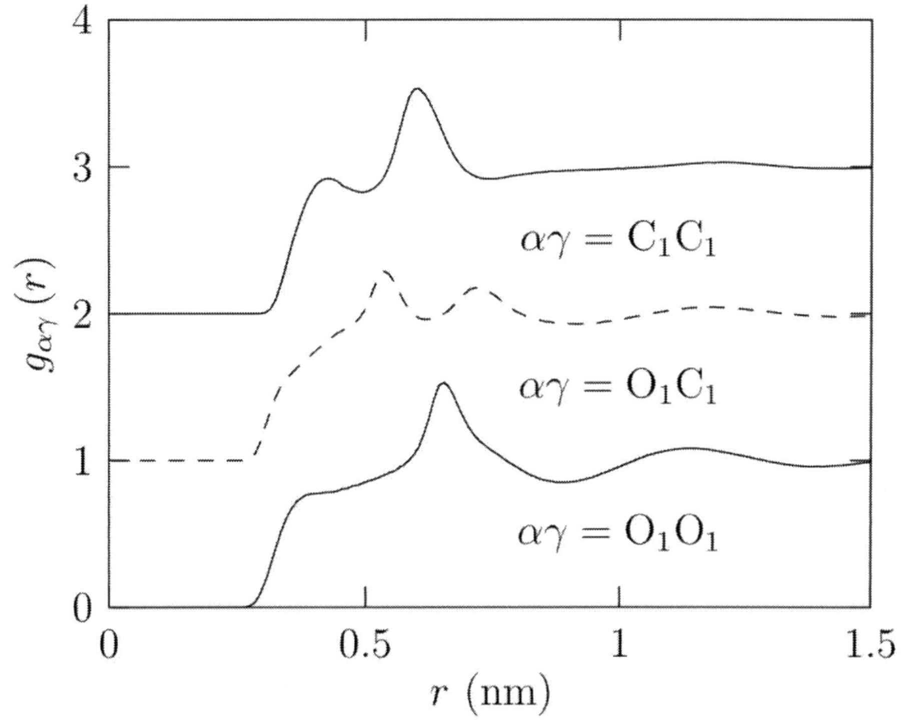

FIG. 2: Observed radial distribution functions for the uniform liquid; see Fig. ??.

at p=1 atm and T = 300K gave 60.1±8.6, about 7% below the experimental value.[15] The analogous result at T = 350K is 49.2±5.4. These results lead to the estimate that $(\partial \ln(\varepsilon/\varepsilon_0)/\partial T)_p \approx -4 \times 10^{-3}\mathrm{K}^{-1}$ which is about 9% too large.[15] Calculations at a series of pressures up to 1 kbar determined the compressibility of this simulation model as 0.26 GPa^{-1}, which is 56% below the experimental compressibility (0.59 GPa^{-1}).[16]

The dependence of the dielectric constant on electric field strength is shown in FIG. 3. The Langevin equation[17] works satisfactorily, and the boundary between low and high-field behavior is found to be about $E_0 \approx 0.07$ V/nm.

A net capacitance can be defined as the ratio of charge for the case considered to the net electrostatic potential change ϕ from well below the lower electrode to well above the upper electrode. Fig. 4 shows the net capacitance of these thin films as it depends on ϕ. This demonstrates that the linear-response dielectric constant does not give the capacitances for Fig. 4. This is due to dielectric saturation and the interfacial structure, which contributes non-negligibly to these electrical characteristics particularly for weak fields. For a substantial

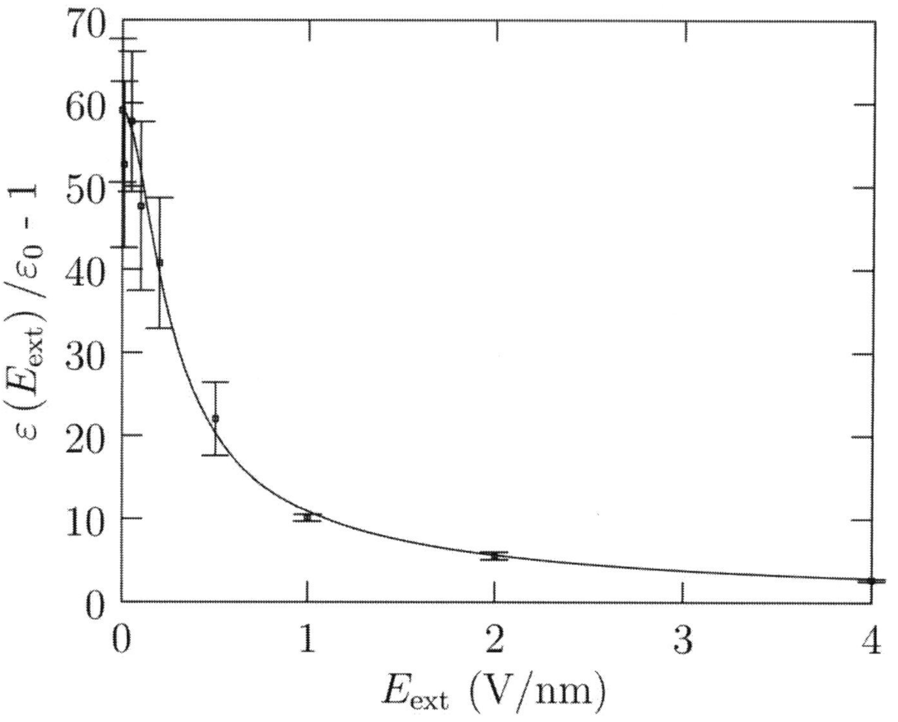

FIG. 3: $\varepsilon\left(E_{\text{ext}}\right)/\varepsilon_0 - 1$ as a function of external electric field strength. The curve is the fitted model $3\left[\frac{\varepsilon(0)}{\varepsilon_0}-1\right] \times \left(\coth(\frac{E_{\text{ext}}}{E_0}) - \frac{E_0}{E_{\text{ext}}}\right) E_0/E_{\text{ext}}$ with $E_0 = 0.07$ (V/nm).[17–20] The error bars indicate a 95% confidence interval.

potential change ϕ over nanometer length-scales, the relevant electric fields are large and a smaller dielectric constant $\varepsilon/\varepsilon_0$ should be used to estimate the capacitance. For example, to obtain a capacitance of 1 μF/cm^2 for the PC thin film between graphite electrodes, a value of about $\varepsilon/\varepsilon_0 \approx 7$ is required. The point of FIG. 4 near $C(\phi) = 1.0\ \mu\text{F/cm}^2$ is $\phi \approx 1.0$ V. This corresponds to the case of total charge on an electrode of $0.5e$, and $E_{\text{ext}} \approx 1$ V/nm. Then consulting FIG. 3, we see that $\varepsilon/\varepsilon_0 \approx 10$, about 40% higher than the required 7. This comparison is not trivial because the intrinsic polarizations of the two distinct interfaces, *e.g.*, on one side the external field is oriented into the liquid film, but on the other side it is oriented out of the film.

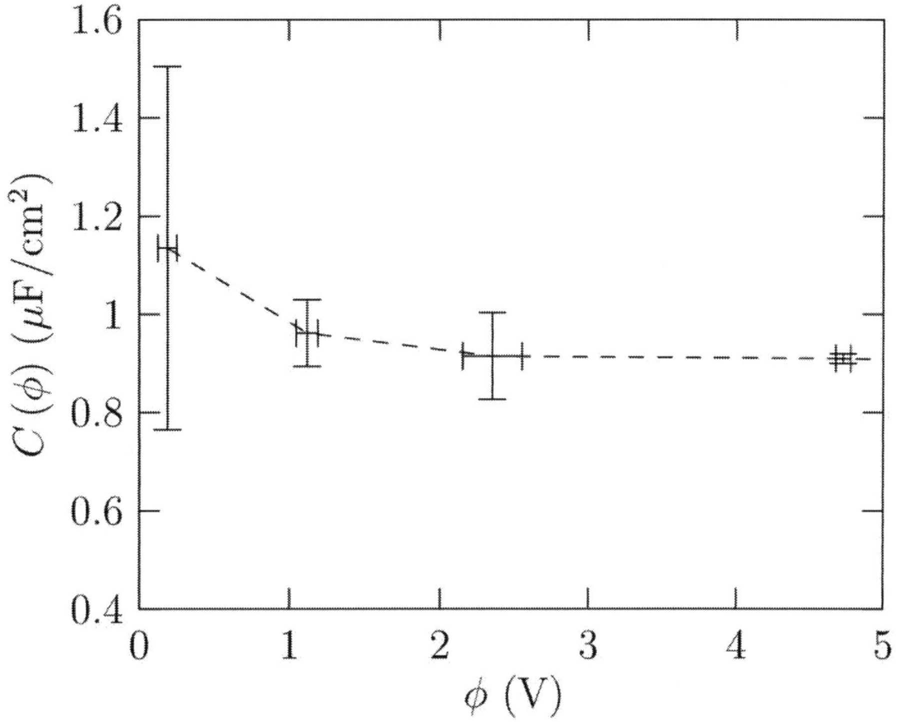

FIG. 4: Net capacitance as it depends on ϕ, the potential difference between the electrodes.

III. CONCLUSIONS

The model of liquid propylene carbonate utilized here produces a bulk dielectric constant in good agreement with experiment, including its temperature dependence. The Langevin equation satisfactorily describes the static non-linear polarization response with the low-field/high-field cross-over parameter of 0.07 V/nm. Strong electric fields within the thin propylene carbonate film reduce the capacitance from the weak-field value that utilizes the bulk liquid linear-response dielectric constant. The non-linear dependence of the dielectric constant on electric field strength permits a reasonable but rough estimate of that reduced capacitance.

---

[1] Futaba, D. N.; Hata, K.; Yamada, T.; Hiraoka, T.; Hayamizu, Y.; Kakudate, Y.; Tanaike, O.; Hatori, H.; Yumura, M.; Iijima, S. *Nature Materials* **2006**, *5*, 987–994.

[2] Yang, L.; Fishbine, B. H.; Migliori, A.; Pratt, L. R. *J. Am. Chem. Soc.* **2009**, 1–4.

[3] Soetens, J. C.; Millot, C.; Maigret, B.; Bako, I. *J. Mol. Liq.* **2001**, *92*, 201 – 216.

[4] Nanbu, N.; Takimoto, K.; Suzuki, K.; Ohtake, M.; Hagiyama, K.; Takehara, M.; Ue, M.; Sasaki, Y. *Chem. Letts.* **2008**, *37*, 476 – 477.

[5] McDonagh, P. M.; Reardon, J. F. *J. Soln. Chem.* **1990**, *19*, 301 – 314.

[6] Jorat, L. J.; Noyel, G. A.; Huck, J. R. *IEEE Transactions on Electrical Insulation* **1991**, *26*, 763 – 9.

[7] Chernyak, Y. *J. Chem. Eng. Data* **2006**, *51*, 416 – 418.

[8] Barthel, J.; Utz, M.; Groß, K.; Gores, H.-J. *J. Sol. Chem.* **1995**, *24*, 1109 – 1123.

[9] Case, D. et al. *University of California, San Francisco.* **2006**.

[10] Wu, Y.; Tepper, H. L.; Voth, G. A. *J. Chem. Phys.* **2006**, *124*, 024503.

[11] Kale, L.; Skeel, R.; Bhandarkar, M.; Brunner, R.; Gursoy, A.; Krawetz, N.; Phillips, J.; Shinozaki, A.; Varadarajan, K.; Schulten, K. *J. Comp. Phys.* **1999**, *151*, 283.

[12] Izaguirre, J. A.; Catarello, D. P.; Wozniak, J. M.; Skeel, R. D. *J. Chem. Phys.* **2001**, *114*, 2090 – 8.

[13] Berendsen, H. J. C.; Postma, J. P. M.; Vangunsteren, W. F.; Dinola, A.; Haak, J. R. *J. Chem. Phys.* **1984**, *81*, 3684 – 90.

[14] Ryckaert, J. P.; Ciccotti, G.; Berendsen, H. J. C. *J. Comput. Phys.* **1977**, *23*, 327–341.

[15] Côte, J.-F.; Brouillette, D.; Desnoyers, J. C.; Rouleau, J.-F.; St-Arnaud, J.-M.; Perron, G. *J. Soln. Chem.* **1996**, *25*, 1163.

[16] Marcus, Y.; Hefter, G. T. *J. Mol. Liq.* **1997**, *73*, 61–74.

[17] Booth, F. *J. Chem. Phys.* **1951**, *19*, 391–394.

[18] Yeh, I.-C.; Berkowitz, M. L. *J. Chem. Phys.* **1999**, *110*, 7935.

[19] Dzubiella, J.; Hansen, J.-P. *J. Chem. Phys.* **2005**, *122*, 234706.

[20] Apol, M.; Amadei, A.; Nola, A. D. *J. Chem. Phys.* **2002**, *116*, 4426 – 4436.

[21] Landau, L. D.; Lifshitz, E. M. *Electrodynamics of Continuous Media*; Pergamon, Oxford, 1975; Course of Theoretical Physics, Volume 8, §6.

[22] Jackson, J. D. *Classical Electrodynamics*; John Wiley & Sons, Inc., New York, 1975; §6.7.

tex

Role of Sedimentation in the Assembly of Colloidal Crystals of Oppositely Charged Particles

Reginald E. Rogers Jr., Chemical Engineering, University of Michigan, Ann Arbor, MI

Michael J. Solomon, Chemical Engineering, University of Michigan, Ann Arbor, MI

Ionic colloidal crystals are long-range ordered structures formed from the finely tuned interaction of charged colloids carrying opposite charges [1]. These structures have been self-assembled under near-equilibrium conditions where the solvent is refractive index and near density-matched to the particles. The allowable range of attractions is apparently limited. For example, if the attraction is too great, ionic gels form rather than crystals. Ionic colloidal crystals hold promise in applications ranging from photonic bandgap to chemical and biological sensing materials. Therefore, a full understanding of the extent to which this type of colloidal crystal structure can be formed would benefit a range of potential applications.

We investigate the role of an applied field, gravity, on ionic colloidal crystal formation through study of their sedimentation. We find that the final structure of the ionic colloidal crystal depends on the sedimentation velocity, initial volume fraction, and interparticle interactions. The strength of the gravitational field relative to the randomizing effects of Brownian motion plays a key role in this process and its effect can be quantified by the dimensionless group, Peclet number. Using poly(diphenyl-dimethyl) siloxane stabilized poly(methyl methacrylate) [DPDM-PMMA] particles (positively charged) mixed with poly-12-hydroxystearic acid (PHSA) stabilized PMMA particles (negatively charged), we evaluate the quality of ionic colloidal crystals with cesium chloride (CsCl) structure formed in a solvent system consisting of cylcohexyl bromide (CHB), decalin and tetrabutylammonium chloride (TBAC) salt added for charge screening. The three-dimensional structure of these crystals is characterized by confocal laser scanning microscopy. By varying the ratio of CHB to decalin, we manipulate the sedimentation rate, and subsequently the Peclet number, of the assembly process. We show that a range of Peclet values exists where ionic colloidal crystallization is observed.

[1] Leunissen, M.E., C.G. Christova, A.P. Hynninen, C.P. Royall, A.I. Campbell, A. Imhof, M. Dijkstra, R. van Roij, and A. van Blaaderen, Ionic colloidal crystals of oppositely charged particles. Nature, 2005. 437(7056): p. 235-240.

See more of this Session: Colloidal Assembly and Fabrication III
See more of this Group/Topical: Engineering Sciences and Fundamentals

Convective Deposition of Binary Suspensions

Pisist Kumnorkaew, Lehigh University, Bethlehem, PA

Alexander L. Weldon, Lehigh University, Bethlehem, PA

James F. Gilchrist, Lehigh University, Bethlehem, PA

We investigate the coupling between the suspension properties and the deposition process during convective deposition of aqueous binary suspensions of colloidal microspheres and nanoparticles. The microsphere monolayer structures formed from this rapid and scalable process have use in a variety of optical, chemical, and biochemical sensing applications. Under ideal conditions, co-deposition of two monosized species creates long range order of the larger species embedded within the nanoparticles when the size ratio of particles is sufficiently large. However, when altering the volume fractions of the two species, instabilities arise that lead to modulation of surface morphology and localized order. At the highest concentration of nanoparticles, the larger species is destabilized via depletion interactions that also strongly interfere with the deposition process through agglomeration and altering the particle-substrate affinity. Direct measurements of deposition are visualized via high speed confocal laser scanning microscopy to generate local dynamic information regarding the deposition as well as the long range structure of assembled microsphere coatings.

See more of this Session: Colloidal Assembly and Fabrication III
See more of this Group/Topical: Engineering Sciences and Fundamentals

Coating Regimes for Controlled Convective Assembly of Monolayered Nanoparticle Films

J. Alex Lee, Department of Chemical Engineering and Materials Science, University of Minnesota, Minneapolis, MN

Damien Brewer, Department of Chemical Engineering and Materials Science, University of Minnesota, Minneapolis, MN

Michael Tsapatsis, Department of Chemical Engineering and Materials Science, University of Minnesota, Minneapolis, MN

Convective assembly with controlled evaporation and controlled substrate withdrawal (i.e. in a dip-coating configuration) is well-accepted as a facile method for deposition or coating of colloidal films. The interplay of bulk particle concentration, substrate withdrawal rate, and to a lesser degree evaporation control (by controlled laminar gas flow similar in configuration to air-knife), are shown to mediate film thickness at the thick multilayer regime. At the thinner film regime, the concept of ``thickness" must be recast in terms of ``coverage" of discrete patterned films with void spaces. Even in situations where particle coverage is in excess of that required for complete single-layer coverage, resultant films are often discrete and patterned. We discuss, therefore, the identification of coating windows for which the method can achieve uninterrupted monolayers with a close-packed structure by tuning bulk particle concentration and substrate withdrawal rate.

See more of this Session: Colloidal Assembly and Fabrication III
See more of this Group/Topical: Engineering Sciences and Fundamentals

Colloidal Self-Assembly Templated by Fluid Interfaces

Bum Jun Park, Department of Chemical Engineering, University of Delaware, Newark, DE

Eric M. Furst, Department of Chemical Engineering, University of Delaware, Newark, DE

We present a method for assembling close-packed two-dimensional colloidal crystals by decreasing the height of a fluid oil-water interface. We show that the self-assembly is driven by lateral capillary interactions caused by the overlap of the deformed interfaces above the particles, and furthermore, that this deformation is a result of the disjoining pressure between the charged colloids and the interface. The rigidity of the self-assembled crystals is determined by the strength of the disjoining pressure, resulting in mechanical properties that range from reversible soft-crystals to solid-like crystals. Once the interface contacts the particles, the particles spontaneously transfer onto the interface due to the high three-phase contact angle. This final interfacial structure is consistent with long-range electrostatic interactions between particles confined to an oil-water interface (i.e. open hexagonal arrays). [1]

[1] B. J. Park, et al., Langmuir, 24:1686-1694, 2008.

See more of this Session: Colloidal Assembly and Fabrication III
See more of this Group/Topical: Engineering Sciences and Fundamentals

Measuring the Elasticity of Semiflexible DNA-Linked Colloidal Chains Via Thermal Fluctuations

Dichuan Li, Chemical and Biomolecular Engineering, Rice University, Houston, TX

Sibani Lisa Biswal, Chemical and Biomolecular Engineering, Rice University, Houston, TX

Flexibility is a crucial parameter in determining the equilibrium configuration of microstructures. In this talk, we will focus on the influence of variable flexibility on the shape and morphology of DNA-linked colloidal chains. These multilink chains are comprised of magnetic colloidal particles which are linked together with DNA.

The flexural rigidity, or bending stiffness, is used to describe the chain's resistance to bending forces. We have previously shown that the properties of the linker molecules greatly influence the overall chain flexibility1. In particular, by adjusting the length of the linker molecule, we can tune the overall flexibility of the chain. We have recently constructed a technique to measure the flexural rigidity of a chain by observing its Brownian motion in an aqueous solvent. The measured flexural rigidities unveil a wealth of information about properties of linker molecules and the linking kinetics.

In this talk we will show that these linked colloidal chains often result is a chain with varying flexural rigidity along its backbone. We will compare these results with those of chain assumed to be a linear homogenous elastic filament2-4. We will also elucidate how the properties of the linker molecules and linking kinetics affect the overall chain elasticity.

1. Biswal SL, Gast AP. Mechanics of semiflexible chains formed by poly(ethylene glycol)-linked paramagnetic particles. Physical Review E 2003; 68: 021402/1-/9.

2. Goubault C, Jop P, Fermigier M et al. Flexible magnetic filaments as micromechanical sensors. Phys Rev Lett 2003; 91: 4.

3. Cebers A. Flexible magnetic filaments. Current Opinion in Colloid & Interface Science 2005; 10: 167-75.

4. Cebers A, Cirulis T. Magnetic elastica. Phys Rev E 2007; 76.

See more of this Session: Colloidal Assembly and Fabrication III
See more of this Group/Topical: Engineering Sciences and Fundamentals

Colloidal Self-Assembly Via DNA Hybridization: Particle Heterogeneity and Nucleation Rates

Marie T. Ung, Department of Chemical and Biomolecular Engineering, University of Pennsylvania, Philadelphia, PA

Raynaldo Scarlett, Department of Chemical and Biomolecular Engineering, University of Pennsylvania, Philadelphia, PA

Talid R. Sinno, Department of Chemical and Biomolecular Engineering, University of Pennsylvania, Philadelphia, PA

John C. Crocker, Department of Chemical and Biomolecular Engineering, University of Pennsylvania, Philadelphia, PA

The crystallization of colloids and nanocolloids due to the specific interaction of DNA hybridization is investigated. Ordered nanocomposites produced using the sequence-specific binding properties of DNA are expected to possess useful optical and electronic properties. We find that the crystal nucleation rate is highly sensitive to sphere-to-sphere variation in DNA density. Earlier experiments crystallizing micron sized colloids would not have nucleated were it not for such variation. Uniformly labeled smaller colloids have a finite classical nucleation rate; variation in labeling leads to a much higher nucleation and correspondingly smaller crystallites. This has been confirmed by simulation at finite rate which predicts that DNA labeled colloids at 980nm should not crystallize. This simulation also confirms that the nucleation barrier decreases with smaller particle size.

See more of this Session: Colloidal Assembly and Fabrication III
See more of this Group/Topical: Engineering Sciences and Fundamentals

Design of Patchy Particles Coated with Ternary Self-Assembled Monolayers

Ines C. Pons, Chemical Engineering, University of Michigan, Ann Arbor, MI

Chetana Singh, Chemical Engineering, University of Michigan, Ann Arbor, MI

Hao Jiang, Chemical Engineering, University of Michigan, Ann Arbor, MI

Sharon C. Glotzer, Chemical Engineering, University of Michigan, Ann Arbor, MI

Recent simulations have studied the formation of patterns in a binary mixture of immiscible surfactants adsorbed on the surface of a spherical nanocolloid[1]. The resulting patterns (patches and stripes) were in good agreement with experimental results [2]. Here we perform dissipative particle dynamics (DPD) simulations to study the patterns obtained by adding a third component to the mixture of surfactants. We present results demonstrating the effect of varying nanocolloid size, stoichiometry of the ternary mixture, degree of surfactant immiscibility, and length difference between surfactants.

[1] C Singh, PK Ghorai, MA Horsch, AM Jackson, RG Larson, F Stellacci and SC Glotzer, "Entropy-Mediated Patterning of Surfactant-Coated Nanoparticles and Surfaces", Physical Review Letters 99, 226106 (2007).

[2] AM Jackson, JW Myerson, F Stellacci, "Spontaneous assembly of subnanometre-ordered domains in the ligand shell of monolayer-protected nanoparticles", Nature Materials, 3, 330-336 (2004).

See more of this Session: Colloidal Assembly and Fabrication III
See more of this Group/Topical: Engineering Sciences and Fundamentals

Rheology of Dilute Suspensions of Fractal Clusters with Arbitrary Structure

Yogesh M. Harshe, Chemistry and applied biosciences, Institute for chemical and bioengineering, ETH Zurich, Zurich, Switzerland

Marco Lattuada, Institute for Chemical and Bioengineering, Department of Chemistry and Applied Biosciences, ETH Zurich, Zurich, 8093, Switzerland

Massimo Morbidelli, Department of Chemistry and Applied Biosciences, Institute for Chemical and Bioengineering, Zurich, Switzerland

Abstract:

The viscosity of dispersions containing particulate solid has been studied in the past, however, specifically only for well-defined geometries of the particles. The well studied geometrical shapes predominantly encompass spheres, and prolate or oblate spheroids. In practice the processing of solid suspensions, and in particular coagulation of dispersions of colloidal particles, leads to the formation of clusters with a great diversity of shapes and sizes. Several works in the literature pointed out that the viscosity of suspension strongly depends on the geometry of the contained clusters, their concentration in the suspension, their orientation with respect to the flow, and the magnitude of the applied shear. Little literature has been published since Einstein's pioneering work on the dependence of viscosity of suspensions on clusters with arbitrarily shape and structure.

In the present work, we have developed a generalized framework to account for the contribution to the viscosity of suspensions of clusters, made of identical spherical particles, with arbitrary shape and structure. A wide database of clusters has been created, which have been generated through a combination of different Monte-Carlo methods, and covers a broad range of fractal dimension values, from 1.8 to 3, as well as aspect ratios, and number of particles. In order to determine their hydrodynamic properties, Stokesian Dynamics was implemented and used to estimate the grand resistance matrix of these clusters, treated as rigid bodies. The grand resistance matrices of several clusters with the same mass and fractal dimension have been averaged, in order to account for the variability in the structure of clusters, and to create an average representative cluster with given properties. These average matrices have been used to compute the clusters' first order contribution to the orientation averaged bulk stress tensor, which was obtained by performing Brownian dynamic simulations on individual clusters, including coupling between their rotational and translational motion. The viscosity of dilute suspension was obtained from the average bulk stress tensor for different flow field configurations, and shear rate values.

Keywords:

Suspension, Stokesian dynamics, rigid body

See more of this Session: Colloidal Hydrodynamics II
See more of this Group/Topical: Engineering Sciences and Fundamentals

Shear-Induced and Brownian Coagulation

Max L. Eggersdorfer, Particle Technology Laboratory, Institute of Process Engineering, Deparment of Mechanical and Process Engineering, ETH Zurich, Zurich, Switzerland

Sotiris E. Pratsinis, Particle Technology Laboratory, Institute of Process Engineering, Department of Mechanical and Process Engineering, ETH Zurich, Zurich, Switzerland

Smoluchowski (1917) developed collision frequencies for these two cases of coagulation. One case describes the collision of particles by Brownian motion (perikinetic coagulation) and the other neglects diffusion as particle collisions are induced by the surrounding fluid flow (orthokinetic coagulation). In both cases interparticle forces are neglected. When both mechanisms are present, the combination affects the collision rate and the resulting particle size distribution.

Swift & Friedlander (1964) concluded that the collision frequencies for Brownian motion and laminar shear are independent and additive for a wide range of conditions. Contradictory Van de Ven & Mason (1977) showed by solving the convective diffusion equation for conditions with a low influence of shear (Pe<<1) that the two mechanisms are coupled and are not simply additive. Shear enhances the coagulation rate proportional to the square root of the Peclet number, which represents the ratio between the convective (inertial) and diffusive terms.

$$Pe = \frac{L \cdot V}{D},$$

where L, V and D are the characteristic length, particle velocity and diffusivity, respectively.

Feke & Schowalter (1985) studied the regime for high Peclet numbers (Pe>>1) where the dominant mechanism is shear. Their analysis of binary collisions also revealed a dependency between the contribution of Brownian motion and shear. They concluded that the combination of both can either increase or decrease the collision frequency. All of these models treat collisions between two particles by calculating the flux of a particle to a reference sphere. The flux is typically computed from the pair probability function, which gives the probability to find two particles at a certain position.

Ernst and Pratsinis (2006) showed by population balance simulations that shear-induced coagulation leads to pseudo self-preserving distributions that end to gelation. Heine and Pratsinis (2007) simulated Brownian coagulation at high concentrations with Langevin dynamics (LD) and showed that classic Smoluchowski equation can not describe highly concentrated systems (particle volume fraction above 1%). They found out that for complete particle coalescence upon collision, the self-preserving size distribution is still obtained for high concentrations.

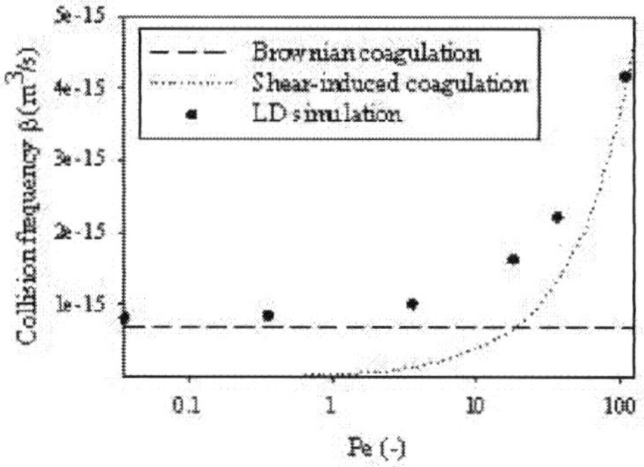

Figure 1. LD simulations for monodisperse particle systems with contributions of shear and Brownian motion.

In the limiting cases, where either peri- or orthokinetic aggregation dominates, the process of coagulation is understood well. The transition regime on the other hand is not well understood.

Here Brownian & shear-induced coagulation are investigated by LD simulations for a multi-particle system of coalescing spheres. Initially all particles are monodisperse. Simulations are done for monodisperse and polydisperse systems. In monodisperse simulations particles are redistributed after a collision. If particles collide in polydisperse simulations, full coalescence takes place. Collision frequencies and particle size distributions are obtained for a wide range of Peclet numbers, including the two limiting cases of pure shear-induced coagulation and pure Brownian coagulation.

Ernst, F. O., & Pratsinis, S. E. (2006). *J. Aerosol Sci.*, 37, 123-142.

Feke, D. L., & Schowalter, W. R. (1985). *J. Colloid Interface Sci.*, 36, 203-214.

Heine, M. C., & Pratsinis, S. E. (2007). *Langmuir*, 19, 9882-9890.

Smoluchowski, M. v., (1917). *Z. Phys. Chem*, 92, 129-&.

Swift, D. L., & Friedlander, S. K. (1964). *J. Colloid Sci.*, 19, 621-647.

van de Ven, T. G. M. & Mason, S. G. (1977). *J. Colloid Polymer Sci.*, 255, 794-804.

See more of this Session: Colloidal Hydrodynamics II
See more of this Group/Topical: Engineering Sciences and Fundamentals

APPLICATION OF THE LATTICE BOLTZMANN METHOD TO MANY PARTICLE DISPERSIONS

Akira Satoh, Akita Prefectural University, Japan
Roy W. CHANTRELL, University of York, U.K.

1. Introduction

For polymeric liquids [1, 2] and particle suspensions [3-5], even nowadays, it is very difficult to simulate the particle motion and the flow field simultaneously with multi-body hydrodynamic interactions among particles or polymers. This difficultly is mainly due to a significant difference in the characteristic times of the motion between solvent molecules and suspended particles (polymers). Since the characteristic time for the motion of solvent molecules is extraordinary shorter than that of suspended particles, the molecular dynamics method [6-8] cannot almost simulate the motion of suspended particles (polymers). An approach to circumvent this difficulty is that solvent molecules are regarded as a uniform liquid and its effect is built into the motion of suspended particles (polymers) macroscopically. This approach is called the Brownian dynamics method [5, 6]. Another approach is to regard solvent molecules as groups or clusters, whose characteristic time is relatively similar to that of suspended particles. A representative method to use the concept of virtual fluid particles is the lattice Boltzmann method [9-11]. The present study attempts to discuss the feasibility of the lattice Boltzmann method as a simulation technique with multi-body hydrodynamic interactions. In concrete, we consider aggregation phenomena in a dispersion composed of magnetic particles in thermodynamic equilibrium in order to investigate whether or not the Brownian motion of dispersed particles is activated on a physically reasonable level.

2. Lattice Boltzmann Method

2.1. BGK Lattice Boltzmann Method

In this method, a simulation region is regarded as a lattice system, as shown in Fig. 1(a), and fluid particles are made to move from site to site on the lattice system. Several lattice models have already been presented, and, in the present study, we employ the D2Q9 model [9-11] shown in Fig. 1(b), since aggregation phenomena in a tow-dimensional system is considered here. The direction of fluid particles in the unit cell is denoted by the notation α, as shown in Fig. 1(b), and α is taken as values of α=0, 1,..., 8. In the BGK model, the particle distribution function in the α-direction depends only on that in the same direction at the neighboring site in the opposite one $\overline{\alpha}$, not on that at the other sites or in the other directions. If an arbitrary lattice point is denoted by $\boldsymbol{r}$ and the particle distribution function in the α-direction at time t is denoted by $f_\alpha(\boldsymbol{r}, t)$, then the particle distribution $f_\alpha(\boldsymbol{r}+\boldsymbol{c}_\alpha\Delta t, t+\Delta t)$ at $(\boldsymbol{r}+\boldsymbol{c}_\alpha\Delta t)$ after the time interval Δt can be obtained from the following equations:

$$\left.\begin{aligned} f_\alpha(\boldsymbol{r}+\boldsymbol{c}_\alpha\Delta t, t+\Delta t) &= \tilde{f}_\alpha(\boldsymbol{r},t) \\ \tilde{f}_\alpha(\boldsymbol{r},t) &= f_\alpha(\boldsymbol{r},t)+\frac{1}{\tau}\{f_\alpha^{(0)}(\boldsymbol{r},t)-f_\alpha(\boldsymbol{r},t)\} \end{aligned}\right\} \qquad (1)$$

in which τ is the relaxation time and $f_\alpha^{(0)}$ is the equilibrium distribution function, expressed

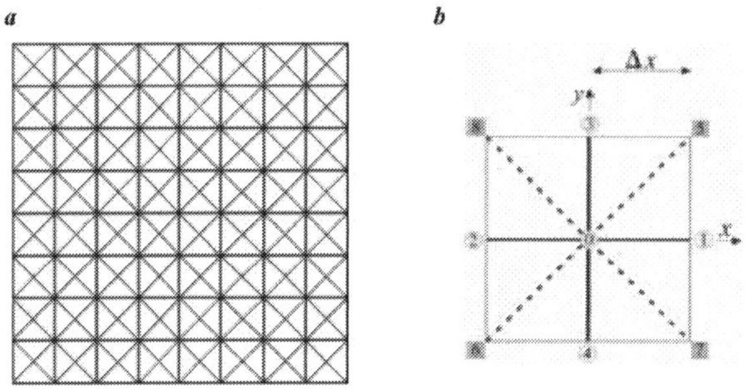

Fig.1 D2Q9 model for two-dimensional lattice Boltzmann simulations: (a) lattice model, and (b) unit cell for D2Q9 model.

with the macroscopic velocity $\boldsymbol{u}$ and the density ρ as

$$f_\alpha^{(0)} = \rho w_\alpha \left\{ 1 + 3\frac{\boldsymbol{c}_\alpha \cdot \boldsymbol{u}}{c^2} - \frac{3u^2}{2c^2} + \frac{9}{2}\cdot\frac{(\boldsymbol{c}_\alpha \cdot \boldsymbol{u})^2}{c^4} \right\} \tag{2}$$

In this equation, w_α is the weight constant, and $\boldsymbol{c}_\alpha$ is the lattice velocity, which can be expressed, in the D2Q9 model, as

$$w_\alpha = \begin{cases} 4/9 & for\ \alpha=0 \\ 1/9 & for\ \alpha=1,2,3,4 \\ 1/36 & for\ \alpha=5,6,7,8 \end{cases} \quad |\boldsymbol{c}_\alpha| = \begin{cases} 0 & for\ \alpha=0 \\ c & for\ \alpha=1,2,3,4 \\ \sqrt{2}c & for\ \alpha=5,6,7,8 \end{cases} \tag{3}$$

In Eq. (3), c is the speed for the movement between the nearest sites, expressed as $c=\Delta x/\Delta t$, and Δx is the length of the unit cell.

2.2. Boundary Conditions

In the present problem, we have to treat the boundary conditions at the outer simulation boundary and at each particle surface. We here consider a two-dimensional system in thermodynamic equilibrium, and the periodic boundary condition [5-8], therefore, is applicable. In this condition, for example, the particle distribution function $f_\alpha(x,y,t)|_{left}$ (α=0, 1, 2, ..., 8) at the left boundary surface in Fig. 1(a) is made to be equal to $f_\alpha(x,y,t)|_{right}$ at the right one. Similar procedure is carried out for the upper and lower boundary surfaces in Fig. 1(a).

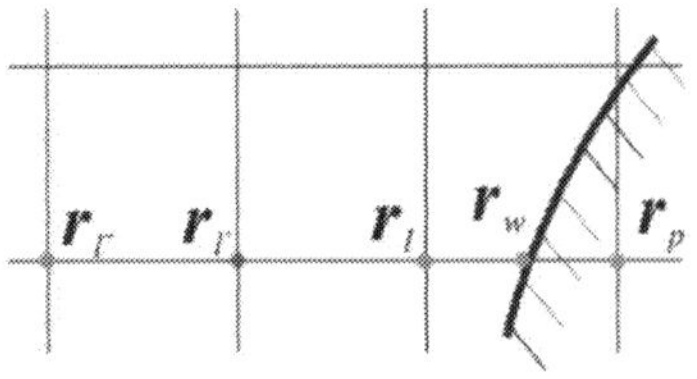

Fig.2 Equilibrium linear YMLS model.

Since dispersed particles move with time, the surfaces of the particles are almost at a position between the two neighboring sites. Hence, it is an important research subject to develop a more accurate boundary model at the surfaces of objects; even nowadays, such studies have being conducted. We treat the modified linear YLMS method [12], which is called the equilibrium linear YLMS method.

We consider the particle distribution function in the α-direction at a certain lattice point (α=2 for Fig. 2). If we use the notation $\boldsymbol{r}_w$ for a point at the particle surface, $\boldsymbol{r}_l$ for the neighboring site in the liquid area, and $\boldsymbol{r}_{l'}$ for the next neighboring site, then the opposite neighboring site $\boldsymbol{r}_p$ becomes inside the particle. This means that $f_2\ (\boldsymbol{r}_l\ ,t+\Delta t)$ cannot be evaluated from Eq. (1). If the particle distribution function $f_2\ (\boldsymbol{r}_w,t+\Delta t)$ at $\boldsymbol{r}_w$ is known, then $f_2\ (\boldsymbol{r}_l\ ,t+\Delta t)$ at $\boldsymbol{r}_l$ can be evaluated from the linear interpolation with $f_2\ (\boldsymbol{r}_{l'}\ ,t+\Delta t)$ at $\boldsymbol{r}_{l'}$ as

$$f_2\left(\boldsymbol{r}_l,t+\Delta t\right)=\frac{\Delta_w}{1+\Delta_w}f_2\left(\boldsymbol{r}_{l'},t+\Delta t\right)+\frac{1}{1+\Delta_w}f_2\left(\boldsymbol{r}_w,t+\Delta t\right) \tag{4}$$

in which Δ_w is expressed as $\Delta_w=|\boldsymbol{r}_l-\boldsymbol{r}_w|/|\boldsymbol{r}_l-\boldsymbol{r}_p|$. In the case of Fig. 2, the direction of $\bar{\alpha}$=2 (i.e., the opposite direction to α=1) is treated, but Eq. (4) is valid for any directions across the particle surface. In order to evaluate $f_2\ (\boldsymbol{r}_l\ ,t+\Delta t)$ from Eq. (4), $f_2\ (\boldsymbol{r}_w,t+\Delta t)$ is necessary, and, in the equilibrium linear YLMS method, the equilibrium distribution shown in Eq. (2) is used as this distribution.

3. Lattice Boltzmann Simulations of Many Particle Dispersions

3.1 Non-dimensionalization

A non-dimensional system is usually treated in simulations. Each quantity is non-dimensionalized by each representative one: for example, time is non-dimensionalized by Δt, distances by Δx, velocities by $c(=\Delta x/\Delta t)$, particle distribution functions by ρ_0, forces by $\rho_0(\Delta x)^2\,\Delta x/(\Delta t)^2$, etc.

The basic equation (1) for the particle distribution function can be expressed in non-dimensional form as

$$\left.\begin{aligned} &f_\alpha^{\,*}(\boldsymbol{r}^*+\boldsymbol{c}_\alpha^{\,*},t^*+1)=\tilde{f}_\alpha^{\,*}(\boldsymbol{r}^*,t^*)\\ &\tilde{f}_\alpha^{\,*}(\boldsymbol{r}^*,t^*)=f_\alpha^{\,*}(\boldsymbol{r}^*,t^*)+\frac{1}{\tau}\{f_\alpha^{(0)*}(\boldsymbol{r}^*,t^*)-f_\alpha^{\,*}(\boldsymbol{r}^*,t^*)\} \end{aligned}\right\} \tag{5}$$

in which

$$f_\alpha^{(0)*}=w_\alpha\rho^*\left\{1+3\boldsymbol{c}_\alpha^{\,*}\cdot\boldsymbol{u}^*+\frac{9}{2}\left(\boldsymbol{c}_\alpha^{\,*}\cdot\boldsymbol{u}^*\right)^2-\frac{3}{2}u^{*2}\right\} \tag{6}$$

$$\left|\boldsymbol{c}_\alpha^{\,*}\right|=\begin{cases}0 & for\ \ \alpha=0\\ 1 & for\ \ \alpha=1,2,3,4\\ \sqrt{2} & for\ \ \alpha=5,6,7,8\end{cases} \tag{7}$$

3.2. Model of Magnetic Particles

Magnetic particles are modeled as a spherical particle with a magnetic dipole in the center, covered with a uniform steric layer. The diameter of the solid part is denoted by D, and the thickness of a steic layer by δ. The force and torque acting on particle i by particle j

and an applied magnetic field can be written in non-dimensional form as

$$\mathbf{F}_{ij}^{(m)*} = -R_m \frac{1}{\tilde{r}_{ij}^{\,4}} \Big[-(\mathbf{n}_i \cdot \mathbf{n}_j)\mathbf{t}_{ij} + 5(\mathbf{n}_i \cdot \mathbf{t}_{ij})$$

$$\times (\mathbf{n}_j \cdot \mathbf{t}_{ij})\mathbf{t}_{ij} - \left\{ (\mathbf{n}_j \cdot \mathbf{t}_{ij})\mathbf{n}_i + (\mathbf{n}_i \cdot \mathbf{t}_{ij})\mathbf{n}_j \right\} \Big] \tag{8}$$

$$\boldsymbol{F}_{ij}^{(V)*} = R_V \boldsymbol{t}_{ij} \ln(\tilde{D}_1 / \tilde{r}_{ij}) \qquad for\ 1 \le \tilde{r}_{ij} \le \tilde{D}_1 \tag{9}$$

$$\boldsymbol{T}_{ij}^{(m)*} = -R_m \frac{D^*}{3} \cdot \frac{1}{\tilde{r}_{ij}^{\,3}} \left\{ \boldsymbol{n}_i \times \boldsymbol{n}_j - 3(\boldsymbol{n}_j \cdot \boldsymbol{t}_{ij})\boldsymbol{n}_i \times \boldsymbol{t}_{ij} \right\} \tag{10}$$

$$\boldsymbol{T}_i^{(H)*} = R_H\, \boldsymbol{n}_i \times \boldsymbol{h} \tag{11}$$

In these equations, $\boldsymbol{n}_i$ is the direction of the magnetic moment $\boldsymbol{m}_i$ of particle i, expressed as $\boldsymbol{n}_i=\boldsymbol{m}_i/m$ ($m=|\boldsymbol{m}_i|$), $\boldsymbol{t}_{ij}$ is the unit vector denoting the direction of the line between $\boldsymbol{r}_i$ and $\boldsymbol{r}_j$, expressed as $\boldsymbol{t}_{ij}=\boldsymbol{r}_{ij}/r_{ij}$, $\boldsymbol{r}_{ij}=\boldsymbol{r}_i-\boldsymbol{r}_j$, $r_{ij}=|\boldsymbol{r}_{ij}|$, $\boldsymbol{H}$ is the applied magnetic field ($H=|\boldsymbol{H}|$), $\boldsymbol{h}=\boldsymbol{H}/H$, k is Boltzmann`s constant, T is the liquid temperature, D_1 is the diameter including the steric layer, expressed as $D_1=D+2\delta$, and t_δ is the steric layer thickness divided by the solid part radius, expressed as $t_\delta=2\delta/D$. The non-dimensional parameters λ, ξ and λ_V are expressed as $\lambda=\mu_0 m^2/(4\pi D^3 kT)$, $\xi=\mu_0 mH/kT$ and $\lambda_V=\pi D^2 n_s/2$, in which μ_0 is the permeability of free space and n_s is the number of surfactant molecules per unit area on the particle surface.

The non-dimensional parameters R_m, R_V and R_H are expressed as

$$\left.\begin{aligned} R_m &= \frac{3kT}{\rho_0(\Delta x)^2(\Delta x/\Delta t)^2} \cdot \frac{1}{D^*}\lambda \\ R_V &= \frac{kT}{\rho_0(\Delta x)^2(\Delta x/\Delta t)^2} \cdot \frac{1}{\delta^*}\lambda_V \\ R_H &= \frac{kT}{\rho_0(\Delta x)^2(\Delta x/\Delta t)^2}\xi \end{aligned}\right\} \tag{12}$$

The equations of motion of magnetic particles are written in non-dimensioned form as

$$\boldsymbol{r}_i^*(t^*+1) = \boldsymbol{r}_i^*(t^*) + \boldsymbol{v}_i^*(t^*) + \frac{2}{\pi\rho_p^* D^{*2}} \boldsymbol{F}_i^*(t^*) \tag{13}$$

$$\boldsymbol{v}_i^*(t^*+1) = \boldsymbol{v}_i^*(t^*) + \frac{4}{\pi\rho_p^* D^{*2}} \boldsymbol{F}_i^*(t^*) \tag{14}$$

$$\boldsymbol{n}_i(t^*+1) = \boldsymbol{n}_i(t^*) + \boldsymbol{\omega}_i(t^*) \times \boldsymbol{n}_i(t^*) \tag{15}$$

$$\boldsymbol{\omega}_i^*(t^*+1) = \boldsymbol{\omega}_i^*(t^*) + \frac{32}{\pi\rho_p^* D^{*4}} \boldsymbol{T}_i^*(t^*) \tag{16}$$

in which, $\tilde{r}_{ij} = r_{ij}/D$, $\tilde{D}_1 = D_1/D$ $(\tilde{\delta} = \delta/D)$.

4. Specification of Quantities for Simulations

The non-dimensional density of magnetic particles, ρ_p^*, is taken as $\rho_p^*=3$, 5, 7 and 10, and, similarly, the non-dimensional particle diameter D^* as $D^*=4$, 6, 8 and 15; it is noted that a small diameter means a more coarse lattice system. The relaxation time τ is 0.7, the volumetric fraction of magnetic particles, ϕ_V, is 0.2, and the number of particles, N_p, is taken as $N_p=100$; we also here consider the case of a fine lattice system such as $D^*=15$, so that a relatively small system such as $N_p=100$ has been treated. In order to compare the present results with those obtained by Monte Carlo and Brownian dynamics simulations [13], the non-dimensional parameter λ_V has been taken as $\lambda_V=150$, λ as $\lambda=0\sim15$, and ξ as $\xi=0\sim20$. The cutoff radius $\tilde{r}_{coff}$ (= r_{coff} /D) is 7.0, and the total number of time steps, N_{timemx}, has been taken as $N_{timemx}=200{,}000$, in which the first ten percent data are discarded, and the remainders are used to conduct averaging procedures.

5. Results and Discussion

5.1 Comparison with Monte Carlo and Brownian Dynamics Results in Aggregate Structures

We discuss the validity of the results of the lattice Boltzmann method, by comparing with those of Monte Carlo and Brownian dynamics simulations. We here focus on the results for $\rho_p^*=5$ and $D^*=8$ for comparison. Figures 3 show the results of aggregate structures obtained by the lattice Boltzmann method and Fig. 3(a) is for $\xi=\lambda=1$, Fig. 3(b) for $\xi=1$ and $\lambda=10$, and Fig. 3(c) for $\xi=20$, $\lambda=10$. Similar results are shown in Figs. 3(d) and 3(e) for Monte Carlo simulations, and, Brownian dynamics results for $\xi=\infty$ are shown in Fig. 3(f).

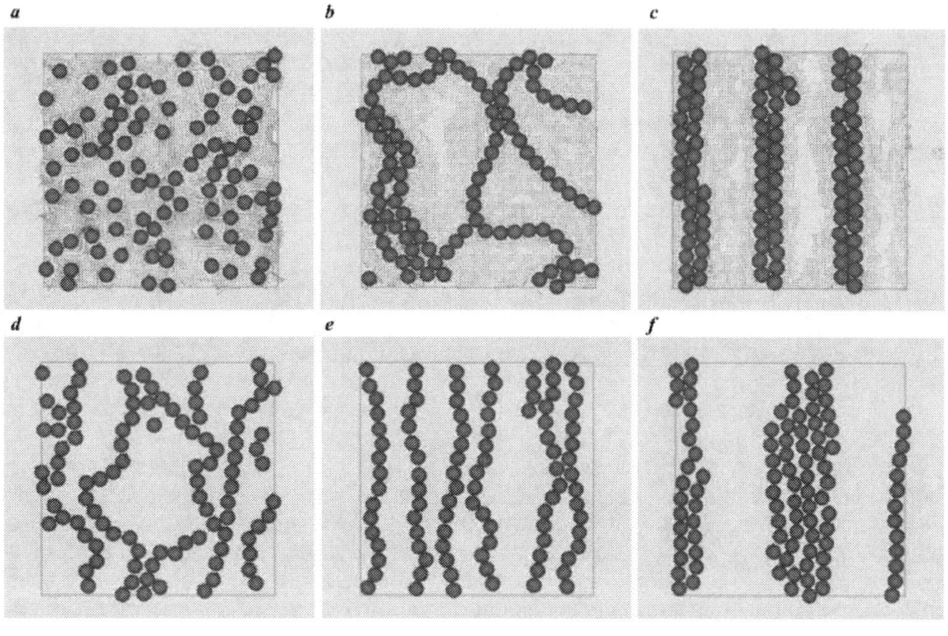

Fig.3 Comparison of lattice Boltzmann (LB) results with Monte Carlo (MC) and Brownian dynamics (BD) ones: (a) $\xi=\lambda=1$, LB ; (b) $\xi=1$ and $\lambda=10$, LB; (c) $\xi=20$ and $\lambda=10$, LB; (d) $\xi=1$ and $\lambda=10$, MC ; (e) $\xi=20$ and $\lambda=10$, MC; (f) $\xi=\infty$ and $\lambda=10$, BD.

First we consider the results in Fig. 3(a) for the lattice Boltzmann method and that for the Monte Carlo method (not shown in a figure), which is obtained for a weak magnetic field of ξ=1. The comparison with each other in the corresponding snapshots clearly shows that qualitative features are in significantly good agreement. That is, for weak magnetic interactions such as λ=1, magnetic particles move solely without significant aggregation, and, as magnetic interactions increase, particles come to aggregate to form clusters. For the case of λ=10, since magnetic particle-particle interactions are much more dominant than the Brownian motion, network-like or necklace-like aggregates are observed to be formed in Figs. 3(b) and 3(d). This good agreement concerning aggregates for λ=10 clearly shows that the Brownian motion of magnetic particles are activated on a physically reasonable level in the lattice Boltzmann method.

Next, we consider the aggregate structures shown in Figs. 3(c), 3(e), and 3(f). Even in this case of a strong magnetic field, the corresponding snapshots show good agreement, but not so significantly as in the previous case of ξ=1. For weak magnetic interactions such as λ=1, the magnetic moments of particles approximately incline in the field direction, but magnetic particles do not aggregate; they move solely. As magnetic interactions increase such as λ=5 and 10, thick chain-like clusters come to be formed, and the present results are in relatively good agreement with those of Brownian dynamics simulations. The internal structures of thick chain-like clusters in Fig. 3(e) are not so compact as in Figs. 3(c) and 3(f), but are relatively loose. These results do not show that the lattice Boltzmann and Brownian dynamics methods are inferior to the Monte Carlo method in reproducing aggregation phenomena, but rather show a weak point in the latter method. That is, the former two methods take into account multi-body hydrodynamic interactions among particles, and, on the other hand, the latter method doesn't. Hence, once long chain-like clusters are formed in the field direction as shown in Figs. 3(c) and 3(f), the mutual effects of magnetic interactions and multi-body hydrodynamic interactions make these loose thick chain-like clusters in more compact formation. We may conclude from these results that the lattice Boltzmann method takes into account multi-body hydrodynamic interactions and can activate the Brownian motion of particles to reproduce physically reasonable aggregate structures.

5.2 Comparison with Monte Carlo and Brownian Dynamics Results in the Pair Correlation Function and Magnetization Curves

Finally, we discuss quantitatively the previous aggregate structures in terms of the pair correlation function and magnetization curves. Figures 4(a), 4(b) and 4(c) show the results of the pair correlation function in the magnetic field direction for the lattice Boltzmann, Monte Carlo and Brownian dynamics methods, respectively. Figures 4(a) and 4(b) are for ξ=20, and Fig. 8(c) is for ξ=∞.

First, we discuss the results in Figs. 4(a), 4(b) and 4(c) for a strong magnetic field case. These results are obtained for relatively large magnetic interactions such as λ=5, 10 and 15, so that, as shown previously, chain-like clusters are formed in the magnetic field direction. Hence, these three simulation methods show good agreement concerning the positions of each peak, which arises at the positions of integer times the particle diameter D_1^*, (strictly, slightly shorter distance than D_1^*). Also, the heights of each peak may be in reasonable agreement among these three simulation methods; as a tendency, Monte Carlo results show lower peaks, then Brownian dynamics ones show slightly higher ones, and lattice Boltzmann simulations indicate higher peak values than the other methods. The previous results concerning aggregate structures say that more stable aggregate formation is reproduced than the Monte Carlo method, since only the present method and the Stokesian dynamics method take into account hydrodynamic interactions, but the Monte Carlo method doesn't. It is, therefore, physically

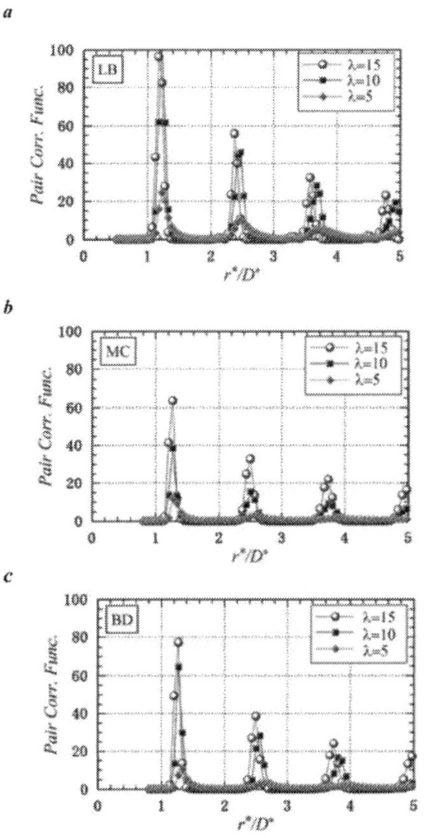

Fig.4 Pair correlation function for (a) ξ=20, LB, (b) ξ=20, MC, and ξ=∞, BD.

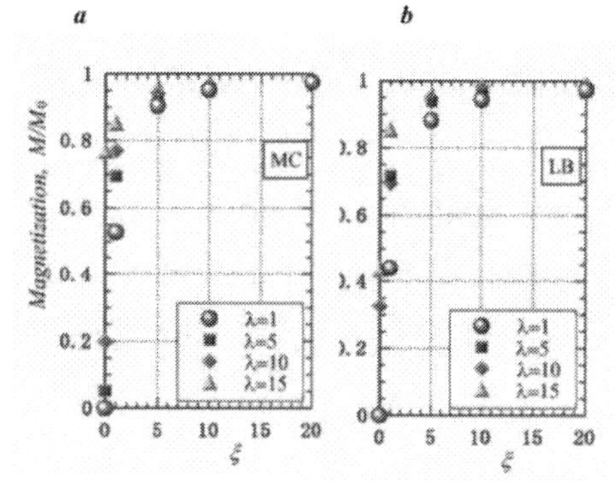

Fig.5 Magnetization curves for (a) lattice Boltzmann method (ρ_p^*=5 and D^*=8) and (b) Monte Carlo method.

reasonable that the former two methods give rise to higher peaks than the latter one. Furthermore, the Brownian dynamics results [13] do not take into account long-range order interactions of hydrodynamics, so that it is quite understandable that the Brownian dynamics results give rise to lower peak values than the present simulation method. We may conclude from these results that the present lattice Boltzmann method gives rise to the most reasonable results.

Finally, we discuss the results of magnetization curves in Fig. 5. Figures 5(a) and 5(b) are for the lattice Boltzmann method and the Monte Carlo method, respectively. The curve of the present method for λ=1 agrees well with that of the Monte Carlo method, and, especially, gives rise to zero at ξ=0, as in the Monte Carlo method. This clearly verifies that the rotational Brownian motion of magnetic particles is activated on a physically reasonable level. Even for large magnetic interactions such as λ=15, the agreement is quite good, and the rotational Brownian motion is seen to be activated reasonably as well. The disagreement at ξ=0 is not due to drawbacks of the present method, but due to relatively small simulation regions used in the present simulations.

6. Conclusion

If relatively coarse lattices are used, physically reasonable aggregate structures can be obtained, but if fine lattices are adopted, the particle Brownian motion cannot be activated properly, which causes invalid aggregate formation of particles. The results concerning the pair correlation function and magnetization curves clearly show that the lattice Boltzmann results are in good agreement with those of Monte Carlo and Brownian dynamics methods. We may conclude from these results that, if relatively coarse lattice systems are used, the translational and rotational Brownian motion of particles are reasonably activated, but if fine lattice systems are adopted, the lattice Boltzmann method cannot give rise to physically reasonable results.

References

[1] R.B. Bird, R.C. Armstrong and O. Hassager (1987), "Dynamics of Polymeric Liquids," *Fluid Mechanics*, Vol.1, John Wiley & Sons, New York.

[2] M. Doi and S.F. Edwards (1986), "The Theory of Polymer Dynamics," Oxford University Press, Oxford.

[3] W.B. Russel, D.A. Saville and W.R. Schowalter (1989), "Colloidal Dispersions," Cambridge University Press, Cambridge.

[4] Theo G. M. van de Ven (1989), "Colloidal Hydrodynamics," Academic Press, New York.

[5] A. Satoh (2003), "Introduction to Molecular-Microsimulation of Colloidal Dispersions," Elsevier Science, Amsterdam.

[6] M.P. Allen and D.J. Tildesley (1987), "Computer Simulation of Liquids," Clarendon Press, Oxford.

[7] S. Kamiyama and A. Satoh (1997), "Molecular Dynamics Simulation," Asakura Shuppan, Tokyo, (in Japanese).

[8] J.M. Haile (1992), "Molecular Dynamics Simulation, Elementary Methods," John Wiley & Sons, New York.

[9] S. Succi (2001), "The Lattice Boltzmann Equation for Fluid Dynamics and Beyond," Clarendon Press, Oxford.

[10] D.H. Rothman and S. Zaleski (1997), "Lattice-Gas Cellular Automata, Simple Models of Complex Hydrodynamics," Cambridge University Press, Cambridge.

[11] B. Chopard and M. Droz (1998), "Cellular Automata Modeling of Physical Systems," Cambridge University Press, Cambridge.

[12] A. Satoh and R. W. Chantrell, (submitted).

[13] A. Satoh, R. W. Chantrell and G. N. Coverdale (1999), "Brownian Dynamics Simulations of Ferromagnetic Colloidal Dispersions in a Simple Shear Flow," *J. Colloid Interface Sci.*, Vol. 209, pp.44-59.

Rheology of Colloidal Suspensions and Gels

M. D. Bybee, Chemical and Biomolecular Engr, University of Illinois at Urbana-Champaign, Urbana, IL

Jonathan Higdon, Chemical and Biomolecular Engineering, University of Illinois, Urbana, IL

The rheology of concentrated colloidal suspensions with attractive and repulsive particle interactions is studied using FLD (fast lubrication dynamics) simulations. The microstructure is determined from dynamic simulations with results presented for varying strengths of attraction, range of attraction and volume fraction. Rheology studies are conducted to evaluate the relaxation modulus for viscoelastic response as well as the frequency dependence of the storage and loss moduli. Step strain /relaxation and oscillatory shear simulations yield consistent results for the moduli. Ensemble averaging and novel noise reduction algorithms are introduced to dela with the extreme. Rheological parameters are computed for a range of strengths of attraction, repulsion, range of attraction and volume fraction. The introduction of repulsive forces reduces the storage modulus by more than order of magnitude compared to attractive gels.

See more of this Session: Colloidal Hydrodynamics II
See more of this Group/Topical: Engineering Sciences and Fundamentals

Numerical Prediction of the Dynamics of Nanoparticles Embedded in a Liquid Crystalline Solvent

Brian T. Gettelfinger, Department of Chemical and Biological Engineering, University of Wisconsin-Madison, Madison, WI

Gary M. Koenig Jr., Department of Chemical and Biological Engineering, University of Wisconsin-Madison, Madison, WI

Jose Moreno-Razo, Department of Chemical and Biological Engineering, University of Wisconsin-Madison, Madison, WI

Nicholas L. Abbott, Department of Chemical and Biological Engineering, University of Wisconsin-Madison, Madison, WI

Juan P. Hernandez-Ortiz, Departamento de Materiales, Universidad Nacional de Colombia, Sede Medellin, Medellin, Colombia

Juan De Pablo, Department of Chemical and Biological Engineering, University of Wisconsin-Madison, Madison, WI

A hierarchical modeling approach has been adopted to examine the structure and dynamics of nanoparticle suspensions in confined liquid crystals. A molecular model and a combination of Monte Carlo and molecular dynamics simulations are used to investigate the defects that arise around the nanoparticles, both at rest and other imposed flow fields, and to explore how such defects influence the aggregation behavior of the particles. The continuum molecular model is solved by resorting to a radial basis function based technique. The validity of the model and our numerical results are established by direct comparison to results of molecular simulations and to experimental mobility data in both the isotropic and nematic phases. The model is then used to examine the response of different types of confinement, surface treatment, and flow field on the aggregation pathways of nanoparticles in liquid crystals.

See more of this Session: Colloidal Hydrodynamics II
See more of this Group/Topical: Engineering Sciences and Fundamentals

Measurements of Structure-Property Relationships in Shear Thickening, Concentrated, near Hard Sphere, Colloidal Dispersions Via 1-2 Plane Flow-SANS

Norman J. Wagner, Department of Chemical Engineering, University of Delaware, Newark, DE

Dennis Kalman, Chemical Engineering, University of Delaware, Newark, DE

Lionel Porcar, NIST Cold Neutron Research Center, National Institute for Standards and Technology, Gaithersburg, MD

In this work we generate structure-property relationships for shearing, concentrated colloidal dispersions through the use of a novel 1-2 plane flow cell sample environment geometry for SANS measurements (Liberatore et al. PRE, 2006), and compare these to theory. Concentrated particle suspensions show widely varying rheological behavior, including shear thinning and shear thickening, even in the absence of particle interactions other than excluded volume (hard spheres). Colloidal suspensions of hard-spheres begin to show significant shear thinning and shear thickening at volume fractions above about 20%; these suspensions show progressively stronger shear thinning and shear thickening with increasing volume fraction. Shear thickening is known to occur via the formation of load bearing hydroclusters from theory, simulations, flow-SANS experiments, and stress-jump rheological measurements, but there are no previous measurements of the structural rearrangements that accompany shear thinning or thickening directly in the plane of shear (the 1-2 or flow-gradient plane). Here, we study model suspensions of near hard-sphere silica particles, approximately 120nm in diameter in a near contrast-matching Newtonian solvent mixture of deuterated ethylene glycol and polyethylene glycol via rheo-SANS in standard 1-3 shear plane (velocity-vorticity) as well as directly in the 1-2 plane (velocity-gradient). Significant anisotropic structural rearrangements under shear are evident. These microstructure changes are compared to the rheological behavior via stress-SANS laws that separate the thermodynamic and hydrodynamic components of the stress which drive shear thinning and shear thickening, respectively. The results are compared against Stokesian Dynamics simulations and theory to provide validation of the mechanisms of shear thinning nad shear thickening in concentrated colloidal dispersions.

See more of this Session: Colloidal Hydrodynamics II
See more of this Group/Topical: Engineering Sciences and Fundamentals

Direct Measurement of Suspension Structure in Pressure Driven Flow

Changbao Gao, Lehigh University, Bethlehem, PA

James F. Gilchrist, Lehigh University, Bethlehem, PA

Many non-Newtonian properties of concentrated suspensions, including shear thinning, normal stress, and shear migration, are attributed to the evolution of structural anisotropy. Investigations of this anisotropy have largely been conducted using simulations such as Stokesian Dynamics and indirect measurement via scattering. Using high speed confocal laser scanning microscopy, we directly investigate the local structure via particle tracking after structural evolution in a pressure driven flow ranging from Péclet number (the ratio of convective to diffusive forces) of $0 < Pe < 2000$ over bulk volume fractions $0.2 \leq \Phi \leq 0.4$. Due to the shear gradient, a single experiment allows both sampling over a range of Péclet numbers and concentrations that vary due to shear migration. Clear structural anisotropy similar to that demonstrated by simulations of Foss and Brady, 2000 and Morris and Katyal, 2002 show strong deviations from homogeneity in radial distribution profiles at high Péclet number and high volume fraction. Moreover, in regions of low Pe located in the center of the pressure driven flow, signatures of colloidal crystallization driven by the high concentration that results from the normal forces generated this anisotropy are apparent.

See more of this Session: Colloidal Hydrodynamics II
See more of this Group/Topical: Engineering Sciences and Fundamentals

Structure of Aggregating Rod Suspensions Under Combined Shear and Electric Fields

Venkat Ganesan, Chemical Engineering, University of Texas at Austin, Austin, TX

Victor Pryamitsyn, The University of Texas at Austin, Austin, TX

Efficient strategies for dispersion of carbon nanotubes in polymeric and solvent matrices constitutes an area of active interest. In this article, we examine from a theoretical perspective the hypothesis that a combination of AC electric and shear fields oriented at an angle may be used to enhance the dispersion of aggregated rod solutions. We present a deterministic and a Smoluchowski equation based analysis of the dynamics of the rod suspensions in a configuration involving combined electric and shear fields. We use this analysis to suggest that a cross-field shear and electric field configuration may result in a ``rotating'' state of the rods which may potential disperse aggregated nanotube suspensions. We test the predictions of our analytical models through Brownian dynamics simulations to analyze the dynamics of a rod suspensions in a cross-field configuration. The results of our simulations display good agreement with our analytical results and serve to delineate the parametric regimes in which the use of a combination of electric and shear fields may enhance the dispersion of aggregated nanotubes.

See more of this Session: Colloidal Hydrodynamics II
See more of this Group/Topical: Engineering Sciences and Fundamentals

Nanoparticle-Organic Hybrid Materials: The Particle That Carries Its Fluid On Its Back

Hsiu-Yu Yu, School of Chemical and Biomolecular Engineering, Cornell University, Ithaca, NY

Donald Koch, School of Chemical and Biomolecular Engineering, Cornell University, Ithaca, NY

Nanoparticle-organic hybrid materials (NOHMs) consist of 10 nm diameter spherical inorganic core particles functionalized with oligomeric organic molecules. Although these systems contain no added solvent, they exhibit fluid behavior with the fluidity provided by the attached oligomers. We will present a theory for the equilibrium structure and transport properties of these materials based on an assumption that the intercore forces are mediated by entropic effects associated with the conformations of the oligomers. The compression of oligomer brushes when two particles approach one another leads to steric repulsion as is typically found in hairy particles. However, the oligomers also lead to entropic attractive forces. The oligomeric fluid is assumed to be incompressible and oligomers (modeled as bead-spring systems) must stretch to fill the space between the cores. This effect prevents the formation of large regions of free volume between the core particles. Because each particle carries its share of the fluid phase, the structure factor at zero wave number is equal to zero, a value indicative of a deficit of one neighboring core particle in the vicinity of any chosen particle. We determine the long-time diffusivity, low shear rate viscosity and high frequency shear modulus for NOHMs in the limiting case where the radius of gyration of the oligomers is large compared with the core radius. In this limiting case, each core experiences weak interactions with the many other cores residing in its neighborhood (within a radius of gyration). Exploiting this limit, the transport properties can be determined in a quasi-analytical manner based on a solution of the non-equilibrium probability density for pairs of particles experiencing an intercore potential of mean force derived from density-functional theory.

See more of this Session: Colloidal Hydrodynamics II
See more of this Group/Topical: Engineering Sciences and Fundamentals

A Pressure-Driven Separation Device with An Integrated Microfluidic Pump

Gwendoline M. Toh-Boyo, Chemistry, University of Wyoming, Laramie, WY

Debashis Dutta, Chemistry, University of Wyoming, Laramie, WY

Fluid and solute samples are commonly moved in microchannels using either electro-osmotic or pressure driven flow. While electroosmotic flow is relatively easier to control in microfluidic channels, pressure-driven flow is often applied in microfluidic applications where electro-osmotic flow is undesirable e.g. liquid chromatography. Steady pressure-driven flow has been realized in microfluidic channels using syringe pumps in this past. Although this method is easy to implement, the difficulty encountered with maintaining and dynamically controlling the pressure generated flows in this design have resulted in poor efficiency and reproducibility of separations. In this work, we report an on chip pressure generation capability that uses a porous membrane made using a sodium silicate precursor. The proposed micro-pump consists of 3 channel segments in a tee geometry one of which has the sodium silicate membrane. Our experimental study has shown that this membrane although has a high electrical conductivity tends to block electro-osmotic flow (EOF) very effectively. In this situation, by applying appropriate electric fields across this design, the EOF generated in one of the open channel segments was blocked by the sodium silicate membrane in the other, to generate a pressure-gradient in the system. The resulting pressure-driven flow could be then effectively maintained and guided into an electric field-free separation channel for performing pressure-driven chromatographic separations. More interestingly, it was shown that for an applied voltage in the system this pressure-driven velocity was independent of channel depth in the microfluidic network. This implies that the proposed strategy for generating pressure-gradients can be equally effective in realizing pressure-driven separations in micro- as well as nanoscale separation channels.

See more of this Session: Microfluidics and Small-Scale Flows III: Technological Advances
See more of this Group/Topical: Engineering Sciences and Fundamentals

Microfluidic Passive Pumping Using Coupled Capillary/Evaporation Effects

Nicholas S. Lynn, Department of Chemical and Biological Engineering, Colorado State University, Fort Collins, CO

David S. Dandy, Department of Chemical and Biological Engineering, Colorado State University, Fort Collins, CO

One of the most important unit operations within microfluidic devices lies in the handling and transport of fluids. The ability to provide temporally stable flows with high precision and repeatability can improve the efficiency and reliability of bioanalytical lab-on-a-chip (LOC) devices. Traditional methods of fluid delivery, such as syringe pumps or electrokinetic phenomena, generally require bulky equipment that can be several orders of magnitude larger in size than the microfluidic network itself. Due to the inherently small size of a microchannel, capillary forces are a convenient method to passively create pressure-driven flow. Although there have been multiple studies concerning LOC devices utilizing capillary forces, none have the ability to create steady-state flows existing for periods longer than O(10) min.

In the present work, a microfluidic passive pumping mechanism is demonstrated that combines the ability to create long term steady state flows from a mm2 footprint. After sample introduction to an inlet reservoir, capillary forces drive fluid through a microfluidic network (μFN) spontaneously such that a small meniscus is created along the corner regions of an outlet reservoir. The small radius of curvature of the outlet reservoir meniscus serves to drive fluid through the system at high differential pressures, where the volumetric flow rate into the reservoir is balanced by the rate of evaporation from the meniscus. The system quickly reaches steady state and is able to provide precise flows through a μFN for periods longer than an hour. Flow rates spanning multiple orders of magnitude are accomplished via control over the dimensions of both the microchannels and outlet reservoirs.

The dynamics of fluid flow in these systems are predicted through use of (1) a geometric model to describe the shape of the outlet reservoir meniscus and (2) a computational fluid dynamics model to calculate the evaporation rate from the meniscus and (3) the Young-Laplace equation to calculate the pressure differential between the inlet and outlet reservoirs. Using this model, we show that fluid flow through μFNs can be accurately predicted to within 15% of laboratory values for systems with a variety of reservoir and microchannel dimensions.

See more of this Session: Microfluidics and Small-Scale Flows III: Technological Advances
See more of this Group/Topical: Engineering Sciences and Fundamentals

Biomolecular Optical Transport in Nanoscale Slot Waveguides

Allen H. J. Yang, Chemical and Biomolecular Engineering, Cornell University, Ithaca, NY

Sean Moore, Mechanical and Aerospace Engineering, Cornell University, Ithaca, NY

Brad Schmidt, Electrical Engineering, Cornell University, Ithaca, NY

Matt Klug, Mechanical and Aerospace Engineering, Cornell University, Ithaca, NY

Michal Lipson, Electrical Engineering, Cornell University, Ithaca, NY

David Erickson, Mechanical and Aerospace Engineering, Cornell University, Ithaca, NY

One of the roadblocks to development of active nanosystems is the ability to controllably deliver nanoscopic matter to and within nanostructures. Optically-driven transport enabled by the integration of rectangular dielectric waveguides with microfluidic devices exploits radiation pressure and gradient forces to trap and propel particles along the waveguide structures. We recently demonstrated an optofluidic device for the trapping and transport of microparticles in the evanescent field of a polymer solid-core waveguide. To the best of our knowledge, researchers have not been able to exploit such waveguide based structures for the trapping and transport of nanoscopic matter, as it would require stronger optical confinement than currently available to overcome random thermal diffusion. Here we demonstrate a new paradigm for nanoscopic transport using sub-wavelength scale planar photonic "slot waveguides". The slot waveguide generates high confinement liquid-core optical modes, creating the necessary optical intensity for nanoparticle trapping. Demonstrated here is the trapping and transport of 75 and 100 nm polystyrene beads flowing in an overlaying microfluidic channel for slots ranging in width from 100 nm to 160 nm. Furthermore, we show that we can trap and transport partially extended λ-DNA in 60 nm slot waveguides. We validate our observations by characterizing the trapping stability using a modified Polanyi-Wigner equation. The architecture demonstrated here is the first step towards an optical nanochannel, bridging the gap between microfluidics and nanofluidics.

Keywords: Optofluidics, DNA trapping, waveguide

See more of this Session: Microfluidics and Small-Scale Flows III: Technological Advances
See more of this Group/Topical: Engineering Sciences and Fundamentals

Handling Solids in Microreactors for Continuous-Flow Synthetic Chemistry

Ryan L. Hartman, Department of Chemical Engineering, Massachusetts Institute of Technology, Cambridge, MA

Nikolay Zaborenko, Department of Chemical Engineering, Massachusetts Institute of Technology, Cambridge, MA

John R. Naber, Department of Chemistry, Massachusetts Institute of Technology, Cambridge, MA

Mahmooda Sultana, Chemical Engineering, Massachusetts Institute of Technology, Cambridge, MA

Stephen L. Buchwald, Department of Chemistry, Massachusetts Institute of Technology, Cambridge, MA

Klavs F. Jensen, Department of Chemical Engineering, Massachusetts Institute of Technology, Cambridge, MA

Microchemical systems catalyze discovery and development of synthetic routes while realizing a more basic understanding of chemistry. Enhanced heat and mass transfer characteristics, safer synthesis of dangerous compounds, isolation of air- and moisture-sensitive chemistries, reduction of hazardous waste, optimized reaction conditions, and rapid experimentation add value to the technology by shortening product development life cycles. Multiphase synthetic chemistry often involves the use of solids, which take the form of products, by-products, reagents, and catalysts. Chemical syntheses forming inorganic salts are difficult to carry out because these solids can clog microchannels. Chemistry and engineering strategies for handling solids must therefore be developed in order to realize complex microchemical synthesis.

We investigate several strategies for handling solids in microchemical systems and apply basic principles to a model coupling chemistry that is widely employed in industry. Microreactor surface modification, reactor design fundamentals, and acoustic forces comprise approaches to overcome this challenge and enable continuous-flow synthesis. These and other results will be presented and discussed.

See more of this Session: Microfluidics and Small-Scale Flows III: Technological Advances
See more of this Group/Topical: Engineering Sciences and Fundamentals

Experimental Visualization of Mass Transfer in Nanofluids Using TIRF Microscopy

Jocelyn Veilleux, Chemical Engineering, McGill University, Montreal, QC, Canada

Sylvain Coulombe, Chemical Engineering, McGill University, Montreal, QC, Canada

Nanofluids - fluids containing suspended nanoparticles at low volume fractions - are presented as an interesting means to improve the cooling capability of conventional heat transfer fluids. Owing to their small size, suspended nanoparticles do not show the major drawbacks associated with suspensions of micrometer or larger size solid particles (erosion, clogging and severe pressure drop). This relative advantage brought many researchers to conduct experimental and numerical investigations to evaluate the cooling capabilities of nanofluids. These investigations have shown an enhancement of the thermal conductivity and, in several cases, this enhancement is anomalously high when compared to continuum model predictions (e.g. Maxwell and Hamilton-Crosser models). However, the reported enhancements show a significant spread and more work is needed to improve our fundamental understanding of such complex systems. Of the many mechanisms that were proposed to explain the anomalous enhancement of nanofluids thermal conductivity, two have emerged of the debate: nanoparticles aggregation and Brownian motion-induced nanoparticles hydrodynamics. The latter mechanism seems more interesting since it could also explain why nanofluids greatly affect mass transport behaviour. Indeed, dye diffusion experiments in suspensions of nanoparticles in water have shown a diffusion pattern that is no longer isotropic and for which the diffusion coefficient is more than one order of magnitude larger than expected. Still, no experimental evidences supporting this suggestion of Brownian motion-induced nanoparticles hydrodynamics have been reported yet.

This project aims at visualizing and characterizing the hydrodynamics of mass transfer in nanofluids using Total Internal Reflection Fluorescence Microscopy (TIRFM). The similarities between heat and mass transfer processes will then be used to infer an explanation for the anomalous enhancement of nanofluids thermal conductivity. TIRFM is chosen over traditional fluorescence and confocal microscopy for its ability to greatly reduce the amount of undesired background fluorescence. Moreover, it probes phenomena occurring very near surfaces (within 200-250 nm), making the technique suitable to study microdevices where nanofluids are likely to find applications. To capture images, TIRFM requires the excitation of fluorescent molecules (or nanoparticles) suspended in the sample fluid by an evanescent wave. In our particular case, TIRFM relies on the fluorescence (emission) tracking of both the host liquid (solution of rhodamine 590 chloride) and the suspended nanoparticles (quantum dots) at two different wavelengths upon laser light stimulation. Filters are used to physically separate the wavelengths before image capture. This experimental setup enables the tracking of individual Brownian nanoparticles (quantum dots) independently of the dye diffusion process, thus making possible the development and validation of a mass transport hydrodynamics model of nanofluids.

In this paper, diffusion of rhodamine 590 chloride in water/alumina nanofluids of various volume fractions is first studied by TIRFM at a magnification of 100X and compared to diffusion in pure water. The injection of the dye solution is realized via a custom-designed PTFE/glass cell that ensures the experimental reproduction of a Heaviside-type initial condition for the concentration profile. TIRFM images are acquired at a rate of 1/3 frame per second, and both qualitative and quantitative analyses are performed. Experimental mass diffusivities are determined via a curve fitting of the acquired concentration versus time profiles at two locations along the PTFE/glass cell. Preliminary results show that the proposed method accurately measures the diffusion coefficient of rhodamine 590 chloride in deionized water, and that the diffusion process in water/alumina nanofluids is affected by the presence of nanoparticles. Current work is focusing on the precise measurement of the diffusion coefficient in nanofluids of various volume fractions, as nanoparticle loading/interaction is expected to influence the diffusion process.

See more of this Session: Microfluidics and Small-Scale Flows III: Technological Advances
See more of this Group/Topical: Engineering Sciences and Fundamentals

Massively Scaled Microfluidic Device for the Flow-Based Formation of 3D Perfusable Scaffold

Lian Leng, Department of Mechanical and Industrial Engineering, University of Toronto, Toronto, ON, Canada

Siavash Aslanbeigi, Division of Engineering Science, University of Toronto, Toronto, ON, Canada

Axel Günther, Department of Mechanical and Industrial Engineering, Institute of Biomaterials and Biomedical Engineering, University of Toronto, Toronto, ON, Canada

Miscible and immiscible liquids have been previously used to define and spatially organize microscale building blocks in planar microchannel networks [1, 2]. Although many applications, including tissue engineering, ultimately require three-dimensional organization, most efforts so far have been limited to planar approaches. The layer-by-layer microfabrication of microstructured materials has been achieved [3], although complex and highly organized three-dimensional microstructures have yet to be fully demonstrated.

We present a microfluidic strategy for the continuous in-flow formation of a soft material with a three-dimensionally organized and tunable microstructure. Our approach is enabled by a massively scaled microfluidic device. A soft lithography process was adapted to consecutively define 300micron wide and 100micron deep microchannels in ten 500micron thin elastomer substrates [4]. The device fabrication was completed by vertically bonding the obtained ten layers. Two fluids, aqueous alginate, A, and calcium chloride, C, entered the device at defined flow rates through separate inlets after which they were distributed into an array of ten by ten microchannels. An alternating configuration of A channels surrounded by C channels was obtained at the device exit where a complex fluid was produced. The three-dimensionally organized fluid was extruded into a water reservoir, in a continuous process. Diffusion of C initiated the polymerization of A, defining the soft material matrix and leaving a vascularized flow network in the areas initially occupied by C. A 10cm long microstructured sample was obtained in less than four minutes. Upon completion of the polymerization process, the microfluidic device was used to perfuse the vascularized sample.

REFERENCES:

1. E. Surenjav, C. Priest, S. Herminghaus, and R. Seeman, "Manipulation of Gel Emulsions by Variable Microchannel Geometry", Lab Chip, 9, 2 (2009)

2. M. T. Tolley, M. Krishnan, D. Erickson, and H. Lipson, "Dynamically Programmable Fluidic Assembly", Appl. Phys. Lett., 93, 254105 (2008)

3. N.W. Choi, M. Cabodi, B. Held, J.P. Gleghorn, L.J. Bonassar, and A.D. Stroock, "Microfluidic Scaffolds for Tissue Engineering", Nature Mat., 6, 11 (2007)

4. M.A. Unger, H.P. Chou, T. Thorsen, A. Scherer, and S.R. Quake, "Monolithic Microfabricated Valves and Pumps by Multilayer Soft Lithography", Science, 288, 5463 (2000)

See more of this Session: Microfluidics and Small-Scale Flows III: Technological Advances
See more of this Group/Topical: Engineering Sciences and Fundamentals

Microfluidic Formation of Ionically Crosslinked Polyamine Gels

Gautam C. Kini, Chemical and Biomolecular Engineering, Rice University, Houston, TX

Justin Tseng-Sing Lai, Department of Biochemistry and Cell-Biology, Rice University, Houston, TX

Michael Wong, Chemical and Biomolecular Engineering and Chemistry, Rice Univvresity, Houston, TX

Sibani Lisa Biswal, Chemical and Biomolecular Engineering, Rice University, Houston, TX

Recent advances in microreaction engineering using microfluidic devices has led to the the generation of polymer structures with a variety of shapes and morphologies. Microfluidic devices allow for the confinement of reagents and precise control of the time of reaction. A current focus area in microfluidic research is on flow and formation of gels in microchannels. Primary applications of gels in microfluidic devices are in forming components for flow control, separations, detection and sensing, geared towards developing micro total analytical (μTAS) systems.[1]

An important approach to incorporating gels within the microchannel is to form the gel structures *in situ*. This route enables gel deposition at precise locations in a microchannel. So far, *in situ* gel formation in microfluidic channels is based on polymerization and electrostatic cross-linking reactions. However, these routes often require additional fabrication steps of flow focusing, geometric confinement and on-line UV/photolithography. Bazargan and Stoeber recently reported on reversible gelation in microchannels, from phase-separation of thermoresponsive pluronics with salt solutions of Na_3PO_4.[2]

In this study, we demonstrate *in situ* polymer gelation from electrostatic cross-linking reactions of cationic poly(allylamine hydrochloride) and citrate anions. This is the first example of *in situ* and in-channel polymer gel formation, which occurs at room temperature and does not require an immiscible phase or a flow focusing geometry. Formation parameters such as charge ratio, flow shear stresses and pH along with existing laminar flow conditions are seen to influence gel morphology. We find that the polyamine exhibits shear-thickening behavior when reacting with the citrate ions to form a viscoelastic gel phase. This gel phase remains stable and intact after cessation of flow. We also show that the extent of polymer gelation and disintegration of gels can be controlled by changing individually, the pH of citric acid and PAH streams. Hence, reversibility of gelation can be built into the system using pH as a variable. A model is proposed based of a phase diagram in charge ratio and flow (shear stress) that describes the mechanism and kinetics of gel-formation.

References:

1. Peterson, D. S., Solid supports for micro analytical systems. Lab on a Chip 2005, 5, (2), 132-139.

2. Bazargan, V.; Stoeber, B., Moving temporary wall in microfluidic devices. Physical Review E 2008, 78, (6).

See more of this Session: Microfluidics and Small-Scale Flows III: Technological Advances
See more of this Group/Topical: Engineering Sciences and Fundamentals

Formation of Supramolecular Hydrogel Microspheres Via Microfluidics

Wanyu Chen, Mechanical Engineering, University of Washington, Seattle, WA

Amy Shen, Mechanical Engineering, University of Washington, Seattle, WA

Yajiang Yang, Chemistry, Huazhong University of Science and Technology, Wuhan, China

Supramolecular hydrogels have attracted increasing interest in recent years due to their fascinating physical and morphological properties and their potential applications in drug delivery, oil recovery, and tissue engineering. Supramolecular hydrogel microspheres are hydrogel particles formed by the self-assembly of hydrogelators in water, through noncovalent interactions. In comparison with the traditional bulk format of the supramolecular hydrogels, hydrogel microspheres provide faster responses to external stimuli. In this paper, we provide a novel strategy to prepare supramolecular hydrogel microspheres with diameters ranging from 15 to 105 microns by using microfluidics. Optical and scanning electron microscopy (SEM) demonstrate that compact entanglement round cage-like aggregates of hydrogelator were formed within the supramolecular hydrogel microsphere, in contrast to loose and less compact aggregates within bulk hydrogel. Thermal analysis (DSC) indicates that supramolecular hydrogel microspheres are more thermally stable and can immobilize more water molecules, owing to compact entangled three-dimensional network structures. This observation is of particular importance for potential drug delivery, bioseparation, and biosensor applications, as it allows unstable biomolecules to be encapsulated inside the hydrogel microspheres.

See more of this Session: Microfluidics and Small-Scale Flows III: Technological Advances
See more of this Group/Topical: Engineering Sciences and Fundamentals

Calcium-Alginate Capsule Formation and Passive Separation in Microfluidic Device

Renee N. Perry, Chemical Engineering, Bucknell University, Lewisburg, PA

Erin Jablonski, Bucknell University, Lewisburg, PA

Janet T. Tesfai, Chemical Engineering, Bucknell University, Lewisburg, PA

Microfluidic devices can be used for many applications, including the formation of well-controlled emulsions. In this study, the capability to continuously create mono-disperse droplets in a microfluidic device is used to form calcium-alginate capsules through chemical crosslinking from aqueous droplets of calcium chloride and sodium alginate suspended in an oil solution. Calcium-alginate capsules have many potential uses, such as immunoisolation of cells, microencapsulation of active drug ingredients, and encapsulation of bitter agents in food or beverage products. Capsule formation is accomplished through fusion of a sodium alginate droplet and a calcium chloride droplet. The high surface tension between the droplet of calcium chloride and sodium alginate necessitates the use of the surfactant sodium dodecyl sulfate (SDS) and a device with a judiciously designed geometry. After creating the capsules, it is necessary to separate them out of the oil solution and into an aqueous solution. A common method of separation is centrifugation, which can damage both the capsules and their contents. The use of a microfluidic device with channel walls of disparate hydrophobicity is used to stabilize co-laminar flow of an oil phase and an aqueous phase. The disparity of hydrophobicity is accomplished by defining one side of the microfluidic device with the hydrogel poly (ethyl glycol) (PEG), which adheres to the glass surface of the device through the use of 3-(trichlorosilyl)-propyl methacrylate (TPM). Due to the difference in surface energy within the channel, the aqueous stream is stabilized near PEG and the oil stream is stabilized near a hydrophobic resin. The hydrophilic capsules travel to the interface of the oil and aqueous phase and are carried out of the device in the aqueous phase. The technique of continuous passive separation using co-laminar flow has shown promising results in separating the calcium-alginate capsules from the oil phase and into the aqueous phase. This unique method of passive separation can be used for other hydrophilic droplets and particles. The method is extended to the encapsulation of cells, molecules and particles of interest within the calcium alginate, and the entire process can be carried out under cell-friendly conditions.

See more of this Session: Microfluidics and Small-Scale Flows III: Technological Advances
See more of this Group/Topical: Engineering Sciences and Fundamentals

HIGH PRESSURE MICROFLUIDICS FOR STUDYING SUPERCRITICAL FLUIDS – LIQUIDS MICRO COFLOWS PROPERTIES

<u>Samuel Marre</u>[1,2], *Cyril Aymonier*[2], *Pascale Subra*[3], *Emmanuel Mignard*[1]

1 CNRS-Rhodia-Université Bordeaux 1 UMR5258 *Laboratoire du Futur*, 178 avenue du docteur Albert Schweitzer, 33608, Pessac, France

2 CNRS, Université de Bordeaux, ICMCB, 87 avenue du docteur Albert Schweitzer, F-33608, Pessac, France

3 Université Bordeaux 1, Laboratoire TREFLE (Transfert, Ecoulements Fluides Energetiques), ENSCBP, 16 avenue Pey-Berland, 33607, Pessac, France

The recent development of microfluidic devices offers several advantages over conventional macroscale processes: enhancement of mass and heat transfers, reproducibility, potential for sensor integration for *in situ* reaction monitoring, rapid screening of parameters, and low reagents consumption during optimization.

Developing microreactors which are pressure and temperature resistant (e.g. $1 < p < 25$ MPa and $0 < T < 300°C$) is of great interest since it addresses part of the limitations of conventional microsystems, such as limited compatible solvents and chemicals and Limited flows regime.

The coupling of microdevices with the specific properties of supercritical fluids –(SCFs) -high diffusivity, gas-like viscosity, liquid-like densities- opens new routes for controlling time and space scales which are crucial for synthesis and provides unique opportunities to 'visualize' supercritical processes compared to the use of blind vessels.

We developed capillary and glass-glass microreactors that allowed for characterizing supercritical fluids and multiphase (liquid / fluid) flows. High speed CCD camera visualization is used to observe, study and screen different flow regimes and phases (liquid, gas, supercritical).
We generated a co-flow using either silica capillary tubings as shown in Figure 1, or a glass – glass microreactor. Dripping and jetting regimes were observed for the first time for fluid - liquid co-flows, and their associated dynamic phase diagrams were plotted. Compressed carbon dioxide and water were used as model binary system and at pressures up to 20 MPa. Although CO_2 can dissolve in water as a result of carbonate formation, it is essentially immiscible with water in the range of temperature investigated in this study (20 – 80°C).
We demonstrated the possibility to tune reversibly the dynamic phase diagram simply by changing the pressure, without neither changing flow rates, nor fluids in use.

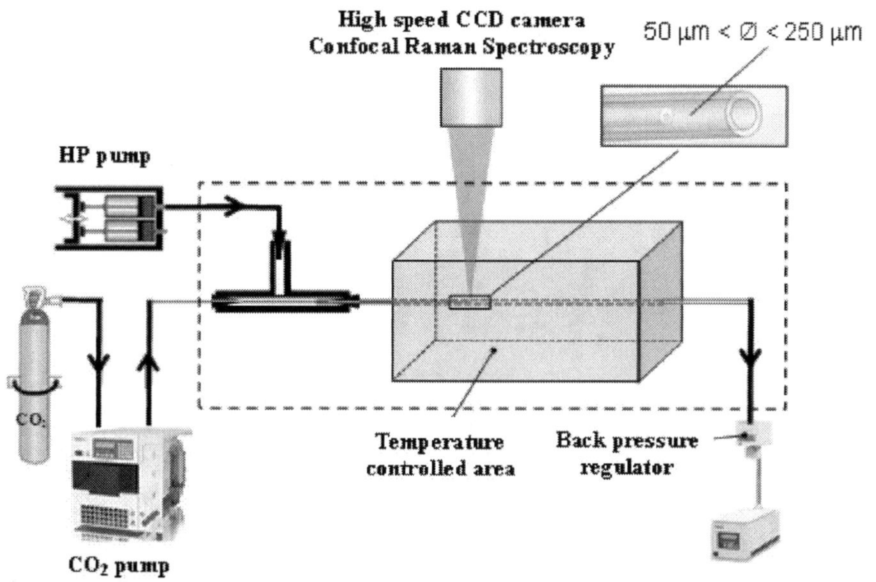

Figure 1. Experimental set-up with capillary microreactor, which can be replaced by glass – glass microreactor

Supercritical CO_2 is the inner fluid and water is the outer fluid. However, it is worth noticing that the reversed system can be investigated, *i.e.* water into CO_2, providing that the silica surface is silanized to become hydrophobic. Varying flow rate of both fluids leads to a (Qint, Qext) phase diagram, as shown in Figure 2:

The large range of values which can be obtained for *(Re, We and Ca)* with compressedCO_2 is of particular interest. Indeed, it is possible to switch from the dripping to the jetting regime playing only with pressure, all other experimental parameters being kept constant. Although dripping to jetting transition have been intensively studied and modeled, none of these models turns out to fit correctly our experimental results (Figure 2 red dotted line). This is understandable, regarding the range of Reynolds number considered here ($10 < Re < 3500$) which turns the flow from laminar to turbulent, and to the fact that $scCO_2$ is a highly dilatable medium, for which every term of the Navier-Stokes equation needs to be taken into account. Therefore, further works need to be done to model these systems.

The large range of values which can be obtained for *(Re, We and Ca)* with compressedCO_2 is of particular interest. Indeed, it is possible to switch from the dripping to the jetting regime playing only with pressure, all other experimental parameters being kept constant. Although dripping to jetting transition have been intensively studied and modeled, none of these models turns out to fit correctly our experimental results (Figure 2 red dotted line). This is understandable, regarding the range of Reynolds number considered here ($10 < Re < 3500$) which turns the flow from laminar to turbulent, and to the fact that $scCO_2$ is a highly dilatable medium, for which every term of the Navier-Stokes equation needs to be taken into account.

Therefore, further works need to be done to model these systems.

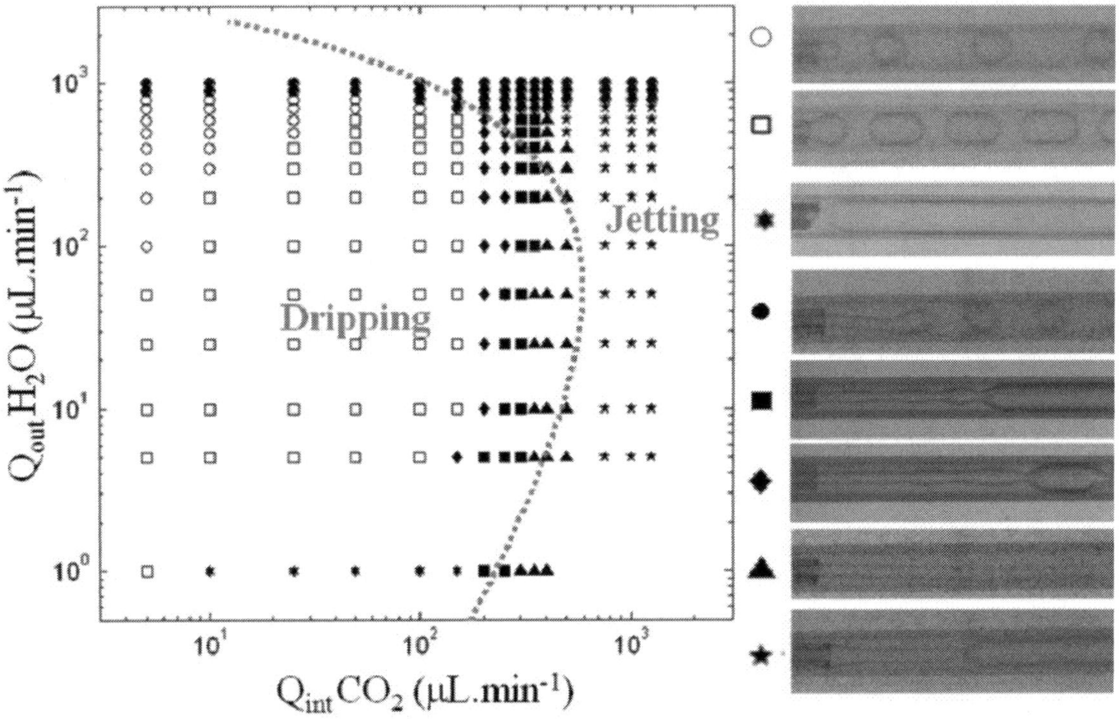

Figure 2. Dynamic phase diagram in the [Q_{int}, Q_{ext}] plane of the $scCO_2$ -liquid water coflowing system. The dripping regime includes spherical droplets (open circle) and slugs (open square). The jets have been observed with various shapes: large straight jet (hexagon), thin jets breaking into droplets (filled circle), wavy jet (filled square), wavy jet breaking into slugs (diamond), thin straight jet (triangle), or more complicated shapes for high Reynolds numbers flows (pentagon). In this configuration, R = 250 μm, r = 75 μm, p = 8 MPa, T = 50°C. The red dotted line corresponds to the dripping to jetting transition that should be obtained when considering a model developed for liquid–liquid micro coflows

To conclude, using transparent microfluidic devices, allows for generating rather easily stable droplets or jets of compressed fluids in immiscible liquids, and opens thus several opportunities in chemistry by using these hydrodynamic structures as microreactors or mixers. Further applications of Supercritical Microfluidics flows approach will be presented for the synthesis and the process of nanomaterials.

Control of Spreading and Drying of a Drug-Solvent Mixture From Marangoni Flows

Marlena Brown, Rutgers University, New Brunswick, NJ

Paul Takhistov, Rutgers University, New Brunswick, NJ

We studied the drying process of dopamine solution drops put by means drop-on-demand tecnique on a wettable solvent cast biopolymeric films. Three stages can be identified. First, the droplet spreads until a maximum radius where evaporation becomes dominant. The second phase is characterized by a constant radius due to competition between spreading and evaporation of the solvent. Finally, the contact line recedes until complete evaporation of the solvent.

During evaporation process we have observed the transformation of a studied system from one phase to another. Furthermore, we have shown that the observed 1st order phase transition can be manipulated via interfacial characteristics of substrate to produce amorphous or crystalline structures upon droplet impingement. In general there exists three regions for which the structure of the splat will exist as either as a structure lacking long range order of the positions of the atoms, a semi crystalline structure with limited long range atomic structure at the peripheries, or an ordered structure where the API exists in a crystalline state. The three distinct structures of the splat observed upon and after impingement are due to differences in wetting phenomena, interaction at the interface of the splat and the substrate, and evaporation kinetics of the drop due to differences in temperature reduction.

See more of this Session: Solid Liquid Interfaces
See more of this Group/Topical: Engineering Sciences and Fundamentals

Adsorption of Charged Colloids in Low Permittivity Solvents

Kwadwo Tettey, Chemical and Biomolecular Engineering, University of Pennsylvania, Philadelphia, PA

Daeyeon Lee, Chemical and Biomolecular Engineering, University of Pennsylvania, Philadelphia, PA

The behavior of charged colloidal particles in aqueous media has been studied extensively over the last century. Several classical theories and equations have been developed to characterize colloid-colloid and colloid-surface interactions in such media. Numerous studies have shown the effects of surfactants and salts on the adsorption of colloidal particles onto various surfaces from aqueous solutions. The behavior of charged colloids in low permittivity solvents, however, is still not well understood. Here, we study the adsorption behavior of charged particles in low permittivity solvents (e.g.hexadecane). Charged colloidal particles in low permittivity solvents are adsorbed onto surfaces with different functionalities. The effects of charge inducing agents on the adsorption of charged colloids in these solvents are also investigated. We use optical and fluorescence microscopy to observe the equilibrium structures formed in these solvents and use Fourier transform analysis to characterize these structures. In addition, a quartz crystal microbalance (QCM) is used to probe the kinetics of particle adsorption in low permittivity environments. With these tools, we show that electrostatic interactions influence the adsorption of colloidal particles from non-aqueous solvents.

See more of this Session: Solid Liquid Interfaces
See more of this Group/Topical: Engineering Sciences and Fundamentals

Wetting by Moving Triple Lines

Neeharika Anantharaju, Mechanical Engineering, Tennessee Technological University, Cookeville, TN

Mahesh V. Panchagnula, Mechanical Engineering, Tennessee Technological University, Cookeville, TN

Srikanth Vedantam, Mechanical Engineering, National University of Singapore, Singapore, Singapore

Wetting hysteresis was experimentally studied on smooth, chemically heterogeneous surfaces using a modified Wilhelmy plate apparatus. The novelty in the current set up lies in its ability to capture the microscopic triple line kinetics in addition to measuring the macroscopic contact angle during the advancing and receding processes. The motion of the triple line is known to play an important role in determining the macroscopic contact angle due to its ability to be pinned at various defect locations on real surfaces. This results in contact angle hysteresis.

Chemical heterogeneity was initially introduced by silanization of a part of smooth glass slide (innately hydrophilic) along its length. The triple line was observed to readily wet the hydrophilic material and hesitant to wet the hydrophobic (silanized) material. This study was extended to one and two-dimensional heterogeneities. The work provides insight into the fabrication of chemically heterogeneous surfaces for desired wetting applications.

See more of this Session: Solid Liquid Interfaces
See more of this Group/Topical: Engineering Sciences and Fundamentals

Characterization of Surface and Frictional Properties of Two-Component Alkyltrichlorosilane Monolayers

Steven G. Vilt, Department of Chemical and Biomolecular Engineering, Vanderbilt University, Nashville, TN

J. Ben Lewis, Department of Chemical and Biomolecular Engineering, Vanderbilt University, Nashville, TN

Jose L. Rivera, Department of Chemical and Biomolecular Engineering, Vanderbilt University, Nashville, TN

Clare McCabe, Department of Chemical and Biomolecular Engineering, Vanderbilt University, Nashville, TN

G. Kane Jennings, Department of Chemical and Biomolecular Engineering, Vanderbilt University, Nashville, TN

To investigate the affect of chain mobility and functional group exposure on friction, we report the tribological properties of two-component monolayers composed of long and short chain alkyltrichlorosilanes on silicon. The tribological properties of these monolayers were measured with a ball-on-flat tribometer at various loads and 0.1 mm/s speeds. The ratio of long and short chain molecules was varied, creating different levels of mobility and average chain cant in the upper liquid region of the film. Once a critical thickness is reached, the tribological properties for the mixed monolayers were indistinguishable from the pure, one-component monolayers although the oleophilicity was much higher for the two-component films. Two-component monolayers created from a combination of shorter perfluoroalkyl silanes and longer alkylsilane, however, showed tribological properties that were dependant on molecular mobility and the applied load. Collectively, the tribological behavior can be explained by the differences in interfacial free energies between hydrocarbon chains near hydrocarbon versus fluorocarbon surfaces, in addition to local repulsion between the hydrocarbon and fluorocarbon chains as quantified through molecular simulations.

See more of this Session: Solid Liquid Interfaces
See more of this Group/Topical: Engineering Sciences and Fundamentals

Interfacial Structure and Interactions in Clay-Polymer Nanocomposites

Qinghua Zeng, School of Materials Science and Engineering, The University of New South Wales, Sydney, Australia

Wen Xu, School of Materials Science and Engineering, The University of New South Wales, Sydney, Australia

Aibing Yu, School of Materials Science and Engineering, The University of New South Wales, Sydney, Australia

Max (Gaoqing) Lu, ARC Centre for Functional Nanomaterials, University of Queensland

Donald R. Paul, Department of Chemical Engineering, The University of Texas at Austin, Austin, TX

In the past decade, polymers reinforced with nanoparticles (e.g., spherical nanoparticles, nanotubes, clay platelets) have been emerged as a new class of materials and received tremendous attention in both scientific and industrial communities. Such polymer nanocomposites have demonstrated novel and often much improved mechanical, thermal, electrical and optical properties unattainable with their macro- and micro-counterparts. Such property enhancement is believed to be closely related to the molecular structure and interactions at the particle-polymer interface. Fundamental understanding of such structure and interactions will help us explore the enhanced properties and develop polymer nanocomposites with controlled structures and properties. In this talk, we will present the recent work toward such fundamental understanding in organoclays and nanocomposites by meaning of molecular modelling.

Keywords: polymer nanocomposites, molecular modelling, nanoparticles, interfacial interaction

References

1. Q. H. Zeng, A. B. Yu, G. Q. (Max) Lu and R. K. Standish. Molecular dynamics simulation of the structural and dynamic properties of organoclay, Journal of Chemical Physics, B 108: 10025-10033 (2004).

2. Q. H. Zeng, A. B. Yu, G. Q. (Max) Lu and D. R. Paul, Clay-based polymer nanocomposites: research and commercial development, Journal of Nanoscience and Nanotechnology, 5: 1574-1592 (2005).

3. Q. H. Zeng, A. B. Yu and G. Q.(Max) Lu, Interfacial interactions and structure of polyurethane intercalated nanocomposite, Nanotechnology, 16: 2757-2763 (2005).

4. D. R. Paul, Q. H. Zeng, A. B. Yu, and G. Q. (Max) Lu, The swelling and molecular packing in organoclays, Journal of Colloid and Interface Science, 292: 462-468 (2005).

5. Q. H. Zeng, A. B. Yu and G. Q. (Max) Lu, Multiscale modelling and simulation of polymer nanocomposites, Progress in Polymer Science, 33: 191-269 (2008).

See more of this Session: Solid Liquid Interfaces
See more of this Group/Topical: Engineering Sciences and Fundamentals

Modeling the Effect of Hydrogen Bonding for Understanding the Dissolution of α-Quartz Surfaces in Water

Adam Arnold Skelton, Chemical Engineering, Vanderbilt University, Nashville, TN

David J. Wesolowski, Chemical sciences division, Oak Ridge National Laboratory, Oak Ridge, TN

Peter T. Cummings, Chemical Engineering, Vanderbilt University, Nashville, TN

The interaction between water and two surfaces relating to different crystallographic faces of α-quartz have been atomistically modeled using the force-field of Lopes et al. (1). Comparisons with X-ray reflectivity experiments (2) have shown agreement between simulation and experiment for the 10-10 and α-quartz surfaces, supporting the accuracy of this model.

The focus of this study is to understand the dissolution of quartz and it is thought that the dissolution process involves both {1} hydrogen bonding between water hydrogen and quartz siloxane oxygen, and {2} close contact of water oxygen to the adjacent silicon (3,4). Different spatial arrangements of atoms on different surfaces/terminations mean that the access of silicon atoms and siloxane oxygens to water is different. The statistics of these close contacts/hydrogen bonds are analyzed and it is considered that atomistic simulations can give us an estimate of the probability of water molecules attacking (i.e. dissolving) quartz surfaces.

Although it has previously been reported that there are two possible terminations for the 10-10 α-quartz surface, only one of these surfaces shows agreement between experiment and simulation. It is postulated that the surface termination that is observed experimentally is the one more resistant to dissolution by water (2). It is suggested that a concerted mechanism involving water, hydroxyls and siloxanes is responsible for the dissolution of the top surface layer thereby exposing the dominant termination. This is supported by the greater number of water-siloxane and hydroxyl-siloxane H-bonds for the less dominant surface than for the more dominant surface.

Step edges for the α-quartz (the dominant surface) and the 10-11 surface have also been simulated in order to further illuminate the mechanism of dissolution. Insight into dissolution has been gained by rationalizing H-bonding statistics with previous AFM results (3), which show that steps form only in particular directions.

(1) P.E.M. Lopes et al . J. Phys. Chem. B 2006, 110, 2782

(2) M.L. Schlegel et al Geochimica et Cosmochimica Acta 2002 66, 3037

(3) L. J. Crisenti, J. D. Kubicki and Susan L. Brantley J. Phys. Chem. A 2006, 110, 198

(4) A. Pelmenschikov, J. Leszczynski and L. G. M. Pettersson, J. Phys. Chem. A 2001, 105, 9528

See more of this Session: Solid Liquid Interfaces
See more of this Group/Topical: Engineering Sciences and Fundamentals

Magnetic Displays - Magnetically Controlled Particle Orientation Enables Tunable Reflectivity

Stephanie B. Bubenhofer, Department of Chemistry and Applied Biosciences, ETH Zurich, 8093 Zurich, Switzerland

Evagelos K. Athanassiou, Department of Chemistry and Applied Biosciences, ETH Zurich, 8093 Zurich, Switzerland

Wendelin J. Stark, Department of Chemistry and Applied Biosciences, ETH Zurich, 8093 Zurich, Switzerland

Controlled orientation of anisotropic micro- and nanostructures in fluids attracts considerable interest in biology and nanotechnology. Electric field induced orientation is common, mostly used for assembling and ordering particles. Magnetic induced orientation was thought to restrict the anisotropic structures to magnetic or magnetically labeled particles. However, recent studies with ferrofluids as forcing agent have shown the possibility of even orienting nonmagnetic microwires by magnetic fields [1]. Here, we present a system with tunable optical properties which is based on the magnetically controlled orientation of aluminum microflakes incorporated in a stabilized, highly superparamagnetic ferrofluid dispersion. By the application of an external magnetic field the orientation of the highly reflecting aluminum flakes could be switched from parallel to vertical, allowing a change in reflectivity. Theoretical investigations have shown that the mechanism of the orientation of the aluminum flakes can be related to a continuum approach of the magnetic fluid. The highly field sensitive change of reflectivity allowed us to demonstrate a possible area of application. A display without the need of background light and a switching time of less than 100 milliseconds was constructed, in which the optical features are visible even under a small angle observation. Besides large screen applications also stationary display (e-paper) or tunable heat isolating windows could be application areas, with the benefit of low energy use [2]. Figure: Switchable Reflectivity. Magnetic fluid film (0.2 mm) between two capton foils (left) reflecting due to a magnetic field parallel to the plane and (right) transparent, due to a perpendicular field. [1] C. Ooi, R.M. Erb, B.B. Yellen, J. Appl. Phys. 103 (2008) 07E910. [2] S.B. Bubenhofer, E.K. Athanassiou, R.N. Grass, F.M. Köhler, M. Rossier, W.J. Stark, in preparation (2009).

See more of this Session: Solid Liquid Interfaces
See more of this Group/Topical: Engineering Sciences and Fundamentals

Interactions Between Carbon-Based Nanoparticles and Lipid Bilayers

Young-Min Ban, Department of Chemical Engineering, University of Florida, Gainesville, FL

Dmitry I. Kopelevich, Department of Chemical Engineering, University of Florida, Gainesville, FL

In this talk, we present results of coarse-grained molecular dynamics (MD) simulations of interactions between dipalmitoylphosphatidylcholine (DPPC) lipid membranes and carbon-based nanoparticles (carbon nanotubes and fullerenes). Understanding these interactions is important for development of biomimetic devices and assessment of potentially harmful effects of nanomaterials on living cells. The considered nanoparticles are hydrophobic and therefore they tend to reside in the interior of the lipid bilayer. It is observed that these nanoparticles may affect several physical properties of the membrane, including its bending modulus, spontaneous curvature, and the lateral pressure profile. Fluctuations of tilt of the lipid molecules are also affected in the direct neighborhood of the nanoparticles. In addition, we observe that the dynamics of the thermal fluctuations of the membrane is significantly affected by the nanoparticles, especially when the latter approach the border between the hydrophilic head-groups and the hydrophobic tail-groups of the lipids.

The information obtained from MD simulations is used to construct a meso-scale model for lipid vesicles with nanoparticles embedded in their membranes. This model is used to assess possibility of the nanoparticle-induced membrane instability. The talk is concluded by a discussion of possible effect of the embedded nanoparticles on biophysical processes in cellular membranes.

See more of this Session: Thermodynamics and Transport in Lipid Bilayers
See more of this Group/Topical: Engineering Sciences and Fundamentals

Phase Behavior Calculations of DOPC Phospholipid Bilayers

Doris M. Grillo, Chemical and Biological Engineering, Northwestern University, Evanston, IL

Igal Szleifer, Department of Biomedical Engineering, Northwestern University, Evanston, IL

Monica Olvera de la Cruz, Northwestern University, Evanston, IL

It has been experimentally shown that the nonspecific adsorption of charged nanoparticles onto the phospholipid bilayers that contain phosphocholine head groups cause a local surface modification at the points where the nanoparticle is adsorbed. When the charged species binds to the phospholipid bilayer surface, its local state of phase can be changed as a result of the reorientation and loss of degrees of freedom of the zwitterionic phosphocholine head group[1]. An accidental shift in the phase transition temperature of a phospholipid bilayer membrane can lead to fatal consequences for living systems. For example, if the presence of a negatively charged nanoparticle can significantly influence the orientation of the DOPC head group and consequently change the integrity of the fluid-like bilayer, vital properties of the bilayer such as the transport of molecules in and out of the cell can be inhibited. Therefore, it is important to understand the theoretical fundamentals of this change of phase state by explicitly taking into account the DOPC head group size, shape and orientation when proposing a molecular model for a lipid bilayer containing phospholipids such as DOPC.[2]

In our theoretical model a self consistent mean-field approach is implemented to the adsorption of a charged nanoparticle onto a DOPC containing lipid bilayer and an extension of the Poisson-Boltzmann equation is derived through the functional minimization of the total free energy. This molecular model takes into account for the first time the size, shape and orientational entropy of the DOPC lipid head group. The results show that these head group considerations are essential in order to properly predict the main chain liquid to gel phase transition of the DOPC containing lipid bilayer.[3]

References

[1] B. Wang, L.zhang, S. Bar and S. Granick, PNAS, vol.105, no. 47 ,18171 (2008)

[2] E. Mbamala, A. Fahr and S. May, Langmuir 22, 5126 (2006)

[3] R. Elliott, M. Schick, and I. Szleifer, Phys. Rev. Letters. 96, 098101 (2006).

See more of this Session: Thermodynamics and Transport in Lipid Bilayers
See more of this Group/Topical: Engineering Sciences and Fundamentals

Effect of Polymer Grafting On the Bilayer Gel to Liquid Crystalline Transition

K. G. Ayappa, Department of Chemical Engineering, Indian Institute of Science, Bangalore, India

Foram M. Thakkar, Chemical Engineering, Indian Institute of Science, Bangalore, India

Grafted polymers on the surface of lipid membranes have potential applications in liposome based drug delivery and supported membrane systems.

The effect of polymer grafting on the phase behaviour of bilayers made up of single tail lipids is investigated using dissipative particle dynamics.

The bilayer is maintained in a tensionless state using a barostat. Simulations are carried out by varying the grafting fraction, defined as the ratio of the number of polymer molecules to the number of lipid molecules, and the length of the lipid tails.

At low grafting fractions the bilayer shows a sharp transition from the gel (L_β) to the liquid crystalline (L_α) phase

This main melting transition temperature is lowered as the grafting fraction is increased and above a critical value of the grafting fraction additional low temperature phases are observed in the main transition.

At low temperatures the L_β phase is transformed to the tilted, $L_{\beta'}$ phase and upon further increase in temperature the interdigitated $L_{\beta I}$ phase is observed. The temperature range over which the intermediate $L_{\beta'}$ and $L_{\beta I}$ phases are observed is a function of the lipid tail length and the grafting fraction.

The presence of the intermediate phases is attributed to the increase in area per head group due to the lateral pressure exerted by the polymer brush. The polymer brush height, area expansion and decrease in the melting temperatures as a function of the grafting fraction were found to follow the scalings predicted by the self consistent mean field theories for grafted polymer membranes. Our study shows that the grafted polymer density can be used to effectively control the temperature range and occurrence of a given bilayer phase.

See more of this Session: Thermodynamics and Transport in Lipid Bilayers
See more of this Group/Topical: Engineering Sciences and Fundamentals

Molecular Dynamics Simulations of Complex Mixed Lipid Bilayers to Model Yeast Membranes

Joseph B. Lim, Department of Chemical and Biomolecular Engineering, University of Maryland, College Park, MD

Jeffery Klauda, Department of Chemical and Biomolecular Engineering, University of Maryland, College Park, MD

There has been growing interest in using molecular dynamics (MD) simulations to investigate membrane-protein systems. However, these simulations have heretofore consisted of no more than three types of lipids, greatly simplifying the compositions of actual membranes. CHARMM-GUI, a web-based graphical user interface for performing various CHARMM functions, has recently been used to generate more complex lipid bilayers composed of six different lipids - cholesterol, DOPC, DPPC, POPA, POPE, and POPS - in order to accurately and realistically model yeast membranes. Four bilayers containing varying amounts of chain saturation, cholesterol concentration, and surface tension were created. These yeast membranes were simulated for a total of 170 ns using the atomistic CHARMM27r force field at 303.15 K. MD simulations with a high concentration of unsaturated chains (73%) showed an increase in surface area per lipid and an decrease in the SCD order parameters for DPPC, as opposed to the membrane created with a high concentration of saturated chains (60-63%). Simulations of the more saturated membranes were in a liquid-ordered state and were in agreement with experimental cholesterol-containing membranes. The unsaturated membrane simulation exhibited a larger average tilt angle of cholesterol with respect to the bilayer normal. Moreover, cholesterol in the unsaturated membrane actually oriented parallel to the bilayer surface for periods of less than a nanosecond. This result supports previous observations of parallel cholesterol existing at the center of polyunsaturated fatty acid membranes, and makes us the first to observe parallel cholesterol in a fully atomistic simulation. Currently, we are using the newly developed atomistic CHARMM35 force field to simulate the yeast bilayers with 25-hydroxysterol and cholesterol. In unsaturated bilayers, oxysterols are believed to be more thermodynamically stable in the parallel orientation compared to cholesterol, so it is our aim to elucidate the dynamics of this phenomenon with our new simulations.

See more of this Session: Thermodynamics and Transport in Lipid Bilayers
See more of this Group/Topical: Engineering Sciences and Fundamentals

Imaging Lipid Distribution within Lipid Membranes Using Nanoscale Secondary Ion Mass Spectrometry

Mary L. Kraft, Chemical & Biomolecular Engineering, University of Illinois at Urbana-Champaign, Urbana, IL

Christopher R. Anderton, Chemistry, University of Illinois at Urbana-Champaign, Urbana, IL

Jessica F. Frisz, Chemistry, University of Illinois at Urbana-Champaign, Urbana, IL

Kaiyan Lou, Chemical & Biomolecular Engineering, University of Illinois at Urbana-Champaign, Urbana, IL

William P. Hanafin, Chemical & Biomolecular Engineering, University of Illinois at Urbana-Champaign, Urbana, IL

Peter K. Weber, Lawrence Livermore National Laboratory, Livermore, CA

Imaging the lateral distributions of cholesterol, specific lipid species, and proteins is in the plasma membrane at the 50 - 200 nm lateral resolution that is relevant to membrane organization is currently a major challenge. Though the distributions of membrane proteins can be unambiguously visualized by immunogold labeling and transmission electron microscopy (TEM), our understanding of how lipids and sterols are organized in the plasma membrane and the forces that drive this organization is poor. We previously demonstrated that the distributions of two isotopically labeled lipids within phase-separated model lipid membranes can be imaged and quantified with a lateral resolution of 100 nm using a nanoscale secondary ion mass spectrometry (nanoSIMS) approach. NanoSIMS performed with the NanoSIMS 50 (Cameca Instruments) reveals the elemental and isotopic composition in a sample with a lateral resolution as high as 50 nm. By incorporating a distinct stable isotope into each lipid species of interest, the isotopically enriched secondary ions produced during nanoSIMS analysis encode for component identity. Here we discuss the application of nanoSIMS to imaging the lipid and cholesterol distribution within more complex biological membranes.

See more of this Session: Thermodynamics and Transport in Lipid Bilayers
See more of this Group/Topical: Engineering Sciences and Fundamentals

Calculation of Partition Coefficients of Chain Anchors in Liquid-Ordered and Liquid-Disordered Phases

Mark Uline, Department of Biomedical Engineering, Northwestern University, Evanston, IL

Gabriel Longo, International School for Advanced Studies (SISSA), Trieste, Italy

Michael Schick, Departement of Physics, Seattle, WA

Igal Szleifer, Department of Biomedical Engineering, Northwestern University, Evanston, IL

One of the most interesting models of biological membranes posits that they are not compositionally homogeneous, but have aggregates rich in cholesterol and saturated lipids that float, like rafts, in a sea of unsaturated lipids [1]. A disagreement centers on if the rafts and the sea are in fact domains of two coexisting phases denoted by liquid-order and liquid-disorder [2]. The liquid-order phase is rich in cholesterol and saturated lipids whose chains are relatively well ordered; the other, liquid-disorder phase, is rich in unsaturated lipids whose chains are not so well ordered. One reason for the interest in this picture is that an inhomogeneous membrane would provide a substrate within which different proteins would be enriched either in rafts or in the sea. Presumably saturated anchors prefer the environment of the raft, or liquid-order phase, while unsaturated and/or otherwise bulky anchors prefer the environment of the sea of unsaturated lipids, the liquid-disorder phase. In order to test this hypothesis on protein anchors, we use a theoretical model membrane that is able to produce coexisting liquid phases [3].

From our theoretical model-membrane ternary system of cholesterol, a lipid with two saturated tails of 16 carbons, such as dipalmitoylphophatidylcholine (DPPC), and a lipid with two mono-unsaturated tails of 18 carbons, such as dioleoylphosphatidylcholine (DOPC) [3], we report calculated partition coefficients for single chains of various lengths and saturation and a few cases of double-chained molecules. The results show that partition coefficients increase with chain length and degree of saturation.

References

[1] O. G. Mouritsen, Life - As a Matter of Fat (Springer-Verlag, Berlin, 2005).

[2] T. McMullen and R. McElhaney, Biochim. Biophys. Acta 1234, 90 (1995).

[3] R. Elliott, M. Schick, and I. Szleifer, Phys. Rev. Letters. 96, 098101 (2006).

See more of this Session: Thermodynamics and Transport in Lipid Bilayers
See more of this Group/Topical: Engineering Sciences and Fundamentals

Molecular Simulations of Liposomal Anti-Cancer Drug Carriers

Emily Boehler, Chemical and Biomolecular Engineering, North Carolina State University, Raleigh, NC

Carol K. Hall, Chemical and Biomolecular Engineering, North Carolina State University, Raleigh, NC

Stavroula Sofou, Polytechnic Intsitute of NYU, Brooklyn, NY

Liposomes have gained much attention recently due to their ability to mediate the intracellular delivery of cancer therapies. Liposomes are artificial vesicles composed of phospholipid bilayer membranes; they can transport both hydrophilic and hydrophobic drugs. Two major goals of liposomal anti-cancer drug delivery are to promote the accumulation and retention of drug-carrying liposomes in tumor cells and to trigger the release of drug molecules inside the cells. A new intermediate resolution model of liposomal drug carriers is used to determine how the properties of the lipid molecules affect their drug trapping efficiencies and to analyze how drug release kinetics changes in response to different external stimuli. The model is essentially a discontinuous version of the "Martini" model for lipids; it is designed for use with discontinuous molecular dynamics (DMD) simulation, a very fast alternative to traditional molecular dynamics simulation. The model is applied to aqueous bilayers composed of two types of phospholipids: those containing phosph-L-serine (PS) head groups and those containing phosphatidylcholine (PC) head groups. Liposomes composed of these species have been found experimentally to phase separate into heterogeneous domains (rafts) as the pH is lowered. Runs are conducted to determine how the tail lengths and mole ratios of the PC and PS phospholipids and the presence of targeting functionalities affect: (1) the formation of heterogeneous domains and membrane permeability, (2) the release of model drug molecules through the leaky liposome surface, and (3) fusion of the liposomal membrane to the endosomal membrane. The results are compared to experimental results found in the Sofou lab.

See more of this Session: Thermodynamics and Transport in Lipid Bilayers
See more of this Group/Topical: Engineering Sciences and Fundamentals

Coarse-Grained Molecular Dynamics Simulations of Transmembrane Assisted Permeation of Gases and Water Across Lipid Membranes

Huajun Yuan, Chemical Engineering, University of Illinois at Chicago, Chicago, IL

Cynthia Jameson, Chemical Engineering, University of Illinois at Chicago, Chicago, IL

Sohail Murad, Chemical Engineering, University of Illinois at Chicago, Chicago, IL

Molecular transport through biological membranes occurs in a range of important chemical and biological processes. We have carried out molecular dynamics simulations to study the transport of a range of gases in a dipalmitoylphosphatidylcholine (DPPC) bilayer membrane. By reducing the degrees of freedom and employing suitable potentials, a coarse-grained model was used to provide direct insight into collective phenomena in biological membranes at large time and length scales.

We also compare the molecular transport across lipid membrane with and without the assistance of transmembrane channels. For assisted transport we have chosen an outer membrane protein A (OmpA) -- a porin from Escherichia coli with a small pore size and high stability. MD simulations studies in such assisted transport are then also compared with available experimental data. The pore characteristics and interaction between lipid and protein were investigated and transport of water molecules within the channel was also studied. The information obtained from this study will lead to better understanding of proteins and advance the use of OmpA for development of biochips and drug delivery systems.

See more of this Session: Thermodynamics and Transport in Lipid Bilayers
See more of this Group/Topical: Engineering Sciences and Fundamentals

Mechanism of Protegrin Dimerization On Lipid Bilayer Membranes: Molecular Dynamics Simulation

Victor Vivcharuk, Chemichal Engineering and Material Science, University of Minnesota, Minneapolis, MN

Yiannis N. Kaznessis, Chemical Engineering and Materials Science, University of Minnesota, Minneapolis, MN

The mechanism of antimicrobial activity of the 18-residue cationic peptide protegrin-1 (PG1) is investigated with atomic-scale molecular dynamics (MD) simulations. In particular, we study the oligomerization of protegrin molecules, because their activity is inseparable from oligomer peptide structures on the membranes of bacteria [1-3]. PG1 is known to form octameric pores in bacterial membranes that lyse the cell by preferentially allowing potassium diffusion [2].

One important question in elucidating the mechanism of antimicrobial activity is whether PG1 forms dimers and then oligomers in water, on the membrane surface, or deep in the membrane core. Literature studies propose different mechanisms and with the simulations we provide some clarity.

MD simulations measuring the free energy difference or the potential of mean force (PMF) between two protegrins under physiological conditions are used to elucidate the efficacy of such interaction in water buffers and in bilayer membranes. The simulated lipid membranes are composed of palmitoylphosphatidylethanolamine (POPE) and palmitoyloleoylphosphatidylglycerol (POPG) lipids, mimicking the bacterial membrane.

More than sixty separate 4 ns-long MD simulations with different orientations of PG1 dimers with anti-parallel and parallel orientations were carried out in different environments. To calculate the PMF we employed a combination of the constrained MD technique and the thermodynamic integration method. The application of the blocking method [4] indicates that our MD trajectories became uncorrelated after ~ 0.8 ns. We calculate the forces on the peptides due to the solvent and quantify the contribution of the effect due to the displacement of ions and water on the binding affinity. Ion contributions are shown to important at all length scales while solvent effects play an important role only in the contact region. The PMF can be used to predict measurable (but currently unknown) binding coefficients. In the presentation, we will reconstruct the minimum free energy path for the peptides to dimerize, oligomerize and form functional pores in lipid bilayers.

[1]. R. M. Murphy, Annu. Rev. Biomed. Eng. 2002, 4, 155-174.

[2]. A. Langham, A. Sayyed-Ahmad, Y Kaznessis, Journal of the American Chemical Society, 2008, 130(13): 4338-4346

[3]. H. Jang, B. Ma, R. Nussinov, BMC Structural Biology 2007, 7:21.

[4]. H. Flyvbjerg, H.G. Petersen, J.Chem.Phys. 1989, 91: 461-466.

See more of this Session: Thermodynamics and Transport in Lipid Bilayers
See more of this Group/Topical: Engineering Sciences and Fundamentals

Effects of Synthetic Linear and Cyclic Peptide Analogs On Liposome Phase Behavior, Transport, and Morphology

Anju Gupta, Chemical Engineering and Department of Biomedical and Pharmaceutical Sciences, University of Rhode Island, Kingston, RI

Geoff Bothun, Chemical Engineering, University of Rhode Island, Kingston, RI

Rob Deluca, Chemical Engineering, University of Rhode Island, Kingston, RI

Guofeng Ye, Biomedical and Pharmaceutical Sciences, University of Rhode Island, Kingston, RI

Deendayal Mandal, Department of Biological and Pharmaceutical Sciences, University of Rhode Island, Kingston, RI

Keykavous Parang, Biomedical and Pharmaceutical Sciences, University of Rhode Island, Kingston, RI

Cell penetrating peptides (CPP) are small cationic peptides that possess the ability to traverse plasma membranes, which makes them candidates as molecular transporters and therapeutic agents. Many CPPs exhibit antimicrobial activity and can disrupt bacterial membranes that are anionic in nature. We have recently reported a series of short novel linear peptide analogues (LPAs), synthesized as Arg-Cn-Arg-Cn-Lys where Cn represents the alkyl linkage separating the cationic residues (n= 4 to 11), as molecular transporters for drug delivery (Ye et al., J Med. Chem. 50, 2007, 3604). To understand LPA and newly-synthesized cyclic peptide analogs (CPA) membrane interaction mechanisms and guide the design of new peptides, we are examining how these peptides disrupt lipid phase behavior, alter membrane permeability, and restructure zwitterionic (DPPC) and mixed zwitterionic/anionic liposomes (DPPC/DPPG and DPPC/DPPS). Isothermal titration calorimetry (ITC) is used to examine the binding of the peptides with the bilayers. For the series, LPA-C4, C7 and C11, LPA-C11 exhibits strong binding to fluid DPPC and DPPC/DPPG bilayers and endothermic effects signifying lipid disordering and distribution, whereas, LPA C4 and C7 displayed weaker exothermic binding. Similarly, differential scanning calorimetry (DSC) shows a decrease in melting temperature and an elimination of the pre-transition in the lipid bilayers with the addition of LPA-C11, while LPA-C4 and C7 showed little effect. While all LPAs bound to the lipids via electrostatic attractions between the cationic residues and the lipid phosphate groups (DPPC, DPPG, and DPPS) and charge attraction between the cationic residues and glycerol headgroups (DPPG and DPPS), our calorimetric studies show that the length of the hydrophobic linkage controls binding and lipid perturbation. Carboxyflurescein permeability studies were conducted to determine if lipid disruption increased membrane leakage. LPA-C11 revealed highest permeability with anionic lipids, consistent with DSC and binding results. Structural and morphological changes of the liposomes were detected by cryogenic transmission electron microscopy (cryo-TEM). In general, LPA-C4 and C7 led to small changes in liposome structure, while LPA-C11 caused liposome fusion and produced large unenclosed liposomes in DPPC and DPPC/DPPG, , For DPPC/DPPS, the addition of all LPAs caused the morphological change from spherical liposomes to disks.. We attribute this to non-ideal mixing between DPPC and DPPS, which was not observed between DPPC and DPPG. Hence, the LPA peptides can bind tightly and restructure both zwitterionic and zwitterionic/anionic membranes, but lead to significant morphological changes when the membrane contains domains rich in zwitterionic or anionic lipids. Currently we are exploring a variety of cyclic peptides consisting of eight alternating amino acids containing electrostatic and hydrophobic residues. We hypothesize that these cyclic peptides will bind strongly to lipid bilayers and can self-assemble within the membrane to function as membrane-spanning channels or pores. A collaborative investigation on the antimicrobial properties possessed by these linear and cyclic peptide analogs will also be presented.

See more of this Session: Biomolecules at Interfaces I

See more of this Group/Topical: Engineering Sciences and Fundamentals

Interaction of Antimicrobial Peptide Ovispirin with Lipid Bilayers

Ann Marie Woys, Department of Chemistry, University of Wisconsin-Madison, Madison, WI

Yu-Shan Lin, Department of Chemistry, University of Wisconsin-Madison, Madison, WI

Allam S. Reddy, Department of Chemical and Biological Engineering, University of Wisconsin-Madison, Madison, WI

James L Skinner, Department of Chemistry, University of Wisconsin-Madison, Madison, WI

Martin Zanni, Department of Chemistry, University of Wisconsin-Madison, Madison, WI

Juan J. De Pablo, Department of Chemical and Biological Engineering, University of Wisconsin, Madison, WI

Antimicrobial peptides kill gram-negative and gram-positive strains of bacteria by interacting with their cell membranes. To understand these interactions at molecular scales it is necessary to develop comprehensive approaches that include novel combinations of experiments and theory. We present one such approach, in which two-dimensional infrared spectroscopy and combined classical-quantum mechanical calculations are used to elucidate the precise nature of interactions between Ovispirin, a 17 residue antimicrobial peptide, and a POPC/POPG mixed bilayer membrane. Classical Monte Carlo and molecular dynamics simulations are used to provide position dependent free-energies for the peptide in the membrane bilayer. A coupled classical-quantum mechanical approach is then used to simulate the 2D IR spectrum corresponding to the classically-generated structures. The results of theoretical predictions are then compared to our own experimental 2D IR measurements. Our results indicate that Ovispirin, in the presence of the membrane, exists in an alpha helical conformation which is adsorbed in an orientation parallel to the bilayer plane. The hydrophobic side chains are buried in the lipid tails, and the polar side chains protrude into the aqueous phase. Our results are consistent with previously published solid-state NMR structures of Ovispirin. Perhaps more importantly, our calculated 2D-IR diagonal line widths for the equilibrium structures obtained in MD our simulations are in quantitative agreement with the corresponding experimental values for the same system.

See more of this Session: Biomolecules at Interfaces I
See more of this Group/Topical: Engineering Sciences and Fundamentals

Characterization of the Structure of Oligopeptide Kinase Substrates Immobilized On Solid Surfaces

Yiqun Bai, Chemical and Biological Engineering, University of Wisconsin- Madison, Madison, WI

Xiaosong Liu, Physical Biosciences Division Lawrence Berkeley National Laboratory, Berkeley, CA

Paul Bertics, Biomolecular Chemistry, University of Wisconsin- Madison, Madison, WI

Franz Himpsel, Physics, University of Wisconsin- Madison, Madison, WI

Nicholas L. Abbott, Department of Chemical and Biological Engineering, University of Wisconsin-Madison, Madison, WI

This presentation addresses two methods for characterization of the structure of oligopeptides immobilized at interfaces. Although many methods exist to measure the functional properties of biomolecules at interfaces (e.g., enzymatic activity), few approaches enable characterization of the structure of the biomolecules. A particular focus is directed to oligopeptides that are substrates of the epidermal growth factor receptor (EGFR). The EGFR is a tyrosine kinase which phosphorylates the oligopeptides, and inhibitors of the tyrosine kinase activity of the EGFR represent an important class of emerging anti-cancer agents. The first method of characterization to be described is based on Near Edge X-ray Absorption Fine Structure Spectroscopy (NEXAFS). We have used NEXAFS to characterize the chemistry of immobilization of the oligopeptides. We have also explored the hypothesis that phosphorylation of the oligopeptides can lead to conformational changes that can be measured by NEXAFS. The second method of characterization involves measurements of the orientational ordering of liquid crystals supported on oligopeptide-decorated surfaces. Our measurements define the sensitivity of the orientational ordering of liquid crystals to the density and structure of the EGFR peptide substrates. We will conclude this presentation by comparing and contrasting the information obtained from these two methods for characterizing the structure and density of biomolecules at surfaces.

See more of this Session: Biomolecules at Interfaces I
See more of this Group/Topical: Engineering Sciences and Fundamentals

Modulating Protein-Surface Interactions Using Mixed Self-Assembled Monolayers

Chrysanty Tedjo, Chemical and Biomolecular Engineering, Vanderbilt University, Nashville, TN

Paul E. Laibinis, Chemical and Biomolecular Engineering, Vanderbilt University, Nashville, TN

The ability of self-assembled monolayers (SAMs) to expose a wide range of functional groups on their surface allows protein-surface interactions to be probed in a controlled manner. With SAMs, protein adsorption studies have often been performed on flat surfaces under static conditions using methods such as ellipsometry, XPS, radio-labeling, and dye quantification. In these approaches, a surface that has been exposed to a protein solution typically undergoes a set of washing procedures before the amount of surface adsorbed protein is analyzed. As a result of procedural differences, there can be lab-to-lab variations in terms of the transition from a non-adsorbing to an adsorbing surface across surfaces of different compositions. More broadly, the role that shear rate has on protein adsorption remains uncertain. For example, the amount of adsorption onto a surface has been reported by different groups to increase, be unaffected, or to decrease with increases in shear rate. Here, we have investigated protein-surface interactions to mixed SAMs formed from Cl3Si(CH2)11(EG)3OCH3 (EG3OMe) and Cl3Si(CH2)7CH3 (C8). These SAMs are covalently anchored to their support and provide a means for creating surfaces of different surface energies. We determined the thicknesses, wettabilities, and surface compositions of the mixes SAMs using various methods of surface characterization. We examined protein-surface interactions onto these mixed SAMs under both static and shear-flow condition. The in-situ measurements of protein adsorption were performed in a flow cell that is integrated with a Total Internal Reflectance Fluorescence (TIRF) system. Using proteins differing in molecular weights and hydrophobicities, we were able to examine the interplay between shear rate, protein characteristics, and surface hydrophobicity and composition on protein adsorption and desorption. The results from this study address observed differences in the threshold conditions for establishing non-fouling behavior toward specific proteins and provide suggestions for broadly defining conditions where levels of protein adsorption may be modulated under flow and static conditions or made to be enhanced for specific proteins over others.

See more of this Session: Biomolecules at Interfaces I
See more of this Group/Topical: Engineering Sciences and Fundamentals

Effect of Hydrophobic Surfactant Proteins SP-B and SP-C On the Phase and Morphology of Protein Deficient Native Surfactant Films

Prajnaparamita Dhar, Chemical Engineering, University of California, Santa Barbara, Goleta, CA

Alan J. Waring, Departments of Medicine and Pediatrics, University of California, Los Angeles, Los Angeles

Joesph A. Zasadzinski, Department of Chemical Engineering, University of California, Santa Barbara, CA

We present alterations in the phase and morphology of a native replacement surfactant film induced either by the removal of the hydrophobic surfactant proteins from a commercially available replacement surfactant, Survanta, or by the systematic addition of bovine hydrophobic surfactant proteins SP-B and SP-C to the protein deficient Survanta lipid film. The surface tension lowering property of a native surfactant film is significantly compromised in the absence of the proteins, as is the ability of the film to undergo reversible collapse. A lack of proteins also causes the characteristic shoulder prevalent in most lung surfactant mixtures to disappear, illustrating the inability of the film to undergo reversible squeeze out by forming "surface associated surfactant reservoirs". Addition of SP-B causes an increase in the amount of material adsorbed from the sub-phase. Further it increases the monolayer stability and the compressibility modulus of film (i.e. decreases the compressibility of the film), thus helping the film achieve a high enough surface pressure during compression as well as quick re-absorption of material during expansion. On the other hand, SP-C plays a dominant role in the formation of bilayer patches containing unsaturated lipids. SP-C also changes the mechanisms of monolayer collapse, and the film collapses via the formation of reversible collapse cracks. However, it is only in the presence of both SP-B and SP-C that the monolayer films are able to perform all the biophysical functions necessary for the proper working of the lung surfactant. Our results explicate the role of the individual proteins in the lipid protein interactions associated with lung surfactant mixtures in a native system. Further, these observations provide conclusive evidence showing that both SP-B and SP-C are equally necessary for the long term survival of air-breathing mammals.

See more of this Session: Biomolecules at Interfaces I
See more of this Group/Topical: Engineering Sciences and Fundamentals

Dramatically Increased Enzymatic Activity of Lysozyme and Myoglobin Adsorbed On the Internal Surface of SBA-15 Mesoporous Silica

Lung-Ching Sang, Chemical Engineering, Rensselaer Polytechnic Institute, Troy, NY

Marc-Olivier Coppens, Isermann Department of Chemical and Biological Engineering, Rensselaer Polytechnic Institute, Troy, NY

When protein molecules are strongly adsorbed on a surface, their activity in catalyzing chemical conversions can be severely altered with respect to free protein molecules in solution, depending on the geometrical and chemical structure of the surface. Local surface curvature plays an important role. For example, on silica nanoparticles and carbon nanotubes, effects of convex surface curvature on protein stability and activity were noted [1,2]. If proteins are similar in size to the pore diameter of mesoporous materials, adsorption on their surface leads to confinement effects that could also alter their activity. Several researchers indeed noted significant, sometimes stabilizing, effects of concave surface curvature [3], although a systematic study requires accurate control over particle morphology and size, and pore morphology and diameter.

By using a series of monodisperse SBA-15 mesoporous silica particles with high surface area, and carefully controlled, unique pore diameter in the range of 5-11 nm, in which large amounts of proteins are strongly adsorbed from buffered solutions, effects of local curvature and protein-surface interactions could be studied in a consistent way. It was verified by multiple washes with buffered solution that the protein would not leach out of the particles. We focused on the activity and stability of two different proteins - lysozyme and myoglobin - after adsorption on SBA-15 with different pore sizes. Lysozyme is an ellipsoidal protein, with dimensions equal to 3nm × 3nm × 4.5nm and myoglobin is a globular protein, with dimensions equal to 3nm × 4nm × 5nm. Effects of the local pore curvature of SBA-15 on the protein structure of the adsorbed protein were studied by comparing the activity and stability of adsorbed and free enzyme, for two representative test reactions including substrates that could diffuse through the pores. The fluorescent and chromogenic products of the test reactions for each case were followed using spectrofluorometry and UV/VIS spectroscopy.

The specific activity of adsorbed lysozyme in pores with a diameter around 5.7 nm pore size was 3 times higher, and in pores of 10.5 nm diameter 2.5 times higher than that of free lysozyme in solution. For adsorbed myoglobin, the activity even increased by a factor of 9 in SBA-15 with pores of 5.7 nm diameter, and a factor of 6 when the pore size was 9 nm, again compared to the free protein. This suggests that the active conformation of each protein has been greatly stabilized by confinement on the concave pore surface of SBA-15. This stabilization is most pronounced in mesoporous SBA-15 with a pore size close to the dimensions of the protein, but leaving room for reactants and products to enter and leave the pores. Heat treatment of protein/SBA-15 samples also showed that the stability of the adsorbed protein was protected.

These results show that the activity and stability of proteins can be significantly enhanced by confining the protein in a mesoporous support with a pore size similar to the protein size. The concave, negative pore curvature of SBA-15 has a remarkable stabilizing effect, higher than that noted for protein adsorption on silica nanoparticles [1] or carbon nanotubes [2], which have a convex, positive surface curvature.

References:

[1]. A.A. Vertegel, R.W. Siegel and J.S. Dordick. Silica nanoparticles size influences the structure and enzymatic activity of adsorbed lysozyme. *Langmuir* **20**, 6800-6807 (2004).

[2]. N.R. Palwai, D.E. Martyn, L.F.F. Neves, Y. Tan, D.E. Resasco and R.G. Harrison. Retention of biological activity and near-infrared absorbance upon adsorption of horseradish peroxidase on single-walled carbon

nanotubes. *Nanotechnology* **18**, 235601 (2007).

[3]. H.H.P. Yiu, and P.A. Wright. Enzymes supported on ordered mesoporous solids: a special case of an inorganic-organic hybrid. *J. Mater. Chem.* **15**, 3690-3700 (2005).

See more of this Session: Biomolecules at Interfaces I
See more of this Group/Topical: Engineering Sciences and Fundamentals

Why Are There Two Types of Collagen IV in Glomerular Basement Membrane?

Michael J. Swickrath, Biomedical Engineering, University of Minnesota, Minneapolis, MN

Kevin D. Dorfman, Department of Chemical Engineering and Materials Science, University of Minnesota, Minneapolis, MN

Yoav Segal, Division of Renal Diseases and Hypertension - Department of Medicine, University of Minnesota, Minneapolis, MN

Victor H. Barocas, Biomedical Engineering, University of Minnesota, Minneapolis, MN

Basement membrane provides, amongst other functions, the mechanical support for tissues in higher organisms. The microstructure of this membrane arises from the self-assembly of its the various components, most notably collagen IV. While virtually all basement membranes contain the α1α1α2 collagen IV protomer, the more interrupted α3α4α5 collagen IV protomer only appears in the specialized basement membranes of organs such as the kidney. We will present Monte Carlo simulation data that suggest a structural role for the α3α4α5 protomers in these more specialized systems, focusing on glomerular filtration in the kidney. In the context of a minimal model of collagen IV, we considered two possible roles for the increased cross-linking in the α3α4α5 protomers: (i) intraprotomeric disulfide bonds increase the protomer stiffness and (ii) interprotomeric disulfide bonding stabilizes lateral interactions. Case (i) tended to produce networks consisting of a large number of small pores, indicating a low membrane permeability. In contrast, we found that case (ii) produces highly connected networks featuring a number of large pores, leading to a high membrane permeability. Importantly, the latter network structures are only achieved when the mole fraction of α3α4α5 protomers in the network reaches physiological levels. As a result, we postulate that the α3α4α5 protomers in the kidney basement membrane impart the requisite combination of mechanical strength and permeability required for glomerular filtration.

See more of this Session: Biomolecules at Interfaces I
See more of this Group/Topical: Engineering Sciences and Fundamentals

Rheological Study of Silica Filled PDMS

Keri Mackenzie, Chemical Engineering, Heriot Watt University, Edinburgh, United Kingdom

Valeria Arrighi, Chemical Engineering, Heriot Watt University, Edinburgh, United Kingdom

Raffaella Ocone, Chemical Engineering, Heriot Watt University, Edinburgh, United Kingdom

The rheological behaviour of silica nano-particles in a series of monodispersed Polydimethylsiloxane (PDMS) samples is studied. It was found that even 1%wt silica had a significant effect on the mobility of the polymer, increasing it by ten fold. At higher weight percent of silica, a further increase in viscosity is observed, however this is not so marked as expected. This finding might be due to the effects of distance between particles and molecular weight.

Unlike pure, low molecular weight PDMS, a Newtonian region is not observed. At increasing shear rate, the viscosities of the particle filled polymers decrease and they begin to approach the viscosity of pure PDMS. This behaviour suggests that the majority of particle-polymer interactions are broken and a further decrease in viscosity should be observed at the point where the onset of shear thinning occurs in pure PDMS.

See more of this Session: Complex Multiphase Flows
See more of this Group/Topical: Engineering Sciences and Fundamentals

Binary Collisions: Suspension Ordering Instead of Hydrodynamic Diffusion

Mauricio Zurita-Gotor, Dept. Ingenieria Aeroespacial y Mecanica de Fluidos, Universidad de Sevilla, Seville, Spain

Jerzy Blawzdziewicz, Mechanical Engineering, Yale University, New Haven, CT

Eligiusz Wajnryb, Institute of Fundamental Technological Research, Polish Academy of Sciences, 00-04 Warszawa, Poland

Binary collisions between rough (or repulsive) particles in dilute suspension flows are assumed to lead to particle diffusion that relaxes suspension non-uniformities. Here we show that binary collisions in wall-bounded shear flow may result in an opposite effect, i.e., development of layered particle ordering. We demonstrate that, depending on particle roughness (or repulsion range), the ordering may either extend through the whole channel width or start from the walls and decay exponentially towards the channel center. The key factor in formation of the layered behavior is existence of the swapping (reversing) binary trajectories [J. Fluid Mech. 592].

See more of this Session: Complex Multiphase Flows
See more of this Group/Topical: Engineering Sciences and Fundamentals

Direct Meso-Scale Simulations of Flexible Fibres in Turbulent Liquid Flow

Jos Derksen, Chemical & Materials Engineering, University of Alberta, Edmonton, AB, Canada

A procedure for direct, meso-scale simulations of flexible fibres immersed in liquid flow is introduced. The fibres are composed of chains of spherical particles connected through ball joints with the bending stiffness of the joints as a variable. The motion of the fibres and the liquid is two-way coupled with full resolution of the solid-liquid interface. First the simulation procedure is validated by means of an analytical solution for sphere doublets in zero-Reynolds simple shear flow. Subsequently we use the numerical method to study inertial flows with fibres, more specifically the interaction of a fibre with isotropic turbulence thereby focussing on the dynamics of bending and stretching of the fibre as a result of the erratic flow structures surrounding it, and on the flow-induced forces and torques in the fibre joints.

See more of this Session: Complex Multiphase Flows
See more of this Group/Topical: Engineering Sciences and Fundamentals

Emulsion Flow through a Packed Granular Material with Multiple Drop Breakup

Alexander Zinchenko, Department of Chemical and Biological Engineering, University of Colorado at Boulder, Boulder, CO

Robert H. Davis, Department of Chemical and Biological Engineering, University of Colorado at Boulder, Boulder, CO

We study the flow of an emulsion of immiscible deformable drops through a randomly packed granular material. This problem is relevant to many applications, including oil filtration through underground reservoirs. Of particular interest is to obtain the pressure gradient - flow rate relationships and delineate the conditions when the emulsion can no longer squeeze through the material due to drop blockage in the pores by capillary forces; it is also intriguing to study the effects of drop breakup in this process. If the emulsion drops were much smaller than the

pores, some effective-medium model could be assumed for the emulsion. However, in most cases, the drops are comparable in size with the particles and pores, and the system has to be handled as a three-phase one, by direct numerical

simulation, with distinct velocities for the drop- and continuous phases. A recent multipole-accelerated boundary-integral algorithm [1] has been extended and systematically used in this work to study emulsion flow

through granular materials with different microstructures. The material skeleton is modeled as a random arrangement of many monodisperse solid spheres rigidly held in a periodic cell in mechanical equilibrium under the action of contact forces. Two extreme cases are of interest (i) frictionless spheres forming a "random close packing" (RCP), with the average density of 0.637 and the coordination number of six, and (ii) highly-frictional particles in a 'random loose packing' (RLP), with the density of packing 0.55 and coordination number of four (within the percolating cluster transmitting the stress). An emulsion of many non-wetting drops flows under a specified volume-averaged pressure gradient. Even with only 50-100 particles and drops in a periodic cell, the problem presents very severe computational difficulties, because of necessary superhigh resolution (~ 10 000 boundary elements per surface) and a large number (~10 000 -50 000) of time steps to evaluate long-time averages for the phase velocities. Multipole acceleration, built in our algorithm, has a two-order-of-magnitude advantage over the standard boundary-integral coding at each time step, and appears to

be, at present, the only way to perform such simulations. Compared to the original algorithm [1], the present version incorporates topological mesh changes on drop surfaces

(with the help of techniques from [2]) to simulate a cascade of multiple drop breakups observed in this problem at sufficiently large capillary numbers. A novel defragmentation algorithm is used to continue simulations

after each pinch-off. Due to geometrical constraints imposed by solid particles, the drop shapes are typically rather compact at breakup, and the daughter drops are comparable in size; small satellite drops are almost never observed. Another significant extension, compared to [1], is ensemble

averaging of the results over many realizations of a granular material. We found that using the empirical Carman-Kozeny correlation for the pure-fluid velocity in the definition for the capillary number significantly reduces the dispersion of data to facilitate ensemble averaging. For drop-to-medium viscosity ratio of unity, drop volume fraction of 20% in the entire space, and nondeformed drop-to-particle size ratio of ~0.5, the average drop- and continuous phase permeabilities are studied as functions of the capillary number, both for RLP and RCP structures; the critical thresholds for squeezing (vs. trapping) to occur are evaluated. The effect of the size ratio is also explored. Generally, the kinematic drop-phase velocity is higher than that for the continuous phase, except for near-critical squeezing conditions. For general viscosity ratios, the boundary-integral iterations for the solid phase are no longer decoupled, which makes the ensemble averaging a formidable task. Nevertheless, in this case, we obtained similar results for individual material realizations.

[1] Zinchenko A.Z., Davis R.H. 2008 Algorithm for direct numerical simulation

of emulsion flow through a granular material. J. Comput. Phys., vol. 227,

pp.7841-7888.

[2] V. Cristini, J. Blawzdziewicz, M. Loewenberg 2001 An adaptive mesh

algorithm for evolving surfaces: simulations of drop breakup and coalescence.

J. Comput. Phys., vol. 168, pp. 445-463.

See more of this Session: Complex Multiphase Flows
See more of this Group/Topical: Engineering Sciences and Fundamentals

Ensemble Averaged Equations of Motion for Power-Law Fluid Flow in Random Fixed Beds of Cylinders and Spheres

John P. Singh, Chemical and Biomolecular Engineering, Cornell University, Ithaca, NY

Donald L. Koch, Chemical and Biomolecular engineering, Cornell University, Ithaca, NY

Darcy's law for the flow of a Newtonian fluid through a porous medium balances the mean pressure gradient with a body force representing the drag exerted on the fixed solid material, yielding a linear relationship between the mean fluid velocity and the pressure drop. A similar force balance applies to non-Newtonian fluid flows in porous media, but the nonlinear dependence of the stress on the strain rate in shear thinning or thickening fluids results in a nonlinear relationship between the pressure gradient and mean fluid velocity. An ensemble average of the equations of motion for a Newtonian fluid over particle configurations in a dilute fixed bed of spheres or cylinders yields Brinkman's equations of motion where the disturbance velocity produced by a test particle is influenced by the Newtonian fluid stress and a body force representing the linear drag on the surrounding particles. A self-consistent calculation of this disturbance velocity and the drag on the test particle yields the permeability of the bed. We consider a similar analysis for a power law fluid where the stress **s** is related to the rate of strain **g** by $\mathbf{s}=m|\mathbf{g}|^{n-1}\mathbf{g}$. In this case, the ensemble averaged momentum equation includes a body force resulting from the nonlinear drag exerted on the surrounding particles, a power-law stress associated with the disturbance velocity of the test particle, and a stress term that is linear with respect to the test particle's disturbance velocity. The latter term results from the interaction of the test particle's velocity disturbance with the random straining motions produced by the neighboring particles. It can be interpreted as arising from the fact that the effective viscosity of the fluid in the far field is influenced more by the shearing motion in the bulk medium than by the disturbance of the test particle. We explore the solutions to these equations using scaling analyses for dilute beds and numerical simulations using the finite element method. In Newtonian flow through random arrays of cylinders, the logarithmic divergence of the disturbance velocity of a cylinder is removed by the Brinkman screening associated with the drag on the surrounding particles leading to a well defined average drag on the cylinders. The effects of particle interactions on the drag in dilute arrays of cylinders and spheres in shear thickening fluids is even more dramatic, where it arrests the algebraic growth of the disturbance velocity with radial position when $n>1$ for cylinders and $n>2$ for spheres. For concentrated fixed beds, we adopt an effective medium theory in which the drag force per unit volume in the medium surrounding a test particle is assumed to be proportional to the local volume fraction of the neighboring particles which is derived from the hard-particle packing.

See more of this Session: Complex Multiphase Flows
See more of this Group/Topical: Engineering Sciences and Fundamentals

Development of a Simulation Method to Study Liquid Phase Maldistribution During the Squeeze Flow of Highly Filled Particulate Pastes

Milan J. Patel, Department of Chemical Engineering & Biotechnology, University of Cambridge, Cambridge, United Kingdom

D. Ian Wilson, Department of Chemical Engineering & Biotechnology, University of Cambridge, Cambridge, United Kingdom

Stuart Blackburn, IRC in Materials Processing, University of Birmingham, Birmingham, United Kingdom

Particulate pastes are widely used to manufacture products such as foods, agrochemicals, pharmaceuticals, ceramics, and solder pastes using common processing techniques such as ram extrusion and injection moulding. These materials are often classified as soft solids and feature a high volume fraction of particulates in a liquid binder. Their rheology is therefore strongly influenced by interparticle contacts, rather than collisions, and exhibits complex yield stress behaviour and wall slip. Furthermore, several types of paste-specific flaw may develop during processing, as discussed by Benbow and Bridgwater (1993, p. 21).

One such flaw is liquid phase maldistribution (or migration), henceforth abbreviated to LPM. When stresses are applied to a paste during ram extrusion, a direct consequence of the multi-phase nature of pastes is that the applied stress is split between the load-bearing particulate network (the solids skeleton) and the liquid present in the gaps (pores) between the particles. These pores form an interconnected network that is permeable to the liquid. If the applied stress varies spatially across the paste, pore liquid pressure gradients develop within the pore network and drive flow of the liquid relative to the solids skeleton. LPM will thus result in an inhomogeneous distribution of liquid. Since the rheological properties of the paste are strongly dependent by liquid content, flow patterns are modified and dead zones may result. In extreme cases of LPM, flow will cease completely, causing damage to the extruder and loss of production.

Detailed and reliable simulation of paste extrusion processes is required to improve design of extrusion dies but can also be used to study flaws such as LPM. The aim of the latter simulations is to link the severity of LPM to the operating conditions and the composition (or formulation) of the paste: the onset of LPM occurs when the relative velocity of the local pore liquid is large compared to the absolute velocity of the solids, so that the liquid has time to redistribute while the paste passes through a high stress region, i.e. the die entry zone in a ram extruder. The relative velocity of the liquid phase to the solids skeleton is a function of the (local) pore liquid pressure gradient and the permeability of the solids skeleton. The latter parameter is determined by the properties of the particles (absolute particle size, size distribution, shape, roughness and deformability), the skeleton (solids volume fraction), liquid (rheology, liquid volume and air volume fractions for unsaturated pastes), and chemical interactions between the solids and the liquid.

Models for paste flow incorporating LPM do exist for ram extrusion, e.g. Rough et al. (2002), and for squeeze flow, e.g. (Poitou and Racineux 2001), Sherwood (2002), Kolenda et al. (2003) and Roussel et al. (2003). However these models are usually one-dimensional. Two-dimensional modelling approaches are well established in the field of soil mechanics and the challenge is to apply these approaches to simulate paste extrusion processes where the strains are much larger than normally arise in soil problems. The squeeze flow test is a relatively simple experimental procedure that shows promise in measuring some of the mechanical constitutive parameters of pastes that are required to calibrate soil mechanics models.

This work investigates the accuracy of this procedure for a model paste and discusses the physical and computational difficulties that arise. The Drucker-Prager constitutive relationship was used to describe the solids skeleton during squeeze flow simulations of a saturated viscoplastic paste comprising glass ballotini and an aqueous glucose liquid binder. The liquid is modelled as Newtonian and incompressible, and Darcy's law is used to represent the interactions between the two phases. The interaction between the paste and the plates was assumed to be frictionless. These simulations were implemented in a two-dimensional axisymmetric finite element model using the ABAQUS/Standard FEM package (Simulia, Rhode Island, USA). A series of simulations have been performed in order to investigate the effect of solids loading and squeeze flow velocity (deformation rate) on paste performance during squeeze flow.

Benbow, J. J. and J. Bridgwater (1993). Paste Flow and Extrusion. Oxford, Clarendon Press: 19-21.

Kolenda, F., P. Retana, et al. (2003). "Identification of rheological parameters by the squeezing test." Powder Technology 130(1-3): 56-62.

Poitou, A. and G. Racineux (2001). "A squeezing experiment showing binder migration in concentrated suspensions." Journal of Rheology 45(3): 609-626.

Rough, S. L., D. I. Wilson, et al. (2002). "A model describing liquid phase migration within an extruding microcrystalline cellulose paste." Chemical Engineering Research and Design 80(7): 701-714.

Roussel, N., C. Lanos, et al. (2003). "Induced heterogeneity in saturated flowing granular media." Powder Technology 138(1): 68-72.

Sherwood, J. D. (2002). "Liquid-solid relative motion during squeeze flow of pastes." Journal of Non-Newtonian Fluid Mechanics 104(1): 1-32.

See more of this Session: Complex Multiphase Flows
See more of this Group/Topical: Engineering Sciences and Fundamentals

Real-Space Characterization of the Evolution of Colloidal Gel Microstructure Under Steady Shear

Bharath Rajaram, University of California, Irvine, CA

Ali Mohraz, Department of Chemical Engineering & Materials Science, University of California, Irvine, Irvine, CA

Understanding the non-linear rheological response of colloidal gels to shear flow is critical in engineering complex fluids and relevant in a wide host of technologies including the food processing and ceramic industries. However, the complex interplay between the deformation of the microstructure due to shear and the subsequent effect on the rheological properties is poorly defined. In this study, the effect of steady shear flow on a density- and refractive index matched, depletion-induced colloidal gel system is directly visualized using confocal microscopy; the three dimensional image slice volumes thus obtained are analyzed using quantitative computational geometry based algorithms. Our results capture, for the first time in real space, local structural information during the evolution of the microstructure in response to applied shear. Three distinct flow regimes are identified from investigations in the flow-gradient plane. Initially, the spatial configuration of the gel changes to a predominant orientation along the extensional axis of flow - for large enough shear rates, this is followed by an unraveling of the gel backbone, forming extended linear domains. The continued application of shear flow ruptures the gel structure, producing disconnected clusters and chains that are translated primarily as a uniform plug along the flow direction. Finally, in the third regime, the steady shear flow induces cluster reaggregation resulting in the separation of the colloids into regularly alternating regions of high and low particle concentrations. Complementary studies in the flow-vorticity plane prior to gel rupture indicate the presence of heterogeneous flow profiles due to local rearrangement of the gel to accommodate the applied strain. Thus, temporal and spatial fluctuations in the velocity are observed in this plane contrary to the uniform velocity profile present in stable suspensions under shear. The results are compared with the response of gels formed by screening surface electrostatic charges and the key differences are highlighted.

See more of this Session: Complex Multiphase Flows
See more of this Group/Topical: Engineering Sciences and Fundamentals

A Cloud of Rigid Rods Sedimenting in Viscous Fluid

Joontaek Park, Chemical Engineering, University of Florida, Gainesville, FL

Jason E. Butler, Chemical Engineering, University of Florida, Gainesville, FL

Bloen Metzger, Polytech-Marseille, Marseille, France

Elisabeth Guazzelli, Aix-Marseille Universit\'e - Polytech'Marseille - IUSTI CNRS UMR 6595, Marseille, France

We investigate the deformation and breakup of a cloud of rigid, rodlike particles falling under gravity through a viscous fluid in the absence of inertia and interfacial tension. Results from dynamic simulations which include hydrodynamic interaction demonstrate that an initially spherical cloud of rigid rods rapidly evolves into a torus which then shatters into secondary droplets. These smaller droplets, in turn, transition into tori and again break apart in a cascade of events that repeats many times. However, the fluctuations in the velocities and positions of the rods accelerates the breakup process as compared to drops of liquids or spherical particles. The simulation results compare well with experimental visualizations of a cloud of sedimenting rods.

See more of this Session: Complex Multiphase Flows
See more of this Group/Topical: Engineering Sciences and Fundamentals

Magnetoviscosity of dilute magnetic fluids in oscillating and rotating magnetic fields

Jorge H. Sánchez[1,2] and Carlos Rinaldi[1,*]

[1]*Department of Chemical Engineering, University of Puerto Rico, Mayaguez Campus, P.O. Box 9000, Mayaguez PR 00681.*

[2]*Facultad de Ingeniería Química, Universidad Pontificia Bolivariana, Medellín, Colombia.*

ABSTRACT

The magnetic field dependent viscosity (magnetoviscosity) of dilute suspensions of spherical magnetic particles suspended in a Newtonian fluid subjected to both magnetic and shear flow fields was studied through Brownian dynamics simulations. Results are presented for the response of these suspensions to oscillating magnetic fields and magnetic fields that are co-rotating or counter-rotating with respect to the fluid vorticity. A decrease to negative values in the intrinsic magnetoviscosity is observed for oscillating and co-rotating magnetic fields whereas an increase is observed for counter-rotating magnetic fields. The frequency corresponding to zero viscosity and the minimum value in the negative viscosity are lower for co-rotating magnetic fields than for oscillating magnetic fields. In the negative magnetoviscosity regions the particles in a co-rotating magnetic field rotate faster than in an oscillating magnetic field. It is estimated that the flow due to co-rotating particles could be strong enough to obtain a negative effective viscosity in dilute suspension. Moreover, it is shown that the commonly accepted constitutive equation for the antisymmetric stress describes well the intrinsic magnetoviscosity of the suspension.

* Corresponding author.
E-mail address: carlos.rinaldi@upr.edu

Thermophoresis of a Poly-Acrylimide Based Polyelectrolyte

Audrey Hammack, Chemistry, University of Texas at Tyler, Tyler, TX

Jennifer Kreft Pearce, Chemistry, University of Texas at Tyler, Tyler, TX

Thermophoresis, the migration of a species due to a temperature gradient, has been studied extensively, both in experiment and in simulation. Research has yielded thermal diffusion coefficients for migrating polyeletrolytes and a plausible mechanism for migration, but these studies are largely limited to DNA. We present an experiment to study a poly-acyrlamide based, charge-bearing, dye-tagged copolymer, characterized by molecular weight and viscosity, in temperature-graded micro-fluidic channels. From this experiment, we present measured thermal diffusion coefficients and compare to the results for DNA. We also provide data from a lattice-Boltzman method simulation to compare with experimental findings.

See more of this Session: Complex Multiphase Flows
See more of this Group/Topical: Engineering Sciences and Fundamentals

Molecular Dynamics Simulation Study of Pure and Surfactant-Laden Water Surfaces: Comparison of Flexible Water Models

Pak K. Yuet, Process Engineering & Applied Science, Dalhousie University, Halifax, NS, Canada

Daniel Blankschtein, Department of Chemical Engineering, Massachusetts Institute of Technology, Cambridge, MA

Surfactants having complex molecular architectures have been used extensively in industrial applications for many years. Many of these so-called high-performance surfactants possess unusual interfacial and surface characteristics, compared to linear alkyl surfactants such as linear alkyl sulfonates and alkyl ethoxylates. To facilitate the design of industrial processes involving the use of these surfactants, it is essential to develop a fundamental understanding of their properties at interfaces, particularly at the air/water interface. Molecular dynamics simulation is a powerful technique for studying these systems at the molecular level. In applying this technique, however, it is important to have a model system that can adequately capture the characteristics of the water surface. Different water models have been shown to result in very different surface tensions, and a survey of the literature has revealed that bond flexibility may play an important role. In addition, it is generally recognized that simulated values of water surface tension are dependent on the simulation parameters, such as the Lennard-Jones cut-off distance, the system size, and the simulation time. Indeed, the differences in simulation parameters used in previous studies of water surfaces make any direct comparison of water models rather difficult.

The primary objectives of this study are to: (i) compare various flexible water models using a common set of simulation parameters, with particular emphasis on their ability to predict the water surface tension, (ii) study the use of these water models with common surfactants, such as sodium dodecyl sulfate and n-octyl glucoside, for simulating the surface pressure of surfactant-laden surfaces, and (iii) develop a better understanding of the relation between surfactant molecular architecture and surface properties. The simulated surface tensions of the flexible three-center (F3C) model and of the flexible SPC/E model at 300 K are 66.8 mN/m and 70.2 mN/m, respectively, in reasonable agreement with the experimental value of 71.7 mN/m. However, preliminary simulation results have indicated that, in the presence of adsorbed surfactants, the simulated surface pressures are considerably smaller than those observed experimentally. Molecular properties such as interfacial dipole moments, the normal and lateral pressure profiles, two-dimensional radial distribution functions, as well as the interactions between surfactant and water molecules are being examined in detail in order to elucidate the key characteristics in the simulation of water surfaces.

See more of this Session: Modeling of Interfacial Systems I
See more of this Group/Topical: Engineering Sciences and Fundamentals

Molecular Dynamics Studies of Friction in Alkylsilane and Hydroxyalkylsilane Coated Mems Devices

Jose L. Rivera, Department of Chemical and Biomolecular Engineering, Vanderbilt University, Nashville, TN

Oleg A. Mazyar, Department of Chemical and Biomolecular Engineering, Vanderbilt University, Nashville, TN

Clare McCabe, Department of Chemical and Biomolecular Engineering, Vanderbilt University, Nashville, TN

G. Kane Jennings, Department of Chemical and Biomolecular Engineering, Vanderbilt University, Nashville, TN

The adhesion and friction properties of the native amorphous silicon oxide layer in silicon-based micro- and nano-electromechanical devices (MEMS/NEMS) have a strong influence on device control, performance and durability. To improve the friction and adhesion properties surfaces coated with alkylsilanes have been proposed and shown to reduce the friction forces as the chain length of the alkylsilanes increases. In an effort to develop surfaces with varying surface energy that can be used in lubrication systems in which a bound lubricant (i.e., alkylsilane) is combined with a mobile layer, alkylsilanes with hydroxyl terminal groups have also been studied. At low normal loads, the friction force of surfaces coated with hydroxylalkylsilanes are higher than the corresponding surfaces coated with akylsilanes, but surfaces coated with hydroxyalkylsilanes show a maximum as a function of normal loads.

We use molecular dynamics simulations to study the friction forces of silica surfaces coated with alkylsilanes, hydroxyalkylsilanes and their mixtures. The systems of interest have been studied over a wide range of normal loads; from 0 to the highest pressure silica can handle before breaking (~3.7 GPa). In agreement with previous studies, friction forces for the surfaces coated with hydroxyalkylsilanes are one order of magnitude higher than surfaces coated with alkylsilanes. We also find the maximum reported for the surfaces coated hydroxyalkylsilanes, but also find that this is a local maximum, after which, the friction forces behave normally and the friction and normal forces are linearly dependent. The surfaces coated with mixtures of alkylsilanes and hydroxyalkylsilanes show a friction behavior dependent on the composition of the surfaces. The friction forces increase linearly as the composition changes from the pure alkylsilane to the pure hydroxyalkylsilane.

See more of this Session: Modeling of Interfacial Systems I
See more of this Group/Topical: Engineering Sciences and Fundamentals

Effect of Nanoscale Patterning in Surfactant-Coated Nanoparticles On Their Interfacial Properties: A Molecular Dynamics Study

Hao Jiang, Chemical Engineering, University of Michigan, Ann Arbor, MI

Chetana Singh, Chemical Engineering, University of Michigan, Ann Arbor, MI

Jeffrey J. Kuna, Materials Science and Engineering, Massachusetts Institute of Technology, Cambridge, MA

Kislon Voitchovsky, Materials Science and Engineering, Massachusetts Institute of Technology, Cambridge, MA

Francesco Stellacci, Materials Science and Engineering, Massachusetts Institute of Technology, Cambridge, MA

Sharon C. Glotzer, Chemical Engineering, University of Michigan, Ann Arbor, MI

Experiments have shown that interfacial properties of substrates covered by mixed surfactant monolayers depend on the pattern formed by the adsorbed surfactants [1,2]. Interestingly, when a nanoparticle (NP) is coated with a mixture of hydrophilic and hydrophobic molecules in a striped pattern, the total work of adhesion of the NP varies non-monotonically with the hydrophilic stripe width [1]. The basis for this non-monotonic behavior is not clear. Here we present results from our systematic molecular dynamics study of interfacial properties of NPs covered with hydrophilic and hydrophobic ligands in aqueous solutions for a variety of surface patterns. Based on detailed statistical analysis of data from our simulations we infer that two competing effects viz. cavitation and enhanced hydrophilicity, lead to the non-monotonic behavior observed in experiments. With an increase in hydrophilic stripe-width accompanied by a decrease in the hydrophobic stripe-width, there is emergence of the cavitation effect [3] that leads to stress in the h-bond network of surrounding water and therefore decreases work of adhesion. However, an increase in hydrophilic stripe-width is also accompanied by an increase in h-bonding between water molecules and the hydrophilic ligands, and therefore leads to an increase in the work of adhesion. These two competing effects, both of which depend on stripe-width, combine in a complex way to result in the non-monotonic dependence of work of adhesion on stripe width. Finally, we provide a brief comparison between the experimental and simulational findings.

1. JJ Kuna, K Voïtchovsky, C Singh, PK Ghorai, H Jiang, S Mwenifumbo, MM Stevens, F Stellacci and SC Glotzer, "On the Role of Nanometer Scale Structure on Interfacial Energy", submitted 2009.

2. A Centrone, E Penzo, M Sharma, JW Myerson, AM Jackson, N Marzari and F Stellacci, "The role of nanostructure in the wetting behavior of mixed-monolayer-protected metal nanoparticles", PNAS, 9886, 105 (29), 2008.

3. D Chandler, "Interfaces and the driving force of hydrophobic assembly", Nature, 640, 437, 2005.

See more of this Session: Modeling of Interfacial Systems I
See more of this Group/Topical: Engineering Sciences and Fundamentals

Dynamics of Water near Mica Surface

Ateeque Malani, Department of Chemical Engineering, Indian Institute of Science, Bangalore, India

K. G. Ayappa, Department of Chemical Engineering, Indian Institute of Science, Bangalore, India

The study related to dynamical response of interfacial water has application in wide range of areas ranging from biology to geology. Recent experimental studies have shown that even for sub-nanometer confinement, the shear viscosity of water between mica surfaces is only three times that of free water. This high fluidity is attributed to the large exchange rate of water in the hydration shell of potassium (K^+) ions.[1] Water near hydroxylated silica surfaces exhibit decoupling in the translational and rotational motions and a monotonic increase in the decoupling was observed toward the surface.[2] Similar decoupling has also been observed for water near biological molecules.[3] In this study, the dynamics of interfacial water near mica surface is evaluated using molecular dynamics simulations. The analysis shows that water in the contact layer has a high residence time near the surface. The K^+ ions have a strong influence on the dynamics of interfacial water. The presence of K^+ ions, gives rise to a non-monotonic decoupling behaviour in the translational and orientational motions toward the mica surface. It was observed that orientational motions are much more suppressed when compared with translational motions. Despite slowing down of water relaxation near the surface, dynamics of water in the hydration shell of K^+ ion is comparable to that observed in bulk water. The analysis also shows the presence of orientational jumps in the contact layer of mica surface. Recently these jumps have been observed in molecular dynamics simulations of bulk water.[4]

References

1. Raviv, U., Laurat, P. and Klein, J. Fluidity of water confined to subnanometer films. *Nature*, **413**, 51-54, 2001.

2. Castrillon, S. R.-V., Giovambattista, N., Aksay, I. A. and Debenedetti, P. G. Effect of surface polarity on the structure and dynamics of water in nanoscale confinement. *J. Phys. Chem. B*, **113**, 1438-1446, 2009.

3. Bagchi, B. Water dynamics in the hydration layer around proteins and micelles. *Chem. Rev.*, **105**, 3197-3219, 2005.

4. Laage, D. and Hynes, J. T. A molecular jump mechanism of water reorientation. *Science*, **311**, 832-835, 2006.

See more of this Session: Modeling of Interfacial Systems I
See more of this Group/Topical: Engineering Sciences and Fundamentals

Aggregation and Disintegration Model of Latex Colloids

Suresh Ahuja and Chieh-Min Cheng, Xerox Corporation, Rochester, NY

ABSTRACT

Hydrodynamic interactions in colloidal dispersions have been analyzed by Stokesian dynamics (SD), the the lattice Boltzmann method (LBM), the finite element method (FEM), diffusion-limited aggregation, direct numerical simulations (DNS), stochastic rotational dynamics (SRD), fluid particle dynamics (FPD), dissipative particle dynamics (DPD) and fluid particle model (FPM). However, quantitatively reliable simulations have not yet been performed successfully for many-particle dispersions due to the complexity of the system. Recently, reliable and efficient numerical method, called the smoothed profile (SP) method to resolve the hydrodynamic interactions acting on solid particles immersed in Newtonian host fluids. In the SP method, the original sharp boundaries between colloids and host fluids are replaced with diffuse interfaces with finite thickness. Dynamical processes occurring in the dispersion colloidal agglomerate in solvents are greatly influenced by coupling between the dispersed microstructures and the global flow. The average size of the agglomerated fragments depends on the shearing rate and the attractive forces between colloidal particles. The aggregation of colloidal particles is formed by the collisions between aggregates, which are influenced by the flow or by the cohesive forces for small dispersion energies. If there are no significant steric or Coulombic interactions, the depth and range of the depletion interaction can be independently controllable. If Coulombic forces are present, adding salt can shrink the size of the Debye double layer, diminishing the repulsion until the colloids aggregate by van der Waals attraction. Sterically stabilized colloids can be cooled from high temperatures, causing the surface grafted polymer layers on the colloid to change conformation, again allowing the colloids to approach close enough to aggregate by the van der Waals attraction. Our colloids of latex went through aggregation and disintegration on shear deformation and were found to behave as elastic solids at low shear rates and require a finite stress to flow. The behavior of shear stresses is analyzed and modeled to dependent on zeta potential, shear rates and fractal floc size and compared with previous models.

Introduction

In recent years, new numerical methods have been developed for modeling physical and chemical phenomena occurring in the mesoscale. The most popular are lattice Boltzmann gas (LBG)), diffusion-limited aggregation, direct numerical simulations (DNS), or other particle methods such as stochastic rotational dynamics (SRD), fluid particle dynamics (FPD), dissipative particle dynamics (DPD), and fluid particle model (FPM). Molecular dynamics (MD) forms the centerpiece from which the other techniques, such as dissipative particle dynamics (DPD). Colloidal particles with an accessible potential minimum will aggregate over time. The origins of interparticle attraction are pervasive in real colloidal dispersions (e.g. dispersion forces, depletion, bridging), making flocculation and aggregation very common [1].

Brownian Dynamics with full hydrodynamic interactions utilizes a mobility matrix which is based on tensor approximations which are exact in the limit of zero Reynolds number and zero particle volume fraction All these methods assume that hydrodynamic interactions are fully developed and that the dynamics of the fluid and of the solved particles can be treated as fully separated. In reality, this is not the case. Hydrodynamic interactions are time dependent due to local stresses at the fluid-particle interfaces. A number of more recent methods attempt to describe the time dependent long-range hydrodynamics properly with the computational effort scaling linearly with the number of particles.

These include recent mesoscopic methods like dissipative particle dynamics, the lattice-Boltzmann method or stochastic rotation dynamics. However, for small Reynolds numbers, the computational gain of these methods is lost due to the additional effort needed to describe the motion of the fluid. Finite element or finite difference methods need a proper meshing of the computational domain which is not trivial for complicated boundary conditions as in the case of dense suspensions. [2]

AGGREGATION DISINTEGRATION (BREAKAGE) MODEL

Competition between attractive van der Waals-Hamaker and repulsive double-layer forces determines the stability or instability of many colloidal systems. The interaction between colloidal particles can be modeled by the Derjaguin-Landau-Verwey-Overbeek (DLVO theory) where DLVO potential, which is the sum of the screened electrostatic repulsion between their double-layers and of the van der Waals-Hamaker attraction [3]. Shear-induced aggregation and Brownian aggregation are two mechanisms that exist simultaneously in sheared system. The two aggregation mechanisms are independent over a wide range of conditions. Therefore, the aggregation rate J (the subscripts ij are dropped to refer to equal-sized spheres) in a system undergoing both shear-induced aggregation and Brownian aggregation can be written as

$$J = \frac{8k_B T n^2}{3\eta W_B}\left[1 + \frac{4}{3\pi}\frac{W_B}{W_S}P_e\right] \text{---} (1)$$

where W_B is the stability ratio introduced into the Brownian aggregation due to the potential barrier in DLVO theory, η is the solvent viscosity, and Pe is referred to as Péclet number, i.e.

$$Pe = \frac{3\pi\eta a^3\gamma}{k_B T} \text{--} (2)$$

W_B is the stability ratio due to dispersion forces (van der Waals-Hamaker attraction), W_S is the stability ratio which includes hydrodynamic interaction. W_B is given by in terms of maximum potential barrier and stability ratio due to dispersion forces.

$$W_B = W_\infty + 0.25\left[\exp\left(\frac{\varphi_{\max}}{k_B T}\right) - 1\right] \text{--} (3)$$

For spheres of identical composition in a medium where the separation distance is much smaller than the particle radius, the van der Waals contribution to the interaction potential can be calculated, to close approximation, as

$$\varphi_{vdW} = -\frac{aA_H}{12D} \text{--} (4)$$

In the limit of constant surface charge density Φe is approximated as

$$\varphi_e \approx 2\pi\varepsilon\varepsilon_0 a\psi_s^{\,2} \ln\left[\frac{1}{1-\exp(-\kappa D)}\right]$$ -- (5) where ε is the dielectric constant of the medium, ε_0 is the permittivity of free space, ψ_S is the particle surface potential, and κ is the Debye parameter. Replacing the surface potential with the more accessible zeta potential, ζ, the total interaction potential becomes[4]

$$\varphi = -\frac{aA_H}{12D} + 2\pi\varepsilon\varepsilon_0 a\zeta^2 \ln\left[\frac{1}{1-\exp(-\kappa D)}\right]$$ --(6)

The bulk property yield stress, τ_y, is proportional to the inter-particle force at the separation distance times the number of particle contacts per unit area. The number of particle contacts scales as ϕ^2 / a^2 where φ is the volume fraction of the solids and a is the particle radius.

$$\tau_y \propto -\frac{\phi^2}{a^2}\frac{\varphi(D_0)}{D_0}$$ ---(7)

Using the interaction potential from DLVO theory and electrostatic potential, Φe given in equations 4,5 and putting in Equation 7, yield stress is derived [5] as

$$\tau_y \propto \frac{\phi^2}{a}\left(\frac{A_H}{12D_0^{\,2}} - \frac{2\pi\varepsilon\varepsilon_0}{D_0}\zeta^2 \ln\left[\frac{1}{1-\exp(-\kappa D_0)}\right]\right)$$ ---
(8)

The above equation shows that at low zeta potential, van der Waals forces dominate. As the zeta potential becomes significant, yield stress increases as the square of zeta potential.

The effective volume fraction of the flocs is given by following equation

$$\phi_f = \left(\frac{d_f}{d_0}\right)^{3-D} \phi_0$$ ---(9)

where d_f, d_0 and ϕ_0 are the diameter of a floc, diameter of a primary particle and volume fraction of primary particles. D is the fractal dimension and is in the range of 1≤ D≤3. When van der Waal forces dominate, yield stress is dependent on the diameter of the floc as

$$\tau_y = \frac{A}{9\pi d_0 h^2}\left(\frac{\phi_0}{0.52}\right)^{\frac{2}{3-D}}$$ ---(10)

where A is Hamaker constant and h is the minimum separation distance between particle surfaces. At the condition of maximum cubic packing, the volume fraction is 0.52.

Hydrodynamic stresses in elongation or shear in either laminar or turbulent regime can affect dispersion stability and breakage of colloidal aggregates in fractal or non-fractal morphology. The balance between clustering (structure-formation) and breakup (structure-failure) is critical enabler in analyzing structural origin of the rheological properties of complex fluids.

Although much is known about aggregation in dilute dispersions, the details of the process in the concentrated regime, which leads to gelation, are poorly understood. This is partly a consequence of the limited ability to measure colloidal level microstructure in concentrated dispersions. Structure development in concentrated, turbid latex dispersions has been investigated by means of fiber-optic quasielastic light scattering (FOQELS). Recent investigations on colloidal gels employed a modified diffusing wave spectroscopy (DWS) technique. Wyss et al. employed this technique to follow the aggregation in aluminum oxide suspensions and demonstrated a correlation with the evolution of the measured shear modulus at a fixed frequency. Further, the elastic plateau moduli deduced from DWS were in good agreement with rheological measurement. However, measurements of the fractal structure of the aggregating dispersion and a quantitative comparison with measured, dilute aggregation were not made. Dissipative particle dynamics (DPD) is one of the mesoscopic techniques, which allows one to model hydrodynamic behavior with thermal fluctuations. This particle-based off-lattice algorithm was inspired by ideas coupling the advantages of the molecular dynamics and lattice-gas methods. A strong background drawn from statistical mechanics has been provided to DPD from which explicit formulas for transport coefficients in terms of the particle interactions can be derived. Since then the DPD model has attracted a great deal of attention from the chemical community. DPD technique can be used for studying various types of phase separation processes in binary or multi-component fluids. By employing two-level models for which colloidal beds are modeled by MD particles and solvent by DPD fluid particles, one can simulate micellar solutions and large colloidal aggregates, which can use as many as 20 million particles. Light scattering methods can track the evolving microstructure in a gelling colloidal dispersion, and that this evolving microstructure manifests itself as an increase in the shear modulus as measured by oscillatory shear rheology. Aggregation in concentrated dispersions rapidly results in gelation, which complicates studies of the aggregation kinetics and structure evolution, but enables the use of rheological methods to probe structure formation. Gelation in aggregating colloidal dispersions is comparable to cross-linking or physically binding monomer segments forming a polymer gel or network [5].

The aggregation may be irreversible or reversible, respectively. In order for the particles to undergo aggregation into the primary minimum, they need to overcome the energy barrier. The higher the energy barrier, the lower is the probability of aggregation. Hence one can consider the process of aggregation as a rate phenomenon which increases with reduction of the energy barrier and ultimately, in the absence
of any barrier, it becomes very fast. The aggregation of particles in a colloidal system can be controlled by three different mechanisms including perikinetic aggregation due to Brownian motion (dominant for particles
with diameters smaller than 1 μm), shear-induced orthokinetic aggregation (dominant for particles in the range of 1-40 μm), and differential sedimentation aggregation (typically important for particles larger than 40 μm)[6]

Acrylic latexes prepared by emulsion polymerization were subjected to steady shear and sinusoidal dynamic deformation. Viscosity of 0.6φ was measured at different shear rates. At the lowest volume fractions the suspensions show shear rate independent behavior. Because the particle-particle separation distance is large compared to the range of steric inter-particle forces, only the Brownian diffusion dominates. The viscosity remains constant with shear rate as the particles can diffuse freely in the suspension. As the volume fraction increases the inter-particle distance becomes comparable to the particle radius and the hydrodynamic and surface interactions dominate. The viscosity increases with increasing volume fraction and changes from one that is independent of shear rate.to shear thinning. The yield stress of latexes obtained by polynomial fits in shear stress shear rate curves is related to the additional force required to break the bonds between particles or polymer chains that hold the network together before flow can begin. In the attractive suspension, there are attractive surface forces between the particles. Only when the applied stress exceeds a critical amount, the yield stress, does the system flow.

Zeta potential was measured by zeta sizer and correlated to yield stress. Model prediction given in Equation 8 shows that yield stress depends on the square of zeta potential.. On power law fitting, we find that the power law exponent is 1,98 compared to 2.0 expected from model prediction.

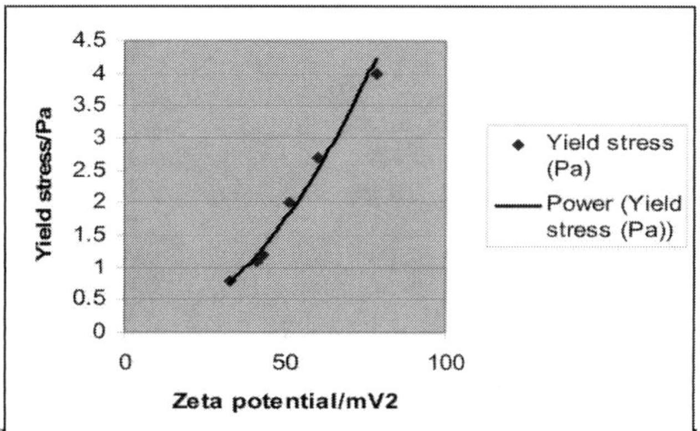

Figure 1 shows dependence of yield stress on zeta potential

Aggregate particle size is related to fractal diameter and fractal dimension. We measured aggregate size of latex . A linear dependence was found as shown in Figure 2. It is consistent with the Equation 8 when zeta potential is significant. Variations in the mechanical strength of aggregates with aggregate cross-sectional area have been modeled. The fractal structures of aggregates were considered in conjunction with fundamental mechanisms of aggregate formation and rupture during the modeling. A quantitative expression of aggregate porosity and characteristic length in terms of aggregate fractal dimensions was derived from fractal dimensional analysis. Predictions from this expression and the Rumpf's equation demonstrated that tensile strengths of aggregates decreased with increasing aggregate cross-sectional area [8]. Aggregation mechanisms can be divided into two different categories: diffusion-limited cluster aggregation (DLCA) and reaction-limited cluster aggregation (RLCA). In DLCA, each collision between particles during aggregation leads to an irreversible bond, whereas in RLCA, particles can escape from a bond site and penetrate deeper into the aggregated structures. Therefore, aggregates formed by RLCA are denser and have a higher fractal dimension than DLCA aggregates. There is a lack of control on the possible microstructures that can be experimentally realized. In order to reach reproducible conditions, the aggregated gels have to be strongly sheared prior to measurement. However, this implies that the microstructure will be governed mainly by the pre-shearing process and not by the particle aggregation kinetics themselves In situ destabilization methods offer the possibility of inducing formation gels in a reproducible way without shearing or mixing [9].

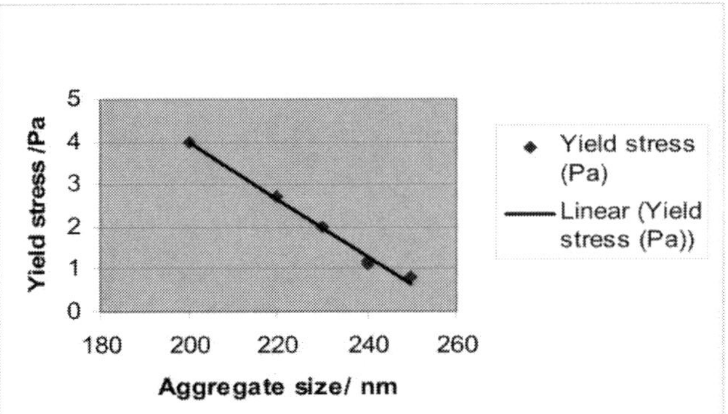

Figure 2 dependence of yield stress on aggregate size.

CONCLUSIONS

Acrylic latex of different aggregate sizes and zeta potential were prepared. The latexes were analyzed by determining their yield stresses. The yield stress dependence on zeta potential and on aggregate size was compared to models on aggregation and disintegration of colloids.

References

1. Witold Dzwinel and David A. Yuen, (2002) Journal of Colloid and Interface Science **247,** pp 463–480
2. J. Harting, M. Harvey, J. Chin, M. Venturoli, and P. V. Coveney.(2005) *Phil. Trans. R. Soc. A*, 363:pp1895–1915,
3. TF. Tadros (2007). General Principles of Colloid Stability and the Role of Surface Forces. I*Colloid Stability: The Role of Surface Forces-Part 1*, ed. TF Tadros. New York: Wiley – VCH
4. W.B. Russel, D.A. Saville and W.R. Schowalter, Colloidal Dispersions, CambridgeUniv. Press, Cambridge, UK, 1989.
5. N. Sakairi , M. Kobayashi, Y. Adachi, (2005) J. Colloid Interface Sci. 283, pp245 -260
6. Stacey L. Elliott, Robert J. Butera, Leo H. Hanusc and Norman J. Wagner (2003) Faraday Discuss., 2003, 123, pp369–383.
7. L Wang, DL Marchisio, RD Vigil, RO Fox. (2005). J. Colloid Interface Sci. 282: pp380-96
8. S. Tang, Y. Ma and C.Shiu (2001), Colloids and Surfaces,180(2)pp7-16
9. Hans M. Wyss, Elena V. Tervoort, and Ludwig J. Gauckler(2005),J. Am. Ceram. Soc., 88 [9] pp2337–2348

An Investigation of the Rapid Cooperative Diffusion of Large Heteroepitaxial Islands

Obioma U. Uche, Sandia National Laboratory, Livermore, CA

Danny Perez, Los Alamos National Laboratory, Los Alamos, NM

Arthur F. Voter, Los Alamos National Laboratory, Los Alamos, NM

John C. Hamilton, Sandia National Laboratory, Livermore, CA

Monolayers of Ag on Cu(001) are well known to reconstruct forming a (10x2) reconstruction with Ag atoms packed hexagonally on the four-fold Cu(001) surface. We have investigated the surface diffusion of hexagonally-packed Ag monolayer islands of various sizes on Cu(001) using computer simulation techniques. Our examination reveals a novel cooperative diffusion mechanism consisting of core glide coupled with the migration of edge vacancies. In particular, analysis of several low-energy pathways reveals the two governing processes: a rapid, glide-centric process and a slower, vacancy-assisted one. It should be noted that this cooperative mechanism was observed for very large islands for which diffusion was orders of magnitude faster than for single Ag atoms at temperatures as low as 175K.

Our findings should provide insight to future experimental research on the size distribution and shapes observed during the growth of thin films in similar systems. Furthermore, these results should have implications for Monte Carlo models of film growth which generally allow only single atom diffusion, thus neglecting such complex, cooperative mechanisms.

See more of this Session: Modeling of Interfacial Systems I
See more of this Group/Topical: Engineering Sciences and Fundamentals

Computational Study of Moving Triple Lines

Neeharika Anantharaju, Mechanical Engineering, Tennessee Technological University, Cookeville, TN

Mahesh V. Panchagnula, Mechanical Engineering, Tennessee Technological University, Cookeville, TN

Srikanth Vedantam, Mechanical Engineering, National University of Singapore, Singapore, Singapore

Wetting of chemically heterogeneous surfaces is modeled using phase field theory. This model studies the one-dimensional kinetic processes involved in wetting a substrate similar to a Wilhelmy technique. The experimental technique provides a complete validation of the phase field theory due to its capability to capture the triple line kinetics in addition to measuring the contact angles during the advancing and receding processes. In the current model, contact angle hysteresis is incorporated by a modified kinetic parameter.

A chemically heterogeneous surface is said to be composed of a predetermined arrangement of two materials. The novelty in the current approach lies in the fact that each of the component materials is constitutively allowed to exhibit hysteresis. We investigate the local shape of the triple line which is known to play an important role in determining the macroscopic contact angle due to its ability to be pinned at various defect locations on real surfaces. We also demonstrate that the shape of the advancing and receding triple line is sensitive to the specific arrangement of the two materials. The work provides insight into the fabrication of chemically heterogeneous surfaces for desired wetting applications.

See more of this Session: Modeling of Interfacial Systems I
See more of this Group/Topical: Engineering Sciences and Fundamentals

LCST Behavior of Lightly Charged Aqueous Polyelectrolytes

David S. Simmons, Chemical Engineering, University of Texas, Austin, Austin, TX

Isaac C. Sanchez, Chemical Engineering, University of Texas, Austin, Austin, TX

Numerous polymers of experimental interest, such as Poly(N-isopropylacrylamide) (PNIPAAm) and PEO, exhibit a lower critical solution temperature (LCST) in aqueous solution. A hydrogen-bonding lattice fluid (HBLF) model employing a combinatorial approach to the hydrogen bonds has previously provided a successful theoretical model for this behavior in the uncharged polymer/water two-component system. However, many systems of experimental interest, such as the human body, contain various electrolytes. The resulting presence of free charges in solution has been experimentally shown to have a significant effect on the LCST of the polymer-water system, presumably through interactions with hydrogen bonds. Furthermore, a common means of modulating the LCST of such systems is the incorporation of a low mole fraction of electrolytic groups into the chain backbone, yielding a lightly charged polyelectrolyte. The authors present an extension of the above HBLF model that addresses these issues by accounting for bound charges on the polymer chain and free charges in solution. The new model allows for the calculation of the LCST of ternary lightly charged polyelectrolyte systems such as Poly(N-isopropylacrylamide-co-acrylamide)-water-acid as a function of temperature, pressure, free salt concentration, pH, and polymer backbone charge concentration.

See more of this Session: Thermodynamics of Polymers II
See more of this Group/Topical: Engineering Sciences and Fundamentals

On the Morphology of Bicontinuous Phases in Blends of Diblock Copolymer and Homopolymer

Juan C. Araque, School of Chemical and Biomolecular Engineering, Cornell University, Ithaca, NY

Francisco Martínez-Veracoechea, School of Chemical and Biomolecular Engineering, Cornell University, Ithaca, NY

Fernando A. Escobedo, School of Chemical and Biomolecular Engineering, Cornell University, Ithaca, NY

We characterize the spontaneous development of morphologically complex phases in binary blends composed of an asymmetric diblock copolymer and a homopolymer with affinity to the minority block. The addition of homopolymer to a well-segregated copolymer melt has a drastic effect in the formation of microstructured bicontinuous phases, i.e., having periodic dividing interfaces without self-intersections. Such phases are of technological relevance for applications in which nanoporous materials with high internal surface area and periodic connectivity are desirable, e.g., templates for high conductivity composites, photonic materials and photovoltaic systems (solar cells). We use particle-based simulations to explore regions of the phase diagram, as function of the homopolymer volume fraction and the homopolymer/copolymer size ratio, where mainly triply periodic phases are predicted to be stable using self-consistent field theory calculations. The stability regions of the phases spontaneously formed in the simulations are in qualitative agreement with those predicted by the theory, and are primarily dominated by interfaces having symmetries corresponding to the diamond (D) and primitive (P) Schwarz surfaces; although other metastable bicontinuous and cocontinuous phases are also observed, some not previously seen before for diblock copolymer melts. At higher volume fractions of homopolymer, the conditions for phase equilibria between the stable bicontinuous phases and a homopolymer-rich phase are investigated with the expanded ensemble formalism. The thermodynamic conditions leading to the stabilization of the D and P phases are correlated to the topological characteristics of the interfaces, and their fluctuations, by following the evolution of geometrical measures during the spinodal decomposition in the particle-based simulations. Some preliminary results are also presented for the formation of bicontinuous phases in composites of diblock copolymers and nanoparticles of different sizes; i.e., for systems where nanoparticles replace the homopolymer in the blends discussed above.

See more of this Session: Thermodynamics of Polymers II
See more of this Group/Topical: Engineering Sciences and Fundamentals

Phase Transition in Polymer Brushes and Its Implication to Surface Forces

Eduardo R. A. Lima, Department of Chemical and Environmental Engineering, University of California, Riverside, Riverside, CA

Jianzhong Wu, Department of Chemical and Environmental Engineering, University of California, Riverside, Riverside, CA

Polymer brushes can be used, e.g., to stabilize colloids, reduce friction between surfaces and to provide lubrication in artificial joints. In this talk, we will present results from the polymer density-functional theory calculations on the structure, phase behavior and colloidal forces at various solution conditions. The implication of theoretical predictions to relevant experiments, in particular, for fabrication of protein-resistant surfaces, will also be discussed.

See more of this Session: Thermodynamics of Polymers II
See more of this Group/Topical: Engineering Sciences and Fundamentals

Application of Interfacial-SAFT (iSAFT) Density Functional Theory in Multiscale Modeling of Inhomogeneous Systems

Zhengzheng Feng, Chemical and Biomolecular Engineering, Rice University, Houston, TX

Chris Emborsky, Chemical and Biomolecular Engineering Department, MS-362, Rice University, Houston, TX

Kenneth R. Cox, Chemical and Biomolecular Engineering, Rice University, Houston, TX

Walter G. Chapman, Chemical and Biomolecular Engineering, Rice University, Houston, TX

Interfacial-SAFT (iSAFT) is a recently developed density functional theory, based on Wertheim's thermodynamic perturbation theory for association, which is computationally simple and thermodynamically consistent in describing structure and thermodynamic properties of polymeric fluids in inhomogeneous environments. In this work, we will present results from iSAFT calculations of the interfacial properties, phase behavior and microstructure of mixtures containing polymer-like molecules and colloidal-like particles. The versatility of the theory allows for a systematic investigation of such physical characteristics as size ratios, polymer chain length, bulk concentrations, intermolecular interactions, and chain heterogeneity. The most recent developments also allows for studying hydrogen bonding and chain branching effects.

See more of this Session: Thermodynamics of Polymers II
See more of this Group/Topical: Engineering Sciences and Fundamentals

Crystaline/Amorphous Transitions in Printed Polymeric Droplets

Phong Huynh, Rutgers University, New Brunswick, NJ

Paul Takhistov, School of Environmental & Biological Sciences, Rutgers University, New Brunswick, NJ

The relationship between structure of a material and cooling rate has been well studied at macro scale. However, this relationship at micro scale is not fully understood. We focus on the solidification of a droplet of polyethylene glycol (PEG) printed onto substrate at different cooling rates. Our data shown that the amorphous PEG appears in small PEG droplet while in the larger droplets, PEG spherulites locates at the center and surrounded by amorphous PEG layer which occupies the thinner, near-contact line region. Phase contrast and DSC techniques have been applied to detail the mechanism of droplet solidification. It was found that the crystal structure appears when the polymer droplet reaches a critical size. A non-steady state heat transfer model has been proposed to interpret this phenomenon.

See more of this Session: Thermodynamics of Polymers II
See more of this Group/Topical: Engineering Sciences and Fundamentals

Nanoparticle/Self-Assembled Monolayer Composite Coatings to Improve MEMS Reliability

Kendall M. Hurst, Department of Chemical Engineering, Auburn University, Auburn, AL

Christopher B. Roberts, Department of Chemical Engineering, Auburn University, Auburn, AL

W. Robert Ashurst, Department of Chemical Engineering, Auburn University, Auburn, AL

Due to persistent reliability issues such as strong adhesion, high friction and structural wear only the most basic microelectromechanical systems (MEMS) are currently used in commercial applications. Several approaches have been examined to overcome the inherent forces which can lead to stiction and friction related problems and device failure. Previous studies have focused on chemically altering silicon-based microstructure surfaces via self-assembled monolayers (SAMs) in order to change the chemical nature of adhesion and friction. While these SAM coatings have provided reduced adhesion and coefficients of friction between two contacting microstructures, SAM formation processes using organic molecular precursors are limited by high reaction sensitivity and low reproducibility. Another approach to overcome inherent attractive forces which cause adhesion and friction is to reduce the real area of contact between two contacting microstructures by increasing the surface roughness. Our previous work has illustrated that ligand-stabilized gold nanoparticles can be conformally deposited onto all surfaces of a micromachine or microdevice using a CO_2-expanded liquid/supercritical drying process. These fundamentally new nanoparticle coatings were shown to effectively increase the surface roughness of polysilicon microstructures, thereby, reducing the adhesion of contacting structures by four orders of magnitude. However, under further examination these gold nanoparticle coatings were not robust as particles were easily translated along the surfaces when contact was made eventually leading to microstructure adhesion. In this work, 3-mercaptopropyl trimethoxysilane (MPTS) and 3-aminopropyl trimethoxysilane (APTMS) self-assembled monolayers are utilized to attach CO_2-expanded liquid deposited gold nanoparticles (AuNPs) to the surface of silicon microstructures in attempt to create more robust nanoparticle coatings. MPTS and APTMS have both been demonstrated to attach free thiol and amino groups to metallic nanoparticles such as gold. The AuNP/SAM composite coatings were tested to determine their effect on microstructure adhesion and coefficients of friction.

See more of this Session: Thin Films, Coatings, and Interfacial Phenomena Using Compressible or Supercritical Fluids
See more of this Group/Topical: Engineering Sciences and Fundamentals

Dynamically Controllable Smart Surfaces Incorporating Reversible Carbon Dioxide Chemistry

Fiaz Mohammed, Chemical & Biomolecular Engineering, Clemson University, Clemson, SC

Christopher L. Kitchens, Chemical & Biomolecular Engineering, Clemson University, Clemson, SC

The use of carbon dioxide with specific amine based functionalities has been studied in the production of ionic liquids, gels and other supramolecualar complexes where the reversible reaction of carbon dioxide (CO2) with free amines results in the formation of carbamate salts. Surfaces which respond to external stimuli with changes in physical structure are a novel concept. The surface properties are tailored to produce specific effects, which is the foundation of a smart surface. In this study, we apply the reversible CO2 "green" chemistry to surfaces in an attempt of creating a smart surface.

See more of this Session: Thin Films, Coatings, and Interfacial Phenomena Using Compressible or Supercritical Fluids
See more of this Group/Topical: Engineering Sciences and Fundamentals

Three Routes to Nanostructured Titania Films for Alternative Energy Applications

Eunyoung You, Department of Chemical Engineering, University of Massachusetts, Amherst, MA

Yu Ying Tang, Polymer Science and Engineering, University of Massachusetts, Amherst, MA

Christos Karanikas, Department of Chemical Engineering, University of Massachusetts, Amherst, MA

Nicholas Hendricks, Polymer Science and Engineering, University of Massachusetts, Amherst, MA

Kenneth R. Carter, Polymer Science and Engineering, University of Massachusetts, Amherst, MA

James J. Watkins, Polymer Science and Engineering, University of Massachusetts, Amherst, MA

Crystalline, nanostructured titania films are of interest for numerous applications including next-generation energy conversion and sensing devices. In this talk we will discuss three approaches to films with defined structures that rely on the unique properties of supercritical fluids for their fabrication and their applications.

The first approach is direct and involves reactive deposition from supercritical CO2 to yield device quality films with exceptional step coverage in high aspect ratio, nano-scale features. We deposit the films via hydrolysis of appropriate precursors using a cold-wall deposition reactor on both inorganic templates and organic substrates patterned by nanoimprint or optical lithography. Removal of the latter by calcination yields hierarchical films with defined pore structures. The second approach involves the infusion and selective condensation polymerization of titania alkoxides within one phase domain of highly ordered, preformed block copolymer templates using supercritical CO2 as the reaction medium followed by removal of the template by calcination. While this approach potentially affords the greatest control of the device nanostructure, it is also the most complex and costly to implement. The final approach, which has recently been developed in our labs, provides a direct spray-on technique in which crystalline nano-scale dendritic metal oxides can be deposited directly onto inorganic as well as thermally sensitive substrates. This approach is low-cost and affords new opportunities for large area applications including photovoltaics.

See more of this Session: Thin Films, Coatings, and Interfacial Phenomena Using Compressible or Supercritical Fluids
See more of this Group/Topical: Engineering Sciences and Fundamentals

Concentration Gradients in Polymer Films Swollen with Supercritical CO2

Bryan D. Vogt, Department of Chemical Enigneering, Arizona State University, Tempe, AZ

The swelling of thin polymer films by supercritical carbon dioxide has been shown to exhibit an anomalous maximum in the sorption isotherms. This swelling maximum occurs when the activity of the fluid phase approaches unity and increases as the critical temperature is approached. Using an in-situ selective reactive templating approach based upon silica condensation within hydrophilic domains of an amphiphilic polymer film, the CO2 distribution within the film can be elucidated from size of the pores generated by the hydrophobic domains as examined using transmission electron microscopy of film cross sections. The origin of the maximum in the swelling of the thin films is demonstrated to be concentration gradients that extend from both interfaces of the film; as the temperature approaches the critical temperature, progressively larger local swelling at both interfaces is found. The impact of both temperature and finite film thickness on the concentration gradients will be discussed.

See more of this Session: Thin Films, Coatings, and Interfacial Phenomena Using Compressible or Supercritical Fluids
See more of this Group/Topical: Engineering Sciences and Fundamentals

Size-Based Separation Processes for Plasmid DNA - Comparison Between Size Exclusion Chromatography, Agarose Gel Electrophoresis, and Membrane Ultrafiltration

David R. Latulippe, Chemical Engineering, Pennsylvania State University, University Park, PA

Andrew L. Zydney, Chemical Engineering, The Pennsylvania State University, University Park, PA

Plasmids are circular pieces of double-stranded DNA that exist in different topological morphologies known as isoforms. The supercoiled isoform is characterized by a 'self-coiling' of the DNA induced by enzyme-generated helical stresses. A nick or break in one strand of the DNA double helix releases the helical stresses and creates the open-circular isoform which lacks any supercoiling. A break in both strands of the DNA double helix creates the linear isoform. The different plasmid isoforms have identical base pair sequences and net charge, but they differ in effective size as determined by static light scattering measurements. Thus, a number of size-based separation techniques have been proposed for both the analysis and purification of plasmid isoforms for gene therapy and DNA-based vaccines. However, there is currently no fundamental understanding of how the morphology of the different plasmid isoforms controls the thermodynamic and transport phenomena involved in these separation processes. The objective of this study was to evaluate the behavior of the supercoiled, open-circular, and linear plasmid isoforms in size exclusion chromatography (SEC), agarose gel electrophoresis (AGE), and membrane ultrafiltration (UF).

Experiments were performed with a 5.76 kbp plasmid supplied in the supercoiled form. Linear and open-circular isoforms were prepared by enzymatic digestion of the supercoiled plasmid. The radius of gyration of the plasmid was evaluated from static light scattering measurements obtained over a range of scattering angles and concentrations. SEC was performed using Sephacryl S-1000 SF resin, with the elution volume evaluated over a range of flow rates and buffer conditions. AGE was performed using 0.7 % horizontal agarose gels, which were stained and then imaged using a UV trans-illuminator. UF was conducted with Ultracel composite regenerated cellulose membranes, data were obtained over a range of filtrate flux using a small laboratory stirred cell.

The radius of gyration of the different plasmid isoforms ranged from 104 nm for the supercoiled isoform to 136 nm for the open-circular and 160 nm for the linear plasmid. The same general behavior was seen in SEC, with the supercoiled isoform having the largest elution volume followed by the open-circular and then the linear plasmid. In contrast, the open-circular isoform migrated the shortest distance in AGE, corresponding to the largest effective size, while the supercoiled plasmid showed the greatest penetration through the gel. This anomalous behavior is likely due to the greater hindrance to electrophoretic motion experienced by the more rigid open-circular plasmid; the linear and supercoiled isoforms have greater flexibility and are thus more easily transported through the agarose gel. The open-circular isoform also had the greatest retention in UF, but in this case the smallest retention was obtained with the linear isoform instead of the supercoiled. The UF data were analyzed using a scaling model based on elongation of the plasmid in the converging flow-field into the membrane pore. Plasmid transmission is predicted to be independent of the plasmid size but to depend critically on the plasmid flexibility, with the greatest transmission of the linear plasmid associated with the increased flexibility of this isoform. The very different behavior observed in SEC, AGE, and UF are directly related to differences in the underlying separation mechanisms and the physical properties of the different isoforms.

See more of this Session: Biomolecules at Interfaces II
See more of this Group/Topical: Engineering Sciences and Fundamentals

Stretched Exponential Kinetics in Thermal Dissociation of Surface-Bound DNA Hairpins

W. Benjamin Rogers, University of Pennsylvania, Philadelphia, PA

John C. Crocker, University of Pennsylvania, Philadelphia, PA

Over the past decade, groups have used a variety of single molecule techniques to study the folding/unfolding kinetics of DNA, RNA, proteins and other biomolecules. In our work, we probe the force dependence of DNA hybridization kinetics by measuring the lifetime distribution of single DNA hairpins under thermal dissociation. We use a scanning line optical tweezers instrument to measure the bound lifetimes of two DNA-coated microspheres under negligible applied tension. The two microspheres share a user-specified potential along the scan direction and are strongly confined in the perpendicular dimensions. The trapping laser intensity is modulated synchronously with a resonant scanner to null all optical contributions to the pair interaction potential near contact. In addition, the laser polarization can be rotated to produce a continuously adjustable optical repulsion, allowing the instrument to double as a force clamp over a modest range of applied tensions. Contrary to expectations for an all-or-none transition, we measure a stretched exponential lifetime distribution for the closed state of a DNA hairpin. We suggest that this unexpected distribution is a result of a slowly varying entropic tension due to the tethered microsphere diffusion and develop a simple model that is capable of explaining both exponential and nonexponential lifetime distributions that have been reported in a number of biophysical experiments.

See more of this Session: Biomolecules at Interfaces II
See more of this Group/Topical: Engineering Sciences and Fundamentals

Performance Analysis of Morpholino Monolayers: Platforms for Label-Free Detection of DNA Molecules

Napoleon Tercero, Chemical Engineering, Columbia University, New York, NY

Rastislav Levicky, Chemical & Biological Engineering, Polytechnic Institute of NYU, Brooklyn, NY

Surface hybridization, a reaction in which nucleic acid molecules in solution react with nucleic acid partners immobilized on a surface, is widely practiced in life science research. In these applications the immobilized partner, or "probe", is typically single-stranded DNA. Because DNA is strongly charged, high salt conditions are required to enable analyte nucleic acids ("targets") in solution to bind with the densely-immobilized DNA probes. High salt, unfortunately, compromises prospects for label-free monitoring or control of the surface hybridization reaction through surface electric fields. In this work Morpholinos, a class of uncharged DNA analogues, are introduced for surface-hybridization applications. Monolayers of Morpholino probes on gold supports can be fabricated with methods similar to those employed with DNA, and are shown to hybridize efficiently and sequence-specifically with target strands. The impact of hybridization on layer organization, as monitored through the surface capacitance, is greatly enhanced for Morpholino relative to DNA probe layers. The molecular mechanisms underlying the diagnostic contrast are identified and further interpreted through comparison to predictions from numerical Poisson-Boltzmann calculations. Interestingly, positive as well as negative capacitive responses (contrast inversion) to hybridization are possible, depending on surface populations of mobile ions as set by the read-out potential. Response sensitivities of Morpholino layers when hybridizing DNA targets are analyzed under conditions spanning an experimentally-relevant range of probe coverage and ionic strength. Quantitative correlation of surface capacitance with target coverage (targets/area) demonstrates sensitivities comparable to or exceeding those of label-free methods such as surface plasmon resonance.

See more of this Session: Biomolecules at Interfaces II
See more of this Group/Topical: Engineering Sciences and Fundamentals

Real Time Detection of DNA Hybridization and Melting On Spotted Oligonucleotide Arrays by Total Internal Reflection Fluorescence

Zhou Xu, Department of Chemical and Biomolecular Engineering, Vanderbilt University, Nashville, TN

Paul E. Laibinis, Department of Chemical and Biomolecular Engineering, Vanderbilt University, Nashville, TN

Complementary base pair interactions between DNA strands form the basis for many methods of DNA detection and identification. DNA microarrays allow the parallel investigation of the ability of a particular sequence to bind to libraries of differing DNA sequences. In most cases, such measurements of DNA-DNA interactions focus on single time results, typically at one temperature, that are performed ex situ. As a result, kinetic and thermodynamic information on the factors that influence DNA binding at surfaces are limited. Our laboratory uses an in situ method for detecting nucleic acid hybridization and melting events in real time at arrayed oligonucleotide surfaces. These measurements are made at the solid-liquid interface using total internal reflection fluorescence (TIRF) imaging. The arrays are constructed by covalently end-immobilizing single-stranded DNA molecules of chosen sequences onto microscopic glass slide surfaces. We have investigated the influences of target DNA concentrations, nucleotide mismatches, and ionic strength on the hybridization kinetics, and used a mathematical model that includes the effects of diffusion and solution-phase depletion in order to obtain estimates for the hybridization rate constants on the surface from our experimental data. The stability of and interactions between DNA duplexes are investigated by measurements of melting curves for hybridized structures formed to the microarrayed surface. TIRF imaging of the microarray through programmed temperature profiles allows direct, concurrent measurements for DNA-DNA interactions. The resulting melting curves of the surface oligos are compared with those predicted from solution phase melting models, showing that single-base mutations are detectable by this in situ method. The talk will examine the abilities of this approach in providing enhanced detection abilities of single- and multiple nucleotide mutations from known gene sequences.

See more of this Session: Biomolecules at Interfaces II
See more of this Group/Topical: Engineering Sciences and Fundamentals

Rational Design of Peptides for DNA Separations

Vikas P. Jain, Department of Chemical Engineering, The City College of New York, New York, NY

Charles Maldarelli, Chemical Engineering, The City College of New York, New York, NY

Raymond Tu, Chemical Engineering, The City College of New York, New York, NY

We have designed a peptide with switchable surface activity, where the folded form (α-helical) of the peptide is amphiphilic and the unfolded form is not. We hypothesize that we can control the secondary structure of the peptide, and, thus, control its surface activity. The primary structure of a peptide is designed in such a way that upon folding the hydrophobic and hydrophilic domains become spatially disjoint, leading to amphiphilic behavior. The ensemble average population of folded states depends on binding to a specific sequence of DNA. The secondary structure has been characterized by using circular dichroism spectropolarimetry, and we show that the peptide has a transient secondary structure as a function of binding. The dynamic behavior of the peptide is characterized by pendant drop/bubble method and modeled accordingly. Here, we hypothesize that only folded peptide adsorbs at the air/water interface, and the dynamic surface tension due to α-helical peptide adsorption is predicted by a diffusive transport model that accounts for peptide folding dynamics and the kinetics of adsorption.

See more of this Session: Biomolecules at Interfaces II
See more of this Group/Topical: Engineering Sciences and Fundamentals

A Tethered-Particle Method Study of DNA Hybridization

Marie T. Ung, Department of Chemical and Biomolecular Engineering, University of Pennsylvania, Philadelphia, PA

John C. Crocker, Department of Chemical and Biomolecular Engineering, University of Pennsylvania, Philadelphia, PA

The tethered-particle method is used to probe the complex process of DNA hybridization. This experiment analyzes tethers formed between PEGylated polystyrene microspheres and PEGylated glass surfaces, to which DNA strands have been grafted. This experiment is performed near the duplex melting temperature, resulting in dynamic binding and unbinding. Single molecular tethers are analyzed using high performance particle tracking with a high-speed camera. By analyzing 2D trajectories of the particle's centroid versus time, we can detect single binding and unbinding events from multiple bridged states. Molecular rates for transitioning between different bridging states are obtained; they indicate that multiple bridged states are less stable than expected for a zero cooperativity model. We hypothesize that this instability may be due to extreme force sensitivity of individual DNA duplexes. Interaction kinetics will also be investigated in the multiple bridge limit.

See more of this Session: Biomolecules at Interfaces II
See more of this Group/Topical: Engineering Sciences and Fundamentals

Chemical Cross-Linking of Polyelectrolyte Nanofilms to Control Mechanical and Cell-Contacting Properties

Jennifer A. Phelps, Yale University, New Haven, CT

Paul R. Van Tassel, Yale University, New Haven, CT

Nanofilms comprised of select polyelectrolytes and constructed by layer-by-layer (LbL) assembly show great promise as surface coatings due to their highly tunable film properties, such as thickness and viscoelasticity, as well as their ability to effectively decouple a material's surface from bulk properties. Nanofilm properties are controlled by physical means such as solution chemistry, pH, and ionic strength, and also by chemical means such as cross-linking of functional groups within the polyelectrolytes. This work focuses on changes in mechanical and cell-contacting properties of nanofilm coatings brought about by different types and extents of chemical cross-linking. EDC-NHS chemistry is employed to form cross-links of controlled extent, both homogeneously throughout the film and in an inhomogeneous manner where increased film rigidity is sought near to the film's surface. Cross-link extent and location are monitored in situ via Fourier transform infrared (FTIR) spectroscopy within the attenuated total reflection (ATR) mode, and film mechanical properties are evaluated using quartz crystal microgravimetry with dissipation monitoring (QCM-D). Cell-contacting properties are investigated using human hepatocellular carcinoma (HepG2) cells. Cross-linking is seen to result insignificantly increased film viscous and elastic moduli, with values for films subjected to non-uniform cross-linking intermediate between fully cross-linked and native films. HepG2 cells adhere strongly to both fully and partially cross-linked films. FTIR spectra confirm the presence of a spatially non-uniform distribution of cross-links, demonstrating the possibility of inhomogeneous films whose bulk and surface properties are effectively decoupled. In conclusion, chemical cross-linking represents a facile and highly effective means of controlling polyelectrolyte nanofilm mechanical and cell contacting properties.

See more of this Session: Biomolecules at Interfaces II
See more of this Group/Topical: Engineering Sciences and Fundamentals

Designing Patchy Particles for Self-Assembly Propensity

Eric Jankowski, Chemical Engineering, University of Michigan, Ann Arbor, MI

Sharon C. Glotzer, Chemical Engineering and Materials Science and Engineering, University of Michigan, Ann Arbor, MI

We utilize a new method, bottom-up building block assembly, to efficiently generate energy- minimizing structures for a family of complex, interacting patchy particles for which *a priori* knowledge of preferred assembled structures is lacking. We introduce the concept of self-assembly propensity and show it depends on the assembly pathways available to the building blocks. By grouping generated configurations into motifs we discern important characteristics for assembling specific patterns and identify building blocks with a propensity to self-assemble a desired configuration. We utilize the assembly pathways to propose strategies for overcoming poor self-assembly propensity.

See more of this Session: Modeling of Interfacial Systems II
See more of this Group/Topical: Engineering Sciences and Fundamentals

Molecular Modeling and Simulation of Core-Shell Particle

María Verónica Carranza, Chemical Engineering Department, Escola Politécnica of the University of Sao Paulo, São Paulo, Brazil

Reinaldo Giudici, Chemical Engineering Department, Escola Politécnica of the University of Sao Paulo, São Paulo, Brazil

Bruno Ramos, Chemical Engineering Department, Escola Politécnica of the University of Sao Paulo, São Paulo, Brazil

In order to develop new materials and composites with designed new properties, it is essential that these properties be predicted before preparation, processing, and experimental characterization. Despite the tremendous advances made in the modeling of the structural, thermal, mechanical and transport properties of materials at the macroscopic level (finite element analysis of complicated structures) there remains tremendous uncertainty about how to predict many critical properties related to performance. The fundamental problem here is that these properties depend on the atomic level interactions and chemistry. Molecular modeling is a powerful tool in the study of atomic structure and interactions at the nanometer length scale, and thus in the prediction of materials properties. The modeling and simulation methods at molecular level usually employ atoms, molecules or their clusters as the basic units considered. Molecular dynamic (MD) is a computer simulation technique that allows one to predict the time evolution of a system of interacting particles and estimate the relevant physical properties (molecular structure and interactions at interface, atomic diffusion coefficient, thermodynamics properties, intercalation kinetics and mechanical properties). Over the last fifteen years, owing to the rapid development of computers, polymeric systems have been the subject of intense study with MD simulations. In this work, it will be discussed the development of this technique as an useful tool to study and predict parameters of an emulsion polymerization process for obtaining core-shell nano-particles of polystyrene/poly(methyl methacrylate).

See more of this Session: Modeling of Interfacial Systems II
See more of this Group/Topical: Engineering Sciences and Fundamentals

DLVO Theory Calculations for Spherical or Cubic Nanoparticles--Some Applications to CuPc Pigments

Jiannan Dong, School of Chemical Engineering, Purdue University, West Lafayette, IN

Yoonjee Park, School of Chemical Engineering, Purdue University, West Lafayette, IN

David .S. Corti, School of Chemical Engineering, Purdue University, West Lafayette, IN

Elias I. Franses, School of Chemical Engineering, Purdue University, West Lafayette, IN

Yan Zhao, Hewlett-Packard Laboratories, Hewlett-Packard Co., Palo Alto, CA

Eric Hanson, Hewlett-Packard Laboratories, Hewlett-Packard Co., Palo Alto, CA

Hou T. Ng, Hewlett-Packard Laboratories, Hewlett-Packard Co., Palo Alto, CA

Even though the DLVO theory has been available for about 60 years, its use is often restrained by some calculation complexities, and the lack of reliable Hamaker constants. A new simple dimensionless group formulation of the key equation for the potential energy vs. interparticle distance is presented. Several key dimensionless groups are identified. The repulsive potential energy maximum is calculated analytically for spheres or cubes which are oriented with their faces parallel to each other, in terms of the dimensionless size Rs (sphere) or l (cube) and Hamaker constant A. The Hamaker constant was calculated with a novel combination of an ab initio and an empirical methodology. With the new formulation one can readily determine the interplay of A, Rs (or l), surface potential (taken to be to the zeta potential) and Debye length (related to the ionic strength) on the DLVO- predicted stability.

The predictions allow easy comparisons with experimental results of the colloidal stability of aqueous dispersions of copper phthalocyanine (CuPc). The comparisons reveal that electrostatic forces play a significant role in the dispersion stabilization. Moreover, as the particles come close together for possible coagulation, other forces start playing key roles in the stability.

The new DLVO equations predict that as Rs (or l) decreases, the dispersion should become less stable. Dispersions of cubes in most cases are predicted to be more stable than spheres of the same size (with 2Rs=l).

See more of this Session: Modeling of Interfacial Systems II
See more of this Group/Topical: Engineering Sciences and Fundamentals

Thermodynamic and Mechanical Properties of Non-Spherical Nematic Colloids: A Computational Study

Shivkumar Bale, Cain Department of Chemical Engineering, Louisiana State University, Baton Rouge, LA

Francisco R. Hung, Cain Department of Chemical Engineering, Louisiana State University, Baton Rouge, LA

We investigate the effects of particle size and shape on the thermodynamic and mechanical properties of non-spherical particles in a nematic liquid crystalline solvent. The particles' inclusion distorts the nematic, which form defect structures around the particles that give rise to anisotropic, strong interparticle interactions (up to several thousands of kT for micron-sized spherical particles). These interactions can be used to bind the particles together into ordered structures [1-4], which have potential applications in colloidal crystals, photonics, optical sensors and nanoscale electronics.

We report numerical calculations for the defect structures and the nematic-mediated potentials of mean force, interparticle interactions and torques for systems containing a few particles in a nematic. Using a mesoscale theory in terms of the tensor order parameter Q of the nematic, we analyze several systems of particles with different shapes (e.g., spherocylindrical, cubic, triangular prisms), and we compare with similar results for spherical particles of comparable sizes. Our results [5] indicate that the nematic forms defect structures consisting of distorted Saturn rings around the non-spherical particles. When the particles are close to each other, the Saturn ring defects can become entangled depending on the size and shape of the particles. Our calculations suggest that the nematic-mediated interparticle interactions are highly directional and strong, up to 5700 kT for the case of nm-sized spherocylinders with their long axis parallel to each other. These interactions can bind the particles together at specific locations, and thus could be used to assemble the particles into ordered structures with different morphologies.

[1] P. Poulin, H. Stark, T. C. Lubensky and D. A. Weitz, Science 275, 1770 (1997); J. C. Loudet, P. Barois and P. Poulin, Nature 407, 611 (2000); J. C. Loudet and P. Poulin, Phys. Rev. Lett. 87, 165503 (2001)

[2] M. Yada, J. Yamamoto and H. Yokoyama, Phys. Rev. Lett. 92, 185501 (2004)

[3] A. B. Nych, U. M. Ognysta, V. M. Pergamenshchik, B. I. Lev, V. G. Nazarenko, I. Muševič, M. Škarabot and O. D. Lavrentovich, Phys. Rev. Lett. 98, 057801 (2007)

[4] I. Muševič, M. Škarabot, U. Tkalec, M. Ravnik and S. Žumer, Science 313, 954 (2006); M. Ravnik, M. Škarabot, S. Žumer, U. Tkalec, I. Poberaj, D. Babič, N. Osterman and I. Muševič, Phys. Rev. Lett. 99, 247801 (2007); U. Tkalec, M. Škarabot, and I. Muševic, Soft Matter 4, 2402 (2008)

[5] F. R. Hung, Phys. Rev. E 79, 021705 (2009); F. R. Hung and S. Bale, Mol. Simulat., in press (2009)

See more of this Session: Modeling of Interfacial Systems II
See more of this Group/Topical: Engineering Sciences and Fundamentals

Linking Atomistic and Mesoscales: Atomistic and Simple Coarse-Grained Models in Application to Biomolecular Problems

Paraskevi Gkeka, School of Engineering, University of Edinburgh, Edinburgh, United Kingdom

Giuseppe Pellicane, Dipartimento di Fisica, Università di Messina, 98166 Messina, Italy

Lev Sarkisov, School of Engineering, University of Edinburgh, Edinburgh, United Kingdom

Computer simulations and molecular modeling provide powerful tools to underpin and explain phenomena occurring on microscopic time and length scales (nanometers, nanoseconds). However for many systems, and in particularly for biological systems, we are interested in effects involving larger scales and taking place over longer periods of times (microseconds). In these situations, simplified, coarse-grained models provide a useful route. The idea of coarse-grained models is to reduce the number of degrees of freedom in the system and thus to make it more computationally tractable. Although, in many cases these models are qualitative, they may provide fundamental insights on the phenomena otherwise unattainable.

In this presentation we will consider several examples of atomistic and coarse-grained models in application to biomolecular problems. The first example considers a particular class of peptides capable of seamless translocation through cell membranes. Due to this property these peptides have been extensively investigated as potential vectors for cellular drug delivery. The actual mechanisms of translocation however vary from system to system and in many cases are unknown. We will construct simple models of these peptides to elucidate these mechanisms and discuss the implications of the model.

In the second example, we employ non-equilibrium molecular dynamics simulation to characterize the effective interactions between lysozyme molecules involved in the formation of two hydrophobic crystal contacts. We show that the effective interactions between crystal contacts do not exceed a few kTs, the range of the attractive part of the potential is less than 4Å and that, within this range, there is a significant depletion of water density between two protein contacts. Our findings highlight the different nature of protein crystallization and protein recognition processes.

See more of this Session: Modeling of Interfacial Systems II
See more of this Group/Topical: Engineering Sciences and Fundamentals

A Brownian Dynamics Study of Protein Deposition On a Solid Surface

Jee-Ching Wang, Department of Chemical and Biological Engineering, Missouri University of Science and Technology, Rolla, MO

Parthasakha Neogi, Department of Chemical and Biological Engineering, Missouri University of Science and Technology, Rolla, MO

When protein molecules deposit at a water-solid interface, they have been observed to form 1-dimensional short strings or 2-dimension dendrites that are one-protein molecule wide. These stringy structures are different from those predicated on random sequential adsorption and indicate the presence of other mechanisms during protein deposition. In this work, we simulate protein deposition on a neutral solid surface with Brownian dynamics simulations. A protein molecule is treated as a charged sphere and interacts with other protein molecules through DLVO theory, which combines the effects of Hamaker (van der Waals) attraction and electrostatic repulsion. The structure of the deposited protein is examined through simulation snapshots, radial distribution function, order parameter, and analysis of coordination number. Depending on the ionic strength and the protein surface charge, the protein-protein interaction can vary from purely attractive to the type with repulsive energy barrier. It is found that protein-protein interactions with repulsive energy barrier lead to slower clustering rate but promote the formation of stringy structures while purely attractive protein-protein interaction leads to faster clustering but more compact structures. The roles of homogeneous and heterogeneous growth mechanisms in protein deposition and cluster formation are also studied.

See more of this Session: Modeling of Interfacial Systems II
See more of this Group/Topical: Engineering Sciences and Fundamentals

Numerical Solution of the Free Boundary Foam Drainage Equation

Dennis A. Siginer, Department of Mechanical Engineering & Department of Mathematics, Petroleum Institute, Abu Dhabi, United Arab Emirates

F. Talay Akyildiz, Department of Mathematics, Petroleum Institute, Abu Dhabi, United Arab Emirates

The numerical solution of the free boundary evolution equation for foams is investigated. The domain with free boundary is mapped to a fixed domain via a change of independent variable; the resulting singular nonlinear partial differential equation is transformed into a nonlinear partial differential equation by a change of dependent variable. After using domain truncation, convergence of the variable step size backward Euler discretization is discussed. Numerical algorithms based on the backward difference and implicit Crank-Nicolson are proposed for solving the field equations and comparative results are presented. Furthermore, it is shown that the reaction-diffusion problem with free drainage represented by a degenerate parabolic partial differential equation is covered by analytic semi-groups set in an abstract Banach space.

See more of this Session: Modeling of Interfacial Systems II
See more of this Group/Topical: Engineering Sciences and Fundamentals

Thermodynamicall Dominant Structures in the Hydration of Aqueous Ions

Safir Merchant, Chemical and Biomolecular Engineering, Johns Hopkins University, Baltimre, MD

Dilip Asthagiri, Department of Chemical and Biomolecular Engineering, Johns Hopkins University, Baltimre, MD

The wealth of structures and processes made possible by biological macromolecules rests in considerable measure on the poly-electrolyte character of the macromolecule. To compensate for the charge on the macromolecule, small inorganic ions are always present in the aqueous medium. However, these ions also participate in ion specific ways in the structure and function of the macromolecule. To understand the basis of ion specific effects, it is necessary to understand how the ions are hydrated and how they influence the solvent matrix. We reformulate the quasichemical theory of solution to study how each stepwise increment in the hydration state of the ion accounts for the thermodynamics of hydration. This molecular *aufbau* approach leads to the surprising insight that only a small subset of water molecules in the first hydration shell of the ion decisively contribute to ion specificity. Further, we find that the distribution of water occupation in a cavity the size and shape of the hydration shell itself is an important element in determining the hydration structure of the ion. These ideas are developed to study the range of ion-specific effects in water and the implications for ion specific effects in biology.

See more of this Session: Thermophysical Properties of Biological Systems
See more of this Group/Topical: Engineering Sciences and Fundamentals

Loop Modeling in Proteins Using a Configurational Bias Approach

Monica Berrondo, Chemical and Biomolecular Engineering, Johns Hopkins University, Baltimore, MD

Jeffrey J. Gray, Chemical and Biomolecular Engineering, Johns Hopkins University, Baltimore, MD

Loop modeling is important in homology modeling, enzyme design, and flexible backbone docking. Despite improvements in protein structure prediction in general, loop modeling remains very challenging, especially due to the difficulties of sampling diversely while maintaining the loop connectivity. Current approaches in loop modeling often only treat loops up to 8 or 12 residues, and the time required to model these loops is long. In this talk, we will present a new loop modeling algorithm based on a configurational bias approach, similar to that used in polymer modeling, where loops are deleted and then re-grown one residue at the time from both ends, alternating between the N- and C-terminus of the loop. The new algorithm reduces the time required to model loops by as much as five-fold. We found that this approach increases the sampling near the native conformation and therefore we are able to find the native conformation without requiring as much sampling (usually near-native structures can be found within the first 100-200 structures). Using the re-growing procedure, preliminary data indicate that for certain loops, the lowest energy structure is as much as 1.2 Å rmsd lower than previously published results for 12-residue loops and much less sampling is needed to achieve these high resolution results. On a small set of 12-residue loops from 10 proteins, the average rmsd of the lowest energy structure was 1.1 Å after only generating 100 structures for each protein. The efficiency and accuracy of the new technique promises to allow more aggressive sampling of flexible loop conformations in applications such as enzyme design and docking of antibodies.

See more of this Session: Thermophysical Properties of Biological Systems
See more of this Group/Topical: Engineering Sciences and Fundamentals

Simulation of Helix Formation and Analysis by Lifson-Roig Theory

Ashish V. Sangwai, Department of Chemical and Biomolecular Engineering, Tulane University, New Orleans, LA

Hank Ashbaugh, Department of Chemical and Biomolecular Engineering, Tulane University, New Orleans, LA

The predilection for short polypeptides to adopt different secondary structural elements, like α-helicies, is encoded in their amino acid sequences. Lifson-Roig is a helix-coil transition theory that predicts the melting curves based on their specific sequences from parameters fitted to fractional helicities determined from circular dichroism experiments. Lifson-Roig theory makes predictions regarding the distribution of helix probabilities along the peptide backbone that is experimentally unavailable. We have performed replica exchange molecular dynamic simulations (REMD) to evaluate helix formation for a large set of polypeptide chains at temperatures ranging from ~250 K to ~350 K. Accurate helical nucleation and propagation probabilities down the simulated peptide sequences were determined. While Lifson-Roig theory accurately reproduces the fractional helicities from simulation, the comparison between the predicted and observed backbone distribution of helical amino acids is only qualitative. These results suggest a moderately longer-range interaction between amino acids in the stabilization of helical structures than assumed by Lifson-Roig.

See more of this Session: Thermophysical Properties of Biological Systems
See more of this Group/Topical: Engineering Sciences and Fundamentals

Atomic-Level Stresses On Proteins Upon Vitrification

Thomas G. Lombardo, Chemical Engineering, Princeton University, Princeton, NJ

Pablo G. Debenedetti, Chemical Engineering, Princeton University, Princeton, NJ

Lyophilization is commonly used in the pharmaceutical and food processing industries to preserve labile biomolecules, but a comprehensive molecular understanding of the process is lacking. This procedure generates both freezing and drying stresses due to the growth of free volume. These stresses, as well as aging in the glass state, can result in undesired molecular mobility which leads to the loss of secondary protein structure [1, 2]. We use molecular dynamics simulations to study the normal and shear stresses acting on individual protein atoms as a result of quenching at constant pressure. We study both isobaric inherent structures and glasses formed via finite-rate quenches. Stresses are calculated on two different proteins, the bee-venom protein melittin and the globular protein ubiquitin, with an atomic-level description of the stress tensor [3]. Protein atoms along the backbone of alpha-helix structures exhibit large increases in the local normal stress after quenching. Many of the atoms which display significant increases belong to carbonyl and amine groups along the protein backbone. These groups are responsible for the formation of hydrogen bonds which stabilize the alpha-helix structure and the increased stress may lead to the degradation of secondary structure.

[1] Webb, S.D., Cleland, J.L., Carpenter, J.F., and Randolph, T.W., *J. Pharmaceut. Sci.*, 92(4), 715, 2003.

[2] Heller, M.C., Caprenter, J.F., and Randolph, T.W., *Biotechnol. Bioeng.*, 63(2), 166, 1999.

[3] Max Born and Kun Huang. *Dynamical Theory of Crystal Lattices.* Clarendon Press, Oxford, 1954.

See more of this Session: Thermophysical Properties of Biological Systems
See more of this Group/Topical: Engineering Sciences and Fundamentals

Large Scale Free Energy and Potential of Mean Force Calculations through Distributed Computing

Michael R. Shirts, Chemical Engineering, University of Virginia, Charlottesville, VA

Recent developments in computational processing power and in theoretical methods are making simulations come increasingly closer to predicting ligand binding affinities and computing accurate, high dimensional potentials of mean force.

This level of precision makes it possible to separate errors caused by poor sampling from the inadequacies of molecular mechanics force fields. These improvements are beginning to change the process of designing drugs with high affinities and specific action from a trial-and-error art to a nanoscale engineering process. However, current computational methods are inefficient and converge relatively slowly, requiring computational resources that far exceed what is currently cost-effective in the pharmaceutical industry. Improvements in methods and increases in available computational power will be necessary to make such calculations feasible in industrial practice.

We present extended ensemble ligand binding simulations for the FKBP-12 system using large scale distributed computing to reach precisions near 1 kcal/mol. Simulations were performed for both AMBER/GAFF and OPLS-AA, two different common force fields used to model biomolecules, in order to understand how much differences in force field can affect results of free energies of binding. Additionally, extended ensemble methods can significantly increase the speed at which simulations sample macromolecular conformations by including Monte Carlo transitions between nonphysical states, and provide some hope for improving efficiency. We will also present data on the comparison between extended ensemble and more traditional free energy calculations.

We also recently presented a minimum variance method to calculate free energies and ensemble averages from multiple equilibrium simulations conducted at different thermodynamic states [1]. This method, based on statistical techniques for correcting biases in sampling, is useful for analyzing both simulations of phase changes and the results of single molecule experiments. This estimator gives the same results as multiple histogram methods [2,3] in the limit of vanishingly small histogram bins, but does not actually require histograms, eliminating bias due to binning.

To solve the problem of efficiently computing averages and free energies with large numbers of states, we derive a modification of the multistate minimum variance method that uses only states with sufficient mutual phase space overlap. For many cases of interest, this drastically decreases the number of states that must be considered with negligible loss of precision.

We demonstrate how these methods can be used to calculate 3-dimensional potentials of mean force from data collected from distributed computing simulations of the ribosome very efficiently.

[1] M. R. Shirts and J. D. Chodera, J. Chem. Phys.129, 124105 (2008)

[2] A. M. Ferrenberg and R. H. Swendsen, Phys. Rev. Lett. 63:1195-1198 (1989)

[3] S. Kumar, D. Bouzida, R. H. Swendsen, P. A. Kollman and J. M. J. Rosenberg, J. Comput. Chem., 13:1011-1021 (1992)

See more of this Session: Thermophysical Properties of Biological Systems
See more of this Group/Topical: Engineering Sciences and Fundamentals

A Hybrid Molecular Mechanics-Elastic Network Model for Studying Zinc Binding to a Zinc-Finger Protein

Dixit Purushottam, Chemical and Biomolecular Engineering, Johns Hopkins University, Baltimre, MD

Dilip Asthagiri, Department of Chemical and Biomolecular Engineering, Johns Hopkins University, Baltimre, MD

It is estimated that nearly half of all proteins require a metal ion as an essential cofactor in the structure and/or function of the molecule. In metalloproteins where the metal acts as a structural element it is not well-understood how the metal ion induces folding and stabilization of the protein. To study such processes requires acknowledging the chemical nature of metal-protein interactions. But a satisfactory description of such chemical interactions requires using computationally expensive, quantum chemical calculations. Here we suggest a way to economize on the quantum chemical calculations by dividing the protein into a metal-binding site that interacts with the rest of the protein (the bulk) that is described within the elastic network framework. Using this approach and the available crystal structure, to within a coupling constant we obtain an effective protein-field on the site particles. Here the value of coupling constant is chosen to reproduce the binding energy distribution of the metal with the site. We use all-atom simulations using classical potentials to obtain the target binding energy distribution, although in principle other metrics or even energy distributions from short all-atom *ab initio* simulations can be used. We use this approach to study Zn(II) binding to a zinc finger protein. We find that without the effective protein-field, the site (in vacuum) is more stable, suggesting that the solvent and the rest of the protein tend to weaken the binding of the metal ion with the site. We present exploratory calculations of the free energy change in replacing Zn(II) with either Co(II) or Fe(II).

See more of this Session: Thermophysical Properties of Biological Systems
See more of this Group/Topical: Engineering Sciences and Fundamentals

Calculations of Absolute Stability of Proteins, Using Molecular Dynamics and Focused Dielectric Constants

Spyridon Vicatos, Chemistry, University of Southern California, Los Angeles, CA

Arieh Warshel, Chemistry, University of Southern California, Los Angeles, CA

Maite Roca, Departament Química Física, Universitat de Valencia, Valencia, Spain

The ability to predict the absolute stability of proteins based on their corresponding sequence and structure is a problem of great fundamental and practical importance. In this work we report an extensive, refinement and validation of our recent approach (Roca et al, FEBS Lett 2007;581(10):2065-2071) for predicting absolute values of protein stability ΔG_{fold}. This approach employs the semimacroscopic protein dipole Langevin dipole method in its linear response approximation version (PDLD/S-LRA) while using optimal values of dielectric constants ε_p and ε_{eff}. The method is optimized and validated on a diverse set of 45 proteins. It is found that optimal values of both dielectric constants are around 40. However the self energy of internal residues and the charge - charge interactions of Lys have to be treated with care, using a somewhat lower values of ε_p and ε_{eff}. The predictions of ΔG_{fold} reported here, have an average error of only 1.8 kcal/mole compared to the observed values, making this method very promising for estimating protein stability. The ability of this method to predict mutations that can stabilize a protein is also investigated. This work provides valuable insight into the complex electrostatic phenomena taking place in folded proteins.

See more of this Session: Thermophysical Properties of Biological Systems
See more of this Group/Topical: Engineering Sciences and Fundamentals

Biomolecular Force Fields for Lysozyme Crystal: Assessed From Molecular Dynamics Simulations

Zhongqiao Hu, Chemical and Biological Engineering, National University of Singapore, Singapore, Singapore

Jianwen Jiang, Chemical land Biological Engineering, National University of Singapore, Singapore, Singapore

Different biomolecular force fields (OPLS-AA, AMBER03 and GROMOS96) in conjunction with SPC, SPC/E and TIP3P water models are assessed for molecular dynamics simulations in a tetragonal lysozyme crystal. The root-mean-square deviations for the CƒÑ atoms of lysozymes are about 0.1 ~ 0.2 nm from OPLS-AA and AMBER03, smaller than 0.4 nm from GROMOS96. All force fields exhibit similar pattern in B-factors, whereas OPLS-AA and AMBER03 accurately reproduce experimental measurements. Despite slight variations, the primary secondary structures are well conserved using different force fields. Water diffusion in the crystal is approximately ten-fold slower than in bulk phase. The directional and average water diffusivities from OPLS-AA and AMBER03 along with SPC/E model match fairly well with experimental data. Compared to GROMOS96, OPLS-AA and AMBER03 predict larger hydrophilic solvent-accessible surface area of lysozyme, more hydrogen bonds between lysozyme and water, and higher percentage of water in hydration shell. SPC, SPC/E and TIP3P water models have similar performance in most energetic and structural properties, but SPC/E outperforms in water diffusion. While all force fields overestimate the mobility and electrical conductivity of NaCl, a combination of OPLS-AA for lysozyme and the Kirkwood-Buff model for ions is superior to others. As attributed to the steric restraints and surface interactions, the mobility and conductivity in the crystal are reduced by one ~ two orders of magnitude from aqueous solution.

See more of this Session: Thermophysical Properties of Biological Systems
See more of this Group/Topical: Engineering Sciences and Fundamentals

Solubility of Fluorinated Compounds in Room-Temperature Ionic Liquids

Mark B. Shiflett, Central Research and Development, DuPont Company, Wilmington, DE

For any applications with a new type of solvents such as room-temperature ionic liquids (RTILs), knowledge of phase behaviors with various chemicals is essential. We have studied vapor-liquid equilibria (VLE) and liquid-liquid equilibria (LLE)) of industrially important fluorinated compounds in a variety of RTILs. Examples of chemicals include: chlorofluorocarbons (CFCs), hydrochlorofluorocarbons (HCFCs), hydrofluorocarbons (HFCs), fluorocarbons (FCs), hydrofluoroethers (HFEs) and fluorinated benzenes. VLE data have been measured with a gravimetric microbalance and successfully correlated with a modified Redlich-Kwong equation of state (EOS). The present EOS predicts partial immiscibilities (VLLE) with lower critical solution temperatures (LCSTs) in the solute-rich side solutions and indicates the Type-V mixture behavior (based on the Konynenburg-Scott classification) for many systems studied here. In order to verify our EOS predictions, we have conducted experiments to show VLLE for some selected binary systems. VLLE data have been obtained using a "volume-mass-measurement" method at an isothermal condition, combined with a cloud-point method. This behavior is quite interesting, and in contrast with ionic liquid solutions of various alcohols and dialcohols, where immiscibility gaps have been well studied experimentally and show upper critical solution temperatures (UCSTs). In this report, we are particularly interested in understanding the global phase behavior of HFCs with ionic liquids and how this provides the basis for developing applications such as absorption cooling and extractive distillation.

See more of this Session: Thermophysical Properties of Ionic Liquids
See more of this Group/Topical: Engineering Sciences and Fundamentals

Experimental Measurements of Ionic Liquid and Water Systems for Absorption Refrigeration

Lindsay E. Ficke, Chemical and Biomolecular Engineering, University of Notre Dame, Notre Dame, IN

Joan F. Brennecke, Chemical and Biomolecular Engineering, University Of Notre Dame, Notre Dame, IN

Absorption refrigeration cycles have many advantages over the traditional vapor-compression cycles. The addition of an ionic liquid working fluid to an absorption refrigeration system has the ability to overcome some of the safety and environmental impacts of the current working fluids in addition to the existing ability to replace electrical energy with low value heat energy. Many ionic liquids are completely miscible with water, which leads our focus to investigating ionic liquids and water for this application.

Thermodynamic measurements and/or predictive modeling must provide the mixture properties and phase behavior in order to evaluate these systems. Vapor-liquid equilibrium is used to determine activity coefficients, accounting for the mixture deviations from ideal behavior. In addition, excess enthalpies are used to determine the strength of interactions between water and the ionic liquid. We have previously presented three 1-ethyl-3-methylimidaolium based ionic liquids and water systems for the absorption refrigeration application. Here, we present a similar evaluation for several new ionic liquid and water systems.

See more of this Session: Thermophysical Properties of Ionic Liquids
See more of this Group/Topical: Engineering Sciences and Fundamentals

Modelling of Gas Solubilities in Certain Ionic Liquids Using a Group Contribution Equation of State

Cor J. Peters, Process and Energy, Delft University of Technology, Faculty of Mechanical, Maritime and Materials Engineering, Delft, Netherlands

Selva Pereda, Process and Energy, Delft University of Technology, Faculty of Mechanical, Maritime and Materials Engineering, Delft, Netherlands

This study presents the results of the thermodynamic modeling of gas solubilities in the homologous [-mim][Tf2N] family using a Group Contribution Equation of State (GC-EoS). The model parameters were estimated on the basis of 1400 experimental data points in the temperature range 278-460 K and pressures up to 45 bar. Results show that the GC-EoS is a suitable model to describe gas solubilities in this family of room temperature ionic liquids. The solubility of various gases like CO2, H2, CH4 and CO were studied. It was found that the IL-family under study has an excellent solvent power for CO2. However, the solubility of H2, CH4 and CO is very limited. Comparison with experimental data show excellent agreement with model predictions.

See more of this Session: Thermophysical Properties of Ionic Liquids
See more of this Group/Topical: Engineering Sciences and Fundamentals

Corresponding States Correlation for Liquid Densities and Gas Solubilities in Ionic Liquids

Jens Abildskov, Chemical and Biochemical Engineering, Technical University of Denmark, 2800 Kgs. Lyngby, Denmark

Martin D. Ellegaard, Department of Chemical and Biochemical Engineering, Technical University of Denmark, 2800 Kgs. Lyngby, Denmark

John P. O'Connell, Chemical Engineering, University of Virginia, Charlottesville, VA

We describe a method for correlating and estimating gas solubilities and liquid densities of ionic liquids based on a corresponding-states form for direct correlation function integrals from fluctuation solution theory. A quantitative description is obtained for volumetric behavior of pure ionic liquids with pressure and temperature. Pure component solubilities and partial molar volumes are treated for hydrogen, carbon monoxide, oxygen and methane in the ionic liquids 1-butyl-3-methyl-imidazolium hexafluorophosphate, [bmim][PF6], and 1-hexyl-3-methyl-imidazolium bis(trifluoromethylsulfonyl)imide, [hmim][Tf2N], including a predictive approach to such systems.

See more of this Session: Thermophysical Properties of Ionic Liquids
See more of this Group/Topical: Engineering Sciences and Fundamentals

Microscopic Structure, Thermodynamic and Transport Properties of Ionic Liquids: Molecular Simulations and Theoretical Predictions

Ioannis G. Economou, Institute of Physical Chemistry, National Center for Scientific Research "Demokritos", GR-15310, Aghia Paraskevi, Greece

Eleni Androulaki, National Center for Scientific Research "Demokritos", GR-15310, Aghia Paraskevi, Greece

Javier Ramos, Department of Macromolecular Physics, Instituto de Estructura de la Materia - CSIC, Madrid, Spain

Georgia - Evangelia Logotheti, National Center for Scientific Research "Demokritos", Aghia Paraskevi, Greece

Andreas Bick, Scienomics SARL, France, France

An atomistic force field developed recently for imidazolium-based [Tf2N] ionic liquids (ILs) based on the combination of density functional theory quantum calculations and the CHARMM force field is used in this work for the determination of microscopic structure, thermodynamic properties and transport properties of IL melts over a wide temperature and pressure range. Molecular Dynamics simulations are performed and results are analyzed for the determination of microscopic phenomena and macroscopic physical properties. A broad range of properties that includes atom pair and center of mass radial distribution functions, mass densities, segmental and chain dynamics, and self-diffusion coefficient are calculated. Comparison against limited literature experimental data reveals that the new force field provides reliable predictions of a broad spectrum of properties. Furthermore, it is shown to be more accurate than earlier atomistic force fields.

Furthermore, calculations from the Perturbed Chain-Polar Statistical Associating Fluid Theory (PC-PSAFT) are presented regarding the phase equilibria of IL - solvent mixtures. Both non-polar and polar solvents are examined. PC-PSAFT provides an accurate correlation of binary and ternary mixture phase equilibria using a single binary interaction parameter.

See more of this Session: Thermophysical Properties of Ionic Liquids
See more of this Group/Topical: Engineering Sciences and Fundamentals

Molecular Dynamics Simulations of CO2 at An Ionic Liquid Interface: Adsorption, Ordering and Interfacial Crossing

Marcos Perez-Blanco, Chemical and Biomolecular Engineering, University of Notre Dame, Notre Dame, IN

Edward J. Maginn, Chemical and Biomolecular Engineering, University of Notre Dame, Notre Dame, IN

Using classical molecular dynamics simulation techniques, the molecular scale characteristics of the gas-liquid interface of the ionic liquid 1-n-butyl-3-methylimidazolium bis(trifluromethylsulfonly)amide ([bmim][Tf2N]) have been studied. The vacuum-liquid surface of the ionic liquid, and the gas liquid interface of the ionic liquid with carbon dioxide, were simulated at a range of temperatures and pressures. The process of carbon dioxide adsorption on the liquid surface and absorption into the bulk liquid was quantified in terms of flux and concentration difference across the interface. Adsorption onto the interface occurs rapidly, in approximately 100 ps, but interfacial crossing and diffusion into the bulk is much slower. At the lower pressures the fraction of CO2 molecules adsorbed is very high, but as the pressure increases a monolayer is formed and a greater fraction of the CO2 molecules stay in the gas phase.

Density profiles with respect to the interfacial normal were computed for each atom in the ionic liquid and CO2. The dependence of the density profiles on the carbon dioxide pressure was evident from comparison of the simulations at different pressures. Interfacial ordering and orientational tendencies were indicated by the relative densities of different atoms as a function of position on the interface normal.

The distribution of angular orientation of selected bond directions, at regular points along the interface normal, was also calculated. This analysis was performed both for the ionic liquid molecules and the CO2 molecules. For the CO2 molecule orientations, a dependence on CO2 pressure is noted.

Residence time distributions of CO2 molecules in the interfacial region were created. It was observed that the rate at which molecules penetrate the surface, relative to the rate at which molecules desorb, is affected by the temperature. The analysis of the residence time distribution seems to indicate that adsorption and desorption occurs on a rapid timescale relative to diffusion through the interface. The time evolution of the density profile of CO2 was studied as it diffused across the interface over the 7 ns time period. Potentials of mean force were also computed for CO2 penetration into the liquid.

See more of this Session: Thermophysical Properties of Ionic Liquids
See more of this Group/Topical: Engineering Sciences and Fundamentals

Binary Vapor-Liquid and Liquid-Liquid Equilibria of Task-Specific Ionic Liquids with Water

Devan E. Kestel, Chemical and Biomolecular Engineering, University Of Notre Dame, Notre Dame, IN

Joan F. Brennecke, Chemical and Biomolecular Engineering, University Of Notre Dame, Notre Dame, IN

Task-specific ionic liquids (TSILs) are ionic liquids containing an appended functional group which chemically complexes with a solute molecule of interest. The TSILs in this study contain either one or two appended amine groups on the cation and/or anion which chemically react with carbon dioxide. The chemical complexation of the TSIL with carbon dioxide provides the potential for further carbon dioxide storage capacity than is possible in ionic liquids which only physically absorb carbon dioxide. In order for a TSIL to be effectively used in an adsorbing /stripping process to capture carbon dioxide, it is also necessary to determine its capacity for absorbing water because this will directly correlate to the amount of energy required to regenerate the TSIL during the stripping phase. Water solubility isotherms were determined experimentally with a Hiden Intelligent Gravimetric Analyser (IGA003) and are reported for various TSILs at 40 C at pressures below saturation (73.849 mbar) along with infinite dilution activity coefficients. In addition, gravimetrically determined liquid-liquid equilibrium values are reported for these TSILs with water at room temperature and atmospheric pressure.

See more of this Session: Thermophysical Properties of Ionic Liquids
See more of this Group/Topical: Engineering Sciences and Fundamentals

Thermodynamic and Transport Properties of Some Novel Task-Specific Ionic Liquids

Kevin N. West, Department of Chemical and Biomolecular Engineering, University of South Alabama, Mobile, AL

Samuel M. Murray, Department of Chemical and Biomolecular Engineering, University of South Alabama, Mobile, AL

James H. Davis Jr., Department of Chemistry, University of South Alabama, Mobile, AL

Ionic liquids are interesting alternative process solvents, as they offer highly tunable solvent properties due to their organic nature, and vanishingly-low vapor pressures which minimize fugitive emissions. One of the most promising ways that ionic liquids are being considered for use is in task-specific applications, where one or more of their properties are design to be exploited to optimize a particular process.

We have synthesized a series of novel task-specific ionic liquids that were designed with properties (solvent properties, melting point, etc.) in mind to optimize their use in specific reactions, separations and materials applications, some of which are significantly above ambient conditions. We have measured the properties of interest, including melting points (and glass-transition temperature when appropriate), densities and viscosities of the pure species as well as the phase behavior, including mixture melting points and cloud points, with other ionic liquids, gases and organic species. Some of these ionic liquids show properties that mimic the biological molecules from which they were derived and others show potential for use in the medical field. Our results indicate that some of the most promising applications may involve mixtures of ionic liquids where multiple properties may be tuned simultaneously to give optimal results for reactions and separations.

See more of this Session: Thermophysical Properties of Ionic Liquids
See more of this Group/Topical: Engineering Sciences and Fundamentals

Transport Properties of Aqueous Solutions in Porous Silica

Lukas Vlcek, Chemical Engineering, Vanderbilt University, Nashville, TN

Eugene Mamontov, Spallation Neutron Source, Oak Ridge National Laboratory, Oak Ridge, TN

David R. Cole, Chemical Sciences Division, Oak Ridge National Laboratory, Oak Ridge, TN

Peter T. Cummings, Department of Chemical and Biomolecular Engineering, Vanderbilt University, Nashville, TN

Nanoconfined water and aqueous solutions exhibit many geologically and technologically important properties which are still not completely understood, such as the mechanism of SiO_2 dissolution or the dynamic transitions of confined aqueous solutions at low temperatures [1].

We present a study of the strucural and transport properties of pure water and aqueous LiCl solutions confined in nanoporous silicates (MCM-41 and SBA-15). The complex dynamics of the system involves fast motions inside pores, slow diffusion of water in bulk silica, as well as surface dissociation of water and SiO_2 dissolution. We evaluate the dynamics of these processes with respect to the pore size, ion concentration, and temperature. Results obtained from simulation studies are compared with available quasi-elastic neutron scattering (QENS) experiments.

Since most of the studied processes are characterized by very long diffusion relaxation times or rare reactive events, typically in the order of nano- to microseconds, they pose a challenge for usual atomistic simulations. To overcome these complications, we use a combination of the reaction ensemble Monte Carlo technique (RxMC) [2] and long GPU accelerated molecular dynamics simulations. The force fields used include common non-reactive models (e.g. SPC/E) for speed, as well as reactive potential models (ReaxFF [3]) for the details of reactive steps.

1) E. Mamontov, D. R. Cole, S. Dai, M. D. Pawel, C. D. Liang, T. Jenkins, G. Gasparovic, and E. Kintzel, Chem. Phys. 352, 117 (2008).

2) J. K. Johnson, A. Z. Panagiotopoulos, and K. E. Gubbins, Mol. Phys. 81, 717 (1994)(2000). W. R. Smith and B. Tříska, J. Chem. Phys. 100, 3019 (1994).

3) A. C. T. van Duin, S. Dasgupta, F. Lorant, and W. A. Goddard III, J. Phys. Chem. A 2001, 105, 9396-9409.

See more of this Session: Transport in Nanostructured Systems
See more of this Group/Topical: Engineering Sciences and Fundamentals

Translocation of Chain Molecules through, Into, and Between Nanopores

Shuang Yang, Department of Chemical and Biochemical Engineering, Rutgers University, Piscataway, NJ

Yang Kan, Department of Chemical and Biochemical Engineering, Rutgers University, Piscataway, NJ

Aleksey Vishnyakov, Department of Chemical and Biochemical Engineering, Rutgers University, Piscataway, NJ

Alexander V. Neimark, Department of Chemical and Biochemical Engineering, Rutgers University, Piscataway, NJ

We study the dynamics of polymer translocation through and into nanopores with two different coarse-grained modeling techniques. The self-consistent field theory (SCFT) represents the polymer chain as a random walk trajectory in an external field fulfilling the Edwards equation, which treats excluded volume effects in a mean field fashion; the process of translocation is described by the Fokker-Plank equation. The dissipative particle dynamics (DPD) represents the polymer chain as a sequence of soft repelling beads linked together by virtual springs in an explicit solvent; the chain movement is directly simulated with Newton equations, which account for random thermal motion, external field, inter-bead forces and friction with explicit solvent. The goal of our work is to understand the interplay of entropic (confinement size) and external field (adsorption potential) effects. Different model set-ups are considered: translocation through a hole in an impermeable membrane under purely diffusive and forced flow conditions, adsorption into a spherical pore from the bulk, and translocation between the pores of different sizes.

See more of this Session: Transport in Nanostructured Systems
See more of this Group/Topical: Engineering Sciences and Fundamentals

Computational Evaluation of Separations with Single-Walled Aluminosilicate Nanotubes

Ji Zang, School of Chemical & Biomolecular Engineering, Georgia Institute of Technology, Atlanta, GA

Shaji Chempath, Theoretical Division, Los Alamos National Lab, Los Alamos, NM

Suchitra Konduri, School of Chemical & Biomolecular Engineering, Georgia Institute of Technology, Atlanta, GA

Sankar Nair, School of Chemical & Biomolecular Engineering, Georgia Institute of Technology, Atlanta, GA

David S. Sholl, School of Chemical & Biomolecular Engineering, Georgia Institute of Technology, Atlanta, GA

Single-walled metal oxide nanotubes are attractive candidates for molecular separations as they present a well-defined solid-state structure, precisely tunable length and diameter, and hydrophilic and functionalizable interiors for obtaining transport and adsorption selectivity. Unlike the situation for carbon nanotubes, little is currently known about molecular separation in metal oxide nanotubes. Recent studies on transport properties of hydrogen-bonding liquids (water, methanol and ethanol) through single-walled aluminosilicate nanotubes have shown that the hydrogen bond network associated with water makes its diffusion behavior different from methanol and ethanol. And mixtures of water and methanol show segregation in the nanotube, with water located closer to the tube wall and the alcohol molecules localized near the center of the nanotube. Here, we study the adsorption and diffusion of water/methanol and CH_4/CO_2 in aluminosilicate nanotubes using Grand Canonical Monte Carlo (GCMC) and molecular dynamics (MD) simulations to investigate the effects of competitive transport of different chemical species. We show that the flexibility of inner surface hydroxyl groups play an important role in adsorption of hydrogen-bonding liquids. Both self- and corrected diffusivity for water (CH_4) are higher than those of methanol (CO_2) at high molecule loadings. By using the adsorption isotherm and transport coefficients of molecules in aluminosilicate nanotubes as input parameters into a continuum model, we examine the possibility of selective transport of molecular species in the present nanotubes. Furthermore, we investigate the effect of organic functionalization of tube interiors on CH_4/CO_2 separation and discuss the possibility of tuning the transport properties of the nanotubes.

See more of this Session: Transport in Nanostructured Systems
See more of this Group/Topical: Engineering Sciences and Fundamentals

Visualization of Frequent Desorption/Re-Adsorption Events for Sorbate at the External Surface of Microporous Crystals by Hyperpolarized Xe-129 2D NMR Exchange Spectroscopy

Amrish R. Menjoge, Department of Chemical Engineering, University of Florida, Gainesville, FL

Clifford R. Bowers, Department of Chemistry, University of Florida, Gainesville, FL

Sergey Vasenkov, Department of Chemical Engineering, University of Florida, Gainesville, FL

Many types of porous membranes and catalyst particles contain microporous crystals, which are interconnected by larger meso- and macropores. Transport of guest molecules through the external surface of microporous crystals can significantly affect the overall rate of transport in membranes and catalysts. Here we report a new experimental methodology that was developed to study how often desorption of sorbate molecules from a microporous crystal is followed by fast re-adsorption into the same crystal. The rates of desorption and re-adsorption processes are also investigated. The experimental approach is based on application of continuous flow hyperpolarized Xe-129 2D NMR exchange spectroscopy (2D EXSY). First results of the study of desorption/re-adsorption events of Xe from/to NaY zeolite crystals will be presented and discussed. The experimental data will be compared with the corresponding results of the dynamic Monte Carlo simulations.

See more of this Session: Transport in Nanostructured Systems
See more of this Group/Topical: Engineering Sciences and Fundamentals

Effective Thermal Conductivity of Graphene Sheet COMPOSITES and Comparison to Carbon Nanotube COMPOSITES

Khoa Bui, Chemical, Biological, and Materials Engineering, University of Oklahoma, Norman, OK

Dimitrios V. Papavassiliou, Chemical, Biological, and Materials Engineering, University of Oklahoma, Norman, OK

By incorporating Carbon nanotubes (CNTs) into a polymer matrix, the effective thermal conductivity of the resulting composites can be increased. For example, it can increase by 27% over pure polymer for the case of epoxy composites [1], or by a factor of almost 4 for the case of high volume fraction single-walled carbon nanotubes in polystyrene [2]. Based, however, on the properties of pure CNTs, one would expect a much higher increase of the effective thermal conductivity of such composite materials, more than an order of magnitude. The presence of resistance to the transfer of heat at the CNT-polymer interface, known as the Kapitza resistance, is the reason for this difference [3]. A lot of research has been done in order to investigate the nature of the Kapitza resistance, but there are not many reports of successfully eliminating or minimizing this effect in CNT composites.

Recently, Balandin et al. [4] reported extremely high values for the thermal conductivity of single-layer graphene sheets (GS) that outperform carbon nanotubes in heat conduction. This result gives rise to the expectation that GS composites could be able to make the unfulfilled promise of CNT composites a reality. In this work, we investigate the effective conductivity of GS composites by means of off-lattice Monte-Carlo algorithms [5]. This method is more efficient than conventional random walk algorithms and faster than molecular dynamics algorithms. We will present the methodology and the validation of the method, which is used to study the effects of GS orientation, dispersion and volume fraction on the effective thermal conductivity of the GS composites. The discussion will include a comparison between theoretical predictions of the value of the thermal resistance at the GS-polymer interface relative to the CNT-polymer interface based on the acoustic mismatch theory and the diffuse mismatch theory, and a comparison between CNT composites and GS composites with similar volume fraction and similar dispersion pattern of the nano-inclusions in the composite matrix.

References

[1] Bryning M. B.; Milkie D. E.; Kikkawa J. M. and Yodh A. G., Appl. Phys. Lett. 2005, 87, 161909.

[2] Peters J. E.; Papavassiliou D.V. and Grady, B. P., Macromolecules, 2008, 41, 7274-7277.

[3] Swartz E. T.; Pohl R. O., Rev. Mod. Phys. 1989, 61, 605.

[4] Balandin, A. A.; Ghosh, S.; Bao, W.; Calizo, I. ; Teweldebrhan, D.; Miao F.; Lau, C. N., Nano Lett. 2008, 7, 902.

[5] Duong H. M.; Papavassiliou D.V.; Mullen K.J.; Maruyama S. Nanotechnology, 2008, 19(6), 065702.

See more of this Session: Transport in Nanostructured Systems
See more of this Group/Topical: Engineering Sciences and Fundamentals

Molecular Simulation of Thermal Transport Across Solid-Fluid Interfaces

Sohail Murad, Chemical Engineering, University of Illinois at Chicago, Chicago, IL

Ishwar K. Puri, Engineering Science and Mechanics, Virginia Tech, Blacksburg, VA

Using molecular simulations, we have investigated heat transfer across the solid-fluid interface of a silica wafer in contact with water vapor. Our results show that the thermal or Kapitza resistance decreases significantly, as the surface becomes more hydrophilic. This is primarily due to increases in adsorption and absorption at the surface, which enhances the intermolecular collision frequency at the interface. Increasing this frequency also reduces the dependence of thermal transport on variations in the interfacial temperature and pressure. Decreasing the density diminishes the intermolecular collision frequency thus increasing the thermal resistance. We have investigated the temperature gradients in the solid phase and their role in the effective thermal conductivity of the solid phase.

See more of this Session: Transport in Nanostructured Systems
See more of this Group/Topical: Engineering Sciences and Fundamentals

Ion Transport in Nanochannels Fabricated by DNA Combing and Imprinting

Nan-Rong Chiou, Chemistry, The Ohio State University, Columbus, OH

Pouyan E. Boukany, Nanoscale Science and Engineering Center for Affordable Nanoengineering of Polymeric Biomedical Devices, The Ohio State University, Columbus,, OH

Orin L. Hemminger, Chemical and Biomolecular Engineering, Ohio State University, Columbus, OH

Bo Yu, Department of Chemical and Biomolecular Engineering, The Ohio State University, Columbus, OH

Ly James Lee, The Ohio State University, Columbus, OH

Nanochannels can precisely manipulate liquids and molecules by providing extremely high levels of spatial confinement and have consequently received a great deal of interest in recent years for both fundamental research and practical applications. We have developed a novel and facile affordable method to fabricate the polymeric nanochannels with the controllable diameter ranging from 10nm to 250nm. Recently, we found that ion conductance is strongly dependent on the size of the channel. The ion transport behavior in the nanochannel is dramatically different from that in the bulk solution. In this study, we will discuss ion transport mechanism in detail occurring at such tiny channels.

See more of this Session: Transport in Nanostructured Systems
See more of this Group/Topical: Engineering Sciences and Fundamentals

Time-Dependent Self-Diffusion of Surfactant Ions Due to Breakup and Reformation of Micelles in Micellar Systems

Amrish R. Menjoge, Chemical Engineering, University of Florida, Gainesville, FL

Sergey Vasenkov, Department of Chemical Engineering, University of Florida, Gainesville, FL

Pulsed field gradient (PFG) NMR technique was used to study diffusion of surfactant ions in the aqueous solutions of anionic surfactant SDS and the mixture of SDS with the cationic surfactant C12TAB. PFG NMR measurements at high magnetic field gradient strength provided separate sets of data on diffusion of SDS and C12TAB ions for a broad range of diffusion times. For small diffusion times at least two surfactant components with different effective diffusivities were observed for each type of surfactant ions. The fast-diffusing component is tentatively assigned to the surfactant ions which experience breakup or reformation of micelles during the diffusion time of the PFG NMR measurements. The slow-diffusing component is assigned to the surfactant ions that do not experience such events during the diffusion time. This assignment was found to be in agreement with the observed patterns of change of the fractions and diffusivities of these components with increasing diffusion time. The fractions of the slow-diffusing component of SDS in the pure SDS sample were observed to be smaller than those in the SDS/C12TAB sample. This observation can be explained by stabilization of micelles due to an addition of the cationic surfactant C12TAB to the aqueous solution of the anionic surfactant SDS.

Acknowledgements

NMR data were obtained at the Advanced Magnetic Resonance Imaging and Spectroscopy (AMRIS) facility in the McKnight Brain Institute of the University of Florida. The surfactant samples were prepared in the group of Professor Dinesh Shah, University of Florida by Dr. Monica A. James-Smith. The authors of this paper are very grateful to Professor Shah for multiple discussions and useful suggestions regarding this work.

See more of this Session: Transport in Nanostructured Systems
See more of this Group/Topical: Engineering Sciences and Fundamentals

Ionic Liquids for the Extractive Distillation of Close-Boiling and Azeotropic Mixtures

Vladimir Mokrushin, Chair of Separation Science & Technology, Friedrich-Alexander University of Erlangen-Nuremberg, Erlangen, Germany

Daniel Assenbaum, Chair of Reaction Engineering, Friedrich-Alexander University of Erlangen-Nuremberg, Erlangen, Germany

Liudmila Mokrushina, Chair of Separation Science & Technology, Friedrich - Alexander University of Erlangen-Nuremberg, Erlangen, Germany

Peter Wasserscheid, Chair of Reaction Engineering, Friedrich-Alexander University of Erlangen-Nuremberg, Erlangen, Germany

Wolfgang Arlt, Chair of Separation Science and Technology, University of Erlangen, Erlangen, Germany

Recently, room temperature ionic liquids (ILs) have been proposed to be used as extractive solvents for the separation of close-boiling substances and azetropic mixtures [Arlt et al. DE 10114734, DE 10136614, WO 2002074718]. Obviously, ILs have an intrinsic advantage over conventional extractive (entrainers) due to their extremely low volatilities and their good thermal stabilities.

This paper presents an extensive study on the feasibility of ionic liquids (IL) for the extractive distillation of close-boiling (alkene/alkane) and azeotropic mixtures (chlorinated alkane/alkane) composed of high-volatile components. Alkenes and alkanes of the same chain length have rather close boiling temperatures, while some of chlorinated alkanes/alkanes mixtures show an azeotropic behaviour (maximum pressure azeotrope). Obviously, both types of mixtures are hardly separable by a simple distillation. Due to their higher polarizability, chlorinated and unsaturated hydrocarbons (such as alkenes, dienes or aromatics) experience a stronger solvation in ionic liquids compared to saturated hydrocarbons. From an engineering point of view, it is a highly relevant question whether this difference is large enough to allow an effective separation technology to be based on these differences. Since alkenes are low boilers (the ene-to-ane separation factor of the pure compounds is slightly higher than 1) and show always higher solubilities in ILs, the task of the entrainer in this case is more difficult than e.g. for the ethanol-water system. In the latter case, water is the high boiler in the mixture of the pure components and stays the high boiler after addition of the ionic liquid [Arlt et al. DE 10114734, DE 10136614, WO 2002074718, Y. Beste et al. DE 10336555]. Thus, a suitable entrainer IL should (i) convert the low-boiling alkene to a high boiler shifting the separation factor below 1, (ii) dissolve the alkene so effectively that the separation factor becomes as low as possible, and additionally (iii) in the case of azetropic mixtures retain the separation factor below 1 at concentrations below and above the azeotropic point.

Obviously, when two or more high-volatile substances dissolve in one solvent, their solubility is not independent of one another, especially at high loads. Experimental studies on the solubility of gas mixtures are however rather tedious and time-consuming. In the present study, a new express experimental procedure for a direct screening of low volatile extractive agents such as ILs is reported. Based on head space chromatography, this method is fast, is applicable to high-volatile mixtures directly, and requires only small amounts of ILs.

More than 20 ILs and their binary mixtures have been screened at ambient temperature and low pressures. Selected ILs have been also studied at elevated temperatures and pressures in an autoclave. The screening results show that every tested IL turns the low boiler alkene to be the high boiler with the ene-to-ane separation factor much below 1 in the whole composition range. In the case of the azeotropic chlorinated alkane/alkane mixtures, IL-entainers are shown to be able to break the azeotrope and thus assure the separation in the whole range of chloroalkane/alkane ratios. By adjusting the cation/anion combination and/or their chemical functionalization, the solubility and separation efficiency can be effectively changed. The gas-mixture solubility increases by elongation of the side chain length in the imidazolium-based cation, the IL separation ability being vice versa decreased. For the IL mixtures studied, individual ILs contribute additively to the quality of the mixed entrainer. The autoclave experiments at elevated temperatures and pressures indicate that the separation ability decreases with growing temperature and loading. Generally, our study proves impressively the great potential of ILs to act as entrainers in the extractive distillation of the close-boiling and azeotropic mixtures composed of high-volatile substances

See more of this Session: Advanced Fluids
See more of this Group/Topical: Engineering Sciences and Fundamentals

Novel Ionic Liquids for CO2 Capture

Surya Sekhar Moganty, Chemical and Biomolecular Engineering, Clarkson University, Potsdam, NY

Pavan Kumar, Department of Chemistry, Clarkson University, Potsdam, NY

Venkat K. Vendra, Chemical and Biomolecular Engineering, Clarkson University, Potsdam, NY

Sitaraman Krishnan, Chemical and Biomolecular Engineering, Clarkson University, Potsdam, NY

Ruth E. Baltus, Chemical and Biomolecular Engineering, Clarkson University, Potsdam, NY

Over the past years, a systematic study of functional groups appended to the imidazolium cations has been investigated in the quest for CO2 philic ionic liquids. [R]mimTf2N ionic liquids have been the primary focus of such studies. The functional groups investigated were alkyl, benzyl, nitrile, ether, hydroxyl, amine, and fluoroalkyl. In this presentation, our research efforts on synthesis and characterization of [R1R2]imTf2N ionic liquids will be presented. Results from fundamental studies of the thermodynamic and transport characteristics of CO¬2 in these novel ionic liquids will be reported. The experimental system used for these measurements involves tracking the decrease in pressure that results from the introduction of target gas into a small closed chamber containing a thin film of ionic liquid. The observed dependency of CO¬2 solubilities on the appended functionalities will be related to the physical properties of the ionic liquids such as glass transition temperature and viscosity. Self diffusion coefficients of imidazolium cation carrying different functional groups were measured at 25C using pulse gradient NMR technique. Effect of functional groups on self diffusivities of cation will also be discussed.

See more of this Session: Advanced Fluids
See more of this Group/Topical: Engineering Sciences and Fundamentals

Hydrophobic Type Interactions Between Ionic Liquids and Inorganic Salt Ions in Aqueous Environments: Molecular Dynamics Simulations

Jindal K. Shah, Department of Chemical and Biomolecular Engineering, University of Notre Dame, Notre Dame, IN

Edward J. Maginn, Department of Chemical and Biomolecular Engineering, University of Notre Dame, Notre Dame, IN

Room temperature ionic liquids (ILs) have been the subject of intense investigation over the last decade or so. Negligible vapor pressure, excellent solvation properties, high thermal and electrochemical stability coupled with the flexibility to design ILs with unique properties make them ideal candidates as solvents in novel product and process development. Liquid-liquid extraction of products from aqueous phase of a biological media is one of the many promising processes that can take advantage of the unique properties of ILs. One of the challenges in development and optimization of such a process is the solubility of ILs in water leading to water contamination problems. Our experimental collaborators have demonstrated that the solubility of ILs in water is dependent on nature and concentration of inorganic salts. In their work, the salting-in/salting-out ability of inorganic salts was found to qualitatively follow a trend known as the Hofmeister series that explains the changes in solubility of the solute in terms of salts as "water structure makers" or "water structure breakers". However, ^{1}H NMR studies conducted by our collaborators suggested that the salting-in/salting-out of ILs was a direct consequence of the unexpected interaction of salt ions with the hydrophobic moieties of the IL cation.

In this talk, we present the results of molecular dynamics simulations of aqueous solutions of the IL, 1-butyl-3-methylimidazolium bis(trifuoromethylsulfonyl)imide [bmim][Tf2N] in the presence of inorganic salts NaCl, Na2SO4 and NaClO4 to obtain molecular level understanding of salting-in/salting-out behavior of ILs. The analysis, in terms of the radial distribution functions, clearly supports the existence of preferential specific interactions between the low electrical charge-density ("apolar moiety") parts of the IL cation and the inorganic salts. Predictions of the Kirkwood-Buff theory of solutions are also in agreement with the salting-in/salting-out behavior observed experimentally.

See more of this Session: Advanced Fluids
See more of this Group/Topical: Engineering Sciences and Fundamentals

Computational Screening of Ionic Liquids for CO_2 Capture Using COSMO-RS

Kazi Z. Sumon, International Test Centre for CO_2 Capture, University of Regina, Regina, SK, Canada

Amr Henni, Faculty of Engineering and Applied Science, University of Regina, Regina, SK, Canada

COSMO-RS method is applied to predict Henry's law coefficient and related thermodynamic quantities for carbon dioxide, methane and ethane in a database of ionic liquids. Effect of temperature and chemical constituents of solvents on the solvation process is explained. Predicted results are compared with experimental data. A short list of potential ionic liquids for CO_2 capture from flue gas and natural gas streams is presented.

See more of this Session: Advanced Fluids
See more of this Group/Topical: Engineering Sciences and Fundamentals

Statistics and Characterization of Ionic Liquids From the Perspective of COSMO-RS Sigma Profiles

Michael Diedenhofen, COSMOlogic GmbH&CoKG, Leverkusen, Germany

Andreas Klamt, COSMOlogic GmbH & Co.KG, Leverkusen, Germany

The Conductorlike Screening Model for Real Solvents (COSMO-RS) has proven to be widely applicable to the qualitative and often quantitative prediction of thermodynamic properties of ionic liquids (IL). In the last years many successful COSMO-RS based screening applications of IL with respect to their suitability as solvents, entrainers, and other purposes have been reported [1-3]. Since the key information entering COSMO-RS is the sigma-profile, i.e. the surface polarity of the compound in a virtual conductor taken from quantum chemical calculations, this entity is very well suited to characterize and classify ionic liquids. In this talk we will present some statistics and resulting insights about similarities and differences of ionic liquids, based on our COSMObaseIL database, which holds the sigma profiles for almost all commonly used IL anions and cations.

1. C. Jork, C. Kristen, D. Pieraccini, A. Stark, C. Chiappe, Y.A. Beste, W. Arlt. Tailor-made ionic liquids. *J. Thermodynamics* **2005**, *37*, 537-558.

2. K. Z. Sumon, E. Z. Hamad, Computational Screening of Ionic Liquids for Aromatic Extraction Using Cosmo-RS, The 2005 Annual Meeting, 2005, 1 November 2005.

3. M. Diedenhofen, F. Eckert, A. Klamt. Prediction of Infinite Dilution activity Coefficients of Organic Compounds in Ionic Liquids Using COSMO-RS. *J. Chem. Eng. Data* **2003**, 48, 475-479.

See more of this Session: Advanced Fluids
See more of this Group/Topical: Engineering Sciences and Fundamentals

Predicting the Phase Behavior of Charged and Polar Fluids Using the SAFT-VR+DE Approach

M. Carolina Dos Ramos, Department of Chemical and Biomolecular Engineering, Vanderbilt University, Nashville, TN

Clare McCabe, Department of Chemical and Biomolecular Engineering, Vanderbilt University, Nashville, TN

Knowledge of the thermodynamic behavior of charged and polar systems is essential for the design and development of many important chemical, biological and environmental processes and has resulted in an increasing demand for predictive tools that accurately describe the thermodynamic behavior of such systems. It is the industrial importance and well known non-ideal thermodynamic behavior of these systems that has lead to increasingly intense academic research and the development of molecular based tools. Traditionally, the description of thermodynamic behavior has been obtained through macroscopic theories that are based on phenomenological equations containing adjustable parameters that lack physical meaning. In this regard, molecular-based equations of state, such as the statistical associating fluid theory (SAFT) [1], which are able to describe the thermodynamic behavior of fluids using adjustable physically based parameters, can be considered a powerful tool to describe the thermodynamics of complex systems. In a recent series of extensions to the SAFT-VR equation of state, McCabe and coworkers [2] have incorporated the different electrostatic interactions (ion-ion, ion-dipole and dipole-dipole) using the mean spherical approximation of Blum *et al.*[3] with the non-primitive model to account for the solvent molecules explicitly in the so-called SAFT-VR+DE equation. In this work, we apply the SAFT-VR+DE approach to deal with pure fluids and mixtures of charged and polar fluids, such as electrolytes, mixed polar solvent electrolytes and ionic liquids.

[1] W. G. Chapman, K. E. Gubbins, G. Jackson, and M. Radosz, *Fluid Phase Equilibria* **52**, 31 (1989); W. G. Chapman, G. Jackson, and K. E. Gubbins, *Molecular Physics* **65** (5), 1057 (1988).

[2] H. Zhao and C. McCabe, *Journal of Chemical Physics* **125**, 104504 (2006); H. Zhao, Y. Ding, and C. McCabe, *Journal of Chemical Physics* **127**, 084514 (2007); H. Zhao, M. C. dos Ramos, and C. McCabe, *Journal of Chemical Physics* **126**, 244503 (2007).

[3] D. Wei and L. Blum, *Journal of Chemical Physics* **87** (5), 2999 (1987); L. Blum and D. Q. Wei, *Journal of Chemical Physics* **87** (1), 555 (1987).

See more of this Session: Advanced Fluids
See more of this Group/Topical: Engineering Sciences and Fundamentals

Hollow Fibers by Electrospinning in Supercritical CO2

Bo Hyun Lee, Chair of Separation Science and Technology, University of Erlangen, Erlangen, Germany

Detlef Freitag, Chair of Separation Science and Technology, University of Erlangen, Erlangen, Germany

Wolfgang Arlt, Chair of Separation Science and Technology, University of Erlangen, Erlangen, Germany

Mark . A McHugh, Dept. of Chemical Engineering, Virginia Commonwealth University, Richmond, VA

Polymer fibers with a diameter of several ten to hundred nanometers have a very large surface area to volume ratio. This enables wide applications of polymer fibers with flexibility in surface functionalities and superior mechanical performance (Xia *et al*., 2004). Especially if polymer fibers offer specific structures such as porous surfaces, hollow, or core-sheath, their application can be widely extended to filters, textiles, sensors, catalysis, or in medicine and pharmacy.

Up to now, hollow fibers have been produced by template processes or using coaxial capillary (Greiner *et al*., 2000, Wendorff *et al*., 2007). The main difficulty of these processes is the limitation of applied materials. Recently, a new method is published by McHugh *et al.* to produce hollow fibers (McHugh *et al*., 2006). This process is based on the solubility of organic solvents into supercritical CO_2. The unique advantage of this process is a one step process without any additives and further processes to form the hollow structure inside of fibers. Supercritical CO_2 dissolves an organic solvent 1-2 order of magnitude faster than conventional liquid antisolvents (McHugh *et al*., 1994). In electrospinning, supercritical CO_2 acts as an antisolvent which is a non-solvent for almost all polymers but miscible with most organic solvents.

A polymer solution with polymer concentrations of 3 to 10 wt% is injected by a HPLC (High pressure liquid chromatography) pump through a thin capillary with an inner diameter of 100 to 1000 μm. The autoclave consists of two stainless steel flanges and one non-conductive plastic vessel. The flanges are connected by a high voltage power supply (0 ~ 30 kV) and act as electrodes. On one of the flanges a collector is placed, to where polymer drops stretch due to the high voltage. The autoclave is pressurized by CO_2 to the desired pressure. During polymer jets are flying to the collector, the organic solvent used will be extracted to the supercritical CO_2.

Our investigation shows that the solubility of organic solvents in CO_2 affects the morphology of fibers during electrospinning. By varying the temperature and pressure, the solubility of organic solvent can be easily modified, which enables the formation of hollow fibers. Hollow fibers by electrospinning in supercritical CO_2 will provide a big potential in a wide range of applications.

Bognitzki, M., Schattka, H., Greiner, A., (2000). Polymer, Metal, and Hybrid Nano- and Mesotubes by Coating Degradable Polymer Template Fibers (TUFT Process). Adv. Mater., 12, 637-640

Dror, Y., Salalha, W., Avrahami, R., Zussman, E., Yarin, A. L., Dersch, R., Greiner, A., Wendorff, J. H., (2007). One-Step Production of Polymeric Microtubes by Co-electrospinning. small, 3, 1064-1073

Li, D., Xia, Y., (2004). Electrospinning of nanofibers: Reinventing the wheel? Adv. Mater., 16, 1151-1170

McHugh, M. A., Krukonis, V. J., (1994). Supercritical fluid extraction, Butterworth-Heinemann, Stoneham

Shen, Z., Thompson, B. E., McHugh, M. A., (2006) Electrospinning in Near-Critical CO_2. Macromolecules, 39, 8553-8555

See more of this Session: Advanced Fluids
See more of this Group/Topical: Engineering Sciences and Fundamentals

Surface Forces against Biofouling

Eduardo R. A. Lima, Department of Chemical and Environmental Engineering, University of California, Riverside, Riverside, CA

Jianzhong Wu, Department of Chemical and Environmental Engineering, University of California, Riverside, Riverside, CA

Polymer-based "non-stick" coatings are promising as the next generation of effective, environmentally-friendly marine antifouling systems by minimizing the adsorption of extracellular polymeric substances (EPS). However, design and development of such systems are often impeded by the poor knowledge of polymer-mediated interactions of biomacromolecules with the protected substrate. In this talk, we will present theoretical predictions of surface forces - electrostatic, van der Waals and hydrophobic interactions - that play an important role in biomacromolecular adsorption. The relevance of theoretical results for practical control of the EPS adsorption will also be discussed.

See more of this Session: Biomolecules at Interfaces III
See more of this Group/Topical: Engineering Sciences and Fundamentals

A Protein Resistant Surface Using Self-Assembly of a Natural Osmolyte with Non-Covalent Forces

Gaurav Anand, Chemical and Biological Engineering, Rensselaer Polytechnic Institute, Troy, NY

Sumanth Jamadagni, Chemical and Biological Engineering, Rensselaer Polytechnic Institute, Troy, NY

Shekhar Garde, Chemical and Biological Engineering, Rensselaer Polytechnic Institute, Troy, NY

Georges Belfort, Chemical and Biological Engineering, Rensselaer Polytechnic Institute, Troy, NY

Preparing protein-resistant surfaces in aqueous solution is challenging, complex and expensive and usually requires surface modification using multi-step wet chemical methods, phase separation hydrophilic from hydrophobic polymers or some form of surface grafting (i.e. photo-induced graft polymerization). We have discovered a simple adsorption and coating process with excellent protein resistance that is fast, relatively stable and inexpensive. It is based on the self-assembly of a natural osmolyte, trimethyl ammonium oxide (TMAO), with non-covalent forces on a hydrophobic support in aqueous solution. The asymmetric character or Janus-like property of TMAO with apolar and polar ends, presents, on adsorption, a mostly oriented self-assembled and stable exposed hydrophilic layer that self-heals. The resultant TMAO coat exhibited similar protein-resistance to a PEG terminated self-assembled monolayer (PEG-SAM). Adhesion measurements using multi molecular force spectroscopy confirm that the TMAO+CH3-SAM surfaces behave like strongly hydrophilic surfaces. The TMAO layer was stable up to 40ºC via washing, remained adsorbed even after 600 min of washing at 24ºC, and can easily be regenerated simply by adding additional TMAO to the solution. The process is simple, self-forming and self-healing in aqueous solution and can be used after device assembly with complex geometries and narrow channels that are normally inaccessible to other covalent surface modification methods. It could also be useful for medical applications since TMAO is naturally present in living cells.

See more of this Session: Biomolecules at Interfaces III
See more of this Group/Topical: Engineering Sciences and Fundamentals

Confocal Microscopy and Competitive Adsorption: a New Look at Polymer-Enhanced Lung Surfactant Adsorption

Ian C. Shieh, Department of Chemical Engineering, University of California, Santa Barbara, CA

Joseph A. Zasadzinski, Department of Chemical Engineering, University of California, Santa Barbara, CA

Lung surfactant (LS) is a mixture of lipids and proteins that lines the air-liquid interface of the alveolar walls. It modulates the surface tension in the lungs which greatly reduces the mechanical work of breathing and also prevents the collapse of the alveoli upon expiration. Blood serum leaking into the alveoli as a result of trauma can lead to LS inhibition, which is one characteristic of acute respiratory distress syndrome (ARDS). One mode of LS inhibition, as observed during ARDS, is the competitive adsorption of albumin to the air-liquid interface of the alveoli which blocks LS from forming a functional monolayer.

The addition of hydrophilic polymers, such as polyethylene glycol (PEG) and chitosan, to the liquid subphase has been shown to enhance interfacial LS adsorption *in vitro*. Optimal amounts of polymer allow LS to form a functional monolayer in the presence of albumin, thus reversing inhibition. Albumin must be at least partially, if not completely, displaced from the air-liquid interface in order for a functional monolayer of LS to form. Imaging of the competitive adsorption process with confocal microscopy has allowed us to better understand the mechanisms behind forming an interfacial LS monolayer under inhibitory conditions. We can simultaneously track LS, polymer, and albumin, as well as separately visualize phenomena occurring at the interface from those occurring in the bulk. As a result of these capabilities, we have studied how changes to various system components affect the transport of LS to the interface and the subsequent displacement of interfacial albumin in order to form a function surfactant monolayer.

See more of this Session: Biomolecules at Interfaces III
See more of this Group/Topical: Engineering Sciences and Fundamentals

Diffusing Colloid Measurements of Protein-Carbohydrate Interactions

Shannon L. Eichmann, Department of Chemical and Biomolecular Engineering, Johns Hopkins University, Baltimore, MD

Michael A. Bevan, Department of Chemical and Biomolecular Engineering, Johns Hopkins University, Baltimore, MD

Specific protein-carbohydrate interactions are critical to numerous processes in biology, medicine, and nanobiotechnologies. As a result, there is great interest in better understanding how to directly measure weak, short-range, and orientation specific protein-carbohydrate interactions on appropriate time, length, and energy scales. In this talk, we present measurements of non-specific and specific interactions between colloids and surfaces with and without conjugated proteins and carbohydrates. Interactions are obtained by novel measurements that analyze evanescent wave scattering and video microscopy measurements of 3D trajectories of colloids above planar surfaces. As a baseline for interpreting specific interactions, results will be presented for non-specific polymer-protein, polymer-carbohydrate, protein-protein, and carbohydrate-carbohydrate interactions. Specific interactions will be presented for colloids conjugated with proteins and several immobilized carbohydrates with different known dissociation constants. Ongoing work is investigating such protein-carbohydrate interactions in the presence of competitive binding with small saccharide ligands. Results show excellent agreement between experiment, theory and Brownian dynamics simulations.

See more of this Session: Biomolecules at Interfaces III
See more of this Group/Topical: Engineering Sciences and Fundamentals

Kinetics and Thermodynamics of Protein Unfolding On Hydrophobic Interaction Chromatography Media Investigated by Global and Local Hydrogen Exchange

Robert E Deitcher, University of Virginia, Charlottesville, VA

Adrian Gospodarek, University of Virginia, Charlottesville, VA

John P. O'Connell, Chemical Engineering, University of Virginia, Charlottesville, VA

Erik J Fernandez, University of Virginia, Charlottesville, VA

It is well known that protein unfolding during hydrophobic interaction chromatography (HIC) can lead to loss of product activity, irreversible binding, and poor selectivity. We utilize hydrogen-deuterium exchange to investigate the conformational state of proteins adsorbed to HIC resins of various ligand types and densities showing a variety of peak shapes in isocratic elution chromatography. Kinetic modeling of global exchange behavior at multiple labeling times reveals equilibrium conformation distributions and unfolding and refolding kinetics on the surface. Exchange behavior of selected systems is also investigated at the peptide level by proteolytic digestion of labeled protein after elution, providing evidence of protein regions where protection is changed by conformational changes upon adsorption. Comparisons are made with computer calculations of local stability of adsorbed proteins.

See more of this Session: Biomolecules at Interfaces III
See more of this Group/Topical: Engineering Sciences and Fundamentals

Adsorption and Bridging Adhesion of MBP with Cytoplasmic Model Myelin Membranes

Younjin Min, Chemical Engineering, University of California, Santa Barbara, Santa Barbara, CA

Kai Kristiansen, Department of Chemical Engineering, University of California at Santa Barbara, Santa Barbara, CA

Dongwoog Lee, Department of Chemical Engineering, University of California at Santa Barbara, Santa Barbara, CA

Joan M. Boggs, The Hospital for Sick Children, Molecular Structure and Function, Toronto, Canada

Cindy Husted, Neuroscience Research Institute, Biological Sciences 2, University of California, Santa Barbara, Santa Barbara, CA

Joesph A. Zasadzinski, Department of Chemical Engineering, University of California, Santa Barbara, CA

Jacob N. Israelachvili, Chemical Engineering, University of California at Santa Barbara, Santa Barbara, CA

The myelin sheath is a multilamellar membrane surrounding the axon of neurons in the nervous system. This sheath acts as an electrical insulator, which allows for faster and more efficient conduction of nerve impulses than unmyelinated nerves. The myelin consists of 8 kinds of lipid molecules on two alternating bilayers (cytoplasmic and extracellular faces of the cell membrane) together with various proteins. Myelin Basic Protein (MBP), which is mainly found on the cytoplasmic face, is believed to have an important role in maintaining myelin structure.

Myelin dysfunctions vary from deterioration of signal transduction to demyelinating diseases such as multiple sclerosis (MS). MS is characterized by the appearance of lesions, reflecting loss of membrane adhesion, swelling across the water gaps, vacuolization, vesiculation, and eventual disintegration of the myelin structure. We want to understand the molecular synergistic interaction responsible for a healthy myelin membrane structure [1].

We have performed Surface Force Apparatus (SFA) measurements on supported cytoplasmic myelin membrane [2], which show that as the bulk MBP concentration increases, the adhesion first increases, and then decreases. Simultaneously, and concurrently, the equilibrium spacing first decreases, and then increases. Thus, optimum adhesion coincides with minimum water gap thickness, which optimizes the electrical shielding properties of the myelin sheath. It appears that hydrophobic and electrostatic interactions are involved both in the adsorption of MBP and the inter-membrane interactions.

References:

[1] Y Hu, I Doudevski, D Wood, M Moscarello, C Husted, C Genain, J A Zasadzinski, and J Israelachvili, "Synergetic interactions of lipids and myelin basic protein", Proceedings of National Academy of Sciences 101 (2004) 13466-13471.

[2] Y Min, K Kristiansen, J M Boggs, C Husted, J A Zasadzinski, and J Israelachvili, "Interaction forces and adhesion of supported myelin lipid bilayers modulated by myelin basic protein", Proceedings of National Academy of Sciences 106 (2009) 3154-3159.

See more of this Session: Biomolecules at Interfaces III
See more of this Group/Topical: Engineering Sciences and Fundamentals

Adjusting the Curvature of Supported Lipid Bilayers by Controlling Surface Interactions

Kai-Wei Liu, Chemical and Biomolecular Engineering, Rice University, Houston, TX

Sibani Lisa Biswal, Chemical and Biomolecular Engineering, Rice University, Houston, TX

Lipid membranes exert forces on their underlying substrates in order to obtain a certain equilibrium curvature. The properties of the substrate can affect many aspects of the lipid membrane structure. In order to further investigate the surface interactions governing the curvatures of lipid bilayers, we use microcantilever surfaces for measuring local mechanical forces applied by the membrane to the substrate. In particular, the surface interactions studied are changes in the electrostatic repulsions or the surface tension of palmitoyl- oleoyl- phosphatidylcholine (POPC) supported lipid bilayers. In one set of experiments, the ratio of positive to neutral charged lipids is adjusted while the curvature of the cantilever is monitored. We find that by increasing the number of positive charges in the lipid membrane, we can increase the curvature of the lipid membrane. Additionally, by introducing cholesterol into the lipid bilayer, we can also increase the curvature of the lipid membrane. This now leads us to create model membranes with a membrane stiffness that can be adjustable. To analyze the origin of microcantilever bending motion induced by a membrane formation, we report a quantitative free energy-based model to describe the artificial membrane system and its transduction into micromechanical motion.

See more of this Session: Biomolecules at Interfaces III
See more of this Group/Topical: Engineering Sciences and Fundamentals

Interfacial Stability: Polymeric Nanoparticles of Itraconazole and Odanacatib

Varun Kumar, Chemical Engineering, Princeton University, Princeton, NJ

Lei Wang, Merck Research Laboratory, West Point

Hsien-Hsin Tung, Merck Research Laboratory, West Point

Robert K. Prud'homme, Chemical Engineering, Princeton University, Princeton, NJ

Polymeric nanoparticles (NPs) are used as vehicles for hydrophobic drug delivery [1,2]. Amphiphilic block copolymers stabilize the hydrophobic drug core during the rapid precipitation process of nanoparticle formation [3]. The stability of formulated nanoparticles is largely governed by Ostwald ripening [4,5].

The successful formulation of itraconazole and odanacatib into nanoparticle form with diameters of 145 nm and 350 nm, respectively, using rapid, block copolymer-directed precipitation is presented [6]. The requirements for interface stability are explained in terms of solute solubility at the nanoparticle/water interface and its dependence on both temperature and solvent concentration. The dependence of stability on particle size was also observed on itraconazole and odanacatib NPs. Odanacatib, even with higher bulk solubility than itraconazole, has a longer time scale for ripening due to the larger sized particles and was verified experimentally [6].

The theory of Ostwald ripening provides the framework for understanding the differences in stability observed for the two compounds. The dynamics of the hydrophobic polymer block plays a major role in long term stability as demonstrated by the behavior of nanoparticles stabilized by Poloxamer vs. polystyrene-b-polyethylene oxide polymers[6]. The instability of NP with Poloxamer arises due to the enhanced drug solubility in Poloxamer at the interface and subsequent adsorption and desorption of Poloxamer micelles at the NP surface [6,7].

1. Allen, C.; Maysinger, D.; Eisenberg, A. Colloids and Surfaces B: Biointerfaces, 1999. 16(1-4):p.3.

2. Kumar, V.; Prud'homme, RK. Journal of Pharmaceutical Sciences, 2008. 97:4904-4914.

3. Johnson, BK.; Prud'homme, RK.Physical Review Letter, 2003. 91:118302.

4. Mullin, JW. Crystallization. 1993.

5. Liu, Y.; Saad, W.; Prud'homme, RK. Physical Review Letter, 2007. 98:036102.

6. Kumar, V.; Wang, L.; Riebe, M.; Tung, HH.; Prud'homme, RK. Molecular Pharmaceutics, 2009. In press.

7. Crisp, MT.; Tucker, CJ.; Rogers, TL.; Williams, RO.; Johnston, KP. Journal of Controlled Release, 2007. 117:351-359.

See more of this Session: Interfacial Phenomena in Pharmaceutics
See more of this Group/Topical: Engineering Sciences and Fundamentals

Surfactant Formulation Principles for Self-Dispersing Aerosol Drug Carriers Based On Marangoni Flow in the Pulmonary Airways

Beautia Dew, Department of Biomedical Engineering, Carnegie Mellon University, Pittsburgh, PA

Markus J. Weygand, Department of Physics, Carnegie Mellon University, Pittsburgh, PA

Chang-You Lin, Department of Physics, Carnegie Mellon University, Pittsburgh, PA

Chiung-Hsuan Chung, Department of Chemical Engineering, Carnegie Mellon University, Pittsburgh, PA

Karthik Peddireddy, Department of Chemical Engineering, Carnegie Mellon University, Pittsburgh, PA

Todd M. Przybycien, Department of Chemical Engineering and Department of Biomedical Engineering, Carnegie Mellon University, Pittsburgh, PA

Joseph M. Pilewski, Division of Pulmonary, Allergy, and Critical Care Medicine, University of Pittsburgh, Pittsburgh, PA

Timothy E. Corcoran, Division of Pulmonary, Allergy, and Critical Care Medicine, University of Pittsburgh, Pittsburgh, PA

Stephen Garoff, Department of Physics, Carnegie Mellon University, Pittsburgh, PA

Robert D. Tilton, Department of Chemical Engineering and Department of Biomedical Engineering, Carnegie Mellon University, Pittsburgh, PA

While inhaled aerosol drugs can deliver substantial doses of medication directly to the lungs, altered patterns of ventilation associated with obstructive lung diseases cause inhaled drugs to deposit non-uniformly. Some lung regions receive very high local doses of medication while other regions go untreated. This presentation concerns recent research suggesting that the addition of surfactant to the aerosol will disperse these drugs more effectively after deposition on the airway surface liquid (ASL). The enhanced dispersion results from the creation of surface tension gradients on the ASL. Surfactants in the deposited droplets adsorb and decrease surface tension locally. Marangoni stresses drive the droplet to spread outward over higher surface tension regions of the surrounding ASL. In this way, drug that initially deposits on mucus obstructions spread over the surface of the obstruction to reach poorly accessible targets. A considerable body of prior research exists concerning the influence of Marangoni stresses on the spreading of exogenous fluids administered in bolus form to the lungs of premature infants being treated for acute respiratory distress syndrome. That research focused on the driving force and initial dynamics of spreading. The current research focuses instead on the factors that determine the extent of spreading, as this is the variable to be optimized for aerosol drug delivery. Model systems are used to isolate the critical surfactant solution characteristics that control the extent of spreading. The relevant subphase for aerosol delivery to the conducting airways of patients with obstructive lung disease is mucus. Pulmonary mucus is an entangled solution of mucin glycoproteins with a variety of other solutes, the type and abundance of which depend on the physiological state of the individual lung. Mucins are surface active and decrease the surface tension of mucus relative to water, and chain entanglements make mucus rheologically complex. Here we use three model subphases that preserve these key features while offering greater simplicity for mechanistic studies: cultured monolayers of human bronchial epithelial cells that express mucus, entangled aqueous mucin solutions, and entangled aqueous solutions of polyacrylamide. We compare surfactants that represent canonical categories: the anionic surfactant sodium dodecyl sulfate (SDS), the cationic surfactant cetyltrimethylammonium bromide (CTAB), and the nonionic polyether surfactant Tyloxapol. These surfactants differ qualitatively in the strength of their interactions with the polymeric subphase and serve to isolate the potential importance of direct surfactant/polymer complexation. Dyes serve as model drugs that can be tracked visually on the subphase. Spreading of dissolved hydrophilic dyes is contrasted with the spreading of surfactant-solubilized hydrophobic dyes. Tracer particles placed on the subphase before droplet deposition allow the motion of the subphase to be tracked independent of the dye solution spreading. Results obtained to date confirm that Marangoni stresses are the principle driving force for surfactant-enhanced spreading. While the hydrophilic dye clearly marks the edges of the spread droplet, the trajectories of surface tracer particles are consistent with migration of surfactant ahead of the spreading droplet. There is strong evidence for a large separation in the time scales of the underlying processes in the spreading. The convective portion of the spreading event is complete before there is significant mixing of the droplet contents and the miscible subphase. Despite the miscibility of the aqueous droplet and subphase, the spreading occurs as if the subphase were immiscible, and the final spread droplet state is captured by a quasi-static interfacial tension balance on a flattened lens of the dosing fluid.

See more of this Session: Interfacial Phenomena in Pharmaceutics
See more of this Group/Topical: Engineering Sciences and Fundamentals

Fabrication of Biodegradable Spheroidal Microparticles Via the Oil-in-Water Emulsion/Solvent Evaporation Method

Michael J. Heslinga, Chemical Engineering, University of Michigan, Ann Arbor, MI

Eric Mastria, Chemical Engineering, University of Michigan, Ann Arbor, MI

Omolola Eniola-Adefeso, Chemical Engineering, University of Michigan, Ann Arbor, MI

Particle shape, in addition to size, is becoming increasingly recognized as important in the design of drug carriers for in vivo use. However, few methods exist for fabricating non-spherical particles from biodegradable polymers. This work describes for the first time the fabrication of biodegradable spheroidal microparticles using the simple oil-in-water emulsion solvent evaporation technique (O/W ESE). Unloaded and paclitaxel-loaded spheroids were fabricated from poly(lactic-co-glycolic acid) (PLGA), and the shape and size of fabricated spheroids were manipulated by controlling fabrication process parameters including stir speed, aqueous and oil phase viscosity, aqueous phase pH, and the polymer molecular weight and end group. The presented data show that high aqueous phase viscosity, basic aqueous phase pH and hydrophilic polymer side chains and end groups are all conditions that favor the formation of spheroidal particles. The described technique is advantageous over methods currently described in the literature in its simplicity in setup, high particle yield and adaptability to a wide range of biodegradable polymers and therapeutics.

See more of this Session: Interfacial Phenomena in Pharmaceutics
See more of this Group/Topical: Engineering Sciences and Fundamentals

Polymer-Surfactant Interactions

Marina Tsianou, Department of Chemical and Biological Engineering, University at Buffalo - The State University of New York, Buffalo, NY

Ankitkumar I. Fajalia, Department of Chemical and Biological Engineering, University at Buffalo - The State University of New York (SUNY), Buffalo, NY

TBA

See more of this Session: Interfacial Phenomena in Pharmaceutics
See more of this Group/Topical: Engineering Sciences and Fundamentals

Lateral Segregation in Polymer Nanocarriers

Rikkert J. Nap, Department of Biomedical Engineering, Northwestern University, Evanston, IL

Igal Szleifer, Department of Biomedical Engineering, Northwestern University, Evanston, IL

Nanocarriers, such as polymeric nanoparticles, micelles and liposomes show great possibilities in biomedical applications, for example for drug delivery or as imaging agents for biodetection and biolabeling. Direct biological applications of most nanoparticles for imaging or pharmaceutics are hindered by the hydrophobicity of the particles. To improve their solubility and biocompatibility the nanoparticles can be tethered with hydrophilic polymers or similarly embedded in the core of a polymer micelle.
Experimental relevant examples of nanocarriers involve nanometer sized micelles made of short diblock copolymers of polyethylene glycol and phospholipids. An important feature of these 'nano' micelles is that the polymers are mobile. They have a liquid-like disorder similar to the positional disorder found in lipid bilayers.
We used a molecular theory, that explicitly include the size, shape, and conformations of all molecules, to investigated the effect of the mobility on the structure of the polymers and on the binding of nanocarriers to a cell surface. The effect of two types of polymers of two different molecular weights is also investigated. A nanocarrier containing several types of polymers can have multiple functions, e.g. simultaneous selective targeting, imaging and drug delivery. In micelles that contain chains of different molecular weights, the long and short polymer chains segregate upon approaching a surface. The short chains are preferentially located near the surface, whereas the longer chains are located at the opposing side of the nanocarrier. The extent of

the segregation is controlled by the total number of polymers, the relative amount and length of short and long polymers. Combining surface mobility with binding of the end-segments to the surface can be used to control the attraction of the nanocariers to the surface. The segregation, allowed by the surface mobility of the polymers and absent for immobile chains, can potentially be used to enhance the strength and selectivity of the binding of nanocarriers to a cell.

See more of this Session: Interfacial Phenomena in Pharmaceutics
See more of this Group/Topical: Engineering Sciences and Fundamentals

The Effect of Interactions with Silica Surface On the Conformation of Antimicrobial Peptide Cecropin P1 Using Molecular Dynamics Simulation

Xiaoyu Wu, Agricultural and Biological Engineering, Purdue University, West Lafayette, IN

Charlene Mello, Molecular Sciences and Engineering Team, Natick Soldier Research, Development & Engineering Center, Natick, MA

Ramanathan Nagarajan, Molecular Sciences and Engineering Team, Natick Soldier Research, Development & Engineering Center, Natick, MA

Ganesan Narsimhan, Purdue University, West Lafayette, IN, U.S.A.

Cecropin P1 is a 31 amino acid peptide found in nematodes from the stomachs of pigs which exhibits antimicrobial activity predominantly against Gram-negative bacteria (with reduced activity against Gram-positive bacteria) through pore formation in cell membrane as a result of its alpha helical structure. An analog of Cecropin P1, known as Cecropin P1 C, where a cysteine residue is attached at the C terminal, is also selected as a model peptide because of a cysteine residue is necessary for the peptides to be attached on a solid surface. It is of interest to relate the loss of antimicrobial activity (if any) as a result of attachment of these peptides to a surface to the conformational changes.

A Molecular Dynamics simulation was performed using AMBER to calculate the secondary structure, potential energy and end-to-end distance for Cecropin P1 and Cecropin P1 C in solution as well as for Cecropin P1 C physically adsorbed onto silica surface or anchored to the surface with a linker molecule. The simulation results showed an equilibrium structure consisting of two α helix regions for Cecropin P1 as well as for Cecropin P1 C with a sharp bend for the latter. Eventhough this could not be compared with NMR structure, the prediction of two α helical regions is consistent with the available structures of other antimicrobial peptides. % α helix of the equilibrium structure is highest (40%) for Cecropin P1 in the presence of 0.12M salt as a result of shielding of electrostatic interactions, followed by Cecropin P1 C (30%) and Cecropin P1 (25%). The conformation of adsorbed Cecropin P1 C on silica surface indicated a α helix content of around 10% as opposed to a value of around 30% in solution. In addition, the adsorbed Cecropin P1 C on silica surface does not exhibit a hairpin bend with the end to end distance being around 4 nm compared to a value of 1.5 to 2 nm in solution. The α helix content and end to end distance of Cecropin P1 C anchored to the silica surface with oligo ethylene oxide spacer molecule were found to be significantly different from those for adsorbed polypeptide.

See more of this Session: Interfacial Phenomena in Pharmaceutics
See more of this Group/Topical: Engineering Sciences and Fundamentals

Electrodeless Electro-Hydrodynamic Printing of Personalized Medicines

Boris Khusid, Chemical, Biological & Pharmaceutical Engineering, New Jersey Institute of Technology, Newark, NJ

Yueyang Shen, Chemical, Biological & Pharmaceutical Engineering, New Jersey Institute of Technology, Newark, NJ

Ezinwa O. Elele, Chemical, Biological & Pharmaceutical Engineering, New Jersey Institute of Technology, Newark, NJ

The authors propose a method appropriate for the drop-on-demand printing of personalized medicines. To eliminate adverse effects of electrochemical reactions at the fluid-electrode interface, the fluid infused into an electrically insulating nozzle to form a pendant drop serves as a floating electrode capacitively coupled to external electrodes. A short voltage pulse is applied to the electrodes to stretch the drop into a liquid bridge that will then break up creating a sessile drop on the substrate. This gentle method for drop deposition has been demonstrated to operate with fluids spanning over three orders of magnitude in viscosity and conductivity. Surprisingly, the formation of a cone-like drop profile was observed to depend not only on the magnitude of electric stress, as assumed by current theories for a Taylor cone, but also on the rate of stress increase. This gentle method for drop deposition has been demonstrated to operate with fluids spanning over three orders of magnitude in viscosity and conductivity.

This work was supported by NSF Engineering Research Center on Structured Organic Particulate Systems.

See more of this Session: Interfacial Phenomena in Pharmaceutics
See more of this Group/Topical: Engineering Sciences and Fundamentals

Insight From Simulations of Laser-Induced Nucleation

Brandon C. Knott, Chemical Engineering, UC Santa Barbara, Santa Barbara, CA

Michael F. Doherty, Chemical Engineering, UC Santa Barbara, Santa Barbara, CA

Baron Peters, Chemical Engineering, UC Santa Barbara, Santa Barbara, CA

Nonphotochemical laser-induced nucleation (NPLIN) has been observed for a variety of supersaturated solutions including urea, glycine, potassium chloride, proteins, and even a liquid crystal former[1,2]. Experiments show that nanosecond pulses of an intense laser (10-1000 MW/cm^2) can reduce the induction time for nucleation from weeks to seconds. The laser frequencies are in the range where the solute molecules are transparent, so many of the proposed mechanisms involve perturbed electronic degrees of freedom that result in altered intermolecular interactions. For example, the optical Kerr hypothesis suggests that the molecules may become partially aligned with their most polarizable axis along the oscillating field direction[1,2]. Another hypothesis due to Isard suggests that the growing nucleus is stabilized due to the difference in dielectric constants of the nucleating phase and the surrounding solution[3]. Simulations using a lattice gas-Potts model are used to test these hypotheses in the first computational investigations of NPLIN. Our results confirm that an optical Kerr aligning mechanism does reduce the critical nucleus size and the free energy barrier to nucleation. However, at the experimental electric field strengths, our model shows that the optical Kerr effect only reduces the free energy barrier by a small fraction of kT which is insufficient to explain the rate enhancements observed in experiments. An important factor that has not been considered in-depth in experiments is the effect of the laser pulse duration. According to the hypotheses in the literature, the nucleation barrier should revert to the unperturbed barrier once the laser pulse ends. For the extremely short pulses used in the experiments, a postcritical nucleus relative to the perturbed critical size may not have time to grow beyond the unperturbed critical nucleus size before the laser turns off. Our simulations show that this dynamical barrier crossing effect may be important for short laser pulses, so we use a Smoluchowski model of the barrier crossing process to derive a threshold pulse duration required to observe a nucleation enhancement. Our results suggest experiments that may be used to test hypotheses based on perturbed molecular interactions and to gain insight into the monomer attachment frequencies for nuclei at the critical size.

References

[1] Matic, et al, Crystal Growth & Design **2005** 5 (4), 1565-1567

[2] Sun, et al, Crystal Growth & Design **2006** 6 (3), 684-689

[3] Alexander and Camp, Crystal Growth & Design **2009** 9 (2), 958-963

See more of this Session: Nucleation and Growth
See more of this Group/Topical: Engineering Sciences and Fundamentals

Simulations of A Vapor Bubble Growth IN Nucleate Boiling

Jinyong Bao, Department of Chemical Engineering, City College of New York and Graduate Center,City University of New York, New York, NY

David Rumschitzki, Chemical Engineering, City College and the Graduate School and University Center (GSUC) of City University of New York, New York, NY

Joel Koplik, Department of Physics, Levich Institute, City College of New York and Graduate Center,City University of New York, New York, NY

Nucleate boiling is recognized as an ideal and effective technique to drastically cool a hot surface applied. It is attractive due to its stable, efficient heat transfer at a relatively low thermal driving force associated with nucleate boiling incipience. It's the dramatic heat transfer enhancements in this regime have attracted intensive interest since the 1930's. However, the problem's inherent complexity has limited the theoretical understanding of the mechanism of nucleate boiling heat transfer enhancement, and its use to predict the rate of heat transfer from a solid to a liquid in contact with it.

In recent years, many numerical simulations of a vapor bubble growth have been based on the micro-layer model, which suggests that the latent heat is involved in vaporizing the micron thick wetting layer that forms beneath a portion of an expanding vapor bubble on a surface that the liquid wets. These simulations are very sensitive to the assumed initial micro-layer thickness and also assumed a uniform, constant metal surface temperature, contrary to the experimental studies. Rarely do these models explicitly recognize the importance of the three-phase contact line at the edge of the vapor bubble on the solid heating surface. Its appearance and the motion of the contact line, we argue from preliminary calculations, is essential to explain the large observed heat enhancement in nucleate boiling; unfortunately, the physics of the contact line and its motion during bubble growth are not well understood. As such, after some preliminary calculations we step down to the molecular scale where forces can be specified accurately and appeal to molecular dynamics to yield the correct physics of the contact line in these situations.

We model the transport process by the quasi-steady heat conduction equation for the growth of a single bubble at the interface of solid and liquid media each of finite extent. We define an axisymmetric, cylindrical region with the bubble at its center and implement an axisymmetric boundary integral equation method to numerically solve the quasistatic problem simulation. This formulation assumes small Reynolds, thermal Peclet, Capillary and Bond numbers. We also apply zero heat flux boundary conditions at the radius of the cylindrical cell in order to mimic the effect of bubble-bubble interactions. Furthermore, we couple the solution of the quasi-static problem with three simple, somewhat ad hoc contact line motion models to simulate the growth of an incipient bubble due to conduction until at least one of our assumptions is no longer valid. These models are (1) the contact line is pinned; (2) the contact line maintains its equilibrium contact angle and (3) a kinematic contact line motion that locally evaporates liquid to advance the contact line by virtue of the heat conducted to the contact line region in the current time step. The effects of different parameters such as conductivity ratio of the liquid and solid, wall superheat and bubble density on bubble growth are examined to reveal the fundamental mechanism. We found that, after an initial, short parameter- and model-dependent growth phase, the volume of the bubble as a function of time approaches an apparently universal time to the 3/2 power, which is in agreement with laser-doppler experimental data and other literatures reported. This stems from our simulation finding that almost 95% of total heat to the vapor bubble transfers to the very small region close to the contact line. Here we first consider the bubble growth when the bubble grows to a size where gravity comes into play, the bubble begins to deform from its spherical shape. If bubble growth nevertheless remains quasi-static, the new shape is still an (equilibrium) solution of Young-Laplace equation for a given bubble volume prescribed by the vaporization during the previous time step and for a contact line position specified by the contact line motion models; in that case, we continue to simulate bubble growth by the above method until it necks.

Since the physics of contact line motion plays an important role in vapor bubble growth and molecular dynamics simulation has the potential to extract the contact line behavior without ad hoc assumptions, with only the molecular interaction physics as an input, we choose this technique to simulate a heterogeneous nucleation and growth of a vapor bubble on a solid surface due to heat transfer from the bottom of the heating solid. Our simulations consider a fluid made of atoms interacting with two body forces of two types, a Lennard-Jones potential to provide an impenetrable atomic core and a finitely extensible nonlinear elastic force to link the atoms into chain molecules. The potential functions for the interaction between solid atoms and interaction between molecules and solid are the same LJ form but with different energy scale parameter. The simulation domain we set for the program of interest is a cuboid, which is composed of the solid wall and fluid region. The solid wall lie on the bottom of simulation box and right above it is the fluid region, which is fully occupied by molecules. The upper boundary of the computation domain is an imaginary perfectly adiabatic wall. The collision of an atom with this boundary is purely elastic. The other 4 vertical boundaries of the domain are applied periodic boundary condition. The velocities of the molecules are scaled at every time step so that we can control the desired temperature.

After achieving thermal equilibrium of all the phases in the domain at a uniform temperature T1, we instantaneously increase the temperature at the bottom (i.e., of the three bottom layers) of the solid to T2 and keep it fixed at that boundary. The heat transfers through solid to the liquid and evaporates the liquid from the layer adjacent to the solid. We use a relative low wettability potential on the solid surface to help the vapor to nucleate by adjusting the attraction parameter between the fluid molecules and those of the solid. Small vapor patches initially appear and disappear randomly in space and time. Finally at some point, one of these patches successfully grows to a stable vapor bubble. We define the contact angle by azimuthally averaging the 2-dimensional density distribution about the center of the bubble. We track the growth of the bubble and the associated change in its contact line and angle and use this as a model for contact line motion during bubble growth. We plug this into continuum conduction calculation described above and compare the physical quantities of interest (e.g., scaling of bubble volume with time, scaling of heat transfer with bubble density and with other physical parameters) that arise from this calculation with those of the previous ad hoc models.

See more of this Session: Nucleation and Growth
See more of this Group/Topical: Engineering Sciences and Fundamentals

Activated Instability of Homogeneous Nucleation in Metastable Fluids

Korosh Torabi, Chemical Engineering, Purdue University, West Lafayette, IN

David S. Corti, Chemical Engineering, Purdue University, West Lafayette, IN

Homogeneous nucleation is the activated process by which the new phase (vapor, liquid or solid) is formed from a bulk metastable fluid (superheated liquid, supercooled vapor or liquid) in the absence of impurities or solid surfaces. According to classical nucleation theory (CNT)[1], if an embryo of the new phase is less than some critical size, the embryo collapses back into the metastable fluid; if the embryo exceeds this critical size, the nucleus "reversibly" grows to macroscopic size. The critical nucleus corresponds to the saddle point in free energy space (if the embryo is not treated as incompressible): a maximum in the radius and a minimum in the number of particles, n, inside the embryo. The free energy surface continues on indefinitely beyond the saddle point, serving to channel the embryo toward the much lower lying minimum corresponding to the new, bulk phase. Though this subsequent growth of nuclei may be rapid, the post-critical embryos nevertheless follow well-defined pathways that describe the reversible change of n and radii. CNT also invokes another feature of the free energy surface, namely that the region about the saddle point is sharply peaked. Hence, only a small area centered about the saddle point describes the most likely transition paths between a pre-critical embryo and the new phase.

Recently, we adapted density-functional theory (DFT) to describe the phenomena of homogeneous nucleation in a metastable fluid comprised of Lennard-Jones particles[2]. In the DFT approach, as opposed to CNT, the non-uniformity of the density throughout the embryo and the surrounding mother-phase is taken into account. In order to generate a free energy surface analogous to the one within CNT, we solve for the density profile that yields the minimum free energy of a system comprised of a spherical embryo of a given radius λ, which contains a given number of particles n surrounded by the metastable mother-phase. The difference between the calculated free energy of the (n, λ) embryo system and the one for the corresponding uniform system is the minimum work of formation of the embryo, $W(n,\lambda)$. Focusing on bubble nucleation within a superheated liquid, the obtained free energy surface reveals some interesting properties not predicted by CNT. For a large enough number of particles inside the bubble, $W(n,\lambda)$ goes through a maximum with respect to the radius, analogous to what is seen in CNT. Increasing the radius further beyond this maximum, however, one reaches a critical radius beyond which no liquid-like density profile can be found that corresponds to the local minimum of the free energy functional. This in turn means there is a limit to the radius of the bubble for a given number of particles, in which we can have a locally stable superheated liquid. In this way our DFT approach, as opposed to CNT, predicts that the free energy surface terminates at a locus of instabilities. Similar results are found for the free energy surface that describes droplet nucleation within a supercooled vapor.

Characterizing the spherical embryo via the radius and number of particles requires the optimization of the free energy functional subject to the n, λ constraint. In addition to the traditional constrained functional optimization method using undetermined Lagrange multipliers, we also employ the number-conserving functional derivative method introduced by Gal [3]. This new approach expedites our numerical calculations considerably and yields equivalent accurate solutions using the method of Lagrange multipliers. Furthermore, we investigate in greater detail the thermodynamic stability of the (n,λ) embryo, particularly at the predicted limits of stability, by studying the proper eigenvalue integral equation that follows from the second functional derivative of the free energy. The developed methodology reveals that the smallest eigenvalue of the second functional derivative operator approaches zero exactly at the same locus of instabilities predicted by our DFT calculations.

We also extend our DFT method to fluids more complex than the relatively simple system of Lennard-Jones particles. We adapt the association model developed by Chapman et al. [4], based on an extension of Wertheim's theory [5], which is able to capture several important features of water-like systems. We demonstrate that homogeneous bubble and droplet nucleation within our water-like system is also described by a free energy surface that is qualitatively similar to the one for the Lennard-Jones fluid. In this way, we demonstrate that the interesting and new properties of the free energy surface, in particular the existence of a locus of instabilities and a relatively flat ridge of transition states all corresponding to the same value of the activation barrier, are not just limited to the Lennard-Jones fluid.

In addition, we perform a series of molecular dynamics (MD) simulations to further validate the predicted instabilities generated by bubbles and droplets beyond or below some critical size. For example, we initialize the superheated liquid in which is contained a spherical bubble of a given n and λ. The (n,λ) embryo-constraint is then relaxed and we allow the liquid to evolve dynamically on its own via additional MD steps. With this procedure, we test if a given configuration described by a given n and λ does in fact lead to a phase change, or whether the bubble collapses rapidly enough to maintain stability, as would be predicted in either case by our DFT stability analysis.

Finally, we also apply the shell particle formulation [6] to rigorously define the volume of an embryo with a given number of particles. Interestingly, we have shown that the thermodynamic properties of a system containing a shell particle can be straightforwardly related to the system that that is not described by a shell particle. In this way, we can straightforwardly calculate the free energy of a shell particle system from our previous DFT results. We discuss how the use of the rigorous shell particle formulation alters some but not all of the properties of the free energy surface of embryo formation.

[1] L. Gunther, Am. J. Phys., 71, 351 (2003).

[2] M. J. Uline and D. S. Corti, Phys. Rev. Letters, 99, 076102 (2007).

[3] T. Gal, Phys. Rev. A, 63, 022506 (2001).

[4] W. G. Chapman, K. E. Gubbins, G. Jackson and M. Radosz, Ind. Eng. Chem. Res., 29, 1709 (1990).

[5] M. S. Wetheim, J. of Stat. Phys., 35, 19 (1983).

[6] D. S. Corti and G. Soto-Campos , J. Chem. Phys., 108, 7959 (1998).

See more of this Session: Nucleation and Growth
See more of this Group/Topical: Engineering Sciences and Fundamentals

Modeling of transient concentration for slow growing conformational polymorphs during batch crystallization.

Lotfi Derdour, San Kiang, Chiajen Lai, Shawn K. Pack, Dimitri Skliar, James Vernille.
Bristol-Myers Squibb Company,
Process Research and Development, One Squibb drive, New Brunswick, NJ 08903, USA
lotfi.derdour@bms.com

Abstract

A simple model to determine the transient concentration during batch crystallization of multiple conformer molecules is presented. The crystallization is assumed to be governed by an advancing step integration mechanism followed by a diffusion period. During the integration period, the concentration of the conformer that crystallizes in solution is considered as the limiting parameter and the low proportion of the right conformer in solution is assumed to be the cause of the observed low nucleation and crystallization rates. This assumption was consolidated by *ab initio* conformation energy calculations which showed that the conformer that crystallizes has a high energy conformation and thus should be present in minor amount in solution. The model is based on the screw dislocation and the birth and spread theories and assumes a constant step advance velocity (CSAV). The prediction of the combined CSAV model / diffusion model provides a good prediction of the transient concentration during the anti-solvent crystallization of a developmental drug. Simulations of different anti-solvent addition scenarios predict a nearly constant and low supersaturation for parabolic profiles and maxima in supersaturation for linear addition rates.

Keywords: conformational polymorphism, slow crystal growth, dislocation, modeling, pharmaceuticals.

Phase Separation During Non-Isothermal Cooling of Polymer- Drug Dispersion

Phong Huynh, Rutgers University, New Brunswick, NJ

Paul Takhistov, School of Environmental & Biological Sciences, Rutgers University, New Brunswick, NJ

Biopolymer can improve the solubility of the poor-soluble drugs. This work focus on how polyethylene glycol 3350 (PEG3350), a polymer widely applied in drug delivery, combines with Ibuprofen (IBU) by hot melt in region I and region II of a eutectic mixture. The experiments demonstrate that the amorphous IBU separates from PEG and locates at the boundary of PEG crystals for mixture in region I while in region II some amorphous IBU is in the PEG crystal the other concentrate at PEG crystals' boundary. Phase diagram of PEG3350-IBU as well as a model is built to integrate the process of phase separation during solidification. Phase contrast optical microscopy technique is applied to analyze the mechanism of non-isothermal solidification of PEG3350-IBU mixture. It was found that the homogeneous nucleation crystallization occurs in region I mixture crystallization while both homogenous and heterogeneous nucleation crystallization appear in the solidification of mixture in region II. Proposed solution of Stephan-type problem of phase separation due to crystallization process is in good agreement with experimental data.

See more of this Session: Nucleation and Growth
See more of this Group/Topical: Engineering Sciences and Fundamentals

Sweating Liquid Marbles

Prasad Bhosale, Department of Chemical Engineering, Tennessee Technological University, Cookeville, TN

Mahesh V. Panchagnula, Dept of Mechanical Engineering, Tennessee Technological University, Cookeville, TN

Liquid marbles have been shown to be an effective means of storing, transporting and creating barrier to diffusive evaporation. We demonstrate a new mechanism for creating micron and submicron sized nanoparticulate liquid marbles that are mechanically and thermally stable. We show that the mechanism for the creation of such marbles is vapor phase diffusion and condensation and that the liquid marble size distribution as well as the mean marble size can be controlled by the heating time and temperature at which the diffusive process is allowed to occur.

See more of this Session: Nucleation and Growth
See more of this Group/Topical: Engineering Sciences and Fundamentals

The Structure and Stability of Alpha-Helical, Orthogonal-Bundle Proteins On Surfaces

Shuai Wei, Department of Chemical Engineering, Brigham Young University, Provo, UT

Thomas A. Knotts IV, Department of Chemical Engineering, Brigham Young University, Provo, UT

The interaction of proteins with surfaces is a key phenomenon in many applications in medicine, biotechnology, and proteomics. Understanding protein/surface interactions is key to improving these technologies, but current understanding of the behavior of proteins on surfaces is lacking. Prevailing theories on the subject, which suggest that proteins should be stabilized when tethered to surfaces, do not explain the experimentally observed fact that proteins are often denatured on surfaces. In an attempt to develop some predictive capabilities with respect to protein/surface interactions, we previously asked the question if the stabilization/destabilization of proteins on surfaces could be correlated to secondary structure and found that no link existed. However, further investigation has revealed that proteins with similar tertiary structure show predictable stabilization patterns. In this presentation, we report how five, alpha-helical, orthogonal-bundle proteins behave on the surface compared to the bulk. By measuring stabilization using melting temperatures and the Gibbs energies of folding, we show that the stability of proteins tethered to surfaces can be correlated to the number of residues in the loop region where the tether is placed and the angle made by the alpha-helices nearest to the tether point. We also show that any destabilization that occurs because of the surface is an enthalpic effect and that surfaces always stabilize proteins entropically.

See more of this Session: Thermophysical Properties of Biological Systems - II
See more of this Group/Topical: Engineering Sciences and Fundamentals

Generalized Structural Polymorphism in Virus Capsid Self-Assembly

Hung D. Nguyen, Department of Chemical Engineering and Materials Science, University of California, Irvine, Irvine, CA

Vijay S. Reddy, Department of Molecular Biology, The Scripps Research Institute, La Jolla, CA

Charles L. Brooks III, Department of Chemistry and Biophysics Program, University of Michigan, Ann Arbor, MI

Self-assembly of viral capsids, which are spherical proteins shells with icosahedral symmetry, holds significant potential for applications in materials science and medicine. However, the complexity of preparing these systems for assembly and the difficulty of quantitative experimental measurements on the assembly process have limited access to critical mechanistic questions that dictate the final product yields and isomorphic forms. Molecular dynamics simulations provide a means of elucidating self-assembly of viral proteins into icosahedral capsids and were the focus of our studies (1-3) by employing our multiscale models for the coat proteins. We examined the spontaneous self-assembly of viral capsids of different sizes containing T = 1, 3, 4, 7, 9, 12, 13, 16, and 19 proteins in each of their 60 icosahedral repeating units (number of proteins per icosahedral capsid = 60xT).

We observed, in addition to icosahedral capsids, a variety of nonicosahedral yet highly ordered and enclosed capsules. Such structural polymorphism was demonstrated to be an inherent property of the coat proteins, independent of the capsid complexity and the elementary kinetic mechanisms. Moreover, there exist two distinctive classes of polymorphic structures: aberrant capsules that are larger than their respective icosahedral capsids when T = 1-7; and capsules that are smaller than their respective icosahedral capsids when T = 7-19. Different kinetic mechanisms responsible for self-assembly of those classes of aberrant structures were deciphered, providing insights into the control of the self-assembly of icosahedral capsids.

References:

1. H.D. Nguyen, V.S. Reddy and C.L. Brooks III, "Deciphering the kinetic mechanism of spontaneous self-assembly of icosahedral capsids," Nano Letters, 7(2) 338-44 (2007)

2. H.D. Nguyen, V.S. Reddy and C.L. Brooks III, "Invariant polymorphism in virus capsid assembly," J. Am. Chem. Soc., 131(7): 2606-14 (2009)

3. H.D. Nguyen and C.L. Brooks III, "Generalized structural polymorphism of self-assembled viral capsids," Nano Letters, 8(12), 4574-81 (2008)

See more of this Session: Thermophysical Properties of Biological Systems - II
See more of this Group/Topical: Engineering Sciences and Fundamentals

Computer Simulation of Protein Aggregation with An Intermediate Resolution Protein Model

Victoria Wagoner, Chemical and Biomolecular Engineering, North Carolina State University, Raleigh, NC

Mookyung Cheon, Chemical and Biomolecular Engineering, North Carolina State University, Raleigh, NC

Carol K. Hall, Chemical and Biomolecular Engineering, North Carolina State University, Raleigh, NC

Assembly of normally soluble proteins into ordered aggregates, known as amyloid fibrils, is a cause or associated symptom of numerous human disorders, including Alzheimer's and the prion diseases. Recent experimental studies have offered tantalizing clues regarding the fibril structure, but our understanding of its assembly is still far from complete. The long term goal of our work is to determine the underlying physical forces responsible for the mis-folding and aggregation of proteins, specifically the beta-amyloid peptide linked to Alzheimer's. Our objective is to extend the intermediate-resolution protein model, PRIME (Protein Intermediate Resolution Model) that we have developed for alanine to the description of all twenty amino acids. PRIME is well suited for modeling protein aggregation because it provides a faithful representation of protein geometry while also capturing the essential features of the forces responsible for protein folding, hydrogen bonding and hydrophobicity.

We extended PRIME to the description of all twenty amino acids (PRIME 20) using a two-pronged approach, focusing on both the protein geometry and energetics. In PRIME 20, each amino acid has a single-sphere side chain with a unique diameter and C-alpha-side chain bond length. Each amino acid is sorted into one of seven classes resulting in twelve interaction types, including both strong and weak attractive, covalent, repulsive, charged, polar and hydrogen bonding. We explored how variations in temperature, concentration, and quench rate affect the formation and structure of several simple peptide sequences known to aggregate. Additionally, we apply PRIME 20 to the beta-amyloid peptide to determine how well it captures this complex aggregation process.

See more of this Session: Thermophysical Properties of Biological Systems - II
See more of this Group/Topical: Engineering Sciences and Fundamentals

Structure, Misfolding and Aggregation of Amylin Protein

Allam S. Reddy, Department of Chemical and Biological Engineering, University of Wisconsin-Madison, Madison, WI

Lu Wang, Department of Chemistry, University of Wisconsin, Madison, WI

Yu-Shan Lin, Department of Chemistry, University of Wisconsin, Madison, WI

Yun L Ling, Department of Chemistry, University of Wisconsin, Madison, WI

Manan Chopra, Department of Chemical and Biological Engineering, University of Wisconsin, Madison, WI

Martin Zanni, Department of Chemistry, University of Wisconsin, Madison, WI

James L Skinner, Department of Chemistry, University of Wisconsin, Madison, WI

Juan J. De Pablo, Department of Chemical and Biological Engineering, University of Wisconsin, Madison, WI

Amyloid deposits of amylin in the pancreas are an important characteristic feature encountered in patients with Type-2 diabetes. Amylin aggregates are considered to be important in the disease pathology and have therefore been studied extensively. However, the secondary structures of the individual peptides have not been clearly identified. In this work we present detailed solution structures of rat and human amylin using a combination of Monte Carlo simulations, molecular dynamics simulations, and 2D IR experimental measurements. Results are also presented regarding the early stages of association of such peptides in solution. A method is presented to determine the free energy of distinct bio-molecular conformations. The human amylin protein is found to adopt a secondary structure that exhibits an alpha-helical segment comprising residues 9-17, and a short anti-parallel beta sheet formed between residues 23-27 and 30-34. The folded structure of the human amylin protein is found to be able to misfold into an anti-parallel beta-hairpin, with the turn region comprising of residues 20-23. In contrast, the rat peptide does not exhibit the propensity to form a beta-hairpin. The results of our calculations are consistent with those of our experimental 2D IR measurements. The predicted structure of rat amylin was found also found to be in agreement with previous NMR results. We then examine the process of association of the human peptide using transition path sampling simulations, and we discuss it in the context of the differences in the secondary structure of the rat and human amylin peptides.

See more of this Session: Thermophysical Properties of Biological Systems - II
See more of this Group/Topical: Engineering Sciences and Fundamentals

Role of Length-Scale Dependent DNA Flexibility in Protein Interactions

Elena Koslover, Biophysics Program, Stanford University, Stanford, CA

Andrew Spakowitz, Chemical Engineering, Stanford University, Stanford, CA

The proteins responsible for reading and packaging the genome must harness the full range of DNA mechanical properties, from bending a rigid DNA strand at short length scales to searching through a meandering, dynamic DNA chain at long length scales. We use the wormlike chain model as a simple description to study the fundamental physics involved at these different regimes. We will present our work on three questions concerning DNA biophysics. First, we look at the effective interaction between DNA bending proteins that arises from elastic stresses in the linker while the DNA is under external tension. Cooperative binding of proteins is required to assemble transcription factor complexes and to condense DNA into chromatin. We demonstrate that the elastic nature of the double helix generates a variable-range oscillatory coupling that may provide a versatile mechanism for tension-mediated regulation of cooperativity and aggregation. We then expand our analytical techniques into practical tools for extracting structural information from single-molecule trapping experiments, with application to nucleosome unwrapping and DNA gyrase mechanics. We consider the relative contributions of elastic bending and entropic fluctuations when measuring DNA kinking at low tensions. Finally, we study the effects of crowding and roadblocks on the competition of timescales for a transcription factor searching through the genome by a combination of one-dimensional sliding and three-dimensional hops.

See more of this Session: Thermophysical Properties of Biological Systems - II
See more of this Group/Topical: Engineering Sciences and Fundamentals

Modeling DNA Hybridization On Surfaces for Improved Microarray Design

Terry J. Schmitt, Department of Chemical Engineering, Brigham Young University, Provo, UT

Thomas A. Knotts IV, Department of Chemical Engineering, Brigham Young University, Provo, UT

DNA microarrays are high throughput, parallel assays for determining which genes are present in a sample and the level of expression of those genes. They have been identified as a key technology in genomic sciences and emergent medical techniques. Despite their abundant use in research laboratories, microarrays have not been used in the clinical setting to the fullest potential due to the difficulty of obtaining reproducible results. Improved understanding of DNA hybridization at the molecular level is key to refining these devices. To this end, we have used a mesoscale model for DNA, which captures both the thermal and mechanical properties of the molecule, to investigate the factors affecting hybridization on surfaces to an extent not previously possible. After briefly describing the model, we will demonstrate how the thermodynamics of hybridization changes in the bulk compared to on a surface, how the crowding of probes on the surface affects hybridization efficiencies, and how targets in solution compete to bind with probes on the surface. We will also demonstrate possible strategies that can be used to enhance hybridization on the surface, including the effect of stretching probe molecules.

See more of this Session: Thermophysical Properties of Biological Systems - II
See more of this Group/Topical: Engineering Sciences and Fundamentals

Validating a DNA Simulation Model through Hairpin Experiments

Margaret C. Linak, Department of Chemical Engineering and Materials Science, University of Minnesota, Minneapolis, MN

Kevin D. Dorfman, Department of Chemical Engineering and Material Science, University of Minnesota, Minneapolis, MN

We will present a comparison between large-scale melting experiments of DNA hairpins in common biological buffers and the predictions of a coarse-grained Brownian Dynamic computation model (M. Kenward and K. D. Dorfman, Journal of Chemical Physics, 130, 1, (2009)). Most computational efforts to date take a physical approach to modeling single stranded DNA. While this is effective for understanding the fundamental properties of self-interacting polymers (with an eye towards the particular interactions that characterize DNA), such studies are rarely related to laboratory experiments in realistic biological conditions. To address this need, we have studied DNA hairpins with AT or GC stems and relatively long non-interacting loops. This system emphasizes the role of stacking and hydrogen bonding energies, which are characteristics of DNA, rather than backbone bending, stiffness and excluded volume interactions, which are generic characteristics of semi-flexible polymers. On the experimental side, we employed a two-color fluorescence assay where (i) the DNA were end-labeled with Cy5 and Iowa Black in order to obtain a FRET signal which provides a metric of the end-to-end probability distribution function and (ii) Sybr Green I was used as an intercalating dye in order to obtain standard melting curves. Seeking not simply the melting temperature, we generated fluorescence intensity versus temperature curves for both the Cy5 and Sybr channels in a variety of common biological buffers including Tris, PBS, and Buffer A. In conjunction with our experiments, we methodically explored the phase space of our computational model by varying the hydrogen bonding parameters, stacking values, and temperature while fixing the DNA's excluded volume, stiff spring, and bending potentials. The comparison of our simulation and experimental results provides a strong test of our model's suitability for capturing the sensitive behavior of simple hairpin systems. Furthermore, our approach and experimental data can be used to validate other similar coarse-grained simulation models.

See more of this Session: Thermophysical Properties of Biological Systems - II
See more of this Group/Topical: Engineering Sciences and Fundamentals

Theoretical and Experimental Investigation of the Impact of Surfaces On Oligonucleotide Melting Temperature: The Effects of Spacer Length and Probe Density

Ayse Bilge Ozel, Department of Chemical Engineering, University of Michigan- Ann Arbor, Ann Arbor, MI

Erdogan Gulari, Department of Chemical Engineering, University of Michigan- Ann Arbor, Ann Arbor, MI

The most commonly used method to predict the melting temperature of an oligonucleotide duplex in solution is the Nearest Neighbor method, which relies on the sequential arrangement of different dinucleotides based on the stabilization effect of the base-stacking interactions between them. It is also the choice of the researchers designing oligonucleotide sequences or probes on the surface for microarray applications. However, the absence of the inclusion of the effects of the surface on oligonucleotide stability and melting temperature leads to an unrealistic representation of the oligonucleotide duplex formation and denaturation in close proximity to the microarray substrate. In this part of the study, semi-empirical models have been developed to look at how these influences, namely electrostatic and entropic blocking, are affected by various system design parameters. Free energy penalty imposed by electrostatic blocking was calculated through the utilization of Electrical Double Layer theory, zeta potential measurements and Surface Partition Model. Entropic blocking free energy penalty term was modeled using a Polymer Physics approach. To further support these findings, various on-surface melting temperature experiments were carried out by utilizing a specially designed integrated real-time microfluidic system. Two design parameters, spacer length and probe density, were examined. Theoretical and experimental results were found to be in agreement with each other and the published trends reached via simulations. At the experimental conditions tested, probe density seems to have more influence in decreasing duplex stability when compared to spacer length. These findings can be used as guidelines in the design of experiments and optimization of hybridization conditions, when spacer length and probe density are considered as design parameters.

See more of this Session: Thermophysical Properties of Biological Systems - II
See more of this Group/Topical: Engineering Sciences and Fundamentals

Computer Simulations of Drug-Carrying Block Copolymer Nanoparticles for Cancer Therapy

Jeff Woodhead, Chemical and Biomolecular Engineering, North Carolina State University, Raleigh, NC

Carol K. Hall, Chemical and Biomolecular Engineering, North Carolina State University, Raleigh, NC

Cancer chemotherapy often requires the untargeted delivery of large amounts of poorly soluble drugs. Block-copolymer nanoparticles are emerging as a viable vehicle for the transportation of chemotherapy drugs to their intended targets and as a means of increasing the solubility of the drugs, thus reducing the dosage that is necessary to fight tumors. We perform discontinuous molecular dynamics (DMD) simulations on systems containing block copolymers and drug molecules in order to understand how block copolymer nanoparticles assemble and what conditions encourage good encapsulation. We will detail the effect of system density, copolymer mole fraction, copolymer composition, and hydrophobic interaction between copolymer head and drug on the encapsulation efficiency and structural phase behavior of the system. We will also explore the internal structure of drug-carrying block copolymer micelles. We find that the presence of drug in the system affects the conditions at which micellization occurs. We also find that the presence of drug affects the nature of the phases that can be discovered, and that a strong head-drug interaction is important in encapsulating drugs. Our results can be used by experimentalists as a framework for optimizing the drug encapsulation process.

See more of this Session: Computational Studies of Self-Assembly
See more of this Group/Topical: Engineering Sciences and Fundamentals

Sequence Effects in the Renaturation of DNA Oligonucleotides: Mechanistic Details and Structure

Edward J. Sambriski, Chemical and Biological Engineering, University of Wisconsin-Madison, Madison, WI

Vanessa Ortiz, Chemical and Biological Engineering, University of Wisconsin-Madison, Madison, WI

Juan J. de Pablo, Chemical and Biological Engineering, University of Wisconsin-Madison, Madison, WI

DNA hybridization plays a central role in biology and, increasingly, in materials science. Yet, there is no precedent for examining the pathways by which specific single stranded DNA sequences interact to assemble into a double helix. A detailed model of DNA is adopted in this work to examine such pathways, and to determine the role of sequence, if any, on DNA hybridization. Free-energy methods are used to extract denaturation profiles and thermodynamic quantities, while transition path sampling (TPS) and forward flux sampling (FFS) calculations are used to extract mechanistic and kinetic details about the renaturation reaction. Results reveal that DNA rehybridization is prompted by a distinct nucleation event involving molecular sites with approximately four bases pairing with partners slightly offset from those involved in ideal duplexation. Nucleation is promoted in regions with repetitive base-pair sequence motifs, which yield multiple possibilities for finding complementary base partners. Repetitive sequences follow a nonspecific pathway to renaturation consistent with a molecular "slithering" mechanism, whereas random sequences favor a restrictive pathway involving the formation of key base-pairs before renaturation fully ensues.

See more of this Session: Computational Studies of Self-Assembly
See more of this Group/Topical: Engineering Sciences and Fundamentals

Self-Assembly of Coarse-Grained Stratum Corneum Lipids

Kevin R. Hadley, Department of Chemical and Biomolecular Engineering, Vanderbilt University, Nashville, TN

Shan Guo, Department of Chemical and Biomolecular Engineering, Vanderbilt University, Nashville, TN

Clare McCabe, Department of Chemical and Biomolecular Engineering, Vanderbilt University, Nashville, TN

The top layer of the skin, known as the stratum corneum, acts as a barrier to foreign agents entering the body and to water leaving the body. Experimentally, while it is known that the lipids of the stratum cornenum are organized in lamellar bilayer structures in which the lipid chains are highly ordered, a detailed picture of the molecular organization of the lipids has not been elucidated. To help understand the molecular level arrangement of the skin lipids both atomistically detailed and coarse-grained molecular models for the three main classes of lipid found in the SC (ceramides, free fatty acids and cholesterol) have been developed for use in molecular dynamics simulations. Coarse-grained models allow molecular simulations to be performed on larger spatial and temporal scales than is possible through atomistic simulation alone and are necessary to study self-assembly in mixed lipid systems. Results will be presented for a number of pure and mixed lipid systems to demonstrate the suitability of the proposed models for describing crystalline and melt lipid phases and future work on studying the complex lipid mixtures seen in the SC will be discussed.

See more of this Session: Computational Studies of Self-Assembly
See more of this Group/Topical: Engineering Sciences and Fundamentals

Self-Assembly of Amphiphilic Molecules Using the i-SAFT Density Functional Theory

Chris Emborsky, Chemical and Biomolecular Engineering Department, MS-362, Rice University, Houston, TX

Zhengzheng Feng, Chemical and Biomolecular Engineering, Rice University, Houston, TX

Kenneth R. Cox, Chemical and Biomolecular Engineering, Rice University, Houston, TX

Walter G. Chapman, Chemical and Biomolecular Engineering Department, MS-362, Rice University, Houston, TX

The effects of an amphiphile (surfactant) on the thermodynamic properties of systems containing chemically dissimilar species (e.g., oil and water) are largely governed by the chemical structure of the amphiphile and physical conditions of the system such as the pressure, temperature, and bulk solubility of the amphiphile. These conditions affect the equilibrium microstructure which in turn defines the interfacial tension. Following justification of the interfacial statistical associating fluid theory (i-SAFT) density functional theory (DFT) approach against molecular simulation data for a model oil-water-surfactant system, a more extensive systematic study is carried out to analyze the role the amphiphile structure and concentration play in determining the equilibrium microstructure and interfacial tension.

See more of this Session: Computational Studies of Self-Assembly
See more of this Group/Topical: Engineering Sciences and Fundamentals

Role of Collective Degrees of Freedom in Formation and Disintegration of Spherical Micelles

Yong Nam Ahn, Department of Chemical Engineering, University of Florida, Gainesville, FL

Gunjan Mohan, Applied Scientific Computing; Coatings Innovation Center, PPG Industries, Allison Park, PA

Dmitry I. Kopelevich, Department of Chemical Engineering, University of Florida, Gainesville, FL

Dynamics of self-assembly and structural transitions in amphiphilic systems play an important role in various processes, ranging from production of nanostructured materials to transport in biological cells. We recently demonstrated [1] that even such simple process as addition of a single surfactant molecule to a micelle involves a complex interplay between micellar microstructure and the monomer configuration. Neglecting cooperative dynamics of these degrees of freedom leads to qualitative discrepancies in predicted surfactant addition rates.

In this talk, we present improvement of our earlier model which enables quantitative predictions of addition/removal rates of surfactants to/from micelles. This is accomplished by a reconstruction of a multi-dimensional free energy landscape parametrized by distance between the micelle and monomer centers of mass, as well as the micellar and monomer configurations. The minimal energy path (MEP) on this landscape is identified. Movement along MEP corresponds to addition and removal of a monomer accompanied by the necessary changes of the micellar and monomer configurations. Analysis of MEP allows us to identify collective degrees of freedom relevant to the monomer addition and removal. However, MEP alone is not sufficient to predict the monomer addition and removal rates since both of these processes involve non-adiabatic stages. Comparable time-scales of several independent degrees of freedom during these non-adiabatic stages imply that the system dynamics cannot be described by a quasi-one-dimensional motion along MEP. Therefore, we obtain solutions of multi-dimensional Langevin equations to correctly describe the non-adiabatic system dynamics.

We conclude this talk by a discussion of extension of the developed methodology to more complex transitions in self-assembled systems.

References

1. G. Mohan and D. I. Kopelevich, "A multiscale model for kinetics of formation and disintegration of spherical micelles", J. Chem. Phys.128, 044905 (2008).

See more of this Session: Computational Studies of Self-Assembly
See more of this Group/Topical: Engineering Sciences and Fundamentals

Computational Studies of Colloidal Dynamics in Entropic Fields

Bryce D. Sturtevant, Chemical Engineering, Purdue University, West Lafayette, IN

David S. Corti, Chemical Engineering, Purdue University, West Lafayette, IN

An important class of inter-particle forces, known as depletion forces, is induced by the presence of other colloidal species and arises solely as a result of entropic considerations. Passive structures etched into the walls of a surface can create entropic force fields of sufficient range and magnitude so that the motion and position of large colloids can be controlled. By providing potentially simple routes for the directed self-assembly of novel mesoscopic structures, the use of entropic force fields is a promising approach to the production of advanced materials. Various issues concerning the feasibility of such methods need to be addressed, however, given that the dynamics of colloidal particles diffusing through an entropic force field is not well known. Since entropic forces become repulsive at intermediate colloidal separations, large repulsive barriers may kinetically stabilize suspensions or prevent deposition. Moreover, the total depletion interaction emerges as a net result of the equilibrium distribution of secondary species and, hence, evolves in time; it remains unclear as to the time scale over which the depletion interaction develops and how the interstitial fluid affects this timescale. We therefore investigate the dynamics of hard-sphere colloids moving above and onto surfaces of various shapes via the use of two mesoscopic simulation techniques, stochastic rotation dynamics (SRD) and dissipative particle dynamics (DPD). While usage of mesoscopic techniques is essential to the computational exploration of colloidal behavior, the effect of coarse-graining, where there is some loss of detail regarding the fluid, raises some issues regarding the fluid that, previously, have not been satisfactorily answered. Although techniques like SRD and DPD have been shown to correctly replicate hydrodynamic effects in a bulk fluid, there has been some uncertainty as to how well these techniques are able to capture fluid effects in the presence of a surface. More specifically, the time scale over which fluid memory effects develop and decay, and, by extension, how these effects correlate with the definition of colloidal local self-diffusion coefficients has not been well understood. We clarify these points and develop a systematic interpretation for defining position-dependent diffusion coefficients. Furthermore, using the coarse-grained techniques, we determine the relative influence of hydrodynamic and entropic effects on particle deposition and agglomeration. Such key issues need to be explored in sufficient detail before an effort can emerge to describe the dynamics and evolution of depletion potentials through a direct appeal to the Fokker-Planck, or probabilistic, description of the underlying stochastic process.

See more of this Session: Computational Studies of Self-Assembly
See more of this Group/Topical: Engineering Sciences and Fundamentals

Thermodynamics and Kinetics of Phase Selection in Binary Colloidal Systems Mediated by DNA Interactions

Raynaldo Scarlett, Chemical and Biomolecular Engineering, University of Pennsylvania, Philadelphia, PA

John Crocker, Department of Chemical and Biomolecular Engineering, University of Pennsylvania, Philadelphia, PA

Talid Sinno, Chemical and Biomolecular Engineering, University of Pennsylvania, Philadelphia, PA

Colloidal self-assembly provides a potential avenue for the design of novel devices with unique optical properties [1]. Colloidal systems also provide useful insights into fundamental mechanisms of phase transitions such as crystal nucleation [2], growth [3] and melting that are otherwise difficult to probe in atomic systems. In particular, micro- and nanoscale particles, functionalized with designer single-stranded DNA brushes, have now been successfully assembled into several crystalline phases, including ordered, binary superlattice structures.

While this early success is promising, further progress requires a rational approach for interaction design, especially if more interesting and useful structures are to be assembled. In this presentation, we apply a comprehensive ensemble of Monte Carlo simulations and free energy calculations to generate a detailed picture of assembly in binary colloidal systems mediated by DNA hybridization. An important aspect of the simulations is that the effective interaction potential produced by the DNA binding has been measured experimentally and modeled with high accuracy, thereby allowing for a close connection between simulation and experiments [4,5].

We show that a range of experimentally observed phenomena pertaining to the growth of binary crystals are explained well by the simulations [6]. In particular, we demonstrate that kinetic factors are important in not only determining the quality (i.e. compositional and morphological order) of a growing crystal, but also its phase. For binary systems in which unlike ("A-B") interactions are dominant, both close-packed (rhcp) and non-close-packed (bcc, or CsCl) superlattice structures are feasible [7], but the competition between them is sensitively dependent on both thermodynamic and kinetic factors. The simulation framework presented here should be useful for finding optimal conditions and DNA designs for experimentally growing crystals with desired structures and low defect densities.

[1] J. D. Joannopoulos, R.D. Meade, and J.N. Winn, Photonic Crystals: Molding the Flow of Light. 1995, Princeton: Princeton University Press.

[2] P.N. Pusey and W.v. Megen, Nature 320, 340 (1986)

[3] U. Gasser, et al., Science 292, 258 (2001)

[4] A.J. Kim, P.L. Biancaniello, and J.C. Crocker, Langmuir. 22, 1991 (2006)

[5] P.L. Biancaniello, A.J. Kim, and J.C. Crocker, Phys. Rev. Lett. 94, 058302 (2005)

[6] A.J. Kim, R.T. Scarlett, et al., Nature Materials 8, 52 (2009)

[7] S. Y. Park, et al., Nature 451, 553 (2008)

See more of this Session: Computational Studies of Self-Assembly
See more of this Group/Topical: Engineering Sciences and Fundamentals

Self-Assembly Propensities and Pathways for Patchy Particles

Eric Jankowski, Chemical Engineering, University of Michigan, Ann Arbor, MI

Sharon C. Glotzer, Chemical Engineering and Materials Science and Engineering, University of Michigan, Ann Arbor, MI

In this work we use a new computational method, bottom-up building block assembly to generate ordered, energy-minimizing configurations for a family of complex interacting patchy particles. We also generate assembly pathways that elucidate ways large clusters can be generated by combing smaller energy-minimizing clusters of patchy particles. We observe common characteristics of assembly pathways that correspond to particles with favorable and unfavorable self-assembly characteristics and demonstrate design strategies for overcoming barriers to self-assembly.

See more of this Session: Computational Studies of Self-Assembly
See more of this Group/Topical: Engineering Sciences and Fundamentals

Molecular Dynamics Simulation of Aggregation Phenomena of TiO2 Nanoparticles in Aqueous and Vacuum Environments

Mozhgan Alimohammadi Zanjani, Chemical Engineering, The Pennsylvania State University, University Park, PA

Kristen A. Fichthorn, Chemical Engineering, Pennsylvania State University, University Park, PA

The aggregation of nanosize clusters is an important phenomenon that occurs in materials synthesis through both colloidal and vapor-phase routes. Particularly interesting is the experimental observation that nanocrystals can also approach one another and merge along specific crystallographic directions, in the oriented attachment mechanism. The ability to direct crystallization through oriented attachment is an exciting prospect that could allow for the creation of new nanostructures with well-defined sizes and shapes. To this end, insight into its origins, which are not currently clear in all cases, would be beneficial.

In this work, we employ molecular dynamics (MD) simulations to understand the origins of oriented attachment in colloidal titanium dioxide (anatase). We use the DL-POLY simulation package to model nanoparticles in the 2-3nm size range, with shapes dictated by the Wulff construction. Our initial studies probe the aggregation (sintering) of two anatase particles in vacuum. We find that oriented attachment occurs in vacuum and is due to electrostatic interactions between under-coordinated O atoms along [101] facet edges and under-coordinated Ti atoms on [001] facets. Oriented attachment is observed experimentally for colloidal nanocrystalline anatase. However, the crystallographic direction is different than what we observe in vacuum, indicating that water plays a role in directing aggregation. In initial efforts to understand the role of water in the aggregation process, we simulate an anatase nanocrystal in water. We characterize the orientation of water around the facets and edges of the nanocrystal, as well as its density distribution. We add surface hydroxyls and hydrogens to probe the role of water dissociation in the hydration of the nanoparticles. Our results indicate that nanoparticle forces in water can significantly differ from those in vacuum.

See more of this Session: Computational Studies of Self-Assembly
See more of this Group/Topical: Engineering Sciences and Fundamentals

New Phase Equilibria Data for Clathrate Hydrates of Tetrahydrofuran + Carbon Dioxide and Tetrahydrofuran + Nitrogen: Experimental Study of Reliability of Literature Data

Amir H. Mohammadi, Centre Energétique et Procédés, CEP-TEP, MINES ParisTech, Fontainebleau, France

Dominique Richon, Centre Energétique et Procédés, CEP-TEP, MINES ParisTech, Fontainebleau, France

Tetrahydrofuran (THF) is an organic water soluble hydrate former, which is used in gas separation, storage and transportation processes using gas hydrate technology. Gas hydrates or clathrate hydrates are solid crystalline compounds, which are formed from combination of water and small guest molecules. Clathrate hydrates can form crystalline structures I (sI), II (sII) and H (sH), where each structure is composed of a certain number of cavities formed by water molecules. The cooperation of two large and small guest molecules are normally required for the formation of structure II with heavy molecules or structure H. Heavy molecules occupy large cavities while small molecule(s) (called help gas) fill the remaining cavities. Tetrahydrofuran can form structure II with or without the presence of a help gas. Most of phase equilibrium data for tetrahydrofuran clathrate hydrates have been reported for the tetrahydrofuran + methane, tetrahydrofuran + carbon dioxide, tetrahydrofuran + nitrogen and tetrahydrofuran + hydrogen systems. A preliminary study shows that some of experimental data reported in the literature are not in good agreement, which require studying their reliability. In this work, we report new phase equilibrium data for clathrate hydrates of tetrahydrofuran + carbon dioxide and tetrahydrofuran + nitrogen. The experimental data were generated using an isochoric pressure-search method. The new equilibrium data are compared with the corresponding experimental data reported in the literature to study the reliability of the literature data.

See more of this Session: Experimental Phase Equilibria and PVT I
See more of this Group/Topical: Engineering Sciences and Fundamentals

Water Solubility in Supercritical Carbon Dioxide and Its Effect On Gas Phase Density

Farshad Tabasinejad, Chemical and Petroleum Engineering, University of Calgary, Calgary, AB, Canada

Y. Barzin, Chemical and Petroleum Engineering, University of Calgary, Calgary, AB, Canada

R. G. Moore, Chemical and Petroleum Engineering, University of Calgary, Calgary, AB, Canada

S. A. Mehta, Chemical and Petroleum Engineering, University of Calgary, Calgary, AB, Canada

K. V. Fraassen, Chemical and Petroleum Engineering, University of Calgary, Calgary, AB, Canada

Phase behavior of CO2-H2O system has been a subject of interest in many fields of science and engineering. CO2 injection based enhanced oil recovery methods, CO2 sequestration in saline aquifers and abandoned hydrocarbon reservoirs, and treatment of waste waters contaminated with dissolved acid gases are some of the subjects dealing with the thermodynamic behavior of the CO2-H2O system. Most of the studies on CO2-H2O system are devoted to the low to moderate pressure and temperature conditions. Only a few studies are dedicated to the high pressure and high temperature conditions.

Due to the paucity of experimental data at such conditions, a series of experiments have been conducted from 423 K to 478 K over a wide range of pressures from 3.8 MPa to 110 MPa. The solubility of water in carbon dioxide as well as the vapor phase density are measured. In order to measure the mass of water and volume of gas in the vapor phase, in each experiment the vapor phase of a L-V equilibrium system is transferred to an equilibrium flash separator equipped with a desiccant and a gasometer. The gas volume is converted to the mass and density based on the ideal behavior of gases at standard conditions.

A comparison between the new generated water solubility in CO2 data and literature data shows that despite the similar trend observed between the Tödheide and Franck[1] data and present data, at the same experimental conditions, lower concentrations of water in the vapor phase are measured in this work. However, the Gillespie and Wilson[2] low pressure data are in good agreement with the new experimental data. There is no consistency between the Takenouchi and Kennedy[3] data and generated data in this work.

The produced data for vapor phase density in this study accurately follow the characteristic trend. Due to the interconnectivity between density data and water content data in the designed experimental procedure, it is believed that the new water content data obtained in this experimental study are reliable.

References:

1. Todheide, K., Franck, E. U., (1963). "Das Zweiphasengebiet und die Kritische Kurve im System Kohlendioxid + Wasser bis zu Drucken von 3500 bar." Z. Physik. Chem. Neue Folge., 37: 387-401.

2. Gillespie, P.C., Wilson, G.M. (1982). "Vapor-liquid and liquid-liquid equilibria: water-methane, water-carbon dioxide, water-hydrogen sulfide, water-n-pentane, water-methane-n-pentane. GPA Research Report RR-48.

3. Takenouchi, S., Kennedy, G. C., (1964). "The Binary System H2O-CO2 at High Temperatures and Pressures." American Journal of Science, 262: 1055-1074.

See more of this Session: Experimental Phase Equilibria and PVT I
See more of this Group/Topical: Engineering Sciences and Fundamentals

High-Pressure Micellar Solutions of Symmetric and Asymmetric Styrene-Diene Diblocks in Compressible near Critical Solvents: MP and CP Respond but MCP Insensitive to Copolymer Molecular Weight, Concentration, and Block Ratio Changes

Winoto Winoto, Chemical & Petroleum Engineering, University of Wyoming, Laramie, WY

Sugata P. Tan, Department of Chemical and Petroleum Engineering, University of Wyoming, Laramie, WY

Youqing Shen, University of Wyoming, Laramie, WY

Maciej Radosz, University of Wyoming, Laramie, WY

Kunlun Hong, Center for Nanophase Materials Science, Oak Ridge National Laboratory, Oak Ridge, TN

Jimmy Mays, Center for Nanophase Materials Sciences, Oak Ridge National Laboratory, Oak Ridge, TN

High-pressure solutions of styrene-diene copolymers in compressible solvents, such as polystyrene-block-polybutadiene and polystyrene-block-polyisoprene in propane or propylene, exhibit a robust micellar region that grows in pressure-temperature coordinates with increasing copolymer concentration, molecular weight, and styrene/diene block ratio. This happens because, while the micellization pressure strongly increases with increasing copolymer concentration, molecular weight, and styrene/diene block ratio, the micellar cloud pressure (the pressure at which the micelles aggregate and precipitate from solution) is largely insensitive to these variables. In other words, neither the block size nor the block ratio nor the copolymer concentration seems to affect much the copolymer separation from solution in the micellar region.

See more of this Session: Experimental Phase Equilibria and PVT I
See more of this Group/Topical: Engineering Sciences and Fundamentals

Vapor-Liquid Equilibria and Saturation Densities of Carbon Dioxide + 1-Pentanol System From 323 to 423 K

Luis A. Galicia-Luna, INSTITUTO POLITECNICO NACIONAL-ESIQIE, Mexico, Mexico

Octavio Elizalde-Solis, Ingeniería Química Petrolera, INSTITUTO POLITECNICO NACIONAL-ESIQIE, Mexico, Mexico

Luis E. Camacho-Camacho, Laboratorio de Termodinámica-ESIQIE, INSTITUTO POLITECNICO NACIONAL-ESIQIE, Mexico, Mexico

Abstract

Information about phase equilibria and volumetric properties is essential in thermodynamics for the process design. Experimental data of both properties obtained in the same experiment is not an easy procedure even at high pressure. The simultaneous determination of these data comes more important for a better correlation and prediction.

The aim of this work is to obtain the vapor-liquid equilibria and saturation densities for the carbon dioxide + 1-pentanol mixture from (323 and 423) K. The apparatus is based on the static-circulation method and has been already tested for other alkanols with lower molecular weight [1]. The vapor-liquid equilibrium was measured in a stainless steel cell connected to a gas chromatograph via a ROLSI sampler-injector as previously reported [2]. This cell is connected to a thermoregulated Anton Paar densimeter DMA 512P for measuring the saturation densities carbon dioxide + 1-pentanol mixture.

References

[1] O. Elizalde-Solis, and L.A. Galicia-Luna, Simultaneous determination of the vapor-liquid equilibria and saturation densities of carbon dioxide + alkanol systems, Proceedings 17th Symposium on Thermophysical Properties - NIST, Boulder, CO-USA, June 21-26, 2009

[2] L.A. Galicia-Luna, A. Ortega-Rodriguez, D. Richon, New apparatus for the fast determination of high-pressure vapor-liquid equilibria of mixtures and of accurate critical pressures, J. Chem. Eng. Data, 45 (2000) 265-271.

* To whom correspondence should be addressed

E-mail: lgalicial@ipn.mx

Tel: (52) 55 5729-6000, Ext. 55133,

Fax: (52) 55 5586-2728

See more of this Session: Experimental Phase Equilibria and PVT I
See more of this Group/Topical: Engineering Sciences and Fundamentals

Compressed Liquid Densities for Binary Mixtures of n-Hexane + n-Decane and n-Heptane + n-Decane From 313 to 363 K

Luis A. Galicia-Luna, INSTITUTO POLITECNICO NACIONAL-ESIQIE, Mexico, Mexico

Octavio Elizalde-Solis, Ingeniería Química Petrolera, INSTITUTO POLITECNICO NACIONAL-ESIQIE, Mexico, Mexico

Rodolfo Quevedo-Nolasco Sr., Laboratorio de Termodinámica, ESIQIE, Politécnico Nacional, Mexico, D.F., Mexico

Jose Domenzain-Gonzalez, INSTITUTO POLITECNICO NACIONAL-ESIQIE, Mexico, Mexico

Abstract

PñT properties in the compressed liquid phase were determined for n-hexane + n-decane and n-heptane + n-decane mixtures at temperatures from 313 to 363 K and pressures up to 25 MPa. A vibrating tube densimeter (VTD) was used to measure these densities. The classical calibration method based on two reference fluids (water and nitrogen) was used for the calibration of the VTD. The experimental densities for each n-alkane were in agreement with those reported in literature as expected. A 6-parameter Equation was used to correlate the mixture densities. This equation represents data within the experimental uncertainty of 0.2 kg/m3.

See more of this Session: Experimental Phase Equilibria and PVT I
See more of this Group/Topical: Engineering Sciences and Fundamentals

PLANNING MISCIBILITY TESTS AND GAS INJECTION PROJECTS FOR FOUR MAJOR KUWAITI RESERVOIRS

Moudi Al-Ajmi[1], Osamah Alomair[2], Adel Elsharkawy[2]
1 Kuwait Oil Company, 2 Kuwait University

ABSTRACT

The escalating oil demand and maturity of most of the giant oil fields in the world, especially in the Middle East, the techniques for improving oil recovery have became more feasible and essential. Kuwait long term strategy is to increase oil production to meet marked demand. Currently, miscible gas injection is considered for enhancing oil production from Kuwaiti oil reservoirs. A key parameter for assessing the applicability of gas injection for a given reservoir is the minimum miscibility pressure (MMP). In this paper various miscibility experiments for planning gas injection projects in major producing fields in Kuwait are discussed. These experiments include swelling tests, slim-tube tests, and core flooding studies. These tests are useful tool for screening of the potential reservoirs for improving their future oil production and for developing suitable EOS for planning gas injection projects of the chosen fields.

INTRODUCTION

Many definitions of miscible displacement have been widely discussed in the literature (Benham et. al., 1965, Stalkup, 1983, Holm, 1987, and Lake, 1989). These definitions clearly convey a consensus that miscibility refers to the absence of an interface between the injected fluids and the reservoir crude oil. The absence of an interface means, in terms of a measurable variable, the value of the interfacial tension between the displacing and displaced fluids is zero.

Miscible displacement is only achieved at pressures greater than a certain minimum. This minimum is called the Minimum Miscibility Pressure (MMP). The MMP of a gas/oil pair is traditionally determined by flooding an oil-saturated slim-tube with a gas at four or five different pressures; the MMP is defined as the lowest pressure at which essentially all oil available for recovery can be displaced by 1.2 of the pore volume of solvent injected. This pressure can be located graphically by the intersection of two lines that define both an immiscible and miscible performance regimes on a plot of recovery versus pressure, or recovery versus composition.

The methods for estimating MMM are classified into experimental and calculation methods. The experimental methods for measuring MMP are; slim tube apparatus, rising bubble apparatus, PVT Cell, variable interfacial tension, and other methods. Slim-tube method is the most common and has been accepted as the standard method to determine MMP. The MMP calculation methods are divided into; correlation, numerical (simulation), and analytical (EOS) methods. Because of their improved speed, analytical methods offer significant promise for developing improved fluid correlations and for use in compositional streamline simulations.

Minimum Miscibility Pressure Experimental Methods

Slim-Tube Test

In this method, the miscibility conditions are determined by conducting the displacements at various pressures or gas enrichment levels in the oil-saturated Slim-tube and monitoring the oil recovery. Then, the oil recovery is plotted against the pressure (Stalkup, 1983). The minimum miscibility pressure is defined as the pressure at which the oil recovery versus pressure curve shows a sharp change in slope. The MMP is traditionally defined as the lowest pressure at which essentially all oil available for recovery can be displaced by 1.2 PV solvent injected. The disadvantages of this method are; it is very time consuming, expensive, does not account for third fluid water, and several points are required to establish the MMP (a minimum of five points are recommended). Furthermore, Slim-tube may give a lower MMP that it actually is because the way the porous medium is packed. On the other hand, Slim-tube test can mimic porous medium and hence are capable of representing multi-contact fluid dispersions mechanism due to mixing in the porous medium.

Rising Bubble Apparatus (RBA)

In the rising bubble experiments, the MMP is inferred from the pressure dependent behavior of rising bubbles. The MMP is determined from the observations of changes in shape and appearance of bubbles of the injected gas as they rise through a thin column of crude oil. The pressure at which a rising gas bubble vanishes in a column of oil is termed as the MMP (Christianson and Haines, 1984). This method is considerably faster with low cost and requires smaller quantities of fluids, compared to Slim-tube. However, the limitations of this technique are that there is some subjectivity associated with the MMP interpretation and that very little quantitative information (changes in composition, interfacial tension, and displacement efficiency) is provided.

Vanishing Interfacial Technique (VIT)
To overcome most of the disadvantages of Slim-tube and RBA approaches to determine MMP, Vanishing Interfacial Technique (VIT) has been developed (Rao, D.N., 1997). Although the experimental set up is quite elaborate, the technique itself is quite simple in terms of its principles. This method is based on the concept that, at miscibility, the value of IFT between the two phases is zero. In this method, the IFT is measured between the injected gas and crude oil at reservoir temperature at varying pressures or enrichment levels of gas phase. The MMP is then determined by extrapolating the plot between IFT and pressure to zero interfacial tension.

Core Flooding
Core flooding tests are used to determine MMP under reservoir conditions. Since an actual reservoir rock is used, the result takes into account "truer" pore throat sizes, and allows the presence of water phases. However, core flood tests are more expensive and time consuming. Moreover, it is often difficult to get good quality and representative core of sufficient length to carry out the tests. Short cores are often inadequate to generate the conditions of multi-contact miscibility.

Minimum Miscibility Pressure Calculation Methods
The calculation methods for MMP can be grouped into three methods; correlation, analytical, and numerical. Brief discussions of each method are listed in the following sections.

Correlation Methods
Any MMP correlation should account for each parameter known to affect the MMP, be based on thermodynamic or physical principles that affect the miscibility of fluids, and be directly related to the multiple contact miscibility process (Enich, et. al., 1985 and 1988). For an initial and quick estimate, operators use correlation currently available in the literature. For screening purposes, they give a fair first guess depending on the data used. Moreover, they are inexpensive and can be detained by simple hand calculation. The success of the correlations is usually limited to the composition range in which these correlations were developed. The MMP correlations fall into two categories: the pure and impure CO_2; while the other category treats MMP's of other gases.

Analytical Method
The analytical method, assuming that appropriate equation of state (EOS) based fluid phase behavior characterization is available. The reliability of any EOS depends on the quality of the data used and the oil composition. The analytical MMP calculation methodology offers a viable alternative for experimental method. It is based on the analytical theory of multi-component gas injection processes. Because of their improved speed, analytical methods offer significant promise for developing improved fluid correlations and for use in compositional streamline simulations. Examples of the analytical methods are; Equation of State Model, Ternary Diagram (Benham et. al., 1959), Multicell Algorithm (Metcalfe et al., 1973, Cook et al., 1969, Zick, 1993, and HØier and Whitson, 2001), Single Mixing Cell Algorithm (Kuo, 1985, Luks et al., 1987, and Wang and Orr, 1998), the Tie-Line Algorithm (Nouar and Flock, 1986, Whitson and Michelsen., 1989, and Wang and Orr., 1997), the C/V Mechanism (Zick, 1986), the Lie Line Algorithm and Method-of Characteristics (MOC) Model (Dindoruk et al., 1997, Johns and Orr, 1996, Wang and Orr, 1998, Jessen, K. and Michelsen, M.L., 1997, Jessen et al.,1998, Wang and Peck,2000, and Yuan and Johns, 2002)

Numerical Method (Simulation)
Among the numerical MMP calculation approaches (third method), one-dimensional compositional simulation is shown to be able to predict MMPs that are consistent with Slim-tube test data, provided that appropriate fluid phase behavior characterization is available and care is taken to account for the effect of numerical dispersion (Zick , 1986, and Stalkup, 1982 and 1987). There are other variations of the compositional simulation approach. For example, the multiple mixing cell method, which has much in common with compositional simulation, can also be used to calculate MMPs accurately, assuming that enough number of cells is used to mitigate the effect of numerical dispersion. The simulation type approaches are generally time consuming, considering the fact that a large number of grid blocks or mixing cells are required to limit the effect of numerical dispersion. In addition, when the number of pressure points at which simulations are performed is not large enough to obtain a reasonably well

defined recovery curve, the numerically calculated MMPs are subject to the visual interpretation of the recovery curve.

OBJECTIVES

The objective of this research is to use various PVT data, miscibility tests and EOS simulations to screen Kuwaiti oil field for possible gas injection. The outcome of this research is to select the optimum gases for injection into major producing oil field in Kuwait. This project is necessary for the following reasons:

- EOR / IOR becomes vital in the petroleum industry
- The successful design and implementation of a miscible gas injection project is dependent upon the accurate determination of MMP, the pressure above which the displacement process becomes multi-contact miscible or the lowest possible, operating pressure at which the gas can miscible displace oil
- If CO2 is selected as one of the injection gases, CO2 emission from power plants and various industries will be used for injection into the selected reservoirs, thus, reduction CO2 emission into the atmosphere.
- Kuwait is planning to increase its future production.
- Oil demand is expected to increase in the near future

EXPERIMENTAL AND RESULTS

Bottom hole and separator samples were taken from four major Kuwaiti reservoirs (Mauddud, Lower Burgan, Zubair, and Minagish Oolite) to study miscibility characteristics of each reservoir. Compositional and PVT analysis were conducted for each sample. In addition, swelling, slim tube displacement, and core displacement test were performed for most of the samples. Finally, simulation studies with 1D-Compositional simulation using ECLIPSE is developed from matched EOS to validate slim-tube test data. The latter is used to study the potential of using other gases for injection into four major reservoirs described in this paper.

Maudad Reservoir (sample1)

Bottom-hole samples as well as separator oil and gas samples are obtained form Mauddud reservoirs. The separator crude oil and gas sample were combined and compared to bottom-hole samples for quality check. Compositional analysis of bottom-hole and recombined samples were very close to be used in all miscibility and swelling test. Mauddud reservoir has a bubble point pressure (BPP) of 1621 psia at 169°F. Other PVT properties such as molecular weight of crude oil, molecular weight of C_{7+}, viscosity, oil formation volume factor, and solution gas oil rate at bubble point pressure are given in **Table 1**. The compositional data of bottom-hole and recombined samples are shown in **Table 2**. Composition of the various gases used in slim tube miscibility test, swelling test, and simulation studies are shown in **Figure 1**. Oil production from Mauddud started in the early of 1960's. The reservoir pressure gradually decline because of the depletion behavior. A field scale water flooding (Inverted 9-spot) in the reservoir was performed in the year 2001. The reservoir pressure is maintained at least 1000 psi above the bubble point pressure. The current reservoir pressure is 3200 psi in the water flood pattern and is considered for miscible gas injection. The reservoir has been screened considering several EOR methods and has indicated that gas injection is the optimum EOR techniques. Therefore it was decided to do laboratory tests and simulation studies to investigate the feasibility of miscible gas injection.

Laboratory Tests

A series of laboratory tests including standard phase behavior (PVT-tests), swelling test, slim tube displacement test, and a core displacement tests were conducted to study the miscibility characteristic of Mauddud reservoir using **Sample 1** for various injection gases. Composition of various gases used in slim tube miscibility tests, swelling tests, core flooding experiments and simulation studies are shown in **Figure 1**. The core flooding experiments were conducted to evaluate the gas displacement efficiency at the current reservoir pressure after water flooding.

Slim Tube Test

The recombined reservoir fluid sample of **Sample 1** was used in the slim tube test as shown in **Figure 1**. Injecting gas RG, which is a synthesis gas that is similar to separator gas of **Sample 1**, has 57% methane, 15% ethane and 12% methane was used in the slim tube test. The gas was injected into slim tube at several pressures. The injection pressures were selected to cover the immiscible and the miscible region behavior of oil recovery during displacement. Oil recovery at each injection pressure after 1.2 pore volume injection fluid is recorded and plotted versus injection pressure **Figure 2**. Slim tube oil recovery indicates no significant increase at pressure higher than 4500 psi which is considered to be the minimum miscibility pressure (MMP). This miscibility pressure is more than 1000 psi above the current average reservoir pressure. Thus, unless the reservoir is pressurized to 4500 psia, RG gas will not be able to achieve miscibility. It is unlikely that RG gas will be considered for gas injection into Mauddud reservoir because of the high miscibility pressure.

Swelling Test

A bottom hole sample of **Sample 1** from Muaddud reservoir was used in the routine PVT tests and swelling test (**Figure 1**). Swelling of **Sample 1** by RG gas was studied at 169°F. The swelling data are shown in **Figure 3**. Liquid factor of total sample volume at various mole% of injected gas at several pressures are conducted. The test indicates that during RG gas injection multiphase behavior occurred and supported by formation of asphaltenes precipitate during the swelling test.

Core Flooding Test

Miscible gas flooding and water alternating gas (WAG) flooding experiment were conducted on core plug taken from well **Sample 1** that has been core in 2003. The test was conducted to study displacement efficiency of gas injection after water flooding. The injection gas used was impure CO_2 having 60% CO_2 and 40% NGL from second stage separator from **Sample 1**. Oil recovery by water flooding and gas injection which followed water flooding is shown in **Figure 4.** Nearly 35% of the oil in the core plug was recovered after 1-PV of water injections. Ultimate oil recovery by water flooding was 40% of original oil in the core plug. Nearly 36% of additional oil was recovered during gas water injection. The results from core flooding experiment indicated that gas injection and WAG are excellent potential EOR for Mauddud reservoir.

Ternary diagram

Equation of state was used to construct the ternary phase diagram for crude oil samples **Sample 1** from Muaddud reservoirs at 169°F. **Figure 5** shows reservoir fluid composition as well as various compositions of injection gases for the ternary diagram at 4500 psi. A tie line drawn from the initial oil composition and tangent to the two- phase envelope shows the minimum composition of the gas required to achieve miscibility with reservoir fluid. At this pressure, the injected RG gas is to the right of the limiting tie line, therefore, multiple contact miscibility develops during displacement as the oil composition occur to the lift of the limiting tie line. Ternary phase diagram for impure CO_2 (60% CO_2 + 40% NGL) also fall to the night of the limiting tie line and achieving miscibility but at lower pressures.

1D-compositional simulation

Slim tube displacement tests were simulated by 1D-compositional simulation using ECLIPSE. A series of 1D slim-tube simulations was performed for **Sample 1** reservoir fluids. The reservoir fluids were characterized using 12-component to describe the reservoir fluids. Various gas injection tests considered to study if miscibility conditions could be achieved at the current reservoir pressure. In these simulation studies a slim tube that is 45 ft long is used. The slim tube were divided into 300 grid blocks, The first simulations test was to investigate the miscibility condition during RG gas injection to compare the simulated result with the measured MMP from slim tube. The simulate studies was also conducted to estimate the MMP for other gases such as pure CO_2, improve CO_2 and other shown in **Table 3**. **Figure 6** shows 1D slim tube displacement of **Sample 1** reservoir fluid with RG gas at various pressures. At 3000 psia which is the current reservoir pressure level, the displacement is immiscible. Simulation of 1D displacement test at 4000 psia shows near miscibility and in the range of 4500 to 5000 psia displacements indicate multiple contact miscibility which is in good agreement with measured value of MMP by slim tube. Measured and simulated results of minimum miscibility pressure for **Sample 1** reservoir fluid for various gases are shown in **Table 3**. The results indicate that CO_2 and impure CO_2 are the best candidate gases to be considered for gas injection into Mauddud reservoir to achieve miscibility given the current reservoir pressure.

Lower Burgan (Sample 2)

One bottom-hole sample was collected from Lower Burgan reservoir. Oil production from Lower Burgan started in late 1950's. The aquifer support of this sandstone reservoir maintained the reservoir pressure at or very close to the initial reservoir pressure of 4,057 psia. The current reservoir pressure is at least 1,000 psia above the bubble point pressure of 2,500-2,800 psia. The potential for application of EOR methods, including gas miscible flooding, is considered in this reservoir.

Laboratory Test

One bottom-hole sample was also collected from lower Burgan as shown in **Table 1**, **Sample 2**. A reservoir condition analysis was performed on the crude oil sample along with conventional PVT studies and a swelling test using CO_2 as injection gas. No slim-tube tests were performed for this reservoir. Compositional analysis of reservoir fluid samples is given reservoir **Table 2**.

Swelling Tests

A bottom hole sample of **Sample 2** from Lower Burgan reservoir was used in the routine PVT tests and swelling test. Swelling of **Sample 2** by CO_2 gas was studied at 195°F. Liquid factor of total sample volume at various mole% of injected gas at several pressures are conducted. The samples has abnormally lower BPP (150 to 400 psia) and higher oil viscosity (40% higher), the crude oil/CO_2 gas pair in the target reservoir showed advantageous properties for the gas flooding in the phase behavior tests including the swelling tests. This is terms of (with reference to 33.6% moles CO_2 gas injection case)The test indicates a high swelling factor in the order of 16% increase in oil volume and a viscosity reduction of around 59% compared to the original fluid viscosity. Furthermore, gas injection processes in Lower Burgan are likely to be subject to multiphase behaviours. This was evidence by the asphaltenes formation during the swelling test (although not quantified).
Results of swelling tests are used to tune the equation of state. Tuned equations of state as well as other miscibility correlations are used to calculate the minimum miscibility pressure (MMP) for Lower Burgan for various gases. The results are shown **Table 3** indicates the following:

1. Pure and impure CO_2 gases have the lowest MMP with Lower Burgan crude oil compared to other gases such as RG, N2 and LG1.
2. The MMP of RG, LG1, and N2 are approximately 2 times CO_2 MMP.
3. More important is that the MMP of CO_2 Lower Burgan crude oil samples can be achieved at the current reservoir pressure, unlike RG, LG1, and N_2.

Zubair Reservoir (Sample 3)

Oil production form Zubair reservoir started in late 1950's. The reservoir pressure gradually decreased in the middle layer (Middle Zubair) in this sandstone reservoir despite of the minor aquifer support. A filed scale peripheral water flooding started in this reservoir in 1999. Within the next decade, the potential for application of gas miscible flooding process is considered. The reservoir pressure is well maintained at relatively high pressure. The gas injection study was conducted to investigate the feasibility of miscible gas injection in this reservoir. It was conducted on fluid samples from Upper Zubair layer which receives moderate to strong aquifer support. The initial reservoir pressure is 4,585 psia and the BBP is 2,660 psia. Its current reservoir pressure is around 4050 to 3,950 psia. The future operating pressure would be in the average of 3,400 psia.

Laboratory Tests

Separator samples consisting of six bottles of stock tank oil along with four bottom-hole samples were taken from the Upper Zubair reservoir. The samples were used for compositional and PVT analysis as well as slim-tube and swelling tests. In addition to the conventional PVT analysis, the recombined oil sample (**Sample 3A**) was used for CO_2 swelling and slim-tube studies to determine CO_2 MMP with Upper Zubair crude. Additionally, the recombined sample, **Sample 3B**, was used to conduct a CO_2 and hydrocarbon lean gas (LG2) swelling tests.

Fluid Composition

The resultant recombined fluid composition (**Sample 3A** and **3B**) consist of over 33% methane (C_1), 8% ethane (C_2), 6% propane (C_3) and over 35% heptanes plus (C7+). These compositions represent a typical fluid type of Upper Zubair reservoir, though **Sample 3B** shows different fluid composition than **Sample 3A**, mainly due to the different methods of recombination.

Recombination Methods and BPP Comparison

Sample 3A has a BPP of 3,025 psia while sample RSS02 has a BPP of 2,473 psia at 195°F. This indicates a fundamental difference between the two recombined samples; the JNOC sample (**Sample 3A**) has oil BPP of 552 psia higher than what was reported by Schlumberger (**Sample 3B**). This is believed to be a difference in the recombination methods which have resulted in 5.66 mole % difference in C_1 content between the two samples; however, the reported compositions for each sample is consistent with the reported bubble point.

Swelling Tests

A bottom hole sample of **Sample 3C** from Lower Burgan reservoir was used in the routine PVT tests and swelling test. Swelling of **Sample 3C** by CO_2 gas was studied at 195°F. The swelling data are shown in **Figure 7**. Liquid factor of total sample volume at various mole% of injected gas at several pressures are conducted. The test indicates a high swelling factor in the order of 16% increase in oil volume and a viscosity reduction of around 59% compared to the original fluid viscosity. Furthermore, gas injection processes in Lower Burgan are likely to be subject to multiphase behaviours. This was evidence by the asphaltenes formation during the swelling test (although not quantified). The swelling tests, specifically on **Sample 3B** had some experimental issues; however, all the swelling tests have showed the following:

1. The saturation pressures are generally lower in the case of CO_2 experiment compared to LG2 test. LB fluid properties are very similar to Upper Zubair reservoir. **Sample 3C** CO_2 swelling test data were compared with Zubair, which showed very good matching.
2. The oil recovery at different pressures is plotted versus pressure. From this plot, the CO_2 MMP pressure was found to be 4,000 psig.
3. CO_2 near-miscible displacement was observed between 3,100-3,200 psig. In the near-miscible process, the recovery is nearly as high as in a miscible displacement even through the fluid flows in porous media as two phases, i.e. gas and liquid.
4. Overall, the comparison shows significant similarity between Lower Zubair and Upper Zubair reservoir fluid/ CO_2 solvent interaction.

Slim Tube Test

The recombined reservoir fluid **Sample 3A** was used in the slim tube test. CO_2 was injected into slim tube at several pressures. The injection pressures were selected to cover the immiscible and the miscible region behavior of oil recovery during displacement. Oil recovery at each injection pressure after 1.2 pore volume injection fluid is recorded and plotted versus injection pressure, **Figure 8.** Slim tube oil recovery increases as a function of increasing the injection pressure. The plot shows two trend of increasing oil recovery versus pressure. The A trend of immiscible displacement which is very sensitive to pressure and the B trend which is less sensitive to pressure. The MMP (4000 psia) is taken as the break-over point of the oil recovery as a function of pressure. This miscibility pressure is close to the current average reservoir pressure of 4050 psia. Thus, if CO_2 is considered for miscible gas injection, unlike other gases there is no need to pressurize the reservoir.

The MMPs of Upper Zubair reservoir with different gases are shown in **Table 3**. Slim-tube column characteristics were not reported, therefore, 1D simulation was not possible to validate the actual results. Never the less, the results of the matched EOS and the "most reliable" correlations indicate the following:

- Like Mauddud and Lower Burgan reservoirs, CO_2 gas is followed by RG have the lowest MMP with Upper Zubair oil compared to N2 and LG1.
- Further testing is required to investigate the MMP of CO_2 with Zubair oil, especially given the uncertainty associated with the experiments.

Minagish oolite (Sample 4)

The Minagish Oolite is a thick lower Cretaceous undersaturated carbonate oil reservoir. It is a mature but relatively undeveloped reservoir. Since discovery in 1959, it has produced 10% of its initial oil in place under a combination of natural depletion, gas re-injection, and aquifer drive. Initial reservoir pressure had declined by about 450 psia prior to the Gulf war in 1990. The well blowouts following the war caused a significant pressure drop of another 700 psia. Following the blowouts, plans were made to re-develop the reservoir and increase its production rate. Since the existing well inventory and the loss of the gas re-injection facility could not sustain the desired plateau rate, additional field development was required including EOR. The original reservoir pressure was 4,750 psia while the current reservoir pressure is ranging from 3,750 to 3,950 psia.

Laboratory Tests

Four bottom-hole samples have taken from the Oolite formation. The samples were submitted PVT laboratory. The reservoir has a bubble point pressure of 1930 psia at 198°F. A good quality crude oil of 33.8 degree API gravity is produced from this reservoir. PVT properties of crude oil such as molecular weight, molecular weight of $C7^+$, asphaltenes content, viscosity, oil formation volume factor and solution gas oil rate at bubble point pressure are given in **Table 1**. The compositions of the bottom hole fluid samples are consistence. A composition representing a typical reservoir fluid type of Oolite formation is shown in Table 2.

Based on the similarity in the compositions of collected reservoir fluid samples, it was decided to commingle two samples for the hydrocarbon gas injection tests, and two fluid samples for the CO_2 gas injection tests. One important observation that can be made when comparing Minagish Oolite sample with the other reservoirs fluid samples (Mauddud, Lower Burgan, and Upper Zubair) is that it is the only sample that showed the presence of Hydrogen Sulfide. In swelling tests both pure CO_2 and hydrocarbon gases were used. However, in slime-tube tests only pure, impure CO_2 and hydrocarbon gas were used. The used synthetic "hydrocarbon gas" composition consists mainly of 65.09% methane, 20.22% ethane, and 14.68% propane.

Swelling Test

A bottom hole sample (**Sample 4**) from Minagish Oolite reservoir was used in the routine PVT tests and swelling test. Swelling by hydrocarbon gas was studied at 189°F. The Oolite hydrocarbon swelling showed a BPP close to Mauddud sample. The hydrocarbon swelling test data showed much of similarity in the swelling factors with Mauddud sample. Oolite CO_2 swelling showed significant similarity with Lower Burgan data.

Slim Tube Test

The reservoir fluid **Sample 4** was used in the slim tube test where pure CO_2 as well as hydrocarbon gas is used in two separate experiments. The gas was injected into slim tube at several pressures. The injection pressures were selected to cover the immiscible and the miscible region behavior of oil recovery during displacement. Oil recovery at four injection pressures after 1.2 pore volume injection fluid is recorded and plotted versus injection pressure **Figure 10**. Oil recovery at 2050 psia and at 2500 psia represents immiscible displacement as indicated by presence of two phase flow through the visual cell and low ultimate oil recovery that is highly dependent on injection pressure. Injection at 3300 psia and 4700 psia show miscible displacement, and little change of oil recovery as a function of increasing the injection pressure. The lack of slim-tube measurements between 2,500 and 3,300 psia is a source of uncertainty in the estimation of CO_2 MMP. The two trends intersect at 2665 psia that could be considered as the MMP. Because of insufficient injection pressure points between 2500 and 3300 psia, the minimum miscibility pressure (MMP) could also be considered in the range of 2650 to 3300 psia. In either interpretation, the CO2 MMP is still below the current reservoir pressure which is favorable for miscible gas injection. In addition to CO_2 , a hydrocarbon gas was used in separate slim tube tests. The slim tube tests using hydrocarbon gas shows an MMP of 3,850 psia which is very close or slightly higher than the current reservoir pressure (3,750-3,950 psia). The produced fluid composition shows an immiscible displacement process at an injection pressure of 3,315 psia. Also the oil recovery produced GOR, and liquid density versus PV injected at 3315 psia support the idea that this test is immiscible or near miscible.

Slim tube displacement tests on **Sample 4** were simulated by 1D-compositional simulation using ECLIPSE. Various gas injection tests were considered to study if miscibility conditions could be achieved at the current reservoir pressure. Among the gases considered are: RG, impure CO_2, N_2, LG and hydrocarbon gas (HG). The estimated MMP for these gases are shown in **Table 3**. The MMPs of Oolite with different gases from the actual experiment, matched EOS, and 1D simulation. The results indicate that pure and impure CO_2 gases have the lowest MMP with Oolite oil compared to HC gas, RG, LG and N2. More importantly is that the MMPs of CO_2 gas with Oolite are achievable with the current reservoir pressure. However, if other gases such as HC gas, RG, LG1, and N2 are considered the reservoir pressured by water flooding which is neither practical nor economical.

SUMMARY AND CONCLUSIONS

Major Kuwaiti oil reservoirs have been under considerations for improving their future oil recovery via applications of several EOR processes. It was found that gas injection is the optimum EOR techniques for many of the oil reservoirs. Therefore it was decided to do laboratory tests and simulation studies to investigate the feasibility of miscible gas injection.

1) Slim- tube tests indicated that MMP for Maudad is 4500 psia. This miscibility pressure is 1000 psia above the current average reservoir pressure. Thus, unless the reservoir is pressurized to 4500 psia, RG gas will not be able to achieve miscibility. Swelling tests also showed the potential of asphaltenes formation during injection of RG gas into Maudad reservoir. It is unlikely that RG gas will be considered for gas injection into Mauddud reservoir because of the high miscibility pressure and the expected asphaltenes formation. Core flooding experiment of Maudad reservoir indicated that gas injection and WAG are excellent potential EOR. Nearly 36% of additional oil was recovered during gas water injection. Simulated slim tube displacement for various gases indicated that CO_2 and impure CO_2 are the best candidate gases to be considered for gas injection into Mauddud reservoir to achieve miscibility given the current reservoir pressure.

2) Swelling tests were conducted of Lower Burgan with CO_2. These tests indicates a high swelling factor in the order of 16% increase in oil volume and a viscosity reduction of around 59% compared to the original fluid viscosity. Furthermore, CO_2 gas injection processes in Lower Burgan are likely to be subject to multiphase behaviours. This was evidence by the asphaltenes formation during the swelling tests.
EOS simulated miscibility tests indicated that Pure and impure CO_2 gases have the lowest MMP with Lower Burgan crude oil compared to other gases such as RG, N2 and LG1. The MMP of RG, LG1, and N2 are approximately 2 times CO_2 MMP. More important is that the MMP of CO_2 Lower Burgan crude oil samples can be achieved at the current reservoir pressure, unlike RG, LG1, and N_2.

3) Separator oil and gas samples from Zubair reservoir are recombined and used to study swelling of this reservoir fluids with CO_2. The swelling test indicated high swelling factor (16%), great viscosity reduction (60% of original viscosity), and the formation of asphaltenes deposits. Slim tube tests showed that the CO_2-MMP of zubair reservoir fluids is 4000 psia which is very close to the current reservoir pressure. EOS simulation of Zubair reservoir with various gases indicated that CO2 is the best candidate gas for injection into this reservoir as it has the lowest MMP and there is no need to pressurize the reservoir.

4) Bottom-hole samples from Managish reservoir were used to study PVT, swelling and slim-tube displacement with pure CO_2 and hydrocarbon gas. The slim tube test has limited injection points. Therefore, the MMP is considered to be in the range of 2650 to 3300 psia for CO_2. CO_2-MMP is below the current reservoir pressure which is favorable for miscible gas injection. Slim-tube tests with hydrocarbon showed that MMP is 3,850 psia which is very close or slightly higher than the current reservoir pressure (3,750-3,950 psia).

5) Compositional simulation studies were used to study miscibility conditions for various gases injection at the current reservoir pressure. The results indicate that pure and impure CO_2 gases have the lowest MMP with Oolite oil compared to HC gas, RG, LG and N2. More importantly is that the MMPs of CO_2 gas with Oolite are achievable with the current reservoir pressure. However, if other gases such as HC gas, RG, LG1, and N2 are considered the reservoir pressured by water flooding which is neither practical nor economical.

REFERENCES

Benham, A. L., Dowden, W. E. and Kunzman, W. J. (1965). Miscible Fluid Displacement-Prediction of Miscibility. *SPE of AIME*. 123.

Stalkup, F. I. (1983). Miscible Displacement. *SPE Monograph Series*. SPE of AIME. 8, 14-17, 22-29.

Holm, L. W. (1987). Miscible Displacement. Petroleum Engineering Handbook, SPE. 45-1.

Lake, L. W. (1989). Enhanced Oil Recovery. *Prentice-Hall Englewood Cliffs, NJ*. 234.

Christiansen, R. L. and Haines, H. K. (1987). Rapid Measurement of Minimum Miscibility Pressure with the Rising-Bubble Apparatus. SPE 13114.

Rao, D.N. (1997). A New Technique of Vanishing Interfacial Tension for Miscibility determination. *Fluid Phase Equilibria*.

Enich, R. M., Holder, G. D., and Mosri, B. I. (1985). An Equation of State Correlation for the Minimum Miscibility Pressure of Carbon Dioxide Floods. SPE 14518.

Enich et al. (1988). A thermodynamic correlation for the MMP in CO_2 flooding of petroleum reservoirs. SPERE, February, 81-92.

Benham, A. L., Dowden, W. E. and Kunzman, W. J. (1959). Miscible Fluid Displacement-Prediction of Miscibility. *Trans. AIME*. 216, 388-398

Metcalfe, R.S., Fussell, D.D. and Shelton, J.L. (1973). A Multicell Equilibrium Separation Model for the Study of Multiple Contact Miscibility in Rich Gas Drives. *SPEJ.* 13, No.3, 147-155.
Cook, A. B. et al. (1969). Effects of Pressure, Temperature, and Type of Oil on Vaporization of Oil During Gas Cycling. RI 7278.
Kuo, S. S. (1985). Prediction of Miscibility for the Enriched Gas Drive Process. SPE 14152.
Luks, K. D., Turek, E. A. and Baker, L. E. (1987). Calculation of Minimum Miscibility Pressure. SPE 14929.
Wang, Y. and Orr, F. M. (1998). Calculation of Minimum Miscibility Pressure. SPE 39683.
Nouar, A. and Flock, D. L. (1986). Prediction of the Minimum Miscibility Pressures of a Vaporizing Gas Drive. SPE 15075.
Whitson, C. H. and Michelsen, M. L. (1989). The Negative Flash. *Fluid Phase Equilibria.* 53,5 I-71.
Wang, Y. and Orr, F. M. (1997). Analytical Calculation of Minimum Miscibility Pressure. *Fluid Phase Equilibria.*
Zick, A. A. (1986). A Combined Condensing/Vaporizing Mechanism in the Displacement of Oil by Enriched Gas. SPE 15493.
Dindoruk, B., Orr, F. M., and Johns, R. T. (1997). Theory of Multicontact Miscible Displacement with Nitrogen. SPE 30771.
Johns, R. T. and Orr, F. M. (1996). Miscible Gas Displacement of Multicomponent Oils. SPE 30798.
Jessen, K. and Michelsen, M.L. (1997). Minimum Miscibility Pressure Calculation. Poster presented at the 1997 IFP Seminar "Production of Reservoir Fluids in Frontier Conditions," Rueil-Malmaison, France, 4–5 December).
Jessen, K., Michelsen, M.L. and Stenby, E. (1998). Effective Algorithm for Calculation of Minimum Miscibility Pressure. SPE 50632.
Wang, Y. and Peck, D. G. (2000). Analytical Calculation of Minimum Miscibility Pressure: Application in a Quantitative Analysis of the Effect of Numerical Dispersion for Different Miscibility Development Mechanisms. SPE 59378.
Yuan, H. and Johns, R. T. (2002). Simplified Method for Calculation Minimum Miscibility Pressure or Enrichment. SPE 77381.
Stalkup, F. I. (1982). Status of Miscible Displacement. SPE 9992.
Stalkup, F.I. (1987). Displacement Behavior of the Condensing/Vaporizing Gas Drive Process. SPE 16715.

Table 1 Characteristic of Reservoir Fluids Used in Miscibility Projects

Reservoir	Reservoir Fluid	T_r, F	BPP, psi	API Gravity	OMW	C_{7+} MW	Wt% Asphalt	Viscosity, cp	OFVF	SGOR
Mauddud	sample 1	169	1621	24.7	154	285	NA	2.77	1.225	342
Lower Burgan	Sample 2	182	2348	32.5	119	260	NA	1.043	1.08	563
Zubair	Sample 3A	195	3.025	33.0	110	256	NA	0.713	-	-
	Sample 3C	195	2473	33.0	112	247	NA	0.72	1.385	683
Minagish Oolite	Sample 4	198	1930	33.8	116	253	2.77	0.871	1.357	589

Tr	Reservoir temperature, °F
BPP	Bubble-point pressure, psia
OMW	Oil molecular weight, Ib/Ib mole
C_{7+} MW	Heptanes-plus molecular weight
Wt% Asphalt	Weight percent asphaltenes
OFVF	Oil formation volume factor, bbl.STB
SGOR	Solution gas-oil ratio, SCF.STB

Table 2 Reservoir Fluid Compotation Used in Miscibility Projects

Component	Sample 1		Sample 2	Sample 3			Sample 4
	Recombined fluid Mole %	Reservoir fluid Mole %	Reservoir fluid Mole %	(A) Recombined fluid 1 Mole %	(B) Recombined fluid 2 Mole %	(C) Reservoir fluid Mole %	Reservoir fluid Mole %
H_2S	0.00	0.00	0.00	0.00	0.00	0.00	2.37
CO_2	0.23	0.27	0.54	0.79	0.79	0.87	5.78
N_2	0.51	0.25	0.30	0.73	0.35	0.45	0.12
C_1	26.03	25.48	33.66	39.35	33.69	40.63	27.25
C_2	7.12	6.96	8.14	8.22	8.52	8.89	9.51
C_3	6.43	6.22	6.72	5.91	7.06	6.70	8.14
$i\text{-}C_4$	1.24	1.17	1.26	1.05	1.24	1.15	1.09
$n\text{-}C_4$	3.96	3.73	3.82	2.79	3.78	3.24	3.96
$i\text{-}C_5$	1.73	1.65	1.67	1.14	1.50	1.23	1.45
$n\text{-}C_5$	2.62	2.68	2.43	1.64	2.26	1.85	2.40
C_6	4.17	3.89	4.41	2.77	3.37	2.92	3.23
C_{7+}	45.96	47.7	37.00	35.62	37.44	32.07	34.7
C_{36+}	7.21	7.35					
Fluid MW	152	154	119	109.73	112.3	102.12	105.04
Fluid Density (gm/cc)	0.763	0.767	0.7842	0.665	0.676	0.639	0.675
Temperature ^{O}(F)	169	169	182	195	195	195	198
C_{7+} MW	290	285	260	256.14	246.56	256.53	237
C_{7+} Density gm/cc	0.910	0.908	0.8902	0.890	0.877	0.877	0.8547
C_{36+} MW	7.21	35 j7					
C_{36+} Density gm/cc	760	730					
Plus Mole%	1.05	1.05		3.12*	20.5**	19.19**	
Plus MW				975*	351.91**	348.02**	
Plus Density					0.917**	0.91**	
Volatile %							
Intermediate %							
MwC_{5+}							

* C_{30+}

** C_{12+}

Table 3 Measured and Calculated MMP for Reservoir Fluids

Reservoir Fluid	Gas	Slim-Tube Test	EOS	1D Simulation	Correlations
Sample 4	Pure CO_2	2,665 (or 3,315)	-	3,315	2,700-3,500
	Impure CO_2	-	2,573-2,999	-	-
	RG	-	3,923-4,423	-	-
	N_2	-	>10,000	-	5400
	LG1	-	7,316	-	5,400
	HC Gas	3,815	-	3,815-4,015	-
Sample 3A	Pure CO_2	4,015	5,644		2,970-(3,750-4,250)
	Impure CO_2	-	3,343		-
	RG	-	4,077		4,650-5,700
	N_2	-	10,285		4,650-5,700
	LG1	-	9,336		4,650-5,700
	LG2	-	6,976		4,650-5,700
Sample 2B	Pure CO_2		5,088		2,970-(3,750-4,250)
	Impure CO_2		2,420-3,140		-
	RG		3,840		4,650-5,700
	N_2		6,011		4,650-5,700
	LG1		8,819		4,650-5,700
	LG2		6,812		4,650-5,700
Sample 2	Pure CO_2	-	-	-	2,750
	Impure CO_2	-	2,422-3,010	-	-
	RG	-	3,957-4,663	-	-
	N_2 and LG1	-	>9,000	-	5,300
Sample 1	Pure CO_2	-	-	2,650-2,900	2,650
	Impure CO_2	-	-	2,764	-
	RG	4,500	4,692	4,500	-
	N_2 and LG1	-	>10,000	-	6,000

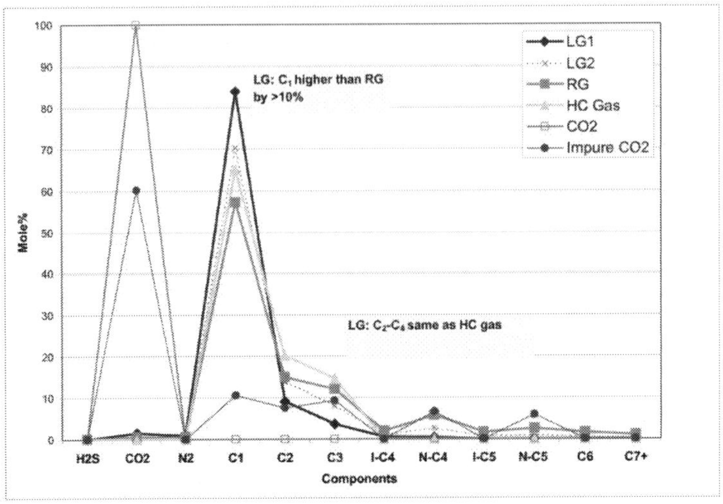

Figure 1 Actual and Hypothetical gases that were used in KOC gas injection experiments and EOS models

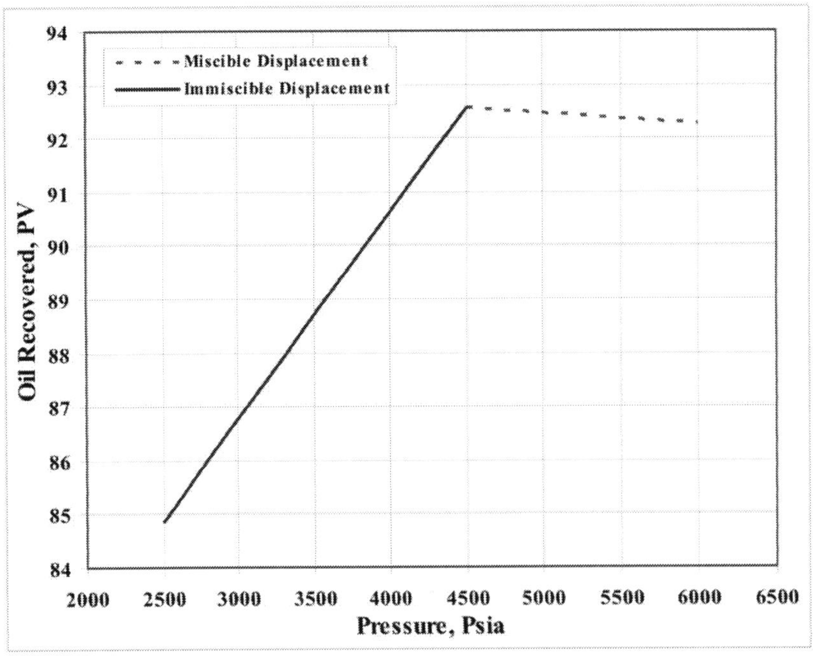

Figure 2 Slim tube tests for sample 1 using RG at 169°F

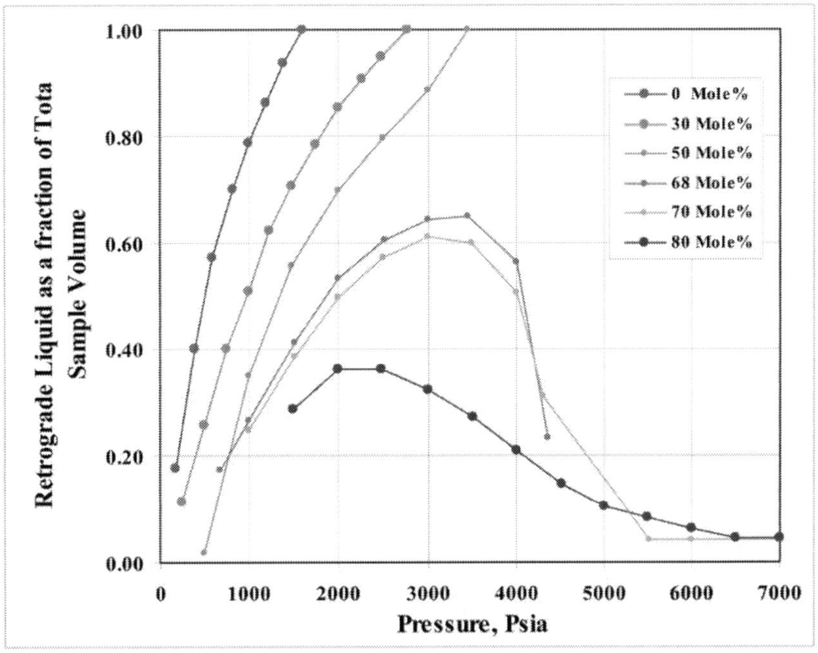

Figure 3 Swelling test for reservoir fluid sample 1 at 169° F using RG

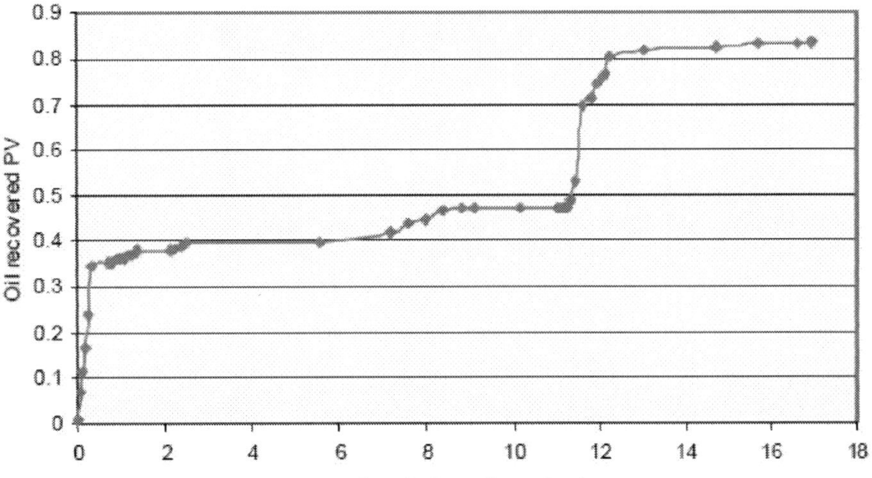

Figure 4 Oil recovery for Sample 1 after RG injection

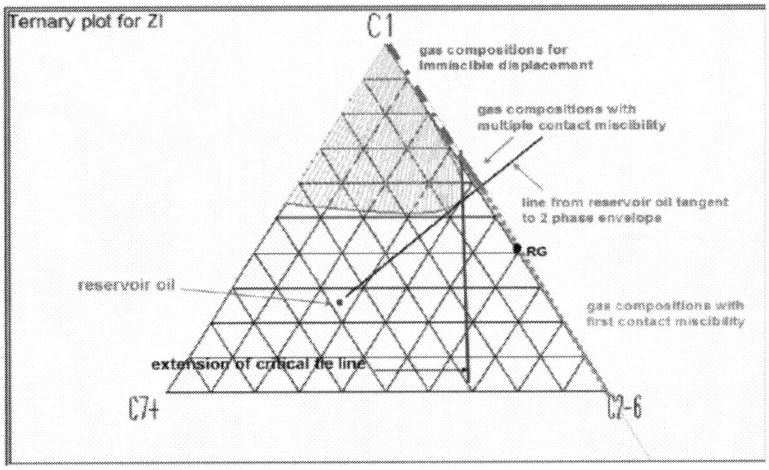

Figure 5 Ternary diagram for Sample 1 and injected RG at 4,500 psia and reservoir temperature of 169° F

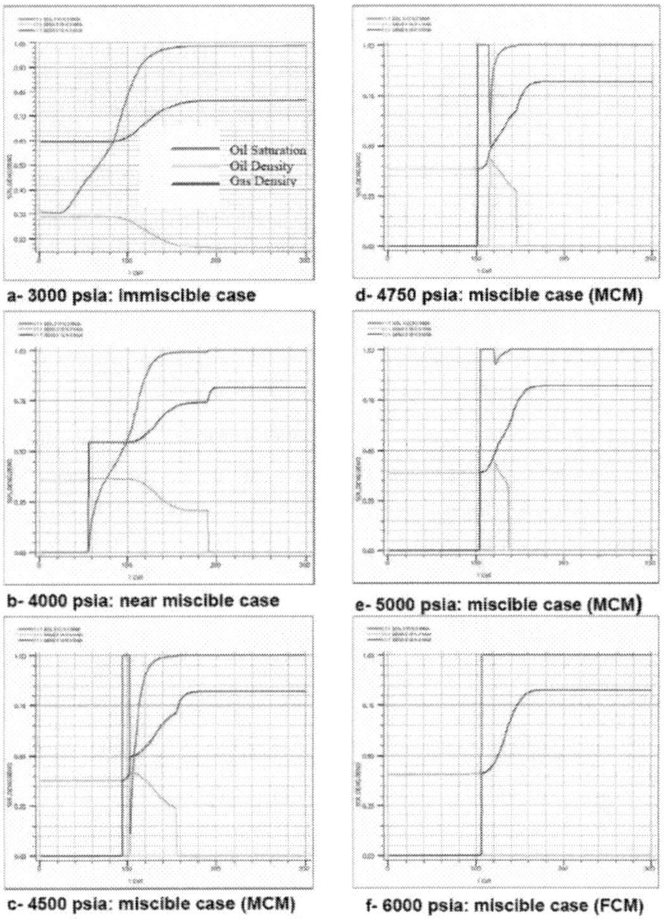

a- 3000 psia: immiscible case

b- 4000 psia: near miscible case

c- 4500 psia: miscible case (MCM)

d- 4750 psia: miscible case (MCM)

e- 5000 psia: miscible case (MCM)

f- 6000 psia: miscible case (FCM)

Figure 6 Slim-tube simulation of Sample 1 reservoir fluid and RG injected gas at 169° F for various injection pressures

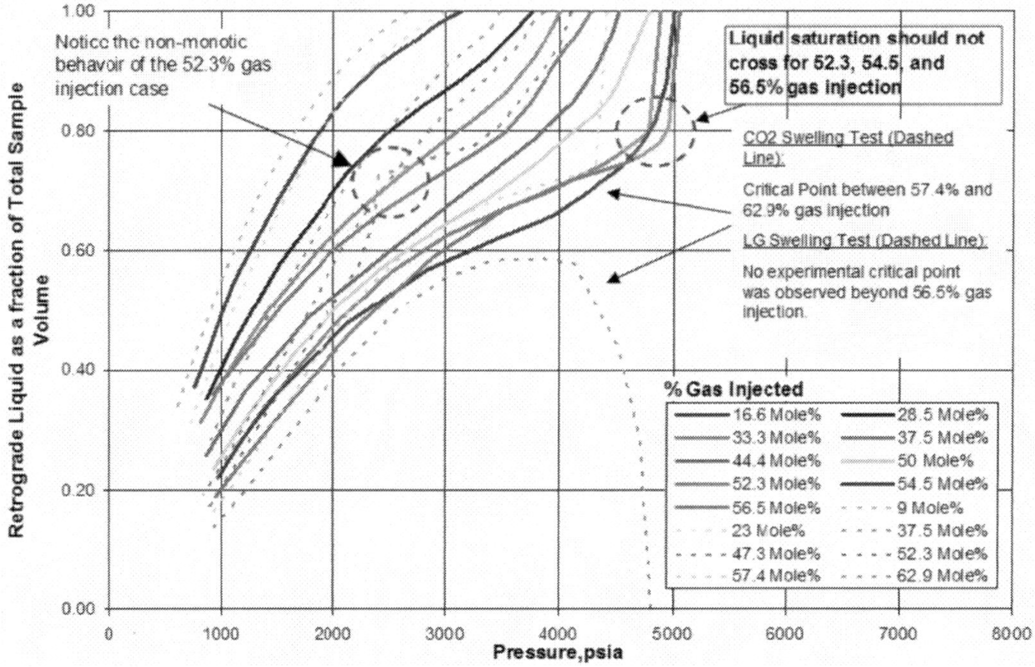

Figure 7 Swelling test for reservoir fluid Sample 3 at 195° F using CO_2

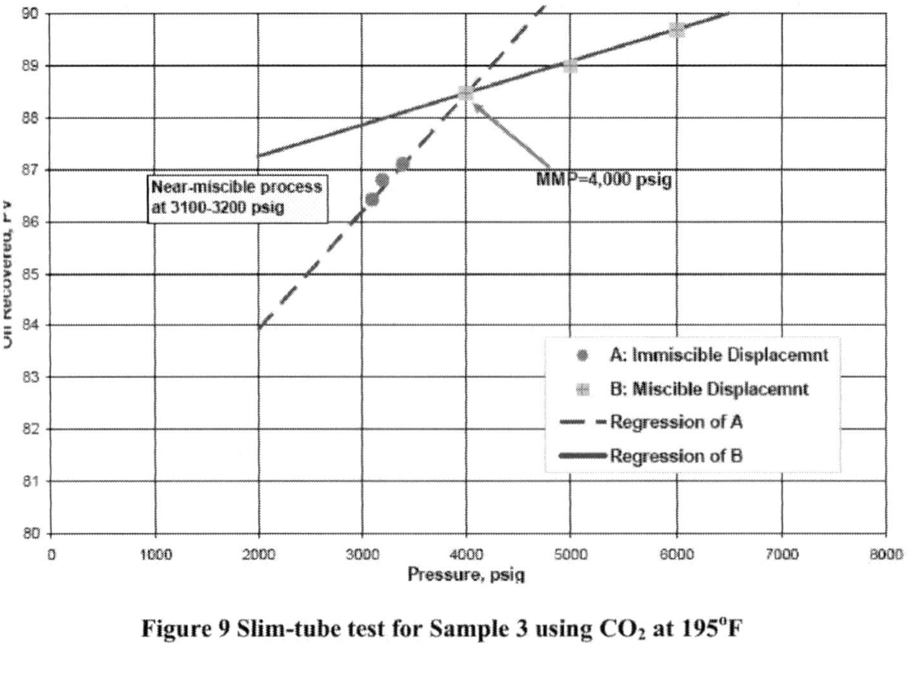

Figure 9 Slim-tube test for Sample 3 using CO_2 at 195°F

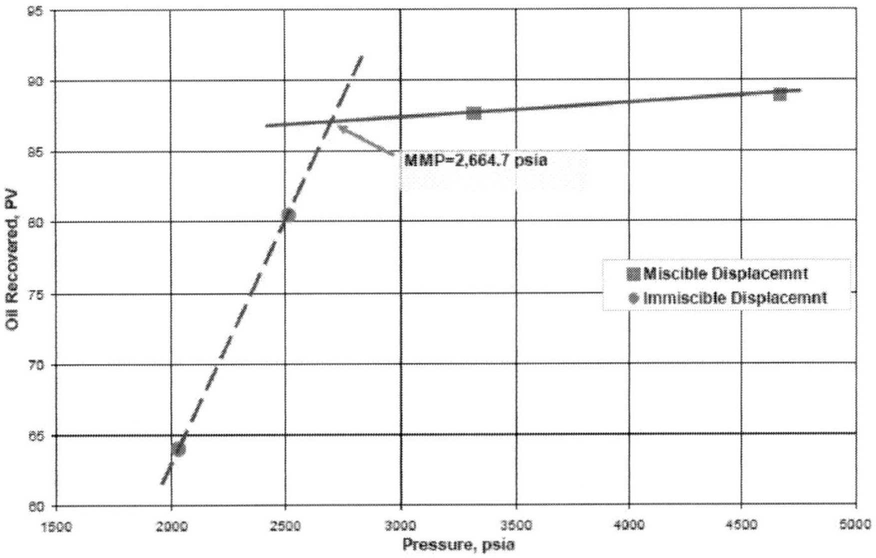

Figure 10a Slim-tube test for Sample 4 using CO_2 at 198°F

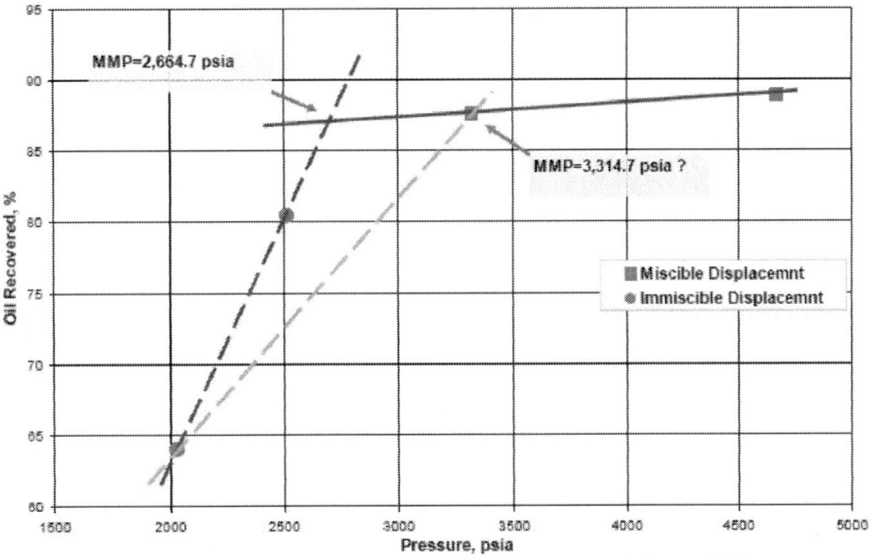

Figure 10b Slim-tube test for Sample 4 using CO_2 at 198°F

Solubilty of Acetylsalicylic Acid in Some Organic Solvents

Gustavo D. Maia,Sciences, Federal University of Alfenas - UNIFAL, Alfenas MG, Brazil

Marco Giulietti, Chemical Engineering, Federal University of Sao Carlos- UFSCar, Sao Carlos-SP, Brazil

The drug known as aspirin is an ordinary medicine that is easily found at any drugstores around the world. Little more than a hundred years ago it was considered to be a miraculous drug that had just appeared at drugstores and pharmacies around Europe. Several similar components had already been produced, though none of them would ever be as successful as the component that is the active principle of aspirin, i.e., the acetylsalicylic acid.

Currently there have been countless reports on the benefits of a small dose of aspirin for the secondary prophylaxis of manifold cardiovascular diseases, such as brain stroke and myocardial infarction due to its capacity to inhibit the thromboxane synthesis, which prevents the formation of both brain and heart thrombi. It is also increasingly evident that aspirin, when regularly used, might play an important role in the prevention of colorectal cancer. Such hypothesis was the topic of an evaluation presented by a working team at the International Agency for Cancer Research in Lyon, France.

Especially in regard to solubility, a few researchers are exclusively concerned with obtaining values for a unique temperature without assessing the solubility in relation to temperature. The solubility is an important parameter to be determined because a series of processes depend upon its recognition as in the event of crystallization. Its study initially faces the problem of lack of information about the behavior of this substance in solution.

The behavior of the acetylsalicylic acid solubility in different temperatures affects, among other aspects, the choice of a crystallization, purification, and storage process. The solubility - or saturation condition - is experimentally determined by heating the suspension and observing at which temperature the entire solid melts. Above this temperature, the solution is said to be sub-saturated, i.e., the entire solute remains dissolved. However, it is possible to cool the same solution at temperatures below the saturation temperature without leading to crystallization. In this case, the solution is known as over-saturated.

Thus, for the determination of acetylsalicylic acid solubility curves and aiming at minimizing the effects of the metastable zones, which could hide the real acetylsalicylic acid solubility, the polythermic method proposed by Nývlt and Giulietti was used.

The temperature dependence of solubility in pure solvents, considering the system as ideal, can be described by a general solubility equation as follows

$$\ln\left(\frac{1}{x_1}\right)=\frac{\Delta h^f}{RT_f}\left(\frac{T_f}{T}-1\right)-\frac{\Delta c_p}{R}\left(\frac{T_f}{T}-1\right)+\frac{\Delta c_p}{R}\ln\frac{T_f}{T} \qquad (1)$$

Where R is the gas constant, T_f is the fusion temperature, T is the study temperature, Δc_p corresponds to the thermal capacity variation between the liquid and the solid phases, Δh^f is the fusion enthalpy, x is the molar fraction - or solubility - and the index 1 stands for the solute. So we can calculate the ideal solubility in any desired temperature range.

In the model proposed by Nývlt and Giulietti, the author considered a binary system where component 1 does not form solid solutions with the solvent. The equilibrium thermodynamic condition at constant pressure is the equality of chemical potentials, μ, of the components in both phases.

Based on this consideration, the authors elaborated an expression that establishes a straight relation between the solubility of a given substance and the system temperature so that the activity coefficient would not explicitly appear in the equation, although it is contained within the adjustable parameters N_1, N_2, N_3

$$\log x = N_1 + \frac{N_2}{T} + N_3 \log(T) \qquad (2)$$

Therefore, this study is aimed at contributing towards the determination of solubility of acetylsalicylic acid in four organic solvents, i.e., ethanol, acetone, propylene glycol, and 2-propanol, in different temperatures, in addition to comparing their solubility values with the expected from an ideal solution. The experimental solubility data were correlated with Nývlt model.

In addition to the industrial relevance, the choice of the solvents was undertaken by taken into account the temperature range in order to obtain an enhanced distribution of temperatures for the solubility study.

It has been noted that the solubility of acetylsalicylic acid is highest in acetone to all studied temperature range until T(K) = 326.3. Propylene glycol presents lower solubility data to all studied temperature range. The solvent, who presents the lowest average percent logarithmic deviation, concerning the ideality of the system, was the ethanol, and the highest value of average percent logarithmic deviation was found in propylene glycol solutions. The experimental solubility data were correlated with Nývlt solubility model. We conclude that Nývlt model was capable to represent properly the experimental data.

See more of this Session: Experimental Phase Equilibria and PVT I
See more of this Group/Topical: Engineering Sciences and Fundamentals

The Importance of Polarizability in the Modeling of Solubility: Quantifying the Effect of An Explicitly Polarizable Solvent On the Solubility of Small Non-Polar Solutes in the Presence of Charged Co-Solutes

Hugh Docherty, Department of Chemical and Biomolecular Engineering, Vanderbilt University, Nashville, TN

Peter J. Dyer, Department of Chemical and Biomolecular Engineering, Vanderbilt University, Nashville, TN

Peter T. Cummings, Department of Chemical and Biomolecular Engineering, Vanderbilt University, Nashville, TN

In recent work,[1] we demonstrated that the addition of polarizability to a simple Lennard-Jones (LJ) model for various small non-polar solute molecules results in quantitative agreement with experiment for the solubility of these molecules in the recent reparameterization of the TIP4P water model (i.e. TIP4P/2005 and TIP4P-Ew). In this work we consider the effect of charged co-solvents (i.e. salting-out) and the advantages of an explicitly polarizable water model, specifically the gaussian charge polarizable model (GCPM) of water of Paricaud *et al.*[2] We will begin by showing that, contrary to current opinion,[3,4] the use of a polarizable solute, together with the TIP4P/2005 and TIP4P-Ew water models, is insufficient to correctly describe the salting-out of small non-polar solutes (Ne, Ar, Kr, Xe and methane) from aqueous solution. We will follow this by demonstrating that these models fail due to an inaccurate solute-water interaction potential and determine whether the highly accurate GCPM water model, coupled with appropriate co-solute and solute models, provides a quantitative description of salting-out.

[1] P.J. Dyer, H. Docherty and P.T. Cummings, J. Chem. Phys. **129**, 02450 (2008)

[2] P. Paricaud, M. Predota, A.A. Chialvo and P.T. Cummings, J. Chem. Phys. **122**, 244511 (2005)

[3] H. Docherty, A. Galindo, E.Sanz and C. Vega, J. Phys. Chem. B. **111**, 8993 (2007)

[4] J. Holzmann, R. Ludwig, A. Geiger and D. Paschek, ChemPhysChem **9**, 2722 (2008)

See more of this Session: Development of Intermolecular Potential Models
See more of this Group/Topical: Engineering Sciences and Fundamentals

Polarizable Water Model for Simulations of Proton Transport

Abhishek Asthana, Chemical Engineering, Brigham Young University, Provo, UT

Dean R. Wheeler, Chemical Engineering, Brigham Young University, Provo, UT

An improved understanding of proton-transport mechanisms in aqueous and hydrophilic systems such as low-temperature fuel cells can be helpful in optimizing such systems. In hydrogen-bonded networks, protons transport by both vehicular and structural diffusion. In particular, structural diffusion which involves the making and breaking of covalent bonds is difficult to model efficiently in large-scale atomistic simulations.

We present here a new first-principles-based polarizable model for water. The model is adapted for simulating proton structural diffusion within molecular dynamics. The model employs charge-equilibration techniques, including an sp-basis parameterization to represent in- and out-of-plane polarization, to describe dynamic electrical response to heterogeneous environments. The quantum aspects of electron correlation and of making and breaking covalent bonds are treated with additional potential expressions. This work demonstrates the nature of collective proton conduction in water under different conditions. Future optimization and improvement of the model are also discussed.

See more of this Session: Development of Intermolecular Potential Models
See more of this Group/Topical: Engineering Sciences and Fundamentals

Implementing Sequence-Dependent Mechanical Properties in a DNA Mesoscale Model

Vanessa Ortiz, Chemical and Biological Engineering, University of Wisconsin-Madison, Madison, WI

Edward J. Sambriski, Chemical and Biological Engineering, University of Wisconsin-Madison, Madison, WI

Juan J. de Pablo, Chemical and Biological Engineering, University of Wisconsin-Madison, Madison, WI

DNA sequence is believed to play an important role in the efficiency of many biologically relevant processes such as nucleosome packing, loop formation and viral encapsidation by influencing the conformational and mechanical properties of DNA. Here we present a DNA model that can reproduce experimentally and computationally observed sequence-dependent mechanical properties. The model accounts for both sequence-dependent base-pair step deformability and base-pairing. The model can capture the de/re-hybridization of DNA and reproduce its melting behavior as a function of CG content, chain length and salt concentration. In addition, the model has also been parameterized to reproduce the atomistic level distributions of bond distances, bend and torsion angles, and in doing so, the model has proved to qualitatively reproduce the sequence-dependent base-pair step deformability. Finally, we look at the effects of this parameterization on the persistence length of short DNA segments.

See more of this Session: Development of Intermolecular Potential Models
See more of this Group/Topical: Engineering Sciences and Fundamentals

Intramolecular Structure of Carbon Dioxide in the Gas, Supercritical, and Solid Phases

J. Ilja Siepmann, University of Minnesota, Minneapolis, MN

Kelly E. Anderson, Department of Chemistry, University of Minnesota, Minneapolis, MN

Recent work has focused attention on possible shifts in the bond angle distribution of carbon dioxide as a consequence of intermolecular interactions in the supercritical phase. To investigate the temperature and phase dependence of the intramolecular structure of carbon dioxide, we performed Feynman path integral Monte Carlo calculations based on a spectroscopically derived analytical potential, first principles molecular dynamics simulations using Kohn-Sham density functional theory, and Monte Carlo simulations employing empirical interaction potentials. Based on various distributions used to characterize the intramolecular structure, we conclude that the aggregation state has a negligible influence on the intramolecular structure, in particular we find that in the classical limit the distributions are remarkably similar for the ideal gas, supercritical, and solid phases. In contrast, an increase in the temperature from 325 to 673 K or inclusion of nuclear quantum effects leads to a significant broadening of the distributions. With respect to the first C-O bond vector, the second bond vector most prefers a collinear arrangement. However, due to the Jacobian factor the maximum in the bond angle distribution at 325 K is shifted to an angle of about 175.7 degree in the classical limit or to 173.0 degree if nuclear quantum effects are included. Nevertheless, an analysis of the temperature dependence of the constant-volume heat capacity demonstrates that carbon dioxide should be viewed as a linear molecule. Gibbs ensemble Monte Carlo simulations are used to explore the influence of flexibility on the vapor-liquid phase equilibria.

See more of this Session: Development of Intermolecular Potential Models
See more of this Group/Topical: Engineering Sciences and Fundamentals

Improving the Lipid Force Field of CHARMM: A Quantum Mechanical and Experimental Approach

Jeffery Klauda, Department of Chemical and Biomolecular Engineering, University of Maryland, College Park, MD

Richard Venable, Lab of Computational Biology, NHLBI, Bethesda, MD

Alexander D. MacKerell, University of Maryland, Department of Pharmaceutical Sciences, Baltimore, MD

Richard W. Pastor, Laboratory of Computational Biology, National Heart, Lung, and Blood Institute, Bethesda, MD

Biological membranes form a barrier to protect the cell from its environment and selectively control the entrance/exit of small molecules. Molecular simulations of these biological membranes require an accurate lipid force field (a major component of the membrane). Previously, extensive *ab initio* quantum mechanical (QM) calculations have been used to improve the aliphatic portion of the CHARMM27 lipid force field. Although this was a significant improvement, the lipid head group required additional modifications to agree with experimental lipid bilayer deuterium order parameters (S_{CD}) and solvation free energies. To improve the solvation free energies, we modified the atomic charges in the carbonyl-glycerol region. The S_{CD}'s for the lipid head group of DPPC were improved by fitting dihedral energy terms to high-level QM calculations and/or to torsional conformational populations that result in agreement with experimental S_{CD}'s. Molecular dynamics (MD) simulations with this new force field, referred to as CHARMM35 (C35), resulted in a significant improvement in the S_{CD}'s and water hydration for DPPC lipid bilayers. The calculated electrostatic profile and lipid bilayer surface tension decreased significantly with the C35 force field. Consequently, the C35 force field resulted in excellent surface areas per lipid with NPT simulations, which is a significant improvement from the C27r force field that required constant area simulations (NPAT) to prevent some bilayers from laterally condensing. MD simulations of other lipid bilayers and monolayers also agreed favorably with experimental densities, monolayer surface tensions, and S_{CD}'s, e.g., DMPC, DOPC, POPC, and POPE. Therefore, CHARMM35 is the pair-wise additive all atom force field of choice for lipids.

See more of this Session: Development of Intermolecular Potential Models
See more of this Group/Topical: Engineering Sciences and Fundamentals

Force Field for Dibenzotiophene and Its Alkyl Derivates

Pedro Navarro-Santos, Chemical Engineering Department, Universidad Michoacana de San Nicolas de Hidalgo, Morelia, Michoacan, Mexico

Jose L. Rivera, Department of Chemical and Biomolecular Engineering, Vanderbilt University, Nashville, TN

Ramon Lopez-Sandoval, Department of Advanced Materials, Instituto Potosino de la Investigacion Cientifica y Tecnologica, San Luis Potosi, SLP, Mexico

Clare McCabe, Department of Chemical and Biomolecular Engineering, Vanderbilt University, Nashville, TN

A force field for dibenzothiophene and its alkyl derivates 4-methyldibnezothiophene and 4,6-dimethyldibenzothiophene has been developed based on ab initio calculations and experimental data. The resulting force field is fully flexible and describes bond, valence angle, and dihedral angle vibrations as well as intermolecular interactions (Lennard-Jones and coulombic). The developed potential was optimized to reproduce the saturation properties of dibenzothiophene and has been validated through molecular dynamics simulations to calculate the orthobaric densities, vapor pressures, surface tension, and vaporization enthalpy in the vapor-liquid region for dibenzothiophene and its alkyl derivates.

See more of this Session: Development of Intermolecular Potential Models
See more of this Group/Topical: Engineering Sciences and Fundamentals

Problems of Molecular Theories of Fluid Mixtures: Intermolecular Potential Models and Combining Rules

Ivo Nezbeda, Chemistry Department, J. E. Purkinje University, Usti n. L., Czech Republic

To predict properties of fluids and fluid mixtures intermolecular potential models are required that, however, are available only for pure fluids. Consequently, certain combining rules must be used, typically the Lorentz-Berthelot (LB) ones. However, no satisfactory agreement with experiment can be achieved unless certain deviations from the LB rules are applied, particularly from the Berthelot rule. Recent studies on mixtures of realistic fluid models have shown that also deviations from the Lorentz rule for the size parameter must be considered. Moreover, we have recently shown that a deviation from the Lorentz rule may even lead to a qualitative change in the behavior of the mixture. On top of this, the use of pure fluid potential models themselves for mixtures must be questioned. Consequently, properties of mixtures are, in principle, unpredictable unless some a priori known experimental result is used.

One possibility to account more appropriately for the cross interactions is to include polarizability. Molecular simulations on polarizable models bring the results to at least qualitative agreement with experimental data but at the expense of considerably increased complexity and CPU time demands. An alternative is to maintain the pure fluid pair-wise additivity and to develop certain rules for the combining rules. Nonetheless, large discrepancies between results obtained for excess properties of the same mixture using different force fields indicate that good performance for pure fluids need not be sufficient for a good description of mixtures. All these findings will be presented and discussed along with an attempt to identify defects in pure fluid models when applied to mixtures.

See more of this Session: Development of Intermolecular Potential Models
See more of this Group/Topical: Engineering Sciences and Fundamentals

Density and Speed of Sound Measurements of Second-Generation Biofuels

Stephanie L. Outcalt, Thermophysical Properties Division, NIST, Boulder, CO

Arno Laesecke, Thermophysical Properties Division, NIST, Boulder, CO

Tara J. Fortin, Thermophysical Properties Division, NIST, Boulder, CO

While the conversion of biomass into ethanol is one of the oldest chemical processes, ethanol has never been optimized as a fuel for transportation but rather as a beverage. Its properties are too different from those of petroleum-based hydrocarbon fuels such as gasoline to be a feasible long-term replacement of the latter. The isomers of butanol (C4H9OH) are much more similar to conventional gasoline but their properties are much less known than those of either ethanol or hydrocarbons. To facilitate their use as fuels, measurements of their thermophysical properties have been started in our laboratory. Results of density and speed of sound measurements for 1- and 2-butanol will be reported in this presentation.

Thoroughly dried samples of the two alcohols were prepared with molecular sieves and confirmed by Karl-Fischer titration. A rapid characterization instrument was used for density and speed of sound measurements at ambient pressure from 278.15 K to 343.15 K. Compressed-liquid densities were measured from 270 K to 470 K at pressures from 0.5 MPa to 50 MPa with a fully automated vibrating-tube densimeter. The calibration and adjustment of the instruments will be discussed in detail because they were extended beyond existing standards.

The density results were correlated by Tait equations for comparisons with literature data. Density data for 1-butanol are available in the literature over the temperature and pressure range of the current measurements. However, they scatter significantly, particularly in the compressed liquid region. Literature data for 2-butanol are relatively scarce. Differences in the thermophysical properties and stability of these butanol isomers, some of the challenges associated with accurate measurements of hygroscopic, polar fluids as well as issues of correlating the measurement results will be discussed in the presentation.

See more of this Session: Experimental Phase Equilibria and PVT II
See more of this Group/Topical: Engineering Sciences and Fundamentals

Determination of critical properties using a dynamic-synthetic apparatus: experimental work and modeling

Chien-Bin Soo[1], Pascal Theveneau[1], Christophe Coquelet[1], Deresh Ramjugernath[2], Dominique Richon[1]

1 : *Mines ParisTech, CEP/TEP, Centre Energétique et Procédés, Rue Saint Honoré, 77305 Fontainebleau, France*
2 : *Thermodynamic Research Unit, School of Chemical Engineering, University KwaZulu Natal, Howard College Campus, Durban, 4041, South Africa*

Critical point data of mixtures are essential for the optimization of petroleum and supercritical extraction processes. These data supply information on fluid behavior at extreme conditions of equilibrium. Due to strong fluctuations of densities at near-critical conditions, it is difficult for most equations of state to satisfactorily predict the phase behavior near the critical region. In addition, the empirical trend of the binary parameters changes after reaching the critical point of the low boiling component, making model predictions less obvious.

In this communication we propose a "dynamic-synthetic" method with an apparatus designed for the visual detection of critical temperatures and pressures. The apparatus is capable of pure component and mixture measurements with the compositions being synthesized for the mixtures. Critical points are reported herein for several pure components and binary mixtures, correlated using various models ranging from polynomial expression to equation of states. Results show that, in the absence of interaction parameters, general disagreements still exist when modeling more complex systems with equation of states.

Keywords: experimental thermodynamics, dynamic-synthetic method, critical points, PC-SAFT, PSRK

Introduction

The calculation of phase behavior at critical regions is often difficult due to the strong fluctuation of densities that occur with changes in temperature and pressure in this region. The prediction of critical points and stability limits, both of which are directly applicable to petroleum and supercritical extraction processes, are usually done using equation of states (EoS). Thus, the inability to perform such calculations with sufficient accuracy would prove detrimental to these processes.

In this communication a "dynamic-synthetic" method-based apparatus has been designed for the visual detection of critical temperature and pressures. Such apparatuses are based on the flow methods first proposed by Roess [1] and later modified by Rosenthal and Teja [2]. The transparent cell allows observation of the critical opalescence as the phase boundary disappears. The apparatus is applicable for pure component measurements, as well as for mixtures of known compositions. Several measurements on hydrocarbons, alcohols and refrigerants have been carried out to investigate the capabilities and limitations of this setup.

Experimental

Material

Table 1 summarizes the purities of the chemicals used in the study. A disadvantage of flow apparatuses is the large volume of materials required; thus an ample supply of chemicals is needed to facilitate the experimental.

Table 1. Purity and supplier of the compounds used in the study

Compound	Purity (% analysis)	Supplier
Propane	≥ 99.95 % $^{v}/_{v}$	Messer-Griesheim
n-Butane	≥ 99.95 % $^{v}/_{v}$	Air Liquide
n-Hexane	≥ 99.0 % GC	Merck
Ethanol	≥ 99.8 % $^{v}/_{v}$	Fluka
1-Propanol	98.0 % GC	Prolabo
Hexafluoropropylene (R-1216)	≥ 99.9 % $^{v}/_{v}$	Pelchem
Hexafluoropropylene oxide	≥ 99.9 % $^{v}/_{v}$	Pelchem
Pentafluoroethane (R-125)	≥ 99.5 % $^{v}/_{v}$	ARKEMA
1,1,1,2 – Tetrafluoroethane (R-134a)	99.95% $^{v}/_{v}$	Dehon
1,1,1,3,3 – Pentafluoropropane (R-245fa)	≥ 99.9 % $^{v}/_{v}$	ARKEMA
1,1,1,3,3 – Pentafluorobutane (R-365mfc)	≥ 99.9 % $^{v}/_{v}$	ARKEMA

Apparatus

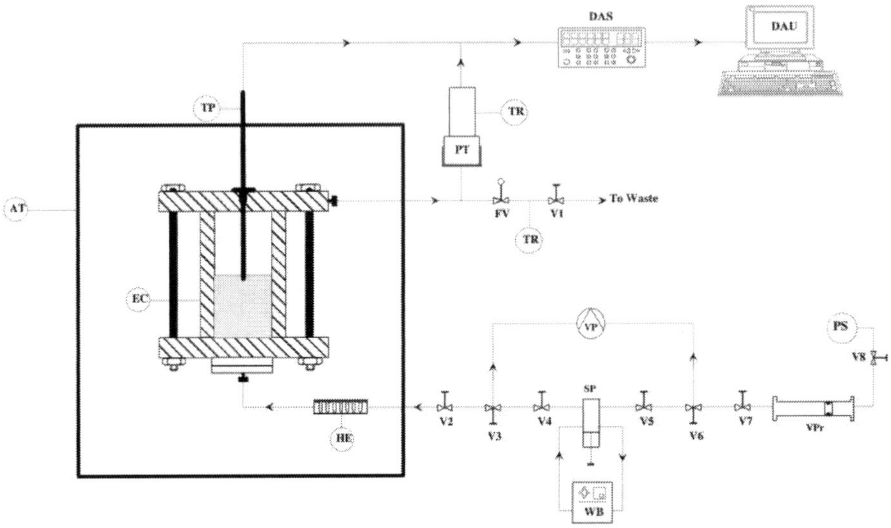

Fig. 1. Flow diagram of the apparatus: AT: Air Thermostat; DAS: Data Acquisition System; DAU: Data Acquisition Unit; EC: Equilibrium Cell; FV: Flow Control Valve; HE: Heat Exchanger; PS: Pressurized source (Nitrogen); PT: Pressure Transducer; SP: Syringe Pump; TP: Platinum Resistance Temperature Probe; TR: Temperature Regulator; V: Valve; VP: Vacuum Pump; VPr: Volumetric Press; WB: Water Bath.

The apparatus is shown schematically in Fig. 1. The circuit is designed to allow complete evacuation of the cell and its lines prior to loading. The sample is prepared in a volumetric press of 100 cm^3 net volume, its composition being accurately determined through weighing on a Mettler Toledo XP2004S balance of ± 0.1 mg maximum uncertainty. Thereafter, it is transferred under pressure to a thermostatic syringe pump (ISCO, Model 260D) capable of delivering its contents at constant flowrate or pressure. The syringe pump is surrounded by an external jacket, within which water, thermally regulated by a water bath (Julabo, Model F32), maintains the pump contents at constant temperature.

The measurement part, a transparent cell, consists of a sapphire tube pressed between two titanium flanges, operating up to 20 MPa and 573 K respectively. The cell is in an air thermostat (France Etuves, Model XU 125) within which a strategically-placed mirror reflects rays of light onto the cell for enhancing the visual access of the critical opalescence. A coiled heat exchanger in the entry line ensures that liquids entering the cell is at the same temperature as the interior of the cell; thus minimizing possible temperature gradients. A 100 Ω platinum resistance thermometer probe (Actifa), placed in the center of the cell, measures the cell temperature. The probe is calibrated against a 25 Ω reference platinum resistance thermometer (TINSLEY Precision Instruments, Type 5187 A), using an oil bath (HAAKE DC 50) for the range 323 – 473 K, and a fluidized alumina bath (Techne, Model FB-08) for the range 423 – 573 K. The 25 Ω reference platinum probe was calibrated by the Laboratoire National d'Essais (Paris) based on the 1990 International Temperature Scale (ITS 90). A Micro-Ohmmeter (Hewlett-Packard, Model 34420 A) was included in the setup. The maximum overall calibration uncertainty is ± 0.02 K in the 50 -170 K range, and ± 0.09 K in the 170 – 300 K range. Pressures are measured with a pressure transducer (Druck, Model PTX 611), which is kept at constant temperature by means of a PID regulator (WEST, model 6100). The pressure transducer is calibrated against a dead weight pressure balance (Desgranges & Huot 5202S, CP 0.3–40MPa, Aubervilliers, France). Pressure calibration uncertainties are estimated to be ± 0.0004 MPa in the 0.1 to 7.1 MPa range, and ± 0.0008 MPa in the 7.1 to 18.1 MPa range. Both the signals from the temperature and pressure probes are transmitted to a data acquisition unit, supplied by Actifa, and displayed on a personal computer via the Ltc10® interface.

The final, flow-monitoring, part consists of two valves, with which variations of the cell volume can be achieved. The valve FV in Fig. 1 is a flow control valve (TOP INDUSTRIE), while V1 is a standard shut-off valve. The exit line emerging from the air thermostat is regulated at a constant temperature by a WEST PID regulator (model 6100). This prevents rapid condensation of the cell contents upon leaving the air thermostat, and in doing so, avoids unexpectedly large drops in pressure. The exit line to the setup is in the form of a stainless steel capillary, with an internal diameter between 100-160 microns.

This experimental setup has the advantage of producing several highly repeatable data points per run. As the binary mixture is prepared separately in a volumetric press by weighing, the uncertainties in molar compositions are very low.

Procedure

Upon charging the ISCO pump with the prepared sample, the contents are pressurized to a single liquid phase, and maintained at a high temperature for a sufficient amount of time. This ensures that diffusion accounts for any remaining unmixed portions within the mixture, as no stirrer is present in the pump. For mixtures containing less soluble components, it may be necessary to induce further mixing through raising and lowering the pump piston continuously.

The cell is first heated to a temperature a few degrees below the expected critical temperature. The pump contents are then transferred to the cell at a constant flowrate, usually at 2 – 3 mL/min. Considerable flashing of the liquid occurs until sufficient pressure is achieved to maintain a liquid phase. The cell is allowed to fill to the halfway mark, at which the flow monitor valves (FV and V1) are opened to maintain the halfway liquid level. The temperature is then slowly raised at a rate of ± 0.5 K/min, during which any changes in the liquid level are compensated by the flow control valves (for coarse adjustments) or the pump flowrate (for finer adjustments). As the cell contents approach its critical temperature, the solution starts to become dark and cloudy, followed by a deep red colour, and finally fades to an orange mist. The sequence of colors does not vary depending on the components. The critical temperature and pressure is recorded at the instant the final orange colour is observed. The temperature is then lowered and the above procedure repeated at least eight times to reduce the bias. The Ltc10® interface refreshes real-time temperature and pressure readings every 0.1s. Standard deviations on experimental temperatures and pressures, resulting from averaging repeated data, are given as 0.5 K and 0.05 MPa respectively.

Results

All experimental data were subjected to the Dixon's Q-test [3] on a 95 % confidence level to remove outliers. The results for the critical points of pure components are graphically summarized in Fig. 2 and compared to various sources [4,5], as well as values obtained through Lydersen's [6] and Joback's [7] methods. For components available on the DIPPR801 [4] database, and from Poling et al. [5], the maximum error upon comparison was 0.35 K and 0.02 MPa for critical temperatures and pressures respectively.

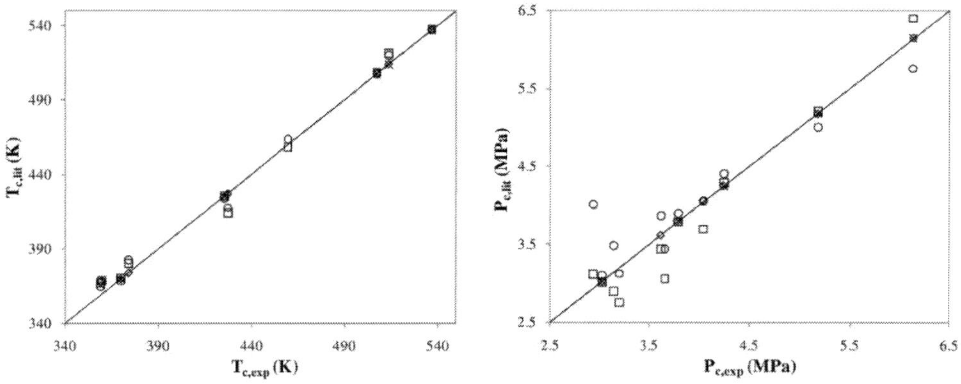

Fig. 2. Literature critical temperatures and pressures compared to our experimental values (x: DIPPR801 [4], ◇: Poling et al. [5], □: Lydersen's method [6], ○: Joback's method [7], — 45° $x = y$ line).

The mixture critical data are presented in Fig. 3, and can be correlated using a simple Redlich-Kister type expression [8],

$$T_{c,mix} = x_1 T_{c,1} + x_2 T_{c,2} + \sum_{i=1}^{n} G_{Ti} x_1 x_2 (2x_1 - 1)^{i-1} \tag{1}$$

$$P_{c,mix} = x_1 P_{c,1} + x_2 P_{c,2} + \sum_{i=1}^{n} G_{Pi} x_1 x_2 (2x_1 - 1)^{i-1} \tag{2}$$

where G_{T1}, G_{T2}, ... G_{Tn} are constants obtained by minimizing the differences between values calculated through the above correlations, and the discrete experimental critical data. The subscript 1 denotes the component with the lower critical temperature, and n, the total number of coefficients employed, is usually sufficient at three or four. Table 2 lists the coefficients obtained for each binary system.

Table 2. Coefficients GT_i and GP_i for the binary systems featured in this work.

	C_3 + n-C_4	C_3 + R134a	n-C_6 + EtOH	EtOH + PrOH
GT_1	17.987	-70.209	-84.232	3.6627
GT_2	4.2069	-10.918	28.991	-0.59662
GT_3	0.71214	-10.918	18.729	0.98173
Max ΔT from Eqn. (1) (K)	± 0.1	± 0.2	± 0.2	± 0.1
GP_1	0.76872	-0.51233	-1.2888	0.26207
GP_2	0.25201	-0.44048	1.3765	0.10324
GP_3	0.11358	1.0880	1.3819x10^{-2}	-0.010051
GP_4		0.36437		
Max ΔP from Eqn. (2) (MPa)	± 0.002	± 0.003	± 0.008	± 0.002

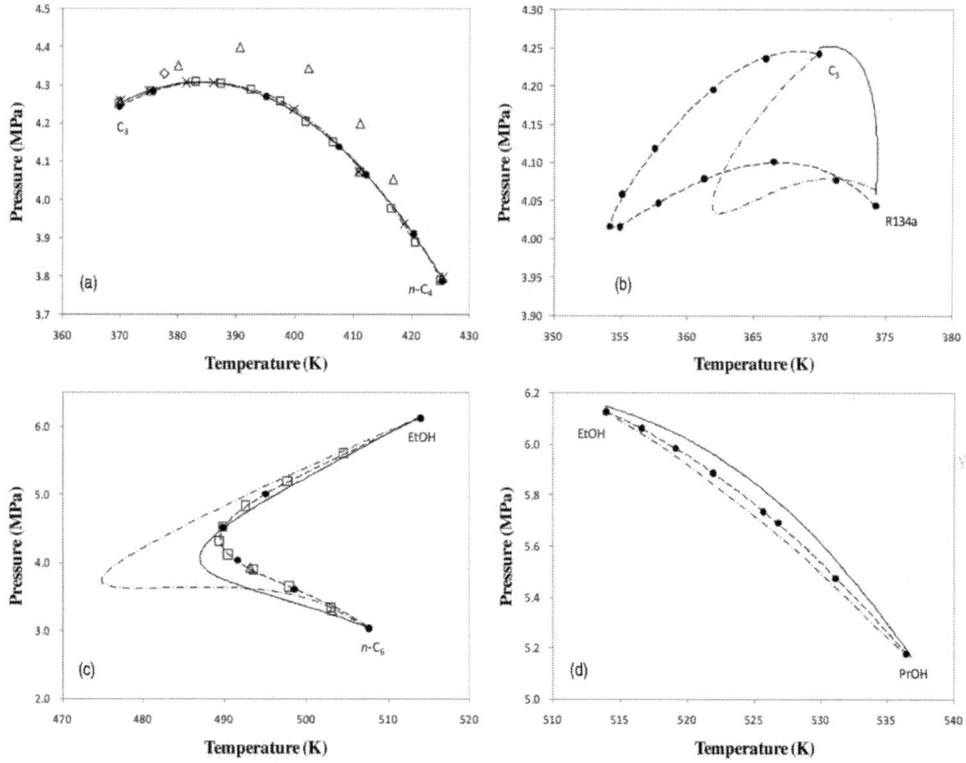

Figs. 3. Critical pressures versus critical temperatures for various binary systems: • Experimental data from this work; — PC-SAFT predictions (k_{ij} = 0) [9]; -•- PSRK predictions [10], -- Redlich-Kister correlation [8] with constants from Table 2. Systems (a): propane + n-butane, □: [11]; Δ: [12]; x: [13]; ◇: [14]. (b): propane + R134a (1,1,1,2-tetrafluoroethane). (c): n-hexane + ethanol, □: [15]; Δ: [16]. (d) ethanol + 1-propanol.

Difficulties were encountered in the binary mixture propane + ethanol due to the high relative volatilities between the two components. This becomes particularly significant between compositions x_{C3} = 0.2 – 0.6, making it currently impossible to maintain a constant global composition in the cell and complete the measurement,

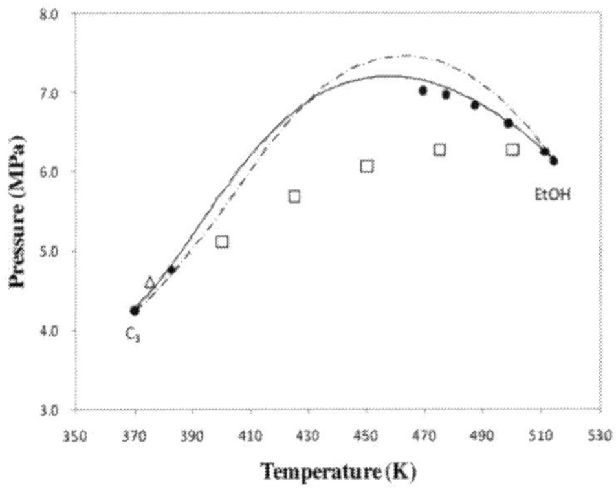

Fig. 4. Critical pressure versus critical temperatures for the binary system propane + ethanol: • Experimental data from this work; — PC-SAFT predictions [9]; –•– PSRK predictions [10], □: [17]; Δ: [18].

A detailed discussion on this matter will be addressed at the conference, along with the predictions from the PC-SAFT [9] and the PSRK model [10] that appears in the results.

Conclusion

A dynamic-synthetic apparatus, capable of temperatures and pressures up to 20 MPa and 573 K, has been developed and commissioned for the measurement of pure component and mixture critical data. The critical points are identified as the onset of the critical opalescence and a sequence of well-defined color changes. The apparatus is capable of producing highly repeatable results, with a maximum standard deviation of 0.5 K and 0.05 MPa on temperature and pressure measurements, respectively.

The critical loci of various hydrocarbon, alcohol and refrigerant binary mixtures have been measured, and treated using a Redlich-Kister expression, the PC-SAFT EoS, and the PSRK predictive model, in an attempt to characterize each model's ability to correlate critical data.

References

1. L.C. Roess, J. Inst. Petrol. Tech. 22 (1936) 665-705.
2. D.J. Rosenthal, A.S. Teja, AIChE J. 35 (1989) 1829-1834.
3. R.B. Dean, W.J. Dixon, Anal. Chem., 23(4) (1951) 636-638.
4. Brigham Young University Thermophysical Properties Laboratory, Design Institute for Physical Properties: Evaluated Standard Thermophysical Property Values, Provo, http://dippr.byu.edu.
5. B. Poling, J. Prausnitz, J. O'Connell, The Properties of Gases and Liquids, 4th ed., McGraw-Hill Book Company, 1988.

6. A.L. Lydersen, Estimation of critical properties of organic compounds by the method of group contributions, University of Wisconsin, Chemical Engineering Department, Eng. Exp. Stn. Rep. 3 (1995).
7. K.G. Joback, S.M. Thesis in chemical engineering, Massachusetts Institute of Technology, Cambridge, Massachusetts, 1984.
8. O. Redlich, A.T. Kister, C.E. Turnquist, Chem. Eng. Prog. Symp. Ser. 48(2) (1952) 49-61.
9. J. Gross, G. Sadowski, Ind. Eng. Chem. Res., 40 (2001) 1244-1260.
10. T. Holderbaum, J. Gmehling, Fluid Phase Equilib., 70 (1991) 251-265.
11. S. Horstmann, K. Fischer, J. Gmehling, Chem. Eng. Technol. 22 (1999) 839 – 842.
12. C.N. Nysewander, B.H. Sage, W.N. Lacey, Ind. Eng. Chem. 32 (1940) 118-123.
13. A. Keglewski, W.B. Kay, J. Phys. Chem. 73 (1969) 3359-3366.
14. H.C. Wiese, J. Jacobs, B.H. Sage, J. Chem. Eng. Data 15 (1970) 82-91
15. P. Sauermann, K. Holzapfel, J. Oprzynski, F. Kohler, W. Poot, T.W. de Loos, Fluid Phase Equilib., 112 (1995) 249-272.
16. J. Seo, J. Lee, H. Kim, Fluid Phase Equilibria, 182 (2001) 199-207.
17. M. Gomez-Nieto, G. Thodos, AIChE J., 24(4) (1978) 672-678.
18. H. Horizoe, T. Tanimoto, I. Yamamoto, Y. Kano , J. Chem. Eng. Jpn., 26 (1993) 482-489.

Volumetric Properties of Hexafluoropropylene and Hexafluoropropylene Oxide. Determination of Phase Diagram and Critical Properties

Christophe Coquelet[1], Ghemina Belaribi[2], Farid Belaribi[2], Alain Valtz[1], Paramespri Naidoo[3], Deresh Ramjugernath[3]

1 Mines ParisTech, Centre Energétique et Procédés, Fontainebleau, France
2 Laboratoire de thermodynamique et modélisation moléculaire, USTHB, Alger, Algérie.
3 School of Chemical Engineering, University of KwaZulu-Natal, Durban, South Africa

Abstract :

Hydrofluoroalkenes, like hexafluoropropylene (R1216) and hexafluoropropylene oxide (HFPO), are being investigated as potential new fluids for refrigeration systems or as fluids for extraction. Considering the different possibilities of utilisation, one of the most interesting properties is the density at and outside saturation. A vibrating tube densitometer technique is used to determine densities for several temperatures between 263 and 383 K and pressures between 0 and 15 MPa. The Forced Path Mechanical Model (Bouchot and Richon, 2001) is used to correlate, at temperature and pressure, the density to the U-tube vibrating period. A sapphire tube cell is also used to determine the pure component vapour pressures.

Using the experimental density values and critical divergence theory concerning density difference between the liquid and vapour phases on the coexistence curve, the critical properties and the phase diagram can be determined.

Introduction

Due to their Global Warming Potential (GWP), hydrofluorocarbons (HFCs) will probably soon be phased out. Historically, they were used because of their zero Ozone Depletion Potential (ODP). Prior to their utilisation research indicated that they led to the depletion of the Earth's ozone layer and they were ultimately banned. HFCs, which do not contain chlorine, pose no threat to the ozone layer, but due to their high stability in the atmosphere, they have a high GWP. This has forced the refrigeration industry (domestic, cars, heat pumps) to consider and find alternate HFCs which have a much lower GWP. In chemistry, researchers have to consider new fluorinated compounds as intermediates for chemical reactions. One solution could be the utilisation of hydrofluoroalkenes, like hexafluoropropene (R1216, CAS Number 116-15-4) or hexafluoropropene oxide (HFPO, CAS Number 428-59-1). The GWP of R1216 is 0.86[1], which is negligible in comparison with the GWP of R134a. Hexafluoropropylene oxide (HFPO) is a versatile fluorointermediate that can be used in the synthesis of fluoromonomers, fluoropolymers and to add fluorine functionality to a variety of organic precursors. R1216 is the key chemical precursor to HFPO which is synthesized in an oxidation process. R1216 is used in the manufacture of fluoropolymers and in other specialty agrochemical and pharmaceutical applications.

In this communication, a complete study of volumetric properties of pure R1216 and HFPO determined using the vibrating tube densitometer technique[2,3] is presented.

Pure component vapor pressures were also measured and critical properties determined from these and density measurements.

Experimental Section

Materials. R1216 and HFPO were supplied by NECSA (South African Nuclear Energy Corporation) with a certified purity greater than 99.99% volume fraction. Gas chromatographic analysis of the samples indicated a single component peak and therefore qualitatively verified the purity.

Vapor pressure apparatus. A classical sapphire tube cell was used for the determination of pure R1216 and HFPO vapor pressures. Temperatures are measured by two Pt100 probes connected to a HP34970A data acquisition unit. These Pt100 probes were periodically calibrated against a 25 Ω reference thermometer (TINSLEY Precision Instrument) certified by the Laboratoire National d'Essais (Paris, France). The resulting uncertainties on temperature measurement are within ± 0.02 K. Pressures were measured using a pressure transducer (Druck PTX611) having a 0 to 4 MPa pressure range. The pressure transducer was calibrated against a dead weight pressure balance (Desgranges & Huot model 5202S) which has a 0.3 to 40 MPa pressure range. The pressure transducer is connected to the HP34970A data acquisition unit. The resulting uncertainties in pressure measurements are within ± 0.0005 MPa.

Vibrating tube densitometer apparatus: A detailed description of a typical vibrating tube density measurement apparatus is given in a previous publication (Bouchot and Richon[2]). The apparatus used in this work uses an Anton Paar DMA 512 vibrating tube. The vibrating tube is made of stainless steel and can be operated at pressures up to 40 MPa. The period of the vibration, τ in seconds, is recorded with a HP53131A data acquisition unit. The uncertainty on the vibrating period values is $\pm 10^{-8}$ s. The temperature of the vibrating tube is controlled by a regulated liquid bath (LAUDA RE206) with stability within ± 0.05 K. The temperature of the remaining parts of the circuit is regulated by a liquid bath controlled by a West type regulator (WEST P6100). Temperatures are measured by two Pt100 probes connected to the HP34970A data acquisition unit. These Pt100 probes are also periodically calibrated against a 25 Ω reference thermometer (TINSLEY Precision Instrument) certified by the Laboratoire National d'Essais (Paris, France). Vacuum was achieved by means of a vacuum pump. Pressures are measured using two pressure transducers (Druck PTX611) with two complementary ranges: (0 to 3) MPa, and (0 to 20) MPa. These sensors were calibrated against a dead weight pressure balance (Desgranges & Huot model 5202S) which has a 0.3 to 40 MPa pressure range and against an electronic balance (fundamental digital pressure standard, Desgranges & Huot model 24610, France) for pressures below 0.3 MPa. The pressure transducers are connected to the HP34970A data acquisition unit.

Experimental Procedure. Details concerning the experimental procedure are fully described in previous papers by some of the present authors [2,3]. The Forced Path Mechanical Calibration model (FPMC method) proposed by Bouchot and Richon[4] is used to convert periods into density values. FPMC parameters were calculated from (PVT) data of a reference fluid (R134a), whose thermodynamic properties are well described by Tillner-Roth and Baehr[5].

Estimation of Uncertainties. The global uncertainty on density data in the vapor and liquid phases are estimated to be ± 0.05%, due to the uncertainties of the mechanical parameters used in the FPMC model. Global temperature uncertainties are estimated to be ± 0.02 K. Global uncertainties on pressure measurements as a result of careful calibration are (± 0.0003 and ± 0.0006) MPa, respectively, for sensor ranges: (3 and 20 MPa).

Experimental results

Vapor pressure Temperature range for measurements were from 253.26 to 358.76 K, and 231.39 to 359.20 K for R1216 and HFPO, respectively. Attempt was made to determine the values of critical temperature and pressure by experimental means. The critical point was expected to be observed with the disappearance of the vapor-liquid interface and the creation of a "fog" in the cell. From this observation, it was determined for R1216 and HFPO that T_C = 358.8 K ± 0.1 K and P_C = 3.129 MPa ± 0.001 MPa, and T_C = 359.3 K ± 0.1 K and P_C = 2.930 MPa ± 0.001 MPa respectively. The measured vapor pressure data were used to fit the parameters of the Frost-Kalkwarf[6] equation (see eq. 1). Average absolute relative deviation is less than 0.2 % and the bias is -0.07%.

$$P = \exp\left(A + \frac{B}{T} + C\ln(T) + D\times10^{-17}\times T^{E}\right) \quad (1)$$

where P (Pa) is the pressure, T (K) is the temperature and A, B, C, D and E are adjustable parameters. The values of each of the parameters are:
A = 51.9463; B = -3799.9997; C = -4.5245; D = 10.1553; and E = 6 for R1216
A = 53.1702; B = -3799.8834; C = -4.7530; D = 12.3069; and E = 6 for HFPO

Densities Figures 1 and 2 show R1216 and HFPO densities respectively, determined at saturation in the 0 to 10 MPa pressure range considering the vapor pressure and the measured densities. The estimated uncertainty is less than 4%.

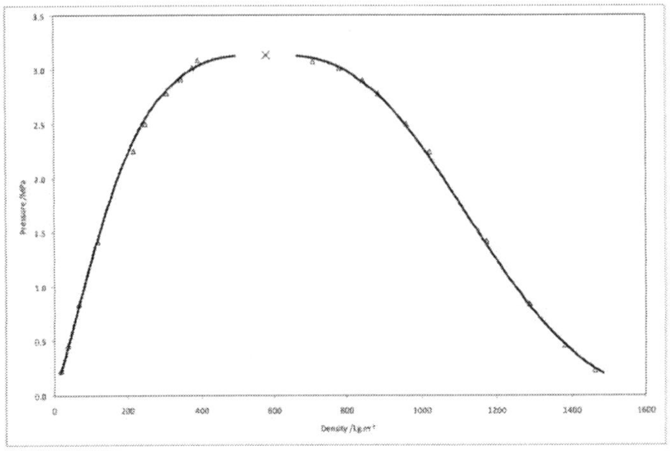

Figure 1: R1216 P-ρ diagram. (Δ): Densities at saturation, (×): critical point. Black line: calculated densities using equations (4) and (5).

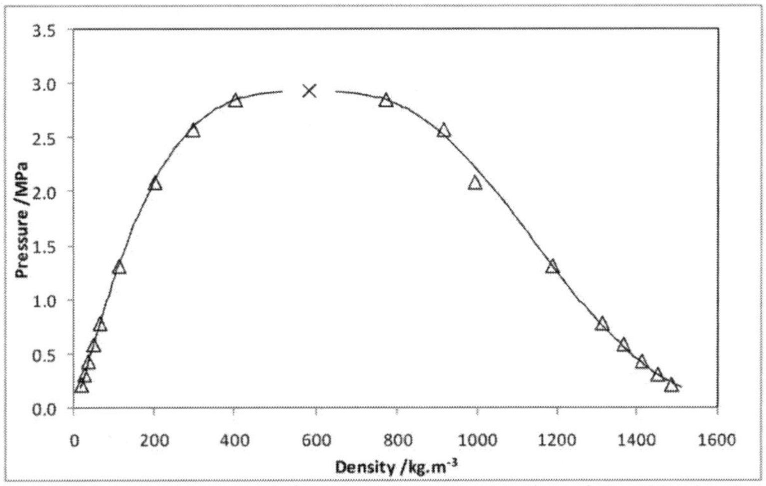

Figure 2: HFPO P-ρ diagram. (Δ): Densities at saturation, (×): critical point. Black line: calculated densities using equations (4) and (5).

Figures 3 and 4 show R1216 and HFPO densities out of saturation respectively (liquid, vapour and supercritical state).

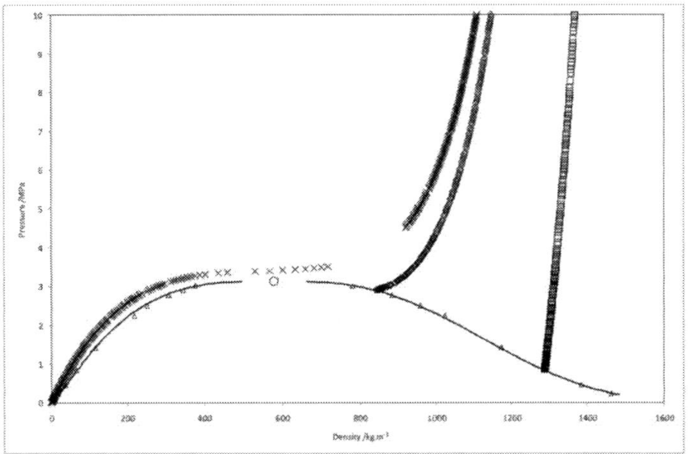

Figure 3: R1216 P-ρ diagram. (Δ): Densities at saturation, Densities out of saturation at: (×):362.90 K, (◊): 355.18 K and (□): 303.28 K, (o): critical point. Solid line: densities at saturation calculated using equations (4) and (5).

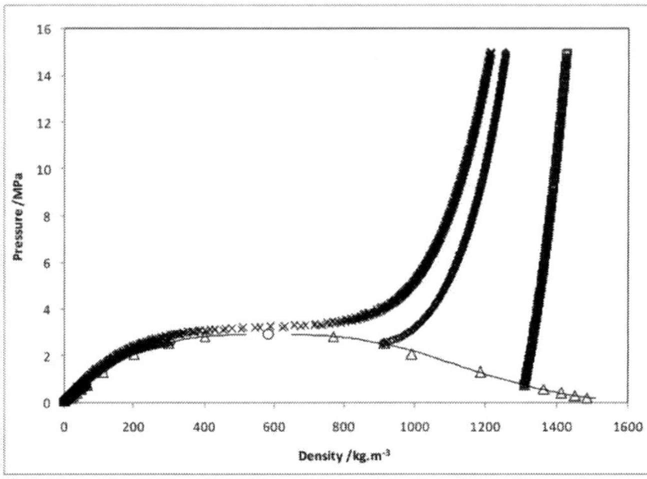

Figure 4: HFPO P-ρ diagram. (Δ): Densities at saturation, Densities out of saturation at: (×): 362.88 K, (◊): 353.06 K and 303.15 K(□): , (o):critical point. Solid line: densities at saturation calculated using equations (4) and (5).

The densities at saturation were used to determine the critical properties of the pure component. Two laws can be used for the determination of critical temperature T_C (K) and critical density ρ_C (kg.m^{-3}). The first is a scaling law directly related to the difference of densities between vapor and liquid phase (eq. 2) and expressed as follows:

$$\rho^L - \rho^V = A(T - T_C)^\beta \tag{2}$$

where β is an universal exponent constant (0.325). It is also assumed that the densities of the coexisting liquid and vapor obey the law of rectilinear diameters (eq. 3) given as:

$$\frac{\rho^L + \rho^V}{2} = B(T - T_C) + \rho_C \tag{3}$$

where ρ^L (kg.m^{-3}) and ρ^V (kg.m^{-3}) are the respective liquid and vapor densities, A (kg.m^{-3}.K$^{-\beta}$) and B (kg.m^{-3}.K) are adjustable parameters. The corresponding values are presented in **Table 1**, along with the critical properties of R1216 and HFPO (including acentric factor).

Table 1: Pure component critical properties obtained

Component	Equations 4 and 5 parameters		Critical properties				
	A	B	T_C /K	P_C/ MPa	ρ_C / kg.m^{-3}	Acentric factor ω	Z_C
R1216	329.47	1.73	358.93	3.1357	579.03	0.3529	0.27226
HFPO	335.12	1.80	359.26	2.9312	584.18	0.3338	0.27737

Figure 1 and 2 show the R1216 and HFPO P-ρ diagram, including their critical points. Using equations 2 and 3, one can define eq. 4 and 5 for the determination of vapor and liquid densities at saturation as follows:

$$\rho^L = B(T - T_C) + \rho_C + \frac{A(T - T_C)^\beta}{2} \quad (4)$$

$$\rho^V = B(T - T_C) + \rho_C - \frac{A(T - T_C)^\beta}{2} \quad (5)$$

Using these two equations, the average absolute relative deviations are 5.9 % and 1% for vapor and liquid density, respectively. The deviation is more important for the vapor density due to its low order of magnitude.

The R1216 and HFPO critical parameters (T_C and P_C) are found to be in very good agreement with those determined through the "vapor pressure" measurements. Concerning, HFP, information obtained from Air Liquide Safety sheets (encyclopedia.airliquide.com/sds/fr/066_AL_FR.pdf) lists T_C = 359.35 K; Dupont de Nemours Safety sheet (www.dupont.com/FluoroIntermediates/en_US/assets/downloads/h96536.pdf) lists T_C = 358.95 K. DIPPR[7] states a value of T_C = 368 K. The value of T_C (358.93 K) obtained in our study is in very good agreement with that listed in the Dupont de Nemours Safety sheet. Concerning HFPO, Dupont de Nemours Safety sheet (www.dupont.com/FluoroIntermediates/en_US/assets/downloads/h88806.pdf) lists T_C = 359.15 K and P_C = 2.896 MPa.

Conclusion

A sapphire tube cell and a vibrating tube densitometer were used to determine the vapour pressures and the densities of pure hexafluoropropylene and hexafluoropropylene oxide. The resultant uncertainties in the measurement of the vapor and liquid densities were ± 0.05% using the FPMC approach. Using these data new values of critical properties were determined and validated through visual measurement on a static cell.

References

1- Acerboni, G.; Buekes, J.A.; Jensen, N.R.; Hjorth, J.; Myhre, G.; Nielsen, C.J. ; Sundet, J.K. *Atmos. Environ.* **2001**, ***35***, 4113-4123.
2- Bouchot, C.; Richon, D. Direct pressure-volume-temperature and vapor-liquid equilibrium measurements with a single equipment using a vibrating tube densimeter up to 393 K and 40 MPa: description of the original apparatus and new data. *Ind. Eng. Chem. Res.* **1998**, *37*, 3295-3304.
3- Galicia-Luna, L. A.; Ortega-Rodriguez, A. New Apparatus for the Fast Determination of High Pressure Vapor-Liquid Equilibria of Mixtures and Accurate Critical Pressures. *J. Chem. Eng. Data* **2000**, *45*, 265-271.
4- Bouchot, C.; Richon, D. An enhanced method to calibrate vibrating tube densimeters. *Fluid Phase Equilib.* **2001**, *191*, 189-208.
5- Tillner-Roth, R.; Baehr, H. D. An international standard formulation for the properties of 1, 1, 1, 2-tetrafluoroethane (HFC-R134a) for temperatures from

170 K to 455 K and pressures up to 70 MPa. *J. Phys. Chem. Ref. Data.* **1994**, *23*, 657-729.
6- Frost A.A.; Kalkwarf, D.R. A semi-empirical equation for the vapor pressure of liquids as a function of temperature *J. Chem. Phys.* **1953**, 21, 264–267.
7- Daubert, T.E.; Danner, R.P.; Sibel, H.M.; Stebbins, C.C. *Physical and thermodynamic properties of pure chemicals*. Data Compilation, Taylor & Francis, Washington, D.C. 1997.

Gas Hydrate Phase Equilibria in the Methane + Carbon dioxide + Water System: Experimental Measurements and Predictions of Dissociation Conditions

Veronica Belandria, Amir H. Mohammadi,* and Dominique Richon

MINES ParisTech, CEP/TEP - Centre Énergétique et Procédés, 35 Rue Saint Honoré, 77305 Fontainebleau, France

ABSTRACT

In this communication, we report experimental dissociation data for clathrate hydrates of methane + carbon dioxide at low, medium and high concentrations of carbon dioxide in the temperature range from 279 to 290 K and pressures up to 13 MPa. An isochoric pressure-search method was used to perform the measurements. The dissociation data generated in this work are compared with the predictions of a thermodynamic model. A discussion is made on the deviations between the experimental and predicted data.

Keywords: Gas hydrate, methane, carbon dioxide, dissociation data.

---

*: Corresponding author, Email: amir-hossein.mohammadi@ensmp.fr Tel.: +(33) 1 64 69 49 70. Fax: +(33) 1 64 69 49 68.

1. Introduction

Carbon dioxide (CO_2) capture and sequestration is subject of major interest within numerous engineering applications, especially in petroleum industry and environmental science [1-3]. One promising approach to capture CO_2 from natural gas is through gas hydrate technology. As hydrates having the capacity to store/separate gases [3-5], the hydrate crystallization process has been proposed for potential applications including CO_2 sequestration and methane recovery from natural gas hydrate in deep sea sediments, and CO_2 capture from flue gases to reduce atmospheric emissions [1-6].

Gas hydrates (or clathrate hydrates) are ice-like crystalline compounds, which are formed through a combination of water and small guest molecules like methane (CH_4), CO_2, *etc.* under suitable conditions of low temperature and high pressure [4]. In a gas hydrate molecule, water form special cavities and guest molecules are trapped inside the cavities. The gas hydrate structures are those of structure I (sI), structure II (sII) and structure H (sH), where each structure is composed of a certain number of cavities formed by water molecules [4]. Because of the difference in chemical affinity between CO_2 and CH_4 in the hydrate structure, when hydrate crystals are formed from a mixture of these two gases CO_2 concentration might be enriched in the hydrate phase while CH_4 would be reduced in the hydrate and increased in the gas phase at equilibrium. The hydrate phase is then dissociated by depressurization or/and heating at ambient temperature and CO_2 can be recovered as a separated gas. This is the principle of gas separation by hydrate crystallization process.

The first experimental study of $CH_4 + CO_2$ hydrates was reported by Unruh and Katz in 1949 [7]. Two decades later, Berecz and Balla-Achs [8] reported that hydrates of CO_2 and CH_4 exhibit a maximum and a minimum in the hydrate isotherms and isobars at 50 mole% CO_2 mole fraction and higher. In 1991, Adisasmito *et al.* [3] confirmed and extended the measurements done by Unruh and Katz [7]. Their data did not show the unusual hydrate instability reported by Berecz and Balla-Achs [8]. Meanwhile, Ohgaki *et al.* [1] reported isothermal phase equilibrium data for the $CH_4 + CO_2$ mixed hydrate system at 280 K. They showed the first experimental evidence that CH_4 is selectively replaced by CO_2 in the hydrate phase. Hydrate formation data of CO_2-rich (> 90 mole%) gas mixtures in water containing small amounts of CH_4 remained to be determined. This was done almost simultaneously by Fan and Guo [9], Seo *et al.* [10] and Hachikubo *et al.* [2].

In this work, we measured and report dissociation data for CH_4 + CO_2 clathrate hydrates at low, medium and high concentrations of CO_2 in the temperature range from 279 to 290 K and at pressures up to 13 MPa. An isochoric pressure-search method [11, 12, 14] was used to estimate the dissociation conditions. The dissociation data generated in this work are compared with the predictions of a thermodynamic model.

2. Experimental setup and method

Table 1 reports the purities and suppliers of the chemicals used in this work. A detailed description of the experimental setup used in this study is provided in the literature [11, 12]. A simplified flow diagram of the experimental setup is shown in figure 1. Briefly, the apparatus is based on the "static–analytic" method with capillary fluid phase sampling [13]. The heart of this apparatus is a cylindrical equilibrium cell made of 316 stainless steel that is suitable for measurements at pressures up to 40 MPa. The cell has a volume of 57.5 cm^3 and two sapphire windows. A magnetic stirrer ensures agitation to facilitate reaching equilibrium. Two platinum resistance thermometers (Pt100) inserted into the vessel were used to measure temperatures and check for equality of temperatures within temperature measurement uncertainties, which is estimated to be less than 0.1 K. This temperature uncertainty estimation results from calibration against a 25-Ω reference platinum resistance thermometer. The pressure in the cell was measured using a pressure transducer (Druck, type PTX611 for pressures up to 20 MPa). Pressure measurement accuracies are estimated to be better than 2 kPa.

An isochoric pressure-search method [11, 12, 14] was followed to determine the hydrate dissociation pressure and temperature conditions. The reliability of the experimental method used has been demonstrated in previous works [11, 12, 15]. Before starting the experiment, the cell was evacuated to eliminate the presence of air. The gas mixture was prepared in the cell by introducing first a certain amount of CO_2, and then CH_4 from the corresponding cylinders through a pressure-regulating valve. After temperature and pressure were stabilized, the valve in the line connecting the vessel and the gas cylinder was closed.

The composition of the gas mixture was analyzed by a gas chromatograph (Varian, CP3800) coupled with a pneumatic Rolsi™ sampler [16]. The operating conditions of the gas chromatograph equipped with a thermal conductivity detector (TCD) are given in Table 2. The analytic column used in this study was a PORAPACK Q model, length: 2m, O.D: 1/8", 80/100 mesh.

Subsequently, 10% (by volume) of the cell was filled with water using a volumetric press. The cell was immersed into the temperature-controlled bath and the temperature was slowly decreased to form the hydrate. Hydrate formation in the cell was detected by a noticeable pressure drop. The temperature was then increased at steps of 0.1 K. At every temperature step, the temperature was kept constant with sufficient time to achieve an equilibrium state in the cell. In this way, a pressure-temperature diagram was obtained for each experimental run, from which the hydrate dissociation point was found [17]. The point at which the slope of pressure-temperature data plots changes sharply is considered to be the point at which all hydrate crystals have dissociated and, hence, is reported as the equilibrium condition [17].

3. Results and discussion

The hydrate dissociation conditions for the methane + carbon dioxide + water system were investigated at different concentrations of carbon dioxide. The compositions of the methane + carbon dioxide gas mixtures are presented in Table 3 and Figure 2. Table 3 also shows the predictions of a thermodynamic model. A detailed description of this model is given elsewhere [18]. The model [18] is briefly based on the equality of fugacity concept, which uses the Valderrama modification of the Patel-Teja equation of state [19] and non-density dependent mixing rules [20] for modeling the fluid phases and the van der Waals and Platteeuw theory [21] is used for modeling the hydrate phase. The absolute deviation (*AD*) in this table was calculated using the following equation:

$$AD(\%) = \left| \frac{p_{\exp} - p_{pred}}{p_{\exp}} \right| \times 100 \qquad (1)$$

where p_{exp} and p_{pred} represent experimental and predicted dissociation pressures, respectively.

We have also compared the experimental and predicted dissociation conditions in Figure 3. As can be observed in this figure and also in Table 3, the predictions of the thermodynamic model [18] generally overestimate the dissociation pressures for the carbon dioxide + methane clathrate hydrates studied in this work. We have assumed both structures I and II for the predictions of the dissociation conditions. Assuming structure I leads to lower deviations between the predicted and experimental data suggesting the stable hydrate structure for the carbon dioxide + methane clathrate hydrates studied in our work is likely structure I. However, the final proof for the stable structure of the methane + carbon dioxide clathrate hydrates requires direct measurements by suitable physical techniques (*e.g.*, NMR, X-ray, or Raman spectroscopy). This preliminary study of the deviations between experimental and predicted data suggests that the existing thermodynamic models may require modifications in order to reliably predict phase behavior of methane + carbon dioxide clathrate hydrates.

4. Conclusions

We reported dissociation data for clathrate hydrates of various methane + carbon dioxide mixtures (Table 3). The gas mixtures were prepared directly in the equilibrium cell using a ROLSI™ sampler, which was connected to a gas chromatograph. An isochoric pressure-search method [11, 12, 14] was used to estimate the dissociation conditions. Our experimental data were compared with the predictions of a thermodynamic model [18] and it was shown that the model over predicts the data. It was suggested that the stable hydrate structure for the carbon dioxide + methane clathrate hydrates is likely structure I. However, direct measurements by suitable physical techniques (*e.g.*, NMR, X-ray, or Raman spectroscopy) are required for the final proof of the stable structure for the methane + carbon dioxide clathrate hydrates.

Acknowledgment

Veronica Belandria wishes to thank *Fundayacucho* of Venezuela for doctoral fellowship. The financial support of *Agence Nationale de Recherche* (ANR) is gratefully acknowledged. The authors also thank Mr Pascal Théveneau for technical support.

References

[1] Ohgaki, K.; Sangawa, H.; Matsubara, T.; Nakano, S. J. Chem. Eng. Jpn. 1996, 29, 478-483.

[2] Hachikubo, A.; Miyamoto, A.; Hyakutake, K.; Abe, K.; Shoji, H. Proceedings of the 4th International Conference on Gas Hydrates, Yokohama, Japan, May 19-23, 2002, p 357.

[3] Adisasmito, S.; Frank, R.; Sloan, D. J. Chem. Eng. Data 1991, 36, 68-71.

[4] Sloan, E. D.; Koh, C. A. Clathrate Hydrates of Natural Gases, third ed., CRC Press, Taylor & Francis Group, Boca Raton, 2007.

[5] Koh, C. A.; Sloan, E. D. AIChE J. 2007, 53, 1636-1643.

[6] Bergeron, S.; Servio, P. Fluid Phase Equilib. 2009, 276, 150-155.

[7] Unruh, C. H.; Katz, D. L., Pet. Trans. AIME 1949, 186, 83-86.

[8] Berecz, E.; Balla-Achs, M. Gas Hydrates (Studies in Inorganic Chemistry Vol. 4), Chapter 3, Elsevier Science Publishers, Amsterdam, Oxford, New York, 1983.

[9] Fan, S. S.; Guo, T. M. J. Chem. Eng. Data 1999, 44, 829-832

[10] Seo, Y.T.; Kang, S.P.; Lee, H.; Lee, C.S.; Sung, W.M. Korean J. Chem. Eng. 2000, 17, 69-667.

[11] Afzal, W.; Mohammadi, A. H. ; Richon, D. J. Chem. Eng. Data 2007, 52, 2053-2055.

[12] Mohammadi, A. H.; Afzal, W.; Richon, D. J. Chem. Eng. Data 2008, 53, 683-686.

[13] Richon, D. Fluid Phase Equilib. 1996, 116, 421-428.

[14] Tohidi, B.; Burgass, R. W.; Danesh, A.; Østergaard, K. K.; Todd, A.C. Ann. N.Y. Acad. Sci. 2000, 912, 924-931.

[15] Mohammadi, A. H.; Belandria, V.; Richon, D. Ind. Eng. Chem. Res. 2009, 48, 5916–5918.

[16] Guilbot, P.; Valtz, A.; Legendre, H.; Richon, D. Analusis 2000, 28, 426-431.

[17] Ohmura, R.; Takeya, S.; Uchida, T.; Ebinuma, T. Ind. Eng. Chem. Res. 2004, 43, 4964-4966.

[18] Heriot-Watt University Hydrate model: http//www.pet.hw.ac.uk/research/hydrate/. (See also: Tohidi, B.; Burgass, R. W.; Danesh, A.; Todd, A.C. SPE 26701. Proc. of the SPE Offshore Europe 93 Conference; 1993, 255-265).

[19] Valderrama, J. O. J. Chem. Eng. Jpn. 1990, 23, 87-91.

[20] Avlonitis, D.; Danesh, A.; Todd, A. C. Fluid Phase Equilib. 1994, 94, 181-216.

[21] van der Waals, J. H.; Platteeuw, J. C. Adv. Chem. Phys. 1959, 2 1-57.

Table 1: Purities and suppliers of chemicals.[a]

chemical	supplier	purity/ (volume %)
Methane	Messer Griesheim	99.995
Carbon dioxide	Air Liquide	99.995

[a] Deionized water was used in all experiments.

Table 2: Operating conditions of the gas chromatograph.

Carrier gas	Helium
Gas rate (ml/min)	25
Detector	TCD
Detector temperature (K)	393
TCD wire temperature (K)	448
Oven temperature (K)	323

Table 3: Experimental (p_{exp}) and predicted (p_{pred}) dissociation conditions using thermodynamic model [18] for carbon dioxide + methane clathrate hydrates in the presence of pure water.

$CO_{2\,(load)}$ / mole fraction	T /K	p_{exp} /MPa	p_{pred} /MPa (assuming structure I)	AD^* /%	p_{pred} /MPa (assuming structure II)	AD^* /%
0.264	284.2	5.29	Model did not converge	-	12.23	131.2
	287.2	9.83	9.94	1.1	18.86	91.9
	289.2	11.62	13.16	13.3	25.51	119.5
0.272	279.1	3.60	3.79	5.3	6.40	77.8
0.490	284.8	5.82	6.47	11.2	17.32	197.6
	289.9	12.41	15.10	21.9	40.36	225.3
0.500	284.9	5.88	Model did not converge	-	17.86	203.7
0.504	279.1	2.96	3.18	7.4	7.53	154.4
0.730	280.6	3.16	3.42	8.2	11.55	265.5
	281.9	4.02	4.03	0.3	15.47	284.8
	289.1	13.06	15.21	16.5	63.88	389.1

*: $AD(\%) = \left|\frac{p_{\exp} - p_{pred}}{p_{\exp}}\right| \times 100$ where p_{exp} and p_{pred} represent experimental and predicted dissociation pressures, respectively.

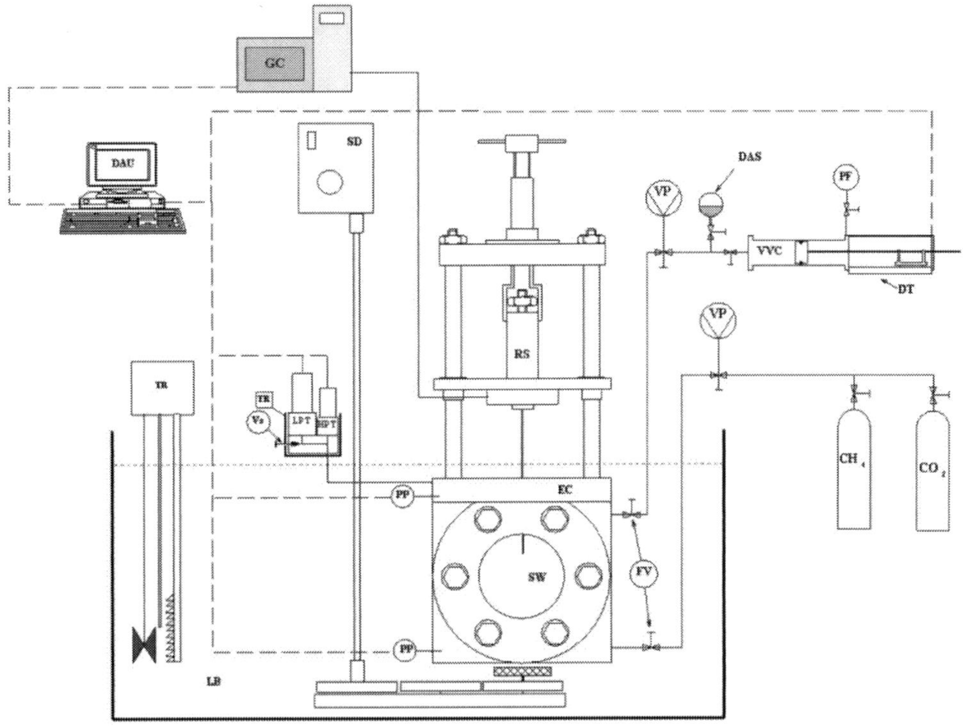

Figure 1: Flow diagram of the experimental setup. DAS: degassed aqueous solution; DAU: data acquisition unit; DT: displacement transducer; EC: equilibrium cell; FV: feeding valve; GC: gas chromatograph; HPT: high pressure transducer; LB: liquid bath; LPT: low pressure transducer (Our of service); PF: pressurizing fluid; PP: platinum probes; RS: Rolsi™ sampler; SD: stirring device; SW: sapphire windows, TR: temperature regulator; VP: vacuum pump; Vs: isolation valve of LPT; VVC: variable volume cell.

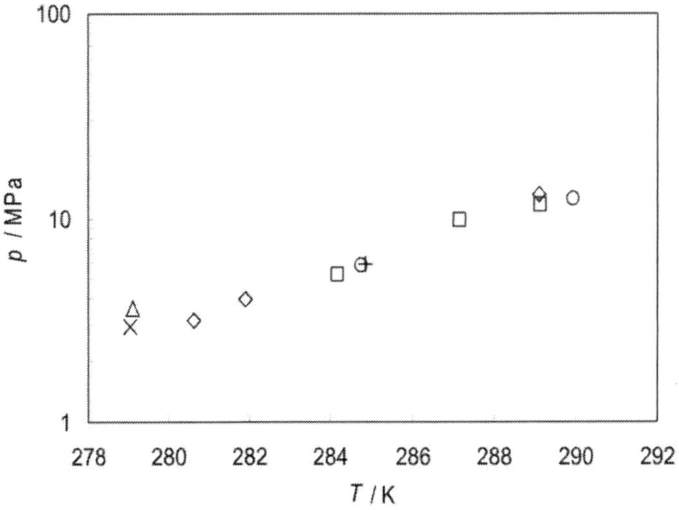

Figure 2: Experimental dissociation data reported in this work for various methane + carbon dioxide clathrate hydrates at different CO_2 load mole fractions: 0.504 (x); 0.272 (△); 0.264 (□); 0.490 (○); 0.500 (+); 0.730 (◇).

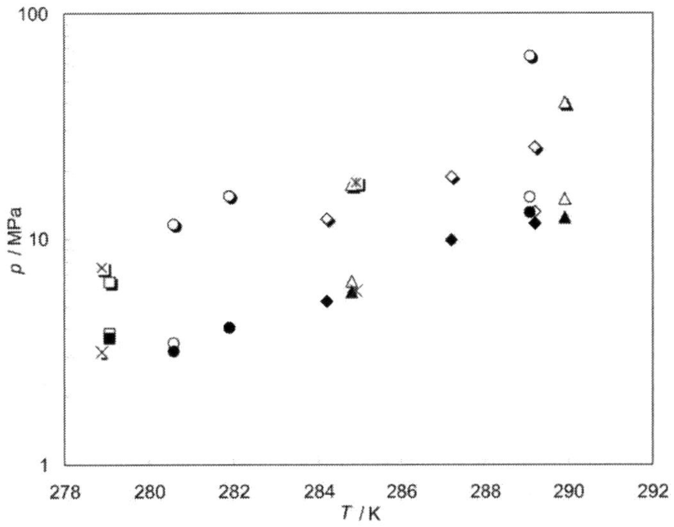

Figure 3: Comparison of the experimental dissociation data reported in this work for various methane + carbon dioxide clathrate hydrates with the predictions of the thermodynamic model [18]. Experimental dissociation conditions at different CO_2 mole fractions loaded, this work: 0.264 (◆); 0.272 (■); 0.490 (▲); 0.500 (*); 0.504 (-); 0.730 (●). Thermodynamic model [18] predictions assuming sI: 0.264 (◇); 0.272 (□); 0.490 (△); 0.500 (model did not converge); 0.504 (x); 0.730 (○).Thermodynamic model [18] predictions assuming sII: 0.264 (◇ shadow); 0.272 (□ shadow); 0.490 (△ shadow); 0.500 (* shadow); 0.504 (x shadow); 0.730 (○ shadow).

Liquid Hydrate Former + Hydrate + Liquid Water Equilibrium Data for the Ethane, Carbon Dioxide or Hydrogen Sulfide + Water System

Amir H. Mohammadi* and Dominique Richon

MINES ParisTech, CEP/TEP - Centre Énergétique et Procédés, 35 Rue Saint Honoré, 77305 Fontainebleau, France.

ABSTRACT

In this work, we report the liquid hydrate former + hydrate + liquid water equilibrium data for the ethane, carbon dioxide or hydrogen sulfide + water system. An isochoric pressure-search method was used to perform the measurements. The new equilibrium data are compared with the experimental data reported in the literature to study the reliability of the literature data.

Keywords: Gas hydrate, ethane, carbon dioxide, hydrogen sulfide, liquid hydrate former + hydrate + liquid water equilibrium, experimental data.

---

*: Corresponding author, Email: amir-hossein.mohammadi@ensmp.fr Tel.: +(33) 1 64 69 49 70. Fax: +(33) 1 64 69 49 68.

1. Introduction

Clathrate hydrates, or gas hydrates, are a group of non-stoechiometric, icelike crystalline compounds formed through a combination of water and suitably sized guest molecule(s) under low-temperatures and elevated pressures. In the clathrate lattice, water molecules form hydrogen-bonded cagelike structures, encapsulating the guest molecule(s), which generally consist of low molecular diameter gases and organic compounds. A concise review of clathrate hydrates is given by Sloan and Koh [1]. Conditions of high pressures and low temperatures as well as presence of water leading to clathrate hydrate formation may be found in oil and gas production, transportation and processing facilities. Formation of gas hydrates can lead to serious operational, economic and safety problems in petroleum industry due to potential blockage of oil and gas equipment [1].

Figure 1 shows a typical pressure-temperature diagram for water + multi-component hydrocarbon systems. As can be seen, the ice-vapor-hydrate (*I-V-H*), liquid water-vapor-hydrate (L_W-*V-H*) and liquid water-liquid hydrocarbon- vapor - hydrate (L_W- L_{HC}-*V-H*) equilibrium occur normally at relatively low and intermediate pressures. However, the L_W-*V-H* and L_W- L_{HC}-*V-H* phase boundaries approach high pressures by increasing temperature. The liquid water- liquid hydrocarbon- hydrate (L_W- L_{HC}- *H*) phase boundary is however stronger function of the system temperature [1]. Most of the existing experimental data and thermodynamic models are limited to low/intermediate pressures (representing *I-V-H*, L_W-*V-H*, L_W- L_{HC}-*V-H* equlibria). At high pressures, the hydrate dissociation pressure exceeds the bubble point pressure of the system and the hydrate phase boundary is reduced to a three-phase curve with only water, liquid hydrocarbon and hydrates in equilibrium (*i.e.*, L_W-L_{HC}-*H* region). In this region, the hydrate dissociation pressure is generally a strong function of the system temperature. [1].

It is the intention of this work to report the L_W-L_{HC}-H equilibrium data for the ethane, carbon dioxide or hydrogen sulfide + water system (From here, the subscript *HC* means liquid ethane, liquid carbon dioxide or liquid hydrogen sulfide). An isochoric pressure-search method [2] was applied to measure equilibrium data. It is expected that this study provides a better understanding of the reliabilities of corresponding equilibrium data reported in the literature.

2. Experimental Section

2.1. Materials

Table 1 reports the purities and suppliers of the materials used in this work.

2.2. Experimental apparatus

Briefly, the main part of the apparatus is a sapphire cylindrical vessel, which can withstand pressures up to 15 MPa. The volume of the vessel is 33.1 cm^3. A stirrer was installed in the vessel to agitate the fluids and hydrate crystals inside it. Two platinum resistance thermometers (Pt100) inserted into the vessel were used to measure temperatures and check for equality of temperatures within temperature measurement uncertainties, which is estimated to be less than 0.1 K. This temperature uncertainty estimation comes from calibration against a 25 Ω reference platinum resistance thermometer. The pressure in the vessel was measured with a DRUCK pressure transducer (Druck, type PTX611 for pressure ranges up to 14 MPa). Pressure measurement uncertainties are estimated to be less than 5 kPa, as a result of calibration against a dead weight balance (Desgranges and Huot, model 520).

2.3. Experimental method

The equilibrium conditions were measured with an isochoric pressure search method [2]. The vessel was immersed into the temperature-controlled bath, and the gas was supplied from a cylinder through a pressure-regulating valve into the vessel. Note that that the vessel was evacuated before introducing any gas. After obtaining temperature and pressure stability (far enough from the hydrate formation region), the valve in the line connecting the vessel and the cylinder was closed. A hand pump (called press) containing liquid water was then used to inject water in the vessel and consequently pressurize the system. Subsequently, temperature was slowly decreased to form the hydrate. Hydrate formation in the vessel was detected by pressure drop. The temperature was then increased with steps of 0.1 K. At every temperature step, temperature was kept constant with sufficient time to achieve a steady equilibrium state in the vessel. In this way, a pressure-temperature diagram was obtained for each experimental run, from which we determined the hydrate dissociation point [3]. If the temperature is increased in the hydrate-forming region, hydrate crystals partially dissociate, thereby substantially increasing the pressure [3]. If the temperature is increased outside the hydrate region, only a smaller increase in the pressure is observed as a result of the change in the equilibrium of the fluids in the vessel [3]. Consequently, the point at which the slope of pressure-temperature data plots changes sharply is considered to be the point at which all hydrate crystals have dissociated and hence reported as the dissociation point [3].

3. Results and conclusion

All the equilibrium data are reported in Tables 2-4 and are plotted in Figures 2-4. We first measured the hydrate dissociation conditions for the ethane + water system. As can be observed in Figure 2*b*, the experimental data measured in this work for the latter system are in acceptable agreement with the experimental data reported in the reference 5. In this figure, the experimental data reported in reference 4 are dispersed. Note that because our vessel can operate up to 15 MPa, therefore we could not check the reliabilities of other experimental data reported in the literature for higher pressures. However, as can be observed in Figure 2*a*, the experimental data reported at higher pressures (references 6 and 7) are apparently reliable and in acceptable agreement with each other. We have done a similar study for the carbon dioxide clathrate hydrates in Figure 3. As can be seen in Figure 3*b*, our experimental data are in acceptable agreement with the experimental data reported in references 8 and 10. The experimental data reported in references 5 and 9 are dispersed. Again, we could not study the reliabilities of other experimental data reported in the literature for higher pressures. However, as can be observed in Figure 3*a*, the experimental data reported at higher pressures (references 6 and 8) are apparently reliable and in acceptable agreement with each other. The experimental data reported in reference 6 truly show a pseudo – retrograde behavior already reported for high pressure conditions [1]. Finally, we repeated this study for clathrate hydrates of hydrogen sulfide, for which limited information is available in the literature [1]. As can be observed in Figure 4, the agreement between our experimental data and those reported in reference 11 are generally acceptable within the temperature and pressure ranges shown in this Figure.

References

[1] Sloan, E. D.; Koh, C. A. *Clathrate Hydrates of Natural Gases*, Third Edition, CRC Press, Taylor & Francis Group, Boca Raton, 2008.

[2] Tohidi, B.; Burgass, R. W.; Danesh, A.; Østergaard, K. K.; Todd, A. C. Improving the Accuracy of Gas Hydrate Dissociation Point Measurements. *Ann. N.Y. Acad. Sci.* **2000**, *912*, 924-931.

[3] Ohmura, R.; Takeya, S.; Uchida, T.; Ebinuma, T. Clathrate Hydrate Formed with Methane and 2-Propanol: Confirmation of Structure II Hydrate Formation. *Ind. Eng. Chem. Res.* **2004**, *43*, 4964-4966.

[4] Roberts, O.L.; Brownscombe, E.R., Howe, L.S. Methane and ethane hydrates. *Oil & Gas J.* **1940**, *39/30*, 37.

[5] Ng, H.J.; Robinson, D.B. Hydrate formation in systems containing methane, ethane, propane, carbon dioxide or hydrogen sulfide in the presence of methanol. *Fluid Phase Equilib.* **1985**, *21*, 145-155.

[6] Nakano, S.; Moritoki, M.; Ohgaki, K. High-Pressure Phase Equilibrium and Raman Microprobe Spectroscopic Studies on the CO_2 Hydrate System. *J. Chem. Eng. Data* **1998**, *43/5*, 807-810.

[7] Morita, K.; Nakano, S.; Ohgaki, K. Structure and stability of ethane hydrate crystal. *Fluid Phase Equilib.* **2000**, *169*, 167-175.

[8] Takenouchi, S.; Kennedy, G. C. *Dissociation Pressures of the Phase* $CO_2.5.75H_2O$. Geol. Notes No. 293; Institute of Geophysics, University of California, Berkeley, 1964; pp 383-390.

[9] Ohgaki, K.; Hamanaka, T. Phase-Behavior of CO_2 Hydrate-Liquid CO_2-H_2O System at High Pressure. Kagaku Kogaku Ronbunshu 1995, 21, 800-803.

[10] Mooijer-van den Heuvel, M. M.; Witteman, R.; Peters, C. J.; Phase behaviour of gas hydrates of carbon dioxide in the presence of tetrahydropyran, cyclobutanone, cyclohexane and methylcyclohexane. *Fluid Phase Equilib.* **2001**, *182*, 97-110.

[11] Selleck, F. T.; Carmichael, L. T.; Sage, B. H. Phase behavior in the hydrogen sulfide - water system. Ind. Eng. Chem. 1952, 44, 2219-2226.

Table 1: Purities and suppliers of materials.[a]

Gas	Supplier	Purity, volume%
Ethane	Messer Griesheim	99.995
Carbon dioxide	Air Liquide	99.995
Hydrogen sulfide	Air Liquide	99.9

[a] Deionized water was used in all experiments.

Table 2: Liquid water – hydrate - liquid ethane equilibrium boundary.

T/K	*p*/MPa
288.0	3.55
288.1	3.98
288.2	5.28
288.3	5.43
288.3	5.67
288.5	7.11
288.7	8.65
288.7	8.99
288.9	9.68

Table 3: Liquid water – hydrate - liquid carbon dioxide equilibrium boundary.

T/K	*p*/MPa
283.2	4.87
283.3	6.11
283.4	7.21
283.4	7.44
283.5	7.97
283.6	8.43
283.6	8.62
283.7	9.28
283.7	9.52

Table 4: Liquid water – hydrate - liquid hydrogen sulfide equilibrium boundary.

T/K	*p*/MPa
302.8	3.49
302.9	4.57
303.0	5.72
303.1	6.71
303.1	6.98
303.2	8.05
303.2	8.24

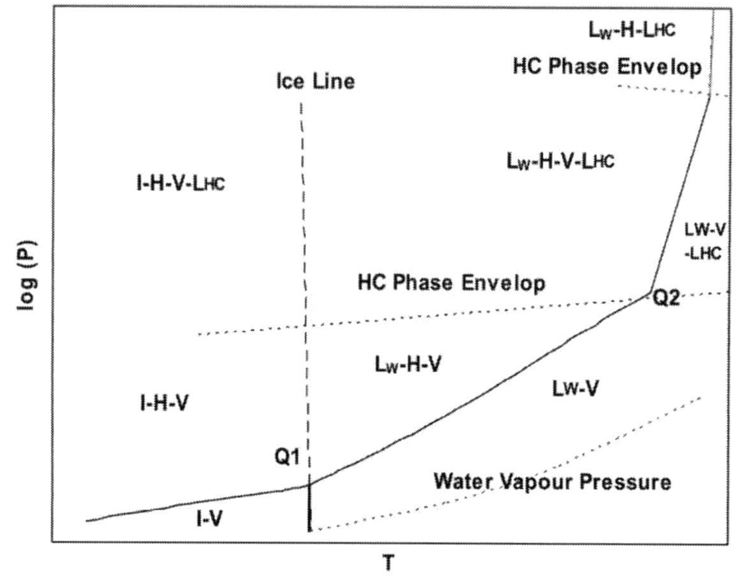

Figure 1: **Typical pressure (*P*) – temperature (*T*) diagram for (free) water – multi-component hydrocarbon systems**. *I*: ice, *H*: hydrate, *V*: vapor, L_{HC}: liquid hydrocarbon (or any hydrate former), L_W: liquid water, Q_1: first quadruple point, Q_2: second quadruple point.

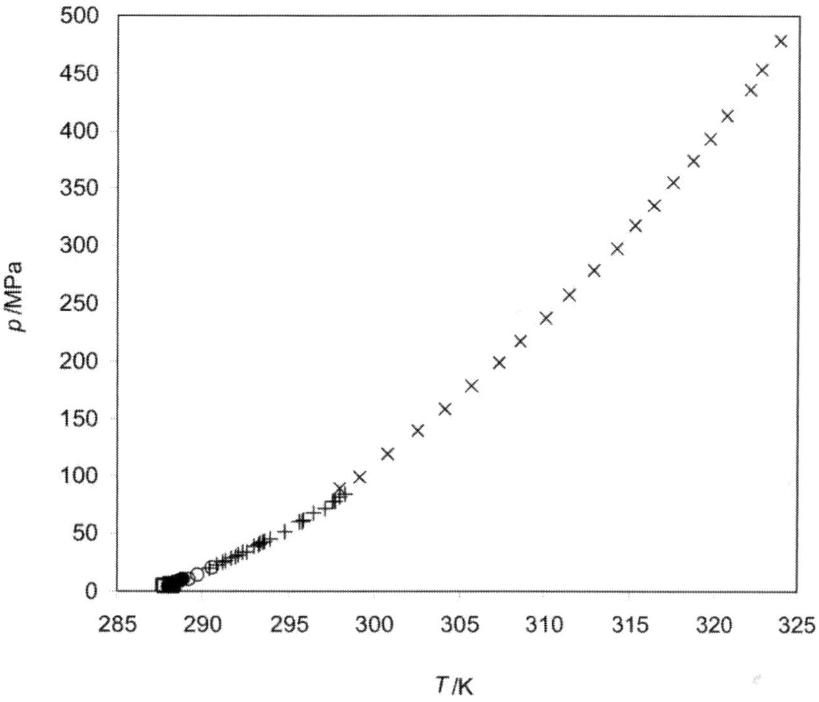

Figure 2*a*: Liquid water – hydrate – liquid ethane phase equilibrium boundary. Symbols represent experimental data. □: [4]; ○: [5]; +: [6]; ×: [7]; ●: This work.

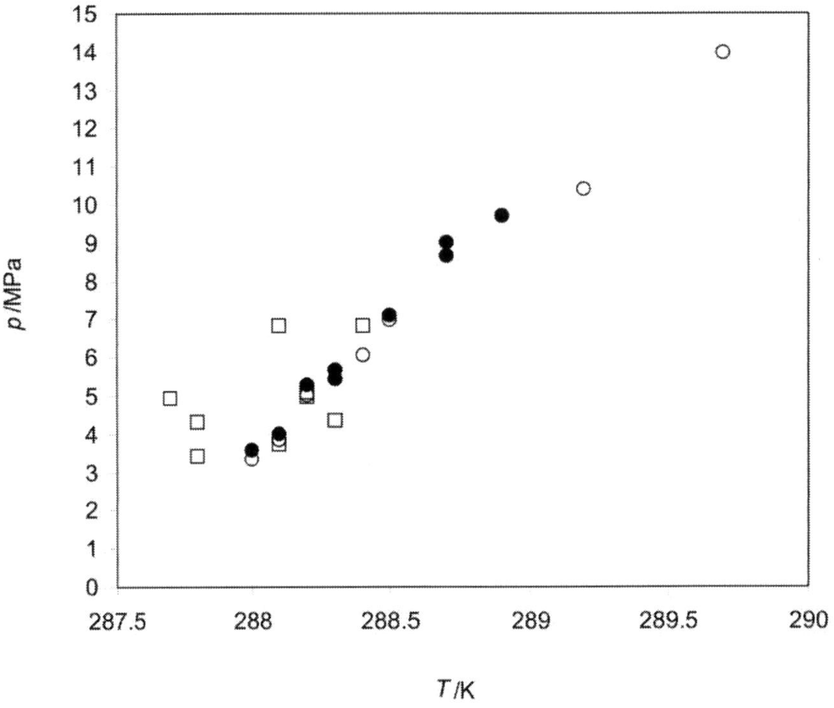

Figure 2*b*: Liquid water – hydrate – liquid ethane equilibrium boundary (Different zoom). Symbols represent experimental data. □: [4]; ○: [5]; ●: This work.

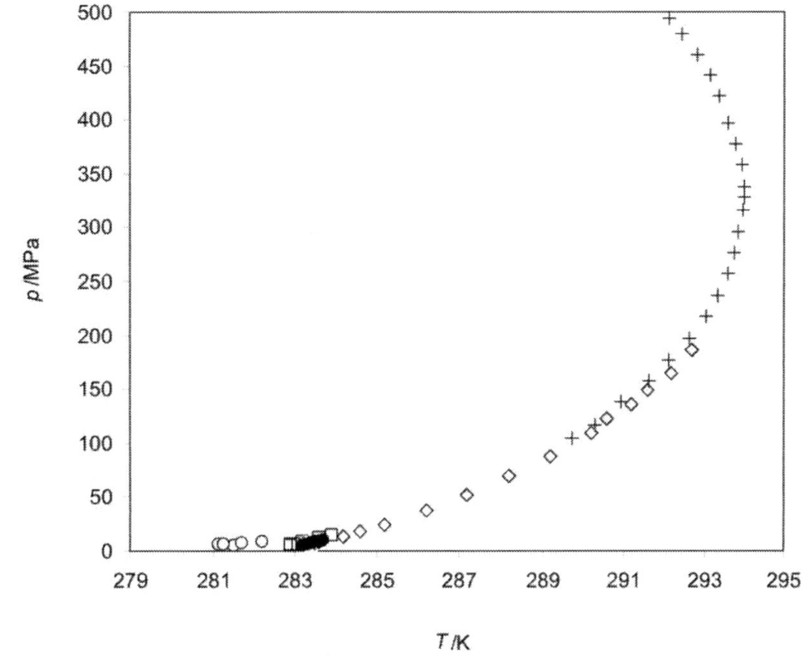

Figure 3*a*: Liquid water – hydrate – liquid carbon dioxide equilibrium boundary. Symbols represent experimental data. ◊: [8]; □: [5]; ○: [9]; +: [6]; ×: [10]; ●: This work.

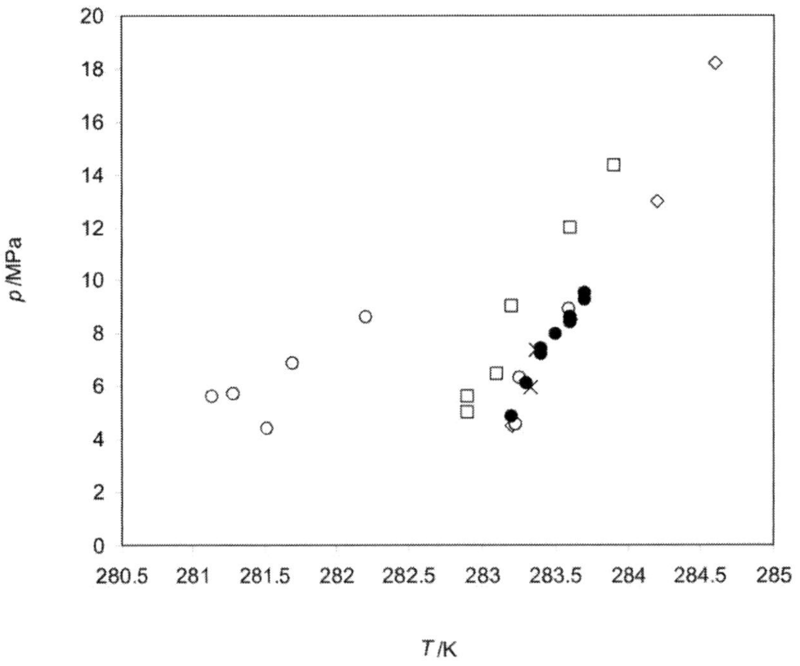

Figure 3*b*: Liquid water – hydrate – liquid carbon dioxide equilibrium boundary (Different zoom). Symbols represent experimental data. ◊: [8]; □: [5]; ○: [9]; ×: [10]; •: This work.

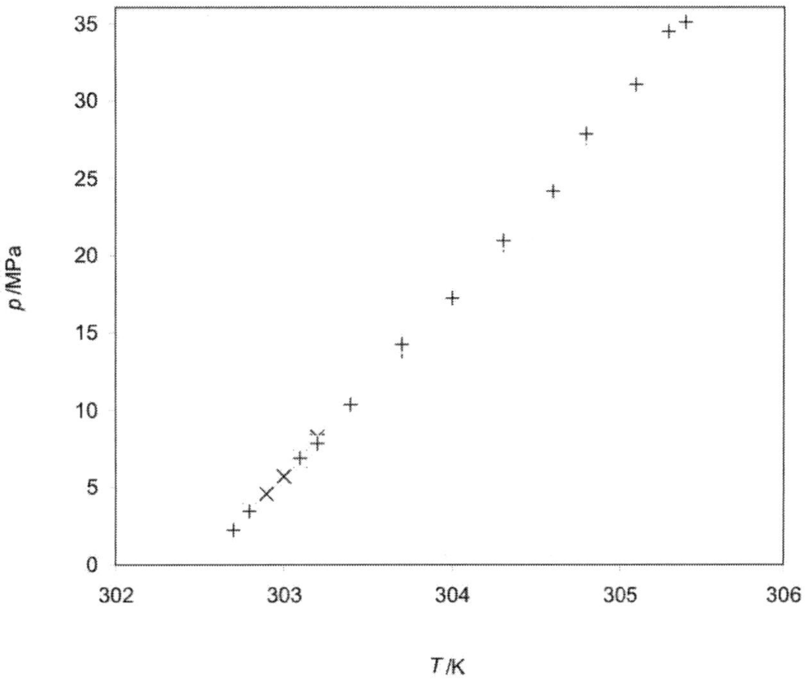

Figure 4: Liquid water – hydrate - liquid hydrogen sulfide equilibrium boundary. Symbols represent experimental data. ×: [11]; +: This work.

AUTHOR INDEX

AUTHOR INDEX

AUTHOR INDEX

AUTHOR INDEX

AUTHOR INDEX

AUTHOR INDEX

AUTHOR INDEX

AUTHOR INDEX

AUTHOR INDEX

AUTHOR INDEX

AUTHOR INDEX

AUTHOR INDEX

AUTHOR INDEX

AUTHOR INDEX